Modern Livestock & Poultry Production

Sixth Edition

Delmar is proud
to support FFA activities

Modern Livestock & Poultry Production

Sixth Edition

James R. Gillespie

DELMAR

THOMSON LEARNING ™

Australia Canada Mexico Singapore Spain United Kingdom United States

DELMAR
THOMSON LEARNING

Modern Livestock & Poultry Production, 6e
James R. Gillespie

Business Unit Director:
Susan L. Simpfenderfer

Acquisitions Editor:
Zina M. Lawrence

Development Editor:
Andrea Edwards Myers

Editorial Assistant:
Elizabeth Gallagher

Executive Marketing Manager:
Donna J. Lewis

Channel Manager:
Nigar Hale

Executive Production Manager:
Wendy A. Troeger

Production Editor:
Kathryn B. Kucharek

Cover Design:
Dutton & Sherman Design

For permission to use material from this text or product,
contact us by
Tel   (800) 730-2214
Fax   (800) 730-2215
www.thomsonrights.com

Library of Congress Cataloging-in-Publication Data

Gillespie, James R.
     Modern livestock & poultry production /
  James R. Gillespie. —6th ed.
        p.   cm.
  Includes index.
  ISBN 0-7668-1607-9
  1. Livestock.   2. Poultry.   I. Title: Modern livestock
& poultry production.   II. Title.

SF61.G5   2000
636—dc21   00-030332

NOTICE TO THE READER

Publisher does not warrant or guarantee any of the products described herein or perform any independent analysis in connection with any of the product information contained herein. Publisher does not assume, and expressly
disclaims, any obligation to obtain and include information other than that provided to it by the manufacturer.

The reader is expressly warned to consider and adopt all safety precautions that might be indicated by the activities herein and to avoid all potential hazards. By following the instructions contained herein, the reader willingly
assumes all risks in connection with such instructions.

The Publisher makes no representation or warranties of any kind, including but not limited to, the warranties of
fitness for particular purpose or merchantability, nor are any such representations implied with respect to the material set forth herein, and the publisher takes no responsibility with respect to such material. The publisher shall
not be liable for any special, consequential, or exemplary damages resulting, in whole or part, from the readers'
use of, or reliance upon, this material.

Contents

v

Preface

Modern Livestock & Poultry Production, Sixth Edition, is designed for vocational-technical students who require competency in all phases and types of livestock production. Its comprehensive, balanced development emphasizes readability, organization, and hands-on activities. The text is based on the most up-to-date information available and is applicable to all areas of the United States.

Section 1 includes a general introduction to the livestock industry—its history, the careers available, and the importance of safety and environmental considerations. Section 2 introduces the student to the topics of feeding and nutrition, while Section 3 provides a sound basis for the understanding and practice of animal breeding. Sections 4 through 11 detail the production of beef cattle, swine, sheep and goats, horses, poultry, rabbits, bison, llamas, alpacas, and ratites. Each section presents information on selection of the stock; feeding, management, and housing; diseases and parasites; and marketing. The section on horses also includes a unit on training and horsemanship.

Every effort has been made to thoroughly update this very successful textbook. The sixth edition of *Modern Livestock & Poultry Production* adds material about raising llamas, alpacas, and ratites to Section 11, Alternate Animals, which already contained information on raising bison. Updates and revisions have been made throughout the text to reflect the most current information available.

The text is designed with the student in mind. Objectives at the beginning of each unit focus attention on the information and skills to be learned, which are then tested in the review section at the end of the unit. A bulleted format for important points, numerous subheads, and unit summaries facilitate the learning process. A list of suggested activities at the end of each unit provides the student and teacher with ideas for the hands-on experiences that are essential for success in the livestock production field. Other outstanding features are the text's many photographs, line drawings, tables, and charts, many new in this edition; an extensive appendix with completely updated tables, a glossary, and a color section featuring scores of photographs to be used for breed identification.

The Instructor's Guide for *Modern Livestock & Poultry Production* provides a list of equipment for teaching the course, a list of topics for which audiovisual aids are available, and suggestions for obtaining information from local resources. A major goal, topics for discussion, and an objective test (with answers) are included for each unit.

ABOUT THE AUTHOR

James R. Gillespie, the author of this text, has extensive training and experience in the field of livestock production and agriculture education. He has received the BS and MS degrees in Agricultural Education from Iowa State University and also holds the degree of Educational Specialist, School Administration, from Western Illinois University. Mr. Gillespie has taught agricultural education at the high school and adult education levels. In addition to other agriculture-related positions, Mr. Gillespie has also been self-employed in farming. Now retired, he was employed by the Illinois State

Board of Education, Department of Adult, Vocational and Technical Education, Program Approval and Evaluation as Regional Vocational Administrator, Region II. Mr. Gillespie is a member of Phi Delta Kappa and the Honor Society of Phi Kappa Phi.

ACKNOWLEDGMENTS

The author and Delmar would like to thank those individuals who reviewed the manuscript and offered suggestions, feedback, and assistance. Their work is greatly appreciated.

Odon Russell
Paxton High School
Paxton, Florida

Cyrus Vernon
Yanceyville, North Carolina

A special thank you is also extended to the contributors of content and illustration.

Jim Arend—Blacksmith, Salem, NY

Mohair Council of America, San Angelo, TX

The Palomino Horse Association, Inc., Jefferson City, MO

United States Department of Agriculture, Veterinary Services, Albany, NY

United States Department of Commerce, Washington, DC

University of Idaho, College of Agriculture, Moscow, ID

Voice of the Tennessee Walking Horse, Louisburg, TN

Walking Horse Report, Louisburg, TN

White Horse Registry, Crabtree, OR

The Livestock Industry

Domestication and Importance of Livestock

Livestock production is an important part of farming in the United States. People are dependent upon livestock production for supplies of food and clothing. The production of livestock involves selection, breeding, feeding, care, and marketing of animals.

Success in raising livestock depends on many factors. Farmers must have knowledge, skill, and patience. They must use the results of research by animal scientists. Research in animal science is carried on by many state universities and the United States Department of Agriculture. Many of the commercial firms that supply and service the animal industry also do research in animal science.

DOMESTICATION OF ANIMALS

Our present civilization has its roots in the domestication of animals. To *domesticate* means to adapt the behavior of an animal to fit the needs of people. The domestication of animals began when early humans had contact with wild animals, which they hunted for food and skins. After a period of time these early humans began to confine some of these animals to ensure a steadier supply of food and clothing. These animals were bred in captivity to replace those that were used. Humans later learned to select animals with certain desirable characteristics to use for breeding purposes. As a result of selective breeding, identifiable breeds began to be developed that would breed true for those characteristics that were determined to be desirable. Before the human race learned to tame and raise animals, it was dependent on hunting and wild plants for food and clothing. With the domestication of animals came the beginnings of a more settled way of life. Domesticated animals supplied a surer source of food and clothing. A better food supply meant an increase in population. More people made it possible to divide the labor within the tribe. Some historians believe that the human race would never have become civilized without the domestication of animals.

Cattle

Modern cattle are descendants of *Bos taurus* and *Bos indicus*. *Bos taurus* are domestic cattle that came from either the Aurochs or the Celtic Shorthorn. The Aurochs were common in Europe. The Celtic Shorthorn were found in the British Isles. *Bos indicus* are the humped cattle found in tropical countries. They are more resistant to some diseases, parasites, and heat than are cattle that came from *Bos taurus*.

Cattle were probably tamed early in the Neolithic (New Stone) Age. The Neolithic Age occurred about 18,000 years ago. Early man used cattle for draft, meat, and milk. Cattle were also a measure of wealth. They are mentioned in records at least 4,000 years old. Various types of cattle were known at that time.

Selection and crossbreeding of cattle for different purposes began early in the history of agriculture. *Selection* means to identify and use for breeding purposes those animals with traits that are considered by the breeder to be desirable. *Crossbreeding* is the mating of animals of different breeds. (See Unit 12 for a discussion of breeding systems.) Selection and crossbreeding resulted in improvement of the animal and the development of different breeds. Most of the improvement has taken place since the middle of the eighteenth century.

Cattle are not native to the United States. Christopher Columbus brought cattle to the New World on his second voyage in 1493. More cattle were brought by Portuguese traders in 1553. The English were the first to bring large numbers of cattle to the United States when they founded the Jamestown colony in 1611. The Spanish type of longhorn cattle arrived in Mexico in 1521. They later spread throughout the western United States as they were brought to Christian missions built by the Spanish.

The number of cattle soon began to grow. Early pioneers took cattle with them as they moved westward. The major growth of the large cattle herds took place in the Great Plains states. This happened because plentiful grazing land was available in this area. The North Central states, which have good supplies of grain, became the main region for finishing cattle (feeding the cattle to market weights) (Figure 1-1).

FIGURE 1-1 Cattle in an Iowa feedlot. Many cattle are finished in Iowa feedlots because of the abundance of corn in this region. *Courtesy of Cobleskill Agricultural & Technical College. Photo by Steven M. Ennis.*

FIGURE 1-2 Illinois is one of the nation's leading hog-producing states. Much of the feed used in raising hogs is raised in the Corn Belt. *Photo by Jim Bodine.*

The early pioneers took dairy cattle as well as beef cattle with them as they moved westward. Before the 1850s most farms had at least one or two dairy cows that provided milk and butter for the family. During the last half of the nineteenth century, however, dairy herds began to increase in size as a market for dairy products began to develop in the growing cities of the United States.

The numbers of dairy cattle in the United States continued to increase until they reached a peak in the middle 1940s. Since that time, the dairy cattle population of the United States has been declining steadily. Milk is still produced in all of the 50 states, but dairying as a major enterprise on the farm tends to be concentrated in several states, as shown in Table 1-4, page 20.

Swine

American breeds of swine come from two wild stocks: the European wild boar (*Sus scrofa*) and the East Indian pig (*Sus vittatus*). Some wild types of piglike animals, which have never been tamed, still exist in certain parts of the world.

The first use of swine for food probably occurred in the Neolithic Age. The first people to tame swine were the Chinese. Written records show that this took place about 4900 B.C. Biblical references to swine occur as early as 1500 B.C. The keeping of swine in Great Britain is mentioned as early as 800 B.C.

Swine were brought to the New World by Columbus on his second voyage in 1493. More swine were brought later by Spanish explorers. The first major increase in the number of swine in the United States occurred in 1539 when the Spanish explorer Hernando DeSoto brought thirteen head of hogs to Florida. As DeSoto moved westward during his exploration, the hogs were taken along. Three years later, by the time the Spanish had reached the upper Mississippi valley, the number of swine had increased to 700 head.

English settlers also brought swine with them to America. The herds grew rapidly in size. Production was soon greater than the local need. Pork and lard were then exported in trade for other products. The main expansion of the swine industry occurred in the Corn Belt states, an area where feed for finishing hogs for market was available (Figure 1-2).

Sheep

Sheep were among the first animals tamed by the human race. They were first tamed during the early Neolithic Age. Sheep are shown on an early Egyptian sculpture dated about 4000 B.C. and are mentioned in the earlier passages of the Bible. Sheep bones have been found in caves and lake dwellings used by early people of Europe.

Wool fabrics have been found in the ruins of Swiss Lake villages. These date back to between 10,000 and 20,000 years ago. The Babylonians used wool for clothing around 4000 B.C. Sheep flocks were important in both Spain and England by 1000 A.D. and, by 1500 A.D., both countries were major sheep-producing areas.

The ancestry of sheep is not as well known as that of other domestic animals. There are more than 200 breeds of sheep in the world. All of the breeds are timid, defenseless, and the least intelligent of the tamed animals. These traits are the result of selection for herding in large bands.

Most present-day sheep probably came from the wild sheep called moufflons and the Asiatic urial. The wild, big-horned sheep of Asia are also the ancestors of some of the present-day breeds.

The only sheep native to North America are the Big Horn or Rocky Mountain sheep. The present-day domestic breeds came from sheep that were imported. Columbus brought sheep to the New World on his second voyage in 1493. Merino sheep were brought to Mexico by Cortez in 1519. British breeds of sheep were imported to Virginia in 1609. Sheep were used by the early colonists mainly for wool production. Importation of breeding stock continued and flock owners selected certain animals to breed that would improve the wool produced.

By 1810, the northeastern part of the United States was the sheep-producing center of the country. However, sheep production gradually moved westward. The number of sheep increased until, by 1840, there were 19 million head of sheep in the United States. The center of the sheep-producing industry moved west when inexpensive rangeland became available (Figure 1-3).

FIGURE 1-3 Sheep grazing on western rangeland convert roughage into wool and meat. *Courtesy of USDA.*

Goats

Goats were first tamed in the Neolithic Age. They may have been the first tamed animals in Western Asia. The goat is believed to be descended from the Pasang or Grecian Ibex. These are species of wild goats found in Asia Minor, Persia, and other countries. Other wild species, such as the Markhors and Tahrs, may be ancestors of some of today's domestic breeds of goats.

Goats are closely related to sheep. The following are the major differences between the two species:

- Sheep have stockier bodies than goats.
- Goats have shorter tails than sheep.
- Goat horns are long and grow upward, backward, and outward; sheep horns are spirally twisted.
- Male goats have beards; male sheep do not.
- Male goats give off a strong odor in the rutting (breeding) season; male sheep do not.
- Goats do not have scent glands in the face and feet; sheep do have these scent glands. [Scent glands are specialized organs that secrete odorous substances (pheromones) that attract females.]

FIGURE 1-4 Angora goats on Texas range. Texas is one of the leading states in Angora goat production. *Courtesy of Texas A&M University.*

- Goats are more intelligent than sheep and have a greater ability to fight and fend for themselves than do sheep.
- Goats can easily return to the wild state; sheep cannot.

Goat remains have been found in the Swiss Lake villages of the Neolithic Age. The Bible refers to the use of mohair from goats.

Early goat importations into the United States came from Switzerland. Records show that milk goats were brought to Virginia and the New England states by Captain John Smith and Lord Delaware. Angora goat flocks in the United States came from Turkey. Most of the increase in milk goat numbers has occurred since 1900. Milk goats are found all over the United States. Many are kept on small farms, and there are few large herds of milk goats. Production of mohair from Angora goats has also increased mainly in the twentieth century. Most Angora goats are found in the western states. Texas has the largest number of Angora goats in the United States (Figure 1-4).

Horses

The horse evolved from a tiny four-toed ancestor called Eohippus (dawn horse). Eohippus was about a foot high and lived in swamps about 58 million years ago. The descendants of Eohippus gradually grew in size. Changes in the feet and skeleton occurred. Eventually the animal was better adapted to the prairie than to the swamp. Eohippus was native to the North American continent. However, it had disappeared entirely before the white man discovered the New World. There were no horses present in the New World when Columbus made his voyages.

The first domestication of the horse seems to have been in Central Asia or Persia before 3000 B.C. The use of the horse spread from there into Europe. Babylonian records show the use of the horse by 2000 B.C. Egyptians were using horses by 1600 B.C. Horses were also used by the ancient Greeks and Romans. The Arabs did not begin to use horses until after about 600 A.D.

Columbus brought horses to the New World on his second voyage in 1493. In 1519, Cortez imported horses into Mexico. The first importation of horses into what is now the United States was by DeSoto in 1539. During DeSoto's explorations throughout the Southeast, many of these horses were left behind. Other Spanish explorers also brought horses with them. These horses spread throughout the western part of what is now the United States. Wild horses found on western ranges are the descendants of these tamed horses.

Saddle horses and draft horses were brought to the United States by the early colonists. For many years, oxen were used as the main draft animal of the colonists. (*Draft* animals are used for pulling loads.) Horses served mainly as pack animals and for riding. The early development of the horse in the United States was primarily associated with riding on plantations. Horse racing also developed into a sport during the 1700s and early 1800s. Heavier draft horses were brought into the colonies by Dutch, Puritan, and Quaker settlers. These horses were used mainly for work.

With the development of other sources of power, the use of draft horses on farms declined. Most horses in the United States today are used for riding and racing (Figure 1-5).

Poultry

Chickens were being raised by the Chinese about 1400 B.C. Poultry were domesticated in India by at least 1000 B.C. Chickens were also known in ancient Egypt. The use of Bantams in the Orient dates back at least 3,000 years.

FIGURE 1-5 Horseback riding is a major recreational activity in the United States today. *Courtesy of B and G Action Photography by Gary Parkin.*

Although poultry and eggs were used for food early in history, poultry raising has only recently become a major commercial enterprise. In the past, most poultry was raised on an individual family basis.

The wild jungle fowl of India (*Gallus gallus*) may have been the early ancestor of most tame chickens. Some other wild species may also have been involved in the development of chickens.

Turkeys are believed to be descended from two wild species. One is found in Mexico and the other in the United States. The turkey was probably tamed by the people originally living in America. Most of the American varieties were probably developed from the species found in the United States.

The wild mallard duck (*Anas boschas*) is thought to be the ancestor of all domestic breeds of ducks. Ducks were tamed at an early date. The Romans referred to ducks more than 2,000 years ago. China has probably raised ducks on a commercial basis longer than other parts of the world.

The goose was probably tamed shortly after the chicken. It was regarded as a sacred bird in Egypt 4,000 years ago. Geese were well distributed over all of Europe 2,000 years ago.

Poultry were brought to the New World by early explorers and colonists. There were poultry in the Jamestown settlement in 1607. Poultry shows, which began to be held around 1850, were important in the development of recognized breeds. In the second half of the nineteenth century, older breeds were perfected and new breeds were developed. The American Poultry Association was formed in 1873. The *American Standard of Perfection* was first published in 1874. The purpose of the Association and the *Standard of Perfection* was to standardize breeds for purposes of showing.

The American poultry industry grew out of the small home flocks raised by early settlers. For a long time, poultry raising was mainly a small enterprise on the farm. However, as the population

FIGURE 1-6 Today, most eggs are produced by caged layers in large confinement flocks. *Courtesy of University of Illinois at Urbana.*

grew, the demand for poultry products increased. Poultry production began to be more specialized in the late 1800s and today is a highly specialized industry. Few small farm flocks remain. Most poultry is raised in large confinement flocks. Much of the poultry industry is concentrated in the southern part of the United States (Figure 1-6).

CLASSIFICATION OF COMMON FARM ANIMALS

A great many species of plants and animals exist on the earth. It is necessary to have a logical system to classify these different species in order to study and describe them. In the past, all living things were divided into two large groups called kingdoms—Plantae (plants) and Animalia (animal). Some forms of life (bacteria, fungi, and protozoa) do not fit easily into either of these kingdoms, however, so other systems of grouping have been developed. Some systems use three large kingdoms; some use four; and some use five. Details about the various grouping systems may be found in most biology textbooks.

A series of terms is used for the scientific name of a member of a kingdom. This is called a hierarchy of categories. Living things are classified as follows: kingdom, phylum (or division), class, order, family, genus, species. Each division of the hierarchy contains one or more groups from the next lower division. For example, a genus contains a group of related species; a family contains a group of related genera; an order contains a group of related families; and so on. This method of grouping continues up the hierarchy so that a kingdom contains all the groups listed below it. Major groups may be divided into subphyla, subclasses, suborders, subfamilies, or subgenera if necessary. Species may also be subdivided into varieties, breeds, or strains for further grouping.

A species is a group of animals with many common traits. When members of this group mate within the group, they will produce fertile young.

In all systems of grouping, one of the large kingdoms is called Animalia. All of the domestic animals found on the farm are included in this kingdom. Domestic farm animals are further classified as phylum Chordata, subphylum Vertebrata, and either class Mammalia (for livestock) or class Aves (for poultry).

Members of class Mammalia, or mammals, share the following characteristics: They have four-chambered hearts; they are warm-blooded, or *homeothermic* (they can maintain their own body temperature). They have a diaphragm that separates the thoracic, or chest, and abdominal cavities; this helps them breathe more efficiently. Their bodies are covered with hair, which provides insulation. The embryo develops in the mother's uterus, and the young are born alive. The mother secretes milk for the young through the mammary glands. There are other traits of the class Mammalia, but these are some of the more important ones shared by farm animals. Common domesticated animals that belong to this class include cattle, sheep, goats, horses, donkeys, swine, bison, alpacas, llamas, and rabbits.

Members of class Aves also have four-chambered hearts and are homeothermic. Instead of hair, their bodies are covered with feathers, which provide insulation. They do not have a diaphragm; instead they have light, hollow bones and an air-sac system attached to the lungs. They lay eggs in which their young develop. They do not secrete milk to nourish the young. These animals have beaks and gizzards instead of teeth. Chickens, ducks, emus, geese, ostrich, rheas, and turkeys all belong to this class.

The scientific name of an animal includes its genus and species names. The genus name is capitalized; the species name is not. In written material, the genus and species names are underlined or italicized.

The classifications of common farm animals are listed in Table 1-1.

TABLE 1-1 Classification of Common Farm Animals.

Common Name	Phylum	Subphylum	Class	Order	Family	Genus	Species
Alpaca	Chordata	Vertebrata	Mammalia	Artiodactyla	Camelidae	*Llama*	*pacos*
Bison	Chordata	Vertebrata	Mammalia	Artiodactyla	Bovidae	*Bison*	*bison*
Cattle	Chordata	Vertebrata	Mammalia	Artiodactyla	Bovidae	*Bos*	*taurus* (most of the domestic breeds)
Cattle	Chordata	Vertebrata	Mammalia	Artiodactyla	Bovidae	*Bos*	*indicus* (humped cattle)
Chicken	Chordata	Vertebrata	Aves	Galliformes	Phasianidae	*Gallus*	*domesticus*
Donkey	Chordata	Vertebrata	Mammalia	Perissodactyla	Equidae	*Equus*	*asinus*
Duck	Chordata	Vertebrata	Aves	Anseriformes	Anatidae	*Anas*	*platyrhyncha*
Emu	Chordata	Vertebrata	Aves	Casuariiformes	Dromiceidae	*Dromiceius*	*novaehollandiae*
Goat	Chordata	Vertebrata	Mammalia	Artiodactyla	Bovidae	*Capra*	*hircus*
Goose	Chordata	Vertebrata	Aves	Anseriformes	Anatidae	*Anser*	*anser*
Horse	Chordata	Vertebrata	Mammalia	Perissodactyla	Equidae	*Equus*	*caballus*
Llama	Chordata	Vertebrata	Mammalia	Artiodactyla	Camelidae	*Llama*	*glama*
Ostrich	Chordata	Vertebrata	Aves	Struthioniformes	Struthionidae	*Struthia*	*camelus*
Rabbit	Chordata	Vertebrata	Mammalia	Lagomorpha	Leporidae	*Oryctolagus*	*cuniculus*
Rhea	Chordata	Vertebrata	Aves	Rheiformes	Rheidae	*Rhea*	*americana*
Sheep	Chordata	Vertebrata	Mammalia	Artiodactyla	Bovidae	*Ovis*	*aries*
Swine	Chordata	Vertebrata	Mammalia	Artiodactyla	Suidae	*Sus*	*scrofa* (evolved from wild hog of Europe)
Swine	Chordata	Vertebrata	Mammalia	Artiodactyla	Suidae	*Sus*	*vittatus* (evolved from wild hog of East India)
Turkey	Chordata	Vertebrata	Aves	Galliformes	Meleagrididae	*Meleagris*	*gallopavo*

FUNCTIONS OF ANIMALS

Some functions of livestock benefit all of society. Other functions are important mainly to individual farms. Taken together, the functions of livestock are a vital part of the total agriculture of a nation. It is useful to understand the functions of livestock when selecting enterprises for a farm. Selecting enterprises involves the choice of what kind of livestock is to be raised. Some functions of livestock, such as the conversion of roughage into food, are factors in this management decision.

Converting Feed into Food

Livestock convert feed grains and roughages into food for human consumption. There is some controversy over the use of feed grains as livestock feed. In the face of world food shortages, it has been suggested that this is not the most efficient use of limited resources.

Nonruminant animals such as swine and poultry are fed large amounts of grain because they cannot use much roughage in their diet. However, about 30 percent of the feed fed to swine and poultry in the United States consists of fish meal, meat and bone meal, milling and fermentation by-products, and tankage. These are feeds that generally cannot be used directly by humans for food.

Ruminants are animals of the suborder Ruminantia of the order Artiodactyla that have a stomach that is divided into several compartments. These animals regurgitate and masticate their feed after they swallow it. Animals in the subdivision Tylopoda have a three-compartment stomach. Typical animals found in this subdivision include camels, llamas, and alpacas. Animals in the subdivision Pecora have a four-compartment stomach. The Pecora are referred to as true ruminants. Typical animals found in this subdivision include cattle, bison, sheep, and goats. The largest of the compartments in true ruminants, the *rumen*, contains microorganisms that allow ruminants to digest many kinds of feed that nonruminant animals cannot use effectively. Antelope, deer, gazelles, and giraffes are examples of other animals in the subdivision Pecora.

Ruminants are important because they have the ability to convert large quantities of materials that cannot be used directly for human food into human food. Almost half of the chemical energy in the major cereal crops such as corn, wheat, and rice is found in parts of the plant, such as the stems, which are not used by humans for food. These crop residues can be converted to human food by ruminants. Waste products from a number of agricultural industries can be used as feed for ruminants. Examples include waste products from fruit and vegetable farming, citrus processing, sugar manufacturing, milling, and cotton ginning. Wood chips, sawdust, and shredded newspaper can also be used as feed for ruminants.

Hay, silage, and pasture is raised on about one third of the cropland in the United States. Approximately 50 percent of the total land area of the continental United States consists of native and natural grasslands and forest lands that produce vegetation that can be grazed by livestock. Forests and rangeland cover a large percent of the land masses of the planet. Only about 11 percent of the world's land area is suitable for the production of foods that can be used directly by humans. About 75 percent of the total energy intake of beef and dairy cattle in the United States consists of roughages and other waste materials that cannot be used directly for human food. The ability of ruminants to convert much of what would not otherwise be available to humans as food adds significantly to the world's total food supply.

In nations with limited grain supplies, ruminants are fed largely on roughages. In the United States and some other nations of the world where grain is plentiful and relatively inexpensive, ruminants are fed rations high in grain during the finishing period. However, during the total lifetime of a beef animal, about 80 percent of the total feed used comes from roughages. Generally, these roughages cannot be used by humans directly for food.

About 80 percent of the human population of the world gets most of its protein, fats (lipids), iron, niacin, and some vitamins (including vitamin B_{12}) from the meat produced by ruminant animals. Because of the kind of land and/or climate where they live, about 14 percent of these people have no other practical source of these nutrients. Food products from ruminants provide about 45 percent of the protein, 32 percent of the fat, 50 percent of the phosphorus, and 77 percent of the calcium found in diets of people living in the United States. About one third of the total amount of food eaten by people in the United States comes from ruminants.

Animal products (ruminant and nonruminant combined) are important sources of nutrients in the average diet in the United States. They provide 35 percent of the energy, 68 percent of the protein, 78 percent of the calcium, 39 percent of the iron, 42 percent of the vitamin A, and 37 to 98 percent of various B vitamins consumed by people in the United States.

The most important livestock sources of protein and energy for human consumption are swine, beef, poultry, and sheep. The efficiency of the major farm animals in converting feed into protein and energy for human use is shown in Figure 1-7.

Clothing

Livestock provide fiber and skins for the production of clothing. The demand for animal fibers for clothing is lower now because of the increased use of synthetic fibers for clothing. However, the use of wool in the United States has been almost constant for the past ten years

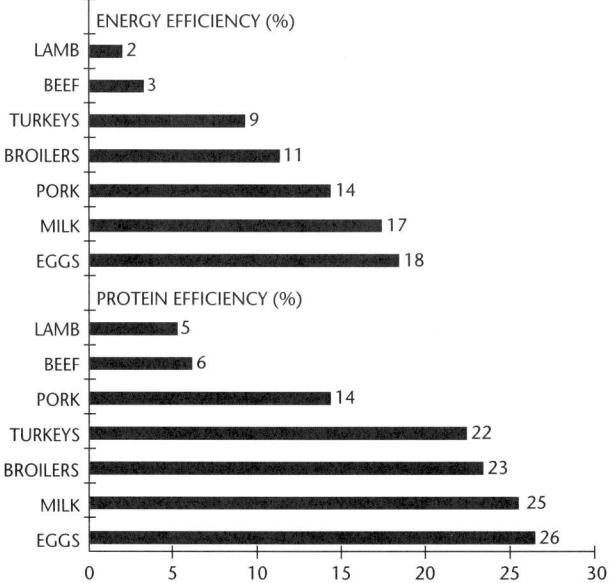

FIGURE 1-7 A comparison of the efficiency of major farm livestock in converting feed calorie intake to food calorie output (energy efficiency) and converting crude protein in feed into edible protein in the form of meat, milk, and eggs (protein efficiency). *Sources:* Vocational Agricultural Service, University of Illinois and National Academy of Sciences.

(Figure 1-8). Most synthetic fibers are oil-based and the price of oil has risen dramatically in recent years; this means that animal fibers continue to be an important resource in human society.

Leather is used for shoes, belts, gloves, and clothing, as well as for other products used by humans. From 5 to 10 percent of the market value of animals comes from the sale of hides. Leather has some characteristics that make it superior to synthetics for the production of clothing. It can allow air to pass through, is more durable, and is warmer than clothing made from synthetics.

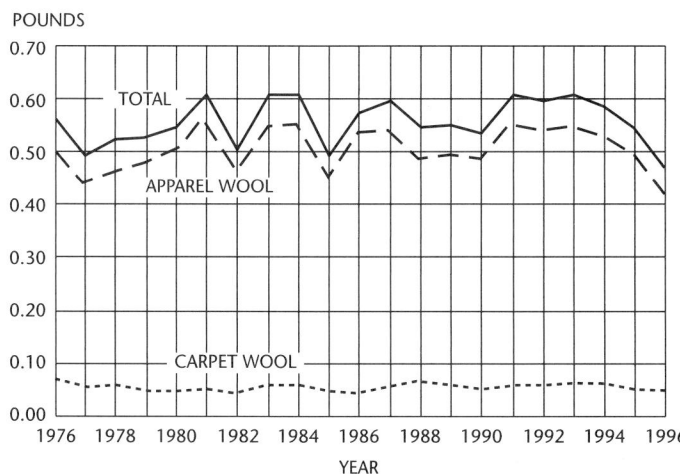

FIGURE 1-8 Consumption per Person for Wool, United States, 1976–1996. *Source:* USDA.

Power

At one time, animals provided much of the power used by the human race. With the development of other sources of power, this use has declined. Very little animal power is now used in the United States. However, in some parts of the world, animals still provide much of the power used by humans.

Recreation

Horseback riding is a major source of recreation for many people today. Racing is also a popular sport. Livestock shows and fairs provide recreation for many people, both as exhibitors and spectators.

Conservation

Livestock help to conserve soil and soil fertility. The grasses and legumes that are used for livestock feed are soil-conserving crops. They form protective covers on the land and help to prevent wind and water erosion. Nutrients are removed from the soil by the crops being grown. When these crops are fed to livestock, about 80 percent of the nutrient value is excreted in the manure. However, by putting the manure back on the soil, the rate of loss of soil fertility can be decreased. The fertilizer value of animal manure and methods of handling manure are discussed in Unit 4.

Animal manure can also be used as a fuel source. In many parts of the world, dried animal manure is burned as a fuel for cooking and to heat homes. Approximately 8 to 12 percent of the world's population depends on dried manure as a fuel source.

Animal manures can be used as a raw material in methane gas digesters. The use of methane gas converters has increased worldwide as a result of the energy crisis brought about by higher oil prices. Fuel for the electricity, cooking, and heating needs of an average United States farm could be supplied by the manure from about 40 cows. Some large farms and feedlots in the United States have built bio-gas plants to utilize the animal manure produced.

Higher feed costs have led to research on the possibility of using animal manures as a supplement in feeds. Recycling animal manure for use as a feed supplement is still in the experimental stage.

Stabilize Farm Economy

Livestock help bring stability to the farm business. Raising livestock makes good use of the resources already available to farmers—land, labor, capital, and management ability—and can increase the farming income. Including livestock in a farm business helps to spread the risks involved in farming over more enterprises. Thus, the farmer is not dependent on only one or two sources of income. In addition, both labor and income are spread more evenly throughout the year.

Concentrate Bulky Feeds

It costs a lot of money to transport bulky feeds such as hay. Livestock convert these bulky feeds into a more concentrated form. This reduces transportation costs to market, a great advantage for farms that produce large amounts of bulky feeds.

By-products

Meat, wool, and leather are not the only products that come from animals. Any product from the animal carcass, other than meat, is called a *by-product*. Thus, wool and leather are by-products from the slaughter of animals. Many other products come from the animal carcass. These include fat, bone, intestine, brain, stomach, blood, and various glands. These by-products are used in the manufacture of many products.

Edible by-products that come from animals include variety meats such as brains, tongue, kidneys, and heart. Oleo stearine, which comes from the fat in the animal carcass, is used in the manufacture of candy and gum. Hooves, horns, bones, and hides produce gelatin, which is used in the production of gelatin desserts, marshmallows, canned meats, and ice cream.

Hides used in the manufacture of leather goods are the most important of the inedible by-products that come from animals. Examples of leather goods made from animal hides include clothing, belts, shoes, purses, furniture, drum heads, and sports equipment.

The inedible fats are used in the production of cosmetics, waxes, soap, lubricants, and printing ink. Bones, horns, and hooves are also used in the production of glue, buttons, bone china, camera film, sandpaper, dice, piano keys, wallpaper, and toothbrushes. Hair from animal hides is used in making brushes, rug padding, house insulation, and upholstering materials for furniture. Artists' paintbrushes are made from the fine hair found in the ears of animals.

Feeds for livestock are made from animal by-products. These include blood meal and meat and bone scraps.

More than 100 drugs used by humans for medical purposes are made from animal by-products. *Insulin* is extracted from the pancreas of animals and is used in the treatment of diabetes. *Cortisone* comes from the adrenal glands and is used for the treatment of rheumatoid arthritis, adrenal insufficiency, some allergies, diseases of the connective tissues, and gout. *Thrombin* comes from the blood of animals and is a coagulant used in surgery to help make blood clot. It is also used in skin-graft operations and for the treatment of ulcers. *Heparin* comes from the lungs and is used to prevent blood clotting during operations. It also helps prevent heart attacks. *Epinephrine* comes from the adrenal glands. It is used for the treatment of some kinds of allergies and to help relieve the symptoms of hay fever and asthma. *Rennet* comes from the stomachs of cattle and is used in cheese making. It also helps babies digest milk. *Corticotropin* (ACTH) comes from the pituitary glands in the brain. It is used for the treatment of some breathing problems, severe allergies, mononucleosis, and leukemia.

By-products are also used in the manufacture of perfumes, fertilizers, candles, lanolin, and glycerine, as well as many other products. Animal by-products make valuable contributions to society.

CONSUMPTION OF LIVESTOCK PRODUCTS

The per capita consumption of major livestock products is shown in Figures 1-9A, 1-9B, 1-9C, 1-9D, and 1-9E. The long-term downward trend in consumption of beef has slowed significantly during the past several years. However, the downward trend in the consumption of all red meat has continued, while the consumption of chicken and turkey has increased. Consumption of eggs has begun to trend slightly upward in recent years and the consumption of dairy products has continued to decline.

Producer organizations have conducted marketing campaigns financed by check-off programs based on producer sales of animals and milk to attempt to increase market share for their particular product. An increased emphasis on producing products that are perceived by the prospective consumer as being safe and healthy has contributed to the slower decline in demand for livestock products. These promotional programs are discussed in more detail in the units related to the marketing of specific livestock products.

Eating habits have also been influenced by more food consumption away from home and an increase in demand for meals that are quick and easy to fix in the home. Marketing methods and price

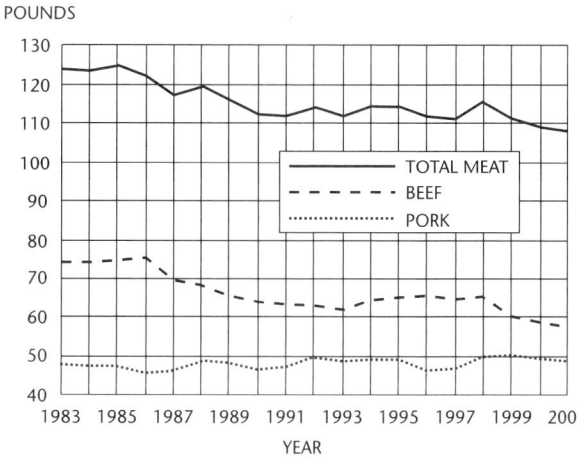

FIGURE 1-9A Consumption (boneless weight) per person for Beef, Pork, and Total Meat, United States, 1983–2001. (Total Meat includes Beef, Pork, Veal, Lamb, and Mutton.) *Sources: Agricultural Statistics,* USDA, 1992, 1998; *USDA Agricultural Baseline Projections to 2008,* USDA, 1999.

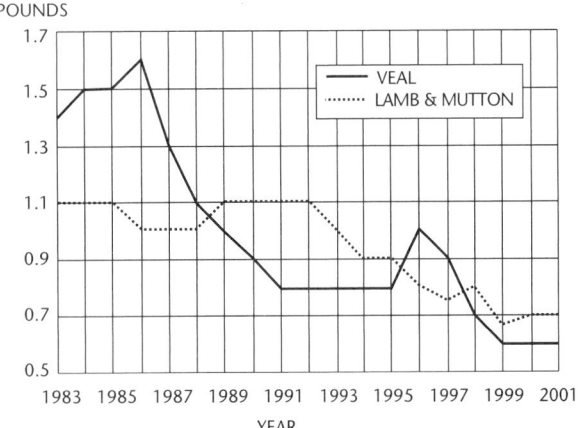

FIGURE 1-9B   Consumption (boneless weight) per person for Veal, Lamb, and Mutton, United States, 1983–2001. *Sources: Agricultural Statistics, USDA, 1992, 1998; USDA Agricultural Baseline Projections to 2008, USDA, 1999.*

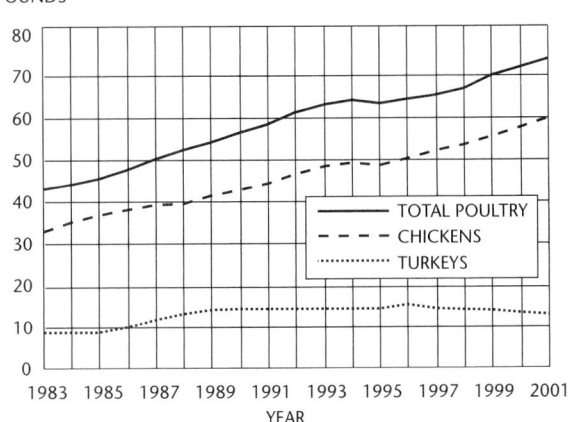

FIGURE 1-9C   Consumption (boneless weight) per person for Chicken, Turkey, and Total Poultry, United States, 1983–2001. *Sources: Agricultural Statistics, USDA, 1992, 1998; USDA Agricultural Baseline Projections to 2008, USDA, 1999.*

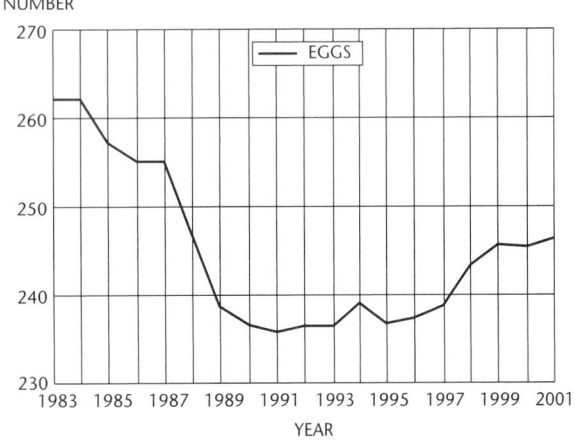

FIGURE 1-9D   Consumption per person for Eggs, United States, 1983–2001. *Sources: Agricultural Statistics, USDA, 1992, 1998; USDA Agricultural Baseline Projections to 2008, USDA, 1999.*

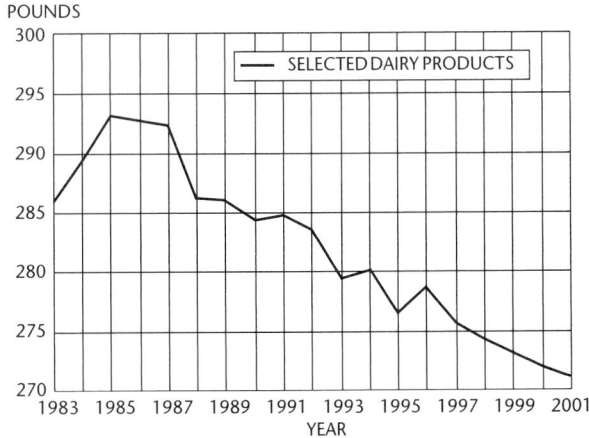

FIGURE 1-9E   Consumption per person for selected Dairy Products, United States, 1983–2001. (Includes Fluid Milk and Cream, Butter, Cheese, Condensed and Evaporated Milk, Ice Cream, Dry Whole Milk, and Non-Fat Dry Milk. Data 1997–2001 projected.) *Sources: Agricultural Statistics, USDA, 1992, 1998.*

competition have also influenced consumer demand for various livestock products.

Consumers are concerned about food safety and nutrition. Food safety concerns are discussed later in this unit. Nutritional concerns about food content include:

- Cholesterol level
- Fat content
- Salt content
- Food additives
- Sugar content
- Artificial coloring

Milk and animal fats are a source of cholesterol in the diet. The use of animal fats has declined significantly since 1967. During this same period, the use of vegetable fats, which are seen as having less cholesterol, has risen rapidly. The per capita consumption of cholesterol in the diet of people in the United States is declining. This decline is related to the decrease in per capita consumption of livestock products that are high in cholesterol.

Producer groups such as the National Pork Producers Council, National Cattlemen's Beef Association, and the Dairy Council, as well as others interested in livestock food products, are taking an active interest in research, promotion, and marketing these products. A greater effort is being made to coordinate meat production among producers, packers, and processors to meet consumer concerns. Grade and yield marketing of livestock helps reduce the amount of fat in meat products by paying the producer a better price for leaner carcasses. Research is being conducted in genetics to develop livestock that produce less fat. The red meat industry is attempting to maintain its share of the market by developing products that are priced competitively and meet the needs of the consumer. The poultry industry is also actively promoting and marketing products both to meet consumer needs and to address concerns related to health and the safety of food products.

SIZE AND SCOPE OF THE LIVESTOCK INDUSTRY IN THE UNITED STATES

Income and Costs

The total value of agricultural products sold from farms in the United States in 1996 was more than 202 billion dollars. The value of livestock and livestock products sold in 1996 was over 92 billion dollars, as shown in Table 1-2. This is 46 percent of the total value of agricultural products sold in 1996. Changes in farm income from livestock and crops for the years 1987 to 1996 are shown in the table.

In 1996, farmers spent over 181 billion dollars for total production expenses. This amount is for both livestock and crop operations (Table 1-3). Feed purchases represented 13.9 percent of this total and livestock purchases were 6.1 percent of the total. The changes in farm production costs from 1987 to 1996 can be seen in the table.

TABLE 1-2 Farm Income: Cash Receipts from Farm Marketings, by Commodities or Commodity Groups, United States, 1987–1996.[1]

Year	Cattle and Calves Million dollars	Hogs Million dollars	Sheep and Lambs Million dollars	Dairy Products Million dollars	Broilers Million dollars	Farm Chickens Million dollars	Chicken Eggs Million dollars	Turkeys Million dollars	Ducks Million dollars	Other Poultry Million dollars	Misc. Other Livestock Million dollars	Total Livestock and Products Million dollars
1987	33,583	10,337	558	17,727	6,177	112	3,208	1,703	11	304	2,276	75,996
1988	36,958	9,221	522	17,632	7,435	95	3,067	1,951	9	311	2,439	79,640
1989	36,429	9,770	487	19,357	8,778	138	3,862	2,235	9	356	2,497	83,918
1990	39,302	11,525	414	20,153	8,365	90	4,010	2,393	8	422	2,537	89,220
1991	38,697	11,036	399	18,007	8,383	67	3,901	2,353	8	441	2,494	85,786
1992	37,272	10,017	460	19,736	9,177	83	3,384	2,397	9	474	2,629	85,637
1993	39,362	10,911	551	19,243	10,416	96	3,779	2,510	8	516	2,779	90,170
1994	36,395	9,883	507	19,935	11,372	78	3,780	2,644	9	563	2,995	88,160
1995	33,983	10,264	559	19,894	11,762	61	3,880	2,882	10	534	3,214	87,042
1996	31,138	12,644	601	22,834	13,906	61	4,757	3,056	11	535	3,371	92,914

Year	Food Grains Million dollars	Feed Crops Million dollars	Cotton Million dollars	Tobacco Million dollars	Oil Crops Million dollars	Vegetables Million dollars	Fruits/ Nuts Million dollars	All Other Crops Million dollars	Total Crops Million dollars
1987	5,790	14,635	4,189	1,816	11,283	9,891	8,056	10,141	65,800
1988	7,469	14,281	4,525	2,069	13,501	9,792	9,032	10,935	71,603
1989	8,247	17,049	5,026	2,410	11,866	11,562	9,151	11,582	76,892
1990	7,480	18,669	5,488	2,733	12,258	11,464	9,416	12,789	80,297
1991	7,325	19,327	5,236	2,881	12,698	11,625	9,923	13,062	82,077
1992	8,467	20,099	5,192	2,958	13,286	11,851	10,179	13,712	85,744
1993	8,180	20,211	5,250	2,948	13,220	13,435	10,284	13,953	87,480
1994	9,545	20,351	6,738	2,656	14,657	13,902	10,335	14,895	93,079
1995	10,417	24,282	6,851	2,548	15,466	14,891	11,074	15,170	100,700
1996	11,550	28,114	7,461	2,796	17,756	14,349	11,714	15,686	109,425

[1]USDA estimates and publishes individual cash receipt values only for major commodities and major producing states. The U.S. receipts for individual commodities, computed as the sum of the reported states, may understate the value of sales for some commodities, with the balance included in the appropriate category labeled "other" or "miscellaneous." The degree of underestimation in some of the minor commodities can be substantial.

Sources: Agricultural Statistics, USDA, 1998.

Leading States in Livestock Production

Table 1-4 shows the ten leading states in several categories of livestock and livestock products production. It also shows the ten leading states in cash receipts from livestock. The ranking of the states shown in the table may change slightly from year to year. However, the states listed tend to be those that are the major producers year after year of the livestock or livestock product category shown.

TABLE 1-3 Expenses: Farm Production Expenses, United States, 1987–2001.

Year	Feed Purchased	Livestock and Poultry Purchased	Seed Purchased	Fertilizer and Lime Purchased	Pesticides Purchased	Fuel and Oil Purchased	Other[1]	Interest	Contract and Hired Labor Expenses	Net Rent to Non-operator Landlord[2]	Capital Consumption	Property Taxes	Total Production Expense
	Million dollars	Million dollars	Million dollars	Million dollars	Million dollars	Million dollars	Million dollars	Million dollars	Million dollars	Million dollars	Million dollars	Million dollars	Million dollars
1987	17,463.2	11,832.1	3,258.9	6,452.5	4,512.2	4,956.6	27,180.7	14,971.9	9,975.5	8,202.9	17,218.5	4,967.9	130,992.9
1988	20,246.2	13,035.8	4,059.5	7,677.8	4,147.7	4,800.0	29,576.8	14,293.9	10,906.7	8,383.9	17,606.2	5,172.9	139,907.4
1989	20,743.6	12,935.1	4,397.2	8,173.9	5,011.5	4,771.5	31,616.5	13,934.9	12,028.8	9,427.4	18,117.5	5,504.5	146,662.4
1990	20,387.8	14,641.5	4,518.9	8,206.3	5,363.2	5,789.7	32,789.8	13,437.2	14,113.4	10,051.7	18,128.5	5,861.9	153,289.9
1991	19,332.8	14,129.1	5,133.4	8,666.1	6,320.5	5,607.5	34,160.6	12,119.1	13,900.0	9,924.4	18,184.2	5,814.7	153,272.4
1992	20,133.0	13,574.2	4,913.4	8,330.7	6,470.6	5,298.4	33,464.8	11,137.5	13,999.7	11,187.5	18,309.5	6,117.0	152,936.3
1993	21,431.2	14,597.3	5,165.0	8,397.5	6,723.3	5,349.8	37,492.0	10,821.5	15,006.3	11,009.1	18,337.8	6,177.0	160,547.8
1994	22,631.2	13,270.1	5,375.7	9,179.7	7,225.0	5,312.0	40,521.4	11,735.2	15,308.6	11,719.9	18,688.2	6,489.9	167,456.9
1995	23,829.3	12,335.3	5,463.3	10,003.0	7,726.5	5,447.7	42,668.7	12,726.4	16,315.8	11,984.0	18,914.3	6,717.2	174,161.5
1996	25,234.5	11,148.1	6,112.1	10,934.2	8,525.1	5,736.3	42,996.3	13,218.3	17,347.9	14,293.1	18,929.5	6,827.8	181,303.2

[1]Includes electricity, repair and maintenance, machine hire and customwork, marketing, storage and transportation, and miscellaneous expenses.
[2]Includes landlord capital consumption.

Source: Agricultural Statistics, USDA, 1998.

TABLE 1-4 Leading States in Livestock and Livestock Product Production.

Beef Cattle and Calves	Fed Cattle Marketed	Swine	Dairy Cattle	Sheep and Lambs	Wool
Texas	Texas	Iowa	California	Texas	Texas
Nebraska	Nebraska	North Carolina	Wisconsin	California	Wyoming
Kansas	Kansas	Minnesota	New York	Wyoming	California
Oklahoma	Colorado	Illinois	Pennsylvania	Colorado	Montana
Missouri	Iowa	Indiana	Minnesota	South Dakota	Colorado
South Dakota	Oklahoma	Nebraska	Texas	Utah	Utah
Iowa	California	Missouri	Michigan	Montana	South Dakota
Colorado	South Dakota	Oklahoma	Idaho	New Mexico	New Mexico
California	Idaho	Ohio	Ohio	Idaho	Idaho
Montana	Minnesota	Kansas	Washington	Iowa	Iowa

Sources: *Agricultural Statistics*, USDA, 1999.
           Cattle, USDA, January 1999.
           Hogs and Pigs, USDA, December 1998.
           Layers and Egg Production, USDA, January 1999.
           Sheep and Goats, USDA, January 1999.
           Wool and Mohair, USDA, March 1999.

Number of Livestock on Farms

Over a period of years livestock and poultry numbers respond to economic conditions and also reflect changes in consumer demand. Table 1-5 shows the trend in livestock numbers in the United States for the period 1989–1999. The increase in numbers for broilers and turkeys reflects an increased consumer demand for food perceived as having a lower cholesterol level. Sheep and goat production have shown the greatest decline in numbers, in part because of lower consumer demand for products from these animals. Beef and swine numbers tend to move in cycles as herds expand and contract primarily in response to economic conditions; however, there has been some weakening in demand for red meat as consumers switch to poultry products that are perceived as being lower in cholesterol.

ANIMAL HEALTH PRODUCTS

The development and sale of animal health products is a growing segment of the livestock industry. The three categories of products included are feed additives, biologicals, and pharmaceuticals. Feed additives are those products used in livestock and poultry production to control or prevent disease, enhance growth, or improve feed efficiency. Biologicals include vaccines, bacterins, and antitoxins. Pharmaceuticals include medicines used in disease control and prevention.

TABLE 1-4 *(Continued)*

Chickens (excluding commercial broilers)	Eggs	Broilers	Turkeys	Goats/ Mohair Production	Leading States in Cash Receipts from Livestock
Ohio	Ohio	Georgia	North Carolina	Texas	Texas
California	California	Arkansas	Minnesota	New Mexico	California
Iowa	Pennsylvania	Alabama	Arkansas	Arizona	Nebraska
Pennsylvania	Iowa	Mississippi	Virginia	Oklahoma	Iowa
Indiana	Indiana	North Carolina	Missouri		Kansas
Georgia	Georgia	Texas	California		North Carolina
Texas	Texas	Maryland	Indiana		Wisconsin
Arkansas	Arkansas	Virginia	South Carolina		Minnesota
Minnesota	Minnesota	Delaware	Pennsylvania		Georgia
North Carolina	Nebraska	Missouri	Ohio		Arkansas

The sale of animal health products is big business, amounting to more than three billion dollars per year. Of this amount, pharmaceutical sales account for a little more than two billion dollars, biologicals total a little more than four hundred million dollars, and feed additives total a little more than six hundred million dollars.

Companies that produce animal health products spend millions of dollars on research and development of new products. Additional millions are spent to defend the continued use of products already approved.

TRENDS IN ANIMAL AGRICULTURE

Consumption and Production

The changes in lifestyle and eating habits of people in the United States will continue to have a significant impact on the production, processing, and marketing of meat, poultry, and dairy products in the twenty-first century. The trend of eating meals in restaurants and fast food businesses will continue to increase. When food is consumed away from home, more of the consumer's expenditure goes toward marketing costs as compared to marketing costs of food consumed at home.

As discussed earlier in this unit, the concern over cholesterol levels in food products has had an impact on consumption patterns for

TABLE 1-5 Trends in Livestock Numbers on Farms, United States, 1989–1999.

	Numbers[a]										
	1989	1990	1991	1992	1993	1994	1995	1996	1997	1998	1999
	Thous.	Thous.	Thous.	Thous.	Thous.	Thous.	Thous.	Thous.	Thous.	Thous.	Thous.
Total all cattle	96,741	95,817	96,394	97,556	99,177	100,988	102,756	103,488	101,656	99,745	98,521
Beef cows	32,488	32,455	32,520	33,007	33,365	34,650	35,156	35,228	34,458	33,885	33,472
Dairy cows	10,138	10,015	9,966	9,728	9,658	9,528	9,487	9,416	9,318	9,199	9,143
Beef heifers	5,325	5,283	5,443	5,643	6,092	6,365	6,475	6,179	6,042	5,764	5,550
Dairy heifers	4,117	4,171	4,093	4,131	4,176	4,144	4,141	4,104	4,058	3,986	4,060
Other heifers	7,631	7,803	8,102	8,048	8,550	9,068	9,275	9,949	10,212	10,051	9,994
Steers	15,431	15,512	15,967	16,424	16,940	17,042	17,463	17,732	17,392	17,189	16,836
Bulls	2,150	2,160	2,196	2,239	2,278	2,307	2,390	2,392	2,350	2,270	2,276
Calves under 500 lbs (227 kg)	19,461	18,418	18,107	18,336	18,118	17,884	18,369	18,488	17,826	17,401	17,190
Hogs[b]	53,788	54,416	57,649	58,202	57,904	59,990	58,264	56,141	61,158	62,206	NA
Sheep and lambs[c]	10,853	11,358	11,174	10,797	10,201	9,742	8,886	8,461	8,024	7,825	7,238
Chickens (for egg production)[d]	357,241	353,179	363,594	371,483	379,640	383,829	384,622	386,974	403,495	NA	NA
Chickens (broiler production)[e]	5,516,521	5,864,150	6,137,150	6,402,490	6,694,310	7,017,540	7,325,670	7,596,760	7,764,200	NA	NA
Turkeys[f]	261,394	282,475	284,910	289,880	287,650	286,585	292,356	302,713	301,251	283,503	NA
Goats (clipped for mohair)[g,h]	2,467	2,000	2,140	1,860	1,970	2,000	1,900	1,600	1,470	1,070	NA

[a]January 1 each year for cattle and sheep.
[b]December 1 each year.
[c]Beginning in 1994 includes new crop lambs.
[d]December 1 each year; does not include commercial broilers.
[e]Broilers are young chickens of the meat-type breeds, raised for the purpose of meat production. These figures are not included in farm production of chickens. Estimates cover the 12-month period, from December 1 of the previous year through November 30. Excludes states that produced less than 500,000 broilers.
[f]Total poults hatched minus death loss of poults and young turkeys during the year.
[g]The number clipped is the sum of goats and kids clipped in the spring and kids clipped in the fall.
[h]1989 includes Texas plus New Mexico, Oklahoma, Arizona, and Michigan; starting in 1990 Texas only.
NA: Data not available.

Sources: *Agricultural Statistics,* USDA, 1998.
  Poultry Production and Value Final Estimates, USDA, 1994–1997.
  Cattle, USDA, January 1999.
  Hogs & Pigs, USDA, March 1999.
  Sheep and Goats, USDA, January 1999.
  Turkeys, USDA, January 1999.

animal products. Per capita consumption of red meats and eggs has decreased in recent years and this trend is expected to continue unless significant changes are made in the saturated fatty acid and cholesterol content of these products. Consumers are demanding leaner meat and less fat in meat and dairy products. The long-term trend of an increase in poultry consumption has resulted in a decrease primarily in beef consumption. It is anticipated that pork consumption will tend to remain fairly constant through the end of the twentieth century.

A trend toward increased efficiency in livestock production is necessary to lower the per unit cost of production. It is anticipated that this increased efficiency will result from larger operations that use more automation.

umber of farms in the United States that raise swine has
de   om approximately one million in the 1970s to about
?   s trend is expected to continue, with only about 5,000 to
s raising hogs by the year 2000. Because hog numbers are
remain about constant during this period, the size of
on will increase dramatically. The swine industry is cur-
g toward large, integrated production systems, much
industry did a few years ago.

n beef production continues to be toward large feed-
hese are expected to be increasingly located outside
hich in the past has fed large numbers of cattle be-
lity of corn. As feeding practices tend to put more
of roughages and other crop by-products for
vailability of grain for finishing cattle will de-

n dairy products currently far outstrips the demand for those pro   s. It will require a further reduction of about 20 percent in dairy cow numbers to more closely balance production with demand. The use of a growth hormone produced in large quantities through biotechnology has the potential to significantly increase milk production per cow. If the use of this technology becomes widespread in the dairy industry, the numbers of dairy cows needed to meet the demand for dairy products will be greatly reduced. It is expected that the size of dairy herds will continue to increase as the number of dairy farms decreases.

The increase in the per capita consumption of poultry is expected to continue. This will cause the trend toward greater numbers of broilers to continue. Currently, the broiler industry is concentrated in the southern states. There is a possibility that there may be some movement of large integrated broiler operations into the Midwest, where feed grains are plentiful and major markets are located. This would reduce the cost of transporting poultry meat into the marketplace.

The trend in sheep numbers has been sharply downward during the past quarter-century. This trend is expected to continue.

Research in livestock production is continuing to improve production efficiency. Areas of research that are having and will continue to have significant impact on the livestock industry include:

- improved twinning rate in beef cattle
- higher embryonic survival rate in swine
- shortening of calving interval for dairy cattle
- improved methods of artificial insemination
- biotechnology, including cloning, superovulation, sex determination, *in vitro* fertilization, and embryo transfer

Another trend in livestock production is increasing the emphasis on disease prevention rather than treatment. It is estimated that livestock producers currently lose about 20 percent of their income from disease, parasites, and toxins. Only about 10 percent of all livestock producers currently use effective disease-prevention management practices. By placing increased emphasis on disease prevention rather than treatment, livestock producers can increase efficiency and profits.

Biotechnology

The use of technology in genetic and reproductive processes in livestock production is in its early stages. This area of research and application is expected to dramatically change many livestock production practices. A discussion of biotechnology in livestock production is found in Unit 11 in this text.

ANIMAL WELFARE AND ANIMAL RIGHTS

The domestication and use of animals for the benefit of humans began many thousands of years ago. Questions about the ethics of animal use date back at least to the time of the ancient Greeks, who held four views regarding the status of animals. The *animists* believed that humans and animals shared and exchanged souls, the *mechanists* believed that neither humans nor animals had souls, the *vitalists* believed that animals had souls but were not as advanced as humans, and the fourth (and largest group) of ancient Greeks believed that animals existed on earth for the benefit and use of humans.

Philosophers since the time of the ancient Greeks have held various views on the status of animals; these views have had little impact on the question of the use of animals until recently. When the majority of the population was directly engaged in farming, most people understood that taking proper care of their animals resulted in greater productivity. As the percent of the population engaged in production agriculture declined during the twentieth century, fewer people had direct contact with animals (except for household pets) and little understanding of the care and management of animals as practiced in modern agriculture. This has provided an opportunity for those opposed to the use of animals to gain many converts to their cause.

There is a significant difference between those people who are concerned about animal welfare and those who believe in animal rights. There are many different groups within these two views, each with their own agendas.

Animal Welfare

Animal welfare supporters emphasize the humane treatment of animals, both in research and production agriculture. They believe that animals can be used to benefit humans. Some animal welfare advocates take the position that there are essential uses of animals (for example, for food and medical research), which they support, and nonessential uses of animals (such as entertainment), which they do not support.

Livestock producers generally support proper feeding and housing, veterinary care, and good management practices because these activities result in more efficient production of meat, milk, eggs, and wool.

The American Animal Welfare Foundation (225 E. 6th St., St. Paul, MN 55101), established in 1991, supports and promotes the humane treatment of animals. Its purpose is to promote responsible animal use and to educate the public about the vital distinction between animal welfare and animal rights. Educational materials and programs available to schools are designed to show how humans benefit from the responsible and humane use of animals.

Animal Rights

The views of animal rights activists vary. Some advocate the total elimination of all animal use by humans (animal liberation), while others recognize that animal use probably is not going to be eliminated and, therefore, work to eliminate animal suffering to the greatest extent possible. If all animal use by humans were to be eliminated, it would have a major impact on society. Those who take a more moderate approach try to achieve their goals by influencing legislation and through public education campaigns.

Some of the livestock production practices that are under attack by animal rights activists include:

■ use of hormones, antibiotics, and additives in animal feeding.
■ caging laying hens.
■ production of veal calves in crates.
■ raising swine in confinement and using farrowing crates for sows.

- management practices such as castration, docking, debeaking, and dehorning.

Other uses of animals to which animal rights activists object include:

- having them as household pets.
- using animals in medical and scientific research (including biotechnology).
- consuming animals' flesh.
- using animal skins (leather and fur) for clothing or other products.
- the use of animal products (including milk and eggs) for food.
- making animals the target of hunting and trapping.
- featuring animals in entertainment activities such as horse and dog racing.

Some militant animal rights activists have been willing to break laws by such actions as stealing research animals and damaging property, to draw attention to their cause. Major targets have included biomedical research facilities, food production and food retail facilities, and retail fur facilities. They have targeted meetings of farm groups with demonstrations protesting the use of animals and animal products. They have also confronted exhibitors at livestock shows with their protests. Investigation by the U.S. Department of Justice and the Department of Agriculture shows that there is an increasing tendency for animal rights violence to be directed at food production facilities and individuals involved in animal production.

Some of the people involved in the animal rights movement are urban dwellers who have limited direct knowledge of livestock production on the farm. Some theologians, philosophers, and human rights activists also have taken up the cause of animal rights.

In recent years, animal rights activists have begun distributing their message to school children by providing free materials to schools under the label of educational material. They also maintain sites on the Internet to distribute their message. Teachers and parents need to carefully review such material to be sure that it is not misleading students about the use of animals in our society. There are a number of Internet sites that provide factual research reports about animal welfare and animal rights.

Legislation

Societies for the prevention of cruelty to animals were organized in both England and the United States in the 1800s. Today there are federal, state, and local laws that address the humane treatment and care of animals. The Animal and Plant Health Inspection Service (APHIS) of the USDA has the responsibility of enforcing federal legislation in this area.

In the United States, the first federal law dealing with the humane treatment of animals was passed in 1873; this legislation mandated that feed and water be provided for farm animals being transported by barge or railroad. Other federal legislation includes:

- Federal Humane Slaughter Act of 1958 (amended in 1978) requires federally inspected slaughter plants to comply with humane methods of slaughter.
- Animal Welfare Act of 1966 addresses the sale, transportation, and handling of dogs and cats used in research institutions.
- Animal Welfare Act amendments of 1970 expand coverage of the 1966 Act to most other warm-blooded animals used in research, animals in zoos and circuses, marine mammals in sea life shows and exhibits, and animals sold in the wholesale pet trade. Retail pet shops, game ranches, livestock shows, rodeos, state and county fairs, and dog and cat shows are not covered by this Act.
- Animal Welfare Act amendment of 1976 extends the 1970 Act to include care and treatment while animals are being transported by common carriers and outlaws animal fighting exhibits (dog or cock fights) unless specifically permitted by state law.
- Horse Protection Act of 1970 (amended in 1976) prohibits the soring of horses; prohibits the transport of sored horses across state lines to compete in shows. (*Soring* is the practice of using chemical or mechanical irritants on the forelegs of the horse. A sored horse lifts its front legs more quickly to relieve the pain. The practice is strictly prohibited by most horse industry organizations and associations.)
- Animal Welfare Act amendment of 1985 provides for the establishment of special committees at all research facilities to oversee animal use; also requires exercise for dogs and provides for the psychological well-being of nonhuman primates at such facilities.
- Animal Enterprise Protection Act of 1992 adds a section to the federal criminal code to deal with vandalism and theft at animal research facilities and threats to research workers.
- Animal Welfare Act amendment of 1993 was passed to help prevent the use of lost or stolen pets in research. It establishes requirements for more documentation from dealers selling animals to research facilities; it specifies that dogs and cats must be held by pounds and animal shelters for at least five days, including a Saturday, before releasing them to dealers.

The Animal Welfare Act, with its amendments, provides only a minimum level of standards for the adequate care and treatment of animals covered by its provisions. Areas addressed in the standards include housing, handling, sanitation, nutrition, water, veterinary care, and protection from extreme weather and temperatures. Farm animals used for food, fiber, or other agricultural purposes are not covered by the Animal Welfare Act.

Livestock producers and others interested in the production and use of animals need to understand that societal concern about the welfare of animals is not going to go away. Decisions about the treatment and use of animals will increasingly be made by society as a whole rather than by individuals. More research is needed on issues that affect the welfare of animals. It is important that decisions about animal welfare and use be made on the basis of scientific information and not emotions.

It is important that those people who produce or work with animals do the best possible job of providing humane treatment for their animals. Animals that are well managed are more efficient producers of meat, milk, eggs, wool, and other products. Livestock producers need to do a good job of letting people know that they care about the welfare of the animals they raise. Students who show livestock need to be aware of the image they are projecting when they are competing. Concern for the welfare of their animals should always be a priority when students are participating in livestock shows.

ANIMAL IDENTIFICATION

A variety of methods are currently used to identify animals. These include ear tags, ear notching, tattoos, electronic collars, electronic ear tags, ear buttons, implants, microchips, and rumen boluses with microchips installed. There are many commercial companies that manufacture animal identification systems and the software packages used to record the data in a computer database.

The livestock industry is in the process of developing a system of unique identifying numbers that can be utilized with one or more of the methods currently in use for livestock identification and that can be used to establish a database of information about individual animals. Computer technology exists that allows embedded microchips to be read and the information secured can then be transferred to a database. Additional information about the animal including place of origin, health records, breeding records, production records, and so on can be added to the database. Standards for security of information and rules for access to database information need to be established.

A unique animal identification coupled with a database of information has been developed by the Holstein Association. The system is known as National Farm Animal Identification and Records (F.A.I.R.). The system is voluntary but many dairy producers are using it as a means of securing better information for managing their dairy herds.

The advantages of developing a system of unique animal identification numbers coupled with a database of information include:

■ provide data for disease control and eradication programs.
■ provide a record of vaccinations and treatments for diseases.

- provide a record of good management practices used by the producer.
- monitor emerging diseases.
- improve quality assurance programs and food safety.
- improve genetics in breeding programs.
- identify animals so they may be returned to their owners in case of theft or loss.
- improve access to international markets (for example, the European Union is developing plans to require identification of meat products that is traceable back to the farm of origin).

The Council on Dairy Cattle Breeding is composed of representatives from artificial insemination organizations, breed associations, and Dairy Herd Improvement Associations. In 1995 the Council began work to develop a better method of identifying dairy cattle in the United States. At that time there were a number of different programs in use that often were not compatible and thus could not exchange information. The Council is working to develop an American ID Number to provide a unique and lifelong ID number for all dairy animals in the United States. The specifications for the American ID number are:

American ID Number

Country code	Alpha	3 characters
ID number	Alphanumeric	12 characters

Additional Database Fields

Species	Alpha	1 character
Breed	Alpha	2 characters
Sex	Alpha	1 character
Registry status	Alpha	2 characters

The ID number is attached to the animal. The International Standards Organization (ISO) assigns a three-letter country code to each nation; this is the code used in the ID number. By using a country code, it is possible to record imported animals into the database without having to renumber a particular animal.

The Council allocates blocks of numbers to organizations that identify dairy animals. The database of numbers allocated to various organizations is available on the Internet at this URL: http://aipl. arsusda.gov/memos/html/usaid.html.

FOOD SAFETY

Despite the fact that the United States has the safest food supply in the world, food safety is becoming a major area of concern among consumers. Much of this concern is fueled by special-interest groups, consumer groups, and the news media. Major issues regarding food safety include:

- Bacteria contamination
- Pesticides in food (generally related to crops—not discussed in this text)
- Drug residues in food (see Unit 7 for withdrawal times before slaughter)
- Irradiation of food
- Genetic engineering (see Unit 11 for general discussion)
- Contamination of food by processors

Consumers are looking for zero health risk in relation to their food. Food safety specialists know there is no such thing as zero risk; they define food safety in terms of risk-benefit. A procedure designed to improve food safety may pose some degree of risk for some consumers. However, the benefits to most consumers may be many times greater than the risk. For example, many additives are used in foods for a variety of purposes, including enhancing flavor, maintaining freshness, and preservation. A small number of people may have an adverse reaction to the additive; most people will not be affected. If the benefits to the majority of consumers far outweigh the potential health hazard to a small number of people, the risk-benefit ratio of using these additives is considered to be acceptable.

Most of the problems with foodborne illness are caused by bacteria (66 percent). Other sources of foodborne illness include chemical (25 percent), viral (5 percent), and parasitic (4 percent).

The most common bacterial causes of foodborne illness are:

- *Salmonella* species (often found in eggs, milk, chickens, beef, and turkey)
- *Campylobacter* (often found in poultry, raw milk, and drinking water)
- *Clostridium botulinum* (lives in the soil, grows in many meats and vegetables; multiplies in improperly processed canned or smoked foods)
- *Staphylococcus aureus* (found on human skin—may enter food supply from improper handling of food by workers)
- *Shigella* (normally found in the intestinal tract of humans—may be transmitted to food by improper sanitation procedures by food handlers)
- *Escherichia coli* (*E. coli*) O157:H7 (a more deadly form of the common *E. coli* bacteria; found in undercooked ground beef—may also be found on other foods such as lettuce, salami, unpasteurized apple cider, and unpasteurized milk)
- *Listeria monocytogenes* (bacteria is more resistant to acidity, salt, nitrite, and heat than many other micro-organisms; often found in

soft cheese, unpasteurized milk, imported seafood products, frozen cooked crab meat, cooked shrimp, and cooked surimi [imitation shellfish]. These bacteria can survive and grow at low temperatures)

■ *Clostridium perfringens* (most often found in meat and meat products; bacteria that survive cooking multiply when food is not kept hot enough [above 140°F (60°C)]. Do not leave prepared foods at room temperature for cooling before storage—refrigerate immediately)

According to the Centers for Disease Control and Prevention (CDC), the four major bacterial pathogens that contaminate meat and poultry products are *Salmonella*, *Campylobacter*, *E. coli* O157:H7, and *Listeria monocytogenes*. The CDC estimates that these cause approximately 4,000 deaths and 5 million illnesses each year in the United States when people eat meat or poultry products that are contaminated.

These bacteria can cause illness ranging from diarrhea, nausea, vomiting, breathing difficulty, and fever to death in extreme cases. Children and the elderly are generally more susceptible and often have more severe reactions to these bacteria than other people. Various species of *Salmonella* are the most common cause of foodborne illness. *Clostridium botulinum* grows in improperly canned foods and produces a toxin that can cause botulism, an illness that often results in paralysis and death. *Escherichia coli* (*E. coli*) is a common cause of diarrhea among travelers.

All fresh-food products contain some bacteria. Proper handling practices such as cooking and holding temperature, personal hygiene of food handlers, and kitchen sanitation will prevent most bacteria problems in food.

Regulation of food additives is the responsibility of the Food and Drug Administration (FDA), with additional review provided by the U.S. Department of Agriculture (USDA). In 1958 the Delaney Clause was added to the Food, Drug, and Cosmetic Act of 1938. The Delaney Clause prohibits the use of any food additive that causes cancer in humans or animals at any dosage. This zero tolerance policy, if applied to all foods, would mean that most foods could not be used. There are many naturally occurring chemicals in foods that can cause cancer at high enough levels or if eaten over a long enough period of time.

The Food Quality Protection Act of 1996 revised the Delaney Clause to set a new standard for carcinogens in food specified as "a reasonable certainty of no harm." This removed the zero tolerance policy relating to food additives and carcinogens that was imposed by the original Delaney Clause.

The USDA's Food Safety and Inspection Service (FSIS) is responsible for verifying that meat and poultry processing plants meet regulatory requirements and take enforcement action when a plant fails to meet these requirements. Historically, the inspection of slaughter plants was based on a visual checking for signs of sick animals or birds before slaughter, and inspecting the carcass for foreign matter, abscesses, or feces contamination. There was no routine microbiological testing done on raw meat or poultry.

In 1993 an outbreak of foodborne illness caused by the *E. coli* O157:H7 bacteria in ground beef showed that the then-current system of plant inspection did not correct the problem of bacteria that are a major cause of foodborne illness. The USDA determined that the meat and poultry inspection system needed to be modernized.

The Sanitation Standard Operating Procedures (SSOPs) along with other pathogen reduction regulations and new concepts for enforcement were introduced in 1997 by the FSIS. These regulations and concepts place the responsibility for food safety on the plants; it also gives plants flexibility in the development and implementation of innovative measures for the production of safe foods. The role of FSIS inspectors and compliance officers is primarily one of verifying that the plant is meeting industry standards and taking enforcement action when necessary. The basis for the current regulations is the belief that if plants properly maintain sanitation and process controls, food products produced by the plant will not be contaminated by dangerous pathogens. The FSIS will not approve the product of the plant if the control systems fail to prevent contamination by pathogens. It is the responsibility of the plants to address any deficiencies found by FSIS inspectors. Failure to correct the deficiencies may result in a suspension of all or part of the plant's operation. Plants have the right to receive notice of alleged violations and file appeals of actions by FSIS.

Plants are required to keep accurate records to verify that their control measures work properly and the product that they produce is safe for human consumption. Criminal prosecution may result if a plant is found to be maintaining false or deceptive records.

Sanitation problems in plants are addressed by the Sanitation Standard Operating Procedures. The focus is on making sure that the plant detects potential sanitation problems such as unclean equipment or poor worker hygiene that may cause contamination of food products by harmful bacteria and then prevents those problems from occurring.

In 1971 a system called Hazard Analysis and Critical Control Points (HACCP) was developed for NASA to monitor the production of food for the space program. Since then some plants have used HACCP on a voluntary basis. A phase-in of mandatory use of HACCP in all slaughter plants began in January 1998 with all large plants (500 or more employees) required to implement a HACCP plan. Small plants (10 or more employees but fewer than 500) were required to implement a HACCP plan by January 1999. All remaining slaughter plants were required to have a HACCP plan in place by January 2000.

The National Advisory Committee on Microbiological Criteria for Foods (NACMCF) defined seven principles for assuring safety in the food supply using HACCP. The HACCP systems used in slaughter plants must be based on those seven principles.

The seven HACCP principles are:

1. The plant conducts a hazard analysis to determine potential food safety hazards and identify preventive measures the plant can use to control those hazards.
2. The plant identifies critical control points (CCP). This is any point, step, or procedure in the process where control can be applied to prevent, eliminate, or reduce a food safety hazard to an acceptable level. Food safety hazards are defined as any biological, chemical, or physical property that may cause a food to be unsafe for human consumption.
3. Critical limits are established for each CCP. A critical limit is defined as the maximum or minimum value to which a physical, biological, or chemical hazard must be controlled at a critical point to prevent, eliminate, or reduce it to an acceptable level.
4. Critical control point monitoring requirements must be established. These activities are necessary to ensure that the process is under control at each CCP. The HACCP plan must list each monitoring procedure and its frequency.
5. Corrective actions must be established whenever monitoring indicates a deviation from an established critical limit. The HACCP plan must identify the corrective actions to be taken if a critical limit is not met. The purpose of corrective actions is to prevent unsafe food from reaching the public.
6. Record keeping procedures must be established by all plants. These records must include its hazard analysis, written HACCP plan, records documenting the monitoring of CCPs, critical limits, verification activities, and the handling of processing deviations.
7. Procedures must be established to verify that the HACCP system is working as planned. Plants are required to validate their own HACCP plans. FSIS does not approve HACCP plans in advance but does review them for conformance with the rules. Verification procedures include review of HACCP plans, CCP records, critical limits, and microbial sampling and analysis. The HACCP plan must include verification tasks to be performed by plant personnel. FSIS inspectors will also perform some verification tasks. Microbial testing to be conducted by both FSIS and the plant is included as one of the verification activities.

Slaughter plants must regularly test animal carcasses for the generic form of *E. coli* to show that their control system for preventing fecal contamination is working. The frequency of sampling is determined by the plant's production volume. Test results must be recorded and be made available to inspectors.

Testing for *Salmonella* contamination is required in slaughter plants and any plants that produce raw ground products. The selection of *Salmonella* as a performance standard was based on its prevalence in raw meat and poultry products and the fact that reliable laboratory tests are available. Also, steps taken to prevent *Salmonella* contamination may also reduce contamination from other bacteria such as *E. coli* O157:H7 and *Campylobacter*. The FSIS has developed a baseline standard based on the prevalence of *Salmonella* in raw products that all plants must meet. The FSIS will continue to collect data and revise its performance standards as necessary. The best way for plants to meet the standards is to develop and implement process controls to prevent product contamination.

Food irradiation is the treatment of food with radioactive isotopes to kill bacteria, insects, and molds that are present in the food. Cobalt-60 is the most widely used source of the short wavelength radiation used for food irradiation. Cesium-137 is sometimes used as the radiation source. The process is carried out in an irradiation facility that confines and directs the energy. The facility has an irradiation chamber and thick concrete walls that confine the energy and protect the workers from stray radiation. The food is moved through the irradiation chamber on a conveyor.

Irradiation does not raise the temperature of the food as it is treated. The process can be used on fresh or frozen foods. The energy waves are not retained by the food; the food does not become radioactive. No significant difference in nutritional quality has been found when irradiated foods are compared to foods processed by other methods. There are a few foods that show undesirable changes when irradiated. An undesirable flavor change occurs in dairy products that are irradiated, and some fruits, such as peaches and nectarines, show a softening of tissues.

Irradiation of food as a method of food preservation began in 1950. Currently 37 countries have approved irradiation for use on a variety of foods. Twenty-four of these countries are using it commercially. There are over 40 irradiation facilities operating in the United States.

Because irradiation is classified as a food additive, the Food and Drug Administration (FDA) must approve its use in the United States. Inspection and monitoring of irradiated meat and poultry and the enforcement of FDA regulations concerning these products are the responsibility of the U.S. Department of Agriculture. All irradiated foods must carry the international symbol called a radura, along with a statement that they have been treated by irradiation (Figure 1-10).

Current FDA approvals for irradiation of food include:

- August 21, 1963—wheat and wheat flour—to control insects
- August 8, 1964—white potatoes—to inhibit sprout development
- July 5, 1983—herbs, spices and vegetable seasonings—decontaminates and controls insects and microorganisms
- June 10, 1985—dry enzyme preparations primarily used in fermentation-type food processes—to control insects and microorganisms
- July 22, 1985—pork—to control the parasite that causes trichinosis
- April 18, 1986—fruits, vegetables and grains—to control insects and inhibit growth and ripening
- May 1, 1990—chicken, turkey, and other fresh or frozen uncooked poultry—control *Salmonella* and other disease-causing bacteria
- December 2, 1997—fresh and frozen red meats such as beef, lamb and pork—slows spoilage and controls disease-causing microorganisms

FIGURE 1-10
The international radura symbol, which must be printed in green.

In 1999 the USDA approved the rules for the irradiation of refrigerated or frozen uncooked meat and some meat products. This action was taken to improve the safety of the food supply in the United States. The rules for irradiation of poultry were changed to make them consistent with the rules for meat. Raw poultry product may now be irradiated before packaging in addition to the treatment of chicken carcasses. Irradiated raw poultry and meat may be used in the manufacture of other products such as sausage or bologna.

Slaughter plants are not required to irradiate their product. Some of the problems involved in the irradiation of meat include uncertainty about consumer acceptance, the volume of product produced each year, and the cost of irradiation compared to other methods of reducing bacteria in foods.

There are several benefits that result from the irradiation of food:

- Preservation of perishable foods—shelf life of foods is extended by destroying organisms that cause spoiling and decomposition.
- Sterilization—makes it possible to store foods for long periods of time without refrigeration.
- Inhibit sprouting—in potatoes, carrots, onions, garlic, and ginger.
- Delay ripening—in bananas, mangos, avocados, and other non-citrus fruits.
- Control insect damage—in grain, fruit, vegetables, dehydrated fruits, spices, and seasonings.
- Control foodborne illness in meat, poultry, and fish—eliminate microorganisms such as *Salmonella* species, *Campylobacter*, *Clostridium botulinum*, *Staphylococcus aureus*, *Shigella*, and *Escherichia coli*.

As is the case with the advent of many new technologies, food irradiation has its opponents. Some people are confused because the

Safe Handling Instructions

This product was prepared from inspected and passed meat and/or poultry. Some food products may contain bacteria that could cause illness if the product is mishandled or cooked improperly. For your protection, follow these safe handling instructions.

 Keep refrigerated or frozen. Thaw in refrigerator or microwave.

 Keep raw meat and poultry separate from other foods. Wash working surfaces (including cutting boards), utensils, and hands after touching raw meat or poultry.

 Cook thoroughly.

 Keep hot foods, hot. Refrigerate leftovers immediately or discard.

FIGURE 1-11   A typical food label showing instructions for safe food handling.

word irradiation sounds like radioactive, thus raising fears about the possibility of cancer or other illnesses. They do not understand that irradiated food does not become radioactive. Another concern is the possibility of toxic radiation products forming in the food. Research done over many years has not shown any health problems resulting from the use of irradiated foods.

The use of food irradiation has been significantly delayed in the United States by the activities of those opposed to this technology. However, recent polls among consumers suggest that a majority is willing to buy irradiated foods, especially in light of the increased safety of the food supply.

Improper handling of meat and poultry in the home is a major cause of foodborne illness. Everything that touches food—bowls, countertops, hands, and utensils—must be kept clean.

- Wash hands thoroughly in hot soapy water before preparing food and after handling raw meat or poultry.
- Juices from raw meat or poultry must not come into contact with other foods.
- Use different plates for cooked foods and raw meat or poultry.
- All utensils that are used on raw meat or poultry must be thoroughly washed in hot soapy water before they are used for cooked foods.
- Counters, cutting boards, and other surfaces where raw meat or poultry was placed must be washed and sanitized before placing cooked foods on them. Surfaces may be sanitized by using a solution of 2–3 teaspoons of household bleach in one quart of warm water. After sanitizing the surface with this mix, rinse with plain hot water.

Bacteria are destroyed by high temperatures (165° to 212°F) when meat and poultry are cooked. Check the internal temperature of meat or poultry in several spots with a meat thermometer, especially if cooking is done in a microwave oven. Internal temperatures of cooked foods vary more when they are prepared in a microwave oven instead of in a conventional oven. Make sure the meat or poultry is thoroughly cooked; there should be no pink meat left internally in the prepared food. Frozen foods usually need about 1½ times the normal cooking time used for thawed foods. Reheat leftover foods at 165°F to kill any bacteria.

Many food retailers now offer food safety pamphlets with specific instructions for safely cooking and storing food. By following the advice in these guides, consumers can prevent many cases of foodborne illnesses. Safe handling instruction labels are now attached to many meats and poultry. A typical label is shown in Figure 1-11. A free information pamphlet, "How To Read The New Food

Label," is available by writing to Consumer Information Center, Department 522A, Pueblo, CO 81009.

Many farm animals are believed to be carriers of the common bacteria that cause foodborne illnesses. It is usually not possible to determine by visual inspection which animals are carrying these bacteria and which are not. Doing simple serum antibody tests or examining cultures from blood or feces samples do not always produce definite proof that the bacteria are present. If the bacteria are present in low numbers, a serum test will not show its presence. Some bacteria live in specific tissues in the animal's body and are not detected by culture tests. Research is needed to develop an accurate and simple test for the presence of bacteria that cause foodborne illnesses that can be used on the farm.

Because of the difficulty in determining which animals are carriers, the possibility of contaminating many uncontaminated carcasses in the packing plant is always present. Ground meat products produced at the plant normally contain meat from several animals. If one of these animals is contaminated, then the entire product becomes contaminated.

More research is needed to determine exactly how these bacteria infect animals and thus get into the food chain. At present, following good management practices on the farm is the best recommendation that can be given. Maintain clean housing, keep animals healthy, and do not use contaminated feed or water.

Closer inspection and regulation of livestock production on the farm by the USDA can be expected as part of an overall program to improve food safety. Tracking of infected animals from the packing plant back to the farm of origin can be expected as a part of this effort. The determination of normal levels of bacteria in livestock is under study and will help establish baselines for inspection and regulation.

Livestock producers have a vital interest in helping to ensure the safety of the food supply. If consumers lose confidence in that safety, livestock producers lose dollars.

ENDANGERED SPECIES

Changes in livestock production during the twentieth century have resulted in a threat of extinction for many species that were once commonly found on the farm. Livestock breeders increasingly select for those factors that increase production, such as faster rate of gain, increased feed efficiency, larger litters, or more milk production. More uniformity within a breed is developed by the use of artificial insemination, embryo transfer, and cloning. There is an increasing emphasis on crossbreeding to produce animals that are considered better able to meet the market demand for their products.

As a result of these changes, many purebred lines of traditional breeds of livestock have almost disappeared. Selection for uniformity, crossbreeding, and the decline in the numbers of many purebred lines has resulted in a loss of genetic diversity within a species. Genetic diversity is important to maintain within a species because it permits adaptation to changing conditions. When genetic diversity is lost, specific breeds within the species are more likely to become extinct.

Modern livestock production is highly dependent upon the use of antimicrobial compounds and anthelmintics to control diseases and parasites (see Unit 7). Natural resistance to diseases and parasites is not usually selected for when breeding animals. The result is a loss of natural resistance to infestations of diseases and parasites. Some disease organisms and parasites develop an immunity to the drugs used for control. Most swine breeds used in the United States are not resistant to scours caused by *Escherichia coli*. Some breeds of Chinese hogs, however, are resistant to this form of scours. These breeds may be of importance in developing a genetic resistance to scours in swine breeds in the United States. Gulf Coast Native sheep, Caribbean hair sheep, and Florida Cracker cattle show a genetic resistance to parasites. These breeds may be of importance in developing this resistance in other sheep and cattle breeds that currently are not resistant to parasite infestations. The lack of breeding stock with a natural resistance to diseases or parasites could result in a serious threat to the food supply.

The preservation of genetic diversity within a species makes it possible to develop new breeds with characteristics that meet changing market needs. Cattle breeds such as the Santa Gertrudis and Senepol were developed in response to a need for cattle that can be raised in hot climates. The Finn sheep has the ability to produce multiple lambs instead of singles or twins. This ability is being developed in the sheep industry in crossbreeding programs.

Maintaining genetic diversity within species also contributes to scientific knowledge and research. Minor breeds often have unusual characteristics that allow study of adaptation, feed use, disease and parasite resistance, and reproductive abilities under a variety of conditions. This knowledge may be essential in the future for the maintenance of an adequate food supply for the world's population.

The American Livestock Breeds Conservancy (Box 477, Pittsboro, NC 27312), founded in 1977, is a nonprofit organization working to prevent the extinction of many breeds of livestock once common in agriculture. A recent census of livestock in the United States and Canada by The American Livestock Breeds Conservancy reveals that many breeds now have low populations and some are very close to extinction.

SUMMARY

The domestication of animals played a vital role in the development of civilization. Domesticated animals provided a more dependable source of food and clothing. Most of the ancestors of present-day farm animals were first tamed in the Neolithic (New Stone) Age. Written records show that animals were being used more than 4,000 years ago. Early explorers brought the various species of farm animals to the United States. These animals spread across the United States as early explorers and colonists moved westward.

Animals have many useful functions. They convert feed into food, are a source of materials for clothing, and provide power and also recreation. They help conserve natural resources and contribute to good farm management by increasing farm profits. Animals concentrate bulky feeds, making them easier to market. Many by-products of animals are also important to society.

Animals and their products contribute billions of dollars to the economy each year. The per person consumption of meat in the United States has been declining in recent years.

The total number of livestock on farms in the United States is several billion. Livestock numbers are expected to continue to increase in the years ahead.

Student Learning Activities

1. Survey the farms in a community, collecting data on size of farm, acres of various crops grown, and kind and size of livestock enterprises on the farms. Prepare a bulletin board display showing a map of the community with the data summarized.
2. Prepare and present an oral report on one of the following topics:
   a. the domestication of animals
   b. the spread of livestock in the United States
   c. the functions of animals
   d. trends in animal agriculture
   e. animal rights issues
   f. food safety
3. Prepare graphs based on Table 1-5 that show trends in livestock numbers in the United States.

Review

1. Why was the domestication of livestock important to the development of civilization?

2. Why did the raising of large cattle herds develop in the Great Plains states whereas the finishing of cattle developed in the North Central states?

3. From what two wild stocks of swine were the American breeds of swine developed?

4. What was the main use of sheep by the early colonists in the United States?

5. Describe the major differences between sheep and goats.

6. What was the main use of the horses brought to the United States by the early colonists?

7. What changes have taken place in the poultry enterprise from colonial times to the present?

8. Name and briefly describe the eight functions of animals in our society.

9. How important economically is the livestock industry as compared to the total farm industry in the United States?

10. Briefly explain the trends in consumption of livestock products per person in the United States from 1971 to 1988.

11. What was the trend in consumption of wool per person in the United States from 1976 to 1996?

12. Briefly explain current trends in animal agriculture.

13. Briefly explain the issue of animal rights.

14. Why are animal by-products important to human society?

Career Opportunities in Animal Science

Objectives

After studying this unit, the student should be able to

- explain the value of a farm background for the individual entering a livestock-related occupation.
- list employment opportunities that require a knowledge of animal science.
- describe the process of choosing an occupation.

EMPLOYMENT IN AGRICULTURE

Agriculture is a basic industry in the United States. The majority of career opportunities in agriculture in the near future are in industries related to agriculture and not in production agriculture itself. More than 200 different careers are available to persons with an interest in agriculture. Many of those careers require a minimum of two years of education beyond high school and many are in the field of Animal Science.

Agriculture and agriculture-related industries provide employment for almost 16 percent of the total work force in the United States. There are five major categories of employment in agriculture-related jobs:

- Farm production and agricultural services. Typical jobs include farmers, hired farm workers, farm managers, and veterinarians— 2.6 percent of the total work force.
- Input suppliers. Typical jobs include employment in wholesale and retail sales of farm equipment; seed, fertilizer, and chemical suppliers; farm machinery manufacturing; and manufacturing and sale of farm chemicals—0.3 percent of the total work force (Figure 2–1).

FIGURE 2-1   This feed mill worker finds knowledge of animal science helpful on his job. *Photo by Jim Bodine.*

- Processing and marketing. Typical jobs include employment in apparel and textile manufacturing; and processing meats, dairy goods, fruits, vegetables, bakery products, and beverages—2.4 percent of the total work force.
- Agricultural wholesale and retail trade establishments. Typical jobs include employment in grocery stores, restaurants, convenience and carry-out stores—10.2 percent of the total work force.
- Indirect agricultural businesses. Typical jobs include employment in chemical and fertilizer mining and manufacturing food processing machinery—0.4 percent of the total work force.

The number of jobs in the first three categories above has been declining during the past 25 years and is expected to continue to decline in the years ahead. These are the jobs that are most closely related to production agriculture. The number of jobs in the last two categories has been increasing during this same period and is expected to continue to offer employment opportunities in the future. These are jobs that are only marginally related to production agriculture.

The number of farms in the United States has been declining since 1936. This trend is expected to continue in the future. The average size of farms has grown steadily larger during this period, while the amount of land in farms has declined only slightly. Approximately 90 percent of the farms in the United States are individually owned and the other 10 percent are owned by partnerships or corporations. The decline in job opportunities in production agriculture and closely related industries is related to the decline in the number of farms.

An expanding population in the United States and the world has resulted in an increased demand for food. This has created more jobs in the businesses that process, package, and market food.

Many job opportunities related to animal science are also available in the fields of agribusiness, communications, science, government, education, and sales. The emergence of new technologies provides additional job opportunities for well-trained people with an interest in animal science. The right combination of skills, education, training, and experience can lead to exciting careers in helping to feed and clothe the people of our nation and the world.

People develop an interest in animal science careers in various ways. High school agriculture programs are one source of information about animal science occupations (Figure 2-2). Courses in animal science in high school can help students learn more about agriculture and develop basic skills. Many colleges also prepare people for various agriculture careers. There are many jobs for men and women with experience and training in animal science.

The person who has lived on a farm and worked with animals already has many skills that are important to job success in agriculture. One who has worked in an agricultural business dealing with animals

FIGURE 2-2   Students in a high school agriculture education class learn about animal science. *Courtesy FFA.*

FIGURE 2-3   In working with livestock, agriculture education students gain many skills that future employers find valuable. *Photo by Michael Dzaman.*

also possesses skills that employers look for in choosing personnel. This experience is a great help to the person who desires a career in animal science (Figure 2-3).

There are many jobs available that relate to animal science and increasingly these jobs require a minimum of a Bachelor's degree. Some of these jobs require education beyond the Bachelor's degree, usually in a specialized area. While the opportunities for employment in production agriculture are limited, many related careers require some knowledge of animal production. Table 2-1 lists some of the career opportunities for people who have a Bachelor of Science degree in animal science.

EMPLOYMENT OPPORTUNITIES FOR COLLEGE GRADUATES IN AGRICULTURE

There is a growing concern among agribusinesses about a future shortage of qualified college graduates with training in the field of agriculture. Current projections indicate that the average annual demand for university graduates with training in agriculture will exceed the annual supply by 5 percent during the next several years.

College graduates in the field of agriculture will find the best job opportunities in basic plant and animal research, food and fiber processing, and agribusiness management and marketing. Figure 2-4 shows the annual percentage of projected openings for college graduates in the field of agriculture. The number of college graduates specializing in the areas of agricultural production and education, communication, and information is expected to exceed the number of annual openings in these areas.

TABLE 2-1 Career Opportunities in Animal Science for Graduates with a B.S. Degree.

Livestock Production
Beef cattle
    Cow/calf operations
    Stocker or grower programs
    Feedlot
Dairy
    Milk production
Swine
    Farrowing operations
    Grower/finisher operations
    Farrow to finish
Sheep
    Ewe flocks
    Wheat pasture growing/finishing
        programs
Poultry
    Broiler production
    Egg production
Horses
    Mare breeding farm
    Training facility

Livestock Feed
Production
Sales
Distribution

Veterinary Medicine
Practice
Research
Product development
Teaching
Inspection

Meat or Dairy Foods
Production
Product development
Quality control
Distribution and marketing

**Livestock Promotion and
    Marketing**
Breed organizations
Livestock publications
Livestock sales / Market reporting

Sales
Feed
Pharmaceuticals
Agricultural chemicals
Livestock supplies

Management
Livestock production enterprises
Sales/marketing companies
Food production/ distribution

Financial Institutions
Banks
Lending agencies

Service Organizations
Extension
Agriculture agents
4-H agents
Teaching (high school, junior
    college or university)

Feed/slaughter inspection
Private consulting

Computer Specialists
Computer programmers
Software distributors
Modeling

**Technology Development and
Application (Biotechnology)**
Laboratory technical support
Animal caretakers
Research scientists
Genetics and animal breeding
Population genetics
Molecular genetics
Genetic engineering
Reproductive management
Endocrinology
Cloning
Embryo technology
Nutrition
Feeding programs
Nutrition/reproduction interactions
Nutrition/health/ immunity
    interactions
Food science
Product development
Food processing
Fermentation

Source: Oklahoma State University.

Scientists, Engineers, and Related Specialists

Some typical animal-science-related occupations found in this category include:

Agricultural Engineer   Food Scientist
Animal Scientist   Geneticist
Biochemist   Microbiologist
Entomologist   Nutritionist
Environmental Engineer   Veterinarian
Food Engineer

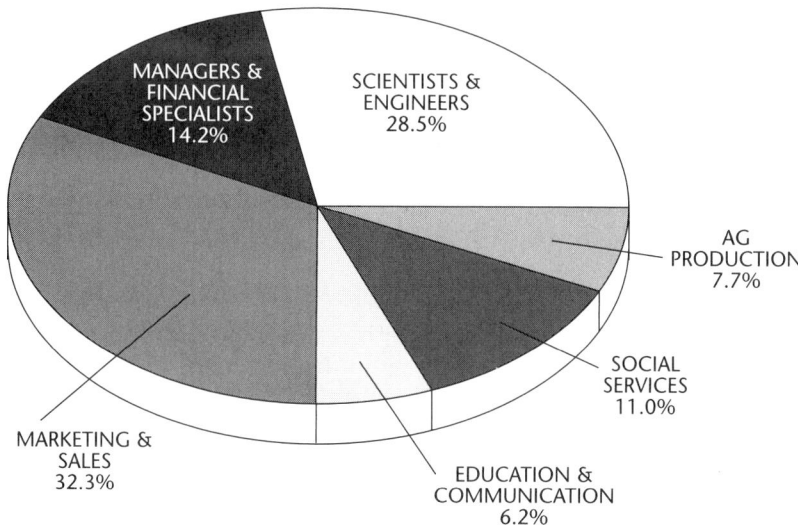

FIGURE 2-4 Employment opportunities for college graduates in the food and agricultural sciences. *Source:* USDA.

Workers in these professions do essential research and development that improves the competitive position of the United States in world markets. Employment opportunities are best for persons with doctoral degrees or postdoctoral experience in molecular genetics, biochemistry, food science, food engineering, nutrition, and environmental science. The annual demand for new veterinarians is expected to be approximately 2,000 during the next several years. Overall, about two thirds of the jobs in this category require advanced degrees for entry-level positions.

Managers and Financial Specialists

Some typical occupations in this category that relate to animal science include:

Business Manager   Food Service Manager

Credit Analyst   Retail Manager

Economist   Wholesale Manager

Financial Analyst

The best job opportunities in this category that relate in some way to animal science are for financial analysts, economists, credit specialists, and restaurant and retail store managers. Food and fiber processing and distribution are expected to provide strong employment possibilities. Opportunities in wholesale and retail businesses that sell feed, agricultural chemicals, and animal health products are expected to be more limited. There is a projected shortage of graduates earning advanced degrees in the occupations included in this group.

Marketing, Merchandising, and Sales Representatives

Some typical occupations related to animal science that are found in this category include:

Food Broker

Grain Merchandiser

Livestock Buyer

Market Analyst

Marketing Manager

Sales Representative

Technical Service Representative

Approximately one third of all the employment opportunities for college graduates in food and agricultural careers are in marketing, merchandising, and sales. In addition to a technical understanding of food and fiber production and/or processing, these workers need good business and communication skills, along with a strong academic preparation. Opportunities in marketing, merchandising, and sales in food and fiber distribution are expanding, while employment opportunities in farm supply and service businesses are decreasing. A baccalaureate degree is required for more than 85 percent of the jobs in this category. People with advanced degrees will find the most employment opportunities in purchasing and buying, market analysis, technical sales, and international trade.

Education, Communication, and Information Specialists

Some typical occupations related to animal science that are found in this category include:

College Faculty Member

Cooperative Extension Agent

Editor

Public Relations Specialist

Reporter

Agriculture Education Teacher

There are some occupations in this group for which the number of college graduates exceeds the number of job opportunities. Other occupations in this group have more job opportunities than the number of college graduates available. Occupations in agricultural public relations and advertising provide the best opportunities for employment. There is a continuing need for well-qualified agriculture education teachers. There is a decreased need for Cooperative Extension Agents. Educators with specializations in nutrition and health may find more opportunities for employment. The need for college faculty members in the fields of basic animal sciences, biochemistry, food services, biotechnology, and agribusiness management specialties is expected to be greater than the need for those in agricultural production management, veterinary medicine, and natural resources.

Social Services Professionals

Some typical occupations related to animal science that are found in this category include:

Dietitian

Food Inspector

Nutrition Counselor

Strong employment opportunities are expected for dietitians and nutritionists. The need for dietitians accounts for more than half of the projected job opportunities in this category.

Agricultural Production Specialists

Some typical occupations related to animal science that are found in this category include:

Farmer	Professional Farm Manager
Feedlot Manager	Rancher

The trend toward larger farms has resulted in fewer job opportunities in production agriculture. Only 28 percent of the farms in the United States account for 87 percent of the total sales of farm products. Technological advances, higher production costs, and unstable markets indicate an increasing need for people in production agriculture to have the business and technical expertise that is provided by a college education.

The major employment opportunities for college graduates in production agriculture are as farm and ranch managers employed by financial institutions and investors. Employment in the fields of financial consulting and production supervision is expected to be available. The opportunities for traditional owner-operators are expected to decrease.

OCCUPATIONS IN ANIMAL SCIENCE

Many occupations are possible in the field of animal science (see Figures 2-5 through 2-8). Some of these occupations are listed here by type of animal. Some general occupations that deal with more than one phase of animal science are also listed.

Descriptions of occupations marked with a * are found in the *Dictionary of Occupational Titles*. Occupational titles that are indented in this list are alternate names for the occupation immediately above the indented titles or may indicate an occupation designated by a field of specialization under the general heading. All indented occupational titles are found under the heading immediately preceding them in the *Dictionary of Occupational Titles*.

FIGURE 2-5   A federal meat inspector at work in one of the more than 3,000 federally inspected meat plants in the United States. *Courtesy of USDA.*

FIGURE 2-6   Sheep shearing is one of the many agriculture-related occupations associated with animal science. *Courtesy of USDA.*

FIGURE 2-7 A farrier is a person who shoes horses. Farriers are a vital part of the nation's horse industry. *Courtesy Kentucky Horse Park.*

General Occupations in Animal Science

agricultural education instructor/FFA advisor

agricultural economist*

agricultural engineer (livestock buildings and equipment)*

agricultural journalist

agricultural news director

animal behaviorist

animal breeder*

animal breeding scientist*

animal caretaker*

– animal attendant

– animal control officer

– farmworker, livestock

animal cytologist*

animal ecologist*

animal health products distributor

animal inspector

animal keeper*

animal nursery worker*

animal nutritionist*

animal physiologist*

animal scientist*

animal taxonomist*

animal trainer*

animal treatment investigator*

– animal control officer

animal science teacher

artificial breeding distributor*

artificial breeding technician*

– breeding technician

artificial insemination herd sire evaluator

artificial insemination herd sire manager

artificial inseminator*

– inseminator

biotechnologist*

biological aide

breed association field worker

breeding researcher

butcher*

central test station manager

central test station worker

county agricultural agent*

– agricultural agent*

– county adviser

– county agent

– extension agent

– extension-service agent

– extension worker

– farm adviser

– farm agent

custom-feed-mill operator*

drug company representative

extension service livestock specialist*

farm broadcaster

farmer, general*

farmhand

farm investment manager

farm loan officer

farm manager*

farmworker, general*

– hired worker

– chore tender

– farm laborer

farmworker, livestock*

– laborer, livestock

– ranch hand, livestock

feed and farm management adviser*

feed mill operator

feed ration developer and analyst

feed research aide*

field sales representative, animal health products

food processing supervisor

4-H club agent*

geneticist*

inspector, grain mill products*

laboratory technician, veterinary*

land bank branch manager

livestock agent*

livestock auctioneer

livestock building dealer

livestock building salesperson

livestock buyer

livestock caretaker

livestock commission agent

livestock equipment dealer

livestock equipment salesperson

livestock farmhand

livestock herder

livestock geneticist

livestock inspector

livestock producer

livestock rancher*

   – livestock breeder

   – livestock farmer

livestock sales representative*

livestock seller

livestock yard attendant*

livestock yard supervisor

magazine writer

manager of retail feed and supply store

market research analyst, agriculture*

marketing analyst

market news analyst

meat cutter*

meat grader

meat inspector*

meat science researcher

muscle biology researcher

nutritional physiologist

nutrition researcher

ova transplant specialist

pharmaceutical company representative

production researcher

reproduction researcher

reproductive physiologist

salesperson, animal feed products

show cattle caretaker, supervisor*

superintendent, grain elevator*

supervisor, animal cruelty investigation*

   – animal humane agent supervisor

supervisor, artificial breeding ranch*

supervisor, feed mill*

supervisor, livestock-yard*

supervisor, stock ranch*

veterinarian*

veterinarian aide*

veterinarian, laboratory animal care*

veterinarian, poultry*

veterinarian's assistant

veterinary anatomist*

veterinary bacteriologist*

veterinary epidemiologist*

veterinary livestock inspector*

veterinary meat inspector*

veterinary parasitologist*

veterinary pathologist*

veterinary pharmacologist*

veterinary physiologist*

veterinary virologist*

veterinary virus-serum inspector*

FIGURE 2-8 A poultry inspector at a poultry-processing plant inspects a chicken before it is slaughtered. Poultry inspectors help assure wholesome poultry on the dinner table. *Courtesy of USDA.*

Beef Occupations

beef breeder
beef farmer
beef farmhand
beef herder
cattle buyer
cattle feeder
cattle rancher*
cowpuncher*
    – puncher

– ranch rider
– rider
feedlot supervisor
feedlot maintenance worker
stock ranch supervisor*
top screw*
    – lead rider
    – ramrod
    – top waddy

 Dairy Occupations

cheesemaker*
    – cheese cooker
dairy barnman
dairy cattle herder
dairy choreman
dairy cow washer
dairy farmer*
dairy farm worker*
    – laborer, dairy farm
dairy helper*
dairy herdsman
dairy herd supervisor (DHIA)
dairy management specialist*
dairy nutrition specialist*
dairy sanitarian
dairy scientist*

dairy technologist*
dairy tester
field contact technician, dairy*
manager, dairy farm*
milker, machine*
    – milking machine operator
milk plant supervisor
milk sampler*
    – sampler
supervisor, dairy farm*
supervisor, dairy processing*
    – butter production supervisor
    – cheese production supervisor
    – instant-powder supervisor
    – pasteurizing supervisor

 Swine Occupations

hog buyer
swine breeder
swine farmer

swine farmhand
swine herder
swine rancher*

 Sheep Occupations

camp tender*
fleece tier*

lamber*
sheep breeder

sheep farmer
sheep farmhand
sheep herder*
  – herder
  – mutton puncher
  – shepherd
sheep shearer*

– sheep clipper
– stock clipper
– wool shearer
supervisor, wool-shearing*
wool buyer
wool-fleece grader*
wool-fleece sorter*

Horse Occupations

barn boss*
  – corral boss
  – hostler
  – lot boss
  – stable manager
equine dentist
farrier
hoof and shoe inspector*
horse breeder
horse exerciser*
horse farmhand
horse herder
horse rancher
horseshoer*

– plater
horse stable attendant
horse trainer*
paddock judge*
race horse trainer*
racing secretary and
   handicapper*
stable attendant*
  – barnworker
  – groom
  – horse tender
  – mule tender
  – stallion keeper

Poultry Occupations

blood tester, fowl*
caponizer*
chick grader*
  – poultry culler
chick sexer*
chicken breeder
egg candler*
farm worker, poultry*
  – helper, chicken farm
  – poultry helper
field service technician, poultry*
grader, dressed poultry*

laborer, poultry farm*
  – laborer, brooder farm
  – laborer, chicken farm
  – laborer, egg-producing farm
  – laborer, fryer farm
  – laborer, pullet farm
  – laborer, turkey farm
laborer, poultry hatcher*
  – hatchery helper
  – incubator helper
manager, poultry farm*
manager, poultry hatchery*

– manager, chicken hatchery
– manager, duck hatchery
– manager, turkey hatchery
poultry breeder*
– chicken fancier
poultry breeding researcher
poultry debeaker*
– debeaker
poultry farmer*
– duck farmer
– poultry farmer, egg
– poultry farmer, meat
– turkey farmer
poultry farmhand
poultry field service technician
poultry geneticist
poultry grader
poultry inseminator*
– artificial insemination technician

poultry nutrition researcher
poultry products sales manager
poultry products technologist
poultry scientist*
poultry tender*
poultry vaccinator*
– chicken vaccinator
supervisor, poultry farm*
– supervisor, brooder farm
– supervisor, egg-producing farm
– supervisor, fryer farm
– supervisor, pullet farm
– supervisor, turkey farm
supervisor, poultry hatchery*
– supervisor, chicken hatchery
– supervisor, turkey hatchery
turkey breeder
turkey producer

Goat Occupations

goat herder*

CHOOSING AN OCCUPATION

Selecting an occupation involves three basic steps. First, one must look at oneself. Second, one should obtain as much information as possible about the occupation in which one is interested. Third, one makes a decision based on the information from steps one and two.

Self-analysis

A person's *ability* is his or her capacity to perform. Grades in school and testing done by guidance counselors can help to determine a person's abilities. It is important to be as realistic as possible when matching abilities to occupations.

Occupations are sometimes chosen on the basis of talents. A *talent* is a natural aptitude a person possesses for performing an activity particularly well. Some people type well, others can sketch or draw, and others get along well with people. Being able to handle animals is an important talent for those considering a job in animal science. Hobbies may be an indication of talents. Other personal qualities should be considered in addition to the talents that a person possesses.

A person's *physical makeup* is his or her health, strength, and stamina. Different occupations have different requirements for physical condition. Limitations due to health problems should be considered when deciding on an occupation.

Previous experience refers to the work a person has done in the past. How well a person has done when working at one job is a good indication of future success on other jobs. Work experience is also a good way to learn more about an occupation in which a person is interested.

Interests are those things that hold a person's attention. People do best when they work at an occupation that interests them. One way to determine one's interests is by taking a test called an *interest inventory*. Guidance counselors are able to give and interpret this kind of test. Interests sometimes change as a person grows older. One must consider whether or not an interest in a particular occupation is temporary or lasting.

Educational aspirations refers to how much education a person desires or can obtain. Many occupations in animal science require education beyond high school. People must decide how long they are willing to study and if it is possible for them to obtain the education needed for a particular occupation.

Attitudes and *values* refer to how a person thinks about life and what things are important to him or her. A person's choice of an occupation should agree with his or her attitudes and values. Many people express their attitudes and values through their work.

Self-concept is how a person sees himself or herself. A positive self-concept is important to job success.

Flexibility, or willingness to change, helps people to make personal adjustments that may be required by an occupation. There is no such thing as a perfect job. All jobs require adjustment. A worker must be able to accept criticism. Flexibility is closely related to success on the job.

Personality is how others see a person. Each person's personality is made up of many different traits. Any given occupation requires certain personality traits for success. Individuals must be sure they have the personality traits necessary for success in a particular job.

Studying an Occupation

When studying an occupation, it helps to have an outline to follow. The following points should be considered when seeking information on a specific job.

The *nature of the work* is what a person has to do on the job. Included are the activities, duties, and responsibilities of the worker. Working conditions are a part of the nature of the work. Working conditions which should be studied include:

a. hours
b. location of work area (indoor or outdoor)
c. physical activity required
d. travel requirements
e. tools or machines used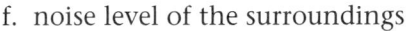
f. noise level of the surroundings
g. presence of dirt and odor
h. possible physical hazards
i. amount of interaction with other people required
j. degree of supervision
k. variety of tasks involved

The mental skills required by a job should also be considered. Some jobs require verbal comprehension, some require reasoning or number ability, while others require spatial ability or mechanical comprehension.

Different jobs require people with different kinds of personalities. Personality requirements may include being sociable, energetic, persuasive, or persistent.

Educational requirements are also an important factor in job selection. Some jobs require only a high school education. For others, a person may need training beyond high school. Some skills can be learned on the job. Sometimes, there are limits on the number of people who are admitted for special training. The cost of the training should also be known.

Some jobs have special entrance requirements, such as a certificate or license. In some fields, a person must belong to a union or a professional association. Other jobs have restrictions that would prevent entry, such as age, height, weight, or physical handicaps.

The demand and supply of workers varies with different occupations and in different parts of the country. Some jobs have a high turnover of workers. It is important to know what the possibility is of obtaining a job in a particular field.

The chances for promotion and the job security available are also points to consider in the choice of an occupation. A person should find out what beginning workers are paid as well as the earnings to be expected after a few years on the job. The opportunity to transfer to another, similar occupation after gaining experience in a particular occupation should be noted. Paid vacations, sick leave policies, pensions or retirement plans, and normal retirement age are other important facts.

Making a Decision

After self-analysis and occupational study, it is time to make a decision. A person must try to match himself or herself as closely as possible to the best occupation. It is a good idea to have several occupations in mind. Most people can succeed at more than one occupation. A person should leave room for a change of career plans in the future.

SOURCES OF INFORMATION ABOUT OCCUPATIONS

Information about occupations is available from a variety of sources. School guidance counselors can provide a great deal of information. Most schools have a copy of the *Dictionary of Occupational Titles,* which gives individual descriptions of more than 20,000 jobs in the United States. Many jobs related to animal science are included.

Agricultural education instructors also have information about jobs in animal science. They can help students explore occupations that interest them. They may be able to give advice on how well a person might fit into a certain occupation.

People who are employed in the occupation in which a person is interested are good sources of information. They know from personal experience the requirements of the job and what it takes to succeed.

Parents may also aid students in making a wise career choice.

State employment services offer help in determining the kinds of jobs available locally. They provide help with testing, counseling, placement, and other services free of charge.

Many books and pamphlets are available that contain information about jobs. A number of these have descriptions of jobs, called *occupational briefs*, which answer many questions about an occupation. Guidance counselors, high school libraries, or agriculture education teachers can supply these books and pamphlets.

Never depend entirely on one source of information about an occupation. Obtain information from as many different sources as possible. The more information a person has, the better the chances of selecting a suitable occupation.

SUMMARY

People with an interest or experience in working with animals have a good chance of success in occupations in animal science. There are many employment opportunities for people who have an interest in animal science.

Agriculture offers many different kinds of jobs and occupations. Livestock production and its related services and supplies, processing, distributing, and marketing are all a part of the agriculture industry. The outlook for job opportunities in agriculture is good.

Three basic steps should be followed in choosing an occupation: (1) self-analysis, (2) studying the occupation one is interested in, and (3) matching personal traits and skills to the occupations of interest.

Many sources of information about occupations are available. School guidance counselors, teachers, people who work in the occupation, and printed material all provide information about various occupations.

Student Learning Activities

1. Prepare a bulletin board display of occupations in animal science.
2. Survey the local community to determine employment opportunities in animal science.
3. Interview a person working in an animal science occupation and present an oral report to the class based on the interview.
4. Invite people in animal science occupations to speak to the class.

Review

1. About what percent of the total labor force in the United States is employed in agriculture-related jobs?
2. What are two ways in which a young person can learn more about a career in animal science?
3. Name five general occupations in animal science.
4. Name three occupations in each of the following areas: (a) beef, (b) swine, (c) sheep, (d) horses, (e) poultry, (f) dairy.
5. List the three basic steps in choosing an occupation.
6. Name five factors that play a part in self-analysis.
7. What types of information should be known about an occupation in order to make a wise career decision?
8. What are three good sources of information about occupations?

Safety in Livestock Production

Objectives

After studying this unit, the student should be able to

- explain the importance of farm safety when working with livestock.
- discuss four types of hazards related to livestock production.
- list the safety practices to be followed when working with beef cattle, hogs, sheep, goats, horses, and poultry.
- develop and use a livestock safety checklist.

Agriculture is one of the most dangerous occupations in the United States. Recent data show a death rate of 21 workers per 100,000. Only mining/quarrying was higher, with a death rate of 25 workers per 100,000. All industries combined show a death rate of 4 workers per 100,000. There are 150,000 disabling accidents involving farm workers. Most fatal farm injuries are caused by machinery. Livestock cause relatively few deaths each year but they are the leading cause of injuries on the farm. Other major causes of non-fatal injuries on farms, in descending order of importance, are machinery (except tractors), hand tools, slips and falls, and tractors. More than 200 children die each year in farm accidents. Most of these deaths result from accidents involving farm machinery. The annual cost of farm accidents in the United States is estimated to be between four and five billion dollars. Table 3-1 shows recent safety statistical data.

Farmers who hire labor are required by law to provide safe and healthy workplaces for their employees. They must inform workers about safety practices. Employees must be told about their rights and responsibilities under the regulations of the Occupational Safety and Health Act (OSHA). Posters have been designed by OSHA for this purpose. OSHA requires farm employers to keep records of work-related

TABLE 3-1   National Safety Statistics—Ranked by Deaths per 100,000 Workers.

Industry	Workers	Deaths	Deaths per 100,000 Workers	Disabling Injuries
Mining/quarrying	600,000	150	25	20,000
Agriculture	3,400,000	710	21	150,000
Construction	6,600,000	1,000	15	350,000
Transportation/ public utilities	6,400,000	750	12	310,000
Manufacturing	18,000,000	610	3	610,000
Government	18,700,000	510	3	580,000
Trade	29,400,000	460	2	880,000
Services	43,400,000	610	1	1,000,000
All Industries combined	126,400,000	4,800	4	3,900,000

Source: Ohio State University Extension.

injuries and illnesses if eleven or more employees work on the farm. Farmers must permit OSHA inspectors to check their farms to see that the law is being followed.

TYPES AND KINDS OF INJURIES

Surveys reported by the National Safety Council show that most of the people hurt by cattle and hogs are males. Almost as many females as males are hurt by horses.

The National Safety Council reports that people in the 45–64 age group are most often hurt in accidents with cattle. Children in the 5–14 age group, and young adults in the 15–25 age group, are most often hurt by horses. Injuries caused by hogs occur most commonly to people in the 25–64 age group.

Injuries from cattle and hogs usually occur in farm buildings or in lots close to the buildings. Accidents involving horses are more common in barnyards, fields, lanes, woods, and along public roads.

Injuries from cows usually occur when they kick or step on people or push them against a hard surface such as the side of a pen. People may be injured by falling when working with cattle. Hogs bite, step on people, or knock them down. The most serious accidents occur with horses and bulls.

Most of the people injured by animals on the farm are farm family members. Fewer than 10 percent of the injuries are to hired help or visitors to the farm.

HUMAN AND ENVIRONMENTAL FACTORS RELATING TO SAFETY

Human error is usually a major factor in the cause of accidents. Being tired, not paying attention, and using poor judgment are frequent causes of accidents that involve animals.

People younger than 25 and older than 64 have more accidents on the farm than people between the ages of 25 and 64. Farming is an occupation in which children are likely to be in the work area. Their curiosity and lack of experience can easily lead them into situations where they may get hurt. As people grow older, they tend to lose some of their strength and agility. They may have poorer balance and failing vision. This may cause them to have more accidents around animals.

Sometimes, workers are not properly instructed in handling animals. This can also result in accidents involving livestock.

A worker who does not feel well may be more likely to have an accident. Sometimes, people are in a hurry to get the job done. This can lead to mistakes in judgment that cause accidents. Long hours of work during a day are often common in farm operations. Being tired increases the chance of having an accident.

Workers who fail to use personal protective equipment in dangerous environments are more likely to be injured than those who dress correctly. There are many dangerous environments involved in livestock operations. These include slippery floors, manure pits, corrals, dusty feed areas, silos, automatic feeding equipment, and confinement livestock buildings (Figure 3-1).

Many confinement livestock buildings have a manure storage pit that is cleaned out only a few times each year. If the building is not properly ventilated, the pit gases can kill workers and livestock. Vent pipes must be installed properly. Improperly installed vent pipes may allow gas fumes to be recycled into the building, which can cause illness and possible death.

A standby source of electrical power is recommended for modern livestock farms. This is especially important for farms with confinement livestock buildings. If pit fans do not operate because of an electrical power failure, a buildup of toxic gases can result. An emergency source of power can be a life-saving measure should this situation occur.

FIGURE 3-1   Proper ventilation of confinement livestock buildings helps to reduce the danger from buildup of toxic concentrations of poisonous gases. However, protective equipment may sometimes be necessary. *Courtesy of University of Illinois at Urbana.*

CHEMICAL SAFETY

The U.S. Environmental Protection Agency (EPA) issued regulations under the Worker Protection Standard for farm chemicals that became effective January 1, 1995. These standards are designed to reduce the health risks associated with pesticide use. They apply to farm employees who work with farm chemicals or related equipment.

Farmers and their immediate families are exempt from the regulations. The standards apply to the use of pesticides for plant production on farms, forests, nurseries, and greenhouses. There are exceptions for the use of pesticides for livestock, pasture, rangeland, structures, gardens, lawns, rights-of-way, and post-harvest applications.

The major provisions of the regulations are:

- Personal protective equipment must be provided to pesticide applicators or handlers.
- Restricted entry to fields treated
  – with highly toxic material: 48 hours.
  – with moderately toxic material: 24 hours.
  – with less toxic material: 12 hours.
- Workers must be told, either in person or by posted signs, that a field has been treated with a pesticide and when it is safe to enter the field after treatment.
- Notices must be posted in easily seen locations showing the date, time, and location of treated fields. The notices must include the brand name, active ingredients, and EPA registration number of the pesticide used.
- Workers must be trained in safe pesticide handling methods at least once every five years.
- Written safety information must be given to workers, and pesticide safety posters must be displayed.
- Water, soap, and towels for washing and decontamination must be provided and located where they are readily accessible to the workers.
- Emergency transportation to a medical facility must be readily available.
- If the worker is applying a Class I (highly toxic) pesticide, he or she must be seen or talked to by the farmer or supervisor every two hours while the work is being done.

The EPA requires that labels on pesticides must provide information that employers need to properly inform their employees about the safe use of the pesticide. Much of the information required under the Worker Protection Standard must be included on the pesticide label.

While the Worker Protection Standard applies only to employees, farmers and their families who work with chemicals should be careful to protect their own health. Following the Worker Protection Standard that protects workers can also protect farmers and family members.

Many chemicals that are used in growing crops or raising livestock are dangerous to people. All workers must be instructed in safe practices relating to the use of these chemicals. Make sure workers who handle farm chemicals read and understand the label instructions on the chemical containers, including the instructions for first aid treatment in case of an accidental spill or ingestion of the chemicals.

The Occupational Safety and Health Administration (OSHA) requires that a Material Safety Data Sheet (MSDS) be available for all chemicals in the workplace. Even if one is not required to have a MSDS available, it is a good safety practice to do so. Any retail outlet that sells hazardous chemicals is required to provide the MSDS for that chemical if a consumer requests one. Anyone using a hazardous chemical should carefully read and understand the MSDS for that chemical before handling it. Be sure the MSDS is current. Check the date it was prepared; it may be wise to contact the manufacturer to make sure it contains the most recent information available.

Information typically found in a MSDS includes:

- The identity of the chemical; the manufacturer's name and address; and an emergency contact number. A non-emergency number for more information may also be included.
- Hazardous ingredients found in the chemical.
- The physical and chemical characteristics of the chemical.
- Fire and explosion potential of the chemical.
- Health hazards posed by the chemical.
- Precautions for safe handling and use of the chemical.
- Procedures for controlling spills of the chemical.
- Control measures for the use of the chemical.

In some parts of the United States, a special permit is required to buy and use some farm chemicals. Information regarding these requirements may be secured from the local Agricultural Extension Service office.

Information about the proper disposal of unused pesticides and empty containers is found on the label of the container. Always follow these instructions to reduce health hazards for workers. Generally, empty containers must be kept in a safe storage area until they can be disposed of properly. Metal, plastic, and glass containers should be rinsed three times, punctured or broken, and then disposed of in a designated landfill or buried at least eighteen inches deep in a location where they will not contaminate surface or ground water. Do not pour unused pesticides down a drain; bury them in a designated landfill. CAUTION: Always check label requirements and local and state disposal regulations before discarding any unused pesticides or empty containers.

Information about chemical regulations may be obtained from the Environmental Protection Agency by calling a toll free number. A regional poison control center may also be reached by calling a toll free number. Current toll free numbers may be secured by calling 1-800-555-1212 and giving the operator the name of the agency desired.

Chemicals, such as pesticides, must be handled with care, and clothing that has been contaminated must be properly washed or disposed of. Some chemicals can enter the body through the skin; this is the most common way dangerous chemicals get into a worker's body. Other dangers include inhaling or swallowing the chemical.

A person who has handled chemicals should always wash his/her hands and face with soap and water. After completing a job that required the use of chemicals, a worker should shower to remove all the chemical from all parts of the body. If an accidental, massive contamination of the body occurs, the worker should immediately take a shower to avoid absorbing any of the chemical into the body.

Care of Clothing Worn While Using Farm Chemicals

The proper care or disposal of clothing worn while using farm chemicals depends upon the level of toxicity of the chemical used. If the label says **CAUTION**, the chemical is slightly toxic and the clothes may be cleaned with one to three machine washings. If the label says **WARNING**, the chemical is moderately toxic and the clothes will need more than three machine washings. If the label says **DANGER POISON**, the chemical is highly toxic and the clothes must be disposed of according to the directions on the label. Clothing that is contaminated with a concentrated chemical should also be disposed of following the label directions.

Guidelines to follow for washing clothes contaminated with farm chemicals include:

- Before washing, store the clothes in a plastic bag; separate from other clothes.
- Handle the contaminated clothes with neoprene or rubber gloves; do not use these gloves for any other purpose.
- Wash the clothes within 8 hours of use.
- Before washing, rinse the clothes by soaking in water in a tub, or hanging them on a line outside and hosing them down with water, or run them through a prewash cycle with agitation in the washing machine.
- Be careful when disposing of the rinse water. If it contains herbicides, do not dispose of it on a garden or lawn area.
- Do not mix contaminated clothes with other clothes in the washing machine.
- Put only a few contaminated clothes in the machine at a time, grouping together those that are contaminated by the same chemical, and set the machine for a full load.
- Use hot water (140° to 150°F) (60° to 65°C) for washing.
- Do not use a suds saver cycle on the washing machine.
- Use a heavy-duty liquid detergent to remove oily chemical residues from emulsifiable concentrate or use a powdered detergent with a phosphate base to remove chemical residues from wettable powders.
- Use a normal 12- to 14-minute wash cycle with two rinses.

- Do not use bleach if the clothing was contaminated with ammonia fertilizers. The ammonia and bleach can combine to form a deadly chlorine gas.
- Increase the amount of detergent by 1.25 times on clothes treated with soil- or water-repellent spray.
- Dry the clothes on a clothesline to avoid contaminating the clothes dryer.
- To remove pesticides from the machine after washing contaminated clothes, run the machine empty through a complete cycle using hot water and detergent.
- After cleaning, wipe the tub with isopropyl alcohol; this will help remove all traces of the chemical.

If any articles of clothing, including shoes, have become badly contaminated with toxic chemicals, they should be burned or buried. It is more difficult to remove oil-based chemicals from synthetic fibers than from clothing with natural fibers such as cotton. It may, therefore, be more likely to be necessary to destroy contaminated synthetic fiber clothing than clothing made from natural fibers. Care must be taken when burning or burying articles of clothing that are contaminated to avoid further contamination of the air or the worker. Bury contaminated clothing at least eighteen inches deep in an area where there is no danger of polluting either surface or subsurface water or dispose of them in a sanitary landfill. Sealing the contaminated clothing in a plastic bag will help reduce the danger of water pollution.

Do not wash and reuse disposable chemical respirators. Washing will not remove chemical contamination from respirators. They are designed to be used once and then thrown away.

Closely woven fabrics are more likely than other types of fabric to become contaminated with chemicals. The close knit fabric acts like a wick to absorb chemicals. It is very difficult to remove chemicals from this type of clothing. Leather gloves and boots also readily absorb chemicals and are almost impossible to decontaminate.

Research at Michigan State University has found that a 100 percent olefin, spun-bonded non-woven material is one of the best fabrics for protective clothing. It is manufactured by DuPont under the trade name Tyvek®. When tested at Michigan State University, Tyvek gave 25 times more protection from chemical penetration than treated chambray.

Standard Tyvek is not totally waterproof, so it is a good practice to use a rubber apron when mixing chemicals or if the possibility exists that the chemicals being used are likely to get the worker wet. Polylaminated Tyvek costs more than standard Tyvek but is totally waterproof. The higher cost may be justified in situations where the worker is more likely to get more of the chemical on his/her clothes.

Disposable coveralls made of either standard or polylaminated

Tyvek are available as protective clothing when handling chemicals on the farm. Tyvek does not breathe and is, therefore, hot to wear. It is, however, a light-weight fabric and is white in color to reflect heat.

Storing Chemicals

Farm chemicals, such as pesticides, should be stored in a safe place where children or others cannot accidentally get into them. Do not use the wrong types of containers, such as soft drink bottles, for chemical storage and be sure the container is properly labeled. The label should contain a warning that the contents are dangerous. Children may not realize that a soft drink bottle contains a dangerous chemical, and they may accidentally drink the contents. National Safety Council data indicates that this type of accident causes many deaths and/or injuries each year.

Chemicals should be kept locked in a cabinet located in an area that does not freeze. Many chemicals used on the farm are damaged or destroyed by freezing. Keep only chemicals in the cabinet. The area around the cabinet should be kept free of other objects and should provide drainage in case of an accidental leak or spill from any of the containers. The area should drain into a safe collection area and not into regular sewer or drainage lines. Do not use the top of the cabinet to store other items that may be accidentally knocked down, possibly upsetting containers and causing a dangerous spill. Proper lighting in the area makes it easier to read the warning labels on the containers.

FIRST AID KITS

First aid kits containing the proper medical supplies should be kept in the home, in livestock buildings, on all major pieces of equipment, and in all vehicles. Several sizes of first aid kits are commercially available. Safety specialists recommend that a first aid kit for farm use should contain the following items:

- various sizes of sterile bandages
- roller bandages: 5 yards long, 1- and 2-inch widths
- various widths of adhesive tape
- triangular bandages for slings
- cotton balls
- bandage scissors
- tweezers
- tongue depressors
- safety pins
- thermometer
- chemical ice bags
- splints: 1/2 inch wide, 1/4 inch thick, 12–15 inches long

- eyewash solution
- tincture of green soap and/or a plastic bottle of soap solution that may be used for cleaning wounds or washing off chemical spills
- one one-ounce bottle of ipecac syrup (it will induce vomiting if a chemical is accidentally swallowed)
- one pint of activated charcoal (can be mixed with water and swallowed to absorb many pesticides)
- one can of evaporated milk and a can opener (milk is an antidote for many chemicals)
- a small container of salt (a solution of 1/2 teaspoon of salt mixed in one quart of water, when drunk, provides first aid for someone in shock)
- a plastic quart container of clean water (in an emergency, water from a stream or pond can be used)
- two empty plastic containers with tight lids for mixing antidotes
- a teaspoon for measuring
- several coins to use in a pay phone to call for help
- a list of names and phone numbers of the nearest doctors, ambulance services, paramedic services, and poison control centers (while these numbers can be secured from the operator, a few minutes saved by being able to dial them directly may mean the difference between life and death in an emergency)

Most commercially available first aid kits will not contain all of the recommended items on the above list. The preparation of a first aid kit containing these items may involve combining a commercial kit with additional items and putting them all in a larger container.

In addition to having a first aid kit readily available, farm workers and family members need to read the labels on chemical containers. These labels contain information on first aid treatment when the chemicals are accidentally swallowed or come into contact with the skin.

HEAT AND HUMIDITY FACTORS RELATING TO SAFETY

High temperature combined with high humidity can be a health hazard for farm workers. Some general guidelines on the relationship between temperature and humidity relating to safety when engaging in physical labor are shown in Figure 3-2. Danger from heat-related problems exists both with lower temperatures and higher humidity and higher temperatures with lower humidity. Under these conditions, workers may be subject to heat exhaustion (sometimes called heat prostration) or to heatstroke. Heatstroke is more dangerous to the worker than heat exhaustion.

The symptoms of heat exhaustion include dizziness, nausea, a feeling of weakness, lowered body temperature, and a cold, clammy feel to the skin. This condition may result in the collapse of the

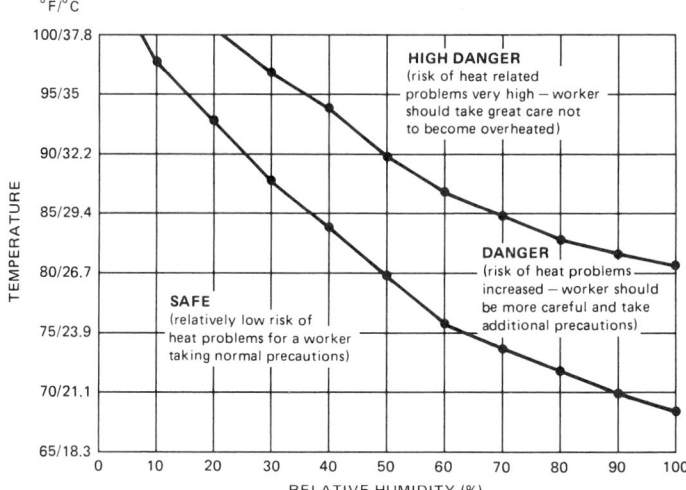

FIGURE 3-2 The relationship of temperature and humidity to safety when engaging in physical labor.

victim. Treatment of heat exhaustion includes rest, drinking fluids, and restoring salt to the body.

Heatstroke occurs when the body cannot get rid of excess heat fast enough by sweating. The average person at rest produces about 70 calories of heat per hour. When engaged in hard physical labor, the rate of heat production may increase by as much as eight times the amount produced at rest. This will result in the production of about 560 calories of heat per hour when the person is working hard.

The symptoms of heatstroke include rapid pulse and a flushed, dry, extremely hot skin. The victim's body temperature can increase to as much as 110°F (43°C) or higher, which causes damage to the central nervous system. If the condition is not treated immediately, the pulse soon becomes weak. The victim may collapse and go into a coma. Heatstroke can result in death if the victim is not treated quickly.

Treatment of heatstroke must be designed to reduce the victim's body temperature as soon as possible. Use ice packs or an ice-water bath to treat heatstroke. If these methods are not immediately available, spraying the victim's body with water will help reduce body temperature until the person can be transported to a facility that can provide better treatment.

People who are engaging in heavy physical work during hot, humid weather can take some precautions to help avoid heat-related problems. Select clothing that is lightweight and light colored. A cotton or cotton blend fabric is preferred because it allows air to pass through. A lightweight khaki is better than heavy denim in hot weather.

Sweating causes a loss of body fluids, which results in weight loss. Any weight loss in excess of five percent of one's normal weight is dangerous when working in hot, humid conditions. A person should drink a lot of water to replace lost body fluids. Sugar-based drinks are not as good as water for replacing body fluids; the sugar-based drink causes the stomach to draw water from other parts of the body to dilute the sugar. This causes a delay in the absorption of water into the body tissues. Eating fruits and vegetables also helps the body to replace electrolytes that are lost in hot weather.

Some people have a greater risk of heat-related problems. Older people are at greater risk because they do not sweat as easily and therefore their bodies do not cool themselves as quickly as those of younger people. People who use alcohol, tranquilizers, or are taking some kinds of medications are also at higher risk from heat-related problems. People who are taking medications and may be working in hot, humid conditions should consult their doctor for precautions to be followed.

HAZARDS IN HANDLING LIVESTOCK

Whenever livestock are handled, there is the possibility of injury to the worker. Loading and unloading operations are particularly dangerous because the animals are excited and confused by what is happening. It is important to have solid facilities for handling livestock. Temporary or makeshift gates, pens, and chutes increase the chances of a worker being injured. Squeeze chutes and headgates should have solid latches to prevent accidental opening and possible injury to the worker (Figure 3-3).

Facilities should be designed so that the worker does not have to enter a small or enclosed area with animals. A mangate or other means of quickly getting out of pens should be provided. Catwalks should be a part of chutes and alleys so that the workers do not have to get into the area with the livestock. A guard rail should be placed on all catwalks that are more than 18 inches off the ground.

Floors must not be slippery or be cluttered with things that might trip the worker. Sharp corners, pinch points, and protrusions should be eliminated from livestock handling facilities.

Lighting should be adequate so that workers can see what they are doing. The National Safety Council recommends at least 10 foot-candles of light in squeeze and loading chute areas. Lighting should be diffused and even. There should be no bright spots that might confuse cattle and cause them to balk.

Beef cattle will seldom attack a person. However, sudden noises may startle them. The cattle may then injure a person by crowding him or her against a hard surface. Never approach cattle from the side or rear. Approach the animal from the front while talking to it. This alerts the animal to the presence of the worker. Always wear boots or hard shoes when working around cattle. Tennis shoes should never be worn around livestock. Cattle kick forward and then to the rear. This is a hazard that must be watched for when working with an

FIGURE 3-3 The use of well-constructed headgates and squeeze chutes reduces the danger to the worker when treating livestock. *Courtesy of University of Illinois at Urbana.*

individual animal. Many of the safety practices outlined for horses also apply to beef cattle.

An understanding of why cattle behave in certain ways helps reduce stress for both the animal and the worker and increases safety. Reducing stress also increases cattle productivity.

The eyes of cattle are located on the sides of their head, which gives them a panoramic view of their surroundings. They have a wider range of peripheral vision than do humans. Cattle have limited depth perception and see things in various shades of black and white. As a result, they are sensitive to movement in their field of vision and also react strongly to contrasting patterns of objects around them. The eyes of sheep, goats, horses, and chickens are also located on the sides of their heads. The visual reaction of these animals is very similar to that of cattle.

Curved chutes that have solid sides and are a uniform color make it easier and safer to move cattle. Anything that creates shadows and contrasting patterns in cattle-working facilities should be avoided, because cattle will slow or stop when they see these things. Cattle sometimes refuse to move across a shadow or bright contrasting patterns on the floor in front of them. Articles of clothing that can move in the wind should not be left hanging on the sides of fences or chutes, because cattle may balk when confronted with these sudden movements.

Cattle react negatively to sudden, loud noises, as well as to high-pitched sounds. When moving cattle, do not yell or make other sudden loud noises. This only tends to confuse the animals and will not get them to move in the desired direction. Cattle should be handled calmly and quietly at all times.

Cattle behavior is patterned as a result of previous experiences. When cattle are handled roughly in working pens, they remember the experience and it becomes very difficult to get them to re-enter the area for further treatment. Avoid the indiscriminate use of electric prods, yelling, punching, and arm waving to move cattle.

To increase safety and make it easier to work with cattle, get them used to being around people when they are on pasture. Handle the cattle carefully in the working pens to avoid unpleasant experiences that they will remember. Don't try to push cattle too hard, and allow them to follow the natural herd leaders. When catching cattle in a headgate, make sure to catch them the first time. It is very difficult to catch them in the headgate after they have been missed on the first attempt.

Cattle hesitate to enter what appears to them to be a dead end. That is one of the main reasons for using curved chutes. Cattle will move more easily into headgates and chutes that are open ahead of them. Also, cattle hesitate to enter a darkened building when it is light outside. Lighting the interior of a truck, especially at night, makes it easier to get cattle to move into the truck.

After treating cattle in a headgate or working pen, allow them to move out of the area at their own speed. Yelling at them or otherwise abusing them to get them to move will make it more difficult to get them to ever enter the area again.

People who work with cattle need to be aware of the "flight zone" concept. The flight zone is an imaginary circle around the animal or the herd. The radius of the circle is fairly small for animals that are used to being around people and much bigger for animals that have not been handled much in the past. The flight zone for cattle generally ranges from 5 to 20 feet. However, cattle that are used to being handled, such as dairy cows, may allow a worker to walk up and touch them without moving away. When the worker enters the animal's or herd's flight zone, the animal or the herd will move away. If the worker stays on the edge of the flight zone, the animals will move in a calm, steady manner.

When moving cattle, the worker should stay in a position where the animal can see the worker. If the worker is directly behind the animal, a sudden, unpredictable movement to one side or the other may result. When the worker is positioned at the edge of the flight zone and about 30 degrees to one side, the animal will move straight ahead; when the worker moves to a position at the edge of the flight zone about 45 degrees to one side of the animal, the animal will turn. Cattle tend to turn in a circle to keep the worker in view. When moving a herd of cattle, the worker should adjust his/her position at the edge of the herd flight zone relative to the lead animal in the herd.

Do not try to head off cattle when they turn and try to go back into an alleyway. A more effective way to handle this situation is to move out of the animal's flight zone when it starts to turn back. This will generally allow the worker to reestablish control over the animal's movements. This is also a safer way to handle the animal, because there is a danger of being knocked down by a charging animal when trying to head it off.

A sow will attack a person if her pigs are hurt or threatened. Never work with small pigs in the same pen as the sow. Hogs will bite or knock a worker down, so care must be taken when moving these animals. A hurdle or solid panel should be used when handling hogs. It is possible to move a hog backward by placing a basket over its head. It will try to back out and can be guided fairly easily to where it is wanted. Small children and visitors should be kept out of hog pens. Do not let children pet hogs through the fence.

An understanding of how hogs behave makes it easier and safer to handle them when they are being moved from one area to another. Unlike cattle, sheep, goats, horses, and chickens, the eyes of hogs are located further forward on their heads. This gives hogs better binocular vision and greater depth perception.

Hogs have a tendency to want to stay in or to return to an area with which they are familiar. When an attempt is made to separate a hog from the herd, that hog will try to return to the herd. However, hogs in a group will follow a leader, trying to maintain visual and physical contact with each other. This makes it easier to separate a group of hogs, rather than one hog, from the herd.

When hogs are handled under artificial light, they will move rather easily from a dark area to a lighted area. However, if hogs are moved from a dark house into bright sunlight, they are likely to balk as they leave the house. When moving hogs at night, it is difficult to get them to move from a lighted house to a dark area outside the house. At night hogs will move more easily into a truck if there is a light in the truck. They will also move more easily through dark alleys or chutes if there is a light at the end of the alley or chute.

Loading chute floors should be made of the same or similar material as the floor the hogs were raised on. Hogs will move through a loading chute more easily if the chute is nearly level or slopes at no more than 25 degrees. Make the outside wall of the loading chute solid so the hogs cannot see outside and become distracted. A loading chute width of 14 to 16 inches reduces the possibility of hogs trying to turn around in the chute.

Sheep and goats, being rather small animals, are generally not considered very dangerous. It is possible to be injured by being butted by a ram or buck. This is a particular hazard for younger children and elderly people.

Poultry are usually not dangerous, but a person may be pecked by a hen or rooster. Geese and gobblers are more likely than chickens to injure a person by pecking them. Female fowl that are setting on a nest hatching eggs can be aggressive. They are liable to attack a person who disturbs them. The wings of the turkey, duck, and goose hen can inflict injury if the setting hen is disturbed. Equipment and dust hazards in poultry facilities are more likely to injure workers than the poultry are.

HORSE SAFETY

Safety with horses is especially important since millions of Americans in both rural and urban areas ride horses. Many of these people are not used to being around animals and must be made aware of basic safety procedures. Serious injury can result from failure to follow safety rules. Horses are variable in temperament. Some are timid and will react violently when frightened. If safety rules are followed when riding or working with horses, these activities can be enjoyable experiences.

Approaching

Horses have good hearing, but they do not see well directly in front or to the rear. Always speak to the horse as you approach. Approach the horse at a 45° angle from the shoulder, never directly from behind. Pet the horse by rubbing its neck or shoulder. Do not reach for the end of its nose. Stay out of kicking range when walking behind the horse. Do not step over the tie rope or walk under it.

Handling

The most important rule to remember is to stay out of kicking range of the horse whenever possible. Another safety measure is to stand close to the horse when working with it. Then, if it kicks, the full force of the kick will not be felt. Work as much as possible from a position near the shoulder of the horse. When working with the horse's tail, take a position near the point of the buttock. Stand to the side and face the rear (Figure 3-4). Do not stand directly behind the horse.

A nervous handler makes the horse nervous. A calm, confident manner that is kind but firm should be used. Let the horse know what is happening. Move slowly when doing things with the horse, such as lifting his feet. A person should learn the peculiarities of his or her horse and tell others who may be working or riding the animal about them.

Use simple methods of restraint. Tying or holding the head is safest when working with the horse. Horses should be tied with about two or three feet of rope. This should be at the height where the lead shank attaches to the halter. Do not leave a halter on a loose horse. The horse might catch his foot in the halter or the halter might catch on a post or other object.

Never tease a horse. If it is necessary to punish the horse, do so at the moment of its disobedience. Never strike a horse around the head.

Protective footwear should always be worn around horses. A horse may step on a person's foot, or there may be nails around the barn that could cause injury. Boots or hard-toed shoes are better footwear than tennis shoes or moccasins. Never go barefooted around horses or barns.

Leading

When leading a horse, walk beside him rather than ahead or behind. Turn the horse to the right, walk around him and keep him on the inside. Horses are usually stronger than people, so it is unwise to try to outpull them. The horse will usually respond to a quick snap on the lead strap if he is properly halter-broken.

The lead strap, halter shank, or reins should not be wrapped around the hand, wrist, or body. Fold the lead strap accordion style in the left hand when leading the horse (Figure 3-5). The right hand should be extended slightly toward the horse. The horse's shoulder will make contact with the elbow first and move a person out of the way.

Be especially careful when leading a horse into a box stall or

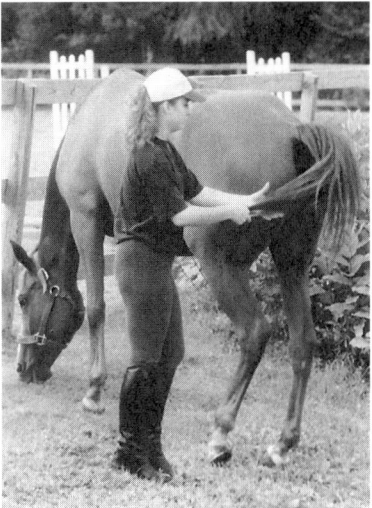

FIGURE 3-4   Stand to one side and face the rear when working with the horse's tail. *Photo by Michael Dzaman.*

FIGURE 3-5   The lead strap is folded accordion style when leading a horse. *Photo by Jim Bodine.*

pasture. Always turn the horse so that he faces the door or gate before releasing him. Otherwise, he may bolt forward when released.

Tying

Horses should be tied with about two feet of rope. They should not be staked out. Be sure the horse is tied far enough away from strange horses so they cannot fight. Long lines and leads should be kept off the ground. This will keep the horse's or rider's feet from being tangled in the lines. The lead shank should be untied before taking the halter off the horse.

Bridling

Do not try to bridle a nervous horse in close quarters. When bridling, stand close to the left side of the horse, just behind the head. In this position, the horse's neck action will push you clear if he throws his head to avoid the bridle.

Saddling

When saddling a horse, a person should stand well back in the clear and reach forward. Do not let the cinch ring strike the off (right) knee of the horse. (See Units 32 and 34 for detailed information on saddles and saddling.) The front cinch of a Western double-rigged saddle should be fastened first. Fasten the rear cinch last. When unsaddling the horse, unfasten the rear cinch first, then the front cinch. Be sure the rear cinch is not so loose that the horse could get his foot caught in it. Make sure the strap connecting the front and rear cinch is secure.

The saddle should be swung into position easily. Do not drop it. Adjust the saddle carefully and check the cinch after walking the horse a few steps. The cinch must be checked again after riding a short distance.

Mounting

Do not mount the horse in a barn or near fences, trees, or other overhanging projections. Horses should be trained to stand perfectly still while the rider is mounting.

Riding

Horses may be frightened by unusual objects or noises. If the horse is frightened, steady him and give him time to calm down. After the horse is calm, ride or lead him past the obstacle. If the horse attempts to run, turn him in a circle and tighten the circle until he stops.

Walk horses up or down hills and go slowly on rough ground or in sand, mud, ice, or snow. Always allow the horse to pick his own way. Try to stay off paved roads. If it is necessary to cross or be on a paved road, slow to a walk or lead the horse.

Do not ride away from another rider who is mounting a horse. Leave a safe distance between riders and do not rush past other riders who are going slower. If it is necessary to pass others, approach slowly and pass on the left side.

Night riding is more dangerous than riding during the day. If the ride is on a road, follow the same rules as for pedestrians. Light-colored clothing should be worn. Riders should carry flashlights and reflectors. Allow the horse more freedom to pick his way. The horse's senses are much keener than the rider's.

Equipment and Clothing

All equipment should be kept in the best possible condition. Replace any strap that is too worn to be safe. Make sure all tack fits the horse. Do not wear spurs on the ground because of the danger of tripping. Clothing should be neat and well fitted so that it will not snag on equipment. Boots and shoes should have heels so the foot will not slip through the stirrup. The horse's feet must be properly trimmed and shod.

Hauling Horses

It is safer for two people to load a horse on a trailer than for one person to attempt it alone. Stand to one side and never directly behind a horse when loading or unloading it from a trailer. The horse should be trained so that it can be sent into the trailer. Do not lead a horse into a trailer unless there is a front exit from the trailer. Be sure the ground around the trailer gives firm footing. Remove all equipment from the horse before loading. Use the halter for leading the horse. Always untie the horse before opening the trailer gate or door when unloading. Make sure the trailer is in good condition.

HAZARDS OF ANIMAL DISEASES

Diseases and parasites that may be transmitted between man and animals are called *zoonoses*. Some of the common zoonoses are rabies, brucellosis, bovine tuberculosis, trichinosis, salmonella, leptospirosis, swine erysipelas, ringworm, tapeworm, and spotted fever. Some of these diseases and parasites are very dangerous to man.

Cleanliness, vaccination, quarantine of sick animals, and avoiding exposure are some of the ways to prevent these diseases. Wear rubber gloves when treating sick animals. A doctor should be called if a person becomes sick after contact with animals.

Cases of human rabies in the United States are extremely rare, with only about two cases per year reported. However, the disease is nearly always fatal once symptoms appear. Because of the seriousness of rabies if it is contracted, people who work around livestock need to take precautions against the possibility of becoming infected by the virus that causes rabies.

The virus is normally transmitted in the saliva of the infected animal. Domestic animals may contract rabies by being bitten by infected wild animals such as skunks, raccoons, or bats. Domestic dogs rarely become infected because of widespread immunization against rabies. Vaccinations must be kept up to date for farm pets such as dogs and cats to reduce the danger of rabies infection.

Workers who treat sick animals should wear rubber gloves to help prevent infection through open cuts on the hands. If the sick animal dies and there is suspicion that rabies might be involved, the head of the animal must be sent to a testing laboratory to determine whether it had rabies. Notify local health officials of a suspected case of rabies. Be careful not to damage the brain tissue. When handling the animal, keep your hands away from its mouth.

Be careful around any wild animal that appears to be acting strangely. Do not pick up or play with wild animals. If there is a suspicion that the animal has rabies, kill it but do not shoot or club it in the head. The brain tissue must be undamaged for laboratory tests. Notify local health officials immediately when a wild animal suspected of having rabies is killed. Preserve the head carefully for laboratory analysis. If you are bitten by any animal, quickly wash the wound with soap and water and have a doctor check the wound.

Workers who come into contact with animals with confirmed cases of rabies must be vaccinated immediately. The vaccine is safe and painless. Five doses are usually given over a four-week period. Injections are given in the hip or arm.

PERSONAL PROTECTIVE EQUIPMENT

Several kinds of personal protective equipment should be used around livestock. Bump caps protect the head around livestock facilities. Respirators should be used around dusty or moldy hay. They should also be used in silos, manure storage areas, and for the use of some pest-control chemicals. Eyes should be protected by goggles from dust, chaff, and chemicals. Glasses should have impact lenses. Protective gloves should be used for certain jobs. Cotton or canvas gloves may be used as hand protection for light work. Leather gloves may be used for heavier work or when working with barbed wire fencing. They increase gripping power and protect the hands when handling rough or abrasive materials. Rubber gloves should be worn around sick animals or when assisting at birth. Safety shoes should always be worn on the farm.

Workers in livestock confinement buildings suffer a high rate of respiratory and other problems. Typical problems include coughing, shortness of breath, scratchy throat, headaches, and watering eyes. Permanent lung damage may result from continued exposure to a contaminated atmosphere in confinement livestock buildings. Some deaths have resulted from toxic gases produced in liquid manure storage pits.

Livestock workers are exposed to three types of atmosphere contamination. Dust and particulate matter come from feed, animal hair, and fecal matter. Pesticides used in treating livestock may cause health problems when inhaled by workers. Toxic and asphyxiating gases are produced in liquid manure storage pits. The four main kinds of dangerous gases produced are ammonia, hydrogen sulfide, methane, and carbon dioxide.

There are two kinds of respiratory protection equipment available for use by workers. Air-purifying respirators use filters to remove the contaminants from the air before it is inhaled. One type of purifying respirator removes particles from the air. The other type removes vapor and gas from the air. A combination filter is available for removing both kinds of contaminants from the air before it is inhaled.

Atmosphere-supplying respirators supply air from a source independent from the surrounding air. One type supplies air through a compressed air line and the other type uses a compressed air cylinder.

Selection of respirator protection is based on the type and concentration of contaminant found in the air. Use only respirators that have been tested and meet the minimum standards of the National Institute for Occupational Safety and Health (NIOSH). Such respirators can be identified by a NIOSH number and label that describes the kind of hazard it protects against.

A coding system has been in use since July 1998 to identify three types of non-powered particulate filter respirators. The "N" code means the respirator is not oil-resistant, the "R" code means it is oil-resistant, and the "P" code means it is oil proof. Efficiency levels of 99.7%, 99%, and 95% are available in all three classes of filters. For example, a filter coded R99 means that it is at least 99% efficient and is resistant to oil.

FACILITIES

There are many hazards in the facilities used for livestock. Slippery steps or floors may cause falls. Electric shock is a possible danger in damp areas. Electrical cords or appliances that may come into contact with water are hazards. Waterers that overflow or faucets that do not turn off completely may cause slippery conditions or electrical shock hazards. Strains can result from lifting heavy loads. Other areas that are hazardous include manure pits, lagoons, livestock confinement buildings, and grain storage areas.

Silo Hazards

As noted earlier, silos are dangerous for several reasons. People can fall from the silo or may be injured by equipment used to fill the silo

FIGURE 3-6 Silos pose several hazards for farm workers. Workers may fall from the silo or be harmed by dangerous gases that build up inside. Always use caution when entering a silo. *Photo by Jim Bodine.*

(Figure 3-6). Nitrogen gases that form from ensiling green material can be deadly. Fermentation of the ensiled material produces mainly carbon dioxide and nitrogen dioxide; hydrogen sulfide and ammonia are produced in smaller quantities. Carbon dioxide is odorless and colorless. Nitrogen dioxide is yellow-brownish to reddish-brown in color and smells like bleach. It is heavier than air and can sink down the unloading chute into the feed room at the base of the silo. The most danger exists during the first twelve to sixty hours after filling the silo. Silo gas can be a danger in an unvented silo for several weeks after the silo is filled. Nitrogen dioxide at low concentration can result in eye irritation and coughing or lung injury. At high concentration it can render a person unconscious and even cause death.

Workers should not enter the silo for at least three weeks after it has been filled. Warning signs of the presence of deadly gas include yellow or brown stains on the sides of the silo chute or feed room wall and dead birds or small animals lying on the floor below the unloading chute. If it is necessary to enter the silo during this period, mechanically vent the silo by running the silo blower for at least 30 minutes before entering. Keep the blower running the entire time the worker is in the silo. Do not depend on respiratory equipment for protection unless it carries its own oxygen supply. Never enter the silo alone during the danger period. Always wear a body harness with a safety lifeline and have another person present and observing during the entire time the worker is in the silo.

The silo room should be ventilated for two weeks after the silo is filled. Keep any doors between the silo room and the barn closed. Do not let children or visitors enter the silo room or the silo during the danger period. Get treatment immediately if exposed to silo gas.

Grain Handling and Storage Hazards

Unloading grain from a storage bin can pose several dangers for workers. Unloading equipment sometimes becomes plugged and a worker must enter the storage facility to correct the situation. The worker may become trapped in the flowing grain if proper precautions are not taken. It only takes four or five seconds for a worker to become submerged to the point of helplessness. Within twenty seconds a worker can become completely covered by the grain.

Another hazard is created when feed and grain bridge over in a storage facility and a hollow space forms near the bottom. If a person falls into this space, he or she may be buried under the feed or grain. This would result in suffocation.

Dust and molds in grain storage areas can cause workers to become sick. Toxic Organic Dust Syndrome can cause symptoms much like those of the flu—coughing, fever, chills, headaches, muscle aches, shortness of breath, and fatigue. Symptoms will last from one to seven days. Farmer's Lung Disease produces similar symptoms but is more serious because it may cause permanent damage. Breathing in large amounts of dust can cause Toxic Organic Dust Syndrome. An allergic reaction to mold spores in the air from moldy grain, hay, or straw causes Farmer's Lung Disease. Symptoms of either illness will become apparent within four to twelve hours after exposure to dust and molds in the air. A person with symptoms of these illnesses should see a doctor for treatment. A blood test can determine which illness is present.

Safety precautions to be followed when unloading grain from a storage facility include:

- Make sure the control circuit on automatic unloading equipment is locked in the off position before entering the storage bin.
- Do not enter the storage bin when the unloading equipment is operating.
- Do not enter a storage bin from the bottom when the material is caked to the sides of the bin or is bridged overhead.
- Use a long pole made of non-conductive material to break up bridged or caked material. Work from the outside of the bin through a door or hatch in the roof. Poles made of conductive material may contact a power line, posing a danger of electrocution.
- Always use safety harness or safety belts that are secured firmly to lines in such a manner that a person will be kept above the material in case he or she falls. Never work alone. Always have at least one other person, who knows what to do in case of an accident, outside the bin and also equipped with a safety harness or belt. (The use of a safety harness or safety belt is mandated by OSHA in facilities subject to OSHA regulations.)
- Be aware that an accumulation of dust in a grain bin can be highly explosive. Do not do anything that might cause an accidental spark that could result in an explosion.
- Be sure the enclosed area is well ventilated before entering and keep it ventilated while working in the area.
- Use a respirator designed for toxic dust/mist that is approved by the National Institute for Occupational Safety and Health. It will have the words "dust/mist" stamped on the mask. Thin paper filters should not be used because they will not filter out the fine dust particles that can penetrate into the lung.

Livestock Confinement Building Hazards

Livestock confinement buildings can have toxic concentrations of gases such as ammonia, methane, hydrogen sulfide, and carbon monoxide. These gases, with the exception of ammonia, are heavier than air and sink to the bottom of manure pits. This forces oxygen out of the area and creates a dangerously lethal atmosphere. Hydrogen sulfide is poisonous to humans and smells like rotten eggs.

It only takes one to three breaths of hydrogen sulfide to kill a person if the gas is present in a high concentration. At a lower concentration it can cause nausea, dizziness, sudden collapse, and severe breathing difficulty.

Workers in livestock confinement buildings must exercise caution because of the danger from toxic gases. Proper ventilation of these buildings is especially important. The greatest danger from lack of ventilation is during the winter when ventilation rates may be reduced below the minimum recommended level in order to conserve heat. The danger from toxic gases is higher when the manure in the storage pit is being agitated. It is safest to remove animals and workers from the building when agitating the manure in the pit. If this is not possible, observe from a safe distance when starting agitation and be prepared to stop the pump if any sign of trouble becomes apparent. Workers should never enter a manure pit during or just after agitation because the dangerous gas levels are at their highest at that time.

A power failure that stops the ventilating fans also poses a hazard both to workers and to the livestock in the building. An alarm system should be installed to warn of power failures and auxiliary power should be available to keep the ventilating system working. Commercial gas monitors are available to measure the level of toxic gases in the building.

Workers should not enter manure pits unless it is absolutely necessary. Before entering the pit, test for oxygen with an oxygen meter and test for toxic gas levels, especially hydrogen sulfide, using an appropriate detector. Continue to monitor the oxygen and gas levels while the worker is in the pit. Make sure the pit is well ventilated, and use self-contained breathing equipment and a safety line attached to the person entering the pit. Always have another worker, also equipped with self-contained breathing equipment and a safety line, available outside the pit to rescue the person going into the pit if it becomes necessary. Never enter the manure pit without being properly equipped, even to rescue another worker; to do so will probably result in the death of the person attempting the rescue as well as the person already in the pit. Have a third person outside the confinement area to provide additional help if needed. This person should also have self-contained breathing equipment available for use. At least one of the workers present to provide assistance should be trained in CPR and first aid.

Carbon monoxide gas may accumulate to dangerous levels in poorly ventilated confinement buildings. Malfunctioning space heaters and unvented radiant heaters in the building are often the source of this gas. Carbon monoxide can cause abortions in sows and humans. It may also cause mental retardation, an outcome that human fetuses are more susceptible to than pig fetuses are. Pregnant women should be careful if they are working in livestock confinement buildings. The use of personal protective equipment to filter air for breathing is recommended for all workers who spend two or more hours per day in these buildings. Minimum protection is provided by disposable paper mask filters; better protection is provided with quarter-face or half-face masks with a screw-on cartridge. An air stream helmet provides the greatest protection.

Even with proper ventilation, dust can be a major problem in livestock confinement buildings. Workers should use a high-quality dust filter that provides protection from 0.3 micron diameter particles. Inexpensive filters generally do not provide this level of protection. The addition of soybean oil or tallow to feed has been shown to reduce the level of dust in confinement buildings. Another danger from dust is the possibility of an explosion and fire. Keeping the building and equipment clean can help reduce the danger of high concentrations of dust.

Provide tight covers for ground-level and below-ground manure storage pits. Keep lagoons and manure storage basins fenced off and posted with signs warning people to keep out.

Any open flames, including cigarettes, should be prohibited in enclosed manure storage areas. If repairs, such as welding, are necessary in the area, be sure it is well ventilated to reduce the risk of explosion and fire. Use only explosion-proof electric motors in livestock confinement buildings. Make sure all lights and electrical wiring are maintained to prevent accidental sparking. Methane gas, which is produced in manure pits, is highly explosive.

Locate first aid supplies and rescue equipment close by the manure storage area. Train all workers in proper first aid procedures for toxic gas problems. Emphasize to workers that they are not to enter manure pits without the proper equipment, even to attempt the rescue of another worker. Move the victim to fresh air if possible. If the victim is not breathing, begin cardiopulmonary resuscitation (CPR) immediately and continue until medical help arrives.

Keep the telephone number of the local fire department or rescue squad readily available by the closest telephone. When phoning for help, be sure to let the fire department or rescue squad know the nature of the problem so they will bring the necessary equipment.

FIRE SAFETY

Fire can be one of the most serious hazards on livestock farms (Figure 3-7). Fires are usually caused by carelessness. Most fires start from electrical equipment, heaters, or careless smoking. Other causes of fires are lightning, arson, and spontaneous combustion.

About 70 percent of all farm fires are caused by some problem in electrical wiring or electrical equipment. Livestock confinement buildings often have high levels of moisture and ammonia gas that can corrode electrical wiring and equipment. In high concentrations, dust can explode if an ignition source such as a spark from electrical

FIGURE 3-7   Farm fires cause serious losses on livestock farms each year. Lack of a good water supply often makes it difficult to fight a farm fire. *Courtesy of USDA.*

wiring is present. Electrical control panels should not be located within the confinement area; put them in an adjacent office or outside the building. Sources of information about the electrical requirements for confinement buildings include the *Agricultural Wiring Handbook* and the National Electrical Code, which provide ratings of materials suitable for use on the farm. Code requirements for home use are not adequate for livestock confinement buildings.

Many things around a farm burn easily. Most buildings are of wood construction, and hay, bedding, and feed are easily ignited. Once a fire starts in a livestock building, it burns and spreads very rapidly.

Safety practices to prevent fires include the following:

- Protect buildings from lightning.
- Store fuels properly.
- Practice good housekeeping by keeping all areas in and around buildings clean and free of debris.
- Do not allow people to smoke in and around buildings.
- Make sure all electrical wiring and equipment is in good condition, and meets code requirements for use in livestock confinement buildings.
- Be careful when using heaters and brooders in livestock buildings.
- Avoid conditions that could cause spontaneous combustion in stored feed.
- When planning new buildings, space them at least 50 feet from other buildings to help slow the spread of fire (spacing of 75 feet will allow better access by fire trucks).
- Use fire-resistant exterior materials (metal siding, asphalt shingles) on new construction.
- Locate fire extinguishers near doorways in all buildings.
- Maintain fire extinguishers properly, with periodic inspection to make sure they are fully charged.

- Provide water outlets and hoses in and around the buildings.
- A pond located near the buildings can provide an emergency supply of water for fire fighting.
- Store combustible chemicals in a safe place.
- Instruct all workers and family members in methods of fire prevention and what to do in case of fire.

People should also know what to do in case a fire starts. When a fire is seen, the first thing to do is call the fire department. Fire extinguishers, water hoses, wet gunnysacks, and shovels and dirt should be used to fight the fire until the fire department arrives. If it is possible, animals should be removed from the buildings. They should be led to a place that is out of the way of the fire fighters.

Classification of Fires

Fires are classified as Class A, Class B, Class C, or Class D. Class A fires are those in which the burning material is wood, paper, textiles, grass, trash, and other similar materials. Water can be used to extinguish Class A fires.

Class B fires are those in which the burning material is grease, gasoline, oils, paints, kerosene, and solvents. Do not use water on a Class B fire because the water can spread the fire. Class B fires must be smothered to extinguish them. Blanketing agents such as carbon dioxide, water-based foam, or a wet blanket may be used to smother a small Class B fire.

Class C fires are those involving burning electrical equipment. Do not use water on a Class C fire unless all power to the area has been cut off. The fire fighter can suffer severe electrical shock if water is used on a Class C fire. High-pressure water fogs can be used on Class C fires.

Class D fires are those involving combustible metals like sodium, potassium, titanium, and magnesium. These fires must be controlled by removing air with a blanket of non-reactive powder like sodium chloride or graphite. Water and carbon dioxide will not control Class D fires because these materials provide a source of oxygen for the burning metal.

Fire Extinguishers

The proper type of fire extinguisher must be used for the different classes of fires. Fire extinguishers are marked with a combination of letters and colors for the class of fire on which they can be used:

- Class A extinguishers are marked with an A in a green triangle.
- Class B extinguishers are marked with a B in a red square.
- Class C extinguishers are marked with a C in a blue circle.
- Class D extinguishers are marked with a D in a yellow five-pointed star.

Class A and Class B fire extinguishers use numbers with the letters. The number with the Class A extinguisher indicates its relative effectiveness for putting out a fire. For example, a 5 would indicate that the extinguisher is five times as effective as an extinguisher marked with a 1. The number on the Class B extinguisher gives an indication of the maximum square foot area of a liquid fire that can be extinguished by the fire extinguisher.

Class C and Class D extinguishers do not use numbers to rate their effectiveness. Select a Class C extinguisher on the basis of the type of construction surrounding the electrical equipment. The nameplate of the Class D extinguisher lists its effectiveness for various metals.

No one type of extinguisher is effective on all types of fires. Some are designated as multipurpose extinguishers and are effective for the classes of fires listed on their labels. Typical multipurpose extinguishers are for Class A, B, and C fires and Class B and C fires. Providing multipurpose fire extinguishers in buildings is recommended to help reduce confusion when a fire occurs.

CHECKLIST OF FARM SAFETY PRACTICES

1. Establish good sanitation, vaccination, and inoculation programs.
2. Plan ahead when working with animals in an enclosed space to provide a way out; have at least two exits from the area.
3. Use proper equipment when handling livestock; make sure all pens, gates, loading chutes, and fences are strong enough for the job and in good repair.
4. Be sure livestock handling is done only by those with enough strength and experience for the job.
5. Use caution when approaching animals to avoid startling them.
6. Teach workers the correct safety measures for handling livestock.
7. Know the animals.
8. Be patient with animals.
9. Do not work with animals when tired.
10. Have enough help available to do the job.
11. Be careful when leading animals and handle lead lines properly.
12. Do not allow horseplay around animals.
13. Keep children and visitors away from animals.
14. Dehorn dangerous animals.
15. Check equipment carefully when riding horses.
16. Do not allow smoking in and around farm buildings and fuel storage and refueling areas; post no smoking signs in these areas.
17. Have working, fully charged ABC type fire extinguishers in barns and other major buildings.

18. Remove all trash and junk in and around buildings to prevent fires and falls.

19. Keep all buildings in good repair.

20. Keep electrical wiring in good condition; check insulation, connections, outlets, and electrical equipment.

21. Use adequate lighting in all buildings.

22. Use proper ventilation in buildings and silos; make sure vents are clear and fans operate properly in all confinement buildings.

23. Keep floors and ramps clean, and free of broken concrete and slippery spots to ensure good footing.

24. Keep a well-maintained first aid kit in all major buildings.

25. Keep emergency telephone numbers posted by each telephone.

26. Have telephones or radios in vehicles and major buildings.

27. Keep entrances to grain, feed, and silage storage areas closed and locked to keep children out.

28. Post warning signs in grain and feed storage areas to warn of the hazard of becoming trapped in flowing grain or feed.

29. Maintain silo and bin ladders in good condition.

30. Shield auger inlets to prevent contact with the auger.

31. Cover loading troughs on augers, elevators, and conveyors with grating.

32. Use caution when moving augers and elevators; check for overhead power lines in the area.

33. Check that the proper shields are in place on all feeding, grinding, and other equipment; DO NOT REMOVE SHIELDS.

34. Use protective equipment such as bump caps, respirators, goggles, and gloves when needed.

35. Store chemicals, fertilizers, medicines, and hardware away from animals and in a room or building that can be locked.

36. Post warning signs at the entrance to areas where chemicals are stored that identify the hazards inside and provide information to fire fighters in case of fire.

37. Mix all chemicals outside or in an open, well-ventilated area in the building.

38. Have first aid equipment and plenty of water available in the area where chemicals are handled.

39. Properly dispose of all chemical containers, following directions on the label.

40. Store only chemicals in the chemical storage area.

41. Carry an ABC type fire extinguisher (minimum size 10 lb) in the combine.

42. Carry an ABC type fire extinguisher (minimum size 5 lb) in the tractor.

43. Maintain ladders and steps on tractors, combines, and other equipment in good repair and free of mud and grease.

44. Keep the operator's platform on combines free of mud, grease, and tools.

45. Keep cab windows and mirrors clean for maximum visibility.

46. Check mufflers and other parts of the exhaust system on tractors, combines, trucks, and other powered equipment to make sure there are no leaks of exhaust fumes.

47. Maintain tires in good condition and properly inflated on all equipment.

48. Make sure all fuel, oil, and hydraulic systems on all equipment is in proper condition.

49. Use reflectors and Slow Moving Vehicle emblems on equipment; make sure they are clean and positioned where they can be easily seen.

50. Make sure all tractors are equipped with roll-over protection cabs or roll bars.

51. Keep all farm ponds fenced to keep children out.

52. Do not carry loaded guns in vehicles, tractors, combines, or other equipment.

53. Keep guns unloaded and locked in a cabinet or gun rack in the home; keep ammunition stored in a location separate from the guns.

SUMMARY

Farming is a dangerous occupation. Many people are injured and some are killed each year when working with livestock. Children and older people are most likely to be injured by livestock.

Many environmental factors play a part in farm accidents. Facilities, especially silos and confinement livestock buildings, should be checked for safety. Protective equipment should be worn when necessary.

It is important that all workers be familiar with the correct handling procedures for various types of livestock. Horse safety is a special problem because so many people are around horses who have little other experience with animals.

Many accidents can be avoided by preventing hazardous situations and knowing safety rules. Good housekeeping on the farm prevents many accidents.

Student Learning Activities

1. Give an oral report on safety practices with livestock.
2. Prepare a bulletin board display of newspaper and magazine stories related to livestock safety.
3. Prepare posters about livestock safety.
4. Survey farms in the community using a livestock safety checklist and formulate recommendations for improvement.
5. Present livestock safety programs to local agricultural groups.
6. Prepare a livestock safety exhibit for display in the community.

Review

1. What is the death rate per 100,000 for workers in agriculture?
2. Why do people under 25 and those over 64 have more accidents on farms than people between 25 and 64 years of age?
3. List five environmental dangers to people working with livestock.
4. Describe the proper procedure for washing clothes that have been contaminated by farm chemicals.
5. Describe safety practices for the proper storage of farm chemicals.
6. List the recommended contents of a first aid kit for use on the farm.
7. What are the symptoms of heat exhaustion?
8. How should heat exhaustion be treated?
9. What are the symptoms of heatstroke?
10. How should heatstroke be treated?
11. Why is water better than sugar-based drinks for replacing lost body fluids?
12. What circumstances might cause some people to be at greater risk from heat-related problems than other people?
13. Why do cattle sometimes refuse to move across a shadow or bright contrasting pattern on the floor in front of them?
14. Based on an understanding of how they react to noise and sudden movement, list several good practices to follow when moving or working with cattle.
15. Describe the "flight zone" concept and describe how it may be used to control the movements of cattle.
16. Describe several practices that make it easier to move hogs from one place to another.
17. Describe how livestock facilities can be designed to prevent accidents.

18. What safety precautions should be followed when approaching a horse?

19. Describe three safety procedures a person should follow when handling a horse.

20. What is the safest way to hold the horse's lead strap?

21. List three safety rules to follow when mounting and riding a horse.

22. Why is it dangerous to enter a silo for a period of time after it has been filled?

23. Describe several safety practices that should be followed to protect workers from the danger of rabies.

24. List three types of atmosphere contamination that might affect people working with livestock.

25. List the four main kinds of dangerous gases found in livestock confinement buildings.

26. Describe the kinds of respiratory protection equipment available for use by workers on farms.

27. List three safety practices that should be followed to reduce the danger from toxic gases in livestock confinement buildings.

28. What safety equipment must a person use when entering the manure pit in a livestock confinement building?

29. List ten safety practices to follow to prevent farm fires.

30. List the four types of fire extinguishers and tell what kinds of fires each may be used on.

31. Name the three most common causes of fires on farms.

32. What should be done if a fire starts on the farm?

33. Describe two ways in which the dangers of animal diseases to humans can be reduced.

34. Name three kinds of personal protective equipment a worker may need when working around livestock.

Livestock and the Environment

Objectives

After studying this unit, the student should be able to

- describe livestock production problems relating to the environment.
- describe methods of handling livestock wastes that reduce environmental pollution and are within the guidelines of current laws and regulations.
- describe the proper way to dispose of dead animals from livestock production operations.
- explain farmer liability under animal trespass laws.

Farmers have become more vulnerable to environmental lawsuits than at any time in the past. Changes in federal and state environmental laws make it easier to take action against farmers and ranchers who knowingly or accidentally damage the environment with chemicals or animal wastes. Lawsuits may be filed by property owners that have suffered damage. In some states, lawsuits may be filed by individuals or groups that have not been directly damaged. Civil and criminal penalties, including heavy fines and imprisonment, may be applied. Those convicted may also be required to pay for cleanup costs, attorney fees, and costs of the prosecution. Farmers need to be aware of changes in environmental laws that affect their current operations or restrict their ability to develop new enterprises or expand existing ones. Insurance carried by farmers may not cover environmental liability. It is wise to develop an environmental compliance plan and keep good written records.

Another area of growing concern for farmers and ranchers is the Endangered Species Act. When a species is declared to be endangered and a particular area is considered to be a critical habitat for that species, then all human activity in that area must be stopped. This includes all farming or ranching activities.

In response to people moving out of cities into rural areas during the late 1970s and early 1980s, all 50 states passed "right to farm" laws. These laws were designed to protect farmers from nuisance lawsuits based on subjective perceptions of people who moved into the area from the city and were unfamiliar with the reality of the sounds and smells of livestock production on farms. The laws protected farmers against nuisance suits as long as the complaints were not based on violations of federal or state laws, negligence in operating the farm, water pollution, or excessive soil erosion.

A challenge to Iowa's right to farm law worked its way through the lower courts and finally reached the state supreme court. In 1998 that court struck down the right to farm law, ruling that the law took valuable property rights away from land owners for the benefit of a few. In 1999, the United States Supreme Court refused to review the Iowa Supreme Court ruling, thus allowing its ruling to stand and bringing into question the status of right to farm laws in all the states.

Property rights of individuals are coming under increasing attack by environmental groups that are urging more government control over land and water use. Some farm groups are working to secure legislation that will help protect private property rights.

Some states are more closely regulating the disposal of medical waste such as needles, syringes, scalpel blades, and blood vials that are used in the treatment of livestock. Before these materials can be disposed of they must be treated to eliminate the possibility that they might transmit infections. Special containers that are puncture and leak resistant are available from veterinarians, waste haulers, hospitals, and local health departments. The medical waste is placed in the special container and then shipped to an approved infectious medical waste facility. State environmental protection agencies may be contacted to determine the location of approved facilities.

Livestock producers must deal with animal wastes, odors, and dead animals in ways that do not harm the environment. They are also legally liable for any damage their livestock may do to other people or their property. Many of these problems require costly solutions. Society must decide whether the benefits are worth the investment.

ENVIRONMENTAL PROBLEMS OF LIVESTOCK PRODUCTION

Changes in Livestock Production

The trend toward larger livestock operations has caused an increase in the concentration of animal wastes on individual farms. The use of large livestock confinement buildings presents special problems related to the disposal of livestock wastes in a manner that is not harmful to the environment or objectionable to others living in the area. Confining cattle in large feedlots results in greater problems in the disposal of livestock wastes (Figure 4-1). Many operators of large feedlots do not have the land on which to spread the manure. This increases the potential for pollution problems.

The Changing Environment of Agriculture

Many people are moving into farm areas to get away from the problems of large cities. Large recreational developments also attract city dwellers to rural areas. Those who live in the cities often find farm odors offensive. They may not realize that odors and livestock wastes are a natural part of livestock production. Farmers must deal with this attitude while still maintaining production.

Handling farm wastes poses different problems than handling waste from cities. The cost for handling these wastes falls on individuals rather than on a whole community. Cities solve their waste problems by building waste disposal plants that may cost several million dollars. However, the cost for each person living in the city may be only $100 to $200. A poultry farm of 200,000 hens, a beef feedlot of 1,200 head, or a 10,500-head hog operation may produce as much waste as a city of 20,000 people. Although several farms in a community may have large feeding operations, they are usually too far apart to use a single disposal plant. Thus, each farm has to bear the cost of taking care of its own waste.

Farmers must develop systems of waste control that are acceptable to others in the changing rural environment. These systems must be a part of their total management plan. They must be affordable and also meet the expectations of nonfarming people who live nearby.

FIGURE 4-1 The cows in this picture are standing on a large pile of manure in a cattle feedlot. *Courtesy of USDA.*

Federal and State Laws

The Federal Water Quality Act of 1965 requires states to have water quality standards. Public hearings must be held before the standards are set up. These standards apply to waters that move from one state to another. They also affect any portion of these waters that are within the borders of a state. The United States Environmental Protection Agency must approve the standards that are set by the state.

Changes in federal law relating to water pollution control are contained in the Federal Water Pollution Control Act of 1972, also known as the Clean Water Act. This law lists national goals for water quality. It establishes cooperation between the federal and state governments to reach these goals. The law prohibits the discharge of any pollutants from a point source into a river or stream without a permit. A large feedlot is an example of a point source of pollution. Thus, waste from a large feedlot cannot be allowed to run into a river or stream. Nonpoint sources of pollution were added to the law in the 1987 amendments to the Federal Water Pollution Control Act. A field that has manure spread on it is an example of a nonpoint source of pollution. The main goal of the 1987 amendments is the development and implementation of programs to control both nonpoint and point sources of pollution. The Act gives to the states the main responsibility for controlling water pollution.

The Refuse Act of 1899 gives the United States Army Corps of Engineers control over some animal waste pollution problems. The Corps can approve or deny an application for permits to let waste run into navigable waters or their tributaries. A permit is required if a confinement feedlot feeds more than 1,000 animal units per year and if there is a direct discharge of waste to the waters. Runoff from natural causes is not considered to be a discharge of waste.

The Solid Waste Disposal Act of 1965 was amended in 1970. This Act sets up federal guidelines for solid waste management. This includes the management of animal manure.

The Federal Clean Air Act establishes national air quality standards. The Act makes the states primarily responsible for making sure that these standards are met, maintained, and enforced. State laws deal with how the standards are to be met. These standards deal with things like dust, grit, organic matter, and open burning as sources of air pollution.

State laws. Most states have set up some type of environmental protection agency. Different states have different names for these agencies, but they all have the job of seeing that the federal and state laws that affect the environment are enforced. Livestock producers are affected by many of the regulations of these agencies.

Many states have laws that deal with nuisances. Nuisances may include odors, dust, chemicals, water pollution, and animal noises. These laws also often control the disposal of dead animals.

Water Pollutants

Water pollution is a major concern when determining the best way to dispose of animal wastes. The Environmental Protection Agency (EPA) monitors six water pollutants that can be measured directly. These are **Biochemical Oxygen Demand (BOD), Fecal Coliform, Fecal Streptococcus, Suspended Solids, Phosphorus,** and **Ammonia.**

Oxygen is required to digest the organic matter in animal manure that is discharged into a river, stream, pond, or lake. Biochemical Oxygen Demand is a measure of the amount of oxygen-demanding organic matter in the water. Fish and other organisms living in the water need oxygen to live. If the BOD level is too high, there may not be enough oxygen left in the water to support life.

There are many types of bacteria found in the intestinal tracts of humans and animals. Two types of these bacteria are Fecal Coliform and Fecal Streptococcus. Water that is contaminated by human or animal waste contains measurable amounts of one or both of these bacteria. Contamination of water by coliform bacteria is undesirable because they can transmit disease to humans and animals.

Some materials like oil, grease, organic matter, and minerals may not dissolve in water. They may also float on the surface of a river, stream, pond, or lake. These materials are called Suspended Solids. They are undesirable because of the odors they may cause. They also give a river, stream, pond, or lake an unsightly appearance.

Algae problems in water may be caused by contamination with phosphorus and ammonia contained in animal waste. These materials provide nutrients that cause the algae in the water to grow by an excessive amount.

HANDLING LIVESTOCK WASTES

Objectives of Waste Management

Animal manure must be handled so that odors, dust, flies, rodents, and other nuisances are controlled. Nitrate problems in water supplies caused by nitrogen in the manure must also be prevented. The system of waste handling must not allow the waste to be dumped into streams, rivers, lakes, or reservoirs. The waste must be in a form that is easily handled and disposed of without causing health and safety hazards to people or animals.

The main objective of manure handling is to prevent surface and ground water pollution. Biological and chemical treatments of animal wastes are too expensive for farmers to use. Generally, the wastes must be held in some way until they can be properly disposed of on the land.

Amount of Waste Produced

The amount of raw manure that an animal produces depends on many factors, including the ration fed and the age of the animal. The data presented in Table 4-1 are based on American Society of Agricultural Engineers (ASAE) research except where otherwise indicated in the table. Data in Table 4-1 that are not based on ASAE research are based on common assumptions and are not proven. The nutrient content data represent guidelines to the approximate fertilizer value of the manure and are not precise. Nutrient content of specific samples of manure may vary by as much as 20 percent above or below the data given. Some factors that affect the nutrient content of manure include (1) length of time in storage, (2) methods of treatment, (3) amount and type of bedding used, and (4) amount of dilution by water entering the system.

About 2 billion tons of manure is produced each year on livestock farms in the United States. Some of this manure is deposited on

FIGURE 4-2   Hogs in a confinement house produce a great deal of manure. *Photo by Steven M. Ennis.*

pastures and rangeland by the animals. A much larger volume of the total is in feedlots, barnyards, and stockpiles. All of this manure has to be taken care of in some way. This is one of the major problems that the livestock producer must solve.

Selecting a System of Manure Handling

An animal feeding operation refers to facilities that house livestock for production purposes. The Environmental Protection Agency (EPA) specifies the conditions that must exist to define an animal feeding operation as one subject to regulation. Livestock must be housed at the location for at least 45 days out of each 12 months and neither crops nor any other plants are normally grown in the area. The 45-day figure is simply the total number of days, not necessarily continuous, during the 12 months. A livestock pasture that is sodded over is not considered to be an animal feeding operation under these regulations. If the farm has two or more kinds of livestock operations that share the same waste disposal system, it is defined as one animal feeding operation for the purposes of EPA regulation.

Choice of a system of manure handling depends primarily on the kind of animal that is being raised. There are many different kinds of facilities that can be used. The farmer must decide how to collect, handle, treat, and dispose of the manure. The ration fed the animals influences the characteristics of the manure produced. The kind of housing and management also affects the kind of system selected (Figure 4-2).

Different sizes and types of feeding operations require different systems. A large operation needs more equipment and automation. A small operation requires less money and more labor. The amount of money the farmer can invest also affects the selection of a system, as does personal preference.

Climate is important when selecting a system. The amount of rainfall and when it occurs, and the amount of evaporation are other factors. Temperature also makes a difference in the kind of system used. The usual direction of the local wind may also influence system selection.

Each farm has characteristics that affect the system selected for that farm. How big the farm is and the soil type must be considered. Whether the land is sloping or level and the kind of crops the farmer grows affect the decision.

Regulations must be taken into account when planning waste handling systems. Federal, state, and local laws must be followed. Zoning may affect the kind of system permitted in an area. Nearby neighbors must be taken into consideration.

Facilities for producing livestock are classified as either unconfined or confined. Unconfined facilities usually make use of pasture or range. In this case, most of the manure is left on the pasture or

TABLE 4-1 Manure Production and Nutrient Content of Manure Produced by Various Species of Animals.

	Weight		Daily Production[a]						Yearly Production		Moisture Content	N		P		K	
	lb	kg	lb	kg	gal	litre	ft³	m³	ton	tonne	%	lb/day	kg/day	lb/day	kg/day	lb/day	kg/day
Beef cattle	500	226.8	30.0	13.6	3.8	14.4	.5	.014	5.5	5.0	88.4	.17	.0771	.056	.0254	.12	.0544
Beef cattle	750	340.2	45.0	20.4	5.6	21.2	.75	.021	8.2	7.4	88.4	.26	.1179	.084	.0381	.19	.0862
Beef cattle	1,000	453.6	60.0	27.2	7.5	28.4	1.00	.028	11.0	10.0	88.4	.34	.1542	.11	.0499	.24	.1089
Beef cattle	1,250	567.0	75.0	34.0	9.4	35.6	1.2	.034	13.7	12.4	88.4	.43	.1950	.14	.0635	.31	.1406
Beef cow[b]			63.0	28.6	7.9	29.9	1.05	.03	11.5	10.4	88.4	.36	.1633	.12	.0544	.26	.1179
Dairy cattle	150	68.0	12.0	5.4	1.5	5.7	.19	.0054	2.2	2.0	87.3	.06	.0272	.01	.0045	.04	.0181
Dairy cattle	250	113.4	20.0	9.1	2.4	9.1	.32	.0091	3.7	3.4	87.3	.1	.0454	.02	.0091	.07	.0318
Dairy cattle	500	226.8	41.0	18.6	5.0	18.9	.66	.019	7.5	6.8	87.3	.2	.0907	.036	.0163	.14	.0635
Dairy cattle	1,000	453.6	82.0	37.2	9.9	37.5	1.32	.037	15.0	13.6	87.3	.41	.1860	.073	.0331	.27	.1225
Dairy cattle	1,400	635.0	115.0	52.2	13.9	52.6	1.85	.052	21.0	19.1	87.3	.57	.2585	.102	.0463	.38	.1724
Pig—nursery	35	15.9	2.3	1.0	.27	1.0	.038	.0011	.4	.4	90.8	.016	.0072	.0052	.0024	.01	.0045
Pig—growing	65	29.5	4.2	1.9	.48	1.8	.07	.002	.8	.7	90.8	.029	.0132	.0098	.0044	.02	.0091
Pig—finishing	150	68.0	9.8	4.4	1.13	4.3	.16	.0045	1.8	1.6	90.8	.068	.0308	.022	.0099	.045	.0204
Pig—finishing	200	90.7	13.0	5.9	1.5	5.7	.22	.0062	2.4	2.2	90.8	.09	.0408	.03	.0136	.059	.0268
Sow—gestating[b]	275	124.7	8.9	4.0	1.1	4.2	.15	.0042	1.6	1.5	90.8	.062	.0281	.021	.0095	.04	.0181
Sow & litter[b]	375	170.1	33.0	15.0	4.0	15.1	.54	.015	6.0	5.4	90.8	.23	.1043	.076	.0345	.15	.0680
Boar[b]	350	158.8	11.0	5.0	1.4	5.3	.19	.0054	2.0	1.8	90.8	.078	.0354	.026	.0118	.051	.0231
Sheep	100	45.4	4.0	1.8	.46	1.7	.062	.0018	.7	.635	75	.045	.0204	.0066	.0029	.032	.0145
Laying hen	4	1.8	.21	.1	.027	.1	.0035	.000099	.038	.034	74.8	.0029	.0013	.0011	.0005	.0012	.0005
Broiler	2	.9	.14	.1	.018	.1	.0024	.000068	.026	.024	74.8	.0024	.0011	.00054	.0002	.00075	.0003
Turkey[b]	15	6.8	.55	.2	.07	.3	.009	.000255	.1	.091	68	.0082	.0037	.0027	.0012	.0025	.0011
Horse	1,000	453.6	45.0	20.4	11.0	41.6	1.5	.042	8.2	7.4	79.1	.27	.1228	.046	.0209	.17	.0771

[a]Daily production based on American Society of Agricultural Engineers (ASAE) data except where otherwise indicated.
[b]Daily production based on common assumptions and not on ASAE data.

FIGURE 4-3   Manure handling is not much of a problem when cattle are on pasture. *Photo by Jim Bodine.*

range by the animal (Figure 4-3). Only small amounts of manure will be deposited in barns and lots. This is a low-cost system from the manure-handling standpoint.

Confined facilities may be open lot, lot and shelter, or a totally enclosed shelter. These types of facilities mean greater costs for handling the manure produced.

Animal manure may be collected and handled as a solid. Other systems collect and handle the manure as a liquid. If the manure is handled as a solid, then bedding may also be handled with the manure. Liquid systems generally cannot handle bedding; however, newspaper bedding usually breaks down completely enough to be useable with these systems. Flushing systems add to the amount of water in the raw manure.

Several types of floors are used, depending on the way in which the manure is to be handled. Housing that uses a solid system may have concrete floors, dirt floors, slotted floors, or solid floors with gutters. Those buildings with liquid systems may use solid concrete floors that are flushed with water. Slotted floors are also used with liquid systems (Figure 4-4).

Liquid systems use pits, lagoons, or storage basins for storing and handling manure. Pits are pumped out and the manure is spread on the land in liquid form. Storage basins may be above or below ground. Above-ground systems are more expensive to build than underground basins. In both systems, a liquid manure pump is used to unload the basin so that the manure can be hauled to the field (Figure 4-5).

Manure pits may be recharged by draining the pit through pipes into a lagoon and then pumping fresh water from the lagoon into the pit. Benefits from recharging the manure pit about once a week include less buildup of dangerous gases and odors, improved feed efficiency, lower death rate of animals in the confinement building, and

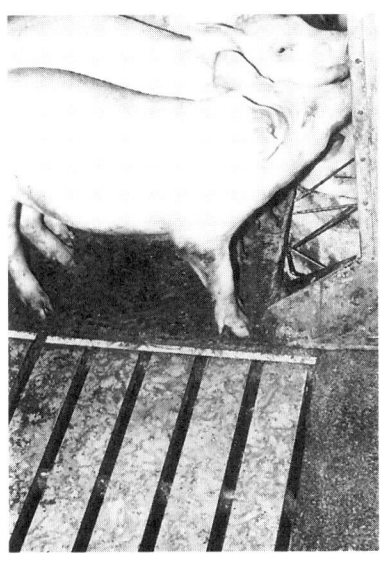

FIGURE 4-4   This swine finishing facility uses slotted floors for waste handling. *Photo by Steven M. Ennis.*

less medication needed for the animals. Recharging systems are used mainly in swine confinement buildings.

To determine the storage space needed for manure in a holding pit or tank, multiply the number of days in the holding period by the daily production of manure. This result is then multiplied by the number of animals that are producing the manure. Water must be added to the manure for proper storage and handling. Increase the storage capacity requirement by about 150 percent to allow for this added water.

The manure in lagoon systems (Figure 4-6) is not unloaded and hauled to the field. Instead, the waste material is broken down by bacteria. Lagoons may be either aerobic or anaerobic. Aerobic systems must have oxygen for the bacteria to work. Aerobic lagoons are shallower and produce less odor but require more area than anaerobic lagoons. They usually require mechanical aerators to control sludge and reduce odors. Anaerobic systems make use of bacteria that work without oxygen. They can handle a larger volume of manure with less cost, labor, and maintenance than aerobic lagoons.

Lagoons must be designed to hold the total amount of manure produced by the livestock, plus all extra water that may be added. Extra water may result from rainfall, feedlot runoff, building wash water, and overflow from livestock waterers. The total design volume needed is about twice the volume needed for just the livestock waste.

Determining the Amount of Livestock Waste to Apply on the Land

The amount of available nitrogen per unit of yield necessary to produce a given crop is called the *agronomic nitrogen rate.* If the recommended

FIGURE 4-5   A wagon used to haul liquid manure to the field for disposal. *Courtesy of Pearson Brothers Company, IL and Van Dale, MN.*

FIGURE 4-6   A lagoon for livestock waste disposal. A diversion terrace keeps rain runoff from filling the lagoon. *Courtesy of USDA.*

TABLE 4-2 Agronomic Nitrogen Rates.

Crop	Pounds of Available Nitrogen
Corn (grain)	1.3/bushel
Corn (silage)	7.5/ton
Barley (straw removed)	1.5/bushel
Grain sorghum (grain)	2.0/100 pounds
Grain sorghum (silage)	7.5/ton
Oats (straw removed)	1.1/bushel
Reed canarygrass	55.0/ton
Rye (straw removed)	2.2/bushel
Sorghum-sudangrass	40.0/ton
Tall fescue	30.0/ton
Wheat (straw removed)	2.3/bushel

agronomic nitrogen rate is followed when applying manure to the land, the crop will be provided with all the nitrogen it needs. If manure is applied at a higher rate than recommended, the excess nitrogen may pollute a water source. The recommended agronomic nitrogen rates for various crops are given in Table 4-2. The agronomic nitrogen rate is shown in pounds of nitrogen per bushel or weight of crop yield.

The nitrogen content of animal wastes varies with the species of animal and the waste storage method used on the farm. The approximate values for nitrogen content of livestock wastes from various sources utilizing different storage methods are given in Table 4-3. These values are only guidelines because there can be a wide variation in the nutrient content of animal wastes. The actual nutrient content of any given sample of livestock waste is affected by the kind of ration fed and the method of collecting and storing the waste. Actual nutrient content can be determined by laboratory analysis.

For a sample calculation, assume that a corn yield of 150 bushels per acre is expected. The livestock waste to be applied comes from a beef feeding operation using pit storage. Table 4-2 shows that 1.3 pounds of nitrogen per bushel of expected yield is required. Therefore, multiply 150 times 1.3 to arrive at 195 pounds of nitrogen required per acre ($150 \times 1.3 = 195$). Table 4-3 shows that waste from a beef feeding operation using pit storage contains from 25 to 50 pounds of nitrogen per 1,000 gallons of waste. For this example, assume the nitrogen content to be 30 pounds per 1,000 gallons of waste. Therefore, 195 pounds of nitrogen required per acre divided by 30 pounds per 1,000 gallons of waste results in 6,500 gallons of livestock waste per acre needed (($195 \div 30) \times 1,000 = 6,500$).

Not all of the nitrogen is available to the plant in the year in which the waste is applied to the soil. It may take from three to five years for all the nitrogen from a given application to become available. More animal waste will need to be applied per year during this period if the amount needed for the expected crop yield is to come from this source. After a three- to five-year period, the amount available each year from the calculated application rate plus that available from applications in previous years will about equal the needs of the crop for the expected yield. An alternative to applying more animal waste during this period is to make up the difference with the application of commercial fertilizer.

Some farmers prefer to use the phosphorus requirements rather than the nitrogen requirements for determining the application rate for animal wastes. It takes less animal waste per acre to meet the phosphorus requirements. The additional nutrients needed are then supplied by adding commercial fertilizer. Tables that contain information on the agronomic rate for phosphorus and the phosphorus content of animal wastes are available from state EPA offices or county Cooperative Extension Service offices.

TABLE 4-3 Nitrogen Content of Livestock Waste.

Species	Nitrogen	
	lb/1000 gal	lb/ton[1]
Beef:		
Pit storage	25–50	
Open lot—runoff	0.5–5.0	
Open lot—solids		10–12
Bedded confinement solids		10–15
Anaerobic lagoon	10–15	
Oxidation ditch	10–25	
Dairy:		
Pit storage	20–40	
Open lot—runoff	0.5–5.0	
Open lot—solids		7–10
Bedded confinement solids		10–15
Anaerobic lagoon	10–15	
Poultry:		25
Swine:		
Pit storage	30–55	
Open lot—runoff	0.5–5.0	
Open lot—solids		10–12
Bedded confinement solids		10–15
Anaerobic lagoon	10–15	
Oxidation ditch	10–25	

[1]At approximately 50% moisture content.

Disposing of the Manure

Most animal waste is eventually spread on the land. All solid handling systems work in this way. Liquid systems, except lagoons, also involve moving the manure to a field to dispose of it.

Manure is valuable as a fertilizer. When commercial fertilizers were inexpensive, the value of animal manure was low. With the increasing cost of chemical fertilizers, animal manure has become more valuable as a fertilizer. The amount of fertilizer value in manure is shown in Table 4-1.

Farmers who have fields to which manure can be hauled are less likely to use lagoons. However, large confinement feeding operations may not have fields available. They often use lagoons to handle the manure.

When livestock wastes are applied to the land, care must be taken to avoid polluting the environment. The following points reflect EPA regulations and should be considered when applying animal wastes to the land:

1. Animal wastes must be incorporated or injected into the soil:
   a. if the percent of slope of the land on which it is being applied

FIGURE 4-7 Research is carried on by the USDA's Agricultural Research Service to find ways of keeping pollution out of rivers and streams. Sampling and recording equipment has been placed in this feedlot catchment pond to aid in this research. *Courtesy of USDA.*

is greater than five percent or more than five tons of soil per acre is lost per year from erosion.

b. if the field is in a 10 year flood plain (the field floods at least one year in ten).

2. Injection or incorporation of the livestock waste will reduce odors.

3. Generally, do not apply livestock wastes on frozen or snow-covered ground. Such application may be done if the land slopes less than five percent or adequate soil erosion control is practiced.

4. Do not apply livestock wastes immediately before or during a rainstorm or to soil saturated with water.

5. Do not apply livestock wastes to grass waterways.

6. Do not apply livestock wastes within 200 feet of surface waters or within 150 feet of a well.

7. Reduce the amount of livestock waste applied to the soil if there is a high water table present or the soil is highly permeable. This will reduce the chances of polluting water supplies.

Feedlot Runoff Control

In an open feedlot, runoff is caused by rainfall or snowmelt. A great deal of manure is carried off the feedlot by the running water. If the runoff gets into a stream or river, it may cause fish kill. Many states have laws to prevent runoff from being channeled into a stream or river (Figure 4-7).

Five ways to prevent runoff are diversion, drainage, debris basins, detention ponds, and disposal. A feedlot operator must plan each of these control measures carefully.

Diversion is preventing surface water from outside the feedlot from getting onto the feedlot. Some feedlots are located at the top of a slope. This prevents much of the water from rain or snowmelt from getting onto the feedlot. If the feedlot is located at the bottom of a slope, a diversion terrace must be built. A *diversion terrace* forces the water to go around the feedlot.

The surface of the feedlot must be properly shaped to divert the runoff water to a drainage channel or pipe in the shortest distance possible. If the surface is graded correctly, water will not collect in ponds in the feedlot. Mounds are sometimes made in feedlots (Figure 4-8). They are constructed with clay soil and packed down. They should be 4 to 5 feet (1.2–1.5 m) high and large enough at the top to let all the animals onto dry ground in wet weather. Feed bunks and waterers should be located so that the animals can get to them on dry ground. There should be no high piles of manure within the pen.

Debris basins are used to catch runoff from the pens. The solids settle to the bottom and the liquids are drained into holding ponds. Debris basins keep about 50 to 85 percent of the solids from reaching the holding ponds. This helps to reduce odor from holding ponds.

FIGURE 4-8 Dirt mounds in the feedlot help keep cattle on drier ground in wet weather. *Courtesy of University of Illinois at Urbana-Champaign.*

Solid manure should be removed regularly from the debris basin. The manure should never be more than 1 foot (0.3 m) deep. A wide, flat channel works best for collecting the debris. It should be about 10 feet (3 m) wide and not more than 3 feet (0.9 m) deep.

A *holding pond* is a temporary storage area for runoff. It is not designed for waste treatment. It should be big enough to hold the runoff from the maximum 24-hour rainfall expected once in 10 years. In drier parts of the United States, the holding pond can be an evaporation area.

Disposal is the final step in controlling runoff from feedlots. The collected water can be used for irrigation of the land, or it may be allowed to evaporate. Holding ponds need to be pumped out fairly often. There must be enough space to hold the runoff from future rains or snowmelts. Local weather bureau records are used to determine how often the pond needs to be pumped.

Gases and Odors from Livestock Wastes

Gases and odors are given off by animal manure. This is caused by anaerobic bacteria breaking down the organic part of the manure. Anaerobic bacteria work when no oxygen is present. The gases produced can be dangerous to people and animals in a confinement facility. Odors may cause people who live close to the farm to take legal action against the farmer.

Gases and odors can be reduced by mixing air with the manure. In liquid manure systems, this is done by installing equipment to force air through the liquid. In solid manure systems, it is more difficult to prevent the gases and odors from forming. In a feedlot, the cattle keep the surface stirred, which allows air to mix with the manure.

It is hard to control odors when hauling manure onto the land. The best way to do this is to mix the manure in the soil as soon as possible after hauling. This can be done by plowing it under or disking it into the soil.

Chemical and bacterial culture odor-control products may be used to control odors from animal manure. These are generally of four types: (1) masking agents, (2) counteractants, (3) deodorants, and (4) digestive deodorants.

Masking agents cover up the odor of wastes with the introduction of another odor. They are considered the most effective of the four types of odor control. However, the masking odor is not always pleasant, and a different odor problem may be created. *Counteractants* attempt to neutralize the odor so no odor remains. These substances are the second most effective type of control. *Deodorants* are chemicals that kill the bacteria that cause the odor. They are not as effective as the first two types. *Digestive deodorants* are bacteria that create a digestive process that eliminates the odor. They are the least effective of the four types. All of these control methods are expensive.

ENVIRONMENT AND NUTRITION

Environmental stress affects the nutrient requirements and intake of animals. Livestock nutrient requirements shown in the older nutrient requirement tables in the Appendix have, for the most part, been established without considering the effects of environmental stress. More recent releases of nutrient requirement publications for beef and swine from the National Academy of Sciences include computer programs that allow the user to include environmental conditions when evaluating diets. As the academy releases additional updates to its nutrient requirements series, it is expected that information will be included to allow the effects of environmental conditions to be considered when evaluating diets.

The environment of animals that are raised in confinement is usually carefully controlled to minimize stress caused by extremes of temperature and humidity. Animals that are raised in less confined environments may be subject to more environmental stress; this needs to be considered when determining their nutritional needs.

The temperature of the air is the primary factor that affects the efficiency of energy use by farm animals; secondary influences include humidity, precipitation, wind, and heat radiation. The combined effect of these factors is referred to as *effective ambient temperature* (EAT). Within limits, animals attempt to compensate for

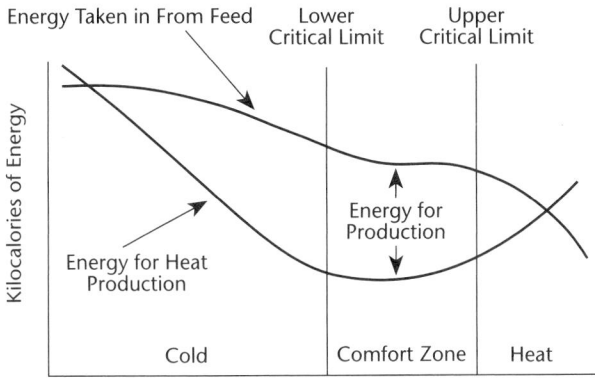

Energy Taken in From Feed   Lower Critical Limit   Upper Critical Limit

Kilocalories of Energy

Energy for Heat Production

Energy for Production

Cold   Comfort Zone   Heat

Effective Ambient Temperature (EAT)

FIGURE 4-9 The general relationship of temperature to energy intake in the feed, energy used for maintenance, and energy available for production.

changes in the EAT by altering feed intake, metabolism, and heat dissipation, (Figure 4-9).

The *comfort zone*, or *thermoneutral zone*, is the range of effective ambient temperatures within which an animal does not have to increase normal metabolic heat production to offset heat loss to the environment. The thermoneutral zone varies with livestock species and may shift up or down as an animal becomes acclimatized to warmer or colder temperatures. For example, as cattle become accustomed to the winter season, their thermoneutral zone may shift downward as much as 27°F (15°C).

The *lower critical temperature* (LCT) is the temperature at which animals will show symptoms of cold stress; feed intake increases, as does metabolic heat production. The *upper critical temperature* (UCT) is the temperature at which animals will show symptoms of heat stress; feed intake is generally lower as animals attempt to reduce the rate of metabolic heat production when the upper critical temperature is reached.

Mature ruminants at high feeding and production levels produce more metabolic heat, have small surface areas relative to total body mass, and have a large amount of insulative tissue. For these reasons, they have significantly lower critical temperatures than do smaller animals such as swine, poultry, or young animals.

Because of their higher metabolic rate, mature ruminants have more difficulty adjusting to high temperatures than to lower temperatures; they become heat stressed more easily than they become cold stressed. The primary method of heat loss in hot environments is through evaporation from the surface of the skin or from the respiratory tract. Providing shade for animals during periods of high temperature will help reduce heat stress.

During periods of high humidity, it is more difficult for animals to lose heat by evaporation. Animals such as cattle (which depend more upon sweating to lose excess heat) are more affected by high humidity than are those such as swine (which do not sweat but lose heat through the respiratory system).

Heat loss from the body by convection and evaporation is affected by the movement of air surrounding the animal. The rate of change in heat transfer is greatest at lower air velocities. *Wind chill index* is a measure of the combined effect of air temperature and speed of air movement. Cold air in motion has a greater adverse impact on animals than cold air that is still. Providing a windbreak for animals during cold weather helps reduce cold stress. Table 4-4 shows the wind chill index for a variety of wind speeds and temperatures.

Precipitation (in the form of rain or wet snow) combined with low temperature and wind can cause animals to lose heat at a rapid rate. The insulation value of an animal's hair, wool, or fur coat is reduced when it is wet or becomes matted by rain or snow, and the animal loses heat more rapidly by conduction. When the hair, wool, or fur coat dries, the animal loses heat by evaporation.

Water intake generally increases as the temperature rises and decreases in colder weather. Cattle will drink more water during cold weather if it is heated; heating water for sheep during cold weather does not appear to increase water intake. Water intake of cattle and sheep tends to decrease as the relative humidity increases. Pregnant and lactating animals have a higher water requirement than nonpregnant and nonlactating animals. Animals fed feeds with a high dry matter content tend to have a higher water requirement than those fed feeds with a higher moisture content.

TABLE 4-4   Wind Chill Index.[1]

Wind Speed (mph)	Temperature (°F)									
	30	20	10	5	0	−5	−10	−20	−30	−40
5	27	16	7	0	−5	−10	−15	−26	−36	−47
10	16	3	−9	−15	−22	−27	−34	−46	−58	−71
15	9	−5	−18	−25	−31	−38	−45	−58	−72	−85
20	4	−10	−24	−31	−39	−46	−53	−67	−81	−95
25	1	−15	−29	−36	−44	−51	−59	−74	−88	−103
30	−2	−18	−33	−41	−49	−56	−64	−79	−93	−109
35	−4	−20	−35	−43	−52	−58	−67	−82	−97	−113
40	−5	−21	−37	−45	−53	−60	−69	−84	−100	−115
45	−6	−22	−38	−46	−54	−62	−70	−85	−102	−117

[1]Wind speed is shown in the left column; actual air temperature is show in the top row. The row and column intersection of actual air temperature and wind speed shows the effective temperature for unprotected skin. The amount of hair, wool, or fur covering that an animal has will influence the wind chill factor for that animal. A heavier covering provides more protection for the animal.

Source: National Weather Service, NOAA, U.S. Department of Commerce.

Feed efficiency is reduced when the temperature is outside the animals' comfort zone. This results in an economic loss for the live-stock producer. The benefit of controlling environmental conditions when raising livestock must be weighed against this economic loss when making management decisions concerning the expenditure of capital assets.

Some nutritional adjustments can be made during cold or hot weather. The energy requirements of the animal during cold weather are higher; however, the protein requirement is about the same as it is when the temperature is in the comfort zone. The percentage of protein in the ration can be reduced during cold weather because feed intake is higher to meet the energy requirements of the animal. Because feed intake is reduced during hot weather, increasing the amount of fat in the ration will help to maintain caloric intake. Fat has a lower heat increment than carbohydrate or protein. The per-centage of protein in the ration may need to be increased in hot weather to meet the needs of the animal.

DISPOSAL OF DEAD ANIMALS

Most states have laws that require the disposal of dead animals within a given period of time after death—usually 24 to 48 hours. It is generally the responsibility of the owner of the animal to dispose of it. This must be done in such a way that no health hazard is cre-ated. Approved methods of disposal vary from state to state. Some of the methods often approved include a licensed disposal plant, bury-ing, disposal pits, burning, and composting.

Diseases may spread from dead animals to people or to other an-imals. Always treat dead animals as though they were diseased. Trucks or other equipment used to haul dead animals should be dis-infected after use. Sometimes there is doubt about the cause of death. If this is the case, a diagnostic laboratory should check the animal for possible diseases.

Dead animals should be hauled in a covered, metal, leakproof ve-hicle. Skinning an animal makes it more difficult to haul the animal and also increases the chance of spreading diseases.

The best place to dispose of dead animals is a rendering or dis-posal plant. There are fewer of these plants today than in previous years. The operating costs have gone up and the prices for products they make have gone down. They also have trouble with liability for air and water pollution.

It is difficult to burn animals. Baby pigs, young chicks, and poults (young turkeys) are about the only animals that are practical to burn. It costs too much to burn large animals. Bad odors are caused by burning animals. Commercial incinerators should be used. Burn-ing dead animals on old brush piles or other open burning is not rec-ommended. In many places this practice is illegal.

When a local disposal plant or sanitary landfill is not available, farmers may bury dead animals. A site should be selected that does not require moving the animal over a public road or over someone else's property. The burial site should be far away from other buildings, public roads, and property boundaries. It should be high enough so that it is above the water table. The site should have good drainage away from water sources such as wells, springs, or streams.

Burying animals requires the use of excavation equipment to dig a large enough pit. Some farmers prepare an excavation in advance. This prevents delay in burying the dead animal.

The carcass should be placed at least 4 feet below the surface and then covered with 4 feet of earth. Covering must be done so that water does not pond in the excavation. The excavation is dug like a trench. The dead animal should be placed at the high end of the trench.

A disposal pit should be rectangular in shape and have vertical sidewalls. If the soil caves in easily, it is necessary to wall the pit with concrete blocks or pressure-treated wood. This increases the cost of the pit. The pit cover may be of concrete, pressure-treated wood, or steel. The cover must be

- watertight.
- able to keep flies and predators out.
- able to keep most of the odor in.
- able to provide an easy way to drop the carcass in.
- constructed in such a way that it can be skidded from an old pit to a new one.

There are several chemicals that help speed the breakdown of the carcass. They also help disinfect and control the spread of diseases. It is not practical to use chemicals on large animals because of the high cost. Always follow the directions and precautions on the label of any chemicals used.

In some areas it is possible to dispose of animal carcasses in sanitary landfills. Check with the operator of the landfill ahead of time to see if it receives dead animals. If it does, the advance notice gives the landfill operator time to prepare for disposing of the animal. In this way, it can be disposed of as soon as it arrives at the landfill.

Some states permit the composting of dead animals, especially poultry and hogs. The poultry industry began the practice of composting dead carcasses in the 1980s. Composting is currently the most used method of dead carcass disposal throughout the poultry industry in the United States. Poultry carcasses will typically decompose in four to six weeks, depending on the size of the bird. Composting is beginning to gain acceptance as a method for disposal of dead animals in the swine industry. The facility needs to be larger when used for swine and the decomposition time is longer than for poultry.

Composting bins are usually constructed on a concrete floor and should have a roof to protect the material in the bin. Sufficient size should be allowed to permit the use of end loaders or a tractor with a loader to handle the compost material. The moisture content of the composting material should be in a range of 40 to 60%. A variety of materials may be used to cover the carcasses in the bin, including sawdust, wood chips, ground corn cobs, wood shavings, or poultry barn litter. The temperature in the compost pile should reach 120–150°F (49–68°C); this temperature level hastens the decomposition process and helps kill harmful microorganisms. A properly constructed compost pile does not give off odors or attract flies and rodents. The animal carcasses are placed on a layer of composting material, then more material is added on top. This layering process continues until the bin is full (typically about five feet [1.5 m]). After an initial decomposition period, the pile is moved to a secondary bin where the composting process is completed. The finished product is safe to spread on fields as fertilizer.

Regulations regarding the disposal of dead animals vary from state to state. Farmers should check state and local regulations to determine which methods are legal in their area. A search of the Internet using the search string "dead animal disposal" in a good search engine will produce information concerning approved methods for disposal of dead animals in various states. Local and state health departments and the Cooperative Extension Service are other good sources of information regarding legal methods for dead animal disposal.

LIVESTOCK LAWS

Animal Trespass

In many states, the owner of an animal is liable for damage an animal does if it strays onto another person's property. Animals that are trespassing may usually be held by the person whose property is damaged. They may be held until the owner makes good on the damage the animal has done. The person holding the animal must notify the owner. If the owner of the animal is not known, public notice must be posted that the animal is being held.

Estray means a domestic animal of unknown ownership that is running at large. A person may take possession of, and use, an estray found on his or her property. However, public notice of possession must be posted. The animal may not be used until the public notice is posted. The notice must describe the animal and tell where an appraisal hearing will take place within fifteen days. If the owner appears, the person holding the animal can collect the cost of keeping, feeding, and advertising the animal. If the estray dies or gets away, the person holding it is not held responsible.

The owner of an estray horse, mule, ass, or a head of cattle has one year in which to claim the animal, and must pay any charges against the animal. The estray becomes the property of the person holding it if the owner does not claim it within one year. The owner of an estray hog, sheep, or goat must appear within three months to claim the animal. Estray laws generally do not apply to poultry.

Animals on Highways

Animals sometimes stray onto public roads. A driver running into such an animal may try to collect damages. Whether or not damages may be collected depends on each individual case. Usually, negligence by the owner of the animal must be proven before damages can be collected.

SUMMARY

Society's growing concern for the environment is affecting livestock production. Farmers must now plan for handling animal wastes within laws and regulations that protect the environment. Larger feeding operations, with more confinement feeding areas, have increased the problems of handling animal wastes. In addition, more people are moving into rural areas and may complain when livestock feeding operations cause odors.

Farmers must select systems of manure handling that do not pollute the environment. The major types of systems use either solid waste management or liquid waste management. Equipment costs can be low if more manual labor is used to handle the manure. On the other hand, high-cost systems reduce the amount of labor required.

Manure is valuable as fertilizer. Thus, the best disposal method is often just to spread the wastes on the land. However, some systems, such as lagoons, do not use land for disposal of animal manure.

Feedlots must be constructed carefully to prevent runoff of the waste into streams and rivers. Proper construction and management of the feedlot also helps to prevent odors from feeding operations.

Gases and odors from livestock wastes can be dangerous and unpleasant. Management of the facility so that the manure is aerated helps to hold down the amount of odor given off. Chemical treatments can be used, but these are expensive.

Dead animals must be disposed of properly. Most states have laws that control the method of disposal. Care must be taken to prevent the spread of diseases.

Farmers are liable for damage their animals do to other people's property. Many states have laws that cover taking possession of stray animals. If an animal strays onto a public road and is involved in an accident, the owner may have to pay for damages if negligence can be proved.

Student Learning Activities

1. Present oral reports on various phases of livestock production as they relate to the environment.
2. Survey the local community to determine what problems farmers have experienced relating to livestock and the environment, and how these problems were solved.
3. Interview farmers to learn their views about current laws and regulations relating to livestock and the environment.
4. Attend local meetings concerning laws and regulations as they affect livestock production.
5. Visit livestock operations designed to conform to current laws and regulations.

Review

1. What changes have occurred in livestock production in recent years that have increased environmental problems?
2. What changes have occurred in the environment of agriculture in recent years that have made the handling of livestock wastes more difficult?
3. List and briefly describe the federal and state laws that affect livestock production as it relates to the environment.
4. From an environmental standpoint, what is the main objective of manure handling?
5. Approximately how many tons of manure are produced each year by livestock on farms in the United States?
6. How does the size of the livestock enterprise affect the system of manure handling used?
7. Describe the operation of a liquid manure-handling system.
8. How valuable is livestock manure as a fertilizer for crop production?
9. List and describe five ways to control runoff from feedlots.
10. Describe how gases and odors from livestock wastes may be controlled.
11. How does the method of livestock raising influence the effect of the environment on nutrition?
12. Define *effective ambient temperature, comfort zone, lower critical temperature*, and *upper critical temperature* and explain their affect on animal nutrition.
13. What is *wind chill index* and how does it affect stress in livestock?
14. How does the temperature affect water intake of animals?

15. What nutritional adjustments may need to be made when the weather is very cold or very hot?

16. List five disposal methods for dead animals.

17. What precautions should be taken when disposing of dead animals?

18. What is the responsibility of the owner of an animal that strays onto another person's property?

19. What is an estray and how should it be handled?

20. Name the six water pollutants that the EPA monitors and briefly discuss each.

21. Name four factors that affect the nutrient content of animal manure.

22. What conditions must exist for the EPA to define an animal feeding operation as subject to regulation?

23. Define agronomic nitrogen rate.

24. Why should the rate of application of livestock waste not exceed the agronomic nitrogen rate?

25. How much livestock waste from a swine feeding operation that uses pit storage should be applied per acre for an expected 130-bushel corn crop?

26. Under what conditions should animal wastes be injected or incorporated into the soil?

27. List four points that should be considered when applying livestock wastes to the land.

Feeding and Nutrition

Digestive Systems and Absorption of Food Nutrients

Knowledge of the different kinds of digestive systems helps in selecting the proper livestock feeds. Understanding the chemical and physical changes that take place after the feed is eaten leads to more efficient livestock feeding.

Digestion is the process of breaking feed down into simple substances that can be absorbed by the body. *Absorption* refers to taking the digested parts of the feed into the bloodstream. *Ruminants* are animals that have a stomach that is divided into several parts (Figure 5-1). Cattle, sheep, and goats are ruminants. Ruminants are cud-chewing animals. *Nonruminants* are animals that have simple, one-compartment stomachs (Figure 5-2). Pigs, horses, and poultry are nonruminants.

DIGESTIVE SYSTEMS

The *digestive system* (or *tract*) consists of the parts of the body involved in chewing and digesting feed. This system also moves the di-

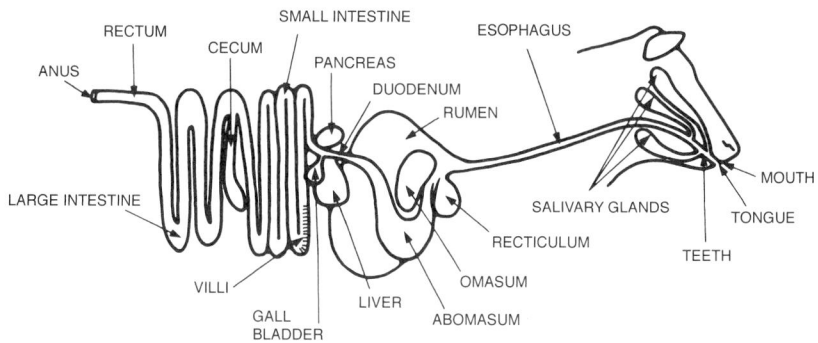

FIGURE 5-1   The digestive tract of the cow, a ruminant animal.

gested feed through the animal's body and absorbs the products of digestion.

The capacities of digestive systems vary greatly among different species of animals (Table 5-1). Within species, the age, breed, and size of animals also affect the capacity of their digestive systems. The digestive systems of ruminant animals are generally larger than those of nonruminants.

There is a great deal of difference among animals in their ability to use various kinds of feed. This difference is mainly the result of differences in their digestive systems. Ruminant animals can digest large quantities of fibrous feeds, such as hay and pasture. Nonruminant animals need a high-energy, low-fiber ration, such as grain. Grains and protein supplements are called *concentrates*.

Cattle and sheep can digest about 44 percent of the roughage they eat. *Roughage* refers to high-fiber feeds, such as hay, silage, and pasture. Horses are able to digest about 39 percent of the roughage in their ration. Swine can digest only about 22 percent of the roughage they eat.

Ruminants can digest large quantities of roughage because of the bacteria present in their digestive systems. These bacteria can produce proteins, B-complex vitamins, and vitamin K. Since ruminant animals produce certain nutrients themselves, the job of balancing their ration is somewhat easier than it is for nonruminants.

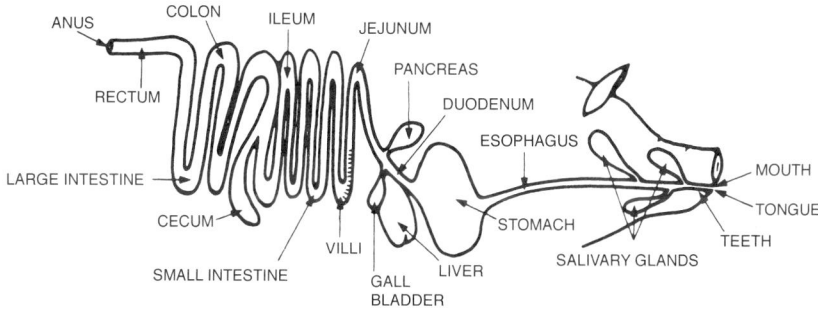

FIGURE 5-2   The digestive tract of the pig, a simple-stomached (nonruminant) animal.

TABLE 5-1 Capacities of the Digestive System of Selected Species (ranges indicate different ages, breeds, and sizes).

Organ/Species	Swine (qts)	Swine (liters)	Horse (qts)	Horse (liters)	Cattle (qts)	Cattle (liters)	Sheep/Goat (qts)	Sheep/Goat (liters)
Rumen					80–192	75.7–181.6	25	23.6
Reticulum					4–12	3.8–11.4	2	1.9
Omasum					8–20	7.6–18.9	1	0.9
Abomasum					8–24	7.6–22.7	4	3.8
Stomach in nonruminants	8	7.57	8–19	7.6–18				
Small intestine	10	9.5	27–67	25.5–63.4	65–69	61.5–65.3	10	9.5
Cecum	1–1.5	0.95–1.4	14–35	13.2–33.1	10	9.5	1	0.9
Large intestine	9–11	8.5–10.4	41–100	38.8–94.6	25–40	23.6–37.8	5–6	4.7–5.7
Total	28–30.5	26.5–28.87	90–221	85.1–209.1	200–367	189.3–347.2	48–49	45.2–46.2

Parts of the Digestive System

The digestive system is made up of a number of parts or *organs*. The system begins at the *mouth,* which is where the food enters the animal's body. The *esophagus* or *gullet* is a tubelike passage from the mouth to the stomach. The *stomach* receives the feed and adds chemicals that help in the digestive process. The food next enters the small intestine. The *small intestine* is a long folded tube attached to the lower end of the stomach. From there, the digested material passes to the large intestine. The *large intestine* is larger in diameter but much shorter in length than the small intestine. The large intestine ends with the *rectum.* Undigested material, called *feces,* is passed from the body through the *anus.*

The digestive system also includes a number of accessory organs. Accessory organs are the teeth, tongue, salivary glands, liver, and pancreas. The first three are located in the mouth. The liver is the largest gland in the body. It is located along the small intestine just past the stomach. The pancreas is located along the upper part of the small intestine.

The digestive systems of most livestock are very similar in terms of the parts they contain. However, there are some differences in the poultry digestive system. Poultry have no teeth. They have a crop and gizzard. They also have two blind pouches, called *ceca,* that are attached to the small intestine. The *cloaca* is an enlarged part connected to the large intestine. Elimination (passing of feces) in poultry is through the *vent.*

THE DIGESTIVE PROCESS

The digestive process is summarized in Table 5-2.

TABLE 5-2 Summary of Digestion.

Location/ Source	Digestive Juice	Enzyme/ Secretion	Action/Function	Comment
Mouth (Salivary glands)	Saliva	Salivary Amylase	Acts on starch/change to maltose.	Of little importance. None in ruminants. Saliva adds moisture to feed. Small amount in poultry.
		Salivary Maltase	Acts on maltose/change to glucose.	
Rumen & Reticulum			Microorganisms act on: protein/nonprotein nitrogen to form essential amino acids. starch/sucrose/cellulose to form volatile fatty acids (mainly acetic, proprionic, butyric), methane, carbon dioxide, and heat. fat to form fatty acids and glycerol. glycerol to form propionic acid.	Synthesize essential amino acids, B complex vitamins, vitamin K.
Omasum			Grinds and squeezes feed/ removes some liquid.	Little digestive action in the omasum.
Stomach/ Abomasum in ruminants/ proventriculus in avian. (Wall of stomach)	Gastric juice	Hydrochloric acid	Stops action of salivary amylase.	
		Pepsin	Acts on protein/change to proteoses, polypeptides, and peptides.	
		Rennin	Acts on milk/curdles the casein.	
		Gastric lipase	Acts on fat/forms fatty acids and gylcerol.	
Gizzard in avian			Grinds and mixes feed.	Digestive juices continue to act on feed.
Small intestine (Pancreas)	Pancreatic juice	Trypsin and Chymotrypsin	Acts on proteins, proteoses, polypeptides, and peptides/produces proteoses, peptones, peptides, and amino acids.	
		Pancreatic amylase	Acts on starch/change to maltose.	Small amounts in ruminants.
		Pancreatic lipase	Acts on fat/forms glycerol, fatty acids, and monoglycerides.	(Continues)

TABLE 5-2 Summary of Digestion. *(Cont.)*

Location/ Source	Digestive Juice	Enzyme/ Secretion	Action/Function	Comment
Small intestine (Pancreas) (Liver)	Bile	Carboxypeptidase	Acts on peptides/forms peptides and amino acids. Acts on fats/forms glycerol and soap.	
(Intestinal wall)	Intestinal juice	Intestinal peptidase (formerly called erepsin)	Acts on remaining proteins, proteoses, peptones, and peptides/produces amino acids.	
		Maltase	Acts on maltose/changes to glucose.	Small amounts in ruminants.
		Sucrase	Acts on sucrose/changes to glucose and fructose.	Small amounts in ruminants.
		Lactase	Acts on lactose/changes to glucose, fructose, and galactose.	Large amounts in young mammals.
		Nucleotidase	Acts on nucleoproteins/forms nucleotides, nucleosides, purines, pyrimidines, phosphoric acid.	
Cecum in horse			Bacterial action digests roughage.	
Large intestine		Cellulase	Acts on cellulose/forms volatile fatty acids. Some digestion continues as material moves from the small intestine to the large intestine.	Mostly in the horse.

Mouth and Esophagus

The chewing action of the mouth and teeth breaks, cuts, and tears up the feed. This increases the surface area of the feed particles which, in turn, helps the chewing and swallowing process. Saliva also stimulates the taste. In ruminants, saliva is important in the chewing of the cud.

In most animals, saliva contains the enzymes, salivary amylase and salivary maltase. *Enzymes* are substances called organic catalysts that speed up the digestive process. Salivary amylase changes some starch to maltose or malt sugar. Salivary maltase changes maltose to glucose.

Ruminants do not chew their food completely when they eat. Roughages, such as hay and silage, and coarse feeds, such as unbroken kernels of corn, are rechewed later. These feeds form ball-like masses in the stomach. The material is then forced back up the esophagus to be chewed again. This is called *rumination* or *chewing the cud.*

The tongue helps direct the feed to the throat for swallowing. The chewed material enters the esophagus. Food is carried down the esophagus by a series of muscle contractions. A valve called the *cardia* is located at the end of the esophagus. This valve prevents food in the stomach from coming back into the esophagus.

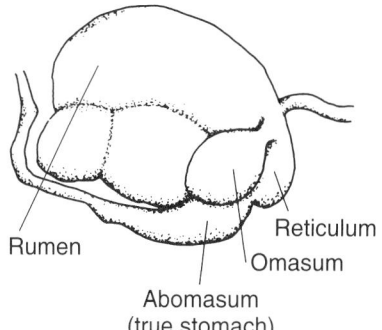

Rumen

Reticulum

Omasum

Abomasum
(true stomach)

FIGURE 5-3   The four divisions of the ruminant stomach.

The ruminant stomach. The four parts of the ruminant stomach are rumen, reticulum, omasum, and abomasum (Figure 5-3). Because of this four-part stomach, digestion in ruminants differs from that in nonruminants.

Ruminants eat rapidly. They do not chew much of their feed before they swallow it. The solid part of the feed goes into the rumen. The liquid part goes into the reticulum, then to the omasum, and on into the abomasum. In the rumen, the feed is mixed and partially broken down by bacteria. A slow churning and mixing action takes place.

When the rumen is full, the animal lies down. The feed is then forced back into the mouth and rumination occurs. Cattle chew their cud about six to eight times per day. A total of five to seven hours each day is spent in rumination.

There is no division between the rumen and the reticulum. Together, they make up about 85 percent of the stomach. They are located on the left side of the middle of the animal. The rumen and reticulum contain millions of microorganisms called bacteria and protozoa. *Bacteria* and *protozoa* are one-celled animals. Muscles in the rumen and reticulum help break the food into smaller particles. This makes it easier for the bacteria to act in the digestive process. Saliva and water are added to the feed to aid in digestion.

It is the bacterial action in the rumen that allows ruminants to use large amounts of roughage. These bacteria can change low-quality protein into the amino acids needed by the animal. *Amino acids* are compounds that contain carbon, hydrogen, oxygen, and nitrogen. They are essential for growth and maintenance of cells. Bacteria also produce many of the vitamins needed by the animal. As the bacteria die, the animal digests them. This process makes protein and vitamins available to the animal.

The kind of ration that is fed affects the growth of the microorganisms in the rumen. The number of microbes in the rumen is reduced when the ration is made up of a large amount of fine-particle material. This effect (of fine-particle material on the rumen) is reduced when animals are fed several times per day. Also, increasing the level of forage in the ration improves the growth rate of microbes in the rumen.

The pH level in the rumen also affects the growth of microbes. Microbes grow best when the pH of the rumen is in the range of 5.5

to 8.0 (optimum pH level is 6.5). A pH level outside of this range reduces the growth rate of the rumen microbes.

Rumen microbes use ammonia and amino acids, energy, and minerals for growth. A shortage of any of these can cause reduced microbe growth in the rumen. Ammonia deficiency may result in reduced efficiency of microbe growth and thus a reduction in the rate and extent of digestion of organic matter in the rumen. This may lead to reduced feed intake by the animal. Providing additional protein sources may stimulate microbe growth in the rumen. The cheapest source of ammonia for rumen microbes is generally some form of non-protein nitrogen such as urea.

Animals sometimes swallow foreign objects, such as wire and nails. These are held in the reticulum. A small magnet is sometimes inserted into the throat of the cow. When it enters the reticulum, it attracts the wire and nails and holds them so that they do not injure the animal.

A large amount of carbon dioxide and methane gas is released by bacterial action in the rumen. These gases must be disposed of through the digestive system. Sometimes the gases form faster than the animal can eliminate them. This may happen when the animal eats large amounts of fresh grass or legumes, and it can cause the animal to bloat.

The omasum is the third part of the ruminant stomach. It makes up about 8 percent of the stomach. The omasum has strong muscles in its walls. The purpose of the omasum is not exactly known. It grinds up a certain amount of feed, but how much is not known. It may also squeeze some of the water out of the feed.

The abomasum is called the *true stomach* of the ruminant. It makes up about 7 percent of the stomach. Feed is mixed with gastric juice in the abomasum. Digestion is carried on here the same as in nonruminant animals.

The nonruminant stomach. When feed enters either the stomach of the nonruminant or the abomasum of the ruminant, gastric juice begins to flow. This fluid comes from glands in the wall of the stomach. The gastric juice contains 0.2 to 0.5 percent hydrocholoric acid. It stops all action of amylase when it mixes with the feed. The gastric juice contains additional enzymes called pepsin, rennin, and gastric lipase. These enzymes act on the feed in the following ways. Pepsin breaks the proteins in the feed into proteoses and peptones. The casein of the milk is curdled by the rennin. Emulsified fats are split by the gastric lipase into glycerol and fatty acids. However, most of the fat entering the stomach is not emulsified. Thus, gastric lipase has little to do in the digestion process.

The muscular walls of the stomach churn and squeeze the feed. Liquids are pushed on into the small intestine. Gastric juice then acts on the solids that remain in the stomach.

Small Intestine

When the partly digested feed leaves the stomach, it is an acid, semi-fluid, gray, pulpy mass. This material is called *chyme*. In the small intestine, the chyme is mixed with three digestive juices: pancreatic juice, bile, and intestinal juice.

Pancreatic juice is secreted (formed and given off) by the pancreas. It contains the enzymes trypsin, pancreatic amylase, pancreatic lipase, and maltase.

Trypsin breaks down proteins not broken down by pepsin. Some of the proteoses and peptones are broken down by trypsin to peptides. Proteoses, peptones, and peptides are combinations of amino acids. Proteoses are the most complex compounds, with peptides being the simplest.

Pancreatic amylase changes starch in the feed into maltose. Pancreatic amylase is more active in this process than salivary amylase because it is found in greater quantities and has a longer time to work on the feed. Sugar and maltose are broken down into even simpler substances. When acted on by maltase, they are changed into a simple sugar called glucose.

Lipase works on fats in the feed. It changes them into fatty acids and glycerol.

Bile. The liver produces a yellowish-green, alkaline, bitter liquid called *bile*. Bile is stored in the gall bladder in all animals except the horse. Bile aids in the digestion of fats and fatty acids. It also helps in the action of the enzyme, lipase. In a final step, fatty acids combine with bile to form soluble bile salts.

Intestinal juice. Glands in the walls of the small intestine produce intestinal juice. This fluid contains peptidase, sucrase, maltase, and lactase, which help in the digestion process. Proteoses and peptones are broken down by peptidase into amino acids. Starches and sugars are broken down by sucrase, maltase, and lactase into the simple sugars, glucose, fructose, and galactose.

Absorption. The wall of the small intestine is lined with many small fingerlike projections called *villi*. These increase the absorption area of the small intestine. Most food nutrients used by the animal are absorbed from the small intestine.

The cecum or "blind gut" is found where the small intestine joins the large intestine. The cecum is a small organ and has little function in most animals, except the horse. In the horse, roughage feeds are

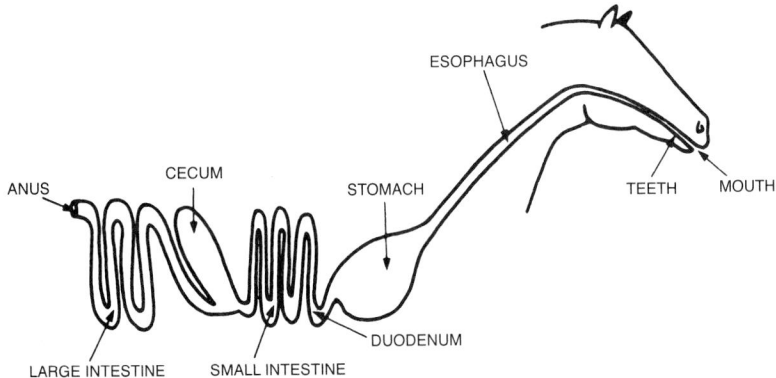

FIGURE 5-4 The digestive tract of the horse, a nonruminant animal. Horses can digest large amounts of roughage because of an organ called the cecum.

digested by bacterial action in the cecum, which explains why the horse can eat large amounts of roughage. The digestive system of the horse is shown in Figure 5-4.

Large Intestine

The large intestine is larger in diameter but much shorter in length than the small intestine. The main function of the large intestine is to absorb water. Material that is not digested and absorbed in the small intestine passes into the large intestine. The enzymes of the small intestine continue to work on the material even after it has passed into the large intestine. Some bacteria also work on the material. The large intestine adds mucus to enable the material to pass through more easily.

Feed materials that are not digested or absorbed are called feces. This material is moved through the large intestine by muscles in the intestinal walls. The undigested part of the feed is passed out of the animal's body through the anus, the opening at the end of the large intestine.

DEVELOPMENT OF THE RUMINANT STOMACH

The abomasum is the only part of the stomach of young ruminant animals that functions. Therefore, young ruminants cannot use roughages in their diet. Milk fed to the animal goes directly to the abomasum. This action continues until the other parts of the stomach have developed.

When the animal is born, the rumen is a very small organ found in the upper left part of the abdomen. After about two months, the rumen moves to its normal position in the mature animal. The reticulum and the omasum grow and develop rapidly during the first two months of the animal's life. By the time the animal is three months old, the rumen has grown large enough to begin to function. The animal can then begin to use more solid feeds and roughage in its diet.

DIGESTION IN POULTRY

Poultry possess certain special digestive organs that are not found in other animals (Figure 5-5). Feed taken in by poultry first goes to the *crop* for storage. Here it is softened by saliva and secretions from the crop wall.

The feed moves from the crop, through the glandular stomach, into the muscular stomach. The large red walls of the muscular stomach are thick, powerful muscles. The muscular stomach is lined with a thick, horny membranelike material called the *epithelium.* Feed particles are crushed and mixed with digestive juices by the *gizzard.*

Two blind pouches, or ceca, are found where the small intestine joins the large intestine. These are about 7 inches long. Their function is not known, but they are usually filled with soft, undigested feed.

The *cloaca* is an enlarged part found where the large intestine joins the vent. Feces from the large intestine are passed out of the body through the vent. Eggs from the oviduct and urine from the kidneys also pass out through the vent.

ABSORPTION OF FEED

Most absorption of digested feed takes place in the small intestine. Some feed is also absorbed from the large intestine. Feed is not broken down enough to be absorbed in the mouth or esophagus. In ruminants, some fatty acids are absorbed from the rumen. The stomach tissue of nonruminants is not suited for absorption.

Villi, the millions of small finger-shaped projections on the wall of the intestine, are the key to absorption because they greatly increase the surface area of the intestine. Each villus (singular of villi) has a network of blood capillaries and a lymph vessel through which nutrients enter the bloodstream.

Digested protein is absorbed in the form of amino acids; some sodium must be present for absorption to properly occur. Digested carbohydrates (starches and sugars) are present as the monosaccharides (simple sugars) glucose, fructose, and galactose. Fiber is in the

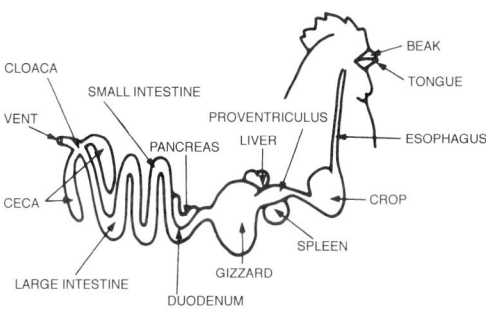

FIGURE 5-5 The digestive tract of the chicken, a nonruminant, is similar to that of other animals, but includes several special organs.

form of short-chained fatty acids. Adenosine triphosphate (ATP) supplies energy for the absorption of monosaccharides and fatty acids. Sodium must also be present for monosaccharide absorption. As the substances are absorbed by the blood capillaries in the villi, they pass through the liver and then into the blood.

Digested fats, in the form of soaps and glycerol, undergo a slightly different process. They are formed again into fats and absorbed by the lymph vessel in the villi. They are then carried through the lymphatic system. They pass through the thoracic duct in the neck into the circulatory system.

Villi also absorb water and dissolved minerals into the bloodstream. Some minerals are present in the form of compounds that are not readily absorbed. Phosphorus from grains, held in the compound phytin, is only partially absorbed by swine and none of it is absorbed in poultry. Ruminants, however, can use this phosphorus because the bacteria in the rumen release enzymes that break phytin down so the phosphorus can be absorbed.

The large intestine also absorbs water and some nutrients directly into the bloodstream. This is done through capillaries in the wall of the intestine. Nutrient absorption in the large intestine is especially important in the horse because much of the microbial digestion of roughage occurs in the cecum.

Most nutrients end their journey in the muscle cells. Some are deposited in the liver. The animal uses the nutrients to replace worn-out cells and to build new ones. Some nutrients are used for energy; others are stored in the form of fat for later use.

METABOLISM

Metabolism is the sum of the processes, both chemical and physical, that are used by living organisms and cells to handle nutrients after they have been absorbed from the digestive system. The metabolic processes include anabolism, catabolism, and oxidation of nutrients. *Anabolism* is the formation and repair of body tissues. *Catabolism* is the breakdown of body tissues into simpler substances and waste products. *Oxidation* of nutrients provides energy for the animal.

SUMMARY

Digestion is the breaking down of feeds into simple substances that can be absorbed into the bloodstream and used by the body cells. Enzymes do most of the job of digestion.

Ruminant animals can use a lot of roughage in their rations. They have a four-part stomach in which bacteria break down the roughage. Nonruminants must have more concentrates, such as grain, in their ration because they have a simple, one-part stomach.

Most digested feed is absorbed from the small intestine of the animal. The small intestine has millions of tiny projections called villi through which nutrients are absorbed and enter the bloodstream.

Student Learning Activities

1. Secure digestive systems of various species of animals from a packing plant. Dissect the systems, identify their parts, and explain their functions.
2. Prepare visuals and give oral reports on various digestive systems, identifying the parts and explaining the functions.
3. Prepare a bulletin board display with pictures of various species classified as ruminant and nonruminant animals. Using pictures or samples of feeds, show typical feeds that are fed to each kind of animal.

Review

1. Define each of the following terms: (a) digestion, (b) absorption, (c) ruminant, (d) nonruminant.
2. Why can ruminants digest large quantities of roughages?
3. Describe the typical digestive system of an animal.
4. How is the digestive system of poultry different from those of other animals?
5. What digestive action occurs in the mouth of an animal?
6. What is meant by "chewing the cud"?
7. List the four parts of the ruminant stomach.
8. What is the function of each part of the ruminant stomach?
9. What digestive action occurs in the small intestine of the animal?
10. What digestive action occurs in the large intestine of the animal?
11. Why are young ruminants unable to use large quantities of roughage?
12. How does digestion in poultry differ from digestion in other animals?
13. How is feed absorbed in an animal's body?
14. What use does an animal make of absorbed nutrients?

Unit 6 *Feed Nutrients*

Objectives

After studying this unit, the student should be able to

- identify the major functions of the basic nutrient groups and identify feeds that are sources of each.
- identify the characteristics of nutrient sources for each basic nutrient group.

A *nutrient* is defined as a chemical element or compound that aids in the support of life. A nutrient becomes a part of the cells of the body. Nutrients are necessary for cells to live, grow, and function properly.

Many different kinds of nutrients are needed by animals. In addition, they must have the right nutrients in the proper balance. Too much of one nutrient and not enough of another may result in unhealthy stock and high feed costs. A lack of one or more nutrients may slow normal growth or production.

Animals differ in the kinds and amounts of nutrients they need. It is the job of animal nutritionists to determine which nutrients animals need. They make recommendations for feeding each kind of animal based on the results of feeding experiments. The farmer who follows these recommendations produces healthy livestock at a profit.

Nutrients are divided into five groups: (1) energy nutrients (carbohydrates, fats, and oils), (2) proteins, (3) vitamins, (4) minerals, and (5) water.

ENERGY NUTRIENTS—CARBOHYDRATES

The main energy nutrients found in animal rations are carbohydrates. Carbohydrates are made up of sugars, starches, cellulose, and lignin. Carbohydrates are chemically composed of carbon, hydrogen, and oxygen.

In a chemical reaction very much like burning, carbohydrates provide energy for the body cells. This energy powers muscular move-

FIGURE 6-1 Ruminant animals can digest large amounts of fiber found in roughages such as hay and pasture. *Photo by Jim Bodine.*

ment such as the heartbeat, walking, breathing, and digestive contractions. Carbohydrates also produce the body heat that helps to keep the animal warm. Extra carbohydrates are stored in the body in the form of fat.

Simple Carbohydrates

The simple carbohydrates are sugars and starches. They are easily digested. Simple carbohydrates are referred to as *nitrogen-free extract (NFE)* and come from the cereal grains such as corn, wheat, oats, barley, rye, and sorghum.

Complex Carbohydrates

The complex carbohydrates, called *fiber,* are cellulose and lignin. These substances are more difficult to digest than simple carbohydrates. Fiber is found mainly in roughages such as hay and pasture plants. Examples are alfalfa, bromegrass, orchard grass, and bluegrass.

Carbohydrate and Fiber Content of Feeds

The dry weight of most grains and roughages ranges from 65 to 80 percent carbohydrates. The *dry weight* of a feed refers to its weight with the moisture content removed. Mature roughages contain more fiber than those harvested when less mature. The mature plant is not easily digested. Hay plants harvested at an early stage of maturity have a higher feed value because they are more easily digested.

Ruminant animals, such as cows, sheep, and goats, can digest large amounts of fiber. A high percent of their ration is roughage (Figure 6-1). Simple-stomached animals, such as swine, cannot digest large amounts of fiber. Their ration must be mostly cereal grains that are more easily digested (Figure 6-2).

FIGURE 6-2 Simple-stomached animals require mainly grain in their ration; they cannot digest large amounts of roughage. *Photo by Jim Bodine.*

A *commercial feed tag* is the label attached to a bag of feed purchased at a grain elevator or feed store. The maximum amount of fiber found in that feed is shown as a percent of the total weight of the bag of feed.

The sources of carbohydrates used for that feed are found in the list of ingredients printed on the tag. Some examples of carbohydrate sources found on a feed tag are cane molasses, ground corn, wheat middlings, and dehydrated alfalfa meal. There are many other sources of carbohydrates used in commercial feeds. The ones used depend on the kind of animal for which the feed is made.

ENERGY NUTRIENTS—FATS AND OILS

Fats and oils are also energy nutrients. They are chemically composed of carbon, hydrogen, and oxygen. They contain more carbon and hydrogen atoms than do carbohydrates. For this reason, the energy value of fats is higher than that of carbohydrates. In fact, fats have 2.25 times the energy value of carbohydrates. *Fats* are solid at body temperature, and *oils* are liquid at body temperature. In animal nutrition, both fats and oils are commonly referred to as *fats*.

Fats are easily digested by animals. They provide energy and body heat. They also carry the fat-soluble vitamins that are in the feed.

Fats come from both vegetable and animal sources. Cereal grains such as corn, oats, and wheat range from 1.8 to 4.4 percent fat. Cereal products used in rations, such as brewer's dried grains, corn gluten meal, distiller's dried grains, and wheat middlings, range from 1.1 to 8.2 percent fat. Animal and vegetable protein concentrates range from 1 to 10.6 percent fat.

The amount of fat in a commercial feed is shown on the feed tag as a percent of the total weight of feed in the bag. The fat content in the feed is expressed as a guaranteed minimum level.

PROTEINS

Proteins are organic compounds made up of amino acids. Amino acids contain carbon, hydrogen, oxygen, and nitrogen. Some amino acids also contain sulphur, phosphorus, and iron.

Proteins supply material to build body tissues. The ligaments, hair, hooves, horns, skin, internal organs, and muscles of the animal body are partially formed from protein. Protein is essential for fetal development in pregnant animals.

If an animal takes in more protein than it needs, the nitrogen is separated and given off in the urine. The material that is left is converted into energy or body fat by the animal.

Ten amino acids are considered to be essential for swine, and thirteen are considered to be essential for poultry (Table 6-1).

TABLE 6-1 Essential and Nonessential Amino Acids for Swine and Poultry.

Essential	Nonessential
Arginine	Alanine
Histidine	Aspartic acid
Isoleucine	Citrulline
Leucine	Cysteine
Lysine	Cystine
Methionine*	Glutamic acid
Phenylalanine**	Glycine
Threonine	Hydroxyglutamic acid
Tryptophan	Hydroxyproline
Valine	Norleucine
	Proline
Additional essential for poultry:	Serine
	Tyrosine
Alanine	
Aspartic acid	
Glycine	
Serine	

*Part can be replaced with Cystine.
**Part can be replaced with Tyrosine.

Nonessential amino acids are needed by animals, but are synthesized in the body from other amino acids and therefore do not have to be provided in the ration for either ruminant or nonruminant animals. Nonruminant animals cannot synthesize the essential amino acids fast enough to meet their needs; therefore, those amino acids must be provided in their rations. Ruminants can generally synthesize the essential amino acids by rumen bacterial action at a rate sufficient to meet their needs.

Recent research suggests that fermentation and bacterial action in the rumen may reduce the quality of protein in the feed. This may result in a deficiency of one or more essential amino acids in ruminant rations. Encapsulating the protein feed reduces exposure to fermentation in the rumen. Heat or chemical treatment with tannic acid or formaldehyde reduces the tendency of the protein to degrade in quality in the rumen. When the high-quality protein is protected, the rumen bacteria are more likely to use non-protein nitrogen sources in fermentation processes. Research is continuing in this phase of animal nutrition.

Sources of Protein

Animal protein sources are considered to be good-quality proteins since they usually contain a good balance of the essential amino acids. Plant protein sources are usually thought of as poor-quality proteins because they often lack some of the essential amino acids.

The amino acid needs of ruminants can be met by feeding proteins from vegetable sources. They can also be met by including urea in the ration. *Urea* is a synthetic nitrogen source that is manufactured from air, water, and carbon. It is then mixed in the ration to provide nitrogen for the making of amino acids in the ruminant's body.

It is necessary to feed simple-stomached animals protein from sources that provide a balance of essential amino acids. This is done to make sure they get the essential amino acids in their feed. If cereal grains are combined in the right combination, they will provide a balanced ration. Soybean oil meal, for example, is one of the most commonly used protein sources for feeding hogs.

A commercial feed tag shows the guaranteed minimum of crude protein contained in that feed, shown as a percent of the weight of the bag of feed. The amount of *crude protein* (total protein) is the amount of ammoniacal nitrogen in the feed multiplied by 6.25. It may contain materials that are not true protein. Plant protein sources often used in commercial feeds include linseed meal, dehulled soybean meal, cottonseed meal, and dehydrated alfalfa meal. Animal protein sources often used are meat meal, fish meal, condensed fish solubles, dried whey, casein, dried milk albumin, and dried skim milk. The synthetic nitrogen source, urea, is also used for ruminant feeds. The list of ingredients on the feed tag shows the protein sources used in the mix.

Not all of the crude protein in a feed is digestible. Thus, the animal cannot use all of it. Some may come from hair, hooves, feathers, or other sources that have a low digestibility. About 60 percent of the crude protein in a roughage ration is considered to be digestible. About 75 percent of the crude protein in a high-concentrate ration is considered to be digestible. *Digestible protein* is approximately the amount of true protein in a feed.

VITAMINS

Vitamins are *trace* organic compounds. In other words, they are needed only in very small amounts by animals. All vitamins contain carbon, but they are not alike chemically.

Vitamins are divided into two groups: fat soluble and water soluble. *Fat-soluble vitamins* can be dissolved in fat. *Water-soluble vitamins* can be dissolved in water.

The fat-soluble vitamins are A, D, E, and K. Vitamin A is associated with healthy eyes, good conception rate, and disease resistance. Vitamin D is associated with good bone development and the mineral balance of the blood. Vitamin E is necessary for normal reproduction and muscle development. Recent research suggests that increasing the level of vitamin E alone or in combination with selenium in the diet may help strengthen the immune system of several animal species. Vitamin K helps the blood to clot and prevents excessive bleeding from injuries.

Some sources of the fat-soluble vitamins are green leafy hay, yellow corn, cod liver and other fish oils, wheat germ oil, and green pastures. Vitamin D is produced in the animal's body if the animal is in direct sunlight part of the day.

The water-soluble vitamins include vitamin C and the B-complex vitamins. The B-complex includes many different vitamins. These are: B_1 (thiamine), riboflavin, niacin, pyridoxine, pantothenic acid, biotin, folic acid, benzoic acid, choline, and vitamin B_{12}.

Vitamin C helps in teeth and bone formation and the prevention of infections. The B-complex vitamins are necessary for chemical reactions in the animal's body. They help improve appetite, growth, and reproduction. The effects of a lack of one of the B-complex vitamins is shown in Figure 6-3.

Sources of Vitamins

Some sources of vitamin C are green pastures and hay. All farm animals seem to produce enough vitamin C in their bodies. Therefore, it does not need to be included as a specific nutrient in their rations. Some sources of the B-complex vitamins are green pastures, cereal grains, green leafy hay, milk, fish solubles, and certain animal proteins.

FIGURE 6-3 These pigs were litter mates. Their rations were the same except for nicotinic acid, one of the B-complex vitamins. The larger pig received enough nicotinic acid, while the smaller pig shows symptoms of deficiency. *Courtesy of USDA.*

Commercial feeds usually include necessary vitamins in the mixture. Vitamins often included in commercial feeds for ruminants are A, D, and E. Vitamins often included in commercial feeds for swine are A, D, E, K, riboflavin, niacin, d-pantothenic acid, B_{12}, and choline chloride.

MINERALS

Minerals are inorganic materials needed in small amounts by animals. Minerals contain no carbon. Thus, if a feed were completely burned, the ash that was left would be the mineral content of the feed.

Minerals are important in animal feeding. They provide material for the growth of bones, teeth, and tissue. Minerals also regulate many of the vital chemical processes of the body. They aid in muscular activities, reproduction, digestion of feed, repair of body tissues, formation of new tissue, and release of energy for body heat. If there is a lack of a certain mineral in an animal's ration, this is called a *deficiency*.

For example, without iron in the blood, oxygen could not be carried to the body cells. A deficiency of iron and copper in baby pig rations will cause anemia. The bones and teeth will not form properly if the animal lacks calcium and phosphorus in its ration.

Recent research suggests that increasing the level of trace minerals in livestock rations above the currently recommended level may improve an animal's immunity to disease. For example, increasing the level of selenium in the diet of calves and swine has increased the antibody response of the immune system. A small increase in the zinc level in the diet has increased the production of white blood cells to fight infections. More research is needed before this practice can be recommended for use on the farm.

Minerals are divided into two groups. *Major minerals* are those needed in large amounts. *Trace minerals* are those needed in small amounts.

Major minerals that are often lacking in animal rations are salt (sodium and chlorine), calcium, and phosphorus. It is important to make sure that these minerals are included in the animal's ration. The ratio of calcium to phosphorus in swine rations should not be greater than 1.5 to 1. The ratio of calcium to phosphorus for ruminants can be as high as 7 to 1.

Trace minerals that are necessary for animals include potassium, sulfur, magnesium, iron, iodine, copper, cobalt, zinc, manganese, boron, molybdenum, fluorine, and selenium. Most of these trace minerals are found in common animal feeds. Usually there are adequate amounts of these minerals in the feed.

In some areas and under certain conditions there may be deficiencies of some trace minerals in the feed available to livestock. These deficiencies may be widespread in a given geographic area or they may occur only in parts of selected fields. Table 6-2 lists trace minerals and where problem areas of deficiencies may occur in the United States. If a trace mineral deficiency in the feed is suspected, it

TABLE 6-2 Trace Mineral Deficient Areas of the United States.

Element	Deficiency Occurrence in the United States
Boron	Throughout the United States on sandy light-textured soils, acid soils, calcareous soils, soils with low amounts of organic matter. Most serious with alfalfa.
Cobalt	Sandy soils in New England states and South Atlantic Coastal Plain, especially in Florida. Legume plants in these areas; grasses and cereal grains in all parts of the United States.
Copper	Most often found on organic soils or very sandy soils in Central wheat belt, New England states, lower Atlantic Coastal Plain. Generally in legumes grown in these areas.
Iodine	Great Lakes states, Dakotas, Montana, Idaho, Washington, Oregon, Nevada, Utah, Colorado, Wyoming, western Nebraska, Southeastern states of the Appalachian Range.
Iron	Alkaline soils of western United States and very sandy soils. Intermountain region in western United States, including northern Nebraska to Kansas, Colorado, western Oklahoma, eastern New Mexico, northwest Texas; also Iowa and California. Often highly localized in specific fields.
Magnesium	Sandy and loamy soils with a high level of available potassium.
Manganese	Great Lakes region and Atlantic Coastal Plain states. Problem is greatest on calcareous soils, peats, mucks, course-textured soils, and poorly drained soils.
Molybdenum	Poorly drained acid soils in the intermountain valleys of western United States and poorly drained acid soils in South Central Florida.
Selenium	Great Lakes states, New England states, upper Appalachian area, Atlantic Coastal plain, Florida. Also, Washington, Oregon, northern two-thirds of California, Nevada, Idaho, western Montana, western Utah. Deficiency may be highly localized in some areas.
Sulphur	Pacific Northwest and some parts of the Great Lakes states.
Zinc	Western United States (generally highly localized on irrigated land), southeastern United States on sandy, well-drained, acid soils, or soils from phosphatic rock parent material. Some localized areas of deficiency in other areas of the United States.

may be wise to have the feed analyzed. When a deficiency is identified, the ration must be properly supplemented with the missing trace minerals.

Some trace minerals, when present in excess quantities, are toxic to livestock. Toxicity may occur when the minerals accumulate in plants grown in the area. Some toxicity problems are discussed in Table 6-3.

Major and trace minerals are usually supplied in commercial feeds. They may be included in a protein supplement or added by the use of a mineral premix. Salt and mineral blocks are often used to provide the additional minerals needed in the ration. Monocalcium phosphate, dicalcium phosphate, ground limestone, steamed bone-meal, and calcium carbonate are usually included in commercial feeds and mineral supplements to provide calcium and phosphorus.

Commercial feed tags show a guaranteed minimum and maximum percent of calcium. The minimum percent of phosphorus is shown on the tag. If salt is added, the minimum and maximum percent of salt is also shown on the feed tag. Major minerals, except salt, are guaranteed in terms of the individual element rather than as compounds. Salt is guaranteed as the compound sodium chloride (NaCl). Trace minerals are guaranteed on the feed tag as a minimum percent of the weight of the feed in the bag.

WATER

Water is so common that its importance as a nutrient is often forgotten. However, water makes up the largest part of most living things. The amount of water in an animal's body varies with the kind of animal, its age, and its condition. In general, the amount of water in animal bodies ranges from 40 to 80 percent. Younger animals have a higher percent of water in their bodies than older animals.

TABLE 6-3 Trace Mineral Toxicity Areas of the United States.

Element	Toxicity Problem Areas in the United States
Boron	San Joaquin valley of California. Semiarid regions with alkaline soils in other areas of the western United States.
Manganese	On poorly drained, acid soils. More likely to affect crop growth than livestock production.
Molybdenum	Forage crops on some alkaline soils in western United States and poorly drained organic soils in South Central Florida. Cattle and sheep most likely to be affected.
Selenium	Some areas of the Plains and Rocky Mountain states; most often on well-drained alkaline soils. Some shrubs and weeds native to semiarid and desert rangelands have an unusual ability to extract selenium from the soil. If grazed, these plants can cause selenium toxicity in livestock. Range grasses and field crops growing in the same areas do not accumulate excessive amounts of selenium in their tissues.

Water has many important functions. It helps to dissolve the nutrients the animal eats. It also helps to control the temperature of the animal's body. Water in the blood acts as a carrier of the nutrients to different parts of the animal's body. Water is necessary for many of the chemical reactions that take place in the body.

A fresh, clean supply of water is necessary for animals to grow and produce profitably. A continuous supply is best for rapid growth and efficient production. If animals do not have a good water supply, they will not make good use of the other nutrients supplied in the ration. Animals can live longer without food than they can without water.

SUMMARY

Nutrients are chemical elements or compounds that aid in the support of life. Animals must have five different groups of nutrients to grow and produce efficiently. Energy nutrients provide energy necessary for movement and production of body heat. Proteins supply material to build body tissues, hooves, horns, hair, and skin. Vitamins help to regulate many of the body's functions. Minerals provide material for bones, teeth, and tissues and help to regulate chemical activity in the body. Water dissolves and carries nutrients, regulates temperature, and is necessary for chemical reactions in the body.

Nutrients are supplied by the grains and forages fed to the animal. Additional nutrients needed by the animal are supplied by commercial feed mixes.

Student Learning Activities

1. Collect samples of nutrient sources from farms and feed stores. Develop an exhibit of these samples showing the approximate percent of each nutrient that comes from each source. Include information about the function of each nutrient in the exhibit.
2. Present a short oral report to the class about new developments in feed nutrients that have been reported in farm magazines and newspapers.
3. Develop a bulletin board exhibit of commercial feed tags. Include information about the functions of the nutrients listed on the tags.
4. Take a field trip to a local grain elevator, feed store, or commercial feed manufacturer. Observe the sources of feed nutrients used in feed mixing, and the method of mixing complete feeds.
5. Take a field trip to a local farm that has facilities for mixing complete feeds. Observe the nutrient sources used and the methods used to obtain a complete feed mix.

Review

1. What is a nutrient?
2. List the five groups of nutrients.
3. What is nitrogen-free extract?
4. What are the four substances that make up carbohydrates?
5. What is the function of carbohydrates?
6. List five feed sources of carbohydrates.
7. What is the function of fats and oils?
8. List five feed sources of fats and oils.
9. What are proteins?
10. What is the function of proteins?
11. Name the ten essential amino acids.
12. What is the difference in value between animal protein sources and plant protein sources?
13. What is urea?
14. List three plant protein sources and three animal protein sources.
15. What is the difference between crude protein and digestible protein?
16. What are vitamins?
17. What is the function of vitamins?
18. Which vitamins are fat soluble and which are water soluble?
19. What are minerals?
20. What is the function of minerals?
21. List the major and trace minerals often needed in animal rations.
22. What is the function of water in an animal's body?

Unit 7

Feed Additives and Hormone Implants

Objectives

After studying this unit, the student should be able to

- discuss the general use and purpose of feed additives and hormone implants.
- describe the proper method of hormone implantation.
- discuss the proper use of feed additives for various species of animals.
- discuss labeling and regulation of feed additives.
- discuss the proper mixing of feed additives in complete rations.
- discuss health issues and concerns relating to the use of feed additives.

Feed additives are materials used in animal rations to improve feed efficiency, promote faster gains, improve animal health, or increase production of animal products. These materials are not generally considered to be nutrients and are used in small amounts in the ration. They are often added to the basic feed mix and require careful handling and mixing. Hormone implants are pelleted synthetic or natural hormones or hormonelike compounds placed under the skin or in the muscle of the animal. Implants are used to lower production costs by improving both rate and efficiency of gain. Feed additives and hormone implants are sometimes called *performance stimulants*.

Feed additives came into common use in livestock feeding in the early 1950s. Since then, a wide variety of feed additives and hormone compounds have been developed, tested, and approved for use with livestock. The use of some of these materials has been discontinued because of toxicity, high cost, lack of proven benefit, or excessive residues in meat and livestock products. However, many feed additives and hormones are still widely used in livestock production. Hormone implants are used mainly in beef cattle operations. The kinds of materials used in livestock production include antimicrobial compounds (such as antibiotics and chemoantibacterials), hormones and hormonelike substances, anthelmintics (dewormers), buffering agents, feed flavors, and bloat preventatives.

KINDS OF FEED ADDITIVES AND HORMONES

Antimicrobial Drugs

Antibiotics and chemoantibacterial (chemotherapeutic) compounds are called antimicrobial drugs because they kill or slow down the growth of some kinds of microorganisms. These compounds are often used as feed additives for livestock and poultry rations.

There are many different kinds of microorganisms (microbes) that live in the bodies of animals. Some of these microbes are beneficial to animals, and some are harmful. When livestock are raised in confinement, there is a greater potential for the spread of harmful microbes among the animals because they are crowded more closely together; the use of antimicrobial drugs helps to keep these harmful microbes under control. The use of these drugs at a lower level in the feed than would be used for treating sick animals is referred to as a *subtherapeutic* level of use. The use of antibiotics at subtherapeutic levels in large cattle feedlots has been declining in recent years. Cattle feeders have become increasingly concerned about the development of resistant strains of bacteria when drugs are used at the subtherapeutic level.

The major difference between antibiotics and chemoantibacterial compounds is the way in which they are produced. Antibiotics are produced by living microorganisms. Chemoantibacterial compounds are made from chemicals. Sometimes an antibiotic and an antibacterial are combined into one compound, called a *chemobiotic*, to combat a problem that is not susceptible to either one individually. In this text, the term antibiotic is used generically to refer to all of these compounds.

There are hundreds of antibiotics. Some of the ones most commonly used in livestock production include chlortetracycline (Aureomycin), neomycin, oxytetracycline (Terramycin), penicillin, streptomycin, and tylosin. Polyether antibiotics are called ionophores and are usually used in the production of ruminant animals. Monensin (Rumensin) and lasalocid sodium (Bovatec) are two commonly used ionophore antibiotics used in beef production. Some common chemoantibacterial compounds used in livestock feeding include carbadox, furazolidone, nitrofurazone, and sulfamethazine.

There is a variation in response to the use of antibiotics among different species and under various feeding conditions. Because antibiotics act to control microbes, there is little or no benefit gained from the feeding of antibiotics when livestock are raised under conditions which are free of harmful microorganisms. Research in livestock nutrition shows several factors that explain why antibiotics increase rate of gain, improve feed efficiency, or improve the general health of animals.

When a substance is referred to as *nutrient-sparing*, it means that the substance allows animals to use available nutrients more effectively. Antibiotics act in several ways to accomplish a nutrient-sparing effect in livestock. Some antibiotics stimulate microbes that are present in the digestive tract to produce more nutrients than they would without the presence of the antibiotic. Some microbes in the digestive tract compete for essential nutrients without improving the performance of the animal; some antibiotics slow down this competition for essential nutrients. An animal must absorb nutrients through the intestinal wall in order to use them in metabolism; some antibiotics help in the development of a thin, healthy intestinal wall, which allows for easier absorption of nutrients.

The rate of metabolism in young, growing animals changes when they are fed antibiotics. When antibiotics are included at low levels in the ration, the daily feed intake is greater and the conversion of feed to meat becomes faster and more efficient as compared with animals that do not have antibiotics included in the ration.

Subclinical diseases are those that are present in the animal's body at levels too low to cause visible effects. Under many conditions of farm feeding, subclinical diseases are present in the animals being fed. When antibiotics are included in the ration, these subclinical diseases are continuously controlled. As a result, the animal is healthier and more vigorous. Because subclinical diseases are controlled, the rate of gain is more uniform among groups of animals that are being fed antibiotics.

Different types of antibiotics vary in the range of microorganisms that they control. Some control many different microorganisms and are thus referred to as *broad-spectrum* antibiotics; those that control only a few microorganisms are called *narrow-spectrum* antibiotics.

Broad-spectrum antibiotics are preferred for use as feed additives. They generally give better results in terms of rate of gain, feed efficiency, and improved animal health. Narrow-spectrum antibiotics are more often used to control a specific disease problem that may be present in the group of animals being fed.

Hormones and Hormonelike Compounds

Hormones are substances produced in the animal's body. Natural hormones are secreted into the body fluids, such as the bloodstream, by various glands in the body. The adrenal cortex, pancreas, pituitary, ovaries, and testes all secrete small amounts of hormones. Hormones regulate many body functions, such as growth, metabolism, and the reproductive cycle. Hormonelike compounds are synthetic substances that act like hormones in the body. Hormones and hormonelike compounds are produced commercially to be used as feed additives, primarily in beef nutrition.

Androgens, progestogens, and estrogens are hormones produced by the sex glands of animals. These hormones increase the rate of protein synthesis and muscle development. They are used in rations to improve feed efficiency and increase the rate of growth. Beef cattle have shown the greatest response to their use either as feed additives or implants.

The first synthetic hormone developed for use in animal feeding was stilbestrol (diethylstilbestrol [DES]). DES is a synthetic estrogen. It was approved by the Federal Food and Drug Administration (FDA) in 1954 as an additive for use in beef cattle finishing rations. The FDA approved DES implants for steers in 1956. Significant increases in rate of gain and feed efficiency in steers were shown to result from the use of DES.

DES was also used to treat women during pregnancy to prevent miscarriages. During the 1960s, medical research indicated that there was some danger that daughters of women who had been treated with DES during pregnancy could develop a rare form of cancer. Research also showed that DES could cause cancer in mice when they were fed massive doses of the drug. In 1972, as a result of these findings, the FDA withdrew approval of DES as a feed additive. This ban was challenged in the courts and DES was again approved for use. In 1979 the FDA again banned the use of DES as a feed additive, and this ban is still in effect today.

The use of hormones and hormonelike compounds in livestock production continues to be controversial. A recent survey of consumers in the United States reveals that 61 percent believe that hormone residues in meat are a serious health hazard; another 32 percent believe that they pose something of a hazard.

The debate over this use of hormones has affected the international markets for meat produced in the United States. Since January 1, 1989, the European Union (EU) has banned the importation of any meat for human consumption that has been treated with anabolic agents. In effect, meat produced with the use of hormones or hormonelike compounds cannot be sold in the EU. Individual meat producers who do not use these compounds in livestock production can still export their products to the EU under individual commercial agreements. The EU has established criteria to monitor these products. However, a problem still exists in monitoring this meat because it is not possible to differentiate between hormones that the animal produces naturally and those that have been implanted.

In 1998 the World Trade Organization (WTO) ruled that the EU ban on meat imported from animals treated with growth hormones is a violation of international trade rules. Following this ruling, the

EU maintained its original ban based on a report that at least one growth hormone used in the United States is carcinogenic. As of the summer of 1999, the issue had not been resolved.

The EU ban is based on consumer concern for human health and safety. However, research by the United States Food and Drug Administration (FDA) has shown that when these compounds are properly used, there are no adverse effects on human health. Researchers at the World Health Organization (WHO) have reached the same conclusion. The amount of these hormones that the human body produces naturally is far greater than any residues found in meat.

Some of the consumer concern is related to the possible carcinogenic effects of these compounds. Research shows that the hormones approved for use in the United States for livestock production would have to be consumed in extremely high dosages to act as carcinogens. When they are used in the proper manner at the approved levels, they pose no danger as carcinogens to humans.

The continuing controversy over the use of these compounds in meat production creates a greater need for livestock producers to exercise care in their use. Careless use provides ammunition for those who would ban these compounds in the United States. If such a ban were imposed, it would result in an estimated 13 percent drop in the amount of beef produced; this could mean annual losses to beef producers of $1 to $3 billion.

The controversy over the use of these compounds has been marked with a great deal of sensationalism and misinformation. Beef producers face a major problem in providing accurate information to the consuming public regarding this issue.

Hormones and hormonelike compounds will be discussed in more detail later in this unit in the section dealing with beef cattle.

Anthelmintics

Anthelmintics (dewormers) are compounds used to control various species of worms that may infest animals. They may be provided to the animal in either feed or water. Some anthelmintics available for use include hygromycin, loxon, phenothiazine, piperzine, thiabendazole, and tramisol.

Some of the more common species of worms that can infest animals at various stages of production include large roundworms, nodular worms, and whipworms. The presence of worms in the animal's system reduces feed efficiency and rate of gain. The level of performance improvement gained from the use of anthelmintics depends on the level of worm infestation found in the animals being treated. Where good management practices (such as rotating pastures or keeping manure cleaned from pens) are followed, there will be less response from the use of anthelmintics. However, if there is an indication of worms in any of the animals in a group, it is safe to assume that all animals in the group have some level of worm infestation. This condition indicates the need to treat the entire group of animals for worm infestation.

Other Feed Additives

A number of substances may be used for specific purposes in livestock feeding. In some cases these are designed to improve growth, rate of gain, or feed efficiency. In other cases they are used for specific purposes.

Coccidiostats are sometimes added to poultry rations to prevent the disease coccidiosis. Some typical coccidiostats used in poultry rations include buquinolate, butynorate, carbarsone, clopidol, decoquinate, ethopabate, furazolidone, lasalocid sodium, momensin, nicarbazin, ormetoprim, oxytetracycline, sulfaquinoxaline, robenidine hydrochloride and zoalene.

The pH level (the acid-base balance) of the fluids in the digestive tract influences the proper digestion of feeds. Sodium bicarbonate and ground limestone are sometimes used in rations of ruminant animals to regulate the pH level or the acid-base balance and thus improve digestion.

The thyroid gland functions to control the rate of metabolism. An additive such as iodinated casein (thyropotein) may be used to increase the amount of the hormone thyroxin in an animal's system. If too much thyroxin is being produced in the body, there are several inhibitors—such as thiourea, thiouracil, and methimazole (tapazole)—that may be used to regulate the amount of thyroxin present.

Ruminant animals may bloat when they eat too much lush, green alfalfa or too much grain. Foaming may occur in the rumen, or a slime layer may build up over the liquid in the rumen that prevents the gases in the rumen from escaping. This causes a noticeable swelling of the animal's midsection. Poloxalene (Bloat Guard) is a bloat preventative that may be added to ruminant feeds when bloat is a problem. It may be used as a precautionary measure when putting animals on pasture in the spring or when starting them on feed in the feedlot. Poloxalene works by breaking up the foam or the slime layer and allowing the gases to escape.

Animals under stress in the feedlot may be calmed by the use of tranquilizers such as hydroxyzine, reserpine, or trifluomeprazine. Enzyme feed additives are available but their cost on a continuous use basis may outweigh the benefits realized from their use. Antioxidants may be added to feeds to prevent rancidity. Sodium bentonite is sometimes used as a pellet binder when feeds are pelleted. Feed flavors are sometimes added to rations to make the ration more palatable or to hide the taste of medication being given in the feed.

Several copper compounds, such as copper sulfate, copper carbonate, copper chloride, and copper oxide, have improved performance when added to swine growing rations. Care must be taken to use the appropriate level of the copper compound and thoroughly mix the feed to prevent toxicity problems. There is no current FDA regulation regarding the use of copper compounds in swine rations. The use of such compounds is becoming popular because they are cheaper than antibiotics. Because copper interferes with the bacterial action in lagoon manure disposal systems, swine producers who use such systems should not use copper compounds in animal rations.

Probiotics are compounds such as yeasts and lactobacilli that change the bacterial population in the digestive tract to a more desirable type. In some cases the use of probiotics has improved animal performance, but the improvement has not been as great as that obtained by the use of antibiotics. Research has shown that probiotics can be used in conjunction with antibiotics in the feeding program.

Some organic acids such as propionic acid are used to slow the development of molds in feed. They are used as preservatives in high-moisture grain.

BEEF CATTLE

The use of antibiotics and hormones as feed additives is one of the most effective management tools available to beef cattle producers. The use of these products significantly increases feed efficiency and rate of gain, and thus increases profits. Table 7-1 gives examples of some feed additives that may be used with beef cattle.

Antibiotics

Cattle being fed a high-energy ration will show a 3- to 5-percent improvement in rate of gain and feed efficiency when a continuous low level (35 to 100 milligrams per head per day) of antibiotic is included in the ration. Daily gain and feed efficiency is even higher for growing cattle being fed low-energy rations.

The improved performance that results from the use of antibiotics generally results from the action of the antibiotics against harmful microorganisms. These microorganisms cause such feedlot disorders as foot rot, liver abscess, respiratory diseases, and shipping fever. The greatest response from the use of antibiotics usually occurs when cattle are under stress or just starting on feed for finishing in the feedlot. However, there is a continued response throughout the feeding period when antibiotics are fed continuously at low levels.

Antibiotics that are available for use with beef cattle include bacitracin methylene disalicylate, chlortetracycline (Aureomycin), erythromycin (Gallimycin), neomycin, oxytetracycline (Terramycin), and tylosin (Tylan).

A high percentage of rations for cattle on feed contain either monensin sodium (Rumensin) or lasalocid sodium (Bovatec). Both are ionophore antibiotics. Monensin sodium either is added to the ration of feedlot cattle at the rate of 20 to 30 grams per ton of complete feed (90% dry matter basis) or may be fed at the rate of 100 to 300 milligrams per head per day. It generally improves feed efficiency, but does not improve the rate of gain. Lasalocid sodium is added to the ration at the rate of 10 to 30 grams per ton of complete feed (90% dry matter basis) to improve feed efficiency. When fed at the rate of 25 to 30 grams per ton of complete feed, it increases the rate of gain.

These compounds affect the fermentation of feed in the rumen, decreasing the proportion of methane gas and increasing the proportion of propionic acid produced. As a result, the conversion of feed energy to growth is improved by a factor of 5 to 10 percent. Feedlot bloat and acidosis are also reduced by these compounds.

Cattle on a high-energy (high-grain) ration show an initial reduction in feed intake of about 20 percent when monensin is first added to the diet. This reduction in feed intake lasts about three to four days. Feed intake then gradually rises to a level of about 10 percent less than the feed intake of cattle that are not being fed monensin. While cattle being fed monensin eat about 2–4 percent less than those not being fed monensin, the rate of gain is about the same. The efficiency of feed use is increased by the addition of monensin to the diet.

Monensin has FDA approval as a feed additive for cattle on pasture that weigh over 400 pounds (181 kg). Cattle on low-energy (pasture or high roughage) diets do not respond in the same manner as those on high-energy diets. Feed intake is not reduced, but the cattle make better use of the energy in the feed. This reduces the daily energy maintenance requirements of the cattle and results in faster gains. Research shows that cattle on pasture with monensin in their diet gain about 16 percent faster than cattle that are not being fed monensin.

Lasalocid sodium (Bovatec) acts in a manner similar to monensin. Research shows that feed efficiency is improved about 8 percent, while rate of gain is increased about 5 percent. A major difference between monensin and lasalocid sodium is the ability of the latter to improve rate of gain. Monensin generally does not improve rate of gain except for cattle on high-roughage rations. Also, lasalocid sodium does not reduce feed intake at the start of the feeding period as much as monensin does. Lasalocid sodium is also less toxic to horses and swine than is monensin.

The claims listed in Table 7-1 are indicative of the purposes of the feed additives, but should not be taken to imply that these are the only effects. In all cases, directions for use and claims on labels should be followed closely. Current approvals for feed additive use are subject to change and the user is cautioned to follow current regulations when using any feed additive. Sources of current information relating to feed additives are given later in this unit.

TABLE 7-1 Examples of Feed Additives That May Be Used with Beef Animals. Other Additives Are Available for Special Purposes and a Number of Combinations May Be Used That Are Not Included in This Table.

Additive (Class of animal)	Amount to Use	Purpose for Use (key at table end)									Withdrawal Before Slaughter
		1	2	3	4	5	6	7	8	9	
Antibiotics:											
Bacitracin (Feedlot cattle)	35 mg/head/day	X									
Bacitracin methylene disalicylate (Feedlot cattle)	70 mg/head/day or 250 mg/head/day for 5 days							X			
Bacitracin zinc (Feedlot cattle)	35–70 mg/head/day	X	X								
Chlortetracycline (Aureomycin)											
(Beef calves)	0.1–0.5 mg/lb of body weight	X	X			X					
(Beef cows)	70–750 mg/head/day	X	X	X		X	X	X		X	
Erythromycin (Gallimycin) (Feedlot cattle)	37 mg/head/day	X	X								
Neomycin (Cattle)	70–140 grams/ton		X			X					
Oxytetracycline (Terramycin)											
(Calves)	200–400 mg/gallon of milk replacer		X			X					
(Calves, birth to 12 weeks)	0.05–0.1 mg/lb of body weight	X	X								
(Calves, in starter feeds or milk replacer)	0.5–5 mg/lb of body weight/day; or 25–75 mg/head/day; or 50–100 grams/ton of dry feed	X	X								5 days if fed in milk replacer or at rate of 2 grams/head in dry feed
(Feedlot cattle)	75 mg/head/day or 0.5–5 mg/lb of body weight/day	X	X		X	X		X		X	
Tylosin (Tylan) (Feedlot cattle)	8–10 grams/ton of feed or 60–90 mg/head/day							X			
Chlortetracycline Sulfamethazine (Aurea S 700) (Feedlot cattle)	350 mg/head/day									X	7 days before slaughter
Antibacterial Agent:											
Ethlenediamine dihydroiodide (EDDI) (Cattle)	400–500 mg/head/day for 2–3 weeks or 50 mg/head/day in feed or salt						X		X		
Ionophores:											
Lasalocid sodium (Bovatec) (Feedlot cattle)	10–30 grams/ton of feed	X	X								
Monensin (Rumensin) (Feedlot cattle)	20–30 grams/ton of feed or 100–300 mg/head/day		X								
(Cattle on pasture)	100–200 mg/head/day	X									
Estrus Suppressant:											
Melengestrol acetate (MGA) (Heifers in feedlot or on pasture)	0.25–0.5 mg/day	X	X								48 hours before slaughter

Key to purposes of use of additives:
- 1 = Rate of gain
- 2 = Feed efficiency
- 3 = Anaplasmosis
- 4 = Bloat
- 5 = Diarrhea
- 6 = Foot rot
- 7 = Liver abscesses
- 8 = Lumpy jaw
- 9 = Respiratory disease

Hormones and Hormonelike Compounds

Many hormone or hormonelike compounds have been developed for use with beef cattle. Some are naturally occurring hormones and others are synthetic substances. In the following discussion, the generic term *hormones* is used to refer to both natural hormones and synthetic hormonelike compounds. While some of these compounds may be added to the feed, it is a common practice to use others as implants in the ear of the animal.

The practice of using hormones in beef production is recognized as one of the most effective management tools available to increase feed efficiency and improve the rate of gain of cattle. Research shows that rate of gain is improved an average of 10 percent, while feed efficiency is improved from 6 to 8 percent through the use of hormones with feedlot cattle.

Hormone feed additive. *Melengestrol acetate (MGA)* is a synthetic hormone similar to progesterone and has FDA approval for inclusion in the feed for beef heifers at the rate of 0.25 to 0.50 milligrams per head per day. It suppresses estrus (prevents the heifer from coming into heat), which reduces the continual mounting seen when heifers are coming into heat in the feedlot. MGA also increases the rate of gain and improves feed efficiency in fattening heifers. Research shows that average daily gain is improved by about 5–11 percent and feed efficiency is improved about 5 percent when MGA is included in the diets of heifers in the feedlot. These results are based on sexually mature, nonpregnant females. MGA gives no response in feed efficiency or rate of gain when fed to steers.

There have been no negative side effects reported with the use of MGA. Sometimes there is a slight enlargement of the mammary glands of the heifers, but this has not been a problem. There have been no reports of any connection between feeding MGA and dark-cutting carcasses.

MGA must be withdrawn from the feed at least 48 hours before the animals are slaughtered. When MGA is withdrawn from the feed more than 72 hours before slaughter, many of the heifers will come into heat. The excitement and activity this causes may result in some loss of weight.

Hormone implants. Table 7-2 lists and summarizes information about some beef cattle implants that are currently approved for use. None of these implants have any restrictions on withdrawal from use before slaughter. All of these implants improve both rate of gain and feed efficiency when they are properly implanted. Combining implants with feed additives improves performance more than using either alone.

Implants stimulate growth hormones in the animal's pituitary gland and change the hormone balance in the animal's body. Implants

TABLE 7-2 Beef Cattle Implants.

Implant (Chemical & dosage)	Approved for	Days Effective
Compudose (Estradiol—24 mg)	Steers—from birth to market Heifers—confined	200
Ralgro (Zeranol—36 mg)	Steers—from birth to market Heifers—from birth to market	70–100
Synovex-S (Progesterone—200 mg + estradiol benzoate—20 mg)	Steers—over 400 pounds (181 kg)	70–100
Synovex-H (Testosterone—200 mg + estradiol benzoate—20 mg)	Heifers—over 400 pounds (181 kg)	70–100
Synovex-C (Progesterone—100 mg + estradiol benzoate—10 mg)	Calves (steers & heifers) 45 days of age to 400 pounds (181 kg)	70–100
Implus-S (Progesterone—200 mg + estradiol benzoate—20 mg)	Steers—over 400 pounds (181 kg)	70–100
Implus-H (Testosterone—200 mg + estradiol benzoate—20 mg)	Heifers—over 400 pounds (181 kg)	70–100
Finaplix-S (Trenbolone acetate—140 mg)	Feeders (steers) over 650 pounds (295 kg)	70–100
Finaplix-H (Trenbolone acetate—200 mg)	Feeders (heifers) over 650 pounds (295 kg)	70–100
Revalor-S (Trenbolone acetate—120 mg + estradiol—24 mg)	Feeders (steers) over 650 pounds (295 kg)	110–120

produce a slightly greater response in steers as compared to heifers; bulls show less response than either steers or heifers. Weight gain response also differs among different ages of cattle. Implants increase weight gains by about 8–15 percent in growing and finishing cattle and by about 8 percent in suckling calves. Feed efficiency in growing and finishing cattle is improved by about 6–10 percent. Rate of gain is higher in implanted suckling heifer calves compared to suckling steer calves.

A feeding trial at Iowa State University revealed that using implants with large-frame steers increased the protein requirements of the ration. Rate of gain increased by 30 percent and feed efficiency improved by 19 percent when the ration contained 12.5–14 percent crude protein compared to a ration containing 9.5–11 percent crude protein.

Variation in response to hormone implants is greatest among cattle on pasture and high-roughage diets. Research indicates that cattle should be gaining about one pound (0.45 kg) per day before there is significant response to the use of implants. Some studies have shown that implanting cull cows on a high concentrate ration with Finaplix-H or Synovex-H will improve rate of gain and feed efficiency. The best response to hormone implants is obtained when cattle are on a high-concentrate finishing diet. Reimplanting before the effective period of the original implant expires will produce improved performance.

Because implants contain an active ingredient that is hormone-like, they can interfere with reproduction or cause complete sterility. Only Synovex-C is approved for use with replacement heifers. No other implants should be used on animals that are to be kept for breeding purposes. The development of the testicles is so severely impaired when implants are used with young bull calves that they cannot be used for reproductive purposes. Research indicates that there is less effect on the sexual development of heifers, but it is still recommended that they not be implanted (except with Synovex-C) if they are to be kept for breeding purposes. However, if they are implanted by mistake during the suckling period, they will probably still be usable for breeding purposes if they do not receive a second implant. Heifers that have been implanted may show a slightly lower conception rate when they are first bred. The closer to puberty that implanting occurs, the more likely it is that reproductive ability will be impaired.

There have been reports of undesirable side effects from implants, such as "buller" steers, high tailheads, and udder development. Buller steers show an unusual amount of sexual activity in the feedlot, which reduces gain and increases labor costs. It is believed that these side effects are probably the result of improper techniques used in implanting.

The proportion of lean meat deposition in the carcass is higher when implants are used on feeder cattle. This may result in fewer carcasses grading Choice. Therefore, cattle need to be fed to slightly higher weights to achieve the same level of marbling reached by cattle that are not implanted.

Ralgro is the commercial name of zeronal, which is a compound derived from the corn mold *Gibberella zeae*. It is approved for use in both steers and slaughter heifers from birth to slaughter. Ralgro is given in a dosage of three 12-milligram pellets (36 mg) for all ages of cattle. No withdrawal time prior to slaughter is required.

Synovex is a combination of natural hormones and is marketed as Synovex-S for steers, Synovex-H for heifers, and Synovex-C for calves of both sexes from 45 days of age to 400 pounds (181 kg). Synovex-S contains 20 milligrams of estradiol benzoate and 200 milligrams of progesterone. Both estradiol benzoate and progesterone are female hormones. Synovex-H contains 20 milligrams of estradiol benzoate and 200 milligrams of testosterone. Testosterone is a male hormone. Synovex-C contains 10 milligrams of estradiol benzoate and 100 milligrams of progesterone. Synovex-S and Synovex-H each contain eight pellets per 220 milligram dose. Synovex-C contains four pellets per 110 milligram dose.

Synovex-S is approved for use with feedlot steers over 400 pounds (181 kg). Synovex-H is approved for use with calves and feedlot heifers over 400 pounds (181 kg). These implants are considered to be effective for 70 to 100 days. No withdrawal time prior to slaughter is required.

Neither Synovex-S nor Synovex-H are approved for use with cattle weighing less than 400 pounds (181 kg). The implant, consisting of eight pellets, is placed in the middle one-third of the ear. The implant site should be at least 1.5 to 2 inches (3.8–5 cm) from the base of the ear.

Synovex-C is approved for use with steer and heifer calves from 45 days of age to 400 pounds (181 kg). The implant is considered to be effective for 70 to 100 days. No withdrawal time prior to slaughter is required.

Implus-S contains 20 milligrams of estradiol benzoate and 200 milligrams of progesterone. It is approved for steers weighing more than 400 pounds (181 kg). *Implus-H* contains 20 milligrams of estradiol benzoate and 200 milligrams of testosterone. It is approved for heifers weighing more than 400 pounds (181 kg). These implants are considered to be effective for 70 to 100 days. No withdrawal time prior to slaughter is required.

Compudose contains 24 milligrams of estradiol 17 β in silicon rubber. The estradiol is released slowly over a 200-day period. Compudose is approved for all ages of steers from birth to slaughter and for confined heifers. The implant is placed in the middle one-third of the ear. No withdrawal time prior to slaughter is required for the Compudose implant. If a second Compudose implant is used, the first implant must be removed before implanting the second one.

Revalor-S contains 120 milligrams of trenbolone acetate (TBA) and 24 milligrams of estradiol. Trenbolone acetate is a synthetic male hormone and estradiol is the female hormone, estrogen. Revalor-S is recommended for use with steers weighing 650 pounds (295 kg) or more. If the feeding period is three to four months, a single implant of Revalor-S at the start of the feeding period is enough. This treatment should give about the same results as implanting with an estrogen product every 70 days. Repeat the implant at about 70 days if the steers are to be fed up to 150 days. For feeding periods up to 200 days, use an estrogen product initially; then implant Revalor-S about 110 days before slaughter.

Implanting procedure. The length of time implants are effective appears to be related to proper technique in applying the implant. If any of the pellets are crushed during implantation, the effective period of the implant is decreased. Some of the undesirable side effects that have been observed with implants appear to also be traceable to improper implantation technique. Improper implanting may also result in some of the animals losing the implant.

Being in too much of a hurry to do the implanting can result in serious economic loss from improperly placed implants. It has been estimated that improper implanting procedures may be causing losses of as much as 20 percent if compared with the loss rate when proper procedures are followed. A little extra time taken during the implanting procedure can pay big dividends for the beef producer.

The following guidelines should be followed to secure the proper placement of the implant:

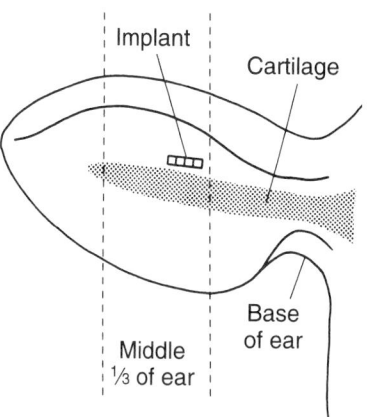

FIGURE 7-1   Proper location for implant.

1. Restrain the animal securely in a headgate or squeeze chute. Use a halter to secure the animal's head to prevent up-and-down head movement.
2. Do not use an instrument with a dull needle as that will make it difficult to penetrate the skin. Make sure the needle is not bent and that it has neither burrs nor rough edges. These conditions often result in crushed pellets.
3. Be sure the needle and the implantation site are both clean. Cotton or a sponge, soaked in alcohol, should be used to clean the needle before injecting another animal. Clean manure and dirt from the implantation site on the ear, using an alcohol-soaked sponge or cotton. Do not use the same cotton or sponge you use to clean the needle. Failure to clean the needle and the implantation site on the ear can result in infection. The infection may encapsulate the implant in connective tissue, rendering it ineffective. Compudose implants are more easily rejected than the other types of implants when infection occurs.
4. Select the proper implanting site on the back surface of the ear. The current recommendation is to place all implants in the middle one-third of the ear—that is, about 1.5 to 2 inches (3.8–5 cm) from the base of the ear (Figure 7-1).
5. With the implanting instrument pointed toward the head and parallel to the ear, lift the loose skin with the point of the needle. Push the needle in, being careful not to go into or through the ear cartilage or hit a vein. Placing the implant in the ear cartilage may cause encapsulation of the implant and result in slower-than-normal absorption of the implant material. Hitting a vein can cause excessive bleeding and possibly result in the loss of the implant.
6. Withdraw the needle slightly before starting the implant. Push the plunger or pull the trigger while slowly withdrawing the needle. Feel for the implant with the opposite hand as the needle is being withdrawn to make sure the pellets have been properly deposited. If the needle is withdrawn too rapidly, the pellets

may be crushed or balled up in one pocket. The pellets should be deposited in a line between the skin and the ear. Crushing the pellets may result in too-rapid absorption and undesirable side effects.

Using Combinations of Feed Additives and Hormones

Antibiotics are sometimes used in combinations in beef production. Chlortetracycline may be used in combination with sulfamethazine for beef cattle at the rate of 350 milligrams of each per head per day for a period of 28 days. This combination is claimed to promote growth and be effective against respiratory diseases. The combination must be withdrawn from use seven days before the animals are sent to slaughter.

A combination of 20 to 30 grams of monensin with 10 grams of tylosin per ton of feed may be used for beef cattle. It is claimed that this combination promotes feed efficiency and is effective against liver abscesses. No withdrawal time prior to slaughter is required.

Several antibiotic combinations are approved for use in calves for the treatment of diarrhea. Oxytetracycline and chlortetracycline may be used at the rate of 50 grams per ton of feed. Fifty grams of oxytetracycline in combination with 35 to 140 grams of neomycin may be used per ton of feed. Another combination of 10 grams of oxytetracycline with 70 to 140 grams of neomycin per ton of feed may also be used for diarrhea. Combinations of oxytetracycline and neomycin may also be used in the milk replacer for calves for the treatment of diarrhea. Label directions for mixing must be carefully followed.

The above discussion is not intended to be all-inclusive regarding the possible use of combinations of antibiotics for beef cattle. Other combinations may be approved for use for specific purposes. Always follow label directions and current regulations when using feed additives.

The ionophore antibiotics (monensin sodium and lasalocid sodium) may be used in combination with hormone implants. Melengestrol acetate (MGA) is also approved for use in combination with hormone implants. The use of these combinations is recommended because of the significant increases in rate of gain and feed efficiency that are obtained. Research indicates that rate of gain is increased about 18 percent and feed efficiency is increased about 12 percent when combinations are used.

Special Purpose Additives

Many beef producers treat all newly purchased cattle for worms, as well as cattle that have just come off of pasture into the feedlot. Coumaphos may be used at the rate of 0.091 grams per 100 pounds (45 kg) of body weight for 6 days to treat gastrointestinal worms. Levamisole phosphate may be used at the rate of 0.08 to 0.8 percent of the ration to treat gastrointestinal and lung worms. Withdrawal period is 48 hours before slaughter. Morantel tartrate may be used at the rate of 0.44 grams per 100 pounds (45 kg) of body weight to treat gastrointestinal worms. Phenothiazine may be used at the rate of 10 grams per 100 pounds (45 kg) of body weight to treat gastrointestinal worms. Thiabendazole may be used at the rate of 3 to 5 grams per 100 pounds (45 kg) of body weight to treat gastrointestinal worms. Thiabendazole must be withdrawn from use three days before slaughter.

Poloxalene (Bloat Guard) may be used at the rate of 1 to 2 grams per 100 pounds (45 kg) of body weight per day for bloat prevention. Rabon may be used at the rate of 0.07 grams per 100 pounds (45 kg) of body weight per day for controlling flies.

The materials discussed above are not intended to be all-inclusive of the kinds of special purpose additives that may be used in beef production. Other products are available for a variety of purposes. Label directions and current regulations must be followed when using any drug in livestock production.

DAIRY CATTLE

The use of feed additives is not as widespread among dairy farmers as it is with beef producers. However, there are a number of additives that may be used with the dairy herd that can increase profits. The returns from the use of these additives varies with the environment and the management practices of individual producers. Table 7-3 gives some examples of feed additives that may be used for dairy animals.

Antibiotics

Antibiotics may be used with replacement heifers that are under 16 weeks of age and are not growing properly. The results of the use of antibiotics with these young heifers is to improve their growth rate and reduce stress from the environment.

Chlortetracycline (Aureomycin) may be used as a feed additive for calves to promote growth, improve feed efficiency, and treat diarrhea. This antibiotic may also be used as a feed additive for non-lactating dairy cows for growth promotion, feed efficiency, and treatment or prevention of diarrhea, liver abscesses, foot rot, respiratory diseases, and anaplasmosis. It may also be used for lactating dairy cows for treatment or prevention of diarrhea, foot rot, and respiratory diseases.

Oxytetracycline (Terramycin) may be used as a feed additive for dairy cows for the treatment of bloat, improved milk production, and treatment of diarrhea. Oxytetracycline/chlortetracycline and oxytetracycline/neomycin combinations may be used for dairy calves in the treatment or prevention of diarrhea.

Monensin may be used to improve feed efficiency and rate of gain in replacement dairy heifers. It may be used from the time the

TABLE 7-3 Examples of Feed Additives That May Be Used with Dairy Animals. Other Additives Are Available for Special Purposes and a Number of Combinations May Be Used That Are Not Included in This Table.

Additive (Class of animal)	Amount to Use	Purpose for Use (key at table end)										Withdrawal Before Slaughter
		1	2	3	4	5	6	7	8	9	10	
Chlortetracycline (Aureomycin) (Dairy cows—nonlactating)	70–750 mg/head/day	X	X	X		X	X	X		X		
Monensin (Rumensin) (Heifers—replacement only— over 400 pounds)	100–200 mg/head/day	X	X									
Neomycin (Cattle)	70–140 grams/ton		X			X						30 days
Oxytetracycline (Terramycin)												
(Calves)	200–400 mg/gallon of milk replacer		X			X						
(Calves, birth to 12 weeks)	0.05–0.1 mg/lb of body weight	X	X									
(Calves, in starter feeds or milk replacer)	0.5–5 mg/lb of body weight/ day; or 25–75 mg/head/day; or 50–100 grams/ton of dry feed	X	X									5 days if fed in milk replacer or at rate of 2 grams/ head in dry feed
(Dairy cattle)	75–100 mg/head/day				X	X					X	
Ethlenediamine dihydroiodide (EDDI) (do not use with lactating cows)	400–500 mg/head/day for 2–3 weeks or 50 mg/head/day in feed or salt						X		X			

Key to purposes of use of additives:

1 = Rate of gain	4 = Bloat	7 = Liver abscesses	9 = Respiratory disease
2 = Feed efficiency	5 = Diarrhea	8 = Lumpy jaw	10 = Milk production
3 = Anaplasmosis	6 = Foot rot		

heifer is 400 pounds (181 kg) to the time of calving. It is not legal to use monensin with lactating dairy cows.

Other Additives

Anhydrous ammonia (NH_3) may be used as a forage preservative for corn silage, low-quality forages, or wet hay. It improves the digestibility of the fiber in the forage and also provides a nonprotein nitrogen source for ruminants. Add 7 pounds (3.2 kg) per ton of wet corn silage and 20 pounds (9 kg) per ton of hay. It may be added to straw or corn stalks at the rate of 40 to 60 pounds (18–27 kg) applied as a gas for one to three weeks.

Buffers such as sodium bicarbonate (baking soda) and sodium sequecarbonate may be added to feed to help preserve a pH of 6.5 to 6.8 in the rumen. This pH level is desirable because it improves feed intake and feed digestibility. Buffers may be fed at the rate of 0.25 to 0.5 pound (0.11–0.22 kg) per head per day. Use a buffer in the feed during the first 120 days after calving if the cows are being fed more than 45 pounds (20 kg) per day of a high-moisture (over 50%) corn silage or are receiving a heavy (more than 2% of body weight) feeding of grain. Buffers may also be used to advantage when cattle are heat stressed, off-feed, or when finely ground forages or grain are being fed.

Isoacids may be fed to lactating dairy cows to improve feed efficiency and the digestion of fiber in the rumen. They are fed at the rate of 0.2 pound (0.09 kg) per head per day. The use of isoacids in the ration may result in from 2 to 6 pounds (0.9–2.7 kg) more milk per day. It may take as long as four weeks from the time feeding of this additive is started until milk production begins to increase.

Propionic acid may be used as a preservative for hay or high-moisture corn. It prevents mold and heating of the hay or corn. It is added at the rate of 0.5 to 1.5 percent of the feed. It is recommended for use when there is danger of spoilage when the hay or corn is too wet. When used at the recommended level, milk fat tests are not lowered.

Propylene glycol may be used at the rate of 0.5 to 1 pound (0.22–0.45 kg) per cow per day to treat ketosis. It is converted to blood sugar thus raising the level of sugar in the bloodstream. It is recommended for use when symptoms of ketosis appear in urine tests or milk ketone tests.

The recent development of a bovine growth hormone has the potential to significantly increase milk production per cow. This development is discussed in detail in Unit 11 of this text.

SHEEP AND GOATS

There are few feed additives available for use with sheep and goats. Table 7-4 gives examples of feed additives that may be used with sheep and goats.

Extensive testing must be done to gain approval for use of drugs as feed additives. The size of the sheep and goat industries are not large enough to make such an intensive effort worthwhile for commercial companies that produce feed additives. Because feed additives do help to prevent disease problems, sheep producers are trying to get a minor species clause approved by the FDA. Such a clause would allow an additive approved for use for one species to be fed to other similar species.

Antibiotics

It is currently a common practice to include a broad-spectrum antibiotic in feeder lamb rations. Most feeder lambs will respond with improved feed efficiency and rate of gain when a feeding level of 25 to 50 milligrams of antibiotic per lamb per day is used. Response is often better when the antibiotic is included in the ration during the first month of the feeding period or when high-energy rations are fed.

TABLE 7-4 Examples of Feed Additives That May Be Used with Sheep and Goats. Other Additives Are Available for Special Purposes That Are Not Shown in This Table.

Additive (Class of animal)	Amount to Use	Purpose for Use (key at table end)						Withdrawal Before Slaughter
		1	2	3	4	5	6	
Chlortetracycline (Aureomycin)								
(Sheep)	20–50 grams/ton	X	X					
(Breeding sheep)	80 mg/head/day			X				
Neomycin (Sheep, goats)	70–140 grams/ton		X		X	X		20 days for sheep and lambs
Oxytetracycline (Terramycin) (Sheep)	10–20 grams/ton	X	X					
	20–100 grams/ton				X	X		
Ethlenediamine dihydroidide (EDDI) (Sheep)	12 mg/head/day						X	

Key to purposes of use of additives:
1 = Rate of gain   3 = Vibrionic abortion   5 = Diarrhea
2 = Feed efficiency   4 = Enterotoxemia   6 = Lumpy jaw

Chlortetracycline (Aureomycin) may be used for sheep to promote growth and improve feed efficiency by as much as 40 percent. When fed at the rate of 60 milligrams per head per day to pregnant ewes from six weeks before lambing to six weeks after lambing, chlortetracycline has reduced lamb deaths by up to 10 percent during the first two weeks after birth. It has also improved the health of the ewe and reduced the incidence of mastitis. When fed at the rate of 80 milligrams per head per day to breeding sheep, chlortetracycline has helped control vibrionic abortion. It is approved for use in sheep at the rate of 20 grams per ton to help prevent enterotoxemia (overeating disease).

Oxytetracycline (Terramycin) may be used with sheep to promote growth and improve feed efficiency. When fed at a higher rate, it is claimed that it is a preventative or treatment for diarrhea and enterotoxemia.

Neomycin may be used with sheep and goats to promote feed efficiency and for the prevention and treatment of diarrhea and enterotoxemia. It must be withdrawn from the feed 20 days prior to slaughter.

Lasalocid sodium may be used to replace antibiotics if a coccidiosis problem exists with feeder lambs. It may be used at the rate of 22 to 33 milligrams per 2.2 pounds (1 kg) of feed. Lasalocid sodium also improves rate of gain and feed efficiency in lambs. It is the only coccidiostat currently approved for use in feedlot lamb diets.

Antibacterial

Ethlenediamine dihydroiodide (EDDI) may be used with sheep at the rate of 12 milligrams per head per day for the prevention or treatment of lumpy jaw.

Special Purpose Additives

Phenothiazine may be used with sheep and goats for the control of gastrointestinal worms. It is used at the rate of 1.25 to 2.5 grams per day or 1 gram per head per day as an additive in the feed or salt.

Thiabendazole may be used with sheep and goats for the control of gastrointestinal worms. It is used at the rate of 2 to 3 grams per 100 pounds (45 kg) of body weight. It must be withdrawn from use 30 days prior to slaughter

Hormone Implant

Ralgro has been approved for use with feedlot lambs. It is not approved for ewes or replacement stock. Ralgro must not be implanted in feedlot lambs within 40 days prior to slaughter.

SWINE

Feed additives are widely used in swine rations. It is important for the swine producer to determine the specific purpose for using any particular feed additive in swine feeding. The economic returns expected from the use of the additive should be greater than the cost of the additive. Table 7-5 gives examples of feed additives that may be used with swine.

Young swine give a greater economic return from the use of antibiotics than any other age group. The use of antibiotics during the breeding, farrowing, and weaning phases of production gives greater returns than at other stages of production. Other factors that affect the economic returns from the use of feed additives include the level of management, the environment in which the hogs are raised, and the history of disease problems on the individual farm.

Antibiotics

Antibiotics that may be used in swine rations for improving rate of gain and feed efficiency include bacitracin methylene disalicylate, bacitracin zinc, bambermycins (Flavomycin), chlortetracycline (Aureomycin), erythromycin, oxytetracycline (Terramycin), penicillin, streptomycin, tylosin, and virginiamycin.

Antibiotics should either not be used at all or used only at low levels when growing gilts unless disease problems are clearly present.

The use of high levels of antibiotics in the ration for 10 days before to 10 days after breeding increases conception rate and litter

TABLE 7-5 Examples of Feed Additives That May Be Used with Swine. Other Additives Are Available for Special Purposes and Combinations That Are Not Shown in this Table.

Additive (Class of animal)	Amount to Use	Purpose for Use (key at table end)					Withdrawal Before Slaughter
		1	2	3	4	5	
Bacitracin methylene disalicylate	10–100 grams/ton of feed	X	X	X			
Bacitracin zinc	10–100 grams/ton of feed	X	X	X	X		
Bambermycins (Flavomycin) (Growing-finishing swine)	2–4 grams/ton of feed	X	X				
Chlortetracycline (Aureomycin)	10–100 grams/ton of feed	X	X		X		
Erythromycin (Young pigs)	10–70 grams/ton of feed	X	X				
Lincomycin	40–100 grams/ton of feed			X			6 days
Neomycin (Young pigs)	200–400 mg/gallon of milk replacer			X	X		20 days
(Older swine)	70–140 grams/ton of feed			X	X		20 days
Oleandomycin (Growing-finishing swine)	5–11.25 grams/ton of feed	X	X				
Oxytetracycline (Terramycin)	7.5–500 grams/ton of feed	X	X		X		5 days if fed at the rate of 500 grams/ton of feed
Penicillin (Procane)	10–50 grams/ton of feed	X	X				
Streptomycin	7.5–75 grams/ton of feed	X	X		X		
Tylosin (Starter feed)	20–100 grams/ton of feed	X	X				
(Grower feed)	10–40 grams/ton of feed	X	X				
Virginiamycin	5–100 grams/ton of feed	X	X	X			
Carbadox	10–50 grams/ton of feed	X	X	X	X		10 weeks
Furazolidone (Sows, baby pigs)	100–300 grams/ton of feed	X		X	X		5 days
Nitrofurazone	500 grams/ton of feed				X		5 days
Roxarsone	0.0025–0.0075% of complete feed	X	X				
Sulfamethazine	100 grams/ton of feed				X		15 days
Sulfathiazole	100 grams/ton of feed				X		7 days
Ethlenediamine dihydroiodide (EDDI)	250–500 mg for 7 days					X	

Key to purposes of use of additives: 1 = Rate of gain   3 = Diarrhea   5 = Respiratory disease
2 = Feed efficiency   4 = Bacterial enteritis

size. There are reasons other than bacterial infections that may cause low conception rates or small litters in sows. A veterinarian should be consulted if serious problems related to conception rates or litter size exist over a period of time.

Antibiotics are generally not needed during the gestation period unless there is a high level of disease or environmental stress present.

Sows and gilts should be treated with an anthelmintic 3 to 4 weeks before farrowing.

The survival rate and the performance of pigs is improved when antibiotics are used in farrowing rations for a period of 7 days before to 14 days after farrowing. Because this is normally a high-stress period, the use of antibiotics to prevent infections during this period can produce significant economic returns.

The use of antibiotics may help prevent lactation problems when sows are nursing their litters. However, there are reasons other than bacterial infections that may cause lactation problems in sows. If problems with poor lactation persist, a veterinarian should be consulted.

Stress and disease risks are high with pigs that have been weaned early or orphaned. Antibiotics should be included in prestarter rations. Starter rations used for pigs weighing up to 30 pounds (11–14 kg) should also include antibiotics because stress and disease risks are also high during this period. Scours is often a problem during the weaning period. It is recommended that a veterinarian should determine the specific bacteria that is causing a scours problem so the proper antibiotic can be selected. This will reduce the cost of treating for scours.

During the growing period when pigs weigh from 25 to 60 pounds (11–27 kg), low levels of antibiotics might be used unless a disease problem exists on the farm. Higher levels of antibiotics should be used for purchased feeder pigs of this weight because of the increased stress of shipping.

During the 60- to 125-pound (27–57 kg) stage of development, the use of antibiotics will result in more economical gains. Higher levels should be used if a disease problem exists on the farm.

The lowest level of response to the use of antibiotics in swine feeding is during the finishing stage of 125 pounds (57 kg) to marketing. If an antibiotic is used, preference should be given to one that does not require a withdrawal period before slaughter

Anthelmintics

An anthelmintic should be used two or three weeks after pigs are weaned. Swine should be treated for worms anytime there appears to be an infestation. A veterinarian should examine a sample of the manure to determine the exact nature of the worm infestation. The recommendation of the veterinarian regarding the anthelmintic to use should be followed.

Piperizine used at the rate of 0.1 to 0.2 percent in the water or 0.2 to 0.4 percent of a complete feed is used to treat large roundworms and nodular worms. Pyrantel tartrate used at the rate of 96 grams per ton of feed may also be used to treat large roundworms and nodular worms. Phenothiazine used at the rate of 5 to 30 grams per ton of feed may be used to treat nodular worms. Levamisole hydrochloride used at the rate of 0.08 percent of a complete feed may be used to treat large roundworms, nodular worms, lungworms, and threadworms. Levamisole must be withdrawn from use 72 hours before slaughter. Dichlorvos may be used at the rate of 0.384 to 0.55 percent of a complete feed to treat large roundworms, nodular worms, whipworms, and thick stomach worms.

Other Additives

Arsanilic acid or sodium arsanilate may be added to swine diets to promote growth and improve feed efficiency. These additives must be withdrawn from the feed five days before slaughter.

Some additives are used in swine production for medicinal purposes. The antibiotics chlortetracycline, oxytetracycline, and tylosin are used for atrophic rhinitis. Chlortetracycline and oxytetracycline are used against leptospirosis. There are no requirements for withdrawal before slaughter for the above-listed additives.

Necrotic enteritis may be treated with carbadox, furazolidone, neomycin, or oxytetracycline. Carbadox must be withdrawn from use 10 weeks before slaughter and should not be fed to pigs weighing more than 75 pounds (34 kg) or receiving less than 15 percent crude protein in a complete ration. Furazolidone must be withdrawn from use five days before slaughter. Neomycin must be withdrawn from use 20 days before slaughter. When oxytetracycline is fed at the rate of 500 grams per ton of feed for 7 to 14 days, it must be withdrawn from use five days before slaughter. There is no requirement for withdrawal of oxytetracycline when it is fed at lower levels in the diet.

Rabon may be used for fly control purposes. It is used at the rate of 0.05 grams per hundred pounds of body weight per day. There is no withdrawal requirement prior to slaughter.

POULTRY

The use of feed additives is a common practice in poultry feeding. They are used primarily to improve feed efficiency and growth in broiler chick and market turkey rations. Some feed additives are also used to improve egg production in laying flocks. Some additives are also used for the prevention or control of diseases. Table 7-6 gives examples of feed additives used in poultry feeding.

Good management practices need to be followed when raising poultry in confinement to reduce disease problems and lessen the dependence on antibiotics for disease control. Access to poultry houses should be restricted and strict clean-up procedures should be followed by all persons who enter the houses. All poultry houses should be thoroughly cleaned and disinfected after the birds are moved out and before bringing new birds into the houses. Disease-carrying predators should be controlled. Avoid overcrowding the birds and make sure the litter is kept dry. Proper ventilation and temperature control in poultry houses also helps to reduce disease problems. If an outbreak of disease does occur, it should be treated promptly and with the proper drugs.

TABLE 7-6 Examples of Feed Additives That May Be Used with Poultry. Other Additives Are Available for Special Purposes and Combinations That Are Not Shown in This Table.

Additive (Class of animal)	Amount to Use	1	2	3	4	5	6	7	8	9	10	11	12	13	14	Withdrawal Before Slaughter
Bacitracin methylene disalicylate (Chickens, turkeys)	4–200 grams/ton of feed	X	X	X			X	X		X						
(Pheasants)	4–50 grams/ton of feed	X	X													
(Quail)	5–20 grams/ton of feed	X	X													
Bacitracin zinc (Chickens, turkeys)	4–500 grams/ton of feed	X	X			X	X	X	X	X						
(Pheasants)	4–50 grams/ton of feed	X	X													
(Quail)	5–20 grams/ton of feed	X	X													
Bambermycins (Broilers, growing turkeys)	1–2 grams/ton of feed	X	X													
(Ducks)	200–400 grams/ton of feed										X					
Chlortetracycline (Aureomycin) (Chickens, turkeys)	10–500 grams/ton of feed	X	X			X			X							
Erythromycin (Chickens, turkeys)	4.6–185 grams/ton of feed	X	X				X		X							24–48 hours
Hygromycin B (Chickens)	8–12 grams/ton of feed								X	X		X				3 days
Lincomycin (Broilers)	2–4 grams/ton of feed	X	X	X												
Neomycin (Poultry)	70–140 grams/ton of feed				X	X										5–14 days
Novobiocin (Chickens, turkeys)	200–350 grams/ton of feed for 5–7 days													X		4 days
Oleandomycin (Broilers, turkeys)	1–2 grams/ton of feed	X	X													
Oxytetracycline (Terramycin) (Chickens)	5–500 grams/ton of feed	X	X						X							3 days at 200 g/T
(Turkeys)	5–200 grams/ ton of feed		X				X							X		3 days at 200 g/T
Penicillin (Procane) (Chickens, turkeys)	2.4–100 grams/ton of feed	X	X		X				X							

(Continues)

TABLE 7-6 *(Cont.)*

Additive (Class of animal)	Amount to Use	Purpose for Use (key at end of table)														Withdrawal Before Slaughter
		1	2	3	4	5	6	7	8	9	10	11	12	13	14	
(Pheasants, Quail)	2.4–50 grams/ton of feed	X	X													
Streptomycin (Chickens, turkeys)	12–180 grams/ton of feed						X	X	X	X						
Tylosin (Chickens)	4–50 grams/ton of feed	X	X													
Virginiamycin (Broilers)	5–15 grams/ton of feed	X	X													
Furazolidone (Chickens, turkeys)	7.5–200 grams/ton of feed	X	X							X					X	5 days
Ipronidazole (Turkeys)	0.00625–0.025% of complete feed	X	X													4–7 days
Roxarsone (Chickens, turkeys)	0.0025–0.005% of complete feed	X	X		X	X										5 days
Sulfaquinoxaline (Chickens, turkeys)	0.015–0.05% of complete feed											X			X	10 days
Ethlenediamine dihydroiodide (EDDI) (Chickens, turkeys)	113 grams/ton of feed for 5–7 days									X						

Key to purposes of use of additives:

1 = Rate of gain	6 = Egg production	11 = Gastrointestinal worms
2 = Feed efficiency	7 = Infectious sinusitis	12 = Breast blisters
3 = Diarrhea	8 = Respiratory disease	13 = General infections
4 = Bacterial enteritis	9 = Blue comb	14 = Fowl typhoid
5 = Egg hatchability	10 = Fowl cholera	

A number of additives may be used to control coccidiosis in poultry. Table 7-7 lists some of these additives with the level of use.

HORSES

Antibiotics are generally not used for low-level feeding over a period of time in horse rations. Chlortetracycline is approved for use with horses under one year of age to promote growth, improve feed efficiency, and reduce stress. It may be fed at the rate of 85 milligrams per head per day. Neomycin may be used with foals at the rate of 200 to 400 milligrams per gallon of milk replacer to combat bacterial enteritis and diarrhea. It may be used with older horses at the rate of 70 to 140 milligrams per ton of feed for the same purposes. Phenothiazine may be used for horses and mules at the rate of 2.5 grams per 100 pounds (45 kg) of body weight for gastrointestinal worm control. It may also be fed at the rate of two grams per head per day for 21 days for the same purpose.

TABLE 7-7 Examples of Feed Additives That May Be Used to Treat Coccidiosis in Poultry.

Additive (Class of animal)	Amount to Use	Withdrawal Before Slaughter
Lasalocid sodium (Broilers)	68–113 grams/ton	3 days
Monensin sodium (Broilers, replacement birds)	90–110 grams/ton	3 days
Oxytetracycline (Chickens)	5–500 grams/ton	3 days at 200 g/ton
Furazolidone (Chickens, turkeys)	7.5–200 grams/ton	5 days
Sulfaquinoxaline (Chickens, turkeys)	0.015–0.05% of total diet	10 days
Buquinolate (Chickens)	0.00825–0.011% of total diet	
Butynorate (Chickens, turkeys)	0.0375–0.07% of total diet	7 days
Carbarsone (Turkeys)	0.0375–0.07% of total diet	5 days
Clopidol (Broilers, replacement birds)	0.0125–0.0250% of total diet	5 days at 0.025%
Decoquinate (Broilers)	0.003% of total diet	
Ethopabate (Chickens)	0.0004% of total diet	
Ormetoprim (Chickens, turkeys)	0.0075% of total diet	5 days
Nicarbazin (Chickens)	0.0125% of total diet	4 days
Robenidine HC (Broilers)	30 grams/ton of feed	5 days
Zoalene (Broilers, replacement birds)	0.004–0.0125% of total diet	
(Turkeys)	0.0125–0.01875% of total diet	

REGULATION OF THE USE OF ADDITIVES

The use of feed additives and hormone implants is strictly regulated in the United States by the Federal Food and Drug Administration (FDA). In Canada, regulation is by the Food Production and Inspection Branch of Agriculture Canada. These regulations change from time to time as new information on effects and dangers becomes available. Many additives may be used only within certain specified levels and for certain species or types of animals. Many additives must also be withdrawn from use in the feed within specified times of marketing the animal.

Current detailed information on feed additive use and regulations may be found in the *Feed Additive Compendium*, published annually by the Miller Publishing Company, 2501 Wayzata Boulevard, Minneapolis, Minnesota 55440. The compendium is updated during the year on a regular basis.

In Canada, information on use and regulations is available in the compendium of *Medicating Ingredient Brochures*, available from the Plant Products Division, Canada Department of Agriculture, Ottawa, Canada.

The *Code of Federal Regulations (CFR)* Title 21 contains official information from the Food and Drug Administration concerning approval of antibiotics and other animal drugs. Revisions of Title 21 are made annually as of April 1 and the *CFR* is updated in individual issues of the *Federal Register*. The *CFR* and the *Federal Register* must be used together to determine the current regulations concerning the use of animal drugs. Title 21 (part 500-599 covering animal drugs, feeds, and related products) is available from the Superintendent of Documents, U.S. Government Printing Office, Washington, D.C. 20402 (inquire for current price). The *Federal Register* is available from the Superintendent of Documents on an annual subscription basis (inquire for current price) and includes monthly issues of the *List of CFR Sections Affected* and *The Federal Register Index*.

Feed Label Requirements

Any feed that contains any level of one or more drugs is defined as a medicated feed. Medicated feeds may be a complete feed mix or may be a premix that is mixed with other feeds to make a complete feed. Premixes contain higher levels of the drug ingredients than do complete mixed medicated feeds.

The FDA requires all manufacturers of medicated feeds to provide information on the label relating to the use of the feed. The word "medicated" must appear under the name of the feed. The specific purpose of the drug or drugs included in the feed must be stated on the label. The name and amounts of all active drug ingredients must be listed. The required withdrawal period (if required) prior to slaughter of the animals must be given. Cautions against misuse must be listed. The directions for the use of the feed must be given.

Medicated feeds that are custom mixed at a local mill must meet the same label requirements as commercial medicated feeds. All labels on medicated feeds are assigned a lot or control number so the batch can be identified at a later date if necessary.

MIXING AND RESIDUE AVOIDANCE

The proper mixing of medicated feeds is important for their safe use. Be sure the drug ingredients are added in the correct proportions and are uniformly mixed in the batch. Failure to do so can have undesirable results—for example, some animals may get too much of the drug and others too little. This reduces the effectiveness of the medicated feed and may result in higher production costs or undesirable side effects.

Drug residues in livestock and livestock products can cause financial losses for farmers. Carcasses of animals may be condemned at slaughter if illegal drug residues are found. Public concern over drug residues in meat and livestock products may result in additional regulation of or restriction on the use of feed additives, which could result in higher production costs for farmers.

The USDA conducts targeted and routine tissue testing in all federally inspected slaughter plants. If evidence of illegal drugs is found in the tissues of a producer's animals, the producer is notified and the animals are held until they are free of the illegal substance. It usually takes two weeks to thirty days to get test results back. This delay causes an economic loss for the farmer. Additional animals cannot be marketed until the tests show them to be free of the illegal drug.

Both horizontal- and vertical-type mixers may be used on the farm for mixing medicated feeds. Regardless of which type of mixer is used, care must be taken to assure a complete mix of the drug ingredients in the complete feed. Proper cleaning of the mixing equipment must be done after each batch of medicated feed is mixed to avoid residue problems in later batches of feed.

A regular procedure should be used when mixing medicated feeds to avoid problems with drug carryover. The following recommendations will help avoid problems:

1. Know the labeled uses, mixing instructions, and withdrawal times for all medications used in the feed.
2. Clean the mixer before use to avoid carryover of drugs from one batch to the next.
3. Do not try to mix more feed in the batch than the mixer can hold.
4. Premix the drugs in a large enough quantity to allow accurate weighing and mixing in the complete feed.
5. Follow the manufacturer's instructions for premixing and adding medications.
6. Use an established order for mixing the medication with the feed. Put in one-half of the supplement, then the medicated and vitamin premixes, then one-fourth of the supplement, then the mineral premix, and then the rest of the supplement. Mix thoroughly; then add the ground grain and continue to mix for another 8 to 10 minutes, or for the amount of time recommended by the equipment manufacturer.
7. Mix all feeds containing medication first, clean the equipment, and then mix the unmedicated feeds. Mix any withdrawal feeds before mixing other nonmedicated feeds.
8. Clean the mixer by putting several hundred pounds of ground corn or soybeans through it. Remove, label, and store this feed for later use in a medicated mix.
9. Keep feed additives stored in a clean, orderly area in their original packages. Control rodents and insects in the storage area. Do not store other chemicals in this area.
10. Read all labels and observe withdrawal times.
11. Keep labeled samples of purchased premixes, supplements, and mixed medicated feeds for at least three months after the livestock to which they were fed have been marketed.
12. Make sure all augers, holding bins, feed wagons, and feeders are

thoroughly cleaned before withdrawal feeds are used. Do not mix withdrawal feeds with medicated feeds in bins or feeders.

KEEPING RECORDS

A good set of up-to-date records on the use of medicated feeds can help the farmer avoid problems with feed contamination and drug residues in livestock and livestock products. The following records are recommended:

1. The date the batch was mixed.
2. The mixing order and the amount of medication added.
3. The mixing time for the batch.
4. The location where the feed is stored.
5. The number, age, and weight of the animals fed from the batch and the amount given per head.
6. The medication that was used, the amount, and the concentration in the batch of feed.
7. The date of cleaning mixers, bins, conveyors, and feeders.

In addition to keeping records, the farmer should have a long-term plan for the use of feed additives. Because microorganisms develop resistance to antibiotics, the farmer should avoid feeding some types of antibiotics so that they might be used in case of an outbreak of disease. Additionally, it is recommended that the farmer periodically change the type of antibiotic being used to reduce the chances of microorganisms building immunity to the additive. This change should be made about once a year. Do not make frequent changes in the type of antibiotic being used.

If good records are kept and a long-term plan for the use of feed additives is followed, it is easier for a veterinarian to properly treat animals if a disease outbreak should occur.

HEALTH CONCERNS

There has been a growing concern in recent years that the continued use of antibiotics in animal agriculture may have an adverse effect on human as well as animal health. This concern centers around the use of antibiotics at subtherapeutic levels in livestock feeding. It is feared that this continuous low-level use of antibiotics might result in the development of resistant strains of microorganisms that could not be effectively treated with antibiotics. Bacterial resistance to drugs has been observed almost from the time antibiotics were first used in animal feeding.

The impact of a ban on the use of antibiotics in animal feeding would vary from species to species, but the overall effect would be to raise the cost of animal products to the consumer. There would be less meat and livestock products produced and the costs of production would be higher. More feed would be required per animal raised and the rate of gain of animals on feed would decrease. There would be an increase in the death loss among animals being raised.

To date, there is no persuasive evidence of animal or human health problems arising after nearly 30 years of using antibiotics in animal feeding. Many of the same antibiotics being used today have been used since the use of antibiotics in animal feeding began, and they are still effective against most microorganisms. There is a need for critical experimental studies on the effect of low-level antibiotic feeding on animal and human health.

The Food and Drug Administration has also expressed concern about the possible carcinogenic (cancer-causing) effects of some feed additives. The 1958 Food Additives Amendments to the Food, Drug and Cosmetic Act of 1938 include prohibitions against adding to food any substance that is a known carcinogen. There is ongoing research into the possible carcinogenic effects of feed additives. If these effects are shown to be present in any additive, it will be withdrawn from use in animal feeding.

SUMMARY

Feed additives are materials that are not considered to be nutrients and are used in small amounts in the ration to improve feed efficiency, promote faster gains, improve animal health, or increase production of animal products. Hormone implants are pelleted synthetic or natural hormones or hormonelike compounds that are used to improve rate and efficiency of gain.

Antibiotics and chemoantibacterial compounds are used to kill or slow the growth of some kinds of microorganisms. The use of these compounds at low levels in the ration over a period of time is a common practice in livestock feeding.

A number of other feed additives may be used in livestock production. Anthelmintics (dewormers) are often used to control various species of worms that may infest animals. Other specialized feed additives are used to prevent or treat various diseases that may affect animals.

Beef producers are major users of feed additives and hormone implants. Feed efficiency and rate of gain are significantly increased by the use of these products. A high percentage of beef cattle rations use either monensin sodium (Rumensin) or lasalocid sodium (Bovatec) either alone or in combination with other feed additives.

Hormone implants improve both rate of gain and feed efficiency when they are properly used. The current recommendation is to place all implants in the middle one-third of the ear, about 1.5 to two inches (3.8–5 cm) from the base of the ear. Implants cannot be used with breeding stock.

The use of feed additives is not as common with dairy animals as it is with beef cattle. Most of the use of feed additives with dairy cattle is with younger animals.

There are few feed additives approved for use with sheep and goats. A broad-spectrum antibiotic is commonly used in feeder lamb rations.

Feed additives are widely used in swine rations. The greatest economic returns from the use of antibiotics occur when they are used for younger animals.

The use of feed additives is a common practice in poultry feeding, especially with broilers and market turkeys.

Feed additives are generally not used with horses. Some antibiotics are approved for use with younger horses to promote growth, improve feed efficiency, and reduce stress.

The use of feed additives and hormone implants is regulated in the United States by the Federal Food and Drug Administration. Their use in Canada is regulated by the Food Production and Inspection Branch of Agriculture Canada. Regulations on the use of additives and hormones change from time to time; therefore, care should be taken to follow current regulations on use of these products.

Any feed that contains any level of one or more drugs is considered a medicated feed. There are strict requirements relating to the labeling of medicated feeds.

It is important that all medicated feeds be properly mixed. Failure to do so can have a severe negative economic impact on livestock producers. A regular procedure should be followed when mixing medicated feeds and careful records should be kept.

There is concern about possible health hazards resulting from the use of feed additives in livestock production. The concern focuses on the possible development of resistant strains of microorganisms that could not be treated with antibiotics. After 30 years of use of antibiotics in animal feeding, there is no persuasive evidence of animal or human health problems resulting from their use.

Student Learning Activities

1. Present a short oral report to the class about new developments with feed additives and/or hormone implants as reported in farm magazines and newspapers.
2. Prepare a bulletin board display of medicated feed tags.
3. Take a field trip to a local farm that has facilities for mixing complete feeds and observe the proper technique for mixing feed additives in the ration.
4. Give a demonstration on the proper technique for making hormone implants in beef cattle.

Review

1. Why are feed additives used in animal rations?

2. Why are hormone implants used in livestock production?

3. What is the major difference between antibiotics and chemoantibacterial compounds?

4. List six antibiotics commonly used in livestock production.

5. List four chemoantibacterial compounds commonly used in livestock production.

6. List four ways in which antibiotics produce the results they do when used in livestock production.

7. What is the difference between broad-spectrum and narrow-spectrum antibiotics? Which type is preferred for use as a feed additive?

8. What is the function of anthelmintics?

9. List four commonly used anthelmintics.

10. When feeding beef cattle a high-energy ration, what is the range of percent improvement in rate of gain and feed efficiency that can be expected from the continuous feeding of a low level of antibiotic?

11. What results can be expected from feeding monensin sodium to feedlot cattle?

12. What results can be expected from feeding lasalocid sodium to feedlot cattle?

13. How do these compounds work to produce the results they do?

14. What is the effect of using melengestrol acetate (MGA) as a feed additive in beef heifer rations?

15. What is the withdrawal period for MGA before slaughter?

16. List and describe the use of hormone implants in beef cattle.

17. Describe the proper procedures for implanting hormone pellets in beef cattle.

18. Briefly explain the use of feed additives with dairy cattle.

19. Why are there few feed additives available for use with sheep and goats?

20. Why is a broad-spectrum antibiotic often used in feeder lamb rations?

21. Briefly explain the results of using chlortetracycline (Aureomycin) in sheep rations.

22. Which age group of swine gives the greatest economic return from the use of antibiotics in the ration?

23. For what two kinds of poultry are feed additives most commonly used?

24. What agency regulates the use of feed additives and hormone implants in the United States? In Canada?

25. What are the sources of current regulations regarding the use of feed additives and hormone implants in the United States? In Canada?

26. List six things that must be on the label of all medicated feeds.

27. Why is the proper mixing of medicated feeds important to livestock producers?

28. How might drug residues in livestock or livestock products cause economic losses for farmers?

29. List the recommended steps for the proper mixing of medicated feeds.

30. List the records that should be kept when using medicated feeds.

31. What is the major concern relating to possible health hazards from the use of feed additives in livestock production?

32. What would be some of the effects of a total ban on the use of antibiotics in livestock production?

Balancing Rations

CLASSIFICATION OF FEEDS

Roughages

Livestock feeds that contain more than 18 percent crude fiber when dry are called *roughages*. Fiber is the hard-to-digest part of the feed. Roughages include hay, silage, pasture, and fodder. There are two general classes of roughages: legume roughages and nonlegume roughages.

Plants that can take nitrogen from the air are called *legumes*. These plants have *nodules* (small swellings or lumps) on their roots that contain bacteria. The bacteria can *fix* the nitrogen from the air in soil and make it available for use by the plant. This is done by combining the free nitrogen with other elements to form nitrogen compounds. All of the clovers, as well as alfalfa, soybeans, trefoil, lespedeza, peas, and beans, are legumes. Many other less-common crops are also legumes. Legumes are usually higher in protein than nonlegume roughages.

Nonlegume roughages cannot use the nitrogen from the air. They are usually lower in protein than the legume roughages. Many common livestock feeds are nonlegume roughages, including corn silage, sorghum silage, fodders, bluegrass, timothy, redtop, bromegrass, orchard grass, fescue, coastal bermudagrass, common bermudagrass, and prairie grasses.

Concentrates

Livestock feeds that contain less than 18 percent crude fiber when dry are called *concentrates*. There are two classes of concentrates: protein supplements and energy feeds.

Protein supplements are livestock feeds that contain 20 percent or more protein. They are divided into two groups based on their source. Those that come from animals or animal by-products are called *animal proteins*. Those that come from plants are called *vegetable proteins*.

Some common animal proteins are tankage, meat scraps, meat and bonemeal, fish meal, dried skimmed milk, dried whole milk, blood meal, and feather meal. (*Tankage* is animal tissues and bones from animal slaughterhouses and rendering plants that are cooked, dried, and ground.) Most animal proteins contain more than 47 percent crude protein. The protein is more variable in quality than protein from vegetable sources. Animal proteins contain a more balanced amount of the essential amino acids than do plant proteins. Thus, animal proteins are sometimes used for balancing rations for swine and poultry.

Some common vegetable proteins are soybean oil meal, cottonseed meal, linseed oil meal, peanut oil meal, corn gluten feed, brewer's dried grains, and distiller's dried grains. Most vegetable proteins contain less than 47 percent crude protein. Soybean oil meal is used more than any of the other protein supplements for livestock rations. Soybean oil meal can supply the necessary amino acids to balance a swine or poultry ration with cereal grains. Vegetable proteins can be used as the only protein supplement for ruminants. Nonruminants, however, may need some animal protein in their ration. Animal proteins give the amino acid balance needed in nonruminant rations when plant protein sources other than soybean oil meal are used.

Commercial protein supplements are made by commercial feed companies. They are mixes of animal and vegetable protein feeds. Each commercial supplement is usually made for one class of animal. Feed companies often mix minerals, vitamins, and antibiotics in their protein supplements. The feed tag on the supplement tells the class of animal for which it is designed. The tag also gives feeding directions and lists the contents of the feed. Feeding directions must always be followed carefully. Feed supplements with antibiotics in them usually must be taken away from the animal for a period of time (withdrawal period) before the animal is sent to market. This practice is required by law. The antibiotic must not be present in the meat when humans eat it.

Livestock feeds with less than 20 percent crude protein are called *energy feeds*. Most of the grains are energy feeds. Some common energy feeds are corn, sorghum grain, oats, barley, rye, wheat, ground ear corn, wheat bran, wheat middlings, dried citrus pulp, dried beet pulp, and dried whey. Corn is the most widely used energy feed. Sorghum grain, oats, and barley are the other commonly used energy feeds.

RATION CHARACTERISTICS

An animal must receive the proper amounts of nutrients in the right proportion to efficiently produce meat, milk, eggs, wool, work, etc. A ration is said to be balanced when it provides the nutrient needs of the animal in the proper proportions. Strictly speaking, a *ration* is the amount of feed given to an animal to meet its needs during a twenty-four-hour period; however, in common practice, the term may refer to feed provided without reference to a time period. A *balanced ration* is one that has all the nutrients the animal needs in the right proportions and amounts. The term *diet* refers to the ration without reference to a specific time period.

A ration must be *palatable* (taste good) in order for the animal to eat it. Moldy feed is often not palatable. Insect and weather damage also lower the palatability of feed. Feed is of no value if the animal will not eat it.

Feed accounts for approximately seventy-five percent of the total cost of raising livestock. To feed livestock profitably it is necessary to develop rations that are as economical as possible. The ration must be palatable and meet the nutritional requirements of the animals. Homegrown feeds are used as much as possible because they are generally less expensive than purchased feeds. Commercial feeds are used when homegrown feeds are not available and also to supply nutrients not provided by homegrown feeds.

Feeds used in rations must not be harmful to the animal's health or lower the quality of the product. Poisonous plants should not be included in diets for livestock. Poisonous plants sometimes grow in hay and pasture fields (Table 8-1). Eradicate these plants before harvesting the hay or allowing animals to graze the pasture. Usually, animals will not eat poisonous plants, but if they are in the hay the animal may not sort them out. If the pasture is sparse, animals may eat poisonous plants that are growing there. Animals get sick every year from eating poisonous plants because farmers do not take care to keep the poisonous plants out of the animals' diet.

It is necessary to balance the intake of roughage and concentrates for the particular species and age of livestock being fed. Ruminants can use more roughage in their diets than nonruminants. Also, younger animals cannot use as much roughage in their diets as can more mature animals. The purpose for which the animal is being fed must also be considered when including roughage in the diet. For example, fattening animals generally should be fed less roughage than breeding animals.

TABLE 8-1 Poisonous Plants.[1]

Common Name(s)	Animals Affected
Arrowgrass	Cattle; sheep
Aster	Sheep
Azalea, western	Sheep
Baccharis	Cattle
Baccharis, eastern; silverling; groundseltree; consumptionweed	Cattle; sheep; poultry
Bitterweed; sneezeweed	Sheep; cattle; horses
Bracken	Horses; cattle, sometimes sheep
Buttercup	All livestock; most commonly cattle
Cherry, wild	Sheep; cattle; horses
Chokecherry	Sheep; cattle
Cocklebur	All livestock; especially hogs; chickens, if seeds ground in feed
Copperweed	Cattle; sheep
Death camas; black snakeroot; crow poison; pink death camas; poison sage; swampgrass; alkaligrass; poison onion	Sheep; cattle; horses
Drymary	Cattle
Dutchman's Breeches	Cattle
Goldenrod	Cattle; sheep; horses
Halogeton	Sheep; occasionally cattle
Hemp; marijuana	Cattle; horses
Henbane	Cattle; sheep; horses
Horsebrush; spring rabbit brush; coal-oil brush	Sheep
Horsetail	Horses
Indian hemp; dogbane; Indian physic	Cattle; horses; sheep
Japanese yew	All livestock
Jimmyweed, rayless goldenrod	All livestock
Jimpson weed; thornapple	All livestock
Larkspur; staggerweed (rocket, azure, tall, dwarf)	Cattle
Laurels (Black sheep and Mountain)	Sheep; goats; cattle; other animals to lesser degree
Locoweed	Cattle; horses; sheep; goats
Lupine; bluebonnet; wild bean; blue pea	Sheep; goats; cattle; hogs; horses
Milkweed (several species)	Cattle; sheep; goats; horses
Nightshade (Black; Deadly) Other species: Horsenettle; bullnettle	All livestock
Oaks	Cattle; sheep; goats; horses; occasionally hogs by the acorns
Oleander	All livestock
Paperflower, greenstem	Sheep
Peganum	Sheep; cattle
Poisonbean	Cattle; sheep; goats
Poison hemlock	Cattle; sheep; goats; horses; hogs; poultry
Poisonvetch	Cattle; sheep
Ragwort; groundsel	Cattle; sheep; horses
Rubberweed (Bitter and Colorado)	All livestock; especially sheep
St. Johnswort; goatweed	Animals with white skin and hair
Snakeroot, white	All livestock
Snakeweed	Cattle; sheep; goats
Spring parsley; wild carrot	Cattle; sheep
Tarweed	Horses; cattle; hogs
Timber milkvetch	Cattle; sheep; goats; horses
Water hemlock	All livestock

[1]Many other plants may be poisonous to livestock under certain conditions. This list is not intended to be all-inclusive but only presents some of the more common poisonous plants that affect livestock.

Micronutrients and feed additives are used in small quantities in the diet. Care must be taken to thoroughly mix these materials to assure uniform distribution in the feed. Failure to do so may result in one animal getting too much of the micronutrient or additive while another animal may get too little. Excessive amounts of some additives may be harmful to an animal. Feed only the recommended amounts of these materials and make sure they are well mixed with the rest of the feed ingredients.

The functions of a ration must be considered when determining the nutrient requirements of livestock. These functions include maintenance, growth, fattening, production, reproduction, and work.

RATION FUNCTIONS

Maintenance

The primary use of the nutrients in a ration is for maintaining the life of the animal. The animal must have energy for the functioning of the heart, for breathing, and for other vital body processes. These activities make up what is called the *basal metabolism* of the animal. Maintenance also includes the use of energy supplied by the ration to keep the animal's body temperature normal. Protein in the body tissues breaks down. Protein from the ration is used to repair these body tissues. Minerals and vitamins are continually being lost from the body and are replaced by those in the ration. Certain fatty acids are needed for good health, and must be supplied by the animal's ration. Water is required for all bodily activities.

If the animal is not being fed enough feed, it may need to use all its ration for maintenance. Thus, it will have none left for other activities, such as growth. Normally, about one-half of the ration fed an animal is used for maintenance. An animal on full feed will use about one-third of its ration for maintenance. *Full feed* means to give an animal all it wants to eat.

Growth

Nutrients in the ration are used for growth only after the maintenance requirements of the animal are met. Animals grow by increases in the size of muscles, bones, organs, and connective tissues. Animals become mature by growing. If they do not grow properly, they will not be productive when they are mature. Animals grow fastest when they are young. The rate of growth slows down as they get older. The larger species of animals usually mature slower than the smaller animals. The growth rate of larger animals is faster than that of smaller animals.

Fattening

Feed nutrients that are not used for maintenance or growth may be used for fattening. Fat is stored in the tissues of the body. Fat stored within the muscles is called *marbling*. Marbling helps make meat juicy and good tasting. The consumer does not want too much fat, however. The object of fattening is to obtain the right amount of fat in the muscle without getting too much fat. Feeds that are high in carbohydrates and fats are used for fattening. They are less expensive than protein feeds.

Production

Cows, swine, horses, sheep, and goats produce milk to feed their young. Dairy goats and cows also produce milk for human use. Chickens produce eggs. Sheep and goats produce wool and mohair. All of this production requires nutrients. The kind of nutrients needed depends on the kind of production.

Reproduction

The proper nutrition is required for reproduction. The animal may become sterile if it does not get an adequate level of nutrients. A sterile animal cannot be bred. Nutrition is extremely important for pregnant animals. Most of the growth of the fetus takes place during the last third of the pregnancy. (The *fetus* is the unborn animal when it is still in the mother's womb.) Animals need additional amounts of nutrients during pregnancy.

Work

Horses do work when they are ridden. Other livestock may be used in some areas to perform work. The energy needed for work comes from carbohydrates, fats, and extra protein in the ration. The other needs of the body are met before nutrients are available for work. The animal may use fat stored in the body for work if the ration does not provide enough nutrients. Animals sweat more when they work. This creates a need for extra salt to make up for that lost by sweating.

BALANCING RATIONS

General Principles

The livestock ration must meet the nutritional needs of the animal. The nutrient allowance figured in the balanced ration should not be more than 3 percent below the animal's requirement, which should be met as closely as possible.

An animal must have a certain amount of dry matter in its ration, or it will be hungry. Its digestive system will not function properly if it does not receive enough dry matter. There is also an upper limit on the total amount of dry matter that an animal can eat. This varies with the kind of animal being fed and its size. The total dry matter in the ration of a full-fed animal should not be more than 3 percent above its need. Total dry matter for animals not on full feed can be considerably above or below their listed needs.

The amount of protein in the ration may be measured by the *total protein (TP)* need of the animal. *Digestible protein (DP)* may also be used as the measure to balance the ration. The essential amino acids must be included when balancing rations for nonruminants. Some feeds being used may be below the average protein content listed in feed composition tables. It is acceptable to allow 5 to 10 percent more protein in the ration than the animal needs. However, too much protein above the animal's needs will raise the cost of the ration.

Four methods are commonly used to measure the energy provided by a ration. Energy may be calculated as *digestible energy (DE)*, *total digestible nutrients (TDN)*, *metabolizable energy (ME)*, and *net energy (NE)* as illustrated by Figure 8-1. The *gross energy* of a feed is measured in a laboratory device called a *bomb calorimeter*. The feed is completely burned (oxidized) in the bomb calorimeter, which contains 25 to 30 atmospheres of oxygen. The gross energy is the total amount of heat released by burning in the bomb calorimeter. *Digestible energy* is the gross energy of a feed minus the energy remaining in the feces of the animal after the feed is digested. *Metabolizable energy*, for ruminants, is the gross energy in the feed eaten minus the energy found in the feces, the energy in the gaseous products of digestion, and the energy in the urine. The energy in the gaseous products of digestion is not considered when determining metabolizable energy for birds and simple-stomached animals.

Net energy is the metabolizable energy minus the heat increment. It is energy used either for maintenance only, or maintenance plus production, or production only. Net energy for maintenance only is written NE_m; for production only it is written NE_p; and for maintenance

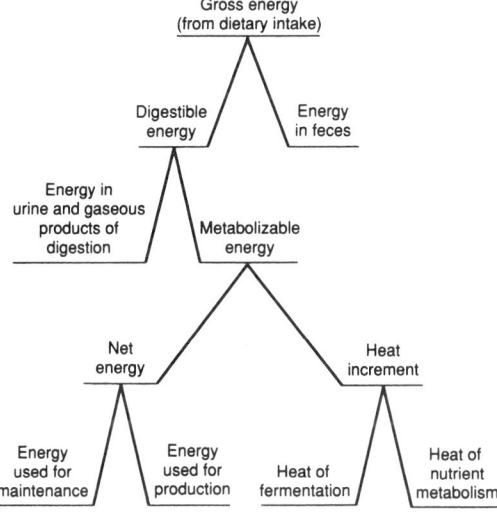

FIGURE 8-1 Utilization of dietary energy by the animal.

plus production it is written NE_{m+p}. Net energy for maintenance is the amount of energy used to keep the animal in energy equilibrium; that is, there is no net gain or loss of energy in the animal's body tissues. The net energy for production is that amount of energy needed by the animal above the amount used for maintenance which is used for work, tissue growth, fat production, fetus growth, or milk, egg, or wool production. The different kinds of net energy are indicated as follows: for work, NE_{work}; for egg production, NE_{egg}; for gain, NE_{gain}; for fetus growth, NE_{preg}; for wool production, NE_{wool}; and so on. The *total digestible nutrients* in a feed is the total of the digestible protein, digestible nitrogen-free extract, digestible crude fiber, and 2.25 times the digestible fat. It gives a measure of the total energy value of a feed when it is fed to an animal. The TDN value of a feed varies with the class of animal to which it is fed. (See feed composition tables in the appendix.) The energy in the ration should not be more than about 5 percent greater than the animal's needs. Animals are limited in the total amount of energy they can use.

Two minerals, calcium and phosphorus, are important in balancing rations. The ratio of calcium to phosphorus should be between 1:1 and 2:1. This ratio is more important than the total amount being fed. The total calcium and phosphorus available in the ration is often more than the animal needs when the other requirements are met.

The other mineral needs of the animal are usually not considered when balancing rations. Trace-mineralized salt will usually meet these needs.

Vitamin A is taken into account when balancing rations. The other vitamins needed are added to the ration without calculating the vitamin content of the feed. The amount needed to meet the minimum daily requirement for the animal is added with a vitamin supplement. The amount of vitamin A in feed will often be more than the animal needs, but this is not harmful. Vitamin deficiencies may occur in cattle and sheep during pregnancy if low-quality legume hay is fed. A vitamin supplement should always be added to a pregnancy ration.

Some feeds are cheaper sources of nutrients than other feeds. Energy feeds should be compared on the basis of price per pound (or kilogram) of energy (TDN, DE, ME, or NE). Protein feeds should be compared in terms of the price per pound (or kilogram) of total protein or digestible protein. The least expensive sources of nutrients should be used as much as possible.

Sampling and Analyzing Feeds

A person must know the nutrient composition of the feeds used to properly balance a ration. Average nutrient composition values can be found in nutrient composition tables such as the one in the appendix of this text. However, some feed, especially forages, can vary considerably from these average values. To get the maximum benefit from the ration, it is wise to have the feeds used analyzed for nutrient content.

Milk cows, beef cows that are nursing calves, and young growing animals require high nutrient levels. Forages fed to these animals should be analyzed for nutrient content. Haylage, silage, and high-moisture grains may vary greatly in moisture and nutrient content. Better rations can be developed if these feeds are also analyzed for moisture and nutrient content.

Growing conditions, such as excessive rainfall or drought, can affect the nutrient content of feeds, especially forages. When unusual growing conditions occur, it is a good idea to have these feeds analyzed. Other indications that a feed should be analyzed include musty odor, unusual amounts of foreign material or pests present in the feed, and a high level of leaf shattering on forages.

The most important tests to be done on feed grains are moisture, protein, and energy content. Test forages for moisture, protein, acid detergent fiber, and neutral detergent fiber. It seldom pays to test for major or trace mineral content. Under some growing conditions it is wise to check forages for nitrate and prussic acid content. Excessive amounts of these substances can be harmful to animals.

Take representative samples of the feed to be analyzed. Take random samples of hay from at least 20 bales. Insert the sampling tube into the center of the bale. About 15 samples from silage and 5 from grain will usually give enough to be representative of the entire lot. Samples of silage or total mixed rations should be taken from the silage feeder or feed mixer as it is being fed. Avoid the top of the feed because it is drier than the entire batch.

Mix the samples from one type of feed and take a subsample from the mixture for analysis. Seal the samples in polyethylene freezer bags; store dry samples in a cool area. Freeze samples that contain more than 15 percent moisture. Send the samples to a testing laboratory as soon as possible. Use a laboratory that is certified by the National Forage Testing Association to ensure the highest accuracy of the test.

A feed analysis report will generally include the following measures:

- Dry matter (DM) (see discussion in next section).
- Crude protein (CP)—the total of both true protein and nonprotein nitrogen.
- Insoluble crude protein (ICP)—the amount of indigestible crude protein in the feed resulting from overheating.
- Adjusted crude protein (ACP)—calculated value, adjusted for insoluble crude protein. If the ICP/CP ratio exceeds 0.1, use this value instead of crude protein when balancing a ration.

- Neutral detergent fiber (NDF)—relatively insoluble material found in the cell wall of plants. May be used to predict feed intake. A low NDF is desirable.
- Acid detergent fiber (ADF)—measures the least digestible part of the feed; includes cellulose, lignin, silica, insoluble crude protein, and ash. A low ADF is desirable.
- Digestible dry matter (DDM)—percent of forage that is digestible.
- Net energy (NE)—an indicator of the true value of a feed. It is the energy left after determining the energy lost through the feces, urine, gas, and heat generated by metabolism. See Figure 8-1 for an explanation of the utilization of dietary energy by the animal.
- Total digestible nutrients (TDN)—the total of the digestible parts of crude fiber, protein, fat, and nitrogen-free extract.
- Dry matter intake (DMI)—estimated maximum consumption of forage dry matter by the animal. It is shown as a percentage of body weight.
- Relative feed value (RFV)—an evaluation of the quality of hay and haylage by combining into one number digestibility and feed intake.

The energy value of forage can also be estimated by plant maturity at harvest and the amount of weather damage to the feed. Late-cut, mature plants are lower in energy than early-cut, immature plants. Weather damage lowers the energy content of forages regardless of the stage of maturity when cut.

Relationship Between 100 Percent Dry Matter Basis and As-fed Basis

Many publications list nutrient requirements and feed composition on a 100 percent dry matter basis. All feeds contain some moisture. The amount varies with the feed, the form of the feed, the stage of growth at which it was harvested, the length of time it was stored, and the conditions under which it was stored. The feed composition table in the appendix of this text shows the average percent of dry matter in the feeds listed.

The term *100 percent dry matter basis* means that the data presented is calculated on the basis of all the moisture removed from the feed. The term *as-fed basis* means the data is calculated on the basis of the average amount of moisture found in the feed as it is used on the farm. The term *air-dry* means the same as the term *as-fed*.

Using the 100 percent dry matter basis makes it easier to compare feeds that have different moisture contents on an as-fed (air-dry) basis. The values, when given on a 100 percent dry matter basis, must be changed to the as-fed basis to find the amounts of feed to actually use. If the feed being used is analyzed, the actual dry matter content is used for this conversion. If no analysis is available, the average dry matter content given in the feed composition table is used. These are averages and actual feeds being used may vary widely from these figures.

The method of converting from one basis to the other is as follows:

Let   a = pounds (kilograms) of feed on 100 percent dry matter basis

    b = pounds (kilograms) of feed on an as-fed (air-dry) basis

    c = the percent of dry matter in the feed

To convert from as-fed (air-dry) basis to 100 percent dry matter basis:

$$a = b \times c$$

That is, the pounds (kilograms) of feed on a 100 percent dry matter basis equals the pounds (kilograms) of feed on an as-fed (air-dry) basis multiplied by the percent of dry matter in the feed.

Example 1: A ration calls for 5.6 pounds (2.6 kilograms) of #2 dent corn on an as-fed basis. The feed composition table shows that corn has 89 percent dry matter. Therefore: $5.6 \times .89 = 5.0$ pounds (or $2.6 \times .89 = 2.3$ kilograms) on a 100 percent dry matter basis. (Amounts have been rounded to the nearest tenth of a pound or kilogram.)

Example 2: A ration calls for 22.4 pounds (10 kilograms) of alfalfa hay on an as-fed basis. The alfalfa has been cut in the mid-bloom stage. The feed composition table shows that mid-bloom alfalfa hay has 89.2 percent dry matter. Therefore: $22.4 \times .892 = 20$ pounds (or $10 \times .892 = 9$ kilograms) on a 100 percent dry matter basis.

Example 3: A ration calls for 21.8 pounds (9.8 kilograms) of corn silage on an as-fed basis. The feed composition table shows that mature corn silage has a dry matter content of 55 percent. Therefore: $21.8 \times .55 = 12$ pounds (or $9.8 \times .55 = 5.4$ kilograms) on a 100 percent dry matter basis.

To convert from 100 percent dry matter basis to as-fed basis:

$$b = \frac{a}{c}$$

That is, the pounds (kilograms) of feed on an as-fed (air-dry) basis equals the pounds (kilograms) of feed on a 100 percent dry matter basis divided by the percent of dry matter in the feed.

Example 1: A ration calls for 5 pounds (2.3 kilograms) of #2 dent corn on a 100 percent dry matter basis. The feed composition table shows that #2 dent corn has 89 percent dry matter. Therefore:

$$\frac{5}{.89} = 5.6 \text{ pounds} \left(\text{or } \frac{2.3}{.89} = 2.6 \text{ kilograms} \right)$$

on an as-fed basis.

Example 2: A ration calls for 20 pounds (9 kilograms) of alfalfa hay on a 100 percent dry matter basis. The alfalfa has been cut in the mid-bloom stage. The feed composition table shows that mid-bloom alfalfa hay has 89.2 percent dry matter. Therefore:

$$\frac{20}{.892} = 22.4 \text{ pounds} \left(\text{or } \frac{9}{.892} = 10 \text{ kilograms} \right)$$

on an as-fed basis.

Example 3: A ration calls for 12 pounds (5.4 kilograms) of corn silage on a 100 percent dry matter basis. The feed composition table shows that mature corn silage has a dry matter content of 55 percent. Therefore:

$$\frac{12}{.55} = 21.8 \text{ pounds} \left(\text{or } \frac{5.4}{.55} = 9.8 \text{ kilograms} \right)$$

on an as-fed basis.

When using nutrient requirement and feed composition tables given on a 100 percent dry matter basis, it is easier to work out the ration on the dry matter basis and then convert the final figures to an as-fed (air-dry) basis.

Rules of Thumb for Balancing Rations

Beef. A maintenance ration for beef cows is primarily roughage. Some supplement may be required, depending on the quality of the roughage fed. The amount of air-dry roughage to feed should equal about 2 percent of the body weight of the animal. For example, if the beef cow weighs 1,213 pounds (550 kilograms), the amount of air-dry roughage to feed for maintenance would be about 24 pounds (11 kilograms). If the ration is being calculated on a 100 percent dry matter basis, 1.8 percent of the animal's body weight is used as the rule of thumb. For a 1,213-pound (550-kilogram) cow this would be about 22 pounds (10 kilograms) of 100 percent dry matter roughage.

Cows nursing calves in drylot should be fed about 50 percent more than dry cows. Silage is substituted at the rate of three (3) parts silage for each one (1) part of dry roughage. Vitamin A supplement may be needed if the roughage is of poor quality.

Fattening rations for beef should be about 2 to 2.5 percent of the animal's body weight, fed as air-dry grain and protein supplement. If the ration is calculated on a 100 percent dry matter basis, use 1.8 to 2.25 percent of the animal's body weight. About 0.5 to 1.0 percent of body weight should be fed as air-dry roughage. On a 100 percent dry matter basis, use 0.45 to 0.9 percent of the animal's body weight. About one part of protein supplement should be fed to each eight to twelve parts of grain. The total ration should have about 10 to 15 percent air-dry roughage. On a 100 percent dry matter basis, about 9 to 13.5 percent of the total ration should be roughage. Rations with a high grain content give the fastest and most efficient weight gains.

Fattening cattle must be fed a mineral supplement. If a high concentrate (grain) ration is fed, a mineral supplement consisting of two parts dicalcium phosphate, two parts limestone, and six parts trace mineralized salt should be fed free choice. *Free choice* means that the supplement is available at all times to the animal. When feeding a high roughage ration, a mineral supplement consisting of two parts trace mineralized salt and one part dicalcium phosphate should be fed free choice. (See Units 15 and 16 for examples of beef rations.)

Swine. Bred sows and gilts that are limit-fed should receive about 3.5 to 4.5 pounds (1.6 to 2.0 kilograms) of air-dry feed in the ration. Feed about 3.15 to 4 pounds (1.44 to 1.8 kilograms) on a 100 percent dry matter basis.

Limit-fed means that the amount of feed given the animal is controlled or limited to less than the animal would eat if given free access to the feed. *Self-fed* means that the animal is given free access to all the feed it will eat.

The ration for bred sows and gilts should contain about 14 percent total (crude) protein. Ground roughage may be added to a self-fed ration to limit the energy intake.

Sows nursing litters should receive about 10 to 15 pounds (4.5 to 6.8 kilograms) of air-dry feed in the ration. Feed 9 to 13.5 pounds (4 to 6 kilograms) when calculating the ration on a 100 percent dry matter basis. The ration should contain about 15 percent total protein.

Growing-finishing pigs are fed according to their size. Fifty-pound (23-kilogram) pigs should be fed about 6.5 percent of their body weight as air-dry feed (5.8 percent on a 100 percent dry matter basis). The total protein in the ration should be about 16 percent. Pigs weighing 100 pounds (45 kilograms) should be fed about 5.5 percent of their body weight on an air-dry basis (5 percent on a 100 percent dry matter basis). The ration should be about 14 percent total protein. Pigs from about 170 pounds (77 kilograms) to market weight should be fed about 4.5 to 3.5 percent of their body weight as air-dry feed (4 to 3 percent on a 100 percent dry matter basis). The ration should have about 13 percent total protein. As the pig becomes larger, the percent of its body weight to be fed as feed in the ration should decrease.

In all cases, the balance of amino acids in the ration is as important as the amount of protein. (See Unit 22 for examples of swine rations).

Sheep. Sheep maintenance rations should have about 3 percent of body weight as air-dry roughage (2.7 percent on a 100 percent dry matter basis). Supplement may be needed to balance the ration. In fattening rations for sheep, about 1.5 to 2 percent of body weight should be fed as air-dry roughage (1.4 to 1.8 percent on a 100 percent dry matter basis). The ration should have about 2 to 3 percent of body weight fed as air-dry grain and protein supplement (1.8 to 2.7 percent on a 100 percent dry matter basis).

Goats. Rules of thumb for goats are similar to those for sheep. Milk goats should receive about 0.5 pound (0.2 kilogram) of air-dry grain (0.45 pound or 0.18 kilogram on a 100 percent dry matter basis) for each pound (kilogram) of milk produced. This is fed in addition to the needs of the animal for maintenance, growth, fetal development, and mohair production.

Horses. Horse rations are based on the amount of work the horse is doing. Table 8-2 gives some guidelines for use in balancing rations for horses. Amounts given in Table 8-2 are on an air-dry (as-fed) basis. Use 90 percent of the amounts given to calculate a ration on a 100 percent dry matter basis.

Poultry. Poultry rations are made up almost entirely of grain and protein supplement. Laying hens need a great deal of calcium for

TABLE 8-2 Rules of Thumb for Feeding Light Horses.

Class	Hours use per day	Pounds/100 lb body wt		Kilograms/100 kg body wt	
		Roughage	Grain	Roughage	Grain
Idle	0	1.5–2.0	0	3.3–4.4	0
Light work	1–3	1.25–1.5	0.5–0.75	2.8–3.3	1.1–1.6
Medium work	3–5	1.0–1.5	0.66–1.0	2.2–3.3	1.4–2.2
Heavy work	5–8	1.0–1.5	1.0–1.4	2.2–3.3	2.2–3.1
Idle mares nursing foals	0	1.5	0.33	3.3	0.7
Growing colts after weaning	0	1.5–2.0	1.0–1.5	3.3–4.4	2.2–3.3
Stallions (breeding season)		0.75–1.5	0.75–1.5	1.6–3.3	1.6–3.3
Pregnant mares		0.5–1.5	0.5–1.5	1.1–3.3	1.1–3.3
Foals before weaning		0.5–0.75	0.5–0.75	1.1–1.6	1.1–1.6

forming the egg shell. A ration for poultry is about 10 percent of body weight, fed as air-dry feed (9 percent if ration is calculated on a 100 percent dry matter basis).

Steps in Balancing a Ration

Step 1. Identify the kind, age, weight, and function of the animal(s) for which the ration is being formulated. In this text, suggested rations and feeding programs are found in the units referring to specific species of animals. These may be used as general guides in formulating rations.

Step 2. Consult a table of nutrient requirements to determine the nutrient needs of the animal(s). These requirements are called feeding standards. Feeding standards are based on average requirements and may not meet the needs under specific feeding conditions. If unusual conditions such as weather stress are present, adjustments in the diet may be needed.

Step 3. Choose the feeds to be used in the ration and consult a feed composition table to determine the nutrient content of the selected feeds. Note that the nutrient content of a feed may be different for different species. Values given in a feed composition table are average values and may not represent the actual composition values of the feeds being used. An analysis of feeds being used is a more accurate method of determining feed composition.

Step 4. Calculate the amounts of each feed to use in the ration. Several methods may be used to do this. The Pearson Square or algebraic equations may be used to balance a ration using two or more feeds. Computer programs may also be used to balance rations.

Step 5. Check the ration formulated against the needs of the animal(s). Be sure it meets the requirements for minerals and vitamins. If there is an excessive amount of a nutrient present, it may be necessary to recalculate the ration to bring it more closely in line with the requirements.

Determining Ration Costs

Check the cost of the nutrients in the ration to determine if this is the most economical ration that is practical to feed. Calculate the cost of the ration per pound (kilogram) or ton (tonne). The daily cost of feeding the animal may also be calculated if a daily consumption rate is known or assumed. In some cases it may be necessary to feed certain nutrients, such as salt or other minerals, on a free-choice basis in addition to the amounts provided in the formulated ration.

The most commonly purchased feeds in rations are the protein supplements. Compare these on the basis of the cost per pound or kilogram of the nutrient content. For example, several protein supplements may be compared as follows:

Percent Protein	Price per Ton	Pounds Protein per Ton	Price per Pound Protein
14%	$186	280	$0.66
16%	200	320	0.625
18%	213	360	0.59
20%	219	400	0.5475

In this example, the 20% protein feed has the lowest cost per pound of protein content although its cost per ton is higher than the other examples.

Other factors to consider when comparing prices include the transportation costs of the feed from the supplier to the farm and the suitability of the particular feed for the class of animal being fed. Feed prices vary over a period of time. Sometimes it is profitable to change feeds used in the diet as prices change. However, the availability of the feed and the effect of a diet change on the performance of the animals being fed must be considered before making changes. Some species of animals react unfavorably when major changes are made in their diet. Changes in diet usually need to be made gradually to avoid a reduction in feed intake, which may result in a reduction in rate of gain or production.

If the livestock feeder knows about how much feed will be needed for a given period of time, it may be profitable to purchase that feed when prices are lower and in larger quantities. The amount of money available and alternate uses of the money must be considered before making this kind of an investment in a feed supply. It may be wise to use an electronic spreadsheet on a computer to make various kinds of projections concerning alternate uses of capital in the farming operation.

Evaluating Diets Using the Computer

The National Research Council published a revised Nutrient Requirements of Beef Cattle (1996) and a revised Nutrient Requirements of Swine (1998) that include computer programs to evaluate diets. Both of these programs generate tables of nutrient requirements based on much more detailed information about management, breed (beef program), and environmental conditions than was previously provided in earlier nutrient requirement publications. Both programs evaluate diets based on detailed user input. It is possible for the user to add feeds and to add different feed analyses to the feed library used in the programs. The programs take into account the many variables that interact in livestock feeding and thus are much more accurate in evaluating diets than other methods described in this text. Because both programs require the user to select feeds and enter estimated amounts to be fed, it is useful to use the ration balancing techniques and the rules of thumb for feeding described in this text to help provide a starting point in these computer programs. It is then possible to quickly refine the inputs to develop a diet to closely match the nutrient requirements of the specific animals being fed.

The methods for balancing rations as outlined in this unit may still be used to gain a general understanding of livestock nutrition and formulating diets. However, the use of the computer programs provided in these revised nutrient requirement publications will allow the user to do a much better job of evaluating diets for a wider range of conditions. The older tables of nutrient requirements are retained in this text for those who wish to use them for learning methods of balancing rations.

The revised nutrient requirement publications may be downloaded without charge from the Internet for those who have access. The download includes the computer programs. The beef program is based on a Lotus 1,2,3 spreadsheet and runs in DOS. The swine program is designed to run in Windows 3.1 or higher, or in the NT operating system. The URL for the National Academy of Sciences where these publications may be downloaded is http://www.nas.edu/. For those who do not have access to the Internet, the publications, including the computer programs, may be ordered in print form from: The National Academy Press, 2101 Constitution Avenue, Washington, DC 20418.

Use of the Pearson Square

It is difficult to balance a ration by trial and error. The Pearson Square is a useful tool for simplifying the balancing of rations. It shows the proportions or percentages of two feeds to be mixed together to give a percent of the needed nutrient.

For example, 2,000 pounds (907 kilograms) of feed is needed to feed a 100-pound (45-kilogram) growing hog. A feeding standards table shows that a 14 percent crude protein ration is needed. Corn and soybean oil meal are selected as feeds. A feed composition table shows that corn has 8.9 percent and soybean oil meal has 45.8 percent crude protein on an as-fed basis. How much corn and soybean oil meal need to be mixed together for 2,000 pounds (907 kilograms) of feed?

Step 1. Draw a square with lines connecting the opposite corners. Write the percent of crude protein needed (14) in the center of the square where the lines cross.

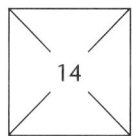

corn 8.9

14

soybean oil
meal 45.8

Step 2. Write the feeds to be used and their crude protein percents at the left-hand corners of the square.

Step 3. Subtract the smaller number from the larger, along the diagonal lines. Write the difference at the opposite end of the diagonals.

$$14 - 8.9 = 5.1$$
$$45.8 - 14 = 31.8$$

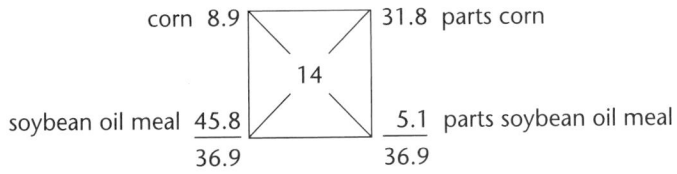

The difference between the percent protein in the soybean oil meal (45.8) and the needed percent protein in the ration (14) is the parts of corn needed (31.8). The difference between the percent protein in the corn (8.9) and the percent protein needed in the ration (14) is the parts of soybean oil meal needed (5.1).

The sum of the numbers on the right equals the difference in the numbers on the left. This fact is used as a check to see if the square is set up correctly.

corn 8.9   31.8 parts corn

14

soybean oil meal 45.8   5.1 parts soybean oil meal

36.9   36.9

Step 4. Divide the parts of each feed by the total parts to find the percent of each feed in the ration.

$$\text{Corn } 31.8 \div 36.9 \times 100 = 86.2\%$$
$$\text{Soybean oil meal } 5.1 \div 36.9 \times 100 = 13.8\%$$

Step 5. It is known that 2,000 pounds (907 kilograms) of the mixture is needed. Thus, the amount of corn needed is 1,724 pounds (782 kilograms). This is found by multiplying the percent of corn in the mix by the total pounds (kilograms) of the mix.

$$2,000 \times 0.862 = 1,724 \text{ pounds}$$
or
$$907 \times 0.862 = 782 \text{ kilograms}$$

The amount of soybean oil meal needed is 276 pounds (125 kilograms). This is found by multiplying the percent of soybean oil meal in the mix by the total pounds (kilograms) of the mix.

$$2,000 \times 0.138 = 276 \text{ pounds}$$
or
$$907 \times 0.138 = 125 \text{ kilograms}$$

(Numbers are rounded off to full pounds or kilograms.)

Step 6. Check the mix to make sure the protein need is met. Multiply the pounds (kilograms) of corn by the percent of protein in the corn ($1{,}724 \times 0.089 = 153$; or $782 \times 0.089 = 69.6$). Multiply the pounds (kilograms) of soybean oil meal by the percent of protein in the soybean oil meal ($276 \times 0.458 = 126$; or $125 \times 0.458 = 57.2$). Add the pounds (kilograms) of protein together. Divide by the total weight of the mix.

$$153 + 126 = 279$$
$$279 \div 2{,}000 \times 100 = 14\%$$
$$69.6 + 57.2 = 126.8$$
$$126.8 \div 907 \times 100 = 14\%$$

The mix is balanced for crude protein content.

Using the Pearson Square to Mix Two Grains with a Supplement

The Pearson Square can be used to find out how much of two grains should be mixed with a supplement. This example is calculated on an as-fed basis. The proportions of each grain to be used must be known or decided upon first.

For example, assume that a 2,000-pound mix of corn, oats, and soybean oil meal is needed. The mix is to contain 16 percent digestible protein. A decision is made to use 3/4 corn and 1/4 oats in the mix. Thus, the proportion of corn to oats is 3:1. How many pounds of corn, oats, and soybean oil meal are needed?

The weighted average percent of protein in the corn and oats is found first. Multiply the proportion of corn (3) by the percent digestible protein in corn (7.1). Do the same for oats. Add the two answers together and divide by the total parts (4). The answer is the weighted average percent of digestible protein in the corn-oats mix.

$$\begin{array}{r} 3 \times 7.1 = 21.3 \\ 1 \times 9.9 = \underline{9.9} \\ 31.2 \end{array}$$

$31.2 \div 4 = 7.8\%$ digestible protein in the corn-oats mix

The Pearson Square is then used to find the pounds of the corn-oats mix and the soybean oil meal needed.

3 parts corn plus 1 part oats 7.8   25.7 parts corn-oats mix

16

soybean oil meal 41.7   8.2 parts soybean oil meal

33.9   33.9 total parts

$$25.7 \div 33.9 \times 100 = 75.8 \text{ corn-oats mix}$$
$$0.758 \times 2,000 = 1,516 \text{ pounds corn-oats mix}$$
$$1,516 \times 3/4 = 1,137 \text{ pounds corn needed}$$
$$1,516 \times 1/4 = 379 \text{ pounds oats needed}$$
$$100 - 75.8 = 24.2\% \text{ soybean oil meal}$$
$$0.242 \times 2,000 = 484 \text{ pounds of soybean oil meal needed}$$

The same method is used to mix two protein supplements with one grain. It is also used to mix two grains and two protein supplements. In each case, the proportions of like feeds to each other (such as the two grains) must be decided upon in advance. The weighted average percent of protein is then found. Finally, the Pearson Square is used to balance the mix. Any of the measures of nutrients in the feed may be used. To balance on energy needs, use TDN, NE, ME, or DE. To balance on protein needs, use total (crude) protein or digestible protein.

Balancing a Swine Ration

Step 1. A ration is needed for a 45-kilogram growing hog.

Step 2. The daily requirements are found in a feeding standards table. They appear as follows:

Feed Intake (kg)	ME (kcal)	Lysine (g)	Ca (g)	P (g)
1.9	6,200	14.3	11.4	9.5

Step 3. Feeds to be used are selected. Their composition is found in a feed composition table. They are listed as follows:

Feed	ME (kcal/kg)	Lysine (%)	Ca (%)	P (%)
Corn	3,420	0.25	0.034	0.33
Soybean oil meal	3,220	2.9	0.38	0.78

Step 4. Use the Pearson Square to find out how much corn and soybean oil meal to mix together to make 1.9 kilograms of feed. A shortage of the amino acid lysine often slows the weight gain of the hog. Therefore, in this example, the ration is balanced on the lysine needs. The other needs are then checked.

The ration must provide 14.3 grams of lysine. This is 0.0143 kilograms. Divide 0.0143 by 1.9 and multiply by 100, which gives .75 as the percent of lysine in the ration. The Pearson Square is set up using the percent of lysine in corn and soybean oil meal.

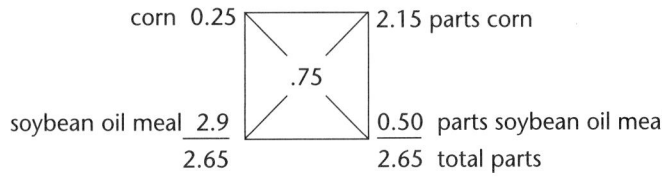

2.15 ÷ 2.65 × 100 = 81.1%

0.811 × 1.9 = 1.54 kg corn needed

0.50 ÷ 2.65 × 100 = 18.9%

0.189 × 1.9 = 0.36 kg soybean oil meal needed

Step 5. The amount of each nutrient in the ration is checked against the needs of the animal (Figure 8-2).

The ration meets or exceeds the needs of the animal for lysine and metabolizable energy. The extra amounts that this ration gives for these nutrients will not harm the animal. They are within the allowable limits for a ration. However, the ration is short on calcium and phosphorus. This need can be met by feeding a mineral supplement. The amounts needed can be added to the mix to meet the needs of the animal.

Feed	Lysine		ME	
	(kg)	(kg)	(kg)	(kg)
Corn	1.54 × .0025 =	0.0039	1.54 × 3,420 =	5,266.8
Soybean oil meal	0.36 × 0.029 =	0.0104	0.36 × 3,220 =	1,159.2
Total provided		0.0143		6,426
Needed by animal		0.0143		6,200
Excess or deficiency		0.0		+226
	Ca		P	
	(g)	(g)	(g)	(g)
Corn	1,540 × 0.00034 =	0.5236	1,540 × 0.0033 =	5.082
Soybean oil meal	360 × 0.0038 =	1.368	360 × 0.0078 =	2.808
Total provided		1.8916		7.89
Needed by animal		11.4		9.5
Excess or deficiency		−9.5084		−1.61

FIGURE 8-2

Balancing a Ration for Beef

Step 1. The ration is for an 800-pound medium-frame steer with an expected daily gain of 2.0 pounds.

Step 2. Find the requirements for the animal in a feeding standards table similar to the tables in the appendix of this text. For this example, determine the energy need in metabolizable energy. Measure the protein need as total protein. Calculate the ration on a 100 percent dry matter basis and convert the final figures to an as-fed basis.

The 800 pound steer in this example has the following requirements:

Dry matter intake (lb) per day	ME (Mcal/lb)	Protein (%)	Ca (%)	P (%)
18.6	1.11	9.2	0.31	0.20

	Total Mcal per day	Pounds protein per day	Pounds Ca per day	Pounds P per day
	20.64	1.71	0.057	0.041

Find the Mcal per day and the pounds per day requirements shown above by multiplying the dry matter intake per day by the Mcal/lb and the percents for the appropriate nutrients.

Step 3.   Select the feeds to be used. Find their composition in a feed composition table. The following feeds are used in this example:

Feed	Dry matter (%)	ME (Mcal/kg)	ME (Mcal/lb)	Protein (%)	Ca (%)	P (%)
Bromegrass hay	89	1.99	0.9	10.0	0.30	0.35
Corn	87	3.00	1.36	9.0	0.07	0.27
Soybean oil meal	90	3.04	1.38	49.9	0.33	0.71

Step 4. Use the rules of thumb outlined earlier in this unit to determine the amount of each feed to use. While this is not as exact as other methods, it will work for general on-farm use. The rule of thumb for the amount of 100 percent dry matter roughage to feed to fattening beef is 0.45 to 0.9 percent of body weight. For this example, 0.9 percent is used:

$$0.009 \times 800 = 7.2 \text{ pounds of hay to feed}$$

Calculate the amount of ME and protein the hay will provide:

$$7.2 \times 0.90 = 6.48 \text{ Mcal of ME from hay}$$
$$7.2 \times 0.10 = 0.72 \text{ pounds protein from hay}$$

Subtract the amount of ME and protein that comes from the hay from the amounts the animal needs:

$$20.64 - 6.48 = 14.16 \text{ Mcal ME from concentrate mix}$$
$$1.71 - 0.72 = 0.99 \text{ pounds protein from concentrate mix}$$

Divide the deficit amount of Mcal in ME by the Mcal in one pound of corn (1.36). Use the Mcal in corn because the largest portion of the concentrate mix is corn. The answer to this calculation gives the pounds of concentrate mix needed.

$$14.16 \div 1.36 = 10.4 \text{ pounds of concentrate mix needed}$$

The pounds of deficit protein is divided by the pounds of concentrate needed and multiplied by 100. The answer is the percentage of protein needed in the concentrate mix.

$0.99 \div 10.4 \times 100 = 9.5\%$ protein in concentrate mix

The Pearson Square is used to determine the amount of corn and soybean oil meal needed.

ground corn 9.0 ⌐‾‾‾‾‾‾‾‾⌐ 40.4 parts of corn

                9.5

soybean oil meal 49.9 ⌐‾‾‾‾‾‾‾⌐ 0.5 parts of soybean oil meal

                40.9   40.9 total parts

$40.4 \div 40.9 \times 100 = 98.8\%$ corn in concentrate mix

$0.988 \times 10.4 = 10.28$ pounds of corn needed

$0.5 \div 40.9 \times 100 = 1.22\%$ soybean oil meal in concentrate mix

$0.0122 \times 10.4 = 0.13$ pounds of soybean oil meal needed

(Answers are rounded.)

Step 5. Check the amount of each nutrient in the ration against the needs of the animal (Figure 8-3).

This ration is exactly balanced for the metabolizable energy and protein needs of the animal. It is slightly deficient in calcium and has a slight excess of phosphorus. While these variations are within the allowable limits for a ration, it might be wise to add a little calcium with a mineral supplement.

The ration was calculated on a 100 percent dry matter basis. The calculated amounts are converted to an as-fed basis by dividing by the percent of dry matter in each feed. The calculations are as follows:

Bromegrass hay	$7.2 \div 0.89 = 8.09$ pounds
Ground ear corn	$10.28 \div 0.87 = 11.82$ pounds
Soybean oil meal	$0.13 \div 0.9 = 0.14$ pounds

The total pounds of feed to be fed is found by adding the pounds of each feed:

hay	8.09
corn	11.82
soybean oil meal	0.14
	20.05 pounds

The animal will eat about 2.5 to 3 percent of its body weight in feed each day. This ration provides 2.51 percent of the body weight in the total ration.

$$20.05 \div 800 \times 100 = 2.51$$

This is within the allowable limits for feeding the animal.

Feed	ME		Protein	
	(Mcal)		(lb)	
Bromegrass hay	$7.2 \times 0.9 =$	6.48	$7.2 \times 0.1 =$	0.72
Corn	$10.28 \times 1.36 =$	13.98	$10.28 \times 0.09 =$	0.93
Soybean oil meal	$0.13 \times 1.38 =$	0.18	$0.13 \times 0.499 =$	0.06
Total provided		20.64		1.71
Needed by animal		20.64		1.71
Excess or deficiency		0.0		0.0

	Ca		P	
	(lb)		(lb)	
Bromegrass hay	$7.2 \times 0.003 =$	0.0216	$7.2 \times 0.0035 =$	0.0252
Corn	$10.28 \times 0.0007 =$	0.0072	$10.28 \times 0.0027 =$	0.0278
Soybean oil meal	$0.13 \times 0.0033 =$	0.0004	$0.13 \times 0.0071 =$	0.0009
Total provided		0.0292		0.0539
Needed by animal		0.057		0.041
Excess or deficiency		–0.0278		+0.0129

FIGURE 8-3

USING ALGEBRAIC EQUATIONS TO BALANCE RATIONS

Algebraic equations may be used instead of the Pearson Square to balance rations. This may be illustrated by using the same problem as the first Pearson Square example shown in this unit. The mix of 2,000 pounds is to be balanced for protein using two feeds. The basic equations are:

X = pounds (or kilograms) of grain needed

Y = pounds (or kilograms) of supplement needed

Equation 1:

$X + Y$ = total pounds (or kilograms) of mix needed

Equation 2:

(percent nutrient in grain) $\times (X)$ + (percent nutrient in supplement) $\times (Y)$ = pounds (or kilograms) of nutrient desired in mix

Place the desired values (express all percents as decimals) in equation 2:

$0.089X + 0.458Y = 280$

(The quantity 280 is found by multiplying the quantity of feed [2,000 lb] by the percent [0.14] [or the amount/kg or lb] of the nutrient desired: $2,000 \times 0.14 = 280$.)

Either X or Y must be canceled by the multiplication of equation 1 by the percentage of nutrient for either X or Y, and the resulting equation 3 is subtracted from equation 2. This example uses the percentage crude protein for corn (0.089), giving *Equation 3:*

$0.089X + 0.089Y = 178$

(The value 178 is found by multiplying 0.089 times 2,000 lb.)

Subtract equation 3 from equation 2:

$$\begin{aligned} 0.089X + 0.458Y &= 280 \\ -0.089X - 0.089Y &= -178 \\ \hline 0.369Y &= 102 \\ Y &= 276 \text{ pounds of soybean meal} \end{aligned}$$

The value of X may be found by substituting the value of Y in equation 1 and solving for X:

$$\begin{aligned} X + 276 &= 2{,}000 \\ X &= 2{,}000 - 276 \\ X &= 1{,}724 \text{ pounds of corn} \end{aligned}$$

Answers in the above example have been rounded to whole numbers. Note that this method gives the same results as the use of the Pearson Square in the first example in this unit.

Algebraic equations may also be used to balance rations using three or more feeds. The same initial step must be taken as when using the Pearson Square, i.e., group similar feeds into two groups and determine the proportions of each to be used in each group. After this is done, the same procedure as outlined above is followed to balance the ration.

Balancing Rations with Simultaneous Algebraic Equations

Simultaneous algebraic equations may be used to balance a ration using two feeds or groups of feeds and balancing for two desired nutrients. Assume that a feed mix is desired for growing-finishing pigs weighing 20–50 kg (44–110 lbs). The mix is to be balanced for lysine and metabolizable energy (ME) requirements using corn and soybean meal. The nutrient and energy composition of the feeds to be used are found in the tables of feed composition in the appendix. When using the tables, note that energy values are different for different classes of livestock. Be sure to secure the value for the type of livestock for which the diet is being formulated. In this example, corn (IFN 4-02-935) and soybean meal (IFN 5-04-604) are to be used. Ground limestone (IFN 6-02-632) and ground defluorinated phosphate (IFN 6-01-780) are used to supply the additional major minerals needed. All values are expressed on an as-fed basis. Lysine is often the first limiting amino acid in swine diets; therefore it is used to balance this example diet. If the lysine requirement is not met, it makes no difference how many of the other essential amino acids are present. The pigs will not grow any faster than the amount of lysine in the diet will permit.

Step 1. Set up the requirements and composition of the feeds:

	Lysine	ME	Calcium	Phosphorus
Req/kg of diet	0.75%	3,260 kcal	0.60%	0.50%
corn	0.25%	3,420 kcal/kg	0.03%	0.29%
SBM	2.90%	3,220 kcal/kg	0.34%	0.70%
Ground limestone			36.07%	0.02%
Defluorinated phosphate			31.65%	13.7%

Step 2. Set up the algebraic equations and solve:

$$X = \text{amount of grain needed per kilogram of diet}$$

$$Y = \text{amount of supplement needed per kilogram of diet}$$

Equation 1:

$$0.25X + 2.90Y = 0.75 \text{ (lysine equation)}$$

Equation 2:

$$3{,}420X + 3{,}220Y = 3{,}260 \text{ (energy equation)}$$

Divide 3,420 by 0.25 to get a factor (3,420/0.25 = 13,680) that is multiplied times equation 1: (13,680) × (0.25X + 2.90Y = 0.75) = 3,420X + 39,672Y = 10,260. The resulting equation is then subtracted from equation 2 to eliminate the X unknown and solve for Y. Alternatively, the Y unknown could be eliminated by dividing 3,420 by 2.90 and then solving for X:

$$
\begin{aligned}
3{,}420X + 3{,}220Y &= 3{,}260 \\
-3{,}420X - 39{,}672Y &= -10{,}260 \\
\hline
-36{,}452Y &= -7{,}000 \\
Y &= 0.192
\end{aligned}
$$

Substitute the value of Y in equation 1 and solve for X:

$$
\begin{aligned}
0.25X + (2.90 \times 0.192) &= 0.75 \\
0.25X + 0.5568 &= 0.75 \\
0.25X &= 0.75 - 0.5569 \\
0.25X &= 0.1931 \\
X &= 0.772413
\end{aligned}
$$

Step 3. The accuracy of the solution may be checked by comparing the computed amounts of lysine and ME provided by this diet with the original requirements:

Lysine requirement

Corn	0.25 × 0.772413	= 0.1931
SBM	2.90 × 0.192	= 0.5569

Total lysine = 0.75 per kilogram of diet

ME requirement
Corn 3,420 × 0.772413 = 2,641.65
SBM 3,220 × 0.192　　=　618.35
　　　　　　　　　　　　Total ME = 3,260 per kilogram of diet

Step 4. Calculate the amount of phosphorus and calcium needed using simultaneous algebraic equations:

　　X = amount of calcium needed per kilogram of diet

　　Y = amount of phosporus needed per kilogram of diet

Determine the amount of calcium and phosphorus supplied by the corn and soybean meal.

　　　　　　　　Calcium:
　　　　　　　　corn: 0.03 × 0.772 = 0.0232
　　　　　　　　SBM: 0.34 × 0.192 = 0.0653

　　　　　　　　Phosphorus:
　　　　　　　　corn: 0.29 × 0.772 = 0.2240
　　　　　　　　SBM: 0.70 × 0.192 = 0.1344

Equation 1:

(percent calcium in ground limestone times X) + (percent calcium in defluorinated phosphate times Y) = (percent calcium required in diet minus [calcium provided in corn plus calcium provided in soybean meal])

　　　　　36.07X + 31.65Y = 0.60 – (0.0232 + 0.0653)
　　　　　36.07X + 31.65Y = 0.5115 (calcium equation)

Equation 2:

(percent phosphorus in ground limestone times X) + (percent phosphorus in defluorinated phosphate times Y) = (percent phosphorus required in diet minus [phosphorus provided in corn plus phosphorus provided in soybean meal])

　　　　　0.02X + 13.70Y = 0.50 – (0.2240 + 0.1344)
　　　　　0.02X + 13.70Y = 0.1416 (phosphorus equation)

Divide 0.2 by 36.07 to get a factor (0.2 ÷ 36.07 = 0.0005545) that is multiplied times equation 1: (0.0005545) × (36.07X + 31.65Y = 0.5115) = 0.02X + 0.01755Y = 0.000284. The resulting equation is then subtracted from equation 2 to eliminate the X unknown and solve for Y.

$$0.02X + 13.70Y = 0.1416$$
$$-0.02X - 0.01755Y = -0.000284$$
$$\overline{13.68245Y = 0.1413}$$
$$Y = 0.010327$$

Substitute the value of Y in equation 1 and solve for X:

$$36.07X + (31.65 \times 0.010327) = 0.5115$$
$$36.07X + 0.32684 = 0.5115$$
$$36.07X = 0.5115 - 0.32684$$
$$36.07X = 0.18466$$
$$X = 0.00512$$

Step 5. Compare the nutrients supplied by the computed diet and the requirements of the animal:

	Lysine	*ME*	*Calcium*	*Phosphorus*
Amount supplied in diet:				
corn	0.1931	2,641.65	0.0232	0.2240
SBM	0.5569	618.35	0.0653	0.1344
Ground limestone			0.1847	0.0001
Defluorinated phosphate			0.3268	0.1415
Total in diet:	0.75	3,260	0.60	0.50
Req/kg of diet:	0.75	3,260	0.60	0.50
Difference:	0.0	0	0	0

The computed ration meets the dietary requirements for lysine, metabolizable energy, calcium, and phosphorus. To provide a completely fortified ration, add 0.25 percent sodium chloride, 0.1 percent trace mineral premix, 0.1 percent vitamin premix, and 0.1 percent antimicrobial premix to the ration. To make the total equal 100 percent, add an amount equal to the difference between the calculated total and 100 to the calculated percent of corn in the diet. The percent of each ingredient in the diet is shown as calculated and as revised:

| | *Percent of each ingredient in the diet* ||
	% as calculated	*Revised % to total 100*
Corn	77.24	78.71
Soybean meal	19.20	19.20
Ground limestone	0.51	0.51
Defluorinated phosphate	1.03	1.03
Sodium chloride	0.25	0.25
Trace mineral premix	0.10	0.10
Vitamin premix	0.10	0.10
Antimicrobial premix	0.10	0.10
Total	98.53	100

USING FIXED INGREDIENTS WHEN FORMULATING DIETS

Feed mixes formulated to provide a complete diet for the animal normally have small amounts of minerals, vitamins, and/or antibiotics added. Generally these total less than 10 percent of the total mix and provide little of the protein or energy needed in the diet. However, these fixed ingredients must be taken into account when formulating diets if the final computed protein and energy needs of the animals are to be met.

The first step in formulating a diet using fixed ingredients is to determine what these ingredients are and how much of each is to be in the final mix. Next, determine if any of these fixed ingredients provide any of the nutrients for which the ration is being balanced. If they do, then these amounts must be calculated and subtracted from the amount to be provided by the major ingredients in the diet. After this is done, then the procedures outlined above may be followed to balance the major ingredients for the mix.

This procedure is demonstrated using algebraic equations. Assume a one ton (2,000 lb) mix is needed to feed finishing hogs weighing 125 pounds (57 kg). The major ingredients selected are corn (IFN 4-02-935) and soybean meal (IFN 5-01-600). The fixed ingredients that provide additional minerals and vitamins do not add either energy or protein to the ration and total 55 pounds. The ration is to be balanced for daily requirements of lysine and ME.

Step 1. Set up the requirements and composition of the feeds:

	Lysine	ME
Daily req. (kg)	0.0122	6,320 kcal
corn (4-02-935)	0.0025	3,300 kcal/kg
SBM (5-04-600)	0.0279	2,972 kcal/kg

Step 2. Set up the algebraic equations and solve:

X = amount of grain needed per day
Y = amount of supplement needed per day

Equation 1:

$$0.0025X + 0.0279Y = 0.0122 \text{ (lysine equation)}$$

Equation 2:

$$3,300X + 2,972Y = 6,320 \text{ (energy equation)}$$

Divide 3,300 by 0.0025 to get a factor that is multiplied times equation 1. The resulting equation is then subtracted from equation 2 to eliminate the X unknown and solve for Y. Alternatively, the Y unknown could be eliminated by dividing 2,972 by 0.0279 and then solving for X:

$$3,300 \div 0.0025 = 1,320,000$$
$$3,300X + 2,972Y = 6,320$$
$$-3,300X - 36,828Y = -16,104$$
$$-33,856Y = -9,784$$
$$Y = 0.289$$

Substitute the value of Y in equation 1 and solve for X:

$$0.0025X + (0.0279 \times 0.289) = 0.0122$$
$$0.0025X + 0.008 = 0.0122$$
$$0.0025X = 0.0122 - 0.008$$
$$0.0025X = 0.0042$$
$$X = 1.68$$

Step 3: The accuracy of the solution may be checked by comparing the computed amounts of lysine and ME provided by this diet with the original requirements:

Lysine requirement (0.0122 kg)

Corn 0.0025×1.68	= 0.0042
SBM 0.0279×0.289	= 0.008
Total lysine	= 0.0122 kg per day

ME requirement (6,320 kcal)

Corn $3,300 \times 1.68$	= 5,544
SBM $2,972 \times 0.289$	= 859
Total ME	= 6,403 kcal per day

Step 4. Determine the amount of corn and soybean meal to mix together to make 1,945 pounds of mix. The amounts of corn and soybean meal needed daily are added together and each amount is divided by the total to determine the percent of each ingredient in the ration. This percentage is then multiplied times 1,945 pounds to determine how many pounds of corn and soybean meal are necessary in the total mix. The balance of the 2,000 pounds is composed of the fixed ingredients previously determined to provide the added minerals and vitamins needed in the ration.

	kg/day	% diet	lb/ton
Corn	1.68	85.3	1,654
SBM	0.289	14.7	286
Total	1.969	100	1,945

SUBSTITUTING SILAGE FOR HAY

Silage replaces hay in a ration at the rate of three (3) parts silage to one (1) part hay. In the beef ration example given in this unit, assume that one-half of the roughage is from silage. The amount of hay and silage used is calculated as follows:

$$7.2 \div 2 = 3.6 \text{ pounds of hay}$$
$$3.6 \times 3 = 10.8 \text{ pounds of silage}$$

The ration is then calculated using the same method as in the example. The ME and protein content of the silage and hay is balanced by adding grain and supplement. The pounds of ME in the hay and silage are added together when finding the pounds of concentrate mix needed. The pounds of protein in the hay and silage are added together when finding the percent of protein needed in the concentrate mix.

USE OF COMPUTERS TO BALANCE RATIONS

Livestock rations can be balanced by the use of a computer. Computer services for balancing rations are offered by many universities and commercial feed companies. A computer can balance a ration quicker than a person can. In addition, a least-cost ration can be calculated. The cost of the computer service must be considered when deciding whether or not to use this service. It may be possible to save money by using a computer service to balance livestock rations.

SUMMARY

Livestock feeds are classified as roughages and concentrates. Roughages have a crude fiber content of more than 18 percent. Concentrates have less than 18 percent crude fiber. Roughages are either legume or nonlegume. Legume roughages can use nitrogen from the air. They are higher in protein content than nonlegume roughages. Concentrates are either energy feeds or protein supplements. Energy feeds are usually grains such as corn or oats. Protein supplements have more than 20 percent protein content. They come from either animal or vegetable sources.

A ration is the amount of feed an animal is given during a 24-hour period. It is balanced if it provides all the nutrients the animal needs for good growth, gain, or production. In addition, the ration must taste good to the animal and it must be economical. The right balance of roughages and concentrates must be in the ration. No harmful materials or excessive amounts of additives should be fed. Micronutrients and additives must be carefully mixed in the right amounts in the ration.

A ration is fed for several purposes. It must provide the nutrients needed for maintenance, growth, pregnancy and, sometimes, work. Many animals are also fed to fatten for market or to produce milk, eggs, or wool. Proper nutrition is also essential for reproduction.

Rations are balanced for the protein and energy needs of the animal. Feeding standards and tables of feed composition are used in balancing rations. The mineral and vitamin needs of the animal are also considered. The price of the feed used is important when calculating least-cost rations.

Certain rules of thumb, which give general guides to use for rations, may be used.

There are five steps in balancing a ration. (1) The kind of animal to be fed is identified. (2) The needs of the animal are found. (3) Feeds are selected and the composition of the feed is found. (4) The amount of each feed to use is calculated. (5) The ration is checked against the needs of the animal to make sure it is balanced. The Pearson Square is a helpful tool to use in balancing rations.

Computers may be used to make balancing rations easier and faster. More nutrients can be considered when using a computer to balance a ration.

Student Learning Activities

1. Calculate balanced rations for livestock on a local farm.
2. Secure prices of feeds from local sources and calculate least-cost, balanced rations for various classes of animals.
3. Prepare an exhibit showing feeds classified as roughages and concentrates.
4. Present an oral report on balancing a ration for livestock on your home farm.
5. Take a field trip to a feed company that uses a computer for balancing rations and observe the method being used.

Review

1. What are roughages?
2. Name the two general classes of roughages.
3. What are the sources of each of these two classes of roughages?
4. What are concentrates?
5. Define the term *ration*.
6. List and briefly explain the six functions of a ration.
7. What are the two common ways to measure the amount of protein in a ration?

8. What are the four common ways to measure the amount of energy provided by a ration?

9. What should be the ratio of calcium to phosphorus in a ration?

10. What are the rules of thumb for balancing rations for (a) beef cattle, (b) swine, (c) sheep and goats, (d) horses, (e) poultry?

11. List and briefly explain the five steps to follow when balancing a ration.

12. Give an example showing how to use the Pearson Square when balancing a ration.

13. Select an animal from the home farm or in the local area and balance a ration for that animal using feeds available locally.

14. How much silage can be substituted for hay in a ration for beef?

15. Show how to convert the amounts in a ration calculated on a dry-matter basis to an as-fed basis.

Animal Breeding

Genetics of Animal Breeding

Objectives

After studying this unit, the student should be able to

- explain how genetics relates to improvement in livestock production.
- describe how cell division occurs.
- diagram and explain how animal characteristics are transmitted.
- diagram and explain sex determination, linkage, crossover, and mutation.

THE IMPORTANCE OF GENETICS

Farm animals today are better than they were one hundred years ago. They produce more meat, milk, eggs, and wool on less feed. Much of this progress in livestock efficiency is the result of the use of genetics. *Genetics* is the study of heredity, or the way in which traits of parents are passed on to offspring. Good breeding programs are based on an application of the principles of genetics.

An Austrian monk named Gregor Johann Mendel is considered to be the founder of the science of genetics. In a period from 1857 to 1865, Mendel did many experiments with garden peas. He proved that certain characteristics, such as color and height, are passed from parent to offspring. Livestock breeders use this fact to select animals for breeding that will produce offspring with desirable characteristics.

Not all differences in animals are caused by genetics. Some are caused by the *environment*, or the conditions under which the animals are raised. This makes the job of selection more difficult. However, methods have been developed that enable farmers to select parent animals with traits that are related to genetics rather than the environment.

SELECTION BASED ON GENETICS

Additive and Nonadditive Gene Effects

Observation of any population of farm animals reveals variation in phenotype and, by inference, variation in genotype. Two factors are responsible for the genetic variation in animals: (1) *additive gene affects*, and (2) *nonadditive gene effects*.

When many different genes are involved in the expression of a trait, that expression is said to be controlled by *additive gene effects*. Individual genes have relatively little effect upon the trait; the effect of each gene is cumulative with very little or no dominance between pairs of alleles. Each member of the gene pair has an equal opportunity to be expressed. Most of the economically important traits of livestock are controlled by additive gene effects. Carcass traits, weight gain, and milk production are examples of traits that have moderate to high heritability and are considered to be greatly influenced by additive gene effects.

Traits that result from additive gene effects are considered to be quantitative. There may be hundreds or even thousands of gene pairs, located on different chromosome pairs, that are involved in the expression of the trait. The environment the animal is raised in often influences the expression of the trait. It is difficult to classify the phenotypes of the animals into distinct categories because they usually follow a continuous distribution. It is hard to identify animals with superior genotypes for quantitative traits.

Nonadditive gene effects control traits by determining how gene pairs act in different combinations with one another. Generally, these traits are readily observable and are controlled by only one or a few pairs of genes. Typically, one of the genes in the pair will be dominant if the animal is heterozygous for the trait being expressed. When combinations of gene pairs give good effects, the offspring will be better than either of its parents. This is sometimes called *hybrid vigor* or *heterosis*.

Traits that result from nonadditive gene effects are considered to be qualitative. The phenotype of these traits can usually be identified easily, there is relatively little environmental effect on these traits, and the genotype usually can be easily determined.

Heritability Estimates

Heritability is the proportion of the total variation (genetic and environmental) that is due to additive gene effects. A *heritability estimate* expresses the likelihood of a trait being passed on from parent to offspring. If a trait has a high heritability, the improvement in the animals' characteristics will be rapid. The improvement is slow for traits with a low heritability, requiring several generations of animals for the desirable trait to become strong. Tables 9-1, 9-2, 9-3, and 42-4 on page 883 list heritability estimates for several species of livestock.

TABLE 9-1 Heritability Estimates for Beef Cattle.

Trait	Heritability (%)
Number born	5
Calving interval (fertility)	10
Percent calf crop	10
Services per conception	10
Conformation score at weaning	25
Cancer eye susceptibility	30
Gain on pasture	30
Weaning weight	30
Yield grade	30
Carcass grade	35
Age at puberty	40
Birth weight	40
Body condition score	40
Carcass—percent lean cuts	40
Conformation score at slaughter	40
Cow maternal ability	40
Efficiency of gain	40
Preweaning gain	40
Yearling frame size	40
Yearling weight	40
Fat thickness	45
Feedlot gain	45
Dressing percent	46
Marbling score	50
Mature weight	50
Scrotal circumference	50
Tenderness	50
Final feedlot weight	60
Retail yield	60
Rib eye area	70

TABLE 9-2 Heritability Estimates for Swine.

Trait	Heritability (%)
Litter survival to weaning	5
Litter size	10
Number farrowed	10
Number pigs weaned	12
Weaning weight (3 weeks)	15
Birth weight	20
Five month weight	25
Number of nipples	25
Conformation	30
Feed efficiency	30
Age at puberty	35
Percent lean cuts	45
Probe backfat (live at 200 lb [99.8 kg])	45
Carcass length	50
Loin muscle area	50
Percent of shoulder	50
Percentage carcass muscle	50
Percent ham	55
Percent fat cuts	60

TABLE 9-3 Heritability Estimates for Sheep.

Trait	Heritability (%)
Number born	13
Conformation score	15
Feed efficiency	20
Fat thickness	25
Milking ability	25
Birth weight	30
Weaning weight	30
Carcass—percent lean cuts	35
Fleece weight	40
Post-weaning daily gain	40
Skin folds	40
Weight of retail cuts	40
Yearling weight	40
Ribeye area	45
Face covering	50
Mature weight	50
Staple length	50

Selecting Breeding Stock

The selection of breeding animals in specific species is discussed in later units in this text. There are computer programs and data bases, developed by universities and breed associations, currently available that can provide information about the breeding value of animals. The use of estimated breeding value and expected progeny difference helps the producer make faster genetic improvement in livestock. Mating systems for livestock are also discussed later in this text.

There are three types of systems that might be used to select breeding animals:

1. Tandem
   - Selection is for one trait at a time; selection for another trait begins when a desired level of performance is achieved in the first.
   - An animal with one desirable trait but other undesirable traits may be kept for breeding purposes.
   - For the most profitable production, emphasis needs to be placed on several traits when selecting breeding stock; tandem selection does not do this.
   - Simple to use but not recommended; it is the least effective of the selection methods.
2. Independent Culling Levels
   - Establishes a performance level for each trait in the selection program that an animal must achieve to be kept for breeding purposes.
   - Selection for the breeding program is based on more than one trait.
   - A disadvantage of this type of selection is that superior performance in one trait cannot offset a trait that does not meet the criteria for selection.
   - Most effective when only a small number of traits are being selected for in the breeding program.
   - This is the second most effective method of selection and is the one most widely used in the livestock industry.
3. Selection Index
   - An index of net merit is established that gives weight to traits based on their economic importance, heritability, and genetic correlations that may exist between the traits.
   - Does not discriminate against a trait with only slightly substandard performance when it is offset by high performance in another trait.
   - Provides more rapid improvement in overall genetic improvement in the breeding group.
   - Extensive records are required to establish the index.
   - Is the most effective method of achieving improvement in genetic merit.

From a practical standpoint, it may be wise for a livestock breeder to use a combination of selection methods in the breeding program. A combination of the independent culling level method and the selection index method may work best for many producers. One or two traits that are particularly critical for the producer may be selected for using the independent culling level method. The selection index might then be used for other traits that are important for that producer.

THE CELL AND CELL DIVISION

An animal's body is made up of millions of cells. Cells are the basic and generally the smallest parts of the body that are capable of sustaining the processes of life (metabolism and reproduction). Figure 9-1 shows the parts of a cell. Most of the cell is made of a material called protoplasm. The *nucleus* contains the hereditary material of the cell, that is, the chromosomes that contain the genes. The nucleus also controls the cells' metabolism, growth, and reproduction. The nucleus is surrounded by the *cytoplasm*. The cytoplasm contains mitochondria, lysosomes, Golgi apparatuses, and ribosomes. The nucleus and cytoplasm are surrounded by the semipermeable *cell membrane*.

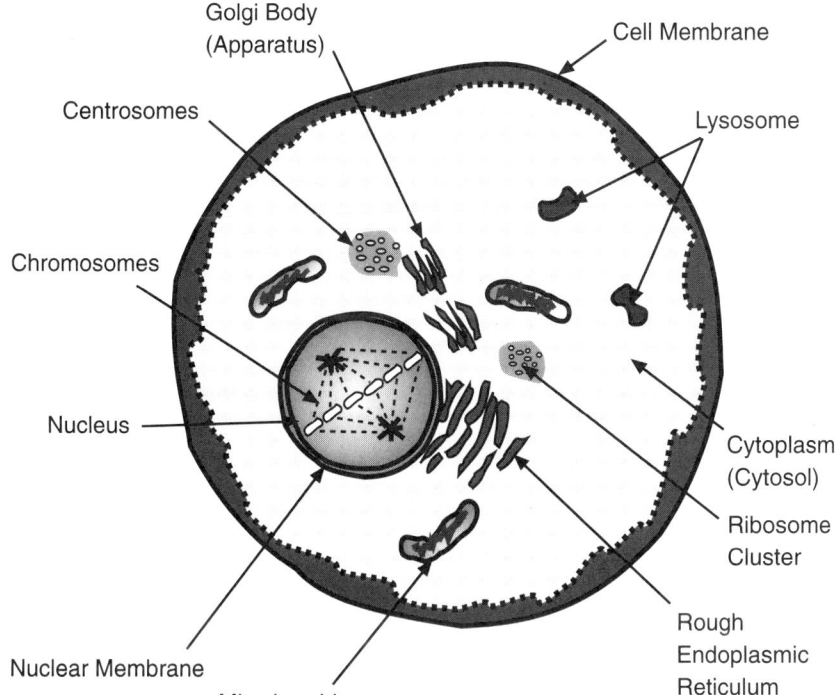

Golgi Body
(Apparatus)

Cell Membrane

Centrosomes

Lysosome

Chromosomes

Nucleus

Cytoplasm
(Cytosol)

Ribosome
Cluster

Nuclear Membrane

Mitochondria

Rough
Endoplasmic
Reticulum

FIGURE 9-1   Parts of the Cell.

Mitosis

Each animal begins as a single cell. This cell divides to make two cells. The cells continue to divide, and groups of cells form specialized tissues and organs in the animal's body. This division of body cells in an animal is called *mitosis*. Mitosis increases the number of body cells, which causes the animal to grow. Old body cells that die are replaced by mitosis. Chromosomes occur in pairs in the nucleus of all body cells except the sperm and ovum. Each parent contributes one-half of the pair. The number of pairs of chromosomes is called the *diploid* number. The diploid number varies from species to species but is constant for each species of animal:

cattle	30	goat	30	chicken	39
swine	19	horse	32	rabbit	22
sheep	27	donkey	31		

During mitosis, the chromosome pairs are duplicated in each daughter cell, so they are exactly like the old cell. Figure 9-2 shows the steps in mitosis. A cell that is not dividing is in the interphase stage. During mitosis there are four typical stages in the division of the cell nucleus. In the order in which they occur these are: prophase, metaphase, anaphase, and telophase.

The ability of body cells to continue to divide throughout the life of the animal is limited. At the end of each chromosome in the nucleus of the cell there is a specific repeating DNA sequence called telomere. The presence of these telomeres is critical for successful cell division. Each time a cell divides, some of the telomere is lost from the end of the chromosome. As the animal ages, the telomeres become shorter and eventually the cells stop dividing. This causes the animal to eventually die of old age, if it does not die from some other cause earlier.

Meiosis

When cells divide by mitosis, the daughter cells contain two of each type of chromosome, that is, they are diploid. The reproductive cells are called gametes. The male gamete is called a sperm cell and the female gamete is called an ovum, or egg, cell. During sexual reproduction two gametes (one sperm and one ovum) unite to form the zygote. If each gamete were diploid, the zygote would have twice as many chromosomes as the parents. Since this does not happen, clearly a mechanism for cell division exists that reduces the number of chromosomes in the gametes by one-half. This specialized type of cell division that occurs in the gametes is called *meiosis*, Figure 9-3.

During meiosis the chromosome pairs are divided in such a manner that each gamete has one of each type of chromosome; the gamete cell has a *haploid* number of chromosomes. The zygote that

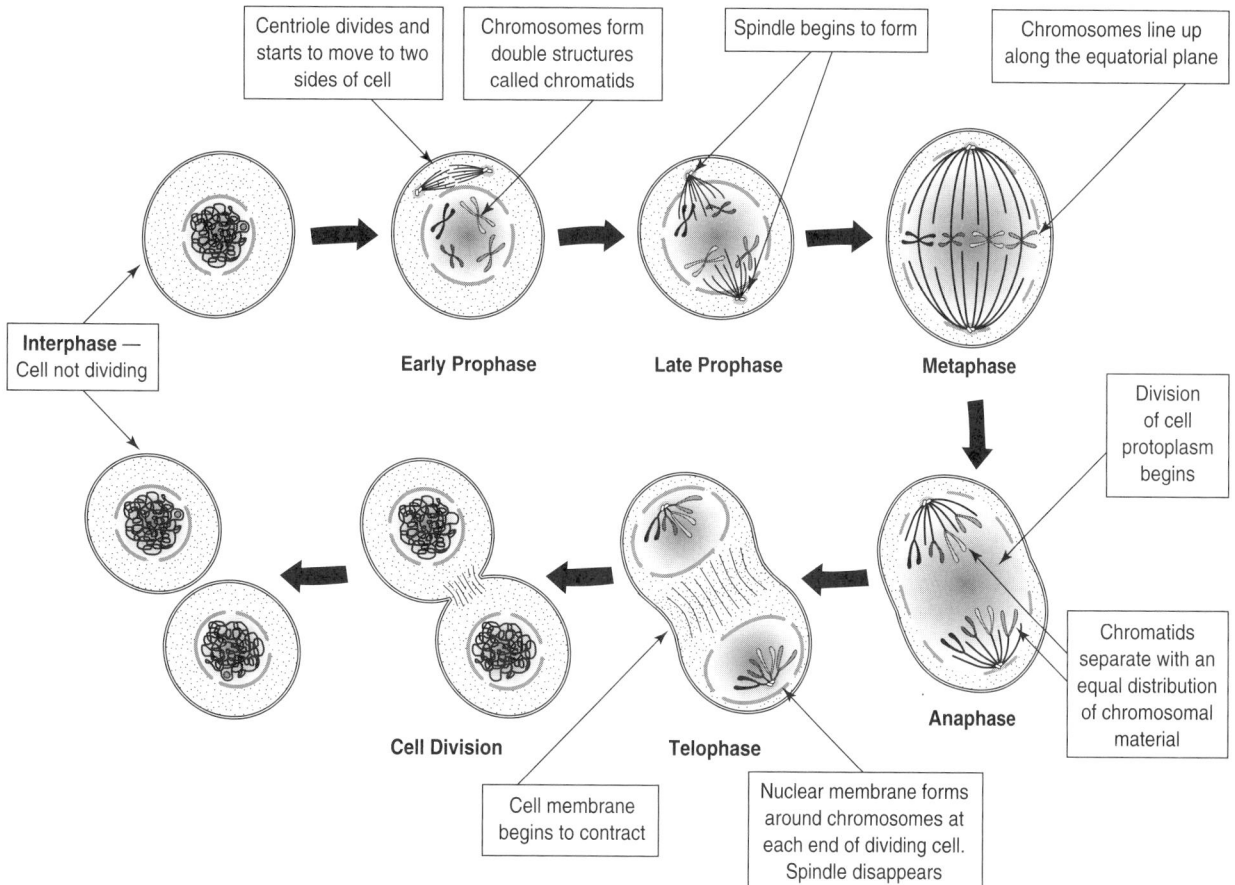

Centriole divides and starts to move to two sides of cell

Chromosomes form double structures called chromatids

Spindle begins to form

Chromosomes line up along the equatorial plane

Interphase — Cell not dividing

Early Prophase

Late Prophase

Metaphase

Division of cell protoplasm begins

Chromatids separate with an equal distribution of chromosomal material

Anaphase

Cell Division

Telophase

Cell membrane begins to contract

Nuclear membrane forms around chromosomes at each end of dividing cell. Spindle disappears

FIGURE 9-2   Cell division (mitosis).

results from the union of the sperm and ovum has a diploid number of chromosomes. One set comes from the sperm and one set comes from the ovum. The chromosome pairs are *homologous*, that is the first chromosome in the sperm matches the first chromosome in the ovum; the rest of the chromosomes in the sperm and ovum match up in a similar manner.

Although both the spermatozoa and the ova are produced by meiosis, there are some differences in the production of each. The production of spermatozoa is called *spermatogenesis*; the production of an ovum is called *oogenesis*, Figure 9-4.

When they reach sexual maturity, male animals begin producing *spermatozoa* (sperm cells) from spermatogonia in the seminiferous tubules in the testes. *Spermatogonia* are the parents of *spermatocytes* (diploid cells that divide by meiosis to produce four spermatids).

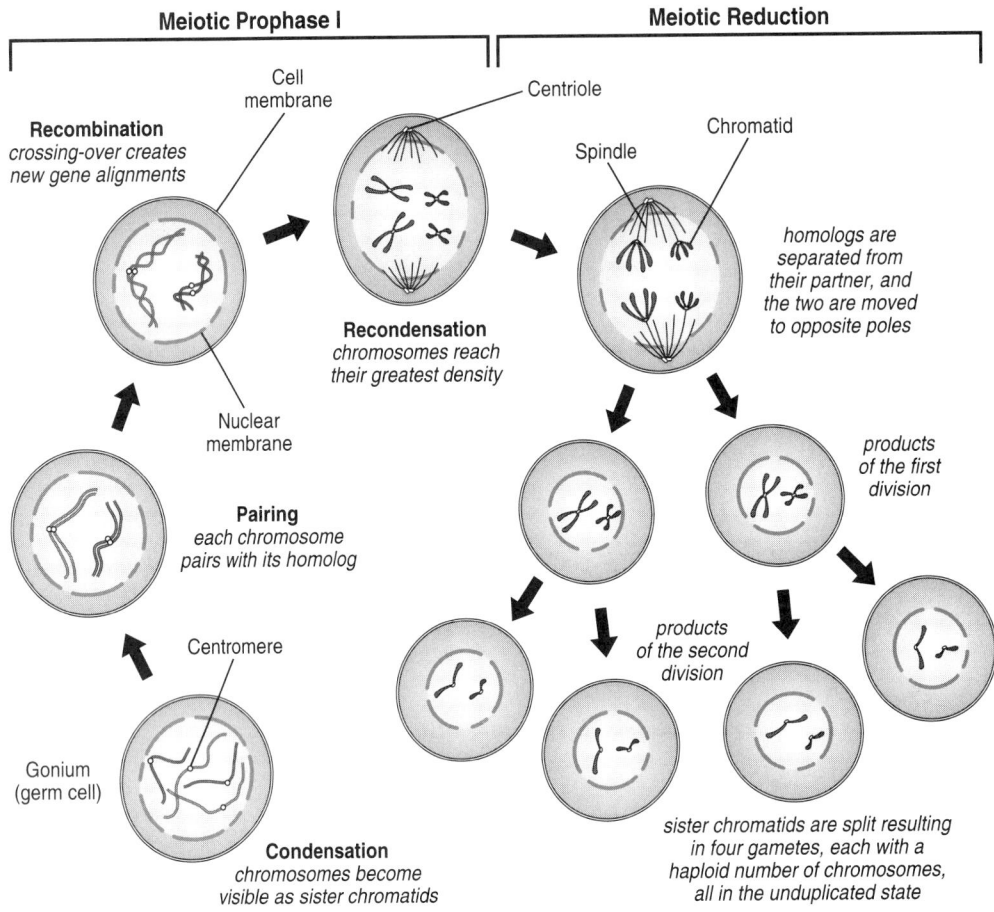

FIGURE 9-3   Cell dividing by meiosis.

A *spermatid* is any of the four haploid cells produced by meiosis that develop into spermatozoa. The first meiotic division acts on a primary spermatocyte to produce two secondary spermatocytes that then divide in the second meiotic division to produce four spermatids. Spermatozoa are small, with only a small amount of cytoplasm in the head that is primarily the nucleus; they develop a long flagellum or tail that gives them a high degree of motility.

At sexual maturity, female animals produce *ova* (egg cells) in the ovaries. (Ova is the plural of ovum.) The first meiotic division acts on a primary oocyte to produce two cells; one is the secondary oocyte and the other is the first polar body. The secondary oocyte is a relatively large body, while the first polar body is quite small. The second meiotic division acting on the secondary oocyte produces one large cell (ootid) and one small cell (second polar body). The ootid develops into the ovum. The first polar body may or may not divide dur-

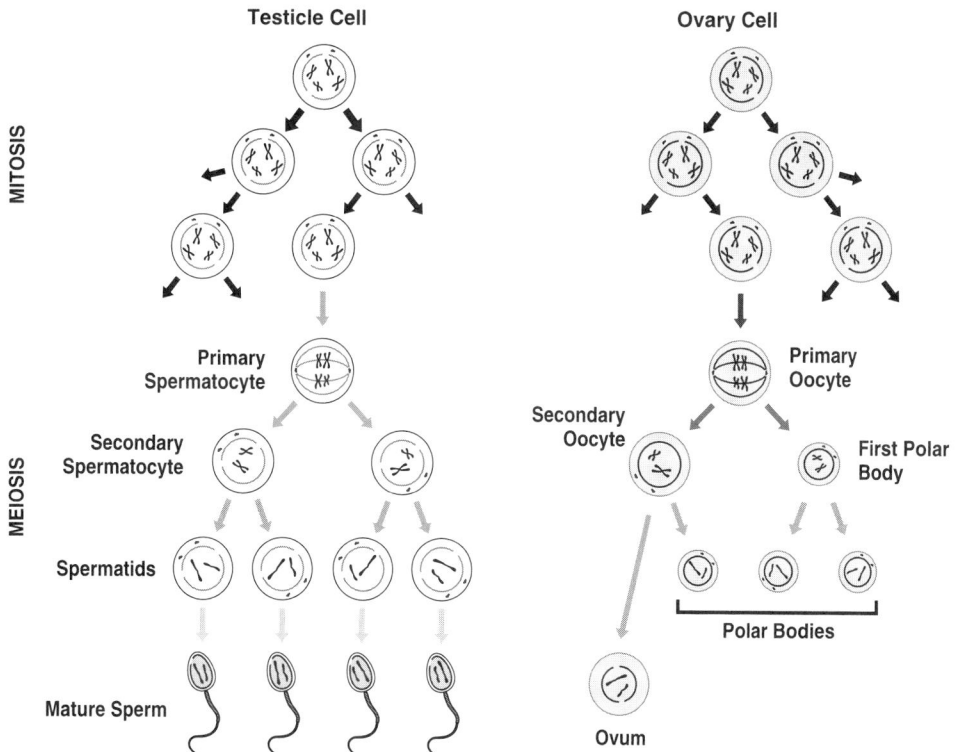

FIGURE 9-4   The steps in production of the sperm (spermatogenesis) and the ovum (oogenesis).

ing the second meiotic division; if it does it produces two second polar bodies. The second polar bodies produced are not functional and are reabsorbed. The single ovum that is produced is large and contains a lot of cytoplasm and stored food; this provides the initial nourishment for the zygote and embryo.

Fertilization

When a sperm cell from the male reaches an egg cell from the female, *fertilization* takes place. The two haploid cells unite to form one complete cell called a zygote. The zygote is diploid; that is, it has a full set of chromosome pairs. This process results in many different possible combinations of traits in the offspring.

TRANSMISSION OF CHARACTERISTICS

Genes

The characteristics of an animal that are inheritable are passed from one generation to the next by genes. *Genes* are located on chromosomes and are composed of DNA. Because chromosomes occur in homologous pairs, the genes that they carry are also paired. The sequence of the locations of the gene pairs are the same in a homologous pair of chromosomes. If a gene pair at a given location is identical, they control a trait in the same way. For example, both may code for the color black. In this case the gene pair is said to be *homozygous*. In some gene pairs each gene codes for a different expression of the same trait. In this case the genes are called *alleles* and the gene pair is said to be heterozygous. For example, one allele may code for black color and the other may code for red color. The same trait is being affected but the alleles are coding for different effects. *Genotype* refers to the combination of genes that an individual possesses.

The genes provide the code for the synthesis of enzymes and other proteins that control the chemical reactions in the body. These chemical reactions ultimately determine the physical characteristics of the animal. The physical appearance of an animal, insofar as its appearance is determined by its genotype, is referred to as its *phenotype*. Environmental conditions may also influence the physical appearance of an animal. For example, the genotype of a beef animal for rate of gain determines a range for that characteristic in which it will fall, but the ration it receives will determine where it actually is in that range.

Some traits are controlled by a single pair of genes; however, most of the traits that are important in animals are controlled by many pairs of genes. For instance, carcass traits, growth rate, and feed efficiency are controlled by many pairs of genes.

The Coding of Genetic Information

The deoxyribonucleic acid (DNA) molecule is shaped like a double helix, Figure 9-5. The sides of the molecule are composed of two strands of deoxyribose (a 5-carbon sugar molecule). The deoxyribose molecules in the side strands are linked together by phosphoric acid. The side strands are linked together by nitrogen-containing bases. Two are purines (adenine [A] and guanine [G]) and two are pyrimidines (thymine [T] and cytosine [C]). A nucleotide is a combination of one of the nitrogenous bases, one phosphate, and one deoxyribose. The nitrogenous base is attached to one side of the deoxyribose and the phosphate to the other. The phosphate is also attached to the next deoxyribose in the chain, forming a polynucleotide chain. Two of these polynucleotide chains twist into a coil, forming the double helix of the DNA molecule.

The nitrogen bases combine in specific combinations as they link across to the other strand in the DNA molecule. Adenine connects only to thymine by two hydrogen bonds; cytosine connects only to guanine by three hydrogen bonds. The possible sequences are AT, TA, CG, and GC. With these sequences, four types of DNA nucleotides can be formed.

DNA ladder separates to form two identical DNA ladders

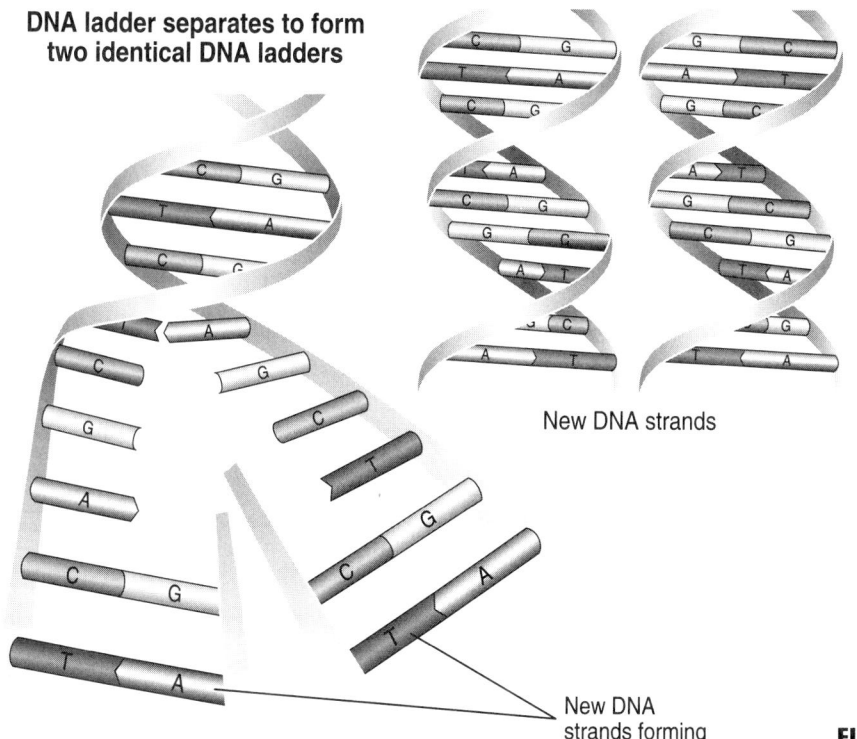

New DNA strands

New DNA strands forming

FIGURE 9-5   The DNA helix.

The factors that distinguish the DNA of one species from that of another are:

- The number of AT and CG pairs.
- The sequence of these pairs.
- Whether the connections are AT, TA, CG, or GC.
- The number of these base pairs that are present (the length of the DNA molecule).

The DNA is duplicated when cells divide. The two strands of the double helix separate and the nitrogenous base attracts its corresponding base. At the same time, the phosphate of one nucleotide is bound to the deoxyribose of the next by an enzyme (DNA polymerase), thus forming a new polynucleotide chain. This results in a new DNA double helix molecule that consists of one old strand and one new strand. The process is called semiconservative replication.

Ribonucleic acid (RNA) regulates protein synthesis in animals. The basic structure of RNA is similar to that of DNA except the nitrogenous base uracil (U) takes the place of thymine, the sugar is ribose instead of deoxyribose, and the binding enzyme during replication is RNA replicase instead of DNA polymerase. The primary function of RNA is carrying the genetic message from DNA for the building of the polypeptide chains that begin the process of protein synthesis.

The basic unit of the genetic code is the *codon*. The sequence of the four nucleotide bases in the DNA or the RNA is the key to the genetic code. Sixty-four different combinations are possible when three bases in specific sequences are used ($4 \times 4 \times 4 = 64$). This provides more than enough codons to identify the amino acids found in protein.

The characteristics of the genetic code are as follows:

- Codons exist as three bases in specific sequences.
- There are multiple codons for the same amino acid.
- Adjacent bases do not overlap to form codons.
- The code is generally nonambiguous under natural conditions.
- The code is colinear, that is, the messenger RNA sequence of codons and the corresponding amino acids of a polypeptide chain are arranged in the same linear sequence.
- Polypeptide chains are initiated by the codon for methionine (AUG).
- The three codons UAA, UAG, and UGA cause both chain termination and chain release; protein release factors are also needed for chain release.
- The code is universal, that is, the same codons code for amino acids similarly in all living organisms.

Dominant and Recessive Genes

A *dominant gene* in a heterozygous pair hides the effect of its allele. The allele that is hidden is called a *recessive gene*. The polled condition in cattle is the result of a dominant gene and is said to be a dominant trait. The horned condition in cattle is a recessive trait. When problems involving genetic inheritance are being worked, the dominant gene is usually represented by a capital letter. The recessive gene is usually represented by a small letter. In the example of polled cattle, the dominant gene is written *P*. The recessive gene is written *p*. (P = polled condition; p = horned condition).

Some other examples of dominant and recessive traits are:

- Black in cattle is dominant over red.
- White face in cattle is dominant to colored face.
- Black in horses is dominant to brown.
- Color in animals is dominant to albinism. (*Albino* animals lack all color.)
- Rose comb in chickens is dominant to single comb.
- Pea comb in chickens is dominant to single comb.
- Barred feather pattern in chickens is dominant to non-barred feather pattern (the dominant gene is also sex linked).
- Normal size in cattle is dominant to "snorter" dwarfism.

Homozygous and Heterozygous Gene Pairs

A *homozygous gene pair* is one that carries two genes for a trait. For example, a polled cow might carry the gene pair PP. A horned cow must carry the gene pair pp. For a cow to have horns, it must carry two recessive genes for the horned trait.

A *heterozygous gene pair* is one that carries two different genes (called *alleles*) that affect a trait. For example, a polled cow might carry the gene pair Pp. This cow is polled (because the P gene is dominant), but carries a recessive gene for the horned trait. If this cow is mated to a bull with a gene pair Pp, some of the calves will be polled and some will have horns.

Six Basic Crosses

There are six basic types of genetic combinations possible when a single gene pair is considered:

- Homozygous × Homozygous (PP × PP) (Both dominant)
- Heterozygous × Heterozygous (Pp × Pp)
- Homozygous (dominant) × Heterozygous (PP × Pp)
- Homozygous (dominant) × Homozygous (recessive) (PP × pp)
- Heterozygous × Homozygous (recessive) (Pp × pp)
- Homozygous (recessive) × Homozygous (recessive) (pp × pp)

It is possible to predict the results of crossing animals with various kinds of genotype. *Genotype* refers to the kinds of gene pairs possessed by the animal. A Punnett square may be used to predict the results of crossing animals with various kinds of genotype. The male gametes are usually shown across the top of the Punnett square and the female gametes along the left side.

Results of crossing are often referred to in the following manner in genetics:

Parental generation	P
First filial generation	F_1
Second filial generation	F_2

Homozygous × Homozygous (PP × PP) (Both dominant). A cross between two polled cattle that are homozygous for the polled trait would be set up as follows:

Male Gametes

	P	P
P	PP	PP
P	PP	PP

(Female Gametes along left side)

- All the F_1 are homozygous for the polled trait with the genotype PP.
- All the F_1 are polled.
- If the parents are homozygous dominant, all the F_1 must be homozygous dominant.

Heterozygous × Heterozygous (Pp × Pp). A cross between two cattle that are heterozygous for the polled trait is set up as follows:

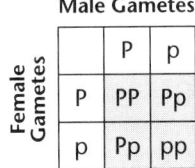

- The F_1 genotypic ratio is 1:2:1 (1 PP, 2 Pp, 1 pp).
- The F_1 phenotypic ratio is 3:1 (3 polled, 1 horned).

Homozygous (dominant) × Heterozygous (PP × Pp). A cross between two cattle, one homozygous and one heterozygous for the polled trait, is set up as follows:

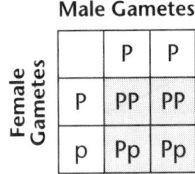

- The F_1 genotypic ratio is 1:1 (2 PP, 2 Pp).
- All the F_1 are polled.

Homozygous (dominant) × Homozygous (recessive) (PP × pp). A cross between two cattle, one homozygous (dominant) and one homozygous (recessive) for the polled trait, is set up as follows:

Male Gametes

		P	P
		P	P
	p	Pp	Pp
	p	Pp	Pp

(Female Gametes)

- All the F_1 are heterozygous with the genotype Pp.
- All the F_1 are polled.

Heterozygous × Homozygous (recessive) (Pp × pp). A cross between two cattle, one heterozygous and one homozygous (recessive) for the polled trait, is set up as follows:

Male Gametes

	P	p
p	Pp	pp
p	Pp	pp

(Female Gametes)

- The F_1 genotypic ratio is 1:1 (2 Pp, 2 pp).
- The F_1 phenotypic ratio is 1:1 (2 polled, 2 horned).

Homozygous × Homozygous (pp × pp) (Both recessive). A cross between two horned cattle, both homozygous (recessive) for the polled trait, is set up as follows:

Male Gametes

	p	p
p	pp	pp
p	pp	pp

(Female Gametes)

- All the F_1 are homozygous for the horned trait with the genotype pp.
- All the F_1 are horned.
- If the parents are homozygous recessive, all the F_1 must be homozygous recessive.

Multiple Gene Pairs

When more than one trait is considered, the possible genotypes and phenotypes increase. For example, if a polled, black cow (PpBb) is crossed with a polled, black bull (PpBb), both animals are heterozygous for the two traits. The Punnett square is set up as follows:

Male Gametes

	PB	Pb	pB	pb
PB	PPBB	PPBb	PpBB	PpBb
Pb	PPBb	PPbb	PpBb	Ppbb
pB	PpBB	PpBb	ppBB	ppBb
pb	PpBb	Ppbb	ppBb	ppbb

(Female Gametes)

- The genotypic ratio of the F_1 is 1:2:2:4:1:1:2:2:1 (1 PPBB, 2 PPBb, 2 PpBB, 4 PpBb, 1 PPbb, 1 ppBB, 2 Ppbb, 2 ppBb, and 1 ppbb).
- The phenotypic ratio of the F_1 is 9:3:3:1 (9 polled, black; 3 polled, red; 3 horned, black; and 1 horned, red).

It is not possible to predict which calf will result from any given mating; the genetic combinations occur by chance. The predicted ratios occur only when a large number of matings occur. The number of offspring necessary in heterozygous matings to approximate the predicated ratios increases rapidly as the number of gene pairs being considered increases, Table 9-4.

Incomplete Dominance

When a heterozygous condition exists for a given trait, one allele is not always dominant over the other. *Incomplete dominance* occurs when the alleles at a gene locus are only partially expressed. This usually produces a phenotype in the offspring that is intermediate between the phenotypes that either of the alleles would express. Some references refer to this phenomenon as blending inheritance or codominance. Codominance, however, is defined differently in some genetics textbooks.

Codominance

Codominance occurs when neither allele in a heterozygous condition dominates the other and both are fully expressed. Note how this definition differs from the definition for incomplete dominance.

The roan color of Shorthorn cattle is an example of codominance. The roan color appears to be an intermediate color between red and white. When the hair coat is examined closely, it is actually

TABLE 9-4 The Number of Offspring Required in Heterozygous Matings to Get the Predicted Ratios and the Number of Genotypes and Phenotypes That Result (where n equals the number of pairs of genes involved).

Number of Pairs of Heterozygous Genes (n)	Number of F_1 Required to Get Predicted Ratio[a] $(4)^n$	Number of F_1 Genotypes $(3)^n$	Number of F_1 Phenotypes $(2)^n$
1	4	3	2
2	16	9	4
3	64	27	8
4	256	81	16
5	1,024	243	32
6	4,096	729	64
7	16,384	2,187	128
8	65,536	6,561	256
9	262,144	19,683	512
10	1,048,576	59,049	1,024
11	4,194,304	177,147	2,048
12	16,777,216	531,441	4,096
13	67,108,864	1,594,323	8,192
14	268,435,456	4,782,969	16,384
15	1,073,741,824	14,348,907	32,768

[a]This is also the total number of possible combinations of gametes.

a mixture of red hairs and white hairs. The individual hairs are not a blend of red and white. To illustrate this trait, let R = red and W = white. If a red animal that is homozygous for red (RR) is mated with a white animal that is homozygous for white (WW), all the F_1 are roan:

Male Gametes

	R	R
W	RW	RW
W	RW	RW

(Female Gametes)

In a mating of the F_1, the F_2 has a genotype ratio of 1:2:1 (1 RR, 2 RW, 1 WW). The phenotype ratio of the F_2 is also 1:2:1 (1 red, 2 roan, 1 white):

Male Gametes

	R	W
R	RR	RW
W	RW	WW

(Female Gametes)

Sex-limited Genes

The phenotypic expression of some genes is determined by the presence or absence of one of the sex hormones; its expression is limited to one sex. These are know as *sex-limited* genes. An example of this is the male and female plumage patterns in chickens. The neck and tail feathers are long, pointed, and curving in male chickens that have the cock-feathering plumage pattern. The hen-feathering and cock-feathering plumage patterns are controlled by a pair of autosomal genes, Table 9-5. The dominant gene H produces the hen-feathering plumage pattern if either sex hormone is present and the recessive gene h produces the cock-feathering plumage pattern if the female sex hormone is absent and hen-feathering if it is present.

Sex-influenced Genes

Some traits are expressed as dominant in one sex but recessive in the other sex; this action is called *sex-influenced* genes. In humans, male

TABLE 9-5 Effect of a Sex-limited Gene on Plumage Patterns in Chickens.

Genotype	Female Phenotype	Male Phenotype
HH	Hen-feathering pattern	Hen-feathering pattern
Hh	Hen-feathering pattern	Hen-feathering pattern
hh	Hen-feathering pattern	Cock-feathering pattern

pattern baldness is an example of sex-influenced genes. Among farm animals, examples of sex-influenced genes include horns in sheep and color spotting in cattle. In sheep, the allele H for horns is dominant in the male; the allele h for polled is dominant in the female. The homozygote HH produces horns in both males and females; the homozygote hh produces the polled trait in both males and females. In cattle, mahogany and white spotting is dominant in males and recessive in females; red and white spotting is dominant in females and recessive in males.

Sex Determination

Mammals. The sex of the offspring is determined at the moment of fertilization. The female mammal has two sex chromosomes in addition to the regular chromosomes. These are shown as XX. Male mammals have only one sex chromosome. The other chromosome of the pair is shown as Y. The female is thus shown as XX. The male is shown as XY. After meiosis, all the egg cells will have an X chromosome, but only one-half the sperm cells will have an X chromosome. The other half of the sperm cells will have a Y chromosome. Thus, the sex of the offspring is determined by the male parent. This can be shown by the use of the Punnett square:

Male Gametes

	X	Y
X	XX	XY
X	XX	XY

Female Gametes

One-half of the offspring are females (XX). One-half of the offspring are males (XY).

Birds. In poultry, the female determines the sex of the offspring. The male carries two sex chromosomes (shown as ZZ). The female carries only one sex chromosome. The other chromosome of the female pair is shown as W. The female is thus shown as ZW.

After meiosis, all the sperm cells carry a Z chromosome. Only one-half of the egg cells carry a Z chromosome; the other half carry a W chromosome. The determination of the sex of the offspring by the female can be shown by use of the Punnett square:

Male Gametes

	Z	Z
Z	ZZ	ZZ
W	ZW	ZW

Female Gametes

As can be seen in the Punnett square, one-half of the offspring are males (ZZ), and one-half of the offspring are females (ZW).

Sex-linked Characteristics

Genes that are carried only on the sex chromosomes are called *sex-linked genes*. An example of a sex-linked trait is the barred color in chickens. Barred color (B) is dominant to black color. The gene for barred color is carried only on the sex chromosome. The results of crossing a barred female (Z^BW) with a black male (Z^bZ^b) is shown as follows. (The color gene is indicated as a superscript on the sex chromosome.)

Male Gametes

	Z^b	Z^b
Z^B	Z^BZ^b	Z^BZ^b
W	Z^bW	Z^bW

Female Gametes

All of the male offspring will be barred and carry a recessive gene for black. All of the females will be black and carry no gene for barred color.

Male Gametes

	Z^B	Z^B
Z^b	Z^BZ^b	Z^BZ^b
W	Z^BW	Z^BW

Female Gametes

More proof that the barred color is sex-linked is found by crossing a black female (Z^bW) with a barred male (Z^BZ^B). All of the offspring are barred.

Linkage

Early studies in genetics were based on the idea that all genes are redistributed in each mating. It was seen, however, that some groups of traits seemed to stay together in the offspring. More work showed that there is a tendency for certain traits to appear in groups in the offspring. This is called *linkage*. Sex-linkage of traits is discussed earlier in this unit. Sometimes, genes that are not on the sex chromosomes also seem to be linked together. This is because they are found on the same chromosome. The closer the genes are located together on a chromosome, the more likely they are to stay together (Figure 9-6).

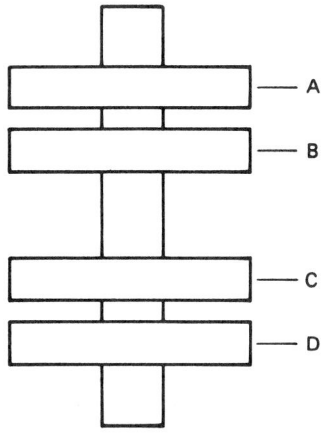

FIGURE 9-6   Gene linkage. Genes A and B will tend to stay together when the chromosomes divide, as will C and D. A and D are not as likely to stay together because they are farther apart.

Crossover

The fact that the predicted results of a mating do not always happen is sometimes the result of *crossover*. During one stage of meiosis the chromosomes line up together. They are very close to each other. Sometimes the chromosomes cross over one another and split (Figure 9-7). This forms new chromosomes with different combinations of genes. The farther apart two genes are on a chromosome, the more likely they are to end up in a new combination.

Mutation

Usually genes are not changed from parent to offspring. However, sometimes something happens that causes a gene to change. When a new trait is shown which did not exist in either parent, this is called *mutation*. Radiation will cause genes to mutate. The new trait is passed on to the offspring of the animal that has the new trait. Some mutations are beneficial, while others are harmful to the animal. Others are not of any importance to the animal. Very few mutations occur. Animal breeders do not depend on mutations to improve animals.

An example of a trait that is believed to be the result of a mutation is the polled Hereford. A cross between horned Herefords resulted in a polled animal. The polled condition is dominant. For the calf to be polled, at least one of its parents would have to be polled. Therefore, the polled calf from horned parents can only be explained as being a mutation.

SUMMARY

Much of the improvement in livestock is the result of using the principles of genetics. The work of Gregor Johann Mendel proved that parents pass their traits to their offspring. The amount of difference between parents and offspring is caused by genetics and the environment. Heritability estimates are used to show how much of the difference in some traits might come from genetics.

An animal's body is made up of millions of cells. Animals grow by cell division. The nucleus of the cell contains chromosomes, which are found in pairs. One chromosome of the pair comes from the father, and one comes from the mother.

Ordinary cell division is called mitosis. In mitosis, each new cell is exactly like the old cell. The reproductive cells are called gametes. Gametes divide by a process called meiosis. In meiosis, the chromosome pairs split, and each chromosome of a pair goes to a different gamete.

The male gamete is called a sperm. The female gamete is called an egg. Fertilization occurs when the sperm cell penetrates an egg cell. The chromosome pairs are formed again when fertilization takes place.

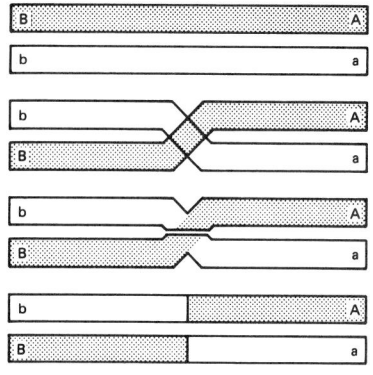

FIGURE 9-7   New combinations of genes are formed when chromosomes cross over and split.

Genes control an animal's traits. Genes are complex molecules found on the chromosomes in pairs. Some traits are controlled by a single pair of genes. Other traits are controlled by combinations of genes. Genes control traits by either additive gene effects or non-additive gene effects. Traits that are highly heritable are controlled by additive gene effects. Traits that are improved by crossbreeding are controlled by nonadditive gene effects.

Some genes are dominant; others are recessive. Dominant genes hide or mask the effect of recessive genes. Some genes are neither dominant nor recessive, and result in incomplete dominance, or a mixture of the two gene effects.

An animal may carry two dominant or two recessive genes for a trait. These are called homozygous pairs. Some animals have a dominant and a recessive gene, forming a heterozygous pair. It is possible to predict the results of mating two animals by using a Punnett square.

The sex of mammals is determined by the male. The female has two sex chromosomes, while the male has only one. One-half of the offspring are male and one-half are female. The sex of poultry is determined by the female. The female chicken has only one sex chromosome. The male chicken has two.

Some characteristics are sex-linked. The genes for those traits are on the sex chromosomes. Some traits tend to always appear together in the offspring. This is because the genes for these traits are on the same chromosomes. However, chromosomes may exchange genes in a process called crossover. This results in new combinations of traits in the offspring. Genes are sometimes changed by mutation, or the effect of outside forces. Radiation, for example, will cause mutation of the genes. Mutations are of little value in improving livestock.

Student Learning Activities

1. Take a field trip to a local livestock producer. Observe the animals, and make a list of traits that you think are (1) due to genetics and (2) due to environment. Suggest traits that should be selected for improving the offspring of the observed animals.

2. Prepare an oral report, including visual aids, on any phase of genetics studied in this unit.

3. Prepare a bulletin board display showing mitosis and meiosis or transmission of traits by mating individuals with a given genetic makeup.

Review

1. What is a heritability estimate and how is it used to improve livestock through breeding?
2. Name the parts of a cell.
3. How many pairs of chromosomes do each of the following animals have: (a) cattle, (b) swine, (c) sheep, (d) goats, (e) horses, (f) chickens?
4. Describe mitosis.
5. Describe meiosis.
6. What is fertilization?
7. Why are genes important in animal breeding?
8. Name and briefly describe the two ways in which genes control inherited traits.
9. Define dominant gene and recessive gene.
10. Define homozygous and heterozygous gene pairs.
11. Demonstrate the use of the Punnett square to predict the traits of the offspring when the male and female carry heterozygous gene pairs of a given trait.
12. Define and give an example of incomplete dominance.
13. How is sex of the offspring determined in mammals?
14. How is sex of the offspring determined in poultry?
15. Define and give an example of a sex-linked characteristic.
16. Define linkage, crossover, and mutation.

Animal Reproduction *Unit* **10**

REPRODUCTION

When organisms multiply, or produce offspring, it is called *reproduction*. Reproduction may be sexual or asexual. *Sexual* reproduction involves the union of a male and a female gamete. *Asexual* reproduction does not involve the gametes. Simple cell division in bacteria is an example of asexual reproduction. All of the common farm animals reproduce by sexual reproduction.

Sexual reproduction begins with the mating of the male and female. This is called *copulation*. The male gamete (sperm) is placed in the reproductive tract of the female. The sperm moves toward the egg cell. Fertilization occurs when the sperm penetrates the egg cell. The new animal, called the *embryo*, begins to grow. It is fed and protected in the female reproductive tract until it is born. *Parturition*, the act of giving birth, is the final step in reproduction.

MALE REPRODUCTIVE SYSTEM

Mammals

The male has special organs for reproduction. The reproductive organs of the bull are shown in Figure 10-1. The reproductive organs of other male mammals are similar to those of the bull.

The *scrotum* is the saclike part of the male reproductive system outside the body cavity that contains the testicles and the epididymis.

221

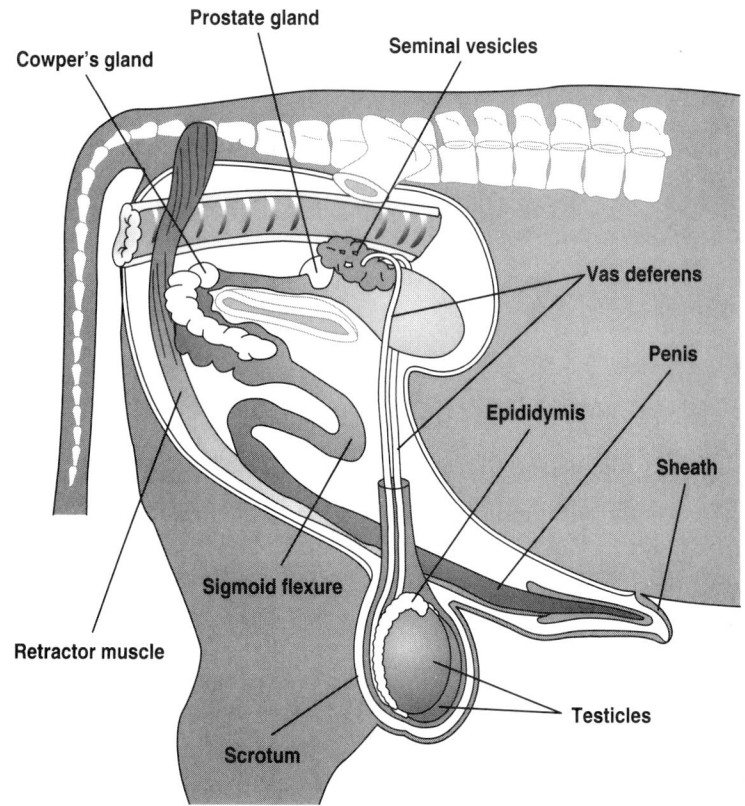

FIGURE 10-1 Reproductive organs of the bull.

The testicles are held in the abdominal cavity of the fetus. After the animal is born the testicles descend into the scrotum. Muscle tissue in the scrotum raises or lowers the testicles in response to the ambient temperature. During cold weather this muscle contracts, raising the testicles closer to the body, and during hot weather it relaxes, allowing the testicles to hang further from the animal's body. The temperature in the scrotum is slightly below the body temperature of the animal, allowing spermatogenesis to occur. Spermatogenesis cannot occur at body temperature. It is reduced if the ambient temperature is too high, producing a temporary reduction in fertility.

If the testicles of an animal are held in the body cavity, the animal is *sterile*, or cannot produce live sperm. A *ridgeling* or *ridgel* is a male in which one or both testicles are held in the body cavity. This is also called *cryptorchidism*, and is an inherited trait. The animal is usually sterile if both testicles are in the body cavity. If one testicle is retained in the body cavity and the other descends into the scrotum, the animal will be fertile. An animal with cryptorchidism (one or both testicles retained) should not be used for breeding.

The *testicles* produce the sperm cells and the male hormone, testosterone. The presence of testosterone maintains the masculine appearance of the animal. A male that is castrated at an early age does not develop the typical masculine appearance of the species and the reproductive organs do not continue to develop. When an adult male is castrated, the reproductive organs diminish in size and lose some of their function.

The *epididymis* is a long, coiled tube that is connected to each testicle. Sperm cells are stored in the epididymis while they mature. Sperm cells that are not moved out of the epididymis by ejaculation during copulation are eventually reabsorbed by the body.

The *vas deferens* is a tube that connects the epididymis with the urethra. Sperm cells move through the vas deferens to the urethra. The vas deferens is inside a protective sheath called the *spermatic cord*.

The *urethra* is the tube that carries urine from the bladder. This tube is found in both male and female mammals. In the male animal, both semen and urine move through the urethra to the end of the penis. The urine is the liquid waste that is collected in the bladder. The semen contains the sperm and other fluids that come from accessory glands. The three accessory glands are the seminal vesicles, the prostate gland, and Cowper's gland.

The *seminal vesicles* open into the urethra. They produce a fluid that protects and transports the sperm.

The *prostate gland* is near the urethra and the bladder. It produces a fluid that is mixed with the seminal fluid.

Cowper's gland produces a fluid that moves down the urethra ahead of the seminal fluid. This fluid cleans and neutralizes the urethra. This helps protect the sperm as they move through the urethra. The mixture of the seminal and prostate fluid and the sperm is called *semen*.

The *penis* deposits the semen within the female reproductive system. The urethra in the penis is surrounded by spongy tissue that fills with blood when the male is sexually aroused. This causes an erection that is necessary for copulation to occur. The *sigmoid flexure* (found in bulls, rams, and boars) and the *retractor muscle* extend the penis from the *sheath*, a tubular fold of skin. After copulation, the blood pressure in the penis subsides and the retractor muscle helps draw the penis back into the sheath. Horses and other mammals do not have a sigmoid flexure. Erection is caused by the blood that fills the spongy tissue when sexual arousal occurs.

Poultry

The reproductive system of the male chicken is shown in Figure 10-2. The *testicles* (which are held within the body cavity) produce the sperm and seminal fluid. The *vas deferens* carries the seminal fluid and sperm cells to the cloaca. The *cloaca* is the enlarged part where the large intestine joins the end of the alimentary canal. The *alimentary canal* is the food-carrying passage that begins at the mouth and

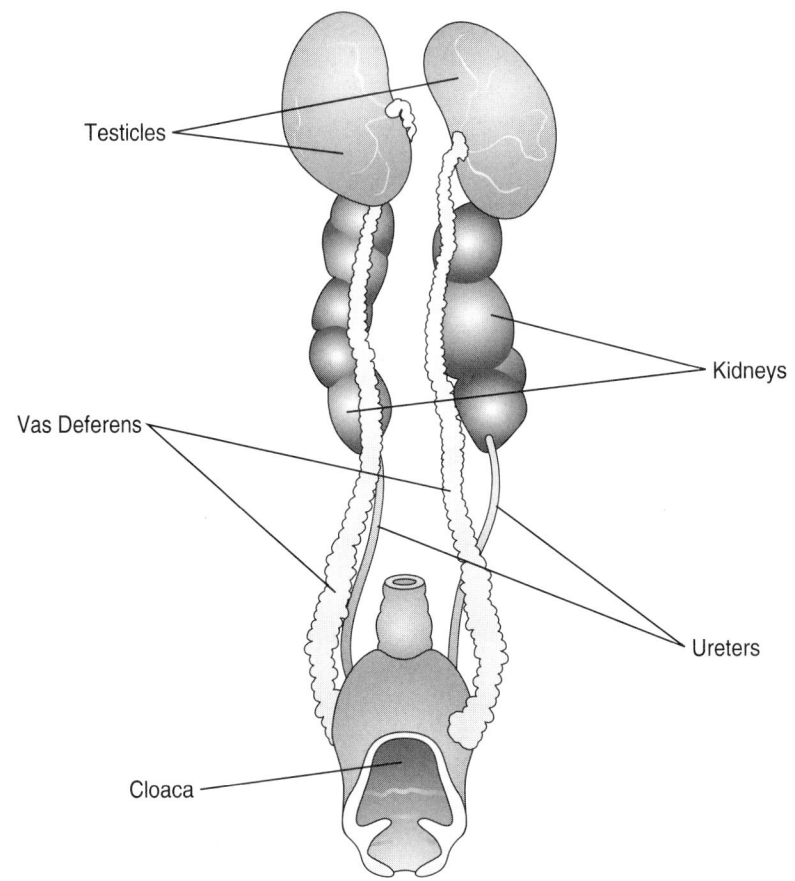

FIGURE 10-2   Reproductive system of the male chicken.

ends at the vent. The *papilla* is the organ in the wall of the cloaca that puts the sperm cells into the hen's reproductive tract.

FEMALE REPRODUCTIVE SYSTEM

Mammals

The female also has special organs for reproduction. They are very different from the male reproductive system. The female produces sex cells in the form of *eggs* or *ova* (singular, *ovum*). The female must also provide the place for the fetus to grow. The fetus is the unborn animal in the later stages of its development. In the early stages of development it is called the *embryo*. The reproductive system of the female cow is shown in Figure 10-3. A top view of the female reproductive system is shown in Figure 10-4. The reproductive systems of other female mammals are similar to that of the cow.

Female farm mammals have two ovaries that produce the ova and two female sex hormones (estrogen and progesterone). There are

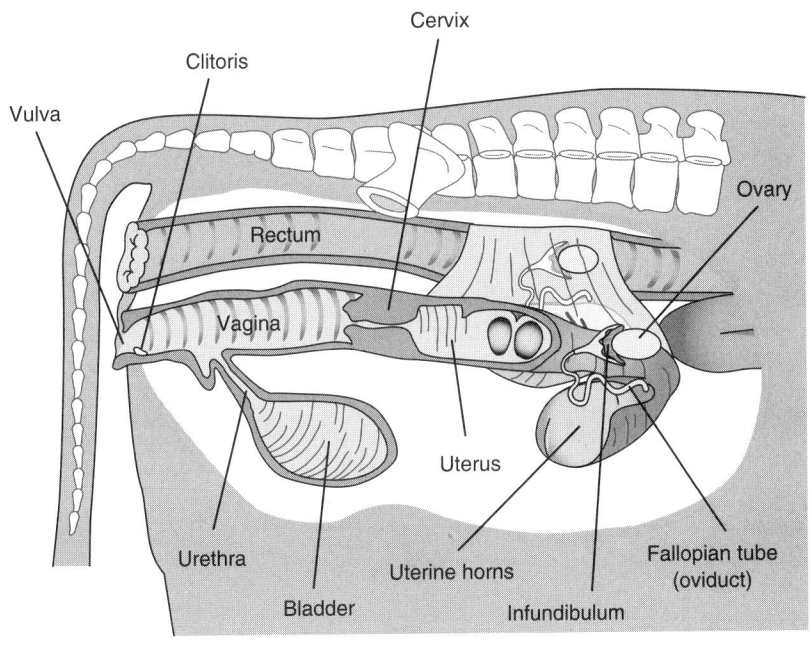

Cervix

Clitoris

Vulva

Ovary

Rectum

Vagina

Uterus

Urethra

Uterine horns

Fallopian tube
(oviduct)

Bladder

Infundibulum

FIGURE 10-3   Reproductive system of female cow, side view.

hundreds of tiny follicles (cavities) on each ovary. In cows, follicles are about the size of the head of a pin. The ova are produced in the follicles. The ovaries also form the *corpora lutea* (singular *corpus luteum*). The function of the corpus luteum is discussed in the section on ovulation.

The *oviducts* are two tubes that carry the ova from the ovaries to the uterus. The oviducts are also called the *Fallopian tubes*. The oviducts are close, but not attached, to the ovaries. The funnel-shaped end of each oviduct that is close to the ovary is called the *infundibulum*. At ovulation the follicle ruptures, releasing an ovum that is caught by the infundibulum. After copulation, sperm move through the uterus to the oviduct. Fertilization of the ovum occurs in the upper end of the oviduct. The zygote (fertilized egg cell) moves to the uterus about two to four days after fertilization.

The *uterus* of mammals is a Y-shaped structure consisting of the body, two uterine horns, and the cervix. The size and shape of the uterus varies among the various species of farm animals. The upper part of the uterus consists of the two uterine horns that develop into the oviducts or Fallopian tubes. Mammals that normally produce litters, such as swine, have relatively large horns and a small body; those that normally produce single offspring or twins, such as cattle and sheep, have smaller horns and a larger body. In all of these species, pregnancy normally occurs in the uterine horns; in horses, pregnancy normally occurs in the body of the uterus. In all species of farm animals, the fetus grows within the uterus, where it remains

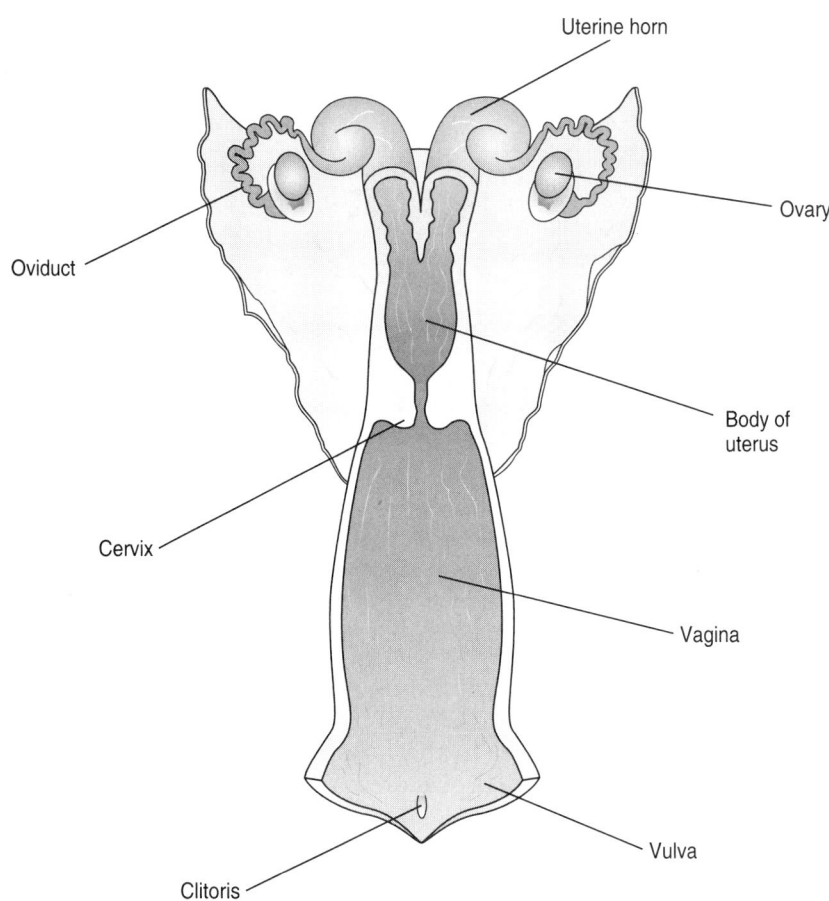

FIGURE 10-4   Reproductive system of female cow, dorsal view.

until parturition. The *cervix* is the lower outlet of the uterus. It is relatively relaxed during estrus to allow the passage of sperm into the uterus; during pregnancy it remains tightly closed to block the entrance of any foreign matter into the uterus.

The *vagina* is the passage between the cervix and the vulva. The lining of the vagina is moist during estrus and dry when the animal is not in estrus. During copulation, the semen is deposited in the vagina. The vagina expands to allow the fetus to pass through at birth.

The *bladder* collects the liquid waste, which is called urine. The urine passes through the urethra to the vagina. The urethra attaches to the floor of the vagina between the cervix and the vulva. It is not a part of the reproductive tract in females.

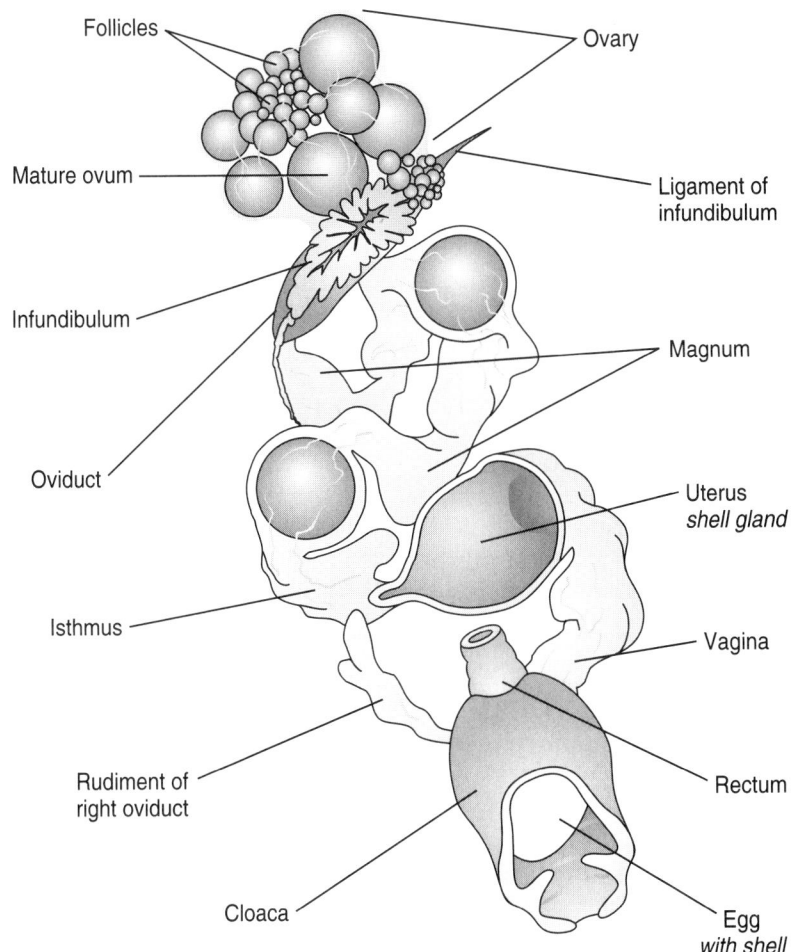

Follicles
Ovary
Mature ovum
Ligament of infundibulum
Infundibulum
Magnum
Oviduct
Uterus *shell gland*
Isthmus
Vagina
Rudiment of right oviduct
Rectum
Cloaca
Egg *with shell*

FIGURE 10-5 Reproductive system of the female chicken.

The *vulva* is the external opening of the reproductive and urinary systems. The exterior, and visible part of the vulva, consists of two folds called the labia majora. The labia minora are two folds located just inside the labia majora. Also located just inside the vulva is the *clitoris*, the sensory and erectile organ of the female. The clitoris develops from the same embryonic tissue as the penis in the male and produces sexual stimulation during copulation.

Poultry

The reproductive system of the female chicken is shown in Figure 10-5. The chicken has two ovaries and two oviducts. The right ovary and oviduct do not function. Only the left ovary and oviduct produce eggs. The ova produced in the ovary develop into egg yolks.

The oviduct of the chicken has five parts. The *funnel* receives the yolk from the ovary. The sperm cells that the chicken receives from the rooster are stored here. The *magnum* secretes the thick white of the egg. It takes about 3 hours for the thick white to be placed around the yolk in the magnum. The yolk and the thick white next moves to the *isthmus*, where two shell membranes are added. It takes about $1\frac{1}{4}$ hours for the shell membranes to be placed around the yolk and thick white.

The thin white and the outer shell are added to the egg in the *uterus*. The egg remains in the uterus about 20 hours. After the egg is completed, it moves to the *vagina*, where it stays for a short time, and then is laid. It takes about 25 to 27 hours for a chicken to produce one egg.

ESTRUS CYCLE

The *estrus*, or heat, period is the time during which the female will accept the male for copulation or breeding. The female mammal begins to have estrus periods when it is old enough to be bred. The estrus cycle begins when a follicle on the ovary begins to develop. The hormone estrogen is produced, and causes the animal to show the signs of estrus.

The signs of estrus in cattle include:

- standing when mounted by another cow (best indicator for time to breed).
- nervousness.
- swelling of the vulva.
- inflamed appearance around the lips of the vulva.
- frequent urination.
- mucus discharge from the vulva.
- trying to mount other cattle (cattle not in estrus may do this).

The signs of estrus in swine include:

- frequent mounting of other sows.
- restless activity.
- swelling of the vulva.
- discharge from the vulva.
- frequent urination.
- occasional loud grunting.

Sheep do not show any visible signs of estrus. The only way to tell if the ewe is in estrus is if she accepts the ram. A teaser ram with an apron to prevent breeding is sometimes used to see if the ewe is in estrus. The apron prevents the ram from completing the act of copulation when mounting the ewe. Most sheep have seasonal estrus periods. They come into estrus only in the fall. Dorset and Tunis breeds, however, come into estrus the year around. Seasonal estrus in sheep seems to be the result of the shorter hours of daylight and the cooler temperatures in the fall.

The signs of estrus in goats include:

- nervousness.
- riding other animals and standing when ridden.
- shaking the tail.
- frequent urination.
- bleating.
- swelling, red appearance of the vulva.
- mucus discharge from the vulva.

Dairy and Angora goats are seasonal breeders just as sheep are. Spanish goats will breed the year around.

The signs of estrus in horses include:

- raised tail.
- relaxation of the vulva.
- frequent urination.
- teasing of other mares.
- apparent desire for company of other horses.
- slight mucus discharge from the vulva.

Horses are partially seasonal in their breeding habits. The breeding season for horses is generally March through May. They have an irregular estrus cycle in the fall and early winter.

Table 10-1 shows the estrus cycles and other breeding information for the common species of farm animals.

OVULATION

Ovulation is the release of the egg cell from the ovary. The number of young that an animal gives birth to at one time is an indication of the number of egg cells released. The time of ovulation is usually near the end of the estrus period. Some animals ovulate after the estrus period ends.

Before ovulation, the egg cell is contained in the follicle. The follicle breaks open, releasing the egg. The egg moves into one of the oviducts. If live sperm are present, the egg may be fertilized. Shortly after ovulation, the corpus luteum forms on the ovary. It releases the hormone progesterone. This hormone causes four things to happen.

- The fertilized egg (embryo) is implanted in the uterus.
- Other eggs are stopped from forming.
- The pregnant condition is maintained.
- The mammary glands begin to develop. The *mammary glands* produce the milk to feed the young when they are born.

TABLE 10-1 Estrus Cycles and Reproductive Traits.[1]

Species	Age at Puberty (months) Range	Length of Cycle (days) Range	Average	Length of Estrus Range	Average	Time of Ovulation	Best Time to Breed	Length of Gestation (days) Range	Ave.
Cattle	4–12 (6–8 more common)	16–24	21	6–35 hours	16–18 hours	20–40 hr from beginning of estrus 10–14 hr after end of estrus	1st, early in estrus; 2nd, 12–20 hr after start of estrus	278–289	283
Swine	4–7	18–24	21	1–5 days Sows: 40–60 hr Gilts: 24–28 hr	3 days	18–60 hr after start of estrus (24 hr before end of estrus)	Sows: last ½ of estrus; Gilts: 2nd day	111–115	114
Sheep	4–8 (16–20 Merino)	14–20	16	1–3 days	30 hr	Near end		144–152	148
Goats	1st autumn	12–30	22	2–3 days	2½ days	Near end	Last ½ of estrus	140–160	151
Horses	10–12	10–37	22	1–37 days	6 days	1 to 2 days before end of estrus	3rd day; 2nd time; if still in heat, 3 days later	310–370	336

[1]The literature reveals a great deal of variation in information concerning estrus cycles and reproductive traits. Information given here is based on most frequently mentioned data.

If the egg is not fertilized, the corpus luteum does not grow. It *atrophies*, or wastes away. This allows another follicle to grow and another estrus period to occur. The time between estrus periods is the amount of time it takes for this to happen.

Animals that have several young at one birth release more than one egg at ovulation. These offspring are said to be *fraternally related*. They each come from a different egg cell. Sometimes one egg cell divides to form two animals. These two animals are *identical*.

FERTILIZATION

Fertilization is the union of the sperm and the egg cells. During copulation, the male animal deposits sperm in the reproductive tract of the female. The sperm move through the reproductive tract of the female until they reach the infundibulum. If an egg cell is present, a sperm cell may penetrate it. Only one sperm cell fertilizes an egg cell. Many millions of sperm cells are present in the reproductive tract as a result of mating. This helps to make sure that at least one sperm will fertilize the egg cell.

Sperm cells cannot live very long in the female reproductive tract. In cattle, for example, sperm cells live only about 24 to 30 hours after mating occurs. The egg cell lives about 12 hours after it is released if it is not fertilized. For fertilization to occur, the animal must be bred to have the live sperm cells and the egg cell present at the same time. If fertilization does not occur, the egg and sperm cells are absorbed by the body. The estrus cycle will repeat itself until the animal does become pregnant.

GESTATION

The *gestation period* is the time during which the animal is pregnant. During pregnancy the fetus develops in the uterus. The fetus is surrounded by a watery fluid enclosed in membranes. Blood vessels in the umbilical cord supply nutrients and oxygen and carry off waste products. The umbilical cord connects from the navel of the fetus to the placenta. The *placenta* lies along the wall of the uterus. Food, oxygen, and wastes are exchanged with the mother through the placenta by a process called *diffusion*.

The fetus grows slowly. Most of its growth is in the last one-third of the gestation period. Early in the growth period, the head, nervous system, and blood vessels develop. The bones and limbs are developed later. The position of the fetus shifts and changes during the gestation period.

PARTURITION

Parturition is the process of giving birth to the new animal. Near the end of the gestation period, the corpus luteum reduces the production of progesterone. There is an increase in the amount of estrogen in the body. This causes the uterine muscles to contract. The contraction of these muscles begins the process of birth.

The first water bag (part of the membranes that surrounded the fetus) soon appears. It gets larger and breaks open. Shortly thereafter, the second water bag, which contains the fetus, appears. The second water bag breaks open and the presentation of the fetus begins. In cattle, normal presentation (position of the fetus at birth) is the front feet first. This is followed by the nose (Figure 10-6). Then the head, shoulders, middle, hips, rear legs and feet appear.

Sometimes the young animal is still enclosed in the membranes when it is born. An attendant should be present at the time of birth to free the animal so that it does not suffocate if this occurs.

FIGURE 10-6 In cattle, a normal presentation at birth is the front feet first, followed by the nose, and then the head. *Photo by Michael Dzaman.*

The umbilical cord is broken at birth. This causes the animal to begin breathing. Because the hormone progesterone has stimulated the mammary glands of the mother, she will normally have milk for the young animal to nurse. This first milk is called *colostrum*. It is rich in antibodies, vitamins, and minerals that the newborn animal needs. *Antibodies* are substances that protect the animal from infections and poisons. It is important that the newborn animal be given colostrum milk during the first 12 to 24 hours after birth. After the first day or so, the mother produces regular milk instead of colostrum.

Several hours after birth the afterbirth is expelled from the uterus. The *afterbirth* consists of the placenta and other membranes that were not expelled earlier when the fetus was born. If the afterbirth is not expelled, it will decay inside the uterus, making the animal sick. If the afterbirth has not been expelled within a few hours, the help of a veterinarian may be needed. Care must be taken to prevent infection when removing the afterbirth from the uterus. The animal may become sterile if proper care is not taken when removing the afterbirth.

Do not allow the animal to eat the afterbirth. Remove it from the pen or area and bury it in lime or burn it. This helps prevent odors and the spread of disease.

REPRODUCTION IN POULTRY

Reproduction in poultry is different in some ways from reproduction in mammals. The young are not carried in the hen's body as they are in mammals. Instead, they develop in the fertilized egg outside of the hen's body.

The process begins with the placing of sperm into the oviduct of the female by the male. The papilla in the cloacal wall of the male deposits the sperm in the cloacal wall of the female. The sperm move up the oviduct to the funnel of the oviduct, where the egg is fertilized. Sperm cells will remain in the oviduct for 2 to 3 weeks after mating. They have full fertilizing ability for about 6 days. After that time, the ability of the sperm to fertilize the egg is decreased. By the tenth day after mating, the sperm have about 50 percent of their fertilizing ability. The fertilizing ability of the sperm is reduced to about 15 percent by the nineteenth day after mating.

After the egg yolk is fertilized, it moves through the reproductive tract of the female. The thick white, shell membranes, thin white, and the shell are added as it moves through the tract. After the egg is laid, the embryo grows inside the shell. It must have the right temperature and humidity to develop properly. During this time, the embryo is fed by the contents of the egg. After incubation, the chicken hatches, or breaks out of the shell.

Incubation of eggs is keeping them at the right temperature and humidity for hatching. An individual hen does this by setting on the eggs in the nest. Commercial hatcheries use mechanical incubators to hatch chickens and other poultry.

The incubation period for chickens is 21 days. For ducks and turkeys it is 28 days. For geese it is 29 to 31 days. Muscovy ducks have an incubation period of 33 to 35 days.

The temperature in the incubator should be 102° to 103°F (38.9°–39.4°C). The relative humidity in the incubator should be 60 percent for the first 18 days. It should be increased to 70 percent during the last 3 days of incubation. The eggs are laid horizontally in the tray. They should be turned twice daily for the first 15 days. For ducks, they should be turned for the first 22 days. This keeps the embryo from sticking to the shell. The incubator should have a small amount of air moving through it. This provides fresh oxygen and removes the carbon dioxide that collects in the bottom of the incubator.

REPRODUCTIVE FAILURES

The general physical condition of the animal has an effect on its ability to reproduce. An animal that is too fat or too thin may not become pregnant when bred. Proper nutrition and exercise will help prevent this problem. Animals in poor physical condition may have trouble giving birth.

There are several infections that affect the reproductive organs of the animal. Some may prevent pregnancy and others may cause abortion. If the animal does become pregnant, it may deliver weak young that may not live. Infection of the uterus or general poor health of the animal may cause difficulty in giving birth. Units on diseases of the various species discuss many of these infections in more detail.

The sexual behavior of animals is affected by the secretion of hormones. When these hormones are not properly secreted, the animal may not be able to reproduce. Sometimes it is necessary to treat the animal with hormones to overcome this problem.

A *cyst* is a swelling containing a fluid or semisolid substance. Cysts may occur in the reproductive organs and cause breeding problems. Some cysts can be removed by surgery or treated with hormones. Not all cysts can be treated.

Cryptorchidism may cause reproductive failure in the male. This is discussed earlier in this unit.

Not all female animals become pregnant when bred. The reasons for this reproductive failure are not always known.

SUMMARY

The mating of the male and female animal begins the process called reproduction. The female gamete is called the ovum or egg cell. The male gamete is called the sperm. The sperm cell is deposited in the female reproductive tract where it moves to the oviduct and fertilizes the egg cell.

The female has an estrus (or heat) cycle during which she will accept the male for mating. The egg cell is released during the heat period.

If the egg cell is fertilized, it moves to the uterus, where it grows into the fetus. Hormones control the release of the egg cell and maintain pregnancy in the animal. When the fetus has reached the right time for birth, the uterine muscles contract, forcing the fetus through the birth canal.

Reproduction in poultry is similar in some ways to reproduction in mammals. The main difference is that the embryo develops in the eggshell outside of the mother's body. Eggs must be kept at the proper temperature and humidity to hatch. This is usually done in an incubator.

Student Learning Activities

1. Take a trip to a packing plant and observe the reproductive organs of male and female livestock.
2. Secure and dissect reproductive organs of male and female animals. Identify each organ by name and record its function.
3. Ask a veterinarian to talk to the class about problems associated with animal reproduction and birth.
4. Using visual aids, give an oral report on reproduction, gestation, or parturition.

Review

1. Name and briefly describe the parts of the reproductive systems of the bull and the cow.
2. Name and briefly describe the parts of the reproductive systems of the male and female chicken.
3. Describe the signs of heat in each of the following: (a) cattle, (b) swine, (c) sheep, (d) goats, (e) horses.
4. Describe ovulation in the mammal.
5. Explain how the egg cell is fertilized in the mammal.
6. Describe what happens during gestation in the mammal.
7. Describe parturition in the mammal.
8. Briefly describe the reproduction process in poultry.

11 *Biotechnology in Livestock Production*

BIOTECHNOLOGY

The human race has always used some form of biotechnology to change living plants and animals or their products for commercial use. Such activities as livestock breeding, crop improvement, and the production of food products are forms of biotechnology.

In the 1970s, laboratory techniques were developed that allowed researchers to identify and manipulate the deoxyribonucleic acid (DNA) that is found in the cells of all living organisms. Information contained in the DNA determines the characteristics of the organism (see Unit 9). The term biotechnology is now usually used to mean manipulating the DNA of living organisms at the molecular and cellular level to produce new commercial applications. One of the things that makes this possible is the fact that DNA from any living organism will function when it is transferred into any other living organism.

Biotechnology is being used in a number of ways:

- genetic engineering
- identification of an individual organism by its DNA sequence (DNA fingerprinting)
- embryo transfer
- semen sexing
- cloning of animals
- rapid diagnosis of infectious diseases
- diagnosis of genetic disorders
- development of vaccines
- identifying genes that control specific traits

There is a need to increase food production to feed an expanding world population. The use of biotechnology has the potential to significantly increase food production and reduce the cost of production, and yet cause less damage to the environment. The United States has led the world in the development and use of biotechnology in agriculture. There is resistance to the use of genetically modified organisms (GMOs) both in the United States and in some other parts of the world. This resistance may have an impact on the export of food products from the United States.

REGULATION AND SAFETY IN THE USE OF BIOTECHNOLOGY

Three federal agencies are involved in the regulation of biotechnology in agriculture and testing for the safety of its products. These are the United States Department of Agriculture (USDA), the Food and Drug Administration (FDA), and the Environmental Protection Agency (EPA). Monitoring of biotechnology within their borders is also done by most state governments.

The USDA is responsible for meat and poultry products, the FDA for all other domestic and imported foods and animal drugs, and the EPA for pesticides. The same standards of safety apply to foods and food ingredients produced by biotechnology that apply to all other foods and food ingredients. Permits must be secured for testing and evaluation of products produced by biotechnology. The product must be shown to be safe for human consumption and must not have any negative impact upon the environment. In the case of animal drugs, rigorous testing is required to show that the meat, milk, or eggs from animals treated with the drugs are safe for human consumption.

There is some controversy regarding the labeling of products produced by biotechnology. Some groups believe that all such products should be labeled as genetically modified. To date, such labeling has not been required by the FDA just because biotechnology is involved in the production of the product. However, in situations where genes for proteins that some people are allergic to are involved, labeling may be used to let people know that a specific protein may be present. Labeling may also be required if the nutrient content of the food is altered.

Biotechnology is being used to develop resistance to insects in some plants. Because the EPA regulates the development and use of pesticides, it is involved in the regulation and safety assurances in the development of plants that are resistant to insects.

The three federal agencies involved in regulation and safety in the use of biotechnology cooperate with each other and with individual states as circumstances require. Typically, states become involved when field testing is being done with plants in their state or products are being moved in or out of the state. State departments of agriculture are the usual agencies involved at the state level. The level of state regulation of biotechnology varies from state to state.

Individuals who want to secure more information regarding the development and testing of products through biotechnology can get copies of permit applications and the results of testing from the USDA.

CLONING

When cells or organisms are genetically identical to each other, they are said to be clones. Some organisms reproduce asexually, which results in new organisms that are genetically identical to the parent. Most organisms that reproduce asexually are single-celled microorganisms such as bacteria. Asexual reproduction usually does not occur in higher, multi-celled organisms. However, cloning does occur naturally in some invertebrates (animals without a backbone). For example, if an earthworm is cut into two pieces, each piece will grow into a new, complete earthworm that is a clone of the original.

Experimental work done on frogs in the 1950s revealed that it was possible to clone vertebrate animals. The nucleus is removed from the egg cell (enucleation) and the nucleus of a body cell of an animal from the same species is placed in the egg cell. This results in a new animal that is genetically identical (a clone) to the animal that contributed the body cell nucleus. In these experiments, embryo body cells were used because the cells of the embryo are relatively unspecialized. As an embryo grows, its body cells specialize to form the various structures of the body. For many years it was generally believed that it was not possible to produce clones from body cells that had matured into specialized structures.

The first successful cloning of a vertebrate organism from mature body cells occurred in Scotland in 1996. Researchers at the Roslin Institute near Edinburgh, Scotland, cloned a sheep (called Dolly) from a mature mammary gland cell of an adult sheep. The technique involved placing the mature mammary gland cell in a solution that stopped its growth by restricting its access to nutrients. Using electricity, the mammary cell was then fused with an enucleated egg cell. After embryo growth began, the egg cell was transplanted into a surrogate ewe to develop into a fetus. More than 200 attempts were needed before this technique resulted in the live birth of a healthy lamb. In addition to the problem of making the technique work, there is an indication that the body cells of the clone are the age of the original donor sheep rather than the chronological age of the clone. (See Unit 9 for a discussion of cell division and the aging of body cells.)

Research is continuing in many laboratories around the world in an effort to develop practicable methods of cloning animals. In 1998, Japanese researchers successfully cloned twins from an adult cow using the same technique that was used in Scotland to clone a sheep. Also in 1998, researchers at the University of Massachusetts cloned six calves using fibroblast cells from a fetus that was 55 days old. A variety of techniques for producing cloned animals are being explored, some of which involve the use of cells that have not yet begun to specialize and others using adult cells.

Interest in cloning animals is high because of the possible benefits that may result. These include:

- testing disease treatments on clones; because the clones are genetically identical any differences in results would be due to the treatment.
- duplicating genetically modified animals to provide organs for human organ transplants.
- duplicating genetically modified animals for the production of pharmaceuticals.
- duplicating animals with desirable genes, such as those that result in faster growth, leaner meat production, or higher milk production.
- cloning animals of endangered species to increase their numbers and prevent the extinction of the species.

GENETIC ENGINEERING

Genetic engineering, also called recombinant DNA (rDNA) technology, is the process of identifying and transferring a gene or genes for a specific trait from one organism to another. *Deoxyribonucleic acid (DNA)* is a long strand of genes. Genes control traits of the organism, such as what it will look like and how its parts function. When doing genetic engineering, researchers must first identify the gene that controls the desired characteristic and locate the gene on the DNA strand. Using a restriction enzyme (one that recognizes a particular sequence of bases on the DNA strand), the DNA is cut at specific points to remove the desired gene. The cut piece of DNA carrying the desired gene is then spliced onto the DNA strand of the vector. The vector may be a virus or the plasmid of a bacterial cell. The vector is then placed in the organism where it will produce the desired action.

The use of genetic engineering can improve livestock performance in a number of ways (Figure 11-1). The potential benefits of genetic engineering in livestock production include:

- developing disease resistant animals.
- developing growth regulators.
- developing new drugs and vaccines.
- specifying the sex of an animal before conception.
- developing animals that use feed more efficiently, grow faster, produce more lean meat, produce more milk, and have greater resistance to disease.
- making the production of pharmaceuticals (those used for humans and those use for animals) easier and less expensive.

Genetic engineering is a complex technology. There are many thousands of genes in plants and animals. Often a characteristic is controlled by a combination of genes rather than by a single gene. Many years of research is sometimes necessary just to identify and locate the gene or genes that control certain characteristics.

In some cases, such as in corn plants, the cell into which the DNA bit is spliced cannot regenerate itself to produce a new corn plant. A process involving tissue culture must be used with corn plants to generate new plants from cells that are used in recombinant DNA research.

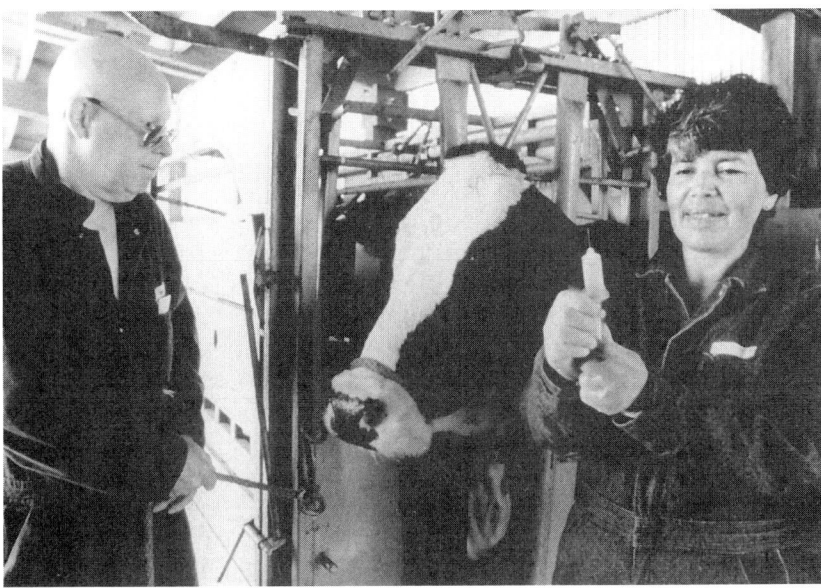

FIGURE 11-1 Cattle performance can be improved with biotechnology. *Courtesy of USDA/ARS.*

BOVINE SOMATOTROPIN (bST)

The ability of dairy cows to increase the production of a bovine growth hormone called bovine somatotropin (bST) has been developed through genetic engineering. Somatotropins are proteins that affect the utilization of energy in the body. Bovine somatotropin causes energy derived from feed to be used for milk production rather than for weight gain, but does not reduce the energy available for body maintenance. It also increases the amount of energy available by improving the breakdown of fat deposits and increasing appetite.

Small amounts of bST are produced naturally by the pituitary gland of the cow. The amount of bST produced by dairy cows has gradually increased over the years as a result of the selection of high-producing cows for breeding stock. However, the only source of bST for experimental work has been the pituitary glands of dead cows. Through the use of genetic engineering, large quantities of bST can be produced. The manufactured hormone is called recombinant bovine somatotropin (rbST). The gene that controls bST production is spliced into the DNA of a bacteria which is then injected into the cow. Because rbST is a protein, it is broken down in the digestive system if ingested orally. That is why treatment must be by injection.

The Food and Drug Administration approved the use of recombinant bovine somatotropin in November 1993. Approval came after almost ten years of testing that determined that milk and meat from cows treated with rbST are safe for human consumption. It became available for use in February 1994.

Because bovine somatotropin is a naturally occurring hormone, trace amounts of it can be found in all milk. Therefore, the Food and Drug Administration has ruled that milk from cows not treated with rbST cannot be labeled "bST free." Such a label would violate the law that requires labeling to be truthful. Stores and dairies are permitted to use labeling that indicates the milk comes from cows not treated with rbST; however, they must be able to document such a claim. The label must also carry a statement such as "No significant difference has been shown between milk derived from rbST-treated cows and non-rbST-treated cows." There is no requirement by the Food and Drug Administration for the mandatory labeling of milk from rbST-treated cows.

The manufacturer of rbST recommends that injections begin nine weeks after the cow has calved; repeat injections are administered every 14 days thereafter. Cows selected for injection should be in good body condition and gaining weight. Some extension dairy specialists recommend that the cow be pregnant before beginning treatment. It is recommended that use of rbST be discontinued about two weeks before the end of the lactation period.

Recombinant bovine somatotropin is sold in pre-filled, single-dose syringes that must be kept refrigerated. The pre-filled syringe is warmed to room temperature before use. The rbST is injected either in the tailhead depression or behind the cow's shoulder. The used syringe is disposed of by placing it in a special container provided by the manufacturer; when filled, the special container is mailed to an approved medical waste disposal facility.

It takes about two or three days for milk production to increase after beginning treatment with rbST. Feed intake should increase two to three weeks after milk production increases. Greater attention to feeding and overall management is necessary when using rbST. Success with rbST depends upon feeding a balanced ration with high-quality forage in sufficient quantity to meet the dry matter intake requirements of the cow. It is not recommended for use when only low-quality forages are available. High feed intake is necessary if the use of rbST is to be effective in increasing milk production.

Current data shows that production may increase by 5–15 pounds (2.2–6.8 kg) per day in well-managed herds. Given the current cost of the product plus the additional feed required, the break-even point is about seven pounds (3.2 kg) of milk per day.

The use of rbST does not overcome problems caused by poor health, poorly balanced rations, inadequate feeding, high somatic cell count, stress from the environment, or breeding problems. Its greatest potential for improvement of milk production is in herds that are already producing above average, have the genetic potential for high milk production, are in good health, and are well managed. There is no indication that the use of rbST causes any health problems or significant increases in mastitis in dairy cows. Because mastitis can be a problem in any high-producing cows, it is important to follow good management practices that reduce the chance for infection.

Consumer resistance to milk from rbST-treated cows may result from the fear of using animal products that contain any kind of additives. Most resistance to the use of rbST has come from consumer advocacy groups that generally oppose the use of all biotechnology in agriculture or the use of animals and animal products for human consumption.

Extensive study and evaluation regarding the safety of milk and meat from animals treated with bST have been done both in the United States and throughout the world. All of these studies have reached the conclusion that the use of bST poses no health problems for humans. A National Institute of Health study concluded that "composition and nutritional value of milk from rbST-treated cows is essentially the same as that of milk from untreated cows" and "meat and milk from rbST-treated cows are as safe as that from untreated cows." The American Medical Association reviewed the safety of bST use and concluded that the milk is safe. They also stated ". . . it is both inappropriate and wrong for special interest groups to play on the health and safety fears of the public to further their own ends." The World Health Organization and the Food and Agriculture Organization of the United Nations concluded that bST use presents no health concerns.

It is estimated that the use of rbST may increase overall milk production in the United States by about two percent. Some opponents of the use of rbST have expressed concern that its use will force small producers out of dairying and cause the loss of jobs in the dairy industry.

A recent analysis by the USDA suggests that the use of rbST may not have as dramatic an effect on the dairy industry as originally believed. The long-term trend in the dairy industry has been toward fewer dairy farms, larger herds, and higher production per cow. The USDA analysis suggests that the use of rbST will accelerate this process, but that it will not fundamentally alter the already existing trend.

Early reports on the adoption rate for the use of bST in dairy herds indicate that about 10 percent of dairy farmers in the United States use bST on at least part of their dairy herd. Early adoption is most widespread in California, where it is estimated that approximately 20 percent of dairy producers are using bST. However, most are not using it on all the cattle in their herds. It is estimated that about eight percent of the dairy cattle in California are actually being treated with bST.

Several characteristics of early adopters have been identified:

- they are generally younger and better educated than the average dairy producer.
- they have larger herds with higher herd production averages.
- they use more intensive management practices, such as using total mixed rations, and keep better records.
- they are often early adopters of other new agricultural technology and tolerate a higher level of risk in their enterprises.

Dairy farmers will need to carefully evaluate their individual dairy enterprises to prepare for the expected effects from the use of rbST. Producers who follow good management practices may find rbST a useful tool.

PORCINE SOMATOTROPIN (pST)

Porcine somatotropin (porcine growth hormone) is a protein produced naturally by the pituitary gland of the pig. Only small amounts are produced and it acts as a growth regulator. The production of somatotropin decreases as the animal becomes mature.

Experimental work done in the early 1970s showed that daily injections of natural pST had an effect on the efficient production of meat in swine. However, the use of natural pST in swine production was not practical because it was not possible to secure sufficient quantities of the growth hormone from natural sources.

The use of recombinant DNA technology has now made it possible to secure pST in larger quantities. Research carried on at several universities has shown that pST affects both growth and carcass characteristics in swine. When somatotropin is released into the bloodstream, it goes to the liver, where it causes the release of somatomedin-C, which is another protein. Somatomedin-C attaches to muscle tissue, causing an increase in cell division. This results in the production of more muscle tissue in the body. Somatotropin also changes the intermediary metabolism of the animal, resulting in a decrease in fat storage and an increase in the accumulation of protein. The combined effects of pST improve the feed efficiency and decrease the amount of fat deposition in the animal's body.

Experimental work to date has shown that the effects of pST are related to the level of dosage used. Too few data are available to draw general conclusions about the level of improvement that might result from the use of pST. However, improvement in feed efficiency of up to 29 percent has been reported with the use of pST. An increase in daily gain of up to 19 percent, an increase in loin eye area of up to 12 percent, and a decrease in back fat of up to 33 percent has also been reported as a result of the limited research that has been conducted.

Recent trials under farm conditions conducted by Iowa State University showed that hogs receiving supplemental porcine somatotropin averaged a 17 percent increase in feed efficiency. Rate of gain increased by six percent. The average backfat was reduced by 14 percent. The hogs in the study were fed to market weights of 240 (108 kg) and 290 pounds (131 kg). Those in the 240-pound group graded higher than the control group. Hogs fed to 290 pounds did not show any decrease in feed efficiency or increase in backfat.

Porcine somatotropin is broken down by enzymes in the small intestine of the pig before it reaches the bloodstream. Therefore, it cannot be given to pigs as a feed additive in the ration. Currently, pST is given to swine as an injection on a daily basis. This is not a practical delivery method for pST in commercial herds. Research is continuing to find an economical method of utilizing pST in swine production.

While pST has great potential for the production of leaner pork with improved feed efficiency, more research is needed before it becomes practical for commercial use. It will also be necessary to demonstrate its safety for human health. Because pST is a natural protein that is broken down in the digestive track of the pig, it is probable that its safety for human health can be easily demonstrated.

OTHER AREAS OF GENETIC ENGINEERING RESEARCH WITH LIVESTOCK

Some additional areas in which research is currently being conducted or considered that involve genetic engineering include:

- identify genetic markers for cattle and swine that make it easier to breed animals for greater resistance to disease and parasites, better-quality meat, improved feed conversion, and higher milk production.
- the gene that determines the horned condition in cattle has been identified. This is the first step in developing naturally hornless cattle, eliminating the need for dehorning. While some cattle have developed the hornless condition through selection, most cattle carry the gene for horns.
- development of a vaccine for Marek's disease in poultry.
- saving and storing valuable embryos and cloning valuable animals.
- reduce undesirable side effects of vaccines.
- develop bacteria for use with non-ruminants (such as swine) to allow them to better use roughages in the ration.
- resistance to leukemia in chickens.
- modifying rumen microorganisms to better utilize low-quality feeds.
- vaccines for controlling foot-and-mouth disease, bluetongue, and swine dysentery.
- manufacturing bovine interferon to help control shipping fever and other diseases in cattle.
- increasing the rate of twinning in cattle.
- resistance to internal parasites.
- accelerating growth rates in response to a specific feed additive.
- a feed additive that activates the uterine gene, improving embryo survival.
- ability to screen progeny to determine growth potential, level of disease resistance, reproductive ability, metabolic efficiency, and nutritional capabilities.
- producing the hormone relaxin for use as a therapeutic agent to help cattle during parturition and reduce calving losses.
- sexing embryos and semen.
- a scours vaccine utilizing cell fusion technology is being commercially produced to help control calf scours.

OPPOSITION TO BIOTECHNOLOGY

There is some opposition to the development and use of genetically engineered products from people who fear the possibility of the release of some new type of uncontrollable disease or other long-term adverse effect on the environment from the use of these products. They claim that there is not enough knowledge regarding these long-term effects. Their solution to these concerns is to attempt to stop the development of genetically altered products in agriculture. Environmental groups have tried to block the testing of genetically engineered agricultural products by securing court orders prohibiting such testing. While these groups have not succeeded in permanently stopping testing, they have caused delays in the development and testing of genetically engineered products.

PATENTS AND GENETIC ENGINEERING

The U.S. Patent Office has recently ruled "non-naturally occurring non-human, multicellular living organisms, including animals, to be patentable subject matter." This ruling means that new animals developed through genetic engineering can be patented. The effect of this ruling is that anyone using an animal that has been patented must pay a royalty to the patent holder. A question has been raised concerning who owns the genes of the offspring of these animals. This question will probably have to be answered through the courts. Some people believe that the ruling of the Patent Office is desirable because it protects the investment of researchers in the development and testing of these animals. The impact of this ruling on the future of the purebred livestock industry is not clear. Some purebred producers believe that there will still be a demand for seedstock that is genetically desirable for use with other genetically engineered products such as growth hormones. It is also possible, however, that the development of genetically engineered animals with superior desirable characteristics will reduce the need for purebred producers. Superior animals may be supplied to farmers from just a few companies, much as hybrid seed for crops is currently produced and distributed.

After the U.S. Patent Office began issuing patents for genetically modified plants, the amount of money spent for research and development in biotechnology increased dramatically. Thousands of patents have now been issued for genetically modified organisms. However, a legal challenge was made in 1999 to the issuance of patents for these organisms. The issue involves seed corn, but the outcome of the case may have a profound impact on the future use of biotechnology in both animal and plant agriculture. It may take several years for this issue to work its way through the courts. It is probable that the final decision will eventually be made in the U.S. Supreme Court.

SLOW PROGRESS IN BIOTECHNOLOGY

New developments in biotechnology have been slow in coming. Making genetic changes in animals, such as gene splicing, has proven to be more difficult than was originally believed. One of the problems faced by researchers in genetic engineering is that many of the performance characteristics that they want to change are controlled by a number of different genes rather than by a single gene. A lack of money for research, together with government regulations, has also had a negative impact on biotechnological developments. Environmental groups have filed lawsuits to try to stop research and testing in this field. Also, many farmers believe that agriculture is already producing more than the market can absorb at a reasonable price for farmers and therefore do not support genetic engineering that might result in greater production.

BIOTECHNOLOGY AND CROP PRODUCTION

The development of genetically engineered plants has proceeded at a much faster pace than has the development of genetically engineered animals. An advantage of using genetic engineering to modify plants is the reduction in time required to produce results. Using regular plant breeding methods requires about seven or eight years to produce a new plant that is only slightly improved. Using genetic engineering allows researchers to significantly alter the final result in about three years. It took only about five years from the time the first genetically engineered plants became commercially available to reach a level of approximately 50 percent of the corn, soybean, and cotton fields planted in the United States to be planted to genetically engineered (transgenic) plants. Two of the major developments that are commercially available are herbicide resistant plants and plants that are resistant to insect pests.

In the agribusiness area, a major impact of this technology is the merger of agricultural chemical companies and seed producing companies. Many millions of dollars are being invested in research and development of transgenic crops. Some developments will be commercially successful and others will not find a place in the market.

Almost anything is possible when developing transgenic plants. Some of the areas under study are:

- frost-resistant varieties.
- salt-resistant varieties.
- drought-resistant varieties.
- plants that produce nutritionally superior food.
- plants that are resistant to viruses, fungi, and other diseases.
- plants that produce higher yields.
- improvement of nitrogen-fixing ability in legumes and development of nitrogen-fixing ability in corn.
- development of corn with significantly higher oil content (5–8% compared to 3.5% for current dent varieties).

These are only a few of the areas in which research is being done in plant improvement. Farmers can expect to see the development of many transgenic plants in the coming years.

EMBRYO TRANSFER

Embryo transfer, especially in the cattle industry, has become a fairly well-established technology in recent years (Figure 11-2). In cattle, the process involves the flushing of embryos from the reproductive

FIGURE 11-2 Computer-enhanced embryo transfer.

track of desirable donor cows and implanting them in other cows that may be of lower quality. The desirable cows can produce many embryos through a process called superovulation.

By utilizing this technology, it is possible to produce many more offspring from desirable cows than would be possible if the cows had to carry each embryo to term before producing another one. The transplanted embryos carry the desirable genetic traits of the donor cow even though they are carried to term in lower-quality host cows.

Other advantages of embryo transfer include:

- proving the productivity of dams and sires more quickly because of the increased number of offspring in a shorter period of time.
- extending the productive life of a female that has been injured and can no longer carry offspring to term.
- ability to import and export desirable genetic traits with embryos when the animals cannot be imported or exported because of potential disease problems.
- faster genetic improvement in the herd.

Because of the current high cost of embryo transfer, its use is generally limited to breeding stock possessing highly desirable genetic traits. When the cost of the procedure is reduced, embryo transfer may become more common for the average cattle raiser or dairy farmer. The current success rate in embryo transfer is about 50 to 60 percent.

Superovulation refers to a process of inducing a cow to produce several oocytes during each estrus cycle. An *oocyte* is a cell that becomes an ovum that may be fertilized by a sperm cell to produce an embryo. Oocytes are contained in the ovaries of the female. Normally, a cow produces one oocyte per estrus cycle. Superovulation is induced by injecting the cow with a hormone that results in the release of several oocytes. Six to eight oocytes are normally produced by this procedure, although as many as 20 have been produced. Hormones that have been used to induce superovulation include gonadotropins, steroids, and prostaglandins. The hormone prostaglandin is currently the one most commonly used in embryo transfer programs.

It is important to make sure that donor cows have proper nutrition and are carefully monitored during their estrus cycles. Most cows will come into heat 21 to 60 days after calving. A hormone injection to induce superovulation is generally done after day 60 following calving. Cows that come into heat shortly after day 21 could be injected for superovulation sooner. However, cows that are used more often in a donor program may need more time after calving before the injection is made. Frequent superovulation may result in the development of cystic ovaries. This is not a serious problem if the condition is properly diagnosed and treated.

After ovulation occurs, sperm is introduced into the reproductive tract by artificial insemination. The fertilized ova develop into embryos in seven days. A saline solution is then used to flush the embryos from the reproductive tract. The embryos may be transferred immediately to another cow or they may be frozen for later use.

Embryos may be removed from the donor cow by surgery. However, this is not a common procedure in current embryo transfer programs. There is a higher level of risk for the donor cow if surgical procedures are used.

Progress has been made in making the process of embryo transfer in cattle simpler. A one-step straw system has been developed which surrounds the collected embryo with a protective agent. The embryo is then frozen in a segment of a plastic straw. The embryo can be thawed in the same straw. It is relatively easy for a technician with a minimum amount of training to make embryo transfers using this technique.

A process for surgically combining two cattle embryos to produce a single calf is being studied. The advantage of using this technique is that it reduces by two years the amount of time necessary to combine the desirable genetic characteristics of four sets of parents.

Another process under study is splitting an embryo to produce two identical calves from one embryo. This process has the potential of increasing the number of calves from cows with desirable characteristics that do not respond to superovulation. The technique is difficult and expensive. To date, it has been used primarily to produce twins for research projects.

Increasing the number of non-identical twins born to cows has been accomplished by implanting two embryos at the same time. In research projects, almost one-half of the cows so implanted have borne twins.

Research is also being conducted in determining the sex of embryos and sexing semen. These practices will permit the purchaser to specify the sex of the purchased embryo and will give the breeder greater control over the sex of the offspring resulting from artificial insemination.

In the United States, the importation and exportation of embryos is regulated by the USDA's Animal and Plant Health Inspection Service (APHIS). A major concern is the possibility of transferring diseases through embryos. Current research suggests that embryos that are properly handled do not transmit diseases. However, research in this area is continuing.

Some countries have very rigid testing requirements for the importation of embryos, while others are less rigid. APHIS is cooperating with the International Embryo Transfer Society to establish international standards for importing and exporting embryos. Breeders in the United States who are interested in exporting embryos should contact the nearest Area Veterinarian-in-Charge. Contact the local USDA office for the location of the Animal and Plant Health Services, Veterinary Services division. Information on current regulations governing embryo exports can be secured in this manner.

SUMMARY

The science of altering genetic and reproductive processes in animals and plants is called *agricultural biotechnology*. Most of the work currently being done in agricultural biotechnology is with genetic engineering and embryo transfer.

Genetic engineering is based on a technology involving recombinant DNA. This involves taking a tiny bit of DNA containing the desired gene from one organism and splicing it onto the DNA strand in another organism. The recipient organism takes on the characteristic controlled by the transferred gene.

Genetic engineering has been used to increase the level of bovine somatotropin in dairy cows, which results in higher milk production. It has also been used with porcine somatotropin to reduce fat and increase lean meat production in swine. Many other areas of research are currently being conducted with genetic engineering in animal and plant science.

Embryo transfer has become an established technology in cattle production. The use of embryo transfer permits the production of many more offspring from genetically desirable animals. Current costs tend to limit its use to highly valuable purebred breeding stock. As costs are lowered, however, the use of embryo transfer may become more common among average cattle and dairy producers.

Student Learning Activities

1. Prepare a bulletin board display of current information on the use of biotechnology in animal and/or plant science.
2. Ask a veterinarian who is familiar with or is using embryo transfer to speak to the class.
3. Interview a veterinarian who is familiar with or is using embryo transfer and prepare a written report or give an oral report to the class.

Review

1. Define agricultural biotechnology.
2. Name the two major areas of research currently being conducted in agricultural biotechnology.
3. Define recombinant DNA technology.
4. Why is genetic engineering often a difficult process?
5. What are somatotropins?
6. What is the potential for increased milk production from the use of bovine somatotropin?
7. Briefly explain some things that must be considered if recombinant bovine somatotropin is used with a dairy herd.
8. Why is there some controversy concerning the use of bovine somatotropin?
9. Describe the mechanism by which porcine somatotropin increases the amount of muscle in a pig's body.
10. Describe the effects of using porcine somatotropin in swine.
11. Why is the use of porcine somatotropin not currently practical for the average pork producer?
12. List five other areas of research currently being conducted in genetic engineering in animal science.
13. Why are some people opposed to the use of biotechnology?
14. What are some of the possible effects of the recent U.S. Patent Office ruling concerning the patentability of genetically engineered animal and plant products?
15. Why is progress in biotechnology slow?
16. List five areas of research currently being conducted in plant biotechnology.
17. List five advantages of embryo transfer.
18. What is the major current disadvantage of embryo transfer?
19. What is superovulation and how is it induced?
20. What is the most common method of removing fertilized embryos from the reproductive tract of the cow?
21. What development has made the process of embryo transfer simpler?
22. What other areas of research relating to embryos are being conducted?
23. What agency in the United States regulates the importation and exportation of embryos?

Unit 12 *Animal Breeding Systems*

Objectives

After studying this unit, the student should be able to

- name and explain common breeding systems used in livestock production.
- explain the effects, advantages, and disadvantages of using various breeding systems.
- identify the factors involved in selecting a breeding system.
- calculate the percent of parental stock in offspring, using various breeding systems.

SYSTEMS OF BREEDING

There are two basic systems of breeding in livestock production: (1) straightbreeding and (2) crossbreeding. Mating animals of the same breed is called *straightbreeding*. Mating animals of different breeds is called *crossbreeding*. Each system has a place in livestock production and is used for a particular purpose. Both have advantages and disadvantages. No one system of breeding is best.

The system of breeding to be used depends on the kind of livestock operation in which the animals are bred. Sometimes, farmers use more than one kind of breeding system. The size of the herd, amount of money available, and goals of the farmer are other factors considered when selecting a system of breeding.

Whenever two animals are mated, either straightbreeding or crossbreeding is used. There are several variations of each system. Straightbreeding includes purebred breeding, inbreeding, outcrossing, and grading up. Several systems of crossbreeding are used. These include two-breed crosses, three-breed crosses, and rotation breeding. These systems are discussed in this unit.

Purebred Breeding

A *purebred* animal is an animal of a particular breed. The animal has the characteristics of the breed to which it belongs. Both parents of a purebred animal must have been purebred. A purebred animal is eligible for registry in the purebred association of that breed if it has no disqualifications. Disqualifications are listed in the rules for registering animals in the breed association. A common disqualification is color markings that purebred breeders regard as undesirable. These are sometimes the result of recessive genes. To register purebred animals, one must be familiar with the rules for registering animals of that breed. Information may be obtained from the breed association.

The ancestors of a purebred animal can be traced all the way back to the original animals accepted for registry in the herd book of the breed association. There is a tendency for purebred animals to be genetically homozygous. Usually, only a small number of animals were originally accepted in the herd book of the breed association. This resulted in some inbreeding and linebreeding in the early history of the breed association. Inbreeding and linebreeding result in greater homozygosity of the genes in a given line of animals.

Purebred animals are not necessarily better than nonpurebred animals. In fact, undesirable recessive characteristics may appear because of the homozygosity of the genes of the parent animals. However, the average purebred animal is generally better than the average nonpurebred animal of the same breed.

The production of purebred animals is a specialized business. Purebred animals provide the foundation stock for crossbreeding to produce market animals. Purebred breeding requires more money for breeding animals than does raising market animals. A purebred breeder usually furnishes foundation stock for other purebred breeders and for those raising market animals. Purebred breeders often show their animals in purebred shows.

Inbreeding

Inbreeding is the mating of related animals. Linebreeding and closebreeding refer to how closely related the animals are that are being mated. The most intensive form of inbreeding is *closebreeding,* in which the animals being mated are very closely related and can be traced back to more than one common ancestor. Examples of closebreeding include sire to daughter, son to dam, or brother to sister (Figure 12-1).

Linebreeding refers to matings of animals that are more distantly related and can be traced back to one common ancestor. Examples are cousins, grandparent to grandoffspring, or half-brother to half-sister (Figure 12-2).

Inbreeding increases the genetic purity of the stock produced. The pairing of the same genes is increased, and the offspring become more genetically homozygous. The result of several generations of

(A represents the male; B the female)	
1st mating	A × B
1st generation	1/2A1/2B
2nd mating	A × 1/2A1/2B
2nd generation	3/4A1/4B

The offspring in the second generation have received 3/4 (75%) of their genetic inheritance from the sire A because he appears closer in the pedigree to the offspring than he does in linebreeding. They have received only 1/4 (25%) of their genetic inheritance from the female B.

FIGURE 12-1   Closebreeding (sire to daughter).

(A represents the male; B & C represent females)

1st matings:	A × B	A × C
1st generation:	1/2A1/2B	1/2A1/2C
2nd mating:	1/2A1/2B × 1/2A1/2C	
2nd generation:	1/2A1/4B1/4C	

The offspring in the second generation have received 1/2 (50%) their genetic inheritance from the sire A because he appears twice in their pedigree. They have received only 1/4 (25%) of their genetic inheritance from each of the females B and C.

FIGURE 12-2 Linebreeding (half-brother to half-sister).

inbreeding is a high degree of genetic purity or homozygosity. Undesirable genes and desirable genes become grouped together in the offspring with greater frequency. This makes the undesirable traits more visible. The breeder can then eliminate animals with these traits from the breeding program. Desirable traits also become more visible. A good program of selection and culling will result in breeding stock with more desirable traits.

Animals with desirable traits that are used for outcrossing usually give better results. It is possible to keep the good traits of an animal in the ancestry of the animals being produced. Inbred animals often transmit desirable genes to their offspring with greater uniformity. The production of inbred lines helps to improve the breed.

Inbreeding requires a carefully planned program of selection and culling. It is expensive because all animals with undesirable traits must be removed from the breeding program. The average animal breeder generally does not find inbreeding a desirable system of breeding to use. It is used more often by universities for experimental work and seedstock breeders that provide animals for crossbreeding in herds producing animals for market.

Outcrossing

Outcrossing is the mating of animals of different families within the same breed. The animals bred are not closely related. The purpose of outcrossing is to bring into the breeding program traits that are desirable but not present in the original animals. Most matings done by purebred breeders are outcrossing. This system is popular because it reduces the chances of undesirable traits appearing in the offspring. The genes for those undesirable traits are still present. However, they are covered up by the outcrossing. Outcrossing is sometimes used in combination with inbreeding programs to bring in traits that are needed.

Linecrossing is mating animals from two different lines of breeding within a breed. The purpose is to bring together desirable traits from different lines of breeding. Some lines cross better than other lines because of different gene combinations. Experience is the best guide in determining the lines to use when linecrossing.

Grading Up

Grading up is the mating of purebred sires to grade females. Most of the animals on farms in the United States are not purebreds. The mating of purebred sires with these grade animals is a good way to improve the quality of animals on the farm. A *grade* animal is any animal not eligible for registry. It does not require as much money since only the purebred sires (or their semen) must be purchased. How quickly the animals are improved depends on the species of animal. Animals with short generations, such as swine, are improved fairly rapidly. Those with longer generations, such as cattle or horses, take longer to produce improvement.

The amount of improvement that results is dependent on the quality of sire selected for the breeding program. Most commercial producers get their purebred sires from purebred breeders. It is important to select the highest quality sire with performance records that the commercial breeder can afford. Offspring of grading up are generally not eligible for registry in the breed association because only one parent is registered. However, some breed associations do permit the offspring of grading up to be registered. Also, new bloodlines were introduced from other breeds in some associations. A few associations are now requiring blood testing as a part of the registration process. A person interested in registering animals should contact the appropriate breed association to determine the current rules for registration.

The greatest percent of improvement comes in the first cross, since fifty percent of the genes of the offspring will be from the purebred sire. Second-generation offspring will be 75 percent purebred. The third generation will be 87.5 percent pure. If the use of a purebred sire continues long enough, the amount of grade breeding left in the offspring will be less than 1 percent (Figure 12-3).

Crossbreeding

Crossbreeding is the mating of two animals from different breeds. The resulting offspring is a hybrid. Crossbreeding usually results in improved traits in the offspring. Dominant genes tend to mask undesirable recessive genes.

Superior traits that result from crossbreeding are called hybrid vigor, or heterosis. Heterosis is measured by the average superiority of the hybrid offspring over the average of the parents. The kind and degree of superiority achieved by crossbreeding varies with different species.

Traits with a high degree of heritability show little improvement from crossbreeding. Those traits with low heritability usually show the greatest improvement as a result of crossbreeding.

(A_1, A_2, A_3, represent purebred sires of a given breed; G represents a grade female)

1st mating:	$A_1 \times G$
1st generation:	$1/2A_1 1/2G$ (50% purebred, 50% grade)
2nd mating:	$A_2 \times 1/2A_1 1/2G$
2nd generation:	$1/2A_2 1/4A_1 1/4G$ (75% purebred, 25% grade)
3rd mating:	$A_3 \times 1/2A_2 1/4A_1 1/4G$
3rd generation:	$1/2A_3 1/4A_2 1/8A_1 1/8G$ (87.5% purebred, 12.5% grade)

FIGURE 12-3   Grading up (purebred sires on grade female).

Animals selected for use in a crossbreeding program must have the desired traits. There will be little or no improvement in the offspring over the parents if animals with undesirable traits are used in a crossbreeding program. Regardless of the crossbreeding system used, the producer must follow a good performance selection program, good management, good nutrition, and good herd health practices to achieve the desired results. Research has shown that well planned crossbreeding programs can increase total productivity in beef herds by 20 to 25 percent.

Beef, swine, and sheep producers usually use crossbreeding for the production of market animals. It is rarely used by dairy producers because they are primarily interested in milk production and the Holstein breed, which is superior to other breeds in this trait, dominates the dairy industry. Poultry producers typically use strains that have been developed from crossing inbred lines.

Crossbreeding systems for beef. The use of crossbreeding in beef cow herds that produce animals for slaughter will generally result in higher profits. Some general considerations regarding a beef crossbreeding program include:

- good recordkeeping is essential.
- calving difficulties may increase when crossing large breed sires with small breed dams.
- there are fewer calving problems if large breed dams are used.
- large breed dams have higher maintenance costs.
- artificial insemination allows access to better bulls.
- to avoid inbreeding, more than one breeding pasture may be required.

Some experimental results of crossbreeding with beef cattle are shown in Table 12-1.

TABLE 12-1 Crossbreeding Effects in Beef Cattle.

Trait	Percent advantage over noncrossbred cattle	
	(2-breed cross)	*(3-breed cross)*
Fertility and percent calf crop	3	7
Weaning weight	5	10
Pounds of calf weaned per cow	6	15
Yearling weight	6	12
Feedlot growth rate	4	5
Carcass traits	0	0
Feed efficiency	0	2

Source: *The F$_1$ Beef Cow and Crossbreeding,* A2556, University of Wisconsin, 1973.

Crossbreeding systems used with beef cattle range from those that are relatively simple to those that are complex. Some typical beef crossbreeding systems include:

Terminal sire crossed with F_1 females. Replacement F_1 (crossbred) females in the herd are purchased and crossed with a terminal bull. All the offspring are marketed.

Rotate herd bull every three or four years. The same breed of bull is used for several years then replaced with a bull of a different breed. Replacement females are selected from the herd.

Two-breed rotation. Bulls from breed A are crossed with cows from breed B. The resulting heifers are bred to bulls from breed B for the duration of their productive life. Replacement heifers chosen from these matings are bred to bulls from breed A. Each succeeding generation of replacement heifers is bred to a bull from the opposite breed used to sire the replacement heifer.

Three-breed rotation. The pattern of breeding is the same as in a two-breed system, except that a bull from a third breed is used in the rotation of sires.

Four- and five-breed rotations. In larger herds, bulls from a fourth or fifth breed may be used in the rotation of sires. This system requires a higher level of management than two- and three-breed systems.

Static terminal sire system. Four breeding groups are needed for this system, as shown in Figure 12-4. The first group (25 percent of the herd) mates bulls (breed A) and cows (breed A) to produce replacement heifers (AA) for group one and group two. The second group (25 percent of the herd) breeds the AA heifers to a bull (breed B) of a different breed, producing crossbred heifers (breed AB). The third group (50 percent of the herd) breeds the AB heifers to a terminal (T) bull

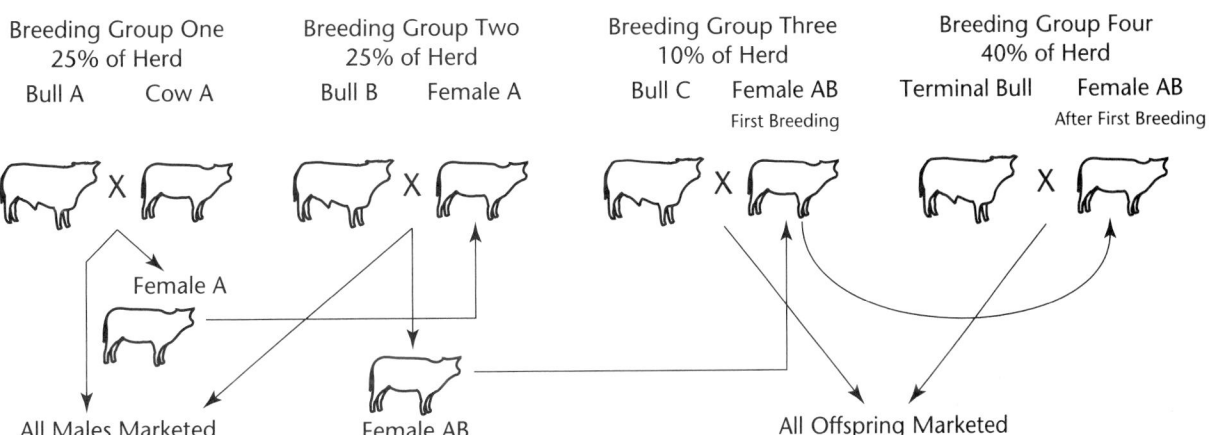

FIGURE 12-4   Static terminal sire system.

selected for ability to transmit a high rate of gain. A sub-group (group four, 10 percent of the herd) of the third group is composed of AB heifers being bred for the first time. These AB heifers are bred to a smaller breed (breed C) bull to reduce first-time calving problems. All the male offspring of groups one and two and all of the offspring of groups three and four are marketed. Any heifers from groups one and two that are not kept for breeding purposes are also marketed.

Rotational-terminal sire system. Two breeding groups are needed for this system. Bulls from breeds A and B are used on a rotating basis on 50 percent of the herd, providing crossbred females for the entire herd. Mature cows in the herd are mated with a terminal bull to produce offspring, all of which are marketed. Replacement females generally come from the matings of bulls A and B with younger cows in the herd.

Composite breeds. The development of a new breed based on crossbreeding with four or more existing breeds of cattle to avoid inbreeding problems. After development, the composite breed is not crossbred with other breeds.

Crossbreeding systems for swine.   Some general considerations related to a swine crossbreeding program include:

- select breeds to use and replacement gilts and boars that meet the objectives of the breeding program.
- select breeds that produce large litters and heavier weaning weights if they are to be used only on the female side of the matings. The white breeds are generally superior in these traits.
- select breeds that have less backfat and higher rate of gain if they are used as terminal boars in the crossbreeding program. The Berkshire, Duroc, Hampshire, Poland China, and Spotted breeds are generally superior in these traits.
- if crossbred sows are used, select those that are at least 50 percent Chester White, Landrace, or Yorkshire. These crosses generally have the superior maternal traits that are desirable.
- select boars that are from sows that rank in the top 25 percent of a herd, as measured by a Sow Productivity Index. Growth rate and backfat thickness are important considerations when selecting boars.
- crossbred boars may be used in a crossbreeding program. Be sure their parents have desirable traits that fit the objectives of the program.

Some experimental results of crossbreeding systems with swine are shown in Table 12-2.

Some typical swine crossbreeding systems include:

Rotational crossbreeding. In *two-breed systems,* a boar from breed A is mated with sows from breed B, producing offspring AB. Selected gilts

TABLE 12-2 Heterosis Advantage for Production Traits in Swine.

| | *Percent advantage over purebred* | | |
Trait	*First cross purebred sow*	*Multiple cross crossbred sow*	*Crossbred boar*
Reproduction			
Conception rate	0.0	8.0	10.0
Pigs born alive	0.5	8.0	0.0
Litter size 21 days	9.0	23.0	0.0
Litter size weaned	10.0	24.0	0.0
Production			
21-day litter weight	10.0	27.0	0.0
Days to 220 lb.	7.5	7.0	0.0
Feed/gain	2.0	1.0	0.0
Carcass composition			
Length	0.3	0.5	0.0
Backfat thickness	−2.0	−2.0	0.0
Loin muscle area	1.0	2.0	0.0
Marbling score	0.3	1.0	0.0

Source: Ahlschwede, W.T., Christians, C.J., Johnson, R.K., and Robison, O.W., *Pork Industry Handbook,* "Crossbreeding Systems for Commercial Pork Production," University of Illinois.

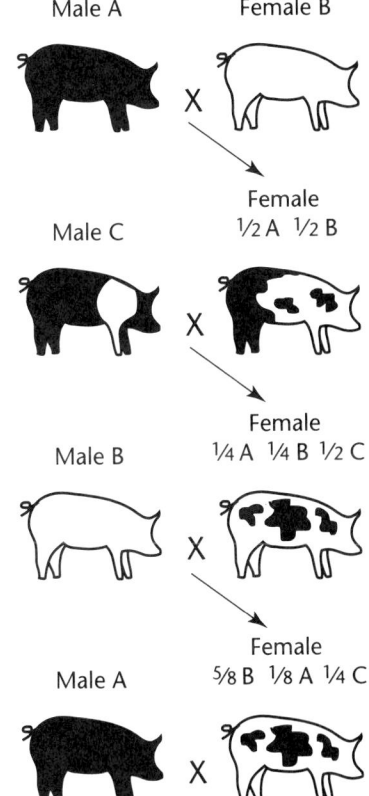

Male A   Female B

Female
½ A ½ B

Male C

Female
¼ A ¼ B ½ C

Male B

Female
⅝ B ⅛ A ¼ C

Male A

A, B, and C represent different breeds. Continue using males of three breeds in sequence. Using males of the three breeds in sequence is called rotating the males; hence, the inclusion of the term *rotation* in the name of this system.

FIGURE 12-5 Three-breed rotation cross.

(AB) are bred to a boar from breed B. Selected gilts from this mating are bred to a boar from breed A. The pattern is repeated, switching back and forth to the breed of the most distantly related boar. *Three-* or *four-breed systems* are more commonly used in swine production. The pattern is the same as that used in the two-breed system except that three or four breeds of boars are used in rotation, as shown in Figure 12-5.

Care must be taken to follow the planned order of breeds used in the rotation, or heterosis will be reduced. Because replacement gilts are selected from within the herd, the chance of bringing disease into the enterprise from purchased breeding stock is greatly reduced.

Terminal crossing system. Crossbred (F_1) females, with superior maternal traits, are bred to boars selected for desirable backfat and rate of gain. All of the offspring go to market. The breeder must either keep a separate herd to produce breeding stock or purchase replacement females. The costs involved in this breeding system are generally higher than in rotational breeding systems. There is some increased health risk if new breeding stock is brought into the herd. Terminal crossing does maintain the maximum advantage of heterosis and breed differences in the breeding system.

Rotaterminal system. The rotational breeding system and the terminal breeding system are combined in this method of crossbreeding. Crossbred females are produced by breeding boars of different breeds in a rotating pattern to crossbred females produced by previous matings in the system. Generally breeds with good maternal traits are used to produce the crossbred females that are bred to terminal boars of other breeds. Terminal boars are selected for desirable backfat and rate of gain. All the offspring produced in the terminal breeding go to market. This system of crossbreeding maintains a high level of heterosis and allows the producer to select breeds with desirable traits. It does require the use of more boars of different breeds.

Crossbreeding systems for sheep. The use of crossbreeding generally increases profits from sheep flocks. Crossbred ewes are hardier, healthier, and produce more milk as compared to non-crossbred ewes. A twelve-year study of crossbred sheep conducted at the Agricultural Experiment Station, University of Idaho, revealed that they produced higher grease (uncleaned), as well as cleaned, fleece weights compared to non-crossbred sheep.

Some typical sheep crossbreeding systems include:

Rotational. The same breeding pattern is used as in beef or swine rotational systems. The lambs are usually kept for flock replacements.

Static. Replacement crossbred ewes for the flock are purchased and bred to a terminal ram. All the lambs are marketed.

Roto-static. A combination of rotational and static crossbreeding systems in which replacement ewes are produced from the flock. These ewes are bred to a terminal ram to produce market lambs. It takes about 25–30 percent of the flock to produce the replacement ewes. The best ewes should be kept for producing replacement ewes.

SUMMARY

The two basic systems of livestock breeding are straightbreeding and crossbreeding. The kind of system used depends on the size of the operation, the amount of money available, and the goals of the farmer.

Purebred animals are eligible for registry in the breed association. They tend to be genetically homozygous. Not all purebred animals are better than nonpurebreds. On the average, however, purebred animals are better than nonpurebreds. Purebreeding requires more money and skill than commercial animal production.

Inbreeding increases the genetic purity of the livestock but generally reduces performance. Both desirable and undesirable traits become more visible. Therefore, inbreeding programs require careful selection and culling of breeding stock. Inbreeding is not commonly used by the average livestock producer. It is usually used by those who do experimental work to improve the breed.

Outcrossing brings genetic traits into the breeding program that tend to hide undesirable traits. Grading up is using purebred sires on grade females. Grading up is a good way for commercial producers to improve grade herds of livestock.

Crossbreeding is the mating of animals from two different breeds. Crossbreeding is used by many commercial producers. It usually results in hybrid vigor. This improves some traits, but has little effect on feed efficiency or carcass traits.

Student Learning Activities

1. Give an oral report on one type of breeding system and explain how a farmer might use it and what results might be expected.
2. Prepare a bulletin board display illustrating matings in a given species and the percent of parental blood in the offspring.
3. Report on breeding systems used on your home farms and why they are used.
4. Interview local farmers and report on the breeding systems they use and the reasons for choosing these systems.

Review

1. Define straightbreeding and crossbreeding.
2. What is purebred breeding?
3. Define and give two examples of inbreeding.
4. Why is inbreeding more commonly used by universities and seed stock producers than by the average livestock producer?
5. Define outcrossing and tell why it might be used in a breeding program.
6. What is linecrossing and why might it be used?
7. What is grading up and why might it be used?
8. Why is crossbreeding used in breeding programs?
9. Name and briefly describe four systems of crossbreeding that might be used in breeding programs.

Beef Cattle

Unit 13 *Breeds of Beef Cattle*

Objectives

After studying this unit, the student should be able to

- describe the characteristics of the beef industry.
- name and describe the various breeds of beef cattle, giving their origin and breed characteristics.
- identify the various breeds of beef cattle by viewing pictures or live animals.

CHARACTERISTICS OF THE BEEF INDUSTRY

About 38 percent of the total income from all livestock and poultry marketing in the United States comes from the beef industry. Marketing of beef accounts for about 18 percent of the total income from all farm marketing (livestock, poultry, and crops) in the United States. The beef industry is the largest single segment of the U.S. agricultural economy. Raising and feeding beef cattle is an important source of income for many farmers.

The annual per capita (per person) consumption of beef in the United States generally increased from 1960 to 1976. Per capita consumption of beef has generally decreased since 1976. Consumption is linked to the supply of beef coming to market, the price of beef, and consumer perception that beef has a high cholesterol level. The concern about cholesterol as a health hazard has caused many people to increase their consumption of poultry and decrease their consumption of beef. Despite this, the total consumption of beef on a boneless, edible weight basis is still higher than the consumption of poultry.

The demand for beef depends on (1) the number of people, (2) income per person, and (3) changes in people's meat preferences. When incomes are high, people tend to buy more beef instead of other kinds of meat. They also want more grain-fed beef instead of vealers (calves grown for veal) and grass-fed cattle. Despite the recent drop in per capita consumption, the demand for beef is expected to continue strong in the future.

Most of the beef eaten in the United States comes from domestic production. About nine percent of the beef supply in the United States comes from imports. About four percent of domestic beef production is exported to the world market.

Small-size herds are typical for beef cow-calf operations, as shown in Figure 13-1. About eighty percent of the cow-calf herds have fewer than 50 head of cows. These smaller herds account for less than one-third of all the beef cows on farms in the United States. Cow-calf herds ranging in size from 50 to 99 head account for a little over 11 percent of the total operations and have about 19 percent of all the beef cows on farms. About one-half of the total number of beef cows on farms are in herds that are over 100 head in size. These herds make up about nine percent of the total number of beef cow-calf herds in the United States.

There is a continuing trend toward fewer small (under 1,000-head capacity) cattle feedlots; however, there has not been any significant increase in the number of large-capacity feedlots. The total number of cattle feedlots in the United States has been declining in recent years. About 95 percent of the total feedlots are small; however, these feedlots feed only about 15 percent of the total cattle coming to market. About one-third of all cattle that are marketed come from a small number of feedlots with over 32,000-head capacities.

The cattle-feeding industry has shifted in recent years from its formerly heavy concentration in the Cornbelt to the Central and Southern Plains states. Most of the high-capacity feedlots are found in Texas, Kansas, Nebraska, and Colorado. There is a trend toward custom feeding cattle in large feedlots, with cow-calf operators retaining ownership of the cattle.

Other changes occurring in the beef industry include:

- efforts to lower unit production cost.
- development of closer coordination of breeding, growing, feeding, marketing, and processing operations, generally through contracts rather than through the vertical integration that is found in the poultry industry.
- greater quality control in production in an effort to reduce the amount of fat in beef and produce carcasses with the characteristics in demand in the marketplace.
- a growing interest in the concept of Integrated Resource Management to increase net income by making the most efficient use of all available resources. Key performance indicators such as bull and cow fertility, calf survival rate, calf growth rate, prices, and production costs are carefully analyzed. A team of specialists including extension personnel, university researchers, animal health company representatives, veterinarians, accountants, and bankers is used to help solve problems that occur in the operation.

The United States is divided into eight cattle-raising regions. Divisions are based on the similarity of conditions in each of the regions.

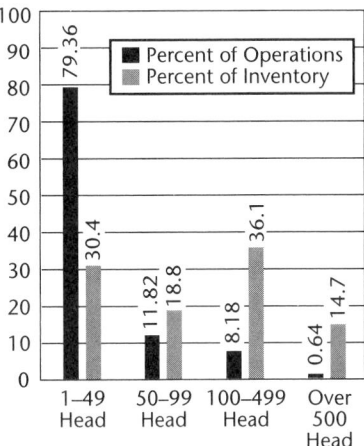

FIGURE 13-1   Beef cow herds: Percent of operations and inventory by size of operations, United States. *Source: Cattle,* USDA, January 1999.

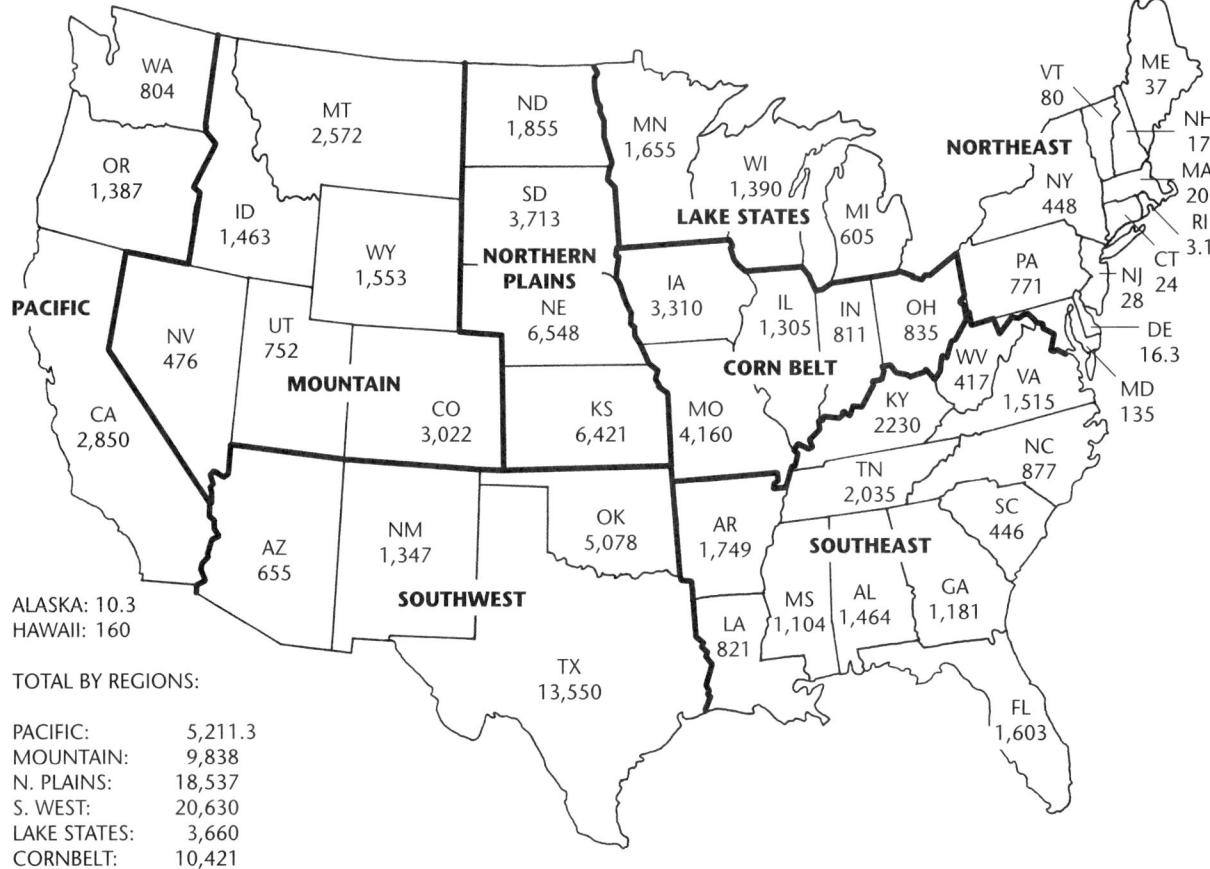

ALASKA: 10.3
HAWAII: 160

TOTAL BY REGIONS:

PACIFIC: 5,211.3
MOUNTAIN: 9,838
N. PLAINS: 18,537
S. WEST: 20,630
LAKE STATES: 3,660
CORNBELT: 10,421
S. EAST: 15,442
N. EAST: 1,579.4
U.S.TOTAL: 85,318.7

Sources: Cattle Raising in the United States, Economic Research Service, USDA Agricultural Economics Report No. 235, January 1973. *Cattle,* USDA, January 1994.

FIGURE 13-2   Beef cattle-raising regions and numbers of beef cattle on farms by state, January 1, 1999 (all classes) (1,000 head).

Figure 13-2 shows the cattle-raising regions and the numbers of cattle on farms in these regions as of January 1, 1999.

Raising beef has the following advantages:

■ Beef use roughages that otherwise would be wasted for feed.
■ Labor requirements may be low.
■ Capital investment can be small.
■ Death losses are usually low.
■ Beef are adapted for use in small operations as well as large ones.
■ There is high demand for meat.

Disadvantages of raising beef include the following factors:

■ Cattle feeding is a high-risk business.
■ Cattle are not efficient converters of concentrated feeds into meat.
■ It takes longer to develop a cattle herd and increase numbers than it does to develop hogs or sheep.
■ The capital investment in modern, efficient feeding operations can be high.

DEVELOPMENT OF BEEF BREEDS

Many modern beef breeds had their origin in Europe. Selection of cattle that later formed a breed was practiced mainly in the British Isles. Most of this selection began in the latter part of the 1700s.

Selection of animals was based on traits that local farmers considered best for their area. The most desirable animals were kept as breeding stock. The breeder culled (removed) from the breeding herd those animals that did not have the desired traits. This increased the gene frequency of the desired traits in the breeding herd. Outside breeding stock was rarely brought into the herds. The number of animals that had the same kind of traits increased. As these animals became more popular, breed societies were formed. A method of registering animals in the breed was established. Eventually, all animals registered in the breed traced their ancestry to the animals originally registered with the breed society.

Newer breeds of beef have been developed during the twentieth century in the United States. Some of these have been developed from crosses of existing breeds or strains of cattle. For example, the Brahman breed is the result of crossing several Zebu breeds from India. Other hybrid breeds have been the result of selecting crossbred animals that possessed desirable traits. These animals were then bred and the offspring selected for the desired traits.

During the late 1960s and 1970s, a number of so-called exotic breeds of cattle were introduced into the United States from Europe. Among these were the Simmental, Limousin, Blonde d'Acquitaine, Chianina, and Maine-Anjou. Other exotic breeds such as the Charolais, although introduced earlier into the United States, gained increased popularity during the 1970s.

There are more than 50 breeds of beef cattle available to producers in the United States. Many of these breeds are registered in small numbers with their respective purebred associations in any given year. In rank order, Angus, Hereford, Polled Hereford, Simmental, Charolais, and Limousin have been the leading beef breeds in the United States based on total numbers registered since 1970. During the 1990s the leading breeds in rank order of registrations have been Angus, Limousin, Simmental, Hereford, Polled Hereford, and Charolais.

SELECTION OF A BREED

Some points that should be considered when selecting a breed are:

- All breeds have both strong and weak traits.
- No one breed is best for all traits.
- Every breed has a wide range of genetic variation.
- The selection of the best animals as breeding stock and the use of good breeding practices are more important than the particular breed selected.

■ The breed selected should be one that seems to produce well in the area where it will be raised.
■ The market demand for a given breed should be determined.
■ Foundation breeding stock should be available at a reasonable cost.

CHARACTERISTICS OF THE BREEDS

Angus

History. The official name of the Angus breed is Aberdeen-Angus. The breed originated in Scotland in the shires of Aberdeen and Angus. The earliest written records of Angus date to the early 1700s. Around 1800, farmers in Scotland began keeping records of pure breeding. In 1862 the first herdbook of the Angus breed was published.

George Grant, of Victoria, Kansas, imported four bulls into the United States in 1873. These were the first recorded importations of Angus into the United States. The American Angus Association was organized in 1883.

During the late 1800s and early 1900s, Angus were found mostly in the midwestern states. The breed has increased in popularity in the range areas. Today, Angus are found in every state in the United States. For many years, Angus have led all other beef breeds in numbers registered.

Description. Angus cattle are black in color (Figures 13-3 A and B). They have a smooth hair coat and are polled. They are an alert and vigorous breed. Angus cattle perform well in the feedlot. They produce a desirable carcass of high-quality, well-marbled meat.

Nearly all Angus are pure for the dominant polled gene. When used in crossbreeding, nearly all the calves are polled. A few Angus carry a recessive gene for the red color. Sometimes, a red calf is born to black parents. The red calf is not eligible for registry in the American Angus Association.

Red Angus

History. Red Angus herds were begun in the United States about 1945. The herds developed from crossing red with red from the Black Angus. Since the red gene is recessive, the offspring of the red crosses are always red.

Red Angus (Figure 13-4) can be registered in the Angus herdbook in Scotland. The American Angus Association registered Red Angus until 1917. The Red Angus Association of America was formed in 1954.

Description. Red Angus are similar to Black Angus except for their color. Since red absorbs less heat than black, the Red Angus can tolerate warmer temperatures somewhat better than Black Angus. In the future, as selection standards change, the Red Angus breed may become less like the parent Black Angus.

FIGURE 13-3A Modern Angus bull. *Courtesy of American Angus Association, MO.*

FIGURE 13-3B Modern Angus cow. *Courtesy of American Angus Association, MO.*

FIGURE 13-4 Red Angus bull. *Courtesy of American Red Angus Magazine, TX.*

Charolais

History. The Charolais (Figures 13-5 A and B) is one of the oldest of the French breeds of beef cattle. It was developed around Charolles in central France. In 1930, two bulls and ten heifers were imported into Mexico. Two more importations were made in 1931 and 1937. The total number of animals imported was 37—eight bulls and 29 cows. In 1936, the King Ranch in Texas imported two bulls from Mexico. These were the first Charolais to be imported into the United States. The breed increased in numbers through breeding Charolais bulls to females of other breeds. New breeding stock was imported into Canada and the United States in 1966. Since then, there have been more importations from Canada, the Bahama Islands, England, Ireland, Japan, and France.

FIGURE 13-5A   Charolais bull. *Courtesy of American-International Charolais Association, MO.*

Description. Charolais cattle are white to light straw color with pink skin. They are a large, heavily muscled breed. Mature bulls weigh 2,000 to 2,500 pounds (907–1,134 kg). Mature cows weigh 1,500 to 1,800 pounds (680–816 kg). Most are naturally horned. Horns are white, slender, and tapered. Naturally polled Charolais may also be registered. Charolais have a high feed efficiency. They are heavily muscled in the round and loin because of generations of selection for this trait.

The Charolais is well adapted to many areas. It is used in many crossbreeding programs.

Charolais are registered with the American-International Charolais Association. The association has an open herdbook. Animals may be registered after five generations of crossing with a Charolais bull.

FIGURE 13-5B   Charolais cows with calves. *Courtesy of American-International Charolais Association, MO.*

Charbray. Charbray cattle are also registered with the Charolais Association. Purebred Charbray are defined as having a minimum of 5/8 to a maximum of 7/8 Charolais breeding. The rest of their inheritance is from registered Brahman or Zebu breeds. Up to 1/32 breeding may be from other breeds. All animals must have originated from registered Charolais and registered Brahman cows or bulls. Charbray-cross bulls are recorded if they are 1/2 to 5/8 Charolais or above 7/8 Charolais. Females are recorded with a minimum of 1/4 Charolais breeding. Charbray-cross females are recorded from mating of registered or recorded Charolais, registered Charbray, or registered Brahman.

Chianina

History. The Chianina (pronounced *Key-a-nee-na*) breed (Figure 13-6) originated in the Chiana Valley in Italy. It is one of the oldest breeds of cattle in Italy and probably one of the oldest in the world. The breed was in existence before the time of the Roman Empire. Chianina semen was first imported into the United States in 1971.

Chianina are used in many crossbreeding programs. The breed association is the American Chianina Association. The association

FIGURE 13-6   Chianina bull. *Courtesy of American Chianina Association, MO.*

does not establish weight or color standards for its members. There are large numbers of quarter-blood and half-blood Chianina in the United States. The color, type, and size vary considerably because of the different kinds of crosses used.

Description. The original Chianina cattle were white with a black switch. The skin pigment is black. They have a high heat tolerance and gentle disposition. Chianina are probably the largest breed of cattle. Mature bulls can grow to 6 feet (1.8 m) at the withers. They can weigh as much as 4,000 pounds (1,814 kg). Mature cows can grow to 5 feet (1.5 m) at the withers. They can weigh as much as 2,400 pounds (1,088 kg).

Chianina are popular in crosses for a number of reasons. They improve the growth rate of the offspring. In addition, they are good foragers and good mothers. They are well adapted to hot and cold climates, rough terrain, and have a high degree of tolerance to insects and diseases. Chianina have fine-textured meat.

Devon

History. Devon cattle originated in southwestern England in the counties of Devon and Somerset. They are sometimes called North Devon to distinguish them from the closely related South Devon breed described later in this unit. It is an old breed of cattle. Some authorities believe they descended from *Bos longifrons*, a small type of aboriginal cattle in Britain.

Devon cattle were brought to the United States by colonists as early as 1623. They were used for milk, beef, ox teams, and leather. The American Devon Cattle Club was established in 1884. Devons are now registered with the Devon Cattle Association, Incorporated. The Devon herdbook was established in 1850.

Description. The color of the Devon is rich red (sometimes called ruby red). They have yellow skin and creamy white, black-tipped horns. Polled Devons trace back to a mutation that occurred in 1915.

Most modern-day Devons are of the beef type. The body is long and moderately deep. The loin and hindquarters carry thick, natural fleshing. The breed is hardy and adaptable to many climates. It has a high heat tolerance.

Devon bulls may be crossed on any breed and the female offspring registered in the Devon Qualified Registry. This permits the breeder to upgrade the herd to registered purebred Devons in four generations. Performance testing and weight-for-age data on the offspring is encouraged by the breed association. The use of artificial insemination is also encouraged to promote superior bulls.

Galloway

History. The Galloway breed (Figures 13-7 A and B) originated in southeast Scotland. The first Galloway cattle were imported into the

FIGURE 13-7A   Galloway bull. *Courtesy of Galloway Cattle Society of America, IL.*

FIGURE 13-7B   Dun Galloway cow. *Courtesy of Galloway Cattle Society of America, IL.*

United States in 1870. Purebreds are registered with the American Galloway Breeders Association.

Description.   Most Galloway cattle are black with soft, wavy hair and a thick undercoat. Other colors include belted, red, dun, and white. They are naturally polled. The Galloway is one of the smallest of the beef breeds.

The breed is noted for its hardiness, carcass quality, and foraging ability. The calves are capable of withstanding more severe weather conditions than many other breeds.

Gelbvieh

History.   The Gelbvieh breed was developed from four yellow breeds of cattle: Glan-Donnersburg, Yellow Franconian, Limburg, and Lahn. These breeds were developed around 1850. They were brought together in the Gelbvieh breed in 1920.

Description.   Gelbvieh cattle are single-colored and vary from cream to reddish yellow. They are of medium weight and size, have good milking ability, and produce a very acceptable carcass.

Hereford

History.   Herefords (Figures 13-8 A and B) originated in the county of Hereford in England. The early breeders selected for a high yield of beef and economical production. Native cattle were bred with white cattle from Flanders. A red bull with a white face was brought into the breeding from Yorkshire in 1750.

The first Herefords to come to the United States were imported by Henry Clay of Kentucky. These cattle were mixed with native cattle of the area. The first purebred breeding herd in the United States was established in 1840 in New York. During the 1870s, large numbers of Herefords were imported and the breed became popular in the United States.

Herefords have been registered by the American Hereford Association since 1881. More Herefords have been registered than cattle of any other breed.

FIGURE 13-8A   American Hereford. *Courtesy of American Hereford Association, MO.*

Description.   Hereford cattle have white faces and red bodies. They have white on the belly, legs, and switch. Herefords are a horned breed. They are docile in nature and easily handled.

The breed is well adapted to the western cattle-raising regions of the United States. They have superior foraging ability, vigor, and hardiness. They produce more calves under adverse conditions than do many other breeds. When Herefords are used in crosses, the white color pattern tends to dominate.

Mature Hereford bulls weigh about 1,840 pounds (834 kg). Mature cows weigh about 1,200 pounds (544 kg). Herefords are popular for their general producing ability.

FIGURE 13-8B   Hereford cattle. *Courtesy of American Hereford Association, MO.*

FIGURE 13-9A   Polled Hereford bull. *Courtesy of American Polled Hereford Association, MO.*

FIGURE 13-9B   Polled Hereford "family." *Courtesy of American Polled Hereford Association, MO.*

FIGURE 13-9C   Polled Hereford cattle. *Courtesy of American Polled Hereford Association, MO.*

Polled Hereford

History. Polled Herefords (Figures 13-9 A, B, and C) originated in Iowa in 1901. Warren Gammon, an Iowa breeder, contacted all Hereford Association members asking if they had naturally polled animals in their herds. He located four bulls and ten cows, which he purchased. These fourteen animals became the foundation of the Polled Hereford breed. Later, other Polled Herefords were found and brought into the breeding. Polled offspring of crosses between Herefords and Polled Herefords have been used to improve the breed.

Description. Polled Herefords have the same traits as Herefords except for the horns. All are descended from purebred horned Herefords. Polled Herefords are eligible for registry in both the American Hereford Association and the American Polled Hereford Association. Many animals are registered in both associations.

Limousin

History. Limousin cattle were named after the province in west-central France where they originated about 7,000 years ago (Figure 13-10). In 1886 a breed association was formed and a herdbook was started. Limousin cattle entered the United States in 1968 when semen was imported from Canada, and registering began in the North American Limousin Foundation.

Description. Limousin cattle have light yellow hair with lighter circles around the eyes and muzzle. The skin is free of pigmentation. The spread of horns is horizontal then forward and upward. The Limousin head is small and short with a broad forehead. The neck is also short. Mature bulls weigh from 2,000 to 2,400 pounds (907–1,088 kg). Mature cows weigh about 1,350 pounds (612 kg). Limousin cattle are noted for their carcass leanness and large loin area.

Maine-Anjou

History. The Maine-Anjou breed (Figure 13-11) originated in France in the 1840s. The breed is the result of crossing English Shorthorns and French Mancelle cows. Maine-Anjou were originally work animals. Through selective breeding, milk producing and beef traits were developed. The Maine-Anjou is today considered an excellent beef producer.

Semen was first imported into the United States from Canada in 1970. The first crossbreds born in the United States were registered in 1972. Animals are registered with the American Maine-Anjou Association.

Description. Maine-Anjou cattle are dark red and white in color. Some animals are roan in color. They have a lightly pigmented skin. They are a horned breed with medium-size horns that curve forward.

They are considered docile and are easily handled. Other traits include a fast growth rate and well-marbled carcass. Mature bulls weigh about 2,750 pounds (1,247 kg).

Murray Grey

History. The Murray Grey breed (Figure 13-12) originated in Australia in 1905. It is the result of Shorthorn-Angus crosses. Semen was imported into the United States from Australia in 1969. The first live Murray Grey cattle were imported in 1972. Only small numbers of purebred bulls and cows have been imported into the United States since that time. The breed is used mainly in crossbreeding programs.

Description. The Murray Grey has a solid-color hair coat that is dark to silver gray. The breed is polled and is considered to be a docile animal. It produces a good-quality carcass on limited grain, and its females have good mothering ability.

The American Murray Grey Association was formed in 1970. All cows with not less than 7/8 Murray Grey breeding may be registered. Bulls with not less than 15/16 Murray Grey breeding may be registered. Both males and females with not less than 1/2 Murray Grey breeding may be recorded.

Scotch Highland

History. The Scotch Highland breed (Figures 13-13 A and B) originated in the Hebrides Islands near Scotland. This breed has been raised in northern Scotland for several centuries. A few of these cattle were imported into the United States in the early 1900s. More were imported in the 1930s. Interest in the breed increased and, in 1948, the American Scotch Highland Breeder's Association was formed. More cattle have been imported since that time. The breed is used in crossbreeding programs. The association is now called the American Highland Cattle Association.

Description. Scotch Highland cattle have a long, coarse outer coat of hair with soft, thick undercoat. The colors are black, brindle, red and light red, dun yellow, and silver. Animals of this breed are hardy and are excellent foragers. They are popular in crossbreeding because they give a winter hardiness to the offspring.

Shorthorn and Polled Shorthorn

History. The Shorthorn breed (Figure 13-14) originated around 1600 in the Tees River Valley of northern England. At that time, they were called Durhams. Major improvement of the breed began in the late 1700s. Shorthorn cattle were imported into Virginia in 1783. The Coates Herdbook was established in 1822 to record Shorthorns. It was the first cattle herdbook and served as a model for other breed

FIGURE 13-10   Limousin feedlot steer. *Courtesy of North American Limousin Foundation, CO.*

FIGURE 13-11   Maine-Anjou bull. *Courtesy of International Maine-Anjou Association, MO.*

FIGURE 13-12   American Murray Grey bull. Murray Grey sire, Michaelong Mesa Grandee, 1975 Australian Murray Grey National Beef Sire of the year. *Courtesy of American Murray Grey Association, MO.*

FIGURE 13-13A Scotch Highland bull. *Courtesy of American Highland Cattle Association, MN.*

FIGURE 13-13B Scotch Highland cow. *Courtesy of American Highland Cattle Association, MN.*

FIGURE 13-14 Shorthorn bull. *Courtesy of American Shorthorn Association, NE.*

herdbooks that followed. Shorthorns were originally a dual-purpose breed. They were bred for both milk and meat production.

The breed was established in the United States on a permanent basis as the result of importations of cattle between 1820 and 1850. The American Shorthorn Herdbook was started in 1846. It was the first beef herdbook to be published in the United States. The American Shorthorn Association was organized in 1872.

Description. Shorthorn cattle are red, white, or roan. They have short horns that curve inward. They are easily handled and have good dispositions. Mature bulls weigh up to 2,400 pounds (1,088 kg). Mature cows weigh up to 1,500 pounds (680 kg). Shorthorns are adaptable to many climates. They have excellent crossing ability with other breeds. Shorthorn cattle have been used in the bloodlines of more than 30 other recognized beef breeds. They are good mothers with excellent milking ability. Shorthorns produce a desirable carcass.

Polled Shorthorns. Polled Shorthorns originated in Minnesota in 1881. They have the same traits as Shorthorns except for being naturally polled. Both horned and polled Shorthorns are registered with the American Shorthorn Association.

Simmental

History. The Simmental breed (Figures 13-15 A and B) originated in the Simmen Valley of Switzerland. It is an old breed, dating back to the Middle Ages. The Simmental Herdbook was established in Switzerland in 1806. It required a performance pedigree for milk and conformation (physical appearance). Meat and carcass traits have since been added to the performance pedigree. About one-half of the cattle in Switzerland are Simmentals. It is the most popular breed of cattle in Europe. In France it is called "Pie Rouge" and in Germany it is known as "Fleckvieh."

Simmentals were first brought into the United States from Canada in 1969. The American Simmental Association was formed in 1968. The herdbook is open to upgrading of beef and dairy stock. All animals must have a performance pedigree to be eligible for registration. The American Simmental Association has no color requirement for registration. Artificial insemination is encouraged. Heifers become purebreds in three topcrosses (7/8 Simmental blood). Bulls are registered as purebreds in four topcrosses (15/16 Simmental blood).

Description. Simmental cattle have white to light straw faces with red to dark red, spotted bodies. They are a horned breed with medium-size horns. The Simmental is a large-bodied animal and is noted for being docile. Mature bulls weigh from 2,300 to 2,600 pounds (1,043–1,179 kg). Mature cows weigh about 1,450 to 1,800 pounds (658–816 kg). They will milk about 9,000 pounds (4,082 kg) of milk per lactation, and the milk will test about 4 percent butterfat.

Simmentals make extremely rapid growth, gaining about 3 pounds (1.4 kg) per day on roughage. They are thickly muscled and produce a carcass without excess fat. They are adaptable to a wide range of climates.

FIGURE 13-15A   American Simmental bull. *Courtesy of American Simmental Association, MT.*

South Devon

History. The South Devon breed was developed in the southwest part of England. It probably originated from large, red cattle that came from Normandy, France. They were brought to England at the time of the Norman invasions. The South Devon is related to the Devon breed. South Devons are dual-purpose cattle (milk and meat). Devons are a single-purpose breed (meat).

The South Devon was first imported into the United States in 1936 and 1947. Only five animals were involved in these importations. In 1969 and 1970, 215 registered cattle were imported by Big Beef Hybrids of Stillwater, Minnesota, for crossbreeding purposes.

FIGURE 13-15B   American Simmental cow. *Courtesy of American Simmental Association, MT.*

Description. The color of the South Devon is a medium-red hair coat that is lighter in the twist. They have soft, curly hair and a very thick hide. The South Devon is a horned breed with average-size horns that curve forward and downward. Mature bulls weigh 2,000 to 2,800 pounds (907–1,270 kg). Mature cows weigh about 1,500 to 1,600 pounds (680–726 kg). It is a very docile breed.

Blonde d'Aquitaine

History. The Blonde d'Aquitaine is a French breed of beef cattle. The breed originated in 1961 when several French breeds were combined. The cattle were originally used for draft, meat, and milk production.

Description. The Blonde d'Aquitaine have heavily muscled bodies with deep chests. The hips are wide and the hindquarters are well developed. Colors are yellow, brown, fawn, or wheat. Mature bulls weigh 2,500 pounds (1,134 kg) and cows weigh 1,500 pounds (680 kg).

Barzona

History. The Barzona breed was developed in Arizona beginning in 1942. The breed was begun by F.N. Bard for use in the desert areas of the southwestern United States. The first cross was between an Africander and a Hereford. Two herds were developed from the females from this cross. Santa Gertrudis bulls were used on one herd; Angus bulls were used on the other. Selection was made for fertility, mothering ability, and gain ability. The herd was closed in 1960; that is, no additional outside breeding stock was used. A cow that results from three

FIGURE 13-16A   Beefmaster bull. *Courtesy of Beefmaster Breeders Universal.*

FIGURE 13-16B   Beefmaster cow. *Courtesy of Beefmaster Breeders Universal.*

FIGURE 13-17   Braford bull. *Courtesy of International Braford Association, TX.*

topcrosses of Barzona registered bulls on a cow of a beef herd is registered as purebred. To be eligible for registration, a bull must be of fourth-generation Barzona breeding. The Barzona Breeders Association of America was formed in 1968.

Description. Barzona cattle are nearly solid red with a little white on the underline and around the head. They are good mothers and are able to subsist on a grass-browse range.

Beefmaster

History. The Beefmaster breed (Figures 13-16 A and B) was begun in 1931 in Texas. Further development of the breed took place in Colorado. The breed is the result of crosses among Herefords, Shorthorns, and Brahmans. The crossing was done with bulls from the three breeds under range conditions. The exact percent of blood of each breed is not known. It is estimated that the breed is about 25 percent Shorthorn, 25 percent Hereford, and 50 percent Brahman.

There are three breed associations: Beefmasters Breeders Universal, Foundation Beefmaster Association, and National Beefmaster Association. Registered cattle either are descendants of the original herd or are from three consecutive topcrosses of Beefmaster breeding.

Each breeder uses a prefix, such as *Smith Beefmaster*, to identify cattle from his herd.

Description. The breed has a variety of colors. Reds and duns are more common than other colors. Some of the cattle are horned and some are polled. Selection has been mainly for good disposition, fertility, gain, conformation, hardiness, and milk production.

Braford

History. Development of the Braford breed (Figure 13-17) began in 1947 by crossing Hereford bulls on Brahman cows at the Adams Ranch near Fort Pierce, Florida. Because Hereford bulls were not well adapted to environmental conditions in southern Florida, crossbred Brahman-Hereford bulls that showed desirable characteristics were used to further develop the foundation herd for the Braford breed. Selection of breeding stock is generally made for those traits of highest economic value. Among these are high fertility, ease of calving, high calf survival rate, milking ability, high weaning rate, growth rate, efficient use of roughages in the diet, and adaptability to the environment.

Description. The color of the Braford is red and shows a Hereford color pattern. The breed is about 5/8 Hereford and 3/8 Brahman. Calves grow rapidly and attain weaning weights of 500 to 800 pounds (227–363 kg) without supplemental feeding. Steers in the feedlot will produce 1,000 pound (454 kg) choice cattle in 12 to 15 months. Mature bulls weigh 1,500 to 2,000 pounds (680–907 kg) and mature cows weigh about 1,150 pounds (522 kg). The breed is noted for its superior maternal ability.

Brahman

History.   The Brahman breed (Figures 13-18 A and B) was developed in the southwestern part of the United States. Between 1854 and 1926 about 266 bulls and 22 females of the *Bos indicus* type of cattle were imported from India. *Bos indicus* cattle have a hump over the shoulders. These cattle are also called Zebu. Several strains of the Zebu cattle were bred to females from several British breeds of cattle. Early breeders selected for hardiness and ability to produce in the climate of the Southwest. Beef conformation and early maturity were also selected for in these early matings.

The major use of the Brahman in the United States is in crossing with other breeds. The resulting hybrids have proven to be desirable beef animals. The Brahman has been used in the development of a number of other newer breeds of beef cattle. The American Brahman Breeders Association was organized in 1924.

Description.   The color of the Brahman is light gray or red to almost black. The most common color is light to medium gray. Red is becoming a popular color with a number of breeders. In addition to the characteristic hump over the shoulders, Brahmans have loose skin (dewlap) under the throat and large drooping ears.

Mature Brahman bulls weigh about 1,600 to 2,200 pounds (726–998 kg). Mature cows weigh from 1,000 to 1,400 pounds (454–635 kg). Brahman cattle have a very high heat tolerance. They are also resistant to disease and insects. They are good mothers and have an excellent ability to forage on poor range. They gain rapidly and produce a quality carcass. However, they do tend to have an unpredictable disposition.

Brangus

History.   The Brangus breed (Figures 13-19 A and B) was developed by crossing Brahman and Angus cattle. Early crossings of these breeds was done at the USDA Experiment Station at Jeanerette, Louisiana. Some of the first crosses were made as early as 1912. The American Brangus Breeders Association was formed in 1949. Since then, the name of the association has been changed to the International Brangus Breeders Association. The Brangus name is a registered trademark. Only animals registered with the breed association can be called Brangus. All present-day Brangus are descendants of foundation animals registered in 1949 or from registered Brahman and Angus cattle that have been enrolled since then.

Brangus cattle are based on foundation stock that is 3/8 Brahman and 5/8 Angus. There are several recognized ways to produce

FIGURE 13-18A   Brahman bull. *Courtesy of American Brahman Breeders Association, TX.*

FIGURE 13-18B   Brahman cow. *Courtesy of American Brahman Breeders Association, TX.*

FIGURE 13-19A   Brangus bull. *Courtesy of International Brangus Breeders Association, TX.*

FIGURE 13-19B   Brangus female. *Courtesy of International Brangus Breeders Association, TX.*

FIGURE 13-20A Red Brangus bull. *Courtesy of American Red Brangus Association, TX.*

FIGURE 13-20B Red Brangus cow. *Courtesy of American Red Brangus Association, TX.*

FIGURE 13-21A American Salers bull. *Courtesy of American Salers Association, CO.*

FIGURE 13-21B American Salers cattle on pasture. *Courtesy of American Salers Association, CO.*

Brangus cattle. A 1/4 Brahman 3/4 Angus crossed with a 1/2 Brahman 1/2 Angus is one way. Another is crossing an animal that is 3/4 Brahman 1/4 Angus with a purebred Angus. The third method is mating of registered Brangus animals.

Description. Brangus cattle are solid black and polled. An inspection is necessary to determine conformation and breed character before the animal may be registered. Brangus are adaptable to different climates. They have good mothering ability, feed efficiency, and produce desirable carcasses.

Red Brangus

History. The Red Brangus breed was developed beginning in 1946 in Texas. Foundation stock came from purebred Angus and Brahman. No percent of blood of either breed is required. However, to be registered, the animal must show traits of both breeds.

Description. Red Brangus are red in color and are a polled breed (Figures 13-20 A and B). Animals that show mostly Brahman and Angus traits are listed as certified Red Brangus. Their offspring may be registered as Brangus if they meet the requirements of the association. In addition to being red in color and polled, they must meet size and conformation requirements. The animals are registered with the American Red Brangus Association. The association was formed in 1956.

Salers

History. The Salers (pronounced *Sa' lair*) breed of cattle are native to the Auvergne region of south central France. Cave drawings, dating back about 7,000 years, found near Salers, France, show cattle believed to be the ancestors of the present breed. Historically, in France, Salers cattle were used for beef, milk, and as draft animals. The first Salers bull was imported into Canada in 1972. Semen from this bull was sold in both Canada and the United States. In 1975 one bull and four heifers were imported directly to the United States. Between 1975 and 1978, 52 heifers and six bulls were imported into the United States and another 100 head were imported into Canada. The present breed in the United States originated with these imports.

The American Salers Association was formed in 1974. The association accepts animals for registration from an upgrading breeding program. The continual use of registered fullblood or purebred Salers sires on herds will produce offspring that are eligible for registry with the association. Current requirements for registration are available from the association.

Description. The Salers is a horned breed. They are a dark mahogany red in color (Figures 13-21 A and B). There is a strain of Salers that are naturally polled and some are black in color. This genetic diversity adds to their value in crossbreeding programs. Salers cows are noted for their ease of calving and good maternal ability. Other desirable characteristics of Salers cattle include good foraging ability on poor range, high weaning weights, and excellent carcass quality that meets current market demand for beef. Salers cattle are also used in cross-breeding programs.

Santa Gertrudis

History. The Santa Gertrudis breed (Figures 13-22 A and B) was developed on the King Ranch in Texas. The breed is the result of crosses of Brahman bulls on Shorthorn cows. Crossbreeding began in 1910 using several different European breeds of beef cows. By 1918, the Brahman-Shorthorn cross showed the most promise. In 1920, a bull named Monkey was born. This bull had 3/8 Brahman and 5/8 Shorthorn blood. Monkey showed outstanding traits and sired more than 150 useful sons. All present-day Santa Gertrudis cattle are descendants of this bull.

The breed association, formed in 1950, is Santa Gertrudis Breeders International. The association has a system of compulsory classification. For registration, an animal must be inspected by an association classifier. A Standard of Excellence was established by the association. An animal that meets the standard is branded with an "S" and is certified as purebred.

Description. The color of the Santa Gertrudis is cherry red. Most of the animals are horned. Some are polled, and these are eligible for registration. They have loose hides with folds of skin on the neck and a sheath or naval flap. The hair grows short and straight in warm climates. It is long in cold climates. Santa Gertrudis cattle are efficient in the feedlot. They produce desirable carcasses with little waste fat. They also resist diseases and insects.

Milking Shorthorn

The Milking Shorthorn (Figure 13-23) has the same background and traits as the Shorthorn breed. It is a dual-purpose breed. Animals have been selected for both meat and milk production. In conformation, the Milking Shorthorn is more angular and less thickly fleshed than the Shorthorn. Both horned and polled Milking Shorthorns are registered in the American Milking Shorthorn Society. The association was formed in 1948. Milking Shorthorns may also be registered in the American Shorthorn Association herdbook.

Red Poll

History. The Red Poll breed originated in eastern England. Development of the breed began in the early 1800s. Native cattle of the shires of Norfolk and Suffolk were crossed to produce the breed. The first Red Polls were imported into the United States in 1873.

FIGURE 13-22A   Santa Gertrudis bull. *Courtesy of Santa Gertrudis Breeders International, TX.*

FIGURE 13-22B   Santa Gertrudis cow. *Courtesy of Santa Gertrudis Breeders International, TX.*

FIGURE 13-23   American Milking Shorthorn bull. *Courtesy of American Milking Shorthorn Society, WI.*

The Red Poll Cattle Club of America was formed in 1883. The Red Poll Cattle Society of Great Britain and Ireland was formed in 1885. The Red Poll Herdbook was established in 1873 in England. Registrations in the United States are in the Red Poll Herdbook, American Series, which is a continuation of the original herdbook. A Gain Register was established in 1960 to recognize preweaning calf gain records. Carcass merit and gain to slaughter is recorded in a Carcass Register. This was established in 1963. Cattle are registered in the American Red Poll Association.

Description. The color of the Red Poll is light red to very dark red. The tail switch contains some natural white. Some white is permitted on the underline. The skin is buff or flesh-colored. Solid black, bluish, or cloudy noses are not permitted for registration. The breed is polled and dual-purpose. They have been bred for both milk and meat production. The emphasis in recent years has been on beef type. The carcass has a high proportion of lean meat. There is little waste fat and the marbling is acceptable.

Hays Converter

History. The Hays Converter breed was developed in Canada by Harry Hays of Calgary, Alberta, Canada. The breed was developed from foundation stock from the Hereford, Brown Swiss, and Holstein cattle breeds. Breed development began in 1957. Animals from the foundation breeds were selected for rate of gain; large, strong frames; sound feet and legs; well-attached udders; good milking ability; high fertility; ease of calving; and winter hardiness. Selection for major traits continued for five generations, with animals showing undesirable traits being culled from the breeding herd.

Description. The color is usually black with white face, feet, and tail. Some are red with white faces. Selection is not made on the basis of color. Bulls weigh about 2,200 pounds (998 kg) and cows weigh about 1,400 pounds (635 kg). Calves reach market weight in 12 to 15 months.

Marchigiana

History. The Marchigiana (pronounced *Mar-key-jahna*) breed (Figure 13-24) originated in Italy near Rome during the fifth century, A.D. Foundation stock originated with crosses of native cattle, including the Chianina. The Italian herdbook for the Marchigiana was established in 1930.

Description. The Marchigiana is a horned breed. Cows are a grayish white. The bulls are somewhat darker in color. The skin, muzzle, and switch are dark in color. Bulls weigh from 2,650 to 3,100 pounds

FIGURE 13-24 Fullblood Marchigiana bull. *Courtesy of Marky Cattle Association, KS.*

(1,202–1,406 kg) while cows weigh 1,400 to 1,800 pounds (635–816 kg). The Marchigiana are used in crossbreeding programs in the United States. Cattle are registered in the American International Marchigiana Society.

Norwegian Red

History. The Norwegian Red breed originated in Norway. They were first imported into the United States in 1973. They are used as dual-purpose cattle in Norway and are the major breed found there.

FIGURE 13-25   Pinzgauer bull. *Courtesy of Pinzgauer Breeders Limited.*

Description. The Norwegian Red are horned and are red or red and white in color. Bulls weigh 2,200 to 2,600 pounds (998–1,179 kg) and cows weigh 1,200 to 1,400 pounds (544–635 kg). They are noted for their good carcasses and feed efficiency. Cattle are registered in the North American Norwegian Red Association.

Pinzgauer

History. The Pinzgauer breed (Figure 13-25) originated in the Alpine regions of Austria, Italy, and Germany.

Description. They are a horned breed and brown in color. There is some white on the top and underlines. Bulls weigh 2,200 to 2,900 pounds (998–1,315 kg) and cows weigh 1,300 to 1,650 pounds (590–748 kg). The Pinzgauer is a hardy breed and is used in crossbreeding programs in the United States. Cattle are registered in the American Pinzgauer Association.

Senepol

History. The Senepol is a Caribbean breed of beef cattle that were originally bred on the island of St. Croix in the United States Virgin Islands. The breed was developed in 1918 from a cross of the Red Poll and the N'Dama. The N'Dama is a humpless longhorn breed from West Africa. A few Senepol were imported into the United States, mainly to Florida and Tennessee, in the late 1970s. Since 1978, the USDA has been conducting crossbreeding experiments with the Senepol at the Subtropical Agricultural Research Station at Brooksville, Florida. The world population of Senepols is estimated at 3,000 to 4,000 head, with 1,500 head of cows registered.

Description. Senepol cattle range in color from tan to dark red, are polled and do not have a hump. They are resistant to ticks and other insects, and tolerate hot weather and a wide range of rainfall conditions. Senepols are similar to the Brahman in adaptation to unfavorable environments but have a much gentler disposition. They are early maturing, and have good mothering characteristics and good milk production. They are seen as having value in the United States in

FIGURE 13-26   Texas Longhorn bull. *Courtesy of Dickinson Livestock.*

crossbreeding programs to produce cattle with desirable characteristics for a subtropical environment. Senepol-Angus crosses have produced carcasses of a slightly higher quality grade than Brahman-Angus crosses.

Texas Longhorn

History.   The Texas Longhorn (Figure 13-26) originated from Spanish Andalusian cattle that were brought to Santa Domingo by Columbus on his second voyage in 1493. In the early 1500s, descendants of these cattle were taken to Mexico by Spanish explorers. As the Spanish continued their explorations, some of these cattle were taken with them into the region that became Texas. Many of the cattle escaped and adapted to the harsh environment of the Southwest. It is estimated that by 1860 about four million descendants of these cattle were running wild in Texas. After the Civil War, cattlemen in Texas began rounding up the Longhorns and moving them along trails to railheads, mainly in Kansas and Missouri. By the 1880s, the Longhorn began to be replaced by European breeds of beef cattle. The Texas Longhorn was almost extinct by 1900. The United States Congress appropriated money in 1927 to establish a federal herd of purebred Longhorns. At that time, 20 cows, three bulls, and four calves were located and established as seed stock at the Wichita Mountains Wildlife Refuge, Cache, Oklahoma. The Texas Longhorn Breeders Association of America was established in 1964. There were about 1,500 Texas Longhorns in existence at that time. Since the 1980s there has been a renewed interest in the Texas Longhorn as genetic seedstock for characteristics desirable in cattle adapted to the subtropical environment. They may also prove to be valuable in breeding programs aimed at producing cattle with meat that is leaner and has lower cholesterol levels.

Description.   The Texas Longhorn has many shadings and combinations of colors. They have horns that curve upward and spread to four feet or more. Their legs are long and their shoulders are large and high. They have a large head with small ears and long hair between their horns. Their neck is short and stocky.

Texas Longhorns are slow maturing, have high fertility, are resistant to many diseases and parasites, and are well adapted to harsh environments. They have the ability to survive on sparse rangeland. They are noted for their easy calving ability, hardiness, and longevity.

Romagnola

History.   The Romagnola (pronounced *Ro-ma-nola*) breed originated in northeastern Italy during the fourth century, A.D. The foundation stock of this breed are the *Bos primigenius podolicus* and the *Bos primigenius nomadicus*. The *Bos primigenius podolicus* (forebears of the modern

Bos taurus) lived in Italy and the *Bos primigenius nomadicus* (forbears of the modern *Bos indicus*) were brought to Italy in the fourth century by Gothic invaders. Several Italian breeds of cattle, all with similar characteristics, were developed from these ancestors. The Romagnola were originally used as draft animals in the fields, which resulted in the development of a muscular body type. After the mechanization of agriculture, the emphasis on breeding and development shifted to beef production. Romagnola cattle were first imported into the United States in the 1970s. Cattle are registered in the American Romagnola Association.

Description. The Romagnola are a horned breed; the females having lyre-shaped horns and the bulls have half-moon-shaped horns. The skin is black pigmented and the hair coat is ivory; bulls have grey hair around the shoulders and eyes. The black skin color is also found on the muzzle, horn tips, tail switch, hoofs, vulva, tip of sheath, and the base of the scrotum. Calves are a light reddish color when born and turn white at about three months of age. The average weight of adult bulls is 2,750 pounds (1,250 kg) and the average weight of adult cows is 1,650 pounds (750 kg). The breed is noted for economical feed conversion with rapid gains, a high dressing percentage, and early maturity.

SUMMARY

The beef industry is an important source of income for farmers. About 18 percent of cash income from crops and livestock comes from beef. The use of beef in the United States has been decreasing in recent years. While many beef producers have small operations, there are also some large feedlots.

Most beef cattle breeds originated in Europe. Some of the newer breeds are the result of crossing breeds from India on European breeds.

Student Learning Activities

1. Present an oral report on the characteristics of the beef industry.
2. On field trips in the community, observe different beef breeds and their characteristics.
3. Conduct a survey of beef producers in the community to determine what beef breeds are used locally.
4. Prepare a bulletin board displaying pictures of the beef breeds.
5. Present an oral report on the history and characteristics of a beef breed to the class.
6. Write to beef breed associations for literature about their breed.

Review

1. How important is the beef industry as compared to the total live-stock production industry?
2. In what part of the United States are most of the large beef feed-lots found?
3. In what part of the United States are most of the smaller beef feedlots found?
4. List the advantages and disadvantages of raising or feeding beef cattle.
5. Describe briefly how modern beef breeds were developed.
6. List seven points that should be considered when selecting a beef breed.
7. Prepare a table that briefly describes the characteristics of each of the beef breeds.
8. Which beef breeds are most common in your area?
9. Which beef breeds were developed in the United States?

Selection and Judging of Beef

TYPES OF BEEF PRODUCTION

There are three main types of beef cattle production systems: (1) cow-calf producers, (2) purebred breeders, and (3) cattle feeders. A farmer may specialize in only one type of operation or combine several kinds of operations. For example, a farmer may produce calves from a cow herd and also feed the calves for slaughter.

Cow-calf Producers

The *cow-calf system* of beef production involves keeping a herd of beef cows. These cows are bred each year to produce calves. The calves are then sold to cattle feeders who feed them to slaughter weights. Most of this type of beef production is done in the western range states and upper Great Plains. Land is used that is not suitable for growing crops. Beef cows are maintained mainly on roughage. Little or no grain is needed for this type of beef production. This type of operation requires less labor and a lower investment in equipment and facilities than other kinds of beef enterprises.

A larger investment in land is usually needed for this type of operation than is necessary for feeding cattle for slaughter. It is difficult to expand or reduce the size of operation quickly. The price received for calves tends to be more closely associated with the supply and demand for calves rather than with the cost of producing them.

Therefore, a producer may not always recover production costs in this type of beef operation.

The cows are usually bred to calve in the spring. Most calves are weaned in the fall and sold as feeders. Sometimes the calves are fed roughage through the winter and sold the next year as yearlings. *Feeder calves* are weaned calves that are under one year of age and are sold to be fed for more growth. *Yearling feeders* are one to two years of age and are sold to be fed to finish for slaughter.

Purebred Breeders

Purebred breeders keep herds of purebred breeding stock. They provide replacement bulls for cow-calf operations. Cow-calf farmers sometimes buy cows or heifers from the purebred breeder to improve the commercial herd. Purebred breeders also sell to other purebred breeders. The purebred breeders are mainly responsible for the genetic improvements that have been made in beef breeds.

A great deal of knowledge and skill are required to raise purebreds, and it should only be attempted by those with experience. The costs are usually higher in this type of cattle business. It takes many years to develop a high-quality herd and achieve success.

Cattle Feeders

The cattle feeder feeds animals for the slaughter market. The objective is to produce finished cattle in the shortest time possible. The operator usually buys feeders or yearlings and finishes them in the feedlot. Some producers feed cattle on pasture for a time and then finish them in the feedlot. There is a trend toward more confinement feedlot finishing of slaughter cattle.

While some roughage can be used in feeder cattle operations, this enterprise requires more grain than cow-calf or purebred production. It usually takes grain to get the quality of finish that is in demand in the marketplace. Feeder operations can easily adjust to changes in feed supplies, operating costs, labor supply, and economic outlook. The cattle feeder can expect a return on investment in four to six months.

The facilities required for confinement feeding of cattle are more expensive than those required for cow-calf operations. Feed costs, labor requirements, and transportation costs are all higher in this type of enterprise. Cattle feeding is a high-risk enterprise because of the fairly large fluctuations in the price of finished cattle.

SELECTION OF BEEF ANIMALS

A beef herd is improved by selecting animals that have the desired traits. The producer must produce what is in demand in the marketplace. Selection is based primarily on conformation and performance records. Pedigree and show ring winnings are of less importance.

Cow-calf producers need to be aware of the relationship between cow size and weaning weights of calves that are produced. In general, for each 100 pound (45 kg) increase in cow size, there is a corresponding increase of 10 to 12 pounds (4.5–5.4 kg) in calf weaning weight. However, experimental work done in Texas has revealed that heavier cows (1,201–1,700 lbs [544.8–771.1 kg]) require 18 percent more energy and 13 percent more protein per day than do medium-weight cows (1,000–1,200 lbs [453.6–544.3 kg]). The average weaning weight of calves from the heavier cows was only 2.4 percent greater than the average weaning weight of calves from the medium-size cows.

In this same experiment, age of the cow was shown to be a factor in calf weaning weights. Twelve-year-old cows produced calves with an average weaning weight that was six percent less than two-year-old cows.

Producers with cow-calf herds need to consider the higher maintenance costs of heavier cows as well as the age of the cows when making selections of breeding stock.

A valuable tool for selecting replacement heifers and herd bulls is the *frame score*. It is used by state and USDA cattle graders in beef performance testing programs. The frame score is a measurement based on observation and height measurements when calves are evaluated at 205 days of age. An estimate of the expected size of the animal when it reaches maturity can be made from its frame score. The use of the frame score in herd sire selection provides a more accurate prediction of expected genetic change in the herd.

Frame scores are made on a 1 through 7 scale (Figure 14-1). The scores represent a range in body types of beef cattle and were developed at the University of Wisconsin. The English breeds of beef cattle are usually covered by body types 1 through 5. Charolais, Simmental, and similar-size cattle of other breeds usually require the use of body types 3 through 7. The age of the animal is a factor in determining its frame score. Many beef performance testing programs use the Missouri system of using a height measurement at the shoulder when determining frame scores.

Muscle conformation scores are also determined for the animal. These scores use a 1 through 7 scale, as follows:

1. an exceptionally thin calf.
2. very light muscled.
3. light muscled.
4. average muscling.
5. heavy muscled.
6. very heavy muscled.
7. double muscled.

A conformation score on a scale of 1 through 17 is also determined for the animal. The low scores indicate inferior animals. Scores

FIGURE 14-1 The seven frame scores used in many beef performance testing programs. *Courtesy of University of Illinois at Urbana-Champaign.*

of 9 through 11 indicate animals that are below average for desirable characteristics. Scores of 12 through 14 indicate animals that are average or slightly below average for some of the desirable characteristics. Scores of 15 through 17 indicate superior animals that are growthy, well balanced, well muscled, and have adequate frame. Superior animals are also structurally sound, have adequate bone, and are highly acceptable in breed and sex character.

Animals with conformation scores in the 14 to 17 range should be considered for herd sires. Replacement heifers should have conformation scores in the 13 to 17 range. Small-framed cattle generally should not be considered for herd replacements even if their conformation score is acceptable.

Conformation

Conformation refers to the appearance of the live animal. It includes the skeletal structure, muscling, fat balance, straightness of the animal's lines, and structural soundness. To describe the conformation of the animal, one must first learn the parts of the animal. The parts of the beef animal are identified in Figure 14-2.

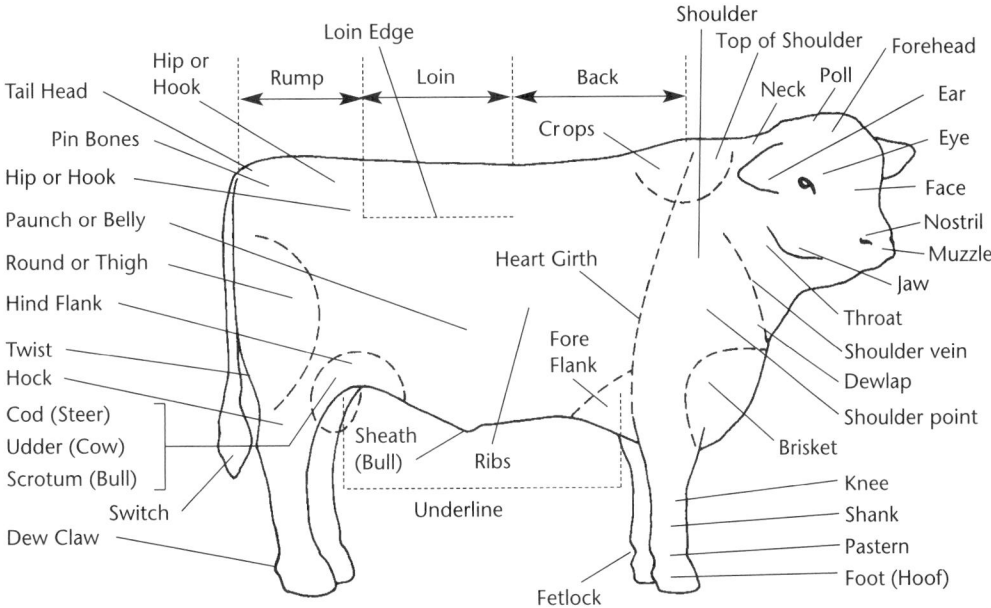

FIGURE 14-2 The body parts of the beef animal.

Desirable conformation of the beef animal includes:

- long, trim, deep-sided body.
- no excess fat on the brisket, foreflank, or hindflank.
- no extra hide around the throat, dewlap, or sheath.
- heavily muscled forearm.
- proper height to the point of the shoulders.
- correct muscling throughout the body.
- maximum development of the round, rump, loin, and rib.

An animal with the proper conformation will produce the maximum amount of high-value cuts. It will have a minimum of less-valuable bone and internal organs. It is necessary to learn the wholesale cuts of the beef animal to properly evaluate the live animal.

Figure 14-3 shows the location of the high- and low-value wholesale cuts of beef. The high-value wholesale cuts come from the round, rump, loin, and rib. The lower-value wholesale cuts come from the chuck, brisket, flank, plate or navel, and shank.

Ultrasonics is the use of high-frequency sound waves to measure fat thickness and loin-eye area. It is a useful tool for selecting meaty animals for breeding purposes. The measurement is made on live animals. Therefore, meaty animals can be identified without killing them. The accuracy and value of the ultrasonic measurement depends on the skill of the operator. Increased use of ultrasonics may be expected as cattle breeders gain confidence in the results.

High-value wholesale cuts

1. Loin
2. Rib
3. Round
4. Rump

Low-value wholesale cuts

5. Chuck
6. Brisket
7. Flank
8. Plate or navel
9. Shank

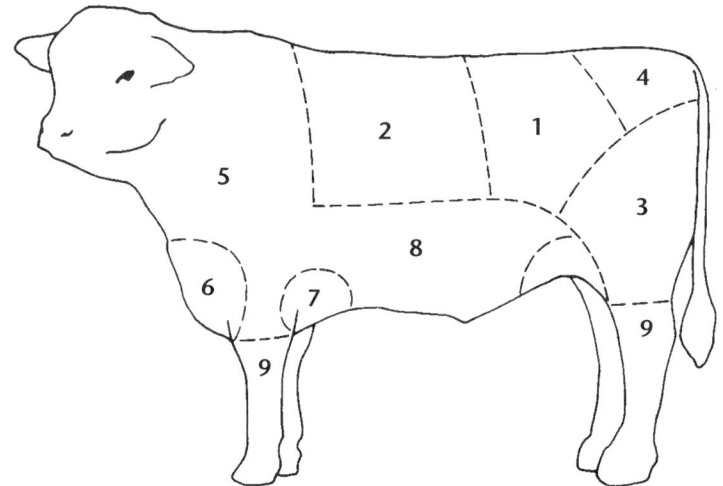

FIGURE 14-3   Location of high- and low-value wholesale cuts of beef.

Performance Records

One of the best ways to select beef animals is on the basis of performance records, of which there are several kinds. Performance testing refers to an animal's own performance in relation to important economic traits. *Production testing* refers to measuring a brood cow's production by the performance of its offspring. (A *brood* animal is kept for breeding.) *Progeny testing* usually refers to the evaluation of a bull by the performance of a number of its offspring. Overall herd evaluation programs are often referred to as performance testing. This may include production and progeny testing. *Performance testing* may be defined as a method of collecting records on beef cattle herds to be used for selecting the most productive animals.

Performance testing is used by both purebred and commercial beef producers. It helps the cattle producers to improve the grade and weight of the calves in the herd more rapidly. Visual evaluation alone is not enough to select the best animals.

Performance testing records are used to:

- cull low-producing cows.
- check on percent calf crop (number of calves weaned divided by number of breeding females in the herd).
- select replacement heifers and bulls.
- measure the productivity of each bull.
- improve herd management.
- improve the grade of calves produced.
- increase the weaning weight of calves produced.
- give buyers more information.
- provide permanent records.

Many states and breed associations have performance testing programs. A producer who desires to do performance testing should contact the local office of the state extension service or the proper breed association for information and forms to use. Setting up and conducting a testing program usually involves three main steps: (1) identifying each cow and calf; (2) recording birth dates of calves; and (3) weighing and grading the calves at weaning time.

Weaning weight is adjusted to a 205-day weight so that all calves in the herd can be compared on an equal basis. The weaning weight is adjusted for the age of the calf, the age of the dam, and the sex of the calf. Most states and breed associations have adopted the recommendations of the National Beef Improvement Federation for the 205-day age basis and the adjustment factors.

The formula for the 205-day weight adjustment is:

$$\frac{\text{Actual weight} - \text{Birth weight}}{\text{Age in days}} \times 205 + \text{Birth weight}$$

If the calf was not weighed at birth, 70 pounds (31.8 kg) may be used as the average birth weight. Some breed associations have established their own standard average birth weight to be used if the calf was not weighed at birth. These standards are shown in Table 14-1.

Calves must be weighed between 160 and 250 days of age to be considered regular age and have 205-day weights and weight ratios calculated for them. If calves are weighed between 120 and 159 days or 251 and 290 days of age, the 205-day weight is calculated but a weight ratio is not calculated for them. These calves are considered to

TABLE 14-1 Breed Standard Birth Weights Used in Performance Testing Programs.

| | Sex of Calf | | | |
| | Females | | Males | |
Breed	(lb)	(kg)	(lb)	(kg)
Angus	65	29.2	75	34.0
Charolais	85	38.6	85	38.6
Chianina	80	36.3	80	36.3
Hereford	70	31.8	75	34.0
Polled Hereford	70	31.8	75	34.0
Limousin	75	34.0	80	36.3
Maine-Anjou	84	38.1	90	40.8
Shorthorn	70	31.8	70	31.8
Simmental	83	37.6	91	41.3

be of irregular age. Performance data for irregular-age calves are not included in sex group summaries, sire summaries, or herd averages.

An adjustment in the 205-day weight of the calf is made for the age of the dam. These adjustments are as follows:

| | Adjustment | |
Age of Dam (years)	Male Calves (lb)	Female Calves (lb)
2	+60	+54
3	+40	+36
4	+20	+18
5 to 10	+ 0	+ 0
11+	+20	+18

The 205-day weight ratio within the sex group is calculated for each regular-age calf. The 205-day weight as adjusted for the age of the dam is divided by the average for the sex group and shown as a percentage. A weight ratio of 90 for a heifer would indicate a calf that is 10 percent below the average for all heifers in the herd. A heifer with a weight ratio of 110 would indicate a calf that is 10 percent above the average for all heifers in the herd.

An adjustment is also made for the sex of the calf. To adjust the 205-day weight to a steer-calf basis, add 5 percent to the weight of a heifer calf and subtract 5 percent from the weight of a bull calf. Adjusting the 205-day weights to a steer-calf basis makes it possible to directly compare heifers and bulls. The information gained from this process can be used for selection of breeding stock as well as in performance testing.

A 205-day adjusted weight ratio is determined for all calves in the herd that are classified as being of regular age. The 205-day weight (adjusted for age of dam and sex of calf) for each regular-age calf is divided by the herd average of 205-day adjusted weights for all regular-age calves. The calves can now be compared on an equal sex basis and each cow's production can be compared to the herd average.

Individual cow productivity is compared by using the 205-day adjusted weight and weight ratio. Individual calf performance within its sex group is compared by using the 205-day weight adjusted for age of dam and the 205-day weight ratio within the sex group.

Performance testing programs also make use of adjusted yearling (365-day) weights. The formula for 365-day adjusted weight is:

205-day weight, adjusted for age of dam
+ (average daily gain on test × 160)

Cattle must be at least 330 days of age and on test at least 140 days in order to calculate a 365-day adjusted weight.

Some producers use a 550-day adjusted weight (approximately 18 months of age) to help in the selection of replacement heifers. The final weight of the animal must be taken at 500 days or later but not before 500 days. The formula for calculating the 550-day adjusted weight is:

$$\frac{(\text{Actual final weight} - \text{Actual weaning weight}) \times 345}{\text{Number of days between weighings}} + \text{205-day weight, adjusted for the age of the dam}$$

Most states and breed associations with testing programs offer computer services to keep and analyze the records. Information provided is for use on the farm where it was gathered. It is not intended for use in comparing one herd with another.

The amount of improvement that can be expected from selecting superior animals depends on the heritability of the traits upon which the selection is based. The performance of an animal depends on genetics and environment. The heritability of a trait is the amount of the performance that comes from genetics. Table 9-1 in Unit 9 shows the heritability estimates for some important traits of beef cattle. Heritability above 40 percent is high; from 20 to 40 percent is medium; below 20 percent is low.

Pedigree

A *pedigree* is the record of the ancestors of an animal. It usually contains only the names of the ancestors. In some cases, it may contain performance data. A person must be familiar with the performance of the individuals in the pedigree if it is to have much value in selection. Only the most recent ancestors are of importance in selection. Ancestors before the fourth generation contribute very little to the current generation.

Sometimes the use of a bloodline becomes a fad. Cattle breeders must be careful not to keep poor animals just because they belong to a given bloodline.

Other Factors in Selection

Only healthy animals should be brought into a herd. The buyer should check the health of an animal carefully before buying it. Females should test negative for brucellosis (Bang's disease), vibriosis, and tuberculosis. It is best to buy females that have been calfhood vaccinated. Check for mange, ringworm, and lumpy jaw. (More information on diseases is included in Unit 17.)

The standards in the show ring are the result of what the consumer wants in the marketplace. Animals that do well in the show ring meet these standards. Therefore, show ring winners are often desirable animals to select for breeding purposes if they meet standards of conformation and performance records.

Selection of the Herd Bull

The selection of the herd bull is one of the most important decisions the cattle breeder makes. Each calf produced receives one-half of its genetic makeup from its sire. When replacement females are selected from the herd, 87.5 percent of the heifer's genetic makeup comes from the last three bulls used in her pedigree.

A major advancement in the improvement of the beef industry has been the development of sire summaries. Most breed associations provide sire summaries that are updated each year. These summaries can be secured directly from individual breed associations.

Some breed associations are now publishing their sire summaries on the Internet. Table 14-2 shows some Internet addresses (URL's) that provide information about some breeds. Entering a breed name

TABLE 14-2. Some Internet Addresses Providing Sire Summaries and Other Breed Information.

Breed or organization	URL	Comments
ABS	www.ABSGLOBAL.com/	Commercial site—provides sire summaries for dairy and beef bulls. Beef: Angus, Charolais, Gelbvieh, Hereford, Limousin, Red Angus, Simmental.
Accelerated	www.accelgen.com/	Commercial site—provides sire summaries for beef and dairy bulls. Beef: Angus, Polled Hereford, Red Angus, Charolais, Simmental. Dairy: Holstein, Brown Swiss, Guernsey, Jersey.
Angus	www.angus.org/	Home page American Angus Association— provides a variety of information including sire summaries for Angus.
Charolais	www.charolaisusa.com/	Provides sire summaries plus much more information about this breed.
Gelbvieh	www.gelbvieh.org/~aga	Goes to home page at //ops.cgsci.colostate.edu/~aga/. Leads to sire summaries at Ag Direct.
Hereford	www.hereford.org/	Home page of American Hereford Association— breed information plus sire summaries.
Limousin	www.agrione.com/LimousinWorld	Home page of magazine *Limousin World*—provides information about breed plus links to many other agricultural pages.
Select Sires	www.selectsires.com/	Commercial site—Provides sire summaries for dairy and beef bulls. Beef: Angus, Red Angus, Charolais, Simmental. Dairy: Holstein.
Simmental	www.simmgene.com/	Home page American Simmental Association— provides breed information plus sire summaries.

in an Internet search engine will usually produce any existing web sites for that breed. Some of these are sites maintained by the breed association, some are sites maintained by universities, and some are sites maintained by people with an interest in the breed (farmers, ranchers, owners, etc.). Always use caution when evaluating information provided by an Internet web site. Consider the source of the information and any agenda the people maintaining the site might have.

Sire summaries provide information on traits that are economically important to cattle producers. Included is information regarding the ability of the bull to transmit growth rate to his offspring. This includes information about expected birth, weaning, and yearling weights. Several breed associations have added carcass information to the sire summaries. Included is information about ribeye area, fat thickness, marbling scores, and carcass weight. It is expected that more breed associations will include carcass information in sire summaries in the future. Carcass data is important when selecting sires to produce animals to meet current market demand for leaner meat.

Many breed associations also include information regarding the daughters of the bulls in the sire summary. This information is referred to as *Maternal Breeding Value* (MBV) and covers ease of calving and ability to wean heavy calves.

The ability of the sire to transmit genetic traits is defined in sire summaries as the *Expected Progeny Difference* (EPD). The EPD is a measure of the degree of difference between the progeny of the bull and the progeny of the average bull of the breed in the trait being measured. It is calculated from data collected on the progeny of the bull. Bulls in a sire summary for a given breed can be directly compared on the basis of their respective EPD values.

A program has been developed by the Meat Animal Research Center, Clay Center, Nebraska, to compare sire EPDs between breeds. This permits producers to select sires for crossbreeding programs that will improve the herd. Current information is available only for sires used in the research. Additional research will provide information for more breeds and additional characteristics.

The EPD is usually given as a plus or minus value; however, some breed associations report EPDs as ratios. An EPD for yearling weight of +65 would show that the progeny of this bull should average 65 pounds more at 365 days of age than the progeny of the average bull of the breed. A minus value would indicate that the progeny of the bull would do poorer in the measured trait than the average bull of the breed.

The data from a sire summary can be used to select herd sires that are superior for traits that are desired. By using a succession of bulls that have high EPDs for selected traits, the producer can develop a superior herd. Care must be taken when selecting traits for a breeding program. There are some undesirable correlations between genetic traits. For example, yearling weight has a positive correlation with birth weight. This could lead to calving difficulties unless sires are selected with high EPD for yearling weight and a moderate or low EPD for birth weight.

Bulls with high EPDs for the selected traits will produce some calves that are not better than the average for the breed. On the other hand, bulls with low EPDs for selected traits will sire some calves that are better than the breed average. However, the overall performance of the progeny of such bulls will be lower than the breed average. The use of sire summaries to select herd sires is a recommended practice for overall herd improvement.

The cow herd and the calves produced need to be evaluated before selecting the herd bull. The areas where the greatest improvement is needed must be found. After the weaknesses of the herd have been identified, a herd bull must be selected that will improve those areas.

Performance record information on the bull must be carefully considered. The bull selected should have desirable conformation, good health, and vigor. Breed traits are a matter of individual preference.

The adjusted weaning weight of the bull should be above the average of the herd. The bull must have plenty of size for age and be structurally correct. The adjusted 365-day weight should be at least 950 to 1,000 pounds (431–454 kg). A bull that has gained rapidly from 8 months to 12 months of age should be selected. Rate of gain is highly heritable.

The sire and dam of the bull should also have records as good producers. The dam should have been rated in the top half of the herd. The sire should have performed well in successfully impregnating cows and in the production of large-framed, fast-gaining calves.

There should be no record of dwarfism in the bull's pedigree. The semen should have been tested for fertility within 30 days of use for breeding. Testicle development should be normal. Both testicles should be present, fully descended, sound, and equal in size.

Measuring the circumference of the scrotum is a valuable method of determining the bull's fitness for breeding. Measurement is taken at the widest area of the scrotum using a tape with a slip cinch. The tape is pulled snug, but not so tight that the scrotum is wrinkled. Bulls with larger scrotums produce more sperm. Research has shown that they also sire heifers that reach puberty at an earlier age. Scrotum size has also been positively correlated with weight gain. English and crossbred bulls should have a minimum scrotum size of 11.8 inches (30 cm) at 12 to 14 months of age. A scrotum size of 13.4 inches (34 cm) at this age is considered very good. Research at Virginia Tech reveals that it is possible to predict scrotum circumference in mature bulls by taking measurements when bulls are between 150 and 350 days of age. If the circumference of the scrotum at that age is less than seven inches (18 cm), it is probable that at maturity the bull will not have developed the minimum scrotum size needed for good breeding performance.

There should be no apparent physical or genetic defects. Bulls that are sickle-hocked, pigeon-toed (feet turn in), or splay-footed (feet turn out), or those with crooked ankles should not be selected. The bull should have a quiet disposition. Bulls that have unruly dispositions tend to produce calves of the same disposition, and calves of that kind do not gain rapidly in the feedlot.

JUDGING BEEF ANIMALS

Judging beef animals consists of comparing one animal to another based on the conformation of the animals. There are usually four animals in a class to be judged. The judge divides the animals into three pairs: a top pair, a middle pair, and a bottom pair. The four animals in the class are then examined and compared to the ideal animal.

A good judge follows a definite procedure. The following is a summary of accepted judging procedures.

First, look at the animals from a distance of about 25 feet. This makes it possible to see all the animals at once. Look over the class for several minutes. Keep in mind the major differences among the animals. Placings should be made on major differences. The judge's first impression of the class is usually best. Look at side, front, and rear views of the animals before making a placing (see Figures 14-4 through 14-6).

As the animals are walked, notice the soundness of feet and legs. Finally, study and handle the animals individually to check the placing. Handling the animals permits comparison of finish and natural fleshing. Feel the animals along the top of the shoulders, ribs, back, loin, rump, and round. Check for smoothness, firmness, and uniformity of finish (Figure 14-7).

Each animal in the class is marked with a number, 1 through 4. The final placing is indicated by listing the numbers of the animals in the order of their placing. For example, if the number 3 animal was placed first, the number 1 animal placed second, the number 4 animal placed third, and the number 2 animal placed fourth, the final placing would be stated 3 1 4 2.

When deciding the placing, begin with either the most or least favorable animal in the class, whichever is easier. Follow this by placing the other animals in the class.

Judging Market Classes

The main points to look for in judging market classes are: (1) type, (2) muscling, (3) finish, (4) carcass merit, (5) yield, (6) quality, (7) balance, (8) style, and (9) smoothness.

Type refers to the general conformation of the animal, and is best determined from the side view. The type most in demand is a thick,

FIGURE 14-4   Look at the animal from the side. *Photo by Jim Bodine.*

FIGURE 14-5   Look at the front view of the animal. *Photo by Jim Bodine.*

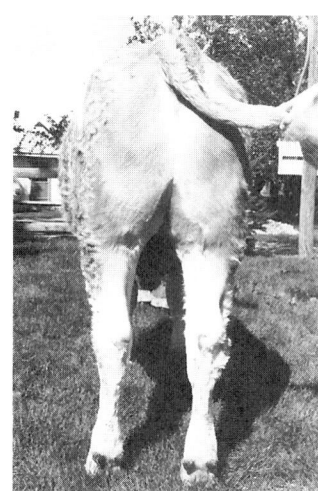
FIGURE 14-6   Look at the rear view of the animal. *Photo by Jim Bodine.*

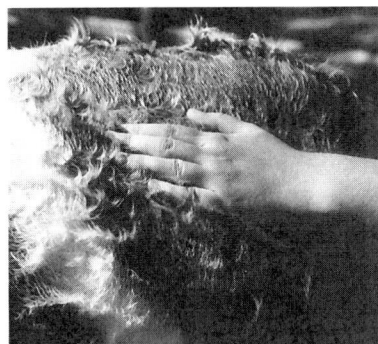

FIGURE 14-7   Handle the animal to check the finish. *Photo by Jim Bodine.*

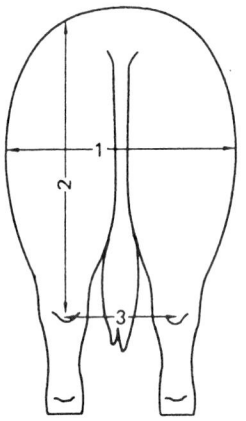

Things to look for:
1. Width of round
2. Depth of round
3. Width between legs

FIGURE 14-8   Things to look for when checking muscling from the rear view of the animal. *Courtesy of University of Illinois at Urbana-Champaign.*

moderately deep-bodied animal. It should have medium length of leg and body, with straight lines and good balance. The current desirable type is one that is not too short-legged. A very low-set steer is just as wrong in type as an extremely tall animal.

Muscling refers to the natural fleshing of the animal. It is an inherited trait. The animal's quarter should be thick, deep, and full. Good width of back, loin, and rump without too much fat also indicates muscling (Figure 14-8).

Finish is the amount of fat cover on the animal. The term *finish* is used only in market classes. A smooth, uniform finish that is not uneven is desirable. When the animal is handled, the finish should be springy. Too hard or too soft a finish is undesirable. The animal should not be too fat or too thin. A 1,000-pound (454-kg) steer should have from 0.3 to 0.6 inch (0.8–1.5 cm) of finish over the loin.

Carcass merit refers to the kind of carcass the animal will produce when slaughtered. A desirable carcass is thick, meaty, and correctly finished. There should be a lot of muscle in the high-priced cuts. Carcass merit must be estimated when judging live animals.

The carcass merit of cattle is expressed by yield grade. There are five yield grades numbered 1 through 5. The fatter the animal, the smaller the percent of yield, since more fat must be trimmed from the retail cuts. *Yield grade* shows the percent of carcass weight in boneless retail cuts from the round, loin, rib, and chuck that have been closely trimmed of fat. The percent of yield for each yield grade is:

Yield grade 1—over 52.4%   Yield grade 4—45.5–47.7%

Yield grade 2—50.1–52.3%   Yield grade 5—less than 45.5%

Yield grade 3—47.8–50.0%

Lighter-weight, lean, heavily muscled cattle are yield grade 1. Heavier, fatter, light-muscled cattle are yield grade 4 or 5. The majority of cattle have a yield grade of 2 or 3.

The USDA uses a dual grading system that includes yield grade and quality grade. Quality grade refers to the amount of marbling and the carcass grade (Prime, Choice, Select, Standard, Commercial, Utility, Cutter, and Canner) of the animal. A more complete description of the quality grades of beef slaughter cattle is found in Unit 19.

Yield is the dressing percent (weight of the chilled carcass compared to the live weight) of the animal. It is found by dividing the chilled carcass weight by the live weight of the animal. This gives a percent of yield. A high yield is desirable. Conformation, finish, quality, and fill affect yield. Other factors affecting yield are the amount of refinement of the head, hide, and bone. An animal that is wasty in the middle will have a lower yield.

Quality refers to the refinement of the head, hide, bone, and hair. A steer should be medium boned. The head should be moderately refined and the hide should be thin and pliable. A soft, fine hair coat is desirable.

Balance refers to the general structure and proportion of the animal's body. A desirable animal is correctly proportioned.

Style means the way the animal shows and the way it carries itself. It should be attractive and alert and show to good advantage (Figure 14-9).

FIGURE 14-9   An animal with style will show to advantage. *Photo by Jim Bodine.*

Smoothness refers to the lack of roughness in finish or bone structure. Rough finish can usually be seen along the top of the animal. Rough bone structure can be seen at the shoulders and hooks. A desirable animal will be smooth rather than rough.

Breeding Classes

Breeding classes are judged for several traits in addition to those listed for market classes. These are (1) condition, (2) size, (3) feet, legs, and bone, (4) breed character, and (5) sex character.

Condition refers to the amount of fat cover the animal has. This is referred to as *finish* in market classes. Always use the term *condition* rather than finish when judging breeding cattle. Muscling is more important than condition when judging breeding cattle. This is because muscling is an inherited trait. Finish depends on the amount of grain the animal has been fed.

Size in relation to age is important when judging an animal. A large animal is better than a small one when both are the same age. Size alone is not the only trait to look for. The other factors related to judging must be kept in mind. Height, by itself, does not mean a larger size. The overall size of the animals must be considered.

Feet, legs, and bone: legs should be strong and straight and have heavy bone. They should be set well out on the corners of the animal. The feet must be large enough to give a good foundation for the animal. Unsoundness in feet and legs makes the animal less valuable for breeding stock.

The breed character is shown in the head and general appearance of the animal. Each breed has its own traits. A good judge must study the traits of the breeds in order to recognize breed character.

Sex character refers to the traits that distinguish the animal as a male or a female. The male should show a heavier development of the forequarters. The neck is thicker and supports a strong, bold head. The overall appearance should be massive and powerful. A female shows more refinement in appearance. The neck and shoulder are lighter. The head and bone show more refinement than those of the male.

Judging Terms

A good livestock judge uses the correct terms to describe the animal being judged. The following is a list of favorable terms used to describe beef cattle. When an animal is being criticized, the opposite of these terms is used. The terms are given in their comparative forms (for example, *thicker* rather than *thick*) as they would be used to compare one animal to another in an actual judging situation.

General terms used for breeding and slaughter cattle:

growthier

heavier muscled

higher quality

more style

more correct set to feet and legs

stronger top

more structurally correct

more even width (uniform width)

more stretch

tighter framed

deeper rib

wider loin

deeper twist

smoother handling

freer, easier in movement

typier

thicker

more nicely balanced

deeper body

smoother top (hooks, tail head, shoulders)

straighter lined

wider, stronger top

bigger framed, more upstanding

longer, leveler rump

more spring of rib

wider back

fuller behind the shoulders

thicker, deeper quarter

thinner hide

Terms and expressions used for slaughter cattle:

more uniformly (desirably, correctly) finished

smoother finished

firmer finished

trimness in brisket, twist, and rear flank

carries more finish over the rump (loin, ribs, back)

thicker fleshed throughout

more finish all down the top

beefier, heavier muscled

deeper, plumper, more bulging round

more natural width and muscling over the back and loin

wider over back and loin

fuller in the rib

wider and fuller in the stifle

thicker, more muscular round

heavier muscled loin

longer, meatier loin

meatier further down the shank

more width through the center of the round

meatier, heavier quartered

wider through the thigh

longer and wider through the rump

larger framed

more size for age

trimmer middled (fronted, brisket)

more refinement of head (hide, bone)

more uniform in width from front to rear

would yield a meatier carcass (more of the high-priced cuts)

would have a more desirable yield grade

would hang up a meatier carcass

would hang up a thicker, heavier muscled carcass

Terms and expressions used for breeding cattle:

typier heifer

more weight for age

larger framed heifer

more size or scale and capacity

larger, more rugged heifer

growthier, beefier heifer

growthier, thicker heifer

more uniform in width from front to rear

deeper, fuller spring of rib

more refined in muscling (female)

moderate thickness of muscling (female)

longer muscling in round (female)

more strength and thickness of muscling down the top (bull)

more bulge and fullness of muscle in the stifle area (bull)

stands more correctly on its legs

stands and walks more correctly

stronger fronted

wider fronted (wider at the chest)

easier walking (more correct on the move)

walks with more style and balance

more feminine head (female)

strong head (bull)

shows more masculinity and ruggedness (bull)

more masculinity about head and neck (bull)

better developed testes (bull)

stronger udder attachment (female)

better balanced udder (female)

more correct udder development (female)

no excessive fat deposits

no excess fat along underline

trimmer, less wasty brisket, rear flanks

more desirable condition

smoother blending of shoulders and rib

smoother through the hooks

neater about the tail head

more style, symmetry, and balance

smoother blending of parts

Oral Reasons

Oral reasons are given to explain the differences between animals that influenced their placings. The differences between two animals are described by using the terms noted earlier and by following these basic guidelines:

Completeness. Mention the main differences; do not spend time on minor details.

Length. Do not speak more than two minutes. Be definite and concise.

Presentation and delivery. When giving reasons, stand 6 to 8 feet from the judge. Look the judge straight in the eye. Hands are kept at the side or behind the back. Stand straight, with feet apart. Reasons are given in a logical order and should be well organized. Speak clearly and not too rapidly. Speak loud enough to be heard, but not too loud. Use complete sentences with correct grammar. Speak with confidence; do not hesitate. Make a concise final statement explaining why the animal at the bottom of the class was so placed.

Accuracy. Only make statements that are true. For example, describe strong points of the animal only if they exist. Grant or admit the good qualities of the other animal in the pair.

Terms. Use the correct terms for the class.
   A set of reasons is organized in the following way:

1. *Introduction*—Give the name of the class and how one placed it.
2. *Top pair*—Give the reasons for placing 1 over 2, using comparative terms. Grant the advantages of 2 over 1, also using comparative terms. Finally, criticize 2 using either descriptive or comparative terms.
3. *Middle pair*—Follow the same outline used for the top pair.
4. *Bottom pair*—Follow the same outline used for the top pair.
5. *Summary*—Give a final statement that explains why the animal at the bottom of the class was so placed. Mention the animal's strong points, if any, and describe its major faults.

Crossbred Steers – Market – Placing 3-2-1-4		
Reason for placing	_Admits or Grants_	_Faults_
3/2 longer rump and loin, heavier muscled, stronger top	2/3 trimmer middled	2 weaker top, rough shoulders
2/1 longer loin and rump, wider through quarter more muscular round, uniformly fleshed over back & fore rib	1/2 better balance, leaner, more style	1 shallow bodied, lacks finish
1/4 more style, trimmer, better balance, stronger typed, better muscle	4/1 more size and length	4 over finished, wasty brisket, rough shoulders

FIGURE 14-10   A sample set of notes for a class of crossbred market steers. Overall placing is shown at the top. Animal 3 was found to be the best in the class.

Taking Notes

Simple notes should be taken while judging the class. They are used to prepare the oral reasons, and are not read to the judge. One way to take notes is to divide the paper into three columns. The first column is headed _Reasons for Placing_. The second is for _Admits_ or _Grants_. The third column lists _Faults_. Use fractions for the top, middle and bottom pairs. For example, 3/2 indicates number 3 animal over number 2 animal. Note what is seen for each pair in the three columns. Figure 14-10 shows a sample set of notes for a market class of crossbred steers.

SUMMARY

The three main kinds of beef cattle production systems are (1) cow-calf producers, (2) purebred breeders, and (3) cattle feeders. Cow-calf producers keep herds of cattle and produce calves for sale to slaughter cattle producers. Purebred producers furnish breeding stock to cow-calf producers. Cattle feeders buy calves or yearlings and feed them to market weights.

Beef animals are selected mainly on the basis of conformation and performance records. Desirable conformation refers to the finish and structure in demand in the marketplace. An animal should produce the maximum amount of the high-priced cuts. Performance records help the producer select animals based on records of previous performance. Some traits are more heritable than others. Cattle producers who use the knowledge of heritability of traits do a better job of improving their beef herds.

Beef animals are judged on the basis of their conformation. A good judge learns the parts of the animal and develops a logical system for evaluating the animal. Presentation of oral reasons, using the proper terms, provides the opportunity to explain why animals were placed in a given way.

Student Learning Activities

1. Using pictures or a live animal, name the parts of the animal.
2. Judge breeding and market classes of beef animals from pictures or live animals. Give oral reasons.
3. Take a field trip to a farm to observe the performance testing procedures being used.
4. Practice performance testing procedures with beef herds on your home farm.
5. Give an oral report to the class on various aspects of selecting beef cattle.
6. Evaluate pedigree information on several beef animals and select the most desirable animals.
7. Attend a beef consignment or dispersal sale and select animals you would buy based on conformation, pedigree, and performance records.
8. If possible, observe the use of ultrasonics in beef selection.
9. Calculate 205-day adjusted weaning weights and adjusted yearling weights.

Review

1. Name the three main types of beef cattle production systems.
2. Briefly describe each system named in question 1.
3. Which type of system is most common in your area? Why?
4. Describe the kind of conformation that is considered desirable in beef cattle.

5. Why is good conformation important when selecting beef breeding stock?

6. Name and briefly describe the kinds of performance records that might be used when selecting beef breeding stock.

7. List nine uses of beef performance testing records.

8. What is a pedigree?

9. How might a pedigree be used when selecting beef breeding stock?

10. Why is selection of the herd bull so important to success in beef production?

11. Briefly describe the traits that are desirable in a beef bull.

12. Describe the general procedure one should follow when judging a class of four animals.

13. List and briefly describe the nine main points to look for when judging a class of market beef animals.

14. Name and briefly describe the five additional points to look for when judging a class of breeding beef animals.

15. What are the five points to remember about procedure when giving a set of oral reasons?

16. Describe the organization of a set of oral reasons.

17. Describe a system for taking notes when preparing to give oral reasons in a judging contest.

Unit *15* *Feeding and Management of the Cow-calf Herd*

Objectives

After studying this unit, the student should be able to

■ plan a feeding program for a cow-calf herd.
■ list and describe approved practices for managing a cow-calf herd.

INTEGRATED RESOURCE MANAGEMENT

Integrated resource management (IRM) is a management tool available to beef producers that uses a team approach to improve the competitiveness, efficiency, and profitability of their beef business. A typical team consists of accountants, bankers, extension personnel, veterinarians, university researchers, soil and water conservationists, and people from allied industries. Team members are all experts in their respective fields. The team analyzes the beef production practices of the farm or ranch to diagnose and solve inefficiencies and make sure the production activities complement each other in a positive manner.

Important components of IRM include setting production and financial goals and accurate record keeping. Both production and financial performance data are gathered and analyzed using the Standard Performance Analysis (SPA). Cattle producers keep accurate financial and production records to be used in the analysis. Typically, the operation is analyzed by its component enterprises. These may include the cow herd, heifer development, feedlot sales, bull sales, pasture use, and forage production. This permits the analysis to pinpoint areas of strength and weakness within the total operation.

The SPA is available as a computer software program to analyze the data. This program was developed at Texas A&M and may be ordered through Cooperative Extension Service offices in those states where beef production is a major component of agriculture. The software has two components: SPA-F for financial information and SPA-P for production information. The SPA-F part of the program helps the operator analyze the economic and financial performance of the operation. It converts cash basis records and inventory changes to an accrual accounting basis. Information returned includes cost per pound of calf weaned, investment per cow, and return on assets. The financial data conforms to the recommendations of the Farm Financial Standards Task Force. The SPA-P part of the program provides information on many production measures, including reproductive and growth performance and feed use. The data may be submitted for inclusion in a national data base for comparison with other beef herds in the region. The SPA may also be completed as handwritten records.

A simpler form of the SPA called SPA-EZ has been developed in an effort to get more cattle producers involved in the program. The simplified form runs in a Microsoft Excel 5.0 format. It provides accurate performance information but in less detail than the full program. The program will determine the annual cost of a cow and the cost per hundredweight of producing a calf. The SPA-EZ may be downloaded from the Internet. There is no charge for the program. The download site may be accessed from a link found at *http://www.beef.org/prodirm/ spa-ez.htm.*

A full SPA analysis that spans several years provides management information that helps the farm or ranch manager

- determine profitability of the enterprise.
- identify components of the business that are meeting or exceeding goals and components that need improvement.
- make better marketing, investment, and production decisions.
- develop goals and track progress toward meeting those goals.
- compare alternative investment opportunities.
- develop employee incentive program.
- monitor and control costs.
- determine how competitive the total business is as well as the individual enterprise on the farm or ranch.
- evaluate the use of resources and identify ways to improve resource use.
- provide information to multiple owners, lenders, and advisors.

Over a recent five-year period, beef producers enrolled in the SPA program in Illinois were able to significantly lower their total costs per cow. This was done by improving the efficiency of the operations after SPA analysis identified areas where cost reduction was possible. Both commercial and purebred beef herds were enrolled in the SPA program. Larger herds had lower total costs per cow than did smaller herds.

The management concepts used in Integrated Resource Management are being increasingly used in other segments of agriculture. Examples include Integrated Pest Management, Integrated Crop Management, and Total Quality Management.

FEEDS

Kinds of Feed

Feeding programs for beef cow-calf herds are based on the use of roughages. Different kinds of roughages are used, depending on where in the United States the beef herd is located. Typical roughages used include pasture, hay, silage, straw, corncobs, and other crop residues. Corn silage is widely used as a roughage feed for beef herds. Alfalfa is the most common roughage in the Midwest and West. Coastal Bermuda grass is more common in the southern coastal states. Native range grasses are utilized to a great extent in the western states.

Less commonly used roughages include oats straw, barley straw, and wheat straw. Cottonseed hulls, peanut hulls, oat hulls, and rice hulls are also used. In the Corn Belt, husklage, corn stover (stalks), corn stover silage, and corncobs may be used. Sorghum and soybean residues also have value for feeding cows.

Grasses often used for pasture or hay include fescue, orchard grass, reed canary grass, smooth bromegrass, Kentucky bluegrass, perennial ryegrass, timothy, redtop, Bahia grass, dallis grass, and Sudan grass. Some of the legumes used include red clover, alsike clover, sercea lespedeza, peanut hay, sweet clover, cowpeas, and soybean hay.

Feed composition tables give the analysis of feeds that can be used for rations for beef herds. Tables of this kind are found in the appendix of this book.

Roughages provide the cheapest source of energy for the cow and the calf. However, adjustments must be made so that the ration meets the needs of cows and calves. If the quality of the roughages is high, little or no supplement is needed. The use of high-quality roughages also increases the weaning weight of calves. Usually, some additional minerals are needed in the ration.

Managing Feed Sources

Forages. Forages should be handled in a way that will keep the labor requirement low. Grazing should be used as much as possible. In some areas, weather does not permit grazing the year-round. In this case, harvesting and storing forages increases the amount available for the beef herd. Crop residues, such as cornstalks, can be used as feed for the beef herd. However, only about 15 to 30 percent of the amount produced is recovered when grazing is the only method of harvesting.

Pasture and hay land. The proper management of pasture and hay land increases the yield of forage. The soil should be tested and the correct type and amount of fertilizer added. If too many cows are fed on a pasture, its quality is reduced. When rotation grazing is used, the carrying capacity of the pasture is increased. *Rotation grazing* is the division of a field by the use of temporary fencing. The cow herd is allowed on only part of the field at a time.

Crop residues. Grazing crop residues reduces feed costs. In northern areas, two acres of cornstalks will carry a pregnant beef cow for about 80 to 100 days. Heavy snow cover reduces the carrying capacity of cornstalk fields (Figure 15-1). The *carrying capacity* of a pasture refers to the number of animals that can be grazed on the pasture during the grazing season.

FIGURE 15-1   This snow cover makes it harder for the cattle to find feed and reduces the carrying capacity of the cornstalk field. *Photo by Jim Bodine.*

Crop residues may be harvested in several ways. Equipment such as balers, forage harvesters, and stackers may be used. Yields of one to two tons per acre may be expected. The moisture content will range from 20 to 55 percent. When stacking or baling, the residues should be below 30 percent moisture content at harvest to prevent spoiling. For corn stover silage, the moisture content should be above 50 percent at harvest and 60 to 70 percent is better. Water should be added if necessary.

Husklage is the material that comes from the corn combine that can be used as a feed. Trailing units are made that will collect this material. It is then dumped in piles and can later be loaded and hauled to a storage site. It can also be chopped for silage or baled. Sometimes it is left in piles in the field for the cattle to eat during the winter.

The digestibility and protein content of harvested crop residues can be increased by treating the residue with anhydrous ammonia (NH_3). Treat the crop residue with an amount of NH_3 equal to 2.5 to 3.0 percent of the weight of the residue.

The material to be treated is sealed under a 6- to 8-mil ultraviolet-resistant clear plastic cover. Black plastic may be used instead of the clear plastic. The ammonia is applied through a hose or pipe, with the amount being controlled with a regulator or gauge. The NH_3 must be applied slowly (one to five minutes per ton of ammonia) to prevent ballooning of the plastic.

The sealed environment must be maintained long enough to complete the treatment of the residue. At low temperatures (under 41°F [5°C]), the environment must be kept sealed longer (more than 8 weeks) than at higher (over 86°F [30°C]) temperatures, where the treatment can be completed in less than one week. The plastic cover can be left on the roughage until it is to be fed. It takes about one day for the free ammonia to dissipate after the plastic is removed.

Treatment with NH_3 can increase the protein content of low-quality roughages by 4 to 6 percentage units. Digestibility may be increased by as much as 10 percentage units. For example, crude protein content might be increased from 4 percent in untreated material to 9 percent in treated material, and digestibility might go from 46 percent in untreated material to 56 percent in treated material. Dry matter intake is also increased when the material is treated with ammonia.

Some problems have been observed when feeding ammonia-treated roughages. Usually these problems resulted when the roughage that was treated was high-quality with a high soluble sugar content, was above 20 percent moisture content when treated, and when an excessive amount of NH_3 (over 3 percent of forage dry matter) was used in the treatment. No problems have been observed from feeding low-quality roughages that have been treated with ammonia.

Experimental work has shown that treating crop residues such as wheat straw, corncobs, and cornstalks with alkaline hydrogen peroxide can increase the digestibility of these materials and release more energy for use by the ruminant animal. Feed intake was also higher with the treated material as compared to untreated material. The result of this experimental work suggests that, with treatment, more crop residue materials might be used in the diet of ruminant animals. This could reduce the use of cereal grains in the production of ruminant animals.

Use of round bales. The use of large round bales (Figure 15-2) to harvest some types of forage can cut labor requirements by up to 60 percent. However, there is about a 20 percent loss of dry matter, energy, and protein content when harvesting this way as compared to using rectangular bales. Research at Pennsylvania State University shows that about 20 to 32 percent more dry matter is needed in the ration when large round bales are fed. Weight gains and cow condition scores are lower when large round bales are fed outside, free choice. Weight gains and cow condition scores are higher with limit-fed rectangular bales that are stored inside.

Large round bales should be baled at 16 to 20 percent moisture to prevent molding. Make the bale tight to prevent loss and store it in a well-drained area.

Losses can be reduced when large round bales are stored inside or under some type of cover. Losses can be as high as 50 percent of the value of the hay if it is not protected during storage. Research at Louisiana State University shows that losses of dry matter in large round bales is approximately 6 percent for inside storage, 15 percent when the bales are stored on wooden platforms and covered with plastic, and 66 percent when the bales are stored unprotected on the ground.

A pole-barn shed can be used for large bale storage. This is probably the best method of storage to prevent losses. Other methods of protection include covering the bales with plastic or completely enclosing them in plastic bags. Equipment is available that puts bales into long plastic tubes. Each tube is 200 feet (61 meters) long and will hold approximately 36 large round bales. If grass hay is enclosed in plastic bags, it is possible to treat it with ammonia to increase its pro-

FIGURE 15-2 Some types of forage can be harvested in large round bales.

tein content. To prevent spoilage on the bottom of the bales, put them on old tires, pallets, crossties, poles, gravel, or other objects that will keep them from making direct contact with the ground. Geotextile materials may be used to create a barrier between the ground and the rock or gravel spread on top. This material may be spun fabric or woven fabric. It is rolled out on the ground and then six inches of rock or gravel is spread on top. This surface becomes very compact over a period of time and provides a good way to reduce waste when large round bales are stored on it.

Some research has been done on using a specially hydrogenated liquid animal fat (tallow) or plant fat (a soy protein isolate) to completely cover the outside of large round bales. After application, the fat hardens into a paraffin-like consistency. No mold or spoilage was found in the bales when the hay was baled at less than 18 percent moisture. Almost no loss in nutrients occurred in bales treated with this covering. Animals will eat the outer layer of hay covered by the material used in the treatment. A tightly wrapped bale, with four-inch twine spacing, gives the best result when using this treatment. The treatment is less successful when bales are loosely wrapped.

The economic returns from protecting large bales in storage are higher in areas with high rainfall. However, some research indicates that there is some loss in feeding value when the bales are unprotected even in areas of low rainfall.

Bales should be placed in such a way that air can circulate around them. If the bales are stored outside, place the ends north and south so they will get more sunlight and dry out quicker after rain.

Access to bales, husklage piles, or stacks of forage is controlled to prevent waste. Feeding gates or electric fencing may be used to control access. A hay feeding rack on skids can also be used. Any system of feeding hay will result in some loss, but the amount of loss can vary widely based on the methods used for feeding and controlling access to the feeding area. When cattle are given free access to the feeding area, losses may be as high as 60 percent. Poor management of forage feeding may increase the total cost of production by as much as 10 percent. Careful management can reduce losses to as little as two percent. Using feed racks, rings, or feeders for any type of forage bales will significantly reduce waste. The quality of the hay also influences the amount of waste; usually more low-quality hay than higher-quality hay is wasted. It may be necessary to limit the amount of hay fed at any one time to no more than a one-day supply in order to reduce waste.

FEEDING DRY, PREGNANT COWS AND HEIFERS

Dry, pregnant cows are fed enough to keep them in good flesh from fall to spring calving. Cows that are of normal weight in the fall should not lose more than 10 percent of their body weight. Thin cows should be fed enough to cause some weight gain during the winter. If cows do not receive enough feed, the calf crop percent is lowered.

Overfeeding should be avoided for several reasons. Cows that are fed too much become too fat. The cost of keeping the cow is more. In addition, cows that are too fat have more trouble calving. Calf losses are higher and milk flow is less.

During the fall grazing season, fall pastures, permanent pastures, and crop residues from small grain fields or cornstalks may be used. During the winter, hay, silage, or harvested crop residues may be fed to the cow herd. If the climate permits, pasture may be used all winter.

Young cows and heifers are still growing and require more feed than mature animals. The amount of feed the cow receives is more important than the kind of feed. The amount of energy the cow needs varies according to size, condition, age, and weather. During cold weather, increase feed or energy intake by one percent for each degree of cold stress.

During the last 30 to 45 days of pregnancy, cows generally need a 10 to 15 percent increase in the protein in the ration, especially if they are being fed stored roughage such as hay. This can be provided with an extra two pounds of high-quality hay per day. Additional protein supplement may be fed if the roughage is low in quality.

The nutrient requirement tables in the appendix of this text may be used to develop rations for the beef breeding herd. Sample rations for wintering dry, pregnant beef cows are given in Table 15-1; sample wintering rations for bred heifers are given in Table 15-2.

Minerals should be fed free choice to the cow herd. Mineral mixes should include calcium, phosphorus, salt, and any trace minerals that are known to be deficient in the area. In areas where grass tetany is a problem, magnesium oxide is included in the mix. (See Unit 17 for a description of this disease.) A good mixture to use is one part trace mineral salt and one part dicalcium phosphate.

Protein supplement is often not needed if high-quality roughages are used. A ration that feeds low-quality roughage, such as range grasses, requires a protein supplement. As much as one-third of the protein in the ration may have to be provided as supplement to cows on range.

Protein blocks are a convenient way to feed protein to the cow herd. The salt content and the hardness of the block control the amount the cow eats. Liquid protein supplements may be fed in lick tanks. Cubed or pelleted protein supplements are handy for hand feeding. However, plenty of bunk space must be provided if they are used. "Boss" cows will get more than their share if bunk space is limited.

Care must be taken to prevent overeating of protein. A hungry or thirsty cow will eat too much protein. Overeating of protein may be partially controlled by feeding plenty of roughage. Make sure there is plenty of water available. This will help flush out excess protein if the cow overeats. Self-feeding of urea is not recommended. Urea may be

TABLE 15-1 Rations for Wintering Dry, Pregnant Beef Cows (wt. 1,000 to 1,100 lb; 453 to 499 kg).

Ration	Amount per Day	
	(lb)	(kg)
1. Legume hay	16–25	7.26–11.3
2. Mixed legume-grass hay (1/3 legume)	18–22	8.2–9.98
3. Legume hay	5–10	2.3–4.5
Straw or low-quality grass hay	10–15	4.5–6.8
4. Legume-grass haylage	30	13.6
5. Corn or grain sorghum silage	35–50	15.9–22.7
Protein supplement (48% Total Protein)	0.5–1	0.23–0.45
6. Legume-grass hay	10	4.5
Straw or cobs	10	4.5
7. Corn or sorghum silage	30	13.6
Legume hay	5	2.3
Straw, low-quality grass hay, cottonseed hulls, ground corncobs, or other low-quality roughage	Unlimited	Unlimited
8. Prairie or grass hay	Unlimited	Unlimited
Protein supplement	0.5–1	0.23–0.45
9. Grass silage	30–40	13.6–18
Straw or low-quality grass hay	Unlimited	Unlimited
10. Grazing crop residue	Unlimited	Unlimited
11. Cornstalk silage	40	18
Legume hay or hay silage	4–5	1.8–2.3
12. Corn silage	30	13.6
Mixed hay	4	1.8
13. Grass silage	25–35	11.3–15.9
Grass or mixed hay	10	4.5
14. Mixed or grass hay	10	4.5
Pea-vine silage	25–35	11.3–15.9
15. Husklage	11	4.99
Alfalfa brome hay	7	3.2
16. Corn stover	12	5.4
Alfalfa brome hay	7	3.2
17. Corn stover	15	6.8
Shelled corn	4	1.8
Protein supplement (35% Total Protein)	1.1	0.5
18. Husklage	15	6.8
Shelled corn	3	1.4
Protein supplement (35% Total Protein)	1.1	0.5
19. Corn silage	43	19.5
20. Soybean stover	20	9
Alfalfa brome hay	7	3.2

Source: Various University Cooperative Extension Service bulletins.

TABLE 15-2 Wintering Rations for Bred Heifers (wt. 800 to 900 lb; 363 to 408 kg).

Ration	Amount to Feed	
	(lb)	**(kg)**
1. Legume-grass hay	20	9
2. Corn silage	50	22.7
Soybean meal	1	0.45
3. Corn silage	25	11
Legume-grass silage	10	4.5
4. Corn silage	45	20.4
Protein supplement (48% Total Protein)	1.5	0.68
5. Legume-grass haylage	35	15.9
6. Legume-grass hay	20	9
Corn (shelled)	5.6	2.5
7. Alfalfa hay	22.8	10.3
Corn (shelled)	3.4	1.5
(Feed vitamin-mineral mix free choice.)		

Source: Various University Cooperative Extension Service bulletins.

toxic to the cow if too much is eaten. The intake of urea must be carefully controlled.

Vitamin A may be needed if poor-quality roughages are fed. Good-quality green roughages usually provide plenty of vitamin A. If the cow has been on good summer pasture, enough vitamin A will be stored in the body to carry the animal for several months in the winter. If the ration is short on vitamin A, 30,000 IU should be fed during the last months of pregnancy.

LACTATION RATIONS

The ration needed for the cow that is nursing a calf depends on how much milk the cow produces. Heavier milk producers have higher requirements than average or low milk producers. The protein requirements for lactation are 160 to 268 percent greater than for dry cows. Energy needs are 36 to 68 percent more. Calcium and phosphorus needs are 100 to 250 percent higher. Vitamin A needs are 18 to 88 percent greater.

Pastures of high quality can usually meet the needs of the lactating cow. Salt and minerals should be provided free choice. When the roughage is of poor quality or limited in amount, some grain should be added to the ration. If there is only a limited amount of legume forage in the ration, protein supplement should also be added.

First-calf heifers (those calving for the first time) require more feed. The heifers are still growing and developing. Weight lost from calving must be regained. Additional feed is necessary if the heifer is to produce enough milk for the calf. The heifer must be in good condition for rebreeding.

When pasture is not available, the lactating cow is fed in drylot. Table 15-3 gives some drylot rations that can be used for lactating cows. Table 15-4 shows sample rations for first-calf heifers in drylot.

TABLE 15-3 Lactating Rations for Cows in Drylot (wt. 1,100 lb; 499 kg).

Ration	(lb)	(kg)
1. Legume-grass hay	10	4.5
Corn silage	40	18
Vitamin A	40,000 IU	
2. Legume-grass hay	30	13.6
3. Legume-grass hay	20	9
Corn	4	1.8
4. Legume-grass silage	74	33.6
5. Legume-grass haylage	50	22.7
6. Corn or grain sorghum silage	60	27.2
Protein supplement (48% Total Protein)	1.5	0.68
7. Mixed hay	20	9
Shelled corn	5	2.3
8. Corn stover	20	9
Shelled corn	5	2.3
Protein supplement (35–40% Total Protein)	2.8	1.3
9. Alfalfa haylage (55% DM)	29.8	13.5
Shelled corn	8	3.6
10. Corn silage	50–55	22.7–24.9
Legume hay	4–5	1.8–2.3

Source: Various University Cooperative Extension Service bulletins.

TABLE 15-4 First-Calf Heifers—Lactation Rations—Drylot.

Ration	(lb)	(kg)
1. Legume-grass hay	25	11.3
Ground shelled corn	3	1.4
2. Corn silage	60	27.2
Soybean meal	1.5	0.68
3. Corn silage	30	13.6
Legume-grass hay	13	5.9
4. Legume-grass silage	65	29.5
Ground shelled corn	3	1.4

Source: Various University Cooperative Extension Service bulletins.

CREEP FEEDING OF CALVES

Creep feeding is a way of providing calves with extra feed. The feed may be grain, commercial creep feed mix, or roughage. It is fed in a feeder in an area cows cannot reach.

The Decision to Creep Feed

Creep feeding may or may not be profitable. The kind of operation, production conditions, and marketing practices influence whether or not creep feeding is economical.

Creep feeding

- produces heavier (30–70 lbs [13.6–31.8 kg]) calves at weaning.
- produces higher grade and more finish at weaning.
- results in calves going on feedlot rations better at weaning time.
- creates less feedlot stress.
- allows cows and calves to stay on poorer-quality pasture for a longer time.

Creep feeding is often used if

- calves are to be sold at weaning.
- calves are to be fed out on high-energy rations.
- cows are milking poorly.
- calves are from first-calf heifers.
- calves were born late in the season.
- calves have above-average inherited growth potential.
- calves were born in the fall.
- calves are to be weaned early (45–90 days).
- calf-feed price ratio is favorable.
- pastures become dry in late summer.
- cows and calves are kept in confinement.

Creep feeding has the following disadvantages:

- If calves are well fed after weaning, the weight advantage from creep feeding is lost.
- When production testing, it is harder to detect differences in inherited gaining ability.
- Replacement heifers may become too fat.
- Non-creep-fed calves usually make faster and more economical gains after weaning as compared to calves that were creep fed before weaning.

Creep feeding is generally not used if

- calves are to be fed through the winter on roughages.
- cows are above-average milk producers.

- the calf-feed price ratio is poor.
- calves are on good pasture.
- heifers are to be kept for herd replacements.
- the milk production of the dam is to be measured.

When and How to Creep Feed

Calves will start to eat grain at about three weeks of age. They eat only small amounts until about six to eight weeks of age. It takes 6 to 9 pounds (2.7–4.0 kg) of feed for each one pound (0.45 kg) of gain. About 280 to 480 pounds (127–218 kg) of feed are required for 40 to 60 pounds (18–27 kg) of gain.

Grains alone will often meet the energy needs of the calf. Milk and pasture provide the protein, minerals, and vitamins the calf needs. Whole oats and cracked corn mixed 50-50 is a good, simple creep feed. If molasses is added to the mix, the feed will taste better, and the calf will eat more. Research at Minnesota reveals that calves prefer rolled shelled corn and linseed meal pellets over whole oats or whole shelled corn. Table 15-5 gives some suggested creep feed rations for self-feeding.

Commercial creep feeds are available. Pelleted mixes are well liked by calves. They are easy to handle and do not blow away in the feeder. The calves eat more and gain more rapidly. Some commercial mixes are medicated. Mixes usually include minerals and vitamins.

Locate the creep feeder near the area where the cows loaf. If salt for the cows is put near the creep feeder, they will gather in the area. This brings the calves to the area, and they will use the creep feeder. Waterers should be located nearby. Locate the creep feeder in the shade, if possible.

The amount of feeder space needed is 4 to 6 inches (10.0–15.2 cm) per calf. Self-feeders or bunks that are portable should be used.

TABLE 15-5 Rations for Creep Feeding Calves—Self-Fed (100 lb [45 kg] mix).

Ration	*(lb)*	*(kg)*
1. Shelled corn	100	45
2. Shelled corn	50	22.7
Oats	50	22.7
3. Ground ear corn	90	40.8
Soybean meal	10	4.5
4. Corn or barley (rolled, cracked, or coarsely ground)	50	22.7
Whole oats	30	13.6
Protein supplement (6.7 to 26% Total Protein)	10	4.5
Molasses (dried or liquid)	10	4.5
5. Shelled corn	90	40.8
Soybean meal	10	4.5

Source: Various University Cooperative Extension Service bulletins.

When feeding for limited energy intake, include a hay self-feeder in the creep. Cover the feeders to protect the feed from the weather.

The opening into the creep must be small enough to keep the cows out. An opening 16 inches (40.6 cm) wide and 36 inches (91.4 cm) high is about right. An adjustable opening is best.

Implants

See Unit 7 for a discussion on the use of implants in beef production.

GROWING REPLACEMENT HEIFERS

At least 15 percent of the cows from the breeding herd are lost each year due to death, breeding failure, or aging. About 30 to 40 percent of the heifers are saved each year to replace the cows that leave the herd. Performance records are the best way to decide which heifers to keep. Only those in the top half in weaning weight should be kept.

Heifer conception rates are lower than cow conception rates. Check the heifers for pregnancy 60 to 90 days after breeding. About 60 to 70 percent of those bred are kept, based on pregnancy and adjusted yearling weight.

Replacement heifers of the British breeds should gain 1.0 to 1.25 pounds (0.45–0.56 kg) per day from weaning to breeding. Larger breeds should gain 1.25 to 1.75 pounds (0.56–0.79 kg) per day.

Puberty is the age at which heifers come in heat. In an efficiently managed cow-calf herd, heifers should reach puberty at 12 to 14 months of age. This goal can be attained with proper selection and feeding of replacement heifers. Generally, heifers reach puberty when they have attained about 65 percent of their mature weight. Heifers of the English breeds should weigh 550 to 625 pounds (249–283 kg) at puberty. Larger breed and crossbred heifers of larger breeds should weigh 675 to 750 pounds (306–340 kg) at puberty. Heifers should be bred according to weight rather than age.

Heifers on good pasture will gain 0.75 to 1.4 pounds (0.34–0.65 kg) per day. When pastures are poor, feed 3 to 5 pounds (1.4–2.3 kg) grain per head per day. At calving time heifers should weigh 900 to 1,050 pounds (408–476 kg).

Feed for heifers must be palatable. Coarse, poor-quality feed is not as good in the ration as better-quality feeds. In regions of cold weather, the heifers need more feed for energy to maintain body heat. Nutrient needs increase 1 percent for each degree of temperature below freezing. Nutrient needs also increase when the heifers are in unprotected areas.

As much as 15 percent of the feed that is fed is wasted. Allowance for waste must be calculated when supplying feed. The amount of feed must be increased as the heifers become heavier. As weight increases, the need for energy is greater. Young heifers do not make good use of non-protein nitrogen sources such as urea.

TABLE 15-6 Rations to Grow Replacement Heifers (450–500 lb; 204–227 kg) (gain 1–1.25 lb [0.45–0.56 kg] daily).

Ration	*(lb)*	*(kg)*
1. Legume-grass hay	10	4.5
Oats	3	1.36
2. Legume-grass haylage	25	11.3
3. Legume-grass hay	10	4.5
Ground ear corn	4	1.8
4. Corn silage	30	13.6
Soybean meal	1.5	.68
5. Corn silage	20	9.1
Legume-grass hay	6	2.7
6. Alfalfa hay	12.5	5.7
Shelled corn or ground ear corn	2.2	1
7. Sorghum silage	43.4	19.6
Protein supplement (35% Total Protein)	1.7	.77
8. Alfalfa hay	4.6	2.1
Corn silage	25.7	11.6
9. Grass hay (brome, orchardgrass, canarygrass)	11.2	5.1
Shelled corn or ground ear corn	3.4	1.5
10. Alfalfa hay	12.5	5.7
Oats	2.6	1.2

(Feed vitamin-mineral mix free choice.)

Source: Various University Cooperative Extension Service bulletins.

Rations for growing replacement heifers that are weaned at 450 to 500 pounds (204–227 kg) are given in Table 15-6. These rations give a daily gain of 1.0 to 1.25 pounds (0.45–0.56 kg). Vitamins and minerals are to be fed free choice with these rations.

GROWING, FEEDING, AND CARE OF BULLS

Wean bulls at six to eight months of age. Feed high-energy rations for about five months after weaning to discover which bulls gain best. Avoid fattening the bulls, however. The best-gaining bulls are used in the herd or kept for sale.

Bulls are full fed until spring. They are then put on pasture to complete growth. Well-grown bulls may be used for breeding at 15 to 18 months of age. Bulls continue to grow slowly until about four years of age.

Bulls fed corn silage are also fed grain at the rate of 1 percent of body weight. When hay or haylage is used, feed grain at the rate of 1.5 percent of the body weight of the bull. Poor-quality roughage must be supplemented with protein. Minerals are fed free choice. Vitamin A is fed at the rate of 30,000 to 50,000 IU per day if the ration is mostly corn silage or limited hay.

FIGURE 15-3 Growing bulls may be self-fed. *Courtesy of University of Illinois at Urbana-Champaign.*

Bulls may be self-fed (Figure 15-3) or hand fed. When self-feeding, use plenty of roughage so the bull will not become too fat or *go off feed* (fail to eat the proper amount of feed).

Table 15-7 lists feed mixtures to use for bulls from weaning to about 700 pounds (318 kg). Table 15-8 gives feed mixtures for bulls over 700 pounds (318 kg). Trace mineralized salt is fed free choice with these rations.

Yearling bulls should gain 1.5 to 2 pounds (0.68–0.90 kg) per day. During the winter, grain may have to be added to the ration for these bulls. The amount of grain required is from 0.5 to 1.0 percent

TABLE 15-7 Feed Mixtures for Weaned Bull Calves (weaning to about 700 lb [318 kg]).

Ration	(lb)	(kg)
1. Corn	1,200	544
Alfalfa-grass hay (ground)	600	272
Soybean meal	200	91
2. Corn	1,200	544
Oats	600	272
Soybean meal	200	91
3. Ground ear corn	1,700	771
Soybean meal	300	136
4. Corn silage	1,860	844
Protein supplement (32–35% Total Protein)	156	71
(Includes vitamin mix and trace minerals)		
5. Shelled corn	460	209
Corn silage	1,431	649
Protein supplement (32–35% Total Protein)	156	71
(Includes vitamin mix and trace minerals)		

Source: Various University Cooperative Extension Service bulletins.

TABLE 15-8 Feed Mixtures for Bulls over 700 lb (318 kg).

Ration	(lb)	(kg)
1. Shelled corn	1,200	544
Alfalfa brome hay	756	343
2. Corn silage	1,940	880
Urea	50	23
Dicalcium Phosphate	10	4.5
3. Shelled corn	440	200
Corn silage	1,500	680
Urea	50	23
Dicalcium Phosphate	10	4.5

(Feed trace mineralized salt free choice.)

Source: Various University Cooperative Extension Service bulletins.

of body weight. If corn silage is included in the ration, grain need not be added.

Two- to four-year-old bulls need more energy and protein than does the cow herd during the winter. These bulls should gain 1.0 to 1.5 pounds (0.45–0.68 kg) per day. A full feed of silage plus 2 pounds (0.90 kg) of 40-percent protein supplement may be used. A good-quality legume hay at the rate of 16 to 20 pounds (7.25–9.0 kg) plus 10 pounds (4.5 kg) of grain per day will also meet the needs of these bulls. Feed a mineral supplement with the ration.

Mature bulls in good condition are fed the same as the cow herd. If the bulls are thin, feed 5 to 6 pounds (2.3–2.7 kg) of grain per day above the amount fed the cow herd. If at least one-half of the feed is legume hay, corn silage, sorghum silage, or grass silage, there will be enough vitamin A. If there is no legume in the ration, feed 1 to 2 pounds (0.45–0.90 kg) of a high-protein supplement. Feed a mineral supplement with the ration.

Bulls lose weight during the breeding season. Enough feed must be fed so that they gain back the weight lost. Additional feed is often needed 6 to 8 weeks before the start of the next breeding season. Bulls that are too fat or too thin have poor fertility. They should be kept in medium flesh and given plenty of exercise. During the breeding season, the bull may need 1 pound (0.45 kg) of protein supplement and 5 pounds (2.3 kg) of grain per day. The amount of feed given is based on the condition of the bull.

Rotate the use of bulls in a large herd during the breeding season. Keep the bull separate from the cow herd when not breeding. On farms where there are no facilities to keep the bull separate, the bull can be run with steers or pregnant cows. Bulls can be dangerous. Care must be taken when handling or working around them.

A lame or long-hooved bull will be a slow or reluctant breeder. If necessary, the hooves should be trimmed several weeks before the breeding season begins. In addition, the bull's semen should be checked for fertility before the breeding season begins. Ninety percent of all breeding problems can be eliminated by giving the bull a physical examination for reproduction soundness before the breeding season.

MANAGEMENT OF THE HERD DURING BREEDING SEASON

The goal of the cow herd owner is a 100 percent calf crop. Careful management helps in reaching this goal. The cow herd should be closely observed during the breeding period. Check for injured or diseased cows or bulls. Watch to see if the bull is servicing (mating with) the cows.

The increased use of sire performance records and scrotal circumference measurements in selecting herd sires has resulted in beef bulls that are more fertile than in the past. These bulls can be mated with more cows than was previously possible. Young bulls with a scrotal circumference of 13.4 inches (34 cm) can mate with 20 to 25 cows with no loss in pregnancy rate. Mature bulls can mate with 25 estrus-synchronized cows or 35 to 40 non-synchronized cows without a decrease in the pregnancy rate. Under range conditions, many ranchers plan to use four bulls per 100 cows. Research in Colorado suggests that the increased pregnancy rate when using four bulls per 100 cows is enough to justify the added cost of the bulls.

The bull should be replaced if a high percent of the cows do not become pregnant after two matings. Check the semen for fertility if it is suspected that the bull is a poor breeder.

Run the bull with the herd for no more than 60 days to maintain a short calving season (40 to 60 days). The majority of the cows should calve early in the season. Begin breeding 20 to 25 days after half of the calves are born. This allows a second and possibly a third heat period for those cows that do not settle (become pregnant) the first time. Older cows come in heat sooner after calving than do first-calf heifers. If yearling heifers are bred 20 days before older cows, they will stay on schedule for the second breeding season.

The conception rate is higher for cows that are gaining weight just before and during the breeding season. Cows that are too fat or too thin are poor breeders. Check the cows for pregnancy 60 to 90 days after the breeding season. Any cows that are still not bred at that time should be sold.

Hot weather lowers the conception rate. Shade, water, and fly protection are necessary during breeding.

The use of a sensing device to determine when cows are coming into heat is being studied by the USDA. The device can also indicate when cows are about to calve.

The device measures the electrical conductivity of the tissue in the vulva and uterine wall. Changes in conductivity indicate when the cow is ovulating or about to calve. The device transmits a signal that may be relayed to a computer in the farm office. Currently the device transmits a weak signal that is difficult to detect.

ARTIFICIAL INSEMINATION

Artificial insemination is the placing of sperm in the female reproductive tract by other than natural means. The breeder uses an inseminating tube to deposit sperm in the cervix and uterus of the cow. Sperm is collected from the bull by any one of several methods. The most common is the use of the artificial vagina. At each ejaculation, a bull produces about 5 cc of semen containing millions of sperm. Five hundred or more cows can be bred from the semen collected in one ejaculation.

An artificial vagina is a tube from 10 to 14 inches (25.4 to 35.5 cm) in length and about 2.5 inches (6.4 cm) in diameter. The outer tube is usually made of some type of heavy rubber, although metal or plastic is sometimes used. There is an inner lining of a thinner rubber. A collection cone and vial is attached to one end of the tube. The other end is open to permit the entrance of the bull's penis. The space between the inner lining and the outer tube is filled with warm water, usually about 140°F (60°C).

The bull is trained to mount either a dummy or a live animal. Cows in heat have been used, although it is becoming common practice to use a steer. The bull's penis is guided into the artificial vagina when he mounts the dummy or the live animal. The bull ejaculates into the artificial vagina and the semen is collected in the vial attached to the device. The technician must be trained and have experience to properly collect usable semen for use in artificial insemination.

The semen is cooled slowly after collection. Rapid cooling kills the semen. The semen can be stored for about one week at 41°F (5°C). Semen that is to be stored for longer periods of time is cooled to –320°F (–195.5°C). Semen stored at this lower temperature will remain fertile for several months.

Semen is placed in the cow's reproductive tract by the rectovaginal method. A plastic glove is placed over the hand and arm of the inseminator. The hand is inserted into the rectum of the cow and the feces removed. The cervix is then grasped through the wall of the rectum. The inseminating tube is inserted through the vagina into the cervix. It is guided by the hand in the rectum. The semen is deposited anywhere from the middle of the cervix to just into the body of the uterus.

The use of artificial insemination with beef cow herds depends on the facilities available. There must be a system that allows close observation of the herd to detect cows in heat. The system must also provide for confining and holding the cow for breeding. It is difficult to use artificial insemination under range conditions. It is more practical for smaller herds.

When using artificial insemination, the following practices should be followed:

1. Check the herd in the early morning and late evening to detect cows in heat.
2. Breed about 12 hours after the cow is first seen in heat.
3. Provide good handling facilities, including a breeding chute.
4. Handle the cows gently.
5. Use a good identification system for the cows.
6. Keep accurate records of date bred, date returned in heat, and calving date.
7. If breeding a purebred herd, check breed association rules for the use of artificial insemination.

Some of the advantages of using artificial insemination are that it

- permits use of superior, performance-tested bulls, in any herd.
- permits easier use of exotic breed bulls.
- improves records for performance testing.
- increases the number of cows that can be bred to superior bulls.
- coordinates well with estrus synchronization programs.
- reduces the spread of disease.

Some of the disadvantages of artificial insemination are that it requires

- a trained inseminator.
- more time and supervision of the herd.
- sterile equipment.
- special handling facilities.

Breeding service catalogs provide information about sires available for artificial insemination. Included is performance data about the animal's sire, dam, and offspring. Predicted performance of offspring may also be available. Careful study of this data will result in the selection of better sires. Unit 14 presents information to be considered when selecting beef-breeding animals.

SYNCHRONIZATION OF ESTRUS

Synchronization of estrus is the use of various compounds (usually hormones) to cause all of the females in a herd to come into heat within a short period of time. The ultimate purpose of synchronization of estrus is to have all of the cows in the herd calve within a short period of time.

The advantages of estrus synchronization include:

- labor involved in detecting estrus in the herd is reduced.
- the scheduling of artificial insemination is easier because most of the females are in heat during a short period of time.
- more uniform calf crops are produced.
- the breeding and calving seasons are reduced.

Some disadvantages of estrus synchronization include:

- sometimes conception rates are low.
- the cost per head is fairly high.
- labor for artificial insemination and calving is concentrated in a short period of time.
- many cows are calving during the same 24 to 36 hours.
- adequate handling facilities must be available.

An estrus synchronization program requires a high level of management. Cows in poor condition will not respond well to treatments used to synchronize estrus. It is important to maintain the proper level of nutrition for the breeding herd and make sure they are in good health. Good facilities are needed for handling the animals and care must be taken to avoid causing the animals stress when handling them. Skilled labor is necessary to help during breeding and calving periods. It is wise to plan a synchronization program with a veterinarian.

WHEN TO BREED HEIFERS

Size is the most important consideration when breeding yearling heifers that are sexually mature. Heifers should weigh 550 to 750 pounds (250–340 kg) when bred. This weight should be from growth, not from fattening.

Age is the second consideration. The goal is to breed the heifers to calve at two years of age. When achieved, this results in an average of one more calf produced during a cow's lifetime.

Calving at two years of age brings cows into production at lower cost. It keeps a higher percent of cows in the herd in production. Fewer replacement heifers are needed each year to maintain a stable herd size.

The conception rate is lower for yearling heifers than for older cows. This lowers the percent of calf crop produced. A longer calving season normally results from breeding yearling heifers. Younger heifers need more help at calving than do older cows.

Good management reduces the calving problems of two-year-old heifers. Heifers are bred earlier so that they will calve about 20 to 30 days before the older cows. This allows more time for heifers to come in heat after the first calving. Heifers require more feed, and should be kept separate from the older cow herd.

Breeding to a bull that sires smaller calves also results in less trouble at calving time. Give heifers additional attention when they are calving. Breed heifers for 40 to 60 days, then check for pregnancy 60 to 90 days later. Heifers that are not bred should be sold.

CALVING

Calving should occur during a 40- to 60-day period. This results in calves that are more uniform in weight and age. Also, it is easier to manage the herd when a shorter calving season is maintained.

Most cow herd operators calve in the spring of the year. Spring calving requires less housing, less feed during the winter, and less labor. Cows with calves do better on pasture. The calves are then ready for the feedlot in the fall when the feed supply is largest.

Calving should start six weeks to three months before pasture season. The exact time depends on the local climate.

Observe the herd closely for signs that the cows are about ready to calve. Good breeding records help the producer plan for calving

FIGURE 15-4 Calving on clean ground helps prevent scours. *Courtesy of University of Illinois at Urbana-Champaign.*

time. When the cow is close to calving, the udder and vulva swell. There is a loosening in the area of the tailhead and pinbones.

An effort should be made to save every calf that is born alive. Calving on clean pasture helps prevent scours (Figure 15-4). Clean, dry sheds or stalls should be used on late winter or early spring pasture in areas where the weather is bad.

Very few older cows need help at calving. About one-half of the two-year-old first-calf heifers will need help at calving. Provide help at calving only if it is necessary.

If the cow has been in labor for two hours or more and there is no sign of progress in giving birth, help is needed. Calving problems are also indicated if: (1) only the tail or head of the calf is visible; (2) more than two feet are visible; (3) the feet are visible beyond the knees, and the head is not visible; (4) the head and only one foot is visible. The experienced cattle raiser can handle some of these problems if the right equipment is available. More serious problems require the help of a veterinarian.

After the calf is born, make sure that it breathes. It may be necessary to clean the mucus from the mouth and nostrils. If necessary, artificial respiration is given by alternate pressure and relaxation on the wall of the chest.

In cold weather, the calf must be kept warm and dry until it is on its feet. The navel cord is disinfected with a 2-percent iodine tincture solution. The calf needs to nurse shortly after birth (Figure 15-5). A weak calf must be helped to nurse. The first milk from the cow, called colostrum, contains nutrients, such as vitamins A and E, and antibodies that the calf requires.

The placenta (afterbirth) is normally expelled within 12 to 24 hours after the calf is born. If it is not, call the veterinarian.

Separate the cows that have calved from those that have not. Provide plenty of clean, fresh water for the cows. Cows with calves need more feed to produce the milk to nurse the calf.

The calf is identified with an ear tag or tattoo. The date of birth should be recorded. Difficulties in calving and the birth weight of the calves are also recorded for performance records.

CASTRATION

Castration is the removal of the testicles of the bull calves. A castrated beef animal is called a *steer*. Calves can be castrated at birth. Bull calves that are going into the feedlot are castrated before they are three to four months of age. Bulls that are castrated after eight months of age have a staggy appearance and bring less profit when sold. (A bull that is *staggy* has the appearance of a mature male.) There are more problems with bleeding and weight loss when older bulls are castrated. Bull calves gain slightly faster to weaning weight. However, this is not

FIGURE 15-5 Nursing calf. *Courtesy of University of Illinois at Urbana-Champaign.*

an advantage. The shrinkage that results from the shock of castration at the older age offsets the initial faster gain.

Bull calves are castrated in several different ways. The most widely used method is the use of a knife. This method should be used only at times of the year when flies are not a problem. Calves should be no more than three to four months of age. Using the knife removes the testicles completely. Because an open wound results, there is some danger from bleeding or infection. The open wound also provides a site for possible screwworm infestation in those parts of the world where the screwworm has not been eradicated.

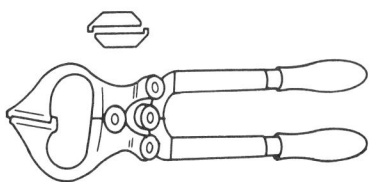

FIGURE 15-6 A type of mechanical pincers used in bloodless castration.

There are two common methods of bloodless castration. One method is the use of pincers called *burdizzo* (Figure 15-6). This tool crushes the cords above the testicles. There is no open wound. However, if the pincers are not applied correctly, the cords may not be completely crushed. This may result in the steer showing signs of stagginess at an older age. This method is a good choice in areas where screwworms are a problem.

Another method is the use of *elastrator bands*. A special instrument is used to place a tight rubber band around the scrotum above the testicles. The blood supply to the testicles is cut off. The testicles waste away from the lack of blood. There is no open wound with this method.

DEHORNING

There are several reasons for dehorning calves. First of all, horned calves often bring less money when sold. If calves are dehorned, less space is needed in feedlots and trucks, and there is less chance of cattle bruising one another. Dehorned cattle cause less damage to facilities.

Calves should be dehorned when they are young. It is easier to handle a young calf, and there is less shock to the animal. If possible, do not dehorn during fly season. If flies are a problem, use a good fly repellant to prevent maggots.

Several methods are used to dehorn calves and cattle. Chemical methods may be used when the calves are under two weeks of age. Liquids, pastes, or caustic sticks are used to apply a chemical to the horn button. Care must be taken to prevent the chemical from touching the skin. The hair is clipped from around the horn button and petroleum is put on the skin. The chemical is then applied to the horn button. The chemical must be dry before the calf is put back with the cow. Keep the calf out of the rain for several days after the chemical is applied. If a paste is used, the hair does not need to be clipped.

When the horns are past the button stage of growth, several other methods are used (Figure 15-7). Calves under 60 days of age may be dehorned with spoons, gouges, or tubes. Larger horns may be removed with Barnes-type dehorners. This type of dehorner has knives that cut off the horns. In range areas, hot irons are often used for dehorning.

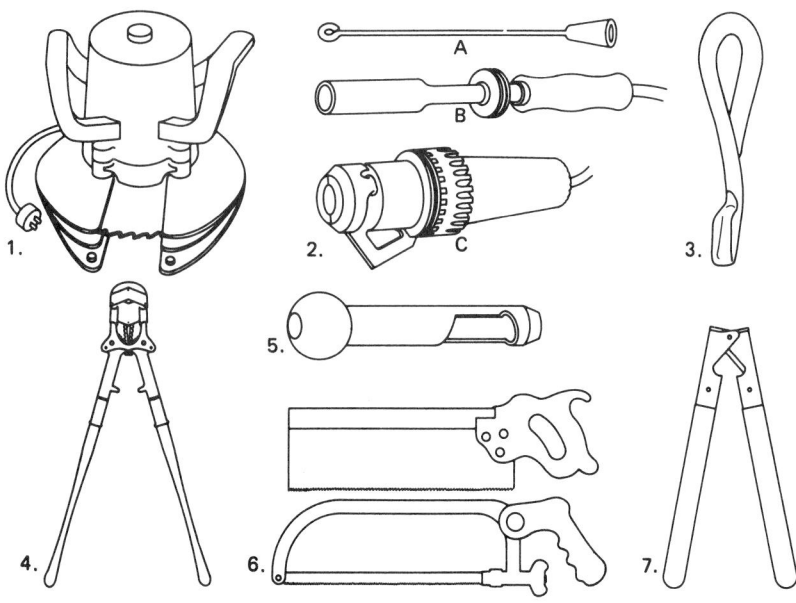

FIGURE 15-7 Instruments used for dehorning cattle.

1. ELECTRIC DEHORNING SAW
2. DEHORNING IRONS B AND C ARE ELECTRICALLY HEATED
3. METAL SPOON OR GOUGE
4. DEHORNING CLIPPER
5. TUBE DEHORNER
6. HAND SAWS
7. MECHANICAL DEHORNER (BARNES-TYPE) FOR CALVES

Hot irons may be used for calves up to four or five months of age. Electrically heated irons are also available, providing a fast and almost bloodless method of dehorning.

Older cattle are dehorned with dehorning clippers or saws. Bleeding is a major problem with older cattle. Use a forceps to pick out the main artery under the cut. The artery is pulled until it breaks, which usually prevents serious bleeding. Animals that have been dehorned by any cutting method should be watched closely. If bleeding continues, pull arteries that have not already been pulled. Another way to stop bleeding is to tie a heavy string around the poll below the dehorned area. Tight bandages may be put on the wound to stop bleeding. If bleeding still continues, call a veterinarian.

BRANDING AND MARKING

Branding and marking cattle is a common practice when herds are large. It is required by law in some western states where cattle are run on rangeland. Most western states require that records of brands be kept on cattle that are slaughtered. Brands are often recorded by county or state governments.

Calves are usually branded before weaning. Hot irons, cold irons used with a commercial branding fluid, and freeze branding are three ways to brand cattle. The hot iron method is the oldest and most commonly used. The cold iron with branding fluid is not used very often. Freeze branding (Figure 15-8) is becoming more widely used.

The calf can be thrown to the ground and held while being branded. It is easier to brand calves in a chute, and this method is becoming more widespread.

The symbols in the brand should be about 4 inches in height. A width of 3/8 inch for the lines of the brand makes it easy to read.

Ear cuts, tattooing, eartagging, and neck chains are other ways of identifying cattle. Cutting the ears is almost as common a method as branding. Earmarks are recorded in brand records and are protected by law. Either ear or both may be cut. Cutting is done in such a way that the mark can be seen from a direct front or rear view.

Ear tattooing is well adapted as a method of marking purebred cattle. It is a more permanent mark than ear cutting. A special instrument is used that has needlelike points (Figure 15-9). The mark is made with indelible ink. No open wound is left.

Ear tags (Figure 15-10) are also widely used on purebred herds. Special clamps are used to clamp tags or buttons in the ear. The identification number is on the tag or button, which may be made of metal or plastic.

Neck chains may be used when the herd owner does not want to make permanent identification marks on the animals. Neck chains

FIGURE 15-8 Cow showing a freeze brand on the side. *Courtesy of American Polled Hereford Association, MO.*

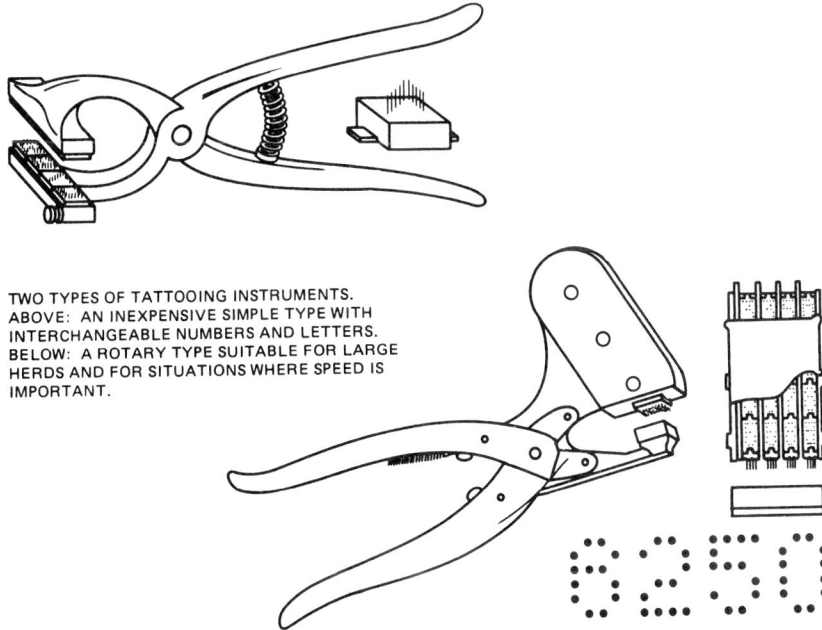

TWO TYPES OF TATTOOING INSTRUMENTS. ABOVE: AN INEXPENSIVE SIMPLE TYPE WITH INTERCHANGEABLE NUMBERS AND LETTERS. BELOW: A ROTARY TYPE SUITABLE FOR LARGE HERDS AND FOR SITUATIONS WHERE SPEED IS IMPORTANT.

FIGURE 15-9 Two types of tattooing instruments.

FIGURE 15-10 Calves identified with ear tags. *Courtesy of Cobleskill Agricultural & Technical College.*

that carry tags with identification numbers on them are usually used with purebred herds. Neck chains are not the best choice when cows are on brushy range. The chains may become caught in the brush and be torn off.

MANAGING WEANED CALVES

The herd owner has several options in managing calves after they are weaned. These include: (1) selling feeder calves, (2) selling yearling feeders, and (3) growing and finishing beef animals. If calves are born in the spring, they weigh about 400 to 500 pounds (181–227 kg) in the fall. Heifer calves weigh about 5 percent less than steer calves at weaning. These calves are sold in the fall as feeder calves.

Calves that have grown to about 650 to 750 pounds (295–340 kg) are sold as yearlings for finishing to market weight. This system uses mostly roughage for feed. If the calves are born in the fall, they are weaned in the spring. They are then fed on pasture for sale as yearling feeders in the fall.

Some herd owners prefer to grow the calves on roughage and then finish them in the feedlot for four to six months. Corn silage or grain and roughage is used for the wintering ration. When the animals are on pasture, little or no grain is fed. The animals are then moved into the feedlot and grain feeding is begun.

PRECONDITIONING CALVES

Preconditioning is the process of preparing calves for the stress of being moved into the feedlot. Most of the procedures involved in preconditioning are accepted as good management practices. They are accomplished before the calves leave the producer's farm.

Calves may be certified as preconditioned if a specific schedule of practices is followed. Preconditioning adds cost to the production of feeder calves. A certificate of preconditioning is shown in Figure 15-11.

Practices involved in preconditioning usually include:

- castration, dehorning, and identification by tattooing or branding.
- maintaining an accurate health record.
- vaccination for brucellosis (heifers only), blackleg/malignant edema, infectious bovine rhinotracheitis (IBR), bovine virus diarrhea (BVD), para-influenza (PI_3), leptospirosis.
- weaning four to six weeks before sale.
- training to eat solid feed from bunk and to drink water from a water tank.
- worming, and treatment for grubs, lice, and mange (if necessary).

Some research shows that feeding calves a preconditioning ration for three to four weeks after weaning and prior to shipping may

```
┌──────────────────────────────────────────────────────────────────┐
│              MISSOURI CERTIFICATE OF PRE-CONDITIONING              │
│                                                                    │
│  This certifies that the following pre-conditioning standards have │
│  been completed on the cattle herein described:                    │
│                                                                    │
│  Identification:              Number Steers _____        │
│                                                                    │
│  Breed:                       Number Heifers _____       │
│                                                                    │
│  Age:                         Total _____                │
│                                                                    │
│  Practices, Treatments & Immunization    Date    Product Used      │
│                                                                    │
│  Weaned                                                            │
│  Castration (knife only)                                           │
│  Dehorning                                                         │
│  Grub Control (if applicable)                                      │
│  Internal Parasite Control                                         │
│  Blackleg - Malignant Edema Vac.                                   │
│  IBR - Vaccine                                                     │
│  Para Influenza - Vaccine                                          │
│  Lepto - Vaccine                                                   │
│  Pasteurella Bacterin                                              │
│  Other                                                             │
│                                                                    │
│  In addition to the above pre-conditioning practices, the heifers  │
│  in this shipment are guaranteed open.                             │
│                                                                    │
│  General Remarks:                                                  │
│                                                                    │
│                                                                    │
│  Certified By                                                      │
└──────────────────────────────────────────────────────────────────┘
```

Information on this certificate is valuable to the purchaser and his veterinarian in making decisions on handling newly purchased cattle. It should reduce the cost of additional vaccinations and treatments.

FIGURE 15-11 A certificate of preconditioning. *Courtesy of University of Missouri, Cooperative Extension Service.*

not be a profitable practice for either the producer or the cattle feeder. Calves that were not weaned before being moved to the feedlot had higher feed efficiency in the feedlot compared to those that were on a preconditioning feeding program. A better feeding alternative appears to be not weaning the calves and limiting the creep feed to one to three pounds per head per day during the last one to two months before shipping the calves to the feedlot. A high energy ration should be used for the calves.

BACKGROUNDING CALVES

Backgrounding is the growing and feeding of calves from weaning until they are ready to enter the feedlot. It may be done by the cow herd owner. Farmers who do not want to own cow herds may buy weaned calves and prepare them for the feedlot. They then sell the calves to others who finish them for market.

Backgrounding is done primarily with roughage rations. Calves are fed for 120 to 150 days. Expected daily gain is 1.5 to 2.0 pounds (0.68–0.90 kg) per day. Calves must not be allowed to become too fat, because fat calves bring lower prices when going into the feedlot for finishing.

SUMMARY

Feeding programs for cow-calf beef herds are based on the use of roughages. Pasture in the summer and silage and hay in the winter are the common feeds used in the ration. The kinds of pasture, silages, and hay used depend on the part of the United States in which the cow-calf herd is located.

Hay can be harvested in rectangular bales or large round bales. Bales may be left in the field for self-feeding or they may be moved to a storage area. More feed is required when using large round bales because of increased losses from this method of feeding.

Dry, pregnant cows are fed to prevent their becoming too fat or too thin during the winter. Younger cows and heifers require more feed than mature cows. Minerals and salt should be fed free choice to cow herds. Protein supplements are seldom needed if the hay quality is good. Vitamin A may be needed when low-quality roughages are fed.

Cows nursing calves require more feed than dry cows. High-quality pastures usually meet the needs of lactating cows.

Creep feeding calves may or may not be profitable. Creep feeding may pay when selling the calves at weaning. If the calves are being kept by the producer and fed roughages through the winter, then creep feeding is probably not economical. Grain is used in the creep feeder. Commercial, pelleted mixes are also available for this use.

It is usually necessary to replace about 15 percent of the cows in the herd each year. The use of performance records to select replacement heifers is recommended. Replacement heifers should weigh about 550 to 750 pounds (250–340 kg) at breeding, depending on the breed. British breeds are bred at the lighter weights.

Bull calves to be kept for breeding purposes are weaned at six to eight months of age. High-energy rations are fed for five months to determine which bulls have the best gaining ability. Bulls that gain best should be selected for breeding. The others should be marketed for slaughter

Well-grown bulls may be used for breeding at 15 to 18 months of age. Bulls that are self-fed require large amounts of roughage so they will not become too fat. Winter rations for two- to four-year-old bulls should contain more energy and protein than cow rations contain. Mature bulls may be fed the same ration as the cow herd.

The beef producer desires a 100 percent calf crop each year. It is essential to make sure that the bull is fertile by testing the semen before the breeding season. Do not run too many cows with the bull when pasture breeding. A bull can breed more cows when pen breeding is used.

Artificial insemination is used in some herds. It requires good facilities and more work than natural breeding. A major advantage is the chance to use superior bulls on the cow herd.

Size is more important than age when determining when to breed heifers that are sexually mature. A good management objective is to have heifers calve at two years of age.

Farmers should aim to have calving occur in a 40- to 60-day period. If this procedure is followed, calves are more uniform in weight and age. Start calving six weeks to three months before pasture season, depending on local climate.

Observe cows closely for signs of calving. If the cow needs help, provide it. Most mature cows need no help in calving. A high percent of heifers do require help.

Castration, dehorning, vaccination, and identification of calves is best done when they are young. There is less shock to the calf at that time. Preconditioning calves involves following a schedule of good management practices.

Backgrounding is the growing of calves on roughages from weaning until they are ready for the feedlot. It is a practice that is flexible. The cow herd owner may want to background calves from the herd before selling them. Some farmers who do not own cow herds may buy calves at weaning and prepare them for the feedlot. These calves are then sold to others for finishing for slaughter.

Student Learning Activities

1. Visit a local beef cow herd, collect data on size of herd, system of production used, and available feeds. Plan a year-round feeding program for that herd.
2. Talk with local beef cow herd operators to find out how they care for the cows and calves at calving time and up to weaning. Report to the class on practices used in the area and recommend any improvements that you feel are needed.
3. Survey beef cow herd operators in the community to determine the system of production used locally. Find out why the farmers use these systems.
4. Prepare and present oral reports on a phase of beef cow-calf herd operations that interests you.
5. Observe and assist in the castration, dehorning, vaccination, and identification marking of calves.
6. Ask a veterinarian to talk to the class about calving problems.
7. Invite a representative of a feed company to talk to the class about beef cow feeding programs.

Review

1. What is the most common kind of roughage used for feeding beef cattle in your area?

2. Why are cow-calf beef feeding programs based on the use of roughages?

3. What kinds of crop residues might be used for feeding cow-calf herds?

4. How can waste be reduced when feeding roughage to cow-calf herds?

5. Why is it important to control the weight gain of dry, pregnant beef cows?

6. Describe a minimum maintenance ration that will meet the requirements of dry, pregnant mature beef cows.

7. Briefly explain the feeding of protein supplement to dry, pregnant beef cows.

8. What minerals and vitamins do dry, pregnant beef cows need and how can they be supplied?

9. What are the protein, energy, mineral, and vitamin requirements of lactating cows as compared to dry cows?

10. Describe how these needs can be met for lactating cows.

11. State the reasons both for and against creep feeding calves.

12. Give a sample ration for creep feeding calves.

13. Briefly explain the use of growth-stimulating implants for beef calves.

14. How much gain per day should replacement heifers of (a) British breeds, and (b) larger breeds have?

15. At what weight should heifers of (a) British breeds, and (b) larger breeds be bred?

16. Give a sample ration for growing replacement heifers.

17. Describe how replacement bulls should be fed.

18. Describe the feeding of mature bulls (a) before the breeding season, and (b) during the breeding season.

19. Briefly explain the number of bulls required for the cow herd during the breeding season.

20. When should breeding begin if a short calving season is desired?

21. Describe the process of artificial insemination.

22. List seven practices that need to be followed if artificial insemination is to be used with a beef breeding herd.

23. List the advantages and disadvantages of using artificial insemination with a beef breeding herd.

24. When should the beef producer plan to have cows calve?

25. Describe the signs that indicate that a cow is about ready to calve.

26. List the signs of trouble that may indicate that a cow will require help in calving.

27. What care should be given the calf immediately after its birth?

28. When should bull calves be castrated?

29. Describe the methods that can be used to castrate bull calves.

30. Why should calves be dehorned?

31. Describe the methods that can be used to dehorn calves.

32. Under what conditions should cattle be branded or marked?

33. Describe some common methods for branding or marking beef cattle.

34. Describe three methods of managing weaned calves.

35. What is preconditioning of calves?

36. List six practices usually followed when preconditioning calves.

37. What is backgrounding of calves?

16 *Feeding and Management of Feeder Cattle*

SYSTEMS OF CATTLE FEEDING

Cow-calf herds are found mostly in western, southwestern, and west north-central parts of the United States. Finishing feeder cattle, on the other hand, is done mostly in the north-central and southern parts of the United States.

Feeding cattle for slaughter requires more grain and protein supplement than that needed for cow-calf herds. That is why many cattle-feeding operations are found in the grain-producing states. There is increased interest in cattle feeding in other parts of the nation.

Types of Feeding Operations

Commercial cattle feedlots are those with a capacity of 1,000 cattle or more. Often, many thousands of cattle are fed in these feedlots at one time. Most or all of the feed needed is purchased rather than grown by the cattle producers. The number of large commercial feedlots is increasing in areas where cow-calf operations are common. Most of the cattle fed in the Plains states, and in Colorado, Arizona, California, and Texas are in commercial feedlots. It is not necessary to move calves a great distance when feedlots are located close to the area where the calves are produced.

Farmer-feeders are farm operators who feed cattle mainly as a way of marketing feed raised on their own farms. Feedlot capacity of these operations is usually less than 1,000 cattle. The size of some of these feeding operations is large enough to make it necessary to purchase additional feed. The majority of the cattle-feeding operations in the central and southern states are of the farmer-feeder type.

FIGURE 16-1   Cattle in the feedlot being finished on a full feed of grain. *Courtesy of University of Illinois at Urbana-Champaign.*

Cattle feeders are in business to make a profit. Homegrown feed is fed with the hope of marketing it at higher prices through the sale of cattle. Feeders must produce the kind of product wanted by the buyers. The current demand is for steers that weigh between 1,000 and 1,250 pounds (454–567 kg) and heifers that weigh between 900 and 1,050 pounds (408–476 kg). Eighty to 90 percent of the cattle fed should grade Choice (be classified as choice quality when marketed).

Types of Finishing

The two general ways to feed cattle for slaughter are: (1) finishing immediately and (2) deferred finishing. The system used depends on the kind of cattle fed, how long they are fed, the feed used, and the market demand.

Finishing immediately is a system in which the feeder cattle are brought to a full feed of grain (Figure 16-1). A full feed of grain means that the cattle are fed mostly grain and smaller amounts of roughage. They are then fed until ready for slaughter. Steer calves are on feed for about 275 days and heifer calves for about 230 days. Yearling steers are on feed for about 175 days, and yearling heifers for about 130 days. Older feeders are on feed for about 100 days. High-quality feeders are best suited to this system. Farms with limited amounts of roughage and plenty of grain are well adapted to this system of cattle feeding. Small amounts of roughage and large amounts of grain are used in this feeding program. Older animals are fed more roughage than calves or yearlings. Heavier animals are particularly well adapted to this system.

There is a trend toward starting calves more slowly on grain. Using this method, calves are wintered on roughages and then put on full feed in the spring. The finished cattle are marketed in the fall.

Deferred finishing systems use more roughage and less grain. Calves are bought in the fall and wintered on roughage. Small amounts of grain are fed. Daily gain is about 1.25 to 1.5 pounds (0.57–0.68 kg). The calves are pastured for 90 to 120 days the next summer. A small amount of grain, or none at all, may be fed on pasture. In the fall, the calves are put in the feedlot for 90 to 120 days. A full feed of grain is fed in the feedlot. Some feeders prefer to put the calves in the feedlot

FIGURE 16-2 Feeder cattle being used to clean up crop residues in a cornfield. *Courtesy of University of Illinois at Urbana-Champaign.*

in the spring. The calves are then fed a full feed of grain for 120 to 150 days.

Yearling feeders may be used to clean up crop residues such as cornfields (Figure 16-2). The yearlings are then wintered on roughage rations. In the spring, the yearlings are finished with a high-grain ration. Finishing is done either on pasture or in drylot. Some feeders use little grain on pasture and finish the yearlings in the feedlot for later marketing. Wintering feeders on winter oat or winter wheat crops before putting them in the feedlot is gaining in popularity in the Southwest.

The deferred system is well suited for the farm that has roughage to market. Corn silage is one of the common feeds used. More beef per acre can be produced with corn silage than with any other crop.

KINDS OF CATTLE TO FEED

There are many choices for the cattle feeder when selecting cattle to feed. Selection is made on the basis of sex, age, weight, and grade of cattle.

Sex

Some heifers must be kept for herd replacements. Therefore, more steers are available for feeding. Steers gain about 10 percent faster than heifers if both are fed for the same length of time. Steers are 10 to 15 percent more efficient in gains. *Efficiency in gain* refers to the amount of feed needed for each pound (kilogram) of gain. The less feed required, the higher the efficiency. Heifers finish at lighter weights than do steers and are usually bought and sold for less money. Heifers are better fit for shorter feeding periods than steers are. Feeder heifers may be pregnant when purchased. This may result in discounts when they are sold for slaughter. The developing fetus is of lower value to the packer than is the rest of the carcass.

Young bulls may be fed for market. At the same age and weight as steers, they gain faster and more efficiently. Bulls produce lean carcasses of about the same quality as steers. However, bulls are not as well accepted in the marketplace. Generally, the feeding of young bulls for market is not recommended.

AGE AND WEIGHT

Feeders are generally divided into three groups—calves, yearlings, and older feeders—based on age and weight. The weight range is 350 to 1,000 pounds (159–454 kg).

Calves are feeders that are less than one year old, usually weighing about 350 to 450 pounds (159–204 kg). They are adapted to many different systems of cattle feeding. Gains are more efficient than they are in older cattle. However, it takes longer to feed calves to slaughter weights. Calves need more grain and less roughage than older cattle when fed immediately for slaughter. Calves are not well adapted to cleaning up crop residues. They do not make good use of low-quality roughage. Death losses are usually higher with calves, and health problems are greater. Since calves are lighter in weight when bought, most of the weight sold is gain. Success in feeding calves depends more on feeding skill than it does on ability to buy and sell.

Yearlings are feeders that are between one and two years old, usually weighing about 550 to 700 pounds (249–318 kg). Yearlings are well adapted to feeding programs using more roughage than that given to calves. They are often used to clean up crop residues. Less time in the feedlot is necessary to finish yearlings for slaughter, and there are fewer health problems.

Older feeders are those that are two years old or older. They weigh about 800 to 1,000 pounds (363–454 kg). These feeders are fed for a short period of time, usually 90 to 100 days. Gains are fast but not as efficient as in younger feeders. Older feeders can make use of more roughage in the ration. Death losses are low. Much of the profit comes from reselling purchased weight. Therefore, more skill in buying and selling is needed for this type of feeder.

Grade

On September 2, 1979, the USDA revised the grades and standards for feeder cattle. This revision came about because of changes in the kinds of cattle being produced and the changes in slaughter cattle and beef carcass grades that became effective in 1976.

Feeder cattle grades are used by the USDA as the basis for reporting market prices of feeder cattle. They are also used as the basis for certifying the grade of feeder cattle that are delivered on future contracts. States use the grades for official feeder cattle grading programs.

The USDA standards for feeder cattle grades apply to cattle that are less than 36 months of age. They may also be used to describe stock cows for market reporting purposes.

Three factors are used to determine the grade of feeder cattle: thriftiness, frame size, and thickness.

Thriftiness refers to the apparent health of the animal and its ability to grow and fatten normally. An unthrifty animal is one that is not expected to grow and fatten normally in its current condition.

Frame size indicates the size of the animal's skeleton (height and body length) in relation to its age. When two animals are the same age, the large-framed one is taller at the withers and hips and has a longer body than one with a smaller frame.

LARGE FRAME

- TALL AND LONG FOR AGE
- HALF INCH OF FAT—
  STEERS, 1200 LBS OR MORE
  HEIFERS, 1000 LBS OR MORE

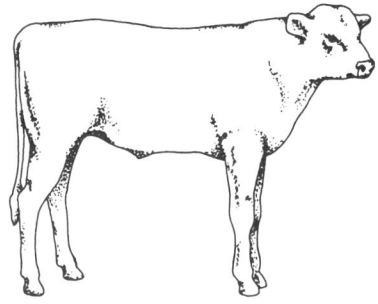

MEDIUM FRAME

- SLIGHTLY TALL AND SLIGHTLY LONG FOR AGE
- HALF INCH OF FAT—
  STEERS, 1000–1200 LBS
  HEIFERS, 850–1000 LBS

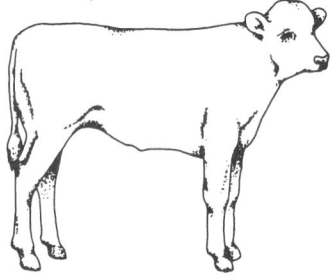

SMALL FRAME

- SMALL FRAME AND SHORTER-BODIED FOR AGE
- HALF INCH OF FAT—
  STEERS, LESS THAN 1000 LBS
  HEIFERS, LESS THAN 850 LBS

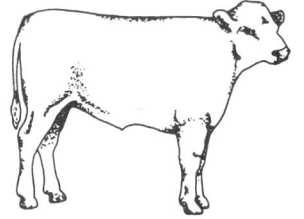

FIGURE 16-3 USDA grades of feeder steers. *Courtesy of Fred L. Williams, Jr., USDA.*

Thickness means the development of the muscle system in relation to the size of the skeleton. When feeder cattle are the same age and frame size, differences in thickness are due to differences in bone structure, muscling, and degree of fatness. A standard degree of fatness—slightly thin—is used when evaluating thickness. Thicker feeder cattle have a higher ratio of muscle to bone when fed to the same degree of fatness as cattle that are not as thick. The thicker cattle will have a higher yield grade at the same degree of fatness as cattle that are not as thick.

There are 10 grades of feeder cattle that result from the use of the above standards: Large Frame, No. 1; Large Frame, No. 2; Large Frame, No. 3; Medium Frame, No. 1; Medium Frame, No. 2; Medium Frame, No. 3; Small Frame, No. 1; Small Frame, No. 2; Small Frame, No. 3; and Inferior. The three frame sizes are shown in Figure 16-3. The three thickness standards are shown in Figure 16-4.

Large frame feeder cattle are thrifty, have large frames, and are tall and long bodied for their age. To produce U.S. Choice carcasses (0.50 inch or 1.27 cm of fat at the twelfth rib), steers would need to be more than 1,200 pounds (544 kg) live weight, and heifers would need to be more than 1,000 pounds (454 kg) live weight.

Medium frame feeder cattle are thrifty, have slightly large frames, and are slightly tall and slightly long bodied for their age. To produce U.S. Choice carcasses, a steer's live weight would need to be 1,000 to 1,200 pounds (454–544 kg), and a heifer's live weight would need to be 850 to 1,000 pounds (385–454 kg).

Small frame feeder cattle are thrifty, have small frames, and are shorter bodied and not as tall as Medium Frame cattle. Small Frame steers would produce U.S. Choice carcasses at live weights under 1,000 pounds (454 kg), and heifers would produce U.S. Choice carcasses at live weights under 850 pounds (385 kg).

No. 1 feeder cattle usually have a high percent of beef breeding. They are thrifty and slightly thick throughout. The forearm and gaskin are slightly thick and full. They show a rounded appearance through the back and loin with moderate width between both the front and rear legs. They usually have a slightly thin covering of fat. No. 1 feeder cattle may have varying degrees of fat.

No. 2 feeder cattle are thrifty and are narrow through the forequarter and the middle part of the rounds. The forearm and gaskin are thin. The back and loin have a sunken appearance. Both front and rear legs are set close together. They usually have a slightly thin covering of fat. No. 2 feeder cattle may have varying degrees of fat.

No. 3 feeder cattle are thrifty. They have less thickness than the minimum requirement for No. 2 feeder cattle.

Inferior feeder cattle are those that are unthrifty. They are not expected to grow or fatten normally. Unthriftiness may be caused by disease, parasites, extreme thinness, or other conditions. These conditions would have to be corrected before the animal could be expected to grow or fatten normally. Cattle that are *double-muscled* (muscular hypertrophy) are also graded Inferior. They cannot be expected to deposit intramuscular fat (marbling) normally. Feeder cattle in the Inferior grade may have any combination of thickness and frame size.

The higher grades of feeder cattle fit best into systems that use a longer feeding period. Less roughage and more grain is fed the higher grades. These cattle finish to higher slaughter grades. Market prices for the better grades are higher in late summer and fall.

The lower grades of feeder cattle are the best choice for systems that utilize shorter feeding periods. More roughage can be profitably used in the ration. These cattle finish at lower slaughter grades. Market prices for the lower grades are higher in the spring and early summer.

BUYING FEEDER CATTLE

Auction markets are the main source for buying feeder cattle. Buying directly from farms or ranches is the second most important source. Only a small percent of feeder cattle are bought through terminal markets (see Unit 19).

Cattle for commercial feedlots are purchased mainly by salaried buyers or order buyers. Some farmer-feeders, themselves, still buy many of the cattle they feed. However, farmer-feeders are increasing their use of order buyers and cattle dealers.

The supply of feeder cattle increases in the fall, with the peak movement of feeder cattle occurring in October. There is a trend toward keeping feedlots full year-round. This has created a demand for feeders at other times of the year. Cow-calf herd operators are adjusting their production to meet this demand.

Feeder and slaughter cattle prices tend to be lower in the fall and winter and higher in the spring and summer. Figure 16-5 shows the seasonal variation in prices of feeder calves and slaughter steers and heifers in the United States. Prices of feeder and slaughter cattle vary greatly from year to year.

Price trends are averages and do not always follow the same pattern every year. A cattle feeder may not be successful in trying to outguess the market. It is best for each cattle feeder to develop a feeding program that is adapted to his or her own feed supply. This program, when followed year after year, results in the greatest overall profit for the cattle feeder.

The profit from feeding cattle is equal to the value of the finished cattle, minus the total costs. Costs include the price of the feeders, feed, labor, veterinary fees, fixed charges for buildings and equipment, interest on investments, and death losses. The cattle buyer must

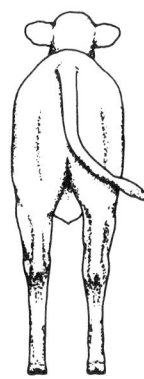

No. 1
- SLIGHTLY THICK THROUGHOUT
- MODERATE WIDTH BETWEEN LEGS
- HIGH PROPORTION OF BEEF BREEDING

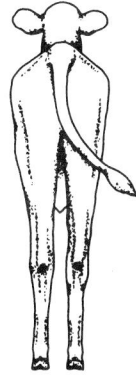

No. 2
- NARROW THROUGHOUT
- LEGS SET CLOSE TOGETHER
- BACK AND LOIN HAVE SUNKEN APPEARANCE

No. 3
- LESS THICKNESS THAN NO. 2
- LEGS CLOSER TOGETHER THAN NO. 2

FIGURE 16-4 USDA grades of feeder steers (Calves). *Courtesy of Fred L. Williams, Jr., USDA.*

% OF ANNUAL
AVERAGE

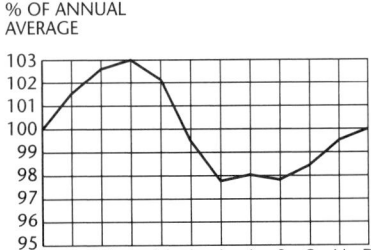

Seasonal variation in prices of slaughter steers and heifers, United States.

% OF ANNUAL
AVERAGE

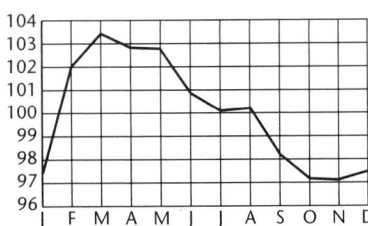

Seasonal variation in prices of feeder calves, United States.

FIGURE 16-5 Seasonal variation in the prices of slaughter steers and heifers and feeder calves in the United States. *Source: USDA.*

make some calculations before deciding on the price to pay for feeder cattle. The *break-even feeder price* is the price paid that just covers all costs without any profit. It may be found by the following formula:

$$\text{Break-even feeder price} = \frac{\text{fed weight}}{\text{feeder weight}} \times \text{expected price of fed cattle}$$
$$- \frac{\text{gained weight}}{\text{feeder weight}} \times \text{cost of gain}$$

Example. Given the following prices and weights, the break-even price is determined as shown.

700-pound feeders

finished cattle weight of 1,050 pounds

expected selling price of finished cattle, $45

cost per hundredweight (cwt) of gain, $35

$$\text{break-even feeder price} = \left(\frac{1,050}{700} \times 45\right) - \left(\frac{350}{700} \times 35\right)$$
$$= (1.5 \times 45) - (0.5 \times 35)$$
$$= 67.50 - 17.50$$
$$= \$50 \text{ per cwt}$$

This calculation tells the farmer that to make a profit the cattle must be purchased at less than $50 per hundredweight.

The profit margin in feeding cattle may be calculated in the following way:

$$\text{cost of feeder} = \frac{\text{feeder weight}}{\text{fed weight}} \times \text{price of feeders per cwt}$$

$$\text{cost of gain} = \frac{\text{gained weight}}{\text{fed weight}} \times \text{cost per cwt of gain}$$

$$\text{total cost} = \text{cost of gain} + \text{cost of feeders}$$

$$\text{margin/cwt} = \text{fed cattle price} - \text{total cost}$$

$$\text{profit per head} = \text{margin/cwt} \times \frac{\text{fed weight}}{100}$$

Example. Given the following prices and weights, the profit margin is determined as shown.

700-pound feeders

$46 per hundredweight of feeder

fed weight of 1,050 pounds

feeding costs of $35 per hundredweight

fed cattle price of $47 per hundredweight

$$\text{cost of feeders} = \frac{700}{1,050} \times 46 = \$30.67$$

$$\text{cost of gain} = \frac{350}{1,050} \times 35 = \$11.67$$

$$\text{total cost} = 11.67 + 30.67 = \$42.34$$

$$\text{margin/cwt} = 47 - 42.34 = \$4.66$$

$$\text{profit per head} = 4.66 \times \frac{1,050}{100} = \$48.93$$

Price is not the only factor to consider when buying feeder cattle. Some general rules to follow when selecting feeders are:

- If possible, buy from herds with performance records. Cattle with good gaining ability are worth more.
- Big-framed, rugged-boned, heavily muscled cattle usually are better gaining cattle.
- Healthy, thin cattle make faster gains than fatter cattle.
- More efficient gains are made by younger, lighter-weight cattle.
- Steers are worth three to four dollars more per hundredweight than heifers.
- Good crossbred cattle gain from 2 to 4 percent better than the average of the parent breeds.

SELECTION OF FEEDS

Roughages

Only limited use is made of roughages in finishing rations. The amount of roughage in the ration varies from 5 to 30 percent. When less than 20 percent of the ration is roughage, the nutrient value of the roughage has little effect on gain. Rate of gain is affected when the percent of roughage in the ration is higher than 20 percent. Better-quality roughages with higher energy content must be fed as the percent of roughage in the ration increases. As the amount of roughage in the ration increases, the rate of gain is slower. More roughages are used in the early part of the feeding period than are used later in the feeding period.

Hay is the most commonly used roughage in finishing rations in some areas of the United States. Legumes make the best kind of hay for this purpose. Legume hays are higher in protein, calcium, and carotene than grass hays. Lower-quality hay requires more supplement to properly balance the ration.

Corn silage is one of the best roughages to use for finishing beef cattle. More beef per acre is produced from corn silage than from any other roughage. It takes about one ton of corn silage to produce 100 pounds (45.4 kg) of beef. The corn silage must be supplemented with protein and minerals. Daily gains of 1½ to 2 pounds (0.68–0.91 kg) can be expected when corn silage is used in the ration. More corn silage is used when corn is high priced to lower the cost of gain. Because cattle gain slower on corn silage, it takes more days in the feedlot to finish the cattle. This, in turn, increases non-feed costs. This is a greater problem for large commercial feeders than it is for smaller farmer-feeders. Wet corn gluten feed may be substituted for corn silage in beef rations. Wet corn gluten feed is high in energy, protein, and phosphorus. It can be combined with crop residues in rations that require little additional supplement to meet the nutritional needs of the animals. Experimental work at the University of Illinois compared wet corn gluten feed combined with (1) corn stalks and (2) fescue hay with corn silage diets for heifers in the feedlot. Rations containing approximately 50 percent wet corn gluten feed, 40 percent forage, and 10 percent supplement on a dry matter basis were compared to a ration containing 90 percent corn silage and 10 percent supplement. Gains on the wet corn gluten feed diets were 10 to 20 percent slower compared to gains on the corn silage diet. Approximately 10 to 20 percent more feed per pound of gain was required on the wet corn gluten feed diets. The relative prices and availability of the feeds need to be considered by the cattle feeder when determining the best diet to use.

Sorghum silages have from 60 to 90 percent of the feeding value of corn silage. The value varies with the amount of grain in the different varieties. Sorghum silage is harvested when the kernels are in the *medium-dough stage*. More mature grain is higher in feed value, but the kernels are harder. When harvested in the more mature stage, the kernels must be broken by grinding so that they are easier to digest. Sorghum silage is low in protein. Protein and mineral supplement must be added to the ration when sorghum silage is fed.

Grass silage is made from grass, legume, or grass-legume mixtures. Grass silages are lower in energy content than silages made from grain crops. Beef cattle make slower gains on grass silage. *Haylage* is low-moisture grass silage. It is more palatable than grass silage and contains more energy and protein. Beef cattle make slightly faster gains on haylage than they do on grass silage. A good-quality grass-legume haylage supplies most of the protein needed by the animal.

Small grains such as oats can be used for silage for finishing beef cattle. These are harvested just after reaching the boot stage. The boot is the upper sheath of the plant. The *boot stage* is that point in growth at which the inflorescence (the flowering part of the plant) expands the boot. From 100 to 200 pounds (45.5–91.0 kg) of carbohydrates are added at the time of ensiling. Corn, molasses, or citrus pulp may be added as the carbohydrate. The protein content is high and the silage is palatable. The quality and nutrient value is increased by the addition of the carbohydrate source.

Lower quality by-product roughages can also be used for finishing. Examples of these kinds of roughages are corncobs, cottonseed hulls, peanut hay, beet tops, oat hulls, and straw. They are usually low in protein, minerals, and vitamins. The ration must be carefully supplemented if they are used. The decision to use low-quality roughages is based on the relative prices of other feeds available.

Pasture

Good-quality pasture lowers the cost of gain for finishing cattle (Figure 16-6). The kinds of pasture discussed in Unit 15 for beef cow-calf herds are also used for finishing beef cattle. When pastures are stocked at the best rates for the type of pasture, rotation grazing produces more beef per acre than continuous grazing. Daily gains per animal are about the same with either method.

Feeding concentrates at the rate of 1 percent of the animal's live weight produces a satisfactory gain, but cattle fed at this rate will generally finish to only Standard or Select grades, rather than Choice or Prime. Grains may be self-fed on pasture to reduce labor costs. However, feeding grain on pasture does increase the total cost of gain.

The protein content of pasture grasses is highest in the spring and lower in the fall. Therefore, more protein supplement is needed in the ration in the fall. The stocking rate must be such that the grasses are kept grazed down. This increases the palatability of the pasture.

Bloating on pasture is prevented by feeding a bloat preventive such as poloxalene. A rate of 2 grams per day per 100 pounds (45.4 kg) of live weight is enough to prevent bloat.

Grain intake must be limited when self-feeding cattle on pasture. Salt or fat may be added to the concentrate mix to limit feed intake on self-feeders. These additives reduce the palatability of the concentrate mix. A level of 10 percent salt in the ration limits the grain intake. Adding fat to the ration may be more profitable depending on the cost of the fat. Limiting the concentrate intake reduces the amount of grain needed by about 50 percent compared to the feed required in drylot.

Little or no protein supplement is needed for grass-legume pastures. Cattle on grasses fertilized with nitrogen or grass in the early

FIGURE 16-6 Good-quality pasture will lower the cost of gain for beef cattle.

stages of growth also need little protein supplement. Cattle on poor-quality pasture or grasses in more mature stages of growth do require protein and minerals added to the ration. Protein supplement is also needed under drought conditions. About 10 to 12 percent total protein in the concentrate ration is enough for most feeding conditions.

Concentrates

Grains usually used for finishing beef cattle are corn, milo (sorghum grain), barley, oats, and wheat. Corn and milo are the most commonly used grains. All other grains are compared to corn for feeding value. The relative prices of the different grains determine which is best to use at any given time. Grains vary in feeding value depending on the variety, how they are processed, and other factors.

High-moisture corn may be treated with ammonia to increase its dry protein content. Experimental work at Purdue University has shown an increase from 9.8 percent to 12.7 percent in dry matter protein content when high-moisture corn was treated at a level of 0.49 percent of the wet corn. The treated corn is brown in color but that does not affect its feed value.

No more than 50 percent of the grain mixture should be composed of wheat. Higher amounts of wheat in the ration tend to decrease feed intake and increase digestive problems and abscessed livers. Wheat should be coarse ground, rolled, or crushed when used in the ration, but it should never be finely ground. The use of a buffer such as sodium bicarbonate in cattle rations with a high wheat content is recommended. Wheat has a feeding value of 100 to 105 percent of the value of corn in beef rations.

Oats are limited to no more than 30 percent of the grain mixture in beef rations. Oats are a good feed for growth and development, but they are not as good for finishing cattle. Oats are lower in energy value and higher in crude fiber content compared to corn.

Milo has about 85 to 95 percent of the nutritional value of corn for beef cattle.

Dry corn gluten feed is a good source of energy and protein for feeder cattle. It is highly digestible and does not limit forage intake. Supplies of dry corn gluten feed may be limited in some areas.

Molasses is added to the ration to increase palatability. It also helps control dust. Molasses may replace up to 10 percent of the grain in the ration.

Methods of Processing Feeds

Grinding is processing a feed through a hammer mill. The feed is beaten with rotating hammers until it is broken up enough to pass through a screen. The diameter of the holes in the screen determines how finely the feed is ground. Different size holes are used for obtaining different degrees of fineness. The feed is dry when it is ground.

High-moisture storage refers to harvesting the grain at a high moisture content (22 to 30 percent for corn and milo) and storing it in a silo.

Rolling refers to processing the grain through a set of smooth rollers that are set close together. The grain is pressed into the form of a flake. Therefore, this process is sometimes called *flaking*. If the grain is heated with steam and then rolled, the process is called *steam flaking*. If the grain is dry, it is called *dry-rolled grain*. *Crimping* is the same as rolling, except the rollers used have corrugated surfaces.

Feeds may be cooked before use, although this practice is seldom profitable. Steam may be used to cook the grain. *Roasting, popping,* and *grain exploding* refer to dry heating of the grain.

Pelleting of feed refers to grinding the feed into small particles and then forming them into a small, hard form called a pellet. *Cubing* is the same process as pelleting, but the cubes are larger in size than the pellets. Individual feeds or complete rations may be pelleted or cubed.

Other methods of feed processing include *wetting, soaking, fermenting,* and *germinating* grains. None of these processing methods is widely used, although they may be used in special cases. However, they are not considered to be practical for general on-farm use.

Grinding generally improves the value of grains fed to beef cattle. However, research conducted at several universities shows that grinding or rolling dry shelled corn has little effect on rate of gain or feed conversion for beef calves or yearlings.

Processing high-moisture shelled corn has shown varied results relating to rate of gain and feed efficiency. Some tests have shown that grinding high-moisture shelled corn before putting it in the silo has slightly reduced feed intake and rate of gain. It may be necessary to include some dry grain in the ration to maintain sufficient dry matter consumption. Animal performance is improved when some dry grain is included in the diet. It may not be profitable to process high-moisture shelled corn when diets containing less than 15 percent roughage are used.

Steam processing both barley and milo increases feed intake and rate of gain. The starch in milo is not as digestible as that in corn. Processing helps to make it more usable for the cattle. Heating and flaking milo can increase rate of gain by 10 percent and improve feed efficiency by 5 percent, as compared to using milo as a dry-rolled grain. Feed requirements are 10 to 18 percent less for high-moisture milo compared to dry ground milo. High-moisture milo has not been shown to improve rate of gain.

High-moisture ground ear corn increases gains by 3 percent. Feed efficiency is improved by 10 percent as compared to dry ground ear corn. High-moisture ground ear corn is a palatable ration that the cattle like. It is easy to start cattle and keep them on feed with this kind of ration.

Pelleting grains gives an increase in feed efficiency of up to 5 percent. Feed efficiency is increased because less feed is wasted when it is pelleted. Complete pelleted rations will slightly increase feed intake.

Grinding forages increases feed intake. Feed consumption is increased because the smaller particles of feed pass more quickly through the digestive system.

Protein Supplements

Natural protein sources often used in finishing rations for beef cattle include soybean, linseed, and cottonseed meals. Most commercial protein supplements for beef cattle are based on soybean meal with some urea included. Urea is a nonprotein nitrogen source that can be utilized by ruminant animals to meet some of their protein requirement.

Lighter-weight (under 600 lb [272 kg]) calves usually gain faster when being started on feed if a natural protein source is used instead of urea. This is because there is a 2- to 3-week adjustment period needed for cattle to become used to urea. If the calves are already eating several pounds of grain per day when started on feed, they will gain just as fast on urea as on a natural protein source.

When rations containing a high level of grain are used, urea may be used to provide half or more of the crude protein equivalent needed by finishing cattle. Yearling feeder cattle can meet all of their protein requirement from urea if the energy level of the diet is high enough. However, performance results have not been as consistent when all of the protein equivalent was from urea, as compared to diets that included natural protein sources.

Urea should be used to supply no more than one-third of the crude protein equivalent needed if a limited-grain ration is being fed. Microorganisms in the rumen must have enough energy available to properly utilize urea. If the diet is low in energy, then less urea should be used.

Urea Fermentation Potential

Research at Iowa State University has been conducted to develop a different system for evaluating protein and urea utilization in cattle. The system accounts for rate of gain and tissue deposition needs, degradability of protein, and urea fermentation potential.

Younger cattle that are still growing require more protein per unit of gain than older cattle that are being fattened. The younger cattle are still developing muscle tissue, which requires more protein. The addition of fat for weight gain in older cattle requires less protein per unit of gain.

Some feeds are broken down more easily by microorganisms in the rumen than other feeds. Protein that is not broken down in the rumen moves on in the digestive tract and is called bypass protein. Metabolizable protein is the combined total of protein from bypass protein and degraded protein that is reformed by rumen microorganisms and is absorbed and used by the animal.

Urea fermentation potential measures the amount of fermentable energy in the ration that is needed to utilize additional urea. It is determined by the total digestible nutrient content of the feed and the amount of ammonia released when the protein is broken down in the rumen by fermentation. High-energy feeds with low protein content tend to have high urea fermentation potential. Low-energy feeds with high protein content tend to have lower urea fermentation potential.

After cattle reach about 700 pounds (317 kg), the bypass protein concept is no longer used for ration formulation. The energy level of a corn/corn silage ration is high enough to allow the use of urea as the major protein source.

Feeding Raw Soybeans to Beef Cattle

When the price of soybeans is low enough, it may be profitable to feed them to beef cattle. The formula for determining when it might be profitable to use raw soybeans in the ration is: (.35 × price/cwt of corn) + (.75 × price/cwt of soybean meal) = price/cwt of whole soybeans. Assume corn is selling for $2.01 per bushel and soybean meal is priced at $139 per ton. Corn is then worth $3.59 per cwt [(2.01/56 = .0359) × 100 = 3.59]. Soybean meal is priced at 6.95/cwt [139/20 = 6.95]. Substituting in the formula: (.35 × 3.59) + (.75 × 6.95) = 6.47. The value of soybeans for feeding purposes is then $3.88 per bushel [(6.47/100) × 60 = $3.88]. If soybeans are selling above this price, it is not profitable to feed them raw; if they are selling below this price, it might be worth considering feeding them raw.

Any processing costs must be added to the calculated value from the above formula. Care must be taken not to feed soybeans that become moldy while in storage. These molds can be toxic to cattle. Beans must be dried to 13 percent moisture or below to prevent molding in storage.

Feeding the beans whole reduces feeding efficiency. The beans should be coarsely crushed when fed. Feed the crushed beans within one week of crushing in hot weather and two weeks in cold weather to avoid spoilage. Start adding the beans to the ration at a low level and gradually increase the amount fed till the desired feeding level is reached. Adding beans too rapidly to the ration can result in diarrhea and lower feeding efficiency. Beans should comprise no more than eight percent of the ration for growing calves and no more than six percent for finishing rations. Do not feed raw beans to young calves with undeveloped rumens. Doing so might cause nutritional muscular dystrophy and abomasal torsion in the calves.

Feeding Wheat Middlings to Beef Cattle

Wheat middlings are a byproduct of the flour milling industry. They provide protein and energy in the ration and contain a good supply of phosphorus and potassium. Some flour mills are pelleting or cubing middlings for sale as cattle feed. Wheat middlings generally compare favorably to other supplements used for cattle feed.

Cattle need to become acclimated to wheat middlings, so they should be added gradually to the ration. There are some problems with transportation and storage of wheat middlings. Mold and spoilage can be problems in storage. Bulk middlings, are more difficult to transport and store than pelleted middlings but many farmers prefer to buy bulk middlings because they cost less than pelleted middlings.

Distiller's Grains as a Protein Source

Research at several universities has been conducted to determine the value of distiller's grains as a replacement for other protein sources in beef cattle rations. Distiller's grains are a by-product of grain alcohol fermentation and are called *stillage*. They are a good source of protein, energy, and phosphorus, but are low in calcium and potassium.

The research shows that wet distiller's corn grains, whether fed as the only protein source or in combination with urea, are equal to soybean meal or corn gluten meal in meeting the total supplementary protein requirement for growing and finishing beef cattle. However, sorghum stillage is not as good as soybean meal as a protein source for beef cattle.

Minerals

Salt, calcium, and phosphorus are the main minerals needed in rations for finishing cattle. Iodized salt is used in iodine-deficient areas. Salt provides the sodium and chlorine the animal needs. Calcium is supplied by limestone or oyster shell flour. Phosphorus is supplied by bonemeal or dicalcium phosphate. If the ration contains high-quality roughages and natural proteins, trace minerals do not give any additional gains. A ration of poor-quality roughages requires the addition of some trace minerals, such as copper and cobalt, to meet the needs of the animal. Urinary calculi (stones caused by precipitation of salts from the urine) are controlled by adding one ounce of technical-grade ammonium chloride to the daily ration.

Vitamins

The vitamins most commonly added to beef cattle rations are A, D, and E. Cattle can produce vitamin K and the B-complex vitamins in the rumen, so these vitamins are usually not necessary in the ration.

Vitamin A is the most important vitamin to add to the ration. A daily intake of 20,000 to 30,000 IU is enough for cattle on feed. When cattle are under stress, the addition of 50,000 IU of vitamin A per head per day may be beneficial.

When the diet contains a high level of a good-quality forage, no additional vitamin A supplementation is needed. Cattle convert carotene to vitamin A at the rate of 400 IU of vitamin A for each milligram of carotene. Good-quality legume forages can contain as much as 20 to 30 milligrams of carotene per pound. Good-quality grass hays can contain as much as 10 to 15 milligrams of carotene per pound.

Cattle that are in the sunlight part of each day do not require additional vitamin D. A shortage of vitamin D may occur in the winter when there is less sunlight. In that case, it should be added to the ration.

Rations that include leafy roughages usually have enough vitamin E. High-grain rations may need 2 to 5 IU of vitamin E added for each pound of feed. Injections of vitamin E for newly arrived cattle help to reduce sickness while starting them on feed.

Additives and Implants

A number of feed additives and hormone implants are available for finishing cattle. Regulations on the use of these products are changing as experimental results become known. Cattle feeders should check current regulations before using any additive or implant. Label instructions must be carefully followed. Feed additives and hormone implants are discussed in more detail in Unit 7 in this text.

FEEDING CATTLE

Feed Intake

Cattle with high feed intake generally have a higher feed efficiency than those with lower feed intake. The amount of feed that cattle will consume is related to (1) the energy level of the ration, (2) the weather, (3) feed palatability, (4) feed processing, and (5) degree of finish on the cattle.

Research shows that cattle will continue to increase feed intake until the energy level of the ration reaches about 55 megacalories per hundredweight net energy for gain (NE_g) on a dry matter basis. This is a level of about 78 percent total digestible nutrients (TDN) in the ration. Increasing the TDN above this level does not increase feed intake and reduces feed efficiency.

Cattle that are switched suddenly from a low-energy ration to a high-energy ration sometimes go off feed (reduce feed intake). The high-energy ration increases the availability of readily fermentable carbohydrates in the rumen; this may cause a low rumen pH level, reduced motility of the digestive tract, and acidosis. These problems can usually be avoided by gradually increasing the energy level in the ration.

Beef cattle generally decrease feed intake in hot weather and increase feed intake in cold weather. At temperatures above 93°F (33.9°C), cattle on full feed will reduce feed intake by 10 to 35 percent. When shade or cooling is available or a low-fiber diet is fed, the reduction in feed intake is less at these high temperatures. When the temperature is between 75°F and 93°F (23.9–33.9°C), feed intake may be reduced 3 to 10 percent. Feed intake is about normal when the temperature is between 59°F and 75°F (15–23.9°C). Feed intake may be increased by 2 to 5 percent when the temperature is between 40°F and 59°F (4.4–15°C). At temperatures of 23–40°F (–5 to 4.4°C), feed intake may be increased by 3 to 8 percent. At 4–23°F (–15.5 to –5°C), feed intake may be increased by 5 to 10 percent. When temperatures are below 4°F (–15.5°C), feed intake may be increased by 8 to 25 percent. Extremely cold temperatures (under –13°F [–25°C]) or storms can cause a temporary decrease in feed intake.

Some management practices that a cattle feeder can follow to increase feed intake during hot weather include:

1. feeding cattle earlier in the morning.
2. feeding more of the ration in the evening.
3. feeding more frequently during the day to reduce spoilage in fermented feeds.
4. treating silage with anhydrous ammonia or other preservatives to increase bunk life.
5. increasing the concentrate level in low-energy diets.
6. feeding a drier ration to increase bunk life of the feed.

Rain can cause a temporary decrease in feed intake of 10 to 30 percent. Mud that is 4 to 8 inches (10–20 cm) deep may reduce feed intake by 5 to 15 percent. Deeper mud (12–24 inches [30.5–61 cm]) may reduce feed intake by 15 to 30 percent. Providing good access to feed and using suitable bedding helps overcome some of the reduction in feed intake caused by muddy conditions.

Palatability of the ration may be reduced by urea, dust, or mold in the feed. Adding molasses or other flavoring agents can help overcome palatability problems caused by the presence of these materials. Adding flavoring agents to the feed when there are no palatability problems present generally will not increase feed intake.

The effect of feed processing on feed intake is discussed earlier in this unit.

Cattle tend to increase feed intake until they reach about 85 to 90 percent of their market weight. Feed intake then levels off for a while and may decline as they get heavier. When feed intake of heavy cattle begins to decrease, it may be an indication that they are about ready for market.

Feeding Management

The time of year that cattle are placed on feed affects feed efficiency and rate of gain. A Kansas study revealed that feedlot performance for steers weighing 700–800 pounds (317–363 kg) was higher when they were placed on feed between March and May. The poorest performance occurred when they were placed on feed in October or November. There was less variation in feedlot performance when the steers weighed 800–900 pounds (363–408 kg) at the start of the feeding period.

While it is difficult to anticipate how much feed cattle will eat from one feeding to the next, the cattle feeder should try to deliver an amount of feed in the feed bunk equal to the amount the cattle will eat before the next feeding. Make sure the feed is fresh and palatable. This improves feed intake and results in fewer digestive problems, especially on high-energy diets. Using low-quality or spoiled feed lowers feed intake. When feed intake is reduced by five percent, rate of gain decreases by about 10 percent.

Cattle should be fed at least twice a day. Increasing the frequency of feeding to four or five times daily may increase feed intake. Increasing the frequency of daily feeding also makes the estimate of how much feed will be eaten between feedings less critical. Cattle should eat all the feed in the bunk each day; however, they should not be out of feed for an extended period of time.

Other good feeding management practices include calculating rations on a dry matter basis, cleaning the feedbunks at least once a week, and following proper mixing procedures to ensure that all ingredients, including additives, are thoroughly mixed into the ration. Proper mixing of the feed improves palatability and decreases the tendency of the cattle to sort through the feed, leaving some portions uneaten.

Starting Cattle on Feed

Many cattle coming into the feedlot have had little or no grain in the growing ration. Some cattle have had grain in a creep feeding ration or have been fed grain on pasture during the growing period. If cattle are put on a full feed of grain too quickly, the microorganisms in the rumen do not have time to adjust to the new ration. This will put the cattle off feed. Cattle should be brought up to a full feed of grain as rapidly as possible without their going off feed.

A good-quality grass hay or first-cutting alfalfa-brome hay makes a good roughage for starting cattle on feed. Cattle will often eat grass hay better than legume hay when starting on feed. Second- or third-cutting alfalfa hay will often cause the cattle to scour (have diarrhea). Grass hay helps to stop scours in cattle. Other roughages that may be used are oat hay, Sudan hay, or green chop.

Corn silage with a protein supplement may be used to start cattle on feed. Hay is provided separately in the ration.

A starting mixture of 80 percent concentrates and 20 percent roughage may be used for cattle that have been creep fed grain or have had grain on grass pasture. The roughage is decreased by 10 percent and the concentrate increased by 10 percent at the end of the first week in the drylot. The ration for the rest of the feeding period may be 90 percent concentrate and 10 percent roughage.

Cattle that have not had any grain before going into the feedlot are fed less concentrate and more roughage to start them on feed. Several methods may be used to bring them to a full feed of grain.

One method is to start the cattle on a mixture of 60 percent concentrates and 40 percent roughages. Over a two-week period, the amount of concentrate is gradually increased while the roughage is decreased. On full feed, they receive 90 percent concentrate and 10 percent roughage in the ration.

Feeding may be started with 90 percent concentrate and 10 percent roughage in the ration. Using this method, the daily feed intake per head is limited to 1 percent of body weight. The amount of feed is gradually increased each day until, at the end of two weeks, the cattle are getting all they will eat.

Another method of starting cattle on feed is to use a ration of 60 to 70 percent roughage with 40 to 30 percent concentrate. The amount of concentrate in the ration is gradually increased over a two-week period. On full feed, the cattle are eating grain at the rate of about 2 percent of body weight.

Cattle to be finished on pasture are started on a ration of 80 percent concentrates and 20 percent roughages fed free choice. The amount of concentrate is increased by 10 percent and the roughage is decreased by 10 percent each week, until the cattle are being fed a ration that is 100 percent concentrates.

The ration should be fortified with vitamin A, antibiotics, and minerals. For the first three weeks, add 50,000 IU of vitamin A per head to the ration on a daily basis. A broad-spectrum antibiotic at the rate of 350 milligrams per head daily for the first three weeks reduces sickness and improves the rate of gain. Minerals should be fed free choice. A mix of 60 percent dicalcium phosphate and 40 percent trace-mineralized salt meets the needs of the cattle.

Rations for Finishing Cattle

The nutrient requirements for feeding cattle are given in the tables in the appendix. The ration is made by using locally available feeds that meet these requirements. Examples and guidelines for balancing rations are found in Unit 8, Balancing Rations.

Table 16-1 gives some examples of rations for cattle on full feed. These examples are only general guidelines. The quality of available

TABLE 16-1 Beef Cattle Rations for Cattle on Full Feed.[1]

Steer Calves—Initial Weight 400–450 pounds (181–204 kg)

Dry Rations	lb/day	kg/day	Silage in Ration	lb/day	kg/day
Grain	11–14	5–6	Grain	10–14	4.5–6.4
High-quality legume hay	5–6	2.3–2.7	Protein supplement	0.5–1.5	0.23–0.68
			High-quality legume hay	2–4	0.9–1.8
			Corn or sorghum silage	8–12	3.6–5.4
Grain	11–14	5–6	Grain	10–14	4.5–6.4
Protein supplement	1.25–1.75	0.57–0.79	Protein supplement	1–1.5	0.45–0.68
Mixed hay or low-quality			Mixed hay or low-quality		
legume hay	4–6	1.8–2.7	legume hay	2–4	0.9–1.8
			Corn or sorghum silage	8–12	3.6–5.4

Steers—Initial Weight 600–750 pounds (272–340 kg)

Dry Rations	lb/day	kg/day	Silage in Ration	lb/day	kg/day
Grain	13–16	5.9–7.3	Grain	12–15	5.4–6.8
High-quality legume hay	6-8	2.7–3.6	Protein supplement	0.75–1.25	0.34–0.57
			High-quality legume hay	3–5	1.4–2.3
			Corn or sorghum silage	14–18	6.4–8.2
Grain	13–16	5.9–7.3	Grain	12–15	5.4–6.8
Protein supplement	1.25–1.75	0.57–0.79	Protein supplement	1–1.5	0.45–0.68
Mixed or low-quality			Mixed hay	2–4	0.9–1.8
legume hay	5–7	2.3–3.2	Corn or sorghum silage	14–18	6.4–8.2
Corn and cob meal	14–17	6.4–7.7	Ground snapped corn	13–17	5.9–7.7
Protein supplement	1–2	0.45–0.9	Protein supplement	2–2.25	0.9–1.1
Mixed hay	4–6	1.8–2.7	Corn or sorghum silage	12–16	5.4–7.3
Ground snapped corn	18–20	8.2–9.1			
Protein supplement	2–2.5	0.9–1.1			
Peanut hay	3–5	1.4–2.3			

Steers—Initial Weight 800–900 pounds (363–408 kg)

Dry Rations	lb/day	kg/day	Silage in Ration	lb/day	kg/day
Grain	16–18	7.3–8.2	Grain	12–16	5.4–7.3
Legume hay	6–8	2.7–3.6	Protein supplement	1.25–1.75	0.57–0.7
			Legume hay	1–3	0.45–1.4
			Corn or sorghum silage	16–20	7.3–9.1
Grain	13–17	5.9–7.7	Grain	12–16	5.4–7.3
Protein supplement	1.5–2	0.68–0.9	Protein supplement	2.5–3.0	1.1–1.4
Mixed hay	7–9	3.2–4.1	Grass hay	3–5	1.4–2.3
			Corn or sorghum silage	16–20	7.3–9.1
Ground snapped corn	14–18	6.4–6.8	Ground snapped corn	15–17	6.8–7.7
Protein supplement	3.0–3.5	1.4–1.6	Protein supplement	2.75–3.25	1.2–1.5
Cottonseed hulls	6–8	2.7–3.6	Corn or sorghum silage	16–18	7.3–8.2

[1]Quantities of feed are averages for entire feeding period. Less feed is required in the early part of the feeding period and more in the later part of the feeding period. More roughage is used early in the feeding period and less in the later part of the feeding period.

Source: USDA, *Finishing Beef Cattle, Farmers Bulletin No. 2196,* May 1973.

feeds and the relative prices of feeds will make it necessary for the feeder to vary the ration slightly from these guides.

Cattle feeders who have fast-growing cattle on high-energy rations may find that increasing the level of protein in the ration will improve feed efficiency and rate of gain. Research at Iowa State University has shown as much as seven percent improvement in feed efficiency and 12 percent improvement in rate of gain by feeding higher levels of protein. The suggested protein levels varies with the current rate of gain:

- less than three pounds (1.4 kg) per day—11 to 11.75% protein.
- three to three and one-quarter pounds (1.4–1.5 kg) per day—12 to 12.5% protein.
- three and one-half pounds (1.6 kg) per day—13 to 13.5% protein.

Response to increased levels of protein varies with the kind of cattle fed, implants used, and overall management.

Research conducted at the University of Nebraska suggests that adjusting the protein level in the ration of feeding cattle during the feeding period may reduce feed costs. The research revealed that making an adjustment in the amount of protein in the ration about every two weeks resulted in reducing protein costs by 15 percent and reducing the amount of nitrogen in the manure by 37 percent.

Rations that are high in crude fiber have given satisfactory gains in experimental trials. These rations contain as much as 70 percent pelleted or cubed alfalfa hay. Gains on high-fiber rations are a little slower than on high-concentrate rations. Feed requirements are about 21 percent less. Fewer animals on high-fiber rations grade USDA Choice as compared to animals on high-concentrate rations.

Using Ultrasonics to Sort Cattle into Feeding Groups

A computer software program developed at Kansas State University measures ultrasonic images of cattle being placed on feed. The computer program determines the length of time necessary to feed the animal to a desired quality and yield grade. Cattle may then be sorted into pens for more efficient feeding since all the animals in a given pen are similar in their feed requirements. A commercial company licenses the software from Kansas State and trains feedlot operators in its use.

Calculating Total Feed Needed

Some general guidelines may be used to calculate the total amount of feed needed to finish cattle. The exact amounts needed vary according to the age and condition of the cattle, kind of feed used, weather conditions, and management practices of the feeder.

Fattening cattle will eat about 2.5 to 3 percent of their body weight in feed each day. Multiplying the average weight of the animal by 2.5 to 3 percent and then multiplying by the number of days in the feeding period gives the amount of feed needed.

Example:

Starting weight of steer	600 lb	272 kg
Length of feeding period	210 days	
Average daily gain	2.4 lb	1.1 kg
Ending weight	1,104 lb	500 kg

Daily ration:

Shelled corn	14.8 lb	6.7 kg
Protein supplement	1.5 lb	0.68 kg
Mixed hay	5 lb	2.27 kg
Feed consumption	2.5 % of average body weight	

$$\frac{\text{start wt.} + \text{end wt.}}{2} \times \left(\begin{array}{l}\text{Feed consumption as percent} \\ \text{of body weight}\end{array}\right) \times \text{days on feed}$$

= total pounds (kg) of feed per head

English

$$\frac{600 + 1,104}{2} \times .025 \times 210 = 4,473 \text{ pounds of feed per head}$$

Metric

$$\frac{272 + 500}{2} \times .025 \times 210 = 2,026.5 \text{ kg of feed per head}$$

Consumption of each feed in ration:

	lb	kg
Corn	$14.8 \times 210 = 3,108$	$6.7 \times 210 = 1,407$
Protein supplement	$1.5 \times 210 = 315$	$0.68 \times 210 = 142.8$
Mixed hay	$5 \times 210 = 1,050$	$2.27 \times 210 = 476.7$
Total feed consumption	4,473	2,026.5

Care of New Feeder Cattle

The most critical time in feeding cattle is the first two or three weeks after they are put into the feedlot for finishing. Important management areas are: feeding, immunization, parasite control, castration and dehorning, treatment of sick cattle, and minimizing stress.

Cattle must be handled carefully to avoid stress. Stressful conditions result when cattle are loaded and transported to the feedlot.

Cattle should have a moderate fill of grass hay and water before loading for shipping. Loading should be done as quickly and quietly as possible to reduce stress. Observe the cattle carefully as they are unloaded to see if any of them show signs of sickness. Isolate sick cattle in separate pens to avoid spreading disease.

Cattle that have been through sale rings, auction markets, and holding yards are more likely to have been exposed to disease. Shipping fever symptoms usually appear within seven to ten days. Cattle that have been in transit longer than that will show signs of shipping fever if they have it.

Newly arrived cattle are put into small lots with plenty of feeder space. This makes it easier for the feeder to observe any cattle that do not go to the bunk to feed. Fresh, palatable feed should be provided several times per day. Dry, dust-free, well-drained lots are recommended.

Most cattle can be started on silage and soybean supplement upon arrival in the feedlot. Sick cattle do better if fed good-quality hay. Plenty of clean, fresh water helps the cattle to recover from shrinkage losses, helps to maintain health, and can be used to give medicines, if needed.

Cattle should be vaccinated as they come off the truck or as soon thereafter as possible. This decreases the amount of handling, thus reducing stress. No one vaccination program is best for all feedlots or groups of cattle. A knowledge of the past history of the cattle helps the operator to plan the vaccination program. About three weeks after arrival in the feedlot, the cattle can be given any vaccinations not given when they arrived.

Three vaccines that should be given to all cattle upon arrival are those for infectious bovine rhinotracheitis (IBR), para-influenza 3 (PI_3), and leptospirosis. Calves that are not castrated or dehorned probably have not been given any vaccines or bacterin. (*Bacterin* is a liquid containing dead or weakened bacteria that causes the animal to build up antibodies that fight the disease caused by the bacteria.) Vaccinations for blackleg and malignant edema should be included in the program for these cattle.

Other vaccines given depend on the history of the cattle and the problems that the feedlot operator has had in the past. Most of these vaccines may be given three weeks after the cattle arrive in the feedlot.

Vitamin A should be included in the ration. A level of 40,000 IU per pound of supplement is adequate. In addition, cattle can be injected with one to two million units of vitamin A as they arrive in the feedlot.

Parasites such as lice, mange, and grubs must be controlled on newly arrived cattle. To eliminate these parasites, cattle may be sprayed or dipped three weeks after they have been placed in the feedlot. Cattle may also need deworming. Dewormers should be used only after the cattle are eating well. Routine application of antibiotics to all cattle entering the feedlot is not recommended.

Castration is not done until the cattle have been in the feedlot for about three weeks. If they need dehorning, this can be done at the same time.

The feedlot operator must watch the cattle closely for signs of sickness. A check should be made at least three times a day. Right after feeding is a good time to spot sick cattle.

Signs of sickness in cattle include:

- moving about slowly.
- rising slowly.
- drooping ears.
- discharges from eyes or nose.
- not eating.
- thin appearance.
- walking stiffly or dragging the hind feet.
- carrying head in abnormal way or drooping head.
- high temperature.
- eyes sunken or dull.
- difficult or rapid breathing, or coughing.
- scours (diarrhea), especially if bloody.

Sick cattle are isolated from the rest of the herd as soon as signs of sickness are observed. A veterinarian can give the most accurate diagnosis and prescribe the best treatment for sick cattle. Calling a veterinarian can help to prevent sickness from spreading to the rest of the cattle.

SUMMARY

Because of the high grain requirements in the finishing ration, most cattle feeding is done in the grain-producing states of the north-central and southern parts of the United States. Large commercial feedlots of more than 1,000 head are found mostly in the Plains states, Colorado, Arizona, Texas, and California. Most of the smaller feedlots of less than 1,000 head are found in the Corn Belt.

Some feeders use more roughage in the ration under a system of deferred finishing. Yearling feeders may be used to clean up crop residues before being put on a full feed.

Steers make better gains than heifers in the feedlot. Calves make more efficient gains than older cattle but take longer to reach market. Crossbred feeder cattle gain faster than the parent breeds. More roughages can be used when feeding lower grades of feeder cattle.

Feeder cattle come from auctions or direct from the farm or ranch where they were born. The peak movement of feeder cattle is in October. Profit from feeding cattle is equal to the value of the finished cattle minus the total cost.

The main feed for feeder cattle is corn. Corn silage is the best roughage that can be used. Common natural protein sources used in feeder cattle rations include soybean, linseed, and cottonseed meal. Urea may be used as a nonprotein source of nitrogen in feeder cattle rations. Cattle need salt, calcium, and phosphorus. It is sometimes necessary to add vitamins A, D, and E to finishing rations.

Cattle with higher feed intake generally have a higher feed efficiency than those with lower feed intake. Feed intake is influenced by energy level in the ration, weather, feed palatability, feed processing, and degree of finish on cattle. The feed intake level begins to decrease when cattle reach 85–90 percent of their market weight.

When starting cattle on feed, the proportion of roughage in the ration is higher at the start and then is gradually decreased. The proportion of concentrate is gradually increased until the cattle are on full feed. Amounts of roughage at the start in the various methods ranges from 10 to 70 percent of the ration. The balance of the ration is concentrate.

New feeder cattle coming on the farm must be watched closely for signs of sickness. Put new cattle in small lots with fresh, palatable feed and water. Vaccination and parasite control programs must be carefully planned and carried out soon after the cattle arrive.

Student Learning Activities

1. Visit local cattle feeders who use different systems of feeding and feed various kinds of cattle. Ask why they follow these feeding programs and prepare an oral report for the class.
2. Prepare a bulletin board displaying pictures of the grades of feeder cattle.
3. Prepare transparencies for an overhead projector and present a report to the class on when to buy feeder cattle and how to calculate possible profit in cattle feeding.
4. Prepare a feeding program for cattle on your home farm.
5. Interview local cattle feeders to learn how they start newly purchased cattle on feed and what management practices are used with cattle on feed. Present a report to the class.
6. Observe various methods of preparing feed for cattle on local farms.
7. Observe and practice vaccination, dehorning, and implanting of cattle on your home farm or a local cattle feeder's farm.

Review

1. Why is more cattle feeding done in the grain-producing states than in other parts of the United States?

2. What is the difference between commercial feedlots and farmer-feeders?

3. Describe (a) finishing immediately, and (b) deferred finishing systems of feeding beef cattle.

4. List and briefly describe four factors that the cattle feeder must consider when selecting cattle for the feedlot.

5. What sources of feeder cattle are available to the feedlot operator?

6. Describe the seasonal variation in prices for feeder cattle and for slaughter cattle.

7. How is the break-even price for feeder cattle determined?

8. How is the profit margin in feeding cattle determined?

9. List six general rules a cattle feeder should follow when selecting feeder cattle.

10. Briefly explain the use of roughages in finishing cattle for the slaughter market.

11. Briefly explain the use of pasture when finishing cattle for the slaughter market.

12. What is the most commonly used grain for finishing cattle for the slaughter market?

13. What is the relative value of some other commonly used grains for finishing cattle?

14. Briefly explain the processing of grain for finishing cattle.

15. How does high-moisture ground ear corn compare to dry ground ear corn for finishing cattle?

16. Name three natural protein sources often used in feeder cattle rations.

17. Briefly explain the use of urea in feeder cattle rations.

18. Name the factors that influence feed intake of feeder cattle and briefly discuss each.

19. List four management practices that a cattle feeder might follow to increase feed intake during hot weather.

20. What minerals are commonly required for finishing cattle?

21. What vitamins should be included in the ration for finishing cattle?

22. Describe how cattle should be started on feed in the feedlot.

23. Give an example of a ration for cattle on full feed in the feedlot.

24. How can stress be minimized as new feeder cattle are started on feed?

25. Name three vaccinations that should be given to cattle upon arrival in the feedlot, if they have not been previously vaccinated.

26. List other management practices that should be followed for cattle as they are brought into the feedlot.

27. List the signs of sickness that a cattle feeder should watch for in cattle in the feedlot.

Unit **17** *Diseases and Parasites of Beef Cattle*

Objectives

After studying this unit, the student should be able to

- explain the importance of maintaining healthy beef cattle.
- identify and recommend prevention and treatment for beef cattle diseases and parasites common to the local area.
- recognize and suggest controls for common nutritional health disorders of beef cattle in the local area.

HERD HEALTH PLAN

The beef producer needs to develop an overall plan for maintaining the health of the beef herd. The key to the success of the health plan is the prevention of problems. Being familiar with the diseases and parasites that affect beef cattle helps farmers plan preventive programs that reduce health problems and increase profits. An important part of the health plan is developing a good working relationship with a veterinarian. Scheduling routine visits by a veterinarian can save money by helping prevent health problems before they become serious.

Characteristics of a good herd health plan include the following practices:

- Work with a veterinarian to develop a herd health program.
- Follow good feeding practices that meet the nutritional needs of the animals.
- Keep good records.
- Vaccinate at the correct time, following all label directions.
- Follow proper procedures for handling and storing vaccines.
- Control parasites.

- Follow good reproductive management procedures.
- Observe the animals to detect signs of disease, correctly diagnose the disease, and treat with the appropriate drugs.

Observing the vital signs (temperature, pulse rate, and respiration rate) in an animal can help in the early detection of health problems. Vital signs will vary with activity and environmental conditions. Normal vital signs in beef cattle are:

- Temperature (normal range is 100.4–102.8°F [38–39.3°C]) (average is 101.5°F [38.6°C]). (Usually temperature is higher in the morning than in the afternoon; younger animals will show a wider range of temperature than mature animals do.)
- Pulse rate (normal range is 60–70 heartbeats per minute).
- Respiration rate (normal range is 10–30 breaths per minute).

Body temperature is taken in the rectum using either a mercury thermometer or a battery-powered digital thermometer. Restrain the animal when inserting the thermometer into the rectum. The pulse rate is taken by finding the artery on the lower edge of the jaw. It may also be taken by finding the artery along the inside of the foreleg or the inside of the hind leg just above the hock. There is another artery that may be used located high on the underside of the tail. Respiration rate is determined by observing the number of times the animal breathes per minute.

It is better to prevent health problems than to try to cure them once they have occurred. Good sanitation programs are essential in preventing diseases and parasites. Feedlots and feed bunks must be kept clean. New additions to herds or to the feedlots should be handled carefully to prevent the spread of diseases or parasites to animals already on the farm or feedlot.

DISEASES

Anthrax

Anthrax is a disease caused by bacteria that may remain in the soil for 40 years or longer. Certain conditions cause the bacteria to become active. Anthrax affects mainly cattle and sheep.

Infection may result from grazing on infected pastures. The bacteria usually enter the animal's body through the mouth. They may enter through the nose or through open wounds. Biting insects, such as horseflies, may spread the disease from one animal to another.

The first sign of anthrax is often the sudden death of the infected animal. Less acute infections show symptoms of high fever, sudden staggering, hard breathing, trembling, and collapse. Death usually occurs within a few hours after these symptoms appear (Figure 17-1).

The carcass of an animal that has died from anthrax should be burned or buried at least six feet deep and covered with quicklime.

FIGURE 17-1   A carcass of a beef animal that has died of anthrax. The carcass bloats soon after death due to rapid decomposition. *Courtesy of USDA.*

Care must be taken not to bury the carcass near wells or streams. In some states, the carcass of an animal infected with anthrax may not be taken to a rendering plant. Care must be taken when handling the carcass of an animal suspected of having anthrax since the disease can be transmitted to people.

Vaccines may be used to control anthrax. In areas where anthrax is a problem, animals should be vaccinated on a yearly basis. Where it is not a common disease, vaccination should be done only upon the advice of a veterinarian.

Bovine Respiratory Syncytial Virus

Bovine Respiratory Syncytial Virus (BRSV) affects the cells lining the respiratory system. As a result, the respiratory system is weakened and becomes more liable to infection from other viruses and bacteria. Nursing and weaned calves are more likely to be affected than are older cattle. Stress from moving or weaning calves increases the changes of infection from this disease. A combination vaccine has been developed for treatment of BRSV, IBR, BVD, and PI_3. Descriptions of the latter three diseases, as well as others, follow.

Bovine Spongiform Encephalopathy (BSE)

Bovine Spongiform Encephalopathy (BSE) is a chronic degenerative disease that affects the central nervous system of cattle. This is one of a class of brain diseases called Transmissible Spongiform Encephalopathies (TSEs) that are relatively rare. Some TSEs affect animals and some affect humans.

TSEs affecting animals include:

- Bovine Spongiform Encephalopathy (BSE)
- Chronic Wasting Disease in deer and elk
- Feline Spongiform Encephalopathy
- Scrapie in sheep and goats
- Transmissible Mink Encephalopathy

TSEs affecting humans include:

- Creutzfeldt-Jakob Disease (CJD) [first identified in the 1920s]
- new variant CJD (nvCJD) [first identified in 1995]
- Fatal Familial Insomnia
- Gerstmann-Straussler-Scheinker Syndrome
- Kuru

BSE is extremely rare in the world and no cases have ever been identified in the United States. It was first diagnosed in Great Britain in 1986 and has been found in a few other countries since then.

There have been no beef imported into the United States from the United Kingdom since 1985. The USDA maintains a constant surveillance program to detect any signs that BSE has infected cattle in the United States. In 1989 the USDA imposed a ban on the importation into the United States of any live ruminant animals or ruminant products from countries with confirmed cases of BSE. In 1997 this ban was extended to include all cattle and sheep and many ruminant products from any part of Europe until such time as the risk from BSE is fully assessed. The Animal and Plant Health Inspection Service of the USDA enforces the current import restrictions.

Cattle that are affected by BSE show symptoms such as nervousness or aggression, muscle twitching, abnormal posture, loss of body weight, decrease in milk production, and difficulty in rising after lying down. There is no treatment and affected animals eventually die. Both beef and dairy cattle are susceptible to BSE.

The cause of BSE is not known; the available evidence indicates that it is neither bacterial nor viral in nature but rather is related to a prion. A prion is a microscopic protein particle that is similar to a virus but lacks nucleic acid. Cattle may contract BSE by ingesting protein in feed that came from an animal protein source that was contaminated by the agent that causes the disease. While a direct link has not been made, there is some suspicion that feeding cattle rendered protein from scrapie infected sheep may be involved in spreading the disease. In 1997 the FDA banned using any mammal-derived protein by-products in cattle feed in the United States.

The incubation period for BSE ranges from two to eight years. The animal usually dies within two weeks to six months after clinical symptoms appear. There is no test to determine if live cattle are affected; the presence of BSE can only be confirmed by postmortem microscopic examination of the brain. The brain tissue of infected animals has a spongy appearance when examined under a microscope. There is no treatment for BSE nor is there a vaccine that can be used to prevent it.

There is no evidence that BSE can be transmitted to humans by direct contact with infected animals or by consuming their meat or dairy products. There also is no evidence that eating meat from BSE infected cattle can cause Creutzfeldt-Jacob Disease (CJD), a human brain disease. In 1995 a new human neurological disease was found in England. It is called new variant Creutzfeldt-Jakob Disease (nvCJD). Research indicates that the same agent that causes BSE in cattle may cause nvCJD in humans. The BSE causative agent has not been found in muscle meat or milk from cattle. It has been found in brain tissue, the spinal cord, corneal tissue, and some other central nervous system tissues of infected animals.

Bovine Virus Diarrhea

Bovine virus diarrhea (BVD) is a common disease throughout the United States. The disease may appear in mild, acute, or chronic forms. BVD spreads by contact and it may be carried on a person's shoes from one herd to another. It can also result from fenceline contact with infected animals.

In the mild form, there are often no symptoms. If symptoms are present, they include fever, coughing, discharge from the nose, slow gains, rapid breathing, and mild diarrhea. Animals that have had the mild form of the disease are immune to further infection.

Animals with the acute form of BVD show symptoms of fever, difficult breathing, discharges from the nose and mouth, and coughing. In addition, ulcers may develop on the mouth and the animal may become lame. Dehydration and weight loss also occur. Diarrhea begins three to seven days after the animal becomes diseased. Pregnant animals may abort if the disease is contracted during the first two months of pregnancy. The fetus may mummify (absorb the fluids in the womb and become hardened and shriveled) if the cow becomes infected with BVD from the 90th to the 120th day of pregnancy. In the later stages of pregnancy, BVD may cause the fetus to suffer brain damage, hairlessness, or underdeveloped lungs.

Chronic cases of BVD result in slow gaining. The hair coat is rough and the animal may become lame.

A modified live virus vaccination is used to prevent BVD. Calves are vaccinated between one day of age and three weeks before weaning. Feeder cattle may be vaccinated upon arrival in the feedlot, or after they have been in the feedlot for two or three weeks. Calves vaccinated before weaning should be vaccinated again when placed in the feedlot.

Pregnant cattle should never be vaccinated. Vaccinate adult cattle only after calving and at least three weeks before breeding. Vaccinate replacement heifers between nine and twelve months of age, but not during the last three weeks before breeding. A single vaccination of older cattle will give immunity for the productive life of the animal. Do not vaccinate animals during a stress period.

There is no cure for BVD. Treatments are given to control diarrhea and secondary infections.

Brucellosis

Brucellosis, a disease caused by a microorganism, causes heavy economic losses in the cattle industry. Federal and state programs of eradication have resulted in the disease being less common than it once was. The germs that cause brucellosis are also dangerous to humans, causing undulant (Malta) fever.

Cattle with brucellosis often abort during the last half of pregnancy. Infected cows may retain the afterbirth (placenta). Other symptoms include sterility in cows and bulls, reduced milk flow in cows, and enlarged testicles in bulls. Calves born to infected cows may be weak.

Brucellosis is spread by infected cattle that are brought into the herd. It may also be picked up by fenceline contact with infected animals. An aborted fetus that carries the *Brucella* organisms may be brought from one farm to another by dogs or other carnivorous animals. Unborn calves may be infected by their mothers and become sources of infection after birth. Cattle can also contract brucellosis by eating or drinking feed or water in which the organism is present. Sniffing or licking an aborted fetus or a calf from a cow with the disease can also spread the disease.

There is no cure for brucellosis. Prevention is accomplished by good management practices. Herd replacements should be purchased only from brucellosis-free herds. All animals new to the herd should be isolated and tested. The herd should be tested periodically (blood or milk tests) to determine whether the disease is present. Infected animals should be disposed of following state and federal guidelines. Calves should be vaccinated between two and six months of age to increase their resistance to the disease. A state that is nearing an eradication stage may not permit calfhood vaccination; calfhood vaccination does not give lifetime protection from brucellosis. Cooperation with state and federal eradication programs helps in controlling the disease.

Blackleg

Blackleg is a disease caused by bacteria that grow only in the absence of oxygen. The disease is most serious when the bacteria lodge in deep wounds. When the bacteria are exposed to air, they form a protective spore (covering) which allows them to live for many years in the soil. The spores enter the animal through the mouth or through wounds. Young cattle are more commonly affected than older cattle.

The first sign of disease may be when one or more animals suddenly die. Before death, the symptoms of blackleg are lameness, swollen muscles, severe depression and, in the early stages, high fever. The animal may be unable to stand. An animal with blackleg is shown in Figure 17-2.

Blackleg is prevented by vaccination. Calves are vaccinated when young and again at weaning. To prevent the spread of the disease, dead animals must be burned or buried.

The disease is treated with massive doses of antibiotics. Treatment is effective only if the disease is diagnosed early. Prevention is more effective and less costly than treatment.

Calf Enteritis (Scours)

Enteritis (scours) is a disease complex (group of diseases) that is most common in the fall, winter, and spring. It is a disease of young calves. Calves over two months of age seldom are affected by the disease.

Symptoms of scours vary. In the acute form, the calf is found in a state of shock. The nose, ears, and legs are cold and the animal may suffer from diarrhea. The calf dies suddenly. The chronic form shows

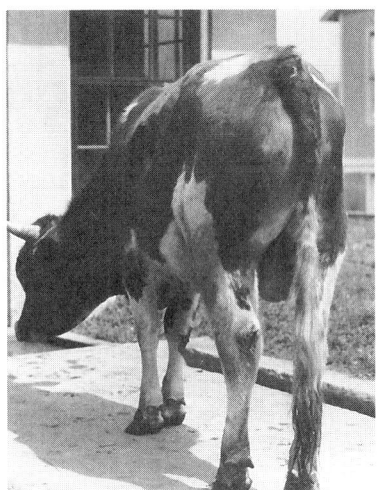

FIGURE 17-2   An animal with blackleg. Note the swelling of the hind leg. *Courtesy of USDA.*

symptoms, including diarrhea, for several days. The calf loses weight and dies after several days if not treated.

The most important factor in control is sanitation. Buildings in which calving occurs must be kept clean. Bedding must be clean and dry. Disinfect buildings and pens where calves are kept. It is important that the calf be fed the first milk (colostrum) from the mother after birth. The colostrum contains antibodies that help prevent scours. Vitamin A helps to control scours. Therefore, supplementing the cow's diet with vitamin A just before calving may help prevent the disease.

The most common types of scours (*E. coli*, Reo virus, and Corona virus) can be controlled by vaccines. The dam is vaccinated at least 30 days before calving; this causes the development of antibodies in the milk. The antibodies are passed on to the calf in the colostrum milk. A combination vaccine for scours and campylobacteriosis (vibriosis) is now available. It can be used 10 weeks before calving.

Extra care must be taken when calves are bucket-fed milk. If they receive milk from dirty buckets they are more likely to become infected with scours. Since too much milk may also lead to scouring, it is good practice to keep the calves a little bit hungry for the first two weeks after birth. A gradual replacement of whole milk with milk replacer starting at about two weeks of age also helps to prevent scours.

Antibiotics and sulfa drugs may be used to treat scours after the symptoms appear. A veterinarian is the best source of information about which drugs are most effective for local conditions.

Foot Rot

Foot rot is caused by a variety of bacteria, fungi, and other organisms that are found in the feedlot. These organisms enter the animal when the skin of the foot is broken in some way. Sharp objects in the feedlot, such as stones, nails, or wire, can cause injuries that make it possible for the organisms to enter the animal's body. Feedlots that are muddy or contain large amounts of manure make the problem worse.

The first noticeable symptom of foot rot is usually lameness. The animal loses its appetite, has a fever, and is depressed. It may not want to stand on its feet or move around. Death may eventually result.

There are no vaccines to prevent foot rot. Sanitation and paved lots help to prevent it. Good drainage and mounds in the feedlots are also helpful in eliminating conditions that encourage the disease. Spreading lime and 5-percent blue vitriol in the areas around water and feed bunks helps in its control.

Massive dosages of penicillin or wide-spectrum antibiotics are given to infected animals. Sulfa drugs also are used to treat foot rot.

Infectious Bovine Rhinotracheitis (IBR, Red Nose, IPV)

Infectious bovine rhinotracheitis is a disease caused by a virus. It occurs in several forms, and is found in cattle throughout the United States and the rest of the world.

Respiratory IBR is the most common form of this disease. Symptoms are fever and a discharge from the nose. The animal may also foam at the mouth, breathe through the mouth, cough, and lose weight. The nose and muzzle become inflamed, which is why this form is sometimes called "red nose." Infection will involve from 15 to 100 percent of the animals. However, death losses from this disease are low.

Genital IBR is known as *infectious pustular vaginitis* (IPV). Symptoms are inflammation in the vagina and swelling of the vulva. Pustules (pus-filled blisters) form in the vagina, and there is a pustular discharge. In the bull, the disease is called *infectious pustular balanoposthitis* (IPB). Symptoms include inflammation of the sheath and penis. Lesions form on the penis and there is a pussy discharge.

Conjunctival IBR is similar to pinkeye. A clear discharge from the eye is the first symptom. The discharge later becomes pussy.

Abortion may be caused by IBR. Abortion may follow the symptoms of respiratory IBR or conjunctival IBR, or it may occur without any other symptoms being observed. Pregnant animals that are exposed to infected animals or vaccinated with modified live virus for IBR may abort.

Encephalitic IBR affects the brain and nervous system of the animal. Symptoms are incoordination, depression, convulsions, and death. Encephalitic IBR affects younger cattle more frequently than older cattle. The symptoms are similar to listeriosis, which is discussed later in the unit.

Because IBR is similar to some other diseases that affect cattle, positive diagnosis can be made only by laboratory tests. A veterinarian should be called if IBR is suspected.

A modified live virus is used to vaccinate for IBR. Calves are vaccinated after they are six months of age. Breeding animals are vaccinated at least two months before breeding if they have not been vaccinated as calves. Pregnant animals should never be vaccinated because vaccination may cause abortion.

Livestock producers must be sure that all new additions to the herd are disease free and have health certificates. Isolate new additions at least 30 days before they come into contact with the rest of the herd. Diseased animals must be isolated at once.

There is no treatment for IBR. Antibiotics and sulfonamides are used to control other infections that may attack the weakened animal.

Johne's Disease (Paratuberculosis)

Johne's Disease is a chronic infection of cattle that causes a thickening of the wall of the intestine. Infected animals have intermittent diarrhea. They lose weight and eventually die. The infected animal does not have a fever. Diagnosis is made with a *Johnin test*, which is done by a veterinarian.

The disease is controlled by checking the health history of animals being brought into the herd. They should not come from herds that have a history of Johne's Disease. If the animal has had chronic diarrhea, suspect the presence of this disease. Do not bring the animal into the herd. There is no effective treatment of infected animals.

Leptospirosis

Leptospirosis is caused by several strains of leptospira bacteria. Infected cattle may not show any symptoms. Acute cases may show a sudden rise in temperature, rapid breathing, loss of appetite, stiffness, bloody urine, jaundice, diarrhea, or abortion. In young cattle, the symptoms are very similar to shipping fever, discussed later in this unit. Death rate ranges from 5 to 15 percent of the infected animals.

Infected animals can spread leptospirosis. Feed that is contaminated from the urine of infected animals can spread the disease. Leptospirosis may be transmitted both from cattle to hogs and from hogs to cattle. Keeping hogs and cattle separate will help control this disease. Good sanitation and isolation for a period of one month of new animals coming into the herd are additional control measures that may be used.

A regular program of vaccination helps control leptospirosis. There are five common strains of the disease. Antibodies from a vaccine for one of the strains do not give protection against other strains. A veterinarian should be consulted to determine the proper vaccine to use for local conditions. Cows should be vaccinated before the breeding season starts and they should be given a booster shot six months later. Annual vaccination is required. Using high levels of antibiotics in the early stages of infection will stop the shedding of the organism. Animals that recover from leptospirosis develop immunity and the future breeding ability of the herd is usually normal.

Listeriosis

Listeriosis, a disease caused by a germ, occurs in three forms. It is spread from animal to animal by contaminated feed or water; by contaminated dust, manure, urine, or saliva; and possibly, by breeding. It is more common in animals that are fed low-quality silage.

Encephalitic listeriosis affects the brain. Initial symptoms include fever, loss of appetite, and dullness. As the disease progresses, the animal may have difficulty standing. The ears, eyelids, or lips droop. The head turns to one side and the animal wanders aimlessly in a circle. The animal may drool. Eventually, it may not be able to stand. Cattle will show these symptoms for 4 to 14 days before death. If treated properly, some will recover.

Abortion is caused from one form of the disease. There may be premature birth of a live fetus or abortion of a dead fetus. Calves that are born live usually die shortly after birth. Most of the abortions occur in the last one-third of the gestation period. The cow usually shows no other symptoms of the disease.

Septicemic listeriosis is more common in older animals. However, it can occur in premature calves. The infection affects the entire body of the animal. Sudden death is a common result of this disease. The animal may become weak, depressed, and lie down much of the time. Hard breathing, slobbering, and a discharge from the nose are other symptoms. The animal dies after several days.

There is no vaccine for control of listeriosis. Sanitation is the best control. Careful disposal of aborted fetuses and afterbirth is another control measure. Feed and water must be kept free from contamination.

Lumpy Jaw (Actinomycosis)

Lumpy jaw is a chronic disease and seldom causes the animal to die. It results in economic loss because the affected body parts are condemned when the animal is slaughtered.

The disease affects the jaw and surrounding bony part of the head. Sometimes it spreads to muscles and other internal organs. Symptoms are tumors or lumps on the jaw. These are filled with a yellow pus that has little odor. The teeth may become loose, making it difficult for the animal to chew. The jawbone becomes spongy and swells, which may result in breathing problems. The animal may lose weight because of its inability to eat.

Surgical treatment by a veterinarian may allow the animal to still be marketed. Complete recovery is usually not possible.

The organism that causes the disease enters the animal's body through wounds. To prevent lumpy jaw, be sure that there are no sharp objects such as wire and barley beards in the pasture or feedlot to cause puncture wounds. Feed having sharp stickers may also cause injuries that allow the organism to enter the body of the animal. Isolate infected animals from the rest of the herd.

Malignant Edema

The symptoms, control, and treatment of malignant edema are much the same as those for blackleg, which is discussed earlier in this unit.

Pinkeye (Infectious Keratitis, Keratoconjunctivitis)

Pinkeye is a disease carried by insects that affects the eyes of the animal. It has both mild and acute forms. In the mild form, the eyeball develops a pinkish color. The cornea (the transparent covering of the iris and pupil) becomes slightly clouded. The symptoms of the acute form are a flowing of tears from the eye and a cloudiness of the cornea. As the infection progresses, the cloudy condition becomes worse and ulcers may develop on the eye. The eye may be damaged so severely that blindness results. The condition may last three to four weeks. If not treated, it will spread to most of the herd.

White-faced cattle and cattle with white pigment around the eye are more likely to be infected with pinkeye. It occurs year-round and may affect any kind of cattle. It is most common during periods of maximum sunlight.

Pinkeye is spread by insects such as face flies, by direct contact with infected animals, by dust, and by tail switching. Cattle that are crowded at feed bunks or waterers may rub heads, which contributes to the spread of the disease. Controlling flies and other insects is one of the main ways to prevent pinkeye.

Vaccinations are now available to control *Moraxella bovis*, the bacteria that is considered to be a main cause of pinkeye. A veterinarian should be consulted for current information on the availability and use of vaccines for pinkeye.

Infected animals should be isolated in a dark place. Antibiotics and sulfa drugs are applied to the infected eye. The medicine must be applied frequently since it is washed out of the eye by the flow of tears. Application should be made at least twice a day. A cloth patch can be used on the affected eye.

A virus form of pinkeye is associated with IBR. The symptoms, treatment, and control of this form are discussed under the section on IBR in this unit.

Shipping Fever (PI₃ Pasteurella, Bovine Respiratory Disease)

Shipping fever is a disease complex that affects the respiratory tract of the animal. It is most common in young cattle at times of stress. A combination of stress, virus infection, and bacterial infection is thought to be the cause.

The stress conditions that contribute to the start of the disease often occur when cattle are moved from the range to the feedlot. Extremes of heat or cold, dust, exhaust fumes, hunger, fright, and rough handling all create stress. This allows the virus and bacteria organisms, which are already present, to attack the respiratory tract of the animal.

The disease may vary from mild to acute. An early symptom is fever. The animal appears depressed, with the head down and the eyes closed. The ears are often drooping. A discharge from the nose develops. The eyes water and the animal loses its appetite. Diarrhea and weight loss may occur. Difficult breathing and coughing follow. Pneumonia may develop, and the animal may die. If it recovers, it will often be a slow-gaining animal.

The disease is prevented by vaccination. Vaccination should be done after the animal is four months of age. The best time to vaccinate is three to four weeks before the animal is exposed to conditions that lead to the disease. Reduction of stress and exposure help to prevent the disease. Good feedlot management and careful handling of new cattle coming into the feedlot also help to reduce shipping fever.

Antibiotics and sulfa drugs are used to treat shipping fever. The treatment must begin as soon as possible after the symptoms are noticed. Treatment after an animal develops pneumonia, should this occur, is of little value.

Trichomoniasis

Trichomoniasis is a venereal disease in cattle caused by a protozoan, *Trichomona fetus*. The organism infects the genital tract of the bull and is transmitted to the cow during breeding. A clean bull can also be infected when breeding an infected cow. The disease can be transmitted through infected semen, even if artificial insemination is used.

Symptoms include abortion early in gestation, low fertility, irregular heat periods, and infection in the uterus. In females, there may be a discharge from the genital tract. An infected bull may not show any symptoms of the disease but is still capable of transmitting it to cows during breeding. The organism is identified by microscopic examination of material from an aborted fetus, the preputial cavity of the bull, or vaginal discharge from the cow.

There is no treatment available for bulls; infected bulls should be slaughtered. Prevent the disease by using a clean bull on cows that have not been exposed to an infected bull. Make sure bulls brought into the herd are free of the disease. Artificial insemination, using semen from clean bulls, will help prevent the spread of the disease. No vaccination is available for trichomoniasis.

Campylobacteriosis

Campylobacteriosis is the new name for a reproductive disease formerly called *vibriosis*. Vaccine labels now carry the new name; however, some manufacturers are also including the trade name "vibrio" on the label to reduce confusion among purchasers.

Campylobacteriosis in cattle is a disease with both an intestinal and venereal form. It is a leading cause of infertility and abortion in the cattle industry, costing many millions of dollars per year.

The intestinal form of vibriosis has little harmful effect on cattle. The venereal form of the disease is more serious. If the organism infects the uterus, there will be some abortion in the herd. However, the number of cows affected is normally small. Cows do not become sterile, and bulls are not affected.

Symptoms of the venereal form include infertility, abortion, and irregular heat periods. In herds that are newly infected, conception rates may drop below 40 percent. The calving season is longer and there are more open cows in the fall. In herds that are chronically infected, the conception rate is lower than normal, usually about 60 to 70 percent. Heifers and new additions to the herd will require repeat breeding or will abort.

Animals should be vaccinated 30 days before breeding. Vaccination must be repeated every year.

The disease is spread from infected bulls to clean cows during breeding. It is possible to treat valuable bulls with antibiotics, but the process is difficult. Infected cows may settle more easily if they are treated with antibiotics. Skipping two heat periods before attempting to breed the cow also improves the conception rate of infected cows. Cows with the disease usually develop immunity and eventually breed again.

The use of artificial insemination (AI) helps to prevent campylobacteriosis. The semen used for AI is treated with antibiotics to eliminate the disease organisms.

Warts

Warts are growths on the skin caused by a virus. They are spread from one animal to another by contact, or by contact with posts, buildings, or other objects warty animals have touched.

Warts may appear singly or in clusters. Some are hard and others are soft. Calves are usually affected on the head, neck, and shoulders. On older cattle, warts often appear on the udder or teats.

Small warts may be clipped off with sterile scissors. Tying a sterile thread around the wart will cause it to slough off in a few days. Applications of glacial acetic acid, iodine tincture, silver nitrate, castor oil, or olive oil may be used to treat warts. Infected animals should be separated from other animals. Disinfecting pens and rubbing posts helps to prevent the spread of warts. Warts can be prevented by vaccination.

Wooden Tongue (Actinobacillosis)

Wooden tongue is a chronic but seldom fatal disease of cattle. It causes economic losses because affected body parts are condemned when the animal is slaughtered.

Symptoms include lesions on the soft tissues of the head. Swelling of the lymph glands of the neck may also occur. These sometimes break open, discharging a creamy pus. Lesions may develop on the tongue. The tongue becomes hard and immobile, and protrudes from the mouth. The animal will drool. As it becomes more difficult for the animal to eat, it takes in less feed and loses weight. Eventually, the animal may die.

The disease is probably spread through contamination of feed by infected animals. Animals should not be fed stemmy hay or be allowed to graze on pastures with plants that might injure the mouth. Surgical treatment of animals with wooden tongue may be necessary. This should be done only by a veterinarian. Infected animals should be isolated from other animals.

Ringworm

Ringworm is a contagious skin disease that may be spread to other animals and humans. Symptoms include round, scaly patches of skin that lack hair and that may appear on any part of the body. The affected area clears up, but the infection spreads to other parts nearby.

Sanitation helps control ringworm. Infected animals should be isolated from the rest of the animals. Iodine tincture or quaternary ammonium compounds may help in the treatment.

EXTERNAL PARASITES

External parasites of beef cattle include flies, lice, mange, mites, and ticks. Each year, there are high losses from these parasites in the United States. Some of these parasites irritate the animals; others are bloodsuckers. They slow down weight gains and, in some cases, damage the hides of the animals. Others carry diseases from one animal to another.

Chemical, biological, mechanical, and cultural control methods are used to reduce losses from parasites. Chemical control methods are the most economical. The use of chemical controls is regulated by state and federal laws. Insecticide registrations and recommendations for use change from year to year. For this reason, specific chemical control agents are not included in this unit. The livestock producer should consult with county extension personnel, university insect control specialists, livestock supply stores, and veterinarians for recommended insecticides for local use on specific parasites.

Carefully read and follow the directions on the labels. Chemicals must be applied only to those animals for which they are intended. Rates of application and withdrawal times must be observed. Keep a record of any chemical used, including the name, rate, dilution, percent of active ingredient, and dates of use.

A program of parasite control includes sanitation. Pens, barns, and feedlots must be kept clean. Manure that is thinly spread outdoors is less likely to provide a place for insect eggs and larvae to develop. Clean up wet litter and areas where seepage occurs. Good drainage in feedlots helps to control insects. Spraying barns with approved chemicals is a recommended practice for controlling insects. Identification of insects and a knowledge of their life cycle is important in the development of control programs.

The use of insecticide-impregnated ear tags has become popular in recent years for controlling some kinds of flies on beef cattle. Not all fly species can be controlled with ear tags. There is also concern that some insects are developing resistance to the insecticides used in the ear tags. The removal of the tags at the end of the fly season appears to help reduce this problem. Some manufacturers are including more than one insecticide in the ear tags to reduce the buildup of resistant insects. The use of sprays or dusts for a quick kill of existing flies at the time of ear tag application is also suggested as a way to prevent a buildup of a resistant population.

Both internal and external parasites may be controlled by using a slow-release bolus placed in the rumen of the animal. The bolus is

FIGURE 17-3 A back rubber may be used to control some external parasites of cattle. *Photo by Jim Bodine.*

used primarily with cattle weighing 275–660 pounds (124.7–299.4 kg) that are on pasture. It is effective for up to 135 days and must be withdrawn 180 days before the animal is slaughtered. The company that developed the bolus claims that treated cattle gain 14 percent faster than untreated cattle.

Flies

Bloodsucking flies include the horn fly, stable fly, horsefly, deerfly, blackfly, and mosquitoes. Other flies that irritate livestock include the screwworm fly, housefly, face fly, and heel fly.

Horn fly. The horn fly is about one-half the size of the housefly. It is gray-black in color. Hundreds of horn flies may cluster on the back, horn, withers, and belly of the animal.

The horn fly lays its eggs in fresh manure. The eggs hatch in one to two days. Maggots become full grown in about five days. Mature maggots pupate (change from larva to pupa form) in manure or soil and emerge as flies in five to seven days.

The horn fly is a problem mainly on pasture. Sanitation methods are of little value in control. Back rubbers containing recommended chemicals (Figure 17-3) are a common control method.

Stable fly. The stable fly is about the same size as the housefly. It is grayish in color, with seven rounded dark spots on top of the abdomen. It stays on the animal long enough to feed and then spends the rest of the day on nearby fences or walls, or in barns.

The stable fly lays its eggs in moist straw, strawy manure, moist feed, or other decaying organic matter. The eggs hatch in one to three days. The maggots reach maturity in eleven to 30 days. The pupal stage takes one to three weeks, at which time adult flies emerge.

Stable flies are more of a problem in feedlots than on pasture. Sanitation is important in control. Insecticides do not give complete control. Spraying areas where the flies rest is also recommended.

Horsefly and deerfly. Horseflies and deerflies are much larger than houseflies. They range from 1/3 inch to 1 1/2 inches in length, depending on the species. They are usually gray, with brown or black colors intermixed. Some have bright green heads or yellowish or reddish-brown bodies. The eyes are usually brilliantly colored.

The eggs are usually laid on vegetation growing in swamps, ponds, or other wet areas. Eggs hatch in four to seven days. The larvae drop into the water or mud. They burrow into mud, moist earth, or decaying organic matter. Some species feed on organic matter. Others feed on insect larvae, snails, earthworms, or other small forms of life. Some species mature in 48 days or less. Others take almost a year to mature.

Mature larvae pupate in drier areas about one to two inches below the surface. The pupal period for most species is two to three weeks.

Other species require one to three months. Adult flies then emerge and live three or four weeks. There is usually one generation per year, although some species may produce two.

Horseflies and deerflies irritate cattle, causing them to gain more slowly. They also feed heavily on blood. Twenty or 30 flies can take almost one-third of a pint of blood in six hours. They transmit a number of diseases, including anthrax and anaplasmosis. (Anaplasmosis is described later in the unit.)

Insecticides do not effectively control horseflies and deerflies although, if applied daily, they may provide some relief. Draining of swampy areas helps control these insects. Moving cattle from pastures in wooded or swampy areas when the flies are attacking them is one way to reduce the problem.

Blackfly (Buffalo Gnats). Blackflies are small, varying in length from 1/25 to 1/5 inch. Colors are orange, brown, and black. They have a humped prothorax.

Eggs are laid on an object in or near flowing water. Eggs hatch in three to 30 days, depending on the species. In northern areas, the winter is spent in the egg stage. Some species survive the winter as larvae. The larval stage varies from ten days to ten weeks, depending on the species. Larvae live underwater. They spin a cocoon that attaches to underwater objects. The pupal stage is from four days to five weeks, depending on the species. Adults emerge from the pupal case, rise to the surface, and fly away.

Mosquitoes. Mosquitoes are small flies with two wings. There are a number of different species and they vary in size. Eggs are laid on water or in low-lying areas that flood. The eggs of most species hatch in two or three days. Some species require a dry period before the eggs hatch. Eggs can remain dormant for many months and then hatch when flooded. Larvae live in the water, and most species must come to the surface to breathe. Larvae usually become pupae within one week. The pupal stage is about two days, after which the adults emerge.

Draining areas where mosquitoes lay their eggs is recommended. Insecticides are more effective when the insects are in the larval stage than when they are adults.

Screwworm fly. The adult screwworm is about twice the size of the common house fly with a bluish-grey or grey body with three dark stripes down its back. It has orange eyes. The larvae are pinkish in color. The adult female mates and then lays up to 400 eggs at one time in open wounds of animals. The female lives for about 31 days and can lay about 2,800 eggs during this time. The eggs hatch in ten to twelve hours and the larvae feed on the flesh of the living animal. They mature in five to seven days, after which they drop off the animal and burrow into the soil to pupate. The pupal stage lasts seven to ten days, after which the adult flies emerge. The adults are ready to mate within three to five days after emergence.

Any open wound on an animal's skin may be a target for screwworm infestation. Multiple infestations may occur in the same wound. If the infestation is not treated, the animal may die in seven to fourteen days. Infested animals may go off feed, exhibit signs of discomfort, and isolate themselves from the rest of the herd. Treatment of an infestation is by topical application of an approved organophosphate insecticide. Continue treatment for two to three days and remove the larvae from the wound with tweezers.

The screwworm was eradicated in the United States by introducing sterile males into the native population. Pupae in a fly population in a production plant are exposed to gamma radiation. The adult flies are sterile. These sterile flies are released from aircraft over areas where the screwworm exists. Over a period of time this resulted in the complete eradication of the native screwworm fly population. This method has been successfully used to eradicate the screwworm from Mexico, Belize, Guatemala, El Salvador, and Honduras. Reinfestation is prevented by establishing a barrier zone of sterile flies between the screwworm free area and the area where it has not yet been eradicated. Efforts are currently ongoing to extend the screwworm free area in Central America, eventually establishing a sterile fly barrier at the Darien Gap between Panama and Colombia.

Since eradication, outbreaks have occurred in Mexico; however, these were successfully contained and eradicated. The screwworm remains a potential threat for livestock in the United States, particularly from animals imported from areas where it has not yet been eradicated. Any suspected infestation should be reported immediately to state or federal animal health authorities.

Housefly. The housefly is about one-fourth inch in length. It is gray with a striped abdomen and brownish-yellow or red markings near the base.

Eggs are laid in manure or other decaying organic matter. The eggs hatch in 8 to 20 hours. The larvae grow to maturity in 5 to 14 days. The pupal stage varies from 3 to 10 days, after which the adult emerges. Sanitation and insecticides are used to control the housefly.

Face fly. The face fly is a little larger than the housefly. It is about the same color, being slightly darker than the housefly. The female feeds around the eyes and nose of cattle. The male is usually not found on animals.

Eggs are laid in fresh cow manure. The eggs hatch in about one day and the larvae mature in three to four days. The mature maggots move to soil and change to the pupae stage. The adult fly emerges in four to six days. Face flies are difficult to control by sanitation. Insecticides are commonly used as a control measure.

Heel fly (cattle grub).   The cattle grub is the larval stage of the heel fly. There are two species of cattle grubs in the United States. The common cattle grub is found in all states except Alaska. The northern cattle grub is found in Canada and northern United States. The cattle grub causes greater losses in the beef industry than any other pest. The major losses come from lower rates of gain and damage to the meat and hides of cattle.

The adult heel fly is about the size of a honey bee, although in some areas it may be larger. The color is similar to that of the honey bee. The body has bands of yellow or orange hair. There are four longitudinal shiny bands on the thorax. Black and orange hairs are found on the legs. The adult heel fly does not feed. It lives for only two or three days on food stored in the abdomen.

The life cycles of the two species are similar. The adult fly lays eggs on the hair of cattle on the hind legs, flanks, or sides. The eggs hatch in three to four days. The larvae burrow into the skin. During the next eight months, the larvae migrate through the body of the animal. When the larvae reach the back of the animal, they create a hole in the skin for breathing. Swellings, called *warbles*, appear on the back of the animal. The larvae, called *grubs*, stay in the back of the animal about one to two months. When they are mature they emerge through the hole and drop to the ground. The grub changes to the pupal form under trash, leaves, or other material on the ground. The pupal stage lasts from 20 to 60 days, depending on the temperature. The adult fly then emerges to mate and lay eggs. There is one generation per year.

Control of cattle grubs is accomplished most effectively with systemic insecticides. A *systemic insecticide* is one that spreads throughout the body of the animal. These insecticides may be applied as wet sprays, dips, backline pour-ons, in mineral mixes, or as feed additives (Figure 17-4). Application is made as soon as possible after heel fly activity has stopped. Early applications are more effective than later ones. In southern states, the application time ranges from late spring to fall. In northern states, the application time ranges from early summer to late fall. Consult veterinarians, county agents, or university specialists for the exact time of application in a given area.

Lice

One species of biting lice, and four species of bloodsucking lice attack beef cattle. Louse eggs are laid on hairs on the animal's body. (*Louse* is the singular of lice.) The eggs hatch in one to two weeks. Young nymphs emerge that mature to adults about one month after hatching.

The louse population is low during the summer and increases in the fall and winter. Cattle with lice have a rough appearance and do not gain at normal rates. Symptoms of lice include cattle rubbing against fences and feed bunks and hair balls left on fences. Bloodsucking lice weaken cattle so that they are more likely to be infected by diseases.

FIGURE 17-4 Cattle grubs may be treated by spraying the animals with a systemic insecticide. *Courtesy of Iowa State University.*

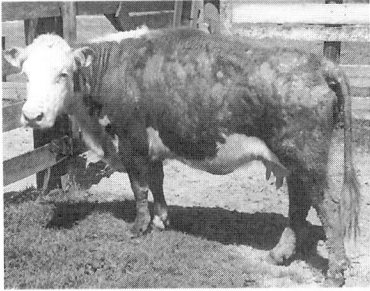

FIGURE 17-5   A beef animal with advanced lesions of common (psoroptic) scabies caused by mites. *Courtesy of USDA.*

Control of lice is by the use of insecticides that may be sprayed on, dusted on, used in a back rubber, or poured on. The best time for treatment is late fall or early winter.

Mites

Several species of mites attack cattle. They live on the skin or burrow into it and cause a condition known as *scab*, *mange*, or *itch*. All of the species that affect cattle are very small in body size. Mite populations are at their lowest in the summer and increase during the winter. Symptoms include the appearance of small pimply areas on the skin that lose hair (Figure 17-5). Infected animals rub, scratch, or lick at the affected areas. Cattle become restless and do not gain well. A heavy infestation may kill the animal.

Treatments that control lice also control mites. Preventive treatment should take place in the fall or early winter.

Ticks

Several species of ticks attack beef cattle. They are serious pests to cattle in some parts of the United States. Ticks are bloodsuckers and also transmit serious diseases among cattle. Cattle that are infested do not gain properly. The tick bites irritate the animals, causing them to rub and scratch at the affected area. This can result in a scabby skin condition or injury.

Ticks are flat, oval shaped, and dark brown or reddish in color. They are common in bushy pastures and wooded areas.

The *spinose ear tick* (Figure 17-6) attaches itself deep in the ear of the animal. This often causes severe irritation, wax buildup, and infection. Infested cattle rub their ears and shake their heads.

Ear ticks are controlled by dipping or treating each infested ear individually. Other ticks are controlled in the same way as lice.

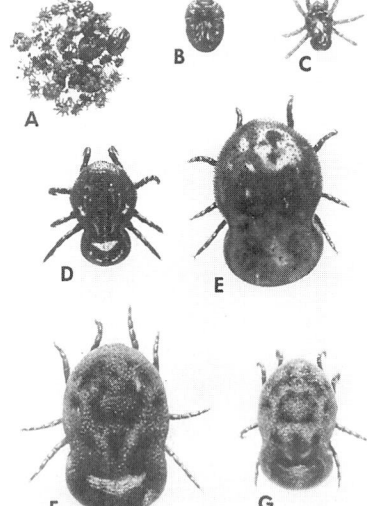

FIGURE 17-6   Ear tick. (a) Ticks and debris collected from ear; (b) engorged larva; (c) young nymph; (d) partially engorged nymph; (e) fully engorged nymph; (f) adult female; (g) adult male. *Courtesy of Texas Agricultural Extension Service, Texas A&M University.*

INTERNAL PARASITES

There are a number of internal parasites that affect cattle. The most common are roundworms, flatworms, coccidia, and anaplasma.

Roundworms

Roundworms found in cattle include stomach worms, nematodirus, threadworm, hookworm, cooperia, nodular worm, whipworm, and lungworm. The stomach worm is the most serious of these worms.

Symptoms of stomach worms include anemia, weakness, constipation, and diarrhea. A condition known as *bottle jaw*, which is a swelling under the jaw, sometimes develops. Worm infestations slow down gains. Symptoms of the other roundworms are similar to those of stomach worms.

Roundworms reproduce rapidly. Good sanitation helps to prevent infestation. Roundworms are treated with one of several chemicals, given as boluses (large pills), drenches, or feed additives.

Flatworms

Flatworms found in cattle include tapeworms and the deer liver fluke. Diarrhea is a symptom of tapeworm infestation. Deer liver fluke symptoms include loss of weight, limping, and weakness in the hind quarters.

Tapeworms have an intermediate host. At one point in their life cycle, they are ingested by free-living soil mites. Feeding cattle on paved lots helps to break the cycle since the mites are not present under this condition.

Snails are the intermediate hosts of deer liver flukes. Pasturing cattle away from streams, ponds, swamps, or other wet areas helps to break the life cycle of the deer liver fluke.

Generally, treatment of flatworms has not been successful. Some new drugs are being tested that may provide effective treatments. Prevention of flatworms is the best control measure.

Coccidia

Coccidia are protozoan organisms that live in the cells of the intestinal lining. They cause irritation of the intestinal wall and bleeding. Symptoms include bloody diarrhea, weakness, and going off feed.

Prevention of *coccidiosis*, the disease caused by coccidia, is based on sanitation. Separate infected animals from the rest of the herd. Sulfa drugs and antibiotics are used to treat coccidiosis.

Anaplasma

Anaplasmosis is a disease caused by protozoan parasites called anaplasma, which destroy red blood cells. It is spread by various biting insects. Symptoms include anemia, weight loss, difficult breathing, abortion, and death. Older cattle are affected more frequently than young cattle.

Reducing insect populations helps to control the spread of anaplasmosis. Some feed additives are used to prevent infestation. Antibiotics are used in its treatment.

NUTRITIONAL HEALTH PROBLEMS

Bloat

Rapid fermentation (breakdown of carbohydrates by enzymes) in the rumen causes too much gas to be produced. The rumen swells and the animal cannot get rid of the gas. This condition is called *bloat*. The major cause of bloat is eating too much green legume too fast. Other feeds can also cause bloat. Some animals will bloat on dry feeds.

One of the main ways to avoid bloat is to prevent animals from overeating legumes in too short a period of time. Feeding grain, dry roughage, or silage before turning the animals onto legume pasture also helps in prevention. Free access to water should be provided at all times.

A stomach tube passed through the mouth helps the animal get rid of the gas. Other treatments include walking the animal on rough ground to make it belch, forcing the animal to drink mineral oil or poloxalene (trade name, Bloat Guard), or inserting a trocar and cannula (a sharp-pointed instrument attached to a tube) into the rumen through the side to allow the gas to escape. The use of the trocar and cannula should only be considered after other methods have failed.

Bovine Pulmonary Emphysema

Symptoms of pulmonary emphysema include panting, coughing, and difficulty in breathing. It occurs among cattle in feedlots. The cause is not known, although the disease may be due to some type of allergy. There is no known prevention or cure. Dust reduction, feeding less high-concentrate feed, and putting cattle on pasture may help relieve the condition.

Brisket Disease

Brisket disease is a heart condition of cattle that occurs at high altitudes. It may be caused by the enlargement of the heart in cattle that cannot adapt to high altitude. Diet may also be a factor.

Symptoms of brisket disease include swelling of the lower neck, brisket, and belly. Cattle are moved to lower altitudes when the symptoms are seen. Limiting water and salt intake and providing a well-balanced ration are also recommended as treatment of the symptoms. Cattle with this condition must be handled with care, or they will die of heart attacks.

Fescue Foot

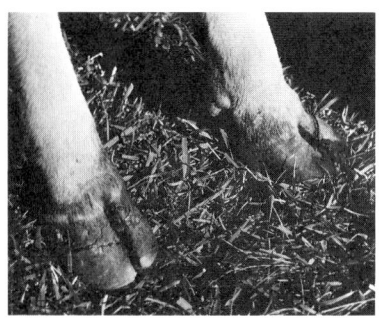

FIGURE 17-7   Splitting of cow hoof caused by fescue foot. *Courtesy of USDA.*

Cattle grazing on pastures of tall fescue sometimes develop a lameness called *fescue foot* (Figure 17-7). The animals shift from one hind foot to the other and sometimes hold one foot off the ground. Swelling develops and the animal may lose one or both of its hind feet.

The exact cause of fescue foot is not known. Possibly, tall fescue may contain a toxic substance. Research is being done to find out if any fungi are involved.

Mixing legumes with fescue helps reduce the number of cases of fescue foot. Feeding hay on pasture and clipping pastures at least once a year also helps. Observe cattle daily for signs of lameness. Remove lame cattle from tall fescue pasture.

Enterotoxemia (Overeating Disease)

Enterotoxemia usually affects cattle on high-concentrate rations. Symptoms include lameness, bloody diarrhea, and bloat. The animal may die in 1 to 24 hours. Vaccinating calves two weeks before putting them on full-feed, high-concentrate rations helps to prevent enterotoxemia. Treatment includes removing concentrates from the diet, feeding roughage, and vaccinating. The animal can gradually be put back on the high-concentrate ration after it is vaccinated.

Fluorosis

Fluorosis occurs in parts of the United States where the fluorine content of the feed or water is too high. It is a poisoning effect that builds up over a period of time. Symptoms include abnormal teeth and bones, stiff joints, diarrhea, and damage to other organs of the body. The animal loses its appetite and becomes thin. Prevention and treatment involves using feeds that do not have a high fluorine content.

Founder

Founder is a swelling of the tissue that attaches the hoof to the foot. It occurs among cattle in the feedlot. Overeating concentrates, sudden changes in ration, drinking too much water, and standing in a stall for long periods cause founder. The animal becomes lame. It shifts its weight from one foot to another and has difficulty in standing. Treatment includes taking the animal off high-concentrate rations and putting wet cold packs on the affected part. Antihistamines are sometimes used in treatment.

Grass Tetany (Hypomagnesemia, Grass Staggers, Winter Tetany, Magnesium Tetany, Wheat Pasture Poisoning)

Grass tetany is found most often in cattle during the lactation period. It sometimes occurs in cattle that are not lactating. It occurs most often when cattle are grazing on grass pastures that are deficient in magnesium. Early symptoms of grass tetany include excitement, loss of coordination, and loss of appetite. The animal may show a trembling of the muscles, convulsions, and coma. Sometimes the animal cannot stand. Death may occur quickly, sometimes within 30 minutes. Animals seldom recover if not treated within 8 to 12 hours.

Prevention includes feeding magnesium in the ration in areas where the soil is deficient in this element. Including legumes in the pasture mix helps to prevent grass tetany. Treatment includes injecting a calcium and magnesium solution into the jugular vein of the animal (Figure 17-8). Cattle with this disease must be handled carefully, since stress may kill the animal. A veterinarian should be called when grass tetany is suspected.

FIGURE 17-8 Injection of calcium-magnesium gluconate into the bloodstream of a cow down with grass tetany. *Courtesy of USDA.*

Hardware Disease (Traumatic Gastritis)

Cattle sometimes pick up sharp metal objects, such as wire, nails, pins, and screws, with their feed. Metal objects collect in the reticulum. When they are sharp, they may puncture the wall of the reticulum causing infection or damage to surrounding organs, such as the heart.

Symptoms of hardware disease include loss of appetite, arched back, fever, stiffness in moving, and less chewing of the cud. The animal may have pain when defecating and when lying down or getting up. The brisket is sometimes flabby and the animal may be bloated.

Hardware disease is prevented by making sure that metal objects do not accidentally become mixed with the feed. Checking for loose wire, nails, and other sharp metal objects in areas where cattle are kept also helps in prevention. A magnet may be placed in the cow's stomach to attract and hold the metal, thus preventing puncture of the wall. The disease is cured by surgically removing the metal.

Nitrate Poisoning

Too much nitrate in feed or water may cause nitrate poisoning in animals. The increase in the use of nitrogen fertilizers has resulted in more frequent occurrences of nitrate poisoning. Nitrate is converted to nitrite in the animal's body. Excess nitrite prevents oxygen from reaching the parts of the body where it is needed.

The affected animal has difficulty breathing and a faster pulse rate. The blood turns brown and the animal may froth at the mouth. Pregnant animals may abort.

Nitrate poisoning can be prevented by analyzing feeds to find the amount of nitrate they contain. If a feed has a high level of nitrate, it may be necessary to add nonnitrate sources of feed to the ration. Silo gases and drainage from silos also contain high nitrate amounts. Keep livestock away from these sources of nitrate poisoning.

Treatment includes the use of antidote tablets that can be secured from a veterinarian. In severe cases, call a veterinarian. Methylene blue is administered in extreme cases. Stop feeding suspect feeds until they have been tested for nitrate content.

Photosensitization

Photosensitization is a skin reaction that occurs in the presence of sunlight after the animal has eaten certain kinds of plants. Symptoms include scratching, rubbing, licking, and tail and head switching. Skin areas turn red and ooze a yellowish fluid. Crusting of the affected areas occurs. Animals lose their appetites and have a fever. Outbreaks occur most often in spring and summer.

There is no drug to prevent photosensitization. Feeding dry roughages and rotation grazing help in prevention. Affected animals are moved to a shaded area and given dry feed. Paint or spray the affected body parts with methylene blue water solution. Sodium thio-sulfate given orally or intravenously may be used as a treatment.

Poisonous Plants

Many kinds of plants can cause injury or poisoning when eaten by cattle. Symptoms vary with the kind of plant involved. Prevention consists of eliminating poisonous plants from pastures and range-land. Treatment is often of little value after the animal is affected.

Urinary Calculi (Water Belly)

Minerals deposited in the urinary organs cause *urinary calculi*, or stones. Steers are affected with this condition more frequently than cows or heifers. Rations high in phosphorus are considered to be a partial cause. Symptoms include restlessness, lying down in a stretched posi-tion, and tail switching. The animal attempts to urinate but only drib-bles. If the urinary tract becomes completely blocked, the bladder will burst. This causes death unless treated.

Prevention includes providing a ration with a calcium-phosphorus ratio of 2:1. Adding salt to the feed, which increases water intake, also helps in prevention. Surgery is the best treatment. Urinary tract re-laxants may be given that help keep the urethra open.

Rumentitis (Liver Abscess Complex)

Cattle on high-concentrate rations are more likely to have liver ab-scesses. *Rumentitis* is caused by microorganisms that are in the soil and manure. Affected cattle do not gain well. No symptoms appear in cat-tle on feed; the first sign of the condition is usually seen when the an-imal is slaughtered and the abscessed liver appears. Infected livers are condemned at slaughter. Feeding antibiotics may help prevent liver abscesses.

White Muscle (Selenium Deficiency)

Cattle fed in areas where there is a deficiency of the trace element se-lenium in the soil may be affected by *white muscle disease*. Muscle dam-age results from a shortage of selenium in the diet. The animal may have trouble walking, breathing, or may die of heart failure. Calves may be born dead or weak. Treatment and prevention consists of giv-ing the animals selenium by injection or orally. *Note:* too much sele-nium in the ration can also be harmful to the animal.

SUMMARY

Profits are reduced when diseases and parasites affect beef cattle. Good management and sanitation programs help to prevent health prob-lems. Planning prevention programs with the help of a veterinarian is recommended.

Many diseases are prevented by vaccination. Antibiotics and sulfa drugs are used to treat some diseases. Buying animals from disease-free herds and isolating new animals help in control programs. Insects spread many diseases, and their control is important in preventing the spread of disease. The symptoms of many diseases are similar. Laboratory tests and the aid of a veterinarian are often needed to identify the disease that is present.

Flies, lice, mites, and ticks are the most common external parasites of cattle. Insecticides are used to control most of these parasites. Label directions must be followed carefully for safe use of insecticides.

Roundworms, flatworms, coccidia, and anaplasma are the most common internal parasites of beef cattle. The stomach worm is the most serious of the internal parasites. Sanitation is the most effective control for internal parasites.

A number of nutritional health problems affect beef cattle. Good management of feeding programs helps to prevent some of these problems. Metal and other foreign objects must be kept out of the feed and feedlot. Trace element deficiencies cause some health problems. The ration should be checked to be sure the necessary trace elements are included. Cattle must be kept away from poisonous plants.

Student Learning Activities

1. Prepare a bulletin board display showing pictures of unsanitary livestock equipment and facilities and contrasting pictures of good sanitary conditions.
2. Prepare and present an oral report on any phase of beef herd health problems.
3. Prepare a bulletin board display showing pictures of animals with diseases or parasites.
4. If possible, practice herd health measures, such as giving injections.
5. Take field trips to observe herd health problems in the local area.
6. Ask a veterinarian to talk to the class about preventive measures for cattle health problems.

Review

1. Why is it important for the cattle producer or feeder to follow practices that help to maintain healthy cattle?
2. Prepare a table that briefly summarizes the symptoms, prevention, and treatment of the common diseases that affect beef cattle.
3. List and briefly describe each of the common external parasites that affect beef cattle.

4. Describe the major steps for control of parasites.

5. List and briefly describe the common internal parasites of beef cattle.

6. List and briefly describe the common nutritional health problems of beef cattle.

Beef Housing and Equipment

After studying this unit, the student should be able to

■ describe the steps in planning for facilities and equipment for beef enterprises.

■ describe the facilities and equipment required for beef enterprises.

PLANNING FOR FACILITIES AND EQUIPMENT

Careful planning of facilities and equipment for beef cattle enterprises is important for the success of the operation. There are many different kinds of beef cattle operations. The kind of facilities and equipment vary with each individual farm.

Careful planning before building can make cattle handling easier. Facilities must be planned to be as safe as possible for the operator and the cattle. Wise planning also helps to save labor.

Factors to consider when planning are the

■ number of cattle in the enterprise.
■ space requirements per head.
■ kind of facilities.
■ location of the facilities.
■ environmental requirements.
■ feed storage and handling methods.
■ amount of land needed.
■ amount of money and labor that is available.
■ opportunity for expansion of the enterprise.
■ coordination of new facilities with existing facilities.

Number of Cattle

Each operator must decide on the number of cattle to be handled in a given operation. This decision is based on several factors, including the availability of feed and labor.

Space Requirements

Facilities are planned according to the number of cattle that are to be handled in the enterprise. It is wise to plan for expansion of the facilities in the future. Build what is needed for the current operation, but allow room for enlarging the facilities if the initial number of cattle increases. Space requirements for different kinds of facilities are given in Table 18-1.

Kind of Facilities

The kind of facilities required depends on the kind of beef enterprise on the farm. Cow-calf operations can be run with limited facilities. The facilities for feedlot operations range from simple to complex.

Facilities can be classified as one of the following types:

- confinement.
- open feedlot.
- open barn and feedlot.
- feeding barn and lot.

Location of Facilities/Environmental Factors

One of the most important decisions to be made is the location of the facilities. Easy access to good roads must be available. Plan for easy movement of cattle to and from pastures. Dust and odors from the facility may be objectionable to neighbors. Therefore, the direction of prevailing winds must be considered. Runoff must be controlled. The facilities should not be located too close to streams, lakes, or ditches. A well-drained site is necessary. Good drainage around feeding areas is also essential. The soil of the chosen location must be able to support large structures, such as silos.

Feed Storage and Handling

Feed storage and handling methods vary from simple, manual systems to complex, automated systems. The system selected must allow for future expansion. Plan so that large equipment will have easy access. An adequate power supply for operation of feeding equipment must be available.

Amount of Land

The amount of land required varies with the size of the operation. For example, a feedlot for 500 head of cattle requires about five acres for the facilities. A 5,000-head facility requires about 35 acres of land. There must be enough land to provide space for lots, alleys, storage, and roadways. Some operators desire enough space for inclusion of an office and feed mill in the facilities. Always allow room for future expansion.

Money and Labor

The amount of money an operator has available affects the size and kind of facility that can be built. Certain types of facilities are more

TABLE 18-1 Space Requirements for Cattle Facilities.

Feedlots	ft²/head	m²/head
Unsurfaced	300–800	27.8–74.3
Partially surfaced	150	13.9
Surfaced with shelter	20 (lot); 30 (barn)	1.8 (lot); 2.8 (barn)
Surfaced no shelter	40–75	3.7–6.9
In sheds or under shade:		
400–800 lb (181–363 kg)	15–20	1.4–1.8
800–1,200 lb (363–544 kg)	20–25	1.8–2.3
over 1,200 lb (544 kg)	25–30	2.3–2.8
Slope of unsurfaced lot	4–6(%)	

	under 600 lb (272 kg) Per Head		600–1,200 lb (272–544 kg) Per Head		Over 1,200 lb (544 kg) Per Head	
Corrals	**ft²**	**m²**	**ft²**	**m²**	**ft²**	**m²**
Holding pens	10–14	0.9–1.3	15–17	1.4–1.6	20	1.9
Sorting pens	8	0.7	12	1.1	15	1.4
Crowding pens	6	0.6	10	0.9	12	1.1
	gal	**L**	**gal**	**L**	**gal**	**L**
Water (per head/day)	8	30.3	10	37.8	12	45.4

Fence	**ft**	**m**
Height (min)	5	1.5
Post depth (min)	2.5	0.76
Post spacing (max)	8	2.4

Working Chute	**in**	**cm**	**in**	**cm**	**in**	**cm**
vertical side-width	18	45.7	22	55.9	26	66
sloping side-width						
bottom	15	38	15	38	16	40.6
top	20	50.8	24	60.9	26	66

	ft	**m**
Length (min)	20	6.1
Fence height (min)	5	1.5
Post depth (min)	3	0.9
Post spacing (max)	6	1.8

Loading Chute		
width	30–42 in	76.2–106.7 cm
length (min)	12 ft	3.6 m
slope (max):		
sloped ramp	3 1/2 in/ft	8.89 cm/30.5 cm
stepped ramp	4-inch riser/18-inch tread	10.2 cm riser/45.7 cm tread

Ramp Height	**in**	**cm**
Gooseneck trailer	15	38.1
Pickup truck	28	71.1
Stock truck	40	101.6
Tractor-trailer	48	121.9
Double-deck	100	254

TABLE 18-1 Space Requirements for Cattle Facilities *(Cont.)*.

	Per Head	
Cold Confinement Buildings	**ft²**	**m²**
Solid floor—bedded	30	9.1
Solid floor—flushing	17–18	5.2–5.5
Totally or partially slotted	17–18	5.2–5.5
Calving pen	100	30.5
Calving space	1 pen/12 cows	

	Lineal Space Per Head			
	One feeding/day		**Two feedings/day**	
Feed Bunks	**in**	**cm**	**in**	**cm**
All animals eat at once:				
400–800 lb (181–363 kg)	18–22	45.7–55.9	9–11	22.8–27.9
800–1,200 lb (363–544 kg)	22–26	55.9–66	11–13	27.9–33.0
over 1,200 lb (544 kg)	26–30	66–76.2	12–15	30.5–38.1
Feed available at all times:				
Hay or silage	9–14	22.8–35.5		
Grain or supplement	3–6	7.6–15.2		
Grain or silage	6	15.2		
Creep or supplement	1 space/5 calves			

Feed Bunk Dimensions	**in**	**cm**
Throat Height		
400–800 lb (181–363 kg)	18	45.7
800–1,200 lb (363–544 kg)	20	50.8
Mature cows/bulls	24	60.9
Width		
Fed from one side	18	45.7
Fed from both sides	48–60	121.9–152.4
	ft	**m**
Concrete apron along bunks	10–12	3.0–3.6
3/4–1 in (1.9–2.5 cm)/ft slope away from bunk		

Water	Per Head/Day			
	Hot weather		**Cold weather**	
40 head/waterer space	**gal**	**L**	**gal**	**L**
Feeders	15–22	56.8–83.3	8–11	50.3–41.6
Cows	18–25	68.1–94.6	9–13	34.1–49.2

Isolation and Sick Pens

Need space for 2 to 5 percent of herd		
Per head	40–50 ft²	3.7–4.6 m²

Mounds (dirt in feedlot)	**ft²**	**m²**
Per head (min)	25	2.3
(Double if windbreak on top of mound—one half each side of windbreak)		
Slope ratio	4:1 to 5:1	

Source: Midwest Plan Service, Iowa State University.

expensive than others. For example, warm confinement barns cost more than other kinds of facilities. (However, research has shown that there is little or no advantage to using warm confinement systems.)

The amount of automation and labor are closely related. Systems with little automation require more labor. If there is a limited labor supply, more highly automated systems should be considered.

Careful planning and layout of the facilities permits future expansion, if desired. New construction should be coordinated with existing facilities.

COW HERD FACILITIES

Generally, beef cow-calf operations require the simplest facilities. Cows can calve on pasture in both spring and fall. Beef cows can be wintered outdoors with a minimum of shelter. Cows can be moved to a pasture or lot close to the farmstead just before calving. Open-front calving barns may be necessary in colder parts of the United States. Portable calf shelters can be used for calves on pasture. Separate feedlot areas should be available for different kinds of cattle, such as mature cows, heifers, bulls, and calves. A corral is needed for any kind of beef cattle operation. A holding pen, working chute, and headgate are essential features of a corral. A good water supply is also necessary. Figure 18-1 shows a plan for a calving barn and corral.

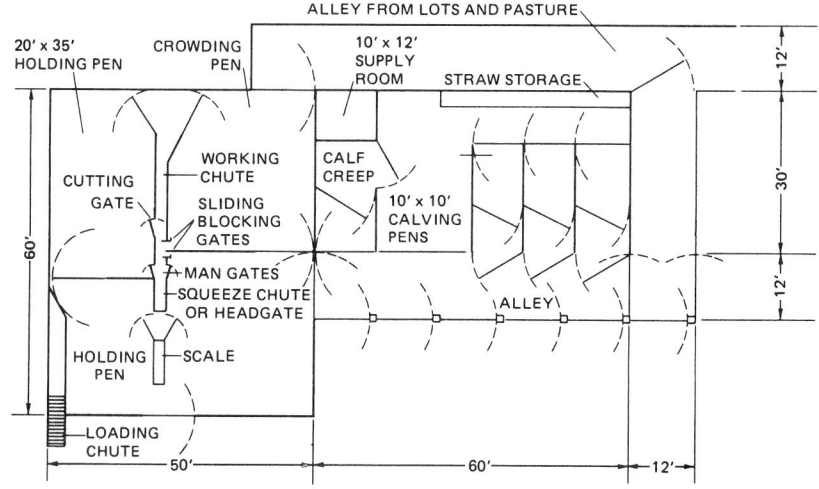

FIGURE 18-1   Plan for a calving barn and corral. *Courtesy of Midwest Plan Service, Iowa State University.*

FEEDLOTS

Confinement Barns

Confinement facilities are of two types: (1) cold and (2) warm.

The cold confinement barn is open on one side. This is usually on the side away from the prevailing winds in the area. The inside temperature is about the same as the outside temperature. Pole-type buildings can be used. A pole-type building is one in which the framework is attached to poles that are set in the ground. Many producers prefer open-span construction because it is easier to clean with mechanical equipment. Open-span construction has no poles or supports in the interior of the building.

Cold confinement barns provide certain advantages over open feedlots or other feeding facilities. First of all, they require less labor. It has also been found that cattle gain better when they have some protection from the weather. The costs of building cold confinement barns are competitive with open lots when the size of the operation is greater than 300 head of cattle. It is usually easier to meet environmental regulations for confinement systems than for other kinds of feeding facilities. Manure handling costs are lower. If slotted floors are used, no bedding is needed. There are fewer problems with flies. Cattle stay cleaner and yield 1 to 2 percent more when slaughtered. It is easier to observe the cattle in a confinement system, especially in bad weather. Cold confinement barns require less land than other kinds of feeding facilities.

Warm confinement barns are closed buildings that are insulated and kept warmer than outside winter temperatures.

They are the most expensive type of cattle-feeding facility to build. Research shows that there is no increase in rate of gain or feed efficiency when cattle are fed in warm confinement barns. This type of facility is, therefore, not recommended for cattle-feeding systems.

Interior specifications. Three types of floor systems are common in confinement barns: solid bedded, slotted, and solid flushing. Solid bedded floors may be either concrete or dirt. They are cheaper than other floor types, but require more bedding and labor.

Slotted floors may be either all slotted or partly slotted. Slats are 6 to 12 inches wide (15.2–30.5 cm) and the slots between the slats are 1¼ to 1¾ inches (3.2–4.4 cm). Partly slotted floors should be at least 40 percent slotted. Animals stay cleaner on slotted floors because manure drops through the slots and is stored in pits under the floor.

Solid flushing floors are usually made of concrete and are designed to slope toward gutters. Manure is flushed with water into the gutters, which empty into a lagoon. Animals do not stay as clean with this type of floor as they do with slotted floors. Larger animals tend to slip on this type of floor. The concrete should have a broom finish to reduce slipping. In addition, more moisture condensation occurs in cold weather with this type of floor than with other types.

Confinement barns have wall heights of 12 to 16 feet (3.6–4.9 m). High walls keep the building cooler in summer and reduce the amount of moisture condensation in the winter. Feed bunks and waterers are placed on opposite sides of the barn. This forces the animals to move around more, thus improving the movement of manure to lower areas on sloping floors or through the slots on slotted floors. Confinement barns must have handling facilities and sick pens. Handling facilities are discussed later in this unit. Sick pens are small pens for isolating sick animals.

Open Feedlots

An open feedlot has no buildings. Protection for cattle is limited to a windbreak fence and sunshades. Several types of windbreak fences may be used to protect cattle from wind and reduce the amount of snow in the feedlot. The fence may be from 4 to 12 feet (1.2–3.7 m) high and may be solid or semi-solid. A solid fence piles the snow higher on the downwind side. A semi-solid fence lets the snow spread out further on the downwind side. The solid fence is better for wind protection; the semi-solid fence gives better snow protection. Local wind speeds and snow amounts need to be considered when determining how far to place windbreak fences from the feedlot. Fifty to 60 feet (15–18 m) is about right for wind speeds of up to 40 miles (64 kilometers) per hour.

Exterior plywood, one-inch boards, or 28-gauge corrugated metal may be used to construct the fence. A semi-solid fence is usually constructed of one-inch boards placed vertically and spaced 2 to 2.5 inches (5–6.4 cm) apart.

Tree windbreaks may be used but they allow the snow to spread further downwind. They must therefore be placed from 100 to 300 feet (30.5–91 m) from the protected area.

Open feedlots are usually not paved. Dirt mounds are used to keep cattle out of the mud. Concrete strips are sometimes installed along feed bunks and waterers. Open feedlots usually require more land area than other kinds of facilities. Good drainage and runoff control are important. Facilities for handling cattle and sick pens are included.

Open Barn and Feedlot

Protection for cattle housed in this kind of facility is provided by the use of an open-front barn (Figure 18-2). Feeding is done in an open lot, and mechanical or fenceline bunks are common. This kind of facility is common in the Midwest where the climate is such that protection of the cattle gives better feed efficiency and rate of gain. This kind of facility is well adapted to smaller feedlots.

FIGURE 18-2 An open-front barn and feedlot. *Photo by Jim Bodine.*

The floor of the barn and the lot are usually unpaved, although some pavement is often installed around the feed bunks and waters. Some operators pave a strip along the open side of the barn extending inward about 4 to 6 feet (1.2–1.8 m) to help control mud. Good drainage and runoff control are needed. Cattle-handling facilities are a part of this kind of facility.

Feeding Barn and Lot

The only major difference between this system and the open barn and feedlot system is that the feed bunks are located inside the barn. Placing the bunks inside has some advantages. The major advantage is that the feed and cattle are protected from the weather. Less feed is lost from the bunks by being windblown or damaged by the weather.

CORRALS

Cattle-handling facilities are an essential part of all beef operations. A corral system provides the following advantages. It

- makes it easier to handle cattle.
- reduces the amount of labor needed to handle cattle.
- saves time when handling cattle.
- reduces stress on the cattle when they are handled.
- reduces injury and weight loss when handling cattle.
- makes cattle handling safer for the workers.
- makes it easier to treat diseased or injured animals.
- makes fly and parasite control easier.

Several pens are necessary in the corral for holding and working cattle. Water must be available in the holding pen. Feed bunks may be needed if the cattle are to be held for any length of time. Access must be provided to the sorting and crowding pens, and to the working chute.

Sorting pens are smaller than holding pens. A corral should have at least two sorting pens. The crowding pen is designed to narrow down to the working chute, thus moving the cattle along. Circular crowding pens make it easier to handle cattle. It is easier to move cattle when they cannot see what lies ahead. Figures 18-3 and 18-4 show plans for two sizes of corrals.

The length of the working chute depends on the number of cattle to be worked. A minimum requirement is that the chute be large enough to hold at least three animals at one time. Some working chutes have slanted sides and are smaller at the bottom than at the top. The chute must be narrow enough so that the cattle cannot turn around. Both straight-line and curved working chutes are used. A walkway should be built along the side of the chute. This makes it easier and safer to work on the cattle. Many operations, such as parasite

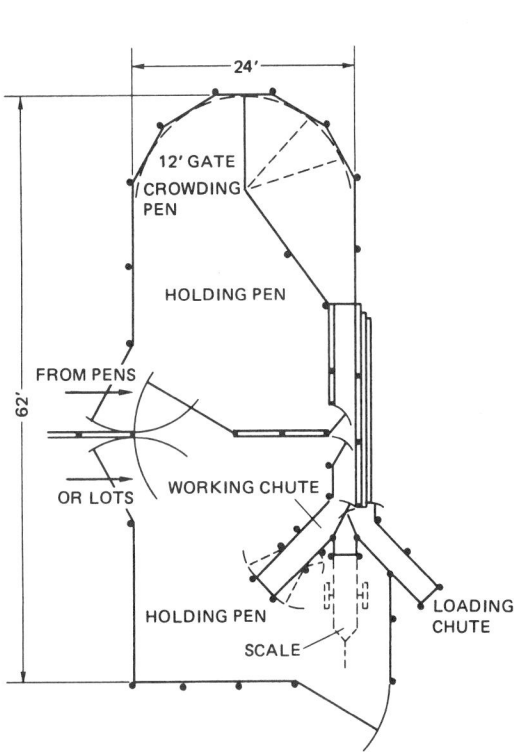

FIGURE 18-3   Plan for a corral for 300 to 1,000 head of cattle. *Courtesy of Midwest Plan Service, Iowa State University.*

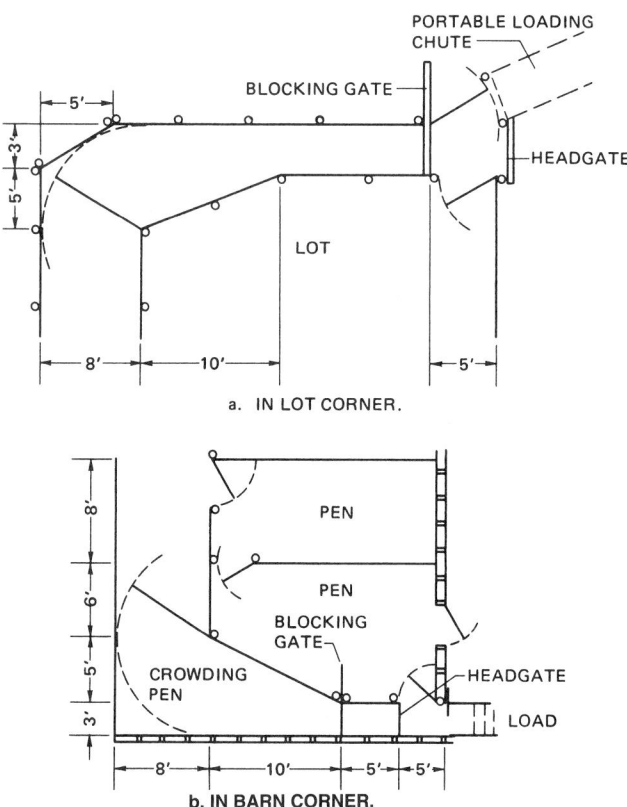

a. IN LOT CORNER.

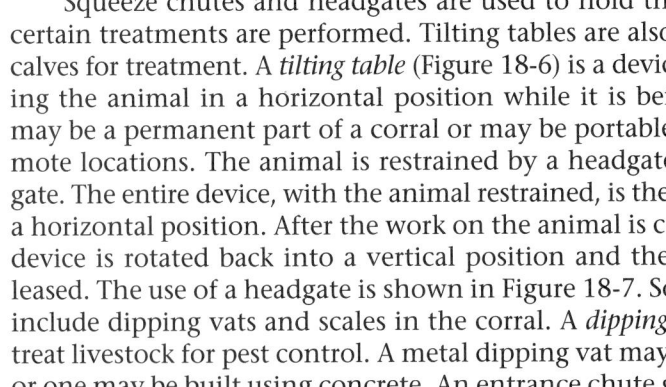

b. IN BARN CORNER.

FIGURE 18-4   Two plans for small herds up to about 75 head. *Courtesy of Midwest Plan Service, Iowa State University.*

FIGURE 18-5   Entrance to a curved alley in a corral. *Courtesy of University of Illinois at Urbana-Champaign.*

control, tattooing, branding, and veterinary treatments, are performed in the working chute. Larger cattle operators use working alleys to connect holding pens, working chute, and feedlot. Figure 18-5 shows the entrance to a curved alley in a corral.

Squeeze chutes and headgates are used to hold the cattle while certain treatments are performed. Tilting tables are also used to hold calves for treatment. A *tilting table* (Figure 18-6) is a device for restraining the animal in a horizontal position while it is being treated. It may be a permanent part of a corral or may be portable for use in remote locations. The animal is restrained by a headgate or a squeeze gate. The entire device, with the animal restrained, is then rotated into a horizontal position. After the work on the animal is completed, the device is rotated back into a vertical position and the animal is released. The use of a headgate is shown in Figure 18-7. Some operators include dipping vats and scales in the corral. A *dipping vat* is used to treat livestock for pest control. A metal dipping vat may be purchased or one may be built using concrete. An entrance chute guides the cattle into the vat, which is filled with the treatment solution. A typical

FIGURE 18-6 Construction and operation of a permanent, wood-frame tilting table for handling calves. *Courtesy of USDA.*

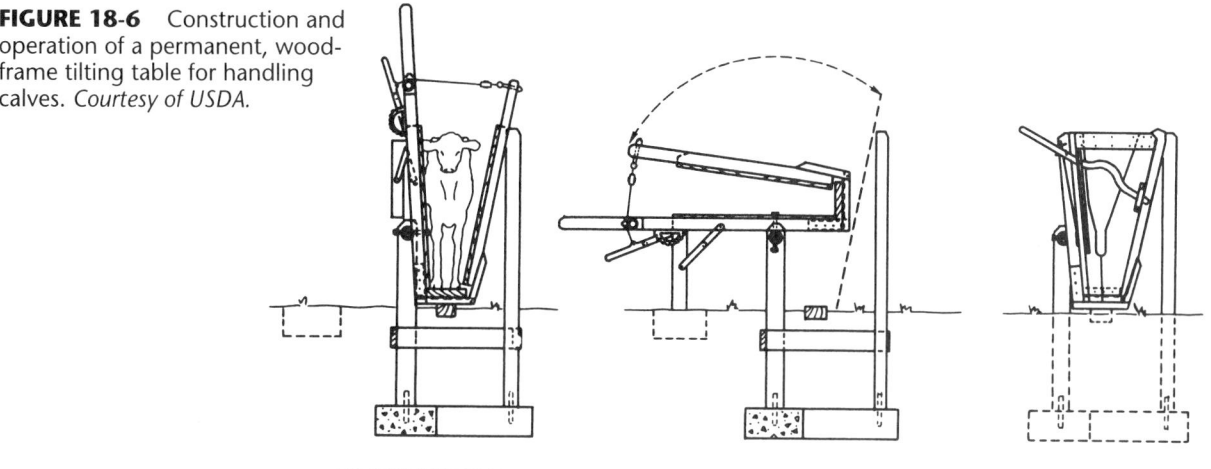

UPRIGHT POSITION

Position the tilting table at the end of a chute equipped with a stop gate at the entrance to the table. After the calf is through the entrance, close the stop gate and apply the self-locking neck yoke to hold the animal snugly and securely so that it won't injure itself. Pull on the tilting lever to lay the calf on its side at a convenient working height.

TABLE POSITION

YOKE END

Any operation or treatment to the animal can now be performed quickly and safely. When finished with the treatment, tilt the table into the upright position, unfasten the neck yoke, and release the calf through the side gate.

vat for cattle will hold about 3,000 gallons (11,356 litres) of treatment solution. It is deep enough so that the animal is completely immersed when it is in the vat. The floor of the entrance end of the vat is a smooth slope that forces the animal into the treatment solution. A typical vat for cattle is about 28 to 32 feet (8.5–9.8 m) in length. The exit end has steps so that the cattle can climb out of the vat into a drip pen. The drip pen provides a place for collecting the treatment solution that drips off the cattle as they come out of the vat. A cover should be placed over the vat when it is not in use. This will keep children out and prevent dilution or contamination of the treatment solution from rain, snow, dust, weeds, or small animals. Large-volume operations are more likely to use dipping vats than smaller operations.

A loading chute should be included in the corral design. Both stepped and sloping ramps are used in loading chutes. Stepped ramps are safer and easier to use than sloping ramps. The sides of the loading chute should be solid so that the cattle cannot see out. Provide a walkway along the sides of the chute. The top should be adjustable for different sizes of trucks.

The materials used to build corrals must have enough strength to hold cattle. Planking or 2" × 6" rough-sawn boards should be used for fencing. Posts and boards are treated with a preservative. Creosote, pentachlorophenol, or copper naphthenate solutions are used for wood treatment. Fence posts should be 5 to 6 inches (12.7–15.2 cm) in top diameter.

FIGURE 18-7 A headgate is used to hold cattle for treatment. *Courtesy of American Polled Hereford Association, MO.*

Posts may be spaced 6 to 10 feet (1.8–3.9 m) apart depending on the weight of the posts and rails. Bolting rails to posts is better than using nails, wire, or lag screws. Sharp or pointed objects should not be allowed to jut into areas through which the cattle move. If the area lacks good drainage, use concrete or limestone to keep cattle out of the mud.

Gate posts should be 6 to 8 inches (15.2–20.3 cm) in diameter. Set gate posts in crushed rock or concrete so that they remain firm. Posts set in light soils must be set deeper than posts set in heavy, gravelly soils. Post depths range from 2½ to 4 feet (0.76–1.2 m). All gate posts should be set at least 4 feet deep (1.2 m).

Gates must be heavy enough to restrain cattle. Construct gates so they swing freely, do not sag, and can be opened easily. No more than 12 inches (30.5 cm) of clearance should be allowed under gates. Gates and fences should be the same height. Gates are usually 10 to 12 feet (3.0–3.6 m) long. Metal gates last longer than wooden gates.

The corral should be in a convenient location. Good access to roads is important. The area where the corral is located should be large enough for trucks to move around it. Choose a well-drained area for the corral. Take advantage of any windbreaks that are available. The corral should be located in such a way that winds do not carry odors to nearby houses.

FEEDING FACILITIES

Feed Storage and Processing

Several kinds of silos are used to store feed. Upright silos include the gas-tight and conventional types. Upright silos (Figure 18-8) cost more but result in lower feed losses than horizontal silos. Gas-tight silos are the most expensive to build, but do the best job of keeping feed. They can be used for many different kinds of feed.

Horizontal silos include the trench, bunker, and stack types. Bunker silos are built aboveground, while trench silos are built below ground level. Stack silos are used for temporary feed storage on the surface of the ground. Horizontal silos require good management to keep feed losses as low as possible.

Horizontal silos are usually the least expensive and best suited to the needs of small feeding operations. Upright silos work well with intermediate-size feeders, while large horizontal silos are the most economical for large feeding operations. The kind of ration used also affects silo choice.

Sealed silos can be used for grain or high-moisture corn. Metal bins can also be used for grain and protein storage. Feed grinding and mixing equipment may be located in a feeding center. Portable grinding and mixing equipment is used on some farms.

Hay may be stored as rectangular bales, small round bales, large round bales, or stacks. Large bales and stacks are handled with special equipment. Ground-level storage requires less labor than overhead

FIGURE 18-8   A conventional upright silo. *Photo by Jim Bodine.*

storage. Well-drained areas are necessary for hay storage. Bales may be stored under some type of shelter to protect the hay from the weather. Large bales stored in the open should be separated by enough space so that the water runs off instead of soaking into the bales. Bales and stacks can be left scattered in the field for feeding. Hay may be self-fed on the ground or in feeders. If fed on the ground, temporary fences are used to reduce waste. A hay self-feeder is shown in Figure 18-9.

Augers, elevators, and other conveying equipment are used for easy handling of grain and protein. Figure 18-10 shows a feed center for a cattle-feeding facility.

Concentrates and roughages may be processed for feeding using a variety of equipment. Portable mixer-grinders can be used for processing both grain and roughage. This equipment can be moved from one location to another. Labor and operating costs are relatively high for portable mixer-grinders.

Electric blender-grinder equipment is lower in operating cost and labor requirement. However, the capacity of this equipment is also relatively low. Concentrates can be measured, ground, and mixed on a continuous basis, but the system does not handle roughage. This type of equipment works well with completely automated systems of feeding concentrates.

Electric batch mill equipment provides accurate control over ration formulation. This type of equipment is relatively expensive and requires careful planning of the operation to ensure an orderly flow of feed through the system. A blender must be added to the system if more than one ingredient is used in the ration.

Hay mills and tub grinders may be used to process roughages for inclusion in the ration when automated feeding systems are used. This type of equipment is relatively expensive.

A well-planned feed center can handle receiving, drying, storage, unloading, elevating, conveying, and processing of feed. Some feed centers are designed to handle the feed in batches. Overhead bins, stationary grinding mills and mixers, and conveyors are included in this type of feed center. In continuous-flow feed centers, the ingredients are stored in overhead bins and fed to an automatic electric mill for processing. The feed is then blended and transported through a conveyor system for feeding.

Feed centers may be located in an area away from the feedlot. Feed is then loaded into a wagon or truck for delivery to the feedlot. If the center is not too far from the feedlot, augers may be used to deliver the feed to the bunks for feeding.

FIGURE 18-9   Hay may be self-fed out of a portable feeder. *Courtesy of University of Illinois at Urbana-Champaign.*

FIGURE 18-10   Metal storage bins are used for grain and protein. *Photo by Jim Bodine.*

Feed Bunks

The kinds of feed bunks used for cattle feeding include:

- portable lot bunks.
- mechanical bunks.
- fenceline bunks.
- covered bunks.

Portable lot bunks are usually made of wood. They are less expensive than some of the other kinds of bunks. Lot bunks range in width from 3 to 5 feet (0.9–1.5 m). They are usually about 14 feet (4.3 m) in length. Both flat-bottomed and V-shaped bottoms are used. Lot bunks may be moved with a forklift or a manure loader.

Mechanical bunks require a large capital investment. However, they save labor by automating cattle feeding. Two types of mechanical bunks are auger and chain-slat conveyor. The newer types are designed to reduce the separation of fine and coarse feed particles. Thus, all cattle get the same ration regardless of where they eat at the bunk. There are a number of different styles of each kind of mechanical bunk. Auger-type feed conveyors, compared to chain-slat conveyors, do a better job of distributing the feed evenly when the supplement is added into the complete ration. The chain-slat conveyor tends to leave the supplement behind. When grain-forage rations are fed, the chain-slat conveyor does a better job of distributing the feed evenly than does the auger type. All mechanical bunks require wiring, electric motors, and electrical controls. Mechanical bunks are wider than other types because the cattle eat from both sides.

Fenceline feed bunks (Figure 18-11) are most commonly used in western United States feedlots and are becoming more common in other areas. They are well adapted to large feedlots. The feed bunks are constructed along outside fences in the feedlot. They are filled from the outside by self-unloading wagons or trucks. The cattle eat from the other side of the bunk.

Fenceline feed bunks are made of wood or concrete, or a combination of the two materials. A concrete apron is usually built to keep the cattle out of the mud. The feed bunk is usually wider at the top than at the bottom. Typical widths are 30 inches (76.2 cm) at the top and 18 inches (45.7 cm) at the bottom. Feeding side height is 18 to 22 inches (45.7–55.9 cm), and the outside height is 30 inches (76.2 cm). Precast, tilt-up concrete fenceline feed bunks are commercially available.

Some feeders are designed with a roof to protect outdoor bunks from the weather. Covered bunks also protect mechanical feeding

FIGURE 18-11 Fenceline feed bunks being filled from a self-unloading wagon. *Courtesy of USDA.*

equipment and provide some shade for cattle. In northern areas, covered bunks should be placed so that the sun can reach the bunk for thawing.

OTHER EQUIPMENT

Cattle gain better when they have a good supply of fresh, clean water. Water should be at 68°F (20°C) for maximum consumption by cattle. Automatic drinking fountains or waterers may be used. Concrete water troughs are also used. A low volume of water pumped continuously to the water trough supplies fresh water. The overflow must be to an area away from the trough. Water must be prevented from freezing in the wintertime. Frost-free hydrants can be used to supply water to a trough.

FIGURE 18-12 Cattle guard gate saves time. *Photo by Jim Bodine.*

Commercial mineral feeders are available. Mineral feeders may also be built out of a metal drum. A weather vane is used on swivel-type feeders to keep out the rain. The weather vane is on top of the feeder. When the wind blows, the vane causes the swivel-mounted feeder to turn so that its opening is facing away from the wind. This protects the contents from rain that would otherwise blow into the opening. Stationary mineral feeders can be built of wood.

In areas of high temperatures, sunshades increase cattle gains. Cattle will use sunshades in any area, but the added cost of the shades may make them impractical in areas of lower temperatures. Permanent sunshades are built on posts set in the ground. Solid roofs of a material such as corrugated metal may be used. However, the area under a solid sunshade will not dry out as well and may create a fly problem. Snow fence or slatted fence makes a practical sunshade. This type of shade can lower temperatures about 10 degrees.

Back rubbers provide treatment for parasite control. Creep feeders, self-feeders, calf shelters, cattle-guard gates, and a cattle treatment stall are other kinds of equipment and facilities that the cattle raiser or feeder may want to include in the facilities. Figure 18-12 shows a cattle-guard gate in a feedlot. A cattle-guard gate allows equipment to be driven into an area without the operator having to stop and open a gate. Cattle will not cross the cattle-guard gate.

SUMMARY

Planning before building increases the value of cattle-raising facilities. Well-planned facilities save labor and make the operation more efficient. Always plan facilities in such a way that future expansion is possible.

Cow-calf herds have the simplest facility requirements. Feedlot operations range widely in their size, complexity, and use of automation.

Warm confinement barns for cattle feeding are not very practical. There is not enough additional gain in this type of housing to justify the added cost. Cold confinement barns provide adequate protection from the weather. Many newer facilities are being built using slotted floors to simplify manure handling. Open feedlots require some kind of windbreak protection for cattle. Dirt mounds are often used to keep cattle out of the mud.

Corrals are an essential part of all kinds of cattle operations. Corrals make it easier and safer to handle cattle. They provide a place to sort and work with the cattle when treatments are needed. A corral should include holding pens, sorting pens, alleyways, a working chute, a squeeze chute, a headgate, and a loading chute. For large operations, a scale is useful. Corrals must be built with materials heavy enough to restrain cattle.

Feed is stored in silos and feed bins. Bunks of various kinds are used for feeding. Portable bunks are the least expensive. Mechanical feed bunks save labor, but require greater initial investment.

Waterers are one of the most important kinds of equipment in the feedlot. Mineral feeders, back rubbers, sunshades, creep feeders, and self-feeders are other kinds of equipment often found in cattle operations.

Student Learning Activities

1. Take field trips to various beef-feeding or beef-herd facilities in the area. Observe and report on the kinds of facilities used.
2. Prepare visuals on housing facilities and/or feed-handling facilities and present an oral report to the class.
3. Prepare a bulletin board display of pictures showing beef facilities and feed-handling equipment.
4. If possible, plan and build facilities on your home farm for a beef project.

Review

1. List ten factors that should be considered when planning facilities for beef cattle.
2. What are the four kinds of facilities commonly used for beef cattle?
3. What factors must be considered when choosing the location of beef facilities?
4. What kind of facilities might be needed for a cow-calf herd operation?

5. Name and briefly describe the two kinds of confinement facilities commonly used for beef operations.

6. Name and briefly describe the three kinds of floors found in beef confinement barns.

7. Describe an open feedlot for beef cattle.

8. List eight reasons for having a cattle corral in a beef operation.

9. Briefly describe the important parts of a cattle corral.

10. Describe the kinds of silos that are used for feed storage in a beef cattle operation.

11. Describe the methods of handling hay for a beef operation.

12. List and briefly describe four kinds of feed bunks used in beef operations.

13. Describe ways in which water can be provided for a beef operation.

14. Under what conditions should sunshades be provided for beef cattle?

Marketing Beef

Objectives

After studying this unit, the student should be able to

- describe the supply and demand cycles for beef and explain how they affect the market price.
- identify and describe the market classes and grades of beef cattle.
- describe the various methods of marketing beef cattle.
- describe how beef cattle should be handled to prevent losses when marketing.
- explain the use of the futures market with beef cattle.

BEEF CHECKOFF AND PROMOTION

The Beef Promotion and Research Act of 1985 established a one dollar per head checkoff for every head of beef sold in the United States. Fifty cents of each dollar goes to the state beef council in the state where the sale occurs. The other fifty cents goes to the Beef Promotion and Research Board for use at the national level. If there is no state council in the state where the sale occurred, all the money goes to the national level. The money collected is used to promote beef as a food, provide consumer information about beef, provide information to the beef industry, promote foreign marketing, and promote research in beef production and utilization.

The checkoff money is generally collected by auctions, packers, and other markets. Animals sold at private sale are also subject to the checkoff, but compliance enforcement is more difficult. Checkoff money is required to be forwarded to the state beef council or the beef board by the 15th of the month following the sale. Penalties for noncompliance include late fees of two percent compounded monthly, a civil penalty of up to $5,000, and a federal penalty of up to $10,000 for each violation. The United States Department of Agriculture is responsible for enforcement of the checkoff.

The activities of the Beef Promotion and Research Board are credited with helping beef maintain a higher share of the meat market than would have otherwise occurred. Research funded by the check-off has addressed such issues as excess carcass fat, quality control, food safety, animal welfare, and environmental concerns.

SUPPLY AND DEMAND

Supply is defined as the amount of a product that producers will offer for sale at a given price at a given time. As prices increase, producers are willing to offer more of the product for sale. As prices go down, less of the product will be offered for sale.

Demand is defined as the amount of a product that buyers will purchase at a given time for a given price. As prices go up there is less demand for a product. As prices go down there tends to be more de-mand for a product.

The combined effects of supply and demand govern the market price of beef cattle. Producers have little effect on the demand for beef. However, their decision to sell or not to sell beef affects the supply. Since cattle raising and feeding is a long-term project, the real influence on supply is made long before the cattle go to market. Management decisions to raise more cattle or fewer cattle eventually affect the price of cattle on the market. Beef cattle numbers tend to run in 9- to 13-year cycles (Figure 19-1).

Seasonal price patterns for feeder calves and slaughter steers and heifers are shown in Unit 16, Figure 16-5. The price of feeder calves tends to be below the yearly average during the fall and winter months. During the spring and summer, the price of feeder calves

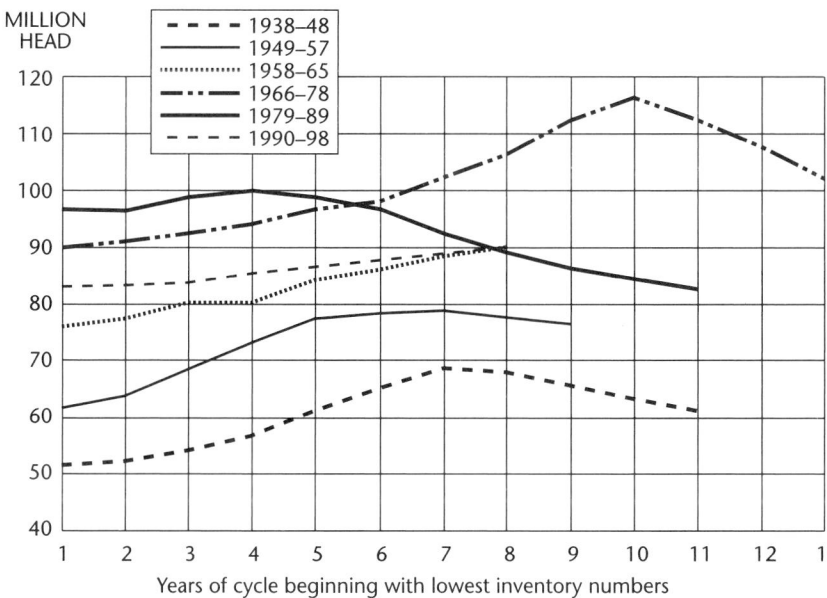

FIGURE 19-1   Inventory of cattle and calves (excluding dairy), by cycles, United States, January 1. *Source:* USDA.

tends to be above the yearly average, which reflects a stronger demand during this period.

The price of slaughter steers and heifers follows a pattern similar to that of feeder calves. The peak prices tend to come in April, after which the prices taper off for the rest of the year.

In recent years, more cattle finishing has been concentrated in large commercial feedlots. This trend has tended to reduce the seasonal variation in fed-cattle prices. Large feedlots are more likely to follow a pattern of year-round placement and marketing of fed cattle. Another factor that has affected the seasonal variation in fed-cattle prices is the tendency to feed cattle for shorter periods of time than was common a few years ago.

The numbers of fed cattle marketed are generally lower during the last quarter of the year. However, an increase in marketing of cows and other nongrain-fed cattle and an increase in the pork supply tends to keep fed-cattle prices lower than the yearly average during this period. Beef marketing tends to be lower during the second quarter of the year; this causes an increase in seasonal prices during this period.

A knowledge of seasonal price changes can help in making management decisions about buying and selling cattle. However, price movements in any given year may be different than the long-term seasonal trends. Current information about supply and demand and other factors that may influence price movement must be carefully evaluated each year when making buying and selling decisions. Federal farm programs, general business conditions, feed supply, and price levels are all factors that may affect seasonal price trends for beef cattle.

Long-term trends show the level of cattle prices over a period of time. General economic conditions have the greatest influence on long-term price trends for cattle. Figure 19-2 shows trends in cattle prices from 1987 to 1997.

METHODS OF MARKETING

Terminal Markets

Terminal markets are also called *central markets* or *public stockyards.* The facilities of the terminal market are owned by a stockyard company. The company charges for the use of the facilities and for feed fed to the cattle while they are in the stockyard. However, title to the livestock does not pass to the stockyard company. In other words, the stockyard never possesses ownership of the cattle.

Terminal markets have two or more commission firms. Cattle are *consigned* to a commission firm by the seller. When goods are consigned, they are not sold to another party, but merely given to another party who acts as a selling agent. The function of the *commission firm,* or selling agent, is to bargain with representatives of the purchaser to get the best price possible for the cattle. Commission firms charge a

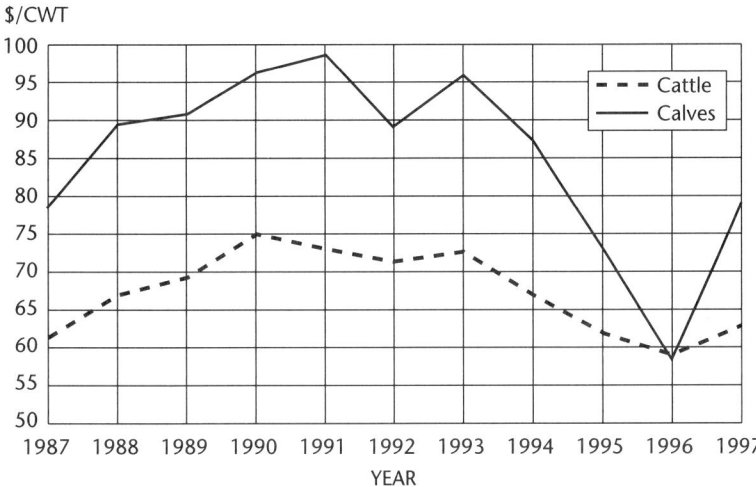

FIGURE 19-2   Yearly average price of cattle and calves, United States, 1987–1997. *Source: Agricultural Statistics,* USDA, 1998 and 1999.

fee, called a *commission,* for their services in marketing the cattle. As noted before, the commission firm does not take title to the cattle. Cattle that are consigned to the firm are sorted into uniform lots for sale.

Costs at a terminal market include charges for yardage, feed, insurance, and selling fees. *Yardage* is the fee charged for the use of stockyard facilities. These costs are deducted from the selling price of the cattle. The seller receives the net amount after the charges are deducted.

Terminal markets exist for both feeder cattle and slaughter cattle. Today, there are about 30 terminal markets compared to about 80 in the 1920s and 1930s. Most terminal markets for feeder cattle are in the western and midwestern states. Slaughter-cattle terminal markets tend to be located near population centers and packing plants.

Commission firm sales of cattle are by private sales. Both buyers and sellers are represented by professionals. Some terminal markets are now using auction sales as a method of selling both feeder and slaughter cattle. However, the use of terminal markets as a method of selling cattle is decreasing. About 12 percent of the packer sources of slaughter cattle is from terminal markets as compared to more than 90 percent in 1925. Larger producers of cattle tend to use other methods of marketing their cattle.

Auction Markets

At *auction markets,* cattle are sold by public bidding, with the animals going to the highest bidder (Figure 19-3). Auction markets are also called *local sale barns* and *community auctions.* They are found in local communities. Auction markets are popular because of their convenience for buyers and the open competition for cattle. Auction markets are of most value to the smaller cattle producer.

FIGURE 19-3   A public cattle auction where cattle are sold to the highest bidder. *Photo by Bill Angell.*

Auction marketing costs include charges for yardage, feed, insurance, brand, and health inspection. Charges are based either on a percent of the selling price or a fixed fee. The costs are paid by the seller of the cattle.

Both feeder and slaughter cattle are sold through auction markets. In some states, cattlemen's associations sponsor consignment sales for feeder cattle. Cooperative auction markets are used in some areas to sell slaughter cattle. Increasing numbers of feeder cattle are being sold through auction markets. The numbers of slaughter cattle being marketed in this way is currently remaining steady. About 18.6 percent of the packers' sources of slaughter cattle come from auction markets, compared to 15.6 percent in 1960.

Direct Selling and Country Markets

Larger cattle producers tend to market their cattle by direct marketing methods. Typically, there are no commission firms or brokers involved in the marketing process.

Contract sales are used to market cattle directly from feeder cattle producers to cattle feeders. The sale is made on the range or farm where the feeder cattle are produced. Large producers of feeder cattle often use contract sales as a method of marketing their cattle.

Cattle dealers buy feeder cattle for cattle feeders. Dealers are either paid on a commission basis, or buy the cattle outright and then resell them at a higher price. Cattle dealers also buy slaughter cattle locally. Usually, they have facilities for handling cattle. These cattle are then shipped on to packing plants for slaughter.

Order buyers buy feeder cattle on order for cattle feeders. Cattle feeders use order buyers because the order buyer knows where the cattle are and is familiar with market conditions. Order buyers also buy slaughter cattle for packing plants. The cattle are bought on the farm and shipped directly to the packing plant. They are weighed at the packing plant and are paid for at prevailing prices for that day.

Increasing numbers of slaughter cattle are being bought on a *grade and yield basis*. Using this system, the value of the animal is determined after it is slaughtered. The carcass is graded, and the carcass weight or yield is determined. Good-quality cattle bring a higher price when marketed in this way. About 23 percent of the cattle and 9 percent of the calves in the United States are marketed on a grade and yield basis. This compares to 14 percent of the cattle and 3 percent of the calves in 1967.

The number of slaughter cattle being marketed by direct marketing methods is increasing. About 69 percent of the packers' source of slaughter cattle is from direct marketing and country dealers as compared to 38 percent in 1960. Increasing numbers of feeder cattle are also being marketed through direct marketing methods.

Electronic Marketing

In some areas of the United States, livestock marketing associations have established a system of marketing utilizing computer technology. This method has been used mainly for marketing feeder cattle, but the potential exists for expansion into marketing other types of cattle.

The system works as a form of auction selling. Each bidder has a computer terminal on-line during the sale. Bids for each lot of cattle are entered by individual bidders on their terminals. When no more bids are received, the lot of cattle is sold and the bidding begins on the next lot.

Bidders can buy the cattle from their offices and do not have to travel to an auction barn. Accurate written descriptions of the cattle and terms of the sale are made available to the bidders several days prior to the beginning of the auction. This information is gathered by graders who work for the livestock marketing association. Information made available includes the number of head, breed or type, sex, estimated average weight, estimated grade, delivery date, weighing location, conditions of sale, flesh condition, feeding and pasture conditions under which the cattle were kept, and information concerning any preconditioning.

Video auctions are growing in popularity as a method of selling cattle. The cattle are videotaped on the ranch or farm where they are located. The videotape is sent to the transmitting station where the auction is located. The auction is telecast for viewing by prospective buyers. Typical information about the cattle includes breed, where they are located, number of head, frame size, estimated weight variance, vaccination and implant information, and current feeding program. Prospective buyers must register with the auction company in advance of the sale. A video auction allows the buyer to see the cattle without moving the cattle to the auction. The purchased cattle are moved directly from the seller's facility to the buyer's facility. This reduces the chances of the cattle having stress and health problems.

A video sales cattle auction company that has been doing video auctions of cattle since 1987 has now expanded its service to the Internet. Potential buyers can see the cattle and make bids online. The service is offered at COW/farms.com, www.farms.com, and www.cattleofferings.com. These URLs also provide links that allow trading in other agricultural supplies and products such as grain, hay, farmland, beef and dairy cattle embryos, chemicals, and livestock feeds. These are commercial sites designed to provide electronic commerce tools to the agricultural community. It is expected that the Internet will continue to increase in importance for livestock producers in terms of both commerce and information in the years ahead.

Packers and Stockyards Act

The *Packers and Stockyards Act* is a federal law that is administered by the United States Department of Agriculture (USDA). All cattle that move across state lines are regulated by this act. The act sets the rules for fair business practices and competition. These rules apply to stockyards, auction markets, packers, market agencies, and dealers who engage in interstate livestock marketing. Individual farmers who buy or sell cattle as a part of their farming operation are not considered to be dealers under the provisions of the act.

Livestock dealers must register with the USDA. They must file a bond based on the volume and type of their business. Packer-buyers who buy only for slaughter do not have to file a bond. Records showing the nature of every transaction must be kept by dealers. Scales must be operated in such a way that give accurate weights. Serially numbered scale tickets are issued. Dealers are forbidden to use any unfair, discriminatory, or deceptive trade practices.

The Packers and Stockyards Act is enforced by representatives of the USDA. Violators of the Act may be warned or ordered to stop violations. If the violation is serious, dealers may be suspended from registration and not allowed to continue to do business. Criminal violations are prosecuted by the Department of Justice.

Purebred Marketing

The marketing of purebred cattle is a specialized business. Purebred cattle are usually sold by private sales or by auction. Purebred producers advertise their herds through breed association magazines and other publications. Selling bulls that have been performance tested also provides advertising for the purebred herd owner. Exhibiting at fairs and shows is another way to advertise purebred cattle. Purebred associations sometimes sponsor consignment sales at which cattle are auctioned. The purebred producer should consign only cattle of the best quality to these sales.

Selecting a Market

The marketplace has two functions. These are to set the value of the cattle and to physically move the cattle from producer to consumer. The individual producer must determine the best market based on price, costs of marketing, and convenience. Many producers select markets mainly on the basis of convenience. Price and costs of marketing should be given more importance than convenience when selecting a market.

Shrinkage

Shrinkage is the loss of weight that occurs as cattle are moved to market. The amount of loss varies from 1 to 5 percent of an animal's weight. The amount of shrinkage is affected by the following factors:

- the distance the cattle are moved. Most loss in weight occurs in the first few miles.
- the amount and kind of feed and water the cattle receive just before shipping.
- the weather.
- the condition of the cattle; thin cattle shrink more than fat cattle.
- how the cattle are handled.
- the sex of the animal.
- the amount of feed, water, and rest the cattle receive during shipping.
- the length of the *fillback period* (time during which cattle are fed) after the cattle reach market.

Losses from shrinkage are absorbed by the seller. A pencil shrink is sometimes used in certain kinds of marketing. In a *pencil shrink,* the amount of shrinkage is estimated and a deduction of weight is made based on the estimate. Tables can be used to determine the net bid when a pencil shrink is being used. For example, if the bid was $40.00/cwt with a pencil shrink of 3 percent, then the net bid is $38.80. If the producer decides to ship the cattle rather than taking the on-farm bid with a pencil shrink, then the cattle must bring a higher price to make up for the shrinkage while being shipped. In this example, if the cattle shrink 3 percent while being shipped, then the market price must be $41.24 to be equal to the bid offered on the farm.

Price Information

If cattle are to be marketed profitably, the producer must be aware of current prices as well as price trends. Live cattle quotations can be obtained from newspapers, radio, television, and by phone. The USDA publishes weekly livestock market news information to which producers can subscribe. Outlook information is available from colleges of agriculture.

Country of Origin Labeling of Meat

A controversy arose in 1999 regarding the labeling of meat products to indicate the country of origin. Those who favored country-of-origin labeling suggested that it would limit competition from imported meat, thus improving sales of meat originating in the United States. Some proponents also framed the argument in terms of the "public's right to know." Several bills were introduced in the U.S. Congress to require country-of-origin labeling of meat. The American Meat Institute (AMI), representing meat and poultry packers and processors, opposed the country-of-origin proposal. The main reason for their opposition was the increased cost to packers, processors, taxpayers (for USDA regulation), and consumers (packer and processor costs are passed on to the final consumer of a product). The AMI claimed that there was no evidence from any research showing that the use of country-of-origin labeling would provide any benefit for producers or consumers of meat and meat products.

The Government Accounting Office and the USDA's Food Safety and Inspection Service both indicated in testimony before a House of Representatives committee that the cost of county-of-origin labeling would be significant.

Clearly it would require a significant amount of paperwork to provide a clear audit trail from the meat in a store back to the actual producer on the farm. Individual animals would have to be accurately identified on the farm and this identification would have to stay with the animal and eventually with its component parts as they are offered for sale to the consumer. This would be necessary to guarantee the accuracy of the label showing the country-of-origin. The issue had still not been settled at the time this text was revised.

Use of Ultrasound to Determine Live Animal Quality

Research is being done at both Iowa State University and Kansas State University to determine the value of ultrasound technology as a means of assessing the probable grade of live cattle. The use of the technology allows the producer to measure the percent of marbling, the amount of fat cover, and the rib eye area in live cattle. Using ultrasound, cattle feeders can sort cattle into separate lots for feeding more uniform groups and determine when cattle are ready for market. Breeders can use the technology to help select breeding stock, especially bulls, with desirable carcass traits. As the cattle industry moves toward a marketing system based on carcass merit, the ability to accurately determine the quality of the carcass in live animals will become more important. Currently, the relatively high cost of the procedure has limited its use. Continued research in the use of ultrasound to determine carcass merit will improve the accuracy and value of this technology.

Effect of Mergers on Marketing

For a number of years there has been an ongoing debate in agriculture regarding mergers of agribusiness corporations into larger corporations. Some observers believe that this trend is putting too much control of the market in the hands of too few people. Whenever the prices paid for livestock drop, the debate becomes more intense. In 1999, there were suggestions that more legislation was needed to control such mergers. There were also those who believe the Justice Department should investigate possible antitrust violations in agriculture. The USDA reports that approximately 80 percent of all steer and heifer slaughter in the United States in 1996 was controlled by the four largest packing companies. This compares to approximately 36 percent of the steer and heifer slaughter that the four largest packing companies controlled in 1980.

A statistical analysis (US Beef Industry: Cattle Cycles, Price Spreads, and Packer Concentration) released in 1999 by the USDA concluded that there was no evidence that packer concentration had any major impact on the market price of live beef cattle. Some farmers, farm organizations, and agricultural economists did not agree with the conclusions of the USDA report. This is another issue that had not been settled at the time this text was revised.

MARKET CLASSES AND GRADES OF BEEF

The purpose of this section is to provide a brief overview of the market classes and grades of beef cattle. The official United States standards for beef cattle may be secured from the USDA, Agricultural Marketing Service, Livestock and Seed Division. The standards are available on the Internet at the URL: www.ams.usda.gov/lsg/stand/st-pubs.htm# Official. This Web site provides official standards for all livestock.

Calves

Beef animals younger than one year of age are called *calves*. After one year of age, beef animals are called *cattle*. Calves are further classed as veal calves (vealers) and slaughter calves.

Veal calves and slaughter calves are generally divided on the basis of the kind of carcass they will produce. The main difference between the two is the color of the lean meat in the carcass. The age and kind of feed the calf has had generally determines whether it is a vealer or a slaughter calf. Vealers usually have had only a milk diet and are under three months of age. Slaughter calves are usually between three and eight months of age and have had feed other than milk in their diet for a period of time.

Light veal calves weigh less than 110 pounds (49.9 kg). *Medium-weight* veal calves weigh from 110 to 180 pounds (49.9–81.7 kg). *Heavy* veal calves weigh more than 180 pounds (81.7 kg). Weight classes for slaughter calves are: *Light*—less than 200 pounds (90.7 kg); *Medium*— 200 to 300 pounds (90.7–136.1 kg); and *Heavy*—more than 300 pounds (136.1 kg). The sex classes of vealers and slaughter calves are steers, heifers, and bulls. All three of these classes are graded on the same standards. The five grades of vealers and slaughter calves are: (1) Prime, (2) Choice, (3) Good, (4) Standard, and (5) Utility.

Slaughter Cattle

Cattle are first divided into feeder or slaughter cattle. This division is based only on their intended use. The class of cattle is based on sex definitions and the grade is based on the apparent carcass merit of the animals.

There are five sex classes of beef cattle. A *steer* is a male that was castrated before reaching sexual maturity and is not showing the

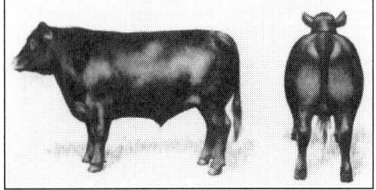

PRIME

CHOICE

SELECT

STANDARD

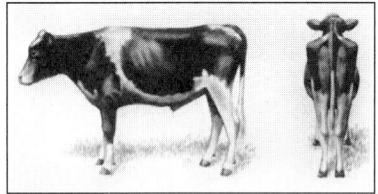

UTILITY

FIGURE 19-4 Slaughter steers—U.S. quality grades. *Courtesy of USDA.*

secondary characteristics of a bull. A *bullock* is a male, usually under 24 months of age, which may be castrated or uncastrated and does show some of the characteristics of a bull. A *heifer* is an immature female that has not had a calf or has not matured as a cow. A *cow* is a female that has had one or more calves. An older female that has not had a calf but has matured is also called a cow. A *bull* is a male, usually over 24 months of age, that has not been castrated. When applying the official USDA standards, any castrated male bovine that shows, or is beginning to show, the mature characteristics of an uncastrated male is considered a bull.

The grades of live slaughter cattle are directly related to the grades their carcasses will produce. As noted in another section of this unit, there is research being conducted on the use of ultrasound technology to help determine the quality of live animals.

The quality grades for steers and heifers are: *Prime, Choice, Select, Standard, Commercial, Utility, Cutter,* and *Canner.* The same quality grades apply to slaughter cows, with the exception that there is no Prime grade for cows. The quality grades for bullocks are Prime, Choice, Select, Standard, and Utility. The Prime, Choice, and Standard grades are generally used only for steers, heifers, and cows that are less than 42 months of age. The Select grade for steers, heifers, and cows is generally limited to animals that are no more than 30 months of age. The Commercial grade is generally applied only to steers, heifers, and cows that are more than 42 months of age. The Utility, Cutter, and Canner grades may be applied to any age of steers, heifers, and cows. The quality grades for bullocks apply only to animals that are no more than 24 months of age.

Five *yield grades* are used to describe slaughter beef. These are numbered *Yield Grade 1* through *Yield Grade 5.* Yield Grade 1 indicates the highest yield of lean meat. Yield Grade 5 indicates the lowest yield of lean meat. The grading of beef carcasses is directly related to live beef grading standards.

Live cattle *quality grades* are based on the amount and distribution of finish on the animal. The firmness and fullness of muscling and maturity of the animal are other factors involved in quality grading. Yield grades in live cattle are based on the muscling in the loin, round, and forearm. Fatness over the back, loin, rump, flank, cod, twist, and brisket is also used in determining yield grade. Fatness is more important than muscling in determining yield grade.

U.S. slaughter quality grades are shown in Figure 19-4. U.S. slaughter yield grades are shown in Figure 19-5.

Carcass Beef

The age of the animal and the apparent sex designation when the animal is slaughtered determines the class of the beef carcass. The five beef carcass classes are (1) steer, (2) bullock, (3) bull, (4) heifer, and (5) cow. The quality and yield grades for carcass beef are the same as those for live slaughter cattle. The amount of marbling in the carcass affects the quality grade of the carcass. *Marbling* refers to the presence and distribution of fat and lean in a cut of meat. When grading beef carcasses, five maturity groups and seven degrees of marbling are used. Figure 19-6 shows the relationship between marbling, maturity, and quality grades of beef. Bullock carcasses are limited to maturity group A. The Select grade of beef is also restricted to maturity group A. Figure 19-7 shows the lower limits of degrees of marbling used in the official United States standards for grades of carcass beef.

The yield grade of a beef carcass is influenced by carcass weight, rib eye area, thickness of fat over the rib eye area, and the amount of kidney, pelvic, and heart fat. A preliminary yield grade for a warm beef carcass is determined by the thickness of fat over the rib eye. Each 0.1 inch (0.25 cm) of fat thickness changes the yield grade by 0.25 of a yield grade. An adjustment is then made for rib eye area and percent of kidney, pelvic, and heart fat to determine the final yield grade. An increase in the amount of these fats decreases the percent of retail cuts from the carcass. Each change of 1 percent of the carcass weight attributed to these fats causes a 0.2 change in the yield grade.

The preliminary yield grade is adjusted for each percent that the kidney, pelvic, and heart fat is more or less than 3.5. For each percent more than 3.5, add 0.2 of a grade to the preliminary grade. For each percent less than 3.5, subtract 0.2 of a grade from the preliminary yield grade.

For example, if the thickness of fat over the rib eye was 0.2 inch (0.5 cm) the preliminary yield grade would be 2.5. Assume the warm carcass weight is 600 pounds (272 kg) and the rib eye area is 12 square inches (77.4 cm^2). The adjustment to the preliminary yield grade is then a minus 0.3. Assume, further, that the percent of kidney, pelvic, and heart fat is 2.5. The adjustment to the preliminary yield grade is then a minus 0.2 of a grade. The final yield grade would then be 2 (2.5 − 0.3 − 0.2 = 2.0).

The higher grades of slaughter beef are usually grain-fed animals that have a high yield of lean cuts with the right amount of marbling. These animals bring higher prices in the market. About 97 percent of the graded beef in the United States grades U.S. Select or higher; 8 percent of the quantity grading Select or higher grades Prime; 83 percent grades Choice; and only 5 percent grades Select. About 53 percent of the commercial beef production in the United States is quality graded.

Research is currently being done on using ultrasound imaging to grade beef carcasses. A hand-held transducer is placed over the ribeye muscle of the carcass as it moves along the line in the packing plant. The information is fed into a laptop computer that calculates the percentage of fat. This allows more accurate grading of the carcass. It is expected that ultrasound imaging will eventually be commonly used for grading beef carcasses.

YIELD GRADE 1

YIELD GRADE 2

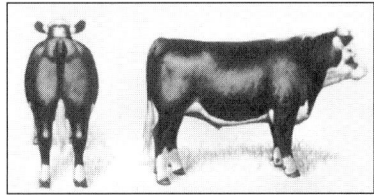

YIELD GRADE 3

YIELD GRADE 4

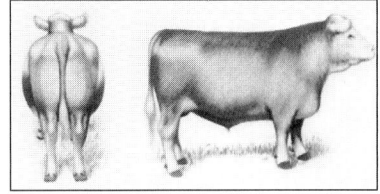

YIELD GRADE 5

FIGURE 19-5 Slaughter steers— U.S. yield grades. *Courtesy of USDA.*

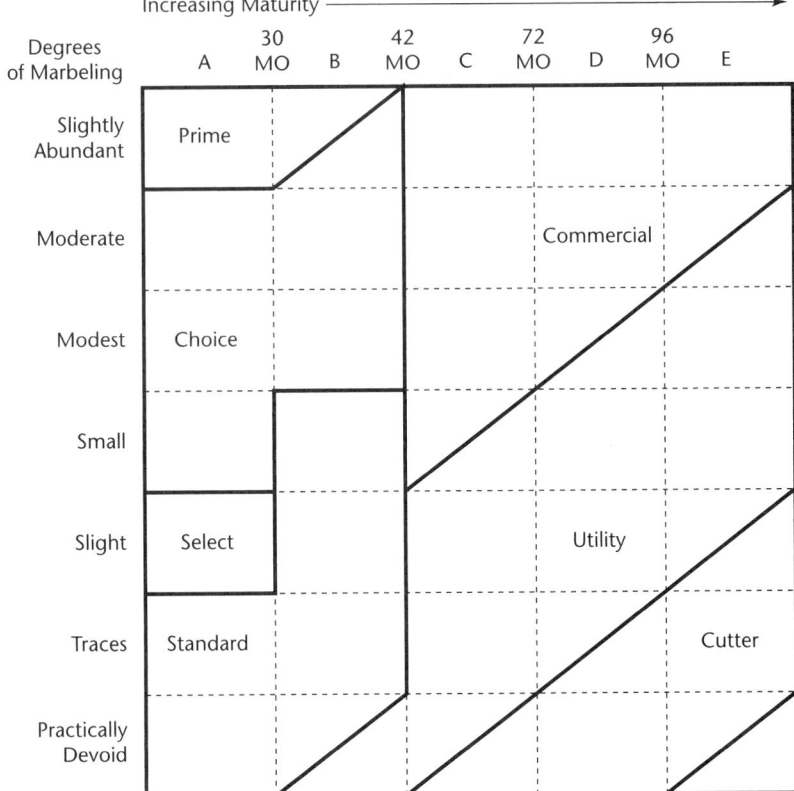

FIGURE 19-6 Relationship between marbling, maturity, and quality. For carcass evaluation and other purposes, three higher degrees of marbling are recognized: slightly abundant, moderate, and very abundant. These are all in the Prime grade, maturity levels A and B. *Source: United States Standards for Grades of Carcass Beef,* USDA, 1997.

Dark cutting beef is a condition in which the lean meat is darker than normal in color. It usually has a gummy or sticky texture. It is believed to be the result of a reduced sugar content in the lean meat at the time of slaughter. The condition is sometimes found when cattle have been subjected to stress conditions just before slaughter.

The condition varies from being barely visible to nearly black in color (*black cutter*). There is little evidence that the condition has any negative effect on the taste of the meat. It is considered in grading beef because it has an effect on consumer acceptance and therefore on the value of the meat.

Depending on how severe the condition is, the grade of carcasses that would otherwise be graded Prime, Choice, or Select may be reduced by one full grade when the condition is present. Beef carcasses that might have otherwise graded Standard or Commercial may be reduced in grade by up to one-half of a grade. Dark cutting beef is not considered when grading Utility, Cutter, and Canner grades.

HANDLING CATTLE PRIOR TO AND DURING MARKETING

Careful handling of cattle just before shipping and during transit to market can save the producer a great deal of money. Losses due to improper handling amount to millions of dollars per year in the United States.

Feed additives and drugs must be removed from the feed for the proper length of time before marketing. Label directions on these additives and drugs must be followed closely.

Cattle can be conditioned before shipping to reduce shrinkage. If they have been on a ration of green feeds or silage, the amount of these feeds should be reduced and the amount of hay increased. Remove the protein from the ration 48 hours before cattle are to be shipped. Reduce the grain in the ration by one-half the day before shipping. Do not feed grain the last 12 hours before shipping. Give the cattle free access to water up to the time of shipping.

If cattle are to be sold on a grade and yield basis all feed can be taken away from them 48 hours before shipping. Allow them all the water they want up to the time of slaughter. Cattle that are to be weighed upon arrival at the plant can be fed the normal ration up to weighing time.

Cattle should be moved slowly and quietly when loading and handling. Do not use clubs or whips or electric prods to force cattle to move. Clubs and whips damage valuable cuts of meat on the carcass. Electric prods overexcite the animals. Canvass slappers can be used with little injury to the cattle. Load only the proper number of cattle on trucks. Overcrowding or underloading increase the chances that cattle will be injured during shipping. Large trucks should be divided into compartments. Protect the cattle from the weather while shipping. Avoid sudden stops and starts with the truck.

CATTLE FUTURES MARKET

The *cattle futures market* is a system of trading in contracts for future delivery of cattle. The contract is a legal document calling for the future delivery of a given commodity. Trading in futures is common with many types of commodities. Trading in contracts means that the contract representing the commodity is bought or sold rather than the actual commodity. Trading in futures is done for both slaughter and feeder cattle. The trading is done at commodity exchanges. The futures market is supervised by the Commodity Exchange Authority of the USDA. The actual buying and selling is supervised by members of the commodity exchange. They charge a fee for their services.

A cattle feeder may use the futures market to hedge on the price of cattle. Hedging provides some protection against price changes. To *hedge,* or protect himself from a future drop in the price of finished cat-

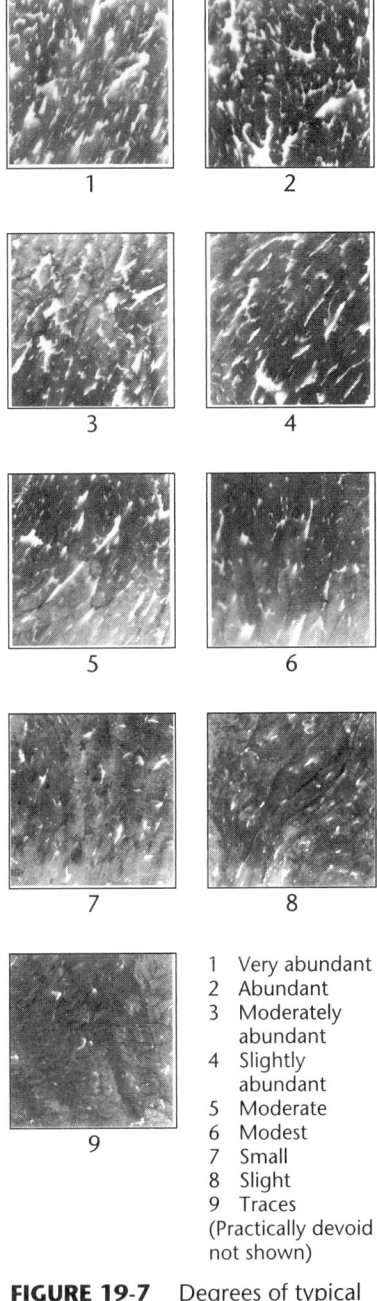

1 Very abundant
2 Abundant
3 Moderately
   abundant
4 Slightly
   abundant
5 Moderate
6 Modest
7 Small
8 Slight
9 Traces
(Practically devoid
not shown)

FIGURE 19-7   Degrees of typical marbling referred to in the official United States standards for grades of carcass beef. *Courtesy of USDA. Illustrations adapted from negatives furnished by New York State College of Agriculture, Cornell University.*

tle, a cattle feeder might sell a contract on the futures market at the same time that he buys feeder cattle. Later, when the finished cattle are sold, the futures contract is bought back. This practice tends to lock in the profits from feeding the cattle.

Because the hedging operation involves opposite actions (buying and selling) in the cash and futures markets, the risk of loss is less if cattle prices go lower than expected. The sale of the futures contract at the start of the hedging period offsets the lower cash price for cattle at the end of the hedging period. However, if cattle prices go higher than expected at the end of the hedging period, the cattle feeder will not realize additional profits because he must buy back a contract at that time or deliver the live cattle to offset the sale of the futures contract at the start of the hedging period. Thus, while hedging reduces risk, it also reduces the opportunity for additional profit that might result from a higher than expected upward movement in the price of live cattle.

There is some risk in using the futures market. The cash price of cattle does not always closely follow the price of futures contracts.

To use the futures market successfully, the cattle feeder must keep good records and understand the pricing system in the market. The futures market tends to bring price stability to beef cattle prices. It also tends to level out seasonal variations in the price of beef. Extensive use of the futures market by cattle feeders tends to take some of the risk out of the beef cattle business.

SUMMARY

Supply and demand govern the price of beef cattle. Beef producers have little effect on the demand for beef. Management decisions of producers do affect the supply of beef in the market. Beef cattle prices vary with the season of the year. There are also long-term trends or cycles in beef cattle prices.

Beef cattle are marketing through terminal markets, auction markets, and by direct selling. The importance of terminal markets has been declining in recent years. Auction markets are found in many local communities. The numbers of cattle marketed in auction markets have not changed much over the years. More cattle are being sold by direct-marketing methods now than there were a few years ago. Larger cattle producers tend to use direct selling rather than other methods of marketing.

The sale of beef cattle that are transported across state lines is regulated by the Packers and Stockyards Act. This act sets forth rules for fair business practices and competition in the marketplace. The USDA administers the act.

Purebred cattle are sold by private sale and by auction. The reputation of the breeder and the methods used for advertising are important factors in the sale of purebred cattle.

Many producers select a marketing method based on convenience. However, price and costs of marketing are more important factors, and should be given first consideration when selecting a marketing method.

Cattle shrink when being shipped to market. Careful handling and management can reduce losses due to shrinkage and damaged carcasses that may occur during shipping.

Cattle are divided into classes based on use, age, weight, and sex. Quality grades are used to describe both feeder and slaughter cattle. Yield grades are also used in describing slaughter cattle. The age of the animal and the amount of marbling in the carcass affect the quality grade. The yield grade is determined by the amount of lean meat that can be cut from the carcass.

The cattle futures market can be used by the cattle producer to reduce price risk. The cattle producer must keep good records and have an understanding of the pricing system to successfully use the cattle futures market.

Student Learning Activities

1. Prepare a bulletin board of charts showing supply and demand cycles, price trends, and local price variations for beef cattle.
2. Take a field trip to various types of markets in the local area. Observe procedures, talk to sellers and buyers, and make a list of the advantages, disadvantages, and costs of each type of market.
3. Survey local beef producers or feeders to determine their marketing practices and why they use the methods they do.
4. Take a field trip to a feedlot or market and grade live cattle. If possible, arrange to see these same cattle after slaughter to observe actual grades.
5. Take a field trip to a packing plant. Grade carcasses of slaughtered animals. Also, observe damage to cattle carcasses from mishandling during shipping.
6. Have a speaker from a packing plant discuss losses from mishandling cattle during shipping.
7. Ask a cattle buyer to speak to the class.
8. Ask a cattle feeder who used the futures market to hedge cattle to explain the procedure to the class.

Review

1. Define the following terms: (a) supply, (b) demand.
2. How do producers affect the supply and demand for beef cattle in the market?

3. List and briefly describe three methods of marketing beef cattle.

4. How does the Packers and Stockyards Act affect the marketing of beef cattle?

5. How are purebred beef cattle marketed?

6. What are three factors that a producer should consider when selecting a market?

7. List eight factors that affect shrinkage when marketing beef cattle.

8. Who absorbs the losses from shrinkage?

9. What sources of price information for beef cattle are available to the producer?

10. List and briefly describe the market classes of beef cattle.

11. What are the factors involved in live cattle quality grades?

12. What are the factors involved in live cattle yield grades?

13. What is the most common grade of beef marketed in the United States?

14. How should cattle be handled during marketing to reduce losses?

15. What is hedging, and why might a cattle feeder use hedging?

Swine

20 *Breeds of Swine*

OVERVIEW OF THE SWINE ENTERPRISE

The Corn Belt states remain the major swine producing region of the United States. About 60 percent of all swine in the United States are raised in this area. The major reason for the popularity of swine in this region is the ready availability of grain, mainly corn, that forms the basis for most swine diets. However, some hogs are raised in every state (Figure 20-1). The leading states in hog production are Iowa, North Carolina, Minnesota, Illinois, Indiana, Nebraska, Missouri, Oklahoma, Ohio, and Kansas. Iowa leads all other states in total hogs produced.

Figure 20-2 shows the percentage of operations and size of swine inventories by the size groups of operations in the United States. Approximately 55 percent of the operations have fewer than 100 head of hogs; however, these operations account for only two percent of the total swine inventory. Twelve percent of the total operations raise 1,000 or more head of hogs, accounting for 77.5 percent of the total swine inventory in the United States. Iowa and North Carolina lead all other states in the number of swine operations with 1,000 or more head of hogs. This reflects an ongoing trend toward large, corporation-owned swine operations. This trend has raised concern among residents and in state legislatures about the environmental impact of large swine operations. Some states have passed or are considering legislation to limit the size of swine operations.

In the past it was a common practice for many farmers to raise hogs when the economic outlook was favorable and to go out of the hog business when returns were not profitable. This practice is not as

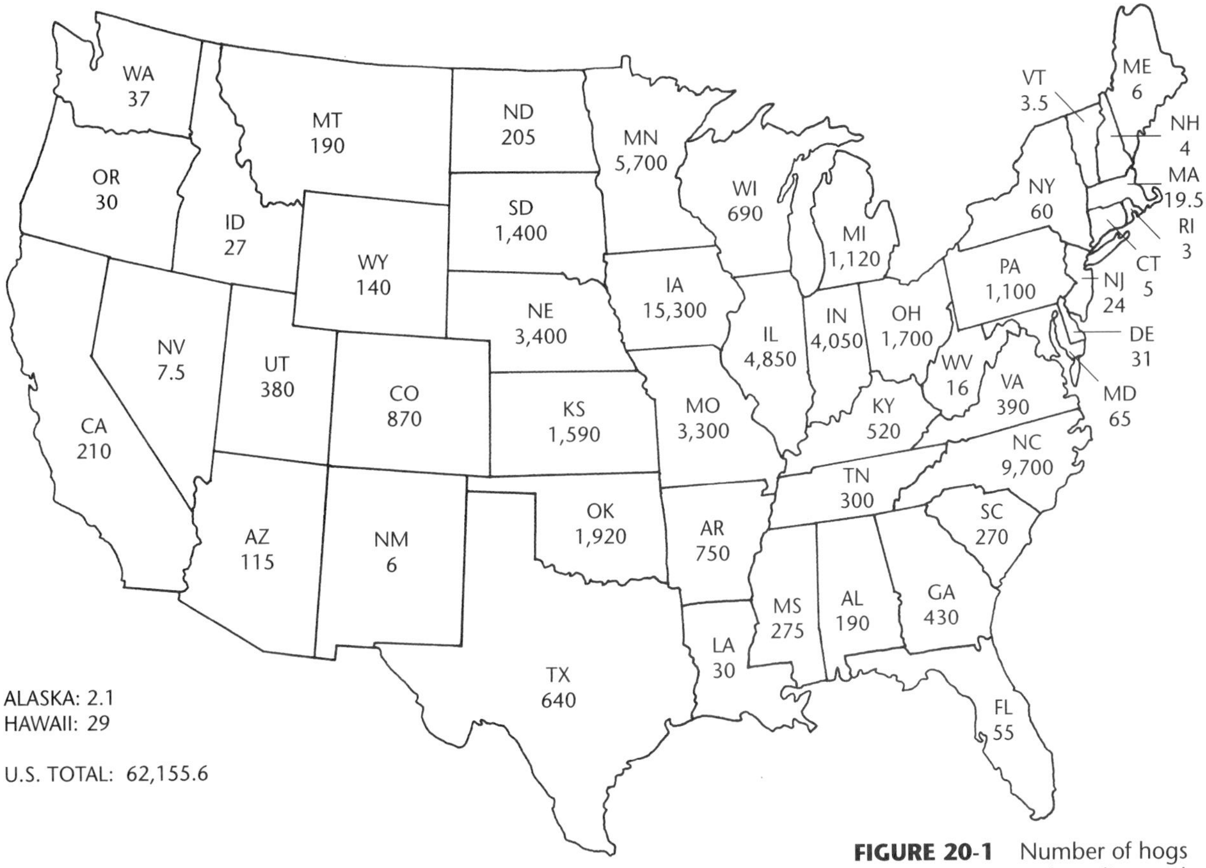

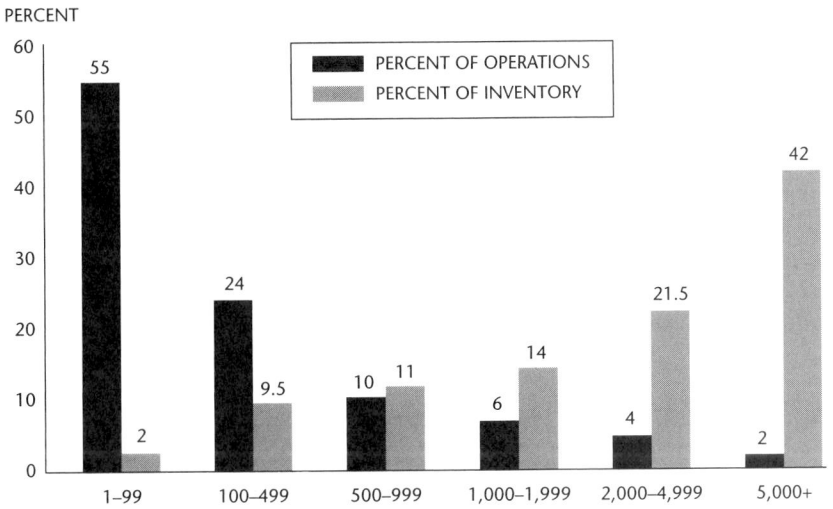

WA 37
OR 30
ID 27
MT 190
ND 205
MN 5,700
WI 690
MI 1,120
NY 60
VT 3.5
ME 6
NH 4
MA 19.5
RI 3
CT 5
NJ 24
DE 31
MD 65
PA 1,100
OH 1,700
IN 4,050
IL 4,850
WV 16
VA 390
KY 520
NV 7.5
UT 380
WY 140
SD 1,400
IA 15,300
NE 3,400
CO 870
KS 1,590
MO 3,300
CA 210
AZ 115
NM 6
OK 1,920
AR 750
TN 300
NC 9,700
SC 270
MS 275
AL 190
GA 430
LA 30
TX 640
FL 55

ALASKA: 2.1
HAWAII: 29

U.S. TOTAL: 62,155.6

FIGURE 20-1 Number of hogs and pigs on farms, breeding and market, December 1, 1998 (1,000 head). *Source*: Hogs And Pigs, USDA, December 1998.

PERCENT

PERCENT OF OPERATIONS
PERCENT OF INVENTORY

1–99: 55, 2
100–499: 24, 9.5
500–999: 10, 11
1,000–1,999: 6, 14
2,000–4,999: 4, 21.5
5,000+: 2, 42

SIZE GROUP OF OPERATIONS—NUMBER OF HEAD

FIGURE 20-2 Swine: Percent of operations and inventory, United States. *Source*: Hogs And Pigs, USDA, December 1998.

common today. The investment required to raise hogs using the current technologies makes it more difficult to be in and out of the business over a period of time. When smaller producers quit raising hogs now, they generally do not return to the business. The producers who stay in the business are making larger investments in facilities and technology to become more efficient and thus improve profitability.

The traditional model for swine production has been a farrow-to-finish enterprise on a farm. There is still a core of farmers in the United States who follow this traditional pattern, especially in the Midwest. A change to more specialization in swine production began in the 1990s. Three different sites, on different farms, are now a common pattern for the swine enterprise. The breeding, gestation, and farrowing phase is located at one site. When pigs are weaned, they are moved to a nursery site where they are fed for 8 to 10 weeks. When they reach 40 to 60 pounds (18–27 kg), they are moved to a finishing facility where they are fed out to market weight.

Production contracts are often used in one or more phases of this production model. The contractor (typically a large operator or corporation) provides the animals, feed, veterinary care, management services, and other necessary inputs. The contractee (typically a smaller farmer) provides labor and facilities in exchange for a guaranteed price per pig delivered. A bonus may be paid for reaching certain production efficiencies. Contracting in swine production has grown most rapidly in those areas where swine production was not a traditional enterprise: the Southeastern (particularly North Carolina) and the Western and Southwestern parts of the United States. Part of the reason for the rapid expansion of contracting in swine production in North Carolina is because the farmers in that state are already familiar with contracting in the poultry industry.

A major change in the swine industry has been the rapid growth of vertical integration. The swine industry is following a pattern similar to that which occurred in the poultry industry several years ago. The same company produces the animals and owns the packing plants that slaughter the animals for market. Five of the ten largest pork producers in the United States in 1999 also owned packing plants. These five companies owned 45 percent of the sows owned by the top fifty producers. It is expected that this trend toward vertical integration will continue despite the fact that some farmers and swine producer organizations are asking state legislators to ban or limit packer ownership of pork production. Some states already have legislation that bans packer ownership of hogs on the farm. These laws are circumvented by changing the ownership of the hogs on paper to make it appear that the company with the packing plant does not own the hogs on the farm. In late 1999 federal legislation was proposed to ban companies with packing plants from owning animals on the farm if they controlled more than a designated percentage of the total production. This issue had still not been settled at the time this revision of the text was being written.

Hogs rank fourth in livestock receipts in the United States. Beef cattle, dairy products, and broilers are the livestock enterprises that have greater total receipts than hogs. Hogs have traditionally been popular as a farm enterprise because the return on investment is generally faster than it is in most other livestock enterprises. The time from breeding to marketing is generally about nine or ten months. The time from farrowing to market is usually five to six months.

Improved technology has led to more pigs weaned per litter, better feed efficiency, and more litters per sow per year. The average number of pigs weaned per litter is now over eight pigs as compared to about seven pigs in 1978. Hogs are efficient converters of feed to meat; they are more efficient than cattle or sheep.

Labor requirements per pound of pork produced are relatively low. The need for manual labor is reduced with the use of automated confinement facilities. The investment in confinement facilities is higher than it is for simpler systems such as raising hogs on pasture. However, few farmers today use pasture raising systems for swine production.

The swine enterprise is profitable about nine years out of ten. That is an average number and does not mean that swine will be profitable in any given year. The management ability of the producer is a factor in determining the profitability of the swine enterprise in any given year. Data from Iowa State University, covering a period of 1984 through 1997 on Iowa farrow-to-finish swine enterprises, revealed that the top one-third of producers made a profit every year. The average one-third of producers made a profit in thirteen out of the fourteen years covered. The low one-third of producers made a profit in only five out of the fourteen years.

The price cycle for hogs is fairly short. The weight range for top butcher prices will vary from year to year depending on supply and demand in the marketplace. The spread in prices for different grades of hogs is small. Hogs have a higher dressing percentage than cattle or sheep; on average it is about 71 percent.

Another recent trend in swine production is the use of multi-year marketing contracts with major packing companies. A typical contract calls for the delivery of specified numbers of hogs of a specific quality on specified dates. This assures the producer an outlet for production and often pays a premium for higher quality hogs. The packer gains an assured supply of quality hogs, which reduces costs. The use of marketing contracts has increased dramatically. In the 1970s only about 2 percent of hogs were market in this manner. In 1999, 59 percent of hogs were marketed through some type of multi-year marketing contract.

SELECTION OF A BREED

It is the goal of the swine producer to raise breeding stock and market pigs that have rapid, efficient growth. These hogs should also yield a high percent of muscle when slaughtered. The modern swine type is described in Unit 21.

There are differences among breeds in the traits that are considered economically important. When selecting breeds to use in crossbreeding programs, consider litter size, growth rate, feed efficiency, carcass length, leanness, and muscle. Data on the major breeds is available as the result of work done at central testing stations. See Tables 20-1, 20-2, and 20-3.

When selecting individual animals, performance test records should be examined. Performance records are sometimes not available for gilts and sows. Selection must often be made on the basis of the boar's performance test record.

Information on each breed is available from the individual breed associations. The location of breeding stock and information on the performance of different bloodlines can be secured from these associations. Most swine breed associations sponsor production registry of breeding stock.

Purebred herds provide breeding stock that can be used in crossbreeding programs to produce market hogs. Purebred animals may be registered in their respective associations if they meet the standards for registration. In rank order, Duroc, Yorkshire, Hampshire, Spotted, Chester White, Landrace, Poland China, and Berkshire have been the leading swine breeds in the United States based on total numbers registered since 1970.

Most swine purebred associations require the following information for the registration of individual hogs:

1. Date farrowed.
2. Number of pigs farrowed.

TABLE 20-1 Comparison of Breed Performance Litter Size.

Breed	Average Litter Size Ratio[a]
Yorkshire	100
Landrace	99
Duroc	91
Chester White	91
Spotted	83
Hampshire	83
Berkshire	77
Poland China	77

[a]Best breed performance is given 100 as compared to each breed.

Source: Data provided in personal communication to the author from Charles J. Christians, Extension Animal Husbandman, University of Minnesota.

TABLE 20-2 Performance of Boars in Central Test Stations.[a]

Breed	Number of Boars Tested	Average Daily Gain lb/day	Feed Efficiency lb feed/lb gain	Backfat inches
Berkshire	349	2.10	2.68	0.83
Chester White	326	2.08	2.59	0.82
Duroc	4,294	2.27	2.49	0.81
Hampshire	1,587	2.17	2.49	0.74
Landrace	685	2.20	2.53	0.78
Poland China	219	2.11	2.68	0.80
Spotted	801	2.09	2.63	0.79
Yorkshire	4,028	2.22	2.49	0.79

[a]Data presented must be used with caution. Boars tested are not a random sample of the breeds and the data may not represent real differences among the breeds.

Source: Ahlschwede, W.T., Christians, C.J., Johnson, R.K., and Robison, O.W., Pork Industry Handbook, "Crossbreeding Systems for Commercial Pork Production," University of Illinois.

TABLE 20-3 Performance of Purebred Barrows Tested at the National Barrow Show.[a]

Breed	Number of Barrows Tested	Average Daily Gain lb/day	Age at 220 Pounds days	10th Rib Backfat inches	Loin Muscle Area square inches
Berkshire	47	1.66	176.40	1.15	4.53
Chester White	66	1.64	175.94	1.26	4.50
Duroc	161	1.73	171.89	1.22	4.72
Hampshire	75	1.60	177.53	0.91	5.22
Landrace	34	1.65	172.00	1.22	4.49
Poland China	47	1.64	173.47	1.12	5.17
Spotted	65	1.68	171.98	1.20	4.72
Yorkshire	93	1.61	178.24	1.15	4.51

[a]Data presented must be used with caution. Barrows tested are not a random sample of the breeds and the data may not represent real differences among the breeds.

Source: Ahlschwede, W.T., Christians, C.J., Johnson, R.K., and Robison, O.W., *Pork Industry Handbook,* "Crossbreeding Systems for Commercial Pork Production," University of Illinois.

3. Number of pigs raised (males and females).
4. Ear notches (must be notched at birth).
5. Name and address of the breeder.
6. Name and address of the owner at farrowing time.
7. Name and registration number of the sire and dam.
8. Litter number.

Ear notches must follow the guidelines for the breed association with which the animal is being registered. Some breed associations require an ear tattoo that shows the year of the pig's birth and the herd it comes from. Litter numbers are assigned by the producer and must not be repeated during the same year.

CHARACTERISTICS OF THE BREEDS

American Landrace

History. The Landrace breed originated in Denmark. In 1934, the first Landrace hogs were imported into the United States for experimental purposes by the USDA. Additional imports of Landrace from the Scandinavian countries occurred in 1954.

Description. Landrace hogs are white in color. They are long bodied and their ears lop forward and down (Figure 20-3). Landrace sows are noted for their mothering ability (Figure 20-4). The breed is also known for its large litters (Figure 20-5).

The American Landrace Association, Inc., the breed association, has its headquarters in West Lafayette, Indiana. The breed association was formed in 1950. Disqualifications for registry are black hair, erect ears, and less than six teats on a side.

FIGURE 20-3   Landrace boar. *Courtesy of American Landrace Association, Inc., IN.*

FIGURE 20-4   Landrace sow. *Courtesy of American Landrace Association, Inc., IN.*

FIGURE 20-5   Landrace sows have large litters and show good mothering ability. *Courtesy of American Landrace Association, Inc., IN.*

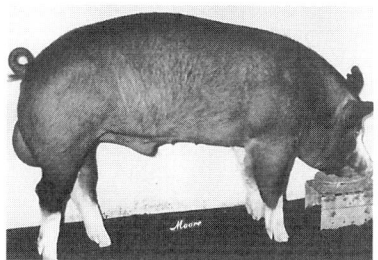

FIGURE 20-6   Berkshire boar. *Courtesy of American Berkshire Association, IN.*

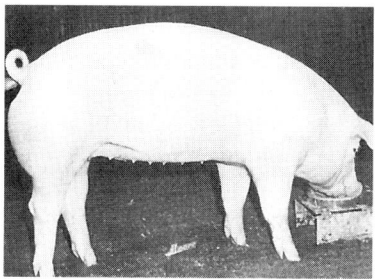

FIGURE 20-7   Chester White gilt (young female). *Courtesy of Chester White Swine Record, IL.*

Berkshire

History. The Berkshire originated in England in and around Berkshire and Wiltshire counties. The development of the breed began during the early and mid 1700s. The first importation of Berkshires into the United States occurred in 1823. Further importations occurred in 1857. All registered Berkshires in the United States trace their ancestry to the 1857 importations.

Description. The Berkshire (Figure 20-6) is a medium-sized hog that produces an acceptable carcass. The animal is black with six white points. Four white points are found on the feet. There is also some white on the face and the tail. The head is slightly dished and the ears are erect.

The American Berkshire Association was formed in 1875. Disqualifications for registration are swirls on the back or sides, large amounts of white hair on the body, or red hair. (A *swirl* is hair growing in a circular pattern from the roots. It usually occurs along the top of the spine. In many breeds it is a disqualification for registry because of the undesirable appearance it gives the animal. However, it does not affect any of the economically important traits of market hogs.) Selection has placed emphasis on fast and efficient growth, meatiness, and good reproduction.

Chester White

History. The Chester White (Figure 20-7) originated in Chester County, Pennsylvania. Additional development of the breed also occurred in Delaware County, Pennsylvania. The original name of the breed was Chester County White. Later the word *County* was dropped from the name.

Yorkshire, Lincolnshire, and Cheshire hogs, all of English origin, were mixed in breeding before 1815. An English white boar, which was called a Bedfordshire or Cumberland, was imported from England sometime between 1815 and 1818. This boar was used on the mixed breeding of the three English breeds mentioned earlier. From these matings, the Chester White breed originated. In 1848, the breed was named Chester County White.

Description. The color of the breed is white (Figure 20-8). The ears droop forward. The breed is noted for its mothering ability. Disqualifications for registry include swirls on the back and sides, or any color other than white.

Several breed associations were formed in the early days of the development of Chester Whites. An effort to combine these associations began around 1911. The present association is the Chester White Swine Record, which was incorporated in 1930. The association headquarters is in Peoria, Illinois.

Duroc

History. The Duroc breed (Figure 20-9) originated from red hogs raised in the Eastern United States before 1865. The New Jersey red hogs were called Jersey Reds. In New York, the red hogs were called Durocs. Some Red Berkshires from Connecticut are also thought to have been included in the early breeding. Intermingling of these breeds resulted in a breed called Duroc-Jersey. The name *Jersey* was later dropped and the breed became known simply as Duroc.

Description. The color of the Duroc is red. Shades vary from light to dark, with a medium cherry being the preferred shade. The Duroc has ears that droop forward. The breed has good mothering ability, growth rate, and feed conversion. It is one of the most popular breeds of hogs in the United States. Disqualifications for registry include swirls on the back and sides, black spots larger than 2 inches in diameter, cryptorchidism (retention of one or both testicles in the body cavity), fewer than six teats on a side, or white hair on the body.

The American Duroc-Jersey Record Association was formed in 1883. Several other associations were later formed to register red hogs. These associations merged to form the United Duroc Swine Registry in 1934. Headquarters of the breed association is located in Peoria, Illinois.

Hampshire

History. The Hampshire breed originated in England. Importations of Hampshires were made into the United States between 1825 and 1835. Major development of the breed occurred in Kentucky, where the belted hogs were known as the Thin Rind.

Description. The Hampshire is black with a white belt that encircles the forepart of the body (Figure 20-10). The forelegs are included in the white belt. To be eligible for registry, the white belt must include no more than two-thirds of the length of the body. White is permitted on the hind legs as long as it does not go above the bottom of the ham or touches the white belt. The Hampshire has erect ears. The breed is noted for its rustling (foraging) ability, muscle, and carcass leanness. It is a popular breed and is used in many crossbreeding programs.

Disqualifications for registry are cryptorchidism, swirls on back or sides, incomplete belt, or white belt more than two-thirds back on the body. Other disqualifications include white on the head (except on the front of the snout), black front legs, white going above the bottom of the ham, or white on the belly extending the full length of the body.

The belted hog was first registered in the American Thin Rind Association. The name of the breed was changed to Hampshire in 1904 and the association was renamed the American Hampshire Record Association. This association was discontinued in 1907, and a new association was formed in Illinois, called the American Hampshire

FIGURE 20-8 Chester White boar. *Courtesy of Swine Genetics.*

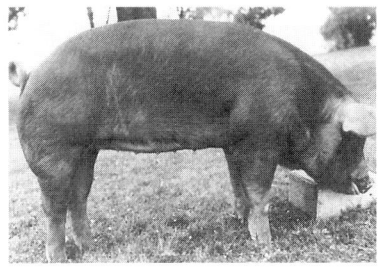

FIGURE 20-9 Duroc gilt. *Courtesy of United Duroc Swine Registry, IL.*

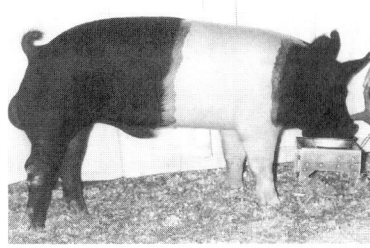

FIGURE 20-10 Hampshire boar. *Courtesy of Hampshire Swine Registry, IN.*

Swine Record Association. The name Hampshire Swine Registry was adopted in 1939. The headquarters for the breed is in West Lafayette, Indiana.

Hereford

History. The Hereford breed was developed in Missouri, Iowa, and Nebraska. Early development of the breed occurred from 1902 to 1925. Foundation stock used in the development of the breed included Duroc and Poland China. Chester White and Hampshire hogs may also have been included in the early breeding.

Description. Herefords are red with a white face. The ears are forward drooping. To be eligible for registry, Hereford hogs must be at least two-thirds red and have some white on the face. Herefords are prolific, good mothers, and have good rustling ability. Disqualifications for registry include no white on the face, less than two-thirds of the body red, swirls on the body, or less than two white feet.

The National Hereford Hog Registry Association was organized in 1934. Hogs from Iowa and Nebraska were selected as foundation stock for original registry. Headquarters for the breed association is in Flandreau, South Dakota.

Pietrain

History. The Pietrain breed originated near the village of Pietrain, Belgium during the early 1950s. The breed gained some popularity in Belgium and was exported to Germany in the early 1960s. Its primary use in Germany is in crossbreeding programs. The breed is also used in the United States in the production of seedstock.

Description. The Pietrain hog is white with black spots. It is medium sized with erect ears, short legs, and muscular hams. The breed has an exceptionally high lean to fat ratio. This makes it desirable in genetic improvement programs. The boars are used on sows of other breeds to improve the carcass qualities of market hogs.

Poland China

History. The Poland China breed (Figure 20-11) originated in the Ohio counties of Butler and Warren. The breed was developed between 1800 and 1850. Russian, Byfield, Big China, Berkshire, and Irish Grazer bloodlines were used in the development of the breed. It is generally believed that no new bloodlines were used in the breeding after 1846.

The breed was originally called the Warren County hog. The name Poland China was officially adopted at the National Swine Breeders Convention in 1872.

Description. The Poland China hog is black with six white points. The white points include the feet, face, and the tip of the tail. The

FIGURE 20-11 Poland China boar. *Courtesy of Poland China Record Association, IL.*

Poland China has forward-drooping ears. Poland Chinas are one of the larger breeds of hogs. These hogs produce carcasses with low back-fat and large loin eyes. They are used in many crossbreeding programs.

Disqualifications for registry include less than six teats on a side, swirls on the upper half of the body, hernia, or cryptorchidism. The absence of any of the white points is not objectionable nor is an occasional splash of white on the body. The breed association is the Poland China Record Association, which is located in Peoria, Illinois.

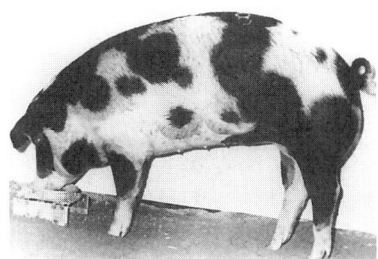

FIGURE 20-12   Spotted Swine gilt. *Courtesy of National Spotted Swine Record, Inc., IL.*

Spotted Breed

History. The Spotted Breed (Figures 20-12 and 20-13) was developed in Indiana. It was created by crossing hogs of Poland China breeding with spotted hogs being grown in the area. Later crosses were made with hogs from England called Gloucester Old Spots.

Description. The color of the Spotted Breed is black and white. At least 20 percent of the body must be either black or white to make the animal eligible for registry. In body type, the Spotted Breed is similar to the Poland China, and also has forward-drooping ears. Breeders strive to produce a large-framed hog with efficient gains and good muscling.

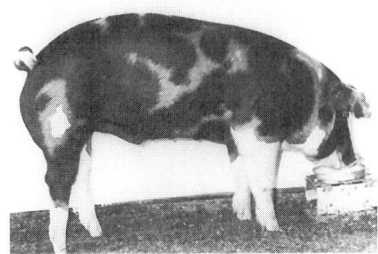

FIGURE 20-13   Spotted Swine barrow (male hog castrated before sexual maturity). *Courtesy of National Spotted Swine Record, Inc., IL.*

The breed association was formed in 1914. At that time, only one recorded parent was necessary for registering an individual hog. The herdbook was closed in 1921. The original name of the breed was the Spotted Poland China. The name was changed to the Spotted Breed in 1960. The herdbook was opened to register purebred Poland Chinas in 1971. This was to provide a broader genetic base. The herdbook was closed again in 1975. (*Closing the herdbook* means that only animals whose parents are registered in the herdbook may be registered.) The breed association is the National Spotted Swine Record and is located in Peoria, Illinois.

Disqualifications for registry include brown or sandy spots, and swirls on any part of the body. Cryptorchids (males with one or both testicles retained) are also disqualified.

Tamworth

History. The Tamworth hog originated in Ireland. Development of the breed took place in England in the counties of Stafford, Warwick, Leicester, and Northhampton. It is one of the oldest of the purebred breeds. Pure breeding began in the early 1800s. The first importations into the United States were in 1881.

Description. The Tamworth is red, with shades varying from light to dark. The ears are erect and it has a long head and snout. The sows are good mothers and have large litters. The breed is noted for its foraging ability. Swirls on the sides and back and inverted teats are disqualifications for registry. The breed association is the Tamworth Swine Association of Winchester, Ohio.

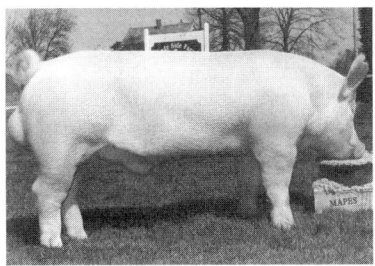

FIGURE 20-14   Yorkshire Boar.
Courtesy of the American Yorkshire Club.

Yorkshire

History. The Yorkshire hog (Figure 20-14) originated in England in the county of Yorkshire, where the breed was called Large White. Importations were made into the United States in the 1800s. At that time, many of the hogs were raised in Minnesota, where they became known as Yorkshires.

Description. The color of the Yorkshire is white. The skin sometimes has black pigmented spots called *freckles*. Hogs with black spots can be registered, but this trait is considered undesirable. The ears are erect and the face slightly dished. The Yorkshire was one of the early bacon-type breeds of hogs. Yorkshires have large litters, high feed efficiency, rapid growth, good mothering ability, and long carcasses. They are often used in crossbreeding programs.

Yorkshires are registered in the American Yorkshire Club of West Lafayette, Indiana. Disqualifications for registry include swirls on the upper third of the body, hair other than white, and blind or inverted teats. Other disqualifications include less than six teats on a side, hernia, and cryptorchidism.

Inbred Breeds

Beginning in 1935, the USDA and several state agricultural experiment stations conducted swine breeding research by crossing purebred lines of hogs. As a result of this work, a number of inbred lines were developed. Later, some private individuals also developed inbred lines by crossing other breeds. Inbreeding programs were followed to fix the traits wanted in the inbred lines. Table 20-4 shows the major breeds developed by these inbreeding programs.

Chinese Pigs

A recent interest has developed in the use of several Chinese breeds of swine in crossbreeding programs. The main reason for this interest is the large litter sizes that are common for these breeds. The most prolific of the Chinese breeds include the Meishan, Fengjing, and Jiaxing Black. These breeds average 15 live pigs per litter at birth and wean 13 pigs per litter. They reach puberty at about three months of age and require 40 percent less feed intake per pound weaned as compared to breeds currently used in the United States.

Research to date indicates that these breeds have slower rates of gain, lower feed efficiency, and poorer carcass characteristics than western breeds. This would indicate that their usefulness in crossbreeding programs would be limited. Research is focusing on trying to determine why they are so prolific. If the reason turns out to be genetic, it may be possible to use genetic engineering techniques to apply the knowledge gained to improve the prolificacy of breeds currently in use in the United States.

TABLE 20-4 Swine Breeds Developed from Inbred Lines.

Breed	Developed By	Where	Date Developed	Date Admitted to Registry	Parent Breeds	Color
Beltsville No. 1	USDA	Beltsville, Maryland	1935	1951	Landrace, Poland China	Black with white markings
Beltsville No. 2	USDA	Beltsville, Maryland	—	1952	Danish Yorkshire, Danish Landrace, Duroc, Hampshire	Light red
C P F	Conner Prairie Farm	Noblesville, Indiana	1956	1964	San Pierre, Beltsville No. 1	Black and white
C P F No. 2	Conner Prairie Farm	Noblesville, Indiana	1959	1964	Yorkshire, Maryland No. 1, Beltsville No. 2	Black and white
Maryland No. 1	USDA, Maryland Ag Exp. Station	Queenstown, Maryland	1941	1951	Landrace, Berkshire	Black and white spotted
Minnesota No. 1	Minnesota Ag Exp. Station, USDA	North Central Station, Minnesota	1936	1946	Danish Landrace, Tamworth	Red
Minnesota No. 2	Minnesota Ag Exp. Station, USDA	Rosemount Exp. Station, Minnesota	1941	1948	Yorkshire, Poland China, San Pierre	Black and white
Minnesota No. 3	Minnesota Ag Exp. Station	Rosemount Exp. Station, Minnesota	1940s	1956	Gloucester Old Spot, Welsh, Large White, C-line Poland, Beltsville No. 2, Minnesota No. 1, Minnesota No. 2, San Pierre	Black and red spotted
Minnesota No. 4	Minnesota Ag Exp. Station	Rosemount Exp. Station, Minnesota	1940s			
Montana No. 1	USDA, Montana Ag Exp. Station	Miles City, Montana	1936	1948	Landrace, Hampshire	Black
Palouse	Washington State Exp. Station, Irrigation Exp. Station	Pullman, Washington, Prosser, Washington	1945	1956	Danish Landrace, Chester White	White
San Pierre	Gerald Johnson	Indiana	—	1953	Berkshire, Chester White	Black and white

SUMMARY

Swine are produced in the areas of the United States where feed grains are raised. This is because the major part of the swine ration is grain. More than one-half of the hogs raised in the United States are raised in the midwestern states. Iowa leads the nation in the number of hogs raised. Hogs are a popular farm enterprise. They are efficient converters of feed to meat. Hogs are profitable about nine years out of ten.

Selection of breeds to use in crossbreeding programs is based on data concerning litter size, growth rate, feed efficiency, carcass length, leanness, and muscle. All major breeders of purebred hogs strive to produce the kind of carcass in demand on the market today.

Most of the breeds of hogs raised in the United States were developed in the United States. Others were developed in Denmark and England. All purebred breeds have associations that register purebred animals. Standards for registration of the individual breeds are set by the breed associations.

Inbred lines of hogs were developed by the USDA and various state experiment stations, as well as by private breeders. These hogs have been used in crossbreeding programs to improve litter size and growth rate.

Student Learning Activities

1. Prepare a bulletin board showing various breeds of swine.
2. Observe the characteristics of the different breeds of swine on farms in the local area.
3. Prepare a list of sources of purebred breeding stock for each of the major breeds that is located within a reasonable distance of the school.
4. Ask local purebred swine breeders to speak to the class about their activities.

Review

1. Why are more than one-half of the hogs raised in the United States raised in the midwestern Corn Belt states?
2. How important is hog production compared to all livestock enterprises in the United States?
3. Why are hogs popular as a livestock enterprise on the farm?
4. What type of hog carcass is in demand in today's market?
5. Prepare a table that lists the common breeds of hogs and briefly describes the traits of each.

Selection and Judging of Swine

SELECTION OF BREEDING STOCK

There has been a major change in the type of market hog in demand today as compared to the typical market hog of the 1980s. Consumers want more lean meat and less fat in their pork. Producers can raise the kind of pork the market demands by using modern production technology and improved breeding programs. Purebred breeders pay careful attention to genetics when selecting bloodlines. Research shows that there are differences among various swine genotypes in their ability to efficiently produce the lean pork the current market demands. The kind of environment and the health management techniques used by the producer also influence the rate and efficiency of lean growth in market hogs.

Practically all market hogs today are produced by some type of crossbreeding program. Swine crossbreeding programs are discussed in more detail in unit 12. Rotational, rotaterminal, and terminal crossbreeding systems are used. Because a terminal crossbreeding system tends to produce leaner pork, it is now used to produce the majority of the market hogs in the United States. Good-quality, leaner, more heavily muscled hogs bring a premium price in the market (Figure 21-1).

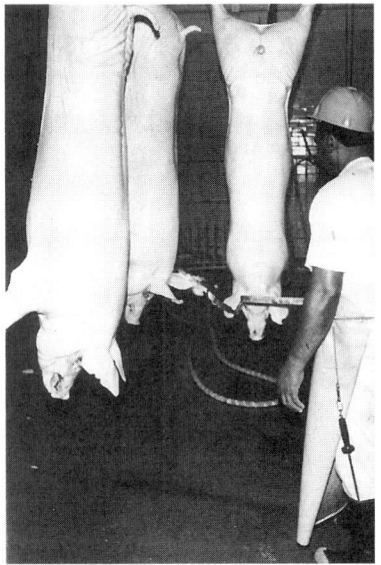

FIGURE 21-1   Good-quality hogs are in demand in today's market. *Courtesy of University of Illinois at Urbana-Champaign.*

Use of Ultrasound

Research is continuing into the use of ultrasound to measure the fat-free lean pork content of live hogs and carcasses. Measuring the amount of fat-free lean pork in live hogs can help producers make decisions about which animals to keep for breeding stock. Packers can use ultrasound measurements to help determine the value of the hog carcass.

Two types of ultrasound machines are currently available:

- A-mode machines use sound waves from a transducer and measure the time required for these waves to reflect back to the machine. This measurement is then converted into a distance measurement.
- Real-time (sometimes called B-Mode) machines use many sound waves at the same time to create a two-dimensional image that can be viewed on a monitor as it is created.

Both types have been tested by researchers at Purdue University. The A-mode machine is less expensive but is not as accurate in measuring the fat-free lean content in either live hogs or carcasses as is the real-time machine. When making genetic selections for lean content in breeding stock, the real-time machine is probably the better choice despite its higher cost. There are certified ultrasound technicians available who travel around the country offering this service to producers who do not want to make the capital investment in the necessary equipment. Depending on the cost of the service and the volume of hogs to be tested, this may be an alternative for some producers.

A list of technicians certified by the National Swine Improvement Federation (NSIF) may be found on the Internet by accessing the home page of the NSIF at the following URL: http://mark.asci.ncsu. edu/nsif/. Information can also be secured by writing to National Swine Improvement Federation, 203 Polk Hall, Box 7621, North Carolina State University, Raleigh, NC 27695-7621.

Parts of the Hog

Live hog.   A diagram of the parts of the live hog is shown in Figure 21-2. To describe or judge a hog, it is necessary to be familiar with the parts of the hog.

Carcass.   The four primal cuts of the hog carcass are: *ham, loin, Boston shoulder (Boston butt)*, and *picnic shoulder*. These four cuts represent the most valuable part of the hog carcass. In a typical 250-pound (113-kg) market hog, these cuts will make up about 44 percent of the live weight and represent about 75 percent of the total value of the animal. Assuming a carcass weight of about 184 pounds (83 kg) from this typical market hog, these cuts represent almost 60 percent of the retail cuts on a semi-boneless basis. These values will vary depending

Plate 1 Angus Bull ⇦
Courtesy of the American Angus Association

Plate 2 Charolais Bull ⇦
Courtesy of the American-International Charolais Association

⇦ **Plate 3 American Salers Bull**
Courtesy of the American Salers Association

Plate 4 Limousin Bull ⇨
Courtesy of the North American Limousin Foundation

Plate 5 American Polled Hereford Bull ⇦
Courtesy of the American Polled Hereford Association

Plate 6 American Simmental Bull ⇦
Courtesy of the American Simmental Association

Plate 7 Ayrshire Cow ⇧
Courtesy of the Ayrshire Breeders' Association

Plate 8 Brown Swiss Cow ⇧
Courtesy of the Brown Swiss Cattle Breeders' Association

Plate 9 Guernsey Cow ⇧
Courtesy of the American Guernsey Association

Plate 10 Holstein-Friesian Cow ⇧
Courtesy of the Holstein Association USA, Inc.

Plate 11 Jersey Cow ⇧
Courtesy of the American Jersey Cattle Club

Plate 12 Milking Shorthorn Cow ⇧
Courtesy of the American Milking Shorthorn Society

Plate 13 Poland China Swine ⬆
Courtesy of the Poland China Record Association

Plate 14 Duroc Swine ⬆
Courtesy of the United Duroc Swine Registry

Plate 15 Chester White Swine ⬆
Courtesy of Swine Genetics

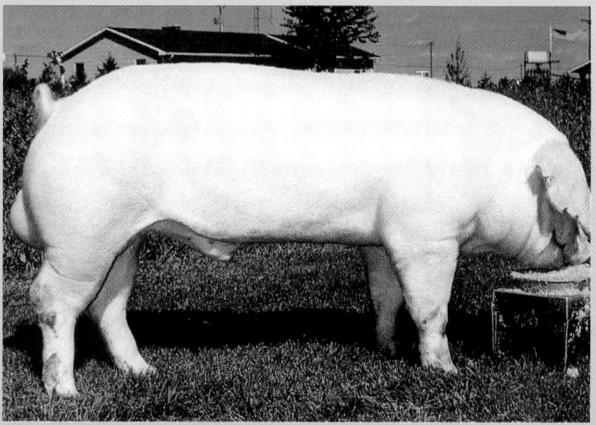

Plate 16 Landrace Swine ⬆
Courtesy of the American Landrace Association

Plate 17 Hereford Swine ⬆
Courtesy of the National Hereford Hog Record Association

Plate 18 Spotted Swine ⬆
Courtesy of Swine Genetics

Plate 19
Columbia Ram
Courtesy of the Columbia Sheep Breeders'

Plate 20
Dorset Ram
Courtesy of the Sheep Breeder Magazine

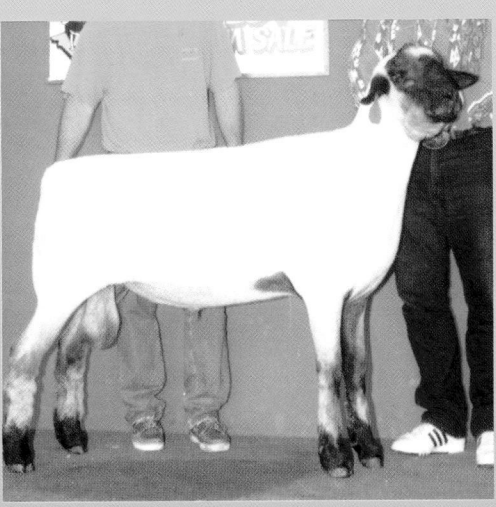

Plate 21
Hampshire Ram
Courtesy of the Sheep Breeder Magazine

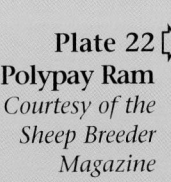

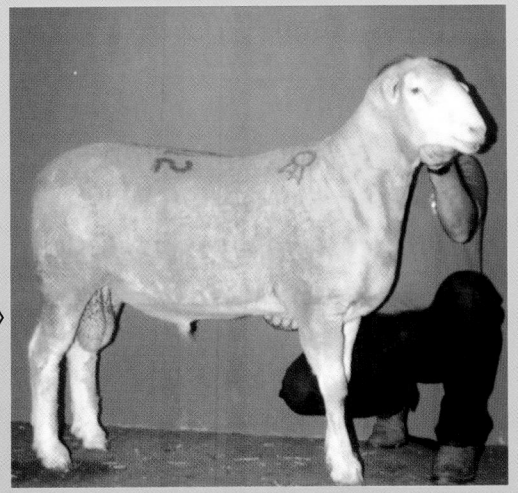

Plate 22
Polypay Ram
Courtesy of the Sheep Breeder Magazine

Plate 23
Rambouillet Ram
Courtesy of the Sheep Breeder Magazine

Plate 24
Suffolk Ram
Courtesy of the Sheep Breeder Magazine

Plate 25 Alpine Goat ⇧
Courtesy of the American Dairy Goat Association

Plate 26 Nubian Goat ⇧
Courtesy of the American Dairy Goat Association

Plate 27 Saanen Goat ⇧
Courtesy of the American Dairy Goat Association

Plate 28 Lamancha Goat ⇧
Courtesy of the American Dairy Goat Association

Plate 29 Toggenburg Goat ⇧
Courtesy of the American Dairy Goat Association

Plate 30 Oberhasli Goat ⇧
Courtesy of the American Dairy Goat Association

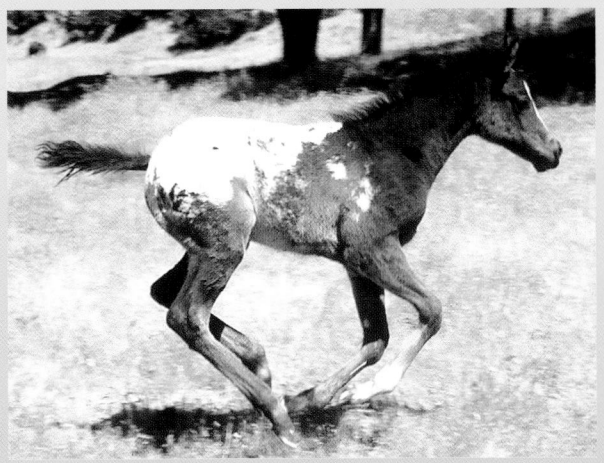

⇧ **Plate 31 Appaloosa**
Courtesy of the Appaloosa Horse Club/ photo by Tom Poulsen

⇧ **Plate 32 Quarter Horse**
Courtesy of the American Quarter Horse Association

◁ **Plate 33 Arabian**
Courtesy of Johnny Johnston

Plate 34 American Paint ▷
Courtesy of Don Shugart

⇧ **Plate 35 Standardbred**
Courtesy of the U.S. Trotting Association

⇧ **Plate 36 Thoroughbred**
Courtesy of the Illinois Racing News

Plate 37
White
Leghorn
Rooster
*Photo by Dr.
Charles Wabeck*

Plate 38
White
Plymouth
Rock Hen
*Photo by Dr.
Charles Wabeck*

Plate 39
Broad Breasted
Large White
Turkey
*Courtesy of the
Minnesota Turkey
Research and
Promotion Council*

Plate 40
Broad Breasted
Bronze Turkey
*Courtesy of Watt
Publishing Co.*

Plate 41
White Pekin
Duck
*Courtesy of
Jurgielewicz
Duck Farm*

Plate 42
Toulouse
Goose
*Photo by Dr.
Charles Wabeck*

Plate 45 Dutch ⇧
Courtesy of the American Rabbit Breeders Association

Plate 46 Checkered Giant ⇧
Courtesy of the American Rabbit Breeders Association

Plate 47 English Lop ⇧
Courtesy of the American Rabbit Breeders Association

Plate 48 New Zealand White ⇧
Courtesy of the American Rabbit Breeders Association

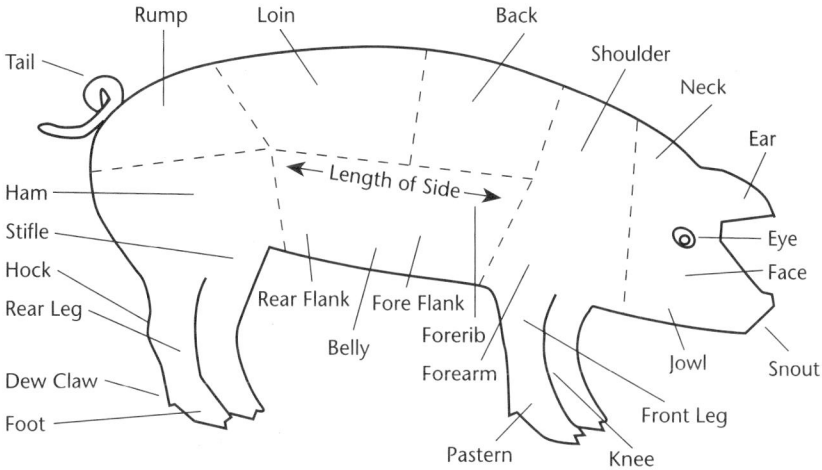

FIGURE 21-2   Parts of a hog.

upon the method of cutting the carcass and the quality of the carcass. Figure 21-3 shows the location of the wholesale cuts of the hog.

Market Hog Description

Improvement in identifying and selecting desirable breeding lines through better genetics and the use of more technology in producing hogs has resulted in significant improvement in the quality of pork produced in the United States. Market hogs are more muscular, meatier, and have less fat as compared to market hogs produced in the past. The typical market hog today is marketed at 250 pounds (113 kg) and produces a carcass weight of 184 pounds (83 kg). It has 0.9 inches (2.3 cm) of backfat at the 10th rib and a loin eye area of 5.2 square inches (33.5 cm^2). The fat-free lean index is 48 percent and it produces 88.6 pounds (40 kg) of lean meat.

The National Pork Producers Council (NPPC) originally described the ideal market pig named Symbol in 1983. Because of significant improvements in the quality of market hogs today, the NPPC has updated its description of the ideal market pig and now calls it Symbol II. This update represents the goals for producers to strive for in the years ahead:[1]

- Producing a 195-pound (88-kg) carcass with desirable muscle quality from a 260-pound (118-kg) live-weight hog marketed at 156 days [164 days for gilts] of age.
- Minimum loin muscle area of 6.5 in^2 (41.9 cm^2) [7.1 in^2 (45.8 cm^2) for gilts] with appropriate color, water holding capacity, and ultimate pH.

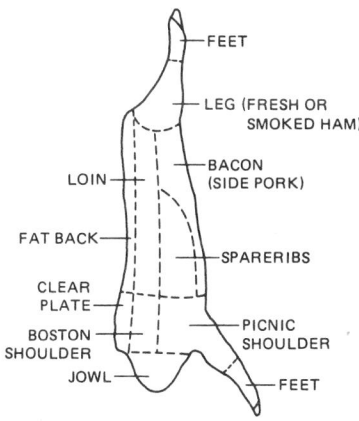

FIGURE 21-3   Wholesale cuts of the hog.

[1]1998–99 Pork Facts, National Pork Producers Council, Des Moines, IA.

■ Intramuscular fat level greater than or equal to 2.9 [2.5 for gilts] percent.
■ High health production system.
■ Free of the stress gene (Porcine Stress Syndrome).
■ Result of a terminal crossbreeding program.
■ From a maternal line capable of weaning 25 pigs per year.
■ Performance from 60 to 260 pounds (27 to 118 kg) on a corn/soy equivalent diet:
  ■ Live-weight feed efficiency of 2.4 (1 pound gain for each 2.4 pounds of feed [0.45 kg gain per 1.1 kg of feed]).
  ■ Fat-free lean gain efficiency of 6.4 [5.9 for gilts] (2.9 kg [2.7 kg for gilts]).
  ■ Fat-free lean gain of 0.78 pounds (0.35 kg) per day.
  ■ Standard reference backfat of 0.8 inch [0.6 inch for gilts].
  ■ Fat-free lean index of 49.8 [52.2 for gilts] percent.
■ Producer has completed NPPC's Environmental Assurance Program and is certified at LEVEL III of the Pork Quality Assurance[SM] (PQA) program.

Genetic Evaluation of Breeding Stock

Both seedstock and market hog producers use performance records for the genetic improvement of hogs. Much progress has been made during the past 20 years in the areas of feed efficiency, growth rate, reproductive efficiency, and carcass quality. This progress has come about because of performance testing and selection programs based on genetic principles that result in the selection of genetically superior breeding stock.

Economic value of improved genetics. Market hog producers who want to improve their herds need to secure their breeding stock from seedstock producers who are consistently doing performance testing and following good selection principles. Seedstock producers using genetic improvement programs must keep good records, do performance testing, and use selection indexes that relate to important economic characteristics to select breeding stock for continued genetic improvement. Superior performance tested boars must be used by seedstock producers to achieve the kind of genetic improvement required to keep pork competitive with other sources of meat in the marketplace.

The major breed associations all have improvement programs in place to take advantage of genetic principles when producing seedstock. A number of companies produce hybrid seedstock using sound genetic principles to improve their breeding stock. These types of breeding programs are expensive and the producers of seedstock expect to make a fair return for their efforts.

Buying better breeding stock will cost the commercial hog producer more money. Using improved breeding stock can significantly improve feed and reproductive efficiency in the commercial herd. Carcass quality can also be improved, producing the kind of market hogs more in demand by packers. Genetically improved breeding stock can return more benefit to the producer than it costs.

Use of performance records. There are many different performance records that might be collected on individual hogs or groups of hogs. When considering improvement programs, those records that have an impact on profitability are the most important. Such records include carcass traits such as loin eye area, length, and muscling score. Other important traits include litter size, litter weight, milking ability, feed efficiency, and number of days to market.

Environmental effects can influence the performance of individual pigs. Typically, records are kept on contemporary groups to help reduce the effect of environmental factors. A contemporary group is a group of animals that are similar in a number of characteristics and have been raised under the same management practices. Age, sex, and breed or cross are often used to set up a contemporary group.

Performance records on close relatives of potential breeding stock are also important. It is possible to secure information about carcass traits after slaughter by keeping records on close relatives. Some central boar testing stations use several animals from the same litter to secure information on traits such as average daily gain, feed efficiency, and carcass quality. The barrows that are litter-mates of the boar on test are slaughtered and carcass information is collected to help determine the breeding value of the boar.

Estimated breeding value (EBV). Estimated breeding values (EBVs) are determined by applying genetic principles to performance records. The EBV is a selection index because it combines information from a number of sources to determine the genetic merit of the individual. Important considerations are the number of records available, heritability of the measured traits, and pedigree information. The performance records of groups of close relatives are considered in determining the EBV of the individual. The performance records of close relatives are more important in determining the EBV than records of more distant relatives. The correlation among genetic traits is also taken into consideration (some traits correlate positively and others negatively). Traits with a high degree of heritability can be expected to be passed on to the offspring of the breeding animal to a greater degree than those traits with a low degree of heritability.

Expected progeny difference (EPD). The EBV is the total estimated genetic value of an animal. Only one-half of the genes of each parent are passed on to the progeny. This is referred to as the Expected Progeny Difference, which is defined as one-half the EBV. In a given population, the average EPD is zero. Individual EPDs are expressed as deviations (+ or –) from the average. The EPD measures how much difference in performance on a given trait one can expect from the progeny of the breeding animal as compared to the average of all animals in the population. The EPDs of potential breeding animals can be compared across herds within the same breed. They cannot be compared between breeds.

Both purebred breeders and commercial companies producing seedstock collect performance records and compute the EPD for their breeding animals. Information from performance records, including records on relatives, may be used in a computer software program called Best Linear Unbiased Predictor (BLUP) to determine the breeding animals with the highest EPDs. This procedure provides a highly accurate measure of the true breeding value of an individual. It also provides information regarding the correlation of multiple traits to determine breeding value.

Sire summaries and EPD values are made available by the various purebred associations, the National Swine Registry, and the National Pork Producers Council. Commercial seedstock companies vary in their policies on making this information available. Links to this information can be found on the Internet from the National Swine Registry home page (URL: www.ansc.purdue.edu/stages/), the National Pork Producers Council home page (URL: www.nppc.org/) and from some breed associations that maintain web pages.

Performance testing programs. There are several genetic evaluation programs that may be used by swine breeders to help determine the breeding value of their hogs.

Central boar testing stations have been used in the United States since 1954 to evaluate boars for genetic merit. Procedures for placing animals on test vary among the stations. Producers who are interested should contact the test station, county extension personnel, or university swine specialists in their area to find out the exact procedures to follow. Some of the test stations test boars for gain, backfat, and feed efficiency. Other test stations test the market progeny of herd boars.

Computer programs called SWINE-EBV are available to analyze performance data collected on the producer's farm. The producer can run the software with the on-site computer so results are quickly available. Data, including data from relatives, is collected on reproduction (number born alive and 21-day litter weight) and post-weaning (days to 230 pounds [104 kg] and backfat thickness) traits and is entered into the computer program. Maternal, general, and terminal sire indexes are computed from the data. Index numbers are ranked according to their economic importance.

The original Swine Testing and Genetic Evaluation System (STAGES) was developed by Purdue University, the USDA Agricultural Research Service, USDA Extension Service, National Association of Swine Records, several purebred associations, and the National Pork Producers Council. This performance and evaluation program was implemented in 1985. A new version of STAGES was implemented in 1998 to provide EPD information for the Yorkshire, Landrace, Duroc, and Hampshire breeds of swine.

Factors used to determine selection indexes for sows and boars include number of pigs born alive, number weaned, 21-day litter weight, growth rate to 250 pounds (113 kg), backfat adjusted to the 250 pound (113 kg) weight, feed conversion, and carcass data. Terminal Sire Indexes (TSI), Maternal Line Indexes (MLI), and Sow Productivity Indexes (SPI) are calculated as well as EPDs for most of the traits that are economically important. An EPD for pounds of lean adjusted to 250 pounds (113 kg) is also calculated.

Animals are evaluated within contemporary groups and breed specific adjustments are made for the EPDs reported. Another feature of the revised STAGES is the use of a rolling base genetic year and the computation of new values every night. This method of computing EPDs provides a higher degree of consistency for the evaluation of each trait from year to year. A better evaluation of the true genetic breeding value for each animal also results from this change. EPDs cannot be compared across breeds.

The National Pork Producers Council has two national genetic evaluation programs for swine. The Terminal Sire Line program was begun in 1993 to evaluate boars from a number of different sire lines. The Maternal Sow Line program was implemented in 1997. These programs are designed to evaluate genetic lines for traits that are important for producing crossbred market hogs. These evaluation programs are designed to test more traits than other evaluation programs currently being used. The programs also test more genetic lines as compared to other programs.

The acid meat gene problem. There is some concern in the U.S. swine industry about a gene that is apparently found only in Hampshire pigs. The gene is also called the Rendement Napole (NP) or Napole gene. This gene has some negative effects on the quality of meat, causing it to be more acidic than normal. This reduces the water holding capacity of the meat, which causes some loss for packers and processors. The gene has a positive effect on growth and carcass traits and improves the eating quality of the meat.

The gene was first observed in Hampshire hogs in France. Most of the research concerning its effects has been done in Europe. The specific gene has not been identified and is found only by testing biopsy samples from the live animal or post-mortem muscle samples. The frequency of the gene in Hampshire populations in the United States is not known. Research is being done at the University of Illinois in cooperation with the National Pork Producers Council and the National Swine Registry to determine how serious the problem actually is in the United States.

Other Considerations in Selecting Breeding Animals

The selection of breeding stock should always begin with a review of performance records and selection indexes. In addition to the genetic evaluation of breeding stock there are other things the producer must consider when selecting breeding animals. These include:

- the type of breeding program used by the purchaser
- the production practices of the seedstock producer
- soundness of the animal
- health of the seedstock herd
- minimum breeding age
- whether to buy or raise replacement gilts

Type of breeding program used by the purchaser. A producer using a rotational crossbreeding program needs to select boars that produce both desirable market hogs and desirable breeding gilts. Such boars should be selected from sows with good maternal characteristics. These boars also need to have good growth and leanness characteristics. Boars being selected for use in a terminal breeding program need to show outstanding growth and carcass traits. One of the reasons for the increased use of terminal crossbreeding programs in the production of market hogs is that it is difficult to make rapid genetic progress when selecting boars for both maternal and market hog characteristics.

Production practices of the seedstock producer. The production of seedstock for breeding programs is done by both individual purebred breeders and large breeding stock companies. The potential buyer needs to determine the reputation of the prospective seller by checking with other people who have purchased seedstock from the seller. Extension specialists and agribusiness people are other sources of information concerning a seedstock producer. The health program followed by the seedstock producer needs to be evaluated. Find out what vaccinations are used and if there have been any disease problems in the herd. Determine whether the producer routinely does performance testing to help assess the genetic value of the seedstock it produces.

Soundness of the animal. Select a boar that has visibly sound reproductive organs. The testicles should be well developed and of equal size. Do not select boars that have umbilical or scrotal hernias or other obvious structural problems. Select boars that are aggressive and show a desire to mate.

Sound underlines are important for both boars and gilts. They should have six or more functional nipples on each side. Check for proper spacing of the nipples and make sure there are no inverted or scarred nipples.

Select gilts and sows that show normal development of the reproductive system. Gilts with small vulvas should not be kept. This is often an indication of internal reproductive defects. Sows that have problems in farrowing should be culled from the herd. Select gilts with strong pasterns and sound feet and legs. Do not purchase breeding stock with obvious defects.

Research conducted at Southern Illinois University suggests that gilts from litters that are predominantly male should not be kept as breeding stock. They do not breed as well and usually have smaller litters. They often have unsound underlines with fewer than six nipples on a side. Research is ongoing to determine what effect male-dominated litters have on other important economic traits in the progeny of gilts from these litters.

Health of the seedstock herd. The health of the herd from which breeding animals are purchased is important. Buy only healthy animals from healthy herds. Herds should be certified brucellosis and pseudorabies free. Breeding animals should be vaccinated for erysipelas and the six strains of leptospirosis. Purchase animals only if they are free of external and internal parasites. Herd health information should be available from the breeder.

Breeding stock should not be either homozygous (nn) or heterozygous (Nn) for the Porcine Stress Syndrome gene. This is an inherited neuromuscular disease. When stressed, animals carrying this recessive gene exhibit symptoms including heavy breathing, tail tremors, splotchy coloring, and occasionally death. Two tests exist, either of which may be used to determine if an animal carries the PSS gene. The most accurate test is a check of the DNA of the animal. The DNA test will identify both homozygous and heterozygous carriers. The other test involves exposing the animal to the anesthetic Halothane. If the animal is homozygous for the PSS gene, it responds within three minutes by showing extreme muscle rigidity. The Halothane test cannot detect animals that are heterozygous for the PSS gene.

Minimum breeding age. Boars must be a minimum of 7½ months of age before they are used for breeding purposes. Breed gilts at the time of their second or third estrus after reaching puberty. Gilts usually have a higher ovulation rate at this time. Replacement boars and gilts should be bought 30 to 60 days before they are to be used. This permits isolation for health checks. Also, they can adjust to the farm and boars can be test mated for breeding performance.

To buy or raise replacement gilts. The increasing use of terminal crossbreeding programs in market hog production means the producer has to make a decision regarding whether to buy or raise replacement gilts. It requires more capital investment to buy replacement gilts. This method may provide the producer with breeding stock of greater genetic value that may offset some of the additional cost. Raising replacement gilts requires additional facilities and a good record keeping system. Producers need to evaluate their own swine enterprise to determine which method of securing replacement gilts works best for them.

SELECTING FEEDER PIGS

Common sources of feeder pigs are pig hatcheries, farmers, auction barns, and dealers who buy and sell feeder pigs. Factors to consider when buying feeder pigs are:

- health
- type
- size
- uniformity

Health. Only healthy pigs should be purchased. Pigs that have visible signs of sickness, such as coughing, infected eyes, rough hair coats, pot bellies, gauntness, or listless appearance, should not be selected. Pigs should be wormed, tail docked, and castrated. Do not buy pigs that show signs of external parasites.

Type. Meaty feeder pigs will produce the kind of carcass in demand on the market. Short, fat pigs will be overfinished when they reach market weight. The same type and conformation that is desirable in breeding hogs should be looked for in feeder pigs.

Size. Feeder pigs usually range from 35 to 80 pounds (15.9–36 kg) in weight. Select pigs that have a good size for their age. Size for age is more important than condition or fatness when selecting feeder pigs.

Uniformity. Uniformity in size, age, condition, and type is desirable in a group of pigs. When these traits are uniform, the pigs will feed out well together. All the pigs in the group will tend to reach market weight at about the same time.

JUDGING HOGS

A hog judge must know the parts of the live hog and the wholesale cuts of the carcass. Refer to Figures 21-2 and 21-3. To judge a class of hogs, look at them from a distance of about 15 feet. To identify them for judging, the hogs in the class have numbers marked on their bodies. Judge them as they move around the ring. Look at each hog and compare it with the ideal hog and the others in the class. Taking notes and giving reasons is done in the same manner as when judging cattle. See Unit 14 to review this procedure.

Using Performance Data in Judging Hogs

Traditionally, judging hogs has been done using only a visual appraisal of the animals in the class, with decisions being based on whether the class was market or breeding hogs. Because of the increased emphasis on genetic improvement in the swine industry, judging contests now often include performance data on the hogs in the class to be used in addition to the traditional visual appraisal in ranking the animals.

When performance data is used in a judging class, the contestants are given printed information about the class. This normally includes the breed and sex of the animals, the planned use of the animals, a description of the production environment, and the plan for marketing the animals or their progeny. Selection priorities may be given with the printed information or the contestants may have to develop their own priorities based upon the information given them. Performance records are provided for each animal. These will include reproduction, growth, and composition records. Records may be provided as individual measurements, may be given as ratios or indexes, or may be Expected Prodigy Differences (EPDs). The most valuable information for judging purposes is the EPDs and the least valuable is individual measurements.

The EPDs are the best records because they combine more information; this makes possible a better evaluation of the true genetic merit of each animal. Individual records do not give enough information to fairly compare the animals in the class. Three indexes are typically used to provide information: (1) Sow Productivity Index, (2) Terminal Sire Index, and (3) Maternal Line Index. All of these indexes are calculated with EPDs. Most indexes use 100 as the average value, therefore an index value above 100 is better than average and a value below 100 is below average. This provides a better comparison of the animals in the class than do individual measurements. When given a choice of performance data, use the EPDs to make preliminary rankings rather than individual measurements or indexes.

When performance records are given, make a preliminary ranking of the class based on that data. Then proceed to rank the class based upon the visual traits as described below. The final ranking of the class must be made on the basis of both performance data and visual appraisal of the animals.

Visual Evaluation of the Class

A number of traits may be visually evaluated when judging hogs. These include body conformation, size, muscling, finish, and reproductive soundness.

Type refers to the conformation of the hog's body. It is judged on the basis of length of side and skeletal size (scale). Body length is related to growth rate and the productivity of the sow. The length of side may be estimated by looking at the distance from a point in the center of the ham to the forepart of the shoulder (Figure 21-4).

A visual evaluation of the amount of muscling on a hog is best seen by looking at the rear view of the animal. In this view, the hog should show a wide back and loin and a deep rump. The ham should be deep, thick, and firm. The chest and shoulders should be wide. Be sure the observed width is due to muscling and not to excess fat. The correct shape across the back is an arc rather than a square (Figure 21-5). There is greater width across the rump and ham than there is across the back. A narrow ham indicates poor muscling.

Finish refers to the amount of fat on the hog. Some fat is desirable, excessive fat is undesirable. The amount of backfat is generally measured at the tenth rib and the last rib section of the loin. Performance data usually includes information about the leanness and muscling of the animals in the class. This information needs to be combined with a visual evaluation. Indications of excessive fat include heavy, wasty jowl; shaky middle, square top; looseness in the ham and crotch; or a roil of fat over the shoulder. If the hog is too wide in the middle, it will not have a desirable yield of lean cuts. This lowers the market value of the hog.

Quality refers to the degree of refinement of head, hair, hide, and bone of the live hog. Moderate refinement is preferred over animals that are too coarse or too refined. A smooth hair coat and hide is desirable; the hide should be free of wrinkles. Pigs with rough skins or roughness over the shoulders are undesirable.

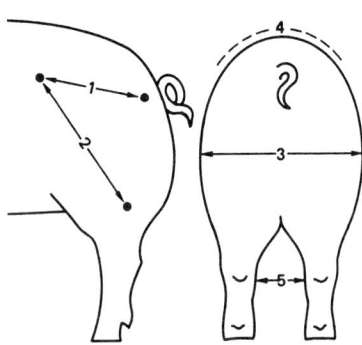

THINGS TO LOOK FOR:
1. LENGTH OF HAM
2. DEPTH OF HAM
3. WIDTH THROUGH CENTER OF HAM
4. CORRECT TURN OVER THE TOP
5. WIDTH BETWEEN HIND LEGS

FIGURE 21-5 Rear and side view of a hog and what to look for in judging. *Courtesy of University of Illinois at Urbana-Champaign.*

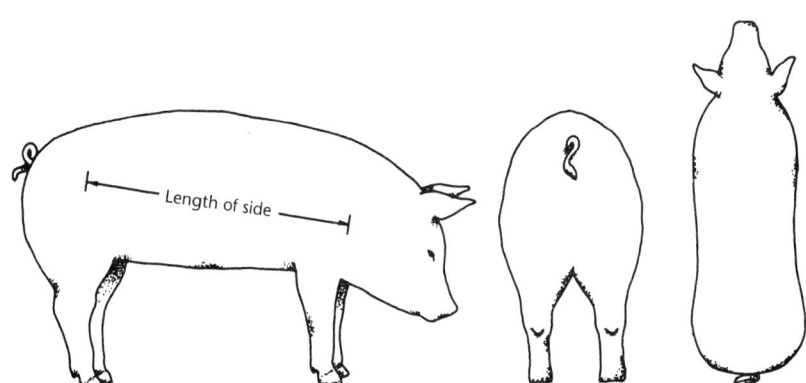

FIGURE 21-4 The length of side gives some indication of type. *Courtesy of University of Illinois at Urbana-Champaign.*

All parts of the hog should be properly proportioned. This is referred to as balance. A well-balanced hog is tight framed and moves well with the correct arch over the back.

Soundness of feet and legs is important. Many hogs are raised on concrete floors and must have good feet and legs to do well under these conditions. Legs must be strong, straight, and set out well on the corners of the hog.

Underline is especially important when evaluating breeding animals. This refers to the mammary development of the hog. There should be a minimum of six nipples per side, none of which are inverted or scarred. Underline is also important in a boar; this trait is passed on to his progeny.

When judging purebred classes, the individuals in the class should show the characteristics of the breed. A good judge must be familiar with breed characteristics. Breed character is often best observed in the head of the animal.

Judging Terms

General terms for market and breeding hogs

Longer, bigger framed

Typier, meatier

Longer and deeper in the ham

Higher quality

Firmer, heavier muscled ham

Squarer rump

Thicker through the rump

Meatier and wider at the loin

More natural thickness down the top

Cleaner, trimmer along the loin edge

Trimmer jowl

Larger skeletal structure

Heavier muscled

Trimmer finished

Smoother

Sounder on front legs

Heavier bone

Nicer turn over the loin

Longer, more correct muscle structure

More correct turn over the top

Trimmer underline

Market hog terms

Longer, larger framed

Heavier muscled

Longer, stretchier side

Firmer finished barrow

Smoother side

More uniform width

Longer rump

Trimmer middled

Typier, meatier barrow

Longer, deeper, fuller in the ham

Cleaner top

More correctly finished

More muscling over the top

Thicker loin

Trimmer in jowl and underline

More uniform arch

Breeding hog terms

Sounder underline	More desirable set to the legs
Stands and walks more correctly	Broodier
More evenly spaced nipples	More breed character
Shows more femininity	Straighter front or hind legs
Wider fronted	More rugged, heavier bone
Meatier gilt	Longer, deeper sided gilt
Roomier-middled gilt	Growthier gilt
Shows more size and scale	Deeper, wider sprung gilt

SUMMARY

Most market hogs produced in the United States today are the result of some type of crossbreeding program. Terminal crosses are most commonly used. Market hogs today are more muscular, produce a larger loin eye area, and have less backfat than those produced 20 years ago.

Performance records are widely used by seedstock producers and market hog producers to evaluate breeding stock. Genetics plays a large role in selection of breeding stock. Expected Progeny Difference is a measure of how much difference in performance on a given trait one can expect from the progeny of a breeding animal as compared to the average of all animals in the population. EPDs can be compared within breeds and across herds but not across breeds. Sire summary and EPD values are available from several sources on the Internet and can be readily accessed by home computer.

In addition to performance records, other criteria to use when selecting breeding stock include soundness of the animal and health of the breeding stock herd.

Feeder pigs must be healthy, meaty hogs to produce the best profits. Select feeder pigs that are healthy and have good size for age. Uniformity within a group is also important

When judging hogs, it is now common practice to include information from performance records combined with visual appraisal to place the class. The best performance records to use for judging hogs are Expected Progeny Differences. Visual appraisal should focus on body conformation, size, muscling, finish, and reproductive soundness. Final placing in a class is made on the basis of both performance data and visual appraisal of the animals.

Student Learning Activities

1. Using a live hog, name the parts of the hog.
2. Judge breeding and market classes of hogs and give oral reasons.

3. Take a field trip to a swine-testing station to observe testing procedures.

4. Practice good selection procedures with a swine project or on the home farm swine herd.

5. Give an oral report to the class on an aspect of swine selection, testing, or judging.

6. Evaluate pedigree information on breeding swine and select the most desirable animals.

7. Attend a purebred swine producers sale and select animals that you would buy, based on type, pedigree, and production records.

8. Observe the use of ultrasonics in swine selection.

Review

1. Describe the use of ultrasound to measure the fat-free lean content of live hogs and carcasses.

2. Name the four primal cuts of the hog carcass.

3. Describe the characteristics of the modern market hog.

4. Describe the goals for market hog characteristics suggested by the National Pork Producers Council.

5. What economic value can be realized by improving genetic lines of breeding stock in the swine industry?

6. What are the most important performance records to be measured in genetic improvement programs?

7. How are Estimated Breeding Values determined?

8. What is Expected Progeny Difference and how is it used in genetic improvement programs?

9. List and briefly describe several performance testing programs currently being used in the swine industry.

10. What is the acid meat gene problem?

11. Name and briefly describe six factors other than performance records that should be considered when selecting breeding animals.

12. What is the minimum breeding age for boars and gilts?

13. How should performance data be used when judging hogs?

14. Name four factors that should be considered when buying feeder pigs.

15. Briefly describe the important visual traits that should be evaluated when judging hogs.

Feeding and Management of Swine

Objectives

After studying this unit, the student should be able to

■ describe different types of swine production.

■ develop feeding programs for the different stages in the life cycle of hogs.

■ describe accepted management practices for the stages in the life cycle of hogs.

The efficient use of resources is the key to profitability in the swine enterprise. To remain competitive, swine producers must select breeding stock that will produce lean hogs and feed efficiently. Using split sex and phase feeding increases feeding efficiency. Improved disease control can be achieved by using an all-in, all-out production cycle.

Two factors that have a major influence on profitability in swine production are the number of pigs weaned per sow per year and feed efficiency. Large corporate farms are currently weaning 21 to 22 pigs per year for each breeding female in their herd. This is the minimum goal that every hog producer needs to strive to achieve. Increasing the number of pigs weaned per litter to more than 9 will help achieve that goal. Females should be bred and managed to produce a minimum of 2.3 litters during each 12 month period. Females that cannot maintain this level of production need to be culled from the breeding herd.

Improved genetics has resulted in hogs that are more efficient in the use of feed. Feed efficiency (that is, the pounds of feed used per hundred pounds of gain) should be in the range of 330 to 350. Feed wastage is a significant factor in determining the amount of feed used. Feeders need to be kept properly adjusted to reduce feed losses. Daily observation is needed to keep feeders properly adjusted.

Table 22-1 shows feed cost per hundredweight of gain based on feeding a sow and litter from farrow to finish. The table assumes a feed conversion of 5.04 bushels of corn and 85 pounds of supplement per hundred pounds of pork produced. These conversion rates are based on recent Illinois Farm Business Record figures and may vary with individual producers. A computer spreadsheet template may be set up using this formula: [(price/bu. corn) × (bushels of corn) + (price/cwt. supplement) × (pounds of supplement)]/100. The formula determines the feed cost per hundredweight of gain at the price intersections of corn and supplement in the table. Different feed cost figures can be obtained by entering different feed conversion figures in the spreadsheet.

TABLE 22-1 Feed Cost per Hundredweight of Gain Based on Feeding a Sow and Litter from Farrow to Finish.

Feed Conversion per CWT of Pork Produced

| Corn (bu.): | 5.04 |
| Protein Supplement (lbs): | 85 |

Price/CWT of Supplement	Price of Corn per Bushel								
	1.50	**1.75**	**2.00**	**2.25**	**2.50**	**2.75**	**3.00**	**3.25**	**3.50**
12.00	17.76	19.02	20.28	21.54	22.80	24.06	25.32	26.58	27.84
12.50	18.19	19.45	20.71	21.97	23.23	24.49	25.75	27.01	28.27
13.00	18.61	19.87	21.13	22.39	23.65	24.91	26.17	27.43	28.69
13.50	19.04	20.30	21.56	22.82	24.08	25.34	26.60	27.86	29.12
14.00	19.46	20.72	21.98	23.24	24.50	25.76	27.02	28.28	29.54
14.50	19.89	21.15	22.41	23.67	24.93	26.19	27.45	28.71	29.97
15.00	20.31	21.57	22.83	24.09	25.35	26.61	27.87	29.13	30.39
15.50	20.74	22.00	23.26	24.52	25.78	27.04	28.30	29.56	30.82
16.00	21.16	22.42	23.68	24.94	26.20	27.46	28.72	29.98	31.24
16.50	21.59	22.85	24.11	25.37	26.63	27.89	29.15	30.41	31.67
17.00	22.01	23.27	24.53	25.79	27.05	28.31	29.57	30.83	32.09
17.50	22.44	23.70	24.96	26.22	27.48	28.74	30.00	31.26	32.52
18.00	22.86	24.12	25.38	26.64	27.90	29.16	30.42	31.68	32.94
18.50	23.29	24.55	25.81	27.07	28.33	29.59	30.85	32.11	33.37
19.00	23.71	24.97	26.23	27.49	28.75	30.01	31.27	32.53	33.79
19.50	24.14	25.40	26.66	27.92	29.18	30.44	31.70	32.96	34.22
20.00	24.56	25.82	27.08	28.34	29.60	30.86	32.12	33.38	34.64
20.50	24.99	26.25	27.51	28.77	30.03	31.29	32.55	33.81	35.07
21.00	25.41	26.67	27.93	29.19	30.45	31.71	32.97	34.23	35.49
21.50	25.84	27.10	28.36	29.62	30.88	32.14	33.40	34.66	35.92
22.00	26.26	27.52	28.78	30.04	31.30	32.56	33.82	35.08	36.34

FIGURE 22-1   One type of swine production is a combination of low-investment housing and pasture. *Photo by Steven M. Ennis.*

TYPES OF SWINE PRODUCTION

Swine production can be divided into two types: purebred and commercial. Commercial production systems can be further divided into three systems: feeder pig production; buying and finishing feeder pigs; and complete sow and litter systems.

The various types of swine production can also be classified according to the kind of housing used. Pasture systems, combinations of pasture and low-investment housing, and high-investment total confinement systems are three types of housing systems used in swine production. The pasture and low-investment housing combination is shown in Figure 22-1.

The characteristics of a pasture management system include:

■ farrowing a smaller number of sows per year.
■ requires enough pasture to be able to rotate pasture use to reduce disease and parasite problems.
■ farrowing only once or twice per year.
■ low investment in buildings.

The characteristics of a confinement management system include:

■ high level of mechanization to reduce labor requirements.
■ high investment in buildings and equipment.
■ multiple farrowings per year with a large number of hogs raised.
■ high level of management ability needed.
■ high degree of control over feeding operation.
■ better year-round working conditions.
■ stringent disease and parasite control program.
■ use of very little high-priced land.

Purebred Production

The production of purebred hogs is a specialized business. Registered purebred hogs make up less than 1 percent of the total hogs raised in the United States. Purebred producers perform an important function in the swine industry by producing the foundation stock used in commercial hog production. Much of the improvement in hog type comes from the work of the purebred producer.

Purebred producers must be excellent managers. They often have higher investments in labor and record keeping than commercial hog producers have. The purebred producer must keep accurate records of the ancestry of the hogs produced. Careful recording of breeding and farrowing dates is essential for the purebred operation. Purebred producers spend a great deal of time advertising, showing, and promoting swine breeds.

In addition to purebred swine producers, seedstock is also produced by large companies that develop lines of breeding stock using genetic principles. Purebred breeders are also using genetic principles to produce improved breeding stock that meets current consumer demand for lean pork. Selection principles and performance testing programs in current use are discussed in Unit 21.

Commercial Production

Most of the pork that is used in the United States is produced by the commercial hog producer. Most commercial hog producers use some type of crossbreeding system to produce hogs for market. Purebred boars are often used on crossbred sows to produce market hogs. Good management is necessary for success in the commercial hog business.

Feeder pig production is an enterprise that produces pigs for sale to feeders, who then feed them to market weights. The feeder pig producer has a breeding herd of sows. The baby pigs are taken care of until they reach weaning weight. A high-producing herd that raises large litters is required. It usually requires an average of 14 to 16 pigs marketed per sow per year to break even in feeder pig production. Health problems must be prevented or carefully treated. The goal of the producer is to raise uniform groups of feeder pigs for sale. Generally, only small investments are required for this type of production. A good manager tries to schedule farrowings so as to have a steady supply of feeder pigs for sale. Less total feed is needed for this type of production system than for the other types.

Buying and finishing feeder pigs is an enterprise in which the operator buys feeder pigs and raises them to market weight. This type of operation requires the least investment and managerial ability of any of the hog production systems. It is possible to feed pigs on pasture or with very limited facilities. However, there is a trend toward investing in more confinement systems. The costs of this type of operation are greater, but hogs gain somewhat more efficiently in confinement systems. The high price of land makes its use as hog pasture questionable.

Buying and finishing feeder pigs does require a high investment to purchase the pigs to be fed. This system is well adapted to the producer who has large amounts of grain for feed. It requires less labor than other systems.

Buying and finishing also has certain disadvantages. The possibilities of health problems are greater in this operation because the purchased feeder pigs may bring diseases to the farm. Market prices of feeder pigs vary a great deal. There is a fairly high risk of not making a profit.

The complete sow and litter system is the most common method of hog production. This operation involves having a breeding herd of sows, farrowing pigs, and caring for and feeding the pigs to market weights. Investment in facilities can be low for pasture systems. Confinement systems, on the other hand, can require very high investments in facilities. The trend is toward more confinement systems, with larger numbers of sows being kept in the producing herd. This system permits spreading the production and, thus, the marketing of pigs, more evenly through the year. This results in an increased potential for profit.

Labor, management, and investment requirements vary considerably with the kind of system used. Pasture systems require more labor, less management, and lower investments. Confinement systems require less labor, more management, and much higher capital investment.

INTEGRATION IN SWINE PRODUCTION

While vertical integration in the poultry industry (see Unit 35) has become commonplace, it has rarely been found in swine production until recently. The practice of contracting with a large company to raise hogs under some type of vertical integration arrangement has become more commonplace in some southern states. Current estimates of the extent of hog contracting in these states range from 20 to 50 percent of their total production. Contracting is now spreading into the Midwest, where most of the pork raised in the United States is still produced.

An alternative to contracting is cooperative ventures between swine enterprises. Cooperative ventures include multipliers, breeding centers, and sow centers. Parent animals for swine producers are provided by multipliers. Breeding centers provide pregnant sows that are returned to the center after the pigs are weaned. Sow centers provide baby pigs to farms where they are fed to market weights.

Under a typical contracting arrangement, the farmer provides the buildings and equipment. The company offering the contract provides the pigs and financing and makes most of the management decisions. These include the rations used, the type of breeding program, the animal health program and veterinary services used, and marketing decisions. Some contractors are offering financing for buildings and training in raising hogs for those not currently in the business. Typically, contracts may be offered for raising feeder pigs or for feeding feeder pigs to market weights.

Under the contracting arrangement, the farmer is generally paid a flat fee per pig raised or marketed. The fee may be adjusted based on the efficiency of the operation as measured by the feed conversion rate.

Contracting reduces the financial investment required by the farmer. However, the farmer loses independence in making management decisions. Control of the swine enterprise is placed in the hands of the company that offers the contract.

Contractors have the advantages of access to large amounts of capital, greater buying and selling power, greater ability to specialize, and the ability to adopt new technology more rapidly.

PORK QUALITY ASSURANCE PROGRAM^SM

In 1989 the National Pork Producers Council implemented the Pork Quality Assurance^SM (PQA) Program. The PQA is a management education program with major emphasis on the swine herd health program. Anyone who raises pork can take part in the program. Producers who are interested in participating in this program do a review of their management practices, especially focusing on the use and handling of animal health products. A series of good management practices is reviewed and a plan is developed for any needed improvements. At the final step in the program, the producer's plans are reviewed and verified by a verifier. This can be a veterinarian, an extension specialist, or an agricultural education instructor. A link to information about this program is found at the National Pork Producers Council home page on the Internet (see Unit 21 for URL).

REDUCING NITROGEN AND PHOSPHORUS EXCRETION

Two problems faced by swine producers are odor and pollution of the environment by excessive amounts of nitrogen and phosphorus in the manure. There is some research that suggests changes in swine diets can help reduce these problems.

Much of the odor problem in confinement operations is related to the release of ammonia from the manure. This is caused by the nitrogen that is in the manure. It is possible to reduce the amount of nitrogen excreted by the pigs by substituting some synthetic lysine for soybean meal in the diet. This lowers the mount of nitrogen excreted and thus reduces the amount of ammonia generated, which in turn reduces the odor from the facility. Use no more than three pounds of synthetic lysine per ton in growing-finishing diets. Replacing too much of the soybean meal with synthetic lysine reduces the amount of other necessary amino acids in the diet and thus has a negative effect on daily gain, feed efficiency, and some carcass traits.

Another method of reducing nitrogen excretion is by using split sex feeding. Barrows have a lower protein requirement than gilts. Feeding the barrows separately from the gilts means they can be fed less protein, thus reducing the amount of nitrogen excreted. Using four to five diets in phase feeding rather than two allows a reduction in the amount of protein fed, and this results in a corresponding reduction in the amount of nitrogen excreted in the manure.

Corn and soybean oil meal, which are the basis of most swine diets, have a fairly high phosphorus content. The problem is that as much as 90 percent of the phosphorus content is in the form of phytic acid, which is not available to the pig. This means that the diet must be supplemented with additional phosphorus, which increases the amount of phosphorus that is released in the manure. This contributes to a problem of excessive amounts of phosphorus being released into the environment when the manure is applied to the land.

Adding the enzyme phytase to the diet results in more of the phytate phosphorus being utilized by the animal. Excretion of phosphorus in the manure has been reduced by as much as 30 percent when the enzyme was added to the diet. Research is also being done on using genetically modified corn that contains less phytate and thus allows the animal to digest more of the phosphorus in the corn. Experimental work with low-phytase corn has shown that its use can result in as much as a 37 percent reduction in the amount of phosphorus excreted in the manure.

Phytic acid binds other minerals such as calcium, zinc, and manganese, reducing their availability to the animal. It appears to also have a negative effect on the digestibility of amino acid. Swine fed the low-phytate corn in their ration developed a larger loin eye muscle, suggesting a better utilization of amino acids in the diet. More research is being conducted to determine the best levels of phytase to be added to the diet to secure optimal performance.

SELECTING FEEDS FOR SWINE

Feed costs range from 55 to 70 percent of the total cost of raising hogs. Combining the right kinds of feed in a well-balanced ration is one of the most important tasks of the hog producer. The nutrient needs of hogs include energy, protein, minerals, vitamins, and water.

Energy Feeds

Corn is the basic energy feed used in hog rations. It is high in digestible carbohydrates, low in fiber, and is palatable. Other feeds that are used as energy sources are compared to corn when determining their feeding value (Table 22-2).

Number two (#2) dent corn contains 8.8 percent protein. However, corn lacks several of the amino acids that are essential for swine nutrition. Lysine and tryptophan are the two amino acids that are not found in corn in large enough amounts. Corn must be supplemented with protein, minerals, and vitamins when fed to hogs. The dry matter in high-moisture corn has the same feeding value as the dry matter in corn at normal moisture levels. High-lysine corn has been developed that may reduce the amount of protein that must be supplemented in swine rations.

TABLE 22-2 The Relative Feeding Value of Various Energy and Protein Feed Sources for Swine.

Feed (as-fed basis)	Relative value % compared to corn	Relative value % compared to soybean meal	Maximum recommended percent of complete diet			
			Gestation	Lactation	Starter	Growing-finishing
Alfalfa meal (dehydrated)	75–85	—	50	10	0	5
Alfalfa meal (sun-cured)	60–70	—	50	10	0	5
Bakery waste	95–110	—	40	40	20	40
Barley	90–100	—	80	80	25	85
Buckwheat	80–90	—	50	0	0	50
Buttermilk, dry	—	75–85	5	5	5	5
Corn (high lysine)	100–105	—	90	90	60	90
Corn (yellow)	100	—	80	80	60	85
Corn gluten meal	—	40–60	5	5	0	5
Corn silage	20–30	—	90	0	0	0
Cottonseed meal (solvent)	—	65–75	5	5	0	5
Fish meal, anchovy	—	140–165	5	5	5	5
Fish meal, menhaden	—	140–165	5	5	5	5
Grain sorghum (milo)	95–100	—	80	80	60	85
Linseed meal	—	55–65	5	5	5	5
Meat & bone meal	—	95–110	10	5	5	5
Molasses	55–65	—	5	5	5	5
Oats	85–95	—	80	10	0	20
Peanut meal, expeller	—	70–80	5	5	0	5
Potatoes	20–25	—	80	0	0	30
Rye	90	—	20	20	0	25
Skim milk, dried	—	95–100	0	0	20	0
Soybean meal, solvent	—	100	25	20	35	22
Soybean meal, solvent, dehulled	—	110–112	22	18	30	20
Soybeans, whole cooked	—	90–100	30	25	40	30
Tankage (meat meal)	—	115–130	10	5	0	5
Triticale	90–95	—	80	80	20	85
Wheat bran	60–65	—	30	5	0	0
Wheat, hard	100–105	—	80	80	60	85
Whey (dried)	135–145	—	5	5	20	5

Whole, shelled corn can be fed free choice. If a protein supplement is to be mixed in the ration, the corn should be ground. Ground ear corn has too much fiber for growing hogs. It can, however, be used in rations for pregnant sows.

Corn coproducts coming from the corn-refining industry may be used in swine feeding. These products have been referred to as by-products in the past. Two products of interest to swine growers are *corn gluten feed* and *corn germ meal.*

Corn gluten feed is not the same product as corn gluten meal. They have different chemical compositions and are produced at different steps in the corn-refining process.

Corn gluten feed has 75 to 80 percent of the energy value of corn grain. It contains 1,120 kilocalories of metabolizable energy per pound and 22 percent crude protein.

Corn gluten feed is available in three forms: wet corn gluten feed; dry, loose corn gluten feed; and dry, pelleted corn gluten feed. The dry, pelleted form is recommended for swine feeding. It is easier to handle and contains more available tryptophan than the other forms.

Dried corn gluten feed can be substituted in the ration for up to 30 percent of the corn in rations for growing-finishing hogs weighing more than 100 pounds (45 kg). Because it is deficient in lysine and tryptophan, corn gluten feed must be properly supplemented with protein. Dried corn gluten feed may substitute for 40 to 50 percent of the corn in gestation rations. A general guideline to follow is that 100 pounds (45 kg) of corn gluten feed is equal to 88 pounds (40 kg) of corn and 12 pounds (5.4 kg) of soybean meal in a swine ration. Corn gluten feed is relatively low in phosphorus; swine rations in which it is used must therefore be supplemented with a minimum of 0.5 percent dicalcium phosphate or 0.1 percent inorganic phosphorus.

Corn germ meal contains 1,360 kilocalories of metabolizable energy per pound and 20 percent crude protein. Experimental work at the University of Illinois indicates that corn germ meal may have limited use in starter diets of young pigs. When fed as the total diet, it proved to be too bulky for sufficient feed intake. No research has been reported on the use of corn germ meal in diets for older pigs. There is relatively little corn germ meal available for feeding because it is usually blended back into corn gluten feed in the corn-refining industry.

Wet corn gluten feed contains 40 percent dry matter and is similar to corn in protein content. It is low in available amino acids, especially lysine and tryptophan. Diets using wet corn gluten feed need to have their amino acid content balanced. Wet corn gluten feed contains only 32 percent of the metabolizable energy of corn.

The use of wet corn gluten feed in swine feeding should be limited to pregnant sows. It is not recommended for growing-finishing pigs. Wet corn gluten feed must be properly balanced for calcium, trace minerals, and vitamins in sow diets. Additional amino acids and energy must also be provided in the diet. For example, a diet of 8 pounds (3.6 kg) of wet corn gluten feed per day needs to be supplemented with at least 1 pound (0.45 kg) of fortified protein and 0.5 pounds (0.22 kg) of corn per day. With a daily consumption of 6 pounds of wet corn gluten feed, sows should also receive 1.25 pounds (0.56 kg) of protein supplement and 1 pound (0.45 kg) of corn per day.

There is a danger of mycotoxins in wet corn gluten feed because the milling process does not remove molds from the corn. It should therefore be fed fresh, because molds will grow easily in the feed.

Barley is a good substitute for corn. In some parts of the United States, more barley than corn is fed to hogs. Barley has a higher fiber content than corn and, therefore, slightly less digestible energy. The protein content of barley is higher than corn. However, like corn, it lacks some of the amino acids required by hogs. Barley must be supplemented with protein, minerals, and vitamins for hog feeding.

Barley is ground medium fine for hog feeding. It may also be rolled or pelleted for hog rations. Barley is not quite as palatable as corn and should be mixed with the protein supplement in the ration. Barley has a relative feeding value of 90 to 100 percent compared to corn. When up to one-third of the corn in the ration is replaced with barley, the hogs will gain as fast as they would on an all-corn ration. If barley is substituted 100 percent for corn, the hogs will show a slightly slower gain.

Barley is sometimes infested with *scab*, a disease that attacks barley. This may make it poisonous to hogs. Scabby barley should never be fed to hogs.

Buckwheat has 80 to 90 percent of the feeding value of corn in swine rations. It contains about 11 percent crude fiber and is not as palatable as corn. Generally, it should be mixed with other grains when fed to hogs. It has 0.65 percent lysine, which is higher than the lysine content of corn. Less protein supplement may be needed when buckwheat is included in the ration. Buckwheat is not recommended for feeding lactating sows or small pigs. It can be used for gestating sows and in growing-finishing rations. While it can be used for up to 100 percent of the corn in growing-finishing rations, this level of feeding will result in a five to ten percent drop in feed efficiency and growth rate. It is recommended that it not be used for more than 50 percent of the ration. Buckwheat contains fagopyrin, which is a photosensitizing agent. This can cause rashes and itching (buckwheat poisoning) when white pigs are exposed to sunlight.

Milo (grain sorghum) is grown widely in the southwestern part of the United States. Milo has a higher protein content than corn. It can replace all of the corn in hog rations. Milo must be supplemented with protein, minerals, and vitamins in hog rations. It has a relative feeding value of 90 to 95 percent compared to corn. Milo should be ground for feeding. Some varieties are unpalatable to hogs and should be mixed with the protein for feeding.

Wheat is equal to or slightly higher in feeding value than corn. Because of its higher protein, lysine, and phosphorus content, its relative feeding value ranges from 100 to 105 percent compared to corn. The energy value of wheat is slightly lower than corn. The relative price of wheat compared to other grains is a determining factor when considering its use in swine rations. Wheat must be coarsely ground when fed to livestock. If ground too fine, it forms a pasty mass in the animal's mouth and reduces feed intake. Feeding problems are reduced when wheat is processed through a roller mill rather than a hammermill.

Oats have a higher protein content than corn, but the quality of the protein is poor. A protein supplement must be used when oats are fed in a hog ration. Oats also have a high fiber content. They have a relative feeding value of 85 to 90 percent compared to corn. Oats should not be substituted for more than 20 percent of the corn in the ration for growing-finishing hogs. If a greater amount of oats is used in the ration, the rate of gain will be slower.

Oats for hog rations should be medium to finely ground. Hulled, rolled oats are an excellent feed for starter rations for baby pigs.

Rye is not a very good feed for hogs. Rye has a relative feeding value of 90 percent compared to corn and is less palatable than other grains. Not more than 25 percent of the grain in the ration should be rye. Rye is harder than corn and should be ground when fed to hogs.

Rye is sometimes infested with a fungus called *ergot*. Ergot will cause abortion in pregnant sows. Rye containing ergot should never be fed to sows. Ergot-infested rye will slow down gains in growing-finishing hogs.

Triticale is a hybrid cereal grain that is the result of a cross between wheat and rye. Triticale has more lysine than corn, but is not as palatable as corn in hog rations. No more than 50 percent of the ration should be made up of triticale. Some varieties of triticale may be infested with ergot. Ergot-infested triticale should not be fed to pregnant sows.

Potatoes may be fed to hogs. They contain mainly carbohydrates, and therefore must be fed with a protein supplement. Heavier hogs make better use of potatoes than younger hogs. It takes about 400 pounds (181 kg) of potatoes to equal the feed value of 100 pounds (45 kg) of corn. They should be fed at the rate of one part potatoes to three parts grain. Potatoes should be cooked before they are fed to hogs.

Bakery wastes may be fed as part of the hog ration. Bakery wastes include stale bread, bread crumbs, cookies, and crackers. The average protein content of these foodstuffs is about 10 percent. Little is known about the amino acid content. A good protein supplement must be fed when bakery wastes are used.

Fats, tallow, and greases provide a high-energy source in hog rations. These substances usually make up less than 5 percent of the ration, depending on the price of fat. They are used to improve the binding qualities of pelleted feed. *Binding quality* refers to how well the feed particles stick together in the pellet. However, carcass quality is decreased if too much fat, tallow, or grease is added to the ration. These substances contain no protein, minerals, or vitamins. Proper nutrient supplements are essential when these substances are part of the ration.

Molasses provides carbohydrate in the ration. It can be substituted for part of the grain. However, molasses should not be more than 5 percent of the ration. Scours may result from overfeeding molasses.

Plant Proteins

Soybean oil meal is available with a 44 percent or 49 percent protein content. The 49-percent protein soybean oil meal is often used in prestarter and starter rations. The two are about equal in value for growing-finishing hogs. The protein quality in soybean oil meal is excellent. Soybean oil meal is the most widely used protein source in hog rations. The amino acid balance is generally good. Soybean oil meal is very palatable to hogs. In fact, they will overeat this substance if it is fed free choice. It must be fed in a completely mixed ration or be mixed with less palatable proteins to prevent overeating. Soybean oil meal may be fed with corn as the only protein supplement. Minerals and vitamins must be added to the ration when soybean oil is not used as the protein source. Other feeds that are used as protein sources are compared to soybean meal when determining their feeding value (see Table 22-2).

Cottonseed meal is 40 to 45 percent protein. However, the protein quality of this material is poor. It is low in lysine and must be fed with other protein sources in hog rations. Cottonseed meal may be fed as 5 percent of the protein in the hog ration. Some cottonseed meal contains gossypol, which is toxic to hogs. If the gossypol has been removed, it may replace up to 50 percent of the soybean oil meal in the ration. Cottonseed meal is low in minerals and fair in vitamin B content. Hogs do not find it very palatable. Do not use cottonseed meal in starter rations.

Linseed meal is 35 to 36 percent protein. Because the protein is of poor quality, linseed meal must be fed with other protein sources in the hog ration. It usually makes up no more than 5 percent of the protein in the ration. Linseed meal contains more calcium than cottonseed meal or soybean oil meal. It has about the same amount of vitamin B as these protein sources. Linseed meal is best fed in combination with animal protein sources. It acts as a laxative in large amounts.

Peanut meal is 47 percent protein. Because the protein is low in several amino acids, peanut meal must be fed with other protein sources. It becomes rancid if stored for more than a few weeks and is low in vitamins and minerals.

Whole soybeans contain about 37 percent protein and can be used to replace soybean oil meal in hog rations. They are higher in energy but lower in protein than the meal. Use six pounds (2.7 kg) of whole cooked soybeans to substitute for five pounds (2.3 kg) of soybean oil meal. The higher energy content of the whole soybeans may increase feed efficiency by about five percent. Do not use raw soybeans in swine growing-finishing rations; they contain an antitrypsin factor that prevents the action of the enzyme trypsin in nonruminants such as swine and poultry. This reduces the availability of tryptophan, an essential amino acid. Heating the soybeans destroys the antitrypsin factor.

Animal Proteins

Tankage and meat scraps contain 50 to 60 percent protein. They have inadequate amounts of the amino acid tryptophan and, therefore, must be used with other protein sources in hog rations. The calcium and phosphorus content of tankage and meat scraps is high. The vitamin content is variable. The B vitamin pantothenic acid is sometimes low. Tankage is not as palatable to hogs as soybean meal. The maximum percent of tankage that should be included in various kinds of complete rations is (1) 10 percent for gestation rations, (2) 5 percent for lactation, grower, and finishing rations, and (3) 0 percent for starter rations.

Meat and bone meal contains 50 percent protein. The amount of bone contained in the mix determines the feeding value. Meat and bone meal is low in lysine when compared with other protein sources. The maximum percent of meat and bone meal that should be included in various kinds of complete rations is (1) 10 percent for gestation rations, and (2) 5 percent for lactation, starter, grower, and finishing rations.

Fish meal is 60 to 70 percent protein. The protein is of excellent quality. Fish meal also is high in minerals and vitamins and very palatable to hogs. It is a good protein source, but usually too expensive to use except in creep rations. The maximum percent of fish meal that should be included in various kinds of complete rations is 5 percent.

Skim milk and buttermilk contain about 33 percent protein when dried. In liquid form, they are worth only about one-tenth as much as dried milk, since the liquid form contains about 90 percent water. The protein quality of skim milk and buttermilk is good, and they are good sources of B vitamins. These milk products are often used in creep rations in dried form. Young pigs cannot consume enough in liquid form to meet their protein needs. The maximum amount of dried skim milk that should be included in starter rations is 20 percent. Dried skim milk should not be included in gestation, lactation, grower, or finishing rations.

Whey, in liquid form, contains only about 1 percent protein. Dried whey contains 13 to 14 percent protein. The protein in whey is of excellent quality. However, liquid whey has a high water content, and hogs cannot consume enough to meet their protein needs. A starter ration may contain up to 20 percent dried whey. Limit gestation, lactation, grower, and finishing rations to no more than 5 percent dried whey. There is some recent research indicating that grower and finishing rations may contain more than 5 percent dried whey without reducing rate of gain or feed efficiency.

Roughages

Alfalfa meal is 13 to 17 percent protein. It has large amounts of vitamins A and B, and is an excellent roughage for hogs. It is also a good source of minerals. Alfalfa meal should be limited to no more than 5 percent of the ration for growing-finishing hogs. It may make up as much as 50 percent of the ration for brood sows. It helps to prevent them from becoming too fat. A maximum of 10 percent of the lactation ration may be alfalfa meal. Do not use alfalfa meal in starter rations.

Alfalfa hay and other hays are generally not used in swine rations except for feeding the breeding herd. Hay must be ground and mixed in the ration for self-feeding sows and gilts. It can be used to make up as much as one-third of the ration.

Silage is most valuable in rations for the breeding herd. Ten to 12 pounds (4.5–5.4 kg) of corn or grass-legume silage can be fed per day to sows and gilts during pregnancy. This must be supplemented with protein and minerals. Moldy silage should never be fed to sows and gilts. (Generally, moldy feed should not be fed to any animal.)

Pasture is also most valuable for feeding the breeding herd. Good-quality pasture supplies the same nutrients as alfalfa meal and hay. Growing-finishing hogs on pasture will not gain as rapidly as those fed in drylot. However, pregnant sows and gilts are given the exercise they need by being fed on pasture. A good-quality pasture supplies enough nutrients so that the amount of concentrates in the ration for the breeding herd may be reduced by up to 40 percent. A balanced ration must be fed when the sows and gilts are on pasture. Pasture is also of value for feeding the herd boar.

Minerals

Four major minerals and six trace minerals are frequently added to hog rations. The major minerals are calcium, phosphorus, sodium, and chlorine. The trace minerals are zinc, iron, copper, selenium, manganese, and iodine.

Salt adds sodium and chlorine to the ration. About 0.5 percent of the ration should consist of salt. The most common calcium source is ground limestone. The ration should contain 0.5 to 0.7 percent calcium. The use of dicalcium phosphate supplies both calcium and phosphorus in the ration. The ration should contain 0.4 to 0.65 percent phosphorus. Other sources of calcium and phosphorus include steamed bone meal and defluorinated rock phosphate.

Feeding too much calcium or phosphorus may reduce the rate of gain for growing-finishing hogs. An excess of calcium will interact with zinc to cause a zinc deficiency. The ratio of calcium to phosphorus in swine diets should be 1.0 to 1.5 calcium to 1.0 total phosphorus in a grain-soybean meal diet.

Trace minerals are often found in commercial protein supplement mixes. Trace-mineralized salt is another source of trace minerals. Special trace mineral premixes are also available which can be added to the ration (Table 22-3).

Iron and copper, which help to prevent anemia, are especially important in baby pig rations. In addition to the iron supplied in the ration, baby pigs should always be given iron shots when they are two to four days old. Zinc is required to prevent parakeratosis. (See Unit 23 for a description of this disease.) Early weaned pigs have a higher zinc requirement than older pigs.

Care must be taken when feeding minerals. Excess minerals in hog rations slow the rate of gain. Minerals should not be added to a ration containing commercial protein supplements unless the feed tag indicates that they are required. A mineral mix can be fed free choice since hogs will not overeat minerals if they are receiving enough minerals in the ration.

Vitamins

Many of the vitamins required by hogs are already present in the feeds used in hog rations. Vitamins that must be added to the ration are: A, D, E, K, riboflavin, niacin, pantothenic acid, choline, and vitamin B_{12}.

Vitamins may be added to the ration as part of complete protein supplements, in mineral-vitamin premixes, or as vitamin premixes. The major differences among these sources is the amount of vitamins they contain and their cost. It is difficult to determine the exact

TABLE 22-3 Suggested Vitamin-Trace Mineral Mix.[1]

Nutrient	Amount per pound of premix[2]	Amount per kilogram of premix	Suggested source
Vitamin A	900,000 IU	1,980,414 IU	Vitamin A palmitate-gelatin coated
Vitamin D	100,000 IU	220,046 IU	Vitamin D3—stabilized
Vitamin E	5,000 IU	11,002 IU	d1-tocopheryl acetate
Vitamin K (Menadione Equivalent)	660 mgs	1,452 mgs	Menadione sodium bisulfite
Riboflavin	1200 mgs	2,641 mgs	Riboflavin
Pantothenic acid	4500 mgs	9,902 mgs	Calcium pantothenate
Niacin	7000 mgs	15,403 mgs	Nicotinamide
Choline chloride	20000 mgs	44,009 mgs	Choline chloride (60%)
Vitamin B_{12}	5 mgs	11 mgs	Vitamin B_{12} in mannitol, (0.1%)
Folic acid	300 mgs	660 mgs	Folic acid
Biotin	40 mgs	88 mgs	D-Biotin
Copper	0.4%	0.4%	$CuSo_4 \cdot 5H_2O$
Iodine	0.008%	0.008%	KIO_4
Iron	4.0%	4.0%	$FeSO_4 \cdot 2H_2O$
Manganese	0.8%	0.8%	$MnSO_4 \cdot H_2O$
Zinc	4.0%	4.0%	ZnO (80% Zn)
Selenium	0.012%	0.012%	$NaSeO_3$ or $NaSeO_4$

[1]Vitamin and trace mineral mixes may be purchased separately. This is advisable if a combination vitamin-trace mineral premix is to be stored longer than three months. Vitamins may lose their potency in the presence of trace minerals if stored for a prolonged period.
[2]Premix is designed to be used at a rate of 5 lb per ton of complete feed for sows and baby pigs and 3 lb per ton of complete feed for growing-finishing swine.

Source: Luce, William G., Hollis, Gilbert R., Mahan, Donald C., and Miller, Elwyn R., *Swine Diets, Pork Industry Handbook,* Cooperative Extension Service, University of Illinois.

amount of vitamins they contain since the feed tags usually do not list the amounts. Past experience with a particular mix is the best guide to follow in selecting a vitamin source.

Complete supplements and mineral-vitamin premixes usually cost more than vitamin premixes. However, if the producer does not have mixing equipment on the farm, it may be best to use complete mixes. Premixes are used in such small amounts per ton that it is difficult to mix them into the ration properly.

Water

Water is one of the most important nutrients in hog rations. Hogs should have plenty of water available at all times (Figure 22-2). The water should be fresh, clean, and no colder than 45°F (7°C). It should be checked periodically for nitrate content. Too much nitrate or nitrite in the water is not good for hogs.

FIGURE 22-2 An automatic watering system provides hogs with fresh, clean water at all times. *Photo by Jim Bodine.*

The estimated daily consumption of water by various classes of hogs that are not undergoing stress is as follows:

Description	Gallons/day
Pigs	
25 lbs	0.5
60 lbs	1.5
100 lbs	1.75
200 lbs	2.5
300 lbs	3.5
Gestating sows	4.5
Sow plus litter	6.0
Nonpregnant gilts	3.2
Pregnant gilts	5.5

Additives

Feed additives increase efficiency in hog production. Additives enable pigs to grow at a faster rate, improve feed conversion, and reduce disease stress. The additives most commonly found in hog rations are anthelmintics, antibiotics, arsenicals, nitrofurans, and sulfa compounds.

Sources of additives include complete protein supplements, complete mixed feeds, and premixes. Premixes must be carefully mixed into the ration for even distribution. Factors to consider when evaluating additive sources include cost, which additives are to be included, and the amounts of additives in the source.

Feed tag instructions must be carefully followed in the use of additives. Withdrawal times must be observed when marketing hogs. These withdrawal times are always listed on the feed tag. Federal regulations govern the required withdrawal times. Penalties are enforced when these regulations are not followed.

FEEDING THE BREEDING HERD

Separate replacement gilts from market hogs when they weigh about 150–200 pounds (68–90 kg). They may be fed a sow diet as shown in Table 22-4. Do not allow gilts to become too fat; they should gain about one pound (0.45 kg) per day before breeding. Breed gilts at about seven to eight months of age when they weigh about 250 to 300 pounds (113 to 136 kg). Boars may be fed the rations shown in Table 22-4; follow the same general rules for weight gain as outlined above for gilts. Pasture may be used in the feeding program for both gilts and boars.

Flushing means to increase the amount of feed fed for a short period of time. A gilt may farrow a bigger litter, depending on her condition, if this practice is followed. When flushing, increase the ration to six to eight pounds (2.7–3.6 kg) about ten days before breeding.

TABLE 22-4 Some Representative Gestation and Lactation Diets for Sows Using Several Different Grain Sources. These Diets May also Be Used for Replacement Gilts.

	Diet number													
	1		**2**		**3**		**4**		**5**		**6**		**7**	
Ingredient	(lb)	(kg)	(lb)	(kg)	(lb)	(kg)	(lb)	(kg)	(lb)	(kg)	(lb)	(kg)	(lb)	(kg)
Corn, yellow	1,627	738	1,253	568	1,302	591	—	—	—	—	—	—	—	—
Barley	—	—	—	—	—	—	—	—	—	—	1,759	798	—	—
Oats	—	—	400	181	—	—	—	—	—	—	—	—	—	—
Sorghum grain	—	—	—	—	—	—	1,617	733	1,469	666	—	—	—	—
Wheat, hard winter	—	—	—	—	—	—	—	—	—	—	—	—	1,565	710
Wheat middlings	—	—	—	—	400	181	—	—	—	—	—	—	—	—
Soybean meal, 44%	295	134	270	122	225	102	306	139	260	118	128	58	165	75
Meat and bone meal, 50%	—	—	—	—	—	—	—	—	—	—	60	27	—	—
Dehydrated alfalfa meal, 17%	—	—	—	—	—	—	—	—	200	91	—	—	200	91
Calcium carbonate	19	9	19	9	25	11	20	9	13	6	16	7	12	5
Dicalcium phosphate	44	20	43	20	33	15	42	19	43	20	22	10	43	20
Salt	10	5	10	5	10	5	10	5	10	5	10	5	10	5
Vitamin-trace mineral mix*	5	2	5	2	5	2	5	2	5	2	5	2	5	2
Total	2,000	907	2,000	907	2,000	907	2,000	907	2,000	907	2,000	907	2,000	907

Calculated Analysis

Protein, %	13.40	13.60	13.70	13.90	13.90	14.40	14.90
Lysine, %	0.62	0.62	0.62	0.62	0.62	0.62	0.62
Tryptophan, %	0.17	0.17	0.17	0.17	0.18	0.18	0.22
Threonine, %	0.51	0.51	0.51	0.48	0.49	0.44	0.50
Methionine + cystine, %	0.54	0.48	0.48	0.42	0.42	0.47	0.54
Calcium, %	0.91	0.90	0.91	0.90	0.91	0.90	0.90
Phosphorus, %	0.70	0.70	0.70	0.70	0.70	0.70	0.71
Metabolizable energy, kcal/lb (kcal/kg)	1,476 (3,254)	1,416 (3,122)	1,441 (3,177)	1,419 (3,128)	1,354 (2,985)	1,338 (2,950)	1,352 (2,981)

*See Table 22-3. It is also recommended that during the gestation period, additional choline (550 grams per ton) be added to the diets. This can be provided by adding 2.5 lb of choline chloride premix containing 50% choline or 2.0 lb of a chloride premix containing 60% choline.

Source: Luce, William G., Hollis, Gilbert R., Mahan, Donald C., and Miller, Elwyn R., *Swine Diets, Pork Industry Handbook,* Cooperative Extension Service, University of Illinois.

Sows that have been on restricted rations before breeding should also be flushed. Gilts and sows should be put back on limited feeding immediately after breeding so they do not get too fat. Limited feeding also helps to reduce fetal death during gestation.

Gilts and sows should not be allowed to become too fat during the gestation period. A gain of 50 to 75 pounds (22–34 kg) is about right for sows. Gilts should gain 70 to 100 pounds (32–45 kg) during gestation. Four to five pounds (1.8–2.3 kg) of feed per day may be hand fed during the first two-thirds of gestation. Increase this to six pounds (2.7 kg) during the last one-third of the gestation period. Rations shown in Table 22-4 may be used for sows and gilts.

Sows and gilts may be fed on pasture during gestation. Alfalfa, ladino, and red clover are good legume pastures to use. Legume pasture may increase estrogen activity, which could impair reproduction. Orchard grass and Kentucky bluegrass are good nonlegumes for hog pastures. Rations shown in Table 22-4 should be fed at the rate of two to three pounds (0.9–1.4 kg) per day on good-quality legume pasture.

Self-feeding may be used during the gestation period if a bulky ration is used. Adding a good-quality ground alfalfa hay to the ration provides the needed bulk. Ground ear corn and oats also add bulk to the ration. Sows and gilts may also be self-fed a high-energy ration. Access to the self-feeder must be limited if this practice is followed. The sows and gilts are prevented from using the self-feeder for two out of three days. Be sure adequate feeder space is available. The hogs do as well as if they were hand fed four pounds (1.8 kg) of a high-energy ration every day.

Corn silage or grass silage may be used in the gestation ration. Silage substitutes for pasture or alfalfa meal in the ration. Feed only as much as the sows or gilts will completely finish eating in two to three hours. For sows, this is about 10 to 15 pounds (4.5–6.8 kg). Gilts will eat about 8 to 12 pounds (3.6–5.4 kg). Silage must be supplemented with 1.0 to 1.5 pounds (0.45–0.68 kg) of protein. Never feed moldy silage (or any other moldy feed); it may cause abortion.

The ration at farrowing time should be bulky. Add wheat bran or ground oats two or three days before farrowing for extra bulk in the ration. At farrowing, about one-third of the ration may be made up of these bulky feeds. Ground legume hay can also be used to add bulk. Adding bulk may help prevent constipation and reduce problems with mastitis-metritis-agalactia (MMA) at farrowing time. Provide a generous supply of fresh water at farrowing time.

During the first few days after farrowing, rations may be limit-fed. Gradually increase the ration until the sow or gilt is on full feed at five to seven days. Some examples of lactation rations are shown in Table 22-4. Lactating sows and gilts will eat from 2.5 to 3 pounds (1.1–1.4 kg) of feed per day per 100 pounds (45 kg) of body weight. The amount of feed fed may also be varied according to the number of pigs being nursed. A rule of thumb is one pound (0.45 kg) of feed for each pig nursed, plus three pounds (1.4 kg) per sow per day of a 15-percent protein ration.

Feed intake may be reduced by as much as 25 percent when the temperature in the farrowing house reaches 80°F (26.6°C) as compared to feed intake at 60°F (15.5°C). Keeping sows cool will help maintain proper feed intake. Adding 10 percent supplemental fat to the diet during hot weather will help maintain the energy level of the diet on reduced feed intake. Wet feeding can also help maintain feed intake during hot weather.

Diets that are high in energy generally reduce feed intake by 5 to 11 percent compared to lower-energy diets. However, the intake of energy is about the same. High-energy diets must be properly supplemented to assure adequate levels of other nutrients when the total feed intake is lower. Feed intake increases with high-fiber diets because of the lower amount of energy. Increasing the protein content of the lactation ration will increase feed intake about 1.1 pound (0.5 kg) for each one-percent increase in protein content. Lactation and gestation rations should contain 0.8 percent calcium to 0.6 percent phosphorous. Feed intake is reduced about 9 percent for each variation of 0.1 percent (calcium) and 0.05 percent (phosphorus) above or below these levels. High levels (above 50 ppb aflatoxin or 4 ppm vomitoxin) of mold toxins also reduce feed intake. Reducing the dust content of feeds by pelleting, adding fat, or using a wet mash will increase feed intake up to 15 percent.

Raw soybeans may be fed to both pregnant and lactating sows. The raw soybeans can replace soybean meal on an equal-weight basis. The crude protein content of the ration is lower, but the lysine content is adequate to meet the requirements of the sows. The advantages of feeding raw soybeans to sows during gestation and lactation include (1) lower feed cost, (2) no processing cost for the beans, and (3) the fat content of the diet is increased without the problem of adding liquid fat to the ration. Raw soybeans have an oil content of 18 percent. Research has shown that feeding raw soybeans to sows does not affect the number of live pigs farrowed or weaned. Pig birth weights are increased but weaning weight is not affected. Feed intake during lactation is not affected and the milk of the sows has a higher fat content.

The herd boar should be kept on limited rations during the breeding season so he will not become too fat. Fat boars are sluggish and make poor breeders. Young boars should be fed enough for moderate weight gains. Feed five to five and one-half pounds (2.3–2.5 kg) per day of a 14 percent crude protein ration to young boars. Mature boars should be fed five to six pounds (2.3–2.7 kg) of a 14 percent crude protein ration per day. During the breeding season feed six to eight pounds (2.7–3.6 kg) of a 16 percent crude protein ration per day. In cold weather provide an additional three and one-half to seven ounces (99–198 g) per day for each degree below 68°F (20°C). Boars can be maintained on four pounds (1.8 kg) of feed when not in use. The same gestation diets recommended for sows may be used for boars (see Table 22-4).

FEEDING BABY PIGS

About one-fourth of the pigs lost before weaning are lost because of poor feeding. Make sure each pig nurses shortly after it is born. Pigs receive disease protection from the colostrum (first) milk of the sow. The sow reaches maximum milk production in three to four weeks. After that, milk production falls off. Pigs must be eating well by then in order to get the nutrients they need.

Baby pigs will start to nibble on a creep feed within a week after they are born, if it is available. Small amounts can be provided in pans and replaced with fresh feed each day. Commercial creep feeds are available. Unless the producer has good mixing equipment, it is probably best to use a commercial creep feed. Baby pigs eat creep feeds better if they are sweetened. Use creep feeds that have the sugar mixed in the pellet rather than sugar-coated pellets. Creep and starter rations are shown in Table 22-5. Be sure that baby pigs are given plenty of clean, fresh water.

TABLE 22-5 Some Representative Baby Pig Diets That May Be Used as Either Creep or Starter Diets. If Postweaning Scours Is a Problem, Substitute 200–400 lb (91–181 kg) of Ground Oats for Corn or Grain Sorghum in Diets 4, 5, or 7 for the First 2–3 Weeks after Weaning.

	Diet number													
	Pigs 10–25 lb. (4.5–11.3 kg)						Pigs 25–40 lb. (11.3–18.1 kg)							
	1		2		3		4		5		6		7	
Ingredient	(lb)	(kg)	(lb)	(kg)	(lb)	(kg)	(lb)	(kg)	(lb)	(kg)	(lb)	(kg)	(lb)	(kg)
Corn, yellow	990	449	1,060	481	1,211	549	1,396	633	625	283	1,159	526	1,279	580
Sorghum grain	—	—	—	—	—	—	—	—	625	283	—	—	—	—
Ground oats	—	—	—	—	—	—	—	—	—	—	200	91	—	—
Soybean meal, 44%	421	191	530	240	390	177	543	246	495	225	530	240	410	186
Fish meal, menhaden	100	45	—	—	100	45	—	—	—	—	—	—	—	—
Dried whey	400	181	—	—	200	91	—	—	200	91	—	—	200	91
Dried skim milk	—	—	200	91	—	—	—	—	—	—	—	—	—	—
Sugar[1]	—	—	100	45	—	—	—	—	—	—	—	—	—	—
Fat	50	23	50	23	50	23	—	—	—	—	50	23	50	23
Lysine, 78% L-Lysine	—	—	—	—	3	1	—	—	—	—	—	—	3	1
Calcium carbonate	7	3	15	7	10	5	15	7	13	6	15	7	13	6
Dicalcium phosphate	20	9	33	15	24	11	34	15	30	14	34	15	33	15
Salt	7	3	7	3	7	3	7	3	7	3	7	3	7	3
Vitamin-trace mineral mix[2]	5	2	5	2	5	2	5	2	5	2	5	2	5	2
Total	2,000	907	2,000	907	2,000	907	2,000	907	2,000	907	2,000	907	2,000	907

Calculated Analysis

Protein, %	19.10		19.50		18.30		17.90		17.60		17.80		15.80	
Lysine, %	1.15		1.15		1.15		0.95		0.95		0.95		0.95	
Tryptophan, %	0.24		0.26		0.21		0.23		0.23		0.23		0.20	
Threonine, %	0.80		0.78		0.76		0.68		0.69		0.67		0.63	
Methionine + cystine, %	0.66		0.64		0.64		0.60		0.56		0.58		0.55	
Calcium, %	0.85		0.86		0.85		0.75		0.75		0.76		0.77	
Phosphorus, %	0.71		0.70		0.70		0.65		0.66		0.65		0.66	
Metabolizable energy, kcal/lb (kcal/kg)	1,516 (3,342)		1,529 (3,371)		1,500 (3,307)		1,478 (3,258)		1,449 (3,194)		1,495 (3,296)		1,520 (3,351)	

[1]Dextrose or hydrolyzed corn starch product.
[2]See Table 22-3.

Source: Luce, William G., Hollis, Gilbert R., Mahan, Donald C., and Miller, Elwyn R., *Swine Diets, Pork Industry Handbook,* Cooperative Extension Service, University of Illinois.

FEEDING GROWING-FINISHING PIGS

Scours (diarrhea) is sometimes a problem during the first two or three weeks after pigs are weaned. This is especially true with early weaned pigs. Replacing 10 to 15 percent of the corn in the ration with ground oats for two or three weeks may help prevent scouring.

When pigs are weaned early, they sometimes lose weight or do not gain weight for the first week or two. The digestive tract of the young pig is immature and only partially functioning. The stomach produces low levels of acid, and the pH level in the stomach and digestive tract is therefore higher (more alkaline). Enzyme activity and utilization of nutrients is more efficient at slightly lower pH levels. Also, bacteria do not grow as readily when the pH level is lower. Research has shown that adding a small amount of fumaric acid to the diet may increase average daily gain and feed efficiency by about 8 percent.

Research has also shown that adding dried whey to the diet for the first two weeks after weaning will help maintain weight gain. A feed containing 20 percent dried whey with soybean meal is recommended. The protein quality of dried whey varies; care must therefore be taken to ensure proper protein content if it is to be included in the diet.

During the first two weeks after weaning, baby pigs have difficulty digesting soybean meal. Processed soybean protein products such as soy protein concentrate, soy flour, and isolated soy protein are more easily digested and do not cause as much intestinal damage in young pigs as does soybean meal. These products are very expensive and are recommended for use only during the first two weeks after weaning.

The use of spray dried porcine plasma (SDPP) during the first two weeks after weaning will increase feed intake and rate of gain in young pigs. Spray dried porcine plasma is also called plasma protein. It is a by-product of blood from pork slaughter plants. SDPP can replace some or all of the dried skim milk in the diet. The maximum recommended level of use is eight to ten percent of the diet. It may be necessary to add more methionine and lactose to the diet when SDPP is used.

Using antibiotics in the starter diet will generally improve feed efficiency by five to ten percent and growth rate by ten to twenty percent. Adding copper sulfate in combination with an antibiotic improves performance more than using either alone. The use of other feed additives does not generally result in significant improvement in rate of gain or feed efficiency for young pigs.

A *phase feeding* program is recommended when pigs are weaned at three weeks of age. Phase feeding is designed to meet the rapidly changing nutritional needs of pigs during the first weeks after early weaning. It helps reduce postweaning growth lag and gets pigs started on a grain and soybean meal diet more quickly.

The Phase I feeding period lasts for seven to ten days for pigs weaned at three weeks of age and three or four days for pigs weaned at four weeks of age. A pelleted diet containing 20 to 22 percent crude protein and 1.45 percent total lysine is used. The diet also contains four to five percent plasma protein, 20 percent food-grade whey, 10 percent food-grade dried skim milk, four to six percent cheese by-product, two to three percent egg protein, and four to six percent soy oil.

The Phase II feeding period follows the Phase I period and lasts for one to two weeks. The diet may be pelleted or meal and contains 18 to 22 percent crude protein and 1.35 percent total lysine. The diet also contains 10 to 15 percent food-grade whey, two and one-half to five percent menhaden fish meal, two to three percent blood meal, and a maximum of eight percent soybean meal.

The Phase III feeding period should start when the pigs weigh about 25 pounds (11 kg) and three to five weeks after they are weaned. A grain-soybean meal diet in pelleted or meal form with 18 to 20 percent crude protein and 1.1 percent total lysine is used. Five to ten percent whey and four to five percent fish meal may be added to the diet. The Phase III feeding period continues until the pigs reach 45 pounds (20 kg).

The nutritional requirements of younger pigs are greater than those of older pigs. Protein content in the ration can be reduced as pigs get older. When pigs are fed on good legume pasture, the protein content of the ration can be reduced about 2 percent. Tables 22-6 and 22-7 show rations for growing-finishing pigs.

The concept of phase feeding has now been extended throughout the finishing period. The reason for dividing the finishing period into more phases is to more nearly match the changing protein needs of the animal. This practice also reduces feed costs and lowers the amount of protein fed. This in turn reduces the amount of nitrogen excreted in the manure, helping to reduce odor and pollution problems.

When the finishing period is divided into four or five feeding phases the amount of protein in the diet is reduced in each progressive phase. The relationship between weight and the percent of crude protein in the diet is approximately:

Weight	*Percent crude protein in diet*
40–65 pounds (18–29 kg)	18–19%
65–95 pounds (29–43 kg)	16–17%
95–140 pounds (43–63 kg)	15–16%
140–195 pounds (63–88 kg)	14–15%
195–255 pounds (88–115 kg)	12–13%

Increasing the number of feeding phases means the producer must closely monitor the growth of the hogs and adjust the diet appropriately. Feed is approximately 60 to 65 percent of the total cost of swine production and about 75 to 80 percent of the variable cost.

TABLE 22-6 Some Representative Diets for Growing Swine (40–125 lb [18–57 kg]) Using Several Different Grain Sources.

| | Diet number | | | | | | | | | | | | | | | |
| | 1 | | 2 | | 3 | | 4 | | 5 | | 6 | | 7 | | 8 | |
Ingredient	(lb)	(kg)	(lb)	(kg)	(lb)	(kg)	(lb)	(kg)	(lb)	(kg)	(lb)	(kg)	(lb)	(kg)	(lb)	(kg)
Corn, yellow	1,555	705	1,368	621	1,228	557	—	—	—	—	804	365	—	—	—	—
Barley	—	—	—	—	—	—	—	—	1,660	753	—	—	829	376	—	—
Oats	—	—	200	91	—	—	—	—	—	—	—	—	—	—	—	—
Sorghum grain	—	—	—	—	—	—	1,549	703	—	—	—	—	—	—	800	363
Wheat, hard winter	—	—	—	—	—	—	—	—	—	—	804	365	829	376	800	363
Wheat middlings	—	—	—	—	400	181	—	—	—	—	—	—	—	—	—	—
Soybean meal, 44%	395	179	383	174	327	148	400	181	293	133	342	155	293	133	350	159
Calcium carbonate	15	7	15	7	21	10	17	8	18	8	16	7	18	8	17	8
Dicalcium phosphate	25	11	24	11	14	6	24	11	19	9	24	11	21	10	23	10
Salt	7	3	7	3	7	3	7	3	7	3	7	3	7	3	7	3
Vitamin-trace mineral mix*	3	1	3	1	3	1	3	1	3	1	3	1	3	1	3	1
Total	2,000	907	2,000	907	2,000	907	2,000	907	2,000	907	2,000	907	2,000	907	2,000	907
Calculated Analysis																
Protein, %	15.30		15.40		15.60		15.70		16.00		15.90		16.30		16.10	
Lysine, %	0.75		0.75		0.75		0.75		0.75		0.75		0.75		0.75	
Tryptophan, %	0.20		0.20		0.20		0.20		0.22		0.21		0.22		0.21	
Threonine, %	0.58		0.58		0.58		0.55		0.55		0.57		0.55		0.55	
Methionine + cystine, %	0.54		0.54		0.52		0.46		0.48		0.56		0.53		0.52	
Calcium, %	0.65		0.65		0.65		0.66		0.65		0.65		0.66		0.66	
Phosphorus, %	0.55		0.55		0.55		0.55		0.55		0.55		0.55		0.55	
Metabolizable energy, kcal/lb (kcal/kg)	1,494 (3,294)		1,464 (3,228)		1,461 (3,221)		1,438 (3,170)		1,360 (2,998)		1,465 (3,230)		1,397 (3,080)		1,437 (3,168)	

*See Table 22-3.

Source: Luce, William G., Hollis, Gilbert R., Mahan, Donald C., and Miller, Elwyn R., Swine Diets, Pork Industry Handbook, Cooperative Extension Service, University of Illinois.

TABLE 22-7 Some Representative Diets for Growing Swine (125 lb [57 kg] to market) Using Several Different Grain Sources.

							Diet number								
	1		2		3		4		5		6		7		8	
Ingredient	(lb)	(kg)	(lb)	(kg)	(lb)	(kg)	(lb)	(kg)	(lb)	(kg)	(lb)	(kg)	(lb)	(kg)	(lb)	(kg)
Corn, yellow	1,662	754	1,473	668	1,329	603	—	—	—	—	902	409	—	—	—	—
Barley	—	—	—	—	—	—	1,770	803	—	—	—	—	—	—	882	400
Oats	—	—	200	91	—	—	—	—	—	—	—	—	—	—	—	—
Sorghum grain	—	—	—	—	—	—	—	—	1,649	748	—	—	852	386	—	—
Wheat, hard winter	—	—	—	—	—	—	—	—	—	—	800	363	851	386	883	401
Wheat middlings	—	—	—	—	400	181	—	—	—	—	—	—	—	—	—	—
Soybean meal, 44%	290	132	280	127	225	102	185	84	304	138	251	114	250	113	190	86
Calcium carbonate	16	7	16	7	19	9	19	9	17	8	16	7	17	8	18	8
Dicalcium phosphate	22	10	21	10	17	8	16	7	20	9	21	10	20	9	17	8
Salt	7	3	7	3	7	3	7	3	7	3	7	3	7	3	7	3
Vitamin-trace mineral mix*	3	1	3	1	3	1	3	1	3	1	3	1	3	1	3	1
Total	2,000	907	2,000	907	2,000	907	2,000	907	2,000	907	2,000	907	2,000	907	2,000	907
Calculated Analysis																
Protein, %	13.40		13.60		13.80		14.30		14.00		14.20		14.50		14.60	
Lysine, %	0.62		0.62		0.62		0.62		0.62		0.62		0.62		0.62	
Tryptophan, %	0.17		0.17		0.17		0.19		0.17		0.19		0.19		0.20	
Threonine, %	0.51		0.51		0.51		0.48		0.48		0.51		0.48		0.48	
Methionine + cystine, %	0.50		0.50		0.48		0.44		0.42		0.53		0.48		0.50	
Calcium, %	0.61		0.60		0.62		0.61		0.61		0.60		0.61		0.60	
Phosphorus, %	0.50		0.50		0.55		0.50		0.50		0.50		0.50		0.50	
Metabolizable energy, kcal/lb (kcal/kg)	1,499	(3,305)	1,469	(3,239)	1,462	(3,223)	1,356	(2,989)	1,442	(3,179)	1,472	(3,245)	1,440	(3,175)	1,398	(3,082)

*See Table 22-3.

Source: Luce, William G., Hollis, Gilbert R., Mahan, Donald C., and Miller, Elwyn R., Swine Diets, Pork Industry Handbook, Cooperative Extension Service, University of Illinois.

The time spent carefully monitoring the feeding program in a swine operation can result in significant savings and thus increase the profitability of the enterprise.

Rations may be fed free choice or as completely mixed feeds. Free choice means the supplement is fed separately from the grain. All the ingredients are mixed together in a completely mixed feed. The pigs receive about the same nutritional effect from either method of feeding. Feed cost for these two methods is about the same. More uniform growth results from a completely mixed feed. Pigs sometimes overeat protein if it is fed free choice. This is especially true if soybean oil meal is the protein used. Overfeeding increases the cost of gain, and therefore should be avoided. On pasture, hogs gain a little faster when fed completely mixed feeds as compared to feeding free choice.

Growing-finishing pigs do not gain as efficiently on barley rations as they do on corn or milo rations. This is mainly because of the lower energy and higher fiber content of barley. Barley that weighs less than 48 pounds (21.8 kg) per bushel has a higher fiber content, which further reduces energy intake and rate of gain. Rate of gain may also be slightly lower on milo rations as compared to corn rations mainly because of the slightly lower energy level in milo.

Nutrient composition tables show the average protein content of corn. However, the crude protein content of corn can vary widely from sample to sample. Laboratory testing of corn for nutrient content can save the farmer money by allowing for more accurate balancing of the ration with protein supplement. A difference of 1 percent in the protein content of corn can mean a difference of 9 percent in the amount of soybean meal needed to balance the ration. Because the protein content of swine rations is critical for efficient growth and feed conversion, the small cost of laboratory testing for nutrient content of corn can be quickly recovered. Agricultural Cooperative Extension Service offices can provide information on the location of testing laboratories.

Molds that produce toxins (poisons) can grow in feed during hot weather. These toxins can reduce the rate of gain. Keeping feed bins and feeders clean and reducing the moisture content of all grains to 14 percent or less can reduce the danger of molds growing in the feed and producing toxins.

Some experimental work has been done on restricted feeding. *Restricted* or *limited feeding* means feeding 75 to 80 percent of full feed. Some improvement in feed efficiency results from restricted feeding. Daily gains are about 0.15 to 0.20 pounds (0.07–0.09 kg) less for each 10 percent reduction in the amount of feed. This increases the time to market by seven to ten days for each 10 percent reduction in feed. If more than a 75 to 80 percent restriction is made, the amount of total feed needed to market increases. Carcass quality is improved by restricted feeding.

FIGURE 22-3 Feeding hogs on full feed requires less labor than restricted feeding. *Photo by Jim Bodine.*

Under most farm conditions, restricted feeding is not yet a recommended practice. Full feeding from weaning to market seems to work better for most producers (Figure 22-3). Restricted feeding should never be used with pigs weighing less than 100 pounds (45 kg). Extra time, labor, and equipment is needed for restricted feeding.

Most swine producers feed growing-finishing barrows and gilts together. Research has shown that the nutritional needs of barrows and gilts are different and overall performance can be improved by feeding them in separate groups. This practice is called *split-sex feeding.*

Gilts need a higher concentration of dietary amino acids than do barrows to maximize lean growth rate. Barrows gain about eight percent faster than gilts; however, gilts generally produce carcasses with less backfat and a larger average loin eye area. Barrows reach their maximum performance as measured by daily gain, feed efficiency, backfat thickness, and loin eye area on a diet of 14 to 15 percent crude protein. Gilts reach their maximum performance as measured by the same traits on a diet of 16 percent crude protein. Gilts have a higher protein and lysine requirement than barrows. Gilts fed a higher protein level in the diet reach market at an earlier age. Split-sex feeding does require more facilities, such as separate feeding areas, separate feed-handling systems, and more feed storage bins. The producer must weigh the economic benefits against the increased requirements to determine the feasibility of using split-sex feeding.

PREPARATION OF FEEDS

Generally, grains used in hog feeding should be ground for most efficient use. Corn, barley, grain sorghum (milo), and oats should be finely ground. Wheat should be coarsely ground.

Pelleting complete feeds improves feed efficiency. Some of this improvement probably results from the lower feed waste that comes from pelleting. Higher-fiber rations are also improved by pelleting. Pelleting results in 4 to 8 percent improvement in rate of gain per ton of feed. Buying a complete pelleted feed may be less expensive than using meal. Most swine producers cannot justify the cost of owning equipment for feed pelleting. In addition, taking home-mixed feeds to a mill to be pelleted is usually not an economical practice.

Liquid or paste feeding reduces feed waste. Rate of gain may increase, but labor costs are generally higher with this method of feeding. Experimental results have not shown a clear advantage for liquid or paste feeding. There is an advantage in wetting complete mixed feeds when limit-feeding hogs. Mix 1.5 parts of water with one part of dry feed for best results.

There are commercial feeders available for use with wet-feeding programs. They are designed to mix the exact proportions of dry grain and water needed in the ration. Feeders made from different materials are available; stainless steel construction will last longer but is more expensive than some other types. Some feeders are made of plastic material, which makes them easier to keep clean. Feeders must be kept in areas that do not freeze, and they must be checked frequently. Better management is needed when following a wet-feeding program.

There is no advantage to cooking, soaking, or fermenting most feeds for hogs when they are full fed. The only exceptions are soybeans and potatoes, which are improved by cooking. Heating corn does not affect its nutritive value.

MANAGEMENT PRACTICES

Prebreeding Management

Producers must decide on the breeding system to be used. Crossbreeding hogs for slaughter is a recommended practice. Crossbred pigs generally grow faster and use feed more efficiently. The sows have larger litters and are better mothers. Crossbreeding systems are discussed in Unit 12.

Multiple farrowing is arranging the breeding program so that groups of sows farrow at regular intervals throughout the year. Multiple farrowing usually results in a higher average price received for hogs on a yearly basis. The chances of selling at better prices are increased as the number of marketings during the year are increased. Other advantages include spreading income more evenly through the year, making more efficient use of facilities, and reducing the investment per pig raised. Multiple farrowing requires better management than other systems. A year-round labor supply is necessary to handle the hogs.

Select replacement gilts at four to five months of age. Separate gilts from finishing hogs and feed separately. Worming of sows and gilts should take place before breeding. They should also be sprayed for external parasites at this time.

The boar should be purchased at least 45–60 days before use. Buy boars from healthy purebred herds that have good performance records. To prevent the spread of disease, isolate the boar from the rest of the herd when he is first brought to the farm. New boars should be treated for internal and external parasites. Semen test the boar or test breed him on a few market gilts before the breeding season begins to be sure he will breed.

The age of the boar is a factor in determining the number of times the boar can mate per day or week. Mating a boar to too many females in a short period of time will deplete the sperm reserve and reduce the boar's sex drive. Table 22-8 shows the current recommendations for the number of services per boar by age.

Conception rate and litter size can be increased by using more than one boar on each female. This is easier to do when using hand-mating or artificial insemination. It can also be done with pen breeding by rotating the boars once a day between pens. Rotating boars from pen to pen also increases the boars' sex drive.

TABLE 22-8 Recommended Maximum Number of Services per Boar, by Age.

	Individual mating system Maximum matings		Pen mating system* Boar to sow ratio
	Daily	Weekly	7–10 day breeding period
Young boar, 8–12 months of age	1	5	1:2 to 4
Mature boar, over 12 months of age	2	7	1:3 to 5

*All sows weaned on the same day.

Artificial insemination has not been widely used in the swine industry in the past except for producers of purebred hogs. With improvements in the technology, there is a trend toward more use of artificial insemination in commercial swine herds. Some advantages of artificial insemination are that it

- increases the ability to bring superior genetics to the herd.
- makes use of the semen from a superior boar to inseminate many more sows than is possible with natural mating.
- reduces risk of disease transmission.
- makes it possible to bring new bloodlines into the herd.

Breeding-Gestation Period

Gilts should be bred when they are seven to eight months of age and weigh 250 to 300 pounds (113–136 kg). Gilts have larger litters if they are bred during their second heat period rather than during their first heat period. Gilts on pasture or in dirt lots begin having heat periods earlier than those confined to concrete feeding floors. Conception rates can be increased if gilts are moved to outside lots by the time they weigh 175 to 200 pounds (79–90 kg). Gilts will begin cycling heat periods earlier if a boar is placed in an adjoining lot, allowing the gilt to see and smell the boar. Boars should be 7½ months of age before being used in a breeding program.

Allowing sows to have fenceline contact with boars stimulates estrus. When using hand-mating or artificial insemination, sows and gilts should be checked for standing heat at least once a day. Checking for standing heat twice a day will increase the conception rate. Gilts should be bred at least twice at 12-hour intervals after standing heat is detected. Breed sows at least twice at 24-hour intervals after standing heat is observed. Breed the first time on the first day of standing heat. Breeding at the above-recommended intervals will increase the conception rate.

Gilts and sows should be kept separate during the gestation period. Boars of the same size and age can be run together during the off-breeding season. Do not allow boars of different ages to run to-gether. The exercise area for a boar should be at least one-quarter acre (0.1 hectare).

Provide shade if animals are on pasture. Avoid overheating the animals and be sure that plenty of fresh water is available. Separate the breeding herd from other hogs on the farm to avoid disease problems.

Common Reproductive Problems

Gilts normally reach puberty before they are 200 days of age. There are several reasons for gilts showing delayed puberty. Delay in reaching puberty is an inherited trait; crossbred gilts generally reach puberty at a younger age than non-crossbred gilts. Duroc and Yorkshire gilts reach puberty later than Landrace and Large White gilts. Age at puberty can be reduced by 20–30 days when gilts are exposed to boars beginning at 140 days of age. Gilts that mature during the period from July through September are more likely to show delayed puberty and anestrus. *Anestrus* is a period of sexual dormancy between two estrus periods. Gilts held in confinement, or in groups of more than 10, also show more delayed estrus and anestrus.

Estrus normally occurs four to 10 days after pigs are weaned from sows. There are several causes of delayed postweaning estrus in sows. A lactation period of less than 21 days reduces the ability of the reproductive tract to recover from pregnancy; also, when young or high-producing sows nurse for more than 30 days, there may be a nutritional drain on the body that can delay postweaning estrus. Increasing energy intake during lactation can help to overcome this problem. When weaning occurs during the period from July through September, sows take longer to come into heat and have more post-weaning anestrus. Sows put in large groups (more than 6) when weaned do not come into heat as readily as those put in smaller groups.

Normally, about 5 percent of the sows and gilts bred will not become pregnant on the first breeding and will come into heat again in 18 to 25 days. There are several reasons why a higher percentage may not become pregnant on the first breeding; these include failure to breed each female at least twice during estrus, low fertility of the boar, and poor sanitation in the breeding area. Gilts should be bred at least twice at 12-hour intervals and sows should be bred at least twice at 24-hour intervals during estrus. Low fertility in the boar may be caused by immaturity, illness or injury, or high temperature during the breeding season. Breeding areas should be cleaned regularly and kept as sanitary as possible.

From 2 to 3 percent of the females will normally come into heat again more than 25 days after breeding. There are several reasons why a higher percentage may do this; matings during the July-through-September period result in more delayed returns to estrus, especially in young sows and gilts. Any disease that causes a fever in sows, enteroviruses, porcine parvovirus (PPV), or mycotoxins in the feed can also cause this condition.

Mycotoxins in feed can cause swelling of the vulva that is not associated with estrus. Feeds should be examined for the presence of mycotoxins. Contaminated feeds should not be fed to the breeding herd.

An abortion rate of 1–2 percent in a swine breeding herd is normal. There are a number of causes of higher abortion rates; among these are diseases including brucellosis, leptospirosis, pseudorabies, porcine parvovirus, or any other disease that causes a fever in sows. Other causes of higher abortion rates include mycotoxins, high levels of carbon monoxide from unvented heaters, and environmental stress. Sows that are bred during the July-to-September period also have more abortions.

A fetal mummification rate of 4–5 percent is normal in swine breeding herds. Several diseases may cause an increase in this rate. These include enteroviruses, porcine parvovirus, and pseudorabies. Vaccination, using older gilts, and culling of females not immune to porcine parvovirus are recommended solutions.

Unusually small litter sizes may result from a number of factors. Several diseases, including enteroviruses, porcine parvovirus, and pseudorabies, may be involved. Other factors include low fertility of the boar, too few matings during estrus, the breed of the sow, breeding gilts too young, and rebreeding sows at the first postweaning estrus. Follow good management practices as outlined in this unit to help overcome this problem.

About 6–8 percent of the pigs will normally be born dead (stillbirths). A higher rate of stillbirths may be caused by several factors. Larger litters normally produce more stillbirths. Older sows have more stillbirths. Overweight sows or gilts, high temperature [above 70–75°F (21–24°C)] in the farrowing house, or carbon monoxide toxicity can also increase the rate of stillbirths in the herd. Leptospirosis, eperythrozoonosis, or porcine parvovirus are also possible causes. Deficiencies of vitamin E or selenium in the diet can cause a higher rate of stillbirths. Reducing the average age of the sow herd, good nutrition during gestation, keeping sows cool in the farrowing house, vaccination, and good disease control can help reduce the incidence of stillbirths in the sow breeding herd.

Farrowing Period

Closely watching the behavior of sows can help the farmer determine when she is about to farrow. A high percent of sows farrow within about six hours after they begin a period of intensive activity. "Intensive activity" is when the sow stands up and lays down more often than once per minute. Sows will start rooting and pawing at the pen floor when they are about ready to begin farrowing.

Farrowing may also be induced by giving an injection of a drug that is commercially available. The injection is given 111 to 113 days after breeding, and the sow usually farrows within 18 to 36 hours later. There are several advantages to having a group of sows farrow within a short period of time: It is easier to even up litter sizes by cross-fostering pigs; labor is more efficiently utilized; and it is also easier to keep a group of sows on a uniform rebreeding schedule when farrowing occurs within a short period of time. The breeding herd can be better managed because the farrowing time is more predictable.

Farrowing facilities must be cleaned and disinfected before the sows are placed in them. Traffic through the farrowing house should be kept at a minimum. Sows must be washed with soap and warm water before they are put in the clean stalls or pens. Sows should be moved into farrowing stalls or pens at least one day before they are due to farrow. Guard rails and artificial heat are used to protect and warm the baby pigs. For newborn pigs, temperature should be 90° to 95°F (32°–35°C) under the heat lamp. Heat lamps are placed 18 inches (45.7 cm) above the pigs. After four or five days, the temperature is lowered to 80° to 85°F (26.7°–29.4°C) by raising the heat lamp.

Many baby pigs can be saved by the operator being present at farrowing time. The sow may need assistance with a difficult farrowing. Pigs that are trapped in the afterbirth can be saved. The mucus must be wiped off and the pigs placed under the heat lamp. Baby pigs must be kept warm and dry.

Needle teeth should be clipped with disinfected clippers. The needle teeth of pigs less than two days old should be clipped at the gum line. Care must be taken not to cut the gum. Clip one-third to one-half of the tooth if the pigs are more than two days old. Avoid injuring the gum. Figure 22-4 shows the method of holding the pig and clipping the needle teeth.

The navel cord should be clipped shortly after the birth of the pig. Cut the cord 1.0 to 1.5 inches (2.5–3.8 cm) from the body. Disinfect with tincture of iodine.

Pigs must be ear notched for identification. Identification is necessary for good record keeping. Accurate records help in selecting replacement animals. Problem litters can also be identified. Identification is useful in other management practices, such as determining rates of gain and feed efficiency. Purebred associations require ear notching for registration of pigs. There are several systems of ear notching. Purebred associations specify which system they require for their records. A standard system is shown in Figure 22-5. The right ear shows the litter number. The left ear identifies the individual pig. (Determination of right and left ear is made by viewing the pig from the rear.)

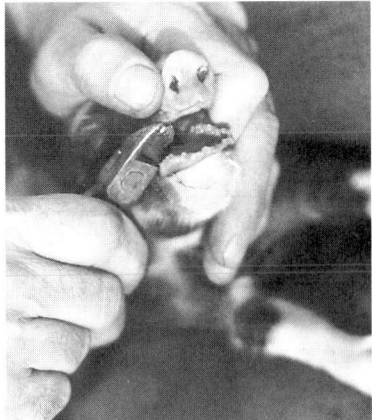

FIGURE 22-4   The needle teeth of baby pigs are clipped to prevent injury to the sow during nursing. *Courtesy of USDA.*

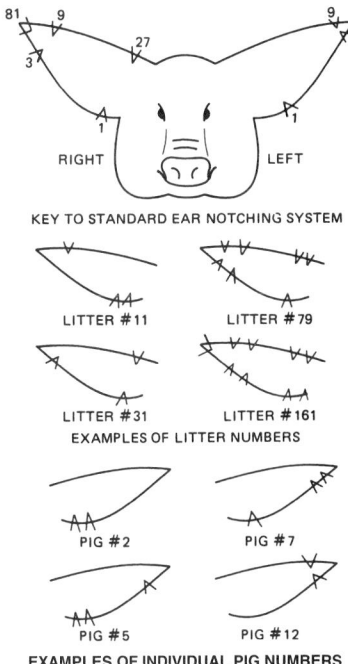

FIGURE 22-5   A standard ear notching system for pigs. *Courtesy of Indiana Cooperative Extension Service, Purdue University.*

An effort should be made to save runt (small) pigs. Extra labor is needed to save these pigs. Using a commercial milk replacer increases their survival rate. A mixture of 1 quart (0.9 litre) milk, 1/2 pint (0.24 litre) Half and Half, and 1 raw egg may also be used. Feed 15 to 20 millilitres once or twice per day. Use a soft plastic tube with a syringe to give the milk replacer or mixture orally. This practice saves about one-half the pigs that otherwise might have died.

Litter sizes should be equalized. Move pigs from large litters to small litters in order to make the litter sizes about equal. This should be done during the first three days after the pigs are farrowed. Be sure that the pigs nurse colostrum milk before moving them. The largest pigs in the litter should be selected for moving. Make sure the sow has the nursing ability and the number of teats necessary for the number of pigs that are in the litter.

Farrowing to Weaning Period

Several important management practices are performed in the period between farrowing and weaning. *Tail docking* is cutting off the pig's tail 1/4 to 1/2 inch (0.6–1.3 cm) from the body. This should be done when the pigs are one to three days old. Side-cutting pliers or a chicken debeaker may be used. Disinfect the tail stub with iodine spray. Disinfect the pliers between use on each pig. Tail docking helps to prevent tail biting among pigs in confinement. Producers of feeder pigs should always tail dock the pigs. Do not dock tails while pigs have scours.

Anemia is prevented by giving iron injections or oral doses of iron. This should be done when the pigs are two to four days old. Injections should be given in the neck or forearm, rather than the ham. Iron-dextran shots are used at the rate of 100 to 150 milligrams per pig. Iron should be given carefully since overdoses may cause shock in the pigs. Repeat the dose at two weeks of age. Iron can also be provided by placing fresh sod in the pen. Iron can also be added to feed or water at the time of the second dose.

Watch the pigs closely for scours. Drugs given orally usually work better than injections in preventing scours. Medication dissolved in water can be used as the pigs become a little older. Sanitation helps in preventing scours. If scours becomes a serious problem, consult a veterinarian.

Male pigs that are raised for slaughter must be castrated. Castration is best done when the pigs are young. There is less stress on the pigs and the job is easier. Boars should be castrated before they are two weeks old. The knife must be clean, sharp, and disinfected. Pig holders are available that make the job a one-person operation. Castration, vaccination, and weaning should not all be done at the same time. This places too much stress on the pigs.

Pigs need to be started on feed as soon as possible. Starting baby pigs on feed is discussed earlier in this unit.

Disease and parasite control is important in reducing losses. Vaccination and worming programs should be tailored to the problems of the individual farm. Consult with a veterinarian for specific recommendations.

There is a trend toward earlier weaning of pigs. Early weaning requires higher levels of management and nutrition. *Early weaning* usually means weaning the pigs before they are five weeks old. The average producer weans pigs sometime between five and eight weeks of age. Pigs should weigh at least 12 pounds (5.4 kg) at the time of weaning. Avoid drafts and great changes in temperature when weaning pigs of any age. Three-week-old pigs require temperatures from 80° to 85°F (26.7°–29.4°C) at weaning. Group pigs according to size at weaning. Groups should contain no more than 30 pigs, if possible.

Weaning to Market

From weaning to market most management centers around feeding and facilities. During this stage hogs are usually raised either in confinement or on pasture. Confinement requires more capital investment, but hogs gain a little faster under confinement conditions. In confinement, fogging systems are often used to cool hogs. Swine facilities are discussed in detail in Unit 24.

Good pasture can reduce the need for protein supplement in the ration. Sample swine rations are given earlier in this unit. Placing a nose-ring in the snout of hogs that are on pasture helps to keep them from rooting in the sod. Use care when placing the nose-ring to avoid injury to the bone structure of the nose. Hogs can be kept from getting too hot on the pasture if shade is provided for them. Control of diseases and parasites is discussed in Unit 23.

Hogs should be grouped in uniform size lots by weight. Groups should be no larger than 50 to 75 head. Weight range should be no more than 20 percent above or below the average weight of the group. Hogs should be marketed at about 230 pounds (104 kg).

Feed accounts for 60 to 65 percent of the expense of raising hogs. Wasted feed reduces feed efficiency. Feed loss can be reduced by making sure feeders are properly adjusted. Self-feeders should be adjusted at least once or twice a week to reduce waste. Controlling rodents in feed storage and feeding areas can also reduce feed loss.

A management practice that can help reduce the incidence of disease in swine herds is *medicated early weaning*. Broad-spectrum antibiotics are used in the sow diet before farrowing and during the lactation period. The pigs are weaned at 10 days or less and moved to a new location. The pigs are given broad-spectrum antibiotics during the first five days after birth. Phase feeding is used with the early weaned pigs. While medicated early weaning shows positive results when used with single-site production systems it is most effective when combined with all-in/all-out multiple-site systems. A major advantage of medicated early weaning is reduced incidence of disease in the herd. Research has shown that average daily gain may be increased by 14 percent and feed efficiency may be increased by nine percent when medicated early weaning is used. Death loss is also reduced using this system. There are increased costs when multiple sites are used. More facilities are needed and pigs and feed must be transported to the other sites. Another concern is the presence of other hog operations near each site used in this system. Depending on the disease problems present in the area, recommendations for distance from other hog-production facilities range from two to ten miles.

Using the *all-in/all-out* method of raising hogs can improve rate of gain and feed efficiency. The incidence of disease in swine herds is reduced using this management method. The all-in/all-out system moves pigs as a group from nursery, through growing and finishing, and to market. Groups consist of pigs farrowed within a short period of time, usually within a two- or three-week period.

Facilities are cleaned by power washing and disinfected between groups of pigs. All bedding, manure, and feed is removed from the facility at the time it is cleaned. Use hot, soapy water and a two percent lye solution for cleaning and rinse with clear water. Disinfectants such as chlorhexidine, chlorine, formaldehyde, or other approved compounds are used after cleaning. A number of commercial products containing approved disinfectants are available. Generally, these disinfectants are effective against a variety of bacteria, fungi, and viruses. Always follow label directions for the use of disinfectants. The facilities are left idle for a short period of time after cleaning and disinfecting before a new group of pigs is brought in.

Dust particles that adversely affect the health of workers in swine confinement buildings are a problem, especially in large operations where workers may spend the entire working day in the building. Research shows that using a light spraying of vegetable oil in the building can significantly reduce the dust and odor level. Daily spraying reduced dust levels by as much as 81 percent and odor intensity by 50 percent. Spraying may be done manually or with automatic equipment. Following this practice does add a small amount to the cost of raising pigs but it may result in a healthier environment for both workers and pigs. More research is needed to determine if the benefits are greater than the costs involved.

FEEDER PIGS

Some producers prefer to sell pigs as feeder pigs. Other producers who do not want to invest in facilities and breeding herds prefer to buy feeder pigs to feed out to market weights. *Feeder pigs* generally are eight to nine weeks of age and average 35 to 50 pounds (15.9–22.7 kg) in weight.

Feeder pig production provides a faster turnover in the volume of pigs handled. Less feed is required for each dollar's worth of pig sold. Labor must be available year-round for feeder pig production. Good sanitation and disease control programs are necessary. Large-volume operators have lower costs per pig produced than do small-volume operators. Net returns are higher for the large-volume producer.

Up to weaning, feeding and management practices are about the same for feeder pig production as for other hog enterprises. Good management and marketing practices are necessary if feeder pig production is to be profitable.

Feeder pigs should be bought from a reliable source. Newly arrived feeder pigs are isolated from other pigs on the farm so that they can be checked for disease. Allow the pigs to rest, keep them cool in hot weather, and allow sufficient space. Sort pigs into uniform lots according to size. The practices for feeding and management to market weights are similar to those discussed earlier in this unit for market hogs.

The same general formulas that are shown in Unit 16 for feeder cattle can be used to determine the value of feeder pigs.

SUMMARY

The two types of swine production are purebred and commercial. Purebred production is a specialized business. Fewer than 1 percent of the hogs raised in the United States are registered purebreds. Purebred producers raise foundation stock used in commercial hog programs.

Commercial hog producers raise hogs to be sold for slaughter. Most commercial producers use some system for crossbreeding. Commercial producers may merely produce feeder pigs, may buy and feed out feeder pigs, or may use a complete sow and litter system.

The most common feeds for hog production are corn and soybean oil meal. Other grains may be substituted for corn. However, most of them do not have the feeding value for hogs that corn does. Other sources of protein may also be used in place of soybean oil meal, although soybean meal oil has the best quality protein. Other sources must be fed in combinations to provide the amino acids required for proper swine nutrition.

Pasture is most valuable in feeding the swine breeding herd. Market hogs raised on pasture gain a little slower than those raised in confinement.

The most important minerals in hog rations are sodium, chlorine, calcium, and phosphorus. Salt, ground limestone, and dicalcium phosphate are the common sources of these minerals. Certain trace minerals are also needed in swine rations. These are usually added by the use of commercial mineral mixes.

Vitamins are added to swine rations by using commercial feeds and vitamin premixes. Water must be clean, fresh, and in adequate quantity for good hog production.

The breeding herd should be fed so that the sows, gilts, and boars do not become too fat. Limiting feed intake is a common way of avoiding overfattening. Self-feeding a bulky ration with ground hay added will also help keep these animals from becoming too fat. Silage may be used in the ration for the breeding herd.

Baby pigs should be started on feed as soon as they will eat. Starter and creep rations should be made available when the pigs are one week old.

Growing-finishing hogs may be fed free choice or on self-feeders. Limited feeding is not a recommended practice for the average hog producer. The protein level in the rations may be reduced as the hogs get older. Pelleted rations increase feed efficiency.

Crossbreeding and multiple farrowing are recommended practices. Breed gilts when they are eight months old and weigh 250 pounds (113.4 kg). Boars should be at least 7½ months old before using them for breeding.

A producer can save more pigs by being present when sows are farrowing. Make sure each pig nurses and is warm and dry. Clip needle teeth, ear mark pigs, disinfect and clip the navel cord, and dock tails during the first day or two after farrowing. Equalize litter sizes and give iron shots during the first few days. Castrate boars before they are two weeks old. Control scours in baby pigs and plan a disease prevention program to keep pigs healthy.

After pigs are weaned, feeding good rations is the most important management practice for successful hog raising. After weaning, pigs should be grouped in uniform lots according to size.

Student Learning Activities

1. Give an oral report on feeding and management practices used on your home farm or on farms of swine producers in the area.

2. Using the practices outlined in this unit, develop a feeding and management program for a swine project on your home farm.

3. Take field trips to observe feeding and management practices in the local area.

4. Ask a major swine producer in the local area to talk to the class about feeding and management practices.

5. Observe and practice good management practices such as clipping needle teeth, tail docking, and castration.

6. Do a community survey of swine producers on feeding and management practices. Report to the class and make suggestions for improvement.

Review

1. What percent of the total hogs raised in the United States are registered purebreds?

2. What is the function of the purebred hog business as it relates to commercial hog production?

3. What is the purpose of production registry and meat certification programs?

4. What type of breeding system is commonly used to produce commercial hogs?

5. Describe the characteristics of a feeder pig production operation.

6. Describe the characteristics of a buying and finishing feeder pig operation.

7. What percent of the total cost of raising hogs is made up of feed costs?

8. Why is corn the basic energy feed used for hog production?

9. Name the two essential amino acids that are especially lacking in corn.

10. Name and briefly describe three energy feeds besides corn that may be used for hog feeding.

11. Describe the characteristics of soybean oil meal as a hog feed.

12. Name and briefly describe three other plant proteins besides soybean oil meal that may be used to feed hogs.

13. Name and describe three animal protein feeds that may be used to feed hogs.

14. Describe the use of roughages to feed hogs.

15. Name four major minerals and six trace minerals and explain how they may be supplied in the ration.

16. How may vitamins be added to the swine ration?

17. Name five feed additives commonly found in hog rations.

18. Describe the feeding of gilts selected for the breeding herd.

19. Describe the practice and purpose of flushing.

20. Describe a feeding program for gilts and sows during the gestation period.

21. What kind of a ration should be fed at farrowing time?

22. Describe how sows and gilts should be fed for the first few days after farrowing.

23. Describe a feeding program for the herd boar.

24. Describe a feeding program for baby pigs.

25. Give a growing-finishing ration for hogs weighing 40 to 125 pounds (18.1–56.7 kg).

26. Give a finishing ration for hogs from 125 pounds (56.7 kg) to market.

27. Compare feeding free choice to feeding completely mixed feeds.

28. Describe feed preparation methods that might be used when feeding hogs.

29. What are the advantages of a multiple farrowing system?

30. At what age should replacement gilts be selected for the breeding herd?

31. How many boars should be provided for the breeding season?

32. How old and how heavy should gilts be before they are bred?

33. List good management practices that should be followed during the breeding-gestation period.

34. Describe management practices that should be followed during the farrowing period.

35. What management practices should be followed during the farrowing to weaning period?

36. What management practices should be followed from weaning to market?

Diseases and Parasites of Swine

Objectives

After studying this unit, the student should be able to

- explain the importance of maintaining good swine health.
- describe the causes, symptoms, prevention, and control of common swine diseases and parasites.
- list good management practices that help to prevent swine diseases and parasites.

The single most important management practice that affects profits from swine raising is maintaining a healthy herd. National averages show that 40 percent of the hogs that are farrowed do not reach market. One-third of the hogs farrowed do not reach weaning. Millions of dollars are lost each year by hog producers because of diseases and parasites. The USDA estimates that the loss due to diseases and parasites averages about $8.00 per hog marketed. Of this amount, diseases cause a loss of $5.00 per pig marketed. Parasites make up the remainder of the loss, or $3.00 per pig marketed.

Herd health problems have become more complex because of certain changes that have taken place in management in recent years. First of all, more hogs are being raised in confinement. In addition, hogs are being fed rations to push them to maximum growth rates. These two management changes create more stress on the pigs that may lead to an increase in herd health problems. Also, because of faster transportation, diseases may be spread farther in a shorter period of time.

DISEASE AND PARASITE PREVENTION

Herd health should be monitored on a continuous basis. Maintain accurate records of average rate of gain, feed intake, feed conversion, death rate, and treatments administered to the swine herd. These records form the basis for early detection of health problems. The use of a veterinary consultant will also help maintain good herd health. Veterinarians can provide diagnostic lab services to help prevent problems before they become serious. A thorough post-mortem examination can accurately identify the cause of an animal's death and helps to prevent spread of a disease through the herd. Blood testing is another health-monitoring technique that can measure exposure to diseases.

Observing the vital signs (temperature, pulse rate, and respiration rate) in an animal can help in the early detection of health problems. Vital signs will vary according to activity and environmental conditions. Normal vital signs in swine are:

- Temperature (normal range): 102.0–103.6°F (38.9–39.8°C) [average is 102.6°F (39.2°C)] (Temperature is usually higher in the morning than in the afternoon; younger animals will show a wider range of temperature than mature animals.)
- Pulse rate (normal range): 60–80 heartbeats per minute.
- Respiration rate (normal range): 8–13 breaths per minute.

Body temperature is taken in the rectum using either a mercury thermometer or a battery-powered digital thermometer. Restrain the animal when inserting the thermometer into the rectum. There is no location on a hog's body where the pulse can be felt directly by finding an artery. The heart must be felt directly over the chest. Respiration rate is determined by observing the number of times the animal breathes per minute.

Many swine producers are participating in the pork industry's Pork Quality Assurance Program. This program emphasizes good management practices in the handling and use of animal health products and encourages producers to annually review herd health programs. This helps reduce drug residue problems in marketed swine, addresses consumer food safety concerns, and improves the export market for pork. The program is funded through a pork checkoff that collects money from producers based on hogs marketed.

A well-planned herd health program is aimed at prevention of diseases and parasites, rather than their treatment. Management practices such as sanitation, isolation of new stock, selecting healthy breeding stock, and proper care of sows and pigs are all part of good herd health maintenance.

The first step in a sanitation program is keeping the facilities clean. High-pressure water cleaning equipment is useful in removing dirt and manure from the hog house. After cleaning with high-pressure water, the facilities should be given a good scrubbing with soapy water. Use of a good disinfectant completes the process. A hot lye solution is an effective disinfectant. Lye solutions kill most germs and viruses.

Caution: Lye solutions are caustic and must be handled carefully. Do not allow the solution to come into contact with the skin or eyes.

Other disinfectants include sodium carbonate, cresol, sodium orthophenylphenate, and iodine. Always use caution when handling any chemical. Read and follow label directions carefully.

Pens for young pigs should receive the most attention when facilities are cleaned and disinfected. Clean pens are also important for older pigs. However, older pigs are not as susceptible to diseases as are young pigs. Proper ventilation and light improve sanitation in the hog house. Bacteria and viruses grow better in dark, damp, poorly ventilated areas. Concrete floors make the cleaning job easier.

Disease organisms may be carried into hog production areas by visitors. Restrict the entry of visitors into hog production facilities; if visitors are allowed to enter the facilities, make sure they are wearing disinfected footwear. A container of disinfectant and a brush for disinfecting shoes should be kept at the entrance for the use of those entering the facilities. It is recommended that clean boots and coveralls be provided to be worn by any off-farm visitors entering hog production facilities. Provide facilities for loading and unloading trucks outside of the hog production area. Dogs, cats, and other animals can carry diseases that may be transmitted to hogs; keep these animals out of the hog production area. A good rat control program will also help reduce the transmission of diseases to hogs.

Other management practices that help control diseases and parasites are discussed in Unit 22. These include isolation of new breeding stock and proper care of sows and pigs at farrowing and during the lactation period. Vaccination and worming programs are adapted to the conditions of the individual farm. A program for a specific farm should be discussed with a local veterinarian. All buildings should be left unoccupied periodically to help prevent a buildup of diseases and parasites.

INFECTIOUS DISEASES

Abscesses

Abscesses, swellings filled with pus, are the result of a bacterial infection that enters the body through the nose or mouth. They are also called *jowl abscesses, cervical abscesses, feeder boils,* or *hog strangles.* The swellings, which vary in size from marbles to baseballs, may appear under the jaw and in the area of the neck. Not all infected hogs show symptoms of abscesses.

Hogs from weaning to market weight are more frequently affected than younger pigs. Affected hogs have a slower rate of gain. Severe cases can cause death. Carcass value is reduced because affected parts are condemned at slaughter.

Abscesses can be prevented by vaccinating hogs with an antibiotic. Other practices that help in prevention are early weaning, buying replacement animals from herds free of abscesses, and good sanitation practices. Feeding antibiotics to young pigs also helps to control abscesses. Treating infected hogs with antibiotics offers some relief from the disease. If penicillin is injected, its use must stop five days before slaughter. (*Note:* the use and withdrawal times for drugs used in treatments are subject to change. Always check current requirements on the label.) Abscesses can be opened and allowed to drain, but this may spread the infection to other animals in the herd.

Actinobacillus Pleuropneumoniae (APP)

Actinobacillus pleuropneumoniae is caused by bacteria. The disease was formerly called Haemophilus pleuropneumonia (HPP). It may appear in both chronic and acute forms. Symptoms of the chronic form include abdominal breathing, high fever (104–107°F [40–41.7°C]), depression, and reluctance to move. Some pigs with the chronic form may die, but most will survive. The surviving pigs will usually have damaged lungs and will not have a normal rate of gain. The first symptom of the acute form is often the sudden death of apparently healthy pigs. The pigs may develop heavy breathing and die within a few minutes. This may occur following a period of stress such as moving the pigs, mixing groups of pigs, weather changes, and poor ventilation in the housing facility. Death can occur in as short a time as 8–12 hours after exposure to the bacteria. The disease can occur in pigs of any age; however, it most commonly occurs in pigs from 40 pounds (18 kg) to market weight. The bacterium is spread through the air. One hundred percent of the pigs in a group can be affected, with death losses reaching 20–40 percent if no treatment measures are taken. A definite diagnosis of the disease is made from cultures of lesions taken in a post-mortem examination.

High levels of penicillin and tetracycline injections are the most effective immediate treatment. A specific antibiotic sensitivity test should be made to identify the proper antibiotic for continued treatment. All pigs in the same facility should be treated with an antibiotic injection to reduce the spread of the disease and death loss. Adding antibiotics to the feed or water is not an effective method of treatment in an acute outbreak. This procedure may be effective as a preventative measure after the initial antibiotic injection is given.

Pigs that have recovered from the disease are carriers. Isolate any additions to the herd for 30 days; serum test at the beginning of the isolation period and again at the end to confirm that they are not carriers. Commercial vaccines are available to help prevent this disease; however, they do not provide complete immunity. Reducing overcrowding and providing good ventilation in the housing facility may help prevent this disease. These measures will also help reduce death losses and improve the chances for successful treatment when an outbreak does occur.

Atrophic Rhinitis

A mild form of *atrophic rhinitis* is caused by a toxin-producing bacterium, *Bordetella bronchiseptica*. A more severe form is caused by the toxin-producing bacterium, *Pasteurella multocida*. The turbinate bones may not grow properly or they may atrophy (shrink) as a result of the infection. The turbinate bones are located in the snout of the pig. They filter and warm the air before it reaches the lungs of the hog. There are other organisms that may cause the turbinate bones to atrophy.

The milder form of the disease generally affects the nasal passages and the tonsils. The more severe form generally affects the nasal passages, tonsils, and lungs. Pigs infected with either form of the disease during the first week of life are more severely affected than those infected when they are older. The mild form has little affect on pigs after they are about nine weeks of age. The more severe form may damage the nasal cavities of pigs up to 16 weeks of age. The more severe form may also damage the liver, kidneys, ends of the long bones, and some components of the blood.

The disease spreads from one pig to another. Litters can be infected by the sow. A herd is usually infected by the introduction of infected breeding stock or feeder pigs. Pigs that appear to be healthy can be carriers of the disease. It is thought that cats, dogs, and rats are also carriers of the bacteria. Other factors that can increase the chances of infection are:

- large herds.
- adding pigs to the herd from several sources.
- large farrowing units.
- large nurseries.
- moving and mixing pigs frequently.
- overcrowding.
- poor ventilation in the housing facility.
- unsanitary conditions.
- continuous flow of pigs through the facilities.
- excessive dust and gases in the facilities.

Pigs are generally infected when young, usually when they are two to three weeks old. They can be infected after weaning. Symptoms usually are seen at about four to twelve weeks of age. These include sneezing, tear flow, nasal discharge, nose bleeding, and twisting of the snout to one side. The atrophy of the turbinate bones leads to a secondary infection of pneumonia. The disease seldom causes death, but rate of gain is slowed, and feed efficiency is lowered.

Vaccines are available for atrophic rhinitis. Vaccinating sows helps reduce the prevalence of atrophic rhinitis in the pigs but does not eliminate the disease. Nursing pigs may be vaccinated, but it takes about two weeks for immunity to develop; by then the pigs may already be infected.

Sulfonamides or oxytetracycline may be included in the sow diet during the last month of gestation to reduce the shedding (discharge) of bacteria. The treatment of nursing pigs with injections of oxytetracycline, or penicillin and streptomycin may reduce the severity of atrophic rhinitis. Medication may be used in the feed or drinking water of weaned pigs.

Good sanitation and management practices must be used with any vaccination and medication program followed. The following management practices are recommended:

- avoid overcrowding pigs.
- keep pigs of different ages separated.
- reduce stress from cold, dampness, and drafts.
- provide adequate ventilation in the housing facilities.
- follow strict sanitation practices.
- follow an all-in/all-out production system.
- use a medicated early weaning program with multiple production sites.
- provide proper care and nutrition at weaning.
- use fewer gilts in relation to older sows in the breeding herd.
- make sure new breeding animals are not infected.
- use Specific Pathogen Free (SPF) breeding stock (SPF hogs come from breeding stock that are surgically removed [generally by cesarean section] from the sows under antiseptic conditions. These hogs are raised in disease-free conditions).

There is no way to eliminate the bacteria after a herd becomes infected. Following good management practices can help reduce the severity of the disease.

Atrophic rhinitis should not be confused with *bull nose*. Pigs with bull nose develop large abscesses on their snouts. These abscesses are filled with a thick pus. Bull nose is associated with unsanitary conditions.

Avian Tuberculosis

Avian tuberculosis is caused by a bacterium. It is spread from chickens to hogs. Few symptoms of the disease appear in market hogs. Symptoms may be noticeable in older hogs that are kept for breeding purposes. The symptoms are a gradual loss in weight or an enlargement of the joints. Diagnosis is made by use of a skin test.

The main preventive measure is to keep chickens and hogs apart. Do not use poultry buildings for hog houses unless they have been disinfected. Do not allow hogs to eat poultry carcasses. There is no cure for avian tuberculosis.

Brucellosis

Brucellosis is caused by bacteria. Abortion early in the gestation period is a common symptom. Sows coming into heat after they were thought to be bred is another symptom. The disease is spread mainly from infected boars. Boars with the disease show swelling of one or both testicles. Blood testing the breeding herd is the only way to determine if the disease is present.

Buying clean breeding stock is the main way to prevent brucellosis. The entire breeding herd should be tested at least once a year. There is no treatment for hogs infected with brucellosis.

Through the efforts of the Cooperative State-Federal Brucellosis Eradication Program there are only a small number of swine herds left that are known to be infected with brucellosis. Under this program, a state is declared brucellosis free when no new cases are reported during a 24-month period and other requirements are met. The USDA is working toward the goal of complete eradication of brucellosis in swine herds in the United States. Most of the remaining known cases of brucellosis in swine herds are in the South and Southeast; the disease has not been a major problem in the Midwest for several years. In 1994 the USDA announced that indemnification payments will be made whenever it is necessary to completely destroy herds infected with or exposed to brucellosis. Prior to this time, only breeding herds were eligible for indemnification payments.

Cholera

At one time, cholera was one of the most serious of all hog diseases in terms of annual losses. A program of cholera eradication has been in effect in the United States for a number of years. The USDA announced early in 1978 that hog cholera has been completely eradicated in the United States. Hog cholera is now classified as an exotic (foreign) disease by the USDA.

Clostridial Diarrhea

Clostridial diarrhea is caused by *Clostridium perfringens* bacteria. This is the same bacteria that cause necrotic enteritis in poultry. The bacteria produce a toxin or endotoxin that damages the cell linings in the intestine. This is a disease of baby pigs, usually under one week of age. Symptoms include listlessness and a yellow, bloodstained, watery diarrhea. The death loss of affected litters is often more than 25 percent. Death usually occurs within a day and a half after the symptoms are first seen.

A chronic, less severe, form of the disease is more difficult to diagnose. The chronic form usually occurs in pigs between five days of age and weaning. Symptoms include a gray-flecked diarrhea and emaciation. These symptoms may be confused with the symptoms of transmissible gastroenteritis, *E. coli* scours, rotavirus, or coccidiosis. A positive diagnosis is made by laboratory analysis of smears taken from infected pigs. Pigs affected by the chronic form have lower feed efficiency and weight gain.

It is necessary to follow a good sanitation program in the farrowing facilities to help prevent this disease. Sanitize the farrowing crates or pens and wash the sows before putting them in the crates or pens.

Drug treatment of newborn pigs may offer limited protection but is generally not practical. Treatment after symptoms appear is generally not effective. Vaccinating sows and gilts at five and two weeks before farrowing may provide some protection against the disease. Where the disease is a major problem, vaccinating at seven weeks and two weeks before farrowing will allow the development of a higher level of antibodies in the colostrum milk.

The best way to control clostridial enteritis is to prevent the transmission of the disease from the sows to the baby pigs. Bacitracin methylene disalicylate may be fed to sows and gilts at the rate of 250 grams per ton of feed for two weeks before farrowing and continuing for three weeks after farrowing. Research has shown that using this treatment significantly reduces the death loss from clostridial enteritis. This treatment also improves the rate of gain in the baby pigs.

Swine Dysentery

Swine dysentery is also known as *vibrionic dysentery, bloody scours,* and *black scours.* It is caused by the bacterium *Serpulina hyodysenteriae.* Swine dysentery most commonly affects pigs from 8 to 14 weeks of age. Any age may be affected, however. In adult swine, the disease is not as severe and may be difficult to diagnose.

Dysentery may be confused with other pig scouring problems. The first symptom is a bloody diarrhea. After a time, the feces become watery and contain a thick mucus. A positive diagnosis of swine dysentery is made by laboratory analysis of feces samples. Weight loss, fever, and refusal to eat are other symptoms. About 20 to 30 percent of the hogs infected die if no treatment is administered. Chronic infection results in slow rates of gain and poor feed efficiency.

The disease spreads from infected hogs to healthy ones. Sows may be carriers of the disease without showing outward symptoms. The bacteria live in the feces of the sow and may be transmitted to nursing pigs. The disease may be spread from farm to farm in feces carried on shoes, boots, or vehicle tires. Field mice are a major reservoir of the bacteria and can spread the disease from farm to farm. Dogs, birds, rats, and flies—even though they are not long-term reservoirs of the bacteria—can spread the disease from farm to farm.

Pigs that are infected with swine dysentery may be treated with drugs added to the drinking water. Because infected pigs do not eat much feed, treatments administered in the feed are of little value. Direct injection of drugs may be used in treatment, but this method is generally not practical. Drugs that are approved for treatment of swine dysentery include:

- Bacitracin methylene disalicylate (1 g/gallon for 7–14 days up to 250 lb [113 kg]; no withdrawal time)
- Carbadox (50 g/ton of feed, up to 75 lb [34 kg]; withdraw 70 days before marketing)
- Gentamicin solution (50 mg/gallon for 3 days; withdraw 3 days before marketing)
- Lincomycin (250 mg/gallon for 5–10 days; withdraw 6 days before marketing)
- Tiamulin (227 mg/gallon for 5 days up to 250 lb [113 kg]; withdrawal 3 days before marketing)
- Tylosin (0.25 g/gallon for 3–10 days; withdraw 2 days before marketing)
- Tylosin injectable (4.0 ml/lb body weight for 3 days; withdraw 14 days before marketing)
- Virginiamycin (25–100 g/ton of feed, used with market swine over 120 lb [54 kg]; or for market swine under 120 lb [54 kg], 100 g/ton for 14 days, followed by 50 g/ton after two weeks; no withdrawal time)

Only healthy hogs should be brought into the herd. Animals should not be purchased from a herd that has had dysentery during the past two years. Isolate newly purchased stock. Avoid spreading the disease from one pen to another. The use of SPF hogs helps break the cycle of dysentery.

Other preventative measures that may be used include maintaining isolated herds, controlling rodents, controlling access to swine facilities, reducing stress, avoiding overcrowding in the facilities, maintaining strict sanitation, and maintaining a clean, dry environment.

Eliminating the disease from the herd may be more cost effective than trying to control it. One method is to totally depopulate the herd, clean and sanitize the facilities, and repopulate with breeding stock free of swine dysentery. A second option is to eradicate the disease without total depopulation of the herd. This involves the use of medications, cleaning and sanitizing the facilities, and eliminating the rodent population. A third option is using two- or three-isolated site production with partial depopulation. It is advisable to work closely with a veterinarian when developing a program for eliminating swine dysentery from the herd.

Edema

Swine edema is caused by toxins produced by a bacterial infection. Stress such as weaning, feed changes, and vaccination appears to make the problem worse. Anywhere from 10 to 35 percent of a herd may be affected. Death losses range from 20 to 100 percent of the pigs affected. Pigs between three to 14 weeks of age are most commonly affected.

Sudden death may be the first symptom seen. The disease may be confused with chemical poisoning. Other symptoms include refusal to eat, convulsions, staggering, and swollen eyelids. There is usually no fever.

Elimination of stress conditions helps to prevent edema. If symptoms appear, take feed away from pigs for 24 hours and then gradually increase to full feed. Sulfa and antibiotics in the feed or drinking water help in prevention and treatment. Observe withdrawal times on the label of the specific drug used.

Eperythrozoonosis

Eperythrozoonosis is caused by a small bacterium-like organism that attaches itself to the membrane of the red blood cell. This results in a destruction of red blood cells in the animal's body. The disease may appear in both chronic and acute forms. Because other infectious and noninfectious problems may produce similar symptoms, it is hard to diagnose this disease, especially in its chronic form. A blood test is available that is useful in identifying infected herds, but it is of limited value in identifying individually infected animals.

Symptoms of eperythrozoonosis include anemia, icterus (yellow discoloration of skin and mucous membranes), slow growth, and reproductive problems. The acute form of the disease is indicated by fever (104–106°F [40–41°C]), anemia, icterus, and an enlarged spleen. Hypoglycemia (low blood sugar) may occur in some hogs; this leads to coma, convulsion, and death. The acute form of the disease is more common in young pigs. The majority of the affected pigs recover, but gain at a slower rate for the rest of their lives.

Sows may be affected by either the chronic or the acute form of the disease. The acute form occurs more often during periods of stress such as farrowing, weaning, or breeding. At farrowing, infected sows may show udder edema (excessive accumulation of serous fluid in the tissue) and a lack of milk. The sow will usually begin to produce normal amounts of milk after a few days.

The chronic form of the disease in sows may be indicated only by persistent anemia. Other indications are decreased fertility, smaller litter size, occasional abortions, increased stillbirth rate, and smaller, weaker pigs at birth.

Growing-finishing pigs may be affected by either the chronic or the acute form of the disease. The acute form is indicated by a yellow discoloration of the abdomen (yellow belly). The chronic form results in slowed growth rate and anemia. Pigs may develop scours after weaning or stop growing during the finishing phase.

Eperythrozoonosis is not spread by direct contact between pigs. The most common method of transmission is by transfer of infected blood from a carrier to non-infected hogs. Infected blood may be transferred by biting insects, blood-contaminated needles, and castration instruments and other surgical equipment.

A veterinarian should be consulted for treatment of the disease. Tetracyclines and arsanilic acid have been effective in suppressing the symptoms, but they do not eliminate the carriers in the herd. The only way to control the disease is to avoid the transfer of infected blood from one animal to another. Control biting insects, lice, and mange to avoid transmission of the disease. Be careful when using needles and surgical equipment to avoid transferring contaminated blood from animal to animal. Be careful when introducing new breeding stock into the herd. Select new breeding stock from herds that have tested negative for eperythrozoonosis.

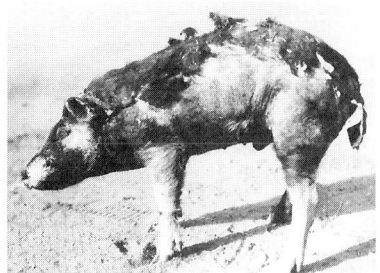

FIGURE 23-1A Skin lesions caused by erysipelas. *Courtesy of USDA.*

Erysipelas

Erysipelas is caused by a bacterium and occurs in acute, mild, and chronic forms. Hogs from weaning to market age are those most affected. Symptoms of the acute form include fever, withdrawal from the herd, lameness, depression, signs of chilling, and sudden death. Red skin lesions in diamond shapes may appear on the skin (Figure 23-1A). The mild form shows the same symptoms as the acute form, only they are less severe. The chronic form may follow recovery from an acute or mild attack. Arthritis is the most common damage resulting from the chronic form (Figure 23-1B).

FIGURE 23-1B Lameness from arthritis resulting from erysipelas. *Courtesy of USDA.*

Prevention can be achieved by vaccination of hogs at six to eight weeks of age. A combination of penicillin and erysipelas serum is used to treat infected hogs.

Exudative Epidermitis (Greasy Pig Disease)

Exudative epidermitis is caused mainly by a common bacterium that lives on the skin of the animal. It enters the animal's body when the skin is broken. Bite wounds, abrasions, lice and mange infections, dirty surgical or injection equipment, or viruses that cause blisters on the skin may create conditions that allow the bacterium to penetrate the skin. Pigs are usually affected during the first five to six weeks of life.

Symptoms of the disease are reddish areas that appear around the eyes, behind the ears, or under the legs. Fluid seeps from these areas and dirt and dander from the pig accumulates in the fluid; this gives the pigs a dirty, greasy appearance. As fluid losses increase, dehydration, lethargy, and lack of appetite result. The average death rate for infected pigs is about 25 percent. In some cases, death may result within a few hours to two days after the initial signs of the disease appear. In other cases, death may be delayed for six to 10 days after the appearance of the initial signs of the disease. Pigs that recover may have a slower rate of gain and poorer feed efficiency because of kidney damage caused by the disease.

A veterinarian should be consulted about the appropriate drugs and control measures to use for this disease. In the early stages of the disease, injections of penicillin or tetracyclines may help. After the disease has had time to fully develop, these antibiotics are of little value. Washing the pigs with a mild soap and water or disinfectant solution may provide some help in treating the disease. Washing should be done daily for three to four days. Remove affected pigs from contact with those that are not affected. Treat unaffected pigs that have been in contact with affected pigs with penicillin or tetracyclines to reduce the chances of their becoming infected.

Control lice and mange to help prevent the disease. Disinfect equipment after treating each pig when giving injections, castrating, clipping needle teeth, or performing other surgical procedures to prevent spreading the bacterium with dirty equipment. Good sanitation procedures in the facilities will help prevent the spread of the disease. Provide a stress-free environment that is warm, dry, and free of drafts. Reduce conditions in the environment that may cause skin abrasions.

Influenza

Swine flu is a respiratory disease caused by a combination of a virus and bacteria. Symptoms include fever, difficult breathing, coughing, going off feed, and weakness. Pigs become ill suddenly, and usually recover in about six days. The death loss is low, but rate of gain is slowed. There is no drug or antibiotic that can be used to cure swine flu. Give sick pigs water and feed and keep them warm. Reduce stress when they are sick.

If swine flu becomes a problem, it may be wise to use a vaccination program to help control the outbreak. Control of the disease can generally be achieved by vaccinating the sows two times per year. After the disease is brought under control, only replacement gilts need to be vaccinated. If vaccinating the sows does not control the disease, the pigs should be vaccinated when they are seven to eight weeks of age.

Leptospirosis

Leptospirosis is a disease caused by bacteria. The most common symptom of leptospirosis is abortion. Infected sows may farrow weak or stillborn pigs. Other symptoms that may appear in chronic cases include a slight fever, diarrhea, and loss of appetite. Convulsions and death may occur in more severe cases. The only sure way to identify leptospirosis is by blood testing the herd.

Feeding antibiotics in the ration helps to prevent the disease. Females should be vaccinated two to three weeks before breeding. A regular program of vaccination helps to prevent leptospirosis.

Mastitis-Metritis-Agalactia (MMA)

The exact cause of *mastitis-metritis-agalactia (MMA)* is not known. It is believed that a bacterial infection, as well as a hormone imbalance, may be involved. Infected sows do not have any milk to nurse their baby pigs. As a result, the pigs scour and die of starvation.

The main symptom of MMA is a lack of milk just before farrowing. The udder is hot and hard. The sow may have a fever, be depressed, and go off feed.

Reducing stress during pregnancy may help in prevention. Make sure the sows have plenty of water and the proper ration. Milk letdown is stimulated by washing the sow's udder with warm water and a mild disinfectant. Observe the sows carefully at farrowing to be sure that milk is being produced and that the pigs are nursing.

If MMA is observed, call a veterinarian. Treatment, which consists of antibiotics and hormones, must begin early. Feeding antibiotics before farrowing does not always prevent the problem.

When the sow has MMA, it is necessary to feed the baby pigs by hand to prevent starvation. Use cow's milk or a milk replacer. Make sure the milk or milk replacer is clean and at body temperature. Adding a tablespoon of corn syrup per pint of milk increases palatability. The baby pigs should be fed at least six times per day. Gradually reduce feeding to three times per day by the end of a week. Pigs should be put on a pig starter ration as soon as possible.

Mycoplasmal Arthritis (PPLO)

Two species of mycoplasma organisms cause arthritis in hogs. (*PPLO* stands for pleuropneumonia-like organisms.) *Arthritis* is a swelling or inflamation of a joint causing difficulty in moving and lameness. Animals with arthritis sometimes have a fever. One form of arthritis may also result in difficulty in breathing.

Prevention of this disease is difficult. Reducing stress may help in prevention. If the problem is chronic, it may be necessary to replace the entire herd. Treatment depends on the form that is diagnosed. Tylosin, lincomycin, penicillin, and cortico steroids are used. Withdrawal times before slaughter are: tylosin (injected), 4 days; lincomycin, 2 days; penicillin, 5 days.

Mycoplasmal Pneumonia

Mycoplasmal pneumonia is also called *swine enzootic pneumonia*. It was once called *virus pig pneumonia* (VPP) because it was thought to be caused by a virus. It is caused by *Mycoplasma hyopneumoniae.* Mycoplasma are very small and pass through ordinary bacteria filters.

Mycoplasmal pneumonia is a chronic disease. Death losses are low. However, feed intake is reduced and rate of gain is slowed, resulting in economic losses. A high percentage (probably 99 percent) of all commercial swine herds in the United States are believed to be affected.

Symptoms of the disease usually appear in pigs that are between six and ten weeks of age or older. Coughing, which may last for one to two months, is the most common symptom, while some symptoms are similar to other diseases that affect hogs. Exact diagnosis must be made by laboratory test. The disease is often accompanied by secondary infections.

The disease is spread from one pig to another, generally in the growing-finishing facilities and usually by direct nose-to-nose contact. It may be spread from the sow to the baby pigs; however, this is not a major method of transmission.

Research has shown that antibiotics are of limited value in treating or controlling this disease. Vaccines are available commercially to prevent mycoplasmal pneumonia. Sanitation and isolation are two preventive measures. Selection of healthy breeding stock is important. The use of SPF pigs also helps to break the cycle. Use of medicated early weaning, an all-in/all-out production system, and a multiple-site system are methods that will help prevent or control the disease. Controlling internal parasites helps reduce the effects of the disease.

Necrotic Enteritis

Necrotic enteritis is a disease caused by bacteria. The disease is sometimes confused with swine dysentery. Symptoms include fever, loss of appetite, and diarrhea. Rate of gain is slowed, and infected animals may die.

Sanitation and good rations help in prevention. Diseased pigs should be isolated from healthy pigs to stop the spread of the disease. The use of bacitracin, chlortetracycline, furazolidone, neomycin sulfate, nitrafurazone, or oxytetracycline is recommended for prevention and treatment of necrotic enteritis. Withdrawal times before slaughter are: bacitracin, none; chlortetracycline (in feed), 7 days; furazolidone, 5 days; neomycin, 20 days; nitrafurazone, 5 days; oxytetracycline, 22 days.

Porcine Reproductive and Respiratory Syndrome (PRRS)

The symptoms of PRRS include late-term fetal death, abortion, weak pigs, and severe respiratory disease in young pigs. At one time it was called Mystery Pig Disease. It is now known to be caused by a virus. There is no treatment for PRRS; however, a vaccine is available. Outbreaks can occur even in herds that have been vaccinated. A genetic test has now been developed that can differentiate between the harmless strain of the virus found in the vaccine and the actual disease-causing virus. Following good management practices to control disease will help reduce the incidence of PRRS in a producer's herd.

Porcine Respiratory Disease Complex (PRDC)

Several infectious agents combined with stress from the environment are responsible for PRDC. Included among the infectious agents are the PRRS virus, swine flu virus, *Mycoplasma hyopneumoniae, Salmonella cholerasuis,* and *Actinobacillus pleuropneumoniae.* This disease complex appears to be more common in multi-site operations.

PRDC usually affects pigs that are 14 to 20 weeks of age. Poor appetite, fever, and coughing are the major symptoms. Rate of gain slows significantly, with an accompanying decline in feed efficiency. Death rate is relatively low at about four to six percent of the affected population. Bringing pigs together from different herds appears to increase the incidence of this disease.

A modified live vaccine is available for use with unbred sows and replacement gilts. Killed PRRS vaccines may be effective when given to sows before farrowing. Antibiotics will help reduce the impact of the disease. Following good herd health management practices will help reduce the incidence of PRDC in producer herds.

Pseudorabies

Pseudorabies is caused by a virus. Outbreaks of this disease have increased in the swine-raising areas of the Midwest in recent years. Pseudorabies affects pigs of all ages. The death losses are highest in baby pigs, often approaching 100 percent of the pigs on a farm.

Symptoms in young pigs include fever, vomiting, tremors, disorientation, incoordination, convulsions, and death. The pigs die within 36 hours after the symptoms appear.

Symptoms in pigs from three weeks to five months of age include fever, going off feed, depression, difficult breathing, vomiting, trembling, and incoordination. The death loss may be as high as 50 to 60 percent in pigs three weeks of age. In older pigs, the death loss may be as little as 5 percent. Pigs that recover have a slower rate of gain.

Symptoms in older hogs include fever, nasal discharge, coughing, and sneezing. Vomiting, diarrhea, or constipation may occur. Sows that are infected may abort their litters, or the pigs may be born dead. Infertility of the sows may also occur.

Pseudorabies may show symptoms much like transmissable gastroenteritis (TGE), leptospirosis, flu, and SMEDI. Laboratory tests are necessary to identify the disease.

Antibiotics are not effective against pseudorabies. Vaccines currently in use do not cure the disease but do limit the transmission of the virus within the herd and prevent symptoms from appearing. The vaccines do not reduce susceptibility to other respiratory diseases that may affect the herd.

The prevention of pseudorabies involves management practices in two areas: (1) breeding, and (2) herd security.

Purchase breeding stock from a certified pseudorabies-free herd. New breeding stock should be isolated for a 60-day period. The isolation facility should be at least 100 feet from other animals. Blood test new breeding stock for pseudorabies before buying the animals. Retest the animals two to three weeks after purchase. Consider the use of artificial insemination or embryo transfer as a means of securing new genetic material in the breeding herd.

Herd security should begin with good fencing that does not allow other animals, including dogs, cats, and wildlife, into the hog production area. Restrict visitor traffic in the area and provide clean boots and coveralls for those who are allowed to visit the facilities; use a disinfectant boot wash at all entry points. Screen open areas to keep birds out. Use an incinerator or deep burial to dispose of dead animals. If they are picked up by truck, locate the pickup area away from the production area. Locate all truck loading and unloading areas away from the production area. Monitor the health of the herd on a continuous basis and keep good records.

Efforts are underway to eradicate swine pseudorabies. Eradication programs should be applied on an area-wide basis. Identify all infected herds in the area and do herd cleanup simultaneously. The least expensive method of herd cleanup is to test the entire herd and market all infected hogs. This method works best when there is a low level of infection in the breeding herds and the virus has not spread. Another method is to remove all young pigs from their mothers and put them in a clean area to avoid any contact with infected hogs. The most expensive method of herd cleanup is to market all hogs in a herd that has some infected animals. Clean the facilities and leave them empty for 30 days before restocking with hogs from certified pseudorabies-free herds.

SMEDI

The term *SMEDI* is formed from the first letters of *stillbirth, mummification, embryonic death,* and *infertility.* It is a disease complex that shows one or more of the following traits: (1) small litters of fewer than four pigs; (2) more than normal repeat breeding; (3) sows that come into heat after being thought bred, but never farrow; (4) dead pigs included in small litters. The dead pigs are often mummified. SMEDI is caused by several viruses.

The only control measures are managerial practices. The breeding herd should be maintained as a closed herd. All sows and gilts should be run together for at least 30 days before breeding. Fenceline contact is allowed with the boar. Do not expose breeding animals to other hogs. Do not allow people, birds, or rodents to enter the area where the breeding herd is kept. Be careful when moving feed trucks and other equipment into the area where the breeding herd is located.

There is no cure for SMEDI. Sows do develop an immunity to the disease. They will normally have good litters the next time they are bred if they have had the disease during the previous gestation.

Streptococcus suis

Streptococcus suis is caused by the bacterium *Strep suis* that lives in the tonsils of pigs. It can cause disease in the brain (meningitis) and in other organs (septicemia). The disease can affect pigs of all ages; however, it is most commonly seen in pigs that have recently been weaned. Meningitis may appear as sudden death or convulsions and then death. More commonly symptoms occur in this order: loss of appetite, reddening of skin, fever, depression, loss of balance, lameness, paralysis, paddling, shaking, and convulsions. Stress appears to be a factor in triggering an outbreak of meningitis.

Septicemia may appear as pneumonia, "fading piglet syndrome," arthritis, and abortion. Pneumonia caused by *Strep suis* most often occurs when the pigs are two to four weeks of age. It can occur during the growing-finishing period. "Fading piglet syndrome" occurs when healthy newborn pigs stop nursing, become listless, are cold to the touch, and die in 12 to 24 hours after birth. *Strep suis* infections are not common in breeding herds but can occur, resulting in a decrease in the farrowing rate.

Carrier pigs can appear healthy and yet have the bacteria in their tonsils or nasal passages. Introducing carriers into noninfected herds usually results in the spread of the disease into the new herd. Dead hog carcasses can harbor the disease and flies can carry it from farm to farm.

Observing the symptoms discussed above does not give a positive diagnosis of the disease. A laboratory analysis of cultured brain tissue from dead animals is the best way to definitely identify *Streptococcus suis*.

Early treatment of affected pigs with injections of penicillin or ampicillin helps prevent death. Tests need to be done to determine the correct antibiotic to use for continued treatment. In some circumstances it may be feasible to treat the entire group when some pigs are identified with the disease.

The key to controlling the disease is to reduce stress in the herd. Avoid overcrowding and drafts and maintain good ventilation in the facilities. Make sure that new breeding stock comes from herds that are free of the disease. Artificial insemination and embryo transfer can be used to bring improved genetic stock from infected herds to noninfected herds. Using SPF pigs and medicated early weaning are other management techniques that will help reduce the incidence of *Streptococcus suis* in a swine herd.

Transmissible Gastroenteritis (TGE)

Transmissible gastroenteritis (TGE) is caused by a virus. Pigs of all ages are affected. The most severe losses occur in pigs under ten days of age. Death loss in baby pigs is almost 100 percent. Older pigs seldom die from the disease. The disease spreads rapidly, especially in farrowing houses. Almost all litters are affected at the same time.

Symptoms in young pigs include vomiting, diarrhea, and death. Death occurs within two to seven days after the symptoms first appear. The feces are whitish, yellowish, or greenish in color. Dehydration and weight loss occur.

Symptoms in feeder pigs and older pigs include going off feed, vomiting, diarrhea, and weight loss. Death loss is low. However, the disease results in slower rates of gain in hogs that recover.

Symptoms in sows include going off feed and having mild diarrhea. Some sows may not show any symptoms. Sows that have the disease and recover have an immunity that lasts about one year. The immunity is passed on to the pigs in their litters. Because of this immunity, sows that recover from TGE should be used to breed another litter.

TGE is very contagious. Control measures consist mainly of good sanitation practices that prevent introduction of the disease. Keep people out of swine areas. Dogs and birds may carry the disease from one area to another, and should also be kept out of swine areas. Disinfect equipment when moving it from one hog area to another. Isolate new stock and do not bring new stock on the farm for one month before farrowing or two weeks after farrowing. Maintain a closed herd. Do not carry manure on shoes or clothing from one hog area to another. Breaking the farrowing cycle for about one month helps to control the disease.

Current vaccinations are not very effective. Some producers try to induce immunity in sows by exposing them to the disease about one month before farrowing. This is a practice that should be done only under the close supervision of a veterinarian.

There is no cure for TGE. Early diagnosis helps reduce losses. Proper disposal of dead pigs helps to prevent the spread of the disease. Dead pigs should be burned, buried, or sent to a rendering plant for disposal. Spread sows out for farrowing if the disease is diagnosed. Antibiotics can only help to reduce the effects of secondary infections.

White Scours (Colibacillosis: Bacterial Enteritis)

White scours is a highly contagious disease caused by several strains of bacteria. It is believed that most baby pig diarrheas are white scours.

The bacteria is present in sows, which are the usual source of infection for baby pigs.

Baby pigs are most commonly affected. The disease has a high death rate. Symptoms include a watery, pale yellowish diarrhea. Pigs become listless and dehydrated, and lose weight. The death loss in older pigs is low, but many pigs that are affected gain at slower rates.

Good sanitation, proper nutrition, and reducing stress help to prevent the disease. Vaccination of the sows is helpful when a specific bacteria is identified as the cause.

Early treatment (within 24 hours) is important if the disease is identified. Tests to identify the bacterial strain are necessary. Treatment is made with the proper antibiotic or sulfa drug after the bacteria strain is identified.

NUTRITIONAL HEALTH PROBLEMS

Anemia

Anemia is a condition caused by a lack of iron in the diet. It affects mainly baby pigs when the sow's milk does not have enough iron for the needs of the nursing pigs. Symptoms appear in pigs from one to two weeks of age. Signs of anemia are poor growth, roughened hair coat, and difficulty in breathing. Sudden death of apparently healthy pigs is also a symptom. Anemia lowers the pig's resistance to other diseases.

Anemia is controlled by giving iron to the baby pigs. Iron may be given as intramuscular injections or by mouth. Providing clean sod in the pen will also help to prevent anemia.

Hypoglycemia

Hypoglycemia is a condition caused by a lack of sugar in the diet. It has a high death rate. Symptoms include shivering, weakness, unsteady gait, dullness, and loss of appetite. The hair coat becomes rough. Diarrhea may develop in some pigs. Death occurs within a day and a half after the symptoms appear.

The main preventive measure is the reduction of stress factors, such as chilling. Be sure that the pigs nurse as soon as possible after they are born. Treat with dextrose solution or dark syrup. Dextrose may be injected or given by mouth. Syrup is given by mouth.

Parakeratosis

Parakeratosis is caused by a lack of zinc in the diet. High calcium content of the diet increases the need for zinc. The disease affects older hogs more frequently than younger ones.

Symptoms include rough, scaly skin, and slower then normal growth. The condition has the appearance of mange. Control and treatment consists of supplying adequate zinc in the ration.

Poisoning

Moldy feed is a common source of poisoning in hogs. Pitch, lead, mercury, pesticides, some plants, salt, and blue-green algae are other possible causes of poisoning.

Do not allow hogs to eat moldy feed or other possibly poisonous substances. Symptoms and treatments of poisoning vary with the type of poison involved. Consult a veterinarian if it is thought that hogs are suffering from poisoning.

Rickets

Rickets is caused by a lack of calcium, phosphorus, or vitamin D in the diet. An improper ratio of calcium to phosphorus in the diet may also cause rickets. Symptoms include slower than normal growth and crooked legs. Control and treatment is accomplished by providing proper amounts of calcium and phosphorus in the correct ratio. Supply vitamin D in the diet or be sure that the pigs are exposed to sunshine.

EXTERNAL PARASITES

Lice

The hog louse is dull, gray-brown in color and about 1/4 inch long. It is a bloodsucking pest. Lice tend to cluster in the ears, on the insides of the legs, and around the folds of skin on the neck of hogs. They cause an extreme irritation when they suck blood from the hog. They are most common in the winter, although they may be present year-round.

Economic losses occur due to reduced rates of gain and lowered carcass value. Infected hogs are more likely to be attacked by other parasites and diseases.

Lice are controlled by insecticides. Sprays and pour-on dusts are available. Observe label directions on limitations for use prior to slaughter.

All new additions to the herd should be treated with an insecticide. Sows should be treated in the fall or winter during the six weeks before farrowing. If an outbreak is detected, treat all the hogs on the farm.

Mange

Mange is caused by a tiny white or yellow mite that bores into the skin. The mite is so small it cannot be seen without magnification. Mites cause severe itching. The hogs are irritated and rub at the affected areas. The areas most often affected are around the eyes, ears, back, and neck. The area becomes inflamed and scabby. The problem is most severe in the fall, winter, and spring. Mites spread rapidly from hog to hog. Insecticides are used to control mange. Sprays or dusts may be used, although dusts are less effective than sprays. Observe label directions on limitations for use prior to slaughter.

INTERNAL PARASITES

Four groups of internal parasites affect hogs: (1) roundworms, (2) tapeworms, (3) flukes, and (4) protozoa. Roundworms cause more damage to hogs than any other internal parasite. The most serious of the roundworms is the large intestinal roundworm (ascarid). Other roundworms include stomach worms, intestinal threadworms, kidney worms, lungworms, nodular worms, whipworms, trichinae, and thorn-headed worms.

Bladder worms, which are incompletely developed tapeworms, are the only tapeworms that affect hogs. Flukes, soft leaflike worms, are not very common in hogs. Protozoa are tiny parasites that can only be seen with a microscope. Coccidia and trichomonads are examples of protozoa that affect hogs. Neither is very common.

Hogs that are infested with internal parasites do not grow or gain weight as fast as other hogs. The infestation makes the hog more likely to have other health problems. A heavy infestation of internal parasites can cause thinness, rough hair coat, weakness, and diarrhea.

Internal parasites cause economic losses because infected hogs make slower gains. Other losses occur because of lower carcass value. Some parts of the carcass may have to be condemned at slaughter because of parasite infestation. Sometimes the entire carcass must be condemned.

Of all the internal parasites previously mentioned, the most common are large roundworms, lungworms, whipworms, nodular worms, and intestinal threadworms. Most worm treatments are aimed at these parasite problems.

Parasites are controlled by good sanitation practices and the proper use of drugs to treat the infestation. Make sure the sows and gilts are clean when put into farrowing crates or pens. A clean farrowing house is necessary. Rotate hog pastures to break the life cycle of parasites. Sows, gilts, and boars should be dewormed one week before breeding and one to two weeks before farrowing. Deworming breeding stock prevents the baby pigs from getting worms from the manure. Deworm young pigs when they are five to six weeks old and again one month later. Deworm feeder pigs when they arrive at the farm and repeat three to four weeks later.

The use of hygromycin B, pyrantel tartrate, or thiabendazole in feeds helps to control several kinds of worms. Common drug treatments for worms are dichlorvos (Atgard), fenbendazole, ivermectin, piperazine, levamisole (Tramisol), hygromycin B, pyrantel tartrate, and thiabendazole. Some of these are given in the feed and others in the water. Check the label for information on the types of worms each drug controls, and instructions for its proper use. Observe label directions for withdrawal times before slaughter.

OTHER HEALTH PROBLEMS

Porcine Stress Syndrome (PSS)

The sudden death of heavily muscled hogs is referred to as *porcine stress syndrome*. Death is the result of a failure in the blood circulation system. It is most common in confinement housing. Genetic factors appear to be involved in the condition. It is believed to be carried on a recessive gene.

Prevention is mainly achieved by selection of breeding stock. Boars and gilts that appear to be more than normally nervous should not be selected for the breeding herd. Signs of PSS are constant movement, tail twitching, and trembling ears. The skin tends to develop splotches when the hog is excited. The splotches are red on white pigs and purple on black pigs. A blood test is available to determine if the pig is stress prone. If a problem develops in a herd, replace the boar with one that is not a carrier of the recessive gene.

Care must be taken with herds in which the problem exists. Avoid stress when handling or moving hogs. Do not crowd hogs in pens, and provide plenty of feeder and waterer space. Do not handle or move hogs in extremely hot weather. Provide plenty of bedding in cold weather. Some hogs will die even when these practices are followed but, in general, death losses will be less severe.

SUMMARY

Disease and parasite problems cause millions of dollars in losses each year in the United States swine industry. Prevention of health problems is the key to reducing losses. Sanitation, selection of healthy breeding stock, and proper management are important factors in preventing losses.

There are many infectious diseases that affect hogs. Vaccinations, sulfa drugs, and antibiotics are useful in the prevention and treatment of many of these diseases. The use of SPF hogs is also helpful in controlling some diseases. The greatest losses from diseases occur among younger pigs.

Nutritional health problems are controlled by proper feeding. Minerals and vitamins are particularly important in the diet. Many health problems are related to a lack of these substances.

Lice and mites are the most common external parasites that affect hogs. Insecticide treatments are used to control these parasites.

The most serious internal parasite of hogs is the large roundworm. Other worms that are major problems are lungworms, nodular worms, whipworms, and intestinal threadworms. A regular program of sanitation and treatment with drugs helps to control these parasites.

Careful selection of breeding stock helps to reduce porcine stress syndrome. Careful handling of hogs in herds where PSS is a problem helps to reduce losses.

Student Learning Activities

1. Prepare a bulletin board display of pictures of unsanitary and sanitary facilities for swine.
2. Prepare and give an oral report on a particular phase of swine herd health management.
3. Prepare a bulletin board display of pictures of animals affected by diseases and parasites.
4. Plan and practice good swine herd health practices with a school swine project or on the home farm.
5. Take field trips to observe swine herd health problems in the local area.
6. Ask a veterinarian to speak to the class about swine herd health management.
7. Conduct a community survey of swine herd health problems. Draw conclusions and present the results to the class.

Review

1. What is the amount of loss caused each year to the hog industry by diseases and parasites?
2. Why have herd health problems increased in recent years in the hog industry?
3. List four management practices that will contribute to good swine herd health.
4. Describe a good sanitation program for the swine herd.
5. Prepare a table listing the infectious diseases of hogs, briefly describing the symptoms, prevention, and treatment.
6. Name the two common external parasites of swine and describe the symptoms, prevention, and treatment.
7. What are the four groups of internal parasites that affect swine?
8. Which is the most serious of the roundworms that affect swine?
9. Describe the kinds of losses caused by internal parasites of swine.
10. Describe how internal parasites of swine may be controlled.
11. Describe porcine stress syndrome (PSS) and tell how it may be prevented and treated.

Unit 24 Swine Housing and Equipment

Objectives

After studying this unit, the student should be able to

- describe facilities required for swine production.
- describe equipment required for swine production.

The kinds of facilities used for swine production are much different today than they were 25 years ago. The trend in swine housing has been away from pasture production systems and toward confinement systems. Today, there is more specialization in hog production. More hogs are being raised per farm, and there are fewer small producers. Electricity, mechanization, and slotted floors have also changed the kind of housing being used. These changes require greater investments in facilities but reduce labor requirements to produce each pig. The amount of net return from swine production depends more on good management than on the kinds of facilities used.

KINDS OF BUILDINGS

Several building systems are possible, based on stages of production. At one time, it was common to use one building from farrowing to finish. Because it does not use space efficiently, this is no longer a common practice. However, the system is still used with feeder pig production. About four farrowings per year can be done with this system when feeder pigs are sold.

The concept of using just two facilities for the swine enterprise is gaining increased interest in recent years. In this system, the pigs are weaned, usually at less than three weeks of age, and moved directly into the facility where they will remain until they reach market weight. Advantages appear to include less stress on the pigs and a saving in costs involved in moving them to a nursery and then to a finishing facility. The major disadvantage is the higher facility cost. A suitable environment for young pigs, as discussed later in this unit, must be initially maintained. More research is needed to determine whether the higher costs associated with wean-to-finish buildings are justified by increased pig performance.

Larger producers with six or more farrowings per year often use housing that is divided into three stages of production: a farrowing to weaning facility, a growing facility, and a finishing facility. The farrowing facility may contain a nursery.

In any of the systems, pasture may be used in part of the program. The use of pasture depends on climate, land availability, the size of the operation, and the value of the land for other uses.

SITE SELECTION AND BUILDING PLACEMENT

Facilities for hog production usually include units for handling hogs from farrowing to finish. This kind of facility requires a large capital investment. Careful planning is necessary to be sure the facility is built in the best location and provides for future expansion.

Odor problems always exist with hog production. Facilities should be located downwind from the residence on the farm. The location in relation to neighbors must also be considered. Protection from wind and snow are other factors to consider in selecting a site.

The facility must have access to an all-weather road. Power and water must be available. Access to these facilities must be considered. Good drainage and runoff control are also necessary. Do not locate buildings in low, wet areas where the drainage is poor. Zoning laws exist in some areas and should be checked. It may also be necessary to secure permits for building.

Methods of handling manure must be considered. Manure handling is usually accomplished by spreading it on a field or using a lagoon. If manure is to be spread on a field, it may be necessary to store manure during part of the year. This must be planned as part of the total facility. Lagoon location must not interfere with future expansion.

VENTILATION, HEATING, AND INSULATION

Hog houses must be well ventilated. A ventilation system has several functions. In the summer, ventilation helps to control the temperature of the air. Hogs produce a great deal of moisture. Ventilation helps to remove from the house the extra moisture that is present in large amounts, especially in the winter. Odors are also removed by ventilation.

Hogs are healthier when they have fresh air. A ventilation system helps to provide the necessary fresh air. The ventilation system also dilutes airborne disease organisms that are present in the hog house.

The amount of ventilation needed for each of these functions varies. If enough ventilation is provided for fresh air and temperature control, the other functions are generally served.

The amount of ventilation required varies with the season. More air movement is needed in the summer than is needed in the winter. Winter ventilation is mainly for moisture and odor control. Summer ventilation is primarily for temperature control and to help keep the hogs cool. It is necessary to provide underfloor ventilation in buildings with slotted floors. This helps to remove the toxic gases and odors that come from the manure.

Warm Confinement Houses versus Cold Confinement Houses

There are two general types of hog houses. *Warm confinement houses* maintain desired temperatures regardless of the outside temperature. *Cold confinement houses* have temperatures that are only slightly warmer (three to ten degrees) than the outside temperature.

It is more important to have warm confinement houses for farrowing and nursery facilities than for growing-finishing facilities. Some producers also use warm confinement houses for growing-finishing hogs. However, hogs of this age do about as well in cold confinement buildings.

Warm confinement houses are more expensive to build. More moisture condensation, fogging, and frost occur in cold confinement houses. However, they are less expensive.

Warm confinement houses use mechanical ventilation systems. The design of the ventilation system depends on the area in which the house is located. Technical details of design and construction for this type of system can be obtained from agricultural engineers, university specialists, or commercial companies.

Cold confinement houses depend on natural ventilation for air movement. Usually, the house is built so one side can be opened to take advantage of prevailing wind direction in the summer. Again, technical details of location and construction design should be obtained for the local area in which the house is to be built.

Baby pigs require temperatures of 80° to 90°F (26.7°–32.2°C). They must be kept dry and free of drafts. Heat lamps, brooders, and underfloor heating systems can be used to provide the conditions necessary for baby pigs. Heat lamps must be no more than 2 feet (0.6 m) above the bedding in a pig brooder. Keep the lamp and cord out of the reach of the sow.

Sows are most comfortable with air temperatures of 50° to 60°F (10°–15.6°C). The body heat of the sows will provide enough heat if the building is well insulated. In cold areas, it may be necessary to provide some additional space heating. This is often done with vented, gas-fired space heaters. There is an increasing interest in the use of solar energy for heating hog houses. Consult with an agricultural engineer for the latest design details.

Weaned pigs are most comfortable at temperatures of 70° to 75°F (21°–23.9°C). Growing-finishing pigs are most comfortable at temperatures of 60° to 70°F (15.6°–21°C). High temperatures are more harmful than low temperatures for the breeding herd.

Energy costs are increasing. Proper insulation helps to reduce the cost of additional heat for hog houses. It keeps the interior surfaces warmer in the winter. This reduces the amount of moisture condensation on the walls and ceiling. Insulation also reduces the amount of animal body heat lost during the winter. This, in turn, helps to reduce heating bills. Fuel for uninsulated houses can cost as much as two to three times as much as that for insulated houses.

Insulation also reduces the amount of cooling necessary in hot climates during the summer. Proper insulation makes the building cooler in the summer. This increases the comfort of the hogs and their rate of gain. It also helps to reduce the cost of cooling the building.

The amount of insulation required varies with the location of the building. Technical advice can be obtained from agricultural engineers and university specialists in the local area.

In hot weather, sows in farrowing houses require cooler temperatures than can be provided by the ventilation system. Some producers use a stream of cool air directed to the sow's nose in each farrowing crate to provide further cooling (Figure 24-1).

Sprinkling systems are also used to help cool hogs. They work best when operated for two to five minutes each hour when the temperature is about 70° to 75°F (21°–23.9°C). A heavy spray should be used to cool the hogs. Fine mists or a fog are not recommended. Such systems are usually temperature and time controlled.

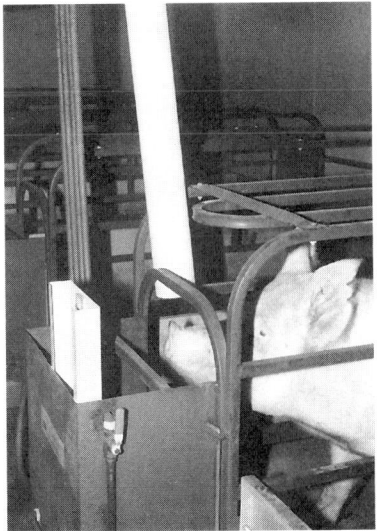

FIGURE 24-1   A stream of cool air directed to the front of the farrowing crate by a drop pipe helps to cool the sow. *Photo by Steven M. Ennis.*

FLOORS

Three general types of floors are used in hog houses: solid, partially slotted, and totally slotted.

Solid floors are usually made of concrete. The cost of solid floors is less than other types of floors. The major disadvantage of solid floors is the difficulty of handling the manure. Floors must be cleaned regularly, which requires a great deal of labor. Few confinement houses use totally solid floors.

Partially slotted floors are a combination of solid and slotted floor. Some hand cleaning of manure is required with partially slotted floors.

Totally slotted floors almost eliminate the handling of manure. However, the cost of slotted floors is more than that of solid floors. Slotted floors are harder on the feet and legs of hogs. Also, it is harder to control the temperature in a slotted-floor building.

A number of different kinds of materials are used for slats. Commercial slats are available in concrete, metal, Fiberglass, and plastic. Metals used include steel, stainless steel, and aluminum. Hardwoods, such as oak, can also be used for the slats. Soft woods should never be used for slats. They wear too fast, warp easily, and can be chewed through by the pigs. Concrete slats generally last the lifetime of the building. Metal slats may corrode if they are allowed to come into contact with wet manure for long periods of time. Metal slats require more support to carry the weight of the hogs.

Narrow slats, with spacing of 3/8 inch (0.9 cm), should be used for baby pigs in farrowing stalls. Wider slats, with spacing of 0.75 to 1.0 inch (1.9–2.54 cm) may be used in finishing houses.

In the farrowing pen, the slats should be placed parallel with the sow. This gives the sow better footing. Some producers use a partially slotted floor in the farrowing crate area. This provides the pigs with a warmer place to sleep.

Manure is handled as a liquid when slotted floors are used. Bedding is not used in liquid manure systems. The bedding will interfere with the operation of some kinds of pumps used in liquid manure-handling systems. Centrifugal pumps without choppers may become clogged by bedding. The chopper-impeller type pumps, however, can handle manure containing chopped bedding. Bedding decomposes slowly in lagoons and increases the odor problem.

BREEDING HERD FACILITIES

Facilities for sows, gilts, and boars may be portable or permanent buildings. Many producers are using total confinement buildings for their breeding herds. The cost for these facilities is higher than that for smaller, portable houses. However, better management of the breeding herd can be achieved in total confinement. Producers with limited capital or rented farms often choose to use more pasture for the breeding herd.

Houses with natural ventilation are adequate for the breeding herd. Boars must be separated from females. If pasture is to be used, locate the buildings so that it is easy to move the animals to pasture. Plan the arrangement of buildings so that it is easy to feed, clean, and water the animals. Boar housing and sow housing should be located close together to make breeding management easier.

Open-shelter housing may be used. Open-front buildings are described in the section on finishing facilities. If solid floors are used, provide about 15 square feet (1.4 m²) per sow. If individual pens are used for boars, provide a pen at least 6 × 8 feet (1.8 × 2.4 m) per pen. Shade, water, and feeding facilities should be available. About 15 to 18 inches (38–45 cm) of feeder space is required per sow.

No bedding is needed in confinement housing if slotted floors are used. Slats measuring 4 to 8 inches (10–20 cm) wide with 1 to 1¼ inch (2.54–3.2 cm) spacing are recommended.

When using a building with an outside apron, provide 11 to 12 square feet (1–1.1 m²) per sow inside and the same amount of space per sow on the apron. Provide 40 square feet (3.7 m²) per boar inside and the same amount per boar on the apron. In confinement systems, provide 40 square feet (3.7 m²) per breeding gilt on solid floors and 24 square feet (2.2 m²) on totally or partly slotted floors. For breeding sows, provide 48 square feet (4.5 m²) per sow on solid floors and 30 square feet (2.8 m²) on totally or partly slotted floors. Up to six gilts or sows can be put in each pen.

Mature boars need 60 square feet (5.6 m²) on solid floors and 40 square feet (3.7 m²) on totally or partly slotted floors. Put one boar per pen.

Gestating gilts should be provided 20 square feet (2.8 m²) per gilt on solid floors and 14 square feet (1.3 m²) per gilt on totally or partly slotted floors. Gestating sows should be provided 24 square feet (2.2 m²) per sow on solid floors and 16 square feet (1.5 m²) per sow on totally or partly slotted floors. Six to twelve gilts or sows can be put in each pen.

Individual breeding pens should be provided. Individual feeding stalls may be used to make it easier to limit-feed each sow.

FARROWING HOUSES

Portable farrowing houses cost less than confinement houses. Producers who farrow fewer than 20 sows per year will find it more economical to use portable houses. Other reasons for using portable housing for farrowing include: (1) farrowing only once or twice per year, (2) limited capital, (3) using rotation pasture, or (4) renting the farm. Individual portable farrowing houses are shown in Figures 24-2 and 24-3.

Confinement housing is more practical for producers who farrow more than 20 sows per year in multiple farrowing. Confinement housing might also be used when: (1) there is a desire to substitute capital for labor, (2) land is too valuable for use as pasture, or (3) a limited amount of space is available.

The main requirements for farrowing housing are that it be warm, dry, and free of drafts. Floors, ventilation, insulation, and heating have been discussed earlier in this unit.

Farrowing usually is done in pens or crates. Pens require more labor in cleaning than do crates. Guard rails and pig brooders help protect the baby pigs in pens. Guard rails are 8 inches (20.3 cm) above the floor and project 6 inches (15.2 cm) from the wall. A 2 × 4 or 3/4-inch pipe bolted to the wall can serve as a guard rail. Pens should be

FIGURE 24-2 Individual, portable A-frame farrowing houses may be used for farrowing on pasture. *Photo by Jim Bodine.*

FIGURE 24-3 A slatted-floor, portable farrowing unit that includes a hog house, farrowing crate, and feeder. *Courtesy of Marting Manufacturing.*

FIGURE 24-4 Farrowing crates will save more pigs than farrowing in pens. *Photo by Steven M. Ennis.*

FIGURE 24-5 This farrowing facility has feed bins located outside the house for automatic feeding. *Photo by Jim Bodine.*

FIGURE 24-6 A portable shade on skids provides shade for raising hogs on pasture. *Photo by Jim Bodine.*

4.5 to 5 feet (1.4–1.5 m) wide and 12 to 14 feet (3.6–4.3 m) long. Partitions between pens should be at least 3 feet (0.9 m) high. Some producers use partially slotted floors in pens. In this case, about one-fourth of the floor area is slotted.

Crates save more pigs than pens (Figure 24-4). In addition, less labor is required when farrowing in crates, and less bedding and floor space is needed. Partially or totally slotted floors are often used with crates. Sows may be fed in the crate or turned out each day for feeding. Crates are 4.5–5.0 feet (1.4–1.5 m) wide and 6.5–8.0 feet (2–2.4 m) long. About two feet (0.6 m) of width is needed for the sow. About 1.5 feet (0.4 m) is needed on each side for the baby pigs.

Automatic feeding and watering is often used in confinement houses. Careful adjustment of the feeders is necessary to prevent waste. The waterers should not leak.

In large confinement houses, an area should be set aside for an office and equipment. Storage space for supplies is also needed. A refrigerator for cold storage may be included. Hot water for washing sows may be provided.

An area for washing the sows before putting them in the pens or farrowing crates should be included. Feed storage is commonly located in large storage bins just outside the confinement house (Figure 24-5). If feeding is not automated, a feed cart makes the job easier.

The nursery unit is often a part of the farrowing facility. When the baby pigs are from a few days to two weeks old, the sow and litter may be moved from the farrowing house to a nursing unit. No more than three sows and litters should be grouped together in one pen. Provide 40 square feet (3.7 m²) per sow and litter. Partially slotted floors are recommended. A pig brooder should be provided at the upper end of the pen.

A pig brooder is a triangular area, usually in one corner of the pen, which is blocked off by boards so the pigs can enter but the sows cannot. It may be open on top with an infrared lamp being hung above the area to provide heat. A hover type brooder is closed on top, usually with a piece of plywood. An incandescent lamp with a reflector is placed on an opening in the cover to provide heat for the pigs. Electric heating cables or hot water pipes can be used under the floor to provide heat in the pig brooder area in place of the lamps. However, these systems are more expensive. Lamps are more commonly used.

GROWING FACILITIES

The *growing period* is from weaning to about 100 pounds (45 kg). Some producers grow pigs on pasture. Portable houses and shade must be provided for pigs on pasture (Figure 24-6). Portable feeders and waterers are also required (Figure 24-7).

Permanent houses are used by some producers for this period. The kind of housing that may be used is similar to that used for finishing houses. Temperature and moisture control are important during this stage of production.

Weaned pigs may be grouped in pens of about 25 head per pen. Provide 3 to 4 square feet (0.3–0.4 m²) per pig. One drinking space should be allowed for each 20 to 25 head. Provide one feeding space for every four pigs. Feeding and watering may be automated to save labor.

Pigs of this size tend to deposit dung over the entire floor of the house. Using partially or totally slotted floors reduces manure-handling problems.

FINISHING FACILITIES

Pigs over 100 pounds (45 kg) grow about as well in cold confinement houses as in warm confinement houses. Three general kinds of houses are used to finish pigs: (1) open front, (2) modified open front, and (3) totally enclosed. Pigs may also be finished on pasture. The same facilities are required for finishing on pasture as for growing (Figure 24-8). One acre of pasture will carry about 50 to 100 pigs.

In confinement housing, temperature and moisture control are the most important considerations. Methods for control are discussed earlier in this unit. The amount of space needed varies with the size of the pigs and the kind of floor used. Do not overcrowd pigs. Problems such as tail-biting, cannibalism, and slower rates of gain result from overcrowding.

Open-front buildings (Figure 24-9) have about one-half of the pen area in front of the building. The open side is placed in such a direction that it protects the pigs in winter and takes advantage of summer wind directions for ventilation. Feeders are placed either under the roof area or in the open part of the pen. Slotted floors may be used in part of the open area for liquid manure handling. The solid part of the floor is sloped toward the front. Bedding or underfloor heat is used in the building. About one-half of the rear of the building should be designed so that it may be opened in summer for ventilation. Drafts are controlled in the winter by constructing a solid partition from floor to ceiling every 40 to 50 feet (12–15 m) the length of the building (Figure 24-10).

A modified open-front house is covered entirely by a roof (Figure 24-11). One side is open. Some provision is often made to partially close the front during the winter.

Totally enclosed houses have no large openings to the outside. Insulation and ventilation must be provided. Floors may be solid, partially slotted, or totally slotted. When manure is stored in a pit under

FIGURE 24-7 A portable feeder for feeding hogs on pasture. *Photo by Jim Bodine.*

FIGURE 24-8 Interior of a swine finishing house showing pens and self-feeders. *Courtesy of University of Illinois at Urbana-Champaign.*

FIGURE 24-9 An open-front swine finishing facility showing pens and slotted floors. *Photo by Jim Bodine.*

FIGURE 24-10 An open-front swine finishing facility. Note the solid partitions that divide the pens inside the building. This helps control drafts in the winter. *Photo by Jim Bodine.*

the building, provision must be made for ventilation in the storage pit.

In confinement buildings, pens are designed to hold from 20 to 30 pigs. Provide 8 to 10 square feet (0.7–0.9 m²) per pig. A drinking space for each 20 to 25 pigs is needed. Provide one self-feeder space for each four pigs. Service alleys are needed in the modified open-front and totally enclosed houses. The service alley should be 4 feet (1.2 m) wide.

Feeding floors made of concrete, with shelters for the hogs, are still used on some farms. However, the trend in newer construction has been away from this type of facility. The open-front system is an adaptation of this type of housing. In moderate and warm climate zones of the United States, a feeding floor with an extension of slotted floor over a lagoon may be used. Shade is provided on one side of the floor. Odor is a problem with this type of feeding floor. Care must be taken in locating the facility. Manure is disposed of in the lagoon and broken down by bacterial action.

FENCING

Pasture should be enclosed with strong, woven-wire fencing at least 3 feet (0.9 m) high. Electric fencing can be used for temporary fences. The wire should be 6 to 8 inches (15–20 cm) off the ground.

Plank or concrete fencing is usually used with open-front or modified open-front houses. Windbreak fences may be made of wood or metal. They should be at least 6 feet (1.8 m) high. Solid windbreak fences provide better wind protection. Fences that are about 80 percent solid are better for snow protection.

FIGURE 24-11 A modified open-front building that is covered entirely by a roof. Feed tanks and loading chute are conveniently located. *Photo by Jim Bodine.*

Gates may be made of wood or metal. Sturdy construction and good hinges are important factors when selecting gates. Stock guards may be used instead of gates for convenience. A stock guard gate is described and pictured in Unit 16.

HANDLING EQUIPMENT

Holding and crowding pens with swinging gates make sorting and moving hogs easier. Alleyways are 20 inches (50.8 cm) wide. Cutting and blocking gates should be located in the alleyways to make sorting easier. The cutting gate swings in the alleyway and allows the operator to direct individual hogs into the desired pens. The blocking gate is also located in the alleyway just ahead of the cutting gate. It is used to stop the flow of hogs through the alleyway while individual hogs are directed into the desired pens by the cutting gate.

A *breeding rack* is used for hand mating. It should be 6 feet (1.8 m) long, 3.5 feet (1.1 m) high, and 32 inches (81 cm) wide. One end serves as a ramp that can be lowered to allow the sow and boar to enter. Shipping, weighing, and ringing crates are other types of equipment that hog producers find useful. Details of construction for these items are available from university agricultural engineering specialists.

Loading chutes may be stationary or portable. Variable-height chutes permit loading into different kinds of trucks. Steps or cleats in the floor of the chute provide firmer footing for the hogs (Figure 24-12).

FIGURE 24-12   A stationary, concrete loading chute. *Photo by Jim Bodine.*

FEEDING AND WATERING EQUIPMENT

Small portable feeders may be used. Larger walk-in feeders and self-feeders are also available. Fenceline feeders are used by some producers. Many different styles are available. Feeders may be constructed on the farm or bought commercially. Hogs are hard on this type of equipment. Sturdiness of construction must be considered when buying any type of feeder. The ease of adjustment is also important, since large amounts of feed are wasted if feeders are difficult to adjust.

Many styles of waterers are available. A barrel or tank on a skid may be used on pasture or in the feedlot. Automatic waterers are popular with confinement systems. Electric heaters may be used to prevent the water from freezing in cold weather. Some waterers have provisions for adding medication to the drinking water.

Feed-handling methods range from the scoop shovel to fully automated equipment. Automation is more expensive, but saves labor. Augers from holding bins are used to move the feed to the feeding areas. Design details are available from agricultural engineers or commercial companies. See Figure 24-13.

FIGURE 24-13   An automated feed-handling system for hog production. *Photo by Jim Bodine.*

OTHER EQUIPMENT

Several small items of equipment are essential for hog production. These include ear notchers, vaccinating syringes, needle teeth clippers, ringing pliers, and castrating knives. These tools are available from local livestock supply stores or through catalogs.

SUMMARY

The trend in swine housing is toward more confinement housing. Housing is needed for farrowing, growing, and finishing hogs. Some producers still use pasture for all or part of the hog production cycle. Portable buildings and equipment are provided for hogs in pasture.

The location of permanent housing requires careful consideration. Odors are the major problem in locating hog houses. The direction of prevailing winds and the location of residences in the area must be considered.

Confinement buildings must have proper ventilation to control temperature and moisture. Additional heat may be required, especially in the farrowing house. Finishing pigs do about as well in open-front buildings as in warm confinement buildings.

The development of slotted floors has reduced the labor needed for manure handling. Most confinement systems use some type of slotted floor.

The plan for a hog enterprise must include fencing, handling equipment, and feeding and watering equipment. Careful planning saves labor. Increased automation also saves labor, but is more expensive.

Student Learning Activities

1. Take field trips to observe various kinds of swine facilities in the local area.
2. Survey the community to determine the types of facilities and equipment commonly used.
3. Construct portable swine equipment in the school shop or on your home farm.
4. Give an oral report on swine housing and equipment.
5. Prepare a bulletin board display of pictures of swine housing and equipment.
6. Plan improved swine facilities for your home farm.

Review

1. What changes have occurred in the hog industry during the past 25 years that have affected the kinds of facilities used?
2. What factors must be considered when selecting the site and building location?
3. What is the function of the ventilation system in a hog house?
4. Describe the differences between warm and cold confinement houses for hogs.
5. What temperature is needed for baby pigs? How can that temperature be provided?
6. What temperatures are needed for (1) sows, (2) weaned pigs, and (3) growing-finishing pigs?
7. Why should insulation be used in hog houses?
8. Describe how hogs can be kept cool in hot weather.
9. Describe the three general types of floors used in hog houses.
10. What kind of facilities are needed for the breeding herd?
11. Under what conditions are portable farrowing houses more practical than confinement housing?
12. Under what conditions are confinement farrowing houses more practical than portable farrowing houses?
13. Describe a farrowing pen.
14. Describe a farrowing crate.
15. What are the advantages of farrowing crates?
16. What kinds of facilities are needed for growing hogs?
17. Describe the three general kinds of houses used for finishing hogs.
18. Describe the kind of fencing needed for hogs.
19. What kinds of handling equipment are necessary for hog production?
20. Describe feeding and watering equipment for hogs.
21. What miscellaneous kinds of equipment are needed for hog production?

Unit 25 Marketing Swine

Objectives

After studying this unit, the student should be able to

- describe three methods of marketing hogs.
- list and describe the grades of market hogs.
- list and describe grades of feeder pigs.

Making the right marketing decisions can mean the difference between profit and loss for the hog producer. Producers must carefully study the kinds of markets that are available. Factors such as the price offered, the costs, and the convenience of the market must be considered. The honesty and reliability of the hog buyer should also be kept in mind.

It is often hard to evaluate different markets on the basis of price. There are many factors that make up the price of hogs in a given market on a given day. The supply of hogs and the demand for hogs can vary from market to market on any given day. The costs at different kinds of markets also vary. The producer must calculate as closely as possible the highest net return that can be obtained for the hogs in each available market. Generally, the market giving the highest net return to the producer is the one to choose.

Producers of high-quality, meaty hogs will generally get better returns by selling on a grade and yield basis. An increasing number of hogs are being marketed on a grade and yield basis. Consumer demand for less fat in red meat has put pressure on the swine industry to produce leaner hogs. Grade and yield marketing provides a price mechanism to reward producers who raise the kind of hogs that are in demand.

A problem with grade and yield marketing is the limited ability of packing plants to evaluate hog carcasses on a large scale. Research is being conducted to develop equipment that can measure the amount of lean meat and fat in individual hog carcasses with a high degree of accuracy and do it at speeds that are compatible with the packing plant slaughter line. One such machine utilizes ultrasound, which measures the depth of fat and muscle by using sound waves; another machine being studied uses electrical conductivity to identify fat and muscle tissue. To be practical, these machines must not only be accurate but must be capable of automatically making measurements on hog carcasses as they move at the rate of 500 head or more per hour on the slaughter line in the packing plant.

Heavy hogs and cull breeding hogs usually sell better at terminal markets or auctions. Many of these types of hogs are also sold by direct-marketing methods.

PORK PROMOTION

The Pork Promotion, Research and Consumer Information Act of 1985 established a National Pork Board that collects funds through an assessment of 45 cents per $100 value of pork sold in the United States. These funds are used to promote the use of pork, develop foreign markets, provide consumer information, and conduct research and producer education programs.

Major advertising campaigns involving television and print media are used to influence consumer demand for pork as a safe, healthy, and nutritious food. Pork exports, especially to the Pacific Rim nations, have increased in recent years. Part of the checkoff funds is used to increase market share in the export market. Research and producer education programs include reviewing the pork production chain from producer to consumer, studying value-based marketing as a method of improving pork quality, and addressing issues of animal care, the environment, and food safety.

Checkoff funds are collected at commercial markets on market hogs and feeder pigs. Seedstock producers and producers who sell farm to farm are responsible for making the proper payments. Anyone with sales of $25 or more per month must make the checkoff payment on a monthly basis. Those with sales of less than $25 per month must make the payment on a quarterly basis. The National Pork Board regularly monitors swine sales nationwide to ensure compliance with the checkoff program. Producers who fail to make the required payments are reported to the USDA, which then takes steps to collect the money. A penalty of 1.5 percent per month and a civil penalty of up to $1,000 per occurrence can be assessed against producers who fail to comply with the checkoff program.

BOAR MEAT

Research is being done to determine the feasibility of feeding boars to market weight instead of castrating them when young to produce barrows. Boars produce more lean meat and have a higher feed efficiency compared to barrows. Boar meat has gained some acceptance on the

European market. Pork producers in the United States may need to develop methods of raising and marketing acceptable boar meat if they are to remain competitive in world markets.

The major objection to boar meat is the odor of the meat. Two compounds, skatole and androsterone, have been identified as the major cause of the odor problem with boar meat. European producers have overcome some of the odor and taste problem by marketing younger, lighter-weight boars. The amount of skatole and androsterone in the animal's body increases with age. More research is needed to determine the right combination of genetics and nutrition that can produce boar meat that is acceptable to the consumer.

KINDS OF MARKETS

Hogs may be sold through: (1) direct marketing, (2) terminal markets, or (3) auction markets. Some hog producers participate in group marketing. Generally, the group markets the hogs through one of the three methods just listed.

Direct Marketing

Direct marketing involves selling to packing plants, order buyers, or country buying stations. There are several kinds of country buying stations. Some are owned by packing plants. Others are independently owned, while still others are cooperatives.

Direct marketing accounts for the majority of the hogs sold in the United States. In this system, the producer deals directly with the buyer. When using direct marketing, the producer must possess selling skill and a knowledge of the markets to be successful in obtaining the best price for the hogs. Generally, the animals are transported shorter distances when sold through direct marketing, and the shrinkage is less.

Terminal Markets

Hogs marketed through terminal markets are consigned to a commission firm. The commission firm deals with the buyer. Commission firms can help the producer select the best time to market hogs. They generally employ people who have a good knowledge of the market.

Less than one percent of the slaughter hogs sold in the United States are marketed through terminal markets. The trend in recent years has been away from terminal markets and toward direct marketing.

There are usually several buyers competing for hogs on the terminal market. This may produce a better price. Prices may vary more widely on the terminal market than in direct marketing. This variation is affected by the number of hogs coming on the market on any given day.

Auction Markets

In some areas, auction markets are an important method of selling hogs. A small percent of the hogs sold in the United States are sold through auction markets. Auction markets are not widely used in the major hog-producing states. Some auction markets are developing new systems to obtain more buyers for the hogs offered. For example, telephone hookups and video auctions permit buyers who are not physically present to bid on the hogs.

Most auction markets have hog sale days once or twice a week. This limits the producer in choosing the day on which to market the hogs. Since auction markets are usually located in the local area, transportation costs and shrinkage are less with this method.

Selling costs involved with terminal and auction markets include commissions, insurance, yardage, and feed costs.

Group Marketing

Some producers use systems of group marketing. Group marketing systems have been established by some major farm organizations. Hog marketing cooperatives also exist in some areas. Some of these groups negotiate contracts with packers to supply hogs. The basic purpose is to obtain higher prices for hogs. The bargaining power of the producer may be increased by using a group marketing method.

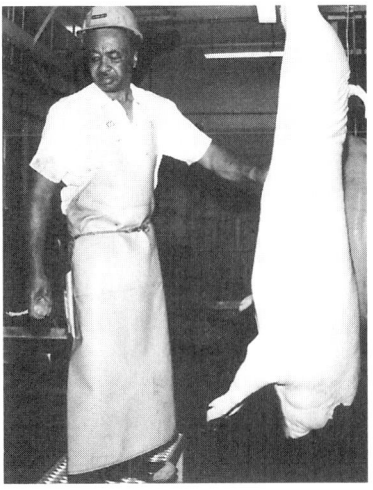

FIGURE 25-1   Hogs sold on the grade and yield basis are evaluated after slaughter at the packing plant. *Courtesy of University of Illinois at Urbana-Champaign.*

PRICING METHODS

Approximately 70 percent of the market hogs sold in the United States are priced on the basis of carcass merit (Figure 25-1 and 25-2). Premium prices are paid for hogs with more lean muscle and less fat. This trend corresponds with the trend toward improved genetics in breeding that results in the production of meatier hogs that meet consumer demand for leaner meat.

Some hogs are still marketed on the basis of the weight of the animal without regard for carcass merit. This method of pricing is rapidly declining.

Another trend in setting hog prices in the market is the use of contracts with producers. More than 60 percent of hogs marketed are now sold under some form of contracting. The formula price contract is the most common type used. The price paid is based on an agreement between the packer and the producer, generally related to a quoted cash price and some formula that sets the final price. Formula pricing generally favors the packer more than the producer, especially when the cash price for hogs is low. Formula pricing does not change the variability found in hog prices because it moves up and down with the variation in the cash price for hogs. This method may increase the average price a producer receives for hogs, typically by one to three dollars per hundredweight.

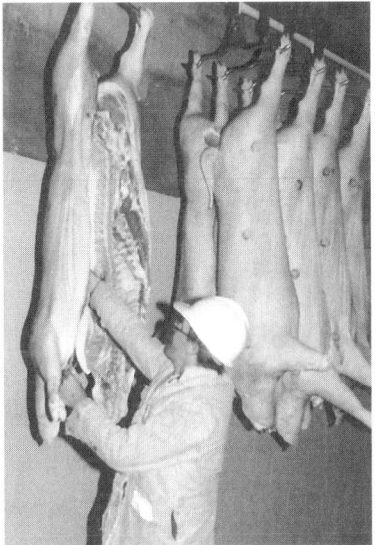

FIGURE 25-2   USDA inspectors check the wholesomeness of all meat produced in plants dealing in interstate commerce or foreign trade. *Courtesy of Utah Agricultural Experiment Station.*

Several other methods of contract pricing are used for marketing hogs. A fixed cash price contract may be tied to the futures market margins. This type of contract may give some advantage to the producer when the cash price of hogs is above the break-even level.

Another type of fixed price contract may be tied to the price of feed or historical price data with an additional amount added for overhead. This type of contract may exist with or without a ledger being maintained. If a ledger is maintained, a debit is recorded to the producer when the cash market price is lower than the contract price; a credit is recorded to the producer when the cash market price is higher than the contract price. Eventually debits and credits usually have to be reconciled so the actual cash price received over a period of time equals the market price over that same time period.

Window contracts provide for some sharing of the price risk between the packer and the producer. A window contract sets a floor for the price to be paid the producer. If the cash market price is lower than this floor, the producer is still paid the amount set by the floor. The packer assumes some of the loss when the price is below the price floor set by the contract and gets some of the gain when the price is above the price floor. Ledgers may or may not be maintained with this type of contract. However, if a ledger is maintained, then the producer may have to eventually repay the debt created by this type of contract when the cash price of hogs goes above the floor set in the window contract. Not all window contracts maintain a ledger.

Some ledger contracts have sunset clauses that in effect forgive the debt accumulated by the producer after the contract expires. When a sunset clause exits, the producer may eventually realize an average price that is higher that the cash market price during the life of the contract. Generally price risk is reduced only by cash contracts tied to futures market margins and contracts without ledgers.

Hogs that are not sold under some type of contract are sold on the cash market. This is referred to as the spot market. Prices on the spot market are determined by supply and demand for hogs on the day they are sold. The use of the spot market to set market hog price is decreasing (Figure 25-3).

MARKET CLASSES AND GRADES[1]

The use (slaughter or feeder), sex, and grade of swine determine their classification. *Slaughter swine* are those that are killed and sold as meat. *Feeder swine* are those that are sold to be fed to higher weights before slaughter.

---

[1]Based on Official United States Standards for Grades of Slaughter Swine, USDA and Feeder Pig Marketing Techniques, Pork Industry Handbook, PIH-72, Cooperative Extension Service, University of Illinois.

FIGURE 25-3   Pork products displayed for sale in a retail store. Consumer demand for pork influences the price of hogs. *Courtesy of Price Chopper Supermarkets.*

The five sex classes of slaughter and feeder swine are barrows, gilts, sows, boars, and stags. The definitions of each class are as follows:

Barrow—A male that was castrated while young. The physical traits of the boar have not developed.

Gilt—A young female that has not farrowed and is not showing any signs of pregnancy.

Sow—An older female that has farrowed or is showing signs of pregnancy.

Boar—A male that has not been castrated.

Stag—A male that was castrated after reaching maturity or beginning to show the physical characteristics of a boar and shows the physical traits of a boar.

The quality of the hogs and pigs is reflected in the grade standards established by the USDA. Grades of slaughter barrows and gilts are based on carcass quality and the yield of the four lean cuts. The four lean cuts are: (1) ham, (2) loin, (3) picnic shoulder, and (4) Boston butt. The quality of the lean is referred to as *acceptable* or *unacceptable*. Acceptable carcasses have bellies that are at least slightly thick overall and not less than 0.6 inches thick at any point. Other factors used to determine quality are the amount and distribution of external finish and the firmness of fat and muscle. Carcasses with acceptable lean quality and firmness of fat are further classed in one of the four top grades, based on the expected yield of the four lean cuts. If the lean quality is unacceptable, the carcass is graded Utility. Animals that will produce carcasses with oily or less than slightly firm fat are also graded Utility.

The five official USDA grades for slaughter barrows and gilts are: (1) *U.S. No. 1,* (2) *U.S. No. 2,* (3) *U.S. No. 3,* (4) *U.S. No. 4,* and (5) *U.S. Utility.* Grades of slaughter sows are discussed later in this unit. Boars and stags are not graded.

The estimated backfat thickness over the last rib and the muscling score are used to determine the official grade of slaughter barrows and gilts. These values are used in a mathematical equation to find the final grade. The equation is: Grade = (4.0 × backfat thickness over last rib, in inches) – (1.0 × muscling score). Muscling scores used in the equation are: thin (inferior) = 1, average = 2, and thick (superior) = 3. If an animal has thin muscling, it cannot be graded U.S. No. 1.

For example, if an animal has an estimated backfat thickness over the last rib of 1.05 inches and thick (superior) muscling the grade is determined as follows: Grade = (4.0 × 1.05) – (1.0 × 3) = 1.2 (U.S. No. 1 grade).

Another way to determine grade is to find a preliminary grade based on backfat thickness over the last rib and then adjust the grade up or down by one grade based on thick (superior) or thin (inferior) muscling. The preliminary grade is based on the following backfat thickness ranges:

Less than 1.00 inch—U.S. No. 1   1.25 to 1.49 inches—U.S. No. 3

1.00 to 1.24 inches—U.S. No. 2   1.50 inches or over—U.S. No. 4

Using the same animal as in the previous example, the preliminary grade is U.S. No. 2 based on the backfat thickness. This is increased one grade to U.S. No. 1 based on the thick (superior) muscling.

An estimate of the fatness and muscling is made by looking at the live animal. Fat deposits develop over the body at different rates, whereas muscle development is fairly uniform. Fat develops more rapidly over the back, at the loin edge, and on the rear flank, shoulder, jowl, and belly. As the animal gets fatter, these areas appear thicker and fuller than other parts of the body. The ham is least affected by increasing fatness, therefore differences in the thickness and fullness of the hams are the best indicators of muscling.

Fatter animals are deeper bodied because of fat deposits in the flank and along the underline. Other signs of fatness are the fullness of the flanks and the thickness and fullness of the jowl.

The three degrees of muscling used when grading slaughter barrows and gilts are thick (superior), average, and thin (inferior). Thick muscled animals with low fat have thicker hams than loins. The loins are full and well-rounded. Animals with thick muscling and a high level of fat are slightly thicker through the hams than through the loins. The back is nearly flat with a slight break into the sides. Average muscled animals with low fat are thicker through the hams than the loins. The loins are full and rounded. The hams and loins have about equal thickness in animals with average muscling and a high level of fat. A thin muscled animal with low fat is slightly thicker through the shoulders and the center of the hams compared to the back. The loins are sloppy and fat. Animals with thin muscling and high fat are wider in the loins than the hams. There is a distinct break from over the loins into the sides.

Slaughter Barrows and Gilts Grade Descriptions

U.S. No. 1 barrows and gilts produce a chilled carcass yielding 60.4 percent, or more, of the four lean cuts. Muscling and backfat combinations that allow hogs to be put into this grade are: average muscling with less than 1.00 inch backfat and thick muscling with less than 1.25 inches of backfat.

U.S. No. 2 barrows and gilts produce a chilled carcass yielding 57.4 to 60.3 percent of the four lean cuts. Muscling and backfat combinations that allow hogs to be put into this grade are: average muscling with 1.00 to 1.24 inches backfat, thick muscling with 1.25 to 1.49 inches of backfat, or thin muscling with less than 1.00 inch of backfat.

U.S. No. 3 barrows and gilts produce a chilled carcass yielding 54.4 to 57.3 percent of the four lean cuts. Muscling and backfat combinations that allow hogs to be put into this grade are: average muscling with 1.25 to 1.49 inches backfat, thick muscling with 1.50 to 1.74 inches of backfat, or thin muscling with 1.00 to 1.24 inches of backfat. Animals with 1.75 inches or more of backfat are not put into this grade.

U.S. No. 4 barrows and gilts produce a chilled carcass yielding less than 54.4 percent of the four lean cuts. Muscling and backfat combinations that allow hogs to be put into this grade are: average muscling with 1.50 inches or more of backfat, thick muscling with 1.75 inches or more of backfat, or thin muscling with 1.25 inches or more of backfat.

U.S. Utility barrows and gilts produce carcasses with unacceptable lean quality or unacceptable belly thickness. Also included are hogs with carcasses that are soft and oily. Muscling and backfat thickness are not considered for hogs in this grade.

Figure 25-4 shows the USDA slaughter swine grades.

Slaughter Sow Grades

The five grades of slaughter sows are (1) *U.S. No. 1*, (2) *U.S. No. 2*, (3) *U.S. No. 3*, (4) *Medium*, and (5) *Cull*. These grades are based on differences in yield of lean and fat cuts and differences in quality of cuts. There is a close relationship between the grades of slaughter sows and

U.S. NO. 1

U.S. NO. 2

U.S. NO. 3

U.S. NO. 4

U.S. UTILITY

FIGURE 25-4 USDA slaughter swine grades. *Courtesy of USDA.*

the grades of sow carcasses. Average backfat thickness is considered when grading sows. The average carcass backfat thickness for each grade of slaughter sows is:

U.S. No. 1—1.5 to 1.9 inches   Medium—1.1 to 1.5 inches

U.S. No. 2—1.9 to 2.3 inches   Cull—Less than 1.1 inches

U.S. No. 3—2.3 or more inches

The standard for each grade includes a description of the live animal and the minimum amount of finish for the grade. There are many combinations of characteristics that qualify an animal for a particular grade.

U.S. No. 1 slaughter sows have just enough finish to produce acceptable palatability in the cuts. These animals have moderate length and are slightly wide in relation to weight. They have a uniform width from top to bottom and front to rear. The back is moderately full and thick with a well-rounded appearance, blending smoothly into the sides. They have moderately long and slightly thick sides. Flanks are slightly thick and full. The fore flank may be slightly deeper than the rear flank. The hams have moderate thickness and fullness with a slightly thick covering of fat. The jowls are moderately thick and full with a trim appearance. Animals in this grade produce U.S. No. 1 carcasses.

U.S. No. 2 slaughter sows have a little more than the minimum finish required to produce acceptable palatability in cuts. These animals are slightly short and are moderately wide in relation to weight. The body is usually wider over the top than at the underline. The shoulders tend to be slightly wider than the hams. The back is full and thick and is slightly flat with a noticeable break into the sides. They have slightly short and moderately thick sides. Flanks are moderately thick and full. The depth of the fore flank is about equal to the rear flank. The hams are thick and full with a moderately thick covering of fat. The jowls are generally thick and full. The neck is short. Animals in this grade produce U.S. No. 2 carcasses.

U.S. No. 3 slaughter sows have much more than the minimum finish required to produce acceptable palatability in the cuts. These animals are short and wide in relation to weight. The body is wider over the top than at the underline. The shoulders tend to be wider than the hams. The back is very full and thick and is nearly flat with a definite break into the sides. They have short, thick sides. Flanks are thick and full. The depth of the fore flank is about equal to the rear flank. The hams are very thick and full with a thick covering of fat. The jowls are generally very thick and full. The neck is short. Animals in this grade produce U.S. No. 3 carcasses.

Medium grade slaughter sows have a little less than the minimum finish required to produce acceptable palatability in the cuts. These animals are long and moderately narrow in relation to weight. The body is usually narrower over the top than at the underline. The shoulders tend to be slightly narrower than the hams. The back is moderately thin and is slightly peaked with a distinct slope toward the sides. They have long and moderately thin sides. Flanks are thin. The depth of the fore flank is greater than the rear flank. The hams are moderately thin and flat. The jowls are generally slightly thin and flat. The neck is long. Animals in this grade produce Medium carcasses.

Cull grade slaughter sows have considerably less finish than that required to produce acceptable palatability in the cuts. These animals are long and narrow in relation to weight. The body is usually narrower over the top than at the underline. The shoulders tend to be narrower than the hams. The back is thin and lacks fullness and is peaked with a definite slope toward the sides. They have very long and thin sides. Flanks are very thin. The depth of the fore flank is considerably greater than the rear flank. The hams are generally thin and flat. The jowls are generally thin and flat. The neck is long. Animals in this grade produce Cull grade carcasses.

Feeder Pig Grades

Feeder pigs are classified in the same grades as slaughter hogs, with the addition of one lower grade. *U.S. Cull* is the lowest grade of feeder pig. Feeder pig grades are used to indicate the expected grade of the pig when it reaches slaughter weight. Unthrifty pigs are classed in either U.S. Utility or U.S. Cull grades. These pigs have failed to grow and gain properly. Poor care or disease can cause a pig to be unthrifty.

The U.S. No. 1 feeder pig has a large frame, thick muscling, and is trim. The legs are set wide apart. The hams are wider than the loins. Feeder pigs in this grade should produce U.S. No. 1 grade carcasses when slaughtered.

The U.S. No. 2 feeder pig has a moderately large frame with moderately thick muscling. The pig is a little fatter than the No. 1. The hams are slightly wider than the loin. The jowl and flank are a little fatter than the No. 1. The side shows less trim. Feeder pigs in this grade should produce U.S. No. 2 grade carcasses when slaughtered.

The U.S. No. 3 feeder pig has a slightly smaller frame with slightly thin muscling. The hams and loin are about the same width. The legs are set fairly close together. The jowl and flank show signs of too much fat. Feeder pigs in this grade should produce U.S. No. 3 grade carcasses when slaughtered.

U.S. NO. 1

U.S. NO. 2

U.S. NO. 3

U.S. NO. 4

U.S. UTILITY

FIGURE 25-5 USDA grades of feeder pigs. *Courtesy of USDA.*

The U.S. No. 4 feeder pig has a small frame with thin muscling. The hams and loin are the same width. The back is flat. The legs are set close together. The jowl and flank are moderately full. The lower ham is beginning to show signs of too much fat. Feeder pigs in this grade should produce U.S. No. 4 grade carcasses when slaughtered.

The U.S. Utility feeder pig shows unthriftiness because of disease or poor care. The skin is wrinkled and the head appears too large for the rest of the body. The pig is rough in appearance. Given good care, pigs in this grade can develop into a higher grade of slaughter hog. Feeder pigs in this grade will produce U.S. Utility grade carcasses when slaughtered unless the unthrifty condition is corrected.

The U.S. Cull feeder pig has a poorer appearance than U.S. Utility. Improper care and disease cause a rough, unthrifty appearance. Pigs in this grade will gain at slower than normal rates and often will not be profitable.

Figure 25-5 shows the USDA grades of feeder pigs.

WEIGHT AND TIME TO SELL

The traditional recommended weight at which to sell slaughter hogs has been 200 to 220 pounds (90.7–99.8 kg). This was based on two conditions related to weight gain: (1) research indicated that feed costs per pound (0.45 kg) of gain increased rapidly above 220 pounds (99.8 kg), and (2) much of the weight added above 220 pounds (99.8 kg) was fat. This section discusses factors that modern hog producers should consider when deciding upon the best weight to market their slaughter hogs.

Recent research indicates that several factors should be considered when deciding on the best weight to market hogs. These factors are: (1) type of hog, (2) hog-feed price ratio, (3) amount of discount for heavier hogs, and (4) the time of year when the hogs are marketed.

Research conducted at several university experiment stations shows that meaty hogs can be fed to heavier weights without a large increase in feed costs. The quality of the hog as measured by the ratio of lean to fat is not decreased appreciably with this type of hog. The experiments show that an increase in feed of 0.70 pounds (0.32 kg) per 100 pounds (45.4 kg) of gain is necessary to go from lighter to heavier market weights. Two studies show a decrease in percent of lean cuts. One shows a decrease of 0.6 percent as hogs were fed from 200 to 250 pounds (90.7–113.4 kg). The other study shows a decrease of 1.3 percent from 220 to 260 pounds (99.8–117.8 kg).

As feed costs increase, the additional returns above feed costs decrease as hogs are fed to heavier weights. As the price of hogs increases, the returns above feed costs increase. It requires careful calculation to determine where the breakeven point occurs as hog-feed price ratios change.

Packers typically discount the price of hogs below 220–230 pounds (99.8–104.3 kg) and above 250–260 pounds (113.4–117.9 kg). This range varies with the number of hogs coming to market and the pricing practices of individual packers. If discounts are high for heavy hogs, it may not pay to feed to the heavier weights.

Part of the decision on feeding to heavier weights depends on the time of year. The price of slaughter hogs changes seasonally based on the supply of pork and consumer demand for pork. Figure 25-6 shows the seasonal variation in the price of slaughter hogs in the United States. Because there are more large commercial swine producers now than there were a few years ago, the supply of hogs coming to market is less variable than it used to be. Large producers tend to distribute farrowings more evenly during the year.

Slaughter hog prices tend to increase from April through July. Prices tend to move down from July to November and then show some recovery through February. They again decline from February to April.

When prices are on the increase, it may pay to feed hogs to heavier weights even though there will be some discount. However, when the price trend is downward, it generally will not pay to feed to heavier market weights. The hog producer should have the flexibility to feed to heavier weights part of the year and to lighter weights at other times of the year.

It must be remembered that the seasonal price trends are averages. Prices in any given year are affected by such factors as the number of hogs being marketed and the demand for pork (Figure 25-7). Hog prices for any given year do not always follow the long-term seasonal price patterns.

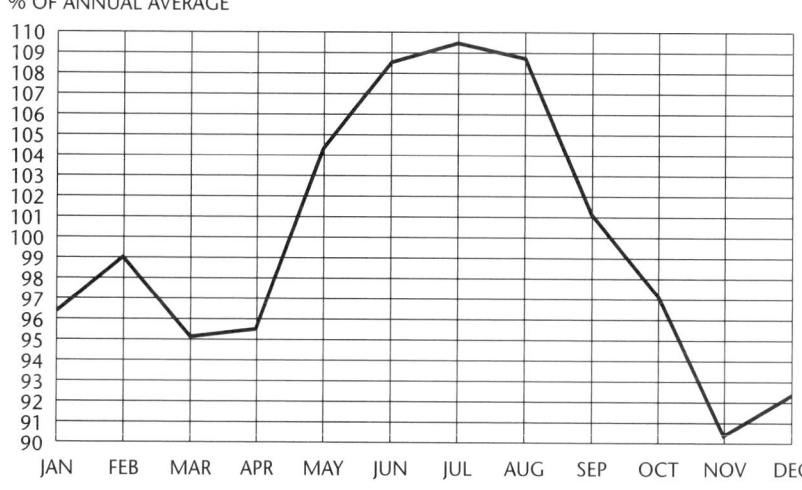

% OF ANNUAL AVERAGE

FIGURE 25-6 Seasonal trend in the price of slaughter barrows and gilts, United States. *Source: USDA.*

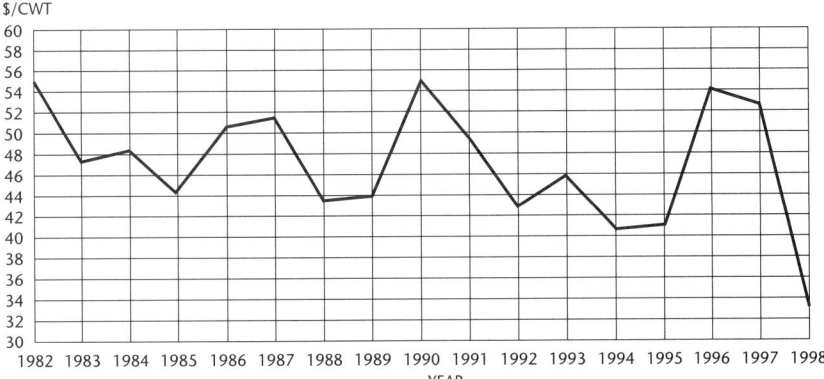

FIGURE 25-7 Annual average price of barrows and gilts, United States, 1982–1998. *Source: USDA.*

SHRINKAGE OF HOGS

Hogs lose weight as they are shipped to market. This weight loss is called *shrinkage*. The distance to market is one of the most important factors in determining the amount of shrinkage. A shrinkage of about 2 percent can be expected, regardless of how close to market the hogs are located. This is caused by the sorting, handling, loading, and hauling that takes place in the first few miles. As distance hauled and time on the road increase, the amount of shrinkage increases. Hogs hauled 150 miles (241.4 km), or more, may shrink as much as 4 percent.

Rough handling increases the amount of shrinkage. Temperatures below 20°F (–6.7°C) or above 60°F (15.5°C) also increase the amount of shrinkage. Careful handling of hogs while sorting and loading reduces losses from shrinkage. It also reduces death losses and the number of damaged carcasses that arrive at market. Keeping the hogs warm in winter and cool in summer while hauling them to market helps to reduce shrinkage and other losses. The wise hog producer reduces stress as much as possible when moving hogs.

FUTURES MARKET AND HEDGING

Live hogs are sometimes traded on the futures market. A *futures contract* establishes a price for live hogs that are to be delivered at some future date. The unit of trading is 30,000 pounds (13,608 kg). This is about 130 to 150 head of market hogs.

One objective of using the futures market is to obtain a higher price for hogs. Futures trading takes place when the futures price is higher than the expected market price of hogs at the time of delivery. The producer runs the risk that the expected market price is not accurate.

Another objective of using the futures market is to reduce the risk of loss if prices go down. By trading on the futures market, the producer locks in the price that will be received for the hogs. If the expected market price is lower than the costs of production, it does not pay to produce the hogs. In such a case, the producer would not be in the futures market. Reducing the risk by locking in the price is called *hedging.*

The producer who wants to hedge on the futures market needs a thorough understanding of the market. Detailed information on using the futures market may be obtained from brokers or from university specialists.

SUMMARY

How to market their hogs is one of the most important decisions that hog producers must make. Markets vary in the price offered, the services given, and the costs charged.

Most hogs in the United States are marketed by direct marketing. Smaller numbers are marketed through terminal markets and by auction market. Some group marketing systems have been established to help producers get higher prices.

Most hogs are sold on the basis of weight produced. Producers of quality meaty hogs can get more for their hogs by selling on a yield and grade basis.

Hogs are classified according to use, sex, weight, and quality. Butcher hogs are graded according to USDA official grading standards. Feeder pigs are also graded by USDA standards. The grades of market hogs are U.S. No. 1, U.S. No. 2, U.S. No. 3, U.S. No. 4, and U.S. Utility. Grading is based on the quality of the lean meat and the percent of four lean cuts that the carcass will produce.

Traditional recommendations have been to market hogs at 200 to 220 pounds (90.7–99.8 kg). Top butcher prices are currently paid for hogs weighing 220 to 260 pounds (99.8–117.9 kg). Meaty hogs can be fed to heavier market weights if the hog feed price ratio is right, the discount for heavy hogs is not too great, and the seasonal price is rising. Hog prices tend to be lower in early spring and higher in the summer.

Shrinkage is increased as hogs are moved greater distances to market. Careful handling to reduce stress when marketing reduces shrinkage and other losses.

The futures market may be used to reduce the risk of loss. A producer using the futures market should consult with experts.

Student Learning Activities

1. Prepare a bulletin board display of charts showing price trends and local prices for hogs.
2. Take a field trip to a local hog market.

3. Ask a manager from a local packing plant to discuss damage to hog carcasses that results from mishandling while marketing.

4. Ask a hog buyer to speak to the class.

5. Ask a hog producer who uses the futures market to hedge to speak to the class on the use of hog futures.

6. Give an oral report on hog marketing.

7. Survey hog producers in the community to determine local marketing practices.

Review

1. Compare direct marketing, terminal markets, and auction markets as methods of marketing hogs.

2. Explain how hogs are priced based on the total weight marketed.

3. Explain how hogs are priced on a yield and grade basis.

4. Describe the use classes of market hogs.

5. Describe the sex classes of market hogs.

6. How is the market grade of a hog determined?

7. Name the four lean cuts of the hog carcass.

8. Describe the five official USDA grades of market hogs.

9. Describe the six official USDA grades of feeder pigs.

10. Name and briefly explain the factors involved when deciding on the best weight to market hogs.

11. Describe the seasonal price trends for butcher hogs.

12. What factors affect the amount of shrinkage as hogs are moved to market?

13. How can the percent of shrinkage be kept low?

14. What are two reasons for a hog producer to use the futures market?

Sheep and Goats

26 *Selection of Sheep*

OVERVIEW OF THE SHEEP ENTERPRISE

Sheep are raised in every state in the United States (Figure 26-1). Range production is concentrated in twelve western states while native, or farm flock production, is found in the rest of the states. Small flocks of fewer than 100 head account for about 90 percent of sheep operations but only about 26 percent of the inventory in the United States (Figure 26-2). Range production accounts for about 70 percent of the total sheep production in the United States; large flocks of 1,000 to 1,500 ewes are common in this area. The production of feeder lambs and wool is located mostly in the southern part of the range area and market lamb production is more in the northern part of the area (Figure 26-3).

In the rest of the United States, farm flocks tend to be small. Sheep production in this area tends to be a secondary enterprise. About half of the sheep numbers in the farm flock area are located in the Corn Belt states. Both wool and market lambs are sold from these flocks. Purebred sheep production is also important in this area.

The number of small sheep enterprises is declining. Most of the lamb feeding is concentrated in large commercial feedlots. The leading states in numbers of sheep and lambs on feed are Colorado, California, Texas, and Wyoming. Large commercial flocks of over 1,000 head account for only two percent of all producers but produce more than 50 percent of the total sheep and lambs in the United States.

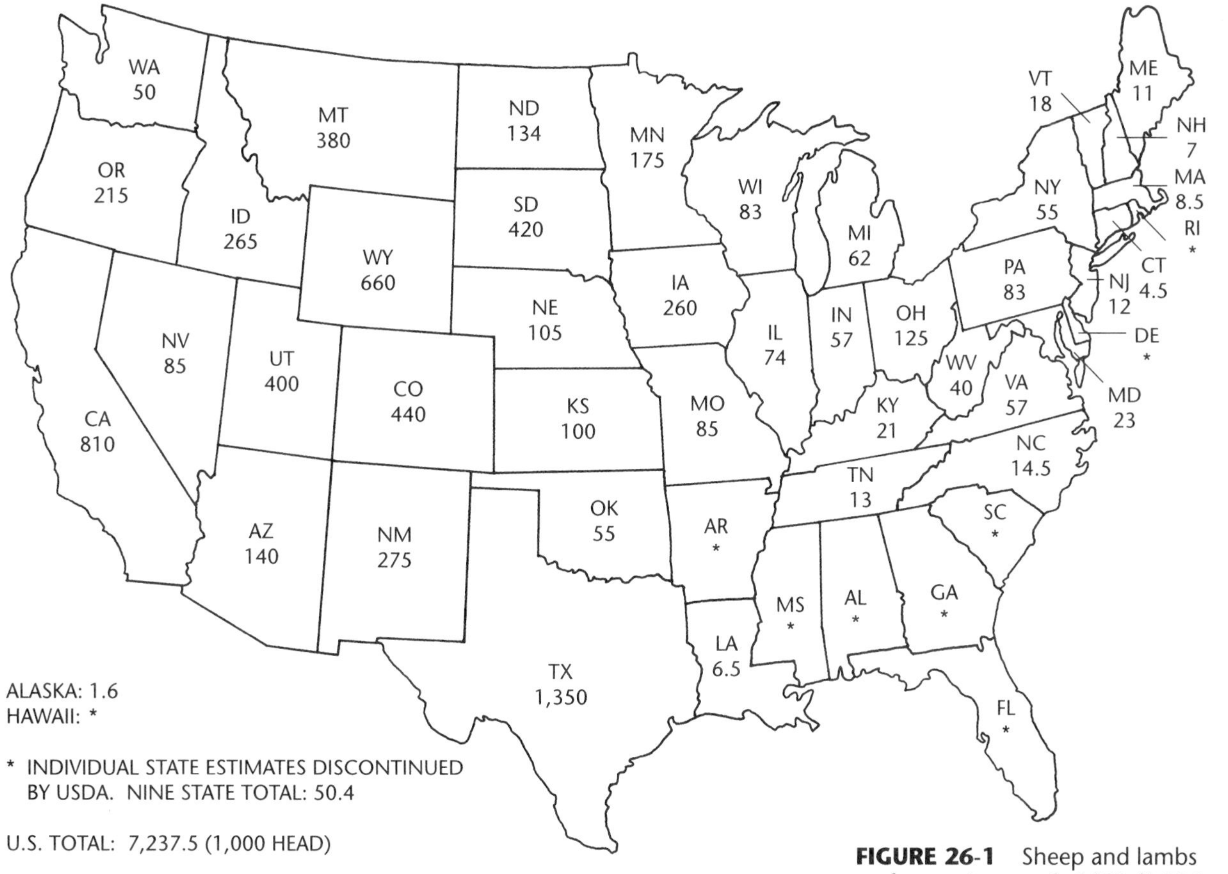

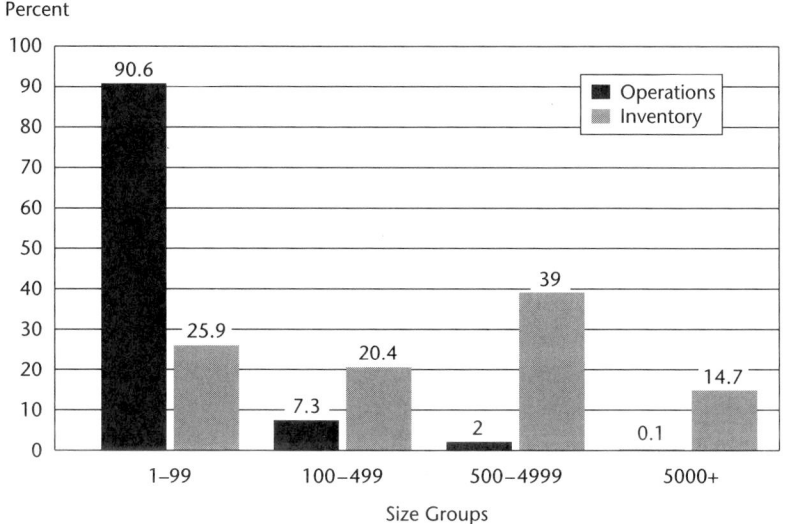

ALASKA: 1.6
HAWAII: *

* INDIVIDUAL STATE ESTIMATES DISCONTINUED
   BY USDA. NINE STATE TOTAL: 50.4

U.S. TOTAL: 7,237.5 (1,000 HEAD)

FIGURE 26-1 Sheep and lambs
on farms—January 1, 1999 (1,000
head). *Source: Sheep and Goats,* USDA,
January 1999.

FIGURE 26-2 Breeding sheep:
Percent of operations and inven-
tory, 1999. *Source: Sheep and
Goats,* USDA, January 1999.

FIGURE 26-3 Sheep being moved on rangeland in Nevada. Sheep do an excellent job of converting roughage into wool and meat. *Courtesy of USDA.*

Several factors have contributed to the decline in the sheep industry in recent years. These include:

- Seasonal demand for lamb meat.
- Low per capita consumption.
- Low wool prices.
- Use of artificial fibers instead of wool in clothing.
- Problems with predators.
- High labor requirement in the sheep enterprise and a lack of suitable labor.
- A lack of improvement in the slaughtering and marketing infrastructure.

Many producers in the western states are now using mixed grazing, that is, raising cattle and sheep on the same land. Mixed grazing is most prevalent in the Northern Plains and Texas. Two-thirds of the sheep producers in these states also raise cattle. The practice of mixed grazing provides more opportunity to increase livestock production than raising cattle or sheep alone.

Sheep make a good second enterprise on a farm because

- wool and lambs provide extra income.
- the initial costs are low.
- the enterprise does not require expensive housing or equipment.
- sheep make use of pasture crops that might otherwise be wasted.
- sheep can be fed on roughages and small amounts of grain.
- market lamb returns compete well with other meat animal enterprises.

The major disadvantages related to the sheep enterprise are that

- dogs and other predatory animals attack sheep.
- sheep are susceptible to internal and external parasites.

■ sheep do not resist diseases or injuries very well.

■ wool prices are quite variable.

CLASSES OF SHEEP

There are a number of ways to classify sheep. The most commonly used classification is by type of wool. The wool type classifications are: (1) fine wool, (2) medium wool, (3) long wool, (4) crossbred wool, (5) carpet wool, and (6) fur sheep.

Fine Wool Breeds

The fine wool breeds are: (1) Rambouillet, (2) American Merino, (3) Delaine Merino, and (4) Debouillet. All of these breeds were developed from the Spanish Merino. They produce a fine wool fiber that has a heavy *yolk,* or oil content. Originally these breeds did not produce good meat carcasses. However, through selection and breeding, the quality of carcass has been improved. These breeds are still used primarily for wool production. They have a strong flocking instinct. A high percent of the sheep in the range areas are of the fine wool breeds. They have the ability to do well on poor-quality rangeland. Fine wool breeds will breed out of season and, thus, can produce lambs in the fall months.

Medium Wool Breeds

The medium wool breeds are: (1) Cheviot, (2) Dorset, (3) Finnish Landrace, (4) Hampshire, (5) Montadale, (6) Oxford, (7) Shropshire, (8) Southdown, (9) Suffolk, and (10) Tunis. Medium wool breeds were originally bred mainly for meat. Wool production is secondary. The fleece is medium in fineness and length. Medium wool breeds are popular in both the range and farm flock production areas. They are used primarily for meat in most programs for meat production.

Long Wool Breeds

The long wool breeds are: (1) Cotswold, (2) Leicester, (3) Lincoln, and (4) Romney. These breeds were developed in England, and are larger than the other breeds of sheep. They produce a long, coarse-fiber wool. They are hardy and prolific, but tend to be late maturing. Depending on slaughter weight, the carcass quality may be poor, carrying too much fat. Excess fat may not be present at slaughter weights of less than 120 pounds (54 kg). These breeds are used mainly in crossbreeding programs.

Crossbred Wool Breeds

The crossbred wool breeds are: (1) Columbia, (2) Corriedale, (3) Panama, (4) Romeldale, (5) Targhee, (6) Tailless, and (7) Southdale.

Most of the crossbred breeds were developed by crossing long wool with fine wool breeds. Some of the newer breeds developed by agricultural experiment stations involved different types of crosses.

Crossbred wool breeds were developed mainly to improve the carcass quality and the length of the wool fiber. These breeds have better banding (flocking) instinct than either the long wool or medium wool breeds. *Banding* (or *flocking*) instinct refers to the tendency of sheep to stay together in a group called a *band* or *flock*. They are well adapted to the western range, but are also popular in the Corn Belt states.

Carpet Wool Breed

There are several breeds of sheep used in countries outside of the United States for the purpose of producing carpet wool. The only breed used in the United States for this purpose is the Black-faced Highland. The fleece is coarse, wiry, and tough. The length of the fiber varies from 1 to 13 inches (2.54–33.0 cm).

Fur Sheep Breed

The only breed of sheep raised mainly for fur in the United States is the Karakul. The fur pelts are taken from the young lambs to make coats. The meat value of the breed is low.

BREEDS OF SHEEP

There are many breeds of sheep available to producers in the United States. Many of the breed associations register only small numbers with their purebred associations in any given year. In rank order, Suffolk, Hampshire, Dorset, Rambouillet, Columbian, Corriedale, Southdown, Shropshire, Polypay, Montadale, and Cheviot have been the leading sheep breeds in the United States based on total numbers registered since 1970.

Fine Wool Breeds

Merino. Merino sheep originated in Spain. They were first imported into the United States in 1801. The three types of Merinos are called *A, B,* and *C*. The A and B types have wrinkled skins, with the A type being more wrinkled than the B type. Wrinkled-skin Merinos are known as *American Merinos*. The C type Merino, called Delaine Merino, has very little wrinkle to the skin. Only the Delaine Merino is popular in the United States.

The fleece of the Delaine Merino is white and grows about 2.5 to 3 inches (6.4–7.6 cm) per year. It is more uniform than the fleece of the American Merino. The rams are horned and the ewes polled. Merinos are medium in size and have angular bodies (Figure 26-4). The

FIGURE 26-4 Merino ram. *Courtesy of American and Delaine Merino Record Association, OH.*

Delaine Merino is the largest of the Merinos. Merinos have a strong banding instinct and are able to do well on poor grazing land.

Rambouillet. The Rambouillet originated in France. The breed was developed from the Spanish Merino. The first importations into the United States took place about 1840. About one-half, or more, of the crossbred sheep in the United States carry some Rambouillet blood. The breed is particularly popular in the western states and is the most popular of the fine wool breeds.

Rambouillets are white in color. They are large and have an angular, blocky body type. They produce a meatier carcass than the Merino, but it is not as good as the carcass of the breeds bred for meat production. There are both horned and polled rams. All ewes are polled. Their fleece grows about 3.5 inches (8.9 cm) per year.

Debouillet. The Debouillet breed originated in the United States. It was developed from crosses of Rambouillet and Delaine Merino. Development began in New Mexico. The Debouillet is adapted to the western range conditions.

The Debouillet is medium-sized and has an angular type body. The color is white. Rams may be horned or polled; ewes are polled. The body is smooth and a long staple wool is produced. *Staple* means the fibers of wool. Long staple wool is the most valuable wool from the fleece.

Medium Wool Breeds

Cheviot. The Cheviot originated in the Cheviot Hills of northern England and southern Scotland. The breed was developed during the 1800s. The first importations into the United States took place about 1838.

The Cheviot is small and has a blocky body type (Figure 26-5). It has a white face and legs and black nostrils. It carries its head erect and presents an alert appearance. The breed is polled. The fleece grows about 4 to 5 inches (10.2–12.7 cm) per year. The breed does not have a strong flocking instinct.

Dorset. The Dorset originated in southern England. The breed was developed in the early 1800s. The first importations into the United States took place about 1887.

The Dorset is of medium size and has a blocky body type (Figures 26-6A and 26-6B). The ears, nose, face, and legs are white. There are both polled and horned strains of Dorset. They produce a medium-coarse fleece. Dorsets shear a light-weight fleece, averaging 7 to 8 pounds (3.2–3.6 kg). They produce a muscular carcass. The ewes will breed out of season, so fall lambs can be produced.

Finnish Landrace (Finnsheep). The Finnsheep originated in Finland. It was first imported into the United States in 1968.

FIGURE 26-5 Cheviot yearling ram. *Courtesy of Sheep Breeder Magazine*

FIGURE 26-6A Polled Dorset ram. *Courtesy of Continental Dorset Club, Inc., IA.*

FIGURE 26-6B Continental Dorset ewes. *Courtesy of Continental Dorset Club, Inc., IA.*

FIGURE 26-7A   Yearling Finn ewe. *Courtesy of Bill Carter, Finnsheep Breeders Association, IN.*

FIGURE 26-7B   Finn ewe with fall born triplets. *Courtesy of Bill Carter, Finnsheep Breeders Association, IN.*

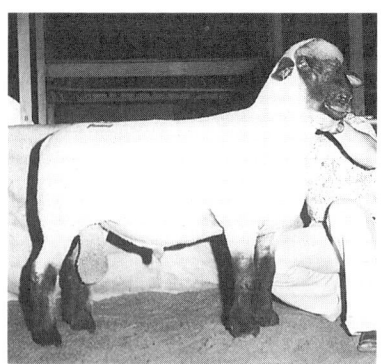

FIGURE 26-8   Hampshire yearling ram. *Courtesy of The Sheep Breeder and Sheepman, MO.*

The Finnsheep is small in size. The ears, nose, face, and legs are white (Figures 26-7A and 26-7B). The breed is generally polled, although some rams may have horns. The Finnsheep produces a medium-coarse fleece that averages 9 to 10 pounds (4–4.5 kg) in weight. The carcass characteristics of the Finnsheep are not as desirable as those possessed by some of the other breeds of sheep.

The Finnsheep is noted for its high lambing rate. Its principal use in the United States is in crossbreeding programs to increase lambing rate. Mature crossbred (1/2 Finnsheep) ewes will commonly produce a 200 to 250 percent lamb crop. Finnsheep crossbred lambs are small at birth but have a high survival rate. When Finnsheep are used in crossbreeding, the lambs produce acceptable carcasses.

Hampshire.   The Hampshire breed originated in southern England. The breed was developed during the late 1700s and early 1800s. It was first imported into the United States before 1840. All of the flocks from these importations disappeared during the Civil War. Later importations occurred around 1881.

Hampshires are large in size and have a blocky body type (Figure 26-8). The face, legs, ears, and nose are black. The breed is polled. Hampshires are medium in maturation rate, are good milkers, and produce lambs that are often ready for market at weaning. They produce a fleece of about 7 to 8 pounds (3.2–3.6 kg) of medium-fine wool. The lamb crop averages about 137 percent. That is, 137 lambs are produced for each 100 ewes. Hampshires are one of the most popular of the medium wool breeds in the United States, especially in the Midwest. They cross well with the fine wool or crossbred breeds to produce market lambs.

Montadale.   The Montadale breed originated in the United States in 1933. Mr. E.H. Mattingly, of St. Louis, Missouri, developed the breed by crossing Columbia ewes with Cheviot rams.

The Montadale is a medium-large breed with a blocky body type (Figure 26-9). The face, ears, and legs are white. There is no wool on the face and legs. The breed is polled. The Montadale produces a good meat carcass. The fleece averages 10 to 12 pounds (4.5–5.4 kg).

Oxford.   The Oxford breed originated in south-central England. The breed was developed about 1836 from crosses of Hampshire, Cotswold, and Southdown. The first importations into the United States occurred about 1846.

The Oxford is one of the largest of the medium wool breeds of sheep (Figure 26-10). It has a blocky body type. The face, ears, and legs are gray to brown. The breed is polled. Wool extends over the poll down to the eyes. The Oxford shears a heavy fleece, averaging 10 to 12 pounds (4.5–5.4 kg). The ewes are prolific and good milkers, and the lambs grow quickly. The breed is used in crossbreeding because of its large size.

Shropshire. The Shropshire breed originated in England around 1860. The first importations into the United States were about 1885. The Shropshire is one of the smaller of the medium wool breeds. It has a blocky body type and is polled (Figure 26-11). The face and legs are dark colored. It has a heavy face covering of wool. Shropshires produce a small carcass. They become too fat when fed to heavy weights. The fleece shears about 9 pounds (4.1 kg). The ewes are good mothers, and the lamb crop is often 150 percent. However, because of their small size, these sheep have lost popularity in the United States.

Southdown. The Southdown breed originated in southern England in the late 1700s. It is one of the oldest sheep breeds in the world, and was used in the formation of most of the medium wool breeds. The first importations into the United States took place about 1803.

The Southdown is small and has a blocky, low-set body type (Figure 26-12). The face, ears, and legs are a gray or mouse brown. The breed is early maturing and polled. They shear a fleece of 5 to 7 pounds (2.3–3.2 kg). Southdowns tend to become overfat at an early age. They have lost popularity in the United States because they are too slow and small in growth.

Suffolk. The Suffolk breed originated in southern England. It was developed in the early 1800s. The first importations into the United States took place about 1888.

The Suffolk is large and has a blocky, muscular body type (Figure 26-13). The face, ears, and legs are black. The breed is polled, and has no wool on the head or legs. The Suffolk shears a fleece of 8 to 10 pounds (3.6–4.5 kg). The lamb crop is often 150 percent or better. The lambs grow rapidly; deposit fat at a slower rate than other breeds; and produce lean, muscular carcasses with desirable yield grades. The Suffolk has gained popularity in the United States, especially in the production of market lambs.

Tunis. The Tunis breed, which originated in North Africa, is an ancient breed of sheep. It was imported into the United States about 1799. It has never been a popular breed in the United States. The color of the face is reddish brown to bright tan. The breed is polled. The Tunis is of medium size and has an angular, blocky body. The wool is rather coarse, and there is none on the face.

Long Wool Breeds

Cotswold. The Cotswold is an old breed of sheep that originated in England. Improvement of the breed occurred during the late 1700s. The first importations into the United States took place about 1832.

The Cotswold is large and has a blocky body type (Figure 26-14). The face and legs are white, although a grayish-white color is not

FIGURE 26-9   Montadale ram lamb. *Courtesy of The Sheep Breeder and Sheepman, MO.*

FIGURE 26-10   Oxford ram. *Courtesy of American Oxford Sheep Association, IL.*

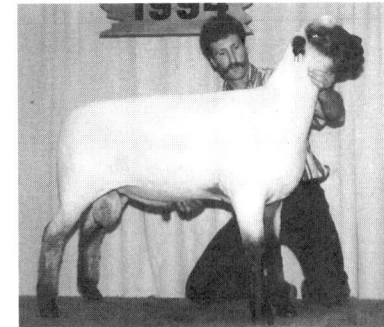

FIGURE 26-11   Shropshire ram lamb. *Courtesy of Sheep Breeder Magazine.*

FIGURE 26-12   Southdown ram. *Courtesy of American Southdown Breeders Association, TX.*

FIGURE 26-13   Suffolk ram lamb. *Courtesy of The Sheep Breeder and Sheepman, MO.*

FIGURE 26-14   Cotswold ram. *Courtesy of The Sheep Breeder and Sheepman, MO.*

objectionable. Dark spots may occur on the face and legs. The wool of this breed grows in long, wavy curls. There is a tuft of wool on the forehead. The wool is coarse and long. The breed is polled. The Cotswold has been used in crossbreeding programs. It is not a popular breed in the United States.

Leicester. The Leicester originated in England. It was developed in the late 1700s. The first importations into the United States took place in the late 1700s.

The Leicester is large and has a blocky body type. The breed is white and polled. There is no wool on the face. The fleece does not give Leicester sheep good protection from the weather. They tend to become wet and chilled. They are only moderately hardy. It is not a popular breed in the United States.

Lincoln. The Lincoln originated in England. It is an old breed, and was first imported into the United States in the late 1700s.

The Lincoln is large and has a blocky body type (Figure 26-15). It is white, polled, and produces a long, coarse fleece. It is a late-maturing breed. The Lincoln produces 12 to 16 pounds (5.4–7.2 kg) of wool. This breed has not been popular in the United States. It has been used mainly for crossbreeding.

Romney. The Romney is also an old breed that originated in southern England. The first importations into the United States took place around 1904.

The Romney is a large, hardy breed with a blocky body type (Figure 26-16). The breed is white and polled. It produces a more compact, finer fleece than the other long wool breeds. For this reason, the Romney is better adapted to wet, marshy areas than are other long wool breeds of sheep. It has been used in crossbreeding programs.

Crossbred Wool Breeds

Columbia. The Columbia originated in the United States in 1912. The breed was developed from a cross of Lincoln rams and Rambouillet ewes. The Bureau of Animal Industry did the primary work in developing the breed in Wyoming and Idaho. The purpose of the cross was to produce a breed that would be better adapted to the intermountainous regions of the West.

The Columbia is a large, blocky breed (Figures 26-17A and 26-17B). It is the largest of the crossbred wool breeds. The face, ears, and legs are white, and the breed is polled. There is no wool on the face. The Columbia is slightly longer legged than other breeds. It shears a 10- to 13-pound (4.5–5.9 kg) fleece and produces a lean market lamb with acceptable leg scores. The Columbia is often used in the Midwest in crossbreeding with black-faced medium wool breeds to produce market lambs.

Corriedale. The Corriedale originated in New Zealand and was developed between 1880 and 1910. It was a result of crossing Lincoln and Leicester rams with Merino ewes. The first importations into the United States took place in 1914. The animals were used in western range conditions.

The Corriedale is a medium-large, blocky breed (Figure 26-18). The face, ears, and legs are white, and the breed is polled. It produces a good-quality wool and acceptable carcass. Corriedales shear a 10- to 15-pound (4.5–6.8 kg) fleece.

Panama. The Panama breed originated in the United States in 1912. It was developed in Idaho from a cross of Rambouillet rams on Lincoln ewes.

The Panama is a large, blocky, polled breed. The color of the face, ears, and legs is white. It is a little smaller than the Columbia and produces a slightly meatier carcass that develops fat deposits at an earlier age.

Romeldale. The Romeldale originated in the United States in 1915. It was developed in California from a cross of Romney rams and Rambouillet ewes.

The Romeldale is a medium-large, blocky breed. The color of the face, ears, and legs is white. The breed is polled. It shears a fleece of 10 to 13 pounds (4.5–5.9 kg). Romeldale has been used for crossbreeding with fine wool ewes to produce lambs adapted to range conditions.

Targhee. The Targhee originated in the United States in 1927. It was developed in Idaho by the USDA from a cross of Rambouillet rams on Corriedale-Lincoln-Rambouillet ewes. A program of interbreeding was then followed based on production performance on the range.

The Targhee is a medium-large, blocky breed, with a white face (Figures 26-19A and 26-19B). The breed is polled and has no wool on the face. It will shear about 10 to 12 pounds (4.5–5.4 kg).

Tailless. The Tailless breed originated in the United States. It was developed by the South Dakota Experiment Station by crossing a fat-rumped, no-tail sheep from Siberia with a number of the improved breeds. The purpose of the cross was to develop a breed that did not require tail docking. The color of the face, ears, and legs is white. The breed is polled and has no wool on the face. Currently, it is not a popular breed in the United States.

Southdale. The Southdale breed originated in the United States. It was developed by the USDA by crossing Southdowns and Corriedales. The purpose of the cross was to produce a breed with improved fleece and meat quality. However, it is not a popular breed, nor is it used much in crossbreeding programs in the United States.

FIGURE 26-15   Lincoln sheep. *Courtesy of Sheep Breeder Magazine.*

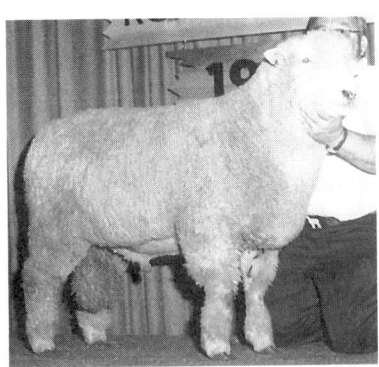

FIGURE 26-16   Romney yearling ram. *Courtesy of The Sheep Breeder and Sheepman, MO.*

FIGURE 26-17A   Columbia ram. *Courtesy of Columbia Sheep Breeders Association of America, OH.*

FIGURE 26-17B Columbia lambs. *Courtesy of Columbia Sheep Breeders Association of America, OH.*

FIGURE 26-18 Corriedale ram. *Courtesy of Sheep Breeder Magazine.*

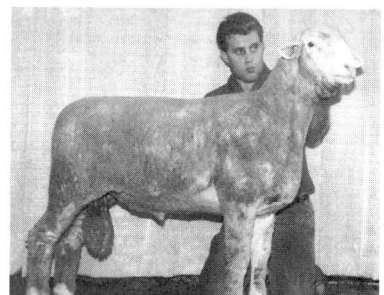

FIGURE 26-19A Targhee sheep. *Courtesy of Sheep Breeder Magazine.*

FIGURE 26-19B Flock of Targhee sheep. *Courtesy of U.S. Targhee Sheep Association.*

Carpet Wool Breed

Black-faced Highland. The Black-faced Highland is an old breed, originating in Scotland. The first importations into the United States took place about 1861.

The breed is small in size. These sheep have a long, coarse outer-coat of the wool with a finer undercoat. The face and legs are black. The rams have large, spirally curved horns. The ewes have small, short, curved horns. The breed can live in rough land areas with little vegetation. It has never been popular in the United States.

Fur Sheep Breed

Karakul. The Karakul originated in Asia and is an ancient breed. The first importations into the United States took place about 1909.

The Karakul is a large, angular-bodied breed (Figures 26-20A and 26-20B). The color of the face, ears, and legs is black or brown. The rams have horns, and the ewes are polled. The lamb pelts of this sheep at birth, and for a few days thereafter, are used as furs. The wool on the mature animal is 6 to 10 inches (15.2–25.4 cm) long. It is coarse, wiry, and of little value. The lambs produce poor carcasses. The Karakul has never been a popular breed in the United States.

SELECTION OF SHEEP

National Sheep Improvement Program

The American Sheep Industry Association supports the National Sheep Improvement Program (NSIP) that is available to all sheep producers. This program provides accurate estimates of the genetic value of breeding sheep. Performance and pedigree information on their sheep is provided by the producer to the NSIP processing center that is located at the offices of the American Polled Hereford Association in Kansas City, Missouri. In return, the producer receives an estimate, known as the flock expected progeny differences (FEPD), of the genetic value of every ewe, ram, and lamb for which information was submitted. The information is based on the individual records plus the records of all relatives of the individual in the flock. Genetic values for reproductive, growth, and fleece traits are currently available.

A Ewe Summary report is provided each year that shows the ewe's pedigree, her performance as a lamb, an update of her FEPD, and the performance of each lamb she has produced. A Ram Summary report provides the same information on rams in the flock. It gives the average performance for the year on his lambs rather than by individuals. A Management Summary report is provided that gives the average performance for the flock for the current year and the previous year. It also reports the number of sheep that left the flock during the year and why they left.

The NSIP currently only reports genetic information on animals within a flock. The report cannot be used to compare animals in different flocks. It is anticipated that as the data base grows it will be possible to provide information that will allow the comparison of animals among different flocks.

Ram Testing Stations

Ram testing stations provide information that allows the comparison of ram lambs from several different flocks. Producers looking for breeding stock can purchase performance-tested rams from ram testing stations.

Ram lambs are usually brought to a station when they are seven to twelve weeks of age. They are on test from eight to twelve weeks. They are weighed at regular intervals, usually every two or three weeks, and the average daily gain is determined. The gain of each ram lamb is compared to the average for its breed group on test. Feed efficiency data is collected at some stations for ram lambs from the same sire.

Ram lambs are evaluated for soundness at the end of the test period. Some stations also provide scrotal circumference measurements, fleece information, and ultrasound fat and loin eye measurements on the test lambs.

Breed Selection

Personal preference is the main factor in selecting a breed of sheep. Other factors that should be considered are

- how well the breed is adapted to the area.
- the market for the product.
- the availability of breeding stock.
- multiple births.

Most of the breeds in the western areas carry a high percent of fine wool breeding. This is of value because of the banding instinct of these breeds. Sheep with a great deal of wool on the face may have more problems with wool blindness in areas where ice, snow, and some types of grasses are grown. Breeds that are more open faced do not have these problems. Some breeds are better suited to live and grow in range areas where the vegetation is not as lush and the ground is rougher.

A purebred breeder must consider the demand for the kind of breed that is produced. Some breeds produce meatier, leaner carcasses than others. If the main market is for meat, then breeds that produce a meaty carcass should be chosen. The market for the kind of wool produced varies from area to area. The breed selected must produce the kind of wool in demand.

FIGURE 26-20A   Two Karakul ewes and a ten-day-old lamb. The ewes are in full fleece. *Courtesy of American Karakul Sheep Registry, WA.*

FIGURE 26-20B   Adult Karakul ewe. *Courtesy of American Karakul Sheep Registry, WA.*

FIGURE 26-21   With the fingers close together and slightly cupped, check for finish over the back. Work from the rear to the front. *Courtesy of Iowa State University.*

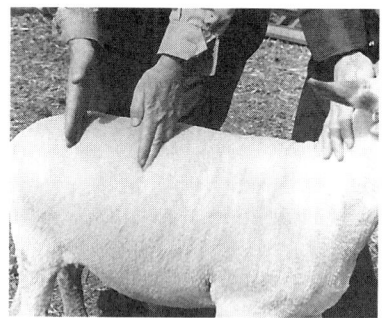

FIGURE 26-22   Check the length of loin. *Courtesy of Iowa State University.*

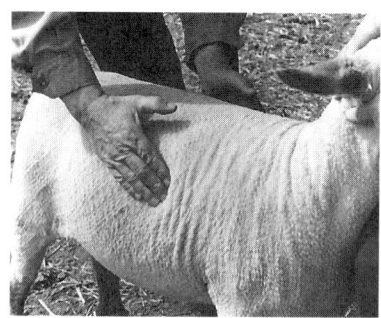

FIGURE 26-23   Check the finish over the ribs. *Courtesy of Iowa State University.*

The producer must consider what kind of breeding stock is available for crossbreeding programs. This depends on the popular purebred breeds in the area and the traits needed in the crossbreeding program. Costs increase if the producer must travel great distances to find the required breed.

Selecting Native or Western Ewes

Ewes that are produced in the western range area are referred to as *western ewes*. Those produced in other parts of the United States are called *native ewes*. Western ewes have a high percent of fine wool breeding. Native ewes are generally of the medium, long, or crossbred wool types.

In the range area, western ewes are generally the desired type to buy. In the farm flock area, either type may be used.

Native ewes generally have a larger lamb crop and produce a more muscular, leaner carcass. Costs are usually less. These ewes are often better adapted to local conditions.

Western ewes are less likely to have parasites. They are often more uniform in size. Western ewes are generally hardier and have longer productive lives.

Selecting Breeding Sheep

All animals selected for the breeding flock must be healthy. Do not select sheep that have diseases or parasites. Some indications of poor health include dark, blue skins; paleness in the lining of the nose and eyelids; lameness; and lack of vigor.

Ewes.   Check the udder for softness and soundness of the teats. Teeth, feet, legs, eyes, and breathing should all be normal. The incisor teeth should meet the dental pad.

The fleece should be heavy. Open-face ewes produce about 15 percent more wool than ewes with a heavy wool covering on the face. Select ewes with uniform fleece. The fiber length should be reasonably long. A dense, compact fleece with bright luster is desired. The yolk should be evenly distributed. Avoid open fleeces with dark-colored fibers.

Large-framed ewes have less difficulty lambing. They are better milkers, are more prolific, and shear heavier fleeces.

Select uniform ewes of correct type. Correct type includes the following traits:

■ generous length, depth, and width of body.
■ medium short, thick neck.
■ wide, deep chest.
■ good spring of rib.
■ depth in the fore and rear flank.

- width and thickness through the loin.
- strong, wide back; tight shoulder.
- long, wide, and level over the rump.
- straight legs with generous width between them.
- strong pasterns, strong feet, medium-sized toes.
- feminine appearance.
- purebred ewes show breed traits.
- all parts of the body blend well together.
- full, plump, muscular legs.
- wide in the dock.
- well-muscled, firm body.

FIGURE 26-24   Check the width of rump. *Courtesy of Iowa State University.*

Well-grown yearling ewes are a better choice than older ewes. Medium, long, and crossbred wool breeds begin to decrease in productivity at six years of age. Fine wool breeds produce for about one year longer. Older ewes are worth less because they have fewer productive years remaining.

Up to the age of four, the front teeth of the lower jaw are a good indication of age. Lambs have small, narrow teeth. At one year of age, the two center incisors are replaced by two broad, large permanent teeth. Each year thereafter, another set of permanent teeth appears until, at four years, the sheep has all its permanent teeth. At this age, the teeth begin to wear down. The sheep begins to lose some of its teeth at about five or six years of age. This condition is called *broken mouth.*

It is necessary to handle sheep to determine the exact conformation and muscling that is present. This is because the wool hides some of these traits from visual inspection. When handling sheep, keep the fingers close together and slightly cupped. Work from rear to front. Press firmly with the finger tips. Check for finish over the back. Feel the shoulders for smoothness. Check length of loin, width and depth of loin, width of rump, and muscling in the leg. The length and quality of the fleece is also checked by handling. Figures 26-21 through 26-26 illustrate the handling of sheep to determine conformation and muscling.

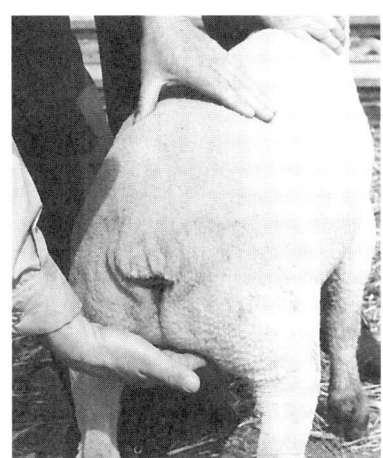

FIGURE 26-25   Check the depth of rump. *Courtesy of Iowa State University.*

Rams. Great care should be exercised in selecting the ram. Select a ram that will bring traits into the flock that are lacking in the ewes. The ram should be large, rugged, muscular, and masculine. He should have plenty of bone. If possible, determine the growth rate of the ram by checking the 90- to 120-day weight. Fast growth rate is 30 percent heritable and is a trait that is passed on to the lambs. Check for two well-developed, pliable testicles. The best indication of potential fertility in a young ram is the circumference of the scrotum. When the ram reaches puberty at five to seven months of age, the average scrotal circumference is 12 to 13 inches (30–33 cm). A ram with below-average scrotal circumference may be late maturing or have low fertility.

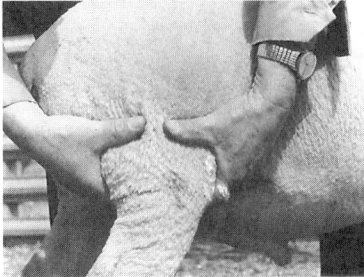

FIGURE 26-26   Check the muscling in the leg area. *Courtesy of Iowa State University.*

Because these traits are inherited, do not select such rams for breeding stock. Select rams with above-average scrotal circumference at puberty. A good-quality ram that costs more to start with will usually be a better investment in the long run. Poor-quality, low-performance rams will not improve the quality of the flock. Purchase rams with superior growth, muscle, soundness, and flesh quality.

JUDGING SHEEP

The general procedures for judging sheep are basically the same as those for judging beef or hogs. However, sheep must be judged on two particular traits—meat and wool.

A judge must be familiar with the parts of the sheep. Figure 26-27 shows the parts of the sheep.

Market Lambs

Market lambs are judged on the basis of type, muscling, finish, carcass merit, yield, quality, balance, style, soundness and smoothness. A number of these traits are described earlier in this unit in the discussion of sheep selection.

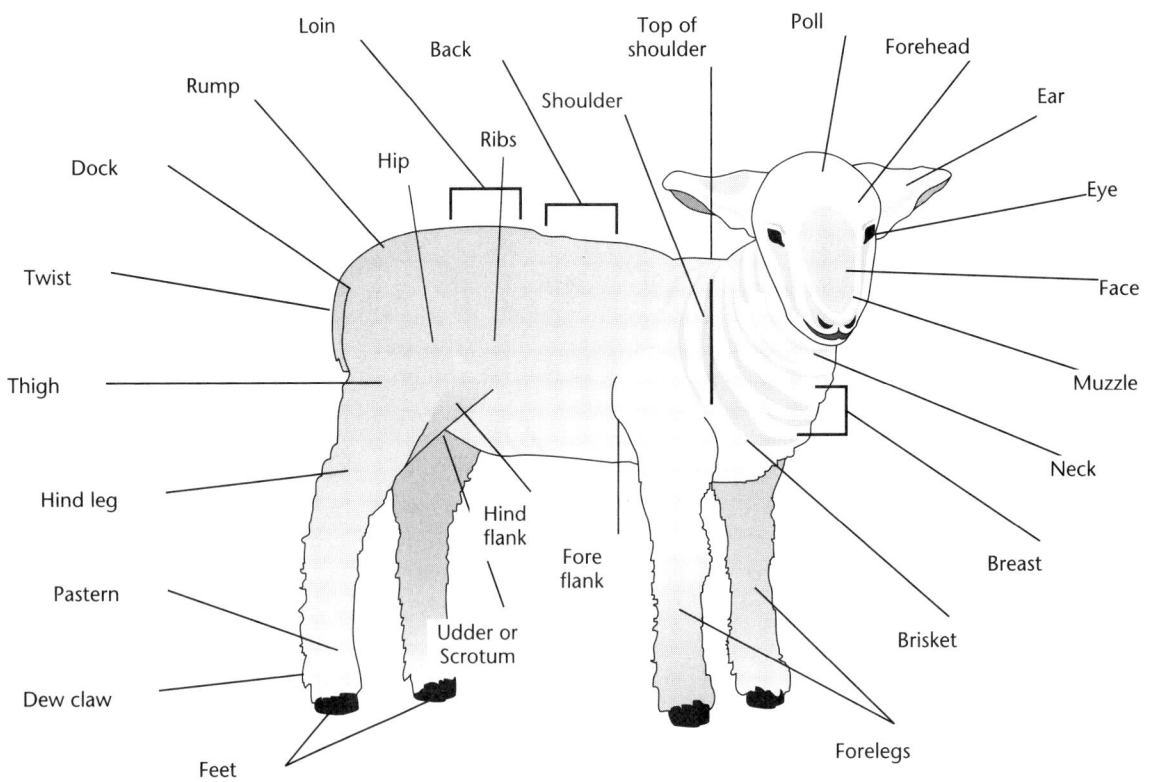

FIGURE 26-27   Parts of a sheep.

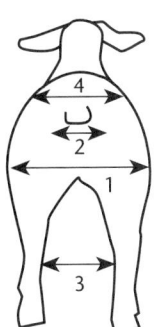 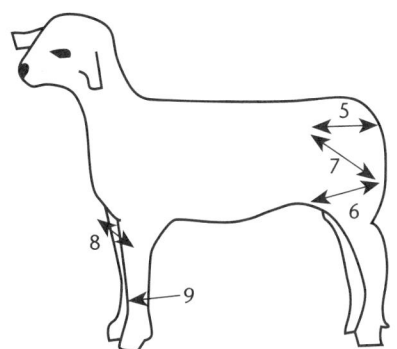

Look for width—
1. at the center of leg
2. at the dock
3. between hind legs
4. over back and loin

Look for—
5. long rump
6. long, bulging stifle
7. depth of leg
8. width at chest
9. heavy bone

FIGURE 26-28   Rear and side view of a meaty lamb and what to look for in judging.

Type is the general build of the sheep. Muscling is shown by the thickness and firmness of the leg, firmness of muscle over the top, and in the shoulders (Figure 26-28). Meaty, heavily muscled lambs are more desirable than those that are overly fat or lightly muscled.

Excessive fat is undesirable. The backfat should measure 0.15 to 0.20 inches (0.38–0.51 cm) at the twelfth rib. Finish is found by handling. Too much finish is shown by a soft, mellow touch. If the backbone and ribs can be felt too readily, the lamb is too thin. The lamb must produce a thick, meaty, correctly finished carcass.

The *pelt* consists of the skin and fleece. A pelt of 3/8 inch (0.95 cm) is desirable. A lamb will have a high dressing percent if it is heavily muscled and trim-middled with a 3/8-inch (0.95-cm) pelt. Too heavy a pelt will decrease dressing percent.

Quality, balance, style, and smoothness are defined and discussed in Unit 14 in the section on judging beef cattle. The same definitions apply to sheep.

Breeding Sheep

The general traits to judge in breeding sheep are discussed earlier in this unit in the section on selection of breeding sheep. Breeding sheep are also judged on condition (finish in market lambs), size, feet, legs, bone, breed and sex character, and fleece.

Notes should be taken and oral reasons given, just as they are with beef and hogs. The judging terms used are similar to those for beef cattle, only using the correct names for the parts of the sheep.

SUMMARY

Sheep are a major farm enterprise in the western range area of the United States. In the rest of the United States, farm flocks tend to be a secondary enterprise.

A common way to classify sheep is by the type of wool. These types are fine wool, long wool, crossbred wool, medium wool, carpet wool, and fur sheep. Fine wool breeding is more common in the western range areas. The medium wool and crossbred wool breeds are more common in the farm flock area. Long wool, carpet wool, and fur sheep are not common in the United States.

Breed selection is mainly a matter of personal preference. However, adaptation of the breed to the area and purpose should also be considered.

Sheep are selected on the basis of health, soundness, fleece, conformation, and age. Sheep are judged on basically the same traits. It is necessary to handle sheep when selecting or judging to determine some of the traits present. Market lambs are judged on the additional trait of finish.

Student Learning Activities

1. Prepare a bulletin board display of pictures of the breeds of sheep.
2. Take a field trip to observe traits of the breeds found on local farms.
3. Survey the local area and prepare a list of purebred breeders that could serve as a source of breeding stock.
4. Name the parts of the sheep using a picture or live sheep.
5. Practice good selection procedures for a sheep project or on your home farm.
6. Give an oral report on breeds, selection, or judging sheep.
7. Attend a purebred sheep sale and select sheep to buy using accepted criteria.

Review

1. What are the two types of sheep production in the United States and where is each located?
2. Briefly describe the characteristics of each type of sheep production.
3. Discuss the advantages and disadvantages of sheep production.
4. Name the six wool type classifications of sheep, list some typical breeds in each, and briefly describe each classification.
5. List and describe the breeds of sheep.
6. What factors should a producer consider when selecting a breed of sheep?
7. Compare native and western ewes.
8. Describe the characteristics that are considered desirable for breeding sheep.

9. Why are well-grown yearling ewes a better choice than older ewes when selecting breeding sheep?

10. Describe how the age of a sheep can be estimated by examining the teeth.

11. Describe how a sheep should be handled to determine conformation and muscling.

12. Describe the characteristics of a desirable breeding ram.

13. Draw a sketch showing the parts of a sheep.

14. Describe the characteristics of a desirable market lamb.

Unit 27

Feeding, Management, and Housing of Sheep

Objectives

After studying this unit, the student should be able to

- describe four systems of raising sheep.
- plan feeding programs for sheep.
- discuss management practices for sheep.
- describe housing and equipment needs for sheep.

SYSTEMS OF RAISING SHEEP

Sheep production may be broadly divided into purebred and commercial producers. A further classification of systems may be made based on the time of lambing, as follows: (1) fall lambs, (2) early spring lambs, (3) late spring lambs, and (4) accelerated lambing. Commercial producers may also buy feeder lambs to finish for market.

Purebred producers produce breeding stock for commercial flocks and for other purebred flocks. A purebred producer must have experience in sheep production. The major goal of the purebred producer is breed improvement. Good production records are necessary.

Commercial producers maintain sheep flocks to produce meat and wool. The size of the flocks varies from small farm flocks to the large flocks found in the western range areas. In the farm flock areas, sheep tend to be a secondary enterprise. In the range areas, sheep are often the major livestock enterprise.

Fall Lambs

Lambs born before December 25th are generally referred to as *fall lambs.* Breeds of sheep that will breed out of season are necessary for this system. Some possible breeds are Rambouillet, Merino, Dorset, Corriedale, and Tunis. Some crosses of these breeds, as well as Hampshires and Suffolks, will also breed out of season.

Fall lambs are marketed from early spring to June. The price for market lambs is usually higher at this time than during the rest of the year. In some areas, lambs are sold at 50 to 90 days of age. They weigh 35 to 60 pounds (15–27 kg) and are called *hothouse lambs.* These lambs go on a specialty market, mainly in New York City and Boston. Another specialty market is that for *Easter lambs.* Easter lambs are sold at weights of 20 to 40 pounds (9–18 kg). This market exists in some large cities.

Some advantages of fall lambing include:

- More favorable weather.
- Better use of equipment.
- Lower feed and labor requirements.
- Feed costs are generally lower.
- Better prices for lambs.
- Lower lamb mortality rate.
- Fits into an accelerated lambing program.

Some disadvantages of fall lambing are:

- More grain is needed.
- It may be difficult to breed the ewes.
- Lambs may pick up parasites from fall pasture.
- Lambing may occur at a busy time of the year.
- Lambs ready for market too early may not bring as good a price as those sold later in the year.
- Lambs have lighter birth weights.
- Lambing percentage is lower.

Early Spring Lambs

Early spring lambs are born in January and February. Breeding must begin about the first of August. Lambs are marketed before the end of June. Creep feeding of lambs is often necessary to prepare them for market on time. Lamb prices are usually better at this time than they are later in the year. There are few problems with parasites in the spring months. Lambing takes place when there is a lighter work load on the farm.

It may be more difficult to get ewes to breed in August. Weather conditions in some areas are more severe at lambing time. This system requires more grain and hay and does not make as much use of pasture. Better housing is needed, and more labor is required to save lambs.

Late Spring Lambs

Late spring lambs are born in March, April, and May. Breeding occurs later in the fall than for early spring lambs. This results in more ewes being settled in a shorter period of time. The lambing season is, therefore, shorter. Most of the feed comes from pasture and roughage. Little grain is required for this system. Producers who do not want to finish out the lambs may put them on the market as stockers and feeders.

Late spring lambing has several disadvantages. Parasite problems are worse with this system. The price for market lambs is not as good at the times these lambs reach market. If pasture is in short supply, additional feed is required to finish the lambs. Care of the lambs may take labor away from other farm tasks that must be done in the late spring and summer.

Accelerated Lambing

Accelerated lambing is a system that produces three lamb crops in two years. Ewes that will breed out of season must be used. It is better to use older ewes for this system. The major purpose is to increase production and, therefore, income, without greatly increasing production costs. To be successful, the producer must be an experienced sheep raiser. Better-than-average management ability is needed.

More labor is required for accelerated lambing. Early weaning of lambs must be practiced. The lambs are finished in the feedlot. Additional feeding is needed for both ewes and lambs.

Synchronized breeding is the forcing of ewes into heat during a short three-to-seven-day period. This is done by using hormone treatments for 14 days before the breeding season. When the hormones are withdrawn, most of the ewes come in heat in one to three days. Synchronized breeding is often used with accelerated lambing programs.

Feeder Lambs

Some farmers prefer to buy weaned lambs and finish them for market. This eliminates the need for a breeding herd. Careful selection of high-quality, healthy lambs is essential for success. This type of production involves a certain amount of risk. If the feeder pays too much, and the market goes down, the chances for profit are less. Lambs may be finished in drylot or grazed on pasture. Good feeding practices are necessary for success.

FEEDING OF SHEEP

Gestation Feeding

Pasture or other roughage is the basic feed for the ewe flock during the gestation period. High-quality hay, silage, or haylage may be used. The hay may be legume, grass, or a mixture of legume and grass. Corn, grass, or legume silages may also be used. Haylage may be legume or grass, or a mixture of the two. Silage must be chopped finer than it is for beef cattle. Stubble and stalk fields may be used for pasture. Native range pastures, permanent pastures, rotation, and temporary pastures may all be used (Figure 27-1).

FIGURE 27-1 A flock of sheep on an Iowa farm pasture. *Photo by Steven M. Ennis.*

Rotating pastures increases the amount of feed that is available to the ewes. Rotation also helps to break the internal parasite cycle. Pastures should be rotated every two to three weeks, depending on their kind and quality. They should not be overstocked. Sheep graze very close to the ground and may kill the vegetation. The stocking rate depends on the kind and quality of pasture used.

Salt, mineral mixes, water, and shade must be provided on pasture. Moving the salt and mineral from place to place will result in more even grazing of the pasture. Supplemental pastures may be needed with native range pastures.

If the pasture is poor, some feeding of hay may be needed. One to two pounds (0.45–0.9 kg) of legume hay will generally meet the ewes' needs. Silage may be substituted for hay at the rate of two to three parts of silage for each one part of hay.

During the last six weeks of the gestation period, the ewe should be fed some concentrate mixture. Corn, grain, sorghum, oats, barley, and bran are often used in concentrate mixes. A protein supplement may be needed, especially if the pasture or hay is of poor quality. Soybean meal, linseed meal, or commercial protein supplements are used. Urea may be used in sheep rations. No more than one-third of the protein should come from urea.

Experimental work in South Dakota suggests that the feeding of an antibiotic may reduce the mortality rate of lambs. Aureomycin fed to ewes at the rate of 60 milligrams per head daily for 80 days, beginning six weeks before lambing, lowered the lamb mortality rate. The mortality rate of the animals receiving antibiotics was 3.9 percent, compared to 14.5 percent for the control group. No increase in rate of gain for the lambs was observed.

Gestating ewes may be self-fed. This practice reduces labor and increases the use of lower-quality roughage. The roughage and concentrate mix is ground and placed in a self-feeder. The main problem is to keep the ewes from becoming too fat. Controlling access to the self-feeder limits the feed intake. The rations used should contain a high percent of roughage.

Lactation Feeding

The amount of grain fed the ewe should be reduced for the first ten days after lambing. A mix of equal parts of oats and bran with hay fed free choice may be used. At about ten days old, the lambs require more milk. Increasing the grain ration at that time will stimulate the ewe's production of milk. Be sure the ewe has access to water.

For about two months after lambing, the ewe requires additional nutrients to produce the milk to feed the lamb. The amount of concentrate in the ration must be increased during this time. Ewes that are not on pasture need about 1.5 to 2.0 pounds (0.7–0.9 kg) of grain daily. Four to six pounds (1.8–2.7 kg) of alfalfa hay should be fed per head per day. If the ewes are on good pasture at lambing, they usually will not need additional grain. Ewes that are nursing twins need more grain than those nursing singles. Heavier ewes need more feed than those of lighter breeds. If low-quality hay or grass hay is used, increase the grain. Some protein and mineral supplement will also be needed with poor-quality hay or grass hay.

After two months, the ration may be reduced to the amounts fed during the last six weeks of gestation. About a week before weaning, reduce the ewe's feed and water. This helps to decrease her milk production so that there will be fewer problems when the lambs are weaned. It also forces the lambs to eat more creep feed.

Feeding the Ram

Before the breeding season, the ram needs only pasture. During the breeding season, feed 1.0 to 1.5 pounds (0.45–0.7 kg) of corn or other grain mix if the ram is thin. Ram lambs require additional grain. During the winter, 3.5 to 5.0 pounds (1.6–2.3 kg) of hay should be fed. Heavier rams require the larger amounts. One pound (0.45 kg) of concentrate mix is enough for any weight of ram. Do not allow the ram to gain too much weight.

Flushing

Flushing is the practice of feeding a ration for ten days to two weeks before breeding and two weeks after breeding that causes the ewe to gain rapidly. Flushing may increase the lamb crop by 10 to 20 percent. If ewes are already fat before breeding, do not flush them or the lambing percentage may be reduced. Flushing may be achieved by putting the ewes on a better-quality pasture. Corn or oats, or a mixture of the two, fed at the rate of 0.5 to 0.75 pounds (0.2–0.3 kg) per head per day is effective.

Feeding Lambs to Weaning

The lamb must receive the colostrum milk as soon as possible after it is born. Colostrum milk is the first milk produced by the ewe. It contains antibodies that help to protect the lamb from infections. It also contains energy, protein, vitamins, and minerals needed by the lamb.

Lambs should be creep fed during the nursing period. A creep feeder is an area that allows only the lambs, and not the ewes, to enter. Lambs will start to eat grain at about ten days to two weeks of age. Feed small amounts of grain and clean out the trough each day. Coarsely grind or crush the grain until the lambs are six weeks of age. Pelleted feed may also be used.

Corn, oats, grain sorghum, and barley are good grains to use. A high-quality legume hay from second or third cutting should be available to the lambs at all times. The creep ration should contain from

14 to 16 percent crude protein. Early weaned lambs should receive 18 percent crude protein in the ration. The addition of molasses increases the palatibility of the ration. Soybean meal, cottonseed meal, linseed meal, or commercial supplements should be included. Urea should not be used for creep feeding lambs. Antibiotics should be included. Lambs on good pasture probably will not benefit greatly from creep feeding.

Lambs are commonly weaned at about three months of age. Early weaning is being practiced by a number of producers. Lambs should be at least nine weeks old and weigh 40 to 50 pounds (18 to 22.7 kg) for early weaning. The milk production of the ewe declines after four weeks. Lambs given high-quality feeds in the creep will gain faster and be ready for market sooner. Early weaning reduces the chances of lambs becoming infested with internal parasites. Early weaned lambs use feed more efficiently, requiring less feed per pound of gain.

Feeding Lambs from Weaning to Market

There is little difference in feeding practices for early weaned, late weaned, or feeder lambs. Use high-quality feeds and change rations slowly. Vaccinate lambs for enterotoxemia (overeating disease) when starting on full feed to protect lambs on high concentrate rations.

Market lambs may be fed out in drylot. Lambs that are eating well and weigh over 40 pounds (18 kg) should not be put on pasture because that slows the rate of gain. Later lambs weaned after July 1 should be drenched before being put on pasture. (A *drench* is a large dose of medicine mixed with liquid and put down the throat of the animal.) Drenching is done to control internal parasites. Several different medicines may be used, including phenothiazine, thiabendazole, haloxon, and tetramizole.

Feeder lambs may be put in the feedlot or they may be fed on pasture for a while and then finished in drylot. Reduce stress for feeder lambs when they arrive on the farm. Allow them to rest, provide grass or legume-grass hay, and water. Isolate sick lambs. Spray or dip to eliminate parasites if necessary. Lightweight lambs make better use of pasture than heavier lambs.

Lambs fed on pasture are grazed until about the first of December, weather permitting. A supplemental ration is then fed at the rate of one pound (0.45 kg) per head per day. This ration is made up of grain and hay balanced to meet the lamb's needs. Around January first, the grain portion of the ration is increased and the hay reduced. If the lambs are put in drylot for finishing, then a protein supplement is added to the ration. If the lambs are fed out on good pasture, the protein supplement is not needed.

Drylot rations use more grains and less roughage. Faster gains will result if the grains make up about 65 percent of the ration and the roughage about 35 percent. Lightweight lambs can use more roughage in the ration. The ration should be about 15 percent protein.

Corn, grain sorghum, barley, wheat, and oats are all popular grain feeds for lambs. Legume or grass-legume hays are commonly used. The kind of hay used depends upon the kind grown in the area. In some areas, cottonseed hulls and peanut hulls are used for roughage. Peanut hay may also be used where it is available. Soybean meal, cottonseed meal, linseed meal, peanut meal, or urea can be used for a protein supplement. The choice depends on the locality. Urea must not be used for more than one-third of the total protein in the ration. It is better to limit the use of urea to the last 25 pounds (11 kg) of feeding before market.

Water, salt, and minerals are needed. Provide clean, fresh water. Salt and minerals may be fed free choice. The calcium-phosphorus ratio should be about 2:1. The copper content of the trace mineral mix should not be too high.

Lambs may be self-fed or hand fed. It takes less labor to self-feed (Figure 27-2). Equipment and processing costs are slightly higher when self-feeding. Self-fed rations can be controlled better if they are ground. There is little advantage to grinding grains if they are to be hand fed. Hand feeding should be done at regular times. It is better to hand feed three times a day rather than twice. Feed only as much as the lamb will eat in 30 minutes.

The concentrate-roughage ratio in the ration should be changed every 7 to 10 days as the lambs become heavier. By the end of the feeding period, the lambs should be receiving 90 percent concentrate and 10 percent roughage.

Feeding Orphan Lambs

Because some ewes refuse to claim their lambs, there are usually orphaned lambs in sheep flocks at lambing time. These lambs can be saved if a ewe that has lost a lamb can be made to accept the orphaned lamb. Blindfolding the ewe is sometimes an effective way to encourage a ewe to accept an orphaned lamb. The ewe may accept the lamb

FIGURE 27-2 Self-feeding lambs requires less labor. Water may be provided with automatic waterers. *Courtesy of Iowa State University.*

if she is put into a headgate and the lamb is allowed to nurse. If all else fails, feed the lamb with a bottle and nipple.

The lamb must receive some colostrum milk if it is to have much of a chance to live. It may be necessary to use colostrum milk from another ewe. Some producers freeze extra colostrum milk so that it is on hand when needed.

Cow's milk may be used, but it does vary somewhat in composition from ewe's milk. Add one tablespoon of corn oil per pint of cow's milk. Warm the milk to body temperature but do not boil it. Keep utensils clean. Feed three ounces (84 g) every three hours. By the end of a week, feed what the lamb will take in five minutes, four times per day. This will be about 0.5 to 0.75 of a pint (0.22–0.35 liter).

A number of good commercial milk replacers are on the market for use in feeding orphan lambs. Follow the directions provided on the package.

Milk dispensers may be used to reduce the labor required. These may be homemade. Commercial milk dispensers are available. Keeping all equipment clean is essential.

Protect the lambs from drafts and cold. Provide dry, well-bedded pens. Heat lamps may be used in cold weather.

Start the lambs on a creep feed as soon as possible. If the lambs are eating well, they can be weaned from the milk at four to six weeks of age. They should be able to be weaned at no later than eight weeks.

Feeding Replacement Ewes

Ewe lambs that are to be kept for flock replacement need to be well fed to be adequately developed at breeding time. Roughage and grain may be used for the ration. Half or more of the ration may be roughage, depending on the size and condition of the ewe. Rations must be adjusted to the growth and condition of the individual flock.

OTHER MANAGEMENT PRACTICES

Breeding

The fine wool breeds and the Dorset are nonseasonal breeders. Some sheep of other breeds will also breed out of season. Generally, the medium wool, long wool, and crossbred wool breeds are seasonal breeders. These breeds normally breed in the fall of the year. The length of the day appears to be the main controlling factor. Rams of all breeds are less seasonal in breeding habit than ewes. The medium wool breed rams are more often affected by periods of temporary sterility. Rams do affect the number of multiple births. This is an inherited trait, so it can be selected for when a ram is being chosen.

Ewes are usually bred to lamb at about two years of age. Since the gestation period is about five months, this means breeding at 18 to 19 months of age. Some producers breed large, well-grown ewe lambs to produce lambs at about one year of age. Ewe lambs born early in the lambing season are more likely to breed as lambs. This practice requires good feeding to produce the growth and earlier sexual maturity that are necessary. Lambs do not have as long a breeding season as older ewes. There are breed differences in the ability to breed early. For example, Rambouillets are slow maturing. Production over the years will be greater if the ewe is bred as a lamb.

Check the ram for fertility before the breeding season. For best breeding results, shear the ram six to eight weeks before the breeding season. Use a marking system to identify when ewes are bred. A marking harness on the ram may be used. It is also possible to use a marking pigment on the ram's brisket. This must be applied to the brisket every second or third day. Paint branding the ewes aids in keeping better records. Use an approved paint branding fluid.

The number of ewes that a ram can breed depends on the age of the ram. When breeding on pasture, a ram lamb can breed about 15 ewes. A yearling ram can breed 25–35 ewes and an aged ram can breed 35–45 ewes. A general rule of thumb is to keep 3 mature rams for each 100 ewes in the breeding flock.

When synchronized breeding is used, hand-mating or limiting the number of matings per ewe is recommended. A vigorous ram can breed five to eight ewes per day when hand-mating is used. Breed the ewe every 12 hours as long as she is in estrus.

Crossbreeding

Crossbreeding is recommended when producing market lambs. While several crossbreeding systems may be used (see Unit 12), a three-breed rotational cross is the most desirable for producing market lambs.

Crossbred lambs have several advantages over straightbred lambs. They

- make more rapid gains.
- are more hardy and vigorous.
- have a lower mortality rate.

The advantages of using crossbred ewes instead of straightbred ewes include:

- greater fertility.
- higher lamb survival rate.
- higher lambing percentage.
- better milk production.

Characteristics to look for when selecting the breed of the ewe to use in a crossbreeding program include:

- early lambing ability.
- high lambing rate.

- greater ease of lambing.
- better maternal instinct.
- higher milk production.
- greater longevity.
- better wool quality and higher quantity of wool produced.
- early sexual maturity.
- greater potential for accelerated lambing.
- good udder soundness.

Characteristics to look for when selecting the breed of the ram to use in a crossbreeding program include:

- rapid growth.
- good carcass quality.
- greater sexual aggressiveness.
- above-average testicle size at puberty.
- high fertility.
- high survival rate of lamb offspring.

Breeds considered particularly desirable for ewe selection include Rambouillet, Merino, Corriedale, Columbia, Targhee, and Polypay. Breeds considered particularly desirable for ram selection include Suffolk, Hampshire, Shropshire, Oxford, Southdown, and Texel. Currently, the most commonly used ram breeds in crossbreeding programs are Suffolk and Hampshire.

Breeds that show desirable characteristics for selection of either ewes or rams for crossbreeding programs include Dorset and Montadale. Finnsheep and Romanov have high lambing rates but poorer fleece quality and quantity. When used in three-breed rotational crosses, Finnsheep and Romanov produce excellent market lambs. Other breeds not mentioned may be used in crossbreeding programs. The producer should give careful consideration to the desired characteristics of both ewe and ram breeds when making selections for breeding stock. Performance records are helpful when making selections for crossbreeding programs.

Gestation Management

Allow pregnant ewes plenty of exercise. Feeding away from the barn forces the ewes to exercise. Watch for signs of pregnancy disease (toxemia or ketosis). Ewes that are too fat or carrying more than one lamb are more likely to develop pregnancy disease.

There are some advantages to shearing the ewes several weeks before lambing. It is more sanitary and the lambs can nurse more easily. The fleece will contain less dirt and manure. Less pen space is required for the ewe. Udder problems can be spotted more easily and it is easier to see when the ewe is close to lambing. Care must be taken not to handle the ewes too roughly. Rough handling may cause

FIGURE 27-3 Pens for lambing must be clean, warm, and dry. *Courtesy of Iowa State University.*

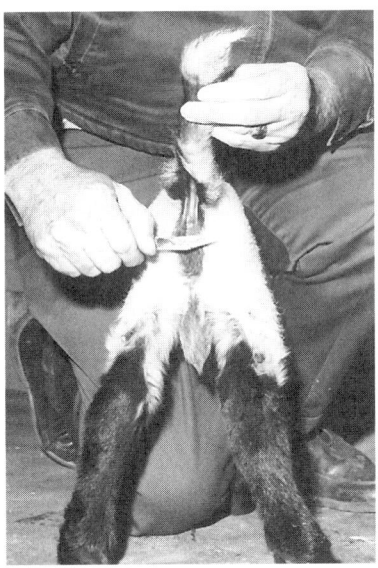

FIGURE 27-4 A lamb may be docked with a pocket knife. *Courtesy of USDA.*

premature lambing. Twice-a-year shearing is recommended in accelerated lambing systems.

If ewes are not sheared, they should be tagged or crutched out. This should be done at least one month before lambing. *Tagging* or *crutching* refers to shearing around the udder, between the legs, and around the dock. The wool around the vulva should be clipped. Ewes with a heavy face covering of wool should be clipped around the eyes.

Lambing Management

If possible, the ewes should not be disturbed during lambing. However, watch the flock carefully, and if a ewe is having difficulty lambing, give assistance.

Pens used for lambing must be clean, warm, and dry (Figure 27-3). When lambing on pasture, the birth should take place in clean pasture to help avoid the spread of parasites.

Be sure that the newborn lamb nurses. If it is weak, help it to nurse. Also check to be sure that the ewe is giving milk. During cold weather, it may be necessary to use heat lamps to keep the lambs warm. The lambs must be dried off after birth to avoid chilling.

Some ewes do not claim their lambs. This problem is less common with older ewes than with ewes lambing for the first time. There is no sure way to persuade the ewe to claim the lamb. Tying the ewe until the lamb nurses sometimes works. Rubbing the ewe's nose and the lamb with the ewe's milk may help. Blindfolding the ewe is sometimes effective.

Disinfect the lamb's navel shortly after it is born with tincture of iodine. Treat sore or irritated eyes twice a day with a saturated solution of boric acid. Some lambs have eyelids that are turned under. This may cause blindness unless it is corrected. If the condition is not too bad, it can be corrected by working the eyelid outward several times a day. When it is severe, it may be necessary to stitch a fold of the eyelid to the skin. Use a sterile needle and silk or nylon thread. This is an inherited defect. Do not select breeding stock that has shown this trait.

Management from Lambing to Weaning

Docking, or cutting off part of the tail, is one of the first management practices performed after lambing. Dock lambs between three and ten days of age. Docked lambs stay cleaner and, therefore, are less likely to get diseases or parasites.

The tail is cut off at the first or second joint or about 1 to 1.5 inches (2.5–3.8 cm) from the body. Docking may be done with a knife, burdizzo, elastrator, emasculator, "all-in-one", electric docker, or hot docking iron (Figures 27-4 and 27-5). Follow good sanitation procedures. Clean and dip all instruments in a disinfectant before use.

Castrate ram lambs when they are young. Many producers dock and castrate lambs at the same time. Castration may be done with a knife, burdizzo, elastrator, "all-in-one", or emasculator. One method of castration is shown in Figure 27-6. Sanitation procedures must be followed. If horses are, or have been on the farm, vaccinate for tetanus.

Ear mark the lambs when they are docked and castrated. Ear marking makes it easier to tell the wethers from the ewes. (A *wether* is a male lamb that has been castrated before reaching sexual maturity.) If a different ear mark is used each year, the sheep can easily be identified as to age. Plastic tags may also be used for ear marking. The lambs should be vaccinated for soremouth at the time of docking, castrating, and ear marking.

Late lambs will do better in the summer if they are sheared. Shear at the time of weaning. A fly repellant is applied to any cuts that occur. In areas where spear and needlegrass are common, the faces, legs, bellies, or the entire lamb may have to be sheared.

Catching and Handling Sheep

When handling sheep, do not cause them to become excited. If they must be handled, confine them in a small area first. Do not handle sheep by the wool. This will cause bruises. Catch sheep around the neck or by the rear flank. Do not pinch the flank. Move sheep by placing the left hand under the jaw and the right hand under the dock. Guide the sheep with the left hand and urge them to move with the right hand.

Spraying and Drenching Sheep

In range areas, sheep are sprayed and drenched as they come out of the shearing pen. Spraying and drenching controls external and internal parasites. Recommended substances for parasite control change. Always follow the current recommendations, which are available from extension specialists, universities, and veterinarians. Sprays and drenches should be changed from time to time to avoid a buildup of resistance in the parasites.

Foot Care

Trimming the feet of sheep helps to prevent problems with foot ailments. Feet should be trimmed twice a year. A sharp knife, pruning shears, or foot rot shears may be used for trimming. Do not trim into the quick. This will cause lameness, bleeding, and create a place for infection to start. Catch and examine any sheep that appear to be lame. Isolate all sheep that show signs of foot infection.

Culling Ewes

Ewes should be culled after the lambs are weaned. Cull ewes with bad udders, broken mouths, or that did not raise a lamb. Ewes that are fat

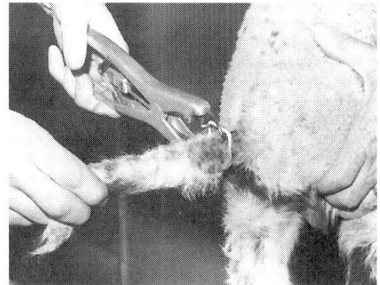

FIGURE 27-5   Using an elastrator to dock a lamb. *Courtesy of USDA.*

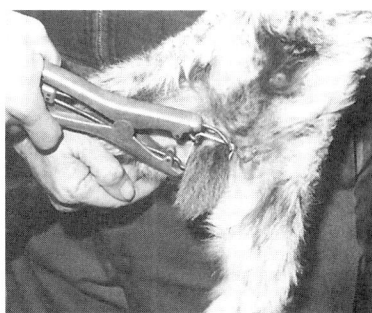

FIGURE 27-6   Castrating a lamb by placing a rubber band around its testicles. The band will cut off circulation and cause the testicles to drop off. *Courtesy of USDA.*

FIGURE 27-7 Professional sheep shearers fleecing sheep. *Courtesy of USDA.*

at weaning time probably did not raise a lamb. Ewes with lumps in the udder should be culled. Also cull ewes with prolapsed uteruses or ruptured abdomens. The condition referred to as *prolapsed uterus* occurs when the uterus protrudes from the vulva. Ewes that have difficulty in lambing may develop prolapsed uterus. An animal with this condition will die if not treated. A veterinarian should be called to administer treatment. Do not keep the ewe.

Records

Identification of individual sheep is necessary for a good system of record keeping. An ear tag designed especially for sheep and goats is commercially available. Birth and weaning weight records are needed. Fleece weight records should be kept. Weights at marketing and carcass evaluations are other important records. Forms are available from extension specialists and universities for keeping records on sheep flocks.

Shearing Sheep

The value of the fleece is lowered by the presence of dirt, manure, hay, straw, burrs, or other foreign matter. Follow management practices all year that will help keep the fleece clean.

Sheep are usually sheared early in the spring. In accelerated lambing systems the sheep are sheared again in July or August. Most shearing is done by custom shearers. Shearing sheep is a specialized occupation that requires special skill (Figure 27-7).

Sheep should be sheared in a clean place. Shear only when the wool is dry. Take sheep off feed for a few hours before shearing. The fleece is removed in one piece. If possible, second cuts should not be made. Do not injure the sheep while shearing them.

Blackfaced and black sheep are separated from the rest of the flock and sheared last. Yearlings are also sheared separately. Tender, coarse, black, and short fleeces are separated from the rest of the clip. If there is a large amount of foreign matter in the neck area, remove this and bag it separately.

A properly rolled and tied fleece is more valuable. The fleece is rolled with the flesh side out. Tie it securely, but not too tightly, using paper twine only. Fleeces are packed in regulation wool sacks. Wool is stored in a clean, dry place until it is marketed. Never use plastic bags to store wool.

PREDATOR LOSS

Losses from animal predators are a major problem for sheep producers in the United States, accounting for almost 40 percent of the total losses from all causes. The economic impact is severe, totaling over 17 million dollars annually in recent years. Nationally, about two-thirds of the total loss from animal predators is caused by coyotes, with dogs accounting for about eleven percent. Other animal predators include mountain lions, bears, foxes, eagles, and bobcats. Losses from coyotes are highest in the western and mountain states. Dogs are the major source of losses in other parts of the United States.

Practical and effective methods of controlling predators depends upon the geographic location of the producer. The use of properly constructed fences is a good method where the use of fences for pasture is common. Electric fences constructed of smooth, high tensile wire may be used. In areas where fences are not commonly used for pastures, the use of guard animals such as dogs, llamas, or donkeys may prove to be effective. Housing the sheep at night may be an effective predator control measure that can be used with small flocks. Sheep producers who are having animal predator problems should contact their local extension office or the USDA Animal Damage Control office for information about how to develop an effective control program.

HOUSING AND EQUIPMENT

It is not necessary or generally practical to provide expensive housing for sheep. Some use has been made of confinement housing systems and slotted floors for sheep. Producers who specialize in sheep production may want to consider such systems. Confinement facilities lend themselves to automated feeding systems. For the average farm flock or for range production, the cost of these systems make their use questionable.

Housing for sheep should provide protection from winter weather. Sheep barns or sheds are often open to the south. The barn must be dry, draft free, and well ventilated. One square foot (0.1 m^2) of window for each 20 square feet (1.9 m^2) of floor area provides adequate light. Provide access to pens for cleaning. Loft floors must be tight to prevent chaff from falling through. Doors should be 8 feet (2.4 m) wide so that sheep do not crowd each other as they go through. Locate feeding equipment for convenience. One type of feeding system is shown in Figure 27-8.

Electricity may be needed in the lambing shed during cold weather. Night lighting, as well as heat lamps, is necessary in cold climates.

Large flocks may justify a shepherd's room in the lambing barn. This can also be used to store other equipment.

Ewes require 12 to 15 square feet ($1.1–1.4 \text{ m}^2$) of floor space. An adjoining lot should provide 50 square feet (4.6 m^2) of space per ewe. A hard-surfaced lot reduces the area needed by 40 percent. Provide 15 to 18 inches (38–46 cm) of space per ewe at hay racks. Two automatic watering cups are adequate for 30 ewes. Waterers and feeders should be 12 to 15 inches (30.5–38 cm) off the floor at the throat. This prevents

FIGURE 27-8 Bunk feeding a farm flock in drylot on an Iowa farm. *Courtesy of Iowa State University.*

the ewes from putting their feet in them. Twelve inches (30.5 cm) of feeder space per lamb is sufficient.

Hinged panels, from 4 to 10 feet (1.2–3 m) in length, are useful for temporary pens. An extension hurdle can be used to crowd sheep into a corner. An *extension hurdle* or *gate* is a portable hurdle that can be lengthened or shortened as needed. It is usually about 3 feet (0.9 m) high and 6 feet (1.8 m) long. A sliding section inside the hurdle can be extended as needed when moving sheep. Maximum extension is about 6 feet (1.8 m). Mineral and salt feeders are also needed. Water may be provided in troughs or with automatic drinking cups.

A dipping tank is practical for larger flocks. Smaller flocks can be sprayed or dusted by hand. Commercial dipping tanks are available or a homemade tank may be used.

Sheep must be fenced in. Fences that are 60 inches (1.5 m) high are considered to be dog-proof. Most sheep can be held by a fence 3 to 4 feet (0.9–1.2 m) high. In range areas, barbed wire fences will hold sheep. A cattle fence can be made sheep-proof by adding three or four barbs, 4 to 5 inches (10–13 cm) apart, on the lower section. Woven wire can also be used.

Sorting and loading chutes may be used in larger flocks. Corrals are also useful for large flocks. Portable shelters and shades are used on pasture. Tilting squeeze, pregnancy testing cradle, blocking stand, weigh crate, and shipping crate are all useful pieces of equipment. A bag holder is needed for the shearing operation. Detailed plans for sheep equipment are available in the *Sheep Handbook Housing and Equipment* from the Midwest Plan Service. This can be secured from cooperating universities or from the Midwest Plan Service, Iowa State University, Ames, Iowa.

Many items of miscellaneous equipment are necessary for sheep production. Docking and castrating equipment; pruning shears; balling gun; syringe; hand shears; tattooing, ear notching, and ear tagging equipment; paper twine; scales; and record books are examples.

SUMMARY

Purebred producers provide breeding stock for commercial and other purebred producers. Commercial producers raise sheep to market meat and wool. Lambs may be born in the fall, early spring, and late spring. Accelerated lambing systems produce three lamb crops in two years.

Roughage is the main feed for sheep. Pasture and good-quality legume or grass-legume hay is used. Silage may be used in place of hay. Corn, grain sorghum, oats, barley, and wheat are grains commonly used in sheep rations. Protein supplements are not needed if good-quality legume pasture or hay is available. Protein supplements to use, if needed, include soybean oil meal, cottonseed meal, linseed meal, and peanut meal.

Do not allow ewes and rams to become too fat. Some breeds of sheep will breed out of season. Others breed only in the fall. Prepare the ewe for breeding by clipping the wool from the vulva, udder, inside of the thigh, and around the eyes. Check the ram for fertility before the breeding season.

Creep feed lambs to get them off to a good start. Save lambs by making sure they nurse and do not become chilled. Dock, castrate, mark, and vaccinate lambs. More concentrate than roughage is used to feed lambs to market weights.

Shear sheep in the spring. Provide a clean place to shear the sheep. Tie the fleece with paper twine only. Store it in a dry place, properly bagged until it is ready for market. Care in handling the fleece at shearing time increases its value.

Expensive housing is not necessary for sheep. Provide dry, well-ventilated housing. Protect sheep from winter weather in cold areas. Use fencing to keep sheep in and dogs out.

Student Learning Activities

1. Give oral reports on feeding, management, and housing of sheep.
2. Develop feeding, management, and housing programs for a sheep project or the home farm, using accepted practices.
3. Take field trips to observe good feeding, management, and housing for sheep.
4. Ask a local sheep producer to speak to the class on feeding, management, and housing for sheep.
5. Observe and practice activities such as docking, castrating, drenching, ear marking, and shearing sheep.
6. Do a community survey of sheep producers to determine feeding, management, and housing practices.
7. Prepare a display of equipment used for sheep production.
8. Construct portable sheep equipment in the school or home shop.

Review

1. What are the functions of the purebred and commercial sheep producers?
2. Name and briefly describe the four systems of sheep production, based on lambing date.
3. Describe a good feeding program for the ewe flock during the gestation period.
4. Describe the feeding program for ewes for the first ten days after lambing.
5. Describe a good feeding program for ewes during the lactation period, starting about ten days after lambing.
6. Describe how ewes should be fed as weaning time approaches.
7. Describe a good feeding program for rams.
8. Describe the practice of flushing.
9. Describe a good feeding program for lambs from birth to weaning.
10. Describe a good feeding program for lambs from weaning to market.
11. How may a producer save more orphan lambs?
12. Describe feeding practices for replacement ewes.
13. What is meant by the term *seasonal breeder*?
14. At what age should ewes be bred for the first time?
15. Describe how ewes should be prepared for breeding.
16. What system is used to identify which ewes are bred during the breeding season?
17. What management practices should be followed with ewes during the gestation period?
18. What management practices should be followed during lambing?
19. Describe the management practices that should be followed with lambs from birth to weaning.
20. Describe how to catch and handle sheep.
21. Why are sheep sprayed and drenched?
22. Describe how to care for a sheep's feet.
23. Describe the kinds of ewes that should be culled from the breeding flock.
24. What kinds of records should be kept on the sheep enterprise?
25. Describe the management practices that produce a more valuable wool clip.
26. Briefly describe the kind of housing that is needed for sheep production.
27. What kind of fencing is used for sheep?
28. Briefly describe other kinds of equipment necessary for sheep production.

Selection, Feeding, and Management of Goats

After studying this unit, the student should be able to

- give a brief description of the goat enterprise.
- list and describe the common breeds of goats.
- select quality breeding stock using generally accepted criteria.
- discuss feeding and management of goats.
- describe housing and equipment required for goat production.

GOAT ENTERPRISES

The following age and sex definitions are used for goats:

- *Doe:* female goat, any age
- *Buck:* male goat, any age
- *Kid:* young goat under one year of age, either sex
- *Yearling:* goat over one year and under two years of age, either sex
- *Wether:* male goat castrated when young

Angora Goats

More than 95 percent of the Angora goat population in the United States is found in Texas. Sixty percent of the world production of mohair comes from Texas. Angora goats also produce meat and help to control brush and weeds in range improvement programs.

Most of the Angora goats in Texas are raised in the Edwards Plateau region, in the south-central part of the state. The Angora goat is best adapted to the dry, mild climate found in this region. Angora goats do not compete with cattle and sheep for pasture. Goats prefer the browse found on the rangeland. *Browse* is the shoots, twigs, and leaves of brush plants found growing on rangeland.

Angora goat flocks range in size from small farm flocks (25–30 head) to large range flocks (several thousand head). There are some purebred breeders who supply bucks to commercial flocks. Purebred flocks are generally small. Commercial flocks concentrate on meat and mohair production. Some producers maintain flocks of wethers mainly for mohair production. These flocks may also be used for controlling brush.

Dairy Goats

The states with the largest number of dairy goats are California and Texas. However, dairy goats are found in every state of the United States. Dairy goats, like Angora goats, can consume browse.

There are few large dairy goat herds in the United States. Most dairy goats are kept in small numbers on farms for family milk production. Dairy goats require less space than dairy cows and are less expensive to raise. In addition, they can be used for meat.

Since very little space is required for dairy goats, they are sometimes kept by people who do not live on farms. However, some areas have zoning restrictions. A person considering keeping a dairy goat should check the local zoning ordinances.

Spanish Goats

The Spanish goat (Figure 28-1) is a mongrel descendant of the milk goat breeds. It is sometimes called a *brush* or *meat goat*, and is used for meat and milk. The main use of Spanish goats is for meat. Spanish goats can live on brush and weeds. They are prolific and can survive with little care. The Spanish goat is more adaptable to varying environments than is the Angora goat. They are most common in Texas.

Cashmere Goats

Cashmere refers to the soft down or winter undercoat of fiber produced by most breeds of goats, except the Angora. There is no true genetic breed of cashmere goats and no breed registry exists for goats that produce the cashmere fiber. Through selective breeding, a cashmere-producing type has been developed. Breeding for this characteristic began in Middle Eastern and Asian countries, resulting in the development of the Kashmir goat found in Kashmir, a region lying partly in India and partly in Pakistan, and in other countries in that part of the world. In recent years, breeders in Australia and New Zealand have been working to improve the production of cashmere from goats. Some breeders in the United States have imported goats from Australia and New Zealand for use on native stocks to improve the cashmere-producing ability of the goats. Good results have been ob-

FIGURE 28-1   Spanish goats. *Courtesy of USDA.*

tained by using the Spanish meat goats as well as the Toggenburg, Saanen, and Nubian dairy breeds.

Most of the world supply of cashmere fiber comes from China, Afghanistan, Iran, Outer Mongolia, India, Australia, and New Zealand. The production of cashmere fiber from goats is a relatively new industry in the United States. The world demand for cashmere normally exceeds the supply.

The growth of cashmere down begins in late June and usually stops in late December. If the fiber is not harvested after growth stops, it will be shed naturally. The fleece shows two types of fiber: a very fine, crimpy down and the longer, outside coarse, straight guard hair. The fine cashmere fiber is separated by combing it out or using a commercial dehairer on the sheared fleece.

Cashmere fiber must be less than 19 microns in diameter; the usual range of diameter size is 16 to 19 microns. The yield of cashmere fiber is in the range of 30–40 percent of the total fleece weight. Solid-color goats are preferred for the production of cashmere fiber. The cashmere from these goats is white, brown, gray, or black. Cashmere from multicolored goats is not as desirable and is classed as white with color or mixed color.

Boer Goats

The Boer goat originated in the Easter Cape Province of South Africa in the early 1900s when farmers began selective breeding for a meat type goat using native goat breeds. During this early development of the Boer goat some crossbreeding occurred with Indian goats and European dairy goats. The name of the breed comes from the Dutch word *boer* meaning "farm"; it was probably used to distinguish the breed from the Angora goats that were also raised in South Africa. Breeders selected for good conformation, rapid growth rate, high fertility, and short hair with red markings around the head and shoulders. Breed standards were established in 1959 with the formation of the Boer Goat Breeder's Association of South Africa.

Current breed standards are a white color with a red head and a white blaze. A few red patches are allowed and a pigmented skin is preferred. The breed is horned and has a Roman nose. Polled individuals will occur occasionally, probably as a result of the European dairy breeds used in the early development of the breed.

Mature males weigh 240–380 pounds (108–172 kg) and mature females weigh 200–265 pounds (90–120 kg). Boer goats in the feedlot have an average daily gain of 0.3–0.4 lbs/day (136–181 g/day).

Males reach puberty at about six months of age and females at about ten to twelve months of age. The breed is prolific, with a kidding rate of 200 percent being common. Because the Boer has an extended breeding season, it is possible to have three kiddings every two years.

Figure 28-2   Pygmy goat—yearling doe. *Courtesy of National Pygmy Goat Association.*

FIGURE 28-3   Pygmy goat—three-month-old buck. *Courtesy of National Pygmy Goat Association.*

Experimental work in Texas using Boer goats in a crossbreeding program with Spanish meat goats indicated that rate of gain increased with an increase in the percentage of Boer breeding. Feed efficiency also increased with the use of Boer breeding as compared to the group with only Spanish meat goat breeding. However, an increase in the percentage of Boer blood in the cross did not significantly increase the feed efficiency in the crossbred groups.

Boer goats were introduced into the United States in 1993. Initially, prices for breeding stock were artificially high. The number of Boer goats in the United States has increased rapidly since their initial introduction and current prices for breeding stock are much lower. The Boer goat has meat production characteristics that are superior to those of the Spanish goat. These characteristics make it valuable for farmers interested in goat meat production.

Pygmy Goats

The Pygmy goat that is found in the United States came from the French Cameroons area of Africa and was originally called the Cameroon Dwarf Goat. Early exports from Africa were sent to zoos in Sweden and Germany. From there exports were made to England, Canada, and the United States.

Mature Pygmy goats (Figures 28-2 and 28-3) are 16 to 23 inches (40–58 cm) at the withers, with the legs and head being relatively short compared to the body length. Genetically polled animals are not accepted for registry in the National Pygmy Goat Association (NPGA). Pygmy goats may be any color; preferred colors are white through gray and black in a grizzled (agouti) pattern. Breed specific standards required by the NPGA include: except for solid black goats, the muzzle, forehead, eyes, and ears are accented in lighter tones than the body; the front and rear hoofs, cannons, crown, dorsal stripe, and martingale are darker than the body color. Goats that are caramel colored must have light vertical stripes on the front sides of darker socks. Some limited random markings are acceptable. These specific breed standards are not required for registry in the American Goat Society (AGS). Female goats may have no beard or one that is sparse or trimmed. Male goats should have a full, long, flowing beard.

Initially, Pygmy goats were primarily exhibited in zoos. Today they are used for meat and milk as well as for 4-H and FFA projects. They are easily handled and because of their small size make good pets.

BREEDS OF GOATS

In rank order, Nubian, Alpine, Saanen, LaMancha, and Toggenburg have been the leading dairy goat breeds based on recent registrations. Relatively few Angora goats are registered with a breed association.

Angora

The Angora goat is an ancient breed. It originated in Turkey in the province of Angora, which is a mountainous region with a dry climate and great extremes in temperature. The first importation of this breed into the United States was in 1849. The Angora goat has been most popular in the southwestern United States.

Angoras are horned, although some polled individuals occur (Figure 28-4). They have long, thin, drooping ears. Angoras are white and open faced. Mature bucks weigh 125 to 175 pounds (56.7–79.4 kg) and mature does weigh 80 to 90 pounds (36.3–40.8 kg).

The fleece of the Angora goat is called *mohair*. There are several types: ringlet, flat, and web. The fleece grows at the rate of 6 to 12 inches (15.2–30.5 cm) per year. Older goats produce a coarser fleece. Therefore, kid fleece is the more valuable. Average mohair production is 6 to 7 pounds (2.7–3.2 kg) per head per year. Selection programs have doubled this production in experimental groups. The fleece is removed twice a year.

Dairy Goats

Although there are a number of breeds of dairy goats, only five breeds are common in the United States. These are: (1) French Alpine, (2) LaMancha, (3) Nubian, (4) Saanen, and (5) Toggenburg.

French Alpine. The French Alpine breed (Figure 28-5) originated in France from Swiss foundation stock. It was first imported into the United States about 1922. The breed ranges in color from pure white to black, with many other varied color patterns. Color shades include fawn, brown, gray buff, and red. French Alpines have erect ears and are short haired. They have no *dewlap* (a fold of flesh under the neck) and may be bearded or not bearded. Some are polled and some have horns. Bucks weigh 170 to 180 pounds (77–81.6 kg) and does weigh 125 to 135 pounds (56.7–61.2 kg). Average milk production is 1,500 to 1,600 pounds (680.4–725.7 kg) per year.

LaMancha. The American LaMancha is a newly developed breed. It was developed by crossing a short-eared Spanish breed with several of the purebred breeds in the United States. LaMancha goats may be any color. They have straight faces with short hair. The breed has two different types of ears: gopher and elf. The *gopher ear* is 1 inch (2.54 cm) or less in length with little or no cartilage. The end is turned up or down. The *elf ear* may be 2 inches (5.1 cm) long. Cartilage shaping is allowed. The end must be turned up or down. Bucks must have the gopher ear to be eligible for registration.

Nubian. The Nubian originated in Africa. The breed as it exists today was developed in England by crossing Nubian bucks with British dairy

FIGURE 28-4 Angora buck. *Courtesy of Dr. Maurice Shelton, Texas A&M University.*

FIGURE 28-5 French Alpine doe. *Courtesy of Laurelwood Acres, CA.*

FIGURE 28-6   Nubian doe. *Courtesy of Laurelwood Acres, CA.*

FIGURE 28-7   Saanen doe. *Courtesy of Laurelwood Acres, CA.*

FIGURE 28-8   Toggenburg doe. *Courtesy of Laurelwood Acres, CA.*

breeds. It was first imported into the United States in about 1896. Nubians may be any color, or combination of colors. Common shades include black, gray, cream, white, tan, and reddish-brown. They have short hair; long, drooping ears; a Roman nose; and no fringe of long hair along the spine (Figure 28-6). Most are polled, although some have horns. The doe is beardless. Bucks weigh 175 to 180 pounds (79.4–81.6 kg) and does weigh 130 to 135 pounds (58.9–61.2 kg). Average milk production is 1,300 to 1,500 pounds (589.7–680.4 kg) per year.

Saanen.   The Saanen (Figure 28-7) originated in Switzerland. It was first imported into the United States about 1904. The color of this breed is white or light cream, with white being preferable. The hair is short and there is a fringe over the spine and thighs. The ears are erect, and bucks have a tuft of hair over the forehead. Saanens are polled, although some may have horns. Both bucks and does are bearded. Bucks weigh 185 pounds (83.9 kg) or more, while does weigh 135 pounds (61.2 kg) or more. Average milk production is 1,800 to 2,000 pounds (816.5–907.2 kg) per year.

Toggenburg.   The Toggenburg originated in Switzerland. It was first imported into the United States about 1893. The color varies from light fawn to dark chocolate. The ears are white with dark spots in the middle. There are two white stripes down the face from the eye to the muzzle (Figure 28-8). The legs and rump are white. Toggenburgs have short- to medium-length hair that lies flat. The ears are erect. They may or may not have *wattles* (a projection of skin hanging from the chin). Toggenburgs are usually polled, although some may have horns. Bucks weigh 150 to 175 pounds (68–79.4 kg), and does weigh 100 to 135 pounds (45.4–61.2 kg). The average milk production is 1,500 to 1,600 pounds (680.4–725.7 kg) per year.

SELECTION OF GOATS

Angora

Angora goats are judged on the basis of (1) body and (2) fleece. These two factors are weighted equally. The body conformation is judged on the basis of breed type, conformation, amount of bone, constitution and vigor, and size and weight for age. The fleece is judged on the basis of fineness, uniformity and completeness of covering, oil content and luster, density, and character of fleece. Breeding animals are further selected on the basis of age and fertility.

Selection of good breeding stock increases the income from Angora goats. Desirable traits are inheritable. However, a period of years is required for a selection program to improve the income from the Angora enterprise. Bucks must be carefully selected to improve the breeding flock. Good management practices must be followed if the selection program is to be effective.

Balance must be maintained between selection for body conformation and fleece. Too much emphasis on fleece tends to reduce the size and vigor of the goat. The yearling buck should weigh at least 80 pounds (36.3 kg). The yearling doe should weigh at least 60 pounds (27.2 kg). The goat should have a wide, deep body with good spring of ribs and wide loin. Strong feet and legs with adequate bone are important. An open face is considered better than too much wool covering on the face.

A long, dense staple (fleece) is desirable. The Angora should produce at least 1 inch (2.54 cm) of growth each month. To determine density, part the fleece and observe how much skin area is exposed.

Kemp fibers are large, chalky white hairs. Do not select goats with large amounts of kemp in the fleece. The fleece should be bright, and uniform in fineness and length. It should cover the body of the goat. A light covering under the jaws, throat, or belly is desirable. The mohair should feel soft to the touch. The type of lock (ringlet, flat, or web) is important to the purebred breeder. It is of less importance to the commercial producer.

Breeding animals older than eight years of age are not as productive as younger goats. The fertility of the doe is an important factor in selection. Does that do not breed regularly should be culled. Does that have multiple births should be kept. Records of fleece weight, length of staple, and percentage of kid crop help the producer do a better job of selection.

Dairy Goats

For dairy goat production, breed choice is a matter of personal preference. Milk production is the main goal of most dairy goat owners. Select individuals that give an indication of being good milk producers.

As with any kind of livestock, it is necessary to be familiar with the parts of the animal in order to evaluate it properly. Figure 28-9 shows the parts of the dairy goat.

Scorecards have been developed by the American Dairy Goat Association for evaluation of type. The scorecard indicates the desirable traits to be used in selecting or judging dairy goats. Figure 28-10 shows the scorecard for evaluating does. Figure 28-11 shows the scorecard for evaluating bucks.

Young people who use dairy goats for 4-H and FFA projects may want to show their animals in dairy goat shows. The ADGA dairy goat showmanship scorecard (Figure 28-12) provides valuable suggestions for showing goats.

The front teeth on the lower front jaw of the goat may be used as a guide in determining the age of the goat (Figure 28-13). Teeth develop

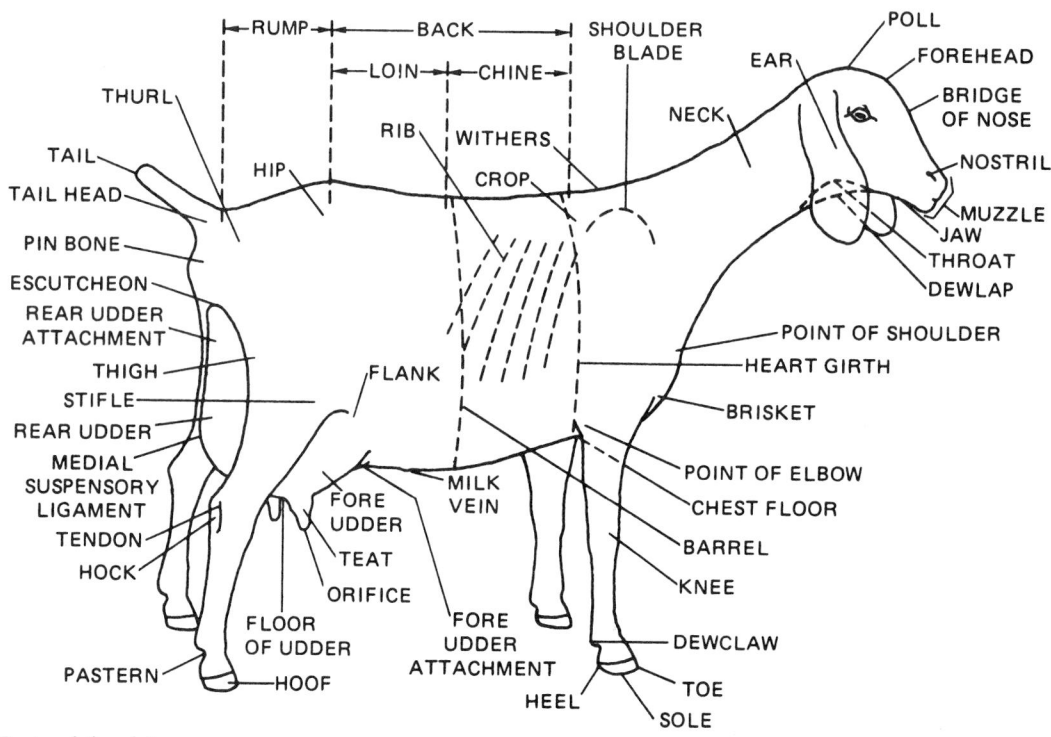

FIGURE 28-9 Parts of the dairy goat. *Courtesy of American Dairy Goat Association, NC.*

in the goat in the same manner as they do in sheep. The number of permanent teeth are an indication of the age of the goat.

Production records and pedigrees give additional information to be used in selection. Official production records from individuals close to the animal in the pedigree are of greater value than those of individuals beyond the fourth generation. Good dairy goats average three to four quarts (2.85–3.8 liters) of milk per day. The normal length of the lactation period is 10 months.

FEEDING GOATS

Angora Goats

The recommended nutrient allowances for Angora goats are given in the appendix of the text. As long as Angora goats are on range with a wide variety of brush, weeds, and grass, additional feeding is not necessary. Generally, additional feed is needed in the winter.

Angora goats need green browse, grass, and weeds to produce good mohair. Guajillo and live oak are rated as excellent browse for Angoras. Other common varieties of brush are rated from good to poor. Black persimmon, mesquite, and white brush are rated among the poorest.

(Ideals of type and breed characteristics must be considered in using this card.)

Based on Order of Observation

1. GENERAL APPEARANCE		30

Attractive individuality revealing vigor; femininity with a harmonious blending and correlation of parts; impressive style and attractive carriage; graceful walk.   **10**
Breed characteristics
Head - medium in length, clean-cut; broad muzzle with large, open nostrils; lean, strong jaw; full, bright eyes; forehead broad between the eyes; ears medium size, alertly carried (except Nubians).
Shoulder blades - set smoothly against the chest wall and withers, forming neat junction with the body.
Back - strong and appearing straight with vertebrae well defined.
Loin - broad, strong, and nearly level.
Rump - long, wide and nearly level.   **8**
   Hips - wide, level with back.
   Thurls - wide apart.
   Pin bones - wide apart, lower than hips, well defined.
   Tail head - slightly above and neatly set between pin bones.
   Tail - symmetrical with body.
Legs - wide apart, squarely set, clean-cut and strong with forelegs straight.
   Hind legs - nearly perpendicular from hock to pastern. When viewed from behind, legs wide apart and nearly straight. Bone flat and flinty; tendons well defined. Pasterns of medium length, strong and springy. Hocks cleanly moulded.   **12**
Feet - short and straight, with deep heel and level sole.

2. DAIRY CHARACTER		20

Animation, angularity, general openness, and freedom from excess tissue, giving due regard to period of lactation.
Neck - long and lean, blending smoothly into shoulders and brisket, clean-cut throat.
Withers - well defined and wedge-shaped with the dorsal process of the vertebrae rising slightly above the shoulder blades.   **20**
Ribs - wide apart; rib bone wide, flat, and long.
Flank - deep, arched, and refined.
Thighs - incurving to flat from the side; apart when viewed from the rear, providing sufficient room for the udder and its attachments.
Skin - fine textured, loose, and pliable. Hair fine.

3. BODY CAPACITY		20

Relatively large in proportion to the size of the animal, providing ample digestive capacity, strength, and vigor.   **12**
Barrel - deep, strongly supported; ribs wide apart and well sprung; depth and width tending to increase toward rear of barrel.
Heart girth - large, resulting from long, well-sprung foreribs; wide chest floor between the front legs, and fullness at the point of elbow.   **8**

4. MAMMARY SYSTEM		30

A capacious, strongly attached, well-carried udder of good quality, indicating heavy production and a long period of usefulness.
Udder - Capacity and Shape - long, wide, and capacious; extended well forward; strongly attached.   **10**
   Rear attachment - high and wide. Halves evenly balanced and symmetrical.   **5**
   Fore attachment - carried well forward, tightly attached without pocket, blending smoothly into body.   **6**
   Texture - soft, pliable, and elastic; free of scar tissue; well collapsed after milking.   **5**
Teats - uniform, of convenient length and size, cylindrical in shape, free from obstructions, well apart, squarely and properly placed, easy to milk.   **4**

TOTAL		100

FIGURE 28-10   ADGA Dairy Goat Scorecard for Does. *Courtesy of American Dairy Goat Association, NC.*

(Ideals of type and breed characteristics must be considered in using this card.)

Based on Order of Observation

1. GENERAL APPEARANCE		45
Attractive individuality revealing vigor, masculinity with a harmonious blending and correlation of parts; impressive style and majestic carriage; graceful and powerful walk.		
Breed Characteristics	10	
Head - medium in length, clean-cut; broad muzzle with large, open nostrils; lean, strong jaw; full, bright eyes; forehead broad between the eyes; ears medium size, alertly carried (except Nubian and LaManchas).	5	
Color - appropriate for breed.		
Shoulder blades - set smoothly against the chest wall and withers, forming neat junction with the body.		
Back - strong and appearing straight with vertebrae well defined.		
Loin - broad, strong and nearly level.		
Rump - long, wide and nearly level.	12	
Hips - wide, level with back.		
Thurls - wide apart.		
Pin bones - wide apart, lower than hips, well defined.		
Tail head - slightly above and neatly set between pin bones.		
Tail - symmetrical with body.		
Legs - wide apart, squarely set, clean-cut and strong with forelegs straight.		
Hind Legs - nearly perpendicular from hock to pastern. When viewed from behind, legs wide apart and nearly straight. Bone strong, flat and flinty; tendons well defined. Pasterns of medium length, strong and springy. Hocks cleanly moulded.	18	
Feet - short and straight, with deep heel and level sole.		
2. DAIRY CHARACTER		30
Animation, angularity, general openness, and freedom from excess tissue.		
Neck - medium length, strong and blending smoothly into shoulders and brisket.		
Withers - well defined and wedge-shaped with the dorsal process of the vertebrae rising slightly above the shoulder blades.		
Ribs - wide apart, rib bone wide, flat and long.		
Flank - deep, arched and refined.		
Thighs - incurving to flat from the side; apart when viewed from rear.		
Skin - fine textured, loose and pliable. Hair fine.		
3. BODY CAPACITY		25
Relatively large in proportion to size of the animal, providing ample digestive capacity, strength and vigor.		
Barrel - deep, strongly supported; ribs wide apart and well sprung; depth and width tending to increase toward rear of barrel.	13	
Heart Girth - Large, resulting from long, well-sprung foreribs; wide chest floor between the front legs, and fullness at the point of elbow.	12	
TOTAL	100	

FIGURE 28-11 ADGA Dairy Goat Scorecard for Bucks. *Courtesy of American Dairy Goat Association, NC.*

Based on Usual Order of Consideration

1. APPEARANCE OF ANIMAL		40
Condition and Thriftiness - showing normal growth—neither too fat nor too thin.	10	
Hair clean and properly groomed.		
Hoofs trimmed and shaped to enable animal to walk and stand naturally.	10	
Neatly disbudded if the animal is not naturally hornless.		
Clipping - entire body if weather has permitted, showing allowance to get a neat coat of hair by show time; neatly trimmed tail and ears.	10	
Cleanliness - as shown by a clean body as free from stains as possible, with special attention to legs, feet, tail area, nose, and ears.	10	

2. APPEARANCE OF EXHIBITOR		10
Clothes and person neat and clean - white costume preferred.	10	

3. SHOWING ANIMAL IN THE RING		50
Leading - enter, leading the animal at a normal walk around the ring in a clockwise direction, walking on the left side, holding the collar with the right hand. Exhibitor should walk as normally and inconspicuously as possible.		
Goat should lead readily and respond quickly.		
Lead equipment should consist of a collar or small link chain, properly fitted.	10	
As the judge studies the animal, the preferred method of leading is to walk alongside on the side away from the judge.		
Lead slowly with animal's head held high enough for impressive style, attractive carriage, and graceful walk.		
Pose and show an animal so it is between the exhibitor and the judge as much as possible. Avoid exaggerated positions, such as crossing behind the goat.		
Stand or kneel where both judge and animal may be observed.		
Pose animal with front feet squarely beneath and hind feet slightly spread. Where possible, face animal upgrade with her front feet on a slight incline. Neither crowd other exhibitors nor leave too much space when leading into a side-by-side position.		
When judge changes placing, lead animal forward out of line, down or up to the place directed then back through the line, finally making a U-turn to get into position.	15	
To step animal ahead - use slight pull on collar. If the animal steps badly out of place, return her to position by leading her forward and making a circle back thru your position in the line.		
When judge is observing the animal, if she moves out of position, replace her as quickly and inconspicuously as possible.		
Be natural. Overshowing, undue fussing, and maneuvering are objectionable.		
Show animal to best advantage, recognizing the conformation faults of the animal you are leading and striving to help overcome them.	15	
Poise, alertness, and courteous attitude are all desired in the show ring. Showmen should keep an eye on their animals and be aware of the position of the judge at all times—but should not stare at the judge. Persons or things outside the ring should not distract the attention of the showmen. Respond rapidly to requests from judges or officials, and be courteous and sportsmanlike at all times, respecting the rights of other exhibitors. The best showmen will show the animals at all times—not themselves—and will continue exhibiting well until the entire class has been placed, the judge has given his reasons, and he has dismissed the class.	10	
TOTAL		100

Suggested Uniform:
    Long-sleeved white shirt, regulation white pants, 4-H or FFA necktie, 4-H or FFA cap (if applicable), with matching shoes and belt in either black, white, or brown.

FIGURE 28-12   ADGA Dairy Goat Showmanship Scorecard. *Courtesy of American Dairy Goat Association, NC.*

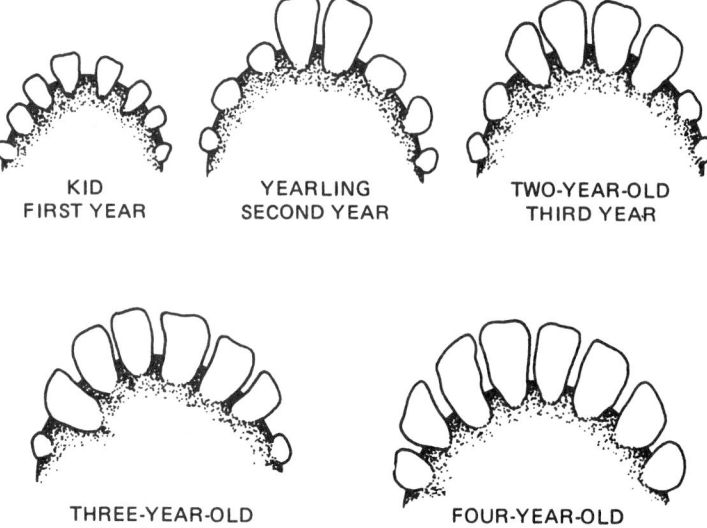

KID
FIRST YEAR

YEARLING
SECOND YEAR

TWO-YEAR-OLD
THIRD YEAR

THREE-YEAR-OLD
FOURTH YEAR

FOUR-YEAR-OLD
FIFTH YEAR

FIGURE 28-13 Determining the age of goats by examining the teeth. *Courtesy of American Dairy Goat Association, NC.*

Supplemental feeding may include pelleted feeds, 20-percent protein range cubes, or shelled yellow corn. Other feeds may be used. The amount to feed depends on the condition of the pasture and the goats. Pregnant does require more feed than dry goats. Protein may be provided by feeding 0.25 to 0.5 pound (0.1–0.2 kg) of cottonseed cake. Yellow corn may be fed at the rate of 0.5 to 1.0 pound (0.2–0.45 kg) per head per day. Goat cubes may be used in place of the cottonseed cake and corn at the rate of 0.5 to 0.75 pound (0.2–0.34 kg) per head per day.

Goats may be self-fed on range, with salt added to the mixture to limit feed intake. Mix seven parts of concentrate to one part of salt if the feeders are located a mile or more from water. If the feeders are placed near water, mix three parts of concentrate to one part of salt to limit feed intake. A common concentrate mix is one part cottonseed meal, three parts ground grain, and one part salt. Alfalfa or other ground roughages may be used in the self-fed mixture. If ground roughage is used, lower the amount of salt in the mix. Moving the feeders from time to time results in better use of the pasture. Salt-controlled feeding should only be used if all other methods of feeding are impractical.

Roughages may be fed as hay. Common hays to use include alfalfa, sorghum, peanut, Sudan, and Johnson grass. At least one pound (0.45 kg) of hay should be fed per head per day. Singed prickly pear and tasajillo may be fed free choice along with 0.25 pound (0.1 kg) of cottonseed cake per head per day.

The feeding of protein blocks during kidding season helps to prevent kids from becoming lost from the does. For best growth, kids should be fed during the winter months.

Dairy Goats

Goats are ruminant animals. They have digestive systems similar to those of cattle. Large amounts of roughage may be used as the basis for feeding goats. Grains are added to supplement low-quality roughages and to provide additional nutrients during growing and lactating periods.

Rations can be made up of grains, pasture, hay, silage, browse, roots, and garden refuse. Commercial protein concentrates may be used as well as home-mixed protein feeds. Commercial dairy mixes will provide the protein needed by the dairy goat. Minerals, vitamins, salt, and water are also essential in the diet of the dairy goat.

Feed good-quality roughages free choice to lactating does. Legumes, grasses, or grass-legume mixtures may be used. The quality is more important than the kind of hay being fed. More protein is needed in the concentrate if grass hays are fed. The crude protein content of the ration should be 12 to 16 percent, depending on the kind and quality of roughage being fed. No additional protein supplement is needed if good-quality legume hay or pasture is available. Silage may be substituted for part of the hay in the ration.

Any of the common grains can be fed. Feed one pound (0.45 kg) of concentrate for each four pounds (1.8 kg) of milk produced daily. Trace minerals and salt should be added to the ration.

Pasture and hay may be fed as the only feed for does during the gestation period. Keep the doe in good flesh. If additional feed is needed, use 0.5 to 1.5 pounds (0.2–0.7 kg) of grain depending on the condition of the doe. Some bran should be fed a few days before freshening. The amount of grain fed at this time should be cut in half. Feed a bran mash for a few days after kidding. Gradually replace the bran mash with grain until does are on a full feed of grain for the lactation period. Do not allow the doe to become too fat during the gestation period.

The buck can be supplied the needed nutrients with pasture or hay. Concentrate may be added to the ration, depending on the condition of the buck and the quality of the pasture or hay. Do not allow the buck to become too fat. During the active breeding season, some additional grain may be fed. The amount needed is one to two pounds (0.45–0.9 kg) daily.

Be sure kids receive the colostrum milk shortly after they are born. Some producers allow the kids to nurse from the does. Others prefer to take the kids away from the does and hand feed the colostrum. Train the kid to drink from a pan or bucket as soon as possible.

Since the does are usually used for milk production, the kids must be fed away from the does. Begin feeding the kids 0.5 pint (0.2 liters) of milk three to four times per day. Cow's milk can be used in place of goat's milk if the change is made gradually. Milk replacers are commercially available. By the time the kid is six weeks old, the amount of milk being fed should be five pints (2.4 liters) per day. Kids begin to eat dry feed during the first three to four weeks. Provide fresh water and high-quality legume hay. Calf starter or rolled grains should be fed. Kids are usually weaned from milk at four to six months of age.

Growing kids are fed mainly on roughages. Low-quality roughages must be supplemented with a 12 to 14 percent protein mix. Grain may be fed at the rate of 0.5 to 1.5 pounds (0.2–0.7 kg) per head daily. Do not allow the kids to become too fat. Adjust the concentrate feed to prevent that from happening.

MANAGEMENT OF GOATS

Angora Goats

Angora goats breed in the fall and kid (give birth) in the spring. It is recommended that three or four bucks be run with each 100 does. Flushing does is a recommended practice. Use a fresh pasture, or feed 0.25 to 0.33 pound (0.1–0.15 kg) of corn per head per day. Keep the does in good condition during gestation to prevent abortion. Only yearling or older bucks should be used for breeding.

Two systems of kidding are in common use. One is to kid does in the pasture. A fresh pasture should be used and the does should be disturbed as little as possible. This system produces hardier goats with less labor.

The other system involves the use of stakes and kidding boxes. The kid is tied to a stake with 15 inches (38 cm) of rope. A box is provided for shelter. The doe is allowed to graze in a nearby pasture and is paired with the kid in the evening. Does and kids must be marked for identification with this system. In general, staking of the kids requires more labor. A modification of this system allows the kids to run free in a large pen. The does are paired with their kids in the evening. Allow kids to go to pasture with the does when they are about one month old. The stake or pen system is commonly used by purebred breeders because it facilitates record keeping.

Newborn kids should be handled as little as possible. Too much handling may cause the doe to disown them. Be sure that the kid nurses soon after it is born. If the doe refuses to claim the kid, put it in a small pen with other unclaimed kids for a few days. Raise orphan kids on a bottle, using cow's milk. Follow procedures outlined for feeding orphaned sheep.

Ear mark kids and vaccinate them for soremouth when they are about one to one and one-half months old. Ear marking is done only for identification. Because plastic tags may pull out in the brush, most producers use a system of ear notching. Figure 28-14 shows the approved system for ear notching registered Angora goats.

Castration is often done in November, December, or January following kidding in the spring. Castrating at the older age produces a

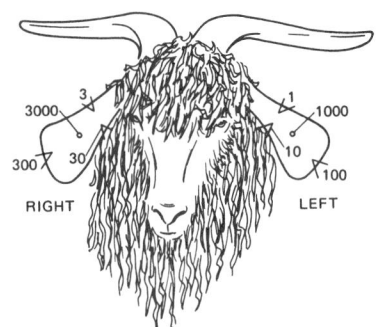

FIGURE 28-14 A system for identification of Angora goats by ear notching.

heavier horn on the wethers. Heavier horns are preferred by buyers. A knife or burdizzo is used for castration. The knife is preferred for younger kids. The burdizzo is best for older animals.

Kids are usually weaned after the fall shearing (Figure 28-15). Move the does away from the kids when weaning. Leave the kids in the familiar pasture.

In Texas, Angora goats are sheared twice a year. Spring shearing time is from February to April, depending on the region. Fall shearing time is mid-July to the end of September. Shearing procedures are similar to those for sheep.

FIGURE 28-15   Shorn Angora goats with their kids. *Courtesy of Dr. Maurice Shelton, Texas A&M University.*

The practice of caping is sometimes followed to protect the goats from the weather after shearing in the spring. In caping, a strip of mohair 3 to 4 inches (7.6–10 cm) wide is left down the goat's back. This should be sheared a month or six weeks after the first shearing. Special goat combs that leave 0.25 to 0.5 inch (0.6–1.3 cm) of stubble on the goats are sometimes used. This stubble provides about the same amount of protection from the weather as does caping. Using the special goat comb eliminates the problem of undesirable staple length that results from caping. Regular combs are generally used for shearing in the fall.

External and internal parasites are controlled by spraying and drenching. Most producers spray and drench goats as they come out of the shearing pen. Follow current recommendations for drenching substances. Always follow directions on the product being used.

Foreign material from pasture being caught in the fleece is a serious problem in mohair production. Needle and speargrass, grass burrs, cockle burrs, horehound, and other plant materials may contaminate the mohair. Moving goats to pastures that contain few if any of these plants for a time before shearing helps to eliminate the problem. Changes in shearing dates will sometimes help reduce plant contamination. Improving pastures by controlling noxious weeds is also recommended.

If oil is applied within 30 days after shearing to control external parasites, it will work out of the fleece before the next shearing. Some producers use oil before shearing to increase fleece weight. Fleeces are discounted by buyers when this practice is followed. The oil is difficult to remove from the fleece and the mohair is left dark and dingy and the fiber luster is harmed. Artificial oils should not be used to increase fleece weight.

Spanish, or Meat-type, Goats

Spanish, or meat-type, goats are similar in many ways to Angora goats. Only the major differences are discussed here.

Spanish goats breed year-round. Most producers follow a twice-a-year breeding system. Does are bred in February and March, and again in September and October. Meat production is the major goal of

Spanish goat enterprises. Selection of breeding animals is done accordingly. Size, body conformation, and rapid growth are important points to consider when making selections.

Supplemental feeding is seldom practiced. Kidding takes place on pasture, and no special attention is given the does at kidding time. Kids are usually marketed at four to five months of age without being marked or castrated. Kids should be vaccinated against soremouth.

Replacement doe kids are selected at weaning time. Doe kids are weaned in drylot in order to teach the does to eat supplemental feed. The replacement does are returned to the breeding flock at about one year of age.

Dairy Goats

Diary goats are seasonal breeders. The breeding season is from September to March. Does are usually bred in September, October, or November and kid in February, March, or April. Only well-grown does should be bred at 10 to 12 months of age. Many producers breed does at 15 to 18 months of age.

Keep the buck in a pen separate from the does during the breeding season. The buck gives off a strong odor during this time, which will affect the odor of the milk. It is generally a good practice to pen the buck separately throughout the year. A buck that does not receive enough exercise may become sterile. Do not overwork the buck during breeding. A year-old buck can breed about 25 does during the season. A mature buck can breed about 50 does during one breeding season. Keep breeding records on every doe.

The gestation period in goats is five months. Watch the does for signs of kidding. The doe becomes more nervous a few hours before kidding. The udder becomes swollen with milk. The doe appears to shrink in the belly. The flanks appear hollow and the tailhead seems to be higher than usual. These changes may occur two to three days before kidding. Do not help the doe with kidding unless she obviously is having difficulty. An experienced producer can provide help with difficult births. The inexperienced producer should obtain help from a veterinarian.

If it is available, put kids on pasture as soon as possible. Provide a pen large enough for the kids to get plenty of exercise. Boxes, ramps, or other elevated objects in the yard provide the kids with running and jumping exercises.

Separate buck kids from doelings by the time they are two to three months of age. The buck kid becomes capable of breeding at the age of three months. Failure to separate the bucks from the does will result in does being bred too young.

Bucks that are not to be kept for breeding should be castrated between one to three weeks of age. Be sure that the testicles have descended into the scrotum before attempting to castrate. A knife, elastrator, or burdizzo may be used to castrate. Castrate before fly season if a knife or elastrator is to be used.

Dehorning should be done when the kids are three to five days old. An electric dehorner, chemicals, or dehorning paste may be used on young kids.

> **Caution:** Dehorning chemicals are caustic and must be used carefully. Pastes are safer than chemicals. Older goats may be dehorned by sawing the horns off. This should only be done in cool weather when flies are not a problem.

Unless goats are kept on hard surfaces they need to have their hooves trimmed from time to time. Trimming prevents foot problems. Goats with untrimmed feet may go lame. Pruning shears may be used to trim the feet. A sharp jackknife or farrier's knife can also be used for trimming.

Goats may be marked for identification by tattooing or ear tagging. Tattooing is done with an instrument sold commercially that is designed for this purpose. The tattoo is placed in the ear or in the soft tissue on one side of the tail. Because ear tags may tear out of the ear on a fence or in brush, ear tagging is not generally recommended. Identification is especially important for purebred producers.

Dairy goats are usually milked on a stand (Figure 28-16). The stand is 1.5 to 2.5 feet (0.46–0.76 m) high. It has a stanchion to hold the doe while she is being milked.

Milk absorbs odors from feed. Any strong-flavored feeds such as silage, onions, or cabbage should be fed only after milking.

The does should be clipped on the udder and flank area. This helps to keep dirt and hair out of the milk, especially if the doe is milked by hand. Does may be milked either by hand or by machine. Goats are milked from the rear or the side.

Washing the udder with warm water and a sanitizer will stimulate milk letdown, as well as promote cleanliness. Use a strip cup to check for mastitis. Milk the first two or three squirts of milk into the strip cup and examine it. Begin milking within two or three minutes after washing the udder. If milking by hand, use a hooded bucket to prevent dirt from getting into the milk.

Milk at regular intervals of 12 hours. This gives the maximum production. The same person should feed, handle, and milk the goat. Following a regular routine gives the most satisfactory results.

Strain the milk through a commercial strainer pad. Cool the milk as quickly as possible after milking. Commercial coolers are available. Running water may also be used to cool the milk. After cooling, keep the milk refrigerated.

Keep milking equipment clean and sanitized. Use a good dairy detergent for cleaning the equipment. Rinse the equipment with warm water as soon as the milking is done. Then scrub it in warm water using a dairy detergent. Brush all surfaces to be sure they are clean.

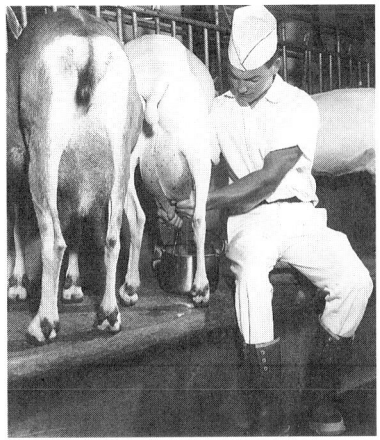

FIGURE 28-16 The most convenient way to milk a dairy goat is from a milking stand. *Courtesy of Laurelwood Acres, CA.*

Rinse again with an acidified rinse to prevent the buildup of milkstone (mineral deposits) on the equipment. Drain all rinse water and store the equipment bottom side up in a place free of dust and flies. Before using the equipment for the next milking, sanitize it with boiling water or rinse it with a dairy sanitizer. Follow directions on the package when using a dairy sanitizer.

Goat's milk is pure white. It has smaller fat globules and a softer curd than cow's milk. Because it is more easily digested, goat's milk is sometimes prescribed by doctors for people who have difficulty digesting cow's milk.

The cream in goat's milk rises more slowly than that in cow's milk. Mechanical separation of cream is recommended. Both hard and soft cheese can be made from goat's milk. Butter made from the cream of goat's milk is white. The composition of goat's milk is similar to that of cow's milk. The fat content is generally higher and the lactose (sugar) content is slightly lower.

HOUSING AND EQUIPMENT

Angora Goats

Angora goats are produced with a minimum of housing and equipment. A shed to protect the goats for four to six weeks after shearing is about the only housing needed. A tight, woven wire fence will keep the goats confined when necessary. Barbed wire fences can be used by spacing the wire about 6 inches (15.2 cm) apart near the bottom. Use additional stays between the posts to hold the wire in place. The same kinds of troughs, feeders, pens, and corrals that are used for sheep can also be used for Angora goats.

Dairy Goats

The amount and kind of housing necessary for dairy goats depends on the number of goats owned and the convenience desired by the producer. A small barn or shed is adequate for a farm with only a few goats. The producer who has a large flock may want more elaborate housing. Two types of housing are commonly used: loose pen and tie stalls. Combinations of housing are also used. The milking does are kept in tie stalls, and loose pens are used for the kids and yearlings.

When using loose housing, provide 15 square feet (1.4 m^2) per goat. A shed, open to the South, may be used. Some provision for shutting the open side in cold weather may be needed in cold climates. More bedding is required for loose housing. Goats must be dehorned if loose housing is used. Loose housing does not cost as much as confinement housing. Figure 28-17 shows a floor plan for loose housing.

Confinement housing is more expensive. However, less bedding is needed. Goats are kept in individual tie stalls or pens. The building

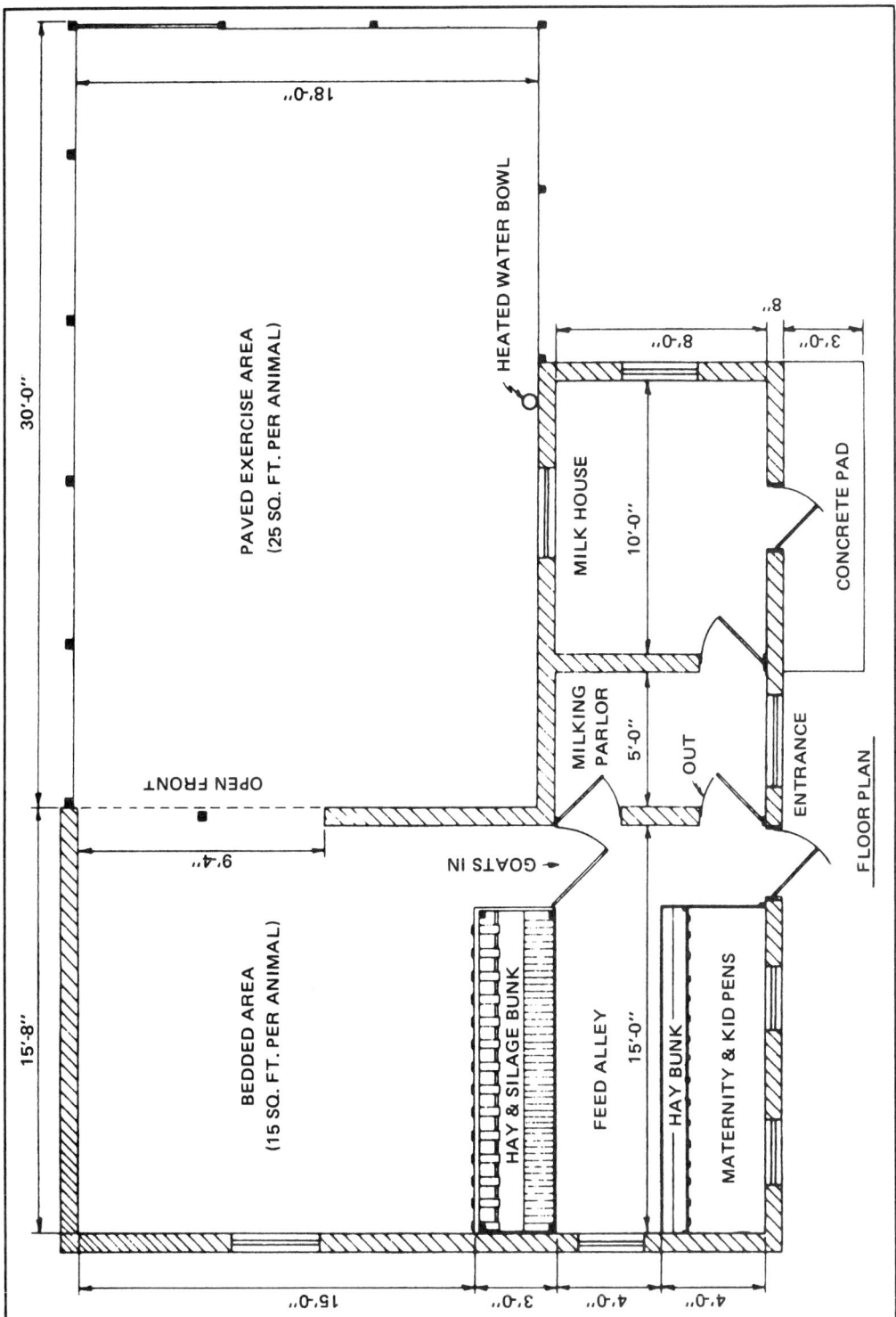

FIGURE 28-17 A floor plan for loose housing of dairy goats. *Courtesy of American Dairy Goat Association, NC.*

may be insulated, and mechanical ventilation is needed. Figure 28-18 shows a floor plan for confinement housing.

Bucks should be housed separately from the rest of the goats. Figure 28-19 shows a plan for a buck barn.

A separate milking area and milk room are recommended. If it is attached to the barn, use double doors to keep dirt, odors, and flies out of the milk area. Figure 28-17 shows the milking parlor and milk room attached to a loose housing barn. Local Board of Health standards may regulate the design and construction of the milking area and milk room.

Equipment needed in the milk room includes:

- a milk cooler.
- a sink for washing equipment.
- hot and cold running water.
- a place to store equipment.
- a milk pail, strainer, brushes.
- an electric milking machine.
- scale for weighing milk.

Woven wire fences are best for goats. The fence should be at least 4.5 to 5 feet (1.4–1.5 m) high. Temporary fencing for dividing a pasture may be electric. Three wires are needed: one 10 inches (25.4 cm), one 20 inches (50.8 cm), and one 40 inches (101.6 cm) from the ground. Tethering is not generally recommended. If it must be done, move the location each day so the goat has feed. Provide shade on pasture or if the goats are tethered.

Feed racks, troughs, hurdles, and other equipment are about the same for goats as for sheep. Figure 28-20 is a plan for a grain feeder. Plans for feed racks appear in Figure 28-21.

SUMMARY

Angora goats are raised for mohair and meat production. Most of the Angora goat production in the United States is found in Texas. Angora goats are fed mainly on browse, with some supplemental feeding being added as needed. Little housing or equipment are required. Angora goats are selected on the basis of body conformation and mohair quality.

Small herds of dairy goats are found in all parts of the United States. Some nonfarming people keep dairy goats to provide milk for the family. There are five breeds of dairy goats commonly found in the United States. These are French Alpine, LaMancha, Nubian, Saanen, and Toggenburg.

Dairy goats are selected for milk production. Housing can be simple or more elaborate, depending on the number of goats owned and the convenience desired. Dairy goats are fed mainly pasture and hay.

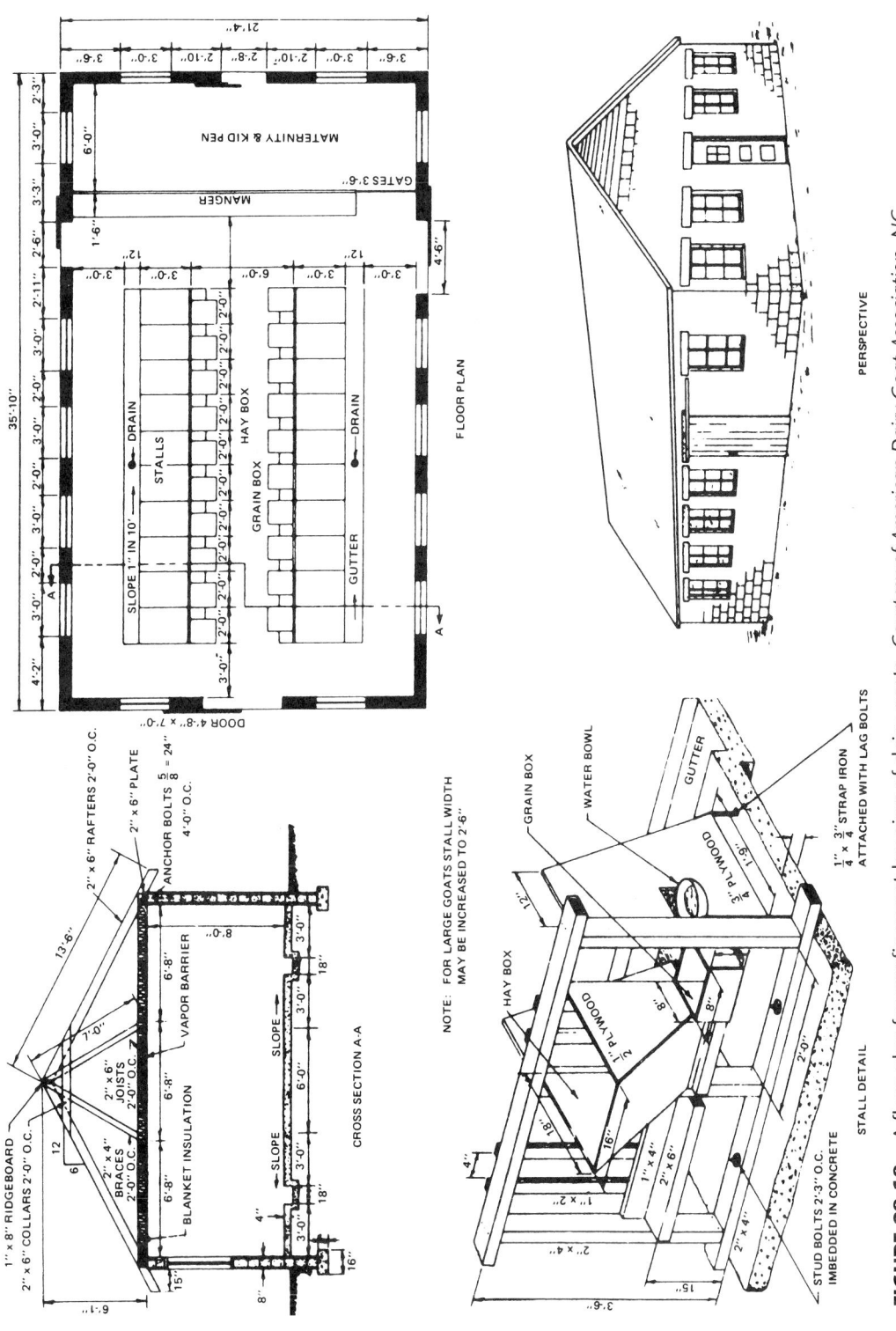

FIGURE 28-18   A floor plan for confinement housing of dairy goats. *Courtesy of American Dairy Goat Association, NC.*

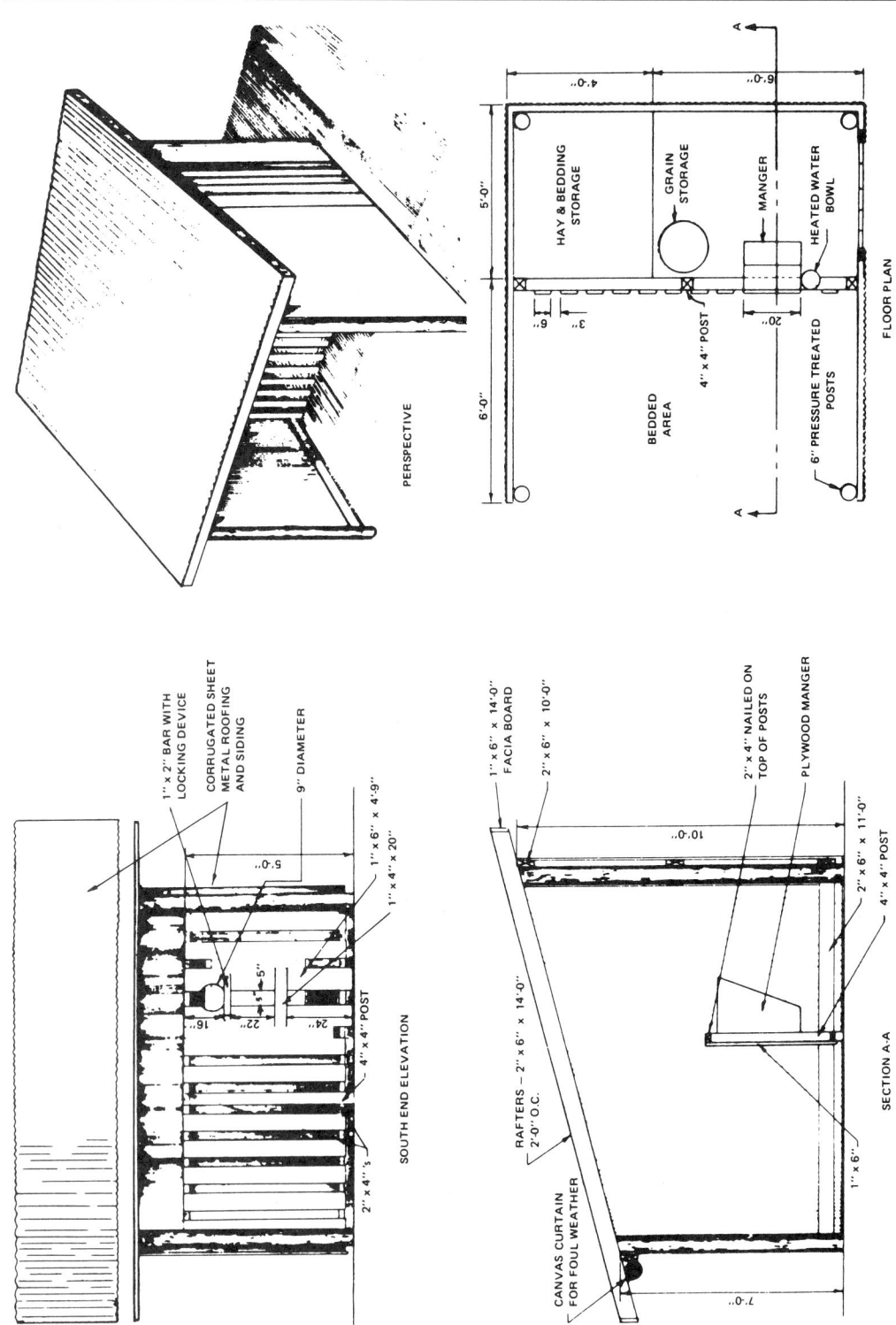

FIGURE 28-19 A floor plan for a buck barn. *Courtesy of American Dairy Goat Association, NC.*

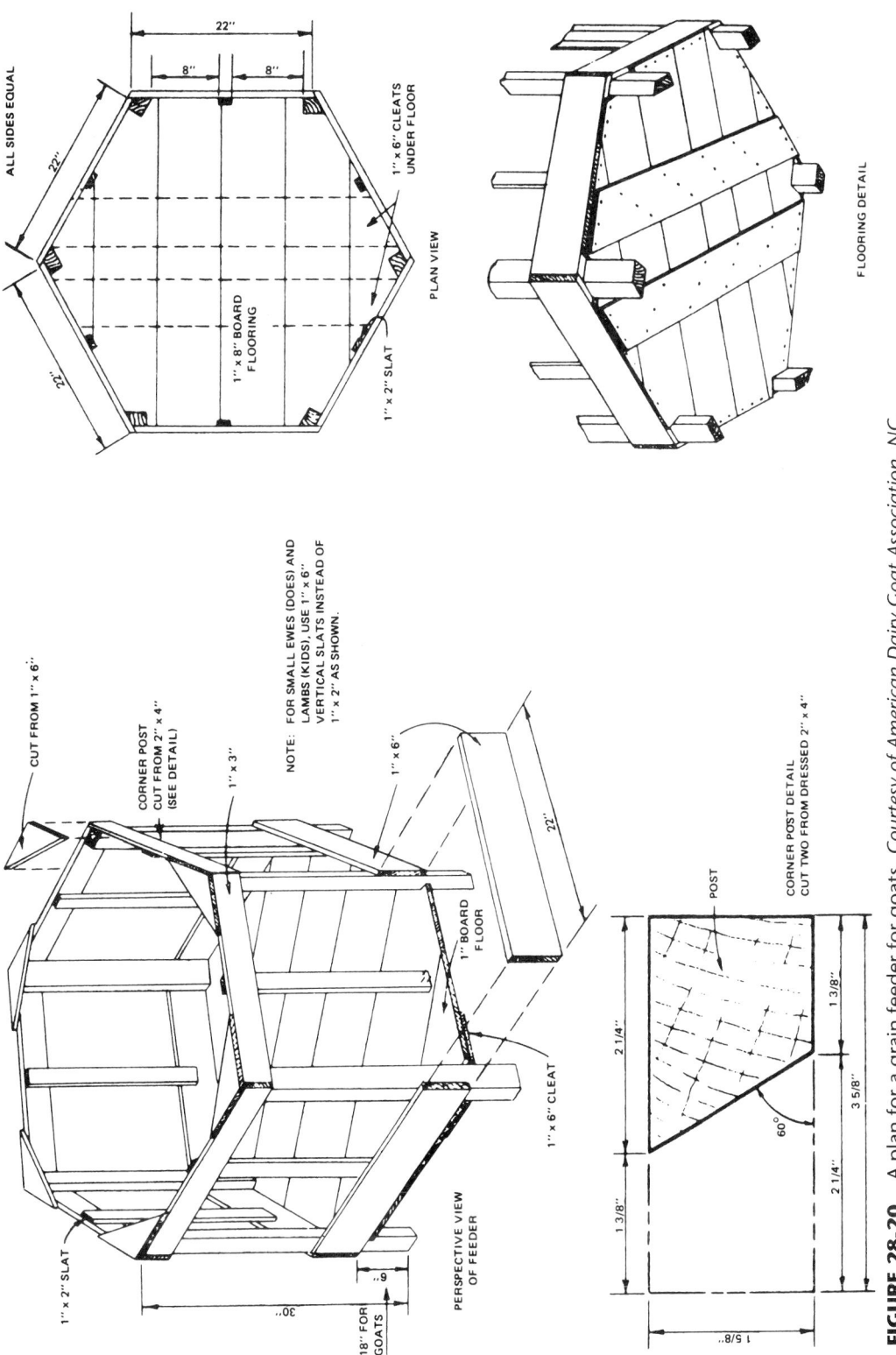

FIGURE 28-20   A plan for a grain feeder for goats. *Courtesy of American Dairy Goat Association, NC.*

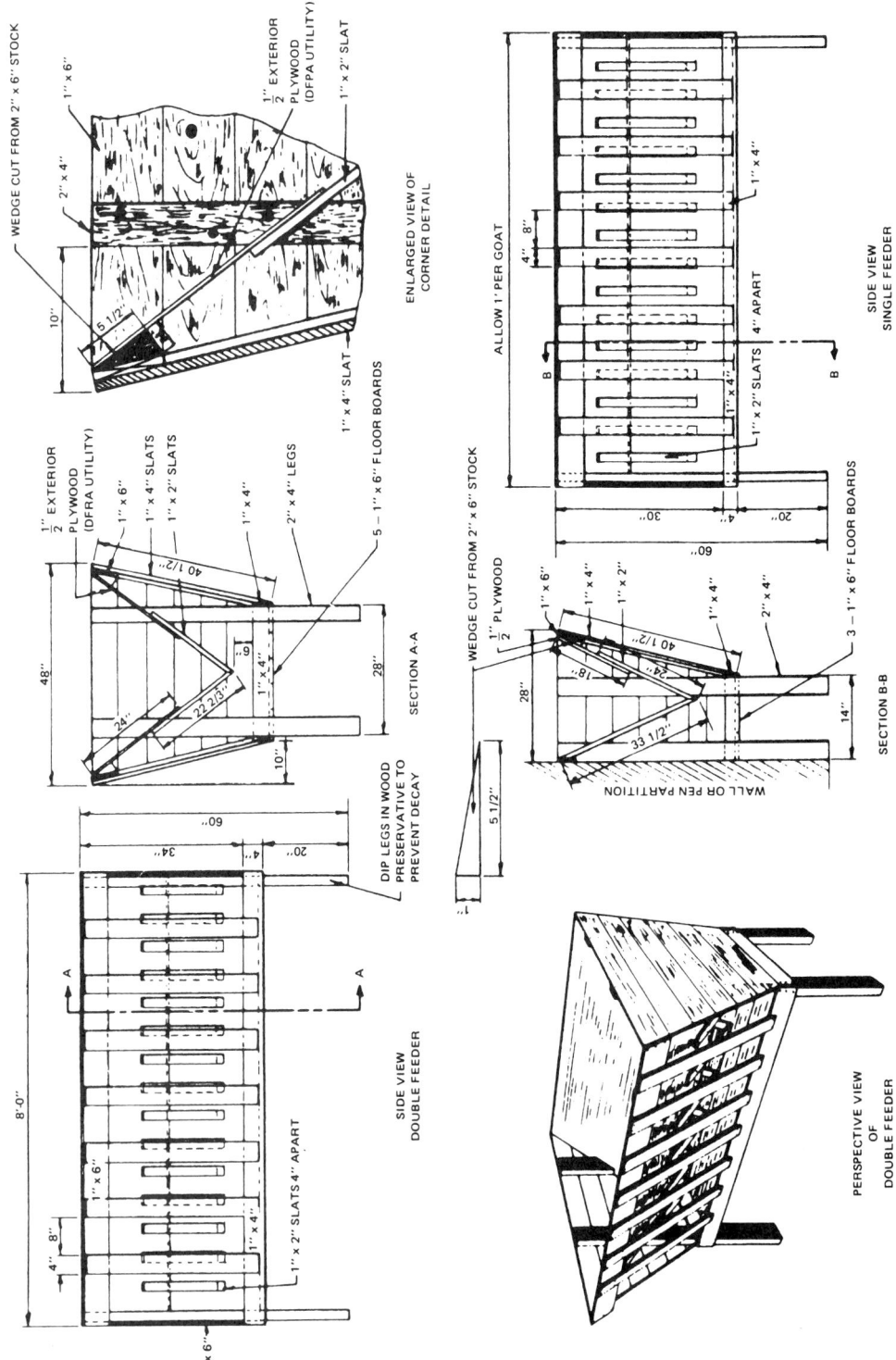

FIGURE 28-21 A plan for feed racks for goats. Courtesy of American Dairy Goat Association, NC.

Some grain and protein supplement is fed the milking does. Both Angora and dairy goats are seasonal breeders. The does are bred in the late fall and kid in the spring.

Spanish goats are raised primarily for meat. Most of this type of production is found in Texas. Spanish goats require little housing and are fed mainly on browse.

Care must be taken when milking dairy goats to keep the milk and the equipment clean. Following a regular routine in milking increases production.

Student Learning Activities

1. Prepare a bulletin board displaying pictures of goat breeds.
2. Name the parts of the goat on a picture or a live animal.
3. Practice good selection procedures for goats when selecting for a project or for home use.
4. Give an oral report on feeding and management of goats.
5. Take field trips to observe good feeding and management practices for goats.
6. Observe and practice dehorning, castration, and marking of goats.
7. Take a field trip to observe facilities and equipment used by goat producers in the area.
8. Prepare a display of equipment used for goat production.
9. Construct portable goat equipment in the school or home shop.
10. Give oral reports on goat facilities and equipment.

Review

1. Why is most of the Angora goat production in the United States found in Texas?
2. Why are small dairy goat herds popular as an enterprise?
3. What is the main use of the Spanish goat?
4. List and briefly describe the breeds of goats.
5. Describe the fleece and desirable body conformation of the Angora goat.
6. Make a sketch showing the parts of the dairy goat.
7. Describe the desirable characteristics of the dairy goat.
8. Describe a good feeding program for Angora goats.
9. Describe a good feeding program for dairy goats.
10. Describe a feeding plan for young kids.
11. Describe a feeding plan for growing kids.

12. List the recommended management practices when breeding Angora goats.
13. Describe two kidding systems in common use with Angora goats.
14. List recommended management practices for Angora goat kids.
15. When are Angora goats sheared?
16. Describe the practice of caping Angora goats.
17. List the management practices that increase the value of the mohair clip.
18. List the recommended management practices for Spanish goats.
19. Describe the recommended management practices for breeding dairy goats.
20. What are the signs that the doe is about to kid?
21. Describe the recommended management practices for caring for dairy goat kids.
22. Describe the recommended practices for milking dairy goats.
23. Describe the housing and equipment needed for an Angora goat enterprise.
24. Describe the kind of housing needed for dairy goats.
25. List the kinds of equipment needed in the milk room.

Diseases and Parasites of Sheep and Goats

Objectives

After studying this unit, the student should be able to

- describe common diseases and parasites of sheep and goats.
- list prevention and control practices for common parasites of sheep and goats.
- outline a program to reduce losses from diseases and parasites.

GENERAL HEALTH RECOMMENDATIONS

A good health program for sheep and goats involves prevention, rather than treatment, of diseases and parasites. Sheep and goats do not respond well to treatment. When they become ill, they frequently do not recover. Often, the value of an individual animal is not high enough to justify the expense of treatment by a veterinarian. This does not mean that the producer should not use the services of a veterinarian. The help of a veterinarian may be needed to plan the use of drugs and to diagnose problems affecting the flock or herd.

Observing the vital signs (temperature, pulse rate, and respiration rate) in an animal can help in the early detection of health problems. Vital signs will vary with activity and environmental conditions. Normal vital signs in sheep and goats are:

- Temperature (sheep, normal range): 100.9–103.8°F (38.3–39.9°C) [average is 102.3°F (39.1°C)] (Temperature is usually higher in the morning than in the afternoon; younger animals will show a wider range of temperature than mature animals.)
- Temperature (goats, normal range): 101.7–105.3°F (38.7–40.7°C) [average is 103.8°F (39.9°C)] (Temperature is usually higher in the

morning than in the afternoon; younger animals will show a wider range of temperature than mature animals.)

- Pulse rate (sheep and goats, normal range): 70–80 heartbeats per minute.
- Respiration rate (sheep and goats, normal range): 12–20 breaths per minute.

Body temperature is taken in the rectum using either a mercury thermometer or a battery-powered digital thermometer. Restrain the animal when inserting the thermometer into the rectum. The pulse rate is taken by finding the artery that runs down the inside of the hind leg. The pulse rate may also be taken by finding the artery that is high up on the inner surface of the thigh just where it emerges from the groin muscle. Respiration rate is determined by observing the number of times the animal breathes per minute.

A program for prevention of health problems includes the following steps:

1. Watch the animals closely for signs of illness.
2. Use the best feeding and management practices possible to prevent problems before they occur.
3. Handle animals with care; avoid stress whenever possible.
4. Follow strict sanitation practices.
5. Treat all wounds with disinfectants.
6. Select only healthy animals for breeding purposes.
7. Isolate and watch newly purchased animals for at least 30 days before putting them with the rest of the animals.
8. Prevent fenceline or other contacts with animals from other farms.
9. Control traffic of trucks, equipment, and people into areas where animals are kept.
10. Isolate sick animals for treatment.
11. When possible, prevent diseases by vaccinating.
12. Control parasites with sprays, dips, dusts, and drenches.
13. Rotate pastures to prevent parasite buildups.
14. Cooperate with a local veterinarian in the prevention and treatment of diseases and parasites.

DISEASES OF SHEEP AND GOATS

A number of diseases are common to sheep, goats, and cattle. While there may be some differences in symptoms, the general description of many of these diseases is very similar. A discussion of the following diseases that may affect sheep and goats, as well as cattle, is found in Unit 17.

actinobacillosis   actinomycosis
  (wooden tongue)   (lumpy jaw)

anthrax   listeriosis
blackleg   malignant edema
bloat   pinkeye
brucellosis   shipping fever
Johne's disease   white muscle disease
leptospirosis

The following diseases generally affect only sheep and/or goats.

Blue Tongue (Sore Muzzle)

Blue tongue is a disease of sheep that is caused by a virus. It occurs mainly in the western United States. It is spread from sheep to sheep by small gnats. Blue tongue weakens a sheep's resistance to other diseases. Death loss (about 5 percent) is usually due to secondary infections such as pneumonia.

Sheep with the disease lose appetite, become sluggish, and have a high fever. The ears, head, muzzle and lips become swollen. The tissues inside the mouth become red and blue. The tongue develops ulcers and the sheep has difficulty eating. Lameness and swelling around the hoof occurs. There is a bad odor, and a discharge from the nose and eyes.

There is no treatment for blue tongue. Treat the secondary infections if the disease occurs. However, a vaccination is available that will prevent the disease. Vaccinate all ewes and bucks at shearing time. Vaccinate replacement lambs at 3½ months of age.

Enterotoxemia (Overeating Disease)

Enterotoxemia is a disease caused by a bacterium. It affects both sheep and goats, with lambs and kids being most often affected. The most common sign of the disease is finding dead lambs or kids. The dead animals have their heads drawn up in an arched and extended position. This is a result of the convulsions that occur before death. If an animal is having convulsions, death will follow quickly. There is no treatment after the symptoms appear.

Enterotoxemia can be controlled through good management, proper feeding, and vaccination. There are several causes of sudden death in lambs. Proper diagnosis of the disease is done through laboratory examination of tissue from animals that have died suddenly. Maintaining a steady intake of feed or milk is the best method of preventing this disease in lambs. Gradually adjust rations for lambs when increasing the concentrate level. Maintain a good source of clean drinking water. Chilling and stress can increase the incidence of enterotoxemia, especially in ewes that are milking heavily and nursing single lambs.

Ewes should be vaccinated four weeks and two weeks before lambing to prevent the disease in nursing lambs. After ewes have been vaccinated, in succeeding years vaccinate them once about two to three weeks before lambing; lambs will secure protection from the disease through the colostrum milk. Vaccinate late-weaned lambs twice before weaning, once late in the nursing period and again two to three weeks later. When practicing early weaning (at 40 days), vaccinate lambs about 10 days before weaning and again about 10 days after weaning. Feeding a continuous low level of chlortetracycline will help protect lambs from enterotoxemia.

If enterotoxemia occurs in feeder lambs, increase the amount of roughage in the ration and raise the level of chlortetracycline to about 200 grams per ton of feed for several days. If the disease occurs in nursing lambs, inject all the lambs with enterotoxemia antiserum; this provides protection for about two to three weeks, after which the lambs should be vaccinated.

Foot Abscess (Bumblefoot)

Foot abscess is a disease that affects the soft tissue of the foot. It is infectious but, unlike foot rot, it is not contagious. It may occur in connection with foot rot, when sheep are in wet or muddy conditions, when the feet have been severely trimmed in wet weather, or when they are put on stubble pasture. Bacteria enter the foot through injuries, causing pus pockets or abscesses. There may be no visible swelling if the infection occurs in the toe or sole of the foot. If it occurs in the heel, there is generally a visible swelling. The disease may affect the joints and tendons. Foot abscess may cause permanent lameness in the animal. Treat the disease by draining the abscesses, applying a medicated dressing, and using systemic antibiotics. Isolate infected sheep on soft, clean footing or slotted floors until healing is completed. Prevent foot abscess by correcting the conditions that can cause foot injuries.

Foot Scald

Foot scald usually affects sheep during periods of extremely wet weather. It is believed to be caused by either the same bacteria or a different strain of the same bacteria that cause foot rot. It is not as severe as foot rot, although it resembles foot rot in its early stages. In its early stage, the skin between the toes becomes inflamed and turns white. As the disease progresses, the rear part of the heel may separate from the hoof. Affected sheep become lame. The forefeet are more often affected than the rear feet. Move affected sheep to a dry area or onto slotted floors.

Treat the sheep with a footbath solution. A five to ten percent copper sulfate or a 10 to 20 percent zinc sulfate solution may be used to treat both foot scald and foot rot. Adding one-half cup of detergent to the zinc sulfate solution increases its effectiveness. Use a zinc sulfate solution daily for seven to ten days for best results. Either of these solutions needs to be in contact with the foot for five to fifteen minutes to be effective in treating either foot scald or foot rot.

Sheep that ingest copper sulfate may develop copper poisoning; use care when handling a copper sulfate solution. Copper sulfate solutions also corrode metal. Zinc sulfate solutions are nonirritating and may be used more often than copper solutions.

Foot Rot

Foot rot is a disease caused by the presence of two different bacteria. Both types must be present for the disease to develop. Foot rot affects both sheep and goats. It is not the same disease as the foot rot that occurs in cattle and is caused by different bacteria.

Foot rot is extremely contagious and may affect the majority of the animals in the flock when an outbreak occurs. It is more likely to occur when animals are on irrigated pastures, wet lowland pastures, or in areas with high rainfall. It is less likely to be a problem when animals are on sandy, well-drained soils and in areas of low rainfall. Foot rot is usually transmitted when animals walk over areas that have been contaminated by infected animals. The disease will spread more rapidly if large numbers of animals are concentrated in a small area.

The death loss from foot rot is low, but affected animals may lose weight. The treatment of the disease requires additional labor, equipment, and materials. It can cause significant economic loss for sheep and goat producers.

Symptoms include lameness, loosening of the hoof wall, and a foul-smelling discharge. It does not form abscesses on the foot. Animals may move around on their knees.

Controlling foot rot involves regular inspection of the animals' feet, proper trimming of feet, keeping animals out of wet areas, keeping bedding dry, and the regular use of a footbath. Footbath solutions as discussed under foot scald may be used. Scattering lime or superphosphate around feed mangers may also help in preventing foot rot.

When an outbreak of foot rot occurs, separate the infected animals from the rest of the flock and put them in a clean, dry area. Trim the feet if necessary. Treat the infected feet in a footbath solution twice daily as outlined above. Severe cases may require the injection of antibiotics. Check the animals and treat them every three days for at least four treatments. Be sure the animals are completely free of foot rot before returning them to the uninfected flock. A vaccination is available for contagious foot rot.

Lamb Dysentery

Lamb dysentery affects sheep and is caused by bacteria. It affects mainly lambs one to five days of age. Death losses can be high. Symptoms include loss of appetite, depression, diarrhea, and sudden death. There is no treatment. Dysentery can be prevented by following strict sanitation practices and lambing in clean, dry housing, and on clean pasture. Vaccinate ewes six and three weeks before lambing.

Mastitis

Mastitis, which affects both sheep and goats, is caused by bacteria or by injury to the udder. The udder becomes swollen, hard, and sore. The affected animal may have a straddling walk. The milk is thick, yellow, and flaky. Often, the lamb or kid cannot nurse. Mastitis is treated by applying hot packs soaked in Epsom salts to the udder four to five times per day. Milk the udder out by hand. Treat with antibiotics. To aid in prevention of mastitis, remove any objects in the barn or pasture that could cause bruises, such as high door sills. Sanitation also helps to prevent the disease. Tag wool from the udder area on sheep, and use proper milking procedures.

Navel Ill

Navel ill affects sheet and goats and is caused by a bacterial infection of the navel. It affects mainly young animals within a few days after birth. Symptoms include fever, loss of appetite, depression, and swelling around the navel. Death may occur quickly. A chronic form results in poor appetite, loss of weight, and lameness in the joints. Antibiotics are used to treat navel ill. Prevention is achieved by good sanitation practices and the use of a disinfectant on the navel at birth.

Pneumonia

Pneumonia is an inflammation of the lungs that affects both sheep and goats. Exposure to cold, damp, drafty conditions may cause pneumonia. It sometimes occurs as a secondary infection with another disease. Parasites in the lungs can also cause pneumonia. Symptoms include difficult breathing, fever, coughing, and loss of appetite. Diarrhea often develops. Some animals die; others recover. In still others, the condition becomes chronic and extends for a period of several weeks. Treat by keeping the animal warm and dry. Use antibiotics when the symptoms are first seen. To prevent pneumonia, keep animals warm and dry, especially after shearing or dipping.

FIGURE 29-1   An advanced stage of scrapie showing wool loss, drooping ears, and staring eyes. *Courtesy of USDA.*

Scrapie

Scrapie (Figure 29-1) is a disease that affects the central nervous system of sheep and goats. The exact nature of the causative agent is not known. It is one of a class of brain diseases called Transmissible Spongiform Encephalopathy (TSE). See the Unit 17 discussion of Bovine Spongiform Encephalopathy (BSE) for more information about TSEs.

There is no cure or vaccine for scrapie and it is always fatal. Scrapie has a long incubation period, typically two to five years, therefore the disease generally affects older animals. Early symptoms include a change in behavior and scratching or rubbing against fixed objects. The animal loses coordination and begins to walk abnormally, typically high stepping with the forelegs, hopping like a rabbit, and swaying of the back end. Other symptoms include weight loss, wool loss, lip smacking, and biting at the feet and legs. Under stress, affected animals will tremble or go into a convulsive like state.

Confirmation of the presence of scrapie in an animal is made after it dies by a microscopic examination of brain tissue. The presence of an abnormal prion protein in the brain tissue confirms the disease.

There appear to be genetic variations among breeds of sheep that are related to their susceptibility to scrapie and the length of the incubation period. The causative agent probably spreads from ewe to lamb by way of the placenta and placental fluids.

In the United States, the USDA's Animal and Plant Health Inspection Service (APHIS) has used a voluntary program since 1992 in an effort to bring scrapie under control. Participants certify the origin of their flocks from scrapie-free flocks. This is a cooperative effort involving producers, veterinarians, state health officials, and APHIS. The interstate movement of sheep from scrapie-infected flocks is restricted. Prior to 1992, the USDA control program emphasized identification and eradication of the disease by quarantining and destroying affected flocks.

Sore Mouth

Sore mouth affects sheep and goats and is caused by a virus. It is more common in younger animals. Symptoms include blisters on the mouth, lips, and nose of the animal. The udder of an older animal affected with the disease will also show blisters. The blisters fill with pus, break and become open sores, and then scab over. The scabs are gray-brown. It is difficult for the animals to nurse or eat, so they lose weight. In areas where it occurs, sore mouth is prevented by vaccination. Treatment after the disease appears is not considered to be practical.

Tetanus

Tetanus affects sheep and goats and is caused by bacteria. The bacteria live in the soil and enter the animal's body through wounds. Symptoms include stiffness, walking with a straddling gait, inability to eat, and rigid jaw and tail. The animal has spasms and dies. There is no treatment after the symptoms appear. Tetanus is prevented by sterilization of docking, castrating, and shearing instruments and the use of disinfectants on open wounds. Vaccination may be recommended for farms where the disease is known to exist.

Vibriosis

Vibriosis affects sheep and is caused by bacteria. Vibriosis in sheep is caused by a different bacterium than the one that causes vibriosis in cattle. The major symptom is abortion. Ewes will again breed normally after abortion. Isolate ewes that abort and burn or bury aborted fetuses. Move ewes that are not affected to a clean area. Antibiotics are used to help control an outbreak. Prevention is by annual vaccination.

NUTRITIONAL PROBLEMS

Some nutritional problems are common to sheep, goats, and cattle. The following nutritional problems that may affect sheep or goats, as well as cattle, are discussed in Unit 17:

bloat

white muscle disease

The following nutritional ailments pose particular problems for sheep and goats.

Constipation and Pinning

Constipation, or difficulty in passing feces, may affect sheep and goats. *Pinning,* which is the pasting of the tail to the anus by the feces, affects mainly very young lambs. These problems may be caused by feeding coarse, dry, or indigestible feed. Overeating may also contribute to the problem. Treat constipation with a warm soapy water enema or feed one to two teaspoonfuls of castor oil. Treat pinning by washing off the accumulated feces. Prevent these problems by proper feeding and docking of the lamb's tail as soon as possible after birth.

Impaction

Impaction occurs when the rumen becomes filled with dry or indigestible feed. Both sheep and goats are subject to this problem. Causes include feeding too much dry feed, sudden changes in feed, overeating feeds when the ration is changed, lack of water, or the presence of any disease that slows proper digestion. Symptoms include poor appetite, bad breath, no cud chewing, constipation, weakness, and in extreme cases, death. Treat by providing plenty of water and taking the feed away from the animal for a few days. Feed a laxative feed for several days. Give the animal fish liver oil, mineral oil, or Epsom salts. Massage the flank to stimulate muscle contractions in the rumen. The problem can be prevented by proper feeding, good pasture, and exercise.

Milk Fever

Milk fever, a disease caused by lack of calcium in the blood, occurs in both sheep and goats. Lambing ewes may be affected. In goats, it may occur shortly after kidding, or it may occur a month or so later in the lactation period. Symptoms include loss of appetite, restlessness, muscle tremors, and difficulty in standing. If milk fever is not treated, the animal will fall into a coma and die. A veterinarian should be called to treat animals with milk fever. Treatment is by injection of calcium. A good ration during gestation may help in prevention of milk fever.

Night Blindness

Night blindness affects both sheep and goats. It is caused by a lack of vitamin A in the diet. Symptoms include the inability to see at night, soreness of the eyes, loss of appetite, weakness, nervousness, and convulsions. Night blindness is treated by adding vitamin A to the diet. Green pasture and green, leafy hay are good sources of vitamin A. Fish liver oil is also a source of vitamin A. The disease can be prevented by feeding a ration that includes sources of vitamin A.

Poisonous Plants

Certain plants are poisonous to both sheep and goats. However, more plants are poisonous to sheep than are poisonous to goats. Symptoms of plant poisoning vary with the type of plant involved. Treatment is often unsuccessful. Prevention includes the elimination of poisonous plants from the pasture. Animals generally will not eat poisonous plants unless they are forced to by lack of other feed. Do not overgraze pastures.

Pregnancy Toxemia (Ketosis)

Pregnancy toxemia affects sheep, especially those carrying twins or triplets. It is a metabolic disorder that usually occurs during the last six weeks of gestation. During this time the fetuses are growing rapidly and the ability of the ewe to take in an adequate amount of feed is reduced. Inadequate carbohydrate metabolism increases the ketone level in the blood.

Symptoms include going off feed, lagging behind the rest of the flock, nervousness, unsteady gait, grinding the teeth, hard breathing, and frequent urination. If untreated, the ewe may die.

Treatment must begin when the symptoms are first observed if it is to be effective. A concentrated source of energy is provided by an administration of oral propylene glycol at the rate of two ounces (59.2 ml), three or four times per day. Do not use molasses or other sugar sources; that treatment may make the condition worse. The most effective treatment is to induce birth when the symptoms appear. Corticosteroids may be used to induce lambing and also to increase carbohydrate metabolism. Treatment should be under the supervision of a veterinarian.

Prevention of pregnancy toxemia may be done with proper nutrition by providing an adequate source of energy in the diet. Avoid stressful conditions and sudden feed changes during the last few weeks of gestation.

Urinary Calculi (Water Belly, Urolithiasis)

Urinary calculi is caused by the formation of small stones, called calculi, in the urinary tract. It generally affects male sheep on high-concentrate rations. The calculi block the urethra, causing retention of the urine. If the condition is not treated quickly, the bladder may rupture, causing death. If the bladder does not rupture, the animal may die of uremic poisoning. While calculi may form in females, they usually do not block the urethra because it is larger than that of the male. Wethers castrated at a young age have a much smaller urethra and penis and are more likely to suffer from blockage of the urethra from smaller calculi. Urinary calculi is more likely to affect western lambs and feeder lambs about four to six weeks after they have been put in the feedlot. Breeding rams are sometimes affected by urinary calculi.

While there are several factors involved in the development of urinary calculi, the main cause is a narrow ratio of calcium to phosphorus in the diet. The ratio of calcium to phosphorus in the diet of feeder lambs should be two parts calcium to one part phosphorus. This ratio may drop as low as 1:1 in high-concentrate diets; this contributes to the development of urinary calculi. Other factors that make the condition worse are cold weather, reduced salt intake, and reduced water intake.

Symptoms include standing with arched backs, depression, low feed intake, straining to urinate, and kicking at the belly. The abdomen may be swollen, especially if the bladder has ruptured; the penis may be swollen if the rupture has occurred in the urethra.

It is better to prevent urinary calculi than to try to treat it after it has occurred. Maintaining the proper ratio of calcium to phosphorus in the diet is the most effective method of prevention. It may be necessary to have a chemical analysis done on the feeds used in the ration to ensure that the calcium-phosphorus ratio is correct. Calcium carbonate (limestone) may be added to the ration to increase the calcium level in the diet. Adding 0.5 percent ammonium chloride to the diet will acidify the urine; this helps to prevent the formation of calculi. Adding ammonium sulfate to the diet will produce the same results. Provide plenty of clean, fresh water at all times. In cold weather the water temperature should be maintained at 45–50°F (7.2–10°C) to increase consumption. If a severe outbreak occurs, it may be necessary to add up to 4 percent salt in the ration. This will increase urine output and dilute the mineral content of the urine. If salt is added to the ration at this level, make sure a good supply of water is available.

If urinary calculi is diagnosed quickly, treatment can be effective. Surgery is required to treat the condition. A veterinarian should be called immediately if the condition is suspected. Recovery is less likely if the bladder ruptures.

EXTERNAL PARASITES

Several external parasites attack sheep and/or goats. External parasites cause losses in the production of wool, mohair, meat, and milk. Sometimes they may cause the death of the animal. A combination of sanitation and the correct use of insecticides helps to control most of the common external parasites of sheep and goats. Recommendations and regulations concerning the use of insecticides are constantly being updated. Producers should contact university entomologists, county extension personnel, or a local veterinarian for current information on permissible control measures.

Lice

Lice (Figure 29-2) are tiny insects that live on animals. Several different kinds of lice attack sheep and goats. Some are bloodsucking; others are biting or chewing lice. They spread rapidly from one animal to another. Lice irritate the animal's body and cause damage to wool. Animals that do a lot of rubbing may be infested with lice. Control is achieved by dipping, spraying, or dusting. Dipping is more practical for large flocks and should be done in warm weather. Smaller flocks should be dusted or sprayed. Dusting is the better method in cold weather.

Sheep Ked

The sheep ked (Figure 29-3) is sometimes called a *sheep tick*. However, it is not a true tick. It is a wingless fly that is about 1/4 inch (0.6 cm) long and has six legs. It is found on both sheep and goats and is a bloodsucker. It is controlled by dipping, spraying, or dusting.

Sheep Bot Fly

The larvae of the sheep bot fly are found in the nasal cavities of sheep and goats. The fly is the adult stage of the insect. It is about 1/2 inch (1.3 cm) long and looks like a honeybee. Infested animals shake their heads, sneeze, have difficulty breathing, and may hold their noses to the ground. They stamp their feet to try to keep the flies away.

Mange and Scab Mites

There are several species of mites that attack sheep and/or goats, causing a condition referred to as *mange, scabies,* or *scab. Mites* are tiny parasites that can hardly be seen with the naked eye. They burrow into the skin, causing irritation. Some cause scabs to form. The sheep scab mite is sometimes called *wet mange.* An animal infested with mites is shown in Figure 29-4. An eradication program against sheep scabies has been in operation in the United States since 1960. It is believed that it is now eradicated in the United States. If any cases appear, they

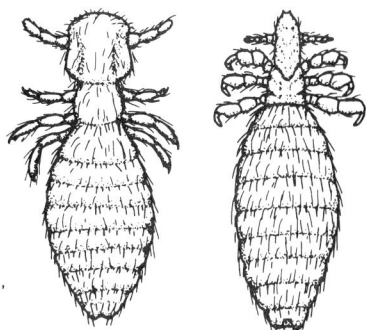

FIGURE 29-2   Two kinds of lice that attack sheep: (left) a biting louse; (right) a sucking louse. *Courtesy of University of Illinois at Urbana-Champaign.*

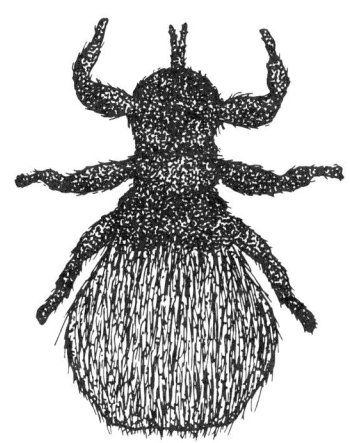

FIGURE 29-3   The adult sheep ked is a type of wingless fly. *Courtesy of University of Illinois at Urbana-Champaign.*

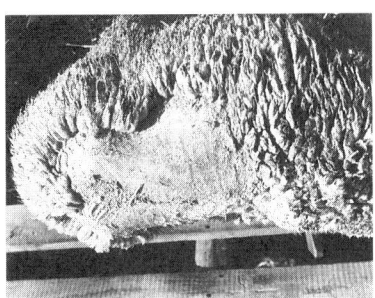

FIGURE 29-4   Sheep head infested with mites. *Courtesy of Iowa State University.*

FIGURE 29-5 Sheep ticks. *Courtesy of Iowa State University.*

must be reported to a veterinarian. The flock is quarantined and control measures are carried out by trained personnel.

Blowflies

Several species of blowflies attack sheep and goats. The screwworm, a maggot of one of these species, was a common pest of sheep and goats in the southwest United States. However, an eradication program has been under way for a period of years. The screwworm has been almost eliminated from the United States. Reinfestations still enter the U.S. from Mexico, where it has not been eradicated. Screwworms enter the animal's body through an open wound and feed on the living flesh in the wound. They breed only in the wounds. Protective smears should be used during the screwworm season. These compounds are placed on open wounds to kill screwworms. Check current recommendations for allowable substances to use.

Wool maggots are caused by other species of blowfly. These do not grow in live flesh. Instead, they live in wool that is wet and matted around wounds. Tagging sheep helps to prevent infestation. Shearing before warm weather and keeping sheep clean also helps to prevent trouble with blowflies. Sprays are available for treating infested animals.

Ticks

Several species of ticks attack sheep and goats. One type of tick is shown in Figure 29-5. The most serious of these is the spinose ear tick. The larvae of this tick lives in the ears of the animal. Ear infections may be caused by the irritation of the tick. The animals lose weight and become listless. Young animals may die from a heavy infestation.

INTERNAL PARASITES

Internal parasites are the most serious health problem of sheep and goats. Economic loss results from loss in weight, lower milk production, poor wool growth, wasted feed, and lower breeding efficiency. Death losses from internal parasites are not high, but some animals do die.

Good management is the key to controlling losses from internal parasites. Overgrazing pastures and failure to rotate pastures are two factors that contribute to problems with internal parasites. A regular program of internal parasite control should be followed. Several anthelmintics are available for the control of internal parasites. *Anthelmintics* are chemical compounds used for worming animals. Current regulations should be followed in the use of these substances. For up-to-date information on chemical control measures, contact county extension personnel, university specialists, or a local veterinarian.

Some of the symptoms of internal parasite infestation are rough hair coat, weight loss, slow gains, loss of appetite, diarrhea, and anemia.

A swelling may be noticed beneath the lower jaw. The animal may have a constant cough. Young animals are affected more severely by internal parasites than are older animals.

Stomach and Intestinal Worms

Worms found in the stomach and intestines of sheep and goats include: (1) common stomach worm, (2) medium stomach worm, (3) bankrupt worm, (4) thread-necked strongyle, (5) nodular worm, (6) hookworm, and (7) tapeworm. The life cycles of all of these worms are similar. The life histories of three of these worms are shown in Figures 29-6 through 29-8. The following general cycle is found in each type of worm. The adult female deposits eggs that pass out of the animal in the manure. The eggs hatch on the ground. The larvae crawl up on blades of grass, which the sheep or goats eat. The larvae become adults inside the body of the animal. The cycle then begins again.

Female worms deposit from several hundred to several thousand eggs per day. The common stomach worm and the medium stomach worm are bloodsuckers. The other worms damage the wall of the intestine, making it more difficult for the animal to absorb nutrients. The common stomach worm is the most serious of all the internal parasites of sheep and goats.

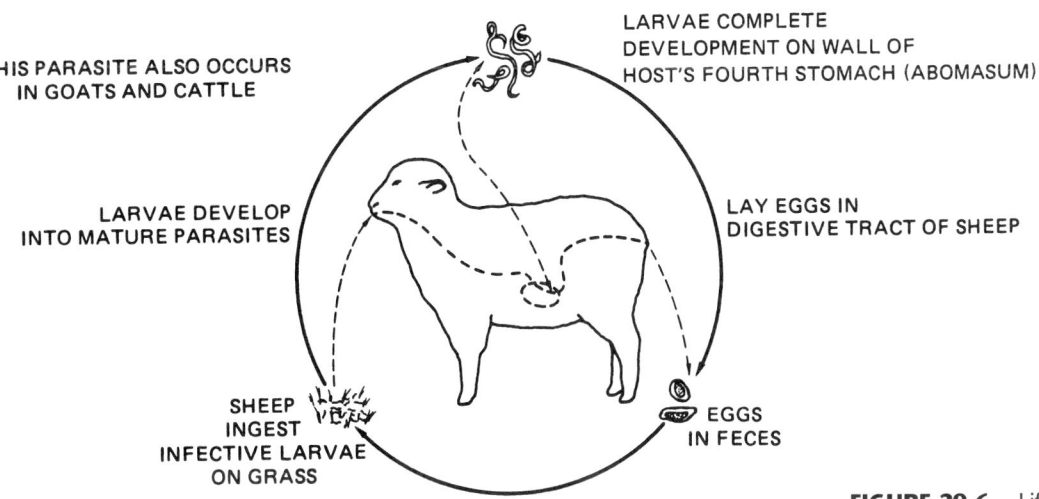

ADULT PARASITE IN FOURTH STOMACH
(ABOMASUM) OF SHEEP

LARVAE COMPLETE
DEVELOPMENT ON WALL OF
HOST'S FOURTH STOMACH (ABOMASUM)

THIS PARASITE ALSO OCCURS
IN GOATS AND CATTLE

LARVAE DEVELOP
INTO MATURE PARASITES

LAY EGGS IN
DIGESTIVE TRACT OF SHEEP

SHEEP
INGEST
INFECTIVE LARVAE
ON GRASS

EGGS
IN FECES

DEPOSITED AND HATCH IN FECES
LARVAE DEVELOP TO INFECTIVE
STAGE AND CONTAMINATE GRASS

FIGURE 29-6   Life history of a common stomach worm of sheep. *Courtesy of University of Illinois at Urbana-Champaign.*

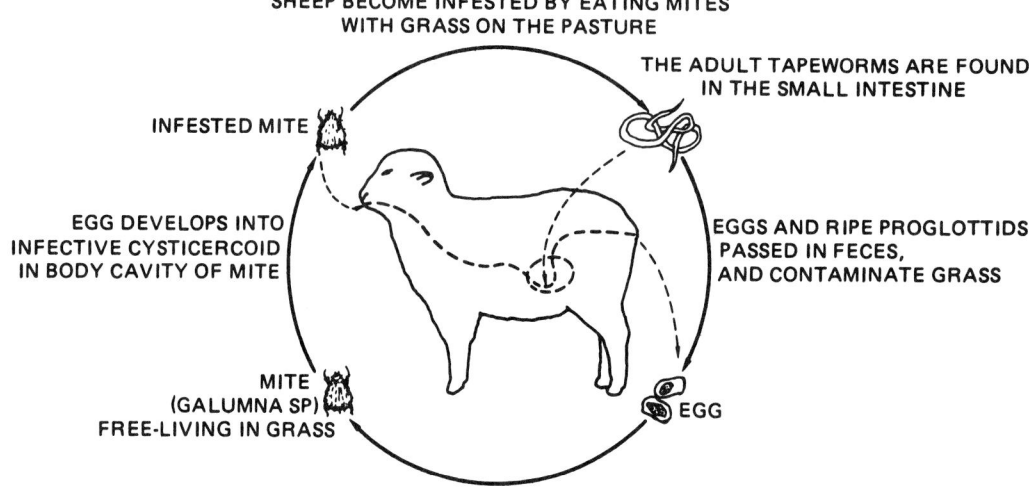

SHEEP BECOME INFESTED BY EATING MITES
WITH GRASS ON THE PASTURE

THE ADULT TAPEWORMS ARE FOUND
IN THE SMALL INTESTINE

INFESTED MITE

EGG DEVELOPS INTO
INFECTIVE CYSTICERCOID
IN BODY CAVITY OF MITE

EGGS AND RIPE PROGLOTTIDS
PASSED IN FECES,
AND CONTAMINATE GRASS

MITE
(GALUMNA SP)
FREE-LIVING IN GRASS

EGG

EGG INGESTED BY MITE. GALUMNA SP

THIS PARASITE IS ALSO FOUND
IN GOATS, CATTLE, AND OTHER
RUMINANTS

FIGURE 29-7   Life history of a tapeworm of sheep. *Courtesy of University of Illinois at Urbana-Champaign.*

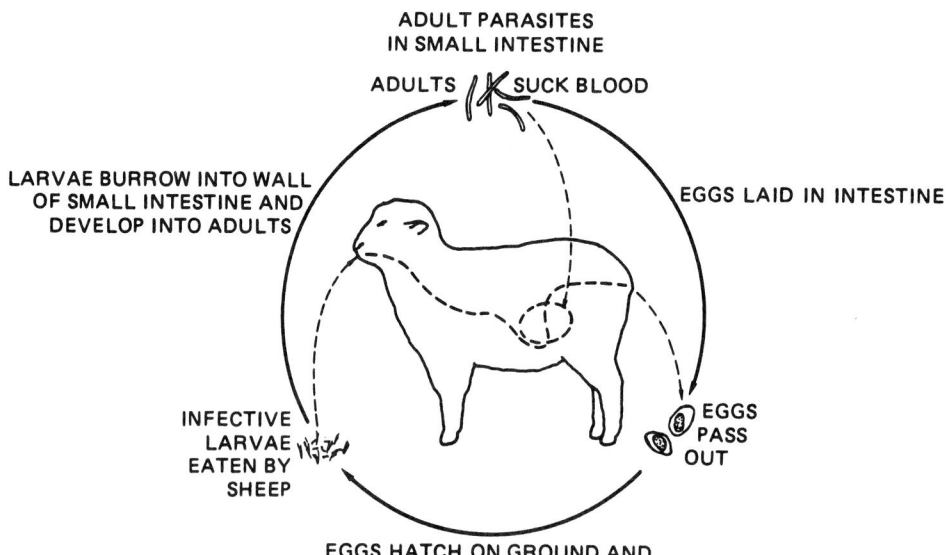

ADULT PARASITES
IN SMALL INTESTINE

ADULTS   SUCK BLOOD

LARVAE BURROW INTO WALL
OF SMALL INTESTINE AND
DEVELOP INTO ADULTS

EGGS LAID IN INTESTINE

INFECTIVE
LARVAE
EATEN BY
SHEEP

EGGS
PASS
OUT

EGGS HATCH ON GROUND AND
INFECTIVE LARVAE DEVELOP
LIFE CYCLE IS DIRECT
NO INTERMEDIATE
HOST IS INVOLVED

THIS PARASITE ALSO OCCURS IN GOATS, CATTLE,
CAMELS, AND VARIOUS ANTELOPES

FIGURE 29-8   Life history of a small intestinal worm of sheep. *Courtesy of Roger Courson, Vocational Agriculture Service, University of Illinois.*

Lungworms

Two species of lungworms infest sheep and goats: the thread lung-worm and the hair lungworm. The adults of both types of worms live in the lungs. Eggs of the thread lungworm hatch in the intestine, and the larvae pass out in the feces. They are then picked up on the feed. The eggs of the hair lungworm hatch in the lungs. The larvae pass out in the feces, and develop in land snails or slugs. The sheep or goats pick up the worms by eating the snails or slugs. The life cycle of the lungworm is shown in Figure 29-9.

Liver Fluke

The liver fluke lives in the liver of the infested sheep or goat. It causes bleeding in the liver. The eggs pass out in the manure. After hatching, the larvae enter a snail. They develop in the snail and then leave the snail and form a cyst (protective covering) that is attached to vegetation. The sheep take in the cyst when eating the vegetation.

Coccidia

Coccidiosis is caused by the small protozoa called *coccidia*. These organisms live in the intestine of the sheep or goat. They cause the cell walls to rupture, and the animal bleeds internally. A bloody diarrhea is one of the symptoms of coccidiosis. The sheep or goat picks up the protozoa organisms from contaminated feed. There are a number of

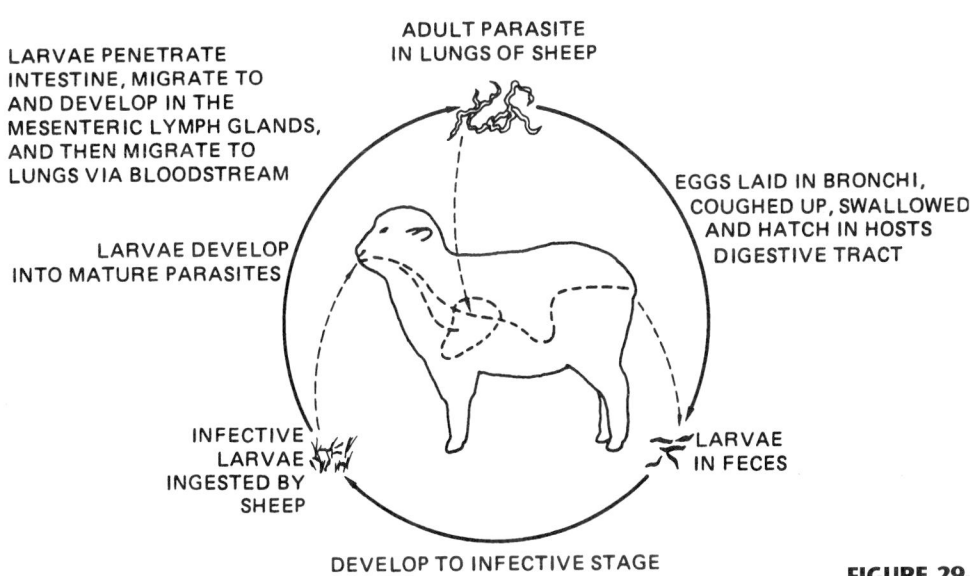

FIGURE 29-9 Life history of a lungworm of sheep. *Courtesy of Roger Courson, Vocational Agriculture Service, University of Illinois.*

species of protozoa that cause coccidiosis. Each kind of livestock is affected by its own type.

Drenching Procedure

Drenching is the oral administration of a liquid medication, generally for the control of internal parasites. It is usually given through the mouth of the animal. A syringe or an automatic drenching gun may be used to drench sheep or goats. The automatic drenching gun is more expensive but saves time. It is more practical when large numbers need to be treated.

The animal's head must be kept in a level position while drenching. Hold the fingers under the muzzle and the thumb over the nose. Lift the upper lip with the thumb and carefully insert the nozzle along the side of the mouth. Administer the dose and release the animal so it can swallow.

> **Caution:** Care must be taken not to injure the animal. If the animal's head is held too high, the material may enter the lungs, causing death. Read and follow the instructions on the label for the type of material being used.

When drenching small flocks, the animals may be penned. The operator then moves among the sheep administering the drench. Each animal is marked as the operation is completed. With larger flocks, a chute should be used. The operator stands in the chute and catches each animal, holding it between his legs as he administers the drench.

SUMMARY

Keeping sheep and goats healthy is best achieved through programs of disease and parasite prevention. Good management practices help to prevent many health problems.

Enterotoxemia is one of the most common diseases of sheep and goats. It can be prevented by a vaccination program. Several nutritional problems affect sheep and goats, many of which can be prevented by proper feeding. Lice and sheep ked are the most common external parasites of sheep and goats. A regular program of treatment helps to control these parasites. Stomach worms are the most serious of the internal parasites that affect sheep and goats. Pasture rotation and proper stocking of pastures controls many internal parasite problems. Internal parasites can be treated by drenching. Care must be taken when drenching sheep and goats to avoid injuring the animals.

Student Learning Activities

1. Prepare and give an oral report on one or more sheep and goat diseases and parasites.
2. Prepare a bulletin board display of pictures of sheep or goats with diseases and/or parasites.
3. Plan and practice good health programs for school sheep or goat projects or for the home farm.
4. Ask a veterinarian to speak to the class about health problems of sheep or goats.
5. Conduct a community survey of health problems of sheep or goats. Develop recommendations that might improve the health of local herds or flocks.

Review

1. List the recommended practices that help to prevent health problems with sheep and goats.
2. List the diseases of sheep and goats, and describe the symptoms, prevention, and treatment of each.
3. List the external parasites that affect sheep and goats, and describe the symptoms, prevention, and treatment of each.
4. How do external parasites affect sheep and goats?
5. How do internal parasites affect sheep and goats?
6. Describe the typical symptoms of internal parasites in sheep and goats.
7. Name and briefly describe the typical life cycles of the worms found in the stomach and intestines of sheep and goats.
8. Name and briefly describe the life cycles of the other common internal parasites of sheep and goats.
9. Describe the procedure for drenching sheep and goats.

30 *Marketing Sheep, Wool, Goats, and Mohair*

Objectives

After studying this unit, the student should be able to

- describe the methods of marketing sheep and wool.
- list the grades of wool and mohair.
- describe the methods of marketing goat meat, milk, and mohair.

MARKETING SHEEP

Promotion of Sheep and Wool

Funding for the promotion of sheep and wool has declined in recent years. The decline in funding began with the phase-out of the National Wool Act that ended incentive payments to producers in 1995. Congress passed the Sheep Promotion, Research and Information Act in 1994 that provided for promotional money to be provided by producer check-off. This is similar to the check-off programs in effect in the beef and swine industries.

Approval of the check-off required a majority vote by producers in a referendum. An initial referendum held in early 1996 was successful; however, the USDA invalidated the results because of perceived errors in conducting the voting. A second referendum held in late 1996 failed to secure the majority of votes needed. As a result there is no check-off to provide money for the promotion of sheep and wool.

In 1997 a group of individuals concerned about the future of the sheep and wool industry met to develop a long-range plan for revitalization of the industry. This plan sets forth a set of goals designed to help the industry regain some of the market share that it has lost in recent years. Goals include:[1]

- eliminate scrapie in the United States
- improve efficiency and decrease costs of production
- provide a consistent supply of quality products on a year-around basis
- increase consumer demand and market share for lamb and wool products
- ensure food safety, animal welfare, and environmental stewardship programs that meet or exceed consumer expectations
- develop a functioning value-based marketing system for lamb and wool
- encourage alliance development to allow better communication and improved risk management

In an attempt to address some of the problems facing the industry, the American Sheep Industry Association established a Sheep Industry Transition Team in early 1998 to help establish a new sheep industry association. As a result the National Sheep Association was formed and developed an organizational structure during 1999. This group is designed to represent range flock producers, farm flock producers, seedstock producers, lamb feeders, packers and processors, wool textile manufacturers, and wool marketing groups.

Types of Markets

Sheep may be marketed through terminal markets, local pools, sale barns, direct to the packer, or by electronic marketing. Electronic marketing of lambs is a new development and it is expected that an increasing number of animals will be marketed in this manner in the years ahead. Direct selling to packers accounts for the largest volume of marketing. The importance of terminal markets has declined in recent years. In 1966, more than 6 million sheep were sold through terminal markets in the United States. This figure has now dropped to fewer than 1 million head. The leading terminal markets include San Angelo, TX; Sioux Falls, SD; Billings, MT; and South St. Paul, MN.

In some areas of the Midwest, auction markets, terminal markets, and local pools are still important ways to market sheep. The producer's choice of a market depends on the kinds of markets available locally, prices paid, numbers of lambs to be sold, and the kind of transportation that is available.

---

[1]Sheep Industry Long Range Plan, August 1997. The entire plan can be accessed on the Internet at: www.sheepusa.org/actplan/texasam. html.

Classes and Grades of Sheep

Sheep are classed according to age, use, sex, and grade. Age classes are lamb, yearling, and sheep. Lambs are further divided by age as hothouse lambs, spring lambs, and lambs. The use classes are slaughter sheep, slaughter lambs, feeder sheep, feeder lambs, breeding sheep, and shearer lambs. The sex classes are ewe, ram, and wether.

Age. *Hothouse lambs* are under three months of age and usually weigh less than 60 pounds (27.2 kg). They are sold in a specialty market between Christmas and Easter. *Spring lambs* are three to seven months of age. They are finished at 70 to 90 pounds (31.8–40.8 kg). Spring lambs are born in the fall and marketed in the spring and early summer. *Lambs* are seven to 12 months of age. The most desirable market weight for lambs is 110 to 130 pounds (49.5–58.9 kg). They are usually milk and grass fed. If they have been fed grain, they are called *fed lambs*. Small-frame lambs may be marketed at 95 to 100 pounds (42.8–45.4 kg).

Yearlings are between one and two years of age. The first pair of permanent incisor teeth are present, but the second pair are not. Sheep are more than two years old.

Use classes. *Slaughter sheep* and *lambs* are marketed for immediate slaughter. *Feeder sheep* and *lambs* require more feeding to finish for slaughter. *Breeding sheep* are western ewes that are sold back to farms and ranches to be bred to produce more lambs. *Shearer lambs* are not finished enough for slaughter. They are shorn and fed to a higher level of finish before slaughter.

Sexes. A *ewe* is a female sheep or lamb. A *ram* is a male sheep or lamb that has not been castrated. A *wether* is a male lamb that was castrated when young. It does not show the sexual traits of the ram.

Grades. Live grades of sheep are based on quality and estimated yield. Quality grades are based on conformation and amount of finish. The four quality grades for lambs and yearlings are: (1) Prime, (2) Choice, (3) Good, and (4) Utility. The four quality grades for slaughter sheep are: (1) Choice, (2) Good, (3) Utility, and (4) Cull. For a description of body conformation and method of handling sheep to determine amount of finish, review Unit 26, Selection of Sheep.

Estimated yield is based on five yield grades indicated by the numbers 1 through 5. Yield grade 1 is the highest yielding, while yield grade 5 is the poorest yielding. The thickness of fat over the ribeye determines the yield grade for all slaughter sheep. The relationship between fat thickness and yield grade is:

- Yield Grade 1   0.00 to 0.15 inch (0.00–0.38 cm) backfat
- Yield Grade 2   0.16 to 0.25 inch (0.41–0.64 cm) backfat

- Yield Grade 3   0.26 to 0.35 inch (0.66–0.89 cm) backfat
- Yield Grade 4   0.36 to 0.45 inch (0.91–1.14 cm) backfat
- Yield Grade 5   0.46 inch (1.15 cm) and greater backfat

Yield grades identify differences in cutability among carcasses. *Cutability* refers to the yield of closely trimmed, boneless retail cuts that come from the major wholesale cuts. The major wholesale cuts are the leg, loin, rib, and shoulder (Figure 30-1). The leg and loin generally yield one-half of the total carcass weight. They represent 75 percent of the value of the carcass. A lack of muscling in the leg or too much finish lowers the estimated yield grade.

Grades for live animals are directly related to the quality and yield grades of the carcasses they will produce. The same grades are used for carcasses as are used for live animals.

The carcass quality grades are based on conformation and quality of the lean. The amount of muscle development gives an indication of conformation. The quality of the lean is determined by the color, firmness of fat and lean, and amount of interior fat deposits. The maturity of the animal affects the standards for judging the quality of the lean.

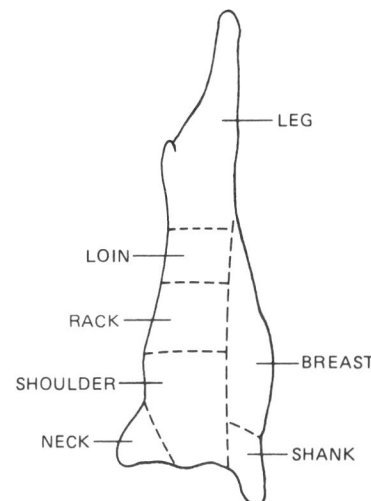

FIGURE 30-1 Wholesale cuts of lamb.

The USDA standards for quality grades of lamb carcasses are briefly described in the following paragraphs. A complete description of standards is available from the USDA.

Minimum qualifications for Prime lamb carcasses are: moderately wide and thick in relation to their length; moderately plump and full legs; moderately wide and thick backs; and moderately thick and full shoulders. Mature lambs have a moderate amount of *feathering* (intermingling of fat with lean) between the ribs and a modest amount of fat streaking within and upon the inside flank muscles, which are a light red color. The lean flesh and external finish are firm; the flanks are moderately full and firm.

Minimum qualifications for Choice lamb carcasses are: slightly wide and thick in relation to their length; slightly plump and full legs; slightly wide and thick backs; slightly thick and full shoulders. Mature lambs have a small amount of feathering between the ribs; a slight amount of fat streaking within and upon the inside flank muscles, which are a moderately light red color; moderately firm lean flesh and external finish; and slightly full and firm flanks.

Minimum qualifications for Good lamb carcasses are: moderately narrow in relation to their length; slightly thin, tapering legs; slightly narrow and thin backs and shoulders. Mature lambs have a slight amount of feathering between the ribs; traces of fat streaking within and upon the inside flank muscles, which are a slightly dark red color; moderately firm lean flesh and external finish; slightly full and firm flanks.

Utility lamb carcasses are: those lambs with characteristics that are inferior to those specified as minimum for the Good grade.

Shrinkage

Sheep shrink as they are moved to market just as do other classes of livestock. The amount of shrinkage depends mainly on the distance traveled and the time in transit. Most of the shrink occurs in the first few miles traveled. Shrinkage of fat lambs ranges from a little under 2 percent to over 8 percent. The average shrinkage is about 3 to 5 percent. Careful handling of sheep while they are being transported to market helps to reduce shrinkage and other losses.

Seasonal Prices

Lamb supplies vary from month to month throughout the year. This contributes to a seasonal variation in the price of lamb. Figure 30-2 shows the seasonal variation in lamb prices as a percent of the yearly average. The variation shown covers prices received by farmers throughout the United States for the years 1982–1998. Lamb prices are normally above the yearly average in February, March, April, May, June, and July, reaching their peak during the months of March, April, and May. They are generally below the yearly average in January, September, October, November, and December. The low point is often reached in November. For the period shown, the price for lamb was at the yearly average generally in August. Prices for any given year may not necessarily follow the long-term trend on a month-by-month basis.

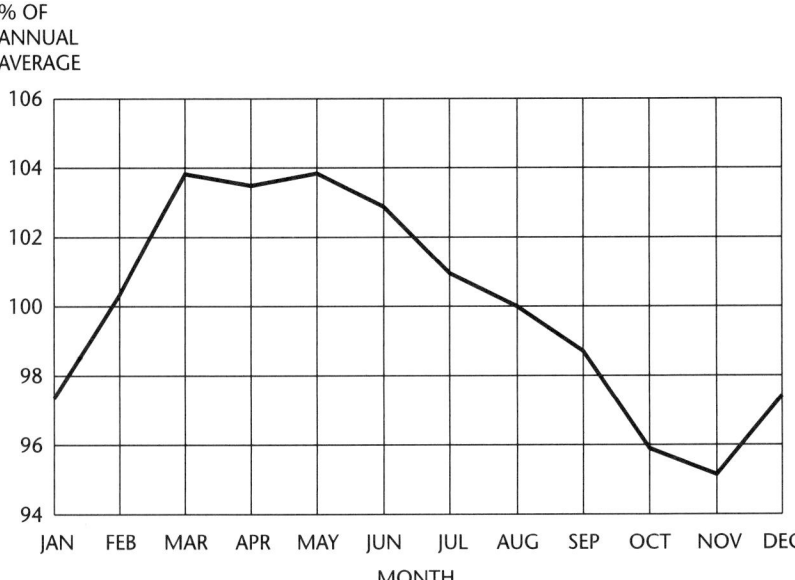

FIGURE 30-2 Seasonal variation in lamb prices received by farmers as a percent of yearly average, United States. *Sources: Annual Price Summary,* USDA, June 1988; *Annual Price Summary,* USDA, July 1998; *Agricultural Prices,* USDA, January– December 1998.

Selling Purebred Sheep

Purebred sheep are sold by private sale or through auction sales sponsored by regional or state sheep associations. Purebred breeders often advertise through magazines that reach sheep producers and by putting up signs at the entrance to the farm.

Sheep from purebred flocks that do not meet the standards for breeding stock are sold for slaughter. The methods for marketing these sheep are the same as for any other commercial sheep producer.

MARKETING WOOL

Wool Markets

Wool is marketed through local buyers, wool pools, cooperatives, warehouse operators, or by direct sale to wool mills. The choices open to the producer depend on the kinds of markets available locally. Local buyers are available in most wool-producing areas. Wool pools and marketing cooperatives have developed in many areas in an effort to obtain a better price for wool. Warehouse sales are common in the range states. Very little wool is marketed directly to wool mills.

Local buyers buy wool on a cash basis. They may or may not pay price differentials for different grades of wool. The producer pays no marketing costs. Prices paid are often lower than those that are paid by other types of markets.

Wool pools grade fleeces by fineness, staple length, color, and cleanness. The wool is sold on a sealed bid basis by grade. Each producer is paid according to the weight and grade of the wool that was consigned. Since wool pools offer a greater volume of wool to prospective buyers, the prices received by the producer are often better.

Wool cooperatives act as marketing agents for producers. A cash advance or partial payment is made to the producer when the wool is delivered to the wool cooperative. The wool is fleece graded at the cooperative warehouse. Final settlement of price is made based on the price received for that grade of wool. Cooperatives also have the advantage of being able to offer larger volumes of wool to the buyer.

Warehouses may buy the wool outright. Others handle the wool on a commission basis. Some warehouses grade the fleeces and market the wool on a graded basis. If the wool is handled on a commission basis, the producer does not receive payment until the wool is sold from the warehouse. Warehouse operators are familiar with the wool market and may be able to obtain better prices for the wool that is consigned to them.

Wool Prices

Wool prices vary seasonally. Prices are usually above the yearly average in March, April, May, June, July, and October, reaching their peak in May. They tend to be below the yearly average in January, February,

August, September, November, and December. The lowest wool prices usually occur in November, December, and January. Figure 30-3 shows the seasonal variation in wool prices received as a percent of the yearly average. Incentive payments that producers received are not included in the data.

The National Wool Act of 1954 established incentive payments for wool and mohair. The purpose was to encourage the production of wool and mohair in the United States. Legislation that ended the incentive payments and repealed the National Wool Act was signed into law in 1993. Incentive payments continued through 1995 during a phase-out period and the final repeal of the Wool Act became effective on December 31, 1995.

The value of the wool fleece is largely determined by the amount of clean wool that it produces. The term *grease* is used to indicate the presence of impurities in the fleece. *Grease wool* refers to the fleece before it is cleaned. *Shrinkage* is the amount of weight lost when the impurities are cleaned out of the fleece. Producing a fleece with less shrinkage increases the value of the fleece. The higher the shrinkage, the lower the grease value of the fleece.

The length, density, and diameter of the wool fiber also affect the value of the wool by determining its grade.

The quality of wool produced in the United States is generally lower than the quality of imported wool. In the United States, wool is usually considered a by-product of the sheep industry, whereas foreign producers tend to take greater care to produce high-quality wool. Wool produced in the United States often contains residues from vegetable matter, urine stains, and dark hairs. The USDA is currently conducting research to find methods of removing impurities from domestic wool without damaging the wool fibers. If the research is successful, it can make domestic wool more competitive in the marketplace.

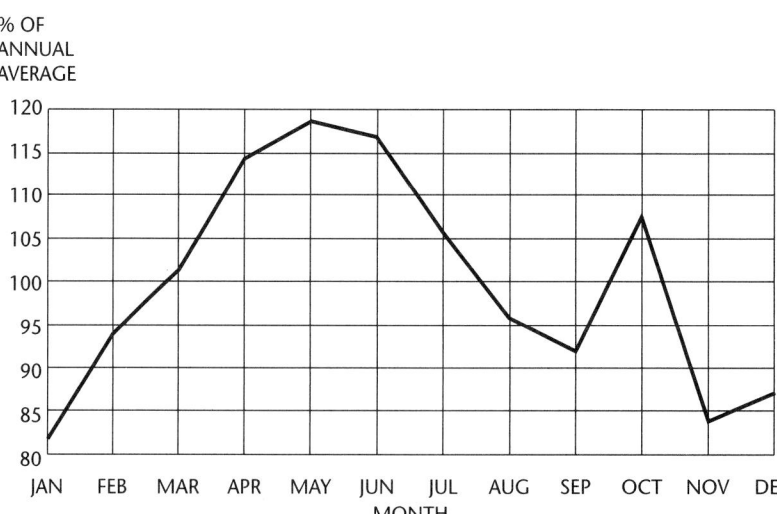

FIGURE 30-3   Seasonal variation in wool prices received by farmers as a percent of yearly average. (Does not include incentive payments.) *Sources: Annual Price Summary,* USDA, June 1988 and July 1993; *Agricultural Prices,* USDA, March 1994.

Wool Grades

Wool grades are based mainly on the diameter of the fiber. Density is the major factor in determining the weight of the fleece. Length of fiber is also considered when grading wool.

Two systems of grading wool have been in common use. These are the American (blood) system and the numerical count (USDA) system.

The American system was originally based on the percent of Merino breeding in the sheep. A 1/2-blood wool came from a sheep with 1/2 of the breeding from a fine wool breed and 1/2 from a coarse wool breed. Today, the system is used only to indicate the diameter of the wool fiber and does not indicate the amount of Merino breeding in the sheep. There are seven grades of wool in the American or blood system. These are: (1) Fine, (2) 1/2 blood, (3) 3/8 blood, (4) 1/4 blood, (5) Low 1/4 blood, (6) Common, and (7) Braid. Fine has the smallest diameter fibers and Braid has the largest.

The numerical count (USDA) system has 16 grades that originally reflected the spinning count. The *spinning count* is the number of hanks of yarn that can be spun from one pound of wool top (partially processed wool). A *hank* of yarn is 560 yards in length. The official USDA grades are: (1) Finer than 80s, (2) 80s, (3) 70s, (4) 64s, (5) 62s, (6) 60s, (7) 58s, (8) 56s, (9) 54s, (10) 50s, (11) 48s, (12) 46s, (13) 44s, (14) 40s, (15) 36s, and (16) Coarser than 36s. Finer than 80s has the smallest diameter fibers and Coarser than 36s has the largest diameter fibers.

A given breed of sheep tends to produce more of one grade of wool than of another. Table 30-1 shows the relationship of breed to the grades produced in the American and numerical count system.

Wool top is wool that has been partially processed.

In addition to grades, wool is also classed according to use. The two major classes of wool are apparel wool and carpet wool. *Apparel wool* is used in the making of clothing. *Carpet wool* is used for making carpets. Carpet wool is coarser than apparel wool. The length of the wool, as well as the fineness, influences the use for which it is suited. Finer grades can be shorter than coarser grades for the same use classification.

Apparel wool is further divided into *worsted* and *woolen* classes. Longer fiber wools are used in the worsted process. These make a relatively strong yarn. Shorter fibers are used in the woolen process, making a weaker yarn.

Wool Futures

A futures market exists for trading wool. Trading is done in grease wool contracts for 6,000 pounds (2,721.6 kg) of clean wool. A producer who

TABLE 30-1 Comparison of Wool Grading Systems and the Grades of Wool Produced by Different Breeds.

American Blood System	USDA Grades	Average fiber diameter (microns)*	Breeds of sheep typically producing grades of wool in range indicated	
Fine	Finer than 80s	17.69 or less	Delaine Merino—	80s or finer
	80s	17.70–19.14	Rambouillet—	64s–70s
	70s	19.15–20.59	Targhee—	62s
	64s	20.60–22.04	Romeldale—	58s–60s
1/2 Blood	62s	22.05–23.49	Corriedale—	60s
	60s	23.50–24.94	Southdown—	56s–60s
3/8 Blood	58s	24.95–26.39	Hampshire, Shropshire—	56s–60s
	56s	26.40–27.84	Suffolk—	54s–58s
1/4 Blood	54s	27.85–29.29	Columbia—	50s–58s
	50s	29.30–30.99	Dorset—	50s–56s
Low 1/4 Blood	48s	31.00–32.69	Cheviot—	48s–50s
	46s	32.70–34.39	Oxford—	46s–50s
Common	44s	34.40–36.19	Tunis, Romney—	44s–48s
Braid	40s	36.20–38.09	Leicester—	40s–46s
	36s	38.10–40.20	Lincoln, Cotswold—	36s–40s
	Coarser than 36s	40.21 or more	Highland—	36s or coarser

*A micron is 1/25,400 of an inch.

Source: USDA.

wishes to use the futures market to hedge should be shearing about 1,200 ewes each year. The futures market may be used to lock in a desirable wool price in anticipation of a drop in the price of wool at the time of shearing.

Trading in futures is done through brokers. There is a commission cost for each contract traded. Producers using the futures market must deposit a margin for each contract traded. A producer who desires to use the futures market should secure reliable advice and help from people familiar with commodities trading.

MARKETING MOHAIR

Over 95 percent of the mohair in the United States is produced in Texas. Most of the mohair is marketed through warehouses. The procedure is the same as that for marketing wool through warehouses. Some of the warehouses grade the mohair. High-quality mohair clips will bring better prices when they are sold on a graded basis. (A *clip* is the wool or mohair produced by a single shearing.) Information gained from graded mohair can be used to improve the breeding program. Animals that produce higher grades of mohair should be kept for breeding purposes. This kind of selection will eventually improve the grade of mohair produced.

Mohair is graded on the basis of the fineness of the fibers. *Kid hair* is finer than *adult hair* and is the most valuable portion of the clip. The coarsest hair comes from bucks and old wethers. It is a good practice for the producer to pack these grades separately at shearing time. Grease mohair grades range from Finer than 40s to Coarser than 18s. The numbers refer to the number of hanks that can be spun from one pound of clean mohair. Average diameters range from under 23.55 microns for the finest grade to over 43.54 microns for the coarsest grade. It is not always profitable to grade mohair clips. The clip must contain over 50 percent 26s to 28s and finer for it to be profitable to grade.

MARKETING GOAT MILK AND MILK PRODUCTS

Most goat milk is used on the farm where it is produced. In certain areas with high dairy goat populations, some milk is sold to consumers. The sale of goat milk must follow local regulations for the marketing of dairy products. Cleanliness and sanitation are essential for the production of a quality product. In some areas, there are evaporating and drying plants to handle goat milk. The milk is then marketed as an evaporated or dried product.

Goat cheese and goat butter are other products made from dairy goat milk. Several kinds of cheeses are made from goat milk. These products are marketed through retail outlets for cheese and butter.

MARKETING GOAT MEAT

Goat meat is called *chevon*. Young kids that weigh 30–40 pounds (13.6–18 kg) are marketed as cabrito. *Cabrito* is Spanish for "little goat." Cabrito is considered a delicacy and is especially good for barbecuing. In some areas, chevon is sold as lamb or mutton. Local regulations in some areas do not permit this marketing practice.

SUMMARY

Sheep are marketed through the same kinds of markets that are used for cattle and hogs. More sheep are sold by direct marketing methods than through terminal markets.

Sheep are classed according to age, use, sex, and grade. A specialty market exists for young lambs called hothouse lambs. The most desirable weight for slaughter lambs is 110 to 130 pounds (49.5–58.9 kg). Slaughter lambs account for the largest volume of sheep going to market.

Sheep are graded on the basis of quality and estimated yield. The body conformation and amount of finish determine the grade. Grades for slaughter lambs and yearlings are Prime, Choice, Good, and Utility. Yield grade 1 gives the highest yield of the major retail cuts. Yield grade 5 is the poorest.

Wool is marketed through local buyers, wool pools, cooperatives, warehouse operators, or direct to mills. The goal of wool pools and cooperatives is to obtain better prices for wool. Warehouses may buy the wool outright or may accept it on a consignment basis. Some warehouses offer grading services.

Wool prices vary with the season. Prices are best in May and lower in the fall and early winter. Wool is graded on the basis of the diameter of the fiber. The finest wool has the smallest diameter and is the most valuable.

Mohair is marketed in much the same manner as wool. Most of the mohair is marketed through warehouses. It is graded on the same basis as wool.

Other goat products include meat, milk, cheeses, and butter. Limited markets exist for these products.

Student Learning Activities

1. Prepare a bulletin board display of charts showing supply and demand cycles, price trends, and local price variations for sheep, goats, wool, and mohair.
2. Take a field trip to local markets for sheep, goats, wool, and mohair.
3. Give an oral report on marketing of sheep, goats, wool, or mohair.
4. Plan and practice good marketing procedures for student projects or for the home farm.
5. Ask a representative from a wool cooperative to speak to the class on wool marketing.
6. Survey local producers to determine local marketing practices with sheep, goats, wool, and mohair.

Review

1. What kinds of markets are used for marketing sheep?
2. Name and briefly describe the age classes of market sheep.
3. Name and briefly describe the use classes of market sheep.
4. Name and briefly describe the sex classes of market sheep.
5. Name and briefly describe the five quality grades of market sheep.
6. List the percents of boneless retail cuts found in each of the five yield grades.
7. What is the average amount of shrinkage when moving sheep to market?

8. Describe the seasonal price pattern for market sheep.
9. Briefly explain the marketing of purebred sheep.
10. Name the kinds of markets available for wool.
11. Describe the seasonal price variations for wool.
12. How is the value of the wool fleece determined?
13. Describe the American (blood) system of grading wool.
14. Describe the numerical count (USDA) system of grading wool.
15. List the typical grades of wool produced by the various breeds of sheep.
16. Name and briefly describe the two use classes of wool.
17. Describe how mohair is graded.
18. What products are marketed from dairy goats?
19. What is a popular use for cabrito?

Horses

Unit 31   *Selection of Horses*

Objectives

After studying this unit, the student should be able to

■ describe the characteristics of the horse industry.

■ describe the common breeds of horses.

■ describe the selection of a horse.

HORSES IN THE UNITED STATES

In the United States, horses are used primarily for recreational purposes. About 75 percent of all horses in the United States are owned for personal pleasure use. Ranching, racing, breeding, and commercial riding make up the other 25 percent.

The three main types of horse enterprises are breeding, training, and boarding stables. Horse breeding farms breed mares and sell the offspring. Some farms specialize in training horses for show or racing. Boarding stables feed and care for horses that are owned by people who do not have facilities for keeping horses. Good management skills, a high level of capital investment, and well-trained workers are required for success in horse enterprises. Breeding, feeding, caring for, and training horses all require a lot of labor.

In 1999, the USDA released its first inventory of equine in the United States. The inventory is based on the 1997 Census of Agriculture plus additional statistical sampling methods to pick up numbers from non-farm commercial operations. The inventory defines equine as including horses, ponies, mules, burros, and donkeys. The January 1, 1999 inventory showed a total of 5.32 million equine in the United States (Figure 31-1). This is a slight (1.3 percent) increase over the previous year. The leading states in horse numbers are Texas, California, and Tennessee. The non-farm population of equine accounted for approximately 39 percent of the total number.

The most popular breeds of horses in the United States, based on purebred registrations, are Quarter Horse, Thoroughbred, Standardbred, Appaloosa, and Arabian. These breeds account for more than 75

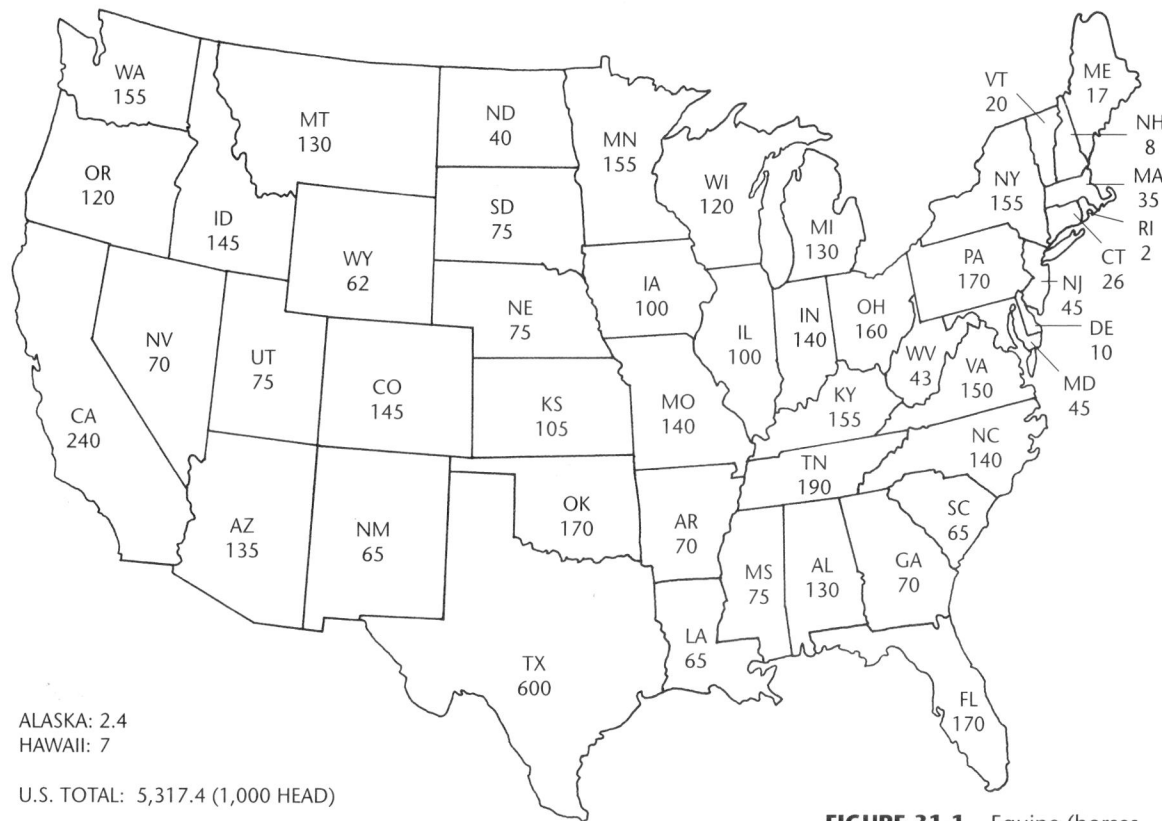

WA
155

MT
130

ND
40

MN
155

VT
20

ME
17

NH
8

OR
120

ID
145

SD
75

WI
120

NY
155

MA
35

RI
2

WY
62

IA
100

MI
130

PA
170

CT
2

NV
70

UT
75

NE
75

IL
100

IN
140

OH
160

NJ
45

DE
10

CA
240

CO
145

KS
105

MO
140

WV
43

KY
155

VA
150

MD
45

NC
140

AZ
135

NM
65

OK
170

AR
70

TN
190

SC
65

MS
75

AL
130

GA
70

TX
600

LA
65

FL
170

ALASKA: 2.4
HAWAII: 7

U.S. TOTAL: 5,317.4 (1,000 HEAD)

FIGURE 31-1 Equine (horses, ponies, mules, burros, donkeys) inventory by states—January 1, 1999 (1,000 head). *Source: Equine, USDA, March 1999.*

percent of all horses registered with breed associations in the United States. The most popular breeds of horses in the world, based on estimates of world numbers, are Thoroughbred, Quarter Horse, Arabian, and Appaloosa. These breeds account for more than 75 percent of all horse numbers worldwide.

The horse industry is big business in the United States, accounting for more than 16 billion dollars of economic activity annually. Horse racing is a major spectator sport and horse shows (Figure 31-2) and rodeos draw large numbers of participants and spectators each year. The USDA equine inventory reported the value of sales of animals for 1998 at more than 1.7 billion dollars.

Horses are beneficial in several ways. They contribute to the economic growth of the nation. They provide people with an opportunity for physical exercise. Working with a horse may provide a release from the tensions of modern society. Owning and caring for a horse contributes to an individual's sense of responsibility. Horse-related events provide an opportunity for members of a family to participate in activities together.

FIGURE 31-2 Judging horses at halter at a horse show. *Photo by Michael Dzaman.*

BREEDS OF LIGHT HORSES AND PONIES

There are many breeds of light horses and ponies. More detailed information about these breeds may be secured by writing to the breed association. Breed associations register purebred horses of their respective breeds.

Light horses are used mainly for riding, driving, and racing. They measure 14-2 to 17 hands at the withers. A *hand* is four inches (10.2 cm). A measure of 14-2 hands is 58 inches (147.3 cm). In the measurement 14-2, the 14 represents the number of hands; the 2 indicates the number of inches. Light horses weigh from 900 to 1,400 pounds (408–635 kg).

Ponies measure under 14-2 hands and weigh from 300 to 900 pounds (136–408 kg). They are used mainly for riding and driving.

American Bashkir Curly

FIGURE 31-3 American Bashkir Curly 10-year-old bay stallion owned by Sunny Martin, Ely, Nevada. First permanent stallion in ABC Registry with five registered Curly get. *Courtesy of American Bashkir Curly Registry, NV.*

The Bashkir Curly (Figure 31-3) originated in Russia. Horses with curly coats have been raised for centuries on the southern slopes of the Ural mountains. This region is known as Bashkiria, giving the breed its name. The hair coat of the Bashkir Curly is long, silky, and curly. The horse sheds the mane hair (and sometimes the tail hair) in the summer. The coat grows back in the winter. This breed can withstand cold climates. Characteristics of the Bashkir Curly include small nostrils and a gentle disposition. All colors are accepted for registration. The breed is used mainly for pleasure riding.

It is not know how the breed came to the United States. Most present-day Bashkir Curlys originated from three horses that were found in the Peter Hanson mountain range in Nevada by Peter Damele in 1898. The American Bashkir Curly Registry was formed in 1971.

American Creme

The American Creme originated in the United States. It is a color breed that may have the type of any of the breeds. A *color breed* is one that is registered on the basis of color. The American Creme is used for pleasure riding, exhibition, and stock horses.

Registration is in the International American Albino Association, Incorporated. They are registered in one of four sections according to the following classifications:

A–Body ivory white, mane white (lighter than body), eyes blue, skin pink.

B–Body cream, mane darker than body, cinnamon buff to ridgeway, eyes dark.

C–Body and mane of the same color, pale cream, eyes blue, skin pink.

D–Body and mane of same color, sooty cream, eyes blue, skin pink.

Combinations of these classifications are also acceptable. The American Creme was recognized as a separate breed for registration in 1970. From 1963 to 1970, cream horses were registered in the National Recording Club. Creme horses so registered can now be transferred to the American Creme Section of the International American Albino Association.

American Gotland

The American Gotland originated in Sweden. The breed is small, ranging from 11 to 13 hands. Colors are bay, black, brown, chestnut, dun, palomino, or roan. Some have leopard or blanket markings. Gotlands were first imported into the United States in 1957. They are used for harness racing and as pleasure horses.

American Mustang

The American Mustang is a descendant of horses brought to the Americas by the Spanish. The breed originated in North Africa. Mustangs may be any color. To be eligible for registry, they must be between 13-2 and 15 hands high. They are used for show and pleasure riding, jumping, endurance trail riding, and as stock horses. The breed association is the American Mustang Association, Incorporated.

American Paint

The American Paint (Figure 31-4) originated in the United States from the descendants of horses brought to the Americas by the Spanish. The Paint horse was popular with the Plains Indians and the early American cowboys. Although the popularity of the breed faded for a time, there has been a renewed interest in the breed in recent years.

FIGURE 31-4   American Paint stallion. *Courtesy of American Paint Horse Association, TX.*

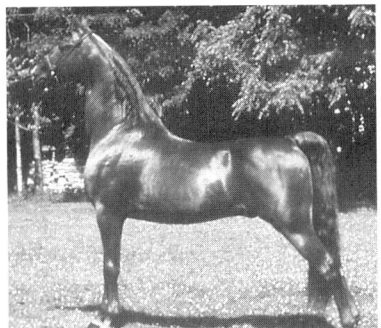

FIGURE 31-5   American Saddle-bred Horse. *Courtesy of American Saddlebred Horse Association, KY.*

FIGURE 31-6   American White. *Courtesy of American Albino Association, Inc., OR.*

The American Paint Stock Horse Association was formed in 1962. This later merged with the American Paint Quarter Horse Association to form the American Paint Horse Association.

The words *paint* and *pinto* both refer to spotted or two-tone horses with white and another color for body markings. The Pinto Horse Association of America, Incorporated, registers horses of all breeds and types. The American Paint Horse Association registers on the basis of registered Paints, Quarter Horses, and Thoroughbreds. The Paint is used for pleasure, show, racing, and stock purposes.

Paint horses have two different color patterns: tobiano and overo. The *tobiano* head is marked in the same way as that of a solid colored horse. The legs are white, at least below the knees and hocks, and there are regular spots on the body. The *overo* has variable color head markings. The white usually does not cross the back between the withers and the tail. One or more legs are dark colored. The body markings are irregular and scattered.

American Saddlebred Horse

The American Saddlebred Horse (Figure 31-5) originated in Kentucky. The average height of the Saddlebred Horse is 15 to 16 hands, and it ranges in weight from 1,000 to 1,200 pounds (453.6–544.3 kg). American Saddlebred Horses may be bay, chestnut, black, or gray. Roan, golden, or palomino animals appear occasionally. The horse may be three- or five-gaited. Horse gaits are described later in this unit. Three-gaited horses perform the walk, trot, and canter; five-gaited horses also have a slow gait (stepping pace, fox trot, or running walk) plus the rack. They are used as pleasure, stock, and harness horses.

Saddlebred horses belonged to the former American Saddle Horse Breeders' Association, formed in Louisville, Kentucky in 1891. They now belong to and are registered in the American Saddlebred Horse Association.

American White

The American White Horse (Figure 31-6) originated in the United States on the White Horse Ranch in Napier, Nebraska. It is a color breed, with many conformation types being represented. The color is snow white with pink skin, and dark eyes. American whites are used as exhibition, pleasure, racing, and light draft horses.

American Whites are registered in the American Albino Association, which was established in 1937. Only Dominant White horses were registered until 1949. Classifications A, B, C, and D were added at that time to register Creme horses. The association now registers American White horses in one section and American Creme horses in another.

Andalusian

The Andalusian originated in Spain. There are few Andalusians in the United States because the Spanish did not permit export of the breed for many years. The colors are bay, white, gray, and occasionally black, roan, or chestnut. They are used for pleasure riding, jumping, exhibition, and bullfighting. The breed association is the American Andalusian Horse Association.

Appaloosa

The Appaloosa (Figure 31-7) originated in the United States from the descendants of horses brought to the Americas by the Spanish. The color patterns of the Appaloosa are variable. Most Appaloosas are white over the hips and loins with dark, round or egg-shaped spots. Occasionally the entire body is mottled with spots. The eye is encircled with white. The hoofs are vertically striped with black and white.

The Nez Perce Indian tribe developed a selective breeding program using these spotted horses. After the Nez Perce Indians lost their war with the United States in 1877, the Appaloosa horses that belonged to them were scattered throughout the west. Interest in the Appaloosa horse increased in the late 1930s. The Appaloosa Horse Club, Incorporated was formed in 1938. Descendants of the Nez Perce horses were the foundation stock for the present breed. The Appaloosa is used for pleasure riding, showing, racing, parades, and as a stock horse.

Appaloosa Pony

The Appaloosa Pony (Figure 31-8) has the same color traits as the Appaloosa Horse. The Pony of the Americas Club is the breed association that registers the pony. The ponies registered show the traits of the Appaloosa, but do not meet the height requirements for horses.

Arabian

The Arabian horse originated in Arabia. It is small to medium in size, ranging from 850 to 1,100 pounds (385.6–499 kg). The colors are mainly bay, gray, or chestnut, with a few being white or black. The skin is always dark and the legs and head often have white markings. The Arabian is used for pleasure riding, racing, and showing, and as a stock horse. The breed associations are the Arabian Horse Registry of America, Incorporated, and the International Arabian Horse Association.

Buckskin

The Buckskin originated in the United States from horses of Spanish descent. The buckskin is a color breed with many different types being

FIGURE 31-7   Appaloosa. *Courtesy of Appaloosa Horse Club, Inc., ID.*

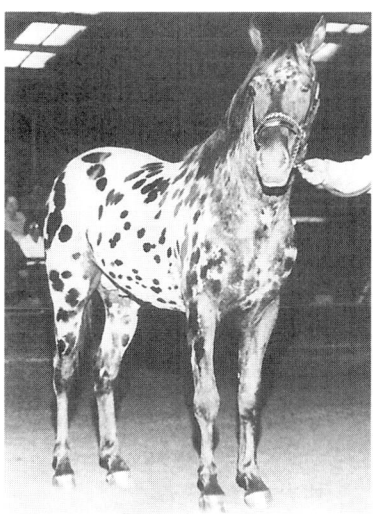

FIGURE 31-8   Appaloosa Pony "Caprice" owned by Bradley and Donna Amick, Alexandria, Indiana. National Grand Champion Stallion at the 1977 Nationals, Lincoln, Illinois, 1977; Indiana Grand Champion Stallion, 1977 Indiana State Show. *Courtesy of National Appaloosa Pony, Inc., IN.*

FIGURE 31-9 Galiceno mare. *Courtesy of Galiceno Horse Breeders Association, TX.*

FIGURE 31-10 Hackney Horse "Peel's Haven" has been grand champion on line many years at the Canadian Exhibition Toronto, Ontario, Canada and at the Royal Winter Fair also held at Toronto. *Courtesy of Art and Brenda Alderman.*

registered. The colors of the Buckskin are buckskin, dun, red dun, or grulla. The American Buckskin is registered in the American Buckskin Registry Association, Incorporated. The International Buckskin is registered in the International Buckskin Horse Association, Incorporated. There are slight differences in the rules for registry in the two associations. The Buckskin is used for pleasure riding, as a stock horse, and for show.

Cleveland Bay

The Cleveland Bay originated in England. The breed is solid bay in color with black legs. It is larger than many of the other light horse breeds, weighing from 1,150 to 1,400 pounds (521.6–635 kg). The Cleveland Bay is used mainly as a general utility horse, for pleasure riding, and driving, and for crossbreeding programs. The breed association is the Cleveland Bay Horse Society of North America.

Connemara Pony

The Connemara Pony originated in Ireland. It was first imported into the United States in 1951. The animals may be bay, black, brown, cream, dun, or gray. Roan or chestnut appear occasionally. The Connemara is used for jumping and for pleasure riding. The breed association is the American Connemara Pony Society.

Galiceno

The Galiceno (Figure 31-9) originated in Spain. It is a small breed, ranging from 12 to 13-2 hands high. The colors are bay, black, chestnut, dun, gray, brown, and palomino. The Galiceno is used mainly for pleasure riding. The breed association is the Galiceno Horse Breeders Association, Incorporated.

Hackney

The Hackney (Figure 31-10) originated in England. It varies in size from 12 to 16 hands and 800 to 1,200 pounds (362.9–544.3 kg). The colors are bay, brown, or chestnut. White markings are common. Roan and black appear occasionally. The Hackney is used mainly as a harness or carriage horse. The breed association is the American Hackney Horse Society.

Missouri Fox Trotting Horse

The Missouri Fox Trotting Horse originated in the Ozark Hills region of Missouri. Its color is sorrel, commonly with white markings. The *fox trot gait* (described later in this unit) is a distinguishing trait. Principal uses of the breed are for pleasure riding, trail riding, and as stock horses. The breed association is the Missouri Fox Trotting Horse Breed Association.

Morab

The Morab horse (Figure 31-11) is a cross between the Morgan and the Arabian. The breed originated in California from foundation stock bred by Martha Doyle Fuller on her ranch at Clovis, California. This foundation stock had been developed since 1955. The Morab may be any color, but without spots on the body. The eyes and skin are black. The main uses of the breed are pleasure riding, showing, endurance riding, and as stock horses.

Morgan

The Morgan originated in the New England states. The breed is descended from a stallion named Justin Morgan that lived in the late 1700s. Bloodlines of the Morgan horse are found in the foundation stock of many of the light horse breeds in the United States.

The color of the Morgan is bay, black, brown, or chestnut. It is used for pleasure riding and as a stock horse. The breed association is the American Morgan Horse Association, Incorporated.

Morocco Spotted Horse

The Morocco Spotted Horse originated in the United States. The breed carries bloodlines of Morocco Barb, Hackney, French Coach Horse, and Saddle Horse breeding. The Morocco Spotted Horse is a color breed, with several breed types of light horses being recognized as eligible for registry. The main colors are black and white or bay and white. Other color combinations are chestnut and white, sorrel and white, blue and white, and palomino and white. Color patterns are either tobiano or overo, or a combination of the two. The tobiano pattern is more common. Stallions must have not less than 10 percent of the secondary color, not counting the face and legs. Mares and geldings must have not less than 3 percent of the secondary color, not counting face and legs. The Morocco Spotted Horse is used for pleasure riding, as a saddle horse, harness horse, and stock horse.

Palomino

The Palomino (Figure 31-12) originated in the United States. Horses of palomino color were found among the descendants of horses brought to the Americas by the Spanish. The color of the breed is golden. It may be three shades lighter or darker than the color of a newly minted gold coin. The palomino color does not breed true in crosses. Various ratios of palomino to other colors result, depending on the cross used. The mane and tail are light colored. The Palomino is used for pleasure riding, and as a stock, harness, and parade horse. There are two breed associations that register Palominos, the Palomino Horse Association, Incorporated, and the Palomino Horse Breeders of America.

FIGURE 31-11   Morab horse. *Courtesy of Morab Horse Registry of America, CA.*

FIGURE 31-12   Palomino Quarter Horse. *Courtesy of Palomino Horse Breeders of America, OK.*

FIGURE 31-13   Pinto. *Courtesy of Pinto Horse Association of America, Inc., TX.*

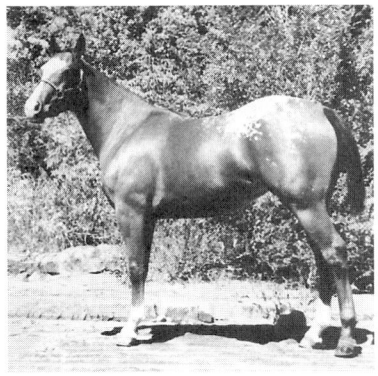

FIGURE 31-14   Pony of the Americas mare. *Courtesy of Pony of the Americas Club, Inc., IN.*

Paso Fino

The Paso Fino originated in the Caribbean area. The colors of this breed are mainly solid bay, chestnut, or black with white markings. Palominos and pintos appear occasionally. The Paso Fino has a paso gait and is used for pleasure riding, showing, parades, and endurance riding. The *paso gait* is four-beat, with the legs on the same side moving together, the hind foot striking the ground an instant before the front foot. There are two breed associations, the American Paso Fino Horse Association and the Paso Fino Horse Association, Incorporated.

Peruvian Paso

The Peruvian Paso originated in Peru from horses that were brought to the Americas by the Spanish. The colors of the breed are bay, black, brown, sorrel, chestnut, dun, buckskin, red dun, grulla, palomino, gray, red roan, and blue roan. Solid colors are preferred, and the skin is dark. The horse is noted for the paso llano and sobreandando gaits. The *paso llano* is an equally spaced, four-beat gait. The feet strike the ground in the following order: (1) left hind foot, (2) left front foot, (3) right hind foot, (4) right front foot. The *sobreandando gait* is usually faster and slightly more lateral than paso llano. The feet strike the ground in the same order as paso llano, but the left hind-left front strike closer together and the right hind-right front strike closer together. The Peruvian Paso is used mainly for pleasure, parade, and endurance riding. There are two breed associations, the American Association of Owners and Breeders of Peruvian Paso Horses and the Peruvian Paso Horse Registry of North America.

Pinto

The Pinto (Figure 31-13) originated in the United States. The color patterns of the Pinto are tobiano and overo. The preferred color distribution is one-half colored and one-half white. Tobiano patterns usually have color on the head, chest, flanks, and some in the tail. The legs are usually white. The overo pattern usually has jagged-edged, white markings on the midsection of the body and neck area. The legs are usually colored rather than white.

There are four types of Pinto horses: stock, pleasure, hunter, and saddle. The Pinto is used as a pleasure, show, stock, and parade horse. The breed association is the Pinto Horse Association of America, Incorporated. The association registers Pinto ponies in a separate registry.

Pony of the Americas

The Pony of the Americas (Figure 31-14) originated in the United States and has appaloosa coloring. Six color patterns are recognized: Snowflake, Frost, Blanket, Leopard, White with black spots on hindquarters, and Marbleized Roan. The breed ranges in size from 11-2 to 13 hands. It is used for pleasure riding and showing by young people. The breed association is the Pony of the Americas Club, Incorporated.

Quarter Horse

The Quarter Horse (Figure 31-15) originated in the United States. During the colonial era, horse racing was a common sport. Since the races seldom were longer than a quarter of a mile, the term *quarter miler* was used to describe these race horses. Horses were bred that could run short distances faster than other breeds. The Quarter Horse was widely used during the westward expansion of the pioneers and on the western ranches.

FIGURE 31-15   American Quarter Horse. *Courtesy of American Quarter Horse Association, TX.*

The colors of the Quarter Horse are bay, black, brown, sorrel, chestnut, dun, buckskin, red dun, grullo, palomino, gray, red roan, and blue roan. Quarter Horses are used as pleasure riding, showing, racing, and stock horses. The breed associations are the American Quarter Horse Association and the National Quarter Horse Registry, Inc.

Racking Horse

The Racking Horse originated in the United States. It is small to medium in size. Average height is 15 hands and average weight is 1,000 pounds (453.6 kg). The colors of the breed include black, white, gray, chestnut, bay, brown, sorrel, roan, and yellow. White markings sometimes appear on the body. The Racking Horse is used for pleasure and show.

FIGURE 31-16   Rangerbred. *Courtesy of Colorado Ranger Horse Association, PA.*

Rangerbred

The Rangerbred (Figure 31-16) originated in the United States from foundation stallions imported from Turkey. These stallions were crossed on Mustang mares. Development of the breed took place in Colorado and, thus, the breed is also known as the Colorado Rangers. Rangerbreds can be any color. Spotted horses are common. The type is similar to the Appaloosa. They are used mainly as stock horses, although they are also shown in the ring. The breed association is the Colorado Ranger Horse Association, Incorporated.

Shetland Pony

The Shetland Pony originated in the Shetland Islands. Shetland Ponies may be any horse color. Both broken and solid color patterns exist. The Shetland Pony is registered in two size classifications: (1) less than 10-3 hands and (2) 10-3 to 11-1 hands. The Shetland Pony is used for pleasure riding by children, and for showing and racing. The breed association is the American Shetland Pony Club.

FIGURE 31-17   "El Mano de Oro," a buckskin Spanish Barb stallion bred by Ilo Belsky, of Eli, Nebraska, and owned by Sally Crouse of Banning, California. *Courtesy of Spanish-Barb Breeders Association, MS.*

FIGURE 31-18   Spanish Mustang. *Courtesy of Southwest Spanish Mustang Association, Inc., OK.*

FIGURE 31-19   Standardbred pacer "Besta Bret." *Courtesy of United States Trotting Association, OH.*

Spanish Barb

The Spanish Barb (Figure 31-17) is descended from horses that were brought to the Americas by the Spaniards. The Barb horse originally came from Africa to Spain. Many of the American light horse breeds carry Barb bloodlines. The most common colors are dun, grullo, sorrel, and roan. It is a small horse, standing 13-3 to 14-1 hands high, and weighing between 800 and 975 pounds (362.9–442.2 kg). Horses currently being registered with the breed association are mainly from three strains: the Romero-McKinley, Belsky, and Buckshot. The breed association is the Spanish-Barb Breeders Association.

Spanish Mustang

The Spanish Mustang (Figure 31-18) originated in the United States from the descendants of horses brought to the Americas by the Spaniards. The Spanish Mustang may have any of the horse colors, with either solid or broken color patterns. The horse is small, standing 13 to 14-2 hands. The Spanish Mustang is used for pleasure riding, and as a stock and pack horse. The breed association is the Southwest Spanish Mustang Association, Incorporated.

Standardbred

The Standardbred originated in the United States. The colors of the breed are mainly bay, black, brown, and chestnut. Other colors that may occur are gray, roan, and dun. The Standardbred was developed as a harness racing horse. Both trotters and pacers have been developed (Figures 31-19 and 31-20). The breed registry is the United States Trotting Association. Standardbreds are also registered in the International Trotting and Pacing Association, Incorporated.

Tennessee Walking Horse

The Tennessee Walking Horse (Figure 31-21) originated in the United States. Common colors include sorrel, chestnut, black, bay, roan, brown, white, gray, and golden. The feet and legs often have white markings. The breed is noted for its running walk gait. The horse is used mainly for pleasure riding and showing. The breed association is the Tennessee Walking Horse Breeders' and Exhibitors' Association.

Thoroughbred

The Thoroughbred originated in England. Development of the breed as a race horse began in the 17th century. Common colors are bay, brown, black, and chestnut. Roan and gray occur occasionally. The face and legs often have white markings. The main use of the Thoroughbred is for racing. However, it is also used in crossbreeding programs. Thoroughbreds are registered with The Jockey Club. Horses with part Thoroughbred breeding may be registered with the American Remount Association, Incorporated, Thoroughbred Half-Bred Registry.

Welsh Pony

The Welsh Pony (Figure 31-22) originated in Wales. The Welsh Pony is a little larger than the Shetland Pony, but smaller than the light horse breeds. Two size classifications are registered: (1) ponies less than 12-2 hands and (2) ponies between 12-2 and 14 hands. The special use of the Welsh Pony is as a riding horse for children. Other uses include pleasure riding, trail riding, harness shows, hunting, and racing. Any of the horse colors, except piebald and skewbald, are accepted. The breed association is the Welsh Pony and Cob Society of America, Incorporated.

Walking Pony

The American Walking Pony (Figure 31-23) originated in the United States. It is a cross between the Welsh Pony and the Tennessee Walking Horse. The colors of either breed are accepted. The pony must be able to do the running walk gait. Its main use is for pleasure riding and showing. The breed association is the American Walking Pony Association.

The American Part-Blooded Horse Registry

Eligible horses that are not purebred may be registered in the American Part-Blooded Horse Registry. This registry was formed in 1939 in response to a demand for registering horses that were not purebred. Half-Bloods, grades, and crosses involving most of the light horse breeds are accepted for registry. Each group has its own register book. The registry certificate authenticates age, breeding, color, and markings. Information and rules for registry may be obtained from the registrar of the association.

BREEDS OF DRAFT HORSES

At one time, the main source of power on the farm was the draft horse. With the mechanization of agriculture, the use and number of draft horses in the United States has decreased. Draft horses are larger, heavier, and more muscular than the light horse breeds. They were selected and bred for the ability to pull heavy loads. The draft horse breeds originated in Europe. There are purebred associations for each of the draft horse breeds in the United States. Few draft horses are registered each year. More information on each breed of draft horse may be secured by contacting the breed association.

FIGURE 31-20   Standardbred trotter "Colonial Charm." *Courtesy of United States Trotting Association, OH.*

FIGURE 31-21   Tennessee Walking Horse. *Courtesy of Tennessee Walking Horse Breeders and Exhibitors Association, TN.*

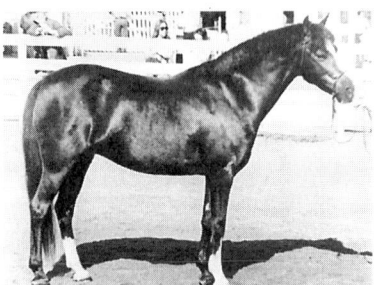

FIGURE 31-22   Welsh Pony. *Courtesy of Welsh Pony & Cob Society of America, Inc., VA.*

FIGURE 31-23 Walking Pony "BT Golden Splendor." *Courtesy of American Walking Pony Association, GA.*

Belgian

The Belgian draft horse originated in Belgium. Common colors of the Belgian are bay, chestnut, and roan, and its average size is 15-2 to 17 hands high. Belgians range in weight from 1,900 to 2,200 pounds (861.8–997.9 kg). The breed association is the Belgian Draft Horse Corporation of America.

Clydesdale

The Clydesdale originated in Scotland. Common colors of the breed are bay and brown, both with white markings. Other colors such as black, chestnut, gray, and roan are sometimes seen. The Clydesdale is somewhat smaller than the Belgian, Percheron, and Shire, standing 16 to 17 hands high. Weights range from 1,700 to 1,900 pounds (771–861.8 kg). The breed association is the Clydesdale Breeders of the United States.

Percheron

The Percheron originated in France. The common colors of the breed are black and gray. Other colors sometimes seen are bay, brown, chestnut, and roan. The Percheron is 16-1 to 16-3 hands high and may weigh from 1,900 to 2,100 pounds (861.8–952.5 kg). The breed association is the Percheron Horse Association of America.

Shire

The Shire originated in England. Common colors are black, brown, bay, gray, and chestnut, frequently with white markings on the face and legs. The Shire is 16 to 17-2 hands high, weighing 1,900 to 2,000 pounds (861.8–907.2 kg). The breed association is The American Shire Horse Association.

Suffolk

The Suffolk originated in England. The only color of the Suffolk is chestnut. Seven shades occur, ranging from dark liver to light golden sorrel. Some white markings are found on the head and legs. Suffolks are the smallest of the draft horse breeds. They range in height from 15-2 to 16-2 hands. The weight range is 1,600 to 1,900 pounds (725.7–861.8 kg). The breed association is The American Suffolk Horse Association.

Donkeys and Mules

Donkey is the common name for the ass. The ass is smaller than the horse, has longer ears, and a short, erect mane. The gestation period is one month longer than that of the horse. The male ass is called a *jack,* and the female ass is called a *jennet.* When a jack is crossed on a mare (female horse) the resulting offspring is called a *mule*. When a stallion (male horse) is crossed on a jennet the resulting offspring is called a *hinny*. The hinny is smaller in size than the mule. Mules are usually sterile; that is, they will not reproduce. The ass and the mule are used mainly as work animals. Miniature donkeys are used as children's pets.

More information about donkeys and mules is available from breed associations: the American Donkey and Mule Society, Incorporated; the Miniature Donkey Registry of the United States; the American Council of Spotted Asses; and the National Miniature Donkey Association.

SELECTION OF HORSES

Definition of Terms

A *foal* is a young horse of either sex up to one year of age. A *filly* is a female less than three years of age. (For Thoroughbreds, fillies include four year olds.) A *colt* is a male less than three years of age. (For Thoroughbreds, colts include four year olds.) A *mare* is a mature female, four years of age and older. (Thoroughbred mares are five years of age or older.) A *stallion* is a mature male four years of age or older. (Thoroughbred stallions are five years of age or older.) A *gelding* is a male that has been castrated.

Use of the Horse

The five general uses of horses are: (1) pleasure, (2) breeding, (3) working stock, (4) show, and (5) sport. Generally, one horse cannot be used in all of these ways. The horse should be selected for its major intended use.

Sources of Horses

Horses can be purchased from breeders, private owners, auctions, and dealers. The most reliable source for obtaining a horse is from a breeder. Horses from this source may be slightly more expensive, but the quality of the animal can be certified. Private owners may be a good source of horses, depending on the reasons they have for selling. Auctions and dealers are less reliable sources since it may be difficult to determine the soundness of the horses offered. The inexperienced person who desires to buy a horse should always obtain the aid of an experienced horseperson when making the purchase.

Age of Horse to Buy

Horses from five to 12 years of age are in the prime of life. If the horse is for a young, inexperienced rider, a horse that is in this age range may be a wise choice. Older horses that are sound are also satisfactory for young riders. Young horses require additional training. The inexperienced rider may be unable to handle a young horse safely.

Sex of Horse to Buy

Stallions are often hard to manage and control. They are not suitable horses for the inexperienced rider. For pleasure riding, a gelding or mare is usually a better choice. A gelding is often very steady and dependable, while mares tend to be more excitable. Of course, a person who desires to raise a foal must purchase a young mare.

Breed Selection

Certain breeds are better adapted to a particular use than others. Thus, the intended use of the horse may help to narrow the choice of breed. Personal preference is also a factor to consider when selecting a breed. A person who is not interested in breeding or extensive showing of horses may prefer to select a good grade or unregistered horse. These horses often make excellent mounts for trail and other pleasure riding. Purebred horses have a greater resale value and can be entered in more horse shows. The demand for a given breed should be considered if the owner's interest is in breeding and raising horses for resale.

Conformation

A person who is evaluating or judging a horse must know the parts of the horse. The parts of the horse are shown in Figure 31-24. The major points to consider when evaluating a horse are shown on a scorecard for judging horses. An example of a scorecard for judging saddle horses is shown in Figure 31-25. The scorecard describes the physical traits that are desirable in a light horse.

Feet and legs. Two of the most important parts of the horse are the feet and legs. The conformation of the legs influences the way the horse moves. Figures 31-26 and 31-27 show the correct and incorrect positions of the front legs and how these influence the way the horse moves. Figures 31-28 and 31-29 show the correct and incorrect positions of the rear legs of the horse.

The feet are just as important as the legs of the horse. A detailed diagram of the hoof is shown in Figure 31-30. The weight of the horse is carried on the wall, bars, and frog. The sole normally does not touch the ground. The hoof must be properly trimmed to keep the horse standing squarely and moving straight. The hoof grows at the rate of about 3/8 to 1/2 inch (0.95–1.27 cm) per month. The hoof should be trimmed every month or six weeks. The hoof must be kept moist. A dry hoof will crack. If possible, it is helpful to allow the horse to stand occasionally in an area where the ground is moist. This helps to keep the

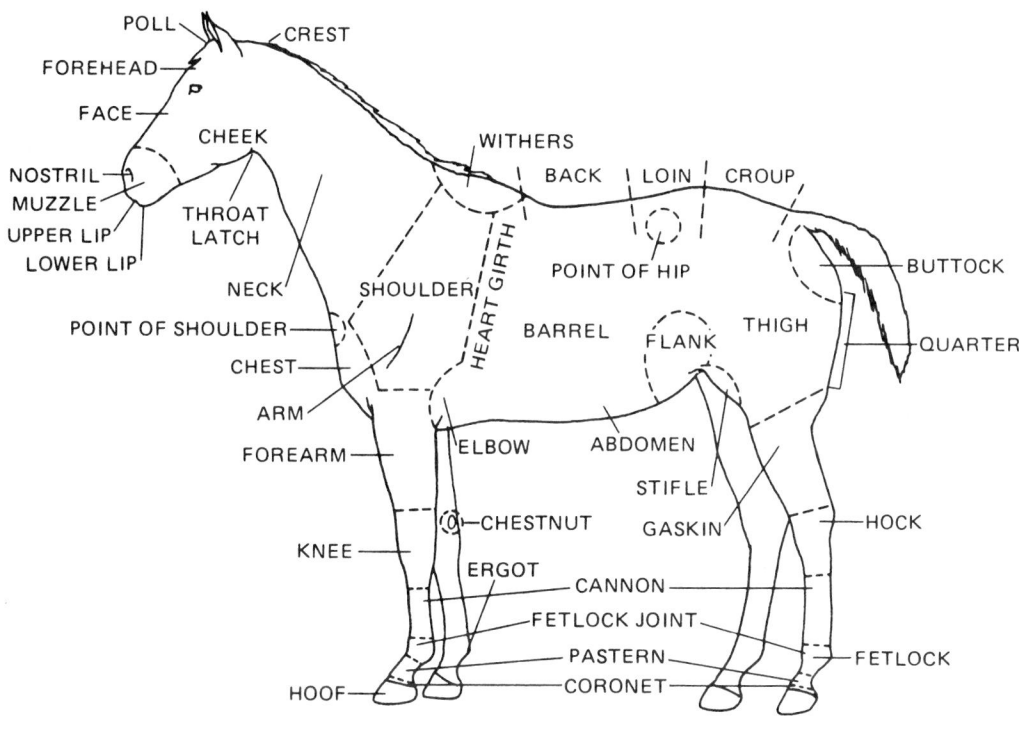

FIGURE 31-24   Parts of a horse.
*Courtesy of Appaloosa Horse Club,
Inc., ID.*

hoof from drying out. A hoof dressing may also be applied to prevent
drying.

The pastern and hoof should form a 45-degree angle with the
ground. Figure 31-31 shows the correct and incorrect angles of the
pastern and hoof. The effect of the angle on the arc of the hoof is also
shown.

Body Colors

The five basic colors of horses are bay, black, brown, chestnut, and
white. *Bay* varies from yellowish tan to bright mahogany. A dark bay
is almost brown. A *black* horse has fine, black hair on the muzzle. A
brown horse is very dark but will have tan or brown hair on the muzzle
or flanks. The *chestnut* color is red with variations from a light yellow
to a dark liver color. Brilliant red-gold and copper shades are also considered chestnut. The *white* horse is pure white and remains the same
color all of its life.

The five major *variations* in horse colors are dun, gray, palomino,
pinto, and roan. *Dun* is a yellow color. The variations are from a
mouse color (grullo) to golden. Dun horses always have a stripe down
their backs. *Gray* is a mixture of black and white hair. The *palomino*
color is golden with a light mane and tail. The color varies from light

SCORE CARD FOR THE LIGHT HORSE

Scale of Points	*Standard or Perfect Score*

General Appearance—12 percent

1. Height ....
2. Weight ....
3. Form (Close but not full made, deep but not broad, symmetrical) 4
4. Quality (Bone clean, dense, fine, yet indicating substance. Tendons and joints sharply defined, hide and hair fine, general refinement, finish) 4
5. Temperament (Active, disposition good, intelligent) 4

Head and Neck—8 percent

6. Head (Size and dimensions in proportion, clear-cut features, straight face line, wide angle in lower jaw) 1
7. Muzzle (Fine, nostrils large, lips thin, trim, even) 1
8. Eyes (Prominent orbit; large, full, bright, clear, lid thin, even curvature) 1
9. Forehead (Broad, full) 1
10. Ears (Medium-size, pointed, set close, carried alert) 1
11. Neck (Long, supple, well crested, not carried too high, throttle well cut out, head well set on) 3

Forehead—22 percent

12. Shoulders (Very long, sloping, yet muscular) 3
13. Arms (Short, muscular, carried well forward) 1
14. Forearm (Long, broad, muscular) 1
15. Knees (Straight, wide, deep, strongly supported) 2
16. Cannons (Short, broad, flat, tendons sharply defined, set well back) 2
17. Fetlocks (Wide, tendons well back, straight, well supported) 2
18. Pasterns (Long, oblique—45 degrees—smooth, strong) 2
19. Feet (Large, round, uniform, straight, slope of wall parallel to slope of pastern, sole concave, bars strong, frog large, elastic, heels wide, full, one-third height of toe, horn dense, smooth, dark color) 5
20. Legs (Direction viewed from in front, a perpendicular line dropped from the point of the shoulder should divide the leg and foot into two lateral halves; viewed from the side, a perpendicular line dropped from the tuberosity of the scapula should pass through the center of the elbow-joint and meet the ground at the center of the foot) 4

Body—12 percent

21. Withers (High, muscular, well finished at top, extending well into back) 3
22. Chest (Medium-wide, deep) 2
23. Ribs (Well sprung, long, close) 2
24. Back (Short, straight, strong, broad) 2
25. Loin (Short, broad, muscular, strongly coupled) 2
26. Flank (Deep, full, long, low underline) 1

Hindquarters—31 percent

27. Hips (Broad, round, smooth) 2
28. Croup (Long, level, round, smooth) 2
29. Tail (Set high, well carried) 2
30. Thighs (Full, muscular) 2
31. Stifles (Broad, full, muscular) 2
32. Gaskins (Broad, muscular) 2
33. Hocks (Straight, wide, point prominent, deep, clean cut, smooth, well supported) 5
34. Cannons (Short, broad, flat, tendons sharply defined, set well back) 2
35. Fetlocks (Wide, tendons well back, straight, well supported) 2
36. Pasterns (Long, oblique—50 degrees—smooth, strong) 2
37. Feet (Large, round—slightly less than in front—uniform, straight, slope of wall parallel to slope of pastern, sole concave, bars strong, frog large and elastic, heels wide, full, one-third height of toe, horn dense, smooth, dark color) 4
38. Legs (Direction viewed from the rear, a perpendicular line dropped from the point of the buttock should divide the leg and foot into lateral halves; viewed from the side, the same line should touch the point of the hock and meet the ground some little distance back of the heel; a perpendicular line dropped from the hipjoint should meet the ground near the center of the foot) 4

Way of Going—15 percent

39. Walk (Rapid, flat footed, in line) 5
40. Trot (Free, straight, smooth, springy, going well off hocks, not extreme knee fold) 5
41. Canter (Slow, collected, either lead, no cross canter) 5

     Total 100

FIGURE 31-25   Score card for the light horse. *Courtesy of Appaloosa Horse Club, Inc., ID.*

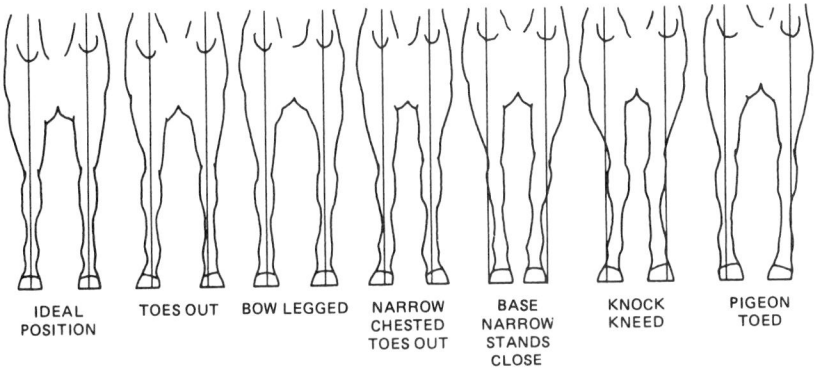

IDEAL POSITION TOES OUT BOW LEGGED NARROW CHESTED TOES OUT BASE NARROW STANDS CLOSE KNOCK KNEED PIGEON TOED

VERTICAL LINE FROM POINT OF SHOULDER SHOULD FALL
IN CENTER OF KNEE, CANNON, PASTERN, AND FOOT

PATH OF THE FEET AS SEEN FROM ABOVE

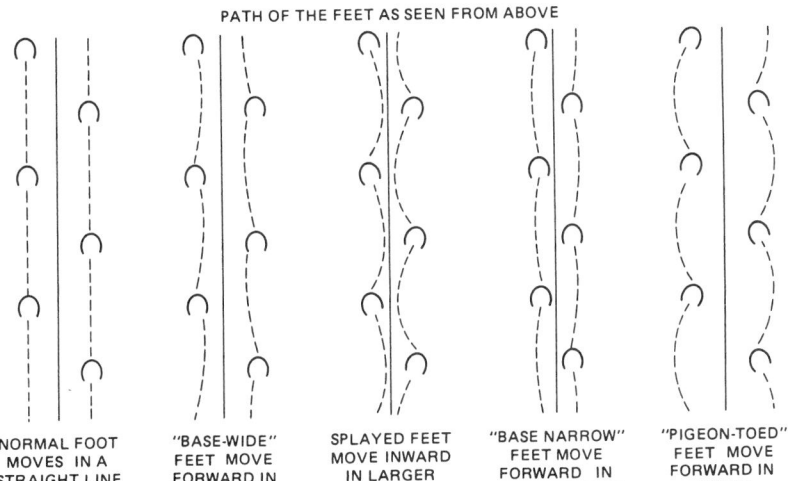

NORMAL FOOT MOVES IN A STRAIGHT LINE "BASE-WIDE" FEET MOVE FORWARD IN INWARD ARCS SPLAYED FEET MOVE INWARD IN LARGER INWARD ARCS "BASE NARROW" FEET MOVE FORWARD IN OUTWARD ARCS "PIGEON-TOED" FEET MOVE FORWARD IN WIDER OUTWARD ARCS

FIGURE 31-26 Correct and incorrect positions of front legs, as shown from the front, and the influence of each on the movement of the horse. *Courtesy of Appaloosa Horse Club, Inc., ID.*

yellow to a bright copper. *Pinto* is sometimes called calico or paint. The horse is spotted with more than one color. *Piebald* is a white and black color combination. *Skewbald* is white with any other color except black. *Roan* horses have white hairs mingled with one or more other hair colors. The *blue roan* is a mixture of black with white hair. The *red roan* is a mixture of bay with white hair. The *strawberry roan* is a mixture of chestnut with white hair.

Face and Leg Markings

Face and leg markings are often used, along with the color of the horse, to identify an individual horse. The common face markings are shown in Figure 31-32. The common leg markings are shown in Figure 31-33.

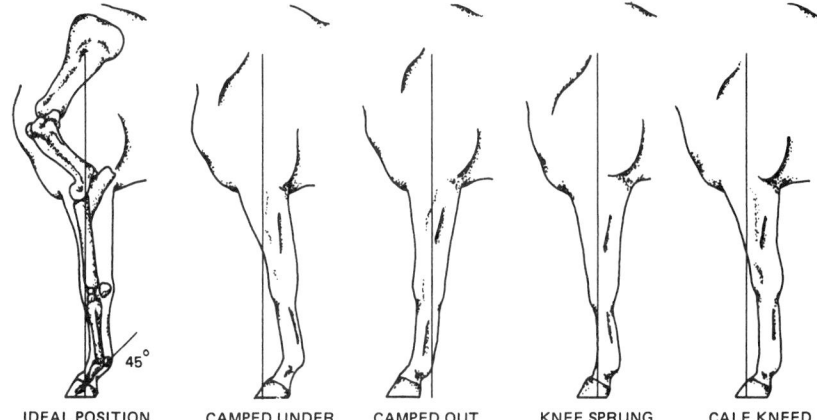

FIGURE 31-27 Correct and incorrect positions of the front legs, as shown from the side. *Courtesy of Appaloosa Horse Club, Inc., ID.*

IDEAL POSITION CAMPED UNDER CAMPED OUT KNEE SPRUNG CALF KNEED

VERTICAL LINE FROM SHOULDER SHOULD FALL THROUGH ELBOW AND CENTER OF FOOT.

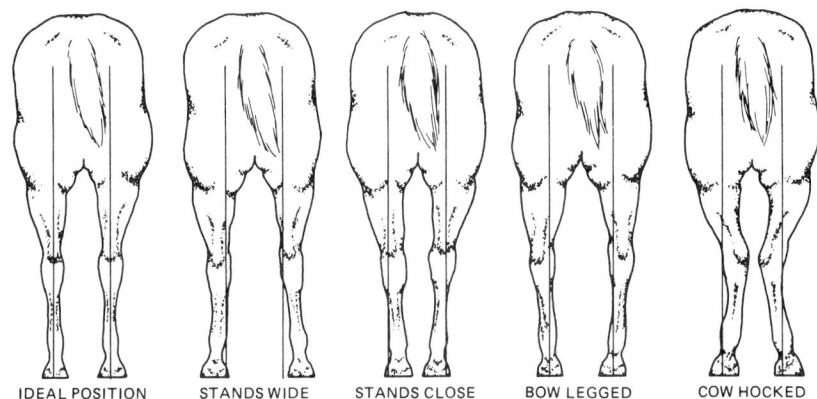

FIGURE 31-28 Correct and incorrect positions of the rear legs, as shown from the rear. *Courtesy of Appaloosa Horse Club, Inc., ID.*

IDEAL POSITION STANDS WIDE STANDS CLOSE BOW LEGGED COW HOCKED

VERTICAL LINE FROM POINT OF BUTTOCK SHOULD FALL IN CENTER OF HOCK, CANNON, PASTERN AND FOOT.

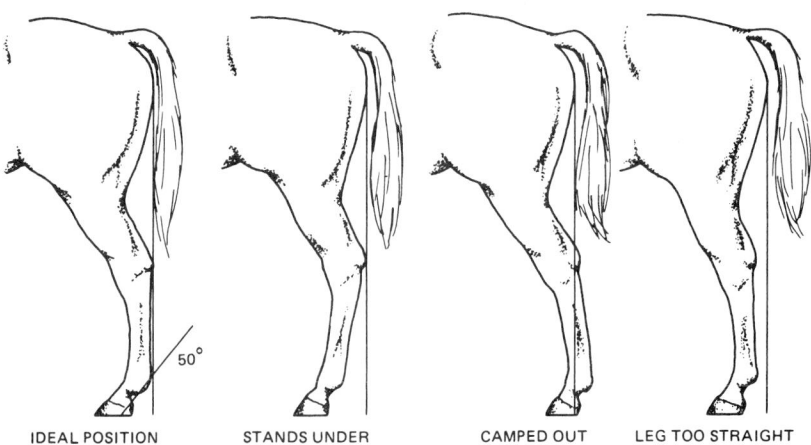

FIGURE 31-29 Correct and incorrect positions of the rear legs, as shown from the side. *Courtesy of Appaloosa Horse Club, Inc., ID.*

IDEAL POSITION STANDS UNDER CAMPED OUT LEG TOO STRAIGHT

VERTICAL LINE FROM POINT OF BUTTOCK SHOULD TOUCH THE REAR EDGE OF CANNON FROM HOCK TO FETLOCK AND MEET THE GROUND BEHIND THE HEEL.

Age of Horses

The approximate age of a horse can be determined by looking at the front teeth. This method is more accurate with younger horses than it is with older horses. A horse that has grazed sandy areas most of its life will appear older than it is because of the wear on the teeth. Figure 31-34 gives some guidelines to use in determining the age of horses by examining their teeth.

Unsoundnesses and Blemishes

An *unsoundness* is a defect that affects the usefulness of the horse. A *blemish* is an imperfection that does not affect the usefulness of the horse. Unsoundnesses are more serious than blemishes. The most serious unsoundnesses are those affecting the feet and legs. The following list defines some common unsoundnesses, blemishes, and faulty conformations. Unsoundnesses and blemishes are identified as follows: U = unsoundness, B = blemish.

Head

1. *Cataract* (U) is an opaqueness or cloudiness of the cornea of the eye.
2. *Blindness* (U) is the partial or complete loss of vision in the eye.
3. *Moon blindness (periodic ophthalmia)* (U) is a cloudiness or inflammation of the eye that occurs at repeated intervals. It may result in permanent blindness.
4. *Poll evil* (U) is an inflammation and swelling of the poll. It often results from bruising.
5. *Roman nose* is faulty conformation.
6. *Parrot mouth* (U) is a condition where the upper jaw overshoots the lower jaw.
7. *Undershot jaw* (U) is a condition in which the upper jaw is shorter than the lower jaw.

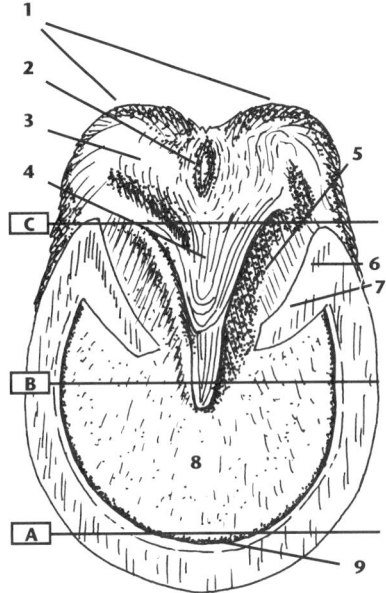

1 Horny bulbs of the heels
2 Middle cleft of the frog
3 Branches of the frog
4 Body of the frog
5 Cleft of the frog
6 Buttress
7 Bars
8 Sole
9 White Line

A to A—Toe
A to B—Side Wall
B to C—Quarter

FIGURE 31-30 The parts of the hoof. *Courtesy of Appaloosa Horse Club, Inc., ID.*

NORMAL FOOT FORMS EVEN ARC IN FLIGHT

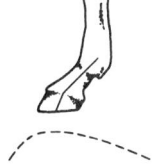

TOO STUBBY – HIGH HEEL AND SHORT TOE CAUSE LENGTHENING OF FIRST HALF OF STRIDE, LONG HEEL TOUCHES GROUND EARLIER WHICH SHORTENS LAST HALF OF STRIDE.

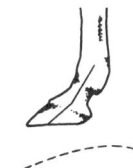

LONG TOE – SHORT HEEL CAUSES SHORTENING OF FIRST HALF OF STRIDE AND LENGTHENING LAST HALF OF STRIDE.

FIGURE 31-31 Correct and incorrect angle of the pastern and hoof and the arc of the hoof at each of the angles. *Courtesy of Appaloosa Horse Club, Inc., ID.*

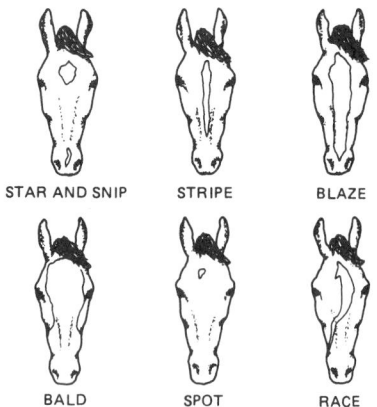

FIGURE 31-32 Common face markings found on horses. *Courtesy of Appaloosa Horse Club, Inc., ID.*

Neck

8. *Ewe neck* is faulty conformation of the neck.

Withers and Shoulders

9. *Fistula of the withers* (U or B) is an inflammation and swelling in the area of the withers. It often results from bruising.
10. *Sweeney* (U) is a decrease in the size of a muscle or group of muscles, usually in the area of the shoulder.

Front legs

11. *Shoe boil* or *capped elbow* (B) is a swelling at the point of the elbow.
12. *Knee sprung* (also called *buck-kneed* or *over in the knee*) is a bending forward of the knee. This is a faulty conformation.
13. *Calf-kneed* (also called *back at the knees*) is a condition in which the knees bend backward. This is a faulty conformation.
14. *Splints* (B) are deposits of bone that occur on the upper, inside part of the cannon bone. Figure 31-35 shows a horse with splints.
15. *Wind-puff (wind-gall, road-puff, road-gall)* (U) is a puffy swelling that occurs on either side of the tendons above the fetlock or knee.
16. *Bowed tendons* (U) are swellings of the tendons on the back of the leg. This may occur on either the front or back legs.

Feet

17. *Ringbone* (U) is a growth of bone on either or both of the bones of the pastern.
18. *Sidebone* (U) is a condition in which the lateral cartilage just above the hoof turns to bone.
19. *Quittor* (U) is a decay of the lateral cartilage resulting in an open sore.
20. *Quarter crack (sand crack)* (B) is a vertical split in the wall of the hoof.
21. *Navicular disease* (U) is an inflammation of the navicular bone and bursa inside the hoof.
22. *Founder* (U) is the inflammation of the sensitive lamina of the hoof. The *laminae* attach the hoof to the fleshy part of the foot.

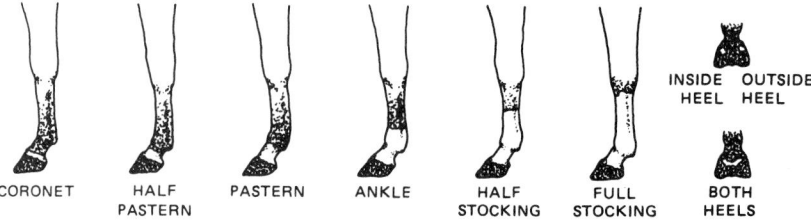

FIGURE 31-33 Common leg markings of horses. *Courtesy of Appaloosa Horse Club, Inc., ID.*

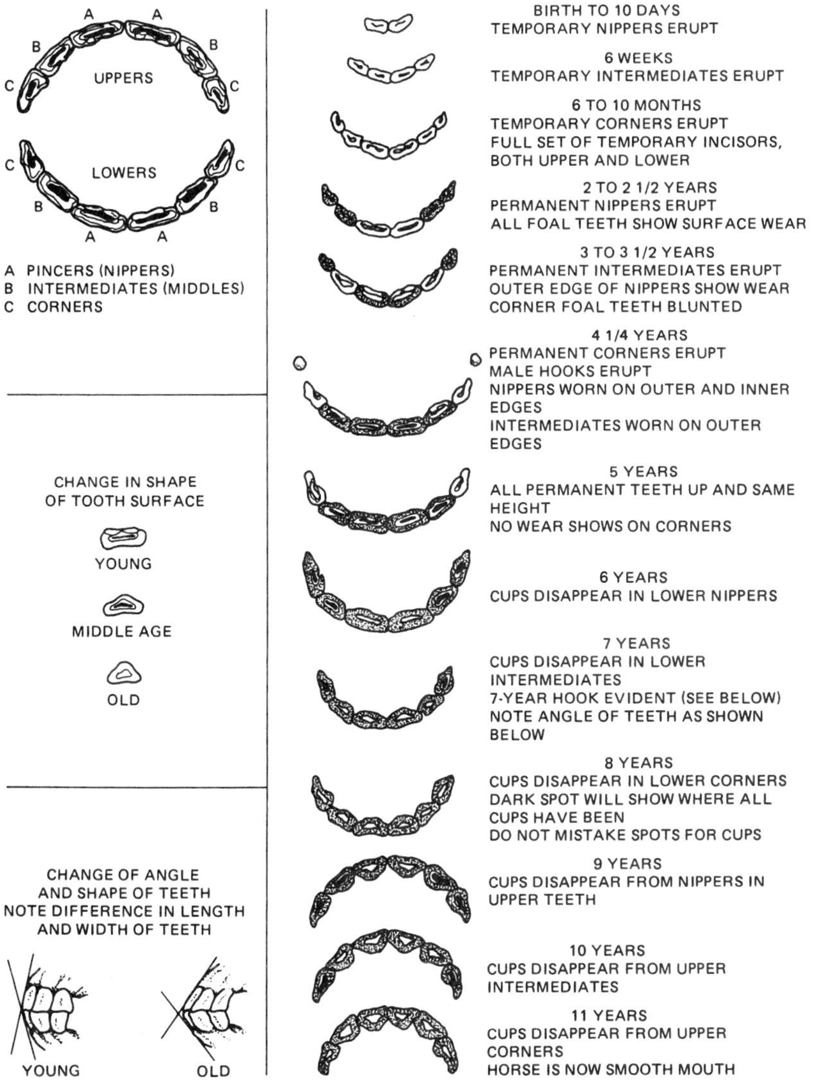

A PINCERS (NIPPERS)
B INTERMEDIATES (MIDDLES)
C CORNERS

UPPERS

LOWERS

CHANGE IN SHAPE
OF TOOTH SURFACE

YOUNG

MIDDLE AGE

OLD

CHANGE OF ANGLE
AND SHAPE OF TEETH
NOTE DIFFERENCE IN LENGTH
AND WIDTH OF TEETH

YOUNG   OLD

BIRTH TO 10 DAYS
TEMPORARY NIPPERS ERUPT

6 WEEKS
TEMPORARY INTERMEDIATES ERUPT

6 TO 10 MONTHS
TEMPORARY CORNERS ERUPT
FULL SET OF TEMPORARY INCISORS,
BOTH UPPER AND LOWER

2 TO 2 1/2 YEARS
PERMANENT NIPPERS ERUPT
ALL FOAL TEETH SHOW SURFACE WEAR

3 TO 3 1/2 YEARS
PERMANENT INTERMEDIATES ERUPT
OUTER EDGE OF NIPPERS SHOW WEAR
CORNER FOAL TEETH BLUNTED

4 1/4 YEARS
PERMANENT CORNERS ERUPT
MALE HOOKS ERUPT
NIPPERS WORN ON OUTER AND INNER
EDGES
INTERMEDIATES WORN ON OUTER
EDGES

5 YEARS
ALL PERMANENT TEETH UP AND SAME
HEIGHT
NO WEAR SHOWS ON CORNERS

6 YEARS
CUPS DISAPPEAR IN LOWER NIPPERS

7 YEARS
CUPS DISAPPEAR IN LOWER
INTERMEDIATES
7-YEAR HOOK EVIDENT (SEE BELOW)
NOTE ANGLE OF TEETH AS SHOWN
BELOW

8 YEARS
CUPS DISAPPEAR IN LOWER CORNERS
DARK SPOT WILL SHOW WHERE ALL
CUPS HAVE BEEN
DO NOT MISTAKE SPOTS FOR CUPS

9 YEARS
CUPS DISAPPEAR FROM NIPPERS IN
UPPER TEETH

10 YEARS
CUPS DISAPPEAR FROM UPPER
INTERMEDIATES

11 YEARS
CUPS DISAPPEAR FROM UPPER
CORNERS
HORSE IS NOW SMOOTH MOUTH

FIGURE 31-34   The number of permanent incisors may be used to determine age to 5 years. After that, use the shape of the biting surface and the angle of the incisors. *Courtesy of Appaloosa Horse Club, Inc., ID.*

23. *Contracted heel* (B) is a condition in which the heel draws in or contracts.
24. *Thrush* (B) is a disease of the frog of the foot. It is caused by filth and may result in lameness.
25. *Corns* (U) are reddish spots on the horny sole. They are usually caused by bruises or improper shoeing. Corns may result in swelling, a pus discharge, and lameness.
26. *Scratches (grease heel)* (U) is an inflammation of the back surfaces of the fetlocks. Scabs form on the area.

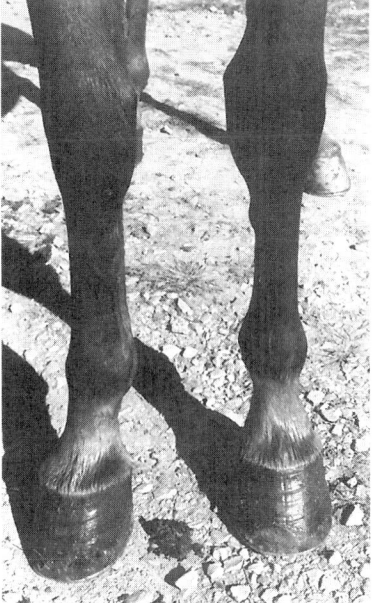

FIGURE 31-35   Horse with splints. *Photo by Jim Bodine.*

Hind Legs

27. *Stifled* (U) is a condition in which the patella (cap) of the stifle joint has been displaced.
28. *Stringhalt* (U) is a condition in which there is a sudden involuntary flexion of one or both hocks. The foot is jerked upward higher than normal.
29. *Thoroughpin* (U) is a puffy swelling just above the hock.
30. *Capped hock* (U or B) is a callus or firm swelling at the point of the hock.
31. *Bog spavin* (U) is a large, soft swelling on the inside and front of the hocks.
32. *Bone spavin (jack spavin)* (U) is a bony growth on the inside of the hock.
33. *Curb* (U) is a hard swelling on the back surface of the rear cannon, about four inches below the point of the hock.
34. *Cocked ankle* (U) is a condition in which the tendons become inflamed and shortened. The fetlock bends forward.
35. *Blood spavin* (B) is a swollen vein over the front and inside of the hock. It does not cause lameness.

Body

36. *Heaves (asthma, broken wind)* (U) is difficulty in breathing. It is caused by lung damage.
37. *Roaring* (U) is difficulty in breathing due to a paralysis of the nerve to the muscles of the larynx. There is a whistling or wheezing sound when the horse breathes in.
38. *Thick wind* (U) is difficulty in breathing in or out due to an obstruction in the respiratory tract.
39. *Rupture (hernia)* (U) is when any internal organ comes through the wall that contains it.
40. *Sway back* is faulty conformation of the back.
41. *Knocked down hip* (U) is a fracture of the external angle of the hip bone. The point of the hip lowers as a result of the fracture.

Vices

Because of long periods of idleness or poor handling, horses sometimes develop bad habits that are called *vices*. The most common vices are cribbing, wind sucking, halter pulling, and kicking.

Cribbing is a behavior in which a horse bites on some part of the feed manger or stall. *Wind sucking* takes place when a horse presses the upper front teeth on some object and pulls back, at the same time sucking air into the stomach. Cribbing and wind sucking often occur together. Fitting a wide strap around the throat so that the larnyx is compressed when pressure is put on the front teeth will help prevent or halt these two vices.

Halter pulling occurs when the horse pulls back against the halter while tied. Using a heavy halter or heavy chain or rope around the neck when tying the horse will help to prevent this vice.

If hoof or shoe marks are on the walls of the stall, the horse may be a *kicker*. Capped hocks and scarred hind legs may also indicate a kicker. Padding the stall, hanging heavy chains from the ceiling, and hanging bags of straw from the ceiling are suggested methods of stopping this vice.

Gaits

The *gait* is the movement of the horse's feet and legs when the horse is in motion. The walk, trot, and gallop are the three natural gaits of the horse. Other gaits include the canter, stepping pace, running walk, fox trot, amble, rack, and pace.

The *walk* is a slow, four-beat gait. Each foot leaves and strikes the ground separately from the other feet. It is the natural gait of the horse.

The *trot* is a fast, two-beat diagonal gait. Opposite front and hind feet leave and strike the ground at the same time.

The *gallop* is a fast, four-beat gait. Each foot strikes the ground separately. The feet strike the ground in the following order: (1) one hind foot, (2) the other hind foot, (3) the diagonal front foot, (4) the other front foot. For a brief moment, all four feet are off the ground. The lead changes when all four feet are off the ground. Most of the drive comes from the hindquarters. The extended gallop is called the *run*.

The *canter* is a slow, three-beat gait. The feet strike the ground in the following order: (1) one hind foot, (2) the other hind foot and the diagonal front foot, (3) the other front foot. The horse may lead from either the right or the left front foot. A western adaptation of a very slow canter is called the *lope*.

The *stepping pace* is a slow, lateral, four-beat gait. The four feet strike the ground separately. The feet strike the ground in the following order: (1) right hind foot, (2) right front foot, (3) left hind foot, (4) left front foot.

The *running walk* is a slow, diagonal, four-beat gait. Each foot leaves and strikes the ground separately from the other feet. The front foot strikes the ground just ahead of the diagonal hind foot. This is a natural gait of the Tennessee Walking Horse.

The *fox trot* is a slow, short, broken trot. The hind foot strikes the ground just ahead of the diagonal front foot.

The *rack* is a fast, even, four-beat gait. The time between each foot striking the ground is the same. The order of the feet striking the ground is the same as in the stepping pace.

The *pace* is a fast, two-beat gait. The front and hind feet on the same side leave and strike the ground at the same time. There is a brief moment when all four feet are off the ground at the same time.

The *amble* is a lateral movement of the horse. It is also called the *traverse* or *side step*. It is not a show gait. The horse moves to one side without going forward or backward.

Pedigree

Pedigree is of greatest importance when selecting race and show horses. It is good to remember that ancestors back of the grandparents have contributed little to the genetic inheritance of the horse that is being considered for purchase. Other factors of selection such as conformation, soundness, training, and so forth, should be considered more important than pedigree.

Records

Breed associations have forms available for keeping records on registered horses. Breeding and production records are kept on both stallions and mares. While forms vary from one association to another, typical information kept for stallions includes the stallion's name and registration number, identifying marks, the mare's owner when bred, date of service, the mare's name, breed, registration number, the number of mares bred, the number of mares settled, the number of live foals, and the performance records of foals. Typical records kept on mares include the mare's name and registration number, the birthdate, owner, breed, sire and dam, identifying marks, number of times bred, number of times settled, number of live foals, and performance record of foals. Health records are often kept with the mare's breeding record. More than one form may be used to keep records on horses. Identification records should be kept for both registered and nonregistered horses in case they are stolen or lost. A record form is generally completed when a registered horse is sold to a new owner. Show and race horses usually have performance records available.

Price

The purchase price of a horse may vary from a small amount to thousands of dollars. Mature, unregistered horses are generally lower in price. Registered horses with potential for showing are higher priced. Horses from outstanding breeding with racing potential will usually cost many thousands of dollars.

Prices of horses vary with the season. Prices are often lower during the fall and winter. The annual costs of keeping a horse must also be considered when thinking about price. Feed, housing, shoeing, and veterinary fees are annual expenses that add to the cost of having a horse. Tack and equipment prices should also be included when determining the cost of the horse to provide a more accurate picture of the cost of ownership. The use to which the horse will be put influences the price of the horse as well as the other costs associated with horse ownership.

SUMMARY

About 75 percent of the horses in the United States are used for personal pleasure riding. Other uses include racing, ranching, breeding, and commercial riding. There are more than 5 million horses in the United States at the present time. More than 16 billion dollars is spent in the horse industry annually.

There are many breeds of light horses and ponies. Most of the horses in the United States are of the light breeds. There are few draft horses in the United States today. Ponies are smaller than light horses. Draft horses are larger than light horses.

Horses should be selected on the basis of conformation, use, age, sex, and soundness. Breed selection is a matter of use and personal preference. Breeders are the best source of horses. Other sources include private owners, auctions, and dealers.

The five basic colors of horses are bay, black, brown, chestnut, and white. Variations of color include dun, gray, palomino, pinto, and roan.

Horses may have a variety of unsoundnesses and blemishes. Unsoundnesses are more serious than blemishes. The most serious unsoundnesses affect the feet and legs of the horse.

The gaits of the horse are the ways in which it moves. The common gaits are walk, trot, gallop, canter, stepping pace, running walk, fox trot, rack, pace, and amble.

Pedigree and price must be considered when selecting a horse. Pedigree is more important for show and race horses. Prices may vary from a small amount to many thousands of dollars.

Student Learning Activities

1. Prepare a bulletin board display of pictures of the breeds of horses.
2. Present an oral report on the origin and traits of the breeds of horses.
3. Survey the community to determine the most popular breeds and numbers of horses locally.
4. Identify the parts of the horse on a picture or on a live horse.
5. On field trips to local farms or stables, observe conformation, unsoundnesses, blemishes, and vices of horses.
6. Determine the age of a live horse by examining the teeth.
7. Give an oral report on horse selection.

Review

1. What percent of the horses in the United States are owned for personal pleasure use?
2. Approximately how many horses are there in the United States?
3. Describe the economic impact of the horse industry in the United States.
4. How popular are horse racing and horse shows in the United States?
5. List several ways in which horses are beneficial to people.
6. Compare light horses and ponies as to size and weight.
7. List and give a brief description of each of the breeds of light horses, ponies, and draft horses.
8. Define the following terms: (a) foal, (b) filly, (c) colt, (d) mare, (e) stallion, (f) gelding.
9. List the five general uses of horses.
10. Briefly explain the various sources from which horses may be secured.
11. A horse of what age is the best to buy? Why?
12. A horse of which sex is usually a better choice for pleasure riding?
13. What factors influence the choice of a horse breed to buy?
14. Draw a sketch of a horse and show the major parts.
15. Briefly describe the physical traits that are desirable in a light horse.
16. Explain the importance of a horse's feet and legs.
17. Name the five basic colors of horses.
18. Describe the five major variations in horse color.
19. Name and describe or sketch the common face markings found on horses.
20. Describe how the age of a horse can be estimated by looking at the teeth.
21. Define the terms *unsoundness* and *blemish*.
22. List and briefly describe some common unsoundnesses and blemishes of horses.
23. Define the term *vice* as it relates to horses.
24. Name and describe some common vices a horse may have.
25. Define the term *gait*.
26. Name and describe the gaits of horses.
27. How important is a pedigree when selecting a horse?
28. How do the prices of horses vary with the season?
29. What factors influence the price of a horse?
30. What costs must a horse owner be prepared to pay, in addition to the original cost of the horse?

Feeding, Management, Housing, and Tack

Objectives

After studying this unit, the student should be able to

- develop a feeding program for horses based on commonly accepted standards.
- describe good management practices for horses.
- describe housing and tack required for horses.

FEEDS FOR HORSES

Horses can utilize large amounts of roughage in the ration. Much of the digestion of the roughage occurs in the large intestine. Bacteria in the large intestine break down the roughage into a form that can be used by the horse. Mature, idle horses can be fed on roughage alone. Horses that are being ridden or otherwise worked require some concentrate added to the ration. Pregnant mares and growing foals also require concentrate in the ration in addition to roughage.

Pastures

The amount of pasture required per horse depends on the use of the pasture, how it is managed, the kind of pasture, and the amount of moisture available. One acre per horse will provide little more than an exercise area. The horse cannot obtain enough feed from one acre to meet its needs. Two acres (for the grazing season) are sufficient if the pasture is well managed and has enough moisture. In drier areas, a greater number of acres per horse is necessary. On native ranges in the western states, two to ten acres per month per horse are required (Figure 32-1).

To produce the greatest amount of feed possible, pastures should be fertilized. Rotation of pastures increases their carrying capacity. In dryland areas, irrigation can be used to improve the yield from the

FIGURE 32-1 This Oregon pasture provides both water and grass for these horses. *Courtesy of USDA.*

FIGURE 32-2 Horses can be hard on pasture. They tend to graze unevenly, resulting in overgrazed and undergrazed areas. *Photo by Jim Bodine.*

pasture. If only horses are grazed, the pasture should be clipped. Grazing cattle with horses reduces the need for clipping. Horses are harder on pastures than cattle. They tend to graze unevenly, resulting in overgrazed and undergrazed areas (Figure 32-2). They also tear up the sod more than cattle do.

Kinds of Pasture

Kentucky bluegrass is considered the best all-around pasture for horses. It is palatable and provides the nutrients needed by horses. It maintains a tough turf that is less subject to damage.

Orchard grass is also an excellent horse pasture. It works well in combination with Kentucky bluegrass or a legume. It stands up well under close grazing.

Tall fescue is not as good a horse pasture as some of the other grasses that are used. It is a good choice for areas of heavy horse traffic because it is not easily damaged. It is not as palatable as other grasses. Horses on tall fescue pasture should receive some grain in the ration. A legume should be seeded with fescue to make a better pasture.

Timothy is palatable for horses. However, it does not stand up well to close grazing and recovers slower than other pasture plants. Rotation grazing should be used with timothy pastures.

Bromegrass is a good pasture for horses. It requires good management practices such as rotation grazing, clipping, and fertilizing.

Reed canary grass is well adapted to areas with a high water table. It makes a good pasture for horses.

Bermuda grass is a good pasture for horses in the southern areas of the United States. It should be seeded with a legume.

Sorghum-Sudan grass crosses, Sudan grass, and pearl millet are not recommended as horse pastures. In some areas of the United States, particularly the Southwest, problems with the disease cystitis syndrome have been reported on these pastures.

Crested wheatgrass, intermediate wheatgrass, and Russian wild rye make good dryland pastures in the western United States. Crested wheatgrass is a good pasture for spring and late fall. Intermediate wheatgrass makes a good early summer pasture. Russian wild rye is a good pasture through the entire grazing season.

Native ranges have a low carrying capacity. To avoid killing out the desired species, rotation grazing is necessary. A given pasture should be left idle once every three years. Native rangelands are fragile. A range badly damaged by overgrazing may take many years to recover.

Legumes that make good horse pastures include alfalfa, white clover, ladino clover, red clover, alsike clover, lespedeza, and birds-foot trefoil. Any of the adapted legumes can be used in a given area. Horses do not bloat, so there is no danger in using legume pastures. The amount of legume in the mix should be about 35 to 40 percent.

Silage

Silage may be used to replace up to one-half of the hay in the horse ration. Corn silage is the best silage to use. Grain sorghum, grass, and grass-legume silages may also be used. Feed high-quality silage that is chopped fine and free of mold. Change the ration slowly when adding silage. Do not feed silage to foals and horses that are being worked hard. It is too bulky for these animals. Legume haylage can be used in place of the silage.

Legume Hay

Legume hay may be made from alfalfa, the clovers, and lespedeza. Legume hays are more palatable than grass hays. They also have a higher protein and mineral content. Alfalfa hay is the best of the legume hays. Red clover is a little lower in protein content than alfalfa hay. Lespedeza hay must be cut at an early stage to be of the best quality. Grass-legume mixtures are often used for horse hay and make an excellent combination for feeding horses.

Grass Hay

Common grass hays are timothy, bromegrass, orchard grass, bermuda grass, prairie hay, and cereal hay. Generally, grass hays do not yield as much feed per acre as legume hays. They tend to be lower in protein, calcium, and vitamins. Timothy hay has long been considered the standard hay for feeding horses. However, it is low in protein and must be fed with a protein supplement or a high protein grain (such as oats). It is better for mature horses than it is for younger, growing horses, stallions, or mares.

Bromegrass is most palatable when harvested in the bloom stage. Orchard grass is not quite as good a horse hay as bromegrass. Several improved varieties of bermudagrass hay are used in the southeastern United States. Prairie hay is not as palatable and is lower in protein than timothy hay. Oat, barley, wheat, and rye hays are often used in the Pacific Coast region of the United States. They should be cut in the soft to stiff dough stage for best quality.

Grain

Oats are considered to be the best grain for horse rations. They are palatable and bulky, which decreases digestive problems. They are higher in protein than corn, but lower in energy value.

Corn is often a better buy on an energy basis than oats. It is especially good for thin horses and those that are worked hard. When corn is fed, care must be taken not to let the horse become too fat. Corn may cause colic, so must be fed with care. A corn-oats mixture makes an excellent grain ration for horses.

Grain sorghum (milo) is best used in a grain mixture. Some varieties are not palatable for horses. Grain sorghum should be cracked or rolled for horse rations.

Barley should be rolled or crushed if fed to horses. It can be substituted for corn in the ration.

Wheat is usually too expensive to feed. It should not make up more than 50 percent of the grain mix. Wheat should be rolled or coarsely ground.

Wheat bran is a bulky, palatable feed for horses. It tends to be slightly laxative, and is often fed to horses in stress conditions.

Cane molasses is used mainly to reduce the dustiness of the ration and to increase palatability. It should not make up more than 4 to 5 percent of the ration. If fed in excess, it acts as a laxative.

Protein Feed

Usually, little protein supplement is needed in horse rations. If at least one-half the roughage is legume, the protein needs of the horse, except in the case of milking mares, will be met. Protein is added to the ration of show horses to improve the hair coat. Protein supplement should be added to the ration if the quality of the roughage is poor.

Soybean meal is an excellent protein supplement for horses. It is high in protein and has a good balance of amino acids.

Cottonseed meal is not as palatable as soybean meal. It is used widely in the Southwest as a protein supplement for horses.

Linseed meal may be too laxative if fed with legume hay. Expeller-type linseed meal (a by-product of an oil extraction process) contains the fatty acid linoleic that is often lacking in horse rations. Linseed meal puts a bloom on the hair coat of the horse.

Alfalfa meal, corn gluten meal, and meat meals, as well as other protein supplements, may be used in horse rations.

Commercial protein supplements are popular with producers who do not want to mix their own rations. Most commercial supplements are developed for a particular feeding program. The directions on the feed tag should be closely followed. Do not use commercial protein feeds that contain urea; horses do not utilize urea efficiently because the cecum is located too far down the digestive tract. In horses, nonprotein nitrogen sources such as urea are converted to protein in the cecum. Too much urea can therefore be toxic to horses.

Pelleted Feeds

The use of complete pelleted rations is gaining in popularity. These rations are carefully balanced by the manufacturer. Convenience is one of their strong points. There is often less waste when pelleted rations are used. Pelleted rations are generally more expensive than roughage and grain rations. Horses tend to eat bedding and chew on wood more when fed pelleted feeds.

Minerals

Horses require salt, calcium, and phosphorus in the ration. The amount of salt needed by the working horse is large. Salt should be fed free choice. Calcium and phosphorus should be fed free choice, separate from the salt. Ground limestone, steamed bone meal, and defluorinated phosphate are common sources of calcium and phosphorus.

Feeding trace-mineralized, iodized salt free choice will supply the salt and trace minerals needed by horses. Check the tag on the trace-mineralized salt for the presence of selenium. Horses need some selenium in their diet, but not all trace-mineralized salt formulations contain this trace mineral.

Mares in early lactation need about twice as much calcium and phosphorus in their diet as do mares that are not nursing foals. Young horses should have a ratio of about 1.1 parts calcium to 1 part phosphorus in their diet. A ratio as high as six parts calcium to one part phosphorus can be tolerated by older horses. Never feed more phosphorus than calcium in a horse's diet.

Vitamins

Horses on good pasture seldom need additional vitamin supplements in the ration. If the hay in the ration is of good quality and at least one-half legume, there probably is no need to add a vitamin supplement to the ration.

When horses are confined in barns, receiving poor-quality hay, or are on drouthy pasture, the diet may not contain enough vitamin A,

D, and E. In these cases, it may be wise to add a vitamin supplement to the diet. The supplement may be mixed in the feed, injected in the muscle, or provided in stabilized mineral blocks. Vitamin K and the B vitamins generally do not need to be added to the ration, as horses can synthesize these in sufficient quantities in the body. However, when horses are under stress, it may be wise to add a small amount of the B vitamins to the diet. Stress conditions may include fast growth, intense training, heavy racing, or breeding. Spent brewer's yeast is a good source of the B vitamins.

Excess vitamins in the diet are not usually toxic to horses. However, horses with liver or kidney problems may not be able to metabolize higher levels of vitamins. The fat soluble vitamins (A, D, E, and K) are more likely to be toxic to horses with liver or kidney problems. Young horses and unborn foals may also have difficulty in metabolizing higher levels of vitamins. It is generally not wise to overfeed vitamins under any conditions.

Water

Horses drink 10 to 12 gallons (37.8–45.4 litres) of water per day. Hard-working horses drink more than this amount. Hot weather increases the need for water. A supply of fresh, clean water should be available at all times. Horses at work must be watered at frequent intervals. However, horses that are hot should be cooled out and allowed to drink only small quantities of water at a time before being allowed to drink their fill. Be sure the water is not too cold when given to hard-working horses.

FEEDING HORSES

Horses are fed according to their size, stage of growth, condition, and amount of work they are performing. Rules of thumb for feeding horses and methods of balancing rations are given in Unit 8. Tables showing the nutrient requirements of horses and the composition of feeds are provided in the appendix of this book.

Horses must be fed and watered regularly, at least twice each day. Adjust the amount and kind of feed according to the condition of the horse. Do not feed moldy or dusty feed. Do not allow a horse that is heated to drink too much water. Water the horse before feeding if water is not available free choice. Hay should be fed before grain. A tired horse should be fed only half of the grain ration and the rest should be fed one hour later. Do not work a horse right after a full feed of grain. When working horses are idle, feed only one-half the grain ration and increase the amount of hay fed.

A guide for feeding horses is given in Table 32-1. Some suggested rations for light horses are given in Table 32-2. In the rations in Table 32-2, milo (grain sorghum), barley, or wheat can be substituted for corn on a pound-for-pound (kilogram-for-kilogram) basis.

Pregnant mares that are healthy, and not too thin or too fat, usually do not need extra grain in their diet if they are fed good-quality hay. Pregnant mares that are older or thin may benefit from the addition of some grain in the diet. If additional grain is added to the ration, do not overfeed the mare or she will become too fat. Mares that are too fat may have more trouble at foaling time.

Feed a small amount of grain along with the hay for the first seven to 10 days after foaling. Feeding too much grain can increase the amount of milk produced; this may cause the foal to scour. After seven to 10 days, slowly increase the amount of grain in the ration.

CARE OF BROOD MARE AND FOAL

Breeding the Mare

Horses have a low conception rate. The national average is 50 to 60 percent. However, application of the basic principles of reproduction can help to increase the conception rate. The basic principles of reproduction are discussed in Unit 10. Information about the estrus cycle and reproductive traits of the horse are also included in that unit.

The mare is more likely to conceive if bred in the months of April, May, or June. This also means that the foal will be dropped the following spring, which is a desirable time of the year for foaling. In the spring, foaling can take place on pasture, which reduces problems with diseases and parasites.

The best age to breed mares for the first time is as three-year-olds, so they will foal when about four years old. A well-grown filly can be bred as a two-year-old, but requires extra care in feeding so that her own body, as well as the developing fetus, will grow properly.

Pasture breeding requires less labor. However, the conception rate is likely to be lower. Hand breeding increases the conception rate.

Brood mares (mares used for reproduction) will usually produce foals until they are 14 to 16 years old. Mares may be rebred during foaling heat, which occurs about nine days after foaling. The practice of rebreeding during the first heat period after foaling heat (about 25 to 30 days after foaling) is becoming more popular. Mares that do not conceive will continue to come into heat about every 21 days.

Mares that are too thin or too fat have lower conception rates. Mares must be given enough exercise and the proper rations before breeding.

Care of Pregnant Mare

The pregnant mare requires a great deal of exercise. Ride or drive the brood mare for a short period of time each day. Provide pasture in which the mare may run. Do not confine the mare to a stall where she will get no exercise.

TABLE 32-1 Horse Feeding Guide.

Suggested concentrate (grain) mixtures[1,2,3] (with all mixtures and for all classes and ages of horses, provide free access in separate containers to (1) plain, loose salt and (2) a mineral mixture containing equal parts of trace mineralized salt,[6] dicalcium phosphate and steamed bone meal.)

Kind of Horse	Daily Allowance	Kind of hay (in season, any good pasture can replace part or all of the hay except for horses at work or in training)	Ration 1	%	Ration 2	%	Ration 3	%
Foals, creep and starting (weighing 100 to 450 lb) (45.4–204.1 kg)	At 1 to 4 months: ½ to ¾ lb grain per 100 lb body weight. (1.1–1.6 kg per 100 kg body weight) At 5 to 6 months: 1 to 1¼ lb grain per 100 lb body weight (2.2–2.8 kg per 100 kg body weight), together with a quantity of hay within same range for each period.	Legume; or legume-grass mixture.	Oats Dried milk by-product[4] Complete Supplement[5]	70 10 20 100	Oats Corn Dried milk by-product[4] Complete Supplement[5]	50 20 10 20 100	Oats Corn Soybean meal (44%) Dried milk by-product[4] Alfalfa meal Molasses (liq.) Limestone Dicalcium phosphate Trace mineral salt[6] Vitamin premix	44.5 18.0 15.0 10.0 5.0 5.0 0.4 1.1 0.5 0.5
			Calculated Nutrient Analysis					
			Protein (%) Calcium (%) Phosphorus (%) Dig. energy kcal/lb kcal/kg	16.40 0.68 0.64 1310 2888		15.80 0.66 0.63 1370 3020		16.00 0.70 0.60 1350 2976
Weanlings (weighing 450–750 lb) (204–340 kg)	1–1½ lb grain per 100 lb body weight (2.2–3.3 kg per 100 kg) together with a quantity of hay in the same range.	Legume-grass mixture, or ½ legume hay, with remainder grass hay.	Oats Complete Supplement[5]	80 20 100	Oats Corn Complete Supplement[5]	50 25 25 100	Oats Corn Soybean meal (44%) Alfalfa meal Molasses (liq.) Limestone Dicalcium phosphate Trace mineral salt[6] Vitamin premix	44.9 23.0 19.5 5.0 5.0 0.3 1.3 0.5 0.5 100.0
			Calculated Nutrient Analysis					
			Protein (%) Calcium (%) Phosphorus (%) Dig. energy kcal/lb kcal/kg	16.20 0.60 0.60 1280 2822				16.50 0.70 0.65 1350 2976

Calculated Nutrient Analysis table and feeding guide (landscape orientation):

Class of horse	Concentrate (grain) mixture and roughage	Roughage / pasture	Complete Supplement[5]
700–1,000 lb (317.5–453.6 kg)	body wt (1.1–2.2 kg and 2.2–3.3 kg per 100 kg)	with remainder grass hay.	
Performance (weighing 900–1,400 lb (408–635 kg))	½–1¾ lb grain and 1–1½ lb hay per 100 lb body weight (1.1–3.9 kg and 2.2–3.3 kg per 100 kg), depending on weight of horse and degree of work expected.		
Pregnant mares (weighing 900–1,400 lb (408–635 kg))	First half: 1½–2 lb hay per 100 lb body weight (3.3–4.4 kg per 100 kg) Last half: ½–1 lb grain and 1–1½ lb hay per 100 lb body weight (1.1–2.2 kg and 2.2–3.3 kg per 100 kg)		
Lactating mares	1–1½ lb grain per 100 lb body weight (2.2–3.3 kg per 100 kg) together with a quantity of hay in the same range.		
Stallions in breeding season (weighing 900–1,400 lb (408–635 kg))	¾–1¼ lb grain per 100 lb body weight (1.6–2.8 kg per 100 kg) together with a quantity of hay within the same range.		
Mature idle horses, stallions, mares, and geldings (weighing 900–1,400 lb (408–635 kg))	1½–1¾ lb hay per 100 lb body weight (3.3–3.9 kg per 100 kg)	Pasture in season; or legume-grass mixture or straight grass hay.	With grass hay, add ½–¾ lb (0.23–0.34 kg) of a high-protein supplement[5] daily.

Complete Supplement[5] composition:

Ingredient	15/100	100
Soybean meal (44%)		9.5
Alfalfa meal		5.0
Molasses (liq.)		5.0
Limestone		0.3
Dicalcium phosphate		0.9
Trace mineral salt[6]		0.5
Vitamin premix		0.5
		100.0

Calculated Nutrient Analysis

Protein (%)	14.10	14.00	14.0
Calcium (%)	0.35	0.44	0.5
Phosphorus (%)	0.47	0.51	0.5
Dig. energy			
kcal/lb	1290	1400	1330
kcal/kg	2844	3086	2932

[1] Grains should be rolled or cracked to increase their bulkiness.
[2] Five to ten percent wheat bran could be substituted for part of the oats or corn.
[3] Five percent linseed meal could be substituted for part of the soybean meal.
[4] Dried whey, dried skim milk, or any similar milk-based product may be used as a source of high protein for the foal.
[5] Any complete supplement containing 30–35 percent crude protein together with approximately 2.5 percent calcium and 1.5 percent phosphorus is suggested.
[6] Any commercial trace mineralized salt for animal use is satisfactory.
[7] The vitamin mixture should provide approximately the following minimum daily requirements per pound (2.2 kg) of final concentrate (grain) ration: vitamin A, 2,000 IU; vitamin D, 200 IU; vitamin E, 20 IU; thiamine, 4.0 mg; niacin, 4.0 mg; riboflavin, 3.0 mg; pantothenic acid, 3.0 mg; and vitamin B₁₂, 5.0 meg.

Source: Jurgens, Marshall H., *Rations For Horses*, AS-387(Rev.), Cooperative Extension Service, Iowa State University, March 1976.

TABLE 32-2 Rations for Light Horses.[1]

A. 1,000 lb (453.6 kg) Mature Idle Horse	(lb)	(kg)
1. Economy Rations:		
a. Pasture when available		
b. Roughages suitable to maintain a beef cow		
c. Timothy hay	16	7.3
Soybean meal	0.5	0.23
2. Standard Rations		
a. Mixed hay	18	8.2
b. Prairie hay	12	5.4
Corn	3	1.4
Soybean Meal	0.5	0.23
c. Timothy hay	12	5.4
Oats	4	1.8
d. Mixed hay	16	7.3
Wheat bran	1	0.45
Molasses	0.5	0.23
3. Conditioning Rations		
a. Timothy	12	5.4
Corn—crackled	3	1.4
Oats—crimped	3	1.4
Wheat bran	1	0.45
b. Mixed hay	12	5.4
Oats—crimped	4	1.8
Corn—cracked	2	0.9
Barley—crimped	1	0.45
Linseed Meal	0.75	0.34

B. 1,000 pound (453.6 kg) Working Horse	Light work (under 3 hrs.)		Medium work (3–5 hrs.)		Heavy work (over 5 hrs.)	
	(lb)	(kg)	(lb)	(kg)	(lb)	(kg)
1. Economy Rations						
Mixed hay	—	—	10	4.5	—	—
Non-legume	—	—	—	—	10	4.5
Grass pasture	free choice		—	—	—	—
Corn	6	2.7	5	2.3	6	2.7
Molasses	—	—	—	—	1	0.45
Oats	—	—	5	2.3	8	3.6
2. Standard Rations						
Mixed hay	12	5.4	10	4.5	—	—
Non-legume	—	—	—	—	10	4.5
Corn—cracked	3	1.4	5	2.3	7	3.2
Barley	—	—	4	1.8	—	—
Oats—crimped	3	1.4	—	—	7	3.2
Wheat bran	—	—	—	—	1	0.45
3. Conditioning Rations						
Mixed hay						
(½ good alfalfa)	12	5.4	—	—	—	—
Non-legume	—	—	10	4.5	10	4.5
Alfalfa leaf meal	—	—	1	0.45	1	0.45
Barley—crimped	1	0.45	—	—	—	—
Corn—cracked	2	0.9	3	1.4	6	2.7
Oats—crimped	3	1.4	5	2.3	6	2.7
Molasses	—	—	1	0.45	1	0.45
Linseed Meal	0.5	0.23	—	—	0.5	0.23
Soybean Meal	—	—	0.5	0.23	—	—
Wheat bran	—	—	1	0.45	1	0.45

TABLE 32-2 Rations for Light Horses. *(Continued)*

	Light Breeding		Heavy Breeding	
C. 1,200 pound (544.3 kg) Stallion	**(lb)**	**(kg)**	**(lb)**	**(kg)**
1. Standard Rations				
Timothy—clover	10	4.5	—	—
Timothy—Lespedeza	—	—	12	5.4
Corn—cracked	2	0.9	—	—
Oats—crimped	4	1.8	9	4.1
Molasses	1	0.45	1	0.45
Soybean Meal	0.5	0.23	1	0.45
Wheat bran	—	—	3	1.4
2. Conditioning Rations				
Timothy—clover hay	10	4.5	—	—
Alfalfa hay	—	—	10	4.5
Corn—cracked	2	0.9	4	1.8
Oats—crimped	4	1.8	8	3.6
Barley—crimped	—	—	3	1.4
Molasses	1	0.45	1	0.45
Linseed Meal	0.5	0.23	0.5	0.23
Wheat bran	1	0.45	1	0.45

	8–9 months				10–11 months			
D. 1,000 pound (453.6 kg) Pregnant Mare	Ration 1		Ration 2		Ration 1		Ration 2	
(0–7 months, same as A above; 8 and 9 months,								
consider rations in B above)	**(lb)**	**(kg)**	**(lb)**	**(kg)**	**(lb)**	**(kg)**	**(lb)**	**(kg)**
Timothy hay	15	6.8	—	—	—	—	—	—
Timothy—clover hay	—	—	15	6.8	—	—	—	—
Non-legume hay	—	—	—	—	15	6.8	—	—
Alfalfa hay	—	—	—	—	—	—	14	6.4
Corn	2	0.9	—	—	—	—	—	—
Corn—cracked	—	—	2	0.9	2	0.9	3	0.9
Oats	1	0.45	—	—	—	—	—	—
Oats—crushed	—	—	1	0.45	2	0.9	3	1.4
Molasses	—	—	1	0.45	0.5	0.23	1	0.45
Soybean Meal	—	—	—	—	0.25	0.11	—	—
Wheat bran	1	0.45	1	0.45	1	0.45	1	0.45

	Ration 1		Ration 2		Ration 3	
E. 1,000 pound (453.6 kg) Lactating Mare	**(lb)**	**(kg)**	**(lb)**	**(kg)**	**(lb)**	**(kg)**
Alfalfa hay	12	5.4	—	—	—	—
Mixed hay	—	—	15	6.8	—	—
Timothy hay	—	—	—	—	15	6.8
Corn—cracked	7	3.2	5	2.3	5	2.3
Oats—crushed	5	2.3	5	2.3	5	2.3
Molasses	—	—	—	—	1	0.45
Soybean Meal	—	—	—	—	0.5	0.23
Wheat bran	—	—	1	0.45	—	—

(Continues)

TABLE 32-2 Rations for Light Horses. *(Continued)*

F. 100–450 pound (45.4–204 kg) Suckling Foals (creep)	Ration 1 (lb)	(kg)	Ration 2 (lb)	(kg)
Alfalfa, good	3–4	1.4–1.8	3–4	1.4–1.8
Corn—cracked	1	0.45	—	—
Oats—crushed	2	0.9	3	1.4
Molasses	—	—	0.5	0.23
Soybean Meal	0.25	0.11	—	—
Linseed Meal	—	—	0.25	0.11
Wheat bran	—	—	1	0.45

G. 500–600 pound (226.8–272 kg) Weanling Foals	Ration 1 (lb)	(kg)	Ration 2 (lb)	(kg)	Ration 3 (lb)	(kg)	Ration 4 (lb)	(kg)
Alfalfa	7	3.2	—	—	—	—	—	—
Alfalfa—Brome	—	—	7	3.2	—	—	—	—
Timothy—Clover	—	—	—	—	7	3.2	9	4.1
Corn—cracked	3	1.4	2	0.9	—	—	1	0.45
Oats—crushed	3	1.4	4	1.8	3	1.4	3	1.4
Molasses	—	—	—	—	0.5	0.23	0.5	0.23
Soybean Meal	—	—	0.5	0.23	0.5	0.23	—	—
Linseed Meal	—	—	—	—	—	—	0.5	0.23
Wheat bran	—	—	—	—	2	0.9	1	0.45

H. 700–800 pound (317.5–362.9 kg) Yearling	Ration 1 (lb)	(kg)	Ration 2 (lb)	(kg)	Ration 3 (lb)	(kg)	Ration 4 (lb)	(kg)
Abundant Pasture	Free choice	—	—	—	—	—	—	—
Mixed hay	—	—	10	4.5	—	—	—	—
Alfalfa—Brome	—	—	—	—	10	4.5	—	—
Timothy—Clover	—	—	—	—	—	—	12	5.4
Corn—cracked	—	—	—	—	2	0.9	3	1.4
Oats—crushed	—	—	3	1.4	2	0.9	—	—
Molasses	—	—	1	0.45	—	—	—	—
Soybean Meal	—	—	0.5	0.23	—	—	0.5	0.23
Wheat bran	—	—	1	0.45	1	0.45	—	—

I. 1,000 pound (453.6 kg) Two Year Old	Ration 1	Ration 2		Ration 3 (lb)	(kg)	Ration 4 (lb)	(kg)
Good Pasture	Free choice	—	—	—	—	—	—
Mixed hay	—	Consider	—	12	5.4	—	—
Non-legume	—	Mature	—	—	—	12	5.4
Corn—cracked	—	Horse	—	2	0.9	—	—
Oats—crushed	—	Rations	—	1	0.45	2	0.9
Molasses	—	Above	—	—	—	1	0.45
Soybean Meal	—	—	—	—	—	0.25	0.11

[1]Rations calculated for horses with mature weight of 1,000–1,200 pounds (453.6–544.3 kg). Adjust for mature weights other than 1,000 lb (453.6 kg) on the following basis:

TABLE 32-2 Rations for Light Horses. *(Continued)*

Weight		
(lb)	*(kg)*	*(% of daily amount in tables)*
400	181.4	50
600	272	68
800	362.9	85
1,000	453.6	100
1,200	544.3	115
1,400	635	130

Source: Bradley, Melvin and Pfander, W.H., *Rations for Light Horses*, Science and Technology Guide 2808, Cooperative Extension Service, University of Missouri, May 1975.

Care at Foaling Time

The mare's udder swells about two to six weeks before foaling time. A week or 10 days before foaling, the muscles over the buttocks appear to shrink. The abdomen drops at this time. The teats fill out to the ends about four to six days before foaling. Wax appears on the ends of the nipples about 12 to 24 hours before foaling. Just before foaling, the mare becomes nervous and restless. Other indications that foaling is near include pawing, lying down and getting up frequently, sweating, lifting the tail, and frequent urinating in small amounts.

Foaling on pasture. In warm weather, a clean pasture away from other livestock is the best place for the mare to foal. However, it is more difficult to assist the mare with foaling on pasture if help is needed.

Foaling in a box stall. A box stall for foaling should be at least 16 × 16 feet (4.9 × 4.9 m) in size. A smooth, well-packed clay floor is preferred. No obstructions, such as feed mangers, should be present in the stall. Isolate the mare from other livestock insofar as possible. Stable the mare in the stall for several days ahead of foaling to allow her to become used to the area. Keep the stall clean and provide large amounts of fresh bedding.

Someone should be nearby the mare when foaling begins, although it is best to remain out of the mare's sight unless she needs help. Mares that are too thin or too fat, and mares foaling for the first time, are more likely to need help at foaling time. If the mare appears to be having difficulty, call a veterinarian for help.

Mares usually foal in about 15 to 30 minutes. A normal presentation of the foal is front feet first with heels down, followed by nose and head, shoulders, middle, hips, and hind legs and feet. If the presentation is not normal, or the mare is taking too much time to foal, call a veterinarian for assistance.

After foaling, be sure the foal is breathing. Remove any mucus from its nose. The navel cord usually breaks 2 to 4 inches (5.1–10.2 cm) from the belly. If it does not break, cut it with clean, dull scissors or scrape it apart with a knife. Dip the navel cord in tincture of iodine to prevent infection.

The foal must nurse shortly after birth to obtain the colostrum milk. Colostrum milk is higher in nutrients and contains antibodies the foal needs.

The foal should have a bowel movement within four to 12 hours after it is born. The feces that are impacted in the bowels during prenatal growth are called *meconium*. If they are not eliminated shortly after birth, they may kill the foal. Feeding colostrum milk usually results in the foal having a bowel movement. If it does not, give the foal an enema. If the foal scours, reduce its milk intake.

Caring for the Suckling Foal

The foal will begin to eat some grain and hay at 10 days to three weeks of age. A creep feeder may be used on pasture. In confinement, provide a low grain box for the foal. Tables 32-1 and 32-2 list several possible rations for use in creep feeding a suckling foal. When well fed during this period, foals reach about one-half their normal mature weight during the first year.

Training the Foal

Training the foal should begin when it is 10 to 14 days old. Begin by putting a well-fitted halter on the foal. After a few days, tie the foal in the stall next to the mare. Be sure the foal does not become tangled in the rope. Keep the foal tied for only 30 to 60 minutes each day. Handle and groom the foal during this time. Pick up its feet to allow it to become used to having its feet handled. Lead the foal with the mare for a few days. Later, lead it by itself at both the walk and the trot. Train the foal to respond to commands to stop and go as they are made. Stand the foal in show position when stopping. It should stand squarely on all four legs with its head up. Use care and patience when working with the foal. Be firm, but do not mistreat it.

Weaning the Foal

Foals can be weaned at four to six months of age. Earlier weaning may be desired if the mare has been rebred during foal heat, is being worked hard, or if either the foal or mare is not doing well. A foal that has had a good creep ration will suffer little setback when weaned. Decreasing the ration of the mare a few days before weaning helps to reduce problems of drying her up.

The foal should be left in the quarters (stall or pasture) and the mare moved to another area at weaning. Do not allow the foal to see the mare for several weeks. Foals may be allowed to run on pasture after several days. Several foals running together may injure each other. To prevent this, only a few should be run together at one time. Make sure that fences are in good repair so that the foals do not become caught in them. Separate timid foals from the rest of the animals. Do not run weaned foals with older horses on pasture.

Castration

Any colt not intended for breeding purposes should be castrated. A veterinarian should perform the castration. Castrating the colt at about one year of age results in a better development of the foreparts of the horse. Castration should be done in the spring before hot weather and flies become a problem. In southern states, time the castration to avoid problems with screwworm infestations.

Breaking

A foal that has been trained at an early age will not need breaking. It is used to being handled. Saddling and harnessing may be done during the winter before it is two years old.

Orphan Foals

An orphan foal may be put on another mare for nursing, or it may be fed milk replacer or cow's milk formula. Feed the foal about one-half pint (0.24 liter) each hour. Larger foals need about one pint (0.47 liter). The feeding interval can be increased to every four hours after four or five days. The quantity may be increased after about one week. Be sure that the foal receives some colostrum milk. Colostrum milk may be frozen and stored for use when needed. Use a bottle and rubber nipple to feed the foal. Switch to a bucket after two weeks. Keep all utensils clean. Start the foal on dry feed as soon as possible.

Drying Up the Mare

When drying up the mare, an oil preparation, such as camphorated oil or a mixture of lard and spirits of camphor, may be rubbed on the udder. The mare should be put on grass hay or low-quality pasture during this time. At first, the udder fills up and becomes tight. Do not milk out the udder at this time. After about five to seven days the udder becomes soft and flabby. At this time, milk out whatever small secretion remains in the udder.

Grooming

Grooming improves the appearance of the horse. It cleans the hair and skin and reduces the chances for skin diseases and parasites. The muscle tone of the horse is also improved.

Equipment commonly used for grooming, as shown in Figure 32-3, includes:

- stiff-bristled cleaning brush.
- smooth-fibered body brush.
- currycomb.
- grooming cloth.
- mane and tail comb.
- clippers or scissors.
- hoof pick.

Follow a pattern when grooming the horse. The *near side* is the horse's left side; the *off side* is the horse's right side. Begin on the near side at the head and work toward the rear. Then follow the same procedure on the off side. Follow good safety procedures when working around the horse.

Use the currycomb on areas caked with dirt and sweat. Do not use the currycomb around the head or below the knees and hocks. A wet sponge or soft brush is used to remove dirt in those areas. Use the currycomb in a circular motion. Finish cleaning the horse with a brush. Always brush in the same direction in which the hair lies (Figure 32-4). Brush in short brisk strokes. The dirt is thrown out of the hair by flicking the brush up at the end of each stroke.

A grooming cloth is used to remove any dust remaining on the horse. Soft linen, terrycloth, or jersey make good grooming cloths.

The mane and tail comb is used to remove dirt and other foreign matter from the mane and tail. The mane is lifted to comb the underside. The roached mane is currently a popular style depending on the breed. A *roached* mane is trimmed down to the crest of the neck, leaving the foretop and a wisp of mane at the withers. Other styles of clipping the mane are used, depending on the type and breed of horse being shown. Remove tangles in the mane and tail using the fingers and a coarse-toothed comb. Trying to pull the tangles out makes them tighter.

The hair may be trimmed from some parts of the body to improve grooming. The mane hair just behind the ears is often trimmed close so that the bridle and halter will fit better (Figure 32-5). Hairs outside the ears may be trimmed to help keep the ears cleaner. Occasionally,

FIGURE 32-3 Horse grooming equipment: (left to right) bot block, shedding blade, sweat scraper, wire hoof brush, hoof pick, currycomb, mane and tail comb, stiff brush, soft brush, ear clippers, mane clippers. *Photo by Jim Bodine.*

the feather is trimmed from the fetlock and pastern. If a shorter tail is desired, hairs of the tail should be pulled, rather than trimmed. Pull mainly from the underside of the tail. The tail of the three-gaited Saddle horse is trimmed closely for a distance of 6 to 8 inches (15.2–20.3 cm) from the base.

CARE OF TEETH

Problems with a horse's teeth can sometimes cause the horse to go off feed even though a good-quality ration is being fed. The horse will often develop an unthrifty appearance. The teeth often wear unevenly and develop sharp edges because the upper molars and premolars normally overlap the bottom teeth. These sharp edges may cut into the horse's cheek while chewing, causing pain. Indications of this problem include weight loss, slobbering, dribbling of feed while eating, and a reluctance to eat. The teeth should be filed with a float. A *float* is a long-handled rasp with a guard to prevent injury to the horse during treatment. Milk teeth sometimes remain in too long and need to be removed. If they are not removed, crooked permanent teeth may result. Regular examination and care of the horse's teeth is recommended.

FOOT CARE

Each foot must be picked up for proper cleaning and inspection. The feet should be inspected every day. Follow a pattern when working with the feet. Start with the near forefoot, then the near hindfoot, next the off forefoot, and finally the off hindfoot. Figure 32-6 shows the proper way to pick up a horse's foot.

Clean the foot from the heel to the toe using the hoof pick. Be sure the depression between the frog and the bars is thoroughly cleaned. This helps to prevent thrush and other infections of the foot. Check for stones, nails, injuries, or loose shoes when cleaning the foot.

A *farrier* is a person who works on horses' feet. The tools used by a farrier are shown in Figures 32-7 and 32-8. Figures 32-9, 32-10, and 32-11 show a farrier preparing a horse's hoof for shoeing.

The hoof should be trimmed every four to six weeks. If the horse is shod, the shoes should be replaced every four to six weeks. Failure to provide proper care for the feet and hooves may result in lameness.

Horses are shod to protect the hooves from wearing too much on hard surfaces. Shoes also help to correct defects in stance and gait. They provide better traction on ice and in mud. Poor hoof structure and growth may be corrected with shoes. Shoes also help to protect the hoof from cracks, corns, and contractions.

Always fit the shoe to the foot. If shoes are left on too long without refitting, the hooves will grow out of proportion. This throws the horse off balance.

FIGURE 32-4 Always brush in the same direction in which the hair lays. *Photo by Michael Dzaman.*

FIGURE 32-5 The mane hair just behind the ears is often trimmed close so the bridle and halter will fit better. *Photo by Michael Dzaman.*

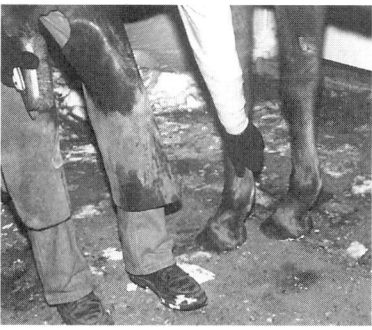 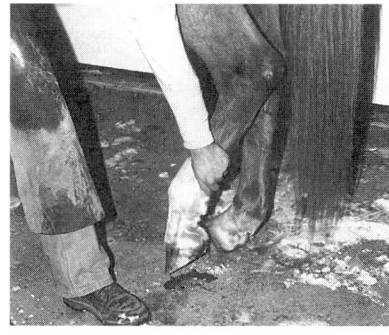

(A) Near forefoot: Slide your left hand down the cannon to the fetlock. Lean with your left shoulder against the horse's shoulder. When the horse shifts weight and relaxes on the foot, pick it up. Reverse for picking up the off forefoot.

(B) Near hindfoot: Grasp the back of the cannon just above the fetlock and lift the foot forward. Reverse sides for picking up the off leg.

FIGURE 32-6   The proper way to pick up a horse's feet. *Photo by Steven M. Ennis.*

FIGURE 32-7   Tools used by a farrier: (left to right) shoeing apron, hoof gauge, hoof knife, hoof nipper, pulloffs, nail clincher, hoop rasp, nailing hammer, clinch bar, turning hammer, anvil. *Photo by Jim Bodine.*

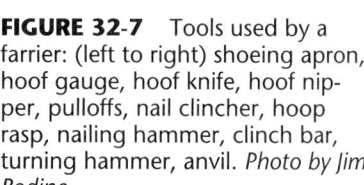

FIGURE 32-8   A forge used by a farrier. *Photo by Jim Bodine.*

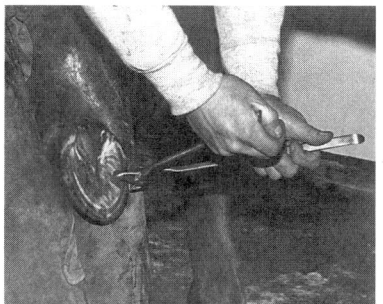

FIGURE 32-9   Use of nippers to cut the horny wall (outer surface) to a proper angle and length to fit the conformation of the animal. *Courtesy of Frank Morton, Farrier. Photo by Steven M. Ennis.*

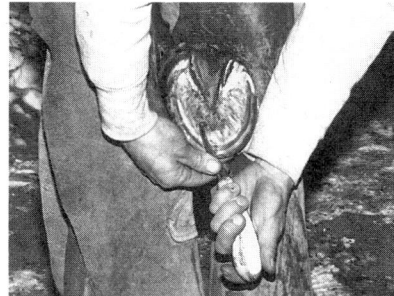

FIGURE 32-10   Use of the hoof knife in conjunction with the nippers to pare the dead and flaky tissue from the sole. *Courtesy of Frank Morton, Farrier. Photo by Steven M. Ennis.*

BUILDINGS FOR HORSES

Horses do not require elaborate or expensive barns. They do need a shelter that will protect them from the cold, storms, sun, and wind. A horse barn or shelter should be located in an area with good drainage. A structure with good ventilation, one without drafts, is important. A space should be provided for storage of feed, bedding, and tack. Figure 32-12 shows a small horse barn that has these features. Plans for horse barns are available from many state cooperative extension services. Figure 32-13 shows plans for a four-stall horse barn. A nine-stall

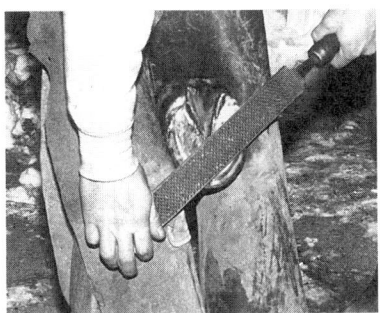

FIGURE 32-11   The rasp is used to prepare a ground-bearing surface by eliminating jagged and sharp corners on the bottom of the hoof wall. *Courtesy of Frank Morton, Farrier. Photo by Steven M. Ennis.*

FIGURE 32-12   A small horse barn with space for feed, bedding, and tack storage. *Photo by Jim Bodine.*

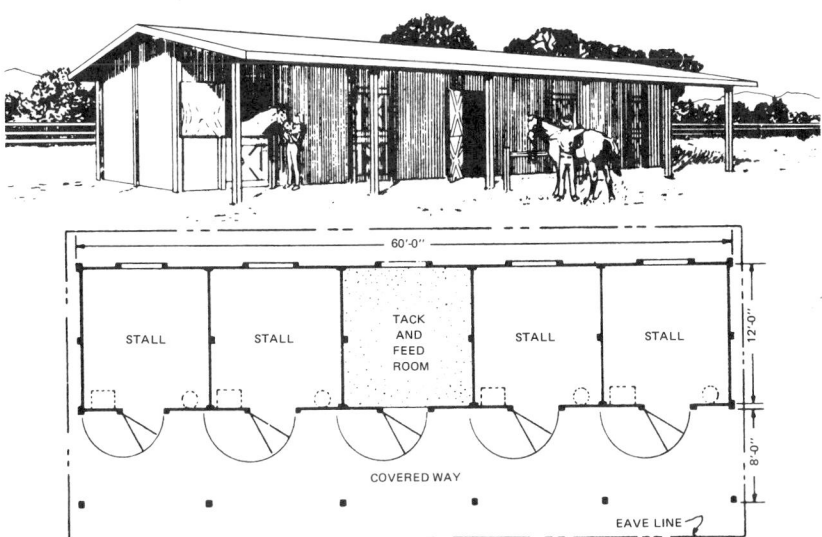

FIGURE 32-13   A four-stall horse barn. The box stalls are 12′ × 12′ (3.7 × 3.7 m). A combination tack and feed room is included. There is an 8′ (2.4 m) covered way on the front. *Courtesy of Clemson University.*

COMFORTABLE QUARTERS FOR FOUR HORSES ARE PROVIDED BY THIS 20′ x 60′ BARN. FOUR 12′ x 12′ BOX STALLS, AS WELL AS A COMBINATION TACK AND FEED ROOM, OPEN TO AN 8′ COVERED WAY.

horse barn is shown in Figure 32-14. A barn for a commercial boarding and riding stable is shown in Figure 32-15.

Small horses such as ponies need box stalls that are 10 × 10 to 10 × 12 feet (3 × 3 to 3 × 3.7 meters). Tie stalls for small horses should be 3 × 6 feet (0.9 × 1.8 meters). Medium-size horses need box stalls that are 10 × 12 to 12 × 12 feet (3 × 3.7 to 3.7 × 3.7 meters). Tie stalls for medium-size horses should be 5 × 9 feet (1.5 × 2.7 meters). Large horses need box stalls that are 12 × 12 to 16 × 16 feet (3.7 × 3.7 to 4.9 × 4.9 meters). Tie stalls for large horses should be 5 × 12 feet (1.5 × 3.7 meters). Mares at foaling time need box stalls that are 16 × 16 to 16 × 20 feet (4.9 × 4.9 to 4.9 × 6.1 meters).

Floors made of clay or wood are better for horses than those made of concrete. They are not as hard on the horses' feet. Straw or wood shavings make good bedding materials for horses. When planning the location of the horse barn, make provisions for cleaning the manure from the barn.

FENCES AND CORRALS

Fences built of wood or poles are the best choice for horses. Woven wire may be used, although the mesh must be small enough to prevent the horse from catching its hoof in it. Barbed wire is not a good material for horse fences. It can cause injury to the horses. A smooth wire or a board may be fastened along the top of a wire fence to help prevent injuries.

CUTAWAY VIEW OF HORSE BARN

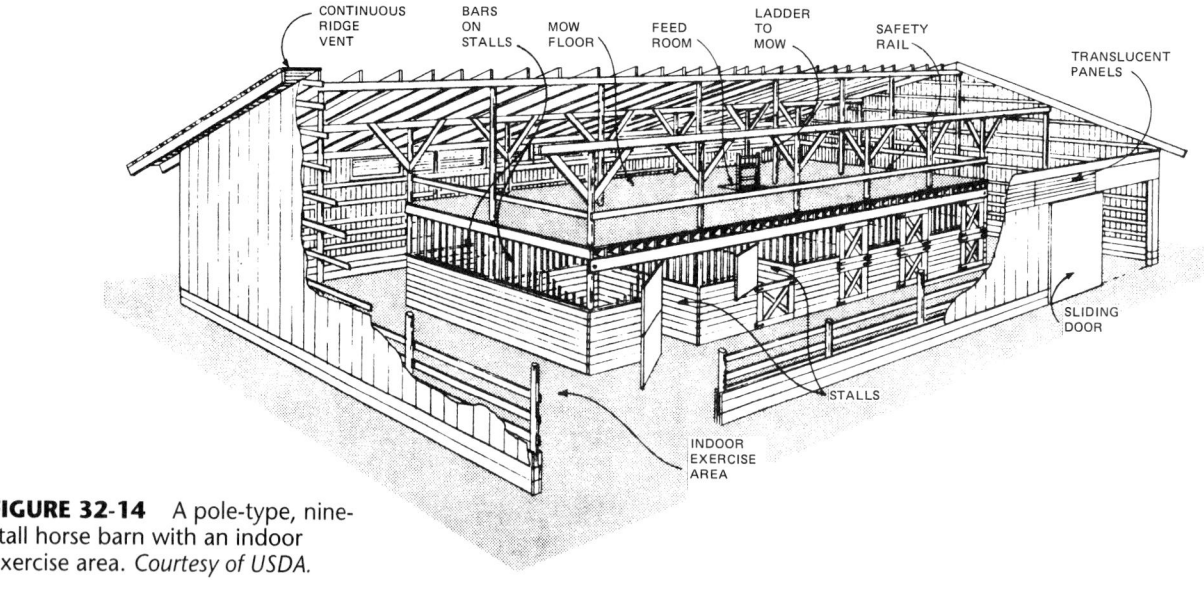

FIGURE 32-14 A pole-type, nine-stall horse barn with an indoor exercise area. *Courtesy of USDA.*

FIGURE 32-15   A barn for a commercial boarding and riding stable. *Photo by Jim Bodine.*

It is convenient to have a corral or paddock near the horse barn. This provides a space for the horses to exercise. Heavy lumber is the best fencing material for a corral or paddock. Place the boards or fencing on the inside of the corral rather than on the outside. If chewing becomes a problem, cover the top board with 1/4-inch chicken mesh wire.

EQUIPMENT FOR FEEDING AND WATERING

A hay rack or manger reduces waste when feeding horses. A grain feeder may be attached to the wall of a stall or placed on a wooden shelf. It should be easily removable for cleaning. Provision should be made for supplying fresh, clean water. A water tank or trough may be used. Measures, such as covering the trough, must be taken to prevent the water from freezing in cold climates. Wooden salt and mineral feeders are also needed. Two separate containers, or one container with two compartments to keep the salt and mineral separate, may be used.

HORSE TRAILERS

Two important considerations in the selection of horse trailers are the safety and protection of the horse. A trailer that is equipped with brakes is safer than one without brakes. Other standard safety equipment includes brake light, taillight, and safety chain on the hitch. These items are required by law in many states.

A padded chest bar helps to prevent injury to the horse. An uneven floor surface gives better footing for the horse. Protect the horse from drafts in the trailer. A small door in the front of the trailer allows a person to leave the trailer after leading a horse inside. Figure 32-16 shows a tandem trailer for transporting horses.

FIGURE 32-16   Loading a horse into a tandem trailer. *Photo by Jim Bodine.*

HORSE TACK

Tack is the equipment used for riding and showing horses. Tack made of good-quality materials and which fits the horse properly should be selected. A good fit is the most important factor to consider in selection of tack. Basic tack consists of the saddle, saddle pad or blanket, bridle, halter, and lead rope. Other kinds of tack required depend on the kind of riding being done.

Tack is expensive and should be given proper care. Leather should be kept clean and oiled. Use saddle soap to clean leather. Several kinds of oils are available for use on leather. Select one that does not rub off on clothing. Use only enough oil to keep the leather soft and pliable. Repair or replace any parts of the tack that become worn or broken.

Saddles

There are many types and styles of saddles. The two most common types are the Western saddle and the English saddle.

Western saddles are large and heavy, but they are more durable than other types of saddles. Western saddles give a comfortable ride. They are designed to be used for work with cattle on western ranges and are used for Western-style riding. A saddle pad or blanket is usually used with the Western saddle. Parts of the Western saddle are shown in Figure 32-17.

English saddles are lighter than Western saddles. They have a flat seat (Figure 32-18). There are many styles, depending on the type of riding

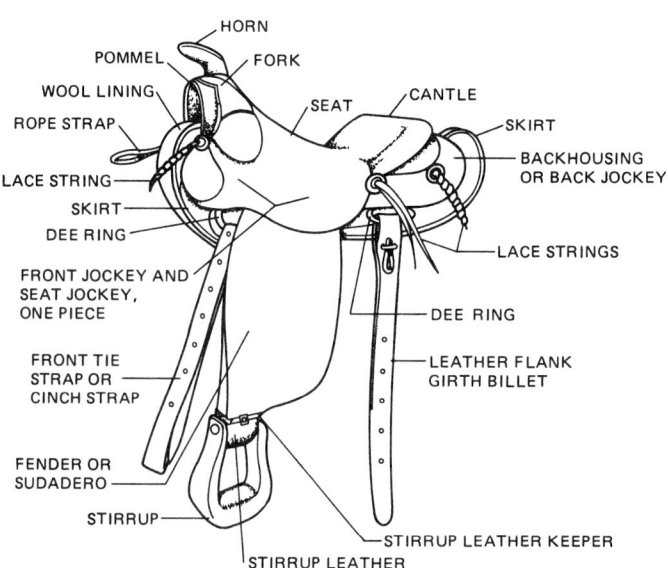

FIGURE 32-17 The parts of the Western saddle. *Courtesy of University of Illinois at Urbana-Champaign.*

being done. They are used for pleasure riding, training, racing, jumping, and polo. Usually, an English saddle is used without a blanket.

Selecting a saddle. The type of saddle selected depends on the style of riding to be done. The saddle must fit both the horse and the rider. It is especially important that the saddle fit the horse at the withers. A poorly fitted saddle will often result in injury to the horse. Stores that sell tack have a wide variety of saddles available (Figure 32-19).

Bridles and Bits

There are many styles of bridles and bits. The style selected depends on the type of riding to be done. The bit must be matched to the style of bridle.

The main purpose of the *bridle* is to hold the bit in the horse's mouth. Common types of bridles are shown in Figure 32-20. The *Weymouth* (A) is a double bridle. A *double bridle* has two bits and two sets of reins. The double bridle is usually used when showing three-and five-gaited horses. The *Pelham bridle* (B) is used for pleasure riding, polo, and hunting. The split-eared bridle (C) may be used on working stock horses. The hackamore (D) is similar to a bridle except it has no bit. Hackamores are often used when breaking young horses because they will not injure the horse's mouth.

The purpose of the *bit* is to help control the horse. Care must be taken not to injure the horse's mouth by improper use of the bit. The bit must be the right length to fit the horse's mouth. The bridle should be adjusted in such a way that the bit just raises the corners of the mouth.

Common English riding bits include the *Weymouth curb bit*, *Pelham curb bit*, *Walking horse bit*, *snaffle bit*, and *Dee race bit*. The curb bit is used in Western show classes. The snaffle bit is the most commonly used bit.

Common Western riding bits include the *hackamore bit*, *roper curbed-cheek bit*, and *spade mouth bit*. These bits are used for working stock and roping horses.

Common driving bits include the *Liverpool bit*, *bar bit*, and *half-cheek snaffle bit*. These bits are used with harness horses.

Halters

Halters are used for tying or leading horses. They are made of rope or leather. When putting a horse on pasture, always remove the halter to prevent it from becoming caught on some object.

Martingales

A *martingale* is a device that prevents the horse from lifting its head too high. It is used on horses that tend to rear. The two types of martingales

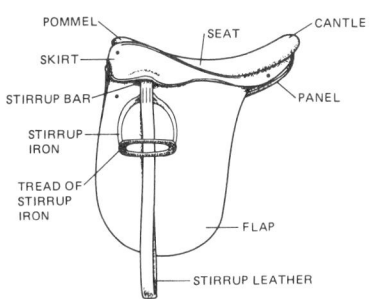
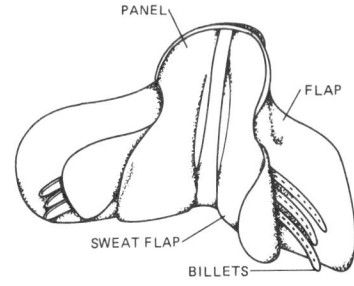

FIGURE 32-18   Parts of the English saddle. *Courtesy of University of Illinois at Urbana-Champaign.*

FIGURE 32-19   A wide variety of saddles are available. *Photos by Michael Dzaman.*

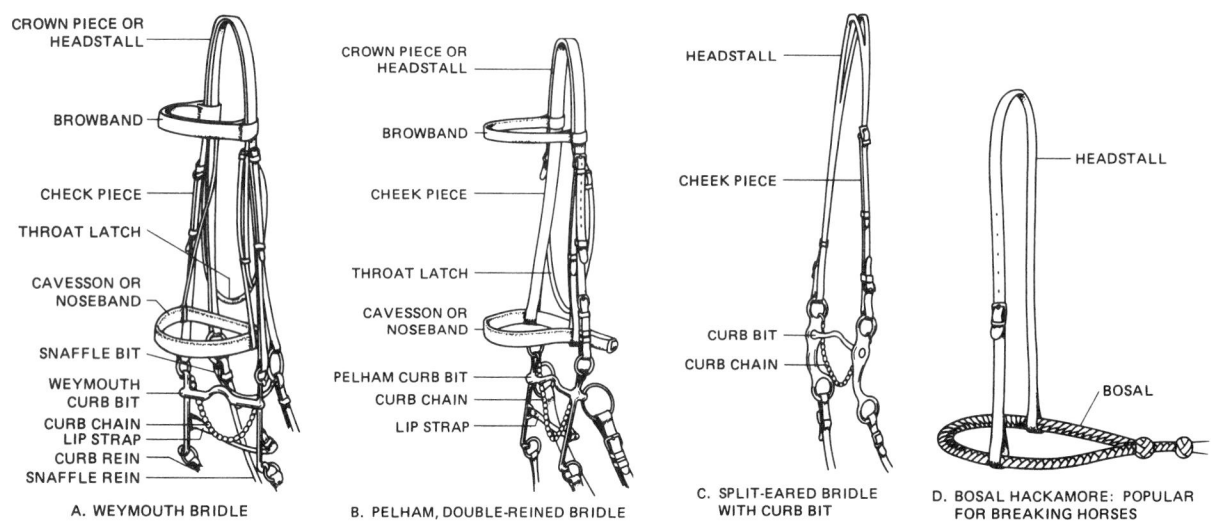

CROWN PIECE OR HEADSTALL
BROWBAND
CHECK PIECE
THROAT LATCH
CAVESSON OR NOSEBAND
SNAFFLE BIT
WEYMOUTH CURB BIT
CURB CHAIN LIP STRAP
CURB REIN
SNAFFLE REIN

A. WEYMOUTH BRIDLE

CROWN PIECE OR HEADSTALL
BROWBAND
CHEEK PIECE
THROAT LATCH
CAVESSON OR NOSEBAND
PELHAM CURB BIT
CURB CHAIN
LIP STRAP

B. PELHAM, DOUBLE-REINED BRIDLE

HEADSTALL
CHEEK PIECE
CURB BIT
CURB CHAIN

C. SPLIT-EARED BRIDLE WITH CURB BIT

HEADSTALL
BOSAL

D. BOSAL HACKAMORE: POPULAR FOR BREAKING HORSES

FIGURE 32-20 A, B, C, D Common types of bridles. *Courtesy of University of Illinois at Urbana-Champaign.*

are the standing and the running (Figure 32-21). The *standing martingale* is attached to the horse's head. The *running martingale* is attached to the reins by two rings.

Riding Clothes

Many different types and styles of riding clothes are designed for various kinds of riding. A good riding boot should be used when pleasure riding. Tennis shoes are not safe for riding or working around horses. They provide neither security in the stirrup nor protection from being stepped on by the horse. If a riding boot is not worn, a good leather shoe with a reinforced toe is the next best choice.

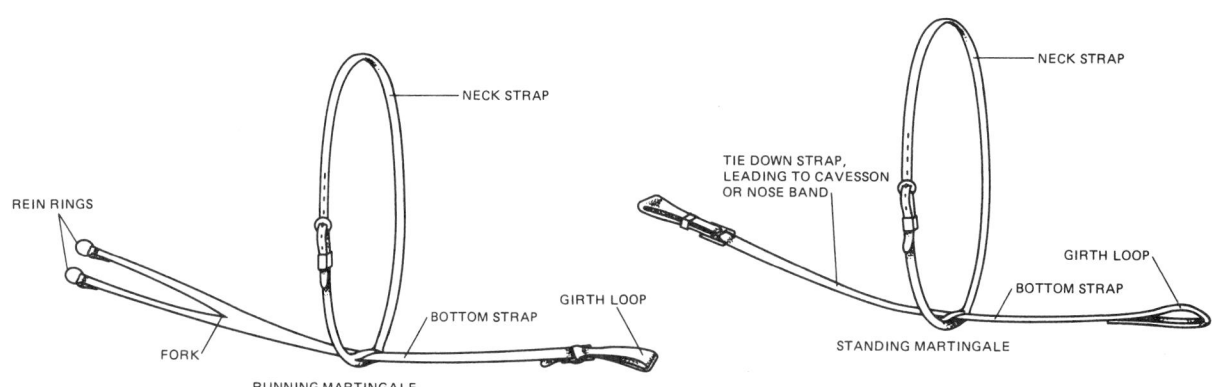

NECK STRAP
REIN RINGS
FORK
BOTTOM STRAP
GIRTH LOOP
RUNNING MARTINGALE

NECK STRAP
TIE DOWN STRAP, LEADING TO CAVESSON OR NOSE BAND
GIRTH LOOP
BOTTOM STRAP
STANDING MARTINGALE

FIGURE 32-21 Two types of martingales. *Courtesy of Appaloosa Horse Club, Inc., ID.*

Harness

A *harness* is used when driving a horse. Figures 32-22 and 32-23 show the parts of the harnesses used for trotting and pacing horses.

Knots

Many different types of knots are used by horseowners. The most common types are shown in Figure 32-24.

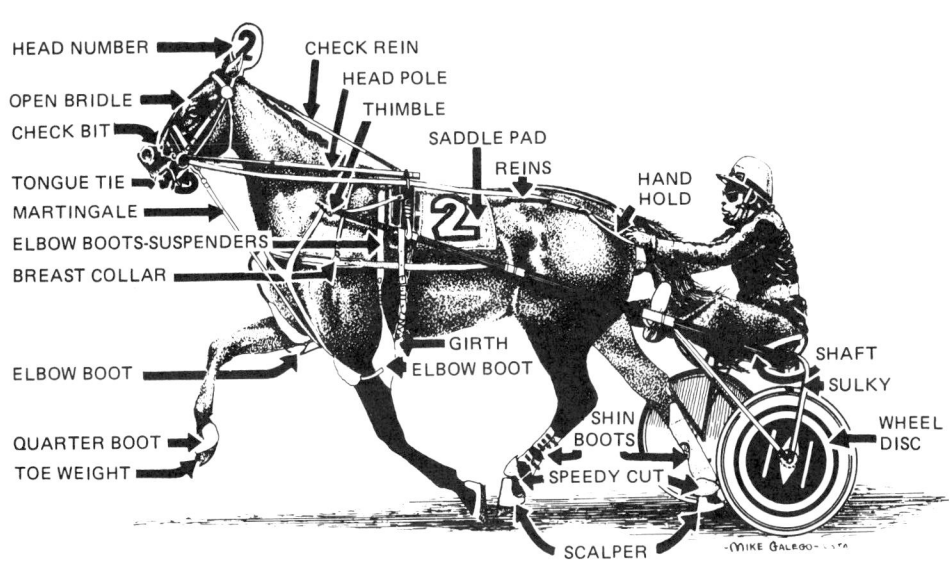

FIGURE 32-22 Equipment commonly worn by the Trotter. *Courtesy of United States Trotting Association, OH.*

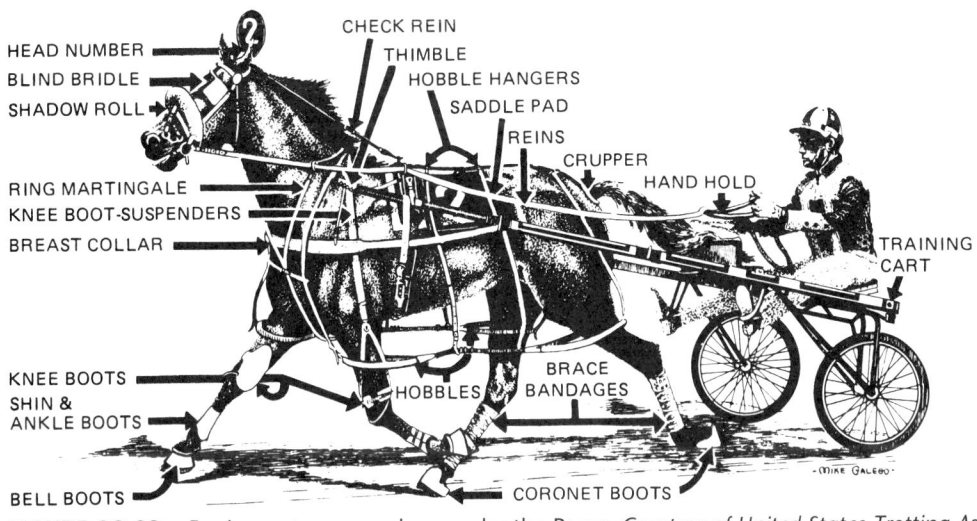

FIGURE 32-23 Equipment commonly worn by the Pacer. *Courtesy of United States Trotting Association, OH.*

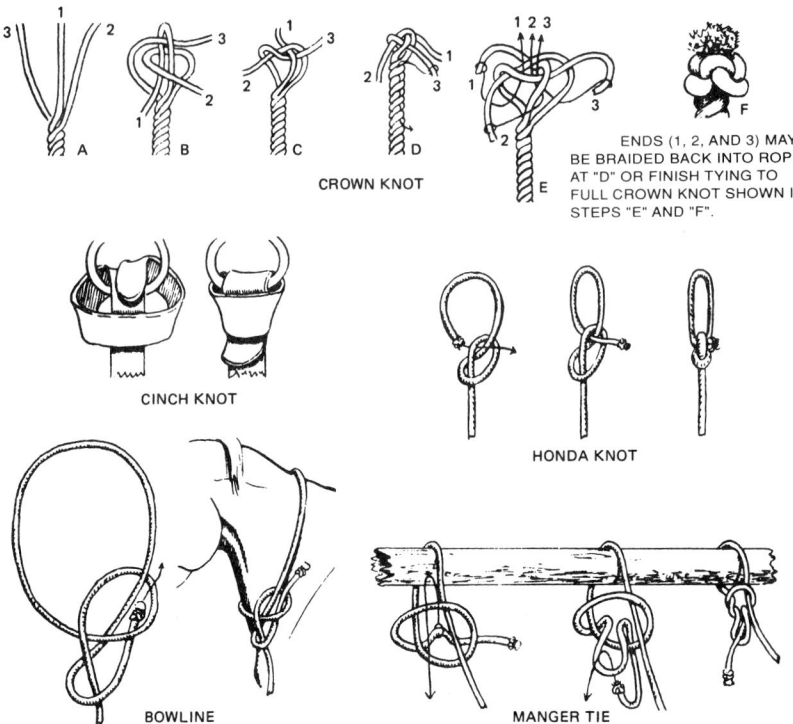

ENDS (1, 2, AND 3) MAY
BE BRAIDED BACK INTO ROPE
AT "D" OR FINISH TYING TO
FULL CROWN KNOT SHOWN IN
STEPS "E" AND "F".

CROWN KNOT

CINCH KNOT

HONDA KNOT

BOWLINE

MANGER TIE

FIGURE 32-24 Knots commonly used by horseowners. *Courtesy of Appaloosa Horse Club, Inc., ID.*

SUMMARY

Mature idle horses can be fed on a ration composed solely of roughage. Horses that are working, pregnant mares, and growing foals need some concentrate in the ration. Grass and legume pastures can provide much of the roughage required by horses. Legume and grass hays are also good sources of roughage for horses. Silage can be used for part of the roughage in the diet.

Oats are the preferred grain for horses. Other grains such as corn, milo, barley, and wheat can also be used in the ration. Usually, little protein supplement is needed in the ration. Soybean meal is an excellent supplement to use, if one is needed. Commercial protein supplements and complete pelleted rations may be used for feeding horses.

Horses require salt, calcium, and phosphorus in the ration. Additional vitamin supplements are seldom necessary. A clean, fresh supply of water should be available for horses.

Horses are fed according to their size, stage of growth, condition, and amount of work they perform. Regular feeding and watering is important.

A mare is more likely to conceive if bred in the spring. Most horseowners prefer to breed mares at three years of age to foal when they are four years old. Pregnant mares need exercise and the proper rations.

A clean pasture is a good place for foaling. An attendant should be present during foaling in case the mare has difficulty giving birth. Treat the navel cord with tincture of iodine and make sure the foal receives the colostrum milk.

The foal is started on dry feeds at 10 days to three weeks of age. Early training of the foal should begin at 10 to 14 days of age. The foal must become accustomed to being handled. Wean the foal at four to six months of age. Colts not intended for breeding are usually castrated when about one year old. This permits better development of the foreparts.

A horse should be groomed regularly. Use a brush to remove dirt from the hair coat and clean the feet carefully. The feet should be trimmed and/or shod every four to six weeks.

Horses do not need expensive shelters. Provide protection from the cold, storms, wind, and sun. Floors made of clay or wood are better for horses than those made of concrete. Fences may be wood or wire. Do not use barbed wire. Watering and feeding equipment should be designed for easy cleaning. Horse trailers should be selected for the safety and protection of the horse.

Tack is expensive and should be kept clean and in good repair. The Western and English saddles are the two most common types in use. Bridles and bits are selected on the basis of the use of the horse. Many different kinds of clothes are used for riding. Select the type based on the kind of riding to be done.

Student Learning Activities

1. Visit horseowners in the area and inquire about the feeding and management practices that are followed.
2. Give an oral report on feeding and/or management practices.
3. Visit local horseowners to observe the housing and equipment used.
4. Prepare a bulletin board display of pictures of tack used with horses.
5. Give an oral report on housing and equipment for horses.
6. Identify the parts of saddles, bridles, and bits.
7. Practice tying knots used by horseowners.

Review

1. What determines how much pasture is needed per horse?
2. How should pastures be maintained to improve their value for horses?

3. List and briefly describe some common pastures used for horses.

4. Describe the place of silage in the feeding of horses.

5. Compare legume and grass hays as feeds for horses.

6. List and briefly explain the grains used for feeding horses.

7. Briefly explain the use of protein feeds for horses.

8. Name and briefly describe some protein feeds used for feeding horses.

9. Briefly explain some advantages and disadvantages of using pelleted feeds for horses.

10. What minerals should be included in horse rations?

11. Briefly explain the use of vitamin supplements in horse rations.

12. How important is water in horse feeding?

13. List some accepted practices to follow when feeding horses.

14. List the management practices to follow when breeding the mare.

15. What practices should be followed in caring for the pregnant mare?

16. What are the signs that indicate the mare is about to foal?

17. What help should be given the mare and foal at foaling time?

18. List a ration that might be used for creep feeding a suckling foal.

19. Describe some possible steps to follow when beginning to train the foal.

20. Describe the practices to follow when weaning the foal.

21. At what age and in which season should a colt be castrated?

22. Describe the practices that help to save orphan foals.

23. List the equipment commonly used for grooming a horse.

24. Describe the procedure to follow when grooming a horse.

25. Describe the care of a horse's feet.

26. Describe the kind of building used for horses.

27. What type of fencing is needed for horses?

28. What equipment is required for feeding and watering horses?

29. Describe some characteristics of a good horse trailer.

30. Define tack and describe how to care for it.

31. Name and describe the two most common kinds of saddles used for riding horses.

32. Name and describe the common kinds of bridles.

33. What are the uses of the halter?

34. What is the use of the martingale?

35. Describe some typical kinds of riding clothes.

36. Name the common kinds of knots used by horseowners.

Diseases and Parasites of Horses

Objectives

After studying this unit, the student should be able to

- identify common diseases and parasites of horses.
- describe prevention measures for diseases and parasites of horses.

PRINCIPLES OF HEALTH CARE

Diseases and parasites cost horseowners hundreds of thousands of dollars every year. A sick horse loses its usefulness to the horseowner. However, inexperienced horseowners may not always recognize horse health problems when they occur. Therefore, some general principles of preventive health care can be followed to help reduce losses from diseases and parasites.

Proper feeding and management help to prevent losses from diseases and parasites. A horse that is properly fed has a higher resistance to diseases and parasites. Dusty or moldy feed should never be used. The mycotoxin *fumonisin* is sometimes found in moldy corn. It is toxic to animals and may cause brain or liver disorders in horses. The amount of water a horse drinks after it has been worked hard should be carefully controlled.

Cleanliness and sanitation also help to prevent horse health problems. Disease organisms and parasites grow in organic waste material. Stalls, barns, and exercise areas must be kept clean, thereby removing or reducing many sources of infection. The use of clean pastures helps to break the life cycle of many parasites that affect horses (Figure 33-1). Do not keep horses in barns that are too warm and humid.

Planning and practicing an immunization and parasite control program prevents many problems before they occur. Planning should be done with the advice of a veterinarian who is familiar with local disease and parasite problems. Horses can be vaccinated against some

FIGURE 33-1 Using clean pastures helps to break the life cycle of many of the parasites that affect horses. *Courtesy of USDA.*

diseases. However, to be effective, vaccination must be done before the horse becomes infected. Periodic examination of the feces by a veterinarian will detect the presence of parasites. Successful treatment depends upon the use of the right drug at the right time. The correct diagnosis of the problem is important to determine the proper treatment to be used. A veterinarian is best qualified to make such a diagnosis. The services of a veterinarian can save the horseowner money that might otherwise be spent on remedies that will not cure the problem.

Proper exercise and good grooming help to keep horses in good health. Horses must have regular exercise to maintain good muscle tone and to prevent stiffness. Good grooming keeps the horse clean and provides an opportunity for the owner to observe any problems that may be developing. Early treatment of horse health problems results in less loss of time and money.

When a disease does occur, isolate the affected horse to prevent the possible spread of the disease. Water and feed containers for the sick horse should be kept separate from other horses. Call a veterinarian to treat the horse.

Watering troughs that are provided for horses at fairs and shows should not be used. Draw water from the tap into a pail rather than dipping it out of a trough.

Do not allow the horse to come in contact with horses that are known to be sick. Access to the area in which the horse is kept should be controlled to prevent the possible spread of diseases by infected horses. Stress lowers the horse's resistance to disease. Therefore, care should be taken to avoid chilling and other causes of stress in horses.

Observing the vital signs (temperature, pulse rate, and respiration rate) in an animal can help in the early detection of health problems. Vital signs will vary with activity and environmental conditions. Normal vital signs in horses are:

- Temperature (normal range): 99–100.8°F (37.2–38.2°C) [average is 100.5°F (38.1°C)] (Temperature is usually higher in the morning than in the afternoon; younger animals will show a wider range of temperature than mature animals.)
- Pulse rate (normal range): 32–44 heartbeats per minute.
- Respiration rate (normal range): 8–16 breaths per minute.

Body temperature is taken in the rectum using either a mercury thermometer or a battery-powered digital thermometer. Restrain the animal when inserting the thermometer into the rectum. The pulse rate is taken by finding the artery on the lower edge of the jaw. It may also be taken by finding the artery along the inside of the forearm where it travels down the bone. Respiration rate is determined by observing the number of times the animal breathes per minute.

DISEASES AND DISORDERS

Anhydrosis

Anhydrosis is a condition in which horses do not sweat normally. The ability of a horse to sweat is important because that is the way body temperature is controlled. Some horses have sweat glands that do not function at all and others function at a less than normal level. Why sweat glands do not function properly is not known. Anhydrosis should be suspected when a horse does not sweat normally during exercise.

Management practices such as riding or working the horse only when it is cool and keeping the horse out of the sun may be necessary. Using fans or air conditioning in the barn will help keep the horse cool. Using a higher fat diet helps because less heat is generated in the digestion of fat as compared to other nutrients. Consultation with a veterinarian may provide some other possible solutions for a specific set of circumstances. One treatment that has shown success is using a thyroid medication. The proper dosage must be used because an excessive amount of thyroid medication may cause other health problems.

ANTHRAX

Anthrax is caused by bacteria. The spores formed by the bacteria can live for many years in the soil. Symptoms include high fever, blood in the feces, rapid breathing, swellings on the body (especially around the neck) and, in the later stages, depression. The horse may bleed from all body openings. The death rate is usually high.

The disease can be prevented by vaccination. Since anthrax occurs only in certain areas of the United States, a local veterinarian should be consulted for information on vaccination.

Anthrax can be transmitted to people. Never open the carcass of a horse that has died of anthrax. Use care in disposing of the carcass. Burn or bury the carcass and cover it with quicklime.

To combat anthrax, isolate sick horses and vaccinate healthy ones. Change pastures, quarantine the area, and practice strict sanitation measures.

Azoturia (Monday-Morning Sickness)

Azoturia may be a nutritional disorder. It develops when a horse is put to work following a period of idleness. The horse becomes stiff, sweats, and has dark-colored urine. The muscles become swollen, tense, and paralyzed.

Azoturia can be prevented by decreasing the amount of grain fed when the horse is idle. The horse should exercise during idle periods and be started back to work slowly.

When symptoms appear, stop the horse from working and do not allow it to move. Use blankets to keep the horse warm and dry. Call a veterinarian for treatment.

Bruises and Swellings

Horses have many chances of becoming bruised. Bleeding may occur under the skin. Apply cold compresses until the bleeding and swelling stop. Then apply heat and liniments to the affected area.

Colic

Colic is not a specific disease, but rather a disease complex encompassing a wide range of conditions that affect the horse's digestive tract. It is usually caused by some type of obstruction that blocks the flow of feed through the intestine, resulting in abdominal pain. The pain occurs when the intestine is distended by an accumulation of gas, fluid, or feed. When it occurs, colic must be treated immediately.

While there are a number of things that can cause colic in horses, the major cause (about 90 percent of the time) is the presence of parasites, usually large strongyles (bloodworms). Colic may be caused by nutritional factors, such as a ration that is too high in energy, sudden changes in feed, feeding a high-fiber ration, or feeding a poor-quality feed. Teeth problems or other mouth disorders that result in improper chewing of the feed may cause colic. If the horse is allowed to drink excessive quantities of cold water before being cooled out after heavy exercise, colic may result. Diseases that cause high fever and reduced intake of feed and water may also cause colic. Feeding an excessive amount of grain can result in colic. Twisting of the intestine is another cause of colic.

Symptoms of colic include severe abdominal pain, uneasiness or restlessness, looking at the flank region, getting up and down, kicking at the belly, sweating, and shifting weight. As the problem continues, the horse may lie down and roll, have an increased pulse and respiration rate, have congested mucous membranes (gums), strain, sweat, and bloat.

It is better to follow good management practices to prevent colic than to have to treat the condition after it occurs. A major preventative measure is to follow a good deworming program to control worms, especially the large strongyles. Check horses' teeth at least twice a year to eliminate chewing problems and other mouth disorders. Feed high-quality rations on a definite schedule. Feeding small amounts of the ration two or three times a day is better than feeding a large amount less frequently. Make changes in the ration slowly, over a seven- to 10-day period. Feeding hay before grain can help prevent colic. Make sure there is a good supply of fresh, clean water and do not allow a horse to drink too much cold water before cooling out immediately after heavy exercise. Make sure there is enough good-quality hay, salt, and minerals for the horses.

If a horse develops colic, call the veterinarian immediately. Colic can be treated satisfactorily if the proper treatment is started quickly. Before the veterinarian arrives, do not allow the horse to eat or lie down and roll; keep horses walking if they want to lie down and roll. If the horse is not trying to lie down and roll, let it stand quietly. The horse may be allowed to drink water. The severe pain may cause the horse to become violent, so the handler must be careful not to get injured. Upon arrival, the veterinarian will proceed with the proper treatment. In some cases, surgery may be necessary.

Distemper (Strangles)

Distemper is caused by a bacterium. It spreads quickly from horse to horse, especially where large numbers are together in one place. Contaminated feed, watering troughs and tack, or direct contact, will spread the disease. Young horses are more likely to get distemper than are older horses.

Symptoms include high fever, loss of appetite, and depression. There is a puslike discharge from the nose. The lymph nodes in the lower jaw and the throat swell.

Isolation of newly arrived animals for two to four weeks and vaccination are methods of prevention. However, because vaccine may cause some problems, it is usually done only on farms with a history of distemper. Consult a veterinarian concerning vaccination.

Antibiotics are used to treat distemper. Abscesses may be drained surgically, if necessary. Give the horse complete rest. Protect it from cold and drafts. Call a veterinarian for diagnosis and treatment.

Encephalomyelitis (Sleeping Sickness)

Encephalomyelitis, a disease that affects the brain, may be caused by any one of several different viruses. The most common forms of the disease are known as Eastern and Western. Outbreaks of Venezuelan encephalomyelitis have occurred in the United States. The viruses are carried by mosquitoes.

The symptoms of the various forms of the disease are similar. High fever, depression, lack of coordination, lack of appetite, drowsiness, drooping ears, and circling are signs of the disease. The horse may die, or it may recover. Death rate may be as high as 90 percent from the Eastern and Venezuelan types. The Western type has a death rate of about 20–30 percent.

Eastern and Western encephalomyelitis are prevented by vaccination. Two injections are given about one to two weeks apart. The vaccination is repeated each year. Venezuelan encephalomyelitis vaccination is given once each year. Controlling mosquitoes helps in preventing the disease.

There are no effective treatments for encephalomyelitis. Call a veterinarian if the disease is suspected.

Equine Abortion

Abortion in mares may be caused by bacteria, viruses, or fungi. Other causes include hormone deficiencies, carrying twins, genetic defects, or other miscellaneous factors. Abortion may occur at various times during pregnancy, depending on the cause.

Virus abortions (rhinopneumonitis and equine arteritis) may be prevented by vaccinations. Bacterial abortions (caused by several different types of bacteria) are best prevented by strict sanitation measures at breeding time. Isolate horses that have aborted. The bedding and aborted fetus should be burned or buried. The area should be disinfected. There are no vaccines for fungi-caused abortions.

Equine Infectious Anemia (Swamp Fever)

Equine infectious anemia is caused by a virus. It is carried from horse to horse by bloodsucking insects. Symptoms include fever, depression, weight loss, weakness, and swelling of the legs. Death often occurs within two to four weeks. Chronic forms cause recurring attacks. Pregnant mares infected with the disease may abort. Horses with the chronic form become carriers of the disease.

Infected horses are destroyed and their carcasses carefully disposed of. Only buy horses that have been tested and found free of the disease. Control all bloodsucking insects. Practice sanitation and sterilize all instruments used on horses after use on each horse. There is no vaccine or treatment for the disease.

Equine Influenza (Flu)

Influenza is caused by viruses. It spreads quickly where large numbers of horses are brought together.

Symptoms include a high temperature, lack of appetite, and a watery nasal discharge. Younger horses are more likely to become infected.

Isolate newly arrived horses and those that have the disease. A vaccine may be used each year to prevent the disease. However, since there are several strains of viruses involved, the horse may be infected by a strain of the disease that was not protected against by the vaccine used.

For treatment, use antibiotics and allow the animal to rest. A veterinarian should be consulted.

Fescue Toxicity (Fescue Foot)

Fescue toxicity is caused by *Acremonium coenophialum,* a fungus that grows inside tall fescue. The fungus produces toxins that inhibit prolactin, a hormone that is essential in the last months of gestation for udder development and colostrum formation. The toxin can also cause lameness, sloughing off the end of the tail, poor weight gain, and an increase in temperature, pulse, and respiration rate.

There is no treatment for fescue toxicity. The animals must be removed from fescue pasture immediately when symptoms are observed. Do not feed hay made from tall fescue to pregnant mares, or use it for bedding. It is recommended that pregnant mares be removed from infected tall fescue pasture at least three months before foaling. A laboratory test is used to determine the presence of the fungus in tall fescue. There are fungus-free varieties of tall fescue available for pasture use.

Founder (Laminitis)

Founder is a nutritional disorder. Common causes are overeating of concentrates, sudden changes in feed, drinking too much water, or standing in a stall for long periods of time. Founder may occur in an acute or chronic form.

Symptoms of the acute form include swelling of the sensitive laminae on one or more feet, lameness, fever, and sweating. Distortion of the hoof is common with the chronic form.

Care in feeding and management help to prevent founder. Use cold applications to treat the acute form. Call a veterinarian for additional treatment. Chronic cases are treated by trimming the hoof and shoeing the horse.

Fractures

Fractures are broken bones. Strain or injury may cause a bone to break. The horse with a fracture usually shows signs of extreme pain when it tries to move. Care in handling horses helps to reduce the chances of fracture. Treatments are seldom practical, except in the case of very valuable animals. Always call a veterinarian for treatment.

Heaves (Broken Wind, Asthma)

Heaves is a nutritional disorder that affects the respiratory system. It often occurs when moldy or dusty hay is fed. It is more common in horses over five years of age.

An affected horse has difficulty in breathing. The air must be forced from the lungs by the abdominal muscles. Other symptoms include a dry cough, nasal discharge, and loss of weight.

Care in selecting feed is the best preventive measure. Never feed moldy or dusty hay. Changing to a pelleted ration may help if the disease has not progressed too far. Putting the horse on pasture may result in some improvement. In advanced cases, there is no effective treatment.

Lameness

Lameness in horses may occur from many different causes. Many of the unsoundnesses of the feet and legs discussed in Unit 31 result in lameness. A veterinarian should be called for the diagnosis and treatment of lameness.

Navel Ill (Joint Ill, Actinobacillosis)

Navel ill is a condition caused by bacteria. It affects newborn foals. The foal refuses to nurse and shows swelling and stiffness in the joints. The animal may have a fever. The foal does not move around. In older foals, there is a loss of appetite and weight loss.

Sanitation and dipping the navel in tincture of iodine at birth help to prevent navel ill. Antibiotics are used in treatment.

Periodic Ophthalmia (Moon Blindness)

The exact cause of moon blindness is not known but it is believed to be some type of infection. The disease affects older horses more often than younger ones. One or both eyes become swollen and the horse keeps its eyes closed. There is a watery discharge from the eye. The cornea may become cloudy. The attack usually clears up in a week to 10 days. The eye may not show much effect or the horse may be blind. Attacks recur at periodic intervals.

Keep the horse at rest in a partially darkened stall. Call a veterinarian for treatment. If left untreated, permanent blindness usually results.

Pneumonia

Pneumonia is caused by bacteria and viruses. Stress, such as chilling, increases the chances of infection. Inhaling dust, smoke, or liquids can also cause pneumonia. It sometimes occurs as a complication of other diseases.

Symptoms include fever, rapid breathing, loss of appetite, and chest pains. Sanitation and prevention of stress will help prevent the disease. There are no vaccines. Prompt treatment by a veterinarian is necessary.

Poisonous Plants

Certain plants found in pastures and hay are poisonous to horses. Symptoms of poisoning vary with the type of plant involved. A few of the plants that are poisonous to horses are bracken, goldenrod, horsetail, locoweed, oleander, ragwort, and tarweed. The best way to control plant poisoning is to remove the plants from pastures and hay fields. Treatment after poisoning occurs is often unsatisfactory.

Rabies (Hydrophobia)

Rabies is a disease caused by a virus. The virus enters the horse's body when it is bitten by an infected dog or wild animal. An affected horse may become quite violent, sometimes attacking other animals or people. The affected animal usually drools saliva. The horse eventually becomes paralyzed and dies.

Rabies can be prevented by vaccinating dogs and controlling wild animals that may carry the disease. Always call a veterinarian if rabies is suspected.

Tetanus (Lockjaw)

Tetanus is caused by bacteria. Bacteria usually enter the animal's body through a puncture wound, but may enter through other types of wounds. The horse becomes nervous and stiff. Muscle spasms and paralysis follow. Death usually occurs in untreated cases. About 30 percent of all horses that are infected recover.

Tetanus is prevented by vaccination. Annual booster shots are given. An unvaccinated horse is given tetanus antitoxin serum if it is injured. Call a veterinarian for treatment of tetanus.

Vesicular Stomatitis

Vesicular stomatitis is caused by a virus. The horse drools saliva and blisters form in the mouth. Provide the infected horse with water and soft feed. There is no vaccination for this disease.

EXTERNAL PARASITES

Common external parasites of horses are (1) flies, (2) lice, (3) mites, (4) ringworm, and (5) ticks. Most of these external parasites also attack beef cattle and other livestock. A discussion of general control measures for external parasites is found in Unit 17. These control measures can be applied to the control of external parasites on horses.

Flies

Biting flies that attack horses include blackflies, deerflies, horn flies, horseflies, mosquitoes, and stable flies. Descriptions of these flies are found in Unit 17.

Horse botflies produce larvae that are parasites of horses. There are three species of botflies. The common botfly and the throat botfly are found throughout the United States. The nose botfly is common in the Midwest and Northwest.

The common botfly is about the size of a honeybee. It has mottled wings and black and yellow hair on the body. The throat botfly is smaller than the common botfly and has no markings on its wings.

The nose botfly is the smallest of the three types. The wings have no markings, and the body hair is dark.

Botflies lay their eggs on the hairs of the horse. The common botfly lays eggs on the horse's front legs. The throat botfly lays eggs on the hairs of the throat and lips. The nose botfly lays eggs on the hairs of the mouth and lips. In all cases, the larvae enter the horse through the mouth and migrate to the stomach. After developing for 10 to 11 months, the larvae pass out of the horse in the feces. The larvae pupate in the soil for about 30 days. Adult flies then emerge and reproduce. The adult flies do not eat. They live from a few days to three weeks. Botfly season in the northern states is from the middle of June until freezing weather. In the southern states, it is from March to December.

Damage by botflies is both direct and indirect. The direct damage is the inflammation caused by the larvae in the stomach. This interferes with digestion. The horse may also be affected with colic from the digestive disturbance caused by the larvae.

The flies cause indirect damage when laying their eggs by frightening and annoying the horses. Horses spend time and effort fighting the flies. They may run to get away from them. Weight loss may result from this excessive activity. Horses that are attacked by the flies while being ridden may run away, possibly endangering the rider. When burrowing in, larvae cause indirect damage by irritating the tongue, gums, and lips. Horses rub against objects to try to relieve the itching and may injure their lips and noses.

Control is accomplished by treating horses with drugs that kill the larvae in the stomach. A veterinarian should be consulted for the current recommended drug treatments. Treatments should be given several times during the fly season and at least once during the winter. In areas where botflies are a problem, community control programs have been effective.

Face flies, houseflies, and screwworm flies attack horses as well as other kinds of animals. Descriptions of these flies are found in Unit 17.

Lice, mites, ringworm, and ticks also attack horses. Descriptions of these external parasites are found in Unit 17. Figure 33-2 shows ticks removed from a horse.

INTERNAL PARASITES

More than 75 different species of internal parasites affect horses. The most important of these are strongyles, ascarids, pinworms, and bots. Bots are the larvae of the botfly and are discussed earlier in this unit in the section on external parasites. Internal parasites are so widespread that all horses are affected by them. Heavy infestations of internal parasites lead to poor physical condition. In extreme cases, infestation

FIGURE 33-2 Winter ticks and eggs taken from a horse. *Courtesy of Iowa State University.*

may cause death. While all ages of horses are affected, young horses are more seriously affected.

Symptoms

Some general indications of internal parasites in horses are:

- weight loss.
- listlessness and poor performance.
- dry, rough hair coat.
- poor appetite.
- bowel problems and colic.
- periodic lameness.
- breathing problems and coughing.
- anemia.
- foals that do not grow well and develop pot bellies.

Diagnosis and Treatment

The only sure diagnosis of internal parasites is by an examination of the horse and the feces by a veterinarian. Worm eggs in the feces will reveal what kind of parasite is affecting the horse. Treatment with the proper drug depends on identification of the parasite.

Treatment for internal parasites is based on the use of drugs. No one drug is effective against all of the different kinds of parasites. Worm medications can be purchased in several different forms that are administered in different ways. Always follow the manufacturer's directions when using any medication.

Life Cycles

The life cycles of strongyles, ascarids, and pinworms are very similar. They follow this general pattern:

1. Eggs are passed out in the feces.
2. Eggs develop to infective stage on vegetation or in litter; or eggs hatch and larvae attach to vegetation.
3. Horse picks up infective eggs or larvae from vegetation or contaminated litter or water.
4. Eggs hatch, larvae migrate through tissues of horse's body.
5. Larvae develop into mature worms and lay eggs.

There are differences in the life cycles of these parasites, especially in the parts of the body to which the larvae migrate.

Large strongyles migrate to the arteries, liver, and gut wall. Adult large strongyles are bloodsuckers. Blood clots may form in the arteries, resulting in complete blockage and death. Large strongyles are considered the most serious of the internal parasites.

Small strongyle larvae migrate to the intestine. There they cause digestive problems. They are not as serious a problem as large strongyles.

Ascarids migrate to the liver and the lungs. Later they are coughed up, reswallowed, and go to the small intestine. They are not bloodsuckers. They are the largest of the worms that affect horses. They may rupture the wall of the small intestine and cause death.

Pinworms travel to the large intestine and do not migrate through other tissues of the body. They cause irritation in the anal region of the horse. The horse rubs the rear quarters to relieve the itching, resulting in the loss of hair from the tail.

Prevention

Sanitation and good management practices are the basis for any parasite prevention program. The eggs of the worms pass out in the feces of the horse. Therefore, proper handling of manure helps in controlling these parasites. Manure should not be spread on horse pastures. In confined areas, pick up the manure at least twice a week. On pastures, use a chain drag to spread horse droppings so the sunlight can destroy eggs and larvae.

Do not overstock pastures. Rotation grazing helps break the life cycle of the worms. Do not graze young horses with older horses. Alternate horses with cattle or sheep on pasture.

Feed hay and grain in mangers or bunks rather than on the ground. Be sure that the water supply is clean.

Keep stalls clean and rebed with clean bedding as often as possible. On earth floors, remove 10 to 12 inches (24.4–30.5 cm) of the surface each year and replace with clean soil.

SUMMARY

The effects of diseases and parasites are very costly for horseowners. Proper feeding and good management help to reduce losses from diseases and parasites. Cleanliness and sanitation are the basis of disease and parasite prevention programs. Proper exercise and grooming help to keep horses in good health.

The most serious diseases of horses are distemper, encephalomyelitis, equine infectious anemia, and equine influenza. Vaccinations are available for some of the diseases that affect horses. A veterinarian should be consulted concerning vaccination programs.

The common external parasites of horses are flies, lice, mites, ringworm, and ticks. The most serious internal parasites are strongyles, ascarids, pinworms, and bots. Insecticides are used to control external parasites. Good management practices designed to break the life cycle of the parasite are used to control internal parasites. A regular worming program using specific drugs should be followed to treat internal parasites.

Student Learning Activities

1. Prepare an oral report on horse health problems.
2. Ask a veterinarian to speak to the class on horse health problems.
3. Survey horseowners in the community concerning common horse health problems and treatments.

Review

1. List the recommended practices for keeping horses in good health.
2. List the diseases of horses, and describe the symptoms of each. Also describe the prevention and treatment of each disease.
3. Name five common external parasites of horses.
4. Describe the life cycle of horse botflies, the damage they cause, and the control measures used for these parasites.
5. Name the four most common internal parasites that affect horses.
6. List the general symptoms of internal parasites of horses.
7. Describe how internal parasites are diagnosed and treated.
8. Describe the life cycles of the internal parasites of horses.
9. List several means of preventing internal parasites in horses.

34 *Training and Horsemanship*

Objectives

After studying this unit, the student should be able to

- explain the basic principles of training a horse.
- describe basic horsemanship procedures.
- list basic procedures for showing a horse.

UNDERSTANDING THE BEHAVIOR OF HORSES

The behavior of horses is based upon survival instincts that have developed over a period of several million years. The ancestors of the horse survived by being alert, by hiding, or by running from danger, and by adapting to changing conditions.

Horses' eyes see independently. Horses can see to the front, side, and rear at the same time. The retina of the eye is arranged so that part of it is closer to the lens than other parts. Horses must raise and lower their heads to bring objects into focus. It is difficult for horses to judge height and distance. Horses' eyes adjust to changing light conditions slower than do human eyes.

It is generally believed that horses are color blind. Horses react quickly to sudden movement. This is why a horse will sometimes shy and throw a rider at a sudden movement of an object along a trail.

Horses have good memories. Therefore, they can be trained and will remember what they learned. Horses remember the rewards and punishments they are given. Reward or punishment must be given immediately following the desired or undesired behavior so that the horse will associate the two.

Horses band together. This is a survival instinct that originated when horses lived only in the wild. When horses ran wild, the horse in the center of the band was safer from attack. The banding instinct also means that horses imitate the behavior of other horses. This has application when riding in a group.

Horses generally have good hearing. They are able to hear tones higher than the human ear can hear. A sound that the rider does not hear may frighten the horse.

Horses have a good sense of smell. A stallion can detect a mare in heat from a great distance if he is downwind of the mare. The colt should be allowed to smell the saddle and blanket the first few times it is saddled. This helps to reassure the colt that these are not dangerous objects.

Horses have several areas on their bodies that are particularly sensitive. These include the mouth, feet, flanks, neck, and shoulders. These sensitive areas are used in training and controlling the horse.

The rider communicates to the horse through the voice, hands, legs, and weight. The horse can learn and remember voice commands. The sensitive mouth responds to the control of the bit. The legs of the rider can direct the horse by applying pressure to sensitive areas of the skin. The horse is able to sense the security or lack of security of the rider. An insecure rider cannot obtain the best performance from a horse.

TRAINING THE YOUNG HORSE

Training a horse requires skill, patience, and careful handling. Handling a foal while it is young makes the job easier. The foal should be handled each day for a short period of time. This helps the foal to overcome its fear.

Haltering

A foal may be halter-broken when it is only a few weeks old. Crowd the foal into a corner of the pen and gently place the halter on its head. If the mare is gentle, it may be used to help crowd the foal into the corner.

Let the foal become used to the feel of the halter by leaving it on for a short period of time. Petting the foal and giving it a small amount of grain will help it to associate the halter with a pleasant experience. Repeat this procedure for a week or two. After the foal has learned to accept the halter, it can be taught to lead.

Leading

To teach the foal to lead, put a loop of rope over the foal's rump. Fasten a lead rope to the halter. Have one end of the rump rope passed through the halter. Pull on the halter rope and the rump rope. This encourages the foal to move forward. The foal may jump forward when the rump rope is pulled. Be alert to avoid being stepped on by the foal.

Working with the foal for about 30 minutes a day for several days will teach it to lead.

Working With the Feet

The foal should be taught to allow its feet to be picked up and handled. This can be done after teaching it to lead. With the foal tied, pick up each foot and put it down. Be gentle with the foal, petting it to keep it calm. Work with the front feet first and then the hind feet. Keep working with the feet until the foal learns to yield its feet without struggling.

Longeing

Longeing is training the horse at the end of a 25- to 30-foot (7.6–9.1 m) line. The horse is worked in a circle. This training may be begun when the horse is a yearling. Some horseowners do not begin this training until the horse is two and one-half years old.

One end of the line is fastened to the halter. The trainer attempts to make the horse move in a circle. Training begins with a small circle. As the horse learns to respond to commands, the circle is enlarged. A light whip may be used to start the horse moving. Never hit the horse hard with the whip. A light touch on the hindquarters is all that should be used.

Teach the horse to circle at a walk. After it has learned this, advance to the trot and slow canter. Work the horse in both directions equally so the horse learns skill in moving both ways in a circle. The faster gaits should be taught only in a large circle.

The horse can be taught to respond to voice commands. Always use the same commands so the horse learns to associate the command with the action.

The longe line can be used to train and exercise the horse (Figure 34-1). It is a good way to exercise a horse at a show.

FIGURE 34-1 Using a longe line to warm up a horse before the show. *Photo by Michael Dzaman.*

Saddling

The horse must be prepared for saddling several days before it is first saddled. This is often accomplished by using a process known as *sacking out*. First, the horse is tied and rubbed with a soft sack or saddle blanket. Rub the sack over the head, neck, back, rump, and legs. The sack is then flipped over and about the body and legs. To further rid the horse of any fear of movement, a soft cotton rope can be placed over the back and pulled back and forth around the body and legs. Walk around the horse slowly twirling a rope. These actions, when repeated for several days, help to prepare the horse for saddling.

When ready to saddle the horse, first let it see and smell the saddle. Slide the saddle blanket on and off the horse several times until it becomes used to it. Before placing the saddle on the horse, the girths and right stirrup are laid over the seat of the saddle. (Refer to Figure 32-17 for the parts of the saddle.) Lift the saddle gently into place. Lower the girth and stirrup on the off side by hand. Do not push them off causing them to hit the side of the horse.

Reach under the horse with the left hand and bring the girth up to the latigo. Slip the latigo through the ring in the girth. Run the latigo up through the Dee ring on the saddle and back through the girth. Draw the girth snug. Walk the horse a few steps and then draw the girth up again. Repeat the operation with the back girth. Do not let the back girth hang too loose or the horse will catch a foot in it. There should be room for a hand between the back girth and the horse. Leave the saddle on for a time to let the horse become used to its feel. The saddling procedure should be repeated for several days before attempting to ride the horse.

Use of Hackamore and Bridle

Some horseowners use a hackamore when starting to ride young horses. The hackamore prevents injury to the horse's mouth. The horse learns to respond to pressure that indicates the direction in which to go.

A bridle with a snaffle bit may also be used to train young horses. The halter is left on and the bridle is slipped over it. Be sure that the reins are even in length and tie them to the saddle horn. The bit should be in contact with the mouth when the head is held naturally. Lead the horse around for a few minutes. Do this for several days to allow the horse to become used to the bit.

Driving

Some trainers *ground drive* the horse before mounting and riding. A halter, hackamore, or bridle may be used. Driving lines should be about 25 feet (7.6 m) long. They are passed through the stirrups and attached to the halter, hackamore, or bit. A lead rope with a helper may be needed the first few times this is done. With the helper leading, the horse is led forward. Use voice commands such as "whoa" and "back" to fit the action. Start and stop frequently. Ground driving helps the young horse relax.

Mounting and Riding

When mounting, turn the left stirrup to receive the foot. Hold the reins in the left hand. The left hand may hold a lock of mane, the bridle, or be placed on the horse's neck. The right hand is placed on the saddle horn. Step up into the left stirrup. Do this several times to get the horse used to the weight. When the horse accepts the weight, put the right foot over slowly and ease into the saddle. Do not allow the right foot or leg to drag over the horse's rump. Take hold of the reins with both hands. Be careful that the horse does not buck. If a horse throws its rider the first time, it will be more difficult to ride. Dismounting is the reverse of mounting.

Allow the horse to walk on its own for a short time. Stop the horse with the voice command "whoa" and a slight pull on the reins. Start and stop the horse for 15 or 20 minutes. Repeat for several days until the horse is used to the rider. At that time, proceed with training the horse to obey other commands. Remember that control of the horse is through voice, hand, leg pressure, and weight. Move from the walk to the trot. Teach the horse to turn by using the reins. Move from the trot to the lope.

After the horse has had several weeks of this training in a small area, it is time to move out to more open spaces. Start at a slow gait and move into a lope. Vary the gaits during each training session. As the horse becomes used to the riding, the time spent each day may be increased.

The horse learns best at a slow gait. After it has learned the basic commands it can be trained to the proper leads. Work the horse in a circle at the lope. Change directions frequently. Moving to the right trains for the right lead and moving the left trains for the left lead. When the horse has learned leads properly it may be trained for various show events. The help of a professional may be needed for advanced training.

HORSEMANSHIP

Horsemanship is the art of riding a horse. It is also called *equitation*. The two general types of equitation are Western and English. The basic principles of horsemanship are similar for both styles of riding. There are differences in tack and clothing. The American Horse Shows Association, breed associations, and individual shows establish rules for judging equitation classes. A person entering a show must become familiar with the rules that govern that particular show.

It is not possible to learn horsemanship by reading about it. The person who wishes to become a competent rider must learn from an instructor. Classes in horsemanship are offered in many localities. Youth groups such as 4-H clubs and FFA chapters may offer instruction and practice in horsemanship for interested members.

In any type of equitation, the rider must master the basic skills of mounting, correct seat, and dismounting. The basic principles are the same for any type of riding.

Mounting

A horse is always mounted from the near side (Figure 34-2). Two body positions are used. When mounting a strange horse, stand by the horse's left shoulder and face a quarter turn to the rear. When mounting a gentle horse, stand by the stirrup fender and face directly across the saddle.

Beginning from either position, the left foot is placed in the stirrup (Figure 34-3). The left hand holds the reins and is placed on the neck of the horse just in front of the withers. The right hand is placed on the saddle horn in Western equitation. In English equitation, the right hand grasps the off side of the cantle. The rider takes one or two hops on the right foot to gain momentum and springs up (Figure 34-4). The right leg is swung over the horse, taking care not to drag across the rump of the horse (Figure 34-5). The weight is shifted to the left leg to maintain balance. The right foot is slipped into the stirrup before the body weight is settled into the saddle (Figure 34-6).

Seat Position

The Western rider sits erect and relaxed in the center of the saddle. Stirrup length should be such that the heels are lower than the toes (Figure 34-7). The balls of the feet are placed on the treads of the stirrups. Keep the toes pointed in the direction of travel, without their being too far in or out. Maintain contact with the saddle with the calves, knees, and thighs. The elbows are kept close to the body. The rein hand is held just above and in front of the saddle. The free hand is held relaxed.

As the horse moves, the rider moves to stay in balance with the horse. It is important to coordinate body rhythm with the movement of the horse. Practice will help the rider learn how to do this.

Dismounting

Dismounting is the reverse of mounting. The left hand holding the reins is placed on the withers. The right hand is placed on the saddle horn or pommel (depending on whether Western or English equitation is used). The rider rises in the stirrups and slips the right foot

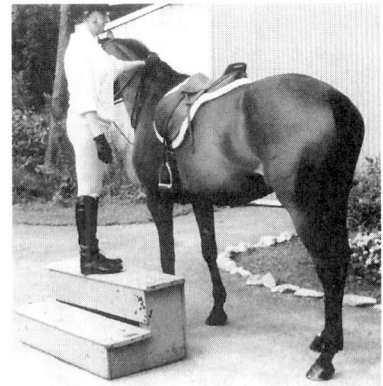

FIGURE 34-2   The horse is mounted from the near (left) side. *Photo by Michael Dzaman.*

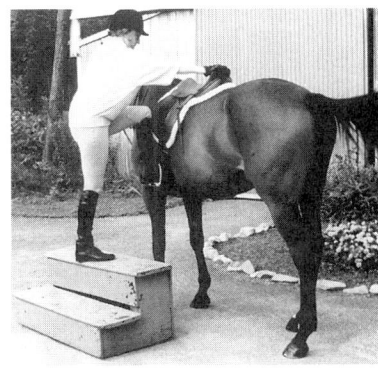

FIGURE 34-3   Begin mounting by placing the left foot in the stirrup. *Photo by Michael Dzaman.*

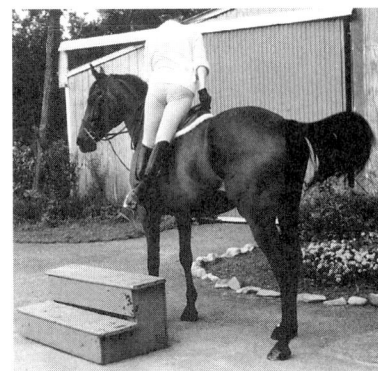

FIGURE 34-4   The rider hops on the right foot to gain momentum and springs up. *Photo by Michael Dzaman.*

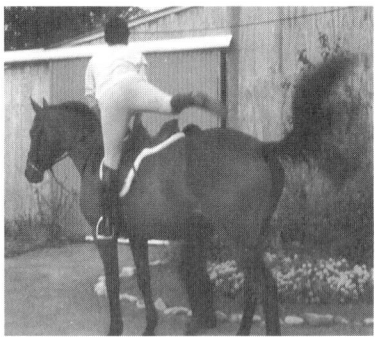

FIGURE 34-5 Swing the right leg over the horse. Do not allow it to drag on the horse's rump. *Photo by Michael Dzaman.*

FIGURE 34-6 The right foot is slipped into the stirrup before the body weight is settled into the saddle. *Photo by Michael Dzaman.*

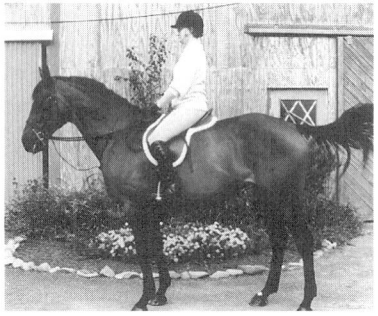

FIGURE 34-7 When in proper seat position, the rider sits relaxed in the center of the saddle. Note that the heels are lower than the toes. *Photo by Michael Dzaman.*

free. The right foot is swung over the horse's back, keeping clear of the rump. In English equitation, the right hand is moved to the cantle of the saddle in preparation for stepping down. It is correct for the rider to either step down or slide down after the right leg is swung over the horse's back. The technique used depends on the size of the horse and the rider. The right leg is kept in close to the near side of the horse. The rider faces slightly forward when the right foot touches the ground. The left foot is removed from the stirrup by pushing down on the heel to slip the foot out of the stirrup. Never roll the left foot on its side to slip it out of the stirrup.

Controlling the Horse

The basic aids used in controlling the horse are the voice, hands, legs, and weight. A well-trained horse will respond to these aids when they are properly used. After training, a horse will respond to light applications of these aids.

When the same words are used consistently during training, the horse learns to associate the words with the action. The horse should be spoken to in a soft, quiet, and firm voice. Yelling at the horse will only make it nervous.

The rider's hands are used to help guide the horse. A light hand is recommended for controlling the horse. The reins are held with a small amount of slack. The correct manner of holding the reins depends on the style of riding. The rider must learn the rules for the style of riding being done.

The rider must maintain body balance to avoid pulling on the reins for bracing. The horse is signaled by using light pulls and slacking on the reins. The reins should never be pulled steadily with great force. This will ruin the horse's mouth. *Neck-reining* is controlling the response of the horse by the weight of the rein against the neck. It takes patience and training to teach a horse to respond to neck-reining.

The rider's legs are used to control forward speed and the movement of the hindquarters. Pressure from the calves and heels is a signal to the horse to move forward. Pressure is also used to change gaits, help stop the horse, and in preparation for backing. Pressure by one leg or the other will get the horse to respond by swinging the hindquarters away from the pressure.

A well-trained horse will respond to slight shifts in the body weight of the rider. The horse shifts its body to balance the weight of the rider. This causes the legs of the horse to be held in place by the weight of the rider or be left free to move because of the absence of weight.

The aids are used to start and stop, move into the gaits, take the correct leads, turn, and back the horse. Instruction and practice are necessary for the rider to learn how to use the aids correctly.

SHOWING AT HALTER

Horses shown at halter are judged for conformation, soundness, and action. The horse must be properly groomed and trained before the show. When showing, the rider must be properly dressed for the class. Clothes must be neat and clean. Fairness and courtesy are always appropriate in the show ring.

The horse is led into the show ring at a brisk walk. The ring steward indicates the direction in which to move. When the horses are lined up, enter the line from the rear. Stand the horse in position evenly with the other horses in the class. The horse will show better if the front feet are slightly higher than the rear feet. Some breeds are shown in a stretched position and others are not. Learn the correct position for the breed of horse being shown.

FIGURE 34-8 Leading the horse while showing at halter. *Photo by Michael Dzaman.*

When in line, stand on the near side of the horse and even with the horse's head. Always leave room on either side of the horse so the judge can see the horses. This also prevents possible injury from other horses in line. Always turn to face the judge as the class is judged. When the judge is inspecting the near side, move to the front of the horse. Never move to the off side of the horse.

The horse is shown at the walk and the trot. Each horse is asked to perform individually by the ring steward.

The horse is led to the indicated line and stopped about one horse length from the person standing at the end of the line. The horse is walked in a brisk manner to the other end of the line. Always lead from the near side and never in front of the horse (Figure 34-8). At the other end of the line, stop about one horse length from the person standing there. When given the signal to return, turn the horse to the right. Pivot the horse on its hindquarters. Trot the horse to the other end of the line. When given the signal, return to the line of horses waiting to be shown. Again, enter the line from the rear.

Various shows and arenas have different classes and patterns of showing. Study the rules for the show being entered. Always watch the ring steward for signals. After entering the ring, continue showing the horse until the class is placed and the judge has given reasons. Leave the ring in the same brisk manner as used when entering. A good horseperson shows good sportsmanship at all times.

EQUITATION CLASSES

Horses can be shown in Western or English equitation. In these classes, the rider is being judged for ability to ride and control the horse. The horse is not being judged as it is in halter classes. The right kind of horse should be used for the class in which it is performing. It must be well trained to perform the maneuvers required in the class.

The rider is judged on a number of points, including position in the saddle, use of the hands, proper tack and dress, and performance of the various gaits appropriate for the class. The rider may be asked to mount and dismount. There are a number of individual tests the rider may be asked to demonstrate.

There are many performance classes in the various styles of riding. A person desiring to enter this type of competition should secure a copy of the official rule book of the American Horse Shows Association. Any given show may also have rules with which the exhibitor must be familiar.

Western Equitation

The rider in Western equitation classes should wear a Western hat and cowboy boots. Spurs may be worn but are not necessary. Chaps are permitted in some shows but not in all; therefore, rules for the show being entered should be followed.

Use a Western saddle in Western equitation shows. Snaffle, curb, spade, and other similar bits are usually allowed in Western shows. Hackamores, martingales, or tie-downs are generally not allowed. Chin straps or curb chains must be at least 1/2-inch wide and lie flat against the jaw. Wire curbs are not allowed. A coiled rope or riata may be attached to the saddle.

The rider maintains an erect position, properly balanced in the center of the saddle and facing the front. The horse is gripped with the inside of the thighs, knees, and calves of the legs. The rider's legs should be nearly parallel to the side of the horse. The upper body of the rider is relaxed but erect. The arms are held with the elbows close to the rider's side. Elbows should not be allowed to flap with the movement of the horse.

The reins may be carried in either hand, but the hand is not changed during the show. Hold the rein hand about three inches above and in front of the saddle horn. Place the other hand on the thigh or carry it in front of the waist, keeping the elbow close to the body.

The reins are usually held between the forefinger and thumb. Some shows permit a finger to be placed between the reins unless a romal is used or the ends of the split reins are held in the hand not used for reining. If the reins are held in the right hand, carry the bight (end of reins) on the right side; carry the bight on the left side if the reins are held in the left hand.

Control of the horse is done mainly by neck-reining with a semi-loose rein. Do not put a strong, constant pressure on the bit.

During Western equitation, the rider maintains a firm seat in the saddle. Do not stand in the stirrups at a trot or lope. Rising and posting the trot is not required in Western equitation as it is in English equitation.

In Western equitation, the horses enter the ring at a walk. During the show the horse is expected to demonstrate a walk, trot (jog), and lope (canter) both clockwise and counterclockwise.

Voice commands should not be heard, nor should leg, feet, and hand cues be noticeable during the competition. Do not slap the horse with the reins.

A well-trained horse will maintain good form during the walk. Judges discount horses that dance, prance, toss the head, and chafe and pull at the bit during the walk.

The trot is a diagonal two-beat gait, which should be performed at moderate speed. If the horse appears to be racing during the trot, it is going too fast; if it appears sluggish, it is going too slow.

The lope or canter is a three-beat gait. It should be performed reasonably slowly, at a restrained gallop. The rider must maintain a firm seat and lean slightly forward with the horse while performing this gait.

Important points that are emphasized in the canter are the ability to execute the correct lead and to change leads easily and quickly. The correct lead is with the foreleg toward the inside of the ring. The fore and hind legs move in unison during the canter.

Begin the canter by lifting the horse slightly with the reins and giving a gentle boot on the fence side of the ring. Do not jerk on the reins, as this may cause the horse to bounce or stop.

During the show, stay on the rail and do not bunch up in a group of riders. Always show courtesy to other riders. Do not hesitate to pass when necessary. Passing is done on the inside of the ring; do not pass on the rail side.

When the judge requests the riders to reverse direction, the turn may be made either toward or away from the rail. In some shows, however, the rules require the turn to be made away from the rail. Riders must be sure to determine the rule for the show in which they are competing. Turning to the inside is more difficult and provides a test of the rider's control of the horse. Be sure to have a firm grip on the reins when executing the turn.

After the gaits are completed, the riders are usually asked to line up the horses in the center of the ring. The horse should stand quietly when lined up. Riders may be asked to back their horses from this line. The horse should back in a straight line when the rider gives a slight pull on the bit. The horse should not pull up or turn its head when backing. Points are lost if the horse swings its head or wrings its tail while backing.

Sometimes riders are asked to dismount and remount their horses. Upon dismounting the rider should hold the reins in the right hand and stand facing the horse.

Riders are sometimes asked to individually demonstrate the figure eight, square stops, and quarter and half turns. The judge will be looking for balance, correct leads, head and tail carriage of the horse, and good form. Additional maneuvers sometimes requested include lope and stop, roll backs, 360-degree turns, riding a serpentine course showing flying changes of lead, and riding without stirrups. Judges may ask riders questions about equitation, tack, or conformation.

English Equitation

Informal clothing, which includes a solid-colored saddle suit (navy, black, brown, beige, or gray), a shirt (white or pastel), a tie (four-in-hand, either solid-colored or striped), jodhpur boots (black or brown), and a soft or hard dark derby, is worn during English equitation. Gloves (white, brown, or black) may be worn but are not required. An alternative to a saddle suit is jodhpurs (without flare or cuff), black or brown jodhpur boots, a shirt, a tie, and a sweater or vest.

Formal attire is required for evening saddle seat equitation and three-gaited classes. This consists of a tuxedo saddle suit, a white stiff-front shirt with a bow tie, and a top hat. Formal jodhpur pants have satin stripes down each leg and formal jodhpur boots are patent leather. Formal attire is never worn in five-gaited classes.

For hunt seat equitation, the rider should wear a conservative plaid or tweed three-button coat; this may be mid-hand or fingertip length with one vent in mid-back. The rider should also wear tan, fawn, black, or brown breeches or flare-top jodhpurs with cuff. The rider's tie may be plaid, striped, or solid-colored and may be stock, rat-catcher, or four-in-hand. Wear black or brown high boots or jodhpur boots and a brown or black hunt cap or hard derby for hunt seat equitation. Black, brown, or tan gloves may also be worn if desired.

Do not wear fancy hats, jockey caps, sleeveless coats, or brightly colored suits in English equitation. Riders with long hair should tie it up or wear a hair net because contestant numbers are worn on the back and must be visible to the judge.

During English equitation, the rider should maintain an upright balanced position that is comfortable; do not lean forward or backward. Look straight ahead between the ears of the horse. Stirrups should be adjusted so the rider's feet hang naturally when mounted. The stirrup should touch the ankle bone.

The rider sits close to the front of the saddle. The ball of the foot is kept evenly on the stirrup bar, with the heels down and toes tilted slightly upward. This position helps the rider maintain a balanced and centered position.

The rider keeps the hands just above the pommel of the saddle at about the waistline. The back of the hands is kept up, with the thumbs close together and the knuckles nearly vertical. The upper arms are kept parallel to the body with the elbows in so the arms do not flap.

In English equitation, the reins are always carried in both hands. If a single-rein bridle is used, the rein comes through the palm and is held between the forefinger and the thumb; when a double-reined bridle is used, the snaffle reins are outside the little fingers and the curb reins are between the little and fourth fingers. An alternative that is acceptable is to hold the curb rein between the middle finger and the fourth finger with two fingers between the snaffle and curb. The reins come out of the palm over the first finger and are held in place by the thumb. If a crop is carried, place it in the hand away from the rail.

When jumping in hunt seat equitation, the rider should lean slightly forward in the seat to help the horse over the jump. The rider's hands are carried closer to the horse's neck when jumping because the horse has a lower head set than the regular saddle horse.

If the show includes hunt seat equitation without jumping, the gaits requested are usually a walk, a trot, and a canter, in both directions in the ring. Riders may be asked to hand gallop and sit (but not post) a slow trot. The body should be kept in a comfortable vertical position for the walk and slow trot; the rider should lean slightly forward at the posting trot, canter, and gallop.

In English equitation, the rider needs a high level of skill because horses used in this type of equitation are often more high-strung and animated than those used in Western equitation. The rider signals the proper gait to the horse by posture, hand signal through the bit, and pressure from the legs. The walk is begun by gathering the reins in the hands above the pommel and urging the horse forward with slight pressure from both legs; the trot is begun by shortening the reins slightly and signalling the horse with a definite pressure from the legs.

During the trot, the rider is expected to post on the correct diagonal (position of the horse's front legs). *Posting* is when the rider rises and sits in the saddle. The rider rises slightly out of the saddle when the horse's shoulder on the fence side rises and sits down when that shoulder falls. Be careful not to rise too far out of the saddle or to sit with a bump.

The canter is usually begun from a walk or from a standing position. In some forms of hunt seat equitation, the canter may be entered from the trot. The canter is a three-beat gait, with the fore and hind legs moving together. It is begun by shortening the reins, turning the horse's head slightly toward the rail, and signalling the horse with the heel behind the girth on the rail side. The correct lead is accomplished when the horse takes the first step with the foreleg toward the center of the ring. If the horse takes the wrong lead, stop and begin again. If the reins are held too tightly, the horse may hop and pound with its front legs; if the reins are held too loosely, the horse may lunge into the canter.

During the competition it is permissible to pass other horses on the inside away from the rail. Do not get bunched up in a group of horses. If this happens, the rider should turn the horse across the ring to the opposite rail. When the judge requests that the horses be reversed, stop and turn the horse toward the rail.

When the horses are lined up, the horse should be positioned squarely with some stretch. Arabians and jumpers may not be required to stretch. When backing the horse, take one or two steps forward and then back the horse in a straight line for about four steps. When backing, the horse is guided by the rider's legs. Riders are usually not required to dismount in English equitation.

Riders may be asked to perform some individual work, such as demonstrating change of diagonals at the trot or showing leads in making a figure eight. The judge may also request that the rider drop the reins so the routine used for picking them up may be checked.

GYMKHANA

Gymkhana is the term used for games on horseback. Many horse shows have gymkhana events (Figure 34-9). Rules, regulations, and diagrams for gymkhana events are available from the American Horse Shows Association, the American Quarter Horse Association, the Appaloosa Horse Club, Inc., and the sponsors of local horse shows.

Some typical events in gymkhana include the following:

- pole bending
- clover-leaf barrel race
- rescue race
- musical chairs
- wheelbarrow race
- keyhole race
- saddling race
- team baton race

Other original events may be included in a local show. Many of these events are timed. The rider and horse are working against the clock and the fastest time wins.

RODEOS

Rodeos are popular horse events in all parts of the United States. A horse must be well trained to participate in rodeo events. Rodeo events require a high degree of skill and experience on the part of the person entering.

Typical rodeo events include the following:

- saddle bronc riding
- bareback bronc riding
- calf and steer roping
- bulldogging (steer wrestling)
- chuck wagon races
- wild horse races
- reining contests
- cutting horse contests

Other events and contests may be included in any given rodeo.

TRAIL RIDING

One of the most popular activities involving the use of horses is trail riding. There are many clubs and groups that organize trail rides. A trail ride may be a one-day event or may last for more than one day.

FIGURE 34-9   Games at horse shows are popular events. *Courtesy of Appaloosa Horse Club, Inc., ID.*

Both the rider and the horse should prepare for trail riding. The rider must develop the endurance needed for trail riding. Short rides each day before the trail ride will help both the rider and the horse prepare for the trail ride. The horse must be properly fed and conditioned, and must become used to the kinds of obstacles that will be found on trail rides.

The trail to be ridden must be carefully selected. Permission must be obtained in advance for crossing private property. A variety of conditions should be provided for on the trail ride. Busy highways should be avoided if at all possible. Provision should be made for rest stops. If the trail ride is to last overnight, then arrangements for camping out must be made. Planning must include measures for handling accidents and emergencies.

Planning for the ride includes determining who will be in charge during the ride. Appropriate equipment and supplies must be secured.

Rules and procedures for each rider to follow must be agreed to before the ride begins. Rules and procedures are based on safety requirements and consideration of others on the trail ride.

SUMMARY

Understanding the reasons for horse behavior is helpful when training and riding horses. Horses have to move their heads to bring objects into focus. This is one reason horses may shy at sudden movements. Horses have good memories, making it possible to train them to respond to commands. Sensitive areas on the body of the horse are used to help control the animal.

Training should begin when the horse is very young. First lessons include haltering, leading, and yielding the feet. When the horse is older, training can begin on the longe line. The horse learns to respond to spoken commands. Prepare a horse for saddling by sacking out. Lift the saddle into place carefully. Be sure that it is properly fastened. Many owners prefer to use a hackamore when beginning to teach a horse to be ridden. A bridle may be used with a snaffle bit. Ground driving may be used to further train the horse before beginning to ride.

Use care when mounting a horse for the first time. Make sure the horse will stand for the weight on its back. Teach the horse at slow gaits first. Later it may be trained at faster gaits.

The two general types of equitation are Western and English. The basic principles are the same for each. The main differences are in the tack and clothes used. Mounting, seat position, and dismounting are basic skills that every rider must have.

The horse is controlled by the voice, hands, legs, and weight of the rider. A well-trained horse will respond quickly to the signals of the rider.

Horses shown at halter are judged on conformation, soundness, and action. Lead the horse from the near side and always face the judge when showing at halter. The horse is shown at the walk and the trot in halter classes.

Equitation classes are judged on the performance of the rider. The rider must have skill in riding and controlling the horse. Position in the saddle, use of the hands, proper tack and dress, and the performance of the various gaits are the main points judged in equitation classes.

Other popular horse events are gymkhana, rodeos, and trail riding. Participation in these events adds to the pleasure of owning a horse.

Student Learning Activities

1. Observe a demonstration on training a young horse.
2. Observe demonstrations on haltering, bridling, saddling, mounting, dismounting, and riding horses.
3. Attend horse shows to observe correct procedures.
4. Prepare an oral report and/or demonstrations on horsemanship and showing horses.

Review

1. What are the survival instincts of horses that have developed over the years that help to explain their behavior?
2. Describe the eyesight of horses.
3. How is a horse's memory used to help train the horse?
4. Name the most sensitive areas on a horse's body.
5. Name four ways a rider communicates commands to a horse.
6. Describe the proper procedures for (a) haltering, (b) leading, (c) working with the feet, (d) longeing, and (e) saddling the horse.
7. Describe how the hackamore and the bridle are used.
8. Describe how ground driving is used when training a horse.
9. Describe the proper way to mount and ride a horse during training.
10. Define the term horsemanship.
11. How may a rider learn horsemanship?
12. Describe the proper way to mount a horse in Western and English equitation.
13. Describe the proper seat position in the saddle.
14. Describe the proper way to dismount from a horse.
15. How does the rider use the basic aids in controlling a horse?

16. Explain how a horse should be shown at halter.
17. Describe the showing of a horse in equitation classes.
18. Name some of the events that are often included in gymkhana.
19. Name some typical events included in a rodeo.
20. List the preparations that are necessary for trail riding.

Section 8

Poultry

Unit 35 Selection of Poultry

Objectives

After studying this unit, the student should be able to

- describe the nature of the poultry industry.
- identify common breeds of poultry.
- explain the selection of poultry for production.

THE POULTRY INDUSTRY

Types of Enterprises

The size of poultry enterprises ranges from small farm flocks to large commercial operations. Most of the poultry raised in the United States is produced in large commercial operations. Regardless of the size of the enterprise, there are three important factors for success with poultry: proper feeding, good management, and sanitation.

The three general types of chicken enterprises are egg production, broiler production, and raising replacement pullets. Most of the turkeys, ducks, and geese in the United States are raised for meat production. Except for hatching purposes, there is little market for turkey, duck, and geese eggs.

In egg production operations, laying hens are kept to produce eggs. Laying hens may be confined in cages or the farmer may use a floor-pen system. Cleaning, grading, and packaging of eggs usually occurs on the farm. When the production cycle is completed, the hens are sold for meat.

Broiler production operations involve raising chickens for meat. High-quality rations are fed to secure rapid, efficient gains. Generally, several flocks of birds are fed and marketed each year. Replacement pullet operations involve raising chicks, which are then sold to egg production or broiler production operations, depending upon the type of bird raised.

Vertical integration is common in the poultry industry. *Vertical integration* means that two or more steps of production, marketing, and processing are linked together. Vertical integration is usually set up by feed manufacturers or poultry processors. They provide the financing needed and have most of the control of management decisions that are made in the production process. Most of the broilers and turkeys in the United States are produced under some type of vertical integration arrangement. More than one-half of the eggs are produced in this manner. Vertical integration has resulted in much larger and more efficient operations in the poultry industry.

Numbers and Trends in Production and Consumption

While the average number of laying hens on farms in the United States has varied since 1967, the general trend has been downward (Table 35-1). The number of farms with laying hens has decreased, while the average size of flock per farm has increased. Much of the egg production in the United States is concentrated in large commercial operations, many of which produce eggs under some type of vertical integration arrangement.

Despite the decrease in the total number of laying hens, annual egg production has increased since 1967. The main reason for this is a significantly higher rate of lay per hen (Table 35-1).

Table 35-1 shows that the consumption of eggs per person has decreased since 1967. Part of the reason for the declining popularity of eggs in the diet may be due to a concern on the part of consumers about the amount of cholesterol in the diet.

The fifteen leading states in numbers of laying hens are Ohio, California, Iowa, Pennsylvania, Indiana, Georgia, Texas, Arkansas, Minnesota, North Carolina, Alabama, Nebraska, Florida, Mississippi, and Missouri (Figure 35-1). The fifteen leading states in egg production are Ohio, California, Pennsylvania, Iowa, Indiana, Georgia, Texas, Arkansas, Minnesota, Nebraska, North Carolina, Florida, Alabama, Missouri, and Mississippi. The rank order in numbers of laying hens and egg production will vary from year to year; however, these fifteen states are generally the major egg producing states in the United States.

The production of broilers in the United States has more than tripled since 1967 (Table 35-2). The per person consumption of poultry meat has also increased significantly since 1967. A major increase in the per capita consumption of poultry meat began in the late 1970s as consumers became increasingly concerned about the level of cholesterol in their diets. Poultry meat is perceived by the consumer as having a lower level of cholesterol than beef or pork.

The fifteen leading states in broiler production are Georgia, Arkansas, Alabama, Mississippi, North Carolina, Texas, Maryland, Virginia, Delaware, Missouri, Oklahoma, South Carolina, Kentucky, Tennessee, and Pennsylvania (Figure 35-2). The rank order of states in numbers of broilers produced will vary from year to year. These fifteen states are usually the major broiler producing states, together producing about 89 percent of the total broiler production in the United States.

TABLE 35-1 Egg Production, Rate of Lay, Number of Layers, and Per Capita Consumption of Eggs, 1967–1998.

| | Egg Production | | Rate of Lay | | Layers | | Per Capita | |
| | Total | Percentage of 1967 | Eggs Per Layer | Percentage of 1967 | Average Number | Percentage of 1967 | Consumption (Eggs) | Percentage of 1967 |
Year	(Million)	(%)	(Number)	(%)	(Thousand)	(%)	(Number)	(%)
1967	69,327	100.0	221	100.0	313,717	100.0	321	100.0
1968	68,156	98.3	220	99.5	309,824	98.8	316	98.4
1969	67,546	97.4	220	100.0	306,886	97.8	310	96.6
1970	68,522	98.8	218	99.1	313,907	100.1	310	96.6
1971	69,649	100.5	223	101.4	312,886	99.7	311	96.9
1972	69,219	99.8	227	103.2	304,504	97.1	303	94.4
1973	66,039	95.3	227	103.2	290,588	92.6	288	89.7
1974	65,620	94.7	230	104.5	284,732	90.8	283	88.2
1975	64,626	93.2	232	105.5	278,101	88.6	276	86.0
1976	64,511	93.1	235	106.8	274,135	87.4	270	84.1
1977	64,602	93.2	235	106.8	274,875	87.6	268	83.5
1978	67,157	96.9	239	108.6	281,544	89.7	273	85.0
1979	69,209	99.8	240	109.1	288,623	92.0	278	86.6
1980	69,686	100.5	242	110.0	287,705	91.7	272	84.7
1981	69,825	100.7	243	110.5	287,774	91.7	266	82.9
1982	69,718	100.6	243	110.5	286,369	91.3	266	82.9
1983	68,169	98.3	247	112.3	276,263	88.1	262	81.6
1984	68,222	98.4	245	111.4	277,960	88.6	262	81.6
1985	68,645	99.0	247	112.3	277,592	88.5	256	79.8
1986	69,106	99.7	248	112.7	279,046	88.9	255	79.4
1987	70,356	101.5	248	112.7	283,872	90.5	255	79.4
1988	69,878	100.8	251	114.1	278,587	88.8	248	77.3
1989	67,503	97.4	250	113.6	270,415	86.2	239	74.5
1990	68,134	98.3	251	114.1	270,946	86.4	236	73.5
1991	69,465	100.2	252	114.5	275,451	87.8	235	73.2
1992	70,749	102.1	254	115.5	278,824	88.9	236	73.5
1993	71,936	103.8	253	115.0	284,770	90.8	236	73.5
1994	73,911	106.6	254	115.5	291,018	92.8	239	74.5
1995	74,258	107.1	253	115.0	293,648	93.6	235	73.2
1996	76,377	110.2	256	116.4	298,270	95.1	237	73.8
1997	77,532	111.8	255	115.9	303,604	96.8	240	74.8
1998[a]	79,717	115.0	256	116.4	312,058	99.5	244	76.0

[a]Preliminary

Sources: Agricultural Statistics, USDA, 1981, 1986, 1988, 1992, 1995, 1999.
    Layers & Egg Production Annual, USDA, January 1999.

Since 1967, the overall trend in both turkey numbers and liveweight produced has been upward (Table 35-3). The greatest increase in turkey production in relation to 1967 has occurred since the late 1970s. As with broilers, this may partially reflect consumer concern regarding cholesterol in the diet. Turkey meat is also perceived by consumers as having less cholesterol than beef or pork. The significant increase in per person consumption of turkey meat, as shown

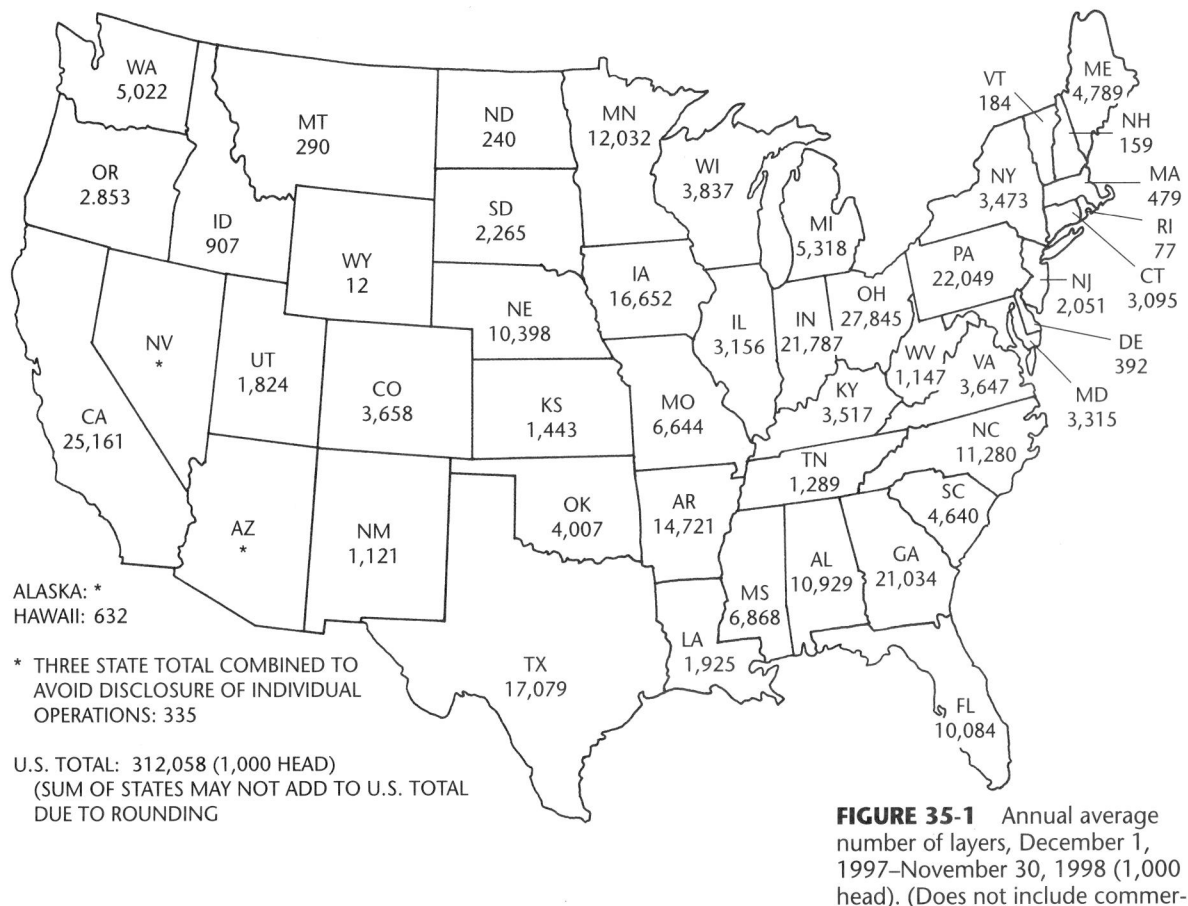

FIGURE 35-1 Annual average number of layers, December 1, 1997–November 30, 1998 (1,000 head). (Does not include commercial broilers.) *Source: Layers & Egg Production Annual,* USDA, January 1999.

in Table 35-3, would appear to indicate that this is the case. The greatest increase in per capita consumption of turkey meat has occurred since the late 1970s.

Turkey production tends not to be concentrated in any one geographical area of the United States (Figure 35-3). The ten leading states in turkey production are North Carolina, Minnesota, Arkansas, Virginia, Missouri, California, Indiana, South Carolina, Pennsylvania, and Ohio. The rank order of turkey producing states will vary from year to year; however, these ten states tend to be the major turkey producing states in the United States.

About 10 million ducks are raised annually in the United States. The raising of ducks for meat is concentrated on Long Island, New York. About 60 percent of the annual production in the United States is on Long Island.

About 1,000,000 geese are raised annually in the United States. The leading states in production are Missouri, Iowa, South Dakota, Minnesota, Wisconsin, Ohio, Indiana, California, and Washington.

TABLE 35-2 Broiler Production and Per Capita Consumption of Chickens, 1967–1998.

	Quantity[b]			Per Capita Consumption[c]		
	Broilers		Percentage of 1967			Percentage of 1967
Year[a]	(Million pounds)	(Million kg)	(%)	(lb)	(kg)	(%)
1967	9,183	4,165	100.0	36.5	16.6	100.0
1968	9,326	4,230	101.6	36.7	16.6	100.5
1969	10,048	4,558	109.4	38.4	17.4	105.2
1970	10,819	4,907	117.8	40.5	18.4	111.0
1971	10,818	4,907	117.8	40.3	18.3	110.4
1972	11,480	5,207	125.0	41.8	19.0	114.5
1973	11,220	5,089	122.2	40.2	18.2	110.1
1974	11,320	5,135	123.3	40.4	18.3	110.7
1975	11,096	5,033	120.8	39.8	18.1	109.0
1976	12,481	5,661	135.9	42.4	19.2	116.2
1977	12,962	5,879	141.2	44.4	20.1	121.6
1978	14,000	6,350	152.5	47.0	21.3	128.8
1979	15,522	7,041	169.0	51.0	23.1	139.7
1980	15,539	7,048	169.2	50.5	22.9	138.4
1981	16,520	7,493	179.9	52.0	23.6	142.5
1982	16,760	7,602	182.5	53.3	24.2	146.0
1983	17,038	7,728	185.5	54.1	24.5	148.2
1984	17,861	8,102	194.5	56.0	25.4	153.4
1985	18,810	8,532	204.8	57.0	25.9	156.2
1986	19,661	8,918	214.1	58.0	26.3	158.9
1987	21,523	9,763	234.4	62.0	28.1	169.9
1988	22,464	10,189	244.6	64.0	29.0	175.3
1989	23,979	10,877	261.1	68.0	30.8	186.3
1990	25,631	11,626	279.1	71.0	32.2	194.5
1991	27,203	12,339	296.2	74.0	33.6	202.7
1992	28,829	13,077	313.9	78.0	35.4	213.7
1993	30,618	13,888	333.4	79.0	35.8	216.4
1994	32,528	14,754	354.2	81.0	36.7	221.9
1995	34,222	15,523	372.7	81.0	36.7	221.9
1996	36,483	16,548	397.3	82.0	37.2	224.7
1997	37,541	17,028	408.8	84.0	38.1	230.1
1998	38,554	17,488	419.8	86.0	39.0	235.6

[a]Computations are made from unrounded data, 1968–69, calendar year; 1970–1998, December previous year through November current year.
[b]Liveweight
[c]Ready to cook basis

Sources: *Agricultural Statistics,* USDA, 1981, 1986, 1988, 1992, 1995, 1999.

Poultry Production and Value, USDA, April 1999.

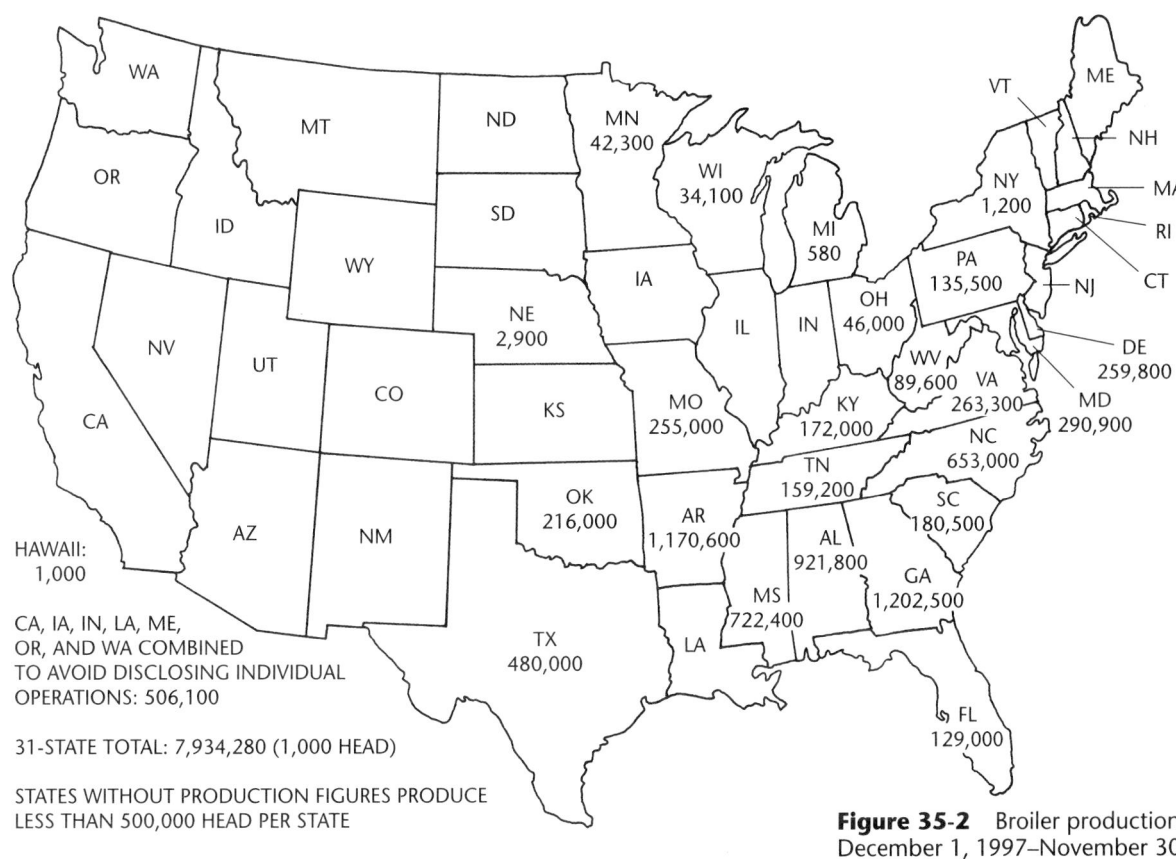

Figure 35-2 Broiler production, December 1, 1997–November 30, 1998 (1,000 head) in 31 major broiler-producing states. *Source: Poultry Production and Value,* USDA, April 1999.

Map labels:

WA
MT
ND
MN 42,300
OR
ID
SD
WI 34,100
MI 580
VT
ME
NH
NY 1,200
MA
RI
PA 135,500
NJ
CT
WY
IA
IL
IN
OH 46,000
NV
UT
CO
NE 2,900
KS
WV 89,600
VA 263,300
DE 259,800
MD 290,900
CA
MO 255,000
KY 172,000
NC 653,000
AZ
NM
OK 216,000
AR 1,170,600
TN 159,200
SC 180,500
HAWAII: 1,000
AL 921,800
GA 1,202,500
MS 722,400
TX 480,000
LA
FL 129,000

CA, IA, IN, LA, ME, OR, AND WA COMBINED TO AVOID DISCLOSING INDIVIDUAL OPERATIONS: 506,100

31-STATE TOTAL: 7,934,280 (1,000 HEAD)

STATES WITHOUT PRODUCTION FIGURES PRODUCE LESS THAN 500,000 HEAD PER STATE

Advantages of Poultry Raising

Raising poultry has a number of advantages, among which are

- high feed efficiency.
- fast return on investment.
- spreading income throughout the year.
- high return compared to feed costs.
- low land requirements.
- adaptability to both small part-time enterprises and large commercial enterprises.
- the operation can be highly mechanized with high output per hour of labor.

Disadvantages of Poultry Raising

There are some problems involved in raising poultry, among which are

- serious problems with diseases and parasites.
- need for a high level of management ability, especially for large commercial flocks.

TABLE 35-3 Turkey Production and Per Capita Consumption, 1967–1998.

Year	Number Raised (1,000 head)	Number Raised Percentage of 1967 (%)	Quantity Turkeys Liveweight (1,000 pounds)	Quantity Turkeys Liveweight (1,000 kg)	Quantity Turkeys Liveweight Percentage of 1967 (%)	Per Capita Consumption (lb)	Per Capita Consumption (kg)[c]	Per Capita Consumption Percentage of 1967 (%)	Average Weight Per Bird (lb)	Average Weight Per Bird (kg)[c]
1967	126,577	100.0	2,343,339	1,062,920	100.0	8.5	3.9	100.0	18.5	8.4
1968	106,709	84.3	2,014,589	913,801	86.0	7.9	3.6	92.9	18.9	8.6
1969	106,736	84.3	2,029,315	920,481	86.6	8.2	3.7	96.5	19.0	8.6
1970	116,139	91.8	2,197,916	996,957	93.8	8.0	3.6	94.1	18.9	8.6
1971	119,657	94.5	2,255,614	1,023,128	96.3	8.3	3.8	97.6	18.9	8.6
1972	128,664	101.6	2,423,618	1,099,334	103.4	8.9	4.0	104.7	18.8	8.5
1973	132,231	104.5	2,451,848	1,112,139	104.6	8.5	3.9	100.0	18.5	8.4
1974	131,909	104.2	2,437,121	1,105,459	104.0	8.8	4.0	103.5	18.5	8.4
1975	124,165	98.1	2,276,504	1,032,604	97.1	8.5	3.9	100.0	18.3	8.3
1976	140,021	110.6	2,606,265	1,182,181	111.2	9.1	4.1	107.1	18.6	8.4
1977	136,390	107.8	2,562,825	1,162,477	109.4	9.1	4.1	107.1	18.8	8.5
1978	138,939	109.8	2,654,788	1,204,191	113.3	9.2	4.2	108.2	19.1	8.7
1979	156,457	123.6	2,957,612	1,341,549	126.2	9.9	4.5	116.5	18.9	8.6
1980	165,243	130.5	3,076,858	1,395,638	131.3	10.5	4.8	123.5	18.6	8.4
1981	170,875	135.0	3,264,463	1,480,734	139.3	10.8	4.9	127.1	19.1	8.7
1982	165,464	130.7	3,175,060	1,440,182	135.5	10.8	4.9	127.1	19.2	8.7
1983	170,723	134.9	3,335,519	1,512,965	142.3	11.3	5.1	132.9	19.5	8.9
1984	171,296	135.3	3,384,393	1,535,134	144.4	11.4	5.2	134.1	19.8	9.0
1985	185,427	146.5	3,703,994	1,680,102	158.1	11.6	5.3	136.5	20.0	9.1
1986	207,232	163.7	4,147,168	1,881,122	177.0	12.9	5.9	151.8	20.0	9.1
1987	240,438	190.0	4,894,858	2,220,268	208.9	14.7	6.7	172.9	20.4	9.2
1988	242,421	191.5	5,059,056	2,294,747	215.9	15.7	7.1	184.7	20.9	9.5
1989	261,394	206.5	5,467,629	2,480,073	233.3	16.6	7.5	195.3	20.9	9.5
1990	282,475	223.2	6,043,155	2,741,127	257.9	17.6	8.0	207.1	21.4	9.7
1991	284,910	225.1	6,114,620	2,773,543	260.9	17.9	8.1	210.6	21.5	9.7
1992	289,880	229.0	6,355,293	2,882,710	271.2	17.9	8.1	210.6	21.9	9.9
1993	287,650	227.3	6,432,577	2,917,765	274.5	17.7	8.0	208.2	22.4	10.1
1994	286,585	226.4	6,540,295	2,966,625	279.1	17.8	8.1	209.4	22.8	10.4
1995	292,356	231.0	6,761,327	3,066,884	288.5	17.9	8.1	210.6	23.1	10.5
1996	302,713	239.2	7,222,834	3,276,220	308.2	18.5	8.4	217.6	23.9	10.8
1997[a]	301,251	238.0	7,225,059	3,277,229	308.3	17.6	8.0	207.1	24.0	10.9
1998[b]	283,503	224.0	7,002,768	3,176,400	298.8	18.1	8.2	212.9	24.7	11.2

[a]Preliminary

[b]Forecast

[c]Conversions rounded to nearest tenth

Sources: Agricultural Statistics, USDA, 1981, 1986, 1988, 1992, 1995, 1999.

   Poultry Production and Value, USDA, April 1999.

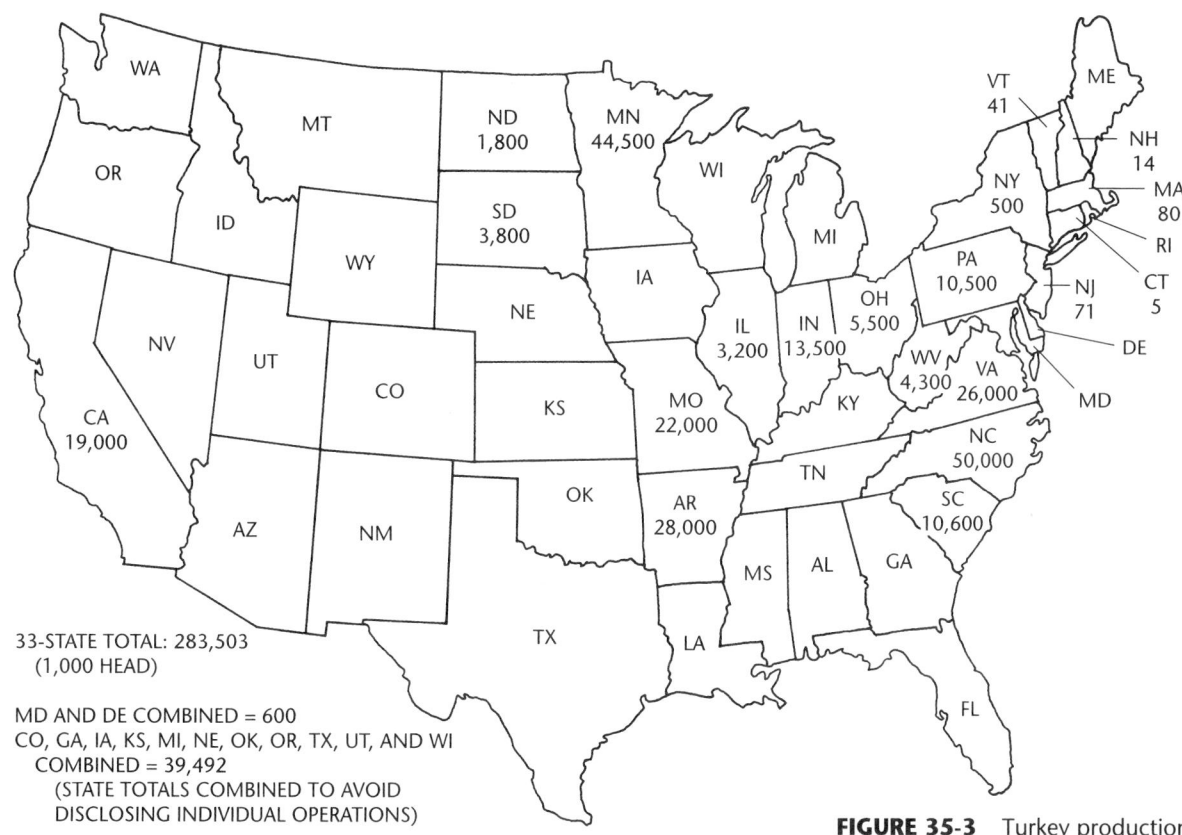

FIGURE 35-3 Turkey production, September 1, 1997–August 31, 1998 (1,000 head) in major turkey-producing states. *Source: Poultry Production and Value,* USDA, April 1999.

- need for large amounts of capital for large operations.
- limitations of zoning on the location of flocks.
- death losses may be high due to predators and stampeding.
- quality of product must be carefully controlled.
- careful marketing is required.
- high volume is needed for an economical enterprise.
- problems of waste disposal and odor.

BREEDS OF POULTRY

Poultry are divided into breeds, varieties, types, and classes. A *breed* is a group of related fowls that breed true to a number of given traits that identify the breed. Breeds are subdivided into *varieties* based on certain traits, such as color of plumage and comb type. *Type* refers to the purpose for which the poultry are bred. The two general types are *egg-type* and *meat-type*. *Class* generally refers to the geographic origin of the poultry. Four classes generally used for chickens in the United States are: (1) Mediterranean, (2) American, (3) English, and (4) Asiatic.

FIGURE 35-4 Rhode Island Red. *Courtesy of Iowa State University.*

FIGURE 35-5 New Hampshire female. *Courtesy of USDA.*

FIGURE 35-6 White Rock. *Courtesy of Iowa State University.*

There are no breed registry associations for poultry such as there are for other farm animals. The American Poultry Association, Incorporated publishes The American Standard of Perfection. This publication lists and describes more than 300 breeds and varieties of poultry. However, very few of these are of commercial importance. Not all commercially important breeds and varieties of poultry are listed in The American Standard of Perfection.

Chickens

Most chickens used for egg and meat production are produced by cross-mating, breed crossing, or inbreeding. Very few pure strains of chickens are used commercially. *Cross-mating* is crossing two or more strains within the same breed. *Breed crossing* is crossing different breeds to get the desired traits. *Inbred chickens* are produced by inbreeding and crossing the inbred lines to get the desired traits.

Most commercial egg-producing flocks are made up of White Leghorn strain crosses because of their superiority in egg production. Hybrid (inbred) crosses are also popular for egg production. The Leghorns and the Hybrids are white egg layers. Brown egg layers are Rhode Island Reds (Figure 35-4), New Hampshires (Figure 35-5), and Barred Plymouth Rocks. These are larger than the Leghorn strains and therefore have higher feed costs per dozen eggs produced. They are not as popular for large commercial flocks. Smaller farm flocks sometimes use these breeds because they produce more meat for home use and are considered to be dual-purpose breeds. However, they are not as efficient in either egg or meat production as types that are bred for one purpose.

Commercial crosses that are bred for meat production are generally used for broiler production operations. A common cross is White Plymouth Rock females (Figure 35-6) mated to Cornish males. These crosses do not have a high rate of egg lay, but are highly efficient as meat producers.

Characteristics of some typical breeds of chickens in four classes are shown in Table 35-4. There are a number of different varieties of most chickens. Color of plumage, type of comb, and size identify the different varieties.

There are a number of different types of comb found in chickens. Some of these are: (1) single, (2) rose, (3) pea, (4) cushion, (5) buttercup, (6) strawberry, and (7) V-shaped. The single comb is the most common type. The other comb types mentioned are mutations. Figure 35-7 shows some of the comb types of chickens.

Bantams are small chickens, usually weighing from 16 to 30 ounces (0.45–0.85 kg) as adults. There are nearly 150 varieties of bantams. Most of them are of the same breeds as the larger chickens. Bantams are generally used for show and as pets. They are not generally bred for egg or meat production.

TABLE 35-4 Characteristics of Some Typical Breeds of Chickens.

Class and Breed	Eggs	Skin	Comb	Eyes	Earlobes	Shanks	Plumage	Comments
Mediterranean:								
Leghorn (white)	White	Yellow	Single	Reddish bay	White	Yellow	White	All three are small in size and are used mainly for egg production—Leghorn most popular.
Minorca (Black)	White	White	Single	Brown	White	Dark Slate	Black	
Andalusian (Blue)	White	White	Single	Reddish bay	White	Dark Slaty Blue	Slaty Blue	
American:								
Plymouth Rock (Barred)	Brown	Yellow	Single	Reddish bay	Red	Yellow	Barred (Sex-linked)	Dual purpose; Used in crosses for sexing chicks at hatching.
Plymouth Rock (White)	Brown	Yellow	Single	Reddish bay	Red	Yellow	White	Primary Use—Broiler
New Hampshire	Brown	Yellow	Single	Reddish bay	Red	Yellow	Red	Primary Use—Broiler
Rhode Island Red	Brown	Yellow	Single	Reddish bay	Red	Yellow	Dark Red	Dual Purpose
English:								
Cornish (White)	Brown	Yellow	Pea	Pearl	Red	Yellow	White	Used in development of male lines for crossbreeding.
Australorp	Tinted	White	Single	Brown	Red	Dark Slate; Bottom feet white	Black	Used in production of crossbreeds.
Asiatic:								
Brahma (Light)	Brown	Yellow	Pea	Reddish bay	Red	Yellow; feathered	Columbian (white & black)	Used in crossbreeding for meat production.

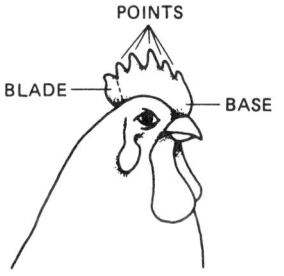

SINGLE

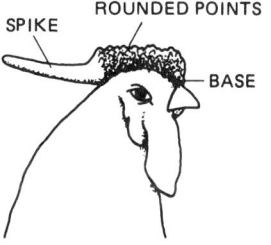

ROSE

PEA

FIGURE 35-7 Comb types found in chickens.

Turkeys

The American Standard of Perfection recognizes only one breed of turkeys. However, a number of varieties of turkeys exist, including the Broad Breasted Bronze, Broad Breasted Large White, Beltsville Small White, Narragansett, Black, Slate, and Bourbon Red. The commercially important varieties are the Broad Breasted Bronze, Broad Breasted Large White, and Beltsville Small White. The Broad Breasted Large White is the most commonly raised meat turkey.

The Broad Breasted Large White was developed from crosses of the Broad Breasted Bronze and the White Holland. The color of the plumage is white. The males have a black beard. Some females have small beards. The shanks, feet, and beak are white to pinkish white, and the throat wattle is red. Since the pinfeathers are white they are not easily seen and therefore do not lower the market grade of the carcass. A *pinfeather* is a feather that is not fully developed. The term generally refers to feathers that are just coming through the skin. Pinfeathers are not easily removed from the carcass. Dark pinfeathers are seen more easily than white pinfeathers. This detracts from the appearance of the carcass and lowers its value. Since market grade is partially associated with value, presence of dark pinfeathers also lowers the market grade of poultry. White turkeys can stand the hot sun better than dark turkeys. The Broad Breasted Large White has the body conformation of the Broad Breasted Bronze, but is slightly smaller when fully grown.

The Broad Breasted Bronze has black plumage and dark-colored pinfeathers. The females have white tips on the black breast feathers. The beard is black. Females normally do not have beards. The shanks and feet are black on young turkeys and change to a pinkish color on the adults. The beak is light at the tip and dark at the base. The Broad Breasted Bronze is the largest of the turkey varieties.

The reproductive ability of the Broad Breasted turkey varieties is not as good as the Beltsville Small White. Fewer eggs with lower fertility and *hatchability* (number of young produced) are produced. Artificial insemination is generally used with all varieties of Broad Breasted turkeys because the heavy males are not good breeders.

The Beltsville Small White (Figure 35-8) was developed by the United States Department of Agriculture. It is similar to the Broad Breasted Large White in color and body type. It averages about 10 pounds (4.5 kg) less in mature body weight than the Broad Breasted varieties.

Ducks

The best breeds of ducks for meat production are the White Pekin, Aylesbury, and Muscovy. Other meat-producing breeds include the

FIGURE 35-8 Beltsville Small White Turkey. *Courtesy of USDA.*

Rouen, Cayuga, Swedish, and Call. The best egg-laying breeds are the Khaki Campbells and Indian Runners.

The White Pekin (Figure 35-9) is the breed generally used in commercial meat production. It reaches a market weight of 7 pounds (3.2 kg) in eight weeks. The adult duck weighs 8 pounds (3.6 kg). White Pekins originated in China and were brought to the United States in the late 1870s. They have white feathers, orange-yellow bills, reddish-yellow shanks and feet, and yellow skin. The eggs are tinted white. White Pekins are not good setters (will not sit on a nest of eggs) and will seldom raise a brood. The White Pekin is nervous and must be handled with care.

The Aylesbury originated in England where it is still the most popular meat duck. They are the same size as the White Pekin. They have white feathers, white skin, flesh-colored bills, and light-orange legs and feet. Their eggs are tinted white. They are not as nervous as the White Pekin. However, they do not set well and, like the White Pekin, seldom raise a brood.

The Muscovy is not related to the other duck breeds. It originated in South America and includes a number of varieties. The White Muscovy is the best meat duck of the Muscovy varieties. The meat is of excellent quality if ducks are marketed before 17 weeks of age. The Muscovy has white feathers and white skin. The adult drakes (males) weigh 10 pounds (4.5 kg) and the adult ducks weigh 7 pounds (3.2 kg). They are good setters and will raise a brood. A cross between the female Muscovy and a drake of the Mallard-type produces a sterile hybrid. The sterile hybrid produces a good meat yield.

The Khaki Campbell (Figure 35-10) originated in England. Ducks of this breed are excellent egg producers, with some strains averaging close to 365 eggs per duck per year. The drake is khaki colored with a brownish-bronze lower back, tail covert, head, and neck. The bill is green and the legs and toes are dark orange. The females are khaki colored with sealbrown heads and necks. The female's bill is greenish-black and the legs and toes are brown. At eight weeks of age they weigh 3.5 to 4 pounds (1.6–1.8 kg). Adult drakes and ducks weigh 4.5 pounds (2.0 kg).

The Indian Runner originated in the East Indies. The three varieties of this breed are White, Penciled, and Fawn and White. The feet and shanks of the three varieties are orange to reddish-orange. The Indian Runner stands erect, and is about the same weight as the Khaki Campbell. It is not as high in egg production as the Khaki Campbell.

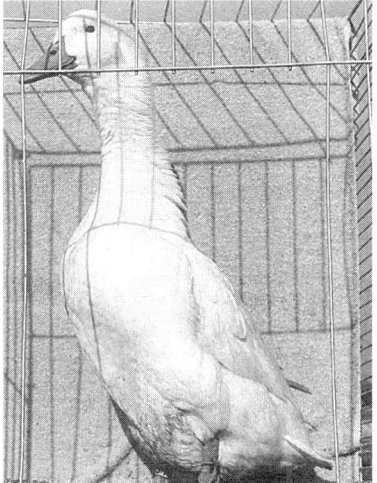

FIGURE 35-9   White Pekin drake. *Courtesy of Dr. C. Darrel Sheraw, Clarion State College.*

FIGURE 35-10   Khaki Campbell drake. *Courtesy of USDA.*

Geese

The five common breeds of geese found in the United States are the Toulouse, Embden, Chinese, Pilgrim, and African. The White Chinese is popular in some areas where it is used for weeding crops such as strawberries, asparagus, sugar beets, mint, and orchards.

The Toulouse originated in France. It is a dark gray with a white abdomen. It has a fold of skin (dewlap) that hangs down from the throat at the upper end of the neck. The Toulouse has a pale orange bill, deep reddish-orange shanks and toes, and dark brown or hazel eyes. The adult gander weighs about 26 pounds (11.8 kg). The adult goose weighs about 20 pounds (9.1 kg).

The Embden is a white breed that originated in Germany. Weights of the geese in this breed are similar to those of the Toulouse.

The Chinese breed exists in two varieties: white and brown. It originated in China. The Chinese has a knob on its beak. Adult ganders weigh about 12 pounds (5.4 kg). The adult goose weighs about 10 pounds (4.5 kg). The Chinese is popular as an exhibition and ornamental breed.

The Pilgrim male is white and the female is gray and white (Figure 35-11). The adult gander weighs about 14 pounds (6.4 kg), and the adult goose weighs about 13 pounds (5.9 kg).

The African is gray with a brown shade. It has a knob on its beak and has a dewlap. The knob and bill are black. The head is light brown and the eyes are dark brown. The adult gander weighs about 20 pounds (9.1 kg) and the adult goose weighs about 18 pounds (8.2 kg).

FIGURE 35-11   Flock of Pilgrim geese. The males are white and the females are gray and white. *Courtesy of USDA.*

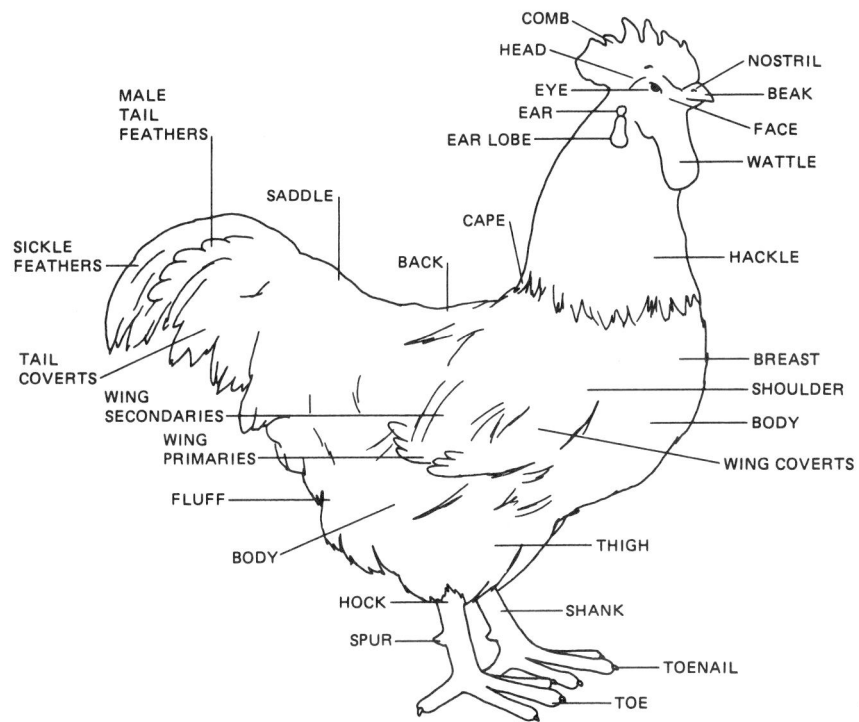

FIGURE 35-12 Parts of a male chicken.

PARTS OF A CHICKEN

To judge, select, or cull chickens, a person must know the names of the parts of the chicken. Study the parts of the chicken shown in Figure 35-12.

SOURCES OF CHICKENS

Commercial hatcheries supply almost all of the chickens produced in the United States. Commercial hatcheries obtain hatching eggs from foundation breeders who raise parent flocks. Foundation breeders usually produce hatching eggs under some type of contract with the hatchery. Large commercial egg and meat producers buy chicks or started pullets from the hatchery.

Small farm flock owners may obtain chickens from (1) a hatchery as day-old chicks, (2) a grower who produces ready-to-lay pullets, or (3) as second year layers from a commercial flock. It is possible, although not too practical, to hatch chicks on the farm.

Day-old chicks should be bought from a reputable hatchery that is U.S. Pullorum-Typhoid Clean. The chicks should be secured from a hatchery that is as close to home as possible. The less distance the chicks have to travel, the lower the loss. Day-old chicks require brooding facilities, which increases the costs of raising them to production age.

Day-old chicks can be bought as straight-run or sexed chicks. *Straight-run* chicks cost less than sexed chicks. Straight-run chicks are about one-half pullets (female) and one-half cockerels (male). *Sexed chicks* are sorted into pullets and cockerels. (Sex-linked characteristics may be used to sort chicks by sex at the hatchery. Sexing chicks by visual inspection is a specialized occupation requiring special training and practice.) The decision as to which to buy depends on how the chickens are to be used. Straight-run chicks will provide some birds for meat. If birds are wanted only for egg production, then pullets should be purchased. If meat production only is desired, cockerels should be purchased.

Started chickens may be available from a grower who specializes in this type of operation. These chickens require less equipment and care than day-old chicks. Started pullets are generally sold at about six to eight weeks of age. Ready-to-lay pullets are sold at about 16 to 20 weeks of age. Started chickens cost more than other chickens because of the feed, care, culling, and management that has already been invested in them.

Most commercial flocks of laying hens are replaced after 12 to 15 months. They go through a molting period and will then lay again. A small farm flock owner could get cheaper chickens with a bred-in potential for high rate of lay from this source.

Farm flocks used for meat production are usually started from day-old chicks. Straight-run chicks are usually bought. Meat-type breeds should be used.

Orders for chicks from hatcheries should be placed several weeks before the birds are to be delivered. Chickens that are hatched in February and March will be in production when egg prices are higher.

CULLING CHICKENS

Culling is the process of removing undesirable chickens from the flock. Culling is generally not done in large commercial egg-laying flocks. These birds have been bred to lay and usually maintain high rates of lay.

Small farm flocks can be improved by a regular culling program. When the chickens arrive on the farm, remove any deformed, weak, or diseased chickens. After laying begins, the flock should be culled regularly to remove those birds that are not laying or are laying at a low rate. Heavy culling should not be necessary during the first eight to nine months of laying.

If the flock is to be kept for a second year of production, culling should take place. However, keeping hens for a second year of production is a questionable practice. Hens usually lay about 20 to 25 percent fewer eggs in the second year of production.

Three things are considered when culling birds for egg production: (1) present production, (2) past production, and (3) rate of production. Culling for these traits is based on the appearance and condition of the body.

The hen should be handled when decisions on culling are being made. Chickens are nervous and must be handled gently. Rest the bird on the palm of one hand with the legs between the fingers. Use the other hand on the back and wings of the hen to steady it while picking it up. The hen is held with the head toward the examiner during examination. With the hen being held on the palm of one hand, use the other hand to examine the body. Keep the hen as quiet as possible during examination. Make sure the hen's feet are on the floor before releasing her after the examination.

To determine whether or not the hen is laying, examine the condition of the comb, wattles, eyes, beak, pubic bones, abdomen, and vent. A laying hen has large, bright red, soft and waxy comb and wattles. The eyes are bright and prominent. The beak is bleaching or bleached. The eyelids and eye ring are bleached. The pubic bones are flexible and spread wide enough for two to four fingers to be placed between them. The abdomen is full, soft, and pliable. The vent is moist, bleached, and enlarged.

A nonlaying hen has a small, pale, scaly, and shrunken comb. The beak is yellow, and the eyes are dull and sunken. The eyelids and eye rings are also yellow. The pelvic bones are stiff and close together with room for less than two fingers between them. The abdomen is full and hard. The vent is dry, puckered, and yellow.

Past production is indicated by the amount of yellow pigment left in the body and the time of molt. *Molting* is the process of losing the feathers from the body and wings. Producing a large number of eggs bleaches the yellow pigment from the hen's body. The beak, eye ring, earlobes, and shanks are bleached white. The feathers are worn and soiled. Molting will occur late in September or October.

The pigment bleaches from the hen's body in a definite order as laying progresses. The pigment leaves the vent first. It becomes fully bleached after about one week of laying. The eye ring is next to bleach, requiring about one week to ten days to fully bleach. The ear lobe is the third part to bleach and requires about one week to 10 days to become fully bleached. The beak bleaches next, starting at the base and progressing to the tip. This takes four to six weeks. The shanks are the last to bleach. The shanks bleach in this order: (1) front of shanks, (2) rear of shanks, (3) tops of toes, and (4) hock joint. It requires four to six months for the shanks to fully bleach.

When the hen stops laying, the pigment returns to the body parts in the same order in which it left. Return of the color takes about one-half of the time required by bleaching.

A hen that has been a poor layer will show yellow pigment in the body parts mentioned above. The molt will come before September.

Molting is a normal process after the hen has been in production. Poor producers usually molt early and may take as many as 20 weeks

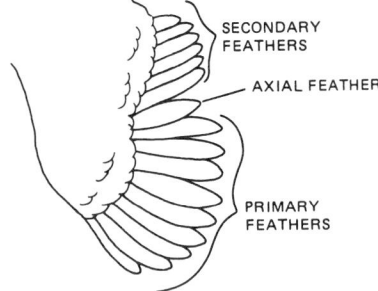

FIGURE 35-13 Location of primary, axial, and secondary feathers on the chicken wing.

to complete the molt. High-producing hens molt late and complete the molt rapidly, some as quickly as six to eight weeks. The hen does not lay eggs during the molt.

Molting occurs in a definite order. The neck feathers fall out first. Then the feathers fall from the back, wings, and body in that order. There are ten primary feathers on the outer part of the wing. These are separated from the secondary feathers on the inner part of the wing by the *axial feather*. Figure 35-13 shows the location of these feathers. The molt begins with the primary feather next to the axial feather. A slow-molting hen sheds one primary feather about every two weeks. A fast-molting hen may shed several primary feathers at the same time. It takes about six weeks for each new feather to become fully grown.

A high rate of lay is indicated by the shape and refinement of the head, the width and depth of the body, the abdominal capacity, the softness and pliability of the abdomen, the thinness of the pubic bones, the thinness and pliability of the skin, and the shape of the shanks.

A hen with a high rate of lay has a moderately deep and broad head. The face is free of wrinkles and the comb and wattles are fine and smooth textured. The hen has large body capacity, the abdomen is soft and pliable, and the pubic bones are flexible and thin. The skin is soft and pliable. The shanks are flat or wedge shaped.

A hen with a poor rate of lay has a long, shallow head and a back that is narrow and tapering. The body capacity is poor and the abdomen is hard. The pubic bones are narrow and thick. The skin is tight and thick. The shanks are round and rough.

SOURCES OF TURKEYS

Turkeys should come from flocks that are tested for and found free of the Arizona and mycoplasma infections, fowl typhoid, paratyphoid, and pullorum disease. Turkey flocks can be started with hatching eggs, day-old poults, or started poults. If the flock is started with older turkeys, the birds should be examined for lice, northern fowl mites, and other external parasites. Isolate the flock for three weeks to make sure it is free of disease and internal parasites. Most turkey growers buy started poults seven to eight weeks of age from producers who specialize in the business of breeding flocks and hatching.

SOURCES OF DUCKS

The production of meat-type breeding stock for ducks is a specialized business. Potential breeders are separated from market flocks at about six to seven weeks of age. One drake should be selected for each six ducks. Trapnesting and progeny testing can be used to select breeding stock that have high fertility, hatchability, and egg production. A *trapnest* is a nest equipped with a door that automatically closes when the bird enters the nest. The bird is held in the nest until released by the farmer. The purpose is to identify the egg laid by a particular bird. Body weight, conformation, and feathering are other important factors in the selection of market ducks.

Eggs must be carefully handled to maintain hatchability. Hatching eggs are stored at a temperature of 55°F (12.8°C) and a relative humidity of 75 percent. Hatching eggs may be stored for two weeks. After the first week they must be turned daily. Eggs must be stored with the small end down.

In commercial flocks, artificial incubators are used to hatch the eggs. Muscovy eggs require 35 days of incubation. Other domestic duck breeds require 28 days in the incubator. Fumigation of the eggs in the incubator will prevent bacteria from entering the egg and killing the developing embryo.

Caution: Do not inhale the formaldehyde fumes used for fumigation. The room containing the incubator must be well ventilated.

Eggs in the incubator are candled after seven or eight days of incubation. Infertile eggs and those with dead embryos are removed. Candling is often repeated after the eggs have been in the incubator for 25 days (32 days for Muscovy eggs).

Ducklings must be protected from chilling after they hatch. Large commercial producers usually take the ducklings from the incubator as they hatch. Small producers may prefer to leave the ducklings in the incubator until the hatch is completed.

A person who is raising only a few ducks may use natural incubation. Muscovy ducks will set on the nest and hatch a brood with little or no trouble. Other breeds do not set eggs very well. Eggs from these breeds may be hatched by setting them under a broody hen. Keep feed and water close to the nest.

SOURCES OF GEESE

Geese may be secured as day-old goslings from a hatchery. They are incubated by artificial means in much the same manner as ducks. Geese, like ducks, can be hatched using natural incubation.

SUMMARY

Most of the poultry raised in the United States is produced in large commercial flocks. Chicken enterprises include egg production, meat production, and raising pullets for replacement. Most of the turkeys, ducks, and geese are raised for meat. Much of the commercial poultry industry is vertically integrated.

The production of eggs in the United States has increased in recent years, while per capita consumption continues to decline. Ohio and California lead all other states in egg production. There has been a major increase in broiler production and per capita consumption of poultry since 1967. Broiler production tends to be concentrated in the southeastern states, with Georgia and Arkansas generally leading the other states in numbers of broilers produced.

Turkey production and per capita consumption has increased significantly since 1967. Turkey production is generally not concentrated in any one part of the United States. North Carolina and Minnesota are often the leading states in numbers of turkeys produced.

About 60 percent of the ducks raised in the United States are raised on Long Island, New York. Geese are raised all over the United States. The leading states are Missouri, Iowa, South Dakota, Minnesota, Wisconsin, Ohio, Indiana, California, and Washington.

Most chickens used for egg production are Leghorn strain crosses. Crossbred chickens are generally used for broiler production. In both cases, the chickens are bred for the type of production desired.

Most of the turkeys raised in the United States are Broad Breasted Large Whites. The White Pekin duck is most commonly used for meat production. The most popular meat breeds of geese are the Toulouse and the Embden.

Most of the poultry is supplied by commercial hatcheries. Poultry may be purchased as day-old, started birds, or ready-to-lay chickens.

Hens are culled to remove unproductive birds from the flock. Culling is based on appearance and condition of the body. Pigmentation and molt are indicators of laying.

Student Learning Activities

1. Prepare and give an oral report on the nature of the poultry industry.
2. Give an oral report on common breeds of poultry.
3. Prepare a bulletin board display of pictures of common breeds of poultry.
4. Survey the local community to determine the common breeds of poultry and the sources of poultry used by local producers.
5. Demonstrate the proper method of culling hens for indications of production.

Review

1. Describe the kinds of enterprises found in the poultry industry.
2. What is meant by the term *vertical integration*?
3. How important is vertical integration in the poultry industry?
4. Describe the trends in production and consumption of poultry products in the United States.

5. Name the ten leading states in the production of eggs.

6. Name the twelve leading states in broiler production.

7. Name the ten leading states in turkey production.

8. Where in the United States is the largest number of ducks raised?

9. Name the leading states in geese production.

10. List the advantages and disadvantages of raising poultry.

11. Name the four classes of chickens generally raised in the United States.

12. What breeding methods are commonly used to produce most of the chickens used for egg and meat production in the United States?

13. What strain crosses are popular for commercial egg producing flocks? Why?

14. What common commercial cross is often used for meat-producing chickens? Why?

15. Name the commercially important varieties of turkeys.

16. Describe each of the varieties named in number 15.

17. Name and describe the common breeds of ducks.

18. Name and describe the common breeds of geese.

19. Draw a sketch of a chicken and name the parts.

20. Compare day-old chicks with started chicks as sources for the producer.

21. What three things are considered when culling chickens for egg production?

22. Describe how a chicken should be handled when culling.

23. Describe the characteristics of a laying and non-laying hen.

24. Describe how a chicken molts.

25. Name the factors a producer should consider when selecting turkey poults.

26. Name the factors a producer should consider when selecting ducks.

36 *Feeding, Management, Housing, and Equipment*

Objectives

After studying this unit, the student should be able to

- describe accepted feeding practices for different kinds of poultry.
- describe approved management practices for different kinds of poultry.
- list housing and equipment required for various kinds of poultry enterprises.

FEEDING POULTRY

Chickens

The cost of feed is about two-thirds of the total cost of producing eggs and meat from chickens. All management practices affect profits in chicken production. Feeding practices must meet the needs of the chickens and still allow room for profit.

Rations for the chickens must supply the protein, carbohydrates, minerals, vitamins, and water that poultry require. In addition, some additives and unidentified growth factors are often used in chicken rations. Tables that give the nutrient requirements of poultry and the composition of common feeds for poultry are found in the appendix of this text.

Most of the energy in poultry diets is supplied by grains, grain by-products, and animal and vegetable fats and oils. Poultry have only a limited ability to use high-fiber feeds such as roughages because of their relative inability to digest the fiber. Grains make up from 50 to 80 percent of the total ration for chickens. Corn is the most commonly used grain in poultry rations. Other grains such as oats, wheat, grain sorghum (milo), and proso millet may be substituted for part of the corn in poultry rations.

When poultry are fed *ad libitum* (given all they will eat), they tend to eat enough to meet their energy requirements. Chickens fed low-energy diets will eat more feed than those fed high-energy diets. Therefore, the amount of required nutrients in a poultry ration must be adjusted in relation to the energy level in the ration in order to ensure that the birds consume the right amount of the needed nutrients. The concentration of nutrients must be increased in high-energy diets because the birds will eat less of the ration per day. The concentration of nutrients should be reduced in low-energy rations because the birds will eat more of these rations per day. High-energy rations usually result in higher efficiency in converting feed to meat or eggs as compared to low-energy rations. Pelleted feeds contain more nutrients per volume of feed; this results in a higher intake of nutrients with the consumption of a given amount of feed as compared to non-pelleted feeds. Adding fat to the ration increases the amount of energy in the diet.

When substituting other grains for corn in poultry diets, care must be taken to ensure that the level of required nutrients in the ration is adjusted to compensate for the difference in energy level and feed intake. Corn has a higher energy level than other grains that can be used in poultry rations. Generally, other grains should be substituted for only part of the corn in the ration because of their lower energy level and higher fiber level. For example, oats or barley should be limited to no more than 10 to 15 percent of the ration.

When preparing poultry feeds, corn should be coarsely ground. Oats, barley, and millet should be finely ground. Wheat and grain sorghum should be processed through a roller mill. If wheat is ground, care must be taken not to grind it too fine.

Fats may be used to increase the energy level of low-energy rations. Animal and vegetable fats should be limited to no more than five to 10 percent of the diet. Fats will increase the palatability of the diet, decrease dustiness, and improve the texture of the feed. Fats are used more often in broiler rations to increase the energy level of the diet. In hot weather, feed containing added fat may become rancid unless it has been properly stabilized.

Proteins supply the essential amino acids needed by chickens. The amino acids most commonly deficient in chicken rations are arginine, glycine, lysine, methionine, and tryptophane. Common sources of protein for chickens include soybean meal, meat scraps, and fish meal. Balancing the ration on the amino acid needs rather than crude protein content assures that the nutritional needs of the chickens are met. This practice may also result in a lower-cost ration.

From 55 to 78 percent of the egg and the chicken's body weight is water. Younger chickens have the highest water content. Chickens need a fresh, clean supply of water at all times.

The daily water consumption of chickens varies with the type of chicken, the temperature, type of diet fed, rate of growth, egg production, and type of watering equipment used. Table 36-1 shows the daily water consumption of chickens at moderate temperatures (68–77°F [20–25°C]). Water consumption will increase at higher temperatures. On very hot days, chickens may consume as much as three times the water used on cooler days.

The mineral requirements of chickens include calcium, phosphorus, manganese, iodine, sodium, chlorine, and zinc. Calcium can be supplied by the addition of oyster shell or limestone. Current recommendations for calcium are that it make up 3.4 percent of the ration for laying hens. It is especially important for laying hens because the formation of the shell of the egg requires a great deal of calcium. Commercial mineral mixes can be used to supply the other minerals needed in the ration.

Vitamins required by chickens include A, D_3, E, K, riboflavin, B_{12}, niacin, pantothenic acid, and choline. Grains, protein sources, and green feeds will supply some of the vitamin needs. However, it is common practice to use a vitamin premix when preparing chicken rations. This ensures that the necessary vitamins are present in the ration.

TABLE 36-1 Daily Water Consumption of Chickens at Various Ages.

Age (weeks)	Broilers gallons/1,000 head/day	Broilers liters/1,000 head/day	White Leghorn Hens gallons/1,000 head/day	White Leghorn Hens liters/1,000 head/day	Brown Hens gallons/1,000 head/day	Brown Hens liters/1,000 head/day
1	8.5	32.1	7.5	28.6	7.5	28.6
2	18.1	68.6	11.3	42.9	15.1	57.1
3	27.4	103.6				
4	37.7	142.9	18.9	71.4	26.4	100.0
5	47.2	178.6				
6	56.6	214.3	26.4	100.0	30.2	114.3
7	66.0	250.0				
8	75.5	285.7	30.2	114.3	34.0	128.6
9						
10			22.6	85.7	37.7	142.9
11						
12			37.7	142.9	41.5	157.1
13						
14			41.5	157.1	41.5	157.1
15						
16			45.3	171.4	45.3	171.4
17						
18			49.1	185.7	49.1	185.7
19						
20			60.4	228.6	56.6	214.3

Source: National Academy of Sciences.

The use of additives in poultry rations is discussed in Unit 7. Proper management practices must be followed whenever feed additives are used in the ration.

Unidentified substances found in dried whey, marine and packinghouse by-products, distillers' solubles, green forages, soybeans, corn, and other natural materials have been found to be beneficial to chickens. These factors stimulate growth, increase reproduction, improve egg quality, or reduce the toxicity of some minerals. Many poultry rations for young birds and breeders contain some of these feeds to add these factors to the diet.

Chickens have no teeth. Grit in the ration helps the gizzard in grinding feed. *Grit* is usually small particles of granite. Research indicates that grit may or may not improve the use of feed with all-mash (completely ground and mixed) rations. It is definitely needed when other types of rations are fed.

Grit comes in small, medium, and large sizes. Feed the small size to chicks, the medium size to growing chickens, and the large size to hens. Young chicks should be fed the grit by sprinkling it on the feed twice a week. For growing chickens and hens, feed grit free-choice in hoppers.

The different kinds of chicken feeds are all available as commercially manufactured feeds. Commercial feeds are available in the form of *mash*, *pellets*, or *crumbles*. Pellet and crumble forms cost more than mash. Chickens may produce slightly better with the pelleted or crumble forms. Pellet and crumble feeds reduce feed waste and the chickens may eat them a little better. The manufacturer's directions for use must be followed closely. Always follow feed tag directions closely when using feeds containing additives.

Four systems that may be used for preparing rations are:

- Use a complete commercial feed.
- Use a commercial protein concentrate and mix with local or homegrown grains.
- Use a commercial vitamin-mineral premix and soybean meal and mix with local or homegrown grain.
- Buy all the individual ingredients and mix the ration.

The choice of system depends on the age of the chickens being fed, the relative costs, the size of the enterprise, and the equipment available for grinding and mixing feeds. Farmers with small flocks generally find it more economical to use a complete commercial feed. Large commercial poultry enterprises generally use some system of mixing either at a local mill or on the farm. For large operations, mixing the feeds usually results in a lower feed cost per ton.

Phase feeding is a system of making specific feeds to be used to meet the changing nutritional requirements of chickens. A number of factors such as rate of lay, egg size, body maintenance needs, and temperature are used to determine the content of the feed. The purpose is to lower feed cost. There are some problems in the use of phase feeding. The procedure for taking into account all of the factors when making up a ration is a complicated one. When more than one age group of chickens is being fed, more feed storage bins and feeders are needed. It requires a high level of management ability on the part of the producer. Phase feeding is more practical for large commercial operations than it is for small farm flocks.

Chickens should be fed only fresh feed. Overfilling feeders results in wasted feed. Hanging feeders should be filled only three-fourths full. Trough feeders should be filled only two-thirds full. Fill feeders in the early morning and refill during the day as needed. Keep the feed clean and never supply moldy or dirty feed. Feeders should be cleaned periodically.

Feed should be stored in a dry place where rats or mice cannot get at it. Feed should be handled and stored in bulk for large operations.

A good manager watches the feed consumption of the flock closely. A drop in feed intake is often the first sign of trouble. Stress, disease outbreak, molting, or other management problems are often first indicated by a drop in feed consumption.

Feed intake is also influenced by the energy level of the ration, level of protein intake, environmental temperature, body size, amount of feather coverage, and rate of growth or egg production.

Feed intake is greater when a high-energy ration is fed and lower when a low-energy ration is fed. Feed intake is also lower at higher environmental temperatures and higher at lower environmental temperatures. Feed intake will change about 1.5 percent for each 1.8°F (1°C) above or below 68° to 70°F (20°–21°C). While temperature is the major factor that affects the level of feed intake, a low intake of protein may cause a higher total feed intake. The other factors mentioned above also have some influence on the level of feed intake.

Feeding for egg production.   Chicks must be given feed and water as soon as they are put into the brooder house. Starter rations should contain 18 percent protein. Complete starter feeds should be used for chicks and fed without grain until they are six weeks old. About 40 pounds (18 kg) of starter mash per 100 chicks is required for the first two weeks. From two to six weeks, 250 to 300 pounds (113–136 kg) of starter mash is needed per 100 chicks. Chick-size grit should be provided when the birds are young and a medium-size given as they become older.

For chicks up to two weeks of age, 100 linear inches (2.4 m) of feeder space is required for 100 chicks. Two hundred linear inches (5 m) of feeder space is needed per 100 chicks from three to six weeks of age. Up to two weeks of age, 25 linear inches (64 cm) of waterer space is needed per 100 chicks. From three to six weeks of age, 50 linear inches (1.3 m) of waterer space is required per 100 chicks.

At six weeks of age, the feed should be changed to a growing ration. From six to 14 weeks the ration should contain 15 percent protein. The ration should contain 12 percent protein from 14 to 20 weeks. Grain may be fed from six weeks on. Grain and mash should be fed in separate feeders. Begin with 10 pounds (4.5 kg) of grain to 100 pounds (45 kg) of mash. Gradually increase the amount of grain until the ration is one-half grain and one-half mash. At 18 to 20 weeks of age change the ration to a laying mash. The change should be made over a two-week period of time.

From 7 to 12 weeks of age, 250 linear inches (6.4 m) of feeder space and 50 linear inches (1.3 m) of waterer space are needed per 100 birds. From 13 to 20 weeks, 300 linear inches (7.6 m) of feeder space and 100 linear inches (2.5 m) of waterer space are needed per 100 birds.

A laying ration should contain 14.5 percent protein. The cost per dozen eggs produced should be used in deciding which is the most economical ration to use. Egg production, body size, health, and temperature all affect the amount of feed a laying hen eats (Figure 36-1). Light breeds eat about 24 pounds (10.9 kg) per 100 hens per day. A hen of the light-weight breeds eats a total of 85 to 90 pounds (38.6–40.8 kg) of feed per year. A heavy-breed hen eats 95 to 115 pounds (40.8–52.2 kg) per year. About two to five pounds (0.9–2.3 kg) of oyster shell and one pound (0.45 kg) of grit per year per hen are required.

Light-weight breeds require 300 linear inches (7.6 m) of feeder space and 50 linear inches (1.3 m) of waterer space. Heavy-weight breeds need 400 linear inches (10.2 m) of feeder space and 100 linear inches (2.5 m) of waterer space per 100 birds.

The ration for the laying flock should not be changed suddenly. This creates a stress in the hens that will result in a drop in egg production. Even changing from a coarse-ground feed to a fine-ground feed may create stress and a drop in production.

FIGURE 36-1 A laying house with automatic waterers, hanging feeders, and slatted floor. *Courtesy of Iowa State University.*

Three systems of feeding are commonly used for laying hens: (1) all mash, (2) mash and grain, and (3) cafeteria.

The *all-mash system* is the use of a complete ground feed. This system is well adapted to use with mechanical feeding systems. It is often used by commercial egg producers.

In the *mash and grain system*, grain and mash are fed separately. A 20- to 26-percent protein supplement is fed with a light grain feeding in the morning. Corn is fed in the evening. Some of the grain may be fed in the litter. This causes the hens to stir up the litter and help keep it drier.

The *cafeteria system* allows the birds to balance their own rations. Grain is fed in one feeder and a 26- to 32-percent protein supplement is fed in another feeder. Feed is kept in the feeders at all times. Older hens may tend to eat too much grain and not enough protein supplement when this system is used. About one-fourth of the feeders should contain protein supplement and three-fourths of the feeders should contain grain.

Feeding efficiency is the number of pounds (kilograms) of feed required to produce a dozen eggs. Records must be kept of the amount of feed used and the egg production in order to calculate feed efficiency. Divide the total pounds (kilograms) of feed fed by the number of dozen eggs produced to calculate how many pounds (kilograms) of feed it took to produce one dozen eggs. Feed efficiency should be about 4.3 pounds (1.95 kg) of feed per dozen eggs. Wasted feed, low rate of lay, health problems, or other management problems lower feed efficiency.

Feeding broilers. For chicks from day-old to six weeks, use a broiler starter ration with at least a 23-percent protein level. Replacement chick starters, designed for egg production type birds, should not be used for broiler starter rations. Broiler starter rations should contain 3 percent added fat and a coccidiostat. The protein level is adjusted to the energy level of the ration. Use chick-sized grit scattered on top of the mash or feed in separate hoppers. At three to six weeks of age, broilers should be fed a ration containing 20 percent protein.

Broilers need 100 linear inches (2.5 m) of feeder space and 25 linear inches (64 cm) of waterer space per 100 birds up to two weeks of age. Provide 300 linear inches (7.6 m) of feeder space and 50 linear inches (1.3 m) of waterer space for birds from three to six weeks of age.

Feed a finishing ration of 18 percent protein to birds from six weeks old to market age (usually, about eight to nine weeks of age). The finishing ration should contain 3 percent or more of added fat. Broiler rations are fed as complete mixed feeds and never as separate grain and protein supplement, as rations for laying hens might be.

Provide 350 linear inches (8.9 m) of feeder space and 75 linear inches (1.9 m) of waterer space per 100 birds from six weeks to market age. Broilers are full fed at all times to obtain the fastest possible gains.

Capons and roasters are fed to heavier weights than broilers. They are fed the same ration as broilers up to six weeks of age. After six weeks, grain is added to the ration in addition to the finishing mash until it is being fed as one-half of the ration.

Vitamins and trace minerals are included in rations in small amounts. Broiler rations must be mixed properly and the quality carefully controlled. This is the responsibility of the feed mill where the mixing is done. Inventory control, assaying of mixed feeds (analysis for content), and assaying of ingredients are methods of insuring proper mixing and quality control.

Feed conversion in broilers refers to the amount of feed needed to produce one pound or one kilogram of live weight. It is found by dividing the total weight of feed fed by the total weight of broilers marketed. Feed conversion after condemnations is found by subtracting the weight of the condemned meat from the weight of the birds marketed before dividing into the total weight of the feed used.

Feed conversion is affected by a number of factors including

- genetic background of the strain being fed.
- type of feed used.
- temperature.
- amount of feed wasted.
- additives used in the feed.
- general management of the operation.

Other factors also affect feed conversion, but those mentioned here are the major ones. Current feed conversion in the broiler industry averages about 1.85 pounds (0.84 kg) of feed per one pound (0.45 kg) of gain for males and 1.95 pounds (0.88 kg) of feed per one pound (0.45 kg) of gain for females.

Feeding the breeding flock. A special mash is made for use with breeder flocks. It contains 14.5 percent protein and is fortified with extra vitamin A, D, B_{12}, riboflavin, pantothenic acid, niacin, and manganese. This mash is more expensive than laying-hen mash. The breeding flock should be started on breeder mash about one month before hatching eggs are to be kept. Breeding flocks may be fed a ration that is one-half mash and one-half grain. The breeding flock is fed in much the same way as is the laying flock.

Replacement pullets for breeding flocks of meat-type chickens tend to grow too fast and develop sexual maturity too early. This results in too many small eggs that are not good for hatching purposes. Egg production tends to be at too low a level at its peak.

Several methods may be used to slow down the rate of growth and sexual maturity. These include limiting feed intake, skip-a-day feeding, low-protein diets, and low-lysine diets.

A *limited-feed intake program* is based on feeding at the rate of about 70 percent of the normal feed level. This program is begun when the birds are from seven to nine weeks of age and is continued until they are 23 weeks old. They are then full fed on a laying mesh.

A *skip-a-day program* involves full feeding every other day. On alternate days, only two pounds (0.9 kg) of grain is fed per 100 birds. This feeding program is followed from the seventh or ninth week to the 23rd week.

A *low-protein diet* is one in which the protein level of the ration is 10 percent. This program is also followed from the seventh or ninth week to the 23rd week.

A *low-lysine diet* will also slow down rate of growth and sexual maturity because of the amino acid imbalance. The diet contains 0.4 to 0.45 percent lysine and 0.6 to 0.7 percent arginine. It is followed from the seventh or ninth week until the 23rd week. The protein level of this diet is 12.5 to 13 percent.

Turkeys

The general principles of feeding turkeys are similar to those for feeding chickens. Major differences are in the protein levels required and the importance of the vitamins biotin and pyridoxine in the turkey diet. Refer to the tables in the appendix for the nutrient requirements of turkeys.

Feeding turkey poults. Poults (young turkeys) must be fed and watered as soon as possible after hatching. When feeding and watering is delayed beyond 36 hours, the poults have difficulty learning to eat. It may be necessary to force feed them to get them started. Start poults on a turkey pre-starter or starter ration. The protein content should be 28 percent. Pre-starters are generally used only for the first week. They contain higher levels of antibiotics, vitamins, amino acids, and energy than do starter feeds. Pre-starters are especially useful if the poults are under unusual stress conditions. From four to eight weeks of age, the starter should contain 26 percent protein and a higher energy level.

Feeding growing turkeys. Turkeys may be moved to range (Figure 36-2) or fed in confinement at eight to 10 weeks of age. Confinement feeding results in faster gains. However, range feeding may result in as much as 10 percent feed savings. A good range contains forage or grain crops. The stocking rate for range varies from 100 to 250 turkeys per acre.

Growing turkeys should be separated by sex because toms have a higher protein requirement than hens. Turkeys generally give better feed conversions when fed complete mixed rations. Pelleting the ration gives the best results. As the turkeys become older, the energy levels of the ration are increased and the protein level is decreased. The tables in the appendix show the amounts needed at various age levels.

FIGURE 36-2 Turkeys may be grown on range. *Courtesy of USDA.*

The daily water consumption of turkeys is shown in Table 36-2. Water consumption will increase in hot weather. For male turkeys at eight to 12 weeks of age and females at eight to 11 weeks of age, the ration should have a protein content of 22 percent. Rations for male turkeys at 12 to 16 weeks of age and female turkeys at 11 to 14 weeks of age should have a protein content of 19 percent. Rations for male turkeys at 16 to 20 weeks of age and female turkeys at 14 to 17 weeks of age should have a protein content of 16.5 percent. Rations for male turkeys at 20 to 24 weeks of age and female turkeys at 17 to 20 weeks of age should have a protein content of 14 percent. Male turkeys are usually marketed at about 24 weeks of age and female turkeys at 20 weeks of age.

Feeding breeding turkeys. Turkeys should be selected at about 16 weeks of age for the breeding flock. A holding diet containing 12 percent protein and of medium energy level is usually fed female breeding turkeys from 16 to 30 weeks of age and males from 16 to 26 weeks of age. A breeding ration containing 14 percent protein is then fed.

TABLE 36-2 Water Consumption of Turkeys at Moderate Temperatures [68–77°F (20–25°C)].

| Age (weeks) | Large White Turkeys | | | |
| | Males | | Females | |
	gallons/1,000 head/day	liters/1,000 head/day	gallons/1,000 head/day	liters/1,000 head/day
1	14.5	55.0	14.5	55.0
2	28.3	107.1	26.0	98.6
3	42.8	162.1	35.1	132.9
4	62.3	235.7	48.1	182.0
5	84.5	320.0	66.0	250.0
6	108.3	410.0	81.1	307.1
7	130.6	494.3	99.6	377.1
8	151.7	574.3	120.0	454.3
9	176.2	667.1	147.2	557.1
10	201.7	763.6	166.0	628.6
11	220.8	835.7	174.4	660.0
12	234.7	888.6	175.9	665.7
13	244.5	925.7	176.6	668.6
14	252.1	954.3	177.4	671.4
15	256.6	971.4	178.1	674.3
16	261.2	988.6	178.9	677.1
17	262.7	994.3	179.6	680.0
18	264.2	1000.0		
19	264.9	1002.9		
20	265.7	1005.7		

Source: National Academy of Sciences.

Males in the breeding flock are fed a breeding ration beginning at about 26 weeks of age. To control the weight of males in the breeding flock, they may be fed a limited diet. Females are normally fed all they will eat.

Feeding Ducks and Geese

Commercial feeds in mash, pelleted, or crumble form are available for ducks and geese. Pelleted or crumble forms are recommended. If a commercial feed for ducks or geese is not available, a chicken feed may be used. However, be sure the chicken feed does not contain a coccidiostat if it is used for ducks or geese.

A starter ration containing 22 percent protein is used for the first two weeks. Feed all that the birds will eat. Geese should be allowed out on pasture during the day, where they will eat grass and bugs. During the first several days, the feed should be placed on rough paper, paper plates, or egg fill flats. Do not place the feed on a smooth surface that could cause the young birds to slip and fall. A smooth surface may cause leg injuries. Supply an insoluble grit in addition to the feed.

Geese may be allowed to forage for feed after they are two weeks old. Under this feeding program, they are fed grain for the last two or three weeks and marketed at about 18 weeks of age. Geese may also be fed a grower ration while they are allowed to forage for feed. Under this feeding program, they may be switched to a high-energy finishing diet and marketed at about 14 weeks of age. Another feeding method is to feed the geese in full confinement and market them at about 10 weeks of age. Geese marketed under this feeding program are called "junior" or "green geese."

Ducks may be raised either in total confinement or in a growing house; this should open onto an exercise area that includes water for swimming or wading. Pelleted feeds are better than mash feeds for ducks. They are usually put on a starter diet containing 22 percent protein for two weeks. They are then fed a grower diet containing 18 percent protein and finished on a diet containing 16 percent protein. Ducks are usually ready for market in seven to eight weeks.

Provide a supply of fresh water that is easily accessible at all times. Ducks and geese consume large amounts of water. The waterers should be designed in such a way that the birds cannot get into them to swim. Water for swimming is not necessary.

Pasture is especially important for geese. They will start to eat grass when only a few days old. Geese can live entirely on pasture after they are five to six weeks of age. A growing ration is recommended, however. Timothy, bluegrass, ladino, white clover, and bromegrass are good pastures. Barley, wheat, and rye make good fall pastures. Geese do not eat alfalfa, sweet clover, or lespedeza.

If the pasture is not of good quality, some additional grain should be fed. Cracked corn, wheat, or milo can be used. Pastures should be rotated. Young goslings should be protected from rain for the first several weeks. Provide shade in hot weather. One acre of pasture will feed about 20 birds.

Ducks will eat some green feed, but are not as good at foraging as geese. Pasture is not necessary for ducks. Farm flocks are usually not confined, however.

A breeder-developer ration is fed to birds being kept as breeders. These birds are given a breeder diet when they are older. Breeder diets contain less energy than grower diets. The breeder diet should be used starting about one month before eggs are to be kept for hatching. Breeder flocks require more calcium, which can be provided by feeding oyster shell.

MANAGING POULTRY

Chickens

Brooding chickens. Small farm flocks are usually brooded on the floor using hover-type brooders (Figure 36-3) or infrared heat lamps. Large commercial flocks are usually brooded in battery brooder units or in wire cages in houses with controlled heat and ventilation.

The brooder house must be prepared for the chicks. Be sure that it is in good repair. Clean and disinfect the house before the chicks arrive. Use clean litter for brooding chicks. Commercial litters such as peat moss or sugarcane pulp may be used. Other materials such as ground corncobs, chopped straw, wood shavings, or sawdust may be used. Battery brooders and wire cages do not use litter.

Three to 4 inches (7.6–10 cm) of litter is placed on the floor. A deep litter system is commonly used. This means that fresh litter is added as needed. It may reach a depth of 8 to 10 inches (20–25 cm). Stir the litter to keep it from packing.

Brooders should be put into operation two days before the chicks arrive. This gives the producer a chance to check the equipment and make sure it is working properly. The temperature under the hover must be 90° to 95°F (32°–35°C) for day-old chicks. Brooder temperatures are measured about 3 inches (7.6 cm) off of the floor and about 3 inches (7.6 cm) inside the outer edge of the hover. Brooder temperatures are reduced about 5°F (2.8°C) each week until the temperature reaches 70° to 75°F (21°–23.9°C). Brooder temperatures must be checked at least twice a day.

The behavior of the chicks gives some indication of proper brooder temperature. If the chicks crowd close together and cheep, they are too cold. If they move out from under the brooder, pant, and hold their wings away from their bodies, they are too hot.

A *hover guard* (chick guard or brooder guard) is used for the first week of brooding to prevent the chicks from wandering away from the heat and becoming chilled. This is a corrugated cardboard or cloth-covered wire about 12 inches (30.5 cm) high. It is placed about two feet

FIGURE 36-3   Broiler chicks being brooded under hover-type brooders in a commercial broiler operation. *Courtesy of USDA.*

(0.9 m) from the hover edge and all the way around it after the chicks are placed underneath. Move it back a short distance each day. Use a wire guard if the chicks are debeaked. When using infrared lamps for brooding, make a circle with the chick guard about 8 feet (2.4 m) in diameter around the heated area. This will keep the chicks confined to the area under the lamp and prevent chilling. With any type of brooder, remove the chick guard after one to two weeks. Blocking the corners of the brooder house with cardboard will prevent the chicks from crowding into the corners and smothering. Chicks may crowd into the corners when they are frightened or if they are cold and have not yet learned to return to the heated area.

Chicks need heat until they are well feathered. This will be at four to eight weeks of age, depending on the season.

Seven to 10 square inches (45–64 cm^2) of space is needed under the hover for each chick. More space is required per chick with electric brooders than with other types. The house should provide a total of 0.75 to 1.0 square feet (0.07–0.09 m^2) of floor space for broiler chicks. Leghorn-type pullet chicks require 1.5 to 2 square feet (0.14–0.18 m^2) of floor space. Heavier breeds need 2.5 to 3 square feet (0.23–0.28 m^2) of floor space.

Roosts are not used for broiler chickens. Using roosts for broiler chicks causes crooked breast bones and breast blisters. Laying hens may use roosts. If they do, then roosts should be provided for the young chickens when they are about four to six weeks of age. About 4 to 6 inches (10–15 cm) of roost space is needed per bird.

A small amount of light should be provided under the brooder. Gas brooders often provide enough light and no additional light is needed. A 7.5- to 10-watt bulb should be used under an electric brooder. Room lights are often used for the first two or three days. After that, no room light is needed at night. Commercial broiler producers often use lights 24 hours a day for four to six weeks. Sometimes they are used until the broilers reach market weight.

A brooder house must provide fresh air. Be sure that there are no drafts on the floor. If the litter is wet and there is a strong odor, the ventilation needs to be improved. This will help prevent respiratory diseases.

Small, trough-type feeders may be used when the chicks are young. As they become older, the feeders should be larger. Mechanical and hanging feeders may be used. Round or trough-type waterers may be used. Automatic waterers may be used when the chicks get older.

Heat lamps (infrared type) may be used for brooding small numbers of chicks. Lamps of this type will brood from 50 to 75 chicks. A chick guard must be installed to prevent drafts and confine the chicks to the heated area. Hang the lamps about 18 inches (45.7 cm) from the litter. The lamps are raised as the chicks grow.

Large commercial operations often use battery or wire cage brooding. Battery brooders are heated with electric heaters. The entire house is heated when wire cages are used. Houses must be well insulated and ventilated.

Day-old chicks in battery brooders require 10 square inches (64 cm²) of space per chick. Space should be increased by 10 to 15 square inches (64–80 cm²) every two weeks. Using plastic or plastic-coated wire floors reduces breast blisters in broilers.

Wire cages are commonly used when brooding commercial-size laying flocks. Leghorn-type chickens need 20 to 30 square inches (129–194 cm²) of floor space for the first seven to eight weeks. They require 45 to 55 square inches (290–355 cm²) from eight to 18 weeks. Heavier breeds require about 25 percent more floor space per chicken.

Debeaking is done to control cannibalism (picking). The causes of cannibalism are not well understood. Crowding, type of ration, overheating, and too much light may be factors in its occurrence. *Debeaking* (Figure 36-4) is the cutting off of one-third to one-half of the upper beak and one-fourth of the lower beak. Debeaking can be done at any age. However, if done too young, it may not be permanent. Debeaking at nine to 12 weeks of age is usually permanent. Birds may be ordered from the hatchery debeaked. Debeaking does not affect the health or growth of the chickens.

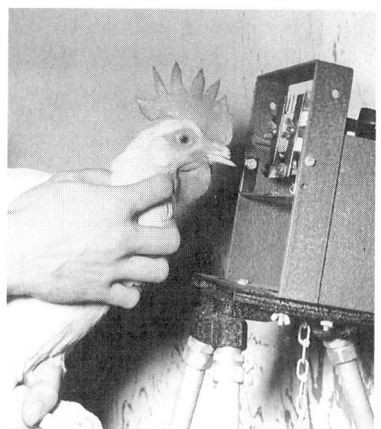

FIGURE 36-4   Debeaking a chicken. *Courtesy of Iowa State University.*

Raising laying pullets. Replacement pullets for laying flocks may be raised in confinement or on range. The trend is toward confinement raising. Less land and labor is needed for confinement systems. Losses from parasites and predators are lower in confinement raising.

Pullets may be raised in the same house that is used for brooding, or they may be moved to a growing house. Pullets should be culled when they are moved to the growing house. They may be grown in cages, using one of two systems: partial cage growing or complete cage growing. *Partial cage growing* is floor brooding for the first six to 10 weeks and then moving the pullets to cages. *Complete cage growing* is brooding and growing entirely in cages.

Light affects the age of sexual maturity and the rate of egg production. The length of the day is the controlling factor. Thus, increasing day length speeds up sexual maturity, and decreasing day length slows down sexual maturity. If pullets reach sexual maturity too early, they will lay small eggs for several months.

Several systems of controlling light during the growing period are used. The basic principle involved is not to grow pullets between 12 and 22 weeks of age under conditions of increasing light. Dark, windowless houses are used in some systems to permit total control of light. Some systems reduce the light gradually during the growing period. Others maintain a constant day length during the growing period. Day length is increased when the pullets are moved into the laying house.

Less feed is required for growing pullets on range. However, more labor is required and there are greater problems with disease and predators. Clean range must be used and must be rotated. One acre can carry

FIGURE 36-5 Shelters are provided for chickens raised on range. *Courtesy of Iowa State University.*

from 300 to 500 birds. Birds of different ages should not be mixed on range. Range shelters and shade are needed (Figure 36-5). Range growing of pullets is better adapted to small flocks than large commercial operations.

Laying flock management. Laying flocks may be housed in open-floor systems or caged systems. Smaller flocks often use open-floor systems. The trend in large commercial flocks is toward using some system of caged layers.

Laying houses must be cleaned and disinfected before pullets are moved in. The laying house should remain empty for at least one week after cleaning before moving the pullets in. The house and equipment should be in good repair and new litter put in the house.

Handle pullets gently when moving them to the laying house. They are easily frightened. Moving them in the dark will excite them less.

Watch the pullets for a few days to make sure they do not pile up and smother. Nests should be cleaned regularly and new nesting material added when needed. Keep waterers and feeders clean. Keep the litter stirred and add new litter when needed. Do not allow the litter to become wet and caked over.

Several types of cages are used for cage laying systems including: single cages, multiple-bird cages, and colony cages. Multiple-bird cages contain two to 10 birds. Colony cages contain 11 to 40 birds.

Single cages may be arranged in single rows, back-to-back, stair-step, or double-deck. *Single row* is two rows of cages separated by an aisle. This system is generally used in narrow houses. The *back-to-back system* uses two rows of single cages placed back-to-back. This permits a sharing of common water and feed troughs. The capacity of the house is increased by the use of this system.

A *stair-step system* is created by offsetting one tier of single cages below another (Figure 36-6). Manure drops to the floor rather than on dropping boards. *Double-deck systems* place one row of cages directly on top of another. Dropping boards are used to catch the manure. Problems with this system include the difficulty of removing the manure and providing enough light to the chickens in the lower cages.

Most multiple-bird cages house two to five birds per cage. Cost per bird is lower than with single cages. Cage arrangements are the same as with single-bird cages. The stair-step arrangement is considered to be the most profitable.

Advantages of colony cages include lower housing and labor costs. Problems include higher death loss, lower production, and more dirty and broken eggs. As more hens are placed in the colony cage, the production level decreases.

Feeding in cages may be done by hand or automatically. Watering may be done by the use of continuous flow troughs, drip nipples, or automatic cups.

FIGURE 36-6 Chickens in cages arranged in stair-step fashion. *Courtesy of University of Illinois at Urbana-Champaign.*

Laying hens require a minimum of 14 hours of light per day. Many controlled lighting systems use a maximum of 17 hours of light per day. Artificial lighting must be used as the length of available natural light decreases with the season. Several types of lighting systems are used. These include all-artificial lights, morning light, evening light, a combination of morning and evening light, and all-night lighting.

Light is increased to 12 hours at 20 weeks of age. It is then increased gradually (15–20 minutes per week) until the maximum desired light is reached. Lights are generally controlled by automatic time clocks.

All-artificial lighting is used in windowless houses. Morning lighting systems turn on lights two to five hours before sunrise and turn off at daybreak. Evening lighting systems turn on in the evening. Dimmers are used to prevent a sudden change from light to dark. Morning-evening systems use lights at each end of the day. All-night lighting is generally used only for second-year laying flocks. Low-wattage bulbs are used in this system.

Regardless of the system used, lights should provide one footcandle of light per bird. Keep the bulbs clean to maintain efficiency of light output. The general principle involved in providing light for laying hens is never to decrease the amount of light while the flock is in production.

Reducing heat stress. Extreme heat is a problem that may affect the growth or egg-laying ability of poultry. Heat stress can also increase the mortality rate in a flock. Birds do not have sweat glands. They depend on evaporation, mainly by panting, to control body temperature. Excessive panting increases the pH of the blood, which reduces the amount of calcium and bicarbonate in the blood. This results in thin-shelled eggs that are more likely to break when handled. Heat stress also reduces appetite, which reduces rate of gain, lowering feed efficiency.

Buildings need to be designed and managed so that a temperature that is as comfortable as possible for poultry is maintained. Poultry are generally comfortable in a range of 45° to 70°F (7°–21°C). Good ventilation is necessary to help keep the interior temperature of the poultry house in the comfort zone for the birds. Circulation fans, fogging nozzles, and evaporative cooling pads are often used to help control the temperature in the poultry house.

Some systems combine natural air flow with circulating fans, while other systems depend entirely upon circulating fans for air movement. If the fans exhaust the air from the building, it is called a negative pressure system. If the fans bring fresh air into the building, it is called a positive pressure system. Tunnel ventilation uses air inlets at one end of the building and exhaust fans at the other end to move air through the building at a minimum velocity of about 350 feet per

FIGURE 36-7 Eggs roll to the front of the cages to be gathered. *Courtesy of University of Illinois at Urbana-Champaign.*

minute. This system does not use a pressure difference between the inside and the outside of the building to move the air. The use of fogging nozzles and evaporative cooling pads are additional options that may be used with circulating fans and tunnel ventilation.

Care and handling of eggs. Eggs lose quality rapidly if they are not handled carefully. Quality is maintained by keeping eggs cool and humid. All equipment used in handling eggs must be kept clean. Open plastic or rubber-coated wire baskets will let the eggs cool more rapidly than closed containers. Egg room temperature should be kept at 65°F (18°C). The humidity should be between 75 and 80 percent. Eggs absorb odors from the environment where they are stored. Never store eggs near anything with a strong odor, such as onions, paint, or kerosene.

In floor systems, eggs should be gathered at least three times a day (Figure 36-7). Caged-layer systems may use some type of automatic egg gathering system such as a conveyor belt. Eggs may also be gathered using mechanized carts. Some research indicates that once-a-day egg gathering in cage systems does not reduce egg quality significantly if the weather is not too hot.

Most of the eggs will be clean if good management practices are followed. If a high number of eggs are dirty, management practices should be reviewed. Dirty eggs must be cleaned. The method of cleaning is often specified by the egg dealer. Eggs may be drycleaned or washed. *Drycleaning* is done by buffing the eggs with fine sandpaper, emery cloth, or steel wool. However, this may increase the number of cracked eggs and requires a great deal of labor.

Egg-washing machines are available and provide the fastest way to clean eggs. Only dirty eggs should be washed. Always follow the equipment manufacturer's directions for the use of the egg washer. Eggs that are washed incorrectly lose quality rapidly.

Containers, fillers, and flats should be precooled before the eggs are packed. This is done by storing them in the egg room for 24 hours before use. Eggs should be shipped and stored prior to sale at 45°F (7°C) or below. Eggs are packed with the large end up. Eggs should not be held more than one week on the farm. Twice-a-week marketing is preferred.

Managing chickens for meat production. Chickens raised for meat production are brooded in the same way as laying chickens. Chickens for meat production are generally raised in confinement, although range can be used. Broilers require 1 square foot (0.09 m^2) of space from two weeks of age to market. Capons and roasters require 2 to 3 square feet (0.18–0.28 m^2) of floor space from 10 to 20 weeks of age. After 20 weeks of age, provide 4 to 5 square feet (0.37–0.46 m^2) of floor space per bird.

Broiler producers usually use lights 24 hours a day. One 60-watt bulb per 200 square feet (18.6 m^2) of floor space provides adequate light. No nests or roosts are used in producing meat-type chickens.

A *capon* is a male chicken that has been surgically castrated. The operation is usually performed when the birds are three to five weeks of age. Capons produce meat that is more tender than that produced by broilers. However, they require more labor and investment to produce. Hormone implants can be used to produce the same results as caponizing.

Breeding flock management. Hatching eggs are produced under agreements with a hatchery. The hatchery generally provides the farmer with the breeding stock having the special bloodlines that the hatchery wants produced.

Ten to 14 cockerels (males) per 100 pullets should be started. These are later culled down to eight to 12 cockerels at 10 weeks of age. Leghorn breeds require one cockerel for each 15 to 17 hens at mating time. Heavier breeds require one cockerel for each 12 to 15 hens.

Brooding is done in the same way as for laying flocks. Raising the chickens requires the management practices outlined for laying flocks. Breeding stock is often revaccinated for Newcastle disease and bronchitis about one month before breeding begins. This gives the newly hatched chicks a temporary immunity to these diseases.

Mating begins at least two weeks before the eggs are to be delivered to the hatchery. Do not put too many birds in one mating pen. This will reduce the matings.

Hatching eggs should be gathered three or four times per day. Gather more often if the weather is very hot or cold. Handle hatching eggs in the same careful way that other eggs are handled. Store hatching eggs at 50° to 60°F (10°–15°C) and at 80 percent humidity. Always pack hatching eggs with the large end up. If eggs have to be held more than one week, tip the crate sharply to prevent the yolk from sticking to the shell.

Hatching eggs must be fumigated within two hours after they are gathered. This will help prevent the spread of egg-borne diseases such as pullorum, fowl typhoid, paratyphoid, paracolon, and navel infection. Formaldehyde or potassium permanganate is used to fumigate the eggs.

Caution: Do not breathe the fumes from fumigation.

Chlorine dioxide may be used as a spray or dip instead of fumigation. It is easier to use and gives better results.

Records. Good records are the key to good management. Records are used to determine the amount of money made from the operation.

They are used as the basis for making management decisions and for selection of breeding stock in breeding programs. Laying flock records must include information on egg production, death losses, amount of feed used, cost of feed, amount of money received from egg sales, and costs of buildings, repairs, and equipment. Meat production operations require records of feed and pounds of birds sold, as well as other production cost figures. From these types of records, the producer can determine profit or less, and what management changes should be made.

Turkeys

Brooding turkeys. The brooder house must be cleaned and disinfected before the poults are placed in it. Old litter is removed and the house is allowed to dry out. Fresh litter is spread on the floor. Shavings, wheat and beardless barley straw, peat moss, shredded cane, rice hulls, processed flax straw, and cedar tow may be used for litter. The litter should be covered with a strong rough-surfaced paper for the first few days. Enclose the area with a poult guard about 1.5 to 3 feet (0.45–0.9 m) from the edge of the brooder hover. The paper cover is taken up after five to six days. Add fresh litter as needed after removing the cover. Keep the litter dry. The poult guard should be 16–18 inches (41–46 cm) high. It may be made of wire or heavy corrugated paper or lightweight aluminum.

Temperature under the brooder for the first two weeks should be 95°F (35°C) for dark poults and 100°F (37.8°C) for light poults. The brooder temperature should be lowered 5°F (2.8°C) each week until heat is no longer needed.

FIGURE 36-8 Turkeys being raised in confinement in Maryland. *Courtesy of USDA.*

Provide ventilation without drafts. Lights are kept on for the first two weeks of brooding. After two weeks, houses with windows need only dim light at night. Windowless houses need about one footcandle of light for 16 hours and one-half footcandle during the other eight hours. Roosts are seldom used when brooding turkeys. Up to eight weeks of age, a minimum of 1 square foot (0.09 m²) of floor space is needed per poult. Larger-type poults require slightly more floor space. Poults in force-ventilated houses may be given slightly less floor space.

Turkeys that are to be raised in confinement (Figures 36-8 and 36-9) should be debeaked at three to five weeks of age. Turkeys that are to be raised on range should not be debeaked unless picking or fighting occurs. *Desnooding* is the removal of the tubular fleshy appendage on the top of the head. Desnooding helps prevent head injuries from picking or fighting. Small- or medium-type turkeys that are to be raised on range should have the flight feathers of one wing clipped to prevent flying. It is not necessary to wing clip confinement-raised turkeys. Toe clipping is done on day-old poults at the hatchery. It prevents scratched and torn backs.

FIGURE 36-9 Turkeys being raised in pole-type house. *Courtesy of USDA.*

Trough-type feeders and waterers can be used for turkeys. Mechanical feeding and watering equipment is available. Feeders and waterers must be cleaned regularly. Moving feeders and waterers helps to prevent wet litter.

Range-growing turkeys. Turkeys can be raised on range (Figure 36-10) although confinement raising is becoming more common. Feed costs are somewhat lower on range. However, losses from diseases, predators, and weather are higher.

Poults can be moved to range at eight weeks of age. Poults should be moved to range only when the weather is good (Figure 36-11). The number of turkeys per acre will vary depending on the type of range. One acre can carry from 125 to 250 birds. On very sandy soil, up to 1,000 turkeys may be put on one acre. One system of range raising involves moving the turkeys to a clean area every one to two weeks (Figure 36-12). This helps to prevent diseases and parasites. Range should be rotated with other crops. A three- or four-year rotation is used. Legumes or grass, or a mixture of the two, may be used for turkey ranges.

Turkeys raised on range require shelter and protection from predators, such as dogs, and wild animals. Double-strand electric fences provide protection against many predators. Six-foot (1.8 m) high, heavy-gauge poultry fencing provides protection from dogs, coyotes, and foxes. Confinement of the turkeys at night also protects against predators.

Confinement-growing turkeys. Poults may be moved to confinement houses when they are six to eight weeks of age. Check the poults the first few nights to be sure they have settled down. Use lights for the first few nights. Be sure that the house is properly ventilated.

Large-type toms require 5.5 square feet (0.5 m^2) of floor space per bird. Hens need 3.5 square feet (0.3 m^2), and mixed flocks need 4.5 square feet (0.4 m^2) of floor space per bird. Softwood shavings or wheat straw are often used for litter.

Managing breeding turkeys. Breeding stock should be blood tested for pullorum disease, fowl typhoid, paratyphoid, infectious sinusitis, and Arizona and Mycoplasma meleagridis infections. Vaccinations are based on the requirements of the hatching egg buyer. (For further information on diseases, see Unit 37).

Low fertility is a problem when natural mating is used. Because of this, artificial insemination is widely practiced with turkey breeding. The first insemination is given when egg production begins. A second insemination is given one week later. This is followed with insemination at two-week intervals. Some of the procedures used are shown in Figures 36-13 through 36-16.

Breeding flocks may be kept on limited range or in confinement. Confinement allows better disease control and management. The labor cost is lower in confinement. Breeding flocks require more floor space than market turkeys.

FIGURE 36-10 Turkeys being raised on range in Arkansas. *Courtesy of USDA.*

FIGURE 36-11 Turkeys being kept in pens on range in Kansas. The portable shades are easily moved from place to place as needed. *Courtesy of USDA.*

FIGURE 36-12 Turkeys on range in Iowa with portable feeders and waterers. *Courtesy of Iowa State University.*

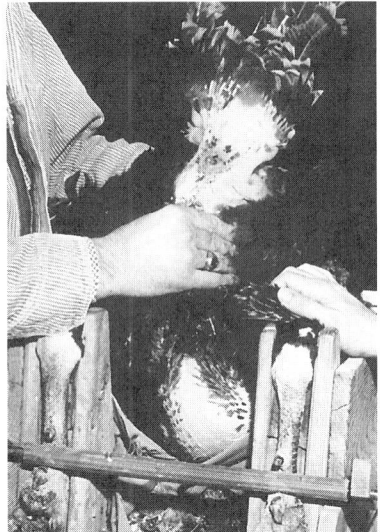

FIGURE 36-13   Stimulating the tom prior to "milking" semen. *Courtesy of University of Missouri, Cooperative Extension Service.*

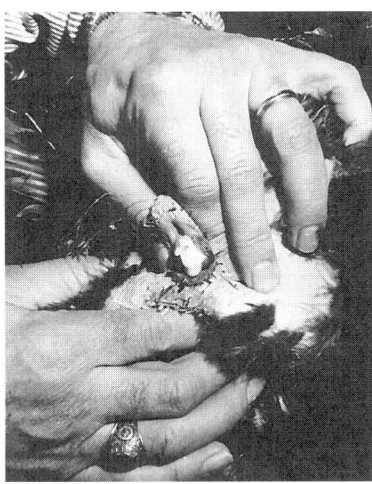

FIGURE 36-14   "Milking" out the semen. *Courtesy of University of Missouri, Cooperative Extension Service.*

Toms should receive artificial light two to three weeks before they are used with the hens. Hens should receive light 30 days before the eggs are needed. A 14-hour day is used for normal spring lighting.

Broodiness of hens is a problem with breeding flocks. *Broodiness* is when a hen stops laying eggs and wants to sit on a nest of eggs to hatch them. Remove broody hens from the flock and confine them in a separate pen. Use slatted floors in the pens. Keep the pens well lighted.

Eggs should be gathered at least every two hours. Eggs are fumigated with formaldehyde gas immediately after gathering. Hold eggs at room temperature for 24 hours. Store eggs at 55° to 60°F (13°–16°C) until they go into the incubator. Store eggs with the large ends up or place them on their sides in trays. Eggs stored over one week must be turned. Dirty eggs lower hatchability. Keep eggs clean by keeping nests clean and gathering often. Eggs that are not too dirty can be washed. Eggs should be delivered to the hatchery at least twice a week.

Ducks and Geese

Brooding. Ducklings and goslings can be brooded naturally or artificially. Broody chicken hens or females of ducks or geese can be used for natural brooding. Young birds not hatched by the female should be placed under her at night. Twelve eggs may be placed under a duck, nine to 10 eggs under a goose, and four to six goose eggs or 10 to 11 duck eggs under a chicken hen. Goose eggs set under a hen should be turned twice a day. A dry shelter is needed for natural brooding.

Brooders used for chickens can be used for artificial brooding of ducks and geese. Ducks and geese grow faster than chickens and require heat for a shorter period of time. Five to six weeks of brooding is usually long enough.

An infrared heat lamp can be used for a small number of eggs. Hang the lamp about 18 inches (45.7 cm) above the floor. One lamp will brood about 30 ducklings or 25 goslings.

Other types of chicken brooders have a capacity of about one-half chick capacity for ducklings and one-third chick capacity for goslings. The hover needs to be 3 to 4 inches (7.6–10 cm) higher than for chicks. Set the brooder at a starting temperature of 90°F (32°C). Reduce the temperature by 5°–10°F (2.8°–5.6°C) each week until a temperature of 70°F (21°C) is reached.

Wire or litter flooring may be used for ducks. Three-quarter inch (1.9 cm) mesh wire placed 4 inches (10 cm) above the floor is used. Wire flooring keeps the ducklings away from manure and moisture. Chopped straw, wood shavings, or peat moss can be used for litter. Litter must be dry and free of mold.

Chopped straw, sawdust, wood shavings, crushed corn cobs, or peat moss make good litters for goslings. Cover the litter with a rough paper for the first four or five days. This prevents the goslings from eating the litter. Smooth paper should not be used. It causes *spraddled legs*, a condition in which the leg is dislocated from the socket where it is attached to the body. Spraddle-legged goslings do not recover. The litter must be kept dry. Wet spots should be removed and fresh litter added as needed.

Rearing. Feeding, watering, and pasture for ducks and geese are discussed earlier in this unit. Ducks and geese should be moved outside as soon as the weather permits. Young ducklings and goslings should not be allowed out in the rain. After they are well-feathered, rain is not a problem.

Pasture for ducks and geese should be fenced. Young birds should be confined with a 2-inch (5 cm) mesh poultry netting. After the birds are four to six weeks of age, an ordinary woven wire fence will hold them. The fence need not be high since the birds seldom fly. Electric fencing has been used successfully with waterfowl.

Birds confined in yards cause a buildup of manure. This must be cleaned periodically, at intervals determined by the number of birds in the yard.

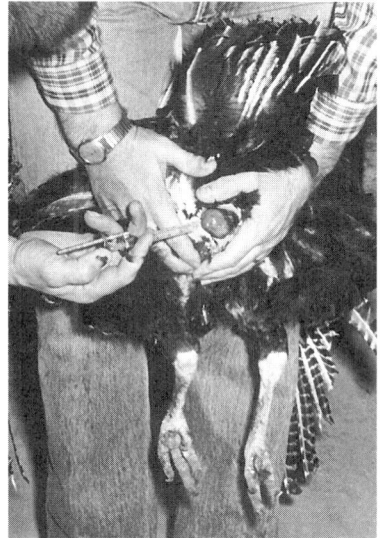

FIGURE 36-15 Eversion of the oviduct prior to insertion of the syringe. *Courtesy of University of Missouri, Cooperative Extension Service.*

HOUSING AND EQUIPMENT

The trend in poultry production has been toward large commercial flocks. With this trend has come an increase in confinement housing for poultry. Initially, pole-type construction with curtain sides was commonly used for confinement operations. This type of construction was relatively economical to build, maintain, and clean. Today, most confinement poultry houses are clear span structures with insulated ceilings and walls. It is easier to provide proper ventilation in this type of housing. There is also a trend toward housing layers in structures that are light-tight. The control of lighting significantly influences egg production. The companies that provide the birds in vertically integrated systems usually specify the type of housing to be used. State colleges of agriculture can provide plans for poultry buildings that are designed for a given geographic area. A person interested in such plans should contact the local Cooperative Extension Service or their state college of agriculture.

Some general characteristics are common to all good poultry houses. Houses should be easy to clean and disinfect. Mechanical cleaning equipment is used in some types of houses. Dampness is a problem in poultry houses. All types of housing must be properly ventilated to help prevent dampness. Insulation should be used in areas where it is needed. Water and electricity are essential for modern poultry operations. Egg, feed, and equipment storage rooms are often included in the house.

Automatic and semi-automatic feeding, watering, and cleaning equipment is common in poultry enterprises. The use of this type of

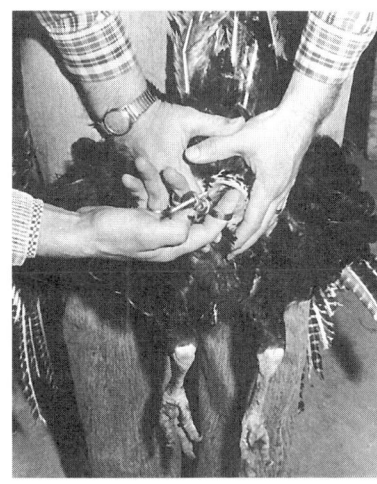

FIGURE 36-16 Syringe of straw is inserted at a depth of 2½ inches (6.35 cm) into the relaxed oviduct. *Courtesy of University of Missouri, Cooperative Extension Service.*

equipment saves time and labor. It also makes it possible for large numbers of birds to be handled in the operation. Bulk feed storage is generally used for large operations.

Poultry raised on range need shelters and shade. Portable feeders and waterers are used on range.

SUMMARY

The greatest single cost of poultry enterprises is feed. Commercially prepared feeds are available for all kinds of poultry. Smaller flock owners generally find it easier to use a complete commercial feed. Large commercial flock owners usually use some system of mixing ingredients to make a complete feed. Poultry require carbohydrates, protein, minerals, vitamins, and water in their rations. Grit is used to help the bird grind the feed in the gizzard. Oyster shell provides calcium for egg production.

Poultry may be brooded on the floor or in cages. Small flocks generally use floor brooding. Large operations may use battery brooder units or wire cages. Proper temperature control is necessary for successful brooding.

Light affects the sexual maturity and egg production rates of poultry. Controlled lighting is generally used in poultry enterprises.

Laying flocks may be handled using floor systems or in cages. Large operations are more likely to use cage systems. Eggs must be carefully handled to maintain quality.

Chickens raised for meat production are usually raised in confinement. Turkeys are raised on range or in confinement.

Ducks and geese can be brooded naturally or artificially. They are easy to raise and do not require expensive housing or equipment.

Automatic or semi-automatic feeding, watering, and cleaning equipment is frequently used in the poultry industry.

Student Learning Activities

1. Give an oral report on feeding, management, housing, or equipment for different kinds of poultry.
2. Prepare a bulletin board display of pictures of housing and equipment used for different kinds of poultry.
3. Visit poultry production enterprises to observe feeding, management, and housing.
4. Survey poultry producers in the local community concerning feeding, management, and housing practices. Prepare a summary of the results.
5. Construct poultry equipment in shop for use with home flocks.

Review

1. What part of the ration supplies the carbohydrates and fats needed by poultry?

2. List some common grains used to feed poultry.

3. Why are fats sometimes added to poultry rations?

4. List the amino acids most commonly lacking in poultry rations.

5. List three common sources of protein for poultry rations.

6. How important is water in the chicken ration?

7. List the minerals commonly required in a chicken ration.

8. What level of calcium should be fed in a chicken ration?

9. List the vitamins commonly required in a chicken ration.

10. How can the poultry producer be sure the ration contains enough vitamins?

11. Briefly explain the use of additives in chicken rations.

12. What are unidentified factors in the chicken ration and what are their effects?

13. Why is grit used in the chicken ration?

14. List three forms in which commercially manufactured chicken feeds may be purchased.

15. List four systems that may be used for preparing chicken rations.

16. Briefly explain the use of phase feeding for chickens.

17. List the good management practices to follow when feeding chickens.

18. Why should the poultry producer keep a close watch on the feed consumption of the flock?

19. Describe the feed that should be used to start chicks.

20. How much feeder and waterer space is needed for chickens up to six weeks of age?

21. Describe the feed that should be used for chickens from six to 12 weeks of age.

22. How much feeder and waterer space is needed for chickens from seven to 20 weeks of age?

23. How much feed per day and per year is needed for light and heavy breeds of laying hens?

24. How much feeder and waterer space is needed for light and heavy breeds of laying hens?

25. Why should the producer avoid sudden changes in the ration for laying hens?

26. Name and describe three systems of feeding that can be used for laying hens.

27. How is feed efficiency determined for laying hens?

28. Describe the kind of starter ration that should be used for broilers.
29. How much feeder and waterer space is needed for broilers?
30. How is feed conversion determined for broilers?
31. Describe the kind of finishing ration that should be used for broilers.
32. What factors affect feed conversion in broilers?
33. Describe a good feeding program for the breeding flock.
34. Describe a good feeding program for turkeys.
35. Describe a good feeding program for ducks and geese.
36. List the recommended practices for brooding chicks.
37. Describe the principles of using artificial light when raising laying pullets.
38. List the recommended practices for managing the laying flock.
39. Describe the kinds of cage systems that can be used with the laying flock.
40. Briefly explain the use of artificial light for the laying flock.
41. List the recommended practices for egg handling.
42. How much space is needed for raising broilers?
43. Briefly explain the use of artificial lights when raising broilers.
44. Describe good management practices for the breeding flock.
45. List the kinds of records that should be kept by a poultry producer.
46. List the recommended practices for brooding turkeys.
47. Describe the growing of turkeys on range.
48. Describe confinement raising of turkeys.
49. List the recommended practices for managing the turkey breeding flock.
50. List the recommended practices for brooding ducks and geese.
51. Describe the practices that should be followed for rearing ducks and geese.
52. Describe the kinds of housing for poultry.
53. List the general characteristics that are common to all good poultry houses.
54. List other kinds of equipment commonly used in growing poultry.

Diseases and Parasites of Poultry

Objectives

After studying this unit, the student should be able to

- establish a disease and parasite control program for poultry.
- identify the symptoms of common poultry diseases.
- identify the symptoms of common parasites of poultry.

MAINTAINING POULTRY HEALTH

A program of preventive management is the best way to control poultry diseases and parasites. Such a program involves sanitation, good management, vaccination, and control of disease outbreaks.

Sanitation

Poultry houses should be completely cleaned and disinfected before new birds are moved in. The following steps will result in a clean house:

1. Take all movable equipment outside of the house. Clean off the manure, and wash and disinfect the equipment. Expose the equipment to sunlight.
2. Clean all of the manure and litter out of the house.
3. Sweep down the walls and ceilings.
4. Scrape and brush the floor clean.
5. Scrub and hose the inside and outside of the house using high pressure. A portable steam cleaner can be used for this operation.
6. Spray the inside of the house with an approved disinfectant. Spray should be applied to all the surfaces of the walls, ceiling, and floor. Do not spray the disinfectant into waterers or feeders. Chlorine, iodine, or quaternary ammonium are good disinfectants for smooth surfaces such as the walls and ceiling. Cresol, phenol, or coal tar-type disinfectants are best for floors, posts, and foundations.

7. Use new, clean, dry, nondusty litter on the floor.
8. Leave the house empty for two weeks to break disease cycles.
9. Lock the door to prevent people from entering and contaminating the clean house.

Insofar as possible, do not allow visitors to enter poultry houses, pens, and yards. Clean coveralls and disinfected rubber footwear should be worn by anyone who must enter the poultry area. Place a foot pan with disinfectant in it at the door, to be used before entering. Replace the disinfectant in the pan frequently.

Use only clean and disinfected equipment. Be cautious about allowing used poultry crates, egg cases, and feed bags to be brought onto the farm. These can spread diseases.

All dead birds must be disposed of promptly. Use of incinerators, composting, or deep burying are recommended for disposal of dead birds. Be sure that disposal methods meet Environmental Protection Agency (EPA) regulations.

Dispose of manure by spreading it thinly on land that is not used for poultry. Do not put poultry on the land where the manure was spread for at least four years.

Eliminate places for pests, such as flies, to breed. Do not pile up manure outside of the poultry house. Control lice and mites inside of the house by using approved chemicals.

Other Health Management Practices

In addition to sanitation, a number of other management practices contribute to good flock health. Among these practices are the following:

1. Buy poultry replacement stock from a reliable, disease-free source.
2. Use day-old chicks and poults.
3. If possible, keep birds of only one age on the farm. Use an all-in, all-out program. (Bring all the birds onto the farm at one time and remove them all at one time.)
4. If it is necessary to keep birds of different ages, separate the flocks by at least 40 feet (12 m).
5. Separate chickens and turkeys. It is best to have only one or the other on the farm.
6. Separate breeder flocks from other poultry. It is best not to have any other poultry on the farm if a breeder flock is kept.
7. Keep pets and flying birds out of the poultry house. Flying birds can be kept out by screening the windows.
8. Provide the proper ventilation in the poultry house.
9. Control rats and mice. Use rat baits and traps as necessary. Make feed bins and storage rooms ratproof. Eliminate places for rats and mice to breed by cleaning up trash and junk.

10. Feed balanced rations to prevent nutritional diseases. Make sure the feed is mixed properly. Provide plenty of fresh, clean water. Keep feeders and waterers clean.
11. Maintain good health records. Records should be kept of vaccinations, disease problems, and medicines used.

Vaccination

Vaccination is not a substitute for good flock health management, but it is helpful in controlling certain diseases. Vaccines are available for Newcastle disease, Marek's disease, infectious bronchitis, fowl pox, epidemic tremors, fowl cholera, laryngotracheitis, infectious bursal disease, erysipelas, and virus hepatitis. Vaccination should be used only in areas where the disease is known to exist. Plan a vaccination program for the specific operation in a specific locality. Some vaccines for certain diseases can only be used with the permission of the state veterinarian. When planning a vaccination program, obtain help from a veterinarian, the Cooperative Extension Service, a hatchery, or feed dealer.

Vaccination causes a stress on poultry. Vaccinate only healthy birds. Read and follow all directions on the vaccine.

Several methods may be used to vaccinate poultry. Individual bird vaccinations are given by injection, intranasally, intraocularly, or through the wing web. *Intranasal vaccination* is placement of the vaccine directly into the nose opening. *Intraocular vaccination* is placement of the vaccine directly into the eye. *Wing web vaccination* is the process of injecting the vaccine into the skin on the underside of the wing web at the elbow. A grooved, double needle instrument is used for wing web vaccination.

Flock treatments are given in the water, by spray, or dust. The method used depends on the disease to be controlled. Individual vaccination causes more stress on the birds than flock treatments. Vaccinations for some diseases can be given in more than one way.

Controlling Disease Outbreaks

It is better to prevent a disease outbreak than to try to control it once it has occurred. Following the sanitation, management, and vaccination suggestions discussed earlier in this unit will help the poultry producer prevent disease outbreaks from occurring.

The poultry flock should be checked daily for signs of disease. A sudden drop in feed and water consumption is often a sign of health problems. Watch the birds to see how they are eating and drinking. If more than 1 percent of the flock is sick, a disease is probably present. Death rate is another sign of disease. During the first three weeks, the normal death rate for chicks is about 2 percent. For turkeys, it is about 3 percent. After three weeks of age, the death rate should not be more than 1 percent per month. A sudden increase in the death rate is an indication of disease.

Most diseases can be accurately diagnosed only in a laboratory. Very few can be accurately diagnosed on the farm. The producer should use the services of a veterinarian or the state diagnostic laboratory to determine which disease is causing the problem. The procedure for collecting needed information and specimens is specified by the laboratory. This procedure should be carefully followed. The recommendations of the veterinarian or laboratory for control of the disease must also be followed for best results.

DISEASES AND DISORDERS

The list of diseases and disorders that affect poultry is extensive. Because of space limitations, only brief descriptions are included in this unit. Positive diagnosis and recommendations for treatment should be obtained from a veterinarian or diagnostic laboratory.

Amyloidosis

Amyloidosis, also called "wooden liver disease," is a common disease of adult ducks. The liver becomes hard and fluids accumulate within the body cavity. The cause of the condition is unknown. Death rate may be as high as 10 percent. There is no known cure.

Aortic Rupture

Aortic rupture affects mainly male turkeys between the ages of eight to 20 weeks. The exact cause is not known, but the disease seems to be related to a high intake of energy feeds. An artery ruptures and the bird bleeds to death internally. Affected birds, which appear to be in good condition, die suddenly. Aortic rupture is prevented by feeding a lower-energy diet and by using a continuous feeding of tranquilizers at a low level.

Arthritis/Synovitis

The *arthritis/synovitis syndrome* is an inflammation of the joints and synovial membranes. The *synovial membrane* is a thin, pliable structure surrounding a joint. The membrane secretes *synovia,* which is a transparent lubricating fluid. The causes include injury, diet, and infection. The three most common forms are infectious synovitis, staphylococci arthritis, and viral arthritis.

Infectious synovitis is caused by a bacterium. It affects chickens and turkeys. The birds become lame, lose weight, and usually have diarrhea. Chickens show swelling of the foot pads and shanks. Turkeys show swelling of the feet and hocks. Death loss from this disease is low. Prevention is through good sanitation, and treatment involves the use of antibiotics.

Staphylococci arthritis is also caused by a bacterium. Affected birds have swollen joints and diarrhea, and show signs of depression. The acute form causes death. Good management helps to prevent the disease. Treatment is with antibiotics.

Viral arthritis is caused by a virus. The birds become lame and show a slow rate of gain. The disease affects only chickens. There is no treatment for the disease.

Aspergillosis (Brooder Pneumonia)

Aspergillosis is caused by a fungus or mold. It affects chickens, ducks, and turkeys. The acute form occurs in young birds and results in a high death rate. A chronic form affects older birds. The symptoms in young birds include loss of appetite, gasping, sleepiness, convulsions, and death. Older birds show a loss of appetite, gasping, and loss of weight. The disease is prevented by using litter that is free of mold. There is no treatment.

Avian Influenza

Avian influenza is caused by several viruses. It affects mainly turkeys and chickens. The birds cough, sneeze, and lose weight. Death rate is low. Good management helps in prevention. Treatment with antibiotics may help reduce losses.

Avian Leukosis

Avian leukosis is a complex of diseases caused by viruses. The most common forms are lymphoid leukosis and Marek's disease. Lymphoid leukosis affects both turkeys and chickens.

Lymphoid leukosis usually affects birds over 16 weeks of age. Sometimes the birds die without showing any symptoms of disease. Usual symptoms are loss of weight, loss of appetite, and diarrhea. The bone form of the disease affects younger birds and is a serious problem in the broiler industry. The affected birds show lameness. Prevention involves the use of resistant strains of birds, sanitation in the hatchery, and keeping the flock isolated. There is no treatment for the disease.

Marek's disease affects mainly young chickens, although it may occur in older birds. It is caused by a virus different from the one that causes lymphoid leukosis. The main symptom is lameness or paralysis of one or both legs. The disease is prevented by vaccination at hatching time. There is no treatment.

Avian Pox (Fowl Pox)

Avian pox is caused by a virus and affects turkeys, chickens, and other birds. Symptoms include wartlike scabs around the head and comb, yellow cankers in the mouth and eyes, and, in turkeys, yellowish-white cankers in the throat. Young birds have a slower rate of growth. Egg production in layers is reduced. The disease is prevented by vaccination. There is no treatment.

Avian Vibrionic Hepatitis

Avian vibrionic hepatitis is caused by a bacteria. It affects chickens in either an acute or chronic form. The acute form causes a higher death loss. The birds show loss of weight, diarrhea, listlessness, shrunken comb, and drop in egg production. Good management and sanitation help in prevention. Drugs are used for treatment.

Blackhead

Blackhead is caused by a protozoan parasite. It affects both chickens and turkeys, but is more serious in turkeys. The disease affects mainly young birds between 6 and 16 weeks of age. Symptoms include droopiness, loss of appetite, darkening of the head, and a watery, yellow diarrhea. Death losses are usually low but may be as high as 90 percent if the disease is not controlled. Diseased chickens often show no signs of the disease.

Brooding on wire or slatted floors helps in prevention. Never house chicken and turkey flocks together. Rotate pastures to prevent the disease. It may be necessary to continuously feed a preventative drug at low levels in the drinking water.

Bluecomb (Transmissible Enteritis; Mud Fever)

Bluecomb in turkeys is caused by a virus and affects all ages. Symptoms in poults include droopiness, dehydration, and gaseous, watery diarrhea. Death loss in poults can be as much as 100 percent. Older birds show a loss of appetite, loss in body weight, diarrhea, and a blueness of the head. Death loss in older birds is not high. Birds that recover may be carriers of the disease.

Sanitation practices help to prevent the disease. If an outbreak has occurred, empty the house and allow it to remain vacant for at least 30 days. Treat by adding antibiotics and molasses to the drinking water. Use one pint (0.47 litres) of molasses to five gallons (19 litres) of water for one day. Follow this treatment with a broad-spectrum antibiotic in the feed or drinking water for five to seven days.

Bluecomb (Pullet Disease)

Bluecomb in chickens is not the same disease as bluecomb in turkeys. The cause is not known but is believed to be a virus. Symptoms include a drop in egg production, loss of appetite, loss of weight, diarrhea, and a darkening of the comb. The crop may become compacted with feed. Sanitation and good management help to prevent the disease. Treatment is with a broad-spectrum antibiotic.

Bumblefoot

Bumblefoot is a disease of the feet that affects turkeys. Hard bumblefoot appears as a hard swelling in the center of the foot. Soft bumblefoot occurs as soft-centered abscesses on the sole of the foot and especially the toes. Bumblefoot can cause lameness and slow growth rate. Male turkeys raised on hard floors and roosts are more likely to be affected by hard bumblefoot. Both male and female turkeys raised on muddy, dirty range or litter are more likely to get soft bumblefoot. Management that eliminates these conditions helps in prevention. There is no treatment.

Botulism (Limberneck; Food Poisoning)

Botulism is the result of the bird eating decayed material that contains a *toxin* (poison) produced by a bacterium. Symptoms include weakness, trembling, paralysis, loose feathers, and dull, closed eyes. Affected birds usually die. Prevention is achieved by eliminating sources of decayed material that might be eaten by the birds. Contaminated water is often a source of the toxin. Be sure that the water supply is clean. Isolate sick birds. An antitoxin may be used to save valuable individual birds.

Cage Fatigue

Cage fatigue appears to be a nutritional problem involving minerals. It commonly affects young pullets that are producing eggs at a high rate. The disease is not common if the diet is properly balanced. The main symptom is paralysis. Proper diet prevents the disease. Treatment is usually not practical.

Coccidiosis

Coccidiosis is caused by a number of protozoan parasites called *coccidia*. Different species of coccidia affect different kinds of poultry. Birds three to eight weeks of age are most frequently affected by this disease. Symptoms include droopiness, huddling of birds, loss of appetite, loss of weight, and diarrhea (which may be bloody). Death loss may be high. The best prevention is the feeding of a coccidiostat in the ration of chicks and poults. Do not feed ducks or geese a chicken feed containing a coccidiostat. Brooding on wire or slat floors will also help to prevent the disease. Drugs may be used to treat affected birds.

Coliform Infections (Colibacillosis)

Coliform infections represent a variety of diseases caused by various strains of the bacteria *Escherichia coli*. The diseases range from severe to mild. Young birds are more commonly affected, although adults may be. The symptoms vary with the type of infection. Death losses may be high in young birds. Some of the symptoms of coliform infections include fever, ruffled feathers, listlessness, difficulty in breathing, and diarrhea. Prevention includes sanitation, fumigation of hatching eggs, and reduction of stress. Drugs for treatment are not very effective but may reduce losses.

Duck Virus Enteritis (Duck Plague)

Duck virus enteritis is caused by a virus. It affects ducks, geese, and other waterfowl. Symptoms include watery diarrhea, nasal discharge, and droopiness. Death loss is high. Sanitation and isolation of birds helps to prevent the disease. There is no treatment.

Degenerative Myopathy

The exact cause of *degenerative myopathy* is not known, but it may be an inherited condition. The inner large breast muscle degenerates and becomes a greenish color. There is a dent in the muscle of one or both sides of the breast. The disease affects mainly turkey breeding hens. Breeding stock that shows signs of the condition should not be used.

Dietary Disorders

A number of diet-related disorders affect poultry. Feeding a well-balanced ration containing the minerals and vitamins needed will prevent most of these.

Endemic Fowl Cholera

Endemic fowl cholera is caused by bacteria. It is a chronic condition that affects mainly chickens. Good management has almost eliminated it from commercial flocks. Symptoms include a nasal discharge, inflammation of the eye, and swelling of the sinus. Prevention is achieved by isolating infected birds and using stock from sources that are free of the disease. If an outbreak occurs, dispose of all birds and clean and disinfect the house. There is no satisfactory treatment.

Epidemic Tremor (Avian Encephalomyelitis)

Epidemic tremor is caused by a virus and affects all ages of chickens. The symptoms usually appear only in young chickens. Chicks become paralyzed and tremble. Death loss is high. Prevention is achieved by vaccinating breeding stock. There is no treatment.

Erysipelas

Erysipelas is caused by a bacterium. In poultry, it is most common in turkeys between four and seven months of age. The first sign of disease is usually several dead birds. Sick birds show weakness, lack of appetite, and yellowish or greenish diarrhea. Death loss can be high. Turkeys can be vaccinated for erysipelas. Treatment is with drugs or a bacterin. Vaccination should be done only if the disease is present in the area.

Fatty Liver Syndrome

The cause of *fatty liver syndrome* is unknown. It affects mainly caged birds on high-energy diets. Symptoms include a drop in egg production, diarrhea, and anemia. Dead birds may be found when no other symptoms have been noticed. Treatment and control measures involve diet adjustment. A nutrition specialist should be consulted.

Fowl Cholera

Fowl cholera is caused by bacteria. It may occur in acute or chronic form. Symptoms of the acute form include purple color of the comb, difficult breathing, watery diarrhea, weakness, droopiness, loss of appetite, loss of weight, lameness, and sudden death. The chronic form shows swollen wattles, earlobes, and joints. Layers will show a drop in egg production. Sanitation helps to prevent the disease. Vaccinate with bacterins in areas where the disease is present. Drugs are used for treatment.

Hemorrhagic Anemia Syndrome

Hemorrhagic anemia syndrome affects chickens of all ages, but mainly those between four and twelve weeks of age. Its cause is unknown. Symptoms include loss of weight, diarrhea, weakness, and anemia. Death losses may be high with the acute form. Sulfa drugs and antibiotics may make the condition worse. Addition of liver solubles to the ration has resulted in some improvement.

Hemorrhagic Enteritis

Hemorrhagic enteritis can occur in turkeys from three weeks to six months of age, but is more common in older birds. Its cause is unknown. Sudden death of several birds on range is often the only symptom. A change in the ration and moving birds to new range may help in treatment. A serum made from the blood of birds that have recovered can be injected to treat the flock.

Hexamitiasis

Hexamitiasis is caused by a protozoan parasite. It does not affect chickens. Turkeys are affected up to 10 weeks of age. Poults huddle together and become listless. There is a rapid weight loss with a watery, foamy, yellowish diarrhea. Sanitation, and brooding on wire or slat floors help to prevent the disease. Broad-spectrum antibiotics have been used with some success in treatment.

Inclusion Body Hepatitis

Inclusion body hepatitis is caused by a virus. The disease is common in commercial chicken flocks. The sudden death of chickens in the flock often is the only symptom. Diseased chickens show symptoms of anemia, yellowing of the skin, depression, and weakness. Death follows these symptoms fairly quickly. There is no vaccine or treatment.

Infectious Bronchitis

Infectious bronchitis is caused by a virus and affects only chickens. Symptoms include difficult breathing, sneezing, gasping, and nasal discharge. Egg production of layers is reduced. Death rate of young chicks is high. Older birds also show loss of appetite and slower growth. Prevention is by vaccination, isolation of flocks, and sanitation. There is no treatment.

Infectious Bursal Disease (Gumboro)

Infectious bursal disease is caused by a virus. It affects young chickens. Symptoms include fever (in the early stages), ruffled feathers, loss of appetite, difficulty in defecation, slight tremors, and dehydration. Birds do not want to move and are unsteady when walking. Vent picking often occurs. Death rate is high. Prevention is by vaccination. There is no treatment.

Infectious Coryza

Infectious coryza is caused by a bacterium. It affects mainly older chickens. Symptoms include swelling around the eyes and wattles, nasal discharge, and swollen sinuses. Egg production in layers is reduced. Good sanitation and management will help in prevention. Drugs are used for treatment.

Laryngotracheitis

Laryngotracheitis is caused by a virus. It affects mainly older chickens. Symptoms include difficult breathing, coughing, and sneezing. Death rate is high. Egg production in layers is reduced. Vaccination is used in areas where the disease exists. The use of the vaccine is restricted. There is no treatment.

Leucocytozoonosis

Leucocytozoonosis is caused by a protozoan parasite. It affects mainly turkey poults and ducklings. Symptoms include lack of appetite, droopiness, weakness, thirst, and difficult breathing. The death rate is high. Sanitation, isolation of brooding, and good management help in prevention. There is no treatment.

Muscle Degeneration

The cause of *muscle degeneration* is not known. Breeding tom turkeys are affected. It usually occurs after stimulatory lighting is started.

Symptoms include neck paralysis and, sometimes, leg paralysis. Birds usually recover after a few days of forced feeding.

Mycoplasmosis

Three species of the bacterium Mycoplasma cause respiratory diseases in poultry. These diseases include chronic respiratory disease/air sac syndrome in chickens, infectious sinusitis in turkeys, airsacculitis in turkeys, and infectious synovitis in chickens and turkeys. Infectious synovitis is discussed earlier in this unit.

Symptoms of *chronic respiratory disease* include nasal discharge, coughing, sneezing, swelling below the eyes, loss of weight, and drop in egg production. Isolation, sanitation, good management, and avoiding stress will help prevent the disease. Treatment is with antibiotics.

Symptoms of *infectious sinusitis* include watery eyes, nasal discharge, swollen sinuses, coughing, and difficult breathing. Prevention and treatment is the same as for chronic respiratory disease.

While there are a number of causes of airsacculitis in turkeys, the bacterium *Mycoplasma meleagridis* is one of the main ones. Breeding flocks as well as young poults may be affected. Yellow deposits are found on the air sacs and sometimes on the lungs of infected birds. Prevention is achieved by treating hatching eggs with antibiotic solutions. There are no drug preventions or controls.

Mycotoxicosis

Fungi or mold growing on feed or litter produce toxins. If these toxins are taken in by the bird, *mycotoxicosis* results. A variety of symptoms may indicate the presence of the disease. Growth rate may be slower, egg production may be reduced, the hatchability of eggs may be lowered, or the bird may die. The condition is best prevented by storing feeds in such a way that the mold does not grow. Do not feed moldy feed or use moldy litter.

Necrotic Enteritis

Necrotic enteritis appears to be caused by toxins produced by bacteria. Sudden death of healthy birds is one of the first indications of the disease. Death rate in a flock may be high. Birds that do not die have slower growth rates and poor feed conversion. Drugs are used in prevention and treatment. Controlling coccidiosis helps to prevent the disease.

Newcastle Disease

Newcastle disease occurs in several forms, each caused by a different virus. Young birds show both breathing problems and signs of nervousness, including difficult breathing, sneezing, gasping, paralysis, and tremors. Adults show the breathing problems and egg production of layers is reduced. Sanitation and vaccination are the methods of prevention. There is no treatment.

New Duck Disease (Infectious Serositis)

New duck disease is caused by bacteria. It is one of the most serious diseases affecting ducklings. Symptoms include sneezing and loss of balance. The death rate is high. Antibiotics are used in treatment.

Omphalitis

Omphalitis appears to be caused by several bacteria. It affects young birds. Symptoms include loss of appetite, diarrhea, and drowsiness. The death rate is high. Good management and sanitation in the hatchery are necessary for prevention. Broad-spectrum antibiotics are used for treatment.

Ornithosis

Ornithosis is caused by a bacterium. In humans it is called *parrot fever*. The disease is not a major problem in the poultry industry, but workers in turkey-processing plants have contracted the disease from infected birds. Symptoms, in turkeys, include coughing, sneezing, and a yellow diarrhea. Isolation of flocks may help in prevention. Antibiotics are used in treating the disease. Affected flocks are quarantined and treated under supervision. Flocks that have had the disease are not kept for breeding purposes.

Poisoning

Birds may be poisoned by eating or drinking any toxic materials. Sprays, disinfectants, nitrates, pesticides, and poisonous plants are a few of the sources of poisoning. Convulsions are a common symptom of poisoning. Birds raised on range are more likely to be poisoned than those raised in confinement. Prevent birds from eating poisonous substances.

Round Heart Disease

The exact cause of *round heart disease* is unknown. It may be the result of a virus, or heredity, or both. It affects mainly young male chickens and turkeys. Stress conditions appear to increase the chances of the disease appearing. The heart becomes enlarged, rounded, and flabby. The liver may be yellow in color. The condition can be prevented by making sure there is adequate heat during brooding and that only disease-free stock are purchased. There is no treatment.

Salmonella and Paracolon Infections

Salmonella and *paracolon* infections are caused by a number of bacteria. Four diseases of poultry that are caused by these bacteria include pullorum disease, fowl typhoid, paratyphoid, and paracolon infections.

Symptoms of *pullorum* include white diarrhea, droopiness, and chilling (Figure 37-1). It affects mainly young birds, and death losses may be very high. Prevent by getting stock from pullorum-free hatcheries. Treatment is not very effective, and the recovered birds become carriers.

Fowl typhoid affects mainly birds 12 weeks of age and older. Symptoms include loss of appetite, thirst, pale comb and wattles, listlessness, and yellow or green diarrhea and sudden death. The death rate may be high. Prevent by securing stock from disease-free flocks and practicing good sanitation on the farm.

Paratyphoid affects mainly young birds. The symptoms are similar to pullorum, as is the death rate. Sanitation in the hatchery and on the farm helps to prevent the disease. Drugs may be used in treatment to lower the death loss.

Paracolon infections are generally known as *Arizona infections*. A large number of bacteria are involved. The symptoms, control, and treatment are similar to paratyphoid. Laboratory diagnosis is needed to identify the disease.

FIGURE 37-1   Chicks sick and dead from pullorum disease. *Courtesy of USDA.*

Trichomoniasis

Trichomoniasis is caused by a protozoan parasite. Turkeys are affected more often than other poultry. Losses are highest among young and growing birds. Symptoms include loss of appetite, loss of weight, and droopiness. Sanitation is the best way to prevent the disease. Affected birds may be treated with copper sulfate added to the drinking water.

Ulcerative Enteritis

The exact cause of *ulcerative enteritis* is not known, but it is suspected that a bacterium is involved. The acute form may cause sudden death. Symptoms of the chronic form include ruffled feathers, humped up posture, and listlessness. A white diarrhea may be present. Antibiotics are used for the prevention and treatment of the disease.

EXTERNAL PARASITES

A number of external parasites are problems in poultry production. Included are chicken mites, northern fowl mites, scaly leg mites, several species of lice, fowl ticks, flies, fleas, and bedbugs. The most serious of these parasites are mites and lice.

Chicken Mite

The chicken mite is also called the *red mite* or the *roost mite*. The chicken mite sucks blood from the birds. Egg production drops and young chickens may die. During the day the chicken mite hides in

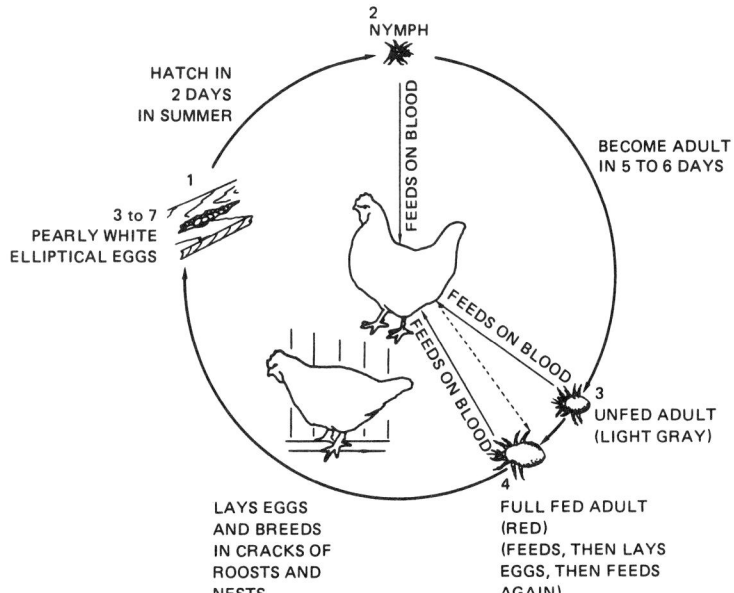

FIGURE 37-2   Life cycle of the common chicken mite. *Courtesy of University of Illinois at Urbana-Champaign.*

cracks and crevices. The life cycle of the common chicken mite is shown in Figure 37-2.

Northern Fowl Mite

The northern fowl mite is a bloodsucker. It spends all of its life cycle on the bird. Sparrows carry these mites and should be kept out of the poultry house. This mite also causes lower egg production and may cause death.

Scaly Leg Mite

This mite causes scaly leg. It lives under the scales on the feet and legs of the bird. Heavy infestations cause a rough appearance and enlargement of the legs.

Poultry Lice

The most common type of poultry louse is the body louse (Figure 37-3). Other lice that attack poultry are the shaft louse (Figure 37-4), fluff louse, wing louse, and the head louse (Figure 37-5). All of these are chewing lice. Infected birds lose weight, egg production drops, and young birds may die.

Fowl Tick

The fowl tick is a dark-colored bloodsucker. It hides during the day in cracks and crevices.

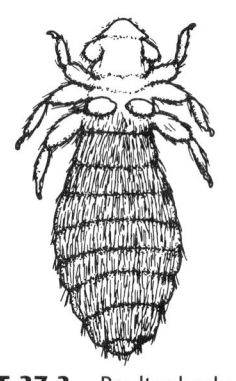

FIGURE 37-3   Poultry body louse. *Courtesy of University of Illinois at Urbana-Champaign.*

Fleas and Bedbugs

Fleas and bedbugs are bloodsuckers, and may become a problem in the poultry house. Fleas stay on the birds. Bedbugs feed on the bird at night and hide in cracks and crevices during the day.

Flies

Flies do not directly attack poultry. However, they may become a nuisance around the poultry house. Heavy fly infestations may cause problems with neighbors.

Control of External Parasites

The basis of a control program is the sanitation procedures outlined earlier in this unit. Insecticides may be used to control these parasites. Since recommendations and regulations on the use of insecticides change from time to time, no specific ones are listed here. Check with poultry specialists, veterinarians, or the Cooperative Extension Service for current recommendations on the use of insecticides.

INTERNAL PARASITES

Internal parasites that affect poultry include large roundworms, crop-worms (capillaria), cecal worms, tapeworms, flukes, gapeworms, and gizzard worms. The life cycles of some of these parasites are illustrated in Figures 37-6 through 37-10. Worm infestations generally cause slow

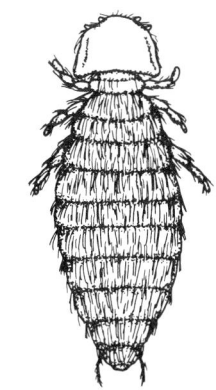

FIGURE 37-4   Poultry shaft louse. *Courtesy of University of Illinois at Urbana-Champaign.*

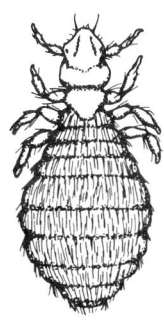

FIGURE 37-5   Poultry head louse. *Courtesy of University of Illinois at Urbana-Champaign.*

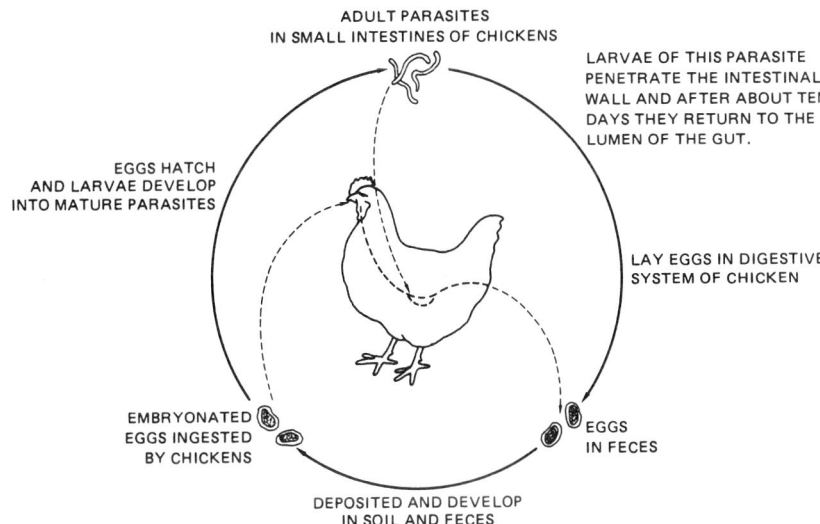

ADULT PARASITES
IN SMALL INTESTINES OF CHICKENS

LARVAE OF THIS PARASITE PENETRATE THE INTESTINAL WALL AND AFTER ABOUT TEN DAYS THEY RETURN TO THE LUMEN OF THE GUT.

EGGS HATCH AND LARVAE DEVELOP INTO MATURE PARASITES

LAY EGGS IN DIGESTIVE SYSTEM OF CHICKEN

EMBRYONATED EGGS INGESTED BY CHICKENS

EGGS IN FECES

DEPOSITED AND DEVELOP IN SOIL AND FECES

THIS PARASITE ALSO OCCURS IN DUCKS, GEESE, PIGEONS, AND TURKEYS.

FIGURE 37-6   Life cycle of large roundworm in poultry. Large roundworms spend their entire life cycle in the small intestine. The eggs must pass through a period of development before they can infest another fowl. *Courtesy of University of Illinois at Urbana-Champaign.*

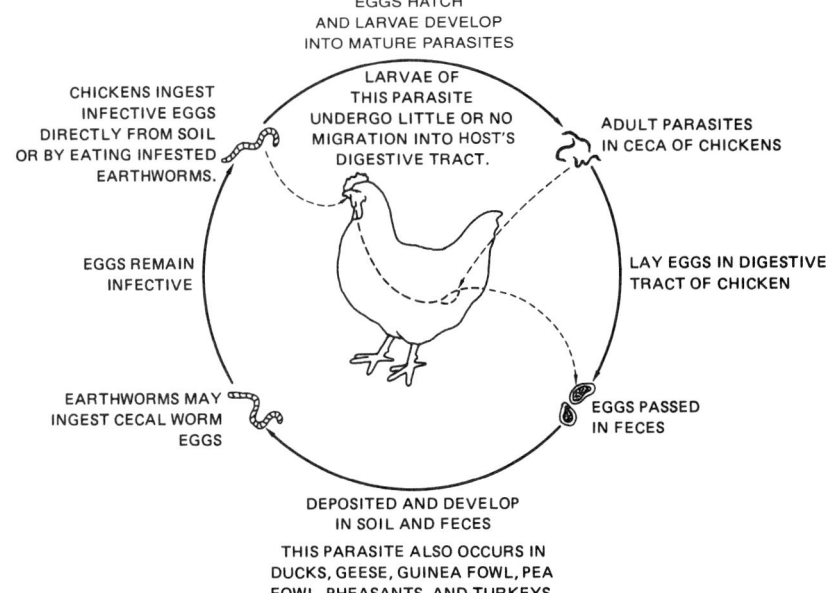

FIGURE 37-7 Life cycle of cecal worm in poultry. Cecal worms are found only in the ceca of chickens. Eggs must pass through a period of 7 to 12 days before infesting another fowl. *Courtesy of University of Illinois at Urbana-Champaign.*

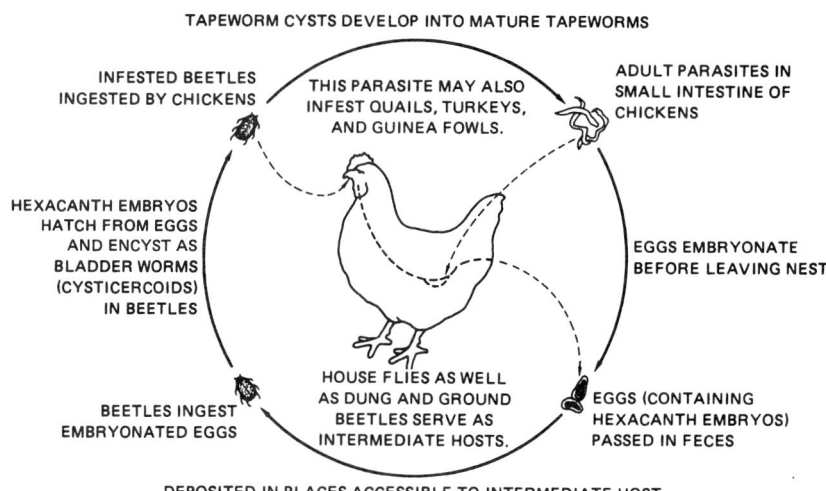

FIGURE 37-8 Life cycle of tapeworm in poultry. Tapeworms have an indirect life cycle—the eggs must develop in an intermediate host, such as flies, beetles, slugs, earthworms, or grasshoppers. *Courtesy of University of Illinois at Urbana-Champaign.*

growth and lower production. Drugs are available for the treatment of some of these worms. There is no known specific treatment for tapeworms.

Control of worms is best achieved by strict sanitation. Rotation of range also helps prevent infestation. Controlling insects and keeping wild birds out of poultry houses is important in control. Birds

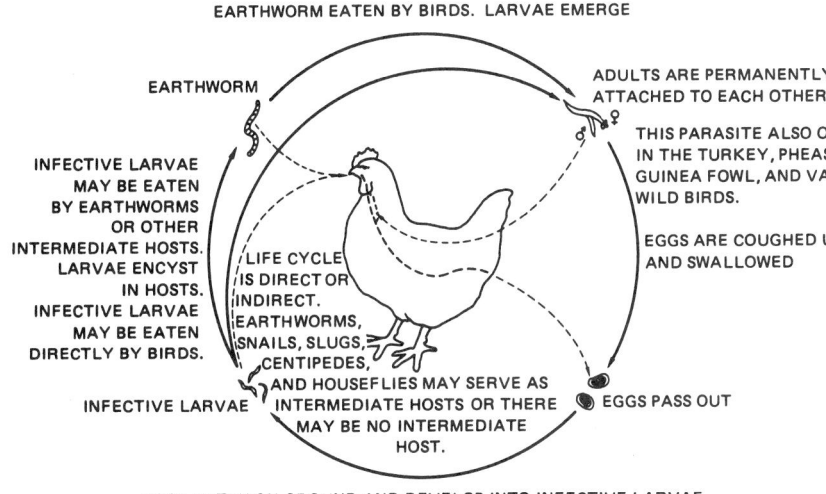

EARTHWORM EATEN BY BIRDS. LARVAE EMERGE

EARTHWORM

ADULTS ARE PERMANENTLY
ATTACHED TO EACH OTHER

THIS PARASITE ALSO OCCURS
IN THE TURKEY, PHEASANT,
GUINEA FOWL, AND VARIOUS
WILD BIRDS.

INFECTIVE LARVAE
MAY BE EATEN
BY EARTHWORMS
OR OTHER
INTERMEDIATE HOSTS.
LARVAE ENCYST
IN HOSTS.
INFECTIVE LARVAE
MAY BE EATEN
DIRECTLY BY BIRDS.

EGGS ARE COUGHED UP
AND SWALLOWED

LIFE CYCLE
IS DIRECT OR
INDIRECT.
EARTHWORMS,
SNAILS, SLUGS,
CENTIPEDES,
AND HOUSEFLIES MAY SERVE AS
INTERMEDIATE HOSTS OR THERE
MAY BE NO INTERMEDIATE
HOST.

INFECTIVE LARVAE

EGGS PASS OUT

EGGS HATCH ON GROUND AND DEVELOP INTO INFECTIVE LARVAE

FIGURE 37-9 Life cycle of gape-worm in poultry. Gapeworms attach themselves to the lining of the trachea. Eggs are present in discharges coughed up by the bird or in the feces. *Courtesy of University of Illinois at Urbana-Champaign.*

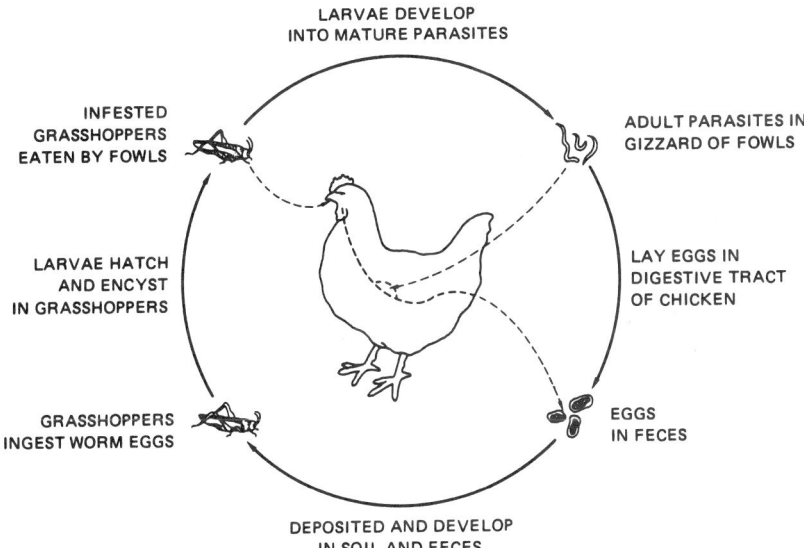

LARVAE DEVELOP
INTO MATURE PARASITES

INFESTED
GRASSHOPPERS
EATEN BY FOWLS

ADULT PARASITES IN
GIZZARD OF FOWLS

LARVAE HATCH
AND ENCYST
IN GRASSHOPPERS

LAY EGGS IN
DIGESTIVE TRACT
OF CHICKEN

GRASSHOPPERS
INGEST WORM EGGS

EGGS
IN FECES

DEPOSITED AND DEVELOP
IN SOIL AND FECES

THIS PARASITE ALSO OCCURS IN TURKEYS.

FIGURE 37-10 Life cycle of gizzard worms in poultry. Gizzard worms have an indirect life cycle. That is, the eggs must develop in an intermediate host such as grasshoppers, weevils, and certain species of beetles. *Courtesy of University of Illinois at Urbana-Champaign.*

raised in confinement and on wire or slat floors seldom have serious infestations of worms. Good management indicates that the flock should be checked periodically for the presence of worms. Current recommendations on the use of drugs for worm control may be secured from poultry specialists, veterinarians, or the Cooperative Extension Service.

SUMMARY

The best way to control poultry diseases and parasites is through a program of prevention. Prevention involves sanitation, good management, vaccination, and control of disease outbreaks.

Sanitation requires complete cleaning and disinfecting of poultry houses and equipment. Isolation of the flock and proper disposal of dead birds and manure are other sanitation practices that should be followed.

All poultry replacement stock should be secured from reliable, disease-free sources. It is best to keep only birds of the same age on the farm. Chickens and turkeys should not be mixed. Feed balanced rations and keep good health records.

Vaccination can be used to prevent some poultry diseases. However, it is not a substitute for good sanitation or management practices. A vaccination program should be planned for a particular operation in a specific area.

The poultry flock should be checked daily for signs of disease. A drop in production or an increase in the death rate may indicate the presence of a disease. Accurate diagnosis of poultry diseases is best done in a laboratory.

Many diseases affect poultry. A large number have similar symptoms and effects. Some affect all kinds of poultry, while others affect only some species.

Mites and lice are the main external parasites that affect poultry. Sanitation and the proper use of insecticides can keep these under control.

The main internal parasites of poultry are worms. Birds raised in confinement or on wire or slat floors have fewer internal parasite problems.

Student Learning Activities

1. Give an oral report on control measures for diseases and parasites or on an individual disease or parasite.
2. Ask a veterinarian to speak to the class on control of poultry diseases and parasites.
3. Survey local poultry producers to determine the disease and parasite problems that are present and the control measures that are used.
4. Practice disease and parasite control measures on the home farm.

Review

1. Outline the steps for a good sanitation program for poultry houses.
2. List other important health management practices for poultry.
3. Briefly explain the vaccination of poultry as part of a good health program.
4. Describe how to control disease outbreaks in a poultry flock.
5. List and briefly describe diseases that affect poultry.
6. List and briefly explain the common external parasites that affect poultry.
7. List and briefly explain the control of internal parasites that affect poultry.

Unit 38 Marketing Poultry and Eggs

Objectives

After studying this unit, the student should be able to

- summarize the production and price trends in poultry and eggs.
- describe the methods of marketing poultry and eggs.

PRICE AND PRODUCTION TRENDS IN POULTRY AND EGGS

Long-Term Production and Price Trends

The quantity of broiler meat produced in the United States has been steadily increasing since 1940. Figures 38-1A and 38-1B illustrate the long-term trends in broiler production and prices. Since the early 1970s the price of broilers has risen steadily.

Figures 38-2A and 38-2B show the long-term trends in egg prices and production. Egg production and prices have shown considerable variation during this period. Production was down sharply in the middle 1970s and has been generally trending upward since then. Egg prices rose abruptly in 1973 and have continued this upward trend, with some annual variation.

Figures 38-3A and 38-3B reflect the long-term trends in turkey production and prices. While there has been some variation in production from year to year, the general trend has been upward. The price of turkeys has shown a greater variation from year to year than the production of turkeys; overall, however, the general trend in price has been upward.

Production and prices tend to follow the laws of supply and demand. It is probable that vertical integration in the poultry industry has also had an effect on the long-term trends in production and prices.

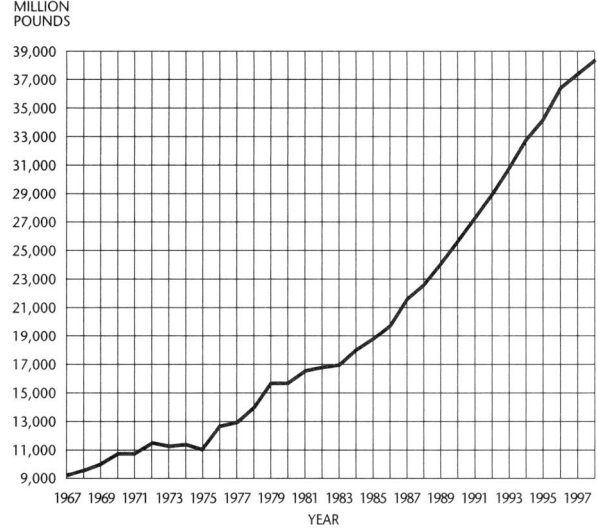

FIGURE 38-1A   Broiler production, 1967–1998. *Sources: Agricultural Statistics,* USDA, 1981, 1986, 1988, 1992, 1995, 1999; *Poultry Production and Value,* USDA, April 1999.

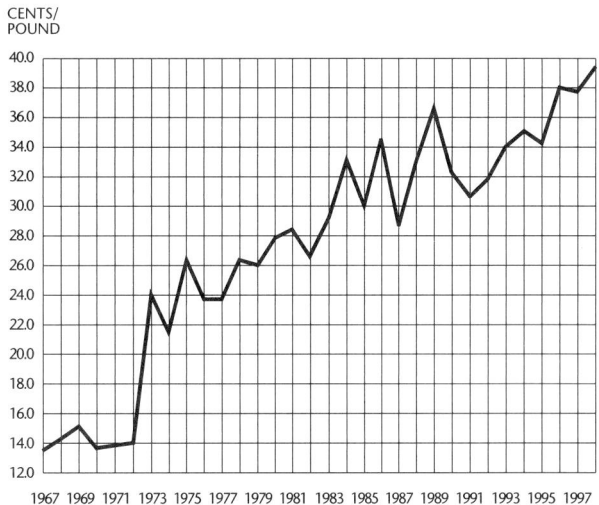

FIGURE 38-1B   Broiler prices, 1967–1998. *Sources: Agricultural Statistics,* USDA, 1981, 1986, 1988, 1992, 1995, 1999; *Poultry Production and Value,* USDA, April 1999.

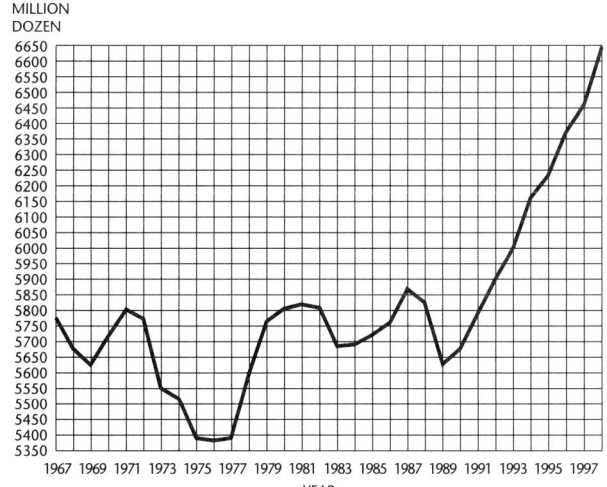

FIGURE 38-2A   Egg production, 1967–1998. *Sources: Agricultural Statistics,* USDA, 1981, 1986, 1988, 1992, 1995, 1999; *Poultry Production and Value,* USDA, April 1999.

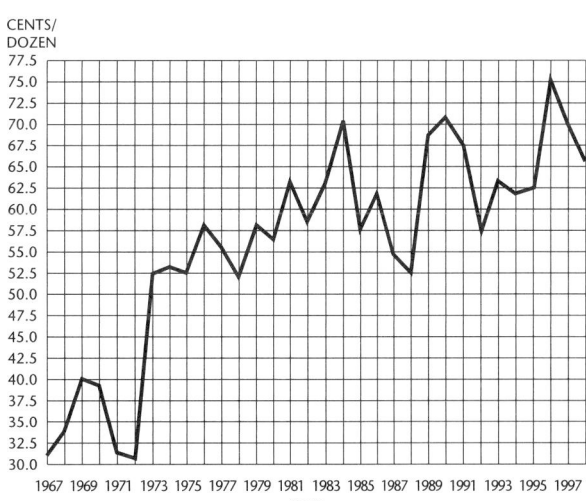

FIGURE 38-2B   Farm prices of eggs, 1967–1998. *Sources: Agricultural Statistics,* USDA, 1981, 1986, 1988, 1992, 1995, 1999; *Poultry Production and Value,* USDA, April 1999.

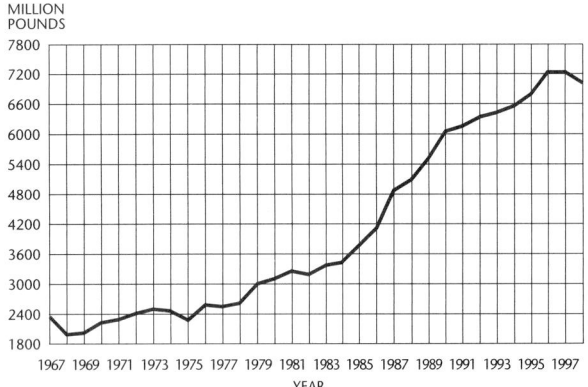

FIGURE 38-3A   Turkey production, 1967–1998.
Sources: Agricultural Statistics, USDA, 1981, 1986, 1988, 1992, 1995, 1999; *Poultry Production and Value,* USDA, April 1999.

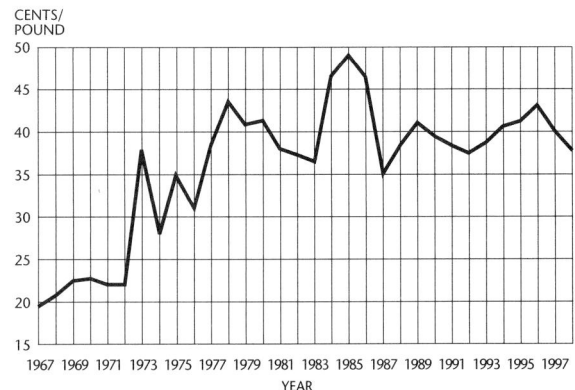

FIGURE 38-3B   Farm prices of turkeys, 1967–1998.
Sources: Agricultural Statistics, USDA, 1981, 1986, 1988, 1992, 1995, 1999; *Poultry Production and Value,* USDA, April 1999.

Seasonal Production and Price Trends

The heavy emphasis on vertical integration in the broiler industry has a tendency to keep seasonal variations in production and prices within a rather narrow range (Figure 38-4).

The seasonal variation in production and prices for eggs is illustrated in Figure 38-5.

Figure 38-6 shows the seasonal variation in production and prices for turkeys. As shown by the graph, the pounds of turkey slaughtered under federal inspection tends to increase during the last half of the

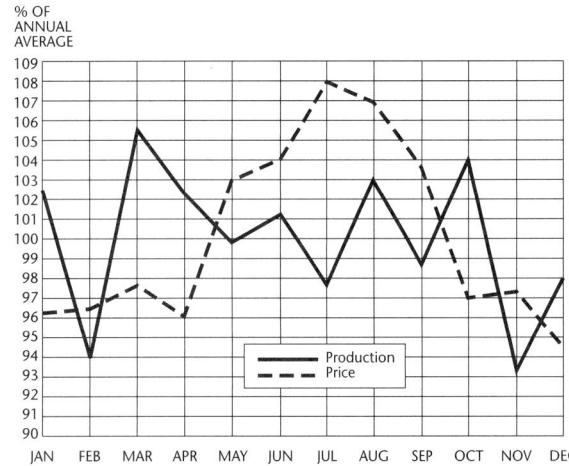

FIGURE 38-4   Seasonal variation in broiler production and prices as a percent of yearly average. *Source:* USDA.

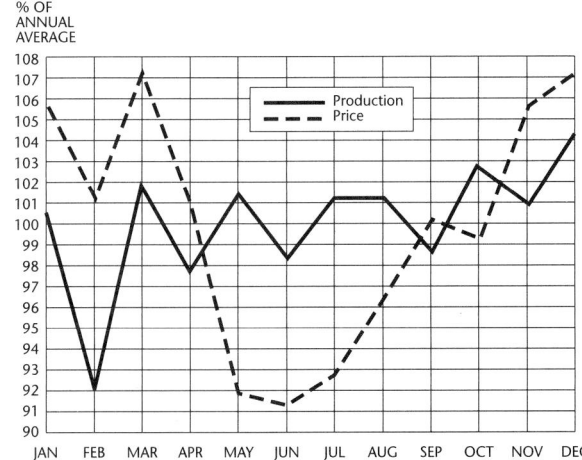

FIGURE 38-5   Seasonal variation in egg production and prices as a percent of yearly average. *Source:* USDA.

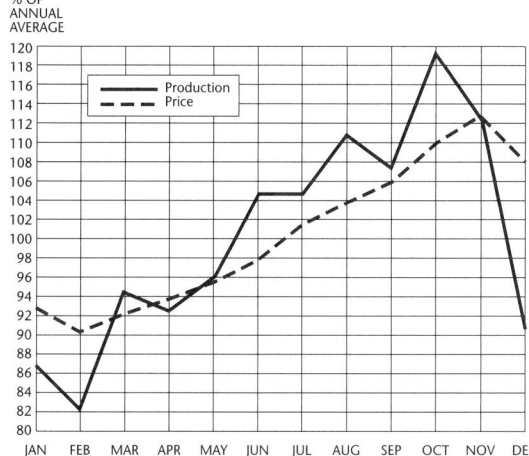

% OF ANNUAL AVERAGE

FIGURE 38-6 Seasonal variation in turkey production and prices as a percent of yearly average. *Source:* USDA.

year. This is the period of highest demand because of the holiday season. Turkey prices are generally stable throughout the year.

MARKETING POULTRY

Broilers and Other Chickens

Vertical integration contract. Approximately 99 percent of the broilers raised in the United States are grown under some type of vertical integration contract. Vertical integration is described in Unit 35. As a result of vertical integration, there are few actual sales of live birds. The birds are owned by the integrator rather than the farmer who feeds them. The contract the farmer has with the integrator spells out the terms under which payment is made for growing the birds.

A contract usually gives a guaranteed minimum payment for the birds. In addition, there is often a bonus clause that provides for payment of an additional amount if the feed conversion is better than the average. Payments are made on the basis of live weight at the time the birds are marketed. Contracts are regulated by the Packers and Stockyards Act. Written contracts are required under this act.

Poultry not under contract. Poultry not produced under contract may be sold live, dressed, or ready-to-cook. Sales may be made to the buyer at the farm, at auction, to dealers, to brokers, or to processing plants. If the poultry is processed on the farm, it may be sold to wholesale dealers or direct to the consumer. However, very little poultry in the United States is sold in this manner.

Preparing birds for market. Chickens should not be fed for 12 hours before slaughter, but should be provided with water. Handle birds carefully to prevent injuries and bruises. Avoid crowding birds in delivery coops. It is best to move birds in the early morning if possible.

Turkeys

About 75 percent of the young turkeys produced in the United States are marketed between August and December. Fryer-roaster turkeys are marketed throughout the year. Most turkeys are marketed through integrated firms, processors, or cooperatives.

Large broad-breasted male turkeys are usually ready for market at 24 weeks, while females are ready at 20 weeks. Small-and-medium-type turkeys are ready at 22 weeks. Fryer-roasters are ready to market at 12 to 13 weeks and light roasters at 17 to 20 weeks.

Preparing birds for market. Turkeys are usually not taken off of feed and water before being moved to market for slaughter. Care must be used when handling turkeys to prevent injuries and bruises that will cause a lowering of the market grade. Turkeys may be handled with less confusion and injury to the birds if they are confined to a small area, such as a chute or pen, when loading them for market. In confinement systems, a small, darkened room works well for catching turkeys.

Ducks

Meat ducks of the Pekin type are usually ready for market at seven to eight weeks of age. Muscovy ducks are ready in 10 to 17 weeks. Ducks are marketed when they have a good finish and are free of pin feathers.

Ducks are taken off of feed eight to 10 hours before they are slaughtered. They are provided with water until they are killed. Use care when catching, handling, and moving ducks to prevent injuries or bruises. Small numbers of ducks may be processed on the farm for direct sale to consumers. Larger producers sell to processing plants that also sell the feathers, feet, heads, and viscera (internal organs) as by-products. Small producers who process their own ducks usually do not find a market for these by-products.

Geese

Geese are usually sold alive off the farm to live-poultry buyers or to processing plants. Most geese are marketed in the fall and winter. The best prices are obtained in large cities around Thanksgiving and Christmastime. Usually, geese are marketed when they weigh 11 to 15 pounds (5–6.8 kg). Geese may be full fed for fast growth. Geese fed in this way are called *green geese* or *junior geese*. They are marketed at about 10 weeks when they weigh 10 to 12 pounds (4.5–5.4 kg).

Goose feathers are valuable in the bedding and clothing industries. Buyers of goose feathers are found in most large cities. The feathers may be sold to a feather processing plant. One pound (0.45 kg) of dry feathers can be produced from three geese.

CLASSES OF READY-TO-COOK POULTRY[1]

Chickens

The USDA sets standards for six classes of ready-to-cook chickens as follows:

1. **Rock Cornish game hen or Cornish game hen.** A young immature chicken of either sex, usually less than five weeks of age, with a ready-to-cook weight of not more than two pounds (0.9 kg).
2. **Broiler or fryer.** A young chicken of either sex, usually less than 10 weeks of age, that is tender-meated with soft, pliable, smooth-textured skin and flexible breastbone cartilage.
3. **Roaster or roasting chicken.** A young chicken, usually less than 12 weeks of age, of either sex, that is tender-meated with soft, pliable, smooth-textured skin and flexible breastbone cartilage.
4. **Capon.** A surgically neutered male chicken, usually less than four months of age, that is tender-meated with soft, pliable, smooth-textured skin.
5. **Hen, fowl, baking chicken, or stewing chicken.** An adult female chicken, usually more than 10 months of age, with meat less tender than that of a roaster or roasting chicken and with a nonflexible breastbone tip.
6. **Cock or rooster.** An adult male chicken with coarse skin, toughened and darkened meat, and nonflexible breastbone tip.

Turkeys

The USDA sets standards for four classes of ready-to-cook turkeys as follows:

1. **Fryer-roaster turkey.** A young, immature turkey of either sex, usually less than 12 weeks of age, that is tender-meated and has soft, pliable, smooth-textured skin and flexible breastbone cartilage. This class may also be labeled "young turkey."
2. **Young turkey.** A turkey generally under six months of age, of either sex, with tender meat and soft, pliable, smooth-textured skin. The breastbone cartilage is a little less flexible than in a fryer-roaster turkey.
3. **Yearling turkey.** A fully matured turkey usually under 15 months of age, of either sex, that is reasonably tender-meated. It has reasonably smooth-textured skin.

---

[1]*Poultry Grading Manual,* USDA, Agricultural Marketing Service, Agriculture Handbook Number 31, April 1998.

4. **Mature or old (hen or tom) turkey.** An adult turkey, usually more than 15 months of age, of either sex, with coarse skin and toughened flesh. Sex designation is optional.

Ducks

The USDA sets standards for three classes of ready-to-cook ducks as follows:

1. **Duckling**. A young duck, usually less than eight weeks of age, of either sex, that is tender-meated and has a soft bill and soft windpipe.
2. **Roaster duck.** A young duck, usually less than 16 weeks of age, of either sex, that is tender-meated and has a bill that is not completely hardened and a windpipe that is easily dented.
3. **Mature duck or old duck.** An adult duck, usually more than six months of age, of either sex, with toughened flesh, a hardened bill, and a hardened windpipe.

Geese

The USDA sets standards for two classes of ready-to-cook geese as follows:

1. **Young goose.** An immature goose of either sex that is tender-meated and has a windpipe that is easily dented.
2. **Mature goose or old goose.** A adult goose of either sex that has toughened flesh and a hardened windpipe.

GRADES OF READY-TO-COOK POULTRY

All poultry slaughtered for human food in the United States must be processed, handled, packaged, and labeled in accordance with Federal law. The applicable laws are the Poultry Products Inspection Act of 1957 and the Wholesome Poultry Products Act of 1968. Before poultry is graded for quality, it must be inspected for wholesomeness and fitness for human food (Figure 38-7). Mandated Federal inspection is paid for by the government. Quality grading is voluntary and is paid for by the processor. Unsound or unwholesome poultry are not eligible for quality grading.

The United States quality grades of poultry are grades A, B, and C. These grades are applied both to the ready-to-cook carcass and to the parts of the carcass. Parts include such things as poultry halves, breast, leg, thigh, drumstick, wing, and tenderloin. No grade standards exist for giblets, detached necks and tails, wing tips, and skin.

The quality grade is determined by an evaluation of the following factors: (1) conformation, (2) fleshing, (3) fat covering, (4) defeathering, (5) exposed flesh, (6) discolorations, (7) disjointed and

FIGURE 38-7 A U.S. Department of Agriculture poultry inspector checks broilers for wholesomeness as they make their way through a cutting and packaging plant. *Courtesy of USDA.*

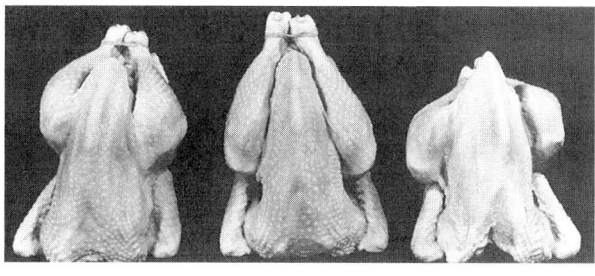

U.S. GRADE A   U.S. GRADE B   U.S. GRADE C

FIGURE 38-8   USDA grades of ready-to-cook young chickens. *Courtesy of USDA.*

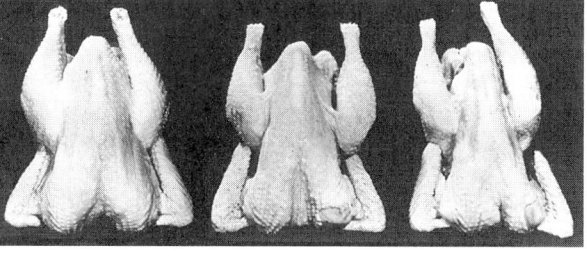

U.S. GRADE A   U.S. GRADE B   U.S. GRADE C

FIGURE 38-9   USDA grades of ready-to-cook stewing chickens. *Courtesy of USDA.*

broken bones, and (8) freezing defects. Grade A is the only grade of poultry that is generally sold at retail. Occasionally grades B and C are sold at retail, but it is more common for these grades to be used in processed poultry products, where they are cut up, chopped, or ground. Figures 38-8 through 38-14 illustrate representative carcasses in the various poultry grades.

The standards for grading ready-to-cook poultry are found in the *USDA Poultry Grading Manual.* These grades apply to all kinds and classes of poultry. The *Poultry Grading Manual*, regulations governing the grading of poultry, and Standards, Grades, and Weight classes for shell eggs can be found on the Internet at links from this URL: www.ams.usda.gov/poultry/pypubs.htm.

MARKETING EGGS

Vertical integration is not as widespread in egg production as it is in broiler production. However, more than half of the eggs produced in the United States are produced under some type of vertical integration in the egg industry. Eggs produced under a vertical integration contract are marketed in much the same manner as broilers. The grower

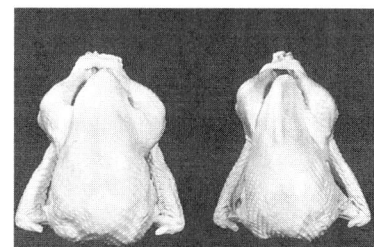

FIGURE 38-10   U.S. Grade A (left) and U.S. Grade B (right) young turkeys. *Courtesy of USDA.*

FIGURE 38-11   Young tom turkey carcass, C Quality. *Courtesy of USDA.*

FIGURE 38-12   Young hen turkey carcass, A Quality. *Courtesy of USDA.*

FIGURE 38-13   Young hen turkey carcass, B Quality. *Courtesy of USDA.*

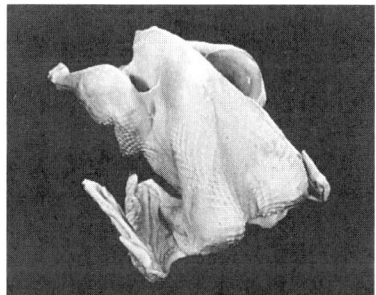

FIGURE 38-14   Young hen turkey carcass, C Quality. *Courtesy of USDA.*

does not own the birds, and the eggs are paid for under the terms of the contract.

Eggs that are not produced under contract are sold to local buyers, produce dealers, or are marketed through cooperatives. Some producers may sell directly to consumers at the farm. Direct sales to consumers are also made through house-to-house routes or at roadside stands. These types of marketing practices require more labor than selling to produce dealers or through cooperatives.

Eggs are highly perishable. They must be handled carefully to prevent spoiling. Egg quality is best preserved by keeping the eggs cool and at the correct humidity after they are laid and until they are used. The proper handling of eggs is described in Unit 36.

CLASSES AND GRADES OF EGGS

The USDA sets standards for weight classes and grades of shell eggs. Grades of eggs are based on these four factors: (1) shell, (2) air cell, (3) white, and (4) yolk. Table 38-1 is a summary of the United States standards for quality of shell eggs. Examples of various grades of eggs are shown in Figures 38-15 through 38-20.

Classes of eggs are based on weight shown as *ounces per dozen*. The weight classes for U.S. Consumer Grades of shell eggs are shown in Table 38-2 and illustrated in Figure 38-21.

The interior quality of the egg is determined by a process called *candling*. The eggs are examined by using a high-intensity light. Eggs are being hand candled in Figure 38-22. Figure 38-23 shows the use of a conveyor belt to mass candle the eggs. This is the most common method of candling.

United States Standards for grades of eggs do not consider the color of the egg. The color of the egg does not affect its nutritive value. However, some markets prefer white eggs, whereas others prefer brown eggs. Therefore, eggs are sometimes sorted by color for the market in which they are to be sold.

Consumer, *wholesale*, and *procurement* grades of eggs are different classifications used in the marketing of eggs. A complete description of these classifications is found in the *United States Standards, Grades, and Weight Classes for Shell Eggs*, which is available from the USDA.

FIGURE 38-15 Grade AA (or Fresh Fancy) egg covers small area; white is thick, stands high; yolk is firm and high. *Courtesy of USDA.*

TABLE 38-1 Summary of United States Standards for Quality of Individual Shell Eggs (based on candled appearance).

Quality Factor	Specifications for Each Quality Factor		
	AA Quality	**A Quality**	**B Quality**
Shell	Clean Unbroken Practically normal	Clean Unbroken Practically normal	Clean to slightly stained* Unbroken Abnormal
Air Cell	1/8 inch or less in depth Unlimited movement and free or bubbly	3/16 inch or less in depth Unlimited movement and free or bubbly	Over 3/16 inch in depth Unlimited movement and free or bubbly
White	Clear Firm	Clear Reasonably firm	Weak and watery Small blood and meat spots present**
Yolk	Outline: slightly defined Practically free from defects	Outline: fairly well defined Practically free from defects	Outline: plainly visible Enlarged and flattened Clearly visible germ development but no blood Other serious defects

Standards of Quality for Eggs with Dirty or Broken Shells

Dirty	Check
Unbroken. Adhering dirt or foreign material, prominent stains, moderate stained areas in excess of B quality	Broken or cracked shell but membranes intact, not leaking***

  *Moderately stained areas permitted (1/32 of surface if localized, or 1/16 if scattered).
 **If they are small (aggregating not more than 1/8 inch in diameter).
***Leaker has broken or cracked shell and membranes, and contents leaking or free to leak.

Source: USDA.

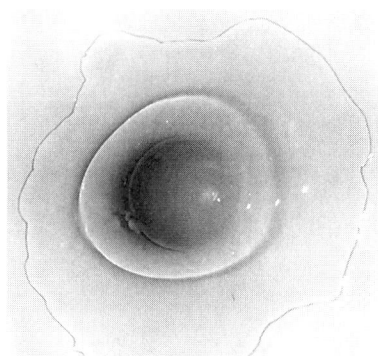

FIGURE 38-16 Top view of Grade AA (or Fresh Fancy) egg. *Courtesy of USDA.*

FIGURE 38-17 Grade A egg covers moderate area; white is reasonably thick, stands fairly high; yolk is firm and high. *Courtesy of USDA.*

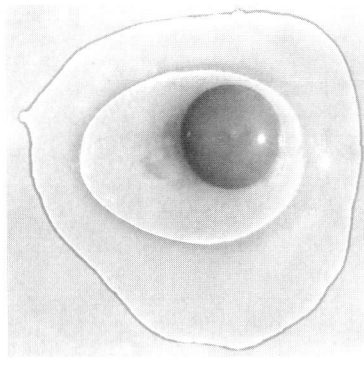

FIGURE 38-18 Top view of Grade A egg. *Courtesy of USDA.*

TABLE 38-2 U.S. Weight Classes for Consumer Grades for Shell Eggs.

Size or Weight Class	Minimum Net Weight per Dozen	Minimum Net Weight per 30 Dozen	Minimum Weight for Individual Eggs at Rate Per Dozen
	Ounces	Pounds	Ounces
Jumbo	30	56	29
Extra large	27	50.5	26
Large	24	45	23
Medium	21	39.5	20
Small	18	34	17
Peewee	15	28	—

Source: United States Standards, Grades, and Weight Classes for Shell Eggs, Agricultural Marketing Service, Poultry Division, United States Department of Agriculture, April 6, 1995.

FIGURE 38-19 Grade B egg covers wide area; has small amount of thick white; yolk is somewhat flattened and enlarged. *Courtesy of USDA.*

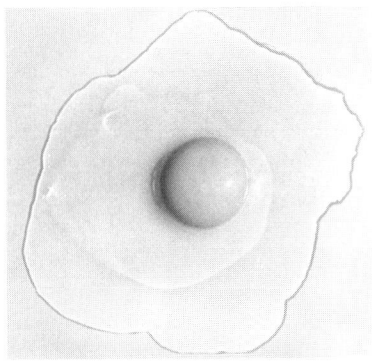

FIGURE 38-20 Top view of Grade B egg. *Courtesy of USDA.*

SUMMARY

The long-term trend in production of broilers and turkeys has been upward. Egg production increased until the early 1970s, but since that time has shown some decrease. Long-term trends in prices show variations resulting from supply and demand and the influence of vertical integration on the market.

Broiler prices vary seasonally within a rather narrow range. Turkey prices tend to be fairly stable throughout the year, showing some increase during the holiday season.

 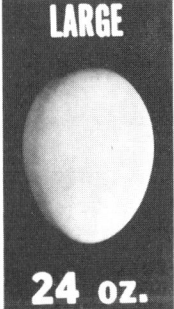 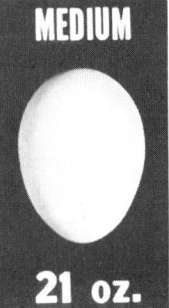  

FIGURE 38-21 U.S. weight classes for eggs. *Courtesy of USDA.*

About 99 percent of the broilers raised in the United States are produced under some type of vertical integration contract. Most of the turkeys in the United States are marketed through integrated firms, processors, or cooperatives. Poultry must be carefully handled during marketing to avoid injuries and bruises.

Ducks and geese may be processed on the farm for direct sales to consumers. Larger producers generally sell to processing plants. Goose feathers are valuable in the bedding and clothing industries.

The USDA has established six classes of ready-to-cook chickens, four classes of ready-to-cook turkeys, three classes of ready-to-cook ducks, and two classes of ready-to-cook geese. Classes are based on age, tenderness of meat, smoothness of skin, and hardness of breastbone cartilage.

The three grades of ready-to-cook poultry are (1) A, (2) B, and (3) C. Standards for grading poultry are found in the *USDA Poultry Grading Manual*.

More than one-half of the eggs produced in the United States are marketed under some type of vertical integration contract. Other eggs are sold to local buyers, produce dealers, cooperatives, or direct to consumers. Eggs must be carefully handled to maintain high quality.

The standards for weight classes and grades of eggs are set by the USDA. Grades are based on (1) shell, (2) air cell, (3) white, and (4) yolk. Weight classes are based on the ounces per dozen eggs.

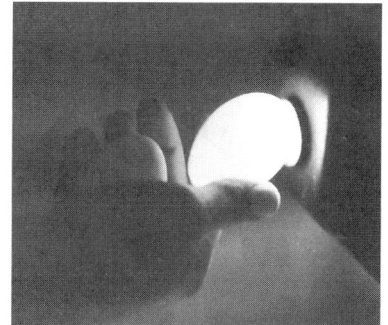

FIGURE 38-22   Hand candling of eggs, using a high-intensity light. *Courtesy of USDA.*

Student Learning Activities

1. Prepare a display of eggs showing the different grades and weight classes.
2. Ask poultry producers to speak to the class on marketing poultry products.
3. Take a field trip to a poultry processing plant to observe inspection and grading procedures.
4. Ask a poultry grader or egg candler to speak to the class.
5. Prepare a bulletin board display of the grades of poultry.

FIGURE 38-23   Mass candling of eggs using a conveyor belt. *Courtesy of USDA.*

Review

1. Describe the long-term production and price trends for broilers, eggs, and turkeys.
2. Describe the seasonal production and price trends for broilers, eggs, and turkeys.
3. Describe the marketing of broilers under contract.
4. List other methods besides contract that are used to market broilers.

5. Describe how broilers should be handled just before and at marketing.
6. List the common methods of marketing turkeys.
7. At what age are turkeys usually marketed?
8. At what age are ducks usually marketed?
9. Describe the marketing of geese.
10. Name and describe the six USDA classes of ready-to-cook chickens.
11. Name and describe the four USDA classes of ready-to-cook turkeys.
12. Name the market classes of ready-to-cook ducks and geese.
13. List the factors that are used for quality grading of ready-to-cook poultry.
14. Describe the grading of ready-to-cook poultry.
15. List and briefly describe the methods of marketing eggs.
16. Briefly describe the United States standards for quality of shell eggs.
17. List the weight classes for U.S. Consumer Grades of shell eggs.
18. How is the interior quality of shell eggs determined?
19. What effect does the color of an egg have on its nutritive value?

Dairy

Breeds of Dairy Cattle

Objectives

After studying this unit, the student should be able to

- describe the characteristics of the dairy enterprise.
- name and describe the breeds of dairy cattle, giving their origin and breed characteristics.
- identify the breeds of dairy cattle by viewing pictures or live animals.

CHARACTERISTICS OF THE DAIRY ENTERPRISE

Dairy cows must be milked at regular intervals, two times per day, seven days per week. Some herds are milked three times per day. The dairy cow is a creature of habit. A change in routine such as a wide variation in milking time or strangers in the milking area can reduce milk production. The modern dairyman uses mechanical equipment to milk, feed, and care for the dairy herd. Milk is moved by pipeline from the milking machine to the bulk tank where it is stored until picked up for market by the milk transport. Few modern dairymen handle milk in cans.

In addition to the regular milking chores, the dairy farmer must raise the crops needed to feed the dairy herd. Corn, small grains, hay, and pasture are raised on the farm to feed the cows. Some of the larger dairy operations buy most or all of the needed feed.

The dairy barn and milk house must be kept clean. In stanchion barns the manure must be removed daily during the winter when the cows are kept inside most of the time. Milking parlors become dirty as the cows pass through for the daily milking. These parlors must be kept washed and clean for the production of milk. Insects must be controlled in and around the dairy buildings.

The dairy farmer must keep records. Milk production records are kept to help in culling cows and in the breeding program. Records of breeding and calving are kept. Crop, animal health, and business records are also necessary.

To be a successful dairy farmer a person must have patience and be willing to work long hours. The capital investment needed to get into dairy farming is high. The average investment is $4,000 to $5,000 per cow. A person must be able to get credit to borrow needed money. Experience on a dairy farm is needed. Experience may be gained by working for a dairy farmer before trying to start for oneself.

The dollar returns on the money invested in the dairy farm may be from six to nine percent. The average labor per cow per year is about 27 hours. The dollar returns per hour of labor vary depending on feed and other costs. Good management is needed to get the best returns per hour of labor invested in the dairy enterprise.

Some of the advantages of the dairy enterprise are:

- Dairy cattle use for feed roughages that might otherwise be wasted.
- Dairying provides a steady income throughout the year.
- Labor is used throughout the year.
- Death losses in the dairy herd are usually low if good management is followed.

Some of the disadvantages of the dairy enterprise are:

- A high capital investment is needed.
- The labor requirement is high and the operator is confined to a regular schedule of milking.
- Training and experience are needed before entering into the dairy business.
- It takes a relatively long time to develop a high-producing dairy herd.
- The demand for dairy products appears to be dropping.

TRENDS IN DAIRY PRODUCTION

The number of milk cows on farms in the United States has shown a steady decline for many years (Figure 39-1). There was a slight increase in dairy cow numbers on farms during the early 1980s, with the numbers again showing a decline since 1985. During the middle 1980s, a Dairy Termination Program administered by the USDA encouraged dairy farmers to sell entire herds for slaughter. The program was designed to reduce an oversupply of milk in the marketplace by reducing the number of producing herds. The effectiveness of the program was modified by an increase in the amount of milk produced per cow in the herds that remained in production.

For more than 40 years, there has been a trend in the United States toward fewer dairy farms, with each having larger herds. In 1954 there were 2,167,000 farms in the United States that reported one or more dairy cows; by 1998, there were only 116,430 farms reporting one or more dairy cows. The average size of dairy herds per farm in 1954 was

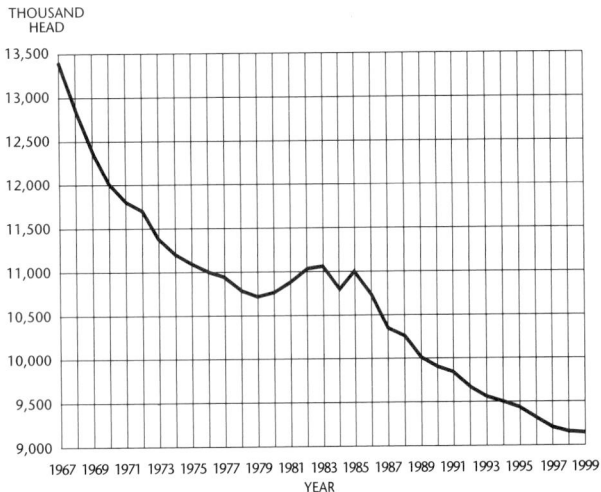

FIGURE 39-1 Trend in milk cow numbers, 1967–1999 (average number during year, excluding heifers not yet fresh). *Sources: Agricultural Statistics,* USDA, 1981, 1986, 1988, 1992, 1995, 1999; *Milk Production,* USDA, February 1999; *Livestock, Dairy, and Poultry,* USDA, April 1999.

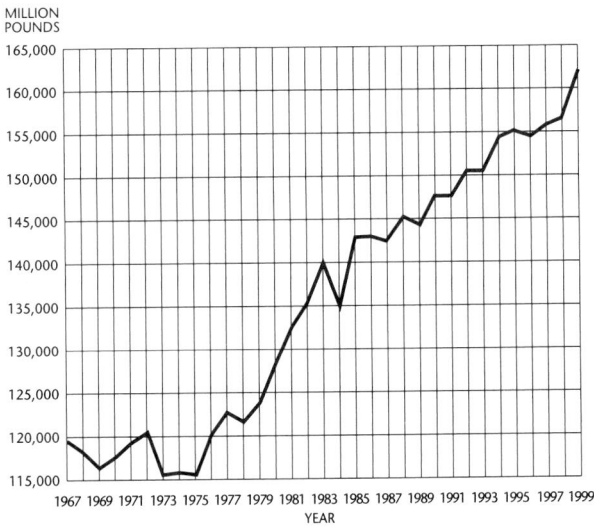

FIGURE 39-2 Total milk production, 1967–1999. *Sources: Agricultural Statistics,* USDA, 1981, 1986, 1988, 1992, 1995, 1999; *Milk Production,* USDA, February 1999; *Livestock, Dairy, and Poultry,* USDA, April 1999.

10 head; by 1998 the average size of dairy herds per farm had increased to 78.5 head.

The total production of milk in the United States, while varying from year to year, has shown a general upward trend (Figure 39-2). This has occurred despite a general downward trend in the number of dairy cows in the United States because of an increase in milk production per cow (Figure 39-3). The production of milkfat per cow has also shown a steady increase (Figure 39-4). This increase in production per cow is the result of better feeding, breeding, and management of the dairy herds.

An indication of better management is the increase in the percentage of all cows that are on test through Dairy Herd Improvement Associations (DHIA). Only 15.6 percent of all milk cows in the United States were on DHIA test in 1967. By 1997 this had increased to 36.8 percent of all milk cows. The average number of cows in DHIA herds increased from 55.7 in 1967 to 124.3 in 1997. Milk and milkfat production per cow is higher in these herds than the average for all herds in the United States (Table 39-1).

Grade A milk is produced on farms that have been certified to meet certain minimum standards. These standards deal with the buildings and conditions under which the milk is produced. The purpose of the standards is to assure that the milk is pure enough to be used for fluid milk consumption.

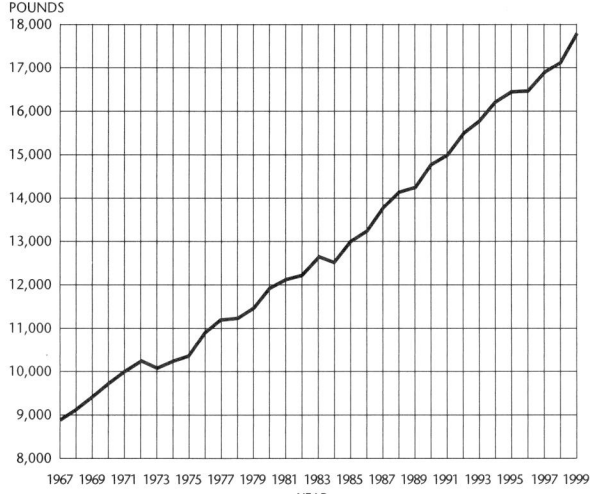

FIGURE 39-3 Milk production per cow, 1967–1999.
Sources: Agricultural Statistics, USDA, 1981, 1986, 1988,
1992, 1995, 1999; *Milk Production,* USDA, February
1999, *Livestock, Dairy, and Poultry,* USDA, April 1999.

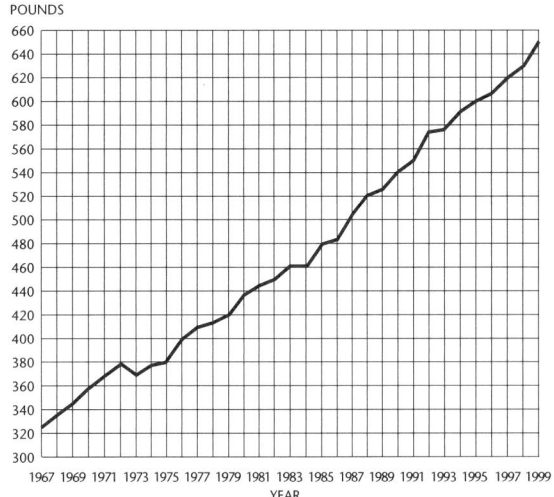

FIGURE 39-4 Production of milkfat per cow,
1967–1999. *Sources: Agricultural Statistics,* USDA, 1981,
1986, 1988, 1992, 1995, 1999; *Milk Production,* USDA,
February 1999; *Livestock, Dairy, and Poultry,* USDA, April
1999.

Grade B milk (manufacturing milk) is produced under conditions
that are less controlled than those for producing Grade A milk. Grade
B milk is used for processing into dairy products such as cheese, butter,
and powdered milk. The processing removes any contamination in
the milk and makes the product safe for human use.

There has been a trend in the United States toward the production of more Grade A milk and less manufacturing grade milk (Grade
B milk). In the late 1940s, 46 percent of the milk produced in the
United States was eligible for fluid use. "Milk eligible for fluid use"
may be interpreted as meaning Grade A milk. Sources of statistical
data do not identify Grade A and Grade B milk but refer instead to
"milk eligible for fluid use" and "manufacturing milk." However, "milk
eligible for fluid use" may be considered to be Grade A milk because
it is assumed that it is either produced under Grade A inspection or
would qualify for Grade A if it were subject to Grade A inspection.
"Manufacturing milk" is considered to be Grade B milk. Between 85
and 90 percent of the milk currently produced in the United States is
eligible for fluid use; that is, it has been produced under conditions
that meet Grade A requirements in most states. Not all Grade A milk
is used for fluid consumption. There is generally an oversupply of milk
eligible for fluid use. This milk is blended with Grade B milk and used
to produce manufactured dairy products. The shift from Grade A to
Grade B milk production is influenced by the price difference between

TABLE 39-1 Comparison of Numbers and Production—DHIA Herds with Average of All Milk Cows in the United States, 1967–1997.

Production of milk and milkfat per cow

Year	Average Number Milk Cows per Year, U.S. in Thousands	Number of Cows on DHIA Test in Thousands	Percent of Total on DHIA Test	Average Number of Cows in DHIA Herds	All Cows in U.S. Milk lb	Milk kg	Milkfat lb	Milkfat kg	% Fat in Milk	Cows in DHIA Herds Milk lb	Milk kg	Milkfat lb	Milkfat kg	% Fat in Milk	DHIA Herds Over All Cows Milk %	Milkfat %
1967	13,415	2,099	15.6	55.7	8,851	4,015	327	148	3.69	12,307	5,582	468	212	3.80	39.0	43.1
1968	12,832	2,132	16.6	57.8	9,135	4,144	335	152	3.67	12,397	5,623	471	214	3.80	35.7	40.6
1969	12,307	2,139	17.4	60.1	9,434	4,279	346	157	3.67	12,553	5,694	476	216	3.79	33.1	37.6
1970	12,000	2,122	17.7	61.9	9,751	4,423	357	162	3.66	12,750	5,783	483	219	3.79	30.8	35.3
1971	11,839	2,245	19.0	67.6	10,015	4,543	367	166	3.66	13,226	5,999	496	225	3.75	32.1	35.1
1972	11,700	2,360	20.2	70.3	10,259	4,653	377	171	3.68	13,287	6,027	499	226	3.76	29.5	32.4
1973	11,413	2,417	21.2	72.9	10,119	4,590	370	168	3.66	13,163	5,971	493	224	3.75	30.1	33.2
1974	11,230	2,433	21.7	75.5	10,293	4,669	377	171	3.67	13,421	6,088	505	229	3.76	30.4	34.0
1975	11,139	2,438	21.9	75.7	10,360	4,699	381	173	3.68	13,632	6,183	511	232	3.75	31.6	34.1
1976	11,032	2,581	23.4	76.5	10,894	4,941	399	181	3.66	14,435	6,548	539	244	3.73	32.5	35.1
1977	10,945	2,704	24.7	77.3	11,206	5,083	410	186	3.65	14,631	6,637	542	246	3.70	30.6	32.2
1978	10,803	2,792	25.8	77.3	11,243	5,100	412	187	3.67	14,644	6,642	542	246	3.70	30.2	31.6
1979	10,734	2,967	27.6	78.9	11,492	5,213	420	191	3.66	14,786	6,707	547	248	3.70	28.7	30.2
1980	10,799	3,197	29.6	80.8	11,891	5,394	435	197	3.65	14,960	6,786	553	251	3.70	25.8	27.1
1981	10,898	3,383	31.0	82.6	12,183	5,526	444	201	3.64	15,134	6,865	558	253	3.68	24.2	25.7
1982	11,011	3,432	31.2	83.6	12,306	5,582	450	204	3.65	15,274	6,928	564	256	3.69	24.1	25.3
1983	11,099	3,447	31.1	84.6	12,622	5,725	461	209	3.66	15,520	7,040	572	259	3.68	23.0	24.1
1984	10,793	3,262	30.2	83.4	12,541	5,688	460	209	3.66	15,587	7,070	577	262	3.70	24.3	25.4
1985	10,981	3,323	30.3	88.2	13,024	5,908	478	217	3.67	16,279	7,384	600	272	3.69	25.0	25.5
1986	10,773	3,103	28.8	89.2	13,285	6,026	487	221	3.67	16,654	7,554	612	278	3.67	25.4	25.7
1987	10,327	3,161	30.6	91.8	13,819	6,268	505	229	3.65	17,008	7,715	625	283	3.68	23.1	23.8
1988	10,224	3,168	31.0	93.1	14,185	6,434	521	236	3.67	17,379	7,883	640	290	3.68	22.5	22.8
1989	10,046	3,213	32.0	95.8	14,323	6,497	528	239	3.68	17,612	7,989	652	296	3.70	23.0	23.5
1990	9,993	3,253	32.6	97.8	14,782	6,705	539	244	3.65	18,031	8,179	662	300	3.67	22.0	22.8
1991	9,826	3,146	32.0	99.9	15,031	6,818	550	249	3.66	18,364	8,330	676	307	3.68	22.2	22.9
1992	9,688	3,138	32.4	103.8	15,570	7,062	573	260	3.68	18,750	8,505	695	315	3.70	20.4	21.3
1993	9,581	3,626	37.8	102.2	15,722	7,131	575	261	3.66	18,719	8,491	690	313	3.68	19.1	20.0
1994	9,494	3,621	38.1	107.0	16,179	7,339	592	269	3.66	19,129	8,677	705	320	3.67	18.2	19.1
1995	9,466	3,527	37.3	111.5	16,405	7,441	601	273	3.66	19,271	8,741	710	322	3.67	17.5	18.1
1996	9,372	3,486	37.2	118.5	16,433	7,454	608	276	3.69	19,192	8,705	713	323	3.70	16.8	17.3
1997	9,252	3,402	36.8	124.3	16,871	7,653	618	280	3.66	19,815	8,988	731	332	3.67	17.5	18.3

Source: *Agricultural Statistics*, USDA, 1981, 1986, 1988, 1992, 1995, 1999.

the two grades of milk. Grade A milk is priced higher than Grade B. When there is little difference in price between the two, there is little reason for a farmer to make the added investment needed to produce Grade A milk. When the price difference is larger, more farmers are willing to invest more money to produce Grade A milk.

Another factor in the drop in the percentage of Grade B milk produced has been a change by many farmers from dairying to other types of farming. Many Grade B dairy producers sold their herds rather than investing more money to meet Grade A standards.

The amount of milk needed in the United States depends upon the population, purchasing power, and per capita consumption. The use of milk substitutes and other drinks also affects the demand for milk. The population of the United States has been rising. Purchasing power has also increased despite inflation. The per capita consumption of fluid milk and dairy products is shown in Figures 39-5A, 39-5B, 39-5C, and 39-5D. In general, the per capita consumption of fluid milk and dairy products has been decreasing for a number of years. Only cheese has shown a steady increase in per capita consumption in recent years. Many substitute products (such as margarine) have come on the market and compete with dairy products. Many times these substitutes are lower priced than dairy products and gain more acceptance from consumers. Substitutes for dairy products may be expected to continue to compete for the consumer's dollar in the marketplace.

FIGURE 39-5A Per capita consumption of fluid milk and cream, 1977–1997. *Source: Agricultural Statistics,* USDA, 1992, 1999.

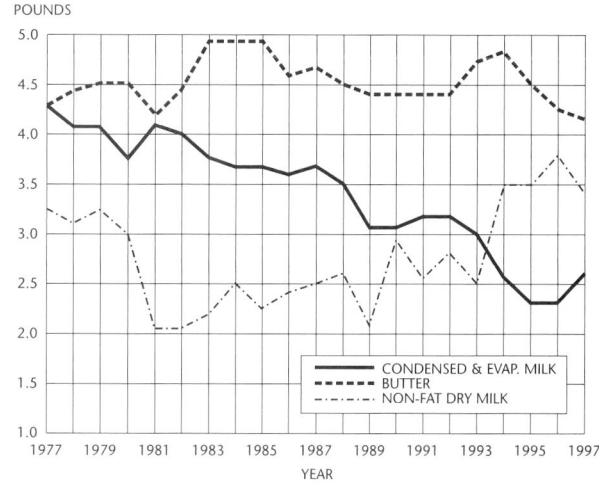

FIGURE 39-5B Per capita consumption of butter, condensed and evaporated milk, and non-fat dry milk, 1977–1997. *Source: Agricultural Statistics,* USDA, 1992, 1999.

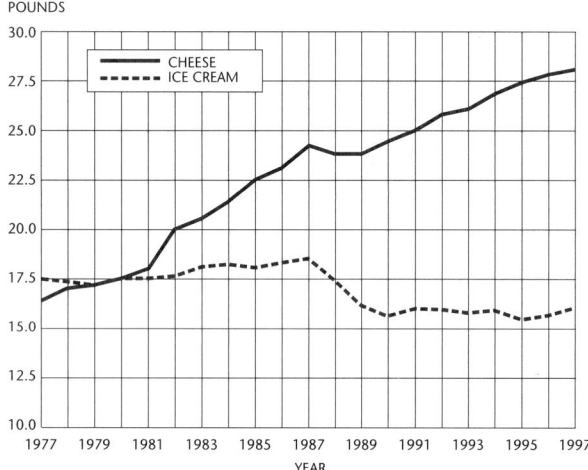

FIGURE 39-5C Per capita consumption of cheese and ice cream, 1977–1997. *Source: Agricultural Statistics,* USDA, 1992, 1999.

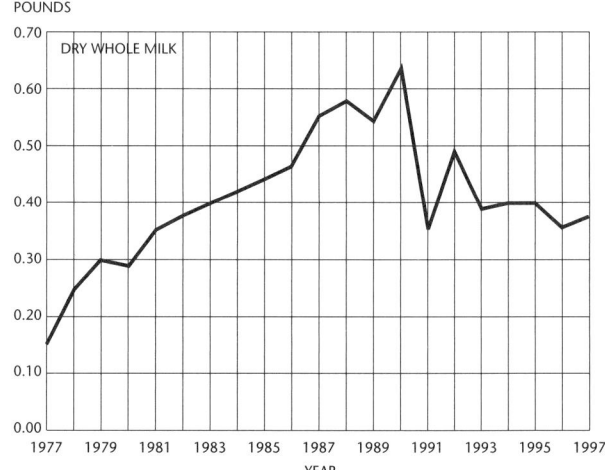

FIGURE 39-5D Per capita consumption of dry whole milk, 1977–1997. *Source: Agricultural Statistics,* USDA, 1992, 1999.

Dairy cows are found in every state in the United States (Figure 39-6). The leading states in dairy cow numbers are California, Wisconsin, New York, Pennsylvania, Minnesota, Texas, Michigan, Idaho, Ohio, and Washington. The rank order of states that lead in dairy cow numbers varies from year to year; however, the states listed above are usually among the top ten.

The production of milk in the United States has been shifting from the midwestern states to the western and southwestern states since the middle 1960s. The rate of this geographic shift has increased since the 1980s. California now produces more milk than Wisconsin. The Midwest produced about 50 percent of the nation's milk supply in 1965; it now produces only about 34 percent. The West and Southwest now produce about 34 percent of the nation's milk supply. The Northeast has shown a smaller decline in milk production, producing about 21 percent of the total supply in 1965 and about 18 percent in recent years.

The increase in milk production in the West and Southwest is attributed to five main factors:

- A long-term population shift to these areas
- More favorable climate
- Economies of scale
- Higher milk production per cow
- A more favorable attitude toward economic growth

Population growth creates a demand for milk and other dairy products. Facilities are cheaper and there are fewer herd health prob-

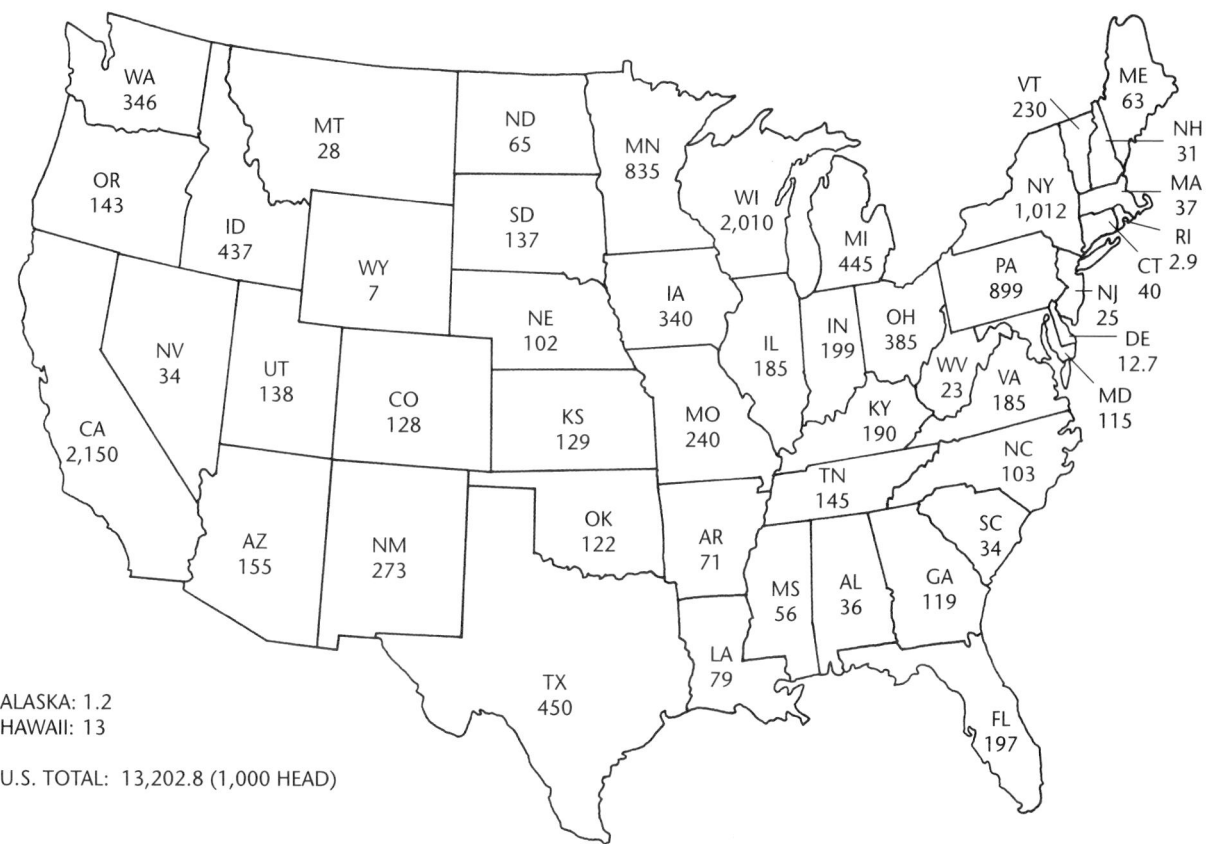

FIGURE 39-6   Milk cows and heifers 500 pounds (227 kg) and over kept for milk cow replacement (1,000 head), January 1, 1999. *Source: Cattle,* USDA, January 1999.

lems where the climate is warmer and dryer. High-quality forages can be produced with irrigated farming, assuring a good feed supply for dairy herds. Almost one-half the dairy herds in California contain over 100 head. Large herd size often results in lower per cow production costs. There has also been active support for economic development on the part of business, bankers, and dairy farmers in the western and southwestern states. This has contributed to the expansion of dairy enterprises in those states.

Dairy herds that exceed 500 head of cows in size account for only two percent of the total number of dairy operations in the United States; however, they account for more than one-fourth of the total inventory of dairy cows (Figure 39-7). Small herds of less than 29 cows make up less than 4 percent of the total inventory but account for almost 31 percent of the total number of operations. More than one-half of all the dairy cows in the United States are found in herds of 100 head or more, making up approximately 18 percent of the total

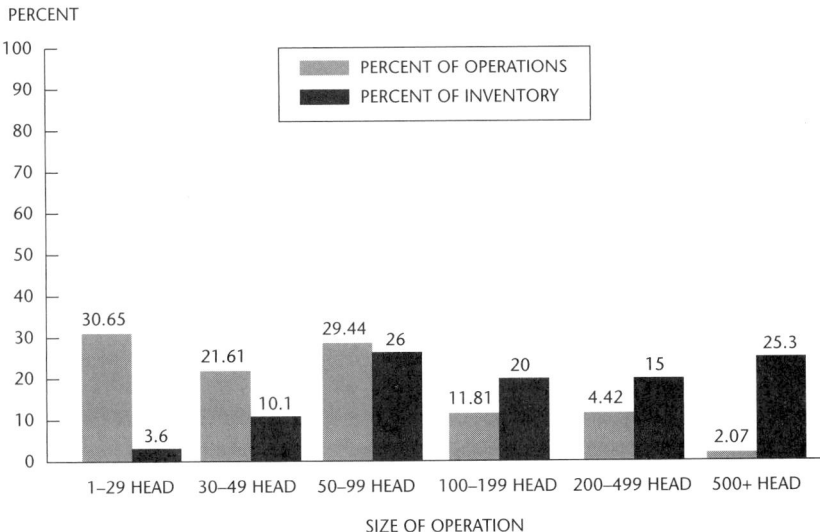

FIGURE 39-7   Milk cows: percent of operations and inventory, United States. *Source: Agricultural Statistics,* USDA, 1999.

number of dairy operations. These statistics reflect a long-term trend in the United States of larger dairy herds and fewer dairy farms.

It is becoming more difficult in many areas to meet environmental regulations when the dairy herd is large. The control of manure runoff is a major concern and has led to a slowdown in the expansion of dairy operations into larger herds. Manure disposal requires large parcels of land and this has limited expansion of some dairy herds in states with high population densities such as California. In recent years there has been some shift of large dairy herds back to the Midwest.

GOVERNMENT INFLUENCE IN DAIRYING

Government at national, state, and local levels influences the dairy enterprise. Examples are the Dairy Herd Improvement Program, milk marketing orders, support prices, import quotas, taxes, zoning ordinances, local health regulations, and animal waste disposal regulations.

THE PUREBRED DAIRY CATTLE ASSOCIATION

The Purebred Dairy Cattle Association (PDCA) was organized in 1940 by the five major dairy breed associations. The PDCA currently includes representatives from the Ayrshire, Brown Swiss, Guernsey, Holstein-Friesian, Jersey, and Milking Shorthorn breed associations. The World Dairy Expo, Madison, Wisconsin currently handles all information about the PDCA. Publications of the PDCA may be obtained at this address: PDCA, Suite 101, 2820 Walton Commons West, Madison, WI 53718.

The PDCA works to improve the dairy industry. Some of the areas in which the PDCA provides leadership include:

- Developing uniform rules for official testing in national production testing programs.
- Developing a unified score card for dairy cattle that is used as a dairy cattle type guide in the United States.
- Developing uniform rules for the artificial insemination of purebred dairy cattle.
- Influencing policies that affect animal health regulations, production testing, cattle exporting, show-ring classification, and show-ring and sales ethics and practices.

The PDCA publishes information in many of the listed areas. These are available to interested persons for a small fee.

BREED SELECTION

There are five major breeds of dairy cattle in the United States—Ayrshire, Brown Swiss, Guernsey, Holstein-Friesian, and Jersey. A sixth breed, The American Milking Shorthorn, is considered a dual-purpose breed. That is, it is used for both beef and milk production. The Milking Shorthorn is described in Unit 13, Breeds of Beef Cattle.

Cattle that are considered dairy breeds are selected for breeding on the basis of their ability to produce large quantities of milk for a long period of time. Other desired traits that are considered in breeding are color and dairy type.

Animals that meet the requirements of the breed association and are recorded in the herdbook of the association are called *registered.* *Grade* animals are those that are the result of mating a registered bull with a native cow or one of mixed breeding. Offspring of animals that are not registered but that have near ancestors that are registered are also called grade.

The name of a breed may be used for either registered or grade animals. The terms *registered* or *grade* should be used in front of the breed name, for example, registered Ayrshire or grade Ayrshire. In the case of grade animals, the breed name used indicates the visible traits of that breed can be easily seen in the animal.

An animal is eligible for registration when the sire and dam are both registered in the association herdbook. The animal must also meet any other qualifications described by the association, such as color, etc. Each breed association sets its own rules for qualifications for registering animals in the association.

Each dairy breed association also has a register of performance. This provides systems of production testing for owners of registered cattle. The rules for performance testing for each breed may be obtained from the respective breed associations.

Several dairy breed associations have established programs to allow owners of grade animals to register their cattle in an official herdbook. The associations benefit from an increased membership base. Cattle owners benefit from the increased value of their cattle. Other benefits include more records for progeny testing bulls, DHIA records may be used in sire summaries, more identified superior bulls become available for use in artificial insemination programs, and crossbred cattle can eventually be upgraded to a registered status. The procedures for these programs vary among associations. Dairy cattle owners interested in these programs should contact the appropriate breed association for more information and current regulations.

It is more important to select individual cows that are high producers than to put too much emphasis on which breed to select. There are some general breed differences that are discussed in more detail as each breed is described.

Some general guidelines for the selection of a breed include:

- Selecting a breed that is common in the area.
- Personal preference.
- Market requirements for the product.

Selecting a breed common in the area increases availability of breeding stock. There will also be a better market for surplus animals.

Personal preference is a matter of individual likes and dislikes. However, the producer should put more emphasis on selecting good individuals than on a particular breed. Breeds do differ in the percent of milkfat produced. Some markets may prefer a lower-testing product while others prefer a higher test. Since market demand may shift, it is not wise to put too much emphasis on a current market demand.

CHARACTERISTICS OF THE BREEDS OF DAIRY CATTLE

Ayrshire

History. The Ayrshire breed originated in the county of Ayr in the southwestern part of Scotland (Figure 39-8). The Ayrshire breed was developed during the last part of the eighteenth century. No known direct importations were used in the development of the breed. Animals were selected that had the desired traits and were mated to produce the color and type wanted by the breeders.

The first importation of Ayrshires into the United States was in 1822. Some importations have taken place since then, mainly from Scotland and Canada. The greatest numbers of Ayrshires are found in the northeastern states in the United States.

The breed association is the Ayrshire Breeders' Association, Brattleboro, VT 05302-1608. It was organized in 1875.

FIGURE 39-8   Ayrshire cow. *Courtesy of Ayrshire Breeders' Association, VT.*

Description. The Ayrshire may be any shade of cherry red. Other colors are mahogany, brown, or white. White may be mixed with red, mahogany, or brown. Each color should be clearly defined. The preferred color is a distinctive red and white. Objectionable colors are black or brindle.

The horns curve up and out. They are of medium length, small at the base, and tapered toward the tips.

Ayrshires have straight lines and well-balanced udders. The udders are attached high behind and extend forward. The teats are medium in size.

Ayrshires are vigorous and strong. They have excellent grazing ability. Mature cows weigh about 1,200 pounds (544 kg). Mature bulls weigh about 1,800 pounds (816 kg).

Ayrshires rank third among the dairy breeds in average milk produced per cow at 11,700 pounds (5,307 kg). They average about four percent milkfat, and rank fourth among the five dairy breeds in average milkfat produced per cow.

FIGURE 39-9   Brown Swiss cow. *Courtesy of Brown Swiss Cattle Breeders' Association of the USA.*

Brown Swiss

History. The Brown Swiss breed originated in Switzerland (Figure 39-9). They are probably one of the oldest of the dairy breeds. They were developed in the Alps. It is believed that no outside breeding was used in the development of the breed after records were kept.

Brown Swiss were first imported into the United States in 1869. Only a small number have ever been imported into the United States. Total importations are believed to be about 25 bulls and 130 cows. All of the approximately 820,000 registered Brown Swiss in the United States are descendants of those importations.

The breed association is the Brown Swiss Cattle Breeders' Association of the U.S.A., P.O. Box 1038, 800 Pleasant Street, Beloit, WI 53511. The association was organized in 1880.

Description. Brown Swiss are solid brown, ranging from light to dark. White and off-color spots are objectionable colors. The nose and tongue are black.

The horns incline forward and slightly upward. They are of medium length and taper toward black tips.

The Brown Swiss are large-framed cattle. Mature cows weigh about 1,500 pounds (680 kg) and bulls about 2,000 pounds (907 kg). The heifers mature more slowly than other dairy breeds. Brown Swiss have a quiet, docile temperament. They are considered to be good grazers.

Brown Swiss are the longest lived of the dairy breeds. They have a high heat tolerance. Bulls of this breed have been used recently in beef crossbreeding programs.

Brown Swiss rank second among the dairy breeds in average milk production per cow at 12,100 pounds (5,488 kg). They average about

FIGURE 39-10   Guernsey cow.
Courtesy of American Guernsey Cattle Club.

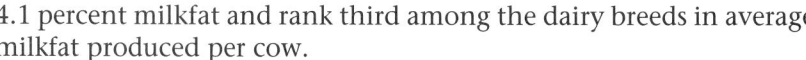

4.1 percent milkfat and rank third among the dairy breeds in average milkfat produced per cow.

Guernsey

History. The Guernsey breed originated on the Isle of Guernsey, which is located in the English Channel off the Coast of France (Figure 39-10). The development of the breed began about 1,000 years ago. Monks on the Isle of Guernsey crossed two breeds of French dairy cattle. These were the Fromond du Leon from Brittany and the Norman Brindles from Normandy. The Guernsey breed was developed through selection of desired traits and elimination of crossbreeding.

Guernseys were first imported into the United States in 1831. Major importations occurred after 1870. Total importations over the years were about 13,000 cattle.

The breed association is the American Guernsey Association, 7614 Slate Ridge Boulevard, Reynoldsburg, OH 43068-0666. The association was organized in 1877. Guernseys rank second in total number of dairy cattle registered in the United States.

Description. The Guernsey may be any shade of fawn with white markings. Black and brindle colors are objectionable. The skin is yellow. A clear or buff muzzle is preferred over smoky or black.

The horns curve outward and to the front. They are medium in length, small and yellow at the base and taper toward the tips.

The Guernsey is an early-maturing breed. They are adaptable and have a gentle behavior. Mature cows will weigh about 1,100 pounds (499 kg) and bulls about 1,800 pounds (816 kg).

Guernseys rank fourth among the dairy breeds in average milk production per cow at 10,600 pounds (4,808 kg). They average about 5 percent milkfat and rank second among the dairy breeds in average milkfat produced per cow. Guernseys produce milk that is golden in color.

Holstein-Friesian

History. The Holstein-Friesian breed originated in the Netherlands (Figure 39-11). Development occurred in the northern province of Friesland and in nearby northern Germany. It is not known when they became a distinct breed. It is believed that selection and breeding that resulted in the Holstein-Friesian breed started about 2,000 years ago. The breed is commonly called Holstein in the United States and Canada. Elsewhere in the world it is called Friesian.

The first importations of Holsteins into the United States were by early Dutch settlers between 1621 and 1664. The first significant importations into the United States were in 1857, 1859, and 1861. Most present-day Holsteins trace their ancestry to importations between 1877 and 1905. There have been no importations since 1905.

FIGURE 39-11   Holstein cow.
Courtesy of Holstein Association USA, Inc.

The first breed association was called the Association of Breeders of Thoroughbred Holstein cattle and was organized in 1871. This association published nine volumes of the Holstein Herdbook. In 1877 the Dutch Friesian Cattle Breeders' Association was organized. They published four volumes of the Dutch Friesian Herdbook. These two associations merged in 1885 to form the Holstein-Friesian Association of America. The herdbook is now called the Holstein-Friesian Herdbook. The association is located at 1 Holstein Place, Brattleboro, VT 05302-0808.

About 90 percent of all dairy cattle in the United States are of Holstein breeding. There are about 1,500,000 registered Holsteins in the United States.

Description. Holsteins are black and white. A recessive gene occasionally causes a red and white color to appear. The switch (tail) has white on it. Solid black or solid white animals are not registered. Off colors include black on the switch, solid black belly, one or more legs encircled with black which touches the hoof at any point, and black and white intermixed to give gray spots.

The horns incline forward and curve inward. They are of medium length and taper toward the tips.

Holsteins are the largest of the dairy breeds. Mature cows weigh about 1,500 pounds (680 kg) and bulls weigh about 2,200 pounds (998 kg). Cows have large udders. Holsteins have excellent grazing ability and a large feed capacity. The cows are generally quiet but the bulls can be mean and dangerous. Holsteins are adaptable to a wide range of conditions.

Holsteins rank first among the dairy breeds in average milk production per cow at 14,500 pounds (6,577 kg). They average about 3.5 percent milkfat and rank fifth among the dairy breeds in average milkfat produced per cow.

Jersey

History. The Jersey breed originated on the Isle of Jersey, which is located in the English Channel off the coast of France (Figure 39-12). It is not known what cattle were the source of the breeding stock that was developed into the Jersey breed. One theory is that the breed developed from early cattle from Normandy and Brittany in France. A law was passed on the Isle of Jersey in 1763 that prohibited the importation of cattle from France. Selection for desired traits developed the breed without further outside breeding after that time.

The earliest importation of Jersey cattle into the United States occurred about 1815. The major importations occurred between 1870 and 1890. The first Jersey cattle registered by the American Jersey Cattle Club were imported in 1850.

FIGURE 39-12   Jersey cow. *Courtesy of American Jersey Cattle Club.*

The breed association is the American Jersey Cattle Club, 6486 East Main Street, Reynoldsburg, OH 43068-2362. The association was organized in 1868.

Description. Jerseys are cream to light fawn to almost black in color. Some animals have white markings. The muzzle is black. The switch and tongue may be black or white.

The Jersey is the smallest of the dairy breeds. Mature cows weigh about 1,000 pounds (453 kg) and bulls weigh about 1,600 pounds (725 kg).

The horns curve inward and are inclined forward. They are of medium length and taper toward the tips.

Jersey cattle have excellent udders that are well attached. They are adaptable and efficient users of feed. Jerseys have excellent grazing ability even on poor pastures. The cows may be somewhat nervous and the bulls can be mean. Jerseys are early maturing and have excellent dairy type.

Jerseys rank fifth among the dairy breeds in average milk production per cow at 10,000 pounds (4,536 kg). They average about 5.4 percent milkfat and rank first among the dairy breeds in average milkfat produced per cow.

SUMMARY

Dairying is an enterprise that requires patience and a willingness to work long hours. Dairy cows must be milked regularly seven days per week. The dairy enterprise requires a large investment of capital. Dairying provides a steady income year-around. A person needs training and experience to be a successful dairy farm operator.

The number of dairy farms and dairy cattle in the United States has been declining in recent years. The average size of the dairy herds has increased. Average production per cow has been increasing. There has been a trend toward the production of more Grade A milk and less manufacturing milk. The per capita consumption of dairy products and fluid milk has been declining.

The five major breeds of dairy cattle in the United States are Ayrshire, Brown Swiss, Guernsey, Holstein-Friesian, and Jersey. Dairy cattle breeds have been developed by selecting for quantity and persistence of milk production. Dairy breed associations register purebred dairy animals. They also keep registers of performance.

More emphasis should be put on selecting good individuals for the herd rather than on which breed to select. There are some differences among dairy breeds in regard to total production per cow and percent of milkfat produced. Holsteins rank first in average production of milk per cow and fifth in average percent of milkfat. Jerseys rank fifth in average production per cow and first in average percent of milkfat.

Student Learning Activities

1. Present an oral report on the characteristics of the dairy enterprise.
2. On field trips in the community, observe different breeds of dairy cattle and their characteristics.
3. Conduct a survey of local dairy producers to determine what dairy breeds are most common in the community.
4. Prepare a bulletin board display of pictures of the dairy breeds.
5. Present an oral report on the history and characteristics of a dairy breed to the class.
6. Write to dairy breed associations for literature about their breed.

Review

1. Describe briefly some of the activities that a dairy farm operator must perform.
2. List the advantages and disadvantages of the dairy enterprise.
3. List and briefly describe some of the important trends in dairy farming.
4. What is Grade A milk? Manufacturing milk?
5. Name the states that lead in dairy cow numbers.
6. Name some of the influences of government in the dairy enterprise.
7. Name the five major breeds of dairy cattle in the United States.
8. What is the difference between registered and grade dairy animals?
9. Prepare a table that briefly describes the characteristics of each of the five major breeds of dairy cattle.
10. Which dairy breeds are most common in your area?

Selecting and Judging Dairy Cattle

After studying this unit, the student should be able to

- select desirable breeding and production animals.
- identify the parts of the dairy animal.
- judge dairy animals.

SELECTING DESIRABLE BREEDING AND PRODUCTION ANIMALS

The selection of desirable dairy animals for breeding and production is based upon the animal's physical appearance, health, milk production records, and pedigree.

The physical appearance (conformation and dairy character) of the animal is referred to as *type*. Type and milk production are closely related. Cows with good dairy type usually produce more milk for a longer period of time. It takes a great deal of practice and study to learn the ideal dairy animal type.

Health records include a history of vaccinations and the general health of the herd from which the animal comes. The apparent health of the animal being considered is also important.

Milk production records that show past performance may or may not be available for the individual animal. If the cow has been in production and in a herd where DHIA records were kept, the record should be available. Such records give some indication of the possible production of the offspring of the cow. Production records for young cows may be used as the basis for predicting future performance. Younger animals that have not been in production will not have production records. Bulls are evaluated on the basis of the production records of their daughters. Young bulls may not have such records available.

Production records should show the pounds (or kilograms) of milk produced, the pounds (or kilograms) of milkfat produced, and the percent of milkfat. To properly evaluate production records, more information is needed. This includes the number of times milked per day, age, feed and care received, and the length of time the cow carried a calf during each lactation.

Production records can only give an estimate of the ability of the cow to transmit high production ability to her daughters. The best indication of a cow's transmitting ability is in the record of her offspring. The best foundation stock is a cow with high production records who also has daughters with high production records.

The pedigree is the record of the animal's ancestors. Pedigrees that are most valuable as a basis for selection give the name, registration number, type rating, production record, and show-ring winnings of each ancestor for three or four generations. Such a record gives a more complete picture of the possible inheritance of type and production than information on only the sire and dam. Pedigrees must be studied carefully. Sometimes they contain misleading information.

JUDGING DAIRY ANIMALS

The Dairy Cow Unified Score Card

Judging dairy animals is a process of comparing the individuals being judged with an ideal dairy type. The ideal dairy type is described in the Dairy Cow Unified Score Card (Figure 40-1).

This score card was developed by the Purebred Dairy Cattle Association (PDCA) to describe the general traits of a good dairy cow. These general traits are the same for all breeds of dairy cattle. Specific breed characteristics are included in the score card. The score card also shows the parts of the dairy cow. A good judge must learn the parts of the dairy animals in order to be able to use the right terms when judging the animals. This score card is used to learn to judge dairy animals and for type classification.

Breed associations use the score card for type classification. The individual's physical appearance is compared to the ideal and a *type classification score*, or number score, is assigned to the animal. Type classification is done by a representative of the breed association at the request of the animal's owner. The official scores for Holsteins are as follows.

Excellent	90–100 points
Very Good	85–89 points
Good Plus	80–84 points
Good	75–79 points
Fair	65–74 points
Poor	50–64 points

DAIRY COW UNIFIED SCORE CARD

Copyrighted by The Purebred Dairy Cattle Association, 1943. Revised, and Copyrighted 1957, 1971, 1982, and 1994.

Breed characteristics should be considered in the application of this score card

MAJOR TRAIT DESCRIPTIONS

	Perfect Score

There are five major classification traits on which a classifier bases a cow's score. Each trait is broken down into body parts to be looked at and ranked.

1) Frame - 15%

15

The skeletal parts of the cow, with the exception of feet and legs, are evaluated. Listed in priority order, the descriptions of the traits to be considered are as follows:

Rump - long and wide throughout with pin bones slightly lower than hip bones. Thurls need to be wide apart and centrally placed between hip bones and pin bones. The tailhead is set slightly above and neatly between pin bones, and the tail is free from coarseness. The vulva is nearly vertical. **Stature** - height, including length in the leg bones. A long bone pattern throughout the body structure is desirable. Height at the withers and hips should be relatively proportionate. **Front End** - adequate constitution with front legs straight, wide apart and squarely placed. Shoulder blades and elbows need to be firmly set against the chest wall. The crops should have adequate fullness. **Back** - straight and strong; the loin - broad, strong, and nearly level. **Breed Characteristics** - overall style and balance. Head should be feminine, clean-cut, slightly dished with broad muzzle, large open nostrils and a strong jaw is desirable.

Rump, Stature, and Front End receive primary consideration when evaluating Frame.

2) Dairy Character - 20%

20

The physical evidence of milking ability is evaluated. Major consideration is given to general openness and angularity while maintaining strength, flatness of bone and freedom from coarseness. Consideration is given to stage of lactation. Listed in priority order, the descriptions of the traits to be considered are as follows:

Ribs - wide apart. Rib bones are wide, flat, deep, and slanted toward the rear. **Thighs** - lean, incurving to flat, and wide apart from the rear. **Withers** - sharp with the chine prominent. **Neck** - long, lean, and blending smoothly into shoulders. A clean-cut throat, dewlap, and brisket are desirable. **Skin** - thin, loose, and pliable.

3) Body Capacity - 10%

10

The volumetric measurement of the capacity of the cow (length x depth x width) is evaluated with age taken into consideration. Listed in priority order the descriptions of the traits to be considered are as follows:

Barrel - long, deep, and wide. Depth and spring of rib increase toward the rear with a deep flank. **Chest** - deep and wide floor with well-sprung fore ribs blending into the shoulders.

The Barrel receives primary consideration when evaluating Body Capacity.

4) Feet and Legs - 15%

15

Feet and rear legs are evaluated. Evidence of mobility is given major consideration. Listed in priority order, the descriptions of the traits to be considered are as follows:

Feet - steep angle and deep heel with short, well-rounded closed toes. **Rear Legs: Rear View** - straight, wide apart with feet squarely placed. **Side View** - a moderate set (angle) to the hock. **Hocks** - cleanly molded, free from coarseness and puffiness with adequate flexibility. **Pasterns** - short and strong with some flexibility.

Slightly more emphasis placed on Feet than on Rear Legs when evaluating this breakdown.

5) Udder - 40%

40

The udder traits are the most heavily weighted. Major consideration is given to the traits that contribute to high milk yield and a long productive life. Listed in priority order, the descriptions of the traits to be considered are as follows:

Udder Depth - moderate depth relative to the hock with adequate capacity and clearance. Consideration is given to lactation number and age.
Teat Placement - squarely placed under each quarter, plumb and properly spaced from side and rear views.
Rear Udder - wide and high, firmly attached with uniform width from top to bottom and slightly rounded to udder floor.
Udder Cleft - evidence of a strong suspensory ligament indicated by adequately defined halving.
Fore Udder - firmly attached with moderate length and ample capacity.
Teats - cylindrical shape and uniform size with medium length and diameter.
Udder Balance and Texture - should exhibit an udder floor that is level as viewed from the side. Quarters should be evenly balanced; soft, pliable and well collapsed after milking.

TOTAL **100**

FIGURE 40-1 The Dairy Cow Unified Score Card. *Courtesy of Purebred Dairy Cattle Association.*

AYRSHIRE   BROWN-SWISS   GUERNSEY

HOLSTEIN-FRIESIAN   JERSEY   MILKING SHORTHORN

BREED CHARACTERISTICS

Except for differences in color, size and head character, all breeds are judged on the same standards as outlined in the Unified Score Card. If any animal is registered by one of the dairy breed associations, no discrimination against color or color pattern is to be made.

AYRSHIRE

Strong and robust, showing constitution and vigor, symmetry, style and balance throughout, and characterized by strongly attached, evenly balanced, well-shaped udder.

HEAD-clean cut, proportionate to body; broad muzzle with large, open nostrils; strong jaw; large, bright eyes; forehead, broad and moderately dished; bridge of nose straight; ears medium size and alertly carried.

COLOR-light to deep cherry red, mahogany, brown, or a combination of any of these colors with white, or white alone, distinctive red and white markings preferred.

SIZE-a mature cow in milk should weigh at least 1200 lbs.

HOLSTEIN

Rugged, feminine qualities in an alert cow possessing Holstein size and vigor.

HEAD-clean cut, proportionate to body; broad muzzle with large, open nostrils; strong jaw; large, bright eyes; forehead, broad and moderately dished; bridge of nose straight; ears medium size and alertly carried.

COLOR-black and white or red and white markings clearly defined.

SIZE-a mature cow in milk should weigh a minimum of 1500 lbs.

MILKING SHORTHORN

Strong and vigorous, but not coarse.

HEAD-clean cut, proportionate to body; broad muzzle with large, open nostrils; strong jaw; large, bright eyes; forehead, broad and moderately dished; bridge of nose straight; ears, medium size and alertly carried.

COLOR-red or white or any combination. (No black markings allowed)

SIZE-a mature cow should weigh 1400 lbs.

BROWN SWISS

Strong and vigorous, but not coarse. Size and ruggedness with quality desired. Extreme refinement undesirable.

HEAD-clean cut, proportionate to body; broad muzzle with large, open nostrils; strong jaw; large, bright eyes; forehead, broad and slightly dished; bridge of nose straight; ears medium size and alertly carried.

COLOR-solid brown varying from very light to dark. Muzzle is black encircled by a mealy colored ring, and the tongue, switch and hooves are black

SIZE-a mature cow in milk should weigh 1500 lbs.

GUERNSEY

Size and strength, with quality and character desired.

HEAD-clean cut, proportionate to body; broad muzzle with large, open nostrils; Strong jaw; large, bright eyes; forehead, broad and slightly dished; bridge of nose straight; ears medium size and alertly carried.

COLOR-a shade of fawn with white markings throughout clearly defined. When other points are equal, clear (buff) muzzle will be favored over a smoky or black muzzle.

SIZE-a mature cow in milk should weigh at least 1150 lbs.

JERSEY

Sharpness with strength indicating productive efficiency.

HEAD-proportionate to stature showing refinement and well chiseled bone structure. Face slightly dished with dark eyes that are well set.

COLOR-some shade of fawn with or without white markings. Muzzle is black encircled by a light colored ring, and the tongue and switch may be either white or black.

SIZE-a mature cow in milk should weigh about 900 lbs.

FACTORS TO BE EVALUATED

The degree of discrimination assigned to each defect is related to its function and heredity. The evaluation of the defect shall be determined by the breeder, the classifier or the judge, based on the guide for discrimination and disqualifications given below.

HORNS
No discrimination for horns.
EYES
1. Blindness in one eye: *Slight discrimination.*
2. Cross or bulging eyes: *Slight discrimination.*
3. Evidence of blindness: *Slight to serious discrimination.*
4. Total blindness: *Disqualification.*
WRY FACE
Slight to serious discrimination.
CROPPED EARS
Slight discrimination.
PARROT JAW
Slight to serious discrimination.
SHOULDERS
Winged: *Slight to serious discrimination.*
TAIL SETTING
Wry tail or other abnormal tail settings: *Slight to serious discrimination.*
CAPPED HIP
No discrimination unless effects mobility.

LEGS AND FEET
1. Lameness - apparently permanent and interfering with normal function: *Disqualification.* Lameness - apparently temporary and not affecting normal function: *Slight discrimination.*
2. Evidence of crampy hind legs: *Serious discrimination.*
3. Evidence of fluid in hocks: *Slight discrimination.*
4. Weak pastern : *Slight to serious discrimination.*
5. Toe out: *Slight discrimination.*
UDDER
1. Lack of defined having: *Slight to serious discrimination.*
2. Udder definately broken away in attachment: *Serious discrimination.*
3. A weak udder attachment: *Slight to serious discrimination.*
4. Blind quarter: *Disqualification.*
5. One or more light quarters, hard spots in udder, obstruction in teat (spider): *Slight to serious discrimination.*

6. Side leak: *Slight discrimination.*
7. Abnormal milk (bloody, clotted, watery): *Possible discrimination.*
LACK OF SIZE
Slight to serious discrimination.
EVIDENCE OF SHARP PRACTICE
(Refer to PDCA Code of Ethics)
1. Animals showing signs of having been tampered with to conceal faults in conformation and to misrepresent the animal's soundness: *Disqualification.*
2. Uncalved heifers showing evidence of having been milked: *Slight to serious discrimination.*
TEMPORARY OR MINOR INJURIES
Blemishes or injuries of a temporary character not affecting animal's usefullness: *Slight to serious discrimination.*
OVERCONDITIONED
Slight to serious discrimination.
FREEMARTIN HEIFERS
Disqualification.

FIGURE 40-1 Continued

The first part of the score card contains five main divisions with the points for a perfect score listed for each. The dairy judge must first become familiar with these divisions and the points for each.

The second part of the score card lists the breed characteristics. The dairy judge must learn these to be able to determine the amount of breed character being shown by the animal being judged.

The third part of the score card is the evaluation of defects. This part helps the dairy judge determine the importance of any defects that the animal being judged may have.

Linear classification is a modification of type classification that utilizes a computer program to score dairy cattle on a number of individual traits. The computer program uses a wider range of scoring for individual traits than is found on the Dairy Cow Unified Score Card and develops an overall type classification score for each animal. Linear classification has not been adopted by all breed associations.

GENERAL DESCRIPTION OF THE SCORE CARD

Dairy character as described on the score card has the highest positive correlation to milk production. Other characteristics on the score card are also important when visually evaluating dairy animals for their ability to efficiently produce milk over a long period of time.

Frame (15 percent)

The frame is defined as the skeletal structure of the cow, except the feet and legs. In priority order the areas considered when evaluating the cow's frame are: rump, stature, front end, back, and breed characteristics. The rump is the highest priority because it is closely related to reproductive efficiency and the support and placement of the udder. The width of the pelvic region affects the ease of calving. The animal should be properly proportioned throughout with a strong and straight topline.

Dairy Character (20 percent)

Dairy character is an indication of milking ability. The priority order for evaluating dairy character characteristics is: ribs, thighs, withers, neck, and skin. Excellent dairy character indicates an animal that is converting feed to milk with maximum efficiency. Animals with poor dairy character are usually coarse and too fat (over conditioned). The ribs should be wide and the thighs should be lean and flat. When viewed from the rear, the thighs should be wide enough to provide plenty of space for the udder attachment.

Body Capacity (10 percent)

The priority order for evaluating body capacity characteristics is: barrel and chest. Good body capacity is needed so the animal can consume the amounts of feed needed for high milk production. Animals with good body capacity can use more roughage in the ration. Adequate body capacity allows proper development of the heart and lungs. Animals with poor body capacity will not be able to maintain high milk production over a long period of time. The age of the animal is taken into consideration when evaluating the length, depth, and width of the body.

Feet and Legs (15 percent)

Feet have a little higher priority than rear legs when evaluating feet and legs. The ability of an animal to reproduce efficiently over a long period of time is closely related to the structure and strength of its feet and legs. Proper placement of the legs improves the ability of the animal to move about with ease. Width in the rear legs provides room for a large udder.

Correct set to the hocks affects the ability of the animal to stand and walk on concrete surfaces over a long period of time. Too much set at the hocks (sickle hock) will cause the legs to weaken as the animal becomes older (Figure 40-2). Legs that are too straight place too much stress on the hocks.

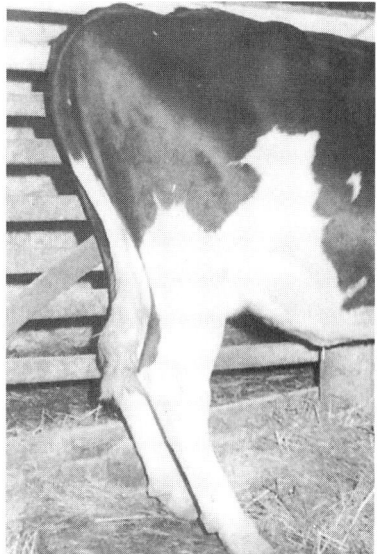

FIGURE 40-2 This leg has too much set (sickle hocked).

Udder (40 percent)

The priority order for evaluating the characteristics of the udder is: udder depth, teat placement, rear udder, udder cleft, fore udder, teats, and balance and texture.

The main purpose of the dairy cow is to produce milk. The udder is the most important part of her body (Figure 40-3). Udders that have poor conformation, are weakly attached, and are poorly balanced do not stand up well under the stress of high production (Figure 40-4).

The size of the udder is generally related to milk-producing capacity. Cows with small udders are usually not high-producing cows.

The udder should be soft, pliable, and elastic. If it is still quite firm and large after milking, it is probably full of fibrous or scar tissue. An udder that is still firm after milking is referred to as *meaty*.

The size and placement of the teats is important for ease of machine milking. Teats that are uneven in size or poorly placed make it more difficult to use the milking machine. Teats should be 1.5 to 2.5 inches (3.8–6.4 cm) long. When the udder is full they should hang straight down.

The mammary veins are blood vessels that circulate blood to the udder. The size of these mammary veins indicates the amount of blood circulation to the udder. Large mammary veins are desirable.

The capacity of the mammary system is reduced by a small udder, deep cuts between the quarters or halves, meaty texture, and small mammary veins. A cow with a poor mammary system is not a good foundation cow for the dairy farmer who wants a high-producing herd.

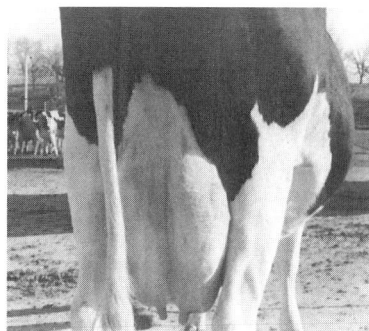

FIGURE 40-3 A well-attached udder, with good width and depth.

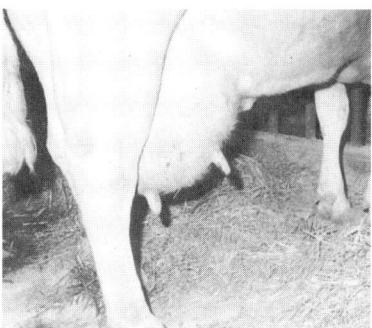

FIGURE 40-4 An udder with poor front attachment and poor teat placement.

FIGURE 40-5 From the side, look at the topline, rump, barrel, heartgirth, shoulders, udder, teats, flank, legs, and neck. *Courtesy of MGM Farms, Robert Miller and Bill McFadden.*

FIGURE 40-6 From the rear, look at the barrel, loin, hips, pinbones, tailhead, udder, and legs.

Steps in Judging

First, view the animal or group of animals from a distance of 20–30 feet (6–9 m). Look at the side, front, and rear of the animals. The animals should be standing on the level or with the front feet slightly higher than the rear.

From the side view, look at the topline, rump, barrel, heartgirth, shoulders, udder, teats, flank, legs, and neck (Figure 40-5). From the rear view, look at the barrel, loin, hips, pinbones, tailhead, udder, and legs (Figure 40-6). From the front view, look at the head, chest, and front legs (Figure 40-7).

Make mental or written notes on each animal as they are viewed on this first inspection. First impressions are usually more accurate and should be carefully noted. Compare each animal with the ideal described in the Dairy Cow Unified Score Card. If there is an easy top or bottom animal or top and bottom pair in the class, make note of it.

The class of animals being judged is usually walked slowly around the judge in a clockwise direction. Look at style, carriage, straightness of legs and topline, udder attachment, and blending of parts while they are walking.

After observing the animals from a distance, move in for a closer inspection. Observe the shape of the withers, quality of hair, mellowness of hide, texture of udder, and development of the mammary veins in this close inspection.

A class of four animals is compared by pairs. That is, the top pair, middle pair, and bottom pair. A procedure for note taking and giving reasons is discussed in Unit 14. This procedure may be followed when judging dairy cattle.

When a pair is close, make the decision on the mammary system. If the mammary system is equal in the pair, use dairy character to determine the placing. If, in a close pair, one cow is dry and the other milking, place the milking cow over the dry cow.

Look for any defects that individual animals may have. Defects are described in the Dairy Cow Unified Score Card. The influence of the defect on the placing ranges from disqualification to slight discrimination depending on the degree of seriousness. For example, permanent lameness will put a cow at the bottom of the class regardless of other traits she may have.

Question class (Type Analysis Questionnaire—TAQ). In some judging contests a different method is being used for judging. This involves asking a series of about 5 to 10 questions about the class. During or after judging, the contestant places the number of the cow that most correctly answers the question beside each question. The official judge determines the correct answers. Scoring is done by giving 5 points for the correct answer, 3 points for the second most correct answer, 1 point for the third most correct answer, and zero points for the wrong answer. If ten questions are used there are a total of 50 points for the class.

Judging dairy heifers. Dairy heifers are judged on much the same points as dairy cows. Heifers do not have as much development as mature cows. Therefore, the judge must visualize how the heifer will develop as she matures. This is especially true for the mammary system. When examining the udder, place emphasis on uniformity of quarters, placement and size of teats, length and width, and rear and fore attachments. A heifer does not have the depth of barrel and dairy temperament of a mature cow.

Springing (pregnant) heifers often carry some surplus fat. This may make them appear to be coarse over the withers. This accumulated fat disappears when the heifer comes into production.

FIGURE 40-7   From the front, look at the head, chest, and front legs.

DAIRY JUDGING TERMS

Comparative terms are used when giving reasons for the placement of a dairy class. The following list contains terms that may be used to compare one animal with another.

General Appearance

More feminine head

More breed character in head and neck

Neck blends more smoothly with the shoulders

Blends smoother at the point of shoulders

Shoulders that blend more smoothly with the body

Tighter through the shoulders

Fuller through the crops

Stronger through the front

Stronger over the back and loins

Stronger and straighter in the loin and chine

More strength in the loin

Weaker in the loin

Longer and more open in the rib

Stronger through the heart

More power and width in the chest

Wider and more nearly level rump

Smoother over the rump

Higher at the thurls

Wider in the thurls

More nearly level from hips to pins

Longer from hips to pins

Wider and more open at the pins

More smoothly blended in at the tail setting

Standing more squarely on the legs

Standing straighter on rear legs

Straighter on rear legs

Stands straighter on legs as she walks

Crooked rear legs

Sickle hocked

Cow hocked

Straighter on the pasterns

Shorter and stronger pasterns

Weaker on the pasterns

Deeper in the heel

Shallow in the heel

Toes out in front

Showing evidence of lameness

A larger cow with more size and scale

More strength and substance

More breed type and character

Lacks the breed character

More style and balance

Shows more style when on the move

Smoother throughout

Dairy Character

Has a more milky appearance

More dairy-like

Cleaner cut head and neck

Longer in the neck

Cleaner at the throat

Longer and thinner in the neck and cleaner in the throat

Sharper over the withers

More open ribbed

More spread to the ribs

Flatter and more open of rib

Sharper in hips and pins

Neater and more refined tailhead

Cleaner over the pin bones

Thinner and cleaner in the thighs

Flatter in the thighs

Cleaner and more refined

More angular throughout

Body Capacity

Cut in behind the shoulders

Smoother behind the shoulders

Fuller in the crops

Narrow chested

Wider on the chest floor

Deeper through the heart

Pinched at the heart

Fuller in the heartgirth

Fuller in the foreribs

Deeper in the fore and rear ribs

A deeper ribbed cow

Deeper and more open ribbed

Has more spring of forerib

Has more spring of rear rib

More open in the rib

Showing greater arch to the ribs

Showing more spring to her rear ribs

Greater spring of rib and depth of body

Narrow body, lacking spring of rib

More capacity and spring in the barrel

Longer and deeper in the barrel

Greater body capacity

Greater feed capacity

Has greater depth of body

Longer bodied

Longer in the body

Deeper bodied

Shows more stretch

Deeper in the flank

Mammary System

More evenly balanced udder

Has a more balanced udder

Udder shows more balance from front to rear

Unbalanced udder

A more uniformly balanced udder that is more level on the floor

A more symmetrical udder with more uniform sized quarters

Udder shows more desirable halving and quartering

Not as badly quartered (or halved)

Udder shows more defined cleavage

Udder is badly quartered (or halved) and hangs forward

Cut up between the quarters

A larger, more capacious udder

Has more quality of udder

More capacity and bloom

More symmetry

More nearly level on the floor

Showing more quality and texture

A softer, more pliable udder

More milky udder with numerous veining

Carries out fuller in the rear quarters

Has more balance of the fore quarters

Has more balance of rear quarters

Carries udder closer to the body

Fore udder attachment blends more smoothly in with the floor of the barrel

Carries her udder closer to the hocks

Udder is more strongly attached both fore and rear

Udder is more firmly attached both fore and rear

Weaker attachment

Has stronger fore udder attachment

Is more firmly attached in the front

Smoother (or stronger) in the fore udder attachment

Carrying farther forward

Loose in front attachment

Weaker fore udder attachment

Higher and wider rear udder attachment

Rear attachment is broken down, allowing the quarter to point out

She is very low and narrow in the rear attachment

Weaker rear udder attachment

Tilted udder

Short front udder

Stronger center support in the udder

Teats more squarely placed

Teats more evenly spaced

Has more uniformly placed teats

Teats too close together

Teats strut

Rear teats too close together

Has more width between rear teats

Teats of more desirable size, shape, and placement

Teats of more uniform size

More desirable in size and shape of teat

Teats hang more nearly plumb

Weaker quarter

Lighter quarter

Underdeveloped rear quarters

SUMMARY

Physical appearance, health, milk production records, and pedigree are important factors to consider when selecting dairy animals for breeding and production.

Dairy animals are judged by comparing the animal being judged with an ideal dairy type. The Dairy Cow Unified Score Card describes the general traits of a good dairy cow. The score card is also used for type classification of dairy cows.

The desirable traits described on the score card are related to the ability of the cow to produce milk over a long period of time. The mammary system is the most important part of the dairy cow.

Follow a definite routine when judging dairy cows. Look at the animal from the side, rear, and front. First impressions are usually more accurate. Use dairy judging terms when giving oral or written reasons.

Student Learning Activities

1. Using pictures or a live animal, name the parts of the dairy cow.
2. Judge a class of dairy cows using pictures or live animals. Give oral reasons.
3. Give an oral report to the class on the selection of desirable breeding cows.
4. Arrange and take a field trip to a local dairy farm to observe type classification of the herd.

Review

1. Name the four things that are considered when selecting dairy animals for breeding and production.
2. Briefly describe each of the four things named in question one.
3. What is the Dairy Cow Unified Score Card?
4. On an outline of a dairy cow provided by the instructor, name the parts of the dairy cow.
5. Name the six type classifications for dairy cows.

6. Briefly describe the breed characteristics for each breed of dairy cow.

7. Name and briefly describe the common defects of the dairy cow.

8. Briefly describe how each of the following is related to the milk producing ability of the dairy cow.
  a. general appearance
  b. dairy character
  c. body capacity
  d. mammary system

9. Describe the steps to follow when judging a dairy cow.

10. Describe how a question class is used when judging dairy cows.

11. How does judging dairy heifers differ from judging dairy cows?

Unit 41 Feeding Dairy Cattle

Objectives

After studying this unit, the student should be able to

- select appropriate feeds for dairy animals.
- selection rations for dairy cows for maximum production.
- select rations for replacement animals for fast, economical growth.
- select an appropriate feeding method.

Feed costs are about 40 to 50 percent of the total cost of producing milk. Cows need to be fed balanced rations to give the most profitable level of production. Milk production of the individual cow is limited by heredity. Differences in milk production among cows are due to about 25 percent heredity and 75 percent environment. Feeding has the most influence on the amount of milk any cow produces. Proper feeding and care allows the cow to produce closer to her potential ability.

Nutrients are used by the dairy animal for the following needs:

- Growth of the immature animal.
- Pregnancy. Needs are small during the first six months and large during the last two or three months of the gestation period.
- Fattening or the regaining of normal body weight lost in lactation and during the dry period.
- Maintenance of the mature animal. Needs vary according to the size of the animal.
- Milk production. Needs vary with the pounds (kilograms) of milk produced and its composition.

Maintenance and milk production needs are the two most important to consider when balancing rations for dairy cows. Cows fed correctly use about one-half of the feed for maintenance and one-half for milk production. Nutrients are used to meet other needs before being used for milk production. The cow will draw on body reserves for her needs if not fed enough feed. Milk production will then go down.

Milk production records are needed to develop a good feeding program. High- and low-producing cows are identified by good records. The feeding program is then adjusted for individual cows or groups of cows. If the cow is not fed enough, milk production goes down. If fed too much, the cow gets too fat and production costs increase.

Dairy cattle are ruminants. (See Unit 5 for a discussion of digestive systems.) The basis for dairy rations is the roughages the cow eats. Dairy cows can use large amounts of roughage such as pasture, hay, haylage, or silage. When these are raised on the dairy farm, they are the cheapest source of nutrients needed by the cow. Feed good-quality roughages and supplement the ration with concentrates (grains and protein feeds).

Moldy feeds may produce toxins that interfere with metabolism, lower the nutritive value of the feed, and cause reproductive problems. Fusarium mold, aflatoxins, and mycotoxins are sometimes found in feeds. If moldy feed must be used, dilute it with feed that does not contain mold; this will reduce the toxic effect of the mold. Proper drying and storing of feeds will reduce mold problems.

Feeding standards for dairy cattle are found in the Appendix of this text. These may be used to help develop balanced rations for dairy animals. (See Unit 8 for methods of balancing rations.)

METHODS OF FEEDING DAIRY COWS

Traditional

Concentrates are fed to cows individually according to milk production. The roughages and concentrates are fed separately. Roughages are usually fed free choice in feed bunks to the entire herd or in mangers in stanchion barns. The concentrate mix is fed individually in mangers in stanchion barns or in the milking parlor during milking (Figures 41-1 and 41-2).

Disadvantages of traditional feeding include:

FIGURE 41-1 Silage topped with concentrate mix may be fed in the stanchion barn. *Photo by Michael Dzaman.*

■ It is hard to measure the amount of forage (roughage) each cow eats. Therefore, it is hard to balance the ration with the right amount of concentrate for each cow.
■ Low producers are often overfed concentrates.
■ High producers are often underfed concentrates.
■ Grain feeding facilities are required in the milking parlor.
■ The level of dust in the milking parlor increases.
■ Cleanup of uneaten grain in the milking parlor is required.
■ Milking in the parlor may be delayed while waiting for the cow to finish eating the grain mix.
■ Feeding in the parlor slows down the milking.
■ Cows do not stand as quietly and defecate more during milking.
■ More labor is necessary.
■ There is less control over the total feeding program.

FIGURE 41-2 Concentrate mix is fed in the milking parlor in feeders located in front of each cow.

- Cost of equipment is higher.
- Careful records of individual production and continual adjustment of concentrate feeding to match production is required.

Advantages of traditional feeding include:

- Less specialized equipment needed.
- Theoretically feeds each cow according to individual needs based on production.
- Permits adjusting concentrate feeding to the stage of lactation.
- Permits challenge feeding of each cow.

A feeding practice that was widely followed in the past was feeding concentrate on a grain-to-milk ratio. That is, a pound of concentrate was fed for so many pounds of milk produced. This is not currently considered to be a good practice. It tends to overfeed the low-producing cow and underfeeds the high-producing cow.

Challenge or Lead Feeding

The practice of feeding higher levels of concentrate to challenge the cow to reach her maximum potential milk production is called *challenge* or *lead feeding*. Challenge feeding gives more concentrates to the cow early in the lactation period and less later in lactation. Dry cows in good condition should get about 1/2 pound (0.22 kg) of concentrate per 100 pounds (45 kg) of body weight. Two to three weeks before calving, increase the amount of concentrate gradually, about 1 pound (0.45 kg) per day, until the cow is eating from 1 to 1 1/2 pounds (0.45–0.68 kg) per 100 pounds (45 kg) of body weight at calving. About three days after calving, increase the concentrate by 1 to 2 pounds (0.45–0.90 kg) per day for two to three weeks after calving. At this time the cow should be getting about 2 pounds (0.90 kg) of concentrate per 100 pounds (45 kg) of body weight. Continue to increase the level of concentrate feeding until milk production levels off. At maximum milk production, hold the level of concentrate feeding constant.

Keep accurate production records. When production drops, decrease the level of concentrate feeding. Decrease by 1 pound (0.45 kg) of concentrate for each 3 pounds (1.36 kg) of milk production drop.

If the cow loses body weight early in the lactation period, she needs more concentrate. Cull cows that do not respond to challenge feeding.

Feeding Total Mixed Rations

A method of feeding that is becoming more popular with modern dairy producers is feeding a total mixed ration. A total mixed ration is one that has all or almost all of the ingredients blended together. This ration is then fed free choice to all the cows in a group.

In large herds the cows are usually divided into several groups. Grouping cows for feeding total mixed rations will be discussed later in this section.

The total mixed ration contains roughages and concentrates combined to meet the energy, protein, mineral, vitamin, and crude fiber needs of the cows. When feeding a total mixed ration free choice, no concentrates are fed in the milking parlor.

Advantages of feeding a total mixed ration include:

- Each cow receives a balanced ration.
- Each cow is challenged to produce to her maximum genetic potential.
- Feeds are used more efficiently.
- NPN (nonprotein nitrogen) is utilized more effectively.
- Fewer cows have digestive problems or go off feed.
- It is not necessary to feed minerals free choice separately from the rest of the ration.
- Rations can be changed easier without affecting consumption.
- Less labor is required for feeding.
- Problems with low milkfat test are reduced.
- Cost of cow housing and feeding facilities are lowered.
- Cows that gain weight early in the lactation period are quickly identified and may be culled.
- Weight gains usually come later in the lactation period.
- It is possible to substitute cheaper grains and urea in the ration because silage tends to mask the taste and dustiness of feed ingredients.
- There are several advantages of not feeding concentrates in the milking parlor:
  - a. Cost of parlor construction and maintenance of feeding equipment is less.
  - b. Cows stand more quietly in the parlor during milking.
  - c. There is less dust and manure in the parlor.
  - d. No concentrate cleanup and no clogging of drain system by concentrates in the parlor.
  - e. It takes less time to milk the cows.
  - f. There is less wasted concentrate.

Disadvantages of feeding a total mixed ration include:

- Special equipment for weighing and mixing the ration is necessary.
- It may be more difficult at first to get cows to enter the milking parlor when they are not fed there.
- It is hard to include hay in the total mixed ration.
- If hay is fed separately, the ration may not be balanced for some cows.
- Low-producing cows tend to get too fat.
- Cows need to be divided into groups for most efficient feeding.

Special Equipment for Feeding Total Mixed Rations

The dairy farmer must have two items of special equipment for successful use of total mixed rations. These are a mixer-blender unit and a weighing device. The mixer-blender unit may be mobile or stationary. It is necessary in order to get the ration completely and uniformly mixed. If the ration is not mixed for a long enough period of time, the blend will not be uniform. If it is mixed for too long a period of time, the blend may be pulverized or shredded. The mixer-blender must be able to handle haylage. Some units cannot handle large amounts of haylage, resulting in equipment failure. Haylage needs to be fine chopped for proper handling in the mixer-blender unit.

The various feeds being used in the ration must be carefully weighed to insure a balanced ration. Weighing devices may be stationary scales or electronic scales on the mixer-blender unit.

Feed Analysis

Analysis of feeds being used in dairy rations is recommended regardless of the type of feeding system used. However, it is especially important for the successful use of total mixed rations. Feed analysis is needed to properly balance the ration and reduce feed costs.

The most important feeds to be analyzed are the roughages used in the ration. Roughages have a greater variation in nutrient content than do grains. However, grains may be analyzed for nutrient content if desired.

Feed testing services are available from both private and state university laboratories in many states. Some feed companies provide feed testing services for their customers. Information on feed testing laboratories may be secured from the Cooperative Agricultural Extension Service, many DHIA supervisors, and agricultural education departments in high schools. The dairy farmer should contact the feed testing laboratory to get the procedures for sampling and packaging feed samples before sending in samples to be tested.

Grouping Cows to Feed Total Mixed Rations

Part of the success of feeding total mixed rations depends on dividing the cow herd into groups. This is done to more nearly match the ration to the needs of the different groups.

Several factors may be considered when deciding on groups for total mixed ration feeding. These are herd size, facilities, and time the cow must wait in the holding area for the milking parlor. Large herds should be divided into several groups. It may be practical to have only one group in a small herd. The kind of facilities available may limit the number of different groups that may be used. If a cow has to be in a holding area for more than two hours before milking, the herd should be divided into smaller groups to reduce this waiting time.

Cows should not be kept away from feed and rest for more than two hours.

The most common way to group cows is according to production level. All dry cows are kept in a separate group. Some dairy farmers prefer to keep first calf heifers in a separate group. Cows with mastitis or other health problems may also be kept in a separate group.

In larger dairy herds, at least three groups should be used. These are high-, medium- and low-producing cows. Higher overall milk production and increased profits will usually result from this type of grouping.

High-producing cows need more concentrate in their ration than do low-producing cows. They are more efficient in converting feed into milk. They are the most profitable cows in the herd. Low-producing cows tend to get too fat when fed at the same level as high-producing cows. Dry cows should be in a separate group from lactating cows.

FIGURE 41-3   A magnetic feeder that permits cows wearing magnets to get extra concentrate.

Magnetic feeders can be used in small herds where two or three groups are not practical (Figure 41-3). This permits high-producing cows equipped with the magnets to get access to additional feed. This creates another group without physically dividing the cows.

Move fresh cows into the high-producing group about three days after calving. Leave them in this group for at least two months to challenge them to reach their maximum potential of production. At the end of two months, move any cow whose production is not high enough to a lower-producing group.

Move cows from one group to another at no more than monthly intervals. This shift is based on DHIA production records for individual cows. Shift cows in small groups rather than one cow at a time. Other factors should be considered when shifting cows from one group to another. These include physical condition of the cow, age, stage of lactation, stage of gestation, and individual cow temperament.

The ration for each group is determined by average production, size of cows, and the average milkfat test for the group. Rations must be adjusted when the quality of the forage being used changes. Analyze forages at least once a month.

Make sure plenty of water is available at all times for each group. If hay is not included in the total mixed ration, it may be fed to the group in a separate manger.

Make sure there is enough bunk space so cows may eat free choice whenever they want to. Keep feed in the manger at all times, especially for the high-producing group.

Several problems may occur when grouping cows. It takes more labor to change groups at milking time and to sort cows when moving them from one group to another. Proper facility design reduces this problem.

When cows are moved from one group to another, some drop in milk production may occur. This is the result of the change in ration and the fighting that might take place. This is usually not a major problem.

If gates are left open, groups may become mixed together. This can be prevented by the farmer taking care in handling the groups.

Automatic Concentrate Feeders

There are three types of automatic concentrate feeders currently in use on dairy farms. These are magnetic, electronic, and transponder. These systems automatically control access to concentrate feed by individual cows. A magnetic or electronic device is attached to each cow that allows access to feed. The transponder controls the amount of feed each cow receives. The purpose is to allow high-producing cows to have more concentrate while limiting the concentrate intake of low-producing cows. Good management is needed to successfully use these feeding systems.

ROUGHAGES FOR DAIRY CATTLE

From 60 to 80 percent of the dry matter in the dairy ration should come from roughages. High-quality roughages can lower the cost of feed for the dairy herd. Roughages used for dairy cattle rations are hay, silage, and pasture.

There can be a wide variation in the nutrient value of forages depending upon the variety, stage of maturity when harvested, moisture content at harvest, growing conditions, method of harvesting, and method of storing. Evaluation of nutrient content should be made for:

- Dry matter (DM) to determine dry matter intake and proper storage.
- Crude protein (CP) and adjusted crude protein (ACP). These will be different if the forage has been heat damaged. Heat damage reduces available protein.
- Acid detergent fiber (ADF). As ADF increases, the energy content of the forage is reduced.
- Neutral detergent fiber (NDF). The level of NDF is a good indicator of forage intake by the animal. Lactating cows should get at least 21 percent of their daily dry matter intake from the NDF in the forage.
- Calcium (Ca) and Phosphorus (P) to assure proper mineral balance in the diet.

Legume and legume-grass mixtures can be compared using a relative feed value (RFV) figure based on the quality of the forage as determined by its acid detergent fiber and neutral detergent fiber content. An ADF content of 41 percent and an NDF content of 53 percent is considered equal to a RFV value of 100. The equation for the relative feed value is: RFV = (digestible dry matter × daily dry matter intake)

÷ 1.29. Forages should have a RFV of 120 or higher for high-producing cows.

Representative samples of the feed are necessary for accurate analysis. Follow the instructions from the testing laboratory for securing, handling, and sending representative samples. Cooperative Extension personnel can provide the locations of feed testing laboratories.

Hay

The feeding value of any hay depends on the kind of hay, stage of maturity when cut, and the harvesting method. Second and third cuttings of hay are generally higher in nutrient value than first cuttings. Early cut hay has a higher feed value than hay that is more mature when harvested.

A high percent of the nutrient value of hay is in the leaves. Harvesting methods that retain more of the leaves results in a higher nutritional value. Windrowing hay when it is too dry causes a high leaf loss. Hay should be windrowed at about 35 to 40 percent moisture content. The use of hay crimpers or conditioners reduces leaf loss during harvesting. The moisture content of the hay should be 20 percent or less when it is baled. This will reduce spoilage and helps maintain nutrient value. Hay that is rain damaged has a lower feeding value compared to similar hay that is not rain damaged. Legume hays contain more protein than grass hays.

Hay has traditionally been handled in a baled form. This requires a lot of hand labor. Some dairy farmers use large, round bales as a method of harvesting hay. This saves labor but usually results in a greater loss of nutritional value. Handling hay in a cubed or wafered form reduces the labor needed. However, machine costs are higher. Experimental results generally show little difference in milk production when hay is fed in the baled, wafered, or cubed form.

Dairy cattle eat much higher quantities of high-quality hay compared to low-quality hay. More concentrate is needed to balance the ration when low-quality hays are used.

Alfalfa is the best hay for dairy cattle use. It yields more protein and TDN per acre than other hay crops. Alfalfa hay should be cut at the one-tenth bloom stage for highest feeding value.

Red clover hay is lower in protein and less palatable compared to alfalfa hay. Sweetclover hay tends to be coarse and stemmy. Sweetclover is better as a pasture rather than a hay for dairy cattle. Lespedeza hay is similar to alfalfa hay in feeding value.

Nonlegume hays such as timothy, millet, oats, prairie grass, Johnsongrass, and Coastal Bermudagrass are generally lower in protein and TDN than alfalfa hay. A high level of fertilization can improve the feeding value of these grass hays.

Mixtures of legume-grass hay are popular in some areas. Alfalfa-bromegrass, alfalfa-timothy, and alfalfa-orchard grass mixtures are high in nutritive value if harvested at the time recommended for alfalfa.

Green Chop

Some dairy farmers harvest forage by chopping it daily and feeding it in bunks. This harvesting method reduces field losses. A major problem is the need to chop the forage each day, especially in bad weather or during times of high labor requirements for other farm operations.

The moisture content of green chop forage varies considerably. As the crop matures, cows generally eat less of the green chop forage. This reduces the amount of energy received from the forage part of the diet.

Silage

Almost any crop can be made into silage. Hay crops, corn, and small grains are common kinds of silage used in dairy rations. Silage has a higher moisture content than dry roughages. Silage fits well with the feeding of total mixed rations. It is in a form that is easily handled by automated equipment. Legume-grass silage may be put into a bunker silo at 60 to 65 percent moisture content.

Haylage is a hay crop made into silage. The first cutting of hay is often harvested as haylage. Both grasses and legumes may be used for haylage. Haylage may be put into a conventional silo at 50 to 60 percent moisture and in an oxygen-limiting silo at 40 to 50 percent moisture. When stored in bunkers, it may be harvested at 75 to 80 percent moisture. This saves the maximum nutrients from standing crop to feeding. Legumes should be cut at one-tenth bloom and grasses at the early heading stage. Harvest sudan grass and sorghum-sudan hybrids before the heads emerge from the boot. Harvest small grains for silage in the boot to early milk stage. Soybeans should be cut for silage when the beans are forming in the pod. As maturity at harvest increases beyond the recommended stage, the palatability and digestibility of the feed decreases.

Grass and legume silages are lower in energy and higher in protein content than corn silage. The amount of dry matter in haylage varies with percent of moisture at the time it is put in the silo. Cows have difficulty in eating enough high-moisture haylage to maintain high milk production. However, when haylage is put in the silo at 50–65 percent moisture content, little difference has been shown experimentally in milk production when compared to cows fed dry hay. Cows apparently use the nutrients in haylage more efficiently.

Putting haylage in the silo at too low a moisture content may cause heat damage. This occurs during the fermentation process in the silo. Heat-damaged haylage provides less energy and protein in the ration. It has a darker color (golden brown to black), caramel odor, and high palatability. Cows like to eat heat-damaged haylage but do not maintain high milk production on it.

If the crop is too dry, water may be added at the silo. Propionic acid may be added to reduce heat damage.

> **Caution:** Propionic acid is corrosive and will cause severe burns. Always use goggles, wear rubber gloves, wash skin immediately upon contact, and clean all equipment right after use.

Some dairy producers prefer to put the forage crop directly into the silo without letting it wilt. Dry ground material (such as grain) or chemical preservatives are added to reduce moisture content and nutrient loss. Materials such as ground grain, dried beet or citrus pulp, dried brewer's grains, hominy, or soybean flakes may be used as preservatives in forage silage. When the forage is direct cut, 100–200 pounds (45–90 kg) of this material may be added per ton of silage.

Chemical preservatives may be added to high-moisture forage silage stored in horizontal silos. Chemicals used as preservatives include propionic acid, sodium metabisulfite, formic acid, or a calcium formate-sodium nitrite mixture. Bacterial cultures or enzyme additives do not show any economic benefit when added to forage silage. No additive or preservative needs to be used if the crop has been wilted to the recommended level before it is put into the silo.

Good management during ensiling helps preserve the best possible nutritive value of the haylage. Cutter knives should be set at 3/8 inch (0.95 cm). Fill the silo as fast as possible and get a good pack on the material. Distribute the material evenly in the silo. Make sure silo walls and doors are tight to keep air out. If possible, put wetter material on top. Wait three to five weeks after filling the silo before starting to feed out of it.

Corn silage is a popular roughage in dairy cattle rations. Some dairy farmers use corn silage as the only roughage for the dairy herd.

Corn silage yields more energy per acre than other forages. It is a highly palatable feed that is easily stored and handled. Corn silage requires less labor to harvest and feed than many other roughages. The quality of corn silage is usually more consistent than other roughages. Good corn silage contains about 50 percent grain on a dry matter basis.

Corn silage contains about 8 percent protein on a dry matter basis. When corn silage is used as the main forage in a dairy ration it must be supplemented with protein. This can be done by feeding a concentrate mix that is higher in protein. Urea or other NPN sources may be added to the corn silage to increase the protein content. A protein supplement like soybean oil meal may be added to the ration. Bacterial cultures or enzyme additives do not add any economic benefits to the use of corn silage.

Compared to other forages, corn silage is low in minerals. Additional minerals must be fed to balance the ration.

Harvest corn silage when the kernels have reached the dent stage. Dry matter should be about 30–35 percent. Set cutter knives on the field chopper at 1/4 to 3/8 inch (0.6–0.95 cm).

Sorghum silage may be made from both forage and grain sorghums. The nutrient content is about the same as corn silage. The crude protein content and energy level are a little lower than corn silage. Sorghum silage is usually a little lower in digestibility than corn silage. Forage sorghum yields about the same per acre as corn silage. Grain sorghum yields per acre are usually a little lower than corn silage.

Dairy cows produce about the same when fed either grain sorghum silage or corn silage. However, cows must eat more grain sorghum silage than corn silage to maintain the same level of production.

Intermediate type sorghum silage yields more per acre than corn silage. It is taller than combine type sorghum. It is grown in some areas that are not well adapted to corn. Milk production is lower when intermediate type sorghum silage is fed.

Coastal Bermudagrass silage has a lower feeding value than corn silage. Storage loss in upright silos is high. Harvesting Coastal Bermudagrass as silage is not recommended.

Small grain silage may be made from small grains such as oats, rye, barley, and wheat. The nutritive value of small grain silages is less than corn silage.

Barley and wheat silage will maintain higher levels of milk production than oat silage. In some areas the use of small grain silage will permit *double cropping* (growing two crops on the same ground in the same year).

Harvest barley, wheat, and oats at the boot to early head stage of growth in order to produce the best yield of energy and protein per acre. These small grains may be harvested by direct-cut without wilting.

Harvest rye for silage at the boot stage of maturity. Cut and wilt it before putting it into the silo.

Growing Austrian winter peas with barley for silage will increase the yield per acre. The protein content of the silage is also increased. This mixture may be direct-cut and put in a conventional upright silo without wilting. If the crop is to be stored in a sealed (air-tight) silo, wilt it to 50–55 percent dry matter before ensiling.

Straw

Straw from small grains (oats, wheat, etc.) is not recommended in rations for cows that are milking. These straws are low in energy, protein, minerals, and vitamins. They mainly add fiber to the ration. Some straw in the ration can be used for dry cows and older heifers. Supplement must be added if straw is used in the ration.

Corn Stover

Corn stover is not recommended in rations for cows that are milking. It may be used in rations for dry cows and heifers. Corn stover is low in protein and carotene. Supplements must be added if corn stover is used in the ration.

Pasture

The effective use of pasture for the dairy herd requires good management. The use of pasture reduces the labor needed for feed and manure handling. Pastures are an excellent source of roughage for dry cows and growing heifers.

Several problems may occur when pastures are used for lactating cows. These include:

- Drop in milk production after cows are on pasture for a time.
- Drop in milkfat test.
- Off flavors in the milk.
- Bloating of cows.
- Reduced grain intake.
- Feces become watery.
- Difficulty in getting cows to come into the milking parlor.

It is hard for lactating cows to get enough dry matter in the ration if pasture is used as the main source of roughage. Pasture has a high moisture content and generally cannot meet the energy needs of high-producing cows. Several management practices may be used to help reduce the problems with lactating cows on pasture.

Limit the grazing time on pasture for lactating cows to 1 to 2 hours per day. Feed dry forage before letting the cows out to pasture. Bring the cows back to the feed lot several hours before milking. Feed silage and/or hay again at this time. Continue feeding the concentrate mixture at the same rate as when the cows are not on pasture. Protein content of the concentrate mix may be reduced some if the pasture is of good quality.

Rotational grazing may be used. Divide the pasture with a temporary electric fence. The high-producing cows are allowed to graze on the fresh pasture. Lower-producing cows, dry cows, and bred heifers are put into the pasture that has been grazed over by the high-producing cows. Clip the pasture regularly to keep weeds down and allow fresh regrowth of the pasture. Pastures need to be well fertilized to provide high-quality forage.

Use pasture as a supplementary roughage. Make sure the lactating cows have access to silage and/or hay to make up for nutritional shortages as pasture quality changes through the grazing season.

Grasses or grass-legume mixtures make good pastures for dairy cattle. Timothy, bromegrass, orchard grass, Coastal Bermudagrass, alfalfa, sweetclover, and ladino clover are examples of crops used for dairy pastures.

Temporary pasture for dairy cattle may be provided by sudan grass, rye, oats, hybrid sorghum, and pearl millet. Prussic acid poisoning may occur on sudan grass if it is grazed before it is 18 inches (45 cm) high. Do not pasture sudan grass after a freeze. The new growth may cause prussic acid poisoning. Rye may cause off flavors in milk. Removing lactating cows from rye pasture several hours before milking will help prevent this problem.

GRAINS FOR DAIRY COWS

Grains are included in the dairy ration mainly for their energy content. Usually, the most limiting factor in milk production is a shortage of energy in the ration. It is important that lactating dairy cows get enough energy in the ration. Grains contain about 70–80 percent total digestible nutrients (TDN).

Grains that are processed before feeding are more digestible. Methods of processing grains include grinding, rolling, crimping, pelleting, or cracking. Grains that are ground too fine may lower digestibility and percent of milkfat. Rumen acidosis may also result from feeding too finely ground grain.

Corn

Corn is the most commonly used grain in dairy cattle rations. It is high in energy value and is palatable. Home-grown corn is usually a cheaper energy source than other grains.

Corn and cob meal contains about 90 percent of the TDN and is higher in fat than shelled corn. The higher fiber content helps keep the percent of milkfat higher. Cows stay on feed better when corn and cob meal is used in the ration. It is usually cheaper to grind ear corn to make corn and cob meal than to shell corn and then grind it.

High-Moisture Corn

Ear or shelled corn may be stored at a higher moisture content in a silo than in a crib. This is called high-moisture corn. High-moisture ground ear corn is stored at 28–32 percent moisture and high-moisture shelled corn at 25–30 percent moisture.

An organic acid (such as propionic or acetic) may be used as a preservative for high-moisture corn. More preservative is needed for high-moisture ear corn than for high-moisture shelled corn.

Ear corn should be ground before being put in the silo. Shelled corn may or may not be ground before ensiling. It should be ground before feeding.

High-moisture corn is a good feed for dairy cattle. On a dry matter basis, it is equal to dry corn in feeding value. More high-moisture corn must be fed to get the same nutrient intake as dry corn.

Oats

Oats are an excellent feed for dairy cows. They are lower in energy value than corn. The protein content is higher than corn. Oats add fiber and bulk to the grain mix. They are a little lower in digestibility than corn. Oats should replace no more than one-half of the corn in the ration.

Barley

Barley has about the same overall energy value for dairy cattle as corn. It is a little higher in protein compared to corn. Barley that is rolled is more palatable than when it is ground. Finely ground barley should make up no more than 50 percent of the grain ration.

Wheat

Wheat is high in energy and protein. Dairy cattle like wheat. Because of price it is not usually used in dairy cattle rations. Wheat should make up no more than 50 percent of the grain ration.

Rye

Rye is not very palatable for dairy cows. It is seldom used in dairy rations. If it is used, it should not be more than one-fourth of the grain in the ration.

PROTEIN SUPPLEMENTS FOR DAIRY COWS

A protein supplement is added to the ration to make up the difference between what the cow needs and what is supplied by the rest of the ration. The amount and quality of the forage in the ration has a great effect on the amount of protein needed in the concentrate mix. Protein supplements are usually the most expensive part of the ration. Therefore, care must be taken to carefully balance the ration to hold the feed cost as low as possible. The quantity of protein is more important than the quality of the protein. Purchase protein for dairy cows on the basis of cost per pound (kg) of protein supplied in the supplement.

Corn Gluten Meal

Corn gluten meal is a by-product from the wet milling of corn for starch and syrup. It may have 40 percent or 60 percent crude protein content; the 60 percent meal is the most common. The energy content of corn gluten meal is slightly lower than corn grain. Because it has a lower palatability, limit the amount of corn gluten meal to no more than five pounds (2.3 kg) per head per day.

Distillers Dried Grains

Distillers dried grains are a by-product of grain fermentation for alcohol production. The nutrient value of distillers dried grains varies depending on the grain used as its source. They contain 23 to 30 percent protein.

Soybean Meal

Soybean meal is the by-product left when oil is extracted from soybeans. Soybean meal is a commonly used protein supplement. It is economical and an excellent source of protein. Soybean meal is found in many commercial dairy protein supplements. It is palatable and highly digestible.

Soybeans

Ground, unprocessed soybeans may be used in grain mixes for dairy cows as an added protein source. If used, they should not make up more than 20 percent of the grain mix. Do not use urea in the ration if soybeans are included.

Processing by roasting or cooking can increase the palatability and stability of the soybeans. Soybeans are usually too high-priced to be used in dairy cow rations.

Sunflower Meal

Sunflower meal supplements range from 28 to 45 percent protein. It is a good source of protein and phosphorus. Because it is less palatable, it should not be topdressed on the feed.

Linseed Meal

Linseed meal is the by-product left when oil is extracted from flax seed. It is a good protein supplement. Linseed meal is usually higher in cost than some other protein supplements. It is sometimes used for fitting show or sale cattle because it adds a shine to the hair coat. Linseed meal is palatable and slightly laxative.

Cottonseed Meal

Cottonseed meal is made from hulled cotton seeds after the oil has been extracted. It is high in protein and may be used as a protein supplement in dairy rations. Cottonseed meal is palatable and slightly constipative.

Urea and Other Nonprotein Nitrogen (NPN)

Urea may be used in dairy rations to supply some of the protein needed by cows. The use of urea usually lowers the cost of the ration. The nitrogen content of urea is 46 percent. It supplies a protein equivalent of 287 percent (46×6.25).

Urea has a bitter taste and must be mixed completely in the ration for cattle to eat it. Too much urea in the ration can be toxic to cattle. Feed no more than 0.4 pound (0.18 kg) of urea per head per day. The concentrate mix should be no more than one percent urea. If urea is added to corn silage at the maximum recommended rate of ten pounds per ton (0.05 percent) then the concentrate mix should contain no more than 0.5 percent urea.

It takes the rumen bacteria some time to adapt to urea in the ration. It is recommended that the amount of urea in the ration be gradually increased over a period of 7–10 days to reach the desired amount in the ration. The amount of urea fed should be limited to 0.2 pound (0.09 kg) per day during early lactation.

Other NPN products may be used as protein substitutes in the ration. Monoammonium phosphate (11% nitrogen, 68.75% crude protein equivalent) and diammonium phosphate (18% nitrogen, 112.5% crude protein equivalent) are two examples of other NPN sources for dairy cattle.

Liquid protein supplements are made by putting a NPN source (usually urea) in a liquid carrier. Molasses is the most common carrier. Other carriers such as fermentation liquors or distillers solubles may be used. Molasses provides an energy source needed for the utilization of the NPN. Phosphoric acid is often added to provide phosphorus and help stabilize the nitrogen.

Most liquid protein supplements contain 32 to 33 percent crude protein equivalent. Liquid protein supplements can be used to provide part of the protein needed in the ration. They may be fed free choice, top dressed (usually on the forage), or blended into a total mixed ration.

There is generally no economic advantage to using liquid protein supplements. The decision to use a dry or a liquid protein supplement depends on the available labor and equipment. Convenience of use is also a factor. The use of liquid protein supplement may be of most value for low-producing cows, dry cows, yearlings, and older heifers when corn silage is the major forage in the ration.

BY-PRODUCTS AND OTHER PROCESSED FEEDS FOR DAIRY CATTLE

Alfalfa Meal

Alfalfa meal is made by grinding alfalfa hay. High-quality hay produces a high-quality meal. It may be pelleted. Because of the grinding, the fiber content will not help in maintaining milkfat test. Alfalfa meal is lower in digestibility than alfalfa hay.

Alfalfa Leaf Meal

Alfalfa leaf meal is made by grinding the leaves of the alfalfa plant. It is higher in protein than alfalfa meal.

Beet Pulp

Beet pulp may be plain or mixed with molasses. It is palatable, bulky, and slightly laxative. Beet pulp is low in protein and high in energy. Up to 30 percent of the ration dry matter may be composed of beet pulp.

Brewer's Grain

Brewer's grain is a by-product of the brewing industry. It may be wet or dry. Wet brewer's grain contains about 80 percent water and therefore requires the feeding of a large amount to get the needed dry matter intake. On a dry matter basis, brewer's grain is higher in protein and lower in net energy compared to corn.

The ration must be gradually adjusted if brewer's grain is to be included. Wet brewer's grain requires special handling and storage.

Citrus Pulp

Citrus pulp has about the same feeding value as beet pulp. It is lower in protein than beet pulp. It is sometimes fed when the price is competitive.

Cottonseed, Whole

Whole cottonseed is high in fat, fiber, and energy but only medium in protein content. The whole cottonseed is white and fuzzy; delinted cottonseed is black and smooth. Do not feed more than seven pounds (3.2 kg) per cow per day.

Corn Gluten Feed

Corn gluten feed is a by-product of the corn wet milling industry. It is medium in energy and protein content but high in fiber. It may be purchased as either a wet or dry feed. Limit the amount of corn gluten feed to not more than 25 percent of the total ration dry matter.

Hominy Feed

Hominy feed is a mixture of the starch part of the corn kernel, the bran, and the corn germ. It is a little higher than ground corn in feeding value. It is palatable and high in energy. Hominy feed may be used in place of ground corn in dairy rations.

Malt Sprouts

Malt sprouts are bitter and need to be mixed with other feeds. They have a medium protein level and only a moderate amount of energy. No more than 20 percent of the grain mix should be composed of malt sprouts.

Molasses

Both cane and beet molasses may be used in dairy rations. They are used mainly to increase the palatability of the ration. Molasses should be no more than 5 to 7 percent of the grain mix. Cane molasses is more commonly used than beet molasses. Molasses supplies energy in the ration. It is low in protein.

Potatoes

Cull potatoes, potato meal, and potato pulp may be used in dairy rations. Potato meal has a high fat content. Potato pulp is palatable and about equal to hominy feed when it is not more than 20 percent of the concentrate mix. Cull potatoes are similar to corn silage in feeding value. Cull potatoes should be chopped when fed. Limit maximum intake to 30 pounds (14 kg) per head per day.

Soybean Hulls

Soybean hulls, soyhulls, and soybean flakes are all by-product feeds from soybean processing. Their fiber content is highly digestible and they may replace starch in the diet. They do not replace forage fiber. These feeds should be limited to no more than 45 percent of the grain ration.

Wheat Bran

Wheat bran adds bulk and fiber to the concentrate mix. It is palatable and slightly laxative. Wheat bran is low in energy and medium in protein content. It should not make up more than 20 to 25 percent of the concentrate mixture.

Wheat Midds

Wheat midds are by-products of wheat milling. They are fine particles of wheat bran, wheat shorts, wheat germ, and other wheat products. They contain a moderate amount of energy and protein. The crude fiber content cannot exceed nine and one-half percent. If used at too high a level in dairy cow diets, they may cause a reduction in milk production. Limit the use of wheat midds to no more than 20 percent of the grain mix.

Whey

Both dried and liquid whey may be fed to dairy cows. Up to 10 percent of the concentrate mix may be dried whey. Dried whey may be added to silage when the silo is being filled. Use 20–100 pounds (9–45 kg) per ton of wet silage.

Liquid whey may be fed free choice. Because of the high water content (90%) cows must consume 15–25 gallons (57–95 litres) per day to get enough dry matter intake. One hundred pounds (45 kg) of liquid whey contains about the same amount of energy as 7.5 pounds (3.4 kg) of most dairy concentrates.

Liquid whey feeding requires adaptation by the cattle. Water must be limited for about four weeks to get the cattle drinking the desired amount of liquid whey. A large amount of liquid in the gutter may be a problem if liquid whey is fed in stanchion barns. A constant supply of fresh liquid whey is necessary. Flies may become a problem when liquid whey is fed.

MINERALS FOR DAIRY CATTLE

Minerals that are needed by dairy cows include calcium, phosphorus, magnesium, potassium, sodium, chloride, sulfur, iodine, iron, copper, cobalt, manganese, zinc, selenium, and molybdenum. The nutrient requirement tables in the Appendix show the required amounts of calcium and phosphorus. The needed amounts of the other minerals have not been clearly established.

Well-balanced rations that include mineral supplements will meet the mineral needs of dairy cows. Steamed bone meal, dicalcium phosphate, and limestone may be used to supply calcium. Dicalcium phosphate, monocalcium phosphate, and monosodium phosphate may be used to supply phosphorus. Salt supplies sodium and chloride. The use of trace mineralized salt will supply many of the minerals needed in small amounts. Commercial mineral mixes fed at the recommended level will normally supply the needed minerals. The ratio of calcium to phosphorus in the ration of lactating cows should be 1.2:1 to 2:1.

Do not feed too much phosphorus in the ration. Doing so increases costs and the amount of pollution caused by manure. The amount of phosphorus in the manure can be reduced by 25 to 30 percent by limiting the amount of supplemental phosphorus fed to not more than 0.38 percent of the diet for cows during the lactation period. Reproduction problems from a lack of phosphorus in the diet do not occur until the level drops to 0.15 to 0.2 percent of the ration. Because of increasing public concern about the environment, it is probable that limitations on the amount of phosphorus that can be applied to the land through manure applications will eventually be imposed by government regulation.

Zinc methionine may be added to the diet at the rate of 4.5 grams per cow per day for cows that have feet and leg problems. Research and field response to this treatment has varied, depending upon forages fed, feeding system used, and herd health. The economic benefit of the treatment must be balanced against its cost.

VITAMIN NEEDS OF DAIRY CATTLE

The nutrient requirement tables in the Appendix give the vitamin A and D requirements for dairy cattle. Some supplementation of vitamins A, D, and E may be needed in dairy rations. Usually there are enough vitamins in the feeds used in dairy rations. Some conditions such as poor-quality forage, high levels of grain feeding, or lack of sunshine may cause a need for additional vitamins.

A commercial vitamin premix may be added to the concentrate mix to supply needed vitamins. Massive doses of vitamins taken over too long a period of time may be toxic.

The addition of six grams of niacin per cow per day to the diet has resulted in increased milk production and feed intake and a decrease in problems with ketosis. The niacin improves rumen digestion and protein synthesis and controls fatty liver. It may be added to the diet beginning two weeks before calving and continued for the first 10 to 12 weeks of lactation. Major benefit is achieved when niacin is added to the diet of cows that are high-producing, are prone to ketosis problems, or are too fat.

WATER NEEDS OF DAIRY CATTLE

Milk is 85 to 87 percent water. Lactating dairy cows require more water in relation to their size than any other farm animal. Dairy cattle suffer quicker from a lack of water than from a lack of any other nutrient. As the air temperature increases, the need for water also increases (Table 41-1). Always keep plenty of fresh water available for dairy cattle.

Keep the water supply free of bacteria and high levels of nitrates and sulphates. A high level of blue-green algae in the water may be toxic to cattle. Do not allow cattle to drink from ponds or lakes that contain a heavy growth of algae.

BODY CONDITION SCORE

Body condition score in dairy cattle refers to the amount of fat the animal is carrying. Scores range from 1 (very thin) to 5 (excessive fat). Body condition scoring is done by observing the amount of depression around the tailhead, the amount of fat covering the pin and pelvic bones, and the amount of fat covering the loin area. The amount of fat covering is determined by feeling with the hands and making as

TABLE 41-1 Water Intake Guideline for Dairy Cattle.

| Weight | | Milk | | Temperature | | | | | |
| | | | | 40°F (4.4°C) or Less | | 60°F (15.5°C) | | 80°F (26.7°C) | |
lb	kg	lb	kg	gal/day	l/day	gal/day	l/day	gal/day	l/day
Heifers									
200	90.7	—	—	2.0	7.6	2.5	9.5	3.3	12.5
400	181.4	—	—	3.7	14.0	4.6	17.4	6.1	23.1
800	362.9	—	—	6.3	23.8	7.9	29.9	10.6	40.1
1,200	544.3	—	—	8.7	32.9	10.8	40.9	14.5	54.9
Dry cows[a]									
1,400	635.0	—	—	9.7	36.7	12.0	45.4	16.2	61.3
1,600	725.7	—	—	10.4	39.4	12.8	48.4	17.3	65.5
Lactating cows[b]									
1,400	635.0	20	9.1	12.0	45.4	14.5	54.9	17.9	67.8
		60	27.2	22.0	83.3	26.1	98.8	24.7	93.5
		80	36.3	27.0	102.2	31.9	120.8	38.7	146.5
		100	45.4	32.0	121.1	37.7	142.7	45.7	173.0

[a]Maintenance and pregnancy.
[b]Maintenance and milk production.

Source: Linn, J.G., Hutjens, M.F., Howard, W.T., Kilmer, L.H., and Otterby, D.E., *Feeding the Dairy Herd,* North Central Regional Extension Publication 346, Agricultural Extension Service, University of Minnesota, Revised 1989.

objective a judgment as possible about the amount of fat covering that is present.

Body condition score 1 is a thin animal showing a deep depression around the tailhead and no fat covering over the rump and loin. Body condition score 2 shows a shallow cavity around the tailhead and a small amount of fat covering the rump and loin areas. Body condition score 3 shows no cavity around the tailhead and fatty tissue over the whole rump and loin area. Body condition score 4 shows folds of fatty tissue over the tailhead and patches of fat over the rump with fairly heavy fat covering over the loin. Body condition score 5 shows the tailhead buried in fatty tissue and heavy fat covering over the rump and loin areas.

For maximum efficiency in milk production, dairy cows must not be too thin or too fat. Using body condition scoring to evaluate the condition of the herd will help the dairy farmer improve feed efficiency and herd health. Thin cows have more health problems; cows that are too fat have more difficulty calving and a higher risk of fatty liver syndrome. Heifers being raised for breeding should also be checked for body condition score. Those that are too fat will have more difficulty calving and may have poorer mammary development, which will reduce their lifetime production potential. Dairy farmers planning to use Bovine Somatotropin (bST) should check their herd for body condition. The use of bST is not recommended for cows that are too thin or too fat.

Cows should be checked for body condition at the following times:

1. *Shortly after calving so feed adjustments may be made if necessary.* At the time of calving a cow should have a body condition score of 3.0 to 3.5. It is normal for the cow to lose some weight during the early part of the lactation period, but the body condition score should not drop below 3.0 to 2.5. A very high-producing cow may drop as low as body condition score 2.0 during this time.

2. *Early in the lactation period, between one to four months.* Body condition score during this part of the lactation period should be around 2.5 to 3.0. Increase the energy level of the ration for cows that are too thin and make sure they have sufficient feed intake to maintain the proper body condition. Cows should reach their peak milk production during this part of the lactation period. If the cow is maintaining a body condition score in the range of 3.0 to 3.5 but does not produce the expected amount of milk, check the ration for protein level, calcium, potassium, phosphorus, and magnesium intake. Make sure there is an adequate water intake. Deficiencies in these nutrients may prevent the cow from reaching her maximum milk production while still maintaining adequate body condition score.

3. *Around the middle of the lactation period or at about four months.* The cow should have a body condition score of about 3.0 during this period. If it appears the animal is becoming too fat (body condition score of 3.5 to 4.0) at this time, energy intake should be reduced. If the animal is becoming too thin, increase the energy intake. Problems often begin earlier in the lactation period, so if a number of animals in the herd are out of body condition at mid-lactation, it is wise to do an earlier check on body condition during the next lactation period.

4. *Check for body condition score again near the end of the lactation period (after eight months).* The cow should be building body reserves at this time in preparation for the next lactation period. The body condition score at this time should be about 3.5. Adjust the ration for energy level if the animal is too thin or too fat. Do not allow the cow to become too fat before drying off in preparation for the next lactation period.

Adjustments are made in the energy, protein, fiber, and acid detergent fiber levels of the ration to correct problems with body condition scores during the lactation and dry periods of the production cycle. The next several sections of this unit discuss various feeding recommendations that will help the dairy farmer better manage the body condition score of the animals in the herd.

FEEDING LACTATING DAIRY COWS

Some general guidelines may be followed when feeding the dairy herd. The total ration should contain from 18 to 19 percent crude protein on a dry matter basis during early lactation. Reduce the crude protein to 13 percent on a dry matter basis late in the lactation period. (See Unit 8 for guidelines based on type of forage.) About 75 to 80 percent of the crude protein is digestible and available for use by the cow.

Protein degradability must be considered when feeding high-producing cows. Some protein is not digested in the rumen but is passed on to the small intestine before it is digested or degraded. This is called "bypass" or "escape" protein. Research indicates that from 33 to 40 percent of the crude protein fed to high producing dairy cows should be bypass protein. Sources of bypass protein include heat-treated whole soybeans or soybean meal, distillers grains, feathermeal, blood meal, and meat and bone meal.

A shortage of energy is usually the most limiting factor in milk production. The total ration should contain 60 to 70 percent total digestible nutrients (TDN). This is equivalent to 0.60 to 0.80 megacalories of net energy per pound of feed. (A megacalorie is 1,000,000 calories.) During early lactation, provide a minimum of 0.78 megacalories per pound (1.72 Mcal/kg) of dry matter as the net energy level for lactation. During the later stage of lactation, this may be reduced to 0.70 megacalories per pound (1.54 Mcal/kg). During the dry period, provide 0.60 megacalories per pound (1.3 Mcal/kg) of dry matter.

The basis of any dairy feeding program is forage. Use high-quality legume forages during early lactation. Feed the cow 1.5 to 2.8 pounds (0.68–1.27 kg) of forage dry matter per 100 pounds (45 kg) of body weight.

Fiber in the ration is needed to maintain milkfat percent in the milk. There should be a minimum of 15 percent crude fiber in the ration. During early lactation, provide a minimum of 18 percent acid detergent fiber (ADF) in the dry matter. During late lactation, increase this to 21 percent or more ADF in the dry matter. The dry matter in the diet should contain a minimum of 21 percent neutral detergent fiber (NDF) from forages.

Recent research at the University of Illinois suggests that the physical form of the fiber in the dairy cow ration during early lactation affects both milk yield and milkfat test as well as the amount of dry matter intake, and the apparent digestibility of the dry matter and neutral detergent fiber. In three groups, all the rations contained 60 percent concentrates, with one ration also containing 40 percent haylage, another 28 percent haylage and 12 percent alfalfa pellets, and the third 12 percent haylage and 28 percent alfalfa pellets. Dry matter intake, milk production, milkfat test, and apparent digestibility of dry matter and neutral detergent fiber were significantly lower for the group receiving 28 percent alfalfa pellets in the ration. This ration was lower in "effective fiber," which reduced the amount of time the cows spent chewing. When cows spend less time chewing, less saliva is produced; this reduces the pH level and affects the ratio of acetate to propionate in the rumen. Lower saliva production also affects the milkfat percentage and the rate of passage of material through the rumen. The results of this research would appear to suggest that care must be taken to provide sufficient levels of "effective fiber" in the dairy ration in addition to the recommended minimum of 15 percent crude fiber.

An increase in milk production and higher feed efficiency can be achieved by properly balancing the amount of neutral detergent fiber and soluble carbohydrates (sugars and starches) in the dairy ration. Too high a level of soluble carbohydrates in the ration may result in rumen acidosis, lower milkfat test, cows going off feed, and laminitis. Feed intake and the energy level of the ration is lowered when the level of neutral detergent fiber is too high. The recommended level of soluble carbohydrates in the ration is 30 to 35 percent, and the recommended level of neutral detergent fiber in the ration is a minimum of 28 percent. An acid detergent fiber level of 19 to 21 percent is recommended. The proper amount of fiber improves rumen digestion and helps maintain the right pH level in the rumen. Soluble carbohydrates are needed for proper microbial growth in the rumen.

The proper balance may be accomplished by combining high soluble carbohydrate sources with lower soluble carbohydrate sources in the ration. Shelled corn has a high soluble carbohydrate level; this may be balanced with beet pulp, oats, or soyhulls, which have lower soluble carbohydrate levels, to prevent milkfat depression, which may occur with a ration that is too high in shelled corn. It is better to use alfalfa forage (lower soluble carbohydrate) rather than corn silage (higher soluble carbohydrate) with high-moisture shelled corn (higher soluble carbohydrate). High-moisture ear corn works well with corn silage in dairy rations because of a better balance of soluble carbohydrates between these two ingredients.

The concept of balancing soluble carbohydrates and neutral detergent fiber in dairy rations is relatively new. Current feed composition tables generally do not provide information on soluble carbohydrate levels in feeds. However, as this concept in dairy feeding becomes more common, it may be expected that information relating to soluble carbohydrate levels will become available.

Both major and trace minerals are needed by the dairy cow. The concentrate mix should contain 0.5 to 1 percent mineralized salt. One percent of the concentrate mix should be a calcium-phosphorus supplement.

Grains and forages may be processed before feeding. If they are, do not grind them too fine. Coarse-to medium-ground feeds are better for dairy cows.

Grain and protein supplements are usually the most expensive part of the ration. Buy feeds on the basis of the least cost per unit of nutrient supplied. Generally, home-grown grains will lower the cost of the ration.

Feed requirements vary with the stage of lactation and gestation. Four feeding phases are identified based on milk production, fat test, dry matter intake, and changes in body weight during lactation. Phase one occurs during the first 70 days of lactation. Milk production is highest during this phase. Phase two is from 70 to 140 days after calving. Milk production is decreasing and dry matter intake is at its highest during this phase. Phase three is from 140 to 305 days after calving. Milk production continues to decrease during this phase. Phase four is the dry period of 45 to 60 days before the next calving and the beginning of a new lactation period. Tables 41-2 and 41-3 show some sample rations for these different feeding periods.

The most critical feeding period is during phase one. Milk production is increasing rapidly at this time. Maximum milk production is reached four to six weeks after calving. Increase the amount of grain in the ration by 1 to 1.5 pounds (0.45–0.68 kg) per day during the first ten weeks of lactation. Keep the fiber level in the ration above 15 percent to keep the rumen working properly. Keep the grain level no higher than 50–55 percent of the total dry matter in the ration. Do not increase the rate of grain feeding too fast or feed more grain than is needed. The cows may go off feed or have problems with acidosis or displaced abomasum if this is done. Too much grain in the diet can also decrease the milkfat percentage. During phase one, the diet must contain 19 percent or more crude protein. Feeding one pound (0.45 kg) of soybean meal or an equivalent commercial supplement for each ten pounds (4.5 kg) of milk produced over 50 pounds (22.7 kg) is recommended. Problems with low peak production and ketosis develop if the nutrient needs of the cow are not met.

Some cows lose weight during early lactation. Research has shown that adding one to 1.5 pounds (0.45–0.68 kg) of fat or oil to the diet during this period can increase milk yield, improve fat test, improve reproductive ability, and improve the general health of the cow. Fats from animal sources are better than unsaturated vegetable fats. Commercial products are available to provide added fats and oils to the ration of dairy cows. The cost of these products as compared to the benefits resulting from their use must be carefully considered. Economical sources of oil include raw soybeans, heat-treated soybeans, raw whole cottonseed, and sunflower seeds.

During phase two, feed the cow to maintain peak milk production as long as possible. The cow reaches her maximum feed intake during this phase and should be maintaining or slightly gaining body weight. Maximum grain intake should be 2.3 percent of body weight and minimum forage intake should be 1.5 percent of body weight. Calculate intake amounts on a dry matter basis. The recommended level of feed intake will maintain rumen function and normal milk-fat test. To keep the rumen functioning at optimum level, use feeds

TABLE 41-2 Example Rations for Lactating Cows [1,350-pound (612-kg) cow, 3.8% fat test].

Item	Phase 1		Phase 2		Phase 3	
	lb	**kg**	**lb**	**kg**	**lb**	**kg**
Milk/day	90	40.8	80	36.3	50	22.7
DM intake/day[a]	49	22.2	51	23.1	38	17.2
Ration #1						
Alfalfa hay (88% DM)–140 RFV, 20% CP	28.00	12.70	34.00	15.42	27.00	12.25
Corn	17.85	8.10	20.40	9.25	13.60	6.17
Oats	3.15	1.43	3.60	1.63	2.40	1.09
Soybean meal—44%	5.00	2.27				
Dicalcium phosphate—18% P	0.50	0.23	0.45	0.20	0.30	0.14
Salt, vitamins, trace minerals	0.30	0.14	0.25	0.11	0.25	0.11
Weight change	−1.50	−0.68			+0.50	+0.23
Ration #2 (corn silage limit fed)						
Alfalfa hay (88% DM)–140 RFV, 20% CP	19.00	8.62	34.00	15.42	23.00	10.43
Corn silage (35% DM)	25.00	11.34	25.00	11.34	25.00	11.34
Corn	15.30	6.94	10.20	4.63	8.50	3.86
Oats	2.70	1.22	1.80	0.82	1.50	0.68
Soybean meal—44%	7.50	3.40	0.30	0.14		
Dicalcium phosphate—18% P	0.45	0.20	0.50	0.23	0.30	0.14
Salt, vitamins, trace minerals	0.30	0.14	0.25	0.11	0.25	0.11
Weight change	−1.20	−0.54			+0.50	+0.23
Ration #3 (hay limit fed)[b]						
Alfalfa-grass hay–113 RFV, 16% CP	10.00	4.54	10.00	4.54	10.00	4.54
Corn silage	41.00	18.60	70.00	31.75	57.00	25.85
Corn	13.60	6.17	9.35	4.24	5.10	2.31
Oats	2.40	1.09	1.65	0.75	0.90	0.41
Soybean meal—44%	11.50	5.22	8.20	3.72	4.50	2.04
Dicalcium phosphate—18% P	0.40	0.18	0.30	0.14	0.25	0.11
Limestone	0.40	0.18	0.30	0.14	0.15	0.07
Salt, vitamins, trace minerals	0.30	0.14	0.25	0.11	0.25	0.11
Weight change	−1.40	-0.64	+0.70	+0.32	+0.50	+0.23
Ration #4						
Alfalfa-grass hay–113 RFV, 16% CP	23.00	10.43	32.00	14.51	24.00	10.89
Corn	18.70	8.48	18.70	8.48	16.15	7.33
Oats	3.30	1.50	3.30	1.50	2.85	1.29
Soybean meal—44%	8.50	3.86	3.50	1.59	1.10	0.50
Dicalcium phosphate—18% P	0.45	0.20	0.40	0.18	0.25	0.11
Limestone	0.20	0.09				
Salt, vitamins, trace minerals	0.30	0.14	0.25	0.11	0.25	0.11
Weight change	−1.90	−0.86			+0.50	+0.23

[a]Estimated average intake during the phase.
[b]Feed amounts may have to be limited during phases 2 and 3 to avoid over-conditioning.

Source: Linn, J.G., Hutjens, M.F., Howard, W.T., Kilmer, L.H., and Otterby, D.E., *Feeding the Dairy Herd,* North Central Regional Extension Publication 346, Agricultural Extension Service, University of Minnesota, Revised 1989.

TABLE 41-3 Example Dry Cow Rations [1,400-pound (635-kg) cow].

	Phase 4	
Ration	**lb**	**kg**
Ration #1 (grass forage)		
Orchard grass hay—12% CP	25.00	11.34
Corn	3.00	1.36
Soybean meal	0.50	0.23
Limestone	0.15	0.07
Trace mineral salt and vitamins	0.10	0.05
Ration #2 (limited legume forage[a])		
Alfalfa hay—RFV 140, 20% CP	12.00	5.44
Corn silage	43.00	19.50
Monosodium phosphate	0.10	0.05
Trace mineral salt and vitamins	0.10	0.05
Ration #3 (limited corn silage)		
Alfalfa-grass hay—RFV 113, 16% CP	21.00	9.53
Corn silage	20.00	9.07
Dicalcium phosphate	0.10	0.05
Trace mineral salt and vitamins	0.10	0.05

[a]Ration contains excess energy as formulated and may over-condition cows in some situations.

Source: Linn, J.G., Hutjens, M.F., Howard, W.T., Kilmer, L.H., and Otterby, D.E., *Feeding the Dairy Herd,* North Central Regional Extension Publication 346, Agricultural Extension Service, University of Minnesota, Revised 1989.

that are high in digestible fiber when grain, on a dry matter basis, makes up 55 to 60 percent of the ration. For best results, feed forages and grain several times a day. Problems that may occur during phase two include lower milk production, low fat test, silent heat, and ketosis.

During the last part of the lactation period (140–305 days after calving) milk production is usually dropping. The nutrient needs of the cow are less. To avoid waste, match grain intake to milk production. Cows that are too thin during this stage of lactation should be fed extra grain. It takes less feed to restore desirable body condition during this period than when the cow is dry.

Younger cows (2- and 3-year-olds) need extra amounts of feed nutrients for continued growth during lactation. If their nutrient needs for growth are not met, they often will not reach their full milk-producing potential. Two-year-olds should get 20 percent more and three-year-olds 10 percent more nutrients.

The feeding sequence, number and size of feedings, and nutrient balance in each feeding can affect total milk production, milkfat test, and general herd health. The feeding sequence is especially important when total mixed rations are not used. Feeding four to eight pounds (1.8–3.6 kg) of forage dry matter approximately 40 to 80 minutes before the grain is fed will increase saliva production and slow the rate at which the grain moves through the rumen, thereby improving digestion. Total mixed rations should be fed at least three to four times per day to maintain an acceptable level of feed intake. Increasing the number of feedings of a total mixed ration above four times daily has resulted in increased milk production for some dairy farmers. Limit the amount of grain fed per cow per feeding to a maximum of five to seven pounds (2.3–3.2 kg). Electronic grain feeders that are currently available can provide from six to 13 feedings per day. If forages are fed separately, feed them several times per day so they stay fresh and palatable and to increase feed intake; clean uneaten feed out of the feed bunks to prevent an accumulation of moldy, unpalatable feed. Nutrients should also be balanced at each feeding to increase milk production. For example, if a feeding consists of a low-energy, high-fiber feed such as grass haylage, then add a high-energy feed with a protein source. To function properly, the rumen microbes need a balance of nutrients at each feeding.

FEEDING DRY COWS

The nutrient needs of dry cows are generally not as high as during lactation. Nutrients are needed for the developing calf and to replace losses in body weight that occurred during lactation. Care must be taken not to overfeed the cow. Overfeeding will result in cows getting too fat during the dry period. Limit the amount of corn silage fed because of its high energy content.

Maintain a total dry matter intake during this period at about two percent of body weight. Roughage intake must be at least one percent of body weight. The amount of grain needed depends on the quality and type of forage fed. Feed no more than one percent of body weight daily; one-half percent is usually enough.

If no grain was fed early in the dry period, some should be fed during the last two weeks before calving. This will prepare the rumen for digesting grain during the lactation period. Begin grain feeding with about 4 pounds (1.8 kg) and slowly increase the amount fed daily.

If retained placentas or metabolic disorders such as milk fever, ketosis, hypocalcemia, and displaced abomasum are a problem in the herd, then feeding a transition ration may be necessary. In addition to adding grain to the diet, the protein level should be raised from 12 percent to 14 or 15 percent. Dry matter intake drops considerably about five days before calving. The ration needs to be adjusted to compensate for the reduced feed intake. Some other additives that may help prevent metabolic disorders are niacin (6 grams/day), anionic salts, and yeast (10–113 grams/day). Anionic salts are not palatable and must be mixed in the feed. These additives are expensive and should only be used if problems are noted in the herd and then only during the last seven to ten days before calving.

Do not feed an excessive amount of calcium and phosphorus. Calcium intake of 1.7 to 2.8 ounces (50–80 grams) and phosphorus intake of 1 to 1.4 ounces (30–40 grams) per cow per day is usually enough. The calcium-phosphorus ratio should be about 2:1. Milk fever problems generally increase when the ration contains more than 0.6 percent calcium and 0.4 percent phosphorus on a dry matter basis.

If the feed sources are of poor quality, add vitamins A and D to the ration. This will increase the calf survival rate and reduce problems with retained placenta and milk fever.

Include trace minerals in the ration. Iodine and cobalt are especially important during the dry period. When urea is fed in the milking ration, begin feeding it about two weeks before calving.

FEEDING HERD REPLACEMENTS

Feeding Calves from Birth to Weaning

Proper feeding of the calf from birth to weaning is critical if death losses are to be held to a minimum. Calves are born with little immunity to disease. Colostrum milk is the first milk secreted by the cow after calving. Colostrum milk, as compared to regular milk, is high in fat, solids not fat, total protein, and antibodies that protect against disease. It is critical that the calf receives colostrum milk within a few hours after being born. The ability of the calf to directly absorb antibodies drops sharply after 24 hours. It is best if the calf gets its first feeding of colostrum milk within 30 minutes after being born. The amount of this first feeding should be equal to 4 to 5 percent of the calf's body weight. The calf should receive an amount of colostrum during the first 24 hours equal to 12 to 15 percent of its body weight. Good-quality colostrum milk is thick and creamy. Colostrum milk that is thin and watery should not be fed to calves because it does not contain the nutrients and antibodies they need. Do not feed newborn calves colostrum milk that is bloody or from cows infected with mastitis. The second milking after calving usually contains about 60 to 70 percent of the antibodies found in the first colostrum milk.

Wash the cow's udder and teats before the calf nurses. The calf is less likely to pick up dirt and germs when the udder and teats are clean.

Colostrum milk may be fed from a bucket or a nipple pail. It is more work to feed the calf this way. However, it is easier to keep track of how much the calf drinks. Feeding too much colostrum milk may cause scours (diarrhea) and increase stress on the calf.

Extra colostrum milk may be frozen and used in case there is none available when other calves are born. (The cow may die, have milk fever, not let down its milk, or may not produce colostrum milk.) Freeze the colostrum milk in a container that will not break or split when frozen. Plastic containers may break when frozen;

half-gallon cardboard milk cartons may be used successfully for freezing colostrum milk, and they provide about the right amount for a newborn calf.

The extra colostrum milk may be stored in large plastic containers at room temperature and allowed to ferment, producing fermented or sour colostrum. The fermentation process produces an acid that prevents the milk from spoiling. Stir the fermented colostrum milk each day. Dilute it half and half with warm water for feeding.

Extra colostrum milk may be fed to older calves, but it must be diluted half and half with warm water. Failure to dilute the colostrum milk when feeding it to older calves may result in scours (diarrhea) or calves going off feed.

Colostrum from several different cows can be mixed together for feeding. This makes a wider range of antibodies available to the calves being fed. This practice may be especially useful if calf diseases are a major problem on the dairy farm.

After three days the calf may be fed whole milk or milk replacer. Feed about five pounds (2.3 kg) to small-breed calves. Large-breed calves are fed about seven pounds (3.2 kg). Feed this amount for three weeks on an early weaning program. Feed for four to five weeks on a liberal milk feeding program. Do not force the young calf to drink more than it will take in 3 to 5 minutes. Overfeeding a liquid diet may cause scours (diarrhea). If the calf develops scours, cut back on the amount of milk or milk replacer fed. When the calf stops scouring, gradually increase feeding to the recommended level.

Use milk replacer of high quality. High-quality milk replacers are made from dairy products rather than plant products. Milk replacers that have the following protein sources are preferred:

- skim milk powder
- buttermilk powder
- dried whole whey
- delactosed whey
- casein
- milk albumin

Plant protein sources are not as digestible as milk protein sources. The following protein sources in milk replacers are considered to be of poorer quality:

- meat solubles
- fish protein concentrate
- soy flour
- distillers dried solubles
- brewer's dried yeast
- oat flour
- wheat flour

A milk replacer should contain at least 20 to 22 percent crude protein if all protein sources are from dairy products. If poorer quality (plant) protein sources are used, they should contain at least 22 to 24 percent crude protein.

A crude fat level of 10 to 20 percent is recommended. Fat reduces scours and provides needed energy. Animal fat sources are better than plant fat sources. Homogenized soy lecithin is a good fat source in milk replacers.

Research at the University of Illinois has shown that adding fat to either whole milk or a milk replacer during the first four weeks of feeding results in a faster rate of gain for calves kept in hutches during cold weather. Fat was added to the ration at the rate of 0.15 pounds (0.7 kg) per day and the calves were fed twice daily at the rate of 9 percent of body weight. At the end of four weeks, the calves getting the additional fat had gained 3.5 pounds (1.6 kg) more than those not receiving any additional fat in the diet.

Lactose (milk sugar) and dextrose are good carbohydrate sources in milk replacers. Starch and sucrose (table sugar) are not good carbohydrate sources for use in milk replacers.

Follow the instructions on the feed tag for mixing the milk replacers with water. A bucket or nipple pail may be used to feed milk or milk replacer. Make sure the pail is clean for each feeding. A dirty pail may increase the chances of the calf getting scours.

Calves are usually fed twice a day in two equal feedings. If the calf is weak it may be advisable to feed it three times a day. Divide the milk or milk replacer into three equal amounts. It is possible to feed calves once a day but the practice is not generally recommended. If once-a-day feeding is done, the amount of liquid in the diet needs to be reduced by about 30 percent; however, the amount of dry matter should remain the same. Calves fed once a day may have more problems with scours.

Calves are usually weaned between four and eight weeks of age. With good management and proper feeding, they may be weaned as early as three weeks of age. However, calves weaned this early may show a reduced growth rate for seven to ten days. By twelve weeks of age, they usually weigh as much as calves weaned at an older age. Calves weaned later than eight weeks may become too fat. Calves should be eating at least one pound (0.45 kg) per 100 pounds (45 kg) of body weight per day of a good-quality calf starter at the time they are weaned.

Calf starter should be fed starting at about three days of age, and no later than 10 to 12 days of age. A good-quality, palatable starter should contain 75 to 80 percent total digestible nutrients (TDN) and 16 to 20 percent crude protein (CP). Some examples of calf starter diets are shown in Table 41-4.

Teach the calf to eat starter by rubbing some on its muzzle or putting a small amount in the milk pail after each milk feeding. The calf will soon start to eat the starter.

Grain starters are fed with forage; complete starters include the forage. Using a complete starter provides better control over nutrient intake and is recommended. Feed a complete starter on a free choice basis, adding some forage when the calves reach three months of age. Feed a grain starter on a free choice basis until daily intake reaches

TABLE 41-4 Examples of Calf Starter Rations.

	Grain Starter Rations, Air Dry Basis								Complete Starter Rations, Air Dry Basis							
	Ration number								Ration number							
	1		2		3		4		1		2		3		4	
Ingredient	lb	kg	lb	kg	lb	kg	lb	kg	lb	kg	lb	kg	lb	kg	lb	kg
Corn	40	18.1	30	13.6	45	20.4	32	14.5	24	10.9	28	12.7	15	6.8	15	6.8
Oats	30	13.6	13	5.9	—	—	15	6.8	35	15.9	—	—	—	—	10	4.5
Wheat Bran	—	—	10	4.5	18	8.2	10	4.5	—	—	—	—	—	—	—	—
Ear Corn	—	—	—	—	—	—	—	—	—	—	—	—	—	—	—	—
Gluten Feed	—	—	—	—	10	4.5	10	4.5	—	—	27	12.2	23	10.4	10	4.5
Distillers Grains	—	—	—	—	15	6.8	10	4.5	—	—	—	—	13	5.9	10	4.5
Beet Pulp	—	—	—	—	—	—	—	—	—	—	10	4.5	13	5.9	10	4.5
Linseed Meal	—	—	10	4.5	—	—	10	4.5	—	—	—	—	—	—	—	—
44% CP Supplement	22.6	10.3	10	4.5	14.8	6.7	5.9	2.7	15	6.8	10.4	4.7	10	4.5	12	5.4
Whey, dried	—	—	10	4.5	—	—	—	—	—	—	—	—	—	—	—	—
Alfalfa	—	—	—	—	—	—	—	—	18.9	8.6	17.9	8.1	19	8.6	16	7.3
Molasses	5	2.3	5	2.3	5	2.3	5	2.3	5	2.3	5	2.3	5	2.3	5	2.3
Mineral 23% Ca and 18% P	0.6	0.3	—	—	—	—	—	—	1.1	0.5	0.7	0.3	0.9	0.4	1	0.5
Feed Limestone or CaCO₃	1.5	0.7	1.7	0.8	1.9	0.9	1.8	0.8	0.7	0.3	0.7	0.3	0.8	0.4	0.7	0.3
Trace Mineral Salt	0.25	0.1	0.25	0.1	0.25	0.1	0.25	0.1	0.3	0.1	0.3	0.1	0.3	0.1	0.3	0.1
Composition, DM basis																
Crude Protein, %	19.7		19.6		19.7		19.6		18.4		19.3		19.4		19.4	
TDN, %	80.7		79.5		81.3		80.0		75.6		78		78		77.4	
NE-Maint. Mcal/lb (Mcal/kg) DM	0.9 (1.98)		0.88 (1.94)		0.9 (1.98)		0.89 (1.96)		0.82 (1.81)		0.85 (1.87)		0.86 (1.90)		0.85 (1.87)	
NE-Growth Mcal/lb (Mcal/kg) DM	0.6 (1.32)		0.59 (1.30)		0.61 (1.34)		0.6 (1.32)		0.54 (1.19)		0.57 (1.26)		0.57 (1.26)		0.56 (1.23)	
ADF, %	7.9		8.3		7.9		10.1		14.2		14.6		14.3		16.1	
NDF, %	16.8		20.4		22.3		25		24.3		29.2		28		30.1	
Calcium, %	0.93		0.95		0.93		0.92		0.82		0.83		0.83		0.85	
Phosphorus, %	0.51		0.59		0.57		0.54		0.51		0.53		0.52		0.52	
Trace Mineral Salt, %	0.28		0.28		0.28		0.28		0.34		0.34		0.34		0.34	
Grain, % in Diet DM	—		—		—		—		91.9		93.1		92.2		95.1	
Forage, % in Diet DM	—		—		—		—		21.4		20.3		21.6		18.1	

Source: Crowley, J., Jorgensen, N., Howard, T., Hoffman, P., and Shaver, R., *Raising Dairy Replacements*, North Central Regional Extension Publication 205, Agricultural Extension Service, University of Wisconsin, March 1991.

four to five pounds (1.8–2.3 kg). Start feeding a good-quality forage with the grain starter about one week before the calves are weaned. Do not feed corn silage or pasture to calves younger than three months of age. The high moisture content of these feeds may reduce feed intake and growth. Low-moisture haylage may be fed if it is kept fresh. Both complete and grain starters are fed until the calves are about four months of age.

Commercial calf starters are usually pelleted; home-mixed starters may be used and should be coarsely ground, rolled, or crushed to improve palatability. Nutrients needed in only small amounts are more uniformly mixed in pelleted form. Commercial calf starters usually have vitamins A, D, and E along with minerals and antibiotics added. If the milk replacer contains antibiotics, do not use a starter with antibiotics. Always follow label instructions when using milk replacers or calf starters that contain antibiotics.

Preventing calf scours will reduce death losses. Some of the causes of calf scours and preventative measures are listed.

Causes	*Preventative Measures*
Overcrowding	Provide 24–28 square feet (2.2–2.6 m²) of bedded area per calf.
Poor ventilation	No direct drafts; minimum of 4 air changes per hour in winter; 15 in summer.
Calves getting wet	Dry bedding; good ventilation; don't spray calves with water.
Overfeeding liquid diet	Follow recommendations for feeding milk or milk replacer.
Low resistance to disease	Provide vitamins A, D, and E orally or by injection right after birth.
Not getting colostrum milk	Make sure calf nurses right after birth.
Dirty feeding pails	Clean completely right after feeding. Store upside down to drain.

Baby calves that get scours dehydrate (lose moisture) rapidly. This often results in death. The calf becomes weak and refuses to drink. The calf may be force fed by putting a tube down the esophagus. Tube feeding equipment can be purchased at feed and livestock supply stores.

A commercial solution may be bought for feeding or a home-mixed solution may be used. A home-mixed solution may be made by mixing the following:

1/2 cup (118 cm³) of light corn syrup

3 teaspoons (15 cm³) of baking soda

4 teaspoons (20 cm³) of salt

1 gallon (3.8 liters) of warm water

Do not force the tube when putting it down the esophagus. Be sure it is in the esophagus and not the trachea. The tube can be felt going down the esophagus by placing a hand on the left side of the calf's neck. After the tube is placed, hold the bag containing the solution above the calf's head. This allows the liquid to drain into the calf's stomach.

The calf may be fed two pounds (0.90 kg) of solution four times per day for two days. On the third day, replace one of the feedings with milk or milk replacer. Replace two of the feedings with milk or milk replacer on the fourth day. Continue to replace with milk or milk replacer until the calf is entirely off the special solution.

Feeding the Calf from Weaning to One Year

Replacement heifers must be fed properly if they are to be ready for breeding at the right time. Heifers that have not been fed to grow properly will not produce as well when they are in the milking herd.

A ration made up of forage fed free choice and limited grain may be fed. The amount of protein supplement needed depends on the amount and quality of the forage fed. Pasture may be used for some of the forage.

Feeding 4 to 5 pounds (1.8–2.3 kg) of grain per day is recommended. Do not allow replacement heifers to become too fat. Reduce the grain intake if the heifer is getting too fat. Table 41-5 shows several grower rations that may be used during this feeding period.

Energy intake levels need to be carefully controlled as heifers grow from three months of age to puberty (nine to 11 months of age). Allowing too much energy intake during this time will result in depressed growth hormone levels, decreased mammary tissue formation, increased fat deposits in the immature mammary gland, and lowered milk yield potential when the heifer comes into production. Research has shown that the effects of overfeeding energy during this period are permanent; that is, the heifer will produce less milk not only during her first lactation, but also through all later lactations. Too little energy intake, however, will result in slower growth and delayed breeding. The effects of a limited underfeeding of energy during this period can be overcome by feeding higher levels of energy during pregnancy and the first lactation. Large breed heifers (Holsteins and Brown Swiss) should gain an average of 1.7 pounds (0.77 kg) per day from three months of age until two months before calving. Guernsey and Ayrshire heifers should gain an average of 1.5 pounds (0.68 kg) and Jersey heifers should gain an average of 1.3 pounds (0.59 kg) per day during this period.

Two ionophores (Lasalocid and Monensin) have been approved for feeding to dairy heifers. Feed efficiency and rate of gain are improved by feeding these ionophores to dairy heifers. Rate of gain is improved by 0.1 to 0.2 pounds (0.045–0.09 kg) per day. Do not use ionophores if more than 50 percent of the diet is a high-energy feed such as corn silage. Non-protein nitrogen sources, such as urea, must be limited to no more than 15 percent of the crude protein equivalent

TABLE 41-5 Some Example Rations for Large Breed Heifers (three months to twelve months of age).

Daily Ration Amounts Shown as Pounds and Kilograms Dry Matter

	Large breed heifers 3–6 months of age, 300 lb (136 kg) average weight								Large breed heifers 7–12 months of age, 600 lb (272 kg) average weight							
	1		2		3		4		1		2		3		4	
Ingredient	lb	kg	lb	kg	lb	kg	lb	kg	lb	kg	lb	kg	lb	kg	lb	kg
Alfalfa—Bud	5.8	2.63	—	—	—	—	—	—	8.7	3.95	—	—	—	—	—	—
Alfalfa—Mid Bloom	—	—	4.8	2.18	—	—	—	—	—	—	7	3.18	12.5	5.67	—	—
Alfalfa—Grass	—	—	—	—	4.3	1.95	—	—	—	—	—	—	—	—	6.1	2.77
Grass Hay	—	—	—	—	—	—	3.5	1.59	—	—	—	—	—	—	—	—
Corn Stalks	—	—	—	—	—	—	—	—	4.2	1.91	—	—	—	—	—	—
Corn Silage	—	—	—	—	—	—	—	—	—	—	6	2.72	—	—	6.1	2.77
Corn, shelled[a]	2.5	1.13	3	1.36	3	1.36	3.3	1.50	2	0.91	1.1	0.50	2.5	1.13	1.2	0.54
44% CP Supplement	0.1	0.05	0.6	0.27	1	0.45	1.4	0.64	—	—	0.6	0.27	—	—	1.2	0.54
Mineral 23% Ca and 18% P	0.03	0.01	0.03	0.01	0.02	0.01	0.09	0.04	0.06	0.03	0.04	0.02	0.04	0.02	0.02	0.01
Feed Limestone or CaCO$_3$	—	—	—	—	0.02	0.01	—	—	—	—	—	—	—	—	—	—
Trace Mineral Salt	0.02	0.01	0.02	0.01	0.02	0.01	0.02	0.01	0.04	0.02	0.04	0.02	0.04	0.02	0.04	0.02
Composition, DM basis																
Dry Matter (DM) intake lb/day (kg/day)	8.5 (3.8)		8.5 (3.8)		8.4 (3.8)		8.3 (3.7)		15 (6.8)		14.8 (6.7)		15.1 (6.8)		14.7 (6.7)	
Crude Protein, %	17.3		16.7		16.6		16.4		14.6		14		15.8		14	
TDN, %	71.6		71.8		72		72.3		65.9		66.4		65.7		66.7	
NE-Maint. Mcal/lb (Mcal/kg) DM	0.76 (1.68)		0.76 (1.68)		0.76 (1.68)		0.77 (1.70)		0.68 (1.50)		0.69 (1.52)		0.68 (1.50)		0.69 (1.52)	
NE-Growth Mcal/lb (Mcal/kg) DM	0.48 (1.06)		0.48 (1.06)		0.48 (1.06)		0.49 (1.08)		0.41 (0.90)		0.42 (0.93)		0.41 (0.90)		0.42 (0.93)	
ADF, %	22		22		22		21		29		28		30		28	
NDF, %	30		31		33		35		43		44		40		46	
Calcium, %	0.95		0.8		0.63		0.63		0.93		0.77		1.06		0.54	
Phosphorus, %	0.35		0.37		0.38		0.38		0.3		0.31		0.31		0.31	
Trace Mineral Salt, %	0.25		0.25		0.25		0.25		0.25		0.25		0.25		0.25	
Grain, % in Diet DM	31		43		49		58		14		12		17		17	
Forage, % in Diet DM	69		57		51		42		86		88		83		83	

[a]Oats, barley or high-energy grain by-products can be used to replace all or part of the corn. Corn or other high-energy feeds, protein supplement, minerals and vitamins may be included in a total grain mix, or dry corn or equivalent amount of dry matter in high-moisture corn may be fed separately and the other required ingredients combined in a complete supplement. Feed high-moisture corn daily to prevent spoilage in bunks.

Source: Crowley, J., Jorgensen, N., Howard, T., Hoffman, P., and Shaver, R., *Raising Dairy Replacements*, North Central Regional Extension Publication 205, Agricultural Extension Service, University of Wisconsin, March 1991.

when an ionophore is fed. Lasalocid may be fed at any age; monensin is approved only for heifers over 400 pounds (181 kg). Both may be fed until calving but may not be fed to milking animals. Follow label directions for amounts to feed.

Feeding Heifers 1 to 2 Years of Age

A good-quality forage is the basis for feeding heifers during this period. Some grain may be needed for proper growth. A gain of 1.6 to 1.8 pounds (0.72–0.8 kg) per day is desirable. A good pasture requires no added forage or grain. Heifers on mature or heavily grazed pasture will need additional feed. Be sure the ration meets the mineral and vitamin needs of the heifer. Table 41-6 shows some rations that may be used.

Feeding during the last two months of gestation is especially important because the condition of the heifer at calving will strongly influence the amount of milk produced during the first lactation. The heifer should be growing more rapidly (up to 2 lb [0.9 kg] per day) during the last two or three months of gestation. Begin feeding grain about six weeks before calving at the rate of one percent of body weight. The exact amount of grain needed depends upon the size and condition of the heifer and the quality of the forage being fed. Be sure the heifer has a balanced ration. Limit the amount of salt in the diet during the last two weeks of gestation. Excess salt can cause udder edema. Heifers that are too fat or too thin have more problems at calving time.

BALANCING RATIONS FOR DAIRY CATTLE

The steps used in balancing rations for dairy cattle are the same as for any other kind of farm animal. A general discussion of balancing rations is given in Unit 8. Nutrient requirements and feed composition tables for dairy cattle are found in the Appendix.

Example Ration Balancing Problem

Balance a ration for a group of cows with an average body weight of 1,300 pounds and an average milk production of 50 pounds per day testing 3.0 percent fat.

Step 1: Find daily requirements.

Daily Requirements	Crude Protein (lb)	NE (Mcal)	Ca (lb)	P (lb)
Maintenance	1.06	9.57	0.046	0.037
Milk production	3.85	14.5	0.12	0.08
Total daily requirements	4.91	24.07	0.166	0.117

The requirements for milk production are found by multiplying the average daily production by the requirement per pound of 3.0 percent

TABLE 41-6 Some Example Rations for Large Breed Heifers (13 months to 22 months of age).

Daily Ration Amounts Shown as Pounds and Kilograms Dry Matter

| Ingredient | Large breed heifers 13–19 months of age, 900 lb (408 kg) average weight | | | | | | | | Large breed heifers 19–22 months of age, 1100 lb (499 kg) average weight | | | | | | | |
| | 1 | | 2 | | 3 | | 4 | | 1 | | 2 | | 3 | | 4 | |
	lb	kg	lb	kg	lb	kg	lb	kg	lb	kg	lb	kg	lb	kg	lb	kg
Alfalfa—Bud	19.9	9.03	—	—	—	—	6	2.72	—	—	—	—	—	—	—	—
Alfalfa—Mid Bloom	—	—	11.2	5.08	—	—	—	—	—	—	25.2	11.43	16	7.26	—	—
Alfalfa—Grass	—	—	—	—	11.9	5.40	4.8	2.18	—	—	—	—	—	—	17.5	7.94
Grass	—	—	—	—	—	—	—	—	20	9.07	—	—	—	—	—	—
Grass Hay	—	—	—	—	—	—	—	—	—	—	—	—	—	—	—	—
Corn Stalks	—	—	—	—	—	—	4.9	2.22	—	—	—	—	—	—	—	—
Corn Silage	—	—	8.8	3.99	8	3.63	4.8	2.18	—	—	—	—	8	3.63	7.7	3.49
Grain-Concentrate	—	—	—	—	—	—	0.6	0.27	—	—	—	—	—	—	—	—
Corn, shelled[a]	—	—	—	—	6	2.72	—	—	2	0.91	—	—	—	—	—	—
44% CP Supplement	—	—	—	—	—	—	—	—	1.6	0.73	—	—	—	—	—	—
Mineral 23% Ca and 18% P	0.02	0.01	0.08	0.04	0.04	0.02	0.09	0.04	—	—	0.04	0.02	0.08	0.04	0.04	0.02
Feed Limestone or CaCO$_3$	—	—	—	—	—	—	—	—	—	—	—	—	—	—	—	—
Trace Mineral Salt	0.05	0.02	0.05	0.02	0.05	0.02	0.05	0.02	0.059	0.03	0.063	0.03	0.06	0.03	0.063	0.03
Composition, DM basis																
Dry Matter (DM) intake lb/day (kg/day)	20 (9.1)		20.1 (9.1)		20.6 (9.3)		21.2 (9.6)		23.7 (10.8)		25.3 (11.5)		24.1 (10.9)		25.3 (11.5)	
Crude Protein, %	19.9		13		12.6		12.3		13.5		18.1		14.1		12.9	
TDN, %	65		65		64		63		64		61		63		61	
NE-Maint. Mcal/lb (Mcal/kg) DM	0.67 (1.48)		0.66 (1.46)		0.65 (1.43)		0.63 (1.39)		0.65 (1.43)		0.61 (1.34)		0.64 (1.41)		0.61 (1.34)	
NE-Growth Mcal/lb (Mcal/kg) DM	0.4 (0.88)		0.4 (0.88)		0.38 (0.84)		0.37 (0.82)		0.38 (0.84)		0.35 (0.77)		0.37 (0.82)		0.35 (0.77)	
ADF, %	30		32		33		33		32		35		32		35	
NDF, %	40		48		52		51		57		47		48		54	
Calcium, %	1.27		0.89		0.66		0.8		0.45		1.23		0.97		0.71	
Phosphorus, %	0.3		0.3		0.3		0.3		0.3		0.29		0.3		0.28	
Trace Mineral Salt, %	0.25		0.25		0.25		0.25		0.25		0.25		0.25		0.25	
Grain, % in Diet DM	0		1		3		3		16		0		1		0	
Forage, % in Diet DM	100		99		97		97		84		100		99		100	

[a]Oats, barley or high-energy grain by-products can be used to replace all or part of the corn. Corn or other high-energy feeds, protein supplement, minerals and vitamins may be included in a total grain mix, or dry corn or equivalent amount of dry matter in high-moisture corn may be fed separately and the other required ingredients combined in a complete supplement. Feed high-moisture corn daily to prevent spoilage in bunks.

Source: Crowley, J., Jorgensen, N., Howard, T., Hoffman, P., and Shaver, R., *Raising Dairy Replacements*, North Central Regional Extension Publication 205, Agricultural Extension Service, University of Wisconsin, March 1991.

milk found in the Appendix table of nutrient requirements for dairy cattle. The requirement per pound of 3.0 percent milk for crude protein is 0.077. Therefore, 50 times 0.077 equals 3.85 pounds of crude protein needed. The same type of calculation is carried out for the other requirements.

Step 2: Calculate nutrients provided by forage fed. Daily forage intake on a dry matter basis is 1.5–2.0 percent of body weight. A 1,300-pound cow will eat from 19.5–26 pounds (dry matter basis) of forage per day ($1,300 \times 0.015 = 19.5$ and $1,300 \times 0.02 = 26$).

Forage intake may need to be limited for high-producing cows so they will eat enough concentrate to meet their needs. If forage intake is limited to less than one percent of body weight, milkfat test may be lowered.

This example will use 1.5 percent of body weight for forage intake. The total amount of forage fed is, therefore, 19.5 pounds. Assume one-half the forage is first-cutting alfalfa-bromegrass hay and the other one-half is corn silage. Assume further that the hay is one-half alfalfa and one-half bromegrass.

All calculations of nutrient content in this example are made on a dry matter basis (Table 41-7). The final ration is then converted to an as-fed basis.

Step 3: Nutrients needed in the concentrate mix.

Subtract the nutrients supplied by the forage from the total needed by the animals (Table 41-8).

Step 4: Amount of concentrate mix required.

An estimate of the pounds of concentrate mix needed is found by dividing the Mcal of NE_L needed by the Mcal supplied by the concentrate mix. Some references use therms for this calculation. One therm equals one megacalorie (Mcal). The average value of 0.84 Mcal per pound is used for calculating on a dry matter basis. Use 0.76 Mcal per pound for calculating on an as-fed basis.

TABLE 41-7 Nutrient Content of Feed.

Feed	lb Fed	Crude Protein (lb)	NE_L (Mcal)	Ca (lb)	P (lb)
Hay					
Alfalfa	4.875	0.84	2.8745	0.0609	0.0112
(ref. #1-00-059)					
Bromegrass	4.875	0.51	3.0956	0.0146	0.0170
(ref. #1-00-887)					
Corn Silage	9.75	0.78	7.0317	0.0263	0.0195
(ref. #3-02-823)					
Total nutrients from forage	19.5	2.13	13.0018	0.1018	0.0477

TABLE 41-8 Calculating Nutrients for Concentrate Mix.

	Crude Protein (lb)	NE_L (Mcal)	Ca (lb)	P (lb)
Total nutrients needed	4.91	24.0700	0.1660	0.1170
From forage	2.13	13.0018	0.1018	0.0477
To be supplied by concentrate mix	2.78	11.0682	0.0642	0.0693

The calculation for this example is 11.0682 divided by 0.84 equals 13.1764 pounds of concentrate mix needed to balance the ration.

Step 5: Percent of protein needed.

The percent of protein needed in the concentrate mix is found by dividing the amount of protein needed by the pounds of concentrate mix needed and multiplying by 100. Therefore, 2.78 divided by 13.1764 times 100 equals 21.09 percent.

Step 6: Use of the Pearson square method to mix grain and supplement.

A simple way to determine how much protein supplement to mix with the grain is to use the Pearson square. The use of the Pearson square is explained in Unit 8.

Assume that ground ear corn (ref. #4-02-849) is to be used with soybean oil meal (ref. #5-04-604) for the concentrate mix. Usually rations are mixed with more than two ingredients. This example uses only two ingredients to simplify the illustration. Methods of calculating a ration using more than one grain or protein supplement are given in Unit 8.

The Pearson square is set up as follows:

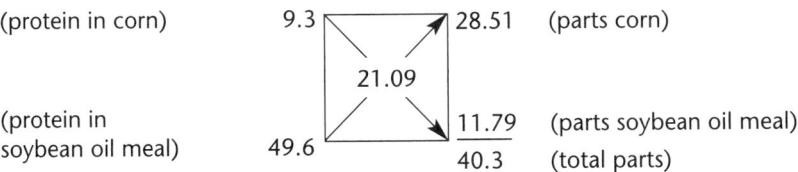

The percent of ground ear corn is 28.51 divided by 40.3 times 100 equals 70.74.

The percent of soybean oil meal is 11.79 divided by 40.3 times 100 equals 29.26.

To convert the percents on a 100 percent dry matter basis to an as fed basis follow these steps:

1. Divide the percent of each feed on a dry matter basis by the percent of dry matter in that feed.
2. Add the results to get a total.

TABLE 41-9 Calculating Percent of Feed for Concentrate Mix.

	% on Dry Matter Basis		% DM			% as Fed
ground ear corn	70.74	÷	87 = 81.3	÷ 114.2 × 100	=	71.2
soybean oil meal	29.26	÷	$89 = \dfrac{32.9}{114.2}$	÷ 114.2 × 100	=	$\dfrac{28.8}{100}$

3. Divide each result in step 1 by the total in step 2 and multiply by 100. The result is the percent of each feed to include when preparing the concentrate mix.

The calculations for this example are shown in Table 41-9. A ton of this mix on an as fed basis contains 1,424 pounds (2,000 × 0.712) ground ear corn and 576 pounds (2,000 × 0.288) soybean oil meal.

The total dry matter intake of cows weighing 1,300 pounds and producing 50 pounds of milk per day is about 2.7 percent of body weight. This is about 35 pounds (1,300 × 0.027) for the cows in this example. The forage dry matter intake is 19.5 pounds. Therefore, the cows will eat 15.5 pounds of the concentrate mix per day.

Step 7: Compare nutrients provided to requirements.

The nutrients provided by this ration are compared to the requirements, as shown in Table 41-10.

The ration provides a little more protein and energy than required. Feeding a little less of the concentrate mix per day will correct this. It is slightly deficient in calcium and phosphorus. Adding a mineral supplement to the mix will provide the needed amounts. It is recommended that 2,000 international units (IU) of vitamin A per pound of mix be added to the ration. This will insure that the cow gets the needed amount of vitamin A.

TABLE 41-10 Nutrients Provided as Compared to Requirements.

	lb Fed (dry matter) (lb)	Crude Protein (lb)	NE_L (Mcal)	Ca (lb)	P (lb)
Forage	19.5	2.13	13.00	0.1018	0.0477
Ground ear corn (15.5 × 0.7074)	10.96	1.02	9.15	0.0055	0.0285
Soybean oil meal (15.5 × 0.2926)	4.54	2.25	3.83	0.0163	0.0340
Nutrients from ration	35.00	5.5	25.98	0.1236	0.1102
Requirements	35.00	4.9	24.07	0.166	0.117
Surplus (deficit)		0.5	1.91	(0.0424)	(0.0492)

Step 8: Determine amount needed on an as fed basis.

The amount of the ration on an as fed basis is found by dividing the amount of feed on a dry matter basis by the percent of dry matter in each feed. The calculations are as follows:

Alfalfa-bromegrass hay (ave. dry matter 89%)
$$9.75 \div 0.89 = 10.96 \text{ pounds}$$

Corn silage
$$9.75 \div 0.35 = 27.86 \text{ pounds}$$

Concentrate mix (ave. dry matter 87.6%)
$$15.5 \div 0.876 = 17.7 \text{ pounds}$$

FEEDING AND REPRODUCTION

Underfeeding or overfeeding energy feeds to developing heifers can lead to reproduction problems. Underfeeding delays the time the heifer reaches first heat. This causes a delay in breeding and shortens the productive life of the cow.

Overfeeding causes first heat to be reached earlier. However, there is a greater chance of the cow having breeding problems later in its life. Overfed cows become sterile at a higher rate than cows fed proper rations. This also shortens the productive life of the cow. Bulls that are too fat have problems with sperm production.

A protein shortage in the ration may cause silent heats or discontinued heats. Protein shortage is more common when corn silage is the main forage fed. Care must be taken to properly balance the ration for protein needs when most of the forage fed is corn silage.

Urea feeding is sometimes blamed for reproductive problems. Studies done at several universities have shown that feeding urea has no apparent effect on calving interval or sterility.

Vitamin A shortage can also cause breeding problems. The addition of vitamin A to the ration will insure that the cow's needs are met.

FEED INVENTORY

A feed inventory is useful when planning current and future feeding programs. A list of the kinds and amounts of feed available and feed needs for the dairy herd is made.

The following management decisions can be made from the feed inventory:

■ Sell extra feed.
■ Buy additional feed needed.
■ Put more feed (forage or corn) in the silo.
■ Make adjustments in the feeding program to fit feeds available.
■ Cull more cows if a feed shortage is apparent.

■ Buy more cows if additional feed is available.

■ Plan future cropping program.

Worksheets and information on capacities of feed storage structures are available from the Cooperative Extension Service in many states.

ON-FARM FEED PROCESSING

Feed processing may be done by commercial mills or on the farm. The amount of feed used per year is a major factor in deciding which alternative to use.

On-farm feed processing requires a large investment in equipment. Additional time is also needed. Care must be taken to be sure trace ingredients are completely mixed in the ration.

The current cost per ton of commercial mixing must be compared to the cost per ton for on-farm feed processing. On-farm feed processing begins to be cheaper than commercial feed processing at about 200 to 250 tons (181–227 tonne) of feed processed per year.

SUMMARY

Feed costs are about 40 to 50 percent of the total cost of producing milk. Feeding balanced rations permits the cow to produce closer to her maximum genetic potential. Dairy rations should be made up of good-quality roughages supplemented with grain and protein supplement.

Dairy cattle have traditionally been fed on an individual basis based on individual milk production. There is a trend toward grouping dairy cows for feeding total mixed rations.

Hay, pasture, green chop, silage, and haylage are generally used for roughage in dairy rations. Alfalfa hay and corn silage are two of the most common roughages used.

Corn is the most commonly used grain for dairy rations. Oats are also an excellent dairy feed. Barley, wheat, and rye are less commonly used.

Soybean oil meal, linseed meal, and cottonseed are often used as protein supplements. Urea may be used as a nonprotein nitrogen source in dairy rations.

A variety of by-product and other processed feeds may be used for dairy cattle feeding. Alfalfa meal, beet pulp, and brewer's grain are examples of by-product and other processed feeds used in dairy rations.

Calcium and phosphorus are two of the most important minerals needed when balancing dairy rations. Salt and trace minerals are also needed.

Vitamins A, D, and E often need to be added to dairy rations. A commercial vitamin pre-mix added to the concentrate will supply needed vitamins.

Lactating dairy cows need more water for their size than any other farm animal. A good supply of fresh water is vital for maximum production.

Lactating cows need to be fed according to the stage of milk production they are in. The first ten weeks after calving is the most critical time for meeting nutritional needs of dairy cows. The most limiting factor in milk production is usually a shortage of energy in the ration.

Newborn calves must get colostrum milk within a short period of time after birth if death losses are to be kept low. Colostrum is the first milk produced by the fresh cow. It contains antibodies that help protect the calf from disease.

Calves should be started on solid feed as early as they will eat it. Proper feeding of replacement heifers to get good growth is essential for high milk production. Poor feeding during the growth period often causes reproductive problems.

Student Learning Activities

1. Prepare a classroom display of roughages and concentrates used locally in dairy rations.
2. Send samples of forages used locally for feeding dairy cattle to a laboratory for analysis.
3. Observe and compare methods used locally for harvesting and storing forages for dairy cattle. Give an oral report to the class.
4. Prepare a classroom display of feed tags from commercial dairy feeds.
5. Visit local dairy farms that use traditional and total mixed ration methods of feeding. Give an oral report to the class.
6. Give an oral report to the class on the feeding of calves from birth to weaning.
7. Calculate a balanced ration for the milking herd on the home farm or for another local dairy farm.

Review

1. Why is it important to feed dairy cows properly?
2. Describe the importance of roughage in dairy rations.
3. List the advantages and disadvantages of traditional dairy feeding methods.
4. List the advantages and disadvantages of total mixed rations for dairy cattle.

5. Why is it especially important to analyze the roughages used in dairy rations?

6. Describe how cows may be grouped for feeding total mixed rations.

7. What problems may occur when grouping cows for feeding total mixed rations?

8. Name three types of automatic concentrate feeders used for dairy cattle and describe their purpose.

9. Describe the use of hay as a roughage for dairy cattle.

10. Briefly explain the practice of using green chop as a roughage for dairy cattle.

11. Name and compare the different kinds of silage often used in dairy rations.

12. Briefly explain the use of pastures for dairy cattle.

13. Name and compare grains commonly used in dairy rations.

14. Name and describe the common protein supplements often used in dairy rations.

15. Briefly explain the use of urea as a nonprotein nitrogen source for dairy cattle.

16. Name and compare by-product and other processed feeds used in dairy rations.

17. Briefly explain the importance of minerals for dairy cattle.

18. List the sources of minerals for dairy cattle.

19. Briefly explain the use of vitamins for dairy cattle.

20. Why is water important for dairy cattle?

21. Briefly describe the five body condition scores used to evaluate dairy cattle.

22. Why is the body condition score of dairy cows and heifers an important concept for dairy farmers to understand and use?

23. When during the lactation cycle should cows be checked for body condition score and what are the appropriate scores for each period?

24. What general guidelines need to be followed when feeding lactating dairy cows?

25. Briefly explain feed requirements for the various stages of lactation for the dairy cow.

26. What are the recommended practices for feeding dry cows?

27. Why is colostrum milk important for newborn calves?

28. Briefly explain feeding the calf from birth to weaning.

29. Describe a good-quality milk replacer.

30. List recommended practices to prevent calf scours.

31. Balance a ration (using local feeds) for a dairy cow weighing 1,400 pounds and producing 60 pounds of 3.5 percent milk daily.

32. Balance a ration (using local feeds) for a mature dry cow weighing 600 kilograms.

33. Balance a ration (using local feeds) for a small breed 300-kilogram growing dairy heifer that is 83 weeks of age and gaining 600 grams per day.

34. Briefly explain the relationship between feeding and reproductive problems in dairy cattle.

35. What management decisions may be made using a feed inventory?

36. Briefly explain on-farm feed processing.

Management of the Dairy Herd

Objectives

After studying this unit, the student should be able to

- describe the use of records in managing the dairy herd.
- cull cows from the dairy herd.
- set goals for the dairy farm.
- manage dry cows.
- raise dairy replacements.
- list and describe other approved dairy management practices.

The data presented in Unit 39 shows a trend toward larger dairy herds on fewer farms. A larger dairy herd does not always mean more net income from dairying. The dairy farmer must follow good management practices to be successful. Selection, feeding, facilities, milking practices, herd health, and marketing are discussed in other units in this section. This unit describes other management practices that are necessary for successful dairy farming.

RECORDS

Good records are needed to serve as the basis for sound management decisions. The dairy farmer must have records to analyze the business. An analysis of the business is needed to make improvements that will increase net income.

The dairy farmer should keep the following kinds of records:

- Production records on individual cows and on the herd
- Feed use records
- Breeding and calving records
- Health records

- Cow identification records
- Financial records (entire farm and enterprise)
- Inventory

DAIRY HERD IMPROVEMENT ASSOCIATIONS (DHIA)

Dairy herd testing is available through the Cooperative Extension Service working with local Dairy Herd Improvement Associations. A DHIA is formed by a group of dairy farmers in the local area to provide production testing services. The DHIA hires a supervisor. Some testing plans require the supervisor to travel to each dairy farm in the association. This person takes and weighs milk samples from each cow in the herd. The milk is tested for fat content and a written report is provided to the dairy farmer.

Other testing plans require the dairy farmer to weigh and take the milk samples. Herds of all types and sizes, both registered and grade, are included in DHIA programs.

The use of DHIA testing has increased over the years. Net income usually increases on farms that use DHIA testing programs compared to those that do not.

There are several kinds of testing programs in DHIA. These include:

- Dairy Herd Improvement (DHI)
- Dairy Herd Improvement Registry (DHIR)
- AM-PM Testing programs
- Owner-Sampler (O-S)
- Weigh-A-Day-A-Month (WADAM)
- Basic Production and Management (BPM)

The type of testing program the dairy farmer decides to use depends on time required, cost, and use of information. These factors vary with the type of testing program. The lowest cost plans are O-S, WADAM, and BPM. The BPM plan requires the least amount of time. All plans provide information for herd management. However, DHI and DHIR are official tests that provide information that may be used in the sale of purebred or surplus cattle.

Dairy Herd Improvement (DHI)

The DHI testing program requires the DHIA supervisor to visit the farm once a month. The supervisor weighs and samples the milk from each cow. Both evening and morning milkings are included. The supervisor also records information on feeding, breeding, calving, and management of the herd.

The records are official because they are made from information collected by the DHIA supervisor. Costs vary from one area to another. DHI records are recognized by the USDA and breed associations.

DHI records may be used to

- identify low-producing cows for culling.
- improve feeding programs.
- identify problem breeders.
- identify cows with chronic mastitis problems.
- develop yearly progress records.
- provide records for the sale of breeding stock.
- establish values if cattle are lost in a disaster.
- give recognition to outstanding dairy operations.
- provide information for research.
- provide information for pedigrees and sale publicity in purebred herds.
- provide information for sire and cow evaluation.

Dairy Herd Improvement Registry (DHIR)

The DHIR testing program is conducted the same way as DHI with some added requirements. These involve enrollment in the program, cow identification, and surprise tests. The dairy farmer must make arrangements with the appropriate breed association to start or stop a DHIR testing program.

Surprise tests are unannounced visits by the surprise test supervisor to test the herd. Surprise tests are done when requested by the breed association or when production levels exceed a preset standard. These standards vary from breed to breed (Table 42-1).

The surprise test is used under the following conditions:

1. Cow has milked more than 90 days and less than 180 days after calving and the cow equals or exceeds either milk or fat production shown in Section A, Table 42-1, on a projected 305-day, twice daily milking, mature equivalent basis. One surprise test is made before the 305th day of lactation.

TABLE 42-1 Production requirements for surprise testing, DHIR program.

	Section A				Section B			
	Milk		Fat		Milk		Fat	
Breed	(lb)	(kg)	(lb)	(kg)	(lb)	(kg)	(lb)	(kg)
Ayrshire	22,500	10,206	800	363	25,000	11,340	875	397
Brown Swiss	23,500	10,659	875	397	26,000	11,340	950	431
Guernsey	20,000	9,072	850	386	22,000	9,979	950	431
Holstein	25,000	11,340	900	408	27,500	12,474	1,000	454
Jersey	17,000	7,711	850	386	19,000	8,618	950	431
Milking Shorthorn	18,000	8,165	700	318	19,000	8,618	800	363

2. Cow has milked more than 180 days after calving and the cow equals or exceeds either milk or fat production shown in Section B, Table 42-1, on a projected 305-day, twice daily milking, mature equivalent basis. Two surprise tests are made before the 305th day of lactation.

DHIR testing costs a little more than DHI testing. The records are recognized by the USDA and the breed associations. More detailed information on specific breed requirements for DHIR may be secured from dairy cattle breed associations.

AM-PM Testing

The AM-PM testing program is conducted by the DHIA supervisor. Only one milking is weighed and sampled on the test day. Alternate morning and evening milkings are weighed and sampled in consecutive test periods. The test is official if the entire herd is milked twice daily and the milking times are recorded. The DHIA supervisor records the beginning and ending times of milking on the test day as well as the two previous milkings.

This testing program is very accurate as compared to DHI testing. The cost is less because the supervisor does not have to return for the second milking on the test day.

Owner-Sampler (O-S)

The owner-sampler testing program requires the dairy farmer to weigh and sample the milk from each cow. This is done once a month. The equipment and forms are supplied by the DHIA.

This testing program provides the same information as DHI, DHIR, and AM-PM testing programs. The results are not official. They are not used in USDA or breed association reports. Owner-sampler testing provides information needed for on-farm herd management. The cost is less than DHI, DHIR, and AM-PM testing.

Weigh-A-Day-A-Month (WADAM)

The Weigh-A-Day-A-Month program involves the dairy farmer taking weights on milk given by each cow. This is done once a month at two consecutive milkings. No milk sample for testing is taken. The breed average or the milk plant fat test is used for the herd.

The cost of WADAM testing is low. Less individual information on each cow is produced by this testing program. The results are not official for USDA or breed association purposes.

Basic Production and Management (BPM)

The Basic Production and Management (BPM) testing program requires the dairy farmer to gather data and report it. Weighing and (if desired) samples are taken at two consecutive milkings once a month. The amount of information from this testing program depends on what the dairy farmer wants. It may be only production with no fat test or samples may be taken and fat tests made. Breeding and feeding records may be included if desired.

No individual body weights, ear tag, or feed information is required for this program. The cost depends on the amount of information the dairy farmer wants. The records are not official.

IDENTIFICATION OF DAIRY ANIMALS

Keeping individual records on dairy cows requires a system of permanent identification. Calves need to be permanently identified shortly after birth. Methods of identification include ear tag, ear badge, neck chain, tattoo, freeze brand, photograph, and ink sketch.

FIGURE 42-1 This cow has both a neck chain and ear tags for identification. *Courtesy of L. DeVere Burton.*

It is recommended that ear tags and neck chains be used together. Neck chains are easier to read in elevated milking parlors (Figure 42-1). It is easier to read ear tags when the cows are eating. Tattoos provide a permanent mark if the ear tag and/or neck chain is lost. The tattoo is placed in the ear of cattle.

A tag inventory record book is recommended. This written record includes the permanent tattoo number, vaccination tag, ear tag, and neck chain number for each cow. Large herd owners may find it useful to keep a location book. This will show by number the groups and pens to which each cow is assigned. This book saves time when it is necessary to locate a given cow.

COMPUTERS FOR COW IDENTIFICATION AND MANAGEMENT

Computer systems that tie together electronic identification of individual cows with various routine management tasks have been developed and are continuing to be improved. Several methods of electronically identifying individual animals include ear tags, neck tags, leg straps, and implanted devices. Ear tags and neck tags are currently the most commonly used methods; most of the devices currently available are too large to be used as implants. Implanting also reduces the range of the device.

Electronic identification units are typically powered by radio signals from a stationary transmitter located at the point where the information is used. This may be a feed dispensing station, in the milking parlor, or some other location, depending upon the management task to be done. The devices may also be battery-powered. The identification unit transmits a signal that identifies the specific animal to which it is attached. The range of the transmitted signal varies from as little as two inches (5 cm) to as much as three feet (0.9 m), with a typical range being six inches (15.2 cm).

The most common use of these systems is for computer-controlled feed dispensing stations. The systems permit control over the feeding of concentrates on an individual basis to cows kept in groups. Upon identifying the individual cow in the feeding station, the computer determines the proper amount of concentrate to be fed, based on the cows' production record that is kept in the computer data base. Some programs even determine the appropriate percent of protein to be included in the concentrate for the individual cow. The computer program keeps track of the amount of feed each cow has received during the day and dispenses the proper amount of feed to avoid feeding too much concentrate at one time. Computer programs currently available typically have the capability to dispense the feed from six to 20 times per day. More complex programs are available that will take into account the amount and quality of forage the animal is receiving and adjust the concentrate mix accordingly, based upon the cow's weight, milk production, and milkfat test.

Research is being conducted on the utilization of electronic identification devices and computer programs for other management tasks. These include estrus detection, mastitis detection, milk production, milkfat test, maximum and average milk flow, milking time, and internal biological changes in the cow's body. Some of these capabilities are already becoming available in commercial systems.

When individual cows are identified in the milking parlor, some systems are capable of accessing a data base that provides the operator with information regarding health needs, time to observe for heat detection, cows that are off feed, and dry cows.

The increased use of electronic identification with computer programs can be expected to improve the management capabilities of dairy farmers. Records collected on the farm can be transmitted electronically to mainframe computers for DHIA, breed association, artificial insemination association, veterinary clinic, or university analysis.

A program developed at Purdue University allows dairy farmers to access data by using a modem attached to their computer. The program is called Direct Access to Records by Telephone (DART). The program replaces the old DHIA method of mailing herd data. DHIA cooperators can download the data for their herd into their own computer and produce printout reports that can be used for herd management. Test supervisors, using a portable computer, can print out reports before leaving the farm. Information can be electronically transmitted to a veterinarian, feed consultant, or breed association from the computer on the farm. The use of computers to maintain and manage herd records is expected to increase over the next several years.

Several dairy herd management software programs are currently available.

STANDARDIZING LACTATION RECORDS

In order to compare production records of dairy cows it is necessary to standardize records to the same basis. The following things are usually considered when standardizing production records:

- Length of lactation
- Number of times milked per day
- Age at calving
- Month of year when calving

The standard length of lactation used for comparing production records is 305 days. Tables of factors for each dairy breed have been developed for adjusting production to the 305-day standard. Factors for both milk and percent milkfat are used.

The standard for number of times milked per day is twice daily (2×). Some dairy farmers milk their cows three times per day (3×). A table of factors is used to convert 3× records to 2×.

A first calf heifer will not produce as much milk as she will when mature. A table of factors has been developed to predict mature equivalent (ME) production for the cow. Age at calving adjustment factors have been developed for different regions of the United States. These are based on DHIA records.

The month of the year during which the cow calves also affects production. Cows calving in the summer months generally produce less milk as compared to cows calving at other times of the year. The adjustment factors for age at calving and month of calving are combined on a regional basis into one table. The factors are different for each breed of dairy cows.

All of the tables of factors have been developed by the USDA from DHIA records. These tables are available from DHIA, the Cooperative Extension Service, and breed associations.

A production record that is reported as 305-2×-ME means that the record has been adjusted to 305 days lactation, twice daily milking, and is a mature equivalent record. If a 305-2×-ME record is used in promoting the sale of breeding stock, the cow's actual production should also be given.

Another adjustment in production records is often made to account for variations in early and late test periods. A cow usually increases milk yield and decreases fat yield during the first two test periods of the lactation. Milk yield usually decreases and fat yield increases during the last test period of the lactation. Tables of factors are used to adjust for these normal changes between tests. Making this adjustment results in a more accurate measure of production.

TABLE 42-2 Why Cows Are Removed from the Milking Herd.

Reason	%
Low production	32.5
Reproduction problems	26.6
Mastitis	10.4
Disease	7.7
Teat or udder injury	7.2
Poor udder conformation	5.0
Accidents and injury	4.0
Poor feet and legs	2.0
Other poor conformation	1.2
Hard to milk or leaks milk	1.9
Poor disposition	0.8
Other miscellaneous reasons	0.7

CULLING

Culling is permanently removing cows from the herd. The most common reason for culling is low milk production. Cows are also culled because of reproduction problems, diseases, udder problems, or other miscellaneous reasons (Table 42-2).

Cows that are poor producers are generally not profitable. Good production records serve as a basis for culling for low production.

The estimated relative producing ability (ERPA) is a prediction of 305-2x-ME production compared to other cows in the herd. This information is available for herds on DHI testing.

The ERPA may be used to help cull low-producing cows from the herd. Cows with low ERPAs will probably be relatively poor producers in future lactations.

Suggestions for deciding which cows to cull include

- those with the lowest ERPA.
- first calf heifers producing 30 percent or more below herd average.
- other cows producing 20 percent or more below herd average.
- those with lowest USDA cow index.
- sell all calves from cows that are low producers (bottom 15 to 20 percent of the herd).
- those that are still not bred 150 days after calving.
- those that have repeated health problems.
- those that show chronic mastitis problems.
- those with poor udders and/or feet and leg problems.
- nervous and/or hard to handle cows.

The dairy farmer must decide when, during the lactation period, a given cow is to be culled. An analysis of production and financial records will show an economic breakeven point for the herd. Both variable and fixed costs are included in finding the breakeven point.

The breakeven point is that production level at which the costs of producing one hundred pounds of milk equal the price received for that milk. A guide that may be used is to cull the cow when her production drops to the breakeven point.

The cost, availability, and estimated productivity of replacement cows must be considered when deciding how many cows to cull. The number of cows that can be handled by the facilities on the farm must also be considered. If the facilities can handle more cows than are available as replacements, then the breakeven point should be calculated on only variable costs. The total fixed costs remain the same regardless of how many cows are in the herd.

GOALS FOR THE DAIRY FARM

A good set of records will tell the dairy farmer what has happened in the business. When planning management changes it is wise to have goals to work toward. DHIA records provide information that may be used when setting herd goals. Goals may be set for production, feeding, reproduction, management, and culling (Table 42-3).

BUDGETS FOR THE DAIRY FARM

Because of the high cost of operating a dairy farm, wise financial planning is essential. A budget is an estimate of income and expenses for a period of time, typically one year. The budget may be compared to current typical figures for dairy farms in the area. Summaries of income and cost figures are available from the Cooperative Extension Service.

Expenses that are much higher or lower than typical figures indicate areas where the dairy farmer needs to carefully examine the business. These wide variations may indicate management problems that need to be corrected.

A budget is usually required by a lender when the dairy farmer is borrowing money for operating expenses or capital investment.

A typical dairy budget may include the following items:

Receipts
milk sales

cull cow sales

calf sales

Expenses—Cash variable
feed purchased

veterinary and medicine

breeding fees

supplies (cleansers, sanitizers, paper towels, inflations, etc.)

DHIA

milk hauling and marketing

bedding purchased

utilities and fuel (dairy share)

hired labor (dairy share)

machinery and equipment repair

building repair

taxes (dairy share)

interest on borrowed capital

Expenses—Cash fixed
property taxes (dairy share)

insurance (dairy share)

miscellaneous (dues, magazines, travel, accounting, legal fees, etc.)

TABLE 42-3 Goals for the Dairy Herd.

Trait	Ideal	High Production (16,000 lb)	Average Production (13,000 lb)	Your Herd Now	Where Do You Want To Be?
General—Yearly					
1. Milk, lb	Increase 1,000 lb			___	___
2. Fat, lb	Increase 40 lb			___	___
3. Value of product	Increase $90			___	___
4. Feed cost/cwt milk	Less than 50% of price received/100 lb milk			___	___
FEEDING					
5. Are all feeds reported accurately?		(See back of barnsheet)		___	___
6. Forage DM/cwt B.W. (Body weight)				___	___
a. Adequate forage supply		1.5–2.0	2.0–2.5	___	___
b. Limited forage supply		1.5	1.5–2.0	___	___
7. Energy Index		95–105	100	___	___
8. Protein Index		110	100	___	___
9. Milk/lb grain D.M. (Dry matter)		2.0–3.0	2.5–4.0	___	___
10. If "herd mix" is used, is the protein correct?		(See printed statement)		___	___
11. Lb grain—dry cows		0–6	0–3	___	___
12. Lb protein—milking cows		(Feed as indicated)		___	___
a. Top-dressing herds		2 lb or less	2 lb or less	___	___
b. Herd-mix herds				___	___
REPRODUCTION					
13. Calving interval		12–13 months		___	___
a. Pregnant		11.3–12.5 months		___	___
b. Possibly pregnant				___	___
14. Calving to 1st breeding (Both pregnant and possibly pregnant)		Less than 75 days		___	___
15. Calving to last bred				___	___
a. Pregnant		Less than 110 days		___	___
b. Possibly pregnant		Less than 85 days		___	___
16. No. cows open 120 days or more				___	___
a. Pregnant		Less than 10%		___	___
b. Possibly pregnant		Less than 10%		___	___
c. Open cows		Less than 10%		___	___

17. No. cows bred-pregnant
    a. 1 time   More than 60%   _____
    b. 2–3 times   Less than 30%   _____
    c. 4+ times   Less than 10%   _____
18. Breeding interval
    a. < 18 days   Less than 5%   _____
    b. 18–24 days   90–100%   _____
    c. > 24 days   Less than 5%   _____

MANAGEMENT

19. Production index—1st lactation   Above 100   _____
20. Age at calving—1st lactation   Less than 2 yr., 2 mo.   _____
21. Peak milk—Holsteins
    a. 1st lactation cows   More than 55 lb   _____
    b. Other cows   More than 70 lb   _____
    c. 1st lactation cows/other cows =   80%   _____
22. Mastitis
    a. Negative   More than 75%   _____
    b. Mild   15–20%   _____
    c. Positive   Less than 10%   _____
23. Length of dry period
    a. < 40 days   Less than 10%   _____
    b. 40–70 days   90–100%   _____
    c. > 70 days   Less than 10%   _____
24. Relationship of "daily milk" and "days in milk"
    a. When DIM is lower than previous month   More milk   _____
    b. When DIM is higher than previous month   Less milk   _____
    c. Persistency measure   Less than 10% drop per month after 50 days   _____
           (Expected value = _____)

CULLING DECISIONS

25. Percent of herd culled   Over 30%   _____
26. Percent of 1st lactation cows culled   Less than 20%   _____
27. Percent of herd culled with a "production index" of under 100   More than 75%   _____

Source: Appleman, R.D., Conlen, B.J., Hutjens, M.F., Mudge, J.W., and Steuernagel, G.R., *Dairy Herd Planning Guide, Dairy Husbandry Fact Sheet No. 16,* Agricultural Extension Service, University of Minnesota, 1977.

Expenses—Non-cash variable

home-grown feeds (market value)

Expenses—Non-cash fixed

depreciation—dairy share (machinery, equipment, buildings, purchased animals)

interest on capital investment—dairy share (equity in buildings, equipment, land, cows, milk base)

The dairy farmer who is faced with making decisions about additional investment in the business must calculate the debt repayment capacity of the business. Information from the dairy budget is useful in making this calculation.

Debt repayment ability is calculated as follows:

Total business cash receipts

   Minus: Business cash operating expenses

   Minus: Family living expenses

   Minus: Interest and principal payments on existing debt

  Balance remaining: Cash available for new investment

A lender will consider the equity the farmer has in the business when determining the risk involved in loaning money for capital investment. Equity is the percent of the business the farmer actually owns. An equity of less than 50 percent is considered very risky. Most lenders prefer that the borrower have at least 60 percent equity. The management ability of the farmer is also considered by the lender when making the decision to lend money.

The dairy farmer with limited capital to invest must often decide which of several alternate investments will be made. The decision must be based on the expected benefits, timing of benefits, and safety of each investment.

The expected benefits are often measured in dollars. The net return from each alternate investment is found by estimating expenses and income produced. Noncash benefits, such as labor saved, are harder to calculate. These are evaluated on an individual preference basis.

It is usually preferable to have the benefits produced early rather than late. Total benefits are usually higher if some net gain is produced shortly after the investment rather than much later.

Evaluating risk is often hard to do. The more difficult it is to calculate the cost of the investment or the benefits received, the higher the risk.

It is important to maintain enough cash flow (available money) to pay operating costs. If the additional investment reduces cash flow to too low a level it may not be a wise investment. Lenders will generally want to evaluate the cash flow situation when deciding about making a loan for capital investment purposes.

Computer programs are now available that will do budget problems and provide information for wise decision making on alternate investments in the farm business. Information on these computer programs is generally available through the Cooperative Extension Service.

BUY OR LEASE DAIRY COWS

An alternative to investing additional capital to buy dairy cows is leasing the cows. It is necessary to carefully prepare budgets on these alternatives before making such a decision.

Things to be considered include available capital, current price of cows, terms of the leasing arrangement, costs of production, and individual goals.

Each situation is different. No general recommendation to buy or to lease can be made.

The dairy farmer who considers leasing dairy cows needs to carefully read all the terms of the lease. Make sure there is no misunderstanding about any of the terms. Deal only with reputable cow leasing firms. Ask others who have leased cows about their experience.

ADJUSTING TO CHANGES IN THE ECONOMY

Production costs and prices received change as the economy moves up or down. The long-range goals of the dairy farmer cannot be changed each time the economy changes. The dairy farmer who is following good management practices will not need to make major changes.

Some minor adjustments may be made to keep costs as low as possible and still maintain high production. For example, the ration may be changed to take advantage of a comparable feed with a lower per nutrient cost as feed prices vary. The use of a computer to calculate least cost rations is helpful.

Careful management of the breeding program is always needed regardless of the economy. Timely breeding and high conception rates are always goals of the efficient manager. It is false economy to use semen from inferior bulls to temporarily lower breeding costs. The dairy farmer is breeding for the future of the herd. Temporary changes in the economy do not change this.

When milk prices drop or feed costs increase, it may be wise to cull inferior cows sooner.

When the economy is moving down, less cash may be available. This can cause cash flow problems. Capital investments may need to be delayed or more carefully planned. Place emphasis on reducing risk and getting maximum benefits sooner.

LABOR MANAGEMENT

It takes about 27 hours of labor per year to take care of one dairy cow and her replacement. In larger dairy herds, labor efficiency slightly reduces the labor required per cow. One person can provide about 3,000 hours of labor per year. Therefore, one person can effectively take care of a herd of 100 to 110 cows. Larger herds need additional labor. In some cases this can be supplied by other members of the family. If family labor is not available, then additional labor must be hired. Large dairy farmers may have several employees with some labor specialization taking place. The dairy farmer who hires labor needs to have skills in managing people. Employees who are dissatisfied can quickly ruin a dairy business.

Some suggestions for effective labor management include:

- Recruit good help by stressing job benefits, interviewing prospective employees, and maintaining a reputation as a good place to work.
- Be as competitive as possible with industry in terms of wages, benefits, hours worked, vacations, etc.
- Put a dollar value on the benefits offered.
- Maintain good personal working relationships with employees. Recognize their needs for satisfactory interpersonal relationships. Communicate effectively with employees.
- Maintain good working conditions. For example, consider straight shifts instead of split shifts for milking in large herds of 200 or more cows.
- Train new employees.
- Use incentives for excellence on the job. Be sure the incentive program is clearly spelled out in writing. It should be based on criteria that can be measured by records.

MANAGING DRY COWS

Dry cows are those that are not producing milk. Most cows need a dry period between lactations. Good management during the dry period increases total herd profits.

Accurate breeding records are needed to determine when the cow is due to calve. The lactation period begins with calving. The average gestation (length of pregnancy) for dairy cows is 283 days. A ten-day variation in gestation is considered normal.

The cow should be dry for 45 to 50 days. The date to begin the dry period is calculated back from the projected date of calving.

Conditioning for the dry period is done during the last few weeks of lactation. USDA research shows that body fat is replaced more efficiently during late lactation than during the dry period. Cows should not be too fat or too thin at the end of the lactation. Weight is controlled by adjusting the grain-to-roughage ratio. Give thin cows a higher percent of grain and fat cows less grain.

There are three ways to dry off the cow.

- Stop milking her.
- Do not milk her out completely the last few days.
- Milk her every other day for several days.

The first method is recommended in most cases. Milk left in the udder causes pressure that stops milk secretion. This helps the drying off process. Do not feed grain or silage for two or three days. Reduce water and forage intake for one or two days. After the feed has been reduced for the recommended time, stop milking the cow.

Cows producing less than 35 pounds (16 kg) of milk per day can be dried off by stopping grain feeding and milking. High-producing cows need to have their production reduced before drying off.

Routine treatment for mastitis at drying off is recommended. Treat the cow for mastitis at the last milking. Use an approved dry cow mastitis treatment and a teat dip. Watch the udder for abnormal swelling for two or three weeks after drying off.

Separate dry cows from the milking herd. They may be grouped with the bred heifers. Allow dry cows to get plenty of exercise. Watch for any health problems that might develop.

Do not overfeed dry cows. Feed mainly good-quality roughage up until the last two or three weeks before calving. Limit body weight gain to no more than 100 pounds (45 kg) from late in the lactation to the next calving. The developing calf will gain about 40 pounds (18 kg) during the last eight weeks of the gestation period.

When dry cows get too fat there are more problems with ketosis, depressed appetite, milk fever, displaced abomasum, and "downer cows." Cows that are too fat have more problems at calving time.

Research shows that it pays to treat dry cows for internal parasites. Follow the advice of a veterinarian on a treatment program.

During the last few days of the dry period watch the cow closely for signs of calving. The muscles around the tailhead relax giving a sunken appearance. The vulva swells. The udder becomes larger and the teats distend. There may be some leaking of milk from the teats.

Check the udder carefully for signs of mastitis. Quarters with lumpy material, watery fluid, or blood may indicate the presence of mastitis. If necessary, treat the quarter for mastitis.

During warm weather the pasture is the best place for calving. A clean, well bedded stall is needed for cold or bad weather. The stall surface should provide good footing. A slippery floor may cause the cow to slip during calving and suffer muscle injury.

A calving pen 12×12 feet (3.6×3.6 m) is large enough. Make sure calving pens are clean, disinfected, and well bedded with fresh, clean bedding.

During calving watch the cow but do not disturb her. Give help at calving only if it becomes apparent that the cow is having unusual difficulty. Young cows, heifers, and cows giving birth to twins or bull calves are more likely to need assistance at calving than older cows.

A cow about to give birth is restless, lying down and getting up frequently. When the water bag (the membranes surrounding the fetus) breaks, the calf should move into the vagina within two hours. This process may take as long as four hours in heifers. Normal birth occurs about one-half to one hour after the calf moves into the vagina. Heifers may take about two hours to give birth. Observe the cow carefully to make sure the presentation of the calf is normal. Normally, the front feet appear first with the head placed on both front legs (anterior presentation). Posterior presentations sometimes occur and are normal if the back legs and tail appear first. If the calf appears in any other presentation, its position must be corrected for the birth to proceed without serious problems. A veterinarian should be called immediately to provide assistance if the presentation of the calf is not normal. Failure to provide proper assistance may result in the death of the calf and the cow.

The placenta (afterbirth) should be expelled within eight hours after the calf is born. Retained placentas occur more often when the birth is difficult, infection is present, or when the cow is too fat. The normal rate of retained placentas in a dairy herd is 5 to 10 percent. A retained placenta rate higher than 10 percent is an indication of poor management, inadequate feeding, or poor sanitation. If the placenta is not expelled within twelve hours after the calf is born, call a veterinarian for assistance.

After calving, watch the cow for signs of milk fever, ketosis, or other health problems. Special health problems of dairy cows are discussed in Unit 44.

Provide fresh water and hay for the cow right after calving. It takes several days to get a cow on full feed after calving. Gradually increase grain feeding. Do not allow too much loss in body weight.

RAISING DAIRY REPLACEMENTS

An important part of successful dairy management is raising replacements for the milking herd. About 30 percent of the average milking herd must be replaced each year. The dairy farmer must either raise or purchase replacements if the herd size is to be maintained or increased.

Approximately 50 percent of the calves born each year are heifers. Calf death losses must be held low. A goal for death loss is under five percent of the calves born. Replacements need to be selected from those animals with the highest potential for milk production.

The advantages of raising dairy herd replacements include:

- Less cost than buying replacements
- Greater control of genetic improvement
- Less chance of bringing disease into the herd
- Use of labor, feed, and facilities that might otherwise not be used
- Provision of herd replacements when needed
- Increased income from sale of extra calves
- Personal satisfaction from herd improvement

Herd Breeding Program

Improving the herd by raising replacements begins with a well-managed herd breeding program. Selecting good sires and dams is the key to genetic improvement. The traits selected in a dairy herd breeding program have different levels of heritability (Table 42-4).

The lower the heritability, the slower the genetic progress in improving that trait. Genetic progress is very slow when heritability is less than 10 to 15 percent. Genetic improvement is slower when selection is made for more than one trait at a time. A breeding program must place emphasis on selection for the most economically valuable traits. Commercial dairy farmers should place emphasis on selecting for milk production.

Genetic evaluation of dairy bulls and cows is based on a USDA-DHIA Animal Model Method. The evaluation values for both bulls and cows are known as Predicted Transmitting Ability (PTA). The PTA value replaces the Predicted Difference value for bulls that was previously used.

Select semen from proven sires with a high PTA for use in the herd breeding program. Younger bulls that have not been completely evaluated may be a good choice if both their sire and dam have high PTA values. Semen from younger bulls is usually not as expensive. Breed no more than 20 to 30 percent of the herd to a young sire that is the offspring of parents with high PTA values. This will give an opportunity to develop records that will prove the genetic value of these younger sires for future breeding programs. Use a proven bull with a high PTA on the rest of the herd.

Due to the high cost of purchasing and maintaining a bull, natural breeding is not recommended. If a herd owner chooses to use natural breeding, always select a bull whose sire and dam both have high PTA values. Less genetic progress in improving desirable traits will be made in a herd where natural breeding is used. More genetic improvement can be gained at a lower cost with the use of artificial insemination.

Keeping good records is an essential part of herd improvement. Records of feeding, reproduction, health, production, and sires used

TABLE 42-4 Heritability of Dairy Traits.

Trait	Heritability (%)
Fat, solids not fat, protein (%)	25
Stature	50
Teat placement	20–31
General appearance	25–29
Milk or fat production	25
Final type classification score	20–30
Mastitis resistance	20–30
Milking qualities	20–30
Rump	25
Feed efficiency	25
Back	23
Fore udder	21
Rear udder	21
Udder support	21
Body capacity (type score)	15–25
Mammary system	15–25
Dairy character	19
Hind legs	15
Longevity	0–15
Front end	12
Feet	11
Head	10
Rate of maturity	0–10
Breeding efficiency	0–10
Disposition	0–10
Udder quality	0

should be kept on every cow in the herd. It is recommended that herds be in a Dairy Herd Improvement program. Breed all heifers to good dairy bulls for their first calving.

The effective herd breeding program is based on the goals of the individual dairy farmer. Some goals that the dairy farmer may wish to achieve include:

- High milk production
- Calving interval of 12 to 13 months
- Reduction of health problems
- Improvement in dairy type
- Cows that are easy to handle (good disposition)
- Ease of milking
- Longevity in the herd
- Production of quality breeding stock

Inbreeding is the mating of close relatives. Avoid inbreeding when managing the dairy herd breeding program. Inbreeding usually results in lower milk production.

Research indicates that it is difficult to predict the influence of a sire on calving difficulty. It is not recommended that calving difficulty scores be considered when breeding milking cows. Sires that are rated below average in calving difficulty should not be used on heifers.

Artificial Insemination

An increasing number of dairy herds in the United States use artificial insemination (AI). The use of AI allows the dairy farmer a wide choice of genetically superior bulls. The risk of disease is less with AI.

The amount of money a dairy farmer should spend for a unit of semen depends on the goals set for the herd. The predicted transmitting ability and the fertility rate of the bull are important for all dairy farmers. The value of the offspring as breeding stock is of importance mainly to the purebred breeder. The commercial dairy farmer can afford to pay more for semen from bulls that will improve the production level of the herd.

Heat Detection

The major problem with AI on the farm is detecting cows in heat. The key to detecting cows in heat is careful and frequent observation of the herd. A good recordkeeping system will help by indicating when to expect a cow to come in heat.

The average cow comes in heat every 21 to 22 days. Observe the cow at least twice a day for signs of heat from day 17 to day 25 after the last heat period. More frequent daily observation will reduce the number of missed heats. Three times per day for 20 minutes each time is recommended.

It is difficult to observe standing heat when cows are confined to stanchions. They must be turned out to pasture or an exercise area to be observed for signs of standing heat. Observe cows for signs of heat at times when feeding or other routine work with the herd is not being done.

- Attempts to mount other cows indicates the cow is coming into heat.
- Restlessness, bawling, excessive walking, drop in feed intake, and decrease in milk production all indicate the cow is coming into heat.
- Slight swelling and reddening of the vulva indicates the cow is coming into heat.
- Pale white or opaque mucus discharge from the vulva indicates the cow is coming into heat.
- Standing when mounted by other cows. (This is the best indication of heat in the dairy cow.)
- An increase of swelling and reddening of the vulva indicates the cow is in "standing heat."
- A thin stream of clear mucus discharge from the vulva indicates the cow is in heat.

Dairy cows stay in standing heat an average of 15 to 18 hours. The range in length of heat is 6 to 36 hours. A bloody discharge from the vulva indicates the heat period is over. Record the date and watch the cow for signs of heat again in 14 to 18 days.

A small percent of dairy cows have quiet heat periods. That is, they do not show the usual signs of heat. Watching for the next heat on the basis of the bloody discharge from the vulva will help in detecting heat in these cows.

Commercial heat detectors are available to help detect heat in a herd of cows. A *heat mount detector* (called a KaMaR) is a device that is glued to the tailhead of each cow. It contains a red dye capsule that releases the dye when the cow is mounted by another animal. It requires four to five seconds of continual pressure to release enough dye to stain the pad red. If the cow is mounted when she is not in heat she will usually not stand long enough for the dye to be released.

This device works best when the herd has access to open areas. It has less value when animals are crowded in holding areas or around feed bunks. The device may be accidentally activated in these areas. Low-hanging tree branches or other low objects may also accidentally activate the device.

Another type of commercial heat detector is the *chin-ball marking device*. It is a special halter with a cone-shaped unit containing a stainless steel ball-bearing attached to the underside of the halter. It contains a marking fluid that is available in several colors. When the cow is mounted, the fluid leaves a visible color on the cow's back.

An implanted computer chip has been developed by researchers at Washington State University that helps determine when a cow is in heat. The computer chip is enclosed in a case that is pressure sensitive and is implanted beneath the skin on the tailhead of the cow. The device records when the cow is mounted. A reader device that may be hand-held or mounted in an area that cows pass through reads the computer chip. The information is fed into a computer that provides a printout showing which cows are in heat. The implant lasts for the lifetime of the cow. Implanting is done by a veterinarian.

A bull (marker bull) that has been surgically modified so it cannot breed the cow may be used with a heat detection device. Surgical modification may involve a vasectomy or removal of the penis. A bull that has had a vasectomy may still spread disease from cow to cow. One with the penis removed will not spread disease. The use of a bull for heat detection means extra feed expense. Also, all bulls are dangerous.

Spade heifers, steers, or estrogen treated cows may be used with these commercial devices for heat detection. There is less expense and danger when using these animals.

The success of these methods still depends on good records, proper animal identification, and careful observation of the herd by the dairy farmer. The use of heat detectors, combined with careful observation, can increase the heat detection rate in the herd. In a recent study, the use of the KaMaR device resulted in an 87 percent heat detection rate compared to only 50 percent when the cows were observed only at milking time.

Estrus or heat synchronization products are available for use with dairy herds. Depending upon the product used, the animal will come into heat within a specified number of hours after the product is administered or the animal is bred at a specified time after the product is removed. These products work well if properly used. However, they will not make up for poor management. The use of estrus synchronization products requires careful planning, proper facilities, and adequate labor. While conception rates with these products are generally satisfactory, research indicates that they are slightly lower than those achieved by breeding with observed estrus.

Time of Breeding

Inseminate the cow from the middle to the last half of the standing heat period. The egg is released from the ovary (ovulation) about 10–14 hours after the end of standing heat. The range in time of ovulation is 3 to 18 hours after heat. Conception is highest when insemination is done 12 to 18 hours before ovulation. Breed cows in the afternoon when they are observed in standing heat that morning. When standing heat is observed in the afternoon, breed the cow the next morning. It is better to breed late (up to 6 hours after standing heat) than to breed in the first half of the heat period.

Isolate the cow in a quiet area away from the rest of the herd at the first signs of standing heat. After breeding, keep the cow quiet and away from the herd until the heat period is over.

About five percent of dairy cows will show signs of heat two to three weeks after breeding even though they are pregnant. If the cow is rebred at this time, do not insert the inseminating tube completely through the cervix. Doing so may cause the cow to abort. Insert the tube no further than mid-cervix when rebreeding cows that may be pregnant.

After the cow is bred, watch her carefully for the next heat period. She may not have settled on the previous breeding. If no more heat periods are seen, have a veterinarian check the cow about two months after the last service to confirm pregnancy.

A herd conception rate of 1.5 to 1.8 services per conception is a desirable goal for the dairy farmer. A herd average of more than two services per conception is an indicator of serious reproductive problems in the herd.

Calving Interval

A 12- to 13-month calving interval is the most profitable. Healthy cows may be bred about 40 days after calving. However, it is recommended that most cows be bred 50 to 60 days after calving. This allows time for the reproductive organs to return to normal after pregnancy and calving. A delay in breeding may be needed for cows that had calving or post-calving problems. Do not breed if any infection is seen in the reproductive tract.

Most cows show signs of heat 34 to 35 days after calving. Cows that have not shown a heat period by 45 days after calving should be checked by a veterinarian. Breeding the cow at the second heat after calving will keep her on schedule for calving interval if she settles. The desirable calving interval can still be maintained if she has to be bred one more time to settle. If additional services are needed to settle the cow, the calving interval will be longer than recommended for maximum profit.

Breeding Heifers

It is recommended that heifers be bred according to size rather than age (Table 42-5).

With proper nutrition, heifers should reach the right size for breeding at about 14 to 15 months of age. Heifers should weigh about 60 percent of their mature weight at the time of breeding. There is a significant increase in lifetime milk production for heifers that calve at 22 to 24 months of age compared to those that calve when they are older.

TABLE 42-5 Recommended Age and Weights for Heifers at Time of Breeding by Breed.

Breed	Minimum Weight		Minimum Age Months
	lb	kg	
Holstein and Brown Swiss	825–875	374–397	14–15
Ayrshire and Guernsey	680–700	308–318	14–15
Jersey	580–600	263–272	14–15
Milking Shorthorn	680–780	308–354	14–15

Freshening Date and Milk Base

The milk base is the amount of milk that may be sold at Class I price from a farm. A dairy farmer's milk base is established during the late summer and early fall months. There is an economic advantage in a high milk base. Therefore, it is a good management practice to breed cows and heifers so they will freshen during this time. Freshening in the fall and winter results in more total milk production from each cow.

Some cows fail to conceive the first few times they are bred. Heat periods are sometimes missed. Therefore, there is a tendency for cows that originally freshened in the fall to, over a period of several years, freshen in the spring.

Breed heifers to freshen in the fall. This will help keep the milk base higher. It is easier to breed heifers for fall freshening than to try to move the freshening date of older cows back to the fall. However, do not delay breeding a heifer for more than two months just to help establish a high milk base.

Care of the Newborn Calf

Check the calf as soon as it is born to be sure it can breath. Wipe any mucous or fetal membrane from its nose. Give artificial respiration by pressing on and releasing the chest wall if the calf is not breathing.

Caution: Some cows will attack a person coming near the calf right after dropping their calf. If the cow shows signs of doing this, fasten her before going near the calf.

Usually, the cow will lick the calf clean right after birth (Figure 42-2). If not, dry the calf with a cloth, towel, or clean burlap sack. Dip the navel cord in a 7 percent tincture of iodine solution to prevent infection. Sometimes there will be bleeding from the navel cord. If so, tie it off with a sterile cotton or linen cord.

A healthy calf will be on its feet within 15 to 20 minutes after it is born (Figure 42-3). It will be nursing within 30 minutes. A weak calf

FIGURE 42-2 A cow will lick the calf clean right after birth. *Photo by Michael Dzaman.*

FIGURE 42-3 A healthy newborn calf will be on its feet within 15–20 minutes after it is born. *Photo by Michael Dzaman.*

must be helped to its feet and be held so it can nurse. The newborn calf must receive colostrum milk if it is going to live. It may be necessary to hand feed a weak calf colostrum milk using a clean nipple bottle. If the calf is too weak to nurse, use a stomach tube to feed the colostrum milk.

Anemia may be prevented by giving the calf an injection of 150 mg of iron dextrin solution within a few hours of birth. An injection of vitamins A, D, and E is also recommended.

Mark the calf for permanent identification before it is removed from its dam. A permanent identification may be made by a freeze brand, tattoo, photograph, or sketch. Some breed associations require a photograph or sketch for identification. Others require a tattoo inside the ear. Registration forms and permanent identification requirements may be secured from the appropriate breed association. Permanent identification in all herds is important for managing the breeding program. Date of birth, sire, and dam are recorded. Sire identification is especially important to aid in developing sire performance information in DHI and DHIR programs.

The adjustment period for both the cow and the calf is shortened by removing the calf from the cow within a few hours after it is born. Continue feeding colostrum from the dam for two or three days. Use a nipple pail or bottle feeder to feed the calf. This will help prevent some digestive problems caused by the calf gulping its milk too fast from an open bucket. It also saves time because it eliminates having to teach the calf to drink from a pail.

Keep all feeding equipment clean and sterile to prevent disease. Wash and sanitize feeding equipment with the same sanitizer strength used to sanitize milking equipment. Clean the equipment after each feeding.

Feeding

See Unit 41 for a discussion on feeding herd replacements.

Dehorning

Horns on dairy cattle do not have any useful purpose. Cows without horns are easier and safer to handle. Horns can cause serious injury to other cows and to people working with the cows. It is recommended that all calves be dehorned at one to two weeks of age. It is easier and less dangerous to dehorn young calves rather than waiting until they are older.

The horn button feels like a hard lump under the skin. An electric dehorner is the best way to destroy the horn growing tissue. It may be necessary to clip the hair around the horn if the horn button is hard to find.

Allow the iron to reach a cherry red heat before using it. Hold the animal tightly and touch the hot iron to the horn button. Hold it in place a few seconds. The skin should show a continuous copper-colored ring around the horn button. If it does not, apply the iron again. The horn button will drop off after several weeks.

A dehorning tube may be used on animals up to four months of age. Several sizes are available. The dehorning tube removes the horn button by cutting it out. Because a small open wound is left, a fly repellent must be used during fly season. The tube must be disinfected between each calf.

Horns may be removed chemically. Caustic potash is used. This is not as neat a method as electric dehorning. Care must be taken not to allow any of the chemical to run down the face or side of the animal's head.

Clip the hair around the horn button. Apply vaseline or grease around the clipped area to keep the chemical on the horn area. Rub the horn button with the stick of caustic potash until a slight bleeding occurs. The horn button will drop off after several weeks.

Keep the calf tied for at least one day after applying the chemical. This will keep the chemical from being rubbed off onto other calves in the area.

> **Caution:** The chemical must not get on the hands or fingers of the person doing the dehorning. It may cause serious burns. Wrap the caustic stick in paper or cloth to hold it. Follow manufacturer's directions for the use of the caustic stick.

Older animals may be dehorned with various kinds of mechanical dehorners and saws. See Unit 15 for illustrations and further discussions of dehorning cattle.

Removing Extra Teats

Extra teats have no useful purpose. They may interfere with milking. Sometimes they leak milk, develop mastitis, or become abscessed. Extra teats are best removed when the calf is two to six weeks of age.

If it is not obvious which are the extra teats, have the calf examined by a veterinarian. Have a veterinarian remove extra teats from older animals.

Lay the calf on the floor and hold it firmly when removing extra teats. Wash and disinfect the area around the teat to be removed. Stretch the teat slightly and cut it off close to the udder with sharp scissors or a knife. Be sure the scissors or knife are clean and sterilized. Use an antiseptic such as iodine on the cut to prevent infection.

A Burdizzo may be used to remove extra teats. This is a type of clamp which, when applied, crushes the blood vessels. The teat will fall off after a period of time.

Raising Young Bulls

In general, follow the same practices and feeding for young bulls as for young heifers. Place a ring in the nose when the bull shows signs of becoming dangerous. As the bull grows, the ring needs to be replaced periodically with larger, heavier rings. A bull is more easily controlled with a staff attached to the ring than by a halter.

Limited use may be made of a well grown bull at one year of age. The bull should be at least two years old before being used heavily. Heavy service is breeding 40 to 50 cows per year.

Bull calves that are not being raised for breeding stock may be sold for veal or fed out for beef. See the beef section of this text for information on raising beef feeders.

Housing

See Unit 45 for a discussion of housing for herd replacements.

Introducing the Heifer to the Milking Herd

About two weeks before calving, introduce the heifer to the milking herd routine. The heifer is easier to manage if she becomes used to the milking parlor or stanchion barn before she freshens. This also allows the heifer to adjust to the feeding program of the milking herd. Handle heifers gently. Heifers that are handled roughly are more difficult to manage in the milking herd.

HOOF TRIMMING

Care of the feet of heifers and cows is often overlooked. Hooves may grow too long and need trimming. If they are not trimmed they will crack and break off. Cows with poor feet often have lower milk production. Bulls may have low-quality semen and refuse to breed cows. Some conformation problems in younger animals may be corrected with proper hoof trimming.

Tools needed for hoof trimming include a hard rubber mallet, a straight wood chisel, a T-handle chisel, rasp, hoof nippers, and a pair of hoof knives.

Stand the animal on a 4 × 8 foot, 3/4-inch sheet of plywood. Confine the animal in a stanchion or tie it securely with a halter.

Trim the front feet first so the animal gets used to the operation. Using the chisel, shorten the toes to the desired length. Do not remove too much as this will cause lameness. Use the rasp to smooth and shape the outer surface.

To trim the sole of the foot, place the animal's knee on a box or a bale of hay or straw. Clean and trim the inner sole with a hoof knife. Use a sharp chisel to level the sole. Do not cut too deeply into the sole.

Check for hoof rot when trimming the feet. Clean out cracks with a hoof knife. Soak cracks with pine tar. If corns are found between the toes, call a veterinarian to treat them.

> **Caution:** Use care when trimming the animal's hind feet. Avoid being kicked by the animal, as serious injury to the person doing the trimming may result.

MAINTAINING MILK PRODUCTION IN HOT WEATHER

Heat stress caused by high temperatures and high humidity will lower milk production and conception rates in dairy cattle. Cows produce the most milk when the temperature ranges from 25° to 65°F (1.7°–21°C). Milk production begins to drop above 80°F (27°C). At 90°F (32°C) the drop in milk production ranges from 3 to 20 percent. The combination of heat and humidity increases stress. When the temperature reaches 100°F (38°C) and the humidity is 20 percent, heat stress begins to be serious and some type of cooling activities need to be used. At 100°F (38°C) and 50 percent humidity, the stress on dairy cattle begins to reach a dangerous level. A combination of 100°F (38°C) and 80 percent humidity can be fatal for dairy cows.

Problems with heat stress are reduced by keeping cows cool. Cows that are outside need access to some type of shade. This can be trees in the pasture or artificial shades made from metal (such as aluminum) sheets. The effectiveness of artificial shade can be improved by painting the top white. Other types of artificial shade may be constructed by suspending snow fence, canvas, or woven plastic fiber from poles.

Provide 6 to 8 feet (1.8–2.4 m) of height under the shade. Make sure that ventilation is adequate. About 40 square feet (3.7 m²) is needed for each cow.

Use fans inside buildings to provide ventilation and reduce the humidity.

Pasture quality usually declines in hot weather. Feed more hay and silage as the pasture quality goes down. Do not put the cows on extremely poor-quality pasture. The cows' energy is used in looking for feed instead of for milk production.

Use the best quality hay or silage available. Low-quality, high-fiber roughage reduces milk production. Maintain high grain rations to help maintain production.

Lush green pasture may lower milkfat percent. Free-choice feeding of hay helps maintain fat test.

Provide plenty of fresh water. The need for water may increase up to 50 percent in hot weather. Feeding dry feed increases the need for water. High-producing cows drink more water than low-producing cows.

Controlling flies helps to prevent a drop in milk production during hot weather. Follow a good fly control program.

Do not force cattle to go long distances to pasture in the hot sun. Avoid forcing cattle to stand in unshaded areas in the hot sun for long periods of time while waiting to be milked. Do not crowd cattle too closely in pens and lots. Water-sprinkling systems in the barnyard or free stall area will help keep cows cool.

It is important to reduce stress as much as possible during hot weather. Too much stress greatly reduces milk production.

SUMMARY

Records are an important management tool on the dairy farm. Dairy Herd Improvement Associations provide many records that help the dairy farmer increase net profits. Several different DHIA programs are available to the dairy farmer. Dairy production records are standardized to a 305-day, twice daily, mature equivalent basis for comparison purposes.

A regular culling program needs to be followed to increase net profits from the dairy herd. Records help determine which cows should be culled from the herd.

A dairy farmer needs to set goals. A budget will help determine how well the goals are being met as well as serve as a basis for management planning. Budgets are needed when the dairy farmer wants to borrow money.

The dairy farmer must manage labor efficiently. Large dairy farms often hire additional labor. Good labor management skills are necessary for success.

Proper feeding and management of dry cows is an important part of dairy farming. Do not allow dry cows to become too fat or too thin. Provide clean quarters for calving.

It is best to raise the replacements needed for the dairy herd. Select for economically important traits when breeding cows. Dairy sires and cows are evaluated for genetic value on predicted transmitting ability. Base a breeding program on the goals for the dairy farm. Dairy farmers make extensive use of artificial insemination. Better sires are available when artificial insemination is used.

Observe cows carefully for standing heat. Breed during the later part of the standing heat period. Dairy cows should calve at 12- to 13-month intervals. Breed cows 50 to 60 days after calving. Breed heifers to calve at 22–24 months of age.

Make sure the newborn calf receives colostrum milk. Mark the calf for permanent identification. Dehorn calves and remove extra teats.

Milk production often drops in hot weather. Proper feeding and keeping cows cool will help maintain production.

Student Learning Activities

1. Talk to dairy farmers in the local area who are enrolled in DHIA programs. Prepare and present an oral report to the class on DHIA programs.
2. Prepare a bulletin board display of various methods of identifying dairy animals.
3. Discuss culling dairy cows with local dairy farmers. Present an oral report to the class.
4. On a field trip to a local dairy farm, observe dehorning and removal of extra teats from dairy calves.
5. On a field trip to a local dairy farm, observe hoof trimming on dairy animals.
6. Prepare and present an oral report to the class on any aspect of dairy herd management.

Review

1. What kinds of records should a dairy farmer keep?
2. Name and briefly describe each kind of DHIA program.
3. What is the value of DHIA programs to the dairy farmer?
4. What methods may be used to identify dairy animals?
5. Why and how are lactation records standardized?
6. What factors are considered when deciding which cows to cull from the milking herd?
7. List some recommended goals for the dairy farmer.
8. Why is a budget of value to the dairy farmer?
9. What items should be included in the budget?
10. How is debt repayment ability calculated?
11. Compare buying to leasing dairy cows.
12. How should the dairy farmer adjust to changes in the economy?
13. Describe an effective labor management plan for the dairy farmer.
14. How long should a cow be dry?
15. Describe three ways to dry off a cow.
16. What are the signs of calving in dairy cows?
17. What are the advantages to the dairy farmer of raising herd replacements?
18. Briefly explain the heritability of traits in dairy cattle.
19. Describe how sires are evaluated for breeding purposes.
20. How are cows evaluated for breeding purposes?
21. Why is artificial insemination widely used on dairy farms?
22. How often does the average dairy cow come in heat?

23. What are the signs of heat in dairy cows?
24. When is the best time to breed a dairy cow during the heat period?
25. What is the most profitable calving interval in dairy cows?
26. How can the proper calving interval be maintained in the dairy herd?

43 *Milking Management*

Objectives

After studying this unit, the student should be able to

- describe the function of the cow's udder.
- describe recommended milking practices.
- describe methods of maintaining milk quality.
- describe the cleaning and sanitizing of milking equipment.
- list and describe off flavors in milk.

FUNCTION OF THE UDDER

The cow's udder is made up of four glands called quarters (Figure 43-1). The udder is attached to the lower abdominal wall by ligaments. Each quarter has a teat that provides an outlet for the milk. A circular muscle (sphincter muscle) at the end of the teat controls the flow of milk.

The udder contains alveoli that manufacture the milk. The raw material for the making of the milk is carried to the alveoli by the blood. The alveoli contain milk cavities. A tubule leads from each alveolus to small ducts that lead to large milk ducts. The large milk ducts empty into a gland cistern. The gland cistern holds about 16 ounces (473 cm³) of milk. Milk passes from the gland cistern through the teat cistern and then through the streak canal at the end of the teat.

A narrow streak canal and a strong sphincter muscle makes the cow harder to milk. A wider streak canal and a weaker sphincter muscle makes the cow easier to milk. If the sphincter muscle is too weak the cow will leak milk when the udder is full.

Between milkings, milk is stored mainly in the milk cavities, tubules, and small ducts. The growth and function of the udder is controlled by hormones.

Milk Let-down

Milk let-down occurs when the cow responds by a conditioned reflex to sensory stimuli such as the washing of the udder. A hormone (oxy-

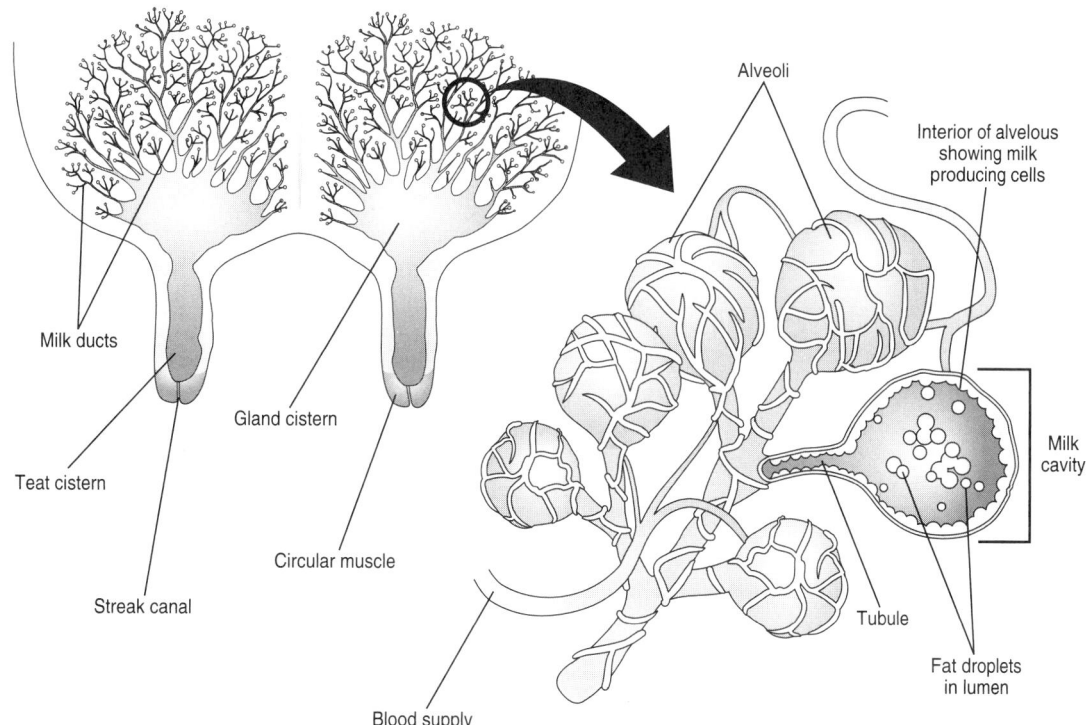

FIGURE 43-1   A side view of the cow's udder showing detail of the alveoli and alveolus.

tocin) is released by the pituitary gland. This hormone is carried by the blood to the udder. It causes muscle-like fibers to contract around the alveoli and ducts. This forces the milk into the larger ducts and gland cistern. Milk let-down must occur before the milk can be effectively removed from the udder. The effect of the oxytocin lasts about 5 to 7 minutes. Milking must be completed in that time to remove most of the milk from the udder.

The udder produces milk all the time. However, as pressure from produced milk increases, the rate of production slows down. Regular milking is important to maintain high production. High-producing cows will stop increasing the amount of milk in the udder after about 8 to 10 hours. In lower-producing cows this takes about 16 to 20 hours.

About 80 percent of the milk is removed from the udder at each milking. If too much milk is left in the udder, the pressure builds up quicker. This causes the cow to dry up sooner.

Gentle washing of the udder helps stimulate milk let-down. Attach the milking machine within one minute of washing. Frightening or hitting the cow will create an emotional disturbance that releases the hormone adrenalin, causing it to interfere with the milk let-down process.

RECOMMENDED MILKING PRACTICES

1. Follow a regular routine. Dairy cows respond with higher production when milked regularly at about the same times each day. Milking interval (time between milkings) should be about equal. The daytime interval can be somewhat shorter than the night interval. For example: 11 hours from morning to evening milking and 13 hours from evening to morning milking.

2. Prepare the cow for milking. Pre-wash extremely dirty udders with a hose or bucket of warm water containing a detergent. Then wash the udder with warm water containing a sanitizing agent such as chlorine or iodine. Use disposable paper towels. Do not dip the towel in the sanitizing solution after it has been removed from the solution. If more solution is needed, pour it on the towel. Dry the udder with a single service dry paper towel. Do not use sponges or rags for washing the udder with the sanitizing solution. Washing the udder helps stimulate milk let-down.

3. Use a strip cup. Milk two or three squirts of milk from each quarter into the strip cup. This stimulates milk let-down. It also removes the first milk, which is usually high in bacteria count. Watch each quarter carefully for abnormal milk.

4. Attach the milking machine within one minute after stimulating milk let-down (Figure 43-2). Be gentle when attaching the teat cups.

5. Remove the milking machine gently. Shut off the vacuum at the claw or break the vacuum seal at the top of the teat cup by pushing down on the top of the liner with a finger. Remove all four teat cups at once. Removing the machine in the wrong way can cause the vacuum to vary at the teat end. This can cause milk droplets to hit against the teat end. These droplets may contain mastitis bacteria and may infect the quarter with mastitis. Leaving the machine on too long may cause damage to the udder.

6. Dip the teats after milking. Use a solution made for teat dipping. Do not use sanitizer made for other purposes. Solutions containing chlorine, iodine, chlorhexidine, or cetyl pyridine chloride appear to be the most effective. The use of a teat dip will help reduce new mastitis infections. It is not a cure for existing infections. Dip at least two-thirds of the teat in the solution. Do this as soon after removing the milking machine as possible. Follow directions on the label of any commercial preparation for teat dipping.

7. Milker's hands must be kept clean. Wash before starting to milk and after handling any infected cow.

8. Do not try to operate too many units. This will result in over-milking and can cause udder damage. Two units per person in a stanchion barn is recommended. Three units per person in a milking parlor are usually enough. A good milker may be able to

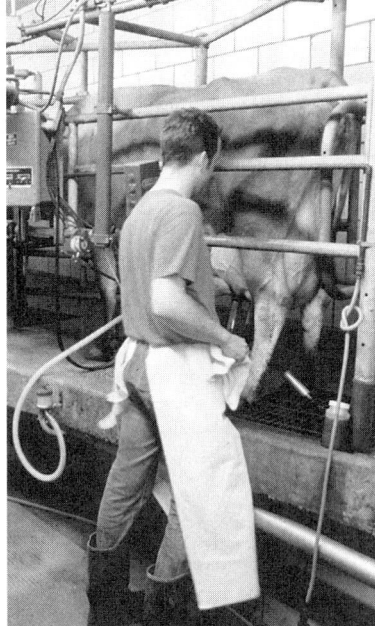

FIGURE 43-2 Attach the milking machine within one minute after stimulating milk let-down. *Photo by Michael Dzaman.*

handle four units in a parlor. A maximum of six units per person is recommended in parlors with automatic milking units.

9. Milking order. Milk heifers, cows in early lactation, and normal cows first. Cows with udder infections are milked last.

Machine Stripping

Machine stripping is the practice of trying to get all of the milk out of the udder before removing the milking machine. This has been a common and recommended practice for many years. About 0.15 to 1.2 pounds (0.07–0.5 kg) of milk are left in the udder when machine stripping is not done. However, research shows that this does not have any significant effect on somatic cell count, milk composition, yield of milk, or bacteria count. The main benefit of not machine stripping cows in large herds is the time saving. This is especially important in large herds. Teat damage is also reduced when machine stripping is not used.

Three-times-per-day Milking

Milking cows three times per day will increase total milk production from 6 to 20 percent as compared to two times per day milking. Milk-fat percent will decrease slightly. Some costs will increase with three-times-per-day milking. There appears to be little effect on herd health. Various studies indicate that an increase of 8 to 10 percent in milk production is needed for three-times-per-day milking to be profitable. A dairy farmer who is considering three-times-per-day milking will need to carefully evaluate the effect on the individual operation before making a final decision.

A major cost when milking three times per day is labor. Some dairy farmers milk three times per day in the winter and two times per day when field work increases. Research shows that greater increases in production from three-times-per-day milking come during the later stages of lactation. Therefore, it is generally recommended that three-times-per-day milking be continued for the entire lactation period.

MILK QUALITY

The production of high-quality milk requires good management practices. The characteristics of high-quality milk include:

- Free of dirt and other sediment
- Low bacteria count
- No chemical contamination
- Low somatic cell count
- No water added
- Has a good flavor

Management practices that will help to produce high-quality milk include:

- Cleanliness of cows, lots, barns, milking parlors, milk houses, and milking equipment
- Keeping cows healthy
- Proper cooling of milk
- Using correct cleaning and sanitizing methods on equipment
- Keeping chemicals out of milk
- Preventing off flavors in milk

Keeping Sediment Out of Milk

Keep the cows clean. Cows that are kept in clean facilities will stay cleaner than cows kept in facilities that are dirty. Keep manure cleaned out and use enough bedding in barns and pens. Use clean, dry bedding.

Make sure there is enough ventilation to reduce humidity and odor problems. Keep loafing and holding areas dry. Paved holding areas are easier to keep clean and dry.

Clipping cows will reduce dirt and bacteria in the milk. It also helps control parasites such as lice and will save time when preparing the cow for milking. Clip cows in the fall and again in late winter. Areas to be clipped include the switch, udder, hind legs, underline, flanks, thighs, thurls, and topline to the poll.

Wash the cow's udder before each milking. This helps keep the milk clean and stimulates milk let-down.

Keep the milking area free of dust. Do not sweep the barn just before milking. Dry feed increases the amount of dust in the air. If the dust from dry feed is a problem, feed at least one hour before milking or after completing milking.

Keep the milking machine units off the floor and out of the manure. If a unit falls to the floor, clean it off before putting it on the cow.

Strain or filter all milk. Strain milk into cans in the milk house, where there is less dust in the air. Do not bang the strainer to force milk through.

Strainers or filters that frequently clog indicate a problem with cleanliness or mastitis. Find and correct the problem. Produce clean milk at the start. Do not depend on the strainer or filter to clean dirty milk.

Keeping Bacteria Count Low

Bacteria are single-celled organisms. High bacteria counts lower milk quality. Several types of bacteria may be found in milk. Some grow in warm or hot milk. Others grow in cold milk.

Problems caused by bacteria in milk include:

- Souring of milk
- Reduced shelf life of milk
- Off flavors in milk
- Ropiness in milk
- Disease transmission through milk

Bacteria counts are lowered by following good management practices, which include:

- Cleanliness in milking area
- Keeping milking area dry
- Reducing dust in milking area
- Clipping cows
- Proper cleaning and sanitizing of equipment
- Washing udders with sanitizing solutions
- Proper milking procedures
- Rapid cooling of milk
- Rinsing and washing equipment properly after milking
- Keeping cows healthy and free of mastitis

CLEANING AND SANITIZING MILKING EQUIPMENT

Two kinds of deposits may be found on milking equipment. Organic deposits come from the fat, protein, and sugar found in milk. Mineral deposits come from inorganic salts (such as calcium, magnesium, and iron) that are in the water and milk.

Milkstone is a combination of the organic and inorganic materials. When equipment is not properly cleaned, milkstone will build up to a point where it can be seen. Deposits on milking equipment provide a place for bacteria to grow and multiply.

Alkaline, chlorinated alkaline, and acid cleaners may be used to remove deposits from milking equipment. Cleaners do not sanitize the equipment. Alkaline and chlorinated alkaline cleaners remove organic deposits. Acid cleaners remove inorganic deposits. Both alkaline and acid cleaners are used in order to remove all types of deposits.

Caution: Do not mix chlorinated alkaline cleaners with acid cleaners. Poisonous chlorine gas is released when these two types are mixed.

Do not use household soaps or detergents to clean dairy equipment. Soaps leave a greasy film on the equipment. Some household cleaners may cause odors or flavors in the milk.

Cleaners must be used at the proper concentration and temperature to be effective. Cleaning solutions need to be kept above 110°F

(43°C) to be effective. Follow the manufacturer's directions for the proper use of cleaners.

Even after proper cleaning, there are still some bacteria on the equipment. A sanitizing solution is used just before milking to kill these bacteria.

Chlorine, iodine, or quaternary ammonium compounds are used to sanitize the equipment. Follow the manufacturer's directions when using any sanitizer.

The method of cleaning and sanitizing varies with the type of equipment. General procedures to follow are listed here. Follow the manufacturer's directions for the type of cleaner and sanitizer used.

Metal Parts of the Milker

Rinse equipment in lukewarm water (100°–120°F; 38°–49°C) as soon as the milking is done. Milk solids that are allowed to dry are hard to remove later. Take the equipment apart. Soak the parts for five minutes in a dairy cleaning solution (120°–130°F; 49°–54°C). Use a hard-bristled brush or plastic sponge to wash the metal parts. Do not use metal or stainless steel sponges. These scratch the surface. Bacteria grow and multiply on scratched surfaces. It is hard to kill these bacteria.

Put the metal parts in an acid rinse (100°–120°F; 38°–49°C). This is especially important if the water is hard. The acid rinse helps prevent a buildup of milkstone. If there is a buildup of milkstone on the metal parts, soak them in an acid cleaner before washing them.

Drain and store the parts upside down. The parts will dry better when stored upside down. Bacteria do not multiply as easily on a dry surface.

Sanitize the parts in a dairy sanitizer just before the next milking. Drain well but do not wash before starting to milk.

Inflations and Rubber Parts

Daily cleaning of these parts begins with a rinse in lukewarm water as soon as milking is done. Use a commercial cleaner to wash the rubber parts. Rinse them in an acid solution. Store them dry or in an acid solution until the next milking.

Inflations last longer if one set is used one week and a second set the following week. Store the inflations when they are not in use in a lye solution or a commercial cleaner made for this purpose. Inflations may be stored dry. Remove them from the lye or cleaner solution after soaking for two hours. Rinse and wash them and let them air dry. Store them in a clean place. Rubber inflations absorb milk fat. This practice removes the milk fat. A lye solution may be made by mixing one-half pound (0.23 kg) of lye in five gallons (18.9 liters) of water.

Caution: Lye solutions are caustic. Store in a crock, plastic pail, or stainless steel pail. Keep all lye solutions out of the reach of children. Wear rubber gloves to protect the hands.

Inspect inflations regularly. Replace any damaged, stretched, soft, spongy, or rough inflations.

Vacuum Lines

Clean vacuum lines on a regular schedule. Vacuum lines can become clogged with dirt, rust, insects, and milk. This reduces the airflow through the line. A reduced airflow causes milking and mastitis problems.

Use a lye solution or a nonfoaming cleaner in hot water.

Caution: Lye solutions are caustic. Handle with care.

The lye solution is prepared by mixing 2 to 3 ounces (59–89 cm^3) of lye per one gallon (3.8 liters) of hot water. Do not put more solution into the system than the trap will hold. If the system has a vacuum reserve tank, use no more solution than one-half the volume of the tank. These precautions are necessary to prevent getting any of the solution into the vacuum pump.

Begin with the stall cock nearest the vacuum pump and allow one quart of solution to be drawn into the line. Use a flexible tube from the stall cock to the pail of solution. Allow air to enter the line before moving to the next stall cock. Repeat the procedure at each stall cock until the last one in the line is reached.

Drain the trap and throw the solution away. If the line is extremely dirty, repeat the procedure.

Rinse the line with hot water. Then wash the line with a hot solution of acid cleaner. Use 3 ounces (89 cm^3) of a nonfoaming cleaner per gallon (3.8 liters) of water for this solution. Start with the stall cock farthest from the vacuum pump and draw some solution through each stall cock.

Rinse the acid cleaner from the line with hot water. Start at the stall cock farthest from the vacuum pump.

Drain the trap and open all the stall cocks. Run the vacuum pump for five minutes to dry the line.

Pipeline Milkers

Pipeline milkers are cleaned in place (CIP) (Figure 43-3). This means they are not taken apart for cleaning. The cleaning solution is circulated through the system. Methods used include: vacuum circulating; pressure or pump circulating; vacuum gravity circulating; and reserve vacuum circulating.

FIGURE 43-3 An automatic pipeline and milking machine washer. *Photo by Michael Dzaman.*

Manufacturer's instructions must be carefully followed for proper cleaning of pipeline milkers. Cleaners made for pipelines must be used. Cleaning requirements are different than those for hand cleaning of milking machines. Stronger solutions that are low foaming are used. Label directions for use must be followed. A general procedure for cleaning pipeline milkers is described here.

As soon as milking is completed, flush the system using a large volume of lukewarm water. Water that is too hot will set the milk film. Cold water will not remove the fat.

Following directions for the specific cleaner used, prepare the cleaning solution. Use hot (160°F; 71°C) water when preparing the solution. Keep the water above 120°F (49°C) during circulation of the solution through the pipeline. A booster heater may be needed to maintain solution temperature.

Circulate the solution as recommended, usually 10 to 20 minutes. Use a lot of clean water to rinse the system, especially if the water is hard. Drain the system and allow it to air dry. Use forced hot air to dry plastic tubing that is over 10 feet (3 m) long. Store plastic tubes in a drying position.

Just before the next milking, circulate a nonfoaming sanitizer through the system. Drain the sanitizer from the system. Do not wash the sanitizer from the system.

Bulk Tank

Bulk tanks are cleaned by hand or by an automatic mechanical system. Always rinse the tank as soon as the milk is pumped out. This is usually done by the milk hauler. Use lukewarm (100°–120°F; 38°–49°C) water to rinse the tank.

Hand cleaning is done using a hard-bristled long-handled brush. All inside surfaces are scrubbed with the washing solution. Mix a chlorinated alkaline or other bulk tank cleaner in hot (120°–160°F; 49°–71°C) water according to the manufacturer's directions. Use a plastic pail that can be placed inside the tank when scrubbing the tank. Take the outlet valve apart and soak it in the wash solution. Rinse the tank using an acid rinse. Just before the next milking, sanitize the tank with an approved bulk tank sanitizing solution. Drain the sanitizer from the tank. Do not rinse the tank after draining the sanitizer.

Mechanical washing is done by spraying the solution with a spray device in the tank. Use a nonfoaming cleaner in mechanical systems.

Follow the manufacturer's directions when preparing the washing solution. Use hot water to mix the solution. Keep the solution above 120°F (49°C) during the wash cycle.

The solution is sprayed for about 10 to 15 minutes on the inside of the tank. Take apart the manhole cover, outlet valve, and gaskets before starting the wash cycle.

Drain the tank after the wash cycle. Brush by hand any parts not contacted by the wash solution. Be sure the outlet valve, tank cover, agitator, calibration rod, and support bridge for the agitator rotor are cleaned. Wash the outside of the tank with the solution drained from the tank.

Rinse the tank with lukewarm water. Then rinse with a foamless acid rinse. Drain the tank completely.

Just before the next milking, rinse the tank with an approved tank sanitizer. Drain the sanitizer from the tank. Do not rinse the tank after draining the sanitizer.

COOLING MILK

Bacteria grow and multiply rapidly in warm milk. Cool milk to 60°F (16°C) within 20 minutes and to 40°F (4°C) within 90 minutes after it is drawn from the cow. Do not allow milk to warm up after it is cool. A bulk tank that fails to cool milk properly must be checked for malfunction.

PREVENTING CHEMICAL CONTAMINATION

Many chemicals and drugs are used by dairy farmers. Improper handling or use of these products can result in milk contamination. Drug and chemical residues are not permitted in milk. Contaminated milk cannot be sold.

Prevent contamination by proper use of chemicals and drugs. Follow label directions for discarding milk when cows are treated for mastitis. Use only sprays and dusts that are approved for dairy barns and cows. Be sure feed is free of pesticide residues. Use only recommended pesticides on forage crops that are to be fed to dairy cows. Always read and follow label directions on any chemical or drug used on the farm.

PREVENTING OFF FLAVORS IN MILK

Milk has a clean, pleasant, and slightly sweet flavor. Off flavors cause dissatisfied customers. Milk sales go down when customers do not like the taste of the milk. The dairy farmer must follow good management practices to prevent off flavors in milk.

Off flavors in milk include:

- Feed flavors
- Rancid
- Flat
- Unclean, cowy, or barny
- Foreign flavors
- Sanitizer, medicinal
- Salty
- High acid (sour), malty
- Oxidized

The most common off flavor in raw milk is caused by feed. Any feed with a strong smell can cause off flavors in the milk. Silage, weeds, grasses, and moldy feeds all cause off flavors. Remove cows from pasture 2 to 4 hours before milking. Feed silage in the barn only after milking instead of before. Be sure there is plenty of ventilation in the barn.

Rancid flavor is strong and bitter. Improper handling of milk can cause rancid flavor. Too much agitation or cooling, warming, and re-cooling milk should be avoided to prevent rancid flavor. Milk is agitated by foaming that is caused by air bubbles in the line. Keep equipment in good operating condition to prevent air entering the line.

A flat flavor in milk is caused by adding water. Be sure equipment is drained before adding milk. Do not add water to milk.

Poor barn ventilation causes unclean, cowy, or barny flavors. Provide good ventilation in the barn and milking area. Make sure all milk-handling equipment is clean.

Kerosene, creosote, paint, and fly spray all cause foreign flavors in milk. Never store these materials in or around the barn or milk house.

Sanitizer and medicinal flavors are caused by improper use of these materials. Follow label directions carefully when using sanitizers and medications. Apply udder medication after milking and wash udders before the next milking. Use teat-dips only after milking.

A salty flavor is caused by cows in late lactation or cows with mastitis. Always milk cows with mastitis separately from the rest of the herd. Do not milk cows more than 10 or 12 months in one lactation.

High acid (sour) and malty flavors are caused by high bacteria counts. Keep equipment, cows, and facilities clean and sanitized. Cool milk rapidly to prevent bacteria growth.

Chemical reactions in the milk cause an oxidized flavor. The milk tastes like wet cardboard. The most common cause is contamination of the milk by small amounts of copper or iron. Lack of green feed in the ration and exposure of the milk to sunlight are other common causes. Do not allow milk to come in contact with copper or iron. The use of stainless steel on all milk contact surfaces reduces the problem. Keep milk away from sunlight.

SUMMARY

The cow's udder is made up of four glands called quarters. Alveoli in the udder manufacture milk. Udder function is controlled by hormones. Milk let-down occurs when the cow responds by a conditioned reflex to sensory stimuli. Gentle washing of the udder stimulates milk let-down.

Follow a regular routine when milking cows. Handle cows gently. Do not overmilk the cows.

Produce high-quality milk by keeping cows and facilities clean, rapidly cooling the milk, and properly cleaning the equipment. Bacteria lower the quality of the milk. Good management practices help to reduce bacteria in the milk.

Mineral and milkstone deposits may be found on milking equipment. Alkaline, chlorinated alkaline, and acid cleaners are used to remove deposits from equipment. Use caution and follow the manufacturer's directions for the use of cleaners and sanitizers.

Off flavors reduce milk quality. There are many causes of off flavors in milk. Good management practices help reduce off flavors.

Student Learning Activities

1. Prepare and present an oral report to the class on any aspect of milking management.
2. On a field trip to a dairy farm, observe proper methods of cleaning and sanitizing milking equipment.
3. Have a dairy plant field person talk to the class on cleaning milking equipment, keeping bacteria count low, and preventing off flavors in milk.
4. Have a local dairy farmer describe proper milking techniques to the class.

Review

1. Describe how the udder of the dairy cow functions.
2. Describe milk let-down in dairy cows.
3. Outline recommended milking procedures for the dairy herd.
4. Briefly explain three-times-per-day milking.
5. What are the characteristics of high-quality milk?
6. What practices help produce high-quality milk?
7. Describe ways to keep sediment out of the milk.
8. List problems caused by high bacteria count in milk.
9. What practices help keep the bacteria count low in milk?
10. What kinds of deposits are found on milking equipment?
11. What kinds of cleaners are used to remove deposits from milk equipment?
12. What kinds of sanitizing solutions are used to sanitize milk equipment?
13. Why must milk equipment be both cleaned and sanitized?
14. Describe the cleaning and sanitizing procedure for each kind of milking equipment.
15. How may chemical contamination of milk be prevented?
16. List and describe the common off flavors in milk.

Unit 44 Dairy Herd Health

Objectives

After studying this unit, the student should be able to

- describe procedures for maintaining herd health.
- describe the proper use of drugs for treating herd health problems.
- describe mastitis and its treatment.
- describe other health problems common to dairy herds.

Dairy cattle are subject to most of the same disease and parasite problems that affect beef cattle. See Unit 17 for a discussion of cattle diseases and parasites.

There are some health considerations that need to be emphasized concerning dairy cattle and they are discussed in this section.

HERD HEALTH PLAN

The dairy farmer needs to develop an overall plan for maintaining the health of the dairy herd. An effective plan puts emphasis on the prevention of problems. The services of a veterinarian should be used on a regular and planned basis. The veterinarian needs to be familiar with the dairy operation on the farm.

A planned program of regular vaccination and herd testing is carried out. Diseases of major concern are brucellosis, Infectious Bovine Rhinotracheitis (IBR), Bovine Virus Diarrhea (BVD), and Parainfluenza-3 (PI$_3$). Symptoms, control, and prevention of these diseases are discussed in Unit 17. Control of other diseases is needed in herds or areas where they are a problem.

Keep health records on all animals in the herd. Make individual physical examinations as needed.

Follow a planned program for mastitis control. Mastitis is one of the major causes of economic losses in dairy herds.

Keep accurate reproduction records. Have a veterinarian examine any cows with breeding or calving problems. Make routine pregnancy examinations.

Follow a planned program of calf health care. Prevent disease by vaccination at the right age. To prevent loss, treat scours and other calf health problems quickly.

Use the services of a veterinarian whenever cows show health problems. Diagnosis and treatment of health problems is a specialized skill. Losses can be high when improper diagnosis and treatment is done.

Herd health problems are reduced by the following management practices:

- Proper feeding of the herd
- Good facilities that are ventilated properly
- Using clean, dry bedding
- Proper cleaning and sanitation of facilities and equipment
- Controlling disease carriers such as flies, birds, and rodents
- Raising the replacements needed for the herd
- Requiring the health records of replacement animals and isolating them from the rest of the herd for 30 days
- Isolating all sick animals from the herd
- Using a veterinarian for quick, accurate diagnosis and treatment of health problems
- Control access to the dairy facilities by posting a sign at the farm entrance informing people that entry is limited and restricted to certain areas to control diseases.
- Require visitors to use some type of protective footwear covering such as plastic boots or use rubber boots and a disinfectant foot dip at the entrance to the facilities.
- Do not allow visitors to have unlimited access to areas where the cattle are kept.
- Have bulk milk pickup and feed delivery points as far from the cows as possible and restrict access by these vehicles to only those designated areas.

Dairy Quality Assurance Program

The Dairy Quality Assurance Program is designed to help dairy farmers produce high-quality milk. Participation is on a voluntary basis. The program was developed by the National Milk Producers Federation and the American Veterinary Medical Association. A producer, in cooperation with a veterinarian, goes through a ten-point checklist of management practices to become certified.

The program identifies several critical control points that help the herd owner produce a quality product. Emphasis is placed on following a preventative health program, including vaccination, housing, nutrition, and sanitation. Producers must use the correct drugs, store them properly, and follow required withdrawal times. The use of drug screening tests and good record keeping are other important aspects of the program. Other management practices, including sanitation, feeding, and milking equipment maintenance, are also reviewed. Producers who complete the certification and follow the management guidelines can reduce the risk of drug violations in the milk by more than 50 percent.

Use of Drugs for Treatment of Dairy Cows and Calves*

- Read drug labels carefully. Hundreds of changes are made in drug labels each year and many of these changes affect the way you should use drugs in treating your valuable animals.
- Use drugs only in the animal species indicated on the label. Drugs meant for one kind of animal can cause adverse drug reactions or illegal drug residues in another species.
- Always make sure you are giving the proper amount of drug for the kind and size of animal you are treating. Overdosing can cause drug residue violations.
- Make sure you are calculating pre-slaughter drug withdrawal and milk discard times accurately. Remember, withdrawal and discard times begin with the last drug administration.
- Always use the correct route of drug administration. Giving oral drugs by injection can cause loss of drug effectiveness. Giving injectable drugs incorrectly can lead to adverse reactions, reduced effectiveness, illegal drug residues, and possibly the death of a fine animal.
- Avoid "double-dosing" your animals. Using the same drug in the feed supply and then by injection can cause illegal residues.
- Keep an accurate record of the drugs you use and identify the animals receiving the drugs. Sending an animal to market too soon after it has been treated or shipping a treated animal because it was not properly identified can be costly mistakes.
- Good drug use records also help when professional animal health care is needed. Veterinarians need to know how much and what kinds of drugs have been given before they can treat animals effectively and safely.
- When injecting animals, select your needles and injection sites with care. Depending on the animal—and sometimes the drug—the wrong needle size, spacing or number of injection sites, or drug amount per site, can result in tissue damage, reduced drug effectiveness, or illegal drug residues.

*Source: Bureau of Veterinary Medicine, *Drug Use Guide: Dairy Cattle and Calves*, D.H.E.W. Pub. No. (FDA 76-6014), May 1976.

■ Remember, feed containing drugs also can cause illegal residues. Make sure you have a reliable source of drug-free feed for your animals to eat during withdrawal periods and that your storage bins and feed troughs are cleaned thoroughly before the withdrawal feed is put in them.

■ For a complete explanation of all the precautions you need to take in using any particular drug or feed medication, first consult the drug label or feed tag. If you have any questions about the proper use of any drug, see your feed dealer or veterinarian.

Mastitis Control

Mastitis is a serious economic problem for dairy farmers. The presence of mastitis in the dairy herd causes losses by

■ lowering milk production from infected cows.
■ increasing the cull rate in the herd.
■ the cost of treatment.
■ loss of infected milk that must be thrown away.
■ increased labor cost to treat infected cows.
■ possible loss of permit to sell milk if infection becomes serious enough.

Mastitis is usually caused by bacteria that get into the udder through the teat opening. The bacteria can enter through an injury to the teat.

Mastitis may be acute or chronic. The symptoms of the acute form include:

■ Inflamed udder
■ Swollen, hot, hard, tender quarter
■ Drop in milk production
■ Abnormal milk (lumpy, stringy, strawcolored, contains blood, yellow clots)
■ Cow goes off feed, shows depression, dull eyes, rough hair coat, chills or fever, constipation
■ Death may result

The symptoms of the chronic form of mastitis include:

■ Abnormal milk (clots, flakes, watery)
■ Slight swelling and hardness of udder that comes and goes
■ Sudden decrease in milk production

The chronic form may not show any symptoms. Therefore, it is often not treated. Sometimes it does not respond to treatment. Chronic mastitis is a more serious economic problem than the acute form. Either form may cause permanent damage to the udder.

Leukocytes are white blood cells that fight infection. The presence of mastitis causes an increase of leukocytes in the milk. Somatic cells are leukocytes and body cells. While all normal milk contains some somatic cells, the goal for the dairy herd should be an average of no more than 150,000 to 200,000 cells per milliliter. Ninety percent of the herd should be below 200,000 cells per milliliter. The somatic cell count can vary considerably from month to month in cows that have a mastitis infection. Daily per head milk losses increases as the somatic cell count increases, ranging from 1.5 pounds (0.68 kg) at 72,000 cells per milliliter to 6.0 pounds (2.7 kg) at somatic cell counts of over one million per milliliter. Somatic cell counts above 500,000 usually indicate a bacteria infection, a cow in late lactation, udder injury, or an old cow. Problem cows should be culled from the herd.

Current regulations state that bulk Grade A milk picked up at the farm cannot have a somatic cell count exceeding 750,000 cells per milliliter. A violation can result in the loss of the Grade A permit. This can result in serious economic loss for the dairy farmer.

Several tests are used to detect high somatic cell counts. The California Mastitis Test (CMT) is a common test to screen the herd for mastitis. It gives an estimate of the somatic cell count and should be used at least once a month.

A small paddle with four cup compartments is used. Milk about one teaspoon of the first milk from a quarter into a cup. Check each quarter separately. A chemical that reacts with the milk is placed in the cups. The presence of leukocytes is shown by the reaction. A slight precipitation shows a low leukocyte count. The development of a heavy gel and a purple color shows a high leukocyte count.

Tests for somatic cell count can be done in a laboratory. Some tests are chemical. Electronic instruments have been developed for giving more accurate somatic cell counts in milk. A direct microscope count may also be made.

A carefully planned and followed mastitis control program is essential for the dairy farm. The control program must be designed to both reduce the number of new infections and effectively treat existing infections in the dairy herd. An additional objective of the control program must be the avoidance of drug residues in the milk.

The following practices should be followed in an effective mastitis control program:

1. Maintain milking equipment in proper operating condition.
2. Practice proper milking procedures.
3. Identify the bacteria causing the infection and determine the extent of the infection in the herd.
4. Promptly treating identified cases of mastitis.
5. Treat all quarters of cows when at drying off time.
6. Cull cows with chronic mastitis problems that do not respond to treatment.

Improperly maintained and operating milking equipment can contribute to a mastitis problem in the herd. The milking vacuum should be relatively stable at 11 to 12 inches of mercury at the claw. The pulsator should maintain 45 to 60 pulsations per minute with a milk/rest (pulsation) ratio of 50/50 to 60/40. (Specific brands of equipment may be designed to operate at different specifications; if so, be sure the equipment is operating within the design specifications.) Check the system while all units are operating. The end of the teat may be damaged if the pulsation ratio and/or rate are too high. This can allow bacteria to enter the teat. Vacuum fluctuation, liner slip, and liner flooding may allow bacteria to enter the teat through the teat canal.

Bacteria can also be transmitted mechanically from cow to cow by the milker claw; this problem can be reduced by properly sanitizing the liners and claw between cows and by milking cows with high somatic cell counts or known infections of mastitis after the other cows in the herd are milked. Care must be taken to keep the sanitizing solution from becoming contaminated when treating the liners and claw between cows. Contaminated sanitizing solution will not prevent the spread of mastitis.

Large herds with a serious mastitis problem may justify the expense of solid state backflushing units, which are now available for use in milking parlors. These units backflush the equipment with a sanitizing solution between each cow. The flush cycle takes from 1 to 3 minutes and uses up to 1.5 gallons (5.7 liters) per cow. This can help reduce the spread of the infection by the equipment.

Proper milking procedure includes washing the udder before applying the milking claw and dipping the teats in a sanitizing solution after milking is completed. Procedures are discussed in detail in Unit 43.

Identifying the bacteria that are causing the mastitis problem is essential for proper treatment. A determination can then be made of the proper antibiotic to use in treating the infection. The use of the California Mastitis Test is discussed earlier in this section. Bacterial culture tests and somatic cell counts taken from four or five daily bulk tank samples will give an indication of the extent of mastitis infection in the herd.

Cows showing symptoms of acute mastitis infection should be treated with intravenous or intramuscular injections of the proper antibiotic. This treatment should always be done under the supervision of a veterinarian. Cases that are not quite as severe may be treated with mastitis infusion tubes in the affected quarters.

Caution: Label directions for period of use and withholding of milk from the market must be followed; failure to do so may result in severe economic loss due to condemnation of entire tanks of milk. Kits are available for testing individual samples of milk before adding them to the bulk tank if there is any doubt about the presence of drug residues.

Treat all quarters of cows being dried off with an approved dry cow mastitis treatment. This practice will help prevent infection during the dry period. If infection is present, the cure rate is higher than if treatment is done during lactation. Treatment also helps damaged tissue regenerate, reduces infection at freshening time, and avoids the problem of drug residues in marketable milk.

Cows with chronic mastitis infection that does not respond to treatment should be culled from the herd; this will eliminate a possible source of infection for other cows.

Displaced Abomasum

Displaced abomasum (DA) is a condition in which the abomasum moves out of place in the abdominal cavity. It is more common in dairy cattle than in beef cattle. The majority of the cases occur shortly after calving.

The symptoms of displaced abomasum include:

- Poor appetite
- Reduced fecal discharge
- Soft or pasty feces
- Diarrhea
- Drop in milk production
- Dull, listless, thin appearance

The kind of ration fed appears to be involved in causing displaced abomasum. Too rapid an increase in grain feeding just before calving increases the chances of DA. Poor-quality, moldy roughage or too much silage in the ration also increases DA.

Do not overfeed silage and concentrates to dry cows. Increase the amount of concentrate in the ration slowly at calving time.

Retained Placenta

Retained placenta is a condition in which the placenta (afterbirth) is not discharged within 12 to 24 hours after calving. It is normal for 10 to 12 percent of dairy cows to have retained placenta. A higher rate indicates a problem that needs attention.

A number of causes may be involved in retained placenta, including:

- Infection in the reproductive tract during pregnancy
- Deficiencies of vitamin A or E, iodine, and selenium
- Calcium to phosphorus ratio in diet out of balance
- Cow too fat (fed too much carbohydrate feeds)
- Stress at calving
- Breeding cow too soon after calving

Good management that prevents the listed causes will help reduce retained placenta. Call a veterinarian to treat a cow with retained placenta.

Ketosis

Ketosis is a nutritional disorder in dairy cows. Blood sugar drops to a low level. It is caused by not feeding enough energy feeds to meet the cow's needs for high milk production. Ketosis usually occurs in the first 6 to 8 weeks after calving.

Symptoms of ketosis include:

- Cow goes off feed shortly after calving
- Drop in milk production
- Loss in body weight
- Cow becomes dull and listless
- Odor of acetone in breath, urine, and milk

Feeding a properly balanced ration will help prevent ketosis. A veterinarian should be called to determine the proper treatment if a cow develops ketosis. Common treatments include injection of glucose into the bloodstream, injection of hormones (cortisone or adrenocorticotrophic hormone), or oral feeding of propylene glycol or sodium propionate. Feeding molasses will not cure ketosis.

Metritis

Metritis is an infection in the uterus. It usually affects a cow within 1 to 10 days after calving. A higher rate of metritis is seen in cows that are too fat at calving time.

Symptoms of metritis include:

- Loss of appetite
- Fever
- Drop in milk production
- Abnormal (thick, cloudy, grey, foul odor) discharge from the vulva
- Standing with back arched
- In severe cases, rapid death

Feeding a properly balanced ration to dry cows helps prevent metritis. Keep the calving area clean and sanitary. Metritis is treated with intrauterine antibiotic drugs. Consult a veterinarian for proper treatment.

Milk Fever

Milk fever (parturient paresis) is caused by a shortage of calcium salts in the blood. It is more common in older, high-producing cows. It usually occurs within a few days after calving.

Symptoms of milk fever include:

- Loss of appetite
- Reduction in quantity of feces passed

- Cow may be excited in early stage
- Staggering
- Cow becomes depressed
- Cold skin, dry muzzle
- Paralysis
- Lies on brisket with head turned back toward side
- In later stages, lies on side with head stretched out
- Bloating
- If not treated, death

Feed a balanced ration to dry cows with the correct calcium-phosphorus ratio. Milk fever is treated by intravenous injection of calcium. Call a veterinarian for treatment.

Internal Parasites

Common internal parasites of cattle are discussed in Unit 17. A regular program of treatment for internal parasites should be followed for the dairy herd. Consult a veterinarian to set up a regular treatment program.

All mature dairy cows should be treated for worms after each lactation. Worm replacement heifers near the end of their pregnancy.

External Parasites

Common external parasites of cattle are discussed in Unit 17. Care must be taken when using insecticides on dairy farms. Insecticides in milk are illegal. Very small amounts can be detected. Use only insecticides approved for dairy animals and facilities. Follow label directions carefully to avoid illegal residues in the milk.

SUMMARY

An effective herd health plan emphasizes prevention of problems. Regular vaccination and herd testing is important. Good management practices help prevent health problems.

Mastitis is the most serious disease that affects dairy cattle. Careful management and proper treatment is needed to keep losses low. Be careful when using drugs to avoid illegal residues in the milk.

A good herd health plan, set up with the assistance of a veterinarian, will increase net profits. Control internal and external parasites.

Student Learning Activities

1. Have a veterinarian speak to the class about mastitis control in dairy herds.
2. Plan a herd health program for a local dairy farm.

3. Prepare and present an oral report to the class on any aspect of dairy herd health.

Review

1. Describe management practices that will help maintain good herd health.

2. What care should be taken when using drugs for treatment of dairy cows and calves?

3. What losses are caused by mastitis?

4. What are the symptoms of acute and chronic mastitis?

5. Describe the California Mastitis Test.

6. How is mastitis treated?

7. Discuss the symptoms, prevention, and treatment of each of the following:
a. displaced abomasum
b. retained placenta
c. ketosis
d. metritis
e. milk fever

8. Briefly explain control of internal and external parasites in the dairy herd.

Unit 45 Dairy Housing and Equipment

Objectives

After studying this unit, the student should be able to

- describe adequate and economical housing for dairy animals.
- describe milking equipment.
- describe milk handling systems.
- describe manure handling systems.

PLANNING THE DAIRY FACILITY

The modern dairy farm requires a high investment in buildings and equipment (Figure 45-1). Careful planning is necessary before building new facilities or remodeling old facilities. Dairy farming is a high labor requirement enterprise. Mechanization can reduce labor requirements but cost more money. Sources of ideas for dairy facilities include farm publications, state universities, Cooperative Extension Service, high school agricultural education departments, other dairy farmers, and manufacturers of dairy equipment. Efficiency, economy, and convenience are important considerations when planning dairy facilities.

Important areas to be considered when planning dairy facilities include the following:

- Location of the facility
- Size of the planned herd
- Laws and regulations that apply to dairy farms
- Source and amount of money available
- Type of milk market available
- Amount of labor available
- Kind of housing system to be used
- Kind of milking system to be used
- Feed handling system to be used

FIGURE 45-1   A modern dairy with free stall housing, milking parlor, and silos. *Photo by Michael Dzaman.*

■ Manure handling system to be used
■ What future expansion might be desired

Location of the Facility

This is a major decision because once the facility is built it cannot be moved. Sometimes the decision is a matter of choosing between re-modeling or building on a new location. Each farm presents a different set of circumstances. While no general recommendation is made, it is often better to build completely new facilities rather than try to re-model old ones. Review all the planning considerations previously listed before making a final decision.

A well-drained location is needed. Provide easy access for vehicles to handle feed, milk, livestock, and manure. Make sure there is a good water supply. Provide room for expansion of the facility. Consider pre-vailing wind direction. Locate the dairy facility downwind from the farm house. Try to avoid a location that will cause odor problems for close neighbors.

Electrical Service

A dairy farm must have a continuous supply of electricity. Make sure the power supply is adequate to meet the needs of the operation. A standby generator is essential in case of power failure.

HOUSING FOR THE MILKING HERD

Many types of housing systems are used for the milking herd. The two most common types are stall (stanchion) barns and free stall barns. Stall barns are more common when herd size is up to 80 cows. Free stall barns are more commonly used with herds larger than 80 cows. Both systems can be mechanized to reduce labor requirements.

Stall Barns

Each cow is confined to an individual stall in a stall barn. The cow is held in a stanchion or a tie stall.

Stanchions have metal yokes that are fastened at the top and bottom. The yoke can pivot from side to side. Some designs have each stanchion released individually. Other designs use a lever to open a group of stanchions at the same time. Less labor is needed for fastening and releasing cows when the lever design is used.

In tie stalls the cow is fastened by a neck chain or strap to the front of the stall. There is a pipe across the front of the stall to prevent the cow from walking forward through the stall. Tie stalls give greater comfort to the cow. More labor is needed to fasten and release each cow from a tie stall.

Advantages of the stall barn include:

- Allows more individual attention to each cow
- Easier to observe and treat each cow
- Better display of breeding stock

Disadvantages of the stall barn include:

- More labor and time needed to fasten and release cows
- Harder to use with a milking parlor
- Harder to keep the area clean
- More bedding needed
- More labor needed for feeding
- Stooping required when milking in stall barns
- Moisture problem is greater, especially in older barns

Stalls must be long enough for cows to lie down with their udders on the platform. There must be enough width for the cow and for the person milking the cow. Teat and udder injuries are more likely to occur when the stall is too small. Stalls that are too long or too short make it hard to keep the cows clean. A minimum gutter width of 16 inches (41 cm) is recommended. Suggested stall sizes for different size cows are shown in Table 45-1.

TABLE 45-1 Platform Size for Cow Stalls When Used with Electric Cow Trainers.

Cow Size	Stanchion Stalls		Tie Stalls	
	Width	Length	Width	Length
Under 1,200 lb (544 kg)	4'-0'' (1.2 m)	5'-6'' (1.67 m)	4'-0'' (1.2 m)	5'-9'' (1.75 m)
1,400 lb (635 kg)	4'-6'' (1.37 m)	5'-9'' (1.75 m)	4'-6'' (1.37 m)	6'-0'' (1.8 m)
Over 1,600 lb (726 kg)	Not Recommended		5'-0'' (1.5 m)	6'-6'' (1.98 m)

Stalls may be arranged with two rows facing either in or out. The recommended arrangement is to have the cows facing out (Figure 45-2). This makes it easier to milk the cows and to install a gutter cleaner. The dairy farmer spends more time working behind the cows as compared to working in front of the cows.

A barn width of 36 feet (11 m) is recommended for larger breeds. Thirty-four feet (10.4 m) may be used for smaller breeds. Figure 45-3 shows a floor plan for a 60-cow stall barn for larger breed cows. Figure 45-4 shows the cross section of the floor plan shown in Figure 45-3.

The use of deep bedding (typically oat straw) has been the traditional way of keeping cows clean in stall barns. Bedding is expensive, takes a lot of labor, is getting harder to find, and does not work well with liquid manure systems.

A bedding chopper is a machine powered by a small gas engine. It chops a bale of bedding material as it is moved down the alley behind the cows. It saves time and labor when using bedding. Poor-quality or rained-on hay may be used for bedding when put through a bedding chopper.

Stall mats made of rubber have been developed as an alternate to deep bedding. Mats that are held in place on top of the concrete tend to get manure, urine, and bedding underneath them. This causes an odor problem. It is hard to clean the area under the mats. Rubber mats can be placed in the fresh concrete during construction. This reduces some of the problems. However, these mats will eventually come loose. Cleaning under them is harder than cleaning under those placed on top of the concrete.

FIGURE 45-2   It is recommended that cows face out in a stanchion barn.

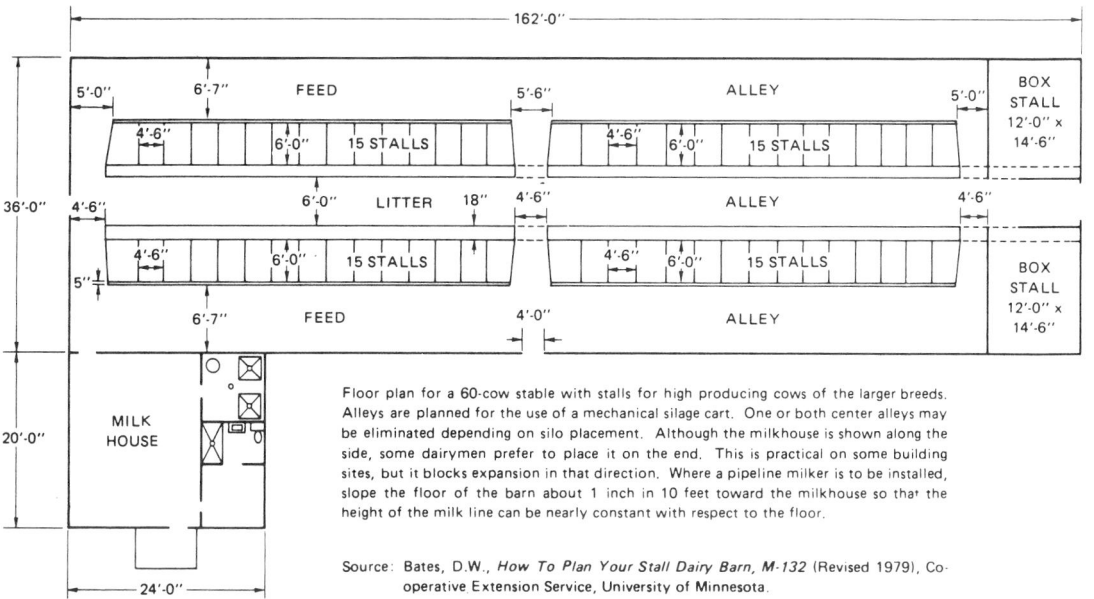

Floor plan for a 60-cow stable with stalls for high producing cows of the larger breeds. Alleys are planned for the use of a mechanical silage cart. One or both center alleys may be eliminated depending on silo placement. Although the milkhouse is shown along the side, some dairymen prefer to place it on the end. This is practical on some building sites, but it blocks expansion in that direction. Where a pipeline milker is to be installed, slope the floor of the barn about 1 inch in 10 feet toward the milkhouse so that the height of the milk line can be nearly constant with respect to the floor.

Source: Bates, D.W., *How To Plan Your Stall Dairy Barn*, M-132 (Revised 1979), Cooperative Extension Service, University of Minnesota.

FIGURE 45-3   Floor plan for a 60-cow stall barn. *Courtesy of Cooperative Extension Service, University of Minnesota.*

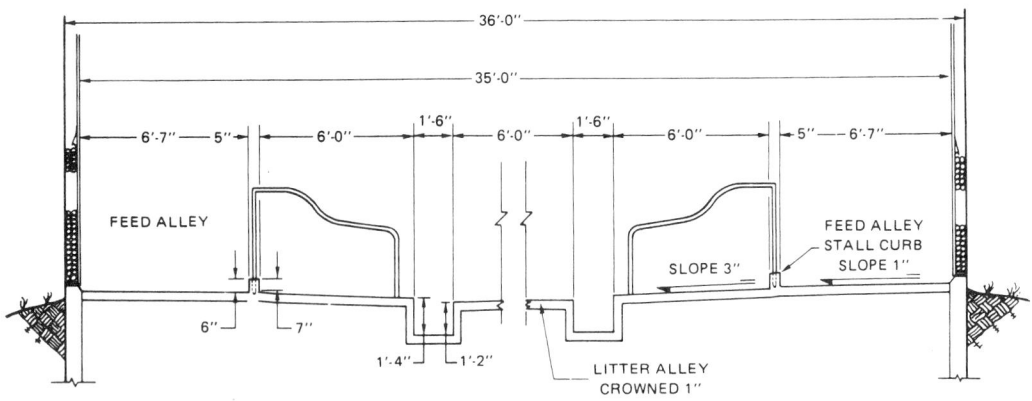

Cross-section of a stable 36 feet wide suitable for use with the floor plan shown in figure 45-3. Note the flat feed alley which also serves as a manger, thus providing more room for a feed cart. A stall curb, 12 inches high, is used for some types of stalls.

Source: Bates, D.W., *How To Plan Your Stall Dairy Barn, M-132* (Revised 1979), Cooperative Extension Service, University of Minnesota.

FIGURE 45-4   Cross section of floor plan shown in Figure 45-3. *Courtesy of Cooperative Extension Service, University of Minnesota.*

There is no completely satisfactory system for the cow platform surface that will keep cows clean. Properly adjusted cow trainers will help force the cow to deposit manure and urine in the gutter. The cow trainer is an electrical device that contacts the cow when she arches her body to defecate or urinate. The electrical charge forces the cow to move back toward the gutter.

Use a finely chopped bedding material on cow mats. In tie stalls use metal tie chains to complete the electric ground for the cow trainer. The rubber mat insulates the cow from the charge. Stanchion stalls provide enough ground through the metal yoke of the stanchion.

Gutter grates are steel bars placed over the gutter. These help keep the cow's tail out of the gutter when she lies down. They also increase the length of the cow bed. The steel bars are installed parallel to the gutter.

Feed carts are used in stall barns to reduce the labor needed to feed the cows. Carts may be homemade or purchased from commercial suppliers. Self-propelled, self-unloading carts are available for use in large stall barns. A separate cart equipped with a scale is often used for feeding grain.

Water cups are used to provide water for the cows in the stalls. There is always some spillage from water cups. Place the cups ahead of the stall curb to prevent wet stall platforms. Provide drainage in the feed manger area to take care of spilled water.

Too many windows in a stall barn make it harder to keep warm in the winter and cool in the summer. They are expensive to install and maintain. Do not depend on windows as part of the ventilation system.

A minimum of 10 foot-candles of light is required by Grade A regulations in all work areas during milking. Place a 100-watt bulb every 10 linear feet (3 m) in the area behind the cows to meet this requirement. Fewer bulbs may be used in the feed alley area. Place alternate lights on different circuits to reduce energy use during times when the cows are not being milked.

Stall barns must be properly insulated and ventilated. Health and moisture problems are common in poorly insulated and poorly ventilated barns.

Follow insulation recommendations for the climate in the local area. Information is available from the Cooperative Extension Service and agriculture engineers at state universities.

Ventilation designs must also conform to local climatic conditions. A common design is to provide enough ventilation capacity for a minimum of four air changes per hour in the winter. Plan for a minimum summer capacity of 30 air changes per hour. Fans controlled by thermostats are used for ventilation capacity above the minimum.

When a manure pit is located beneath the barn it must be ventilated continuously. Failure to do so will create a serious health hazard.

Fresh air must be brought into the building without causing drafts. Provide many small inlets for fresh air. Slot and ceiling intake systems are the most commonly used. Some moisture condensation is normal at the point of fresh air entry.

Provision must be made for electric power failure. Automatic warning systems and a standby generator are recommended.

Remodeling

Older barns may be remodeled to improve stall sizes and reduce labor requirements. Many old barns are not worth remodeling. Whether or not to remodel is an individual decision that must be evaluated in each case. Follow recommendations for stall and gutter widths. Other dimensions may be changed to fit the existing building. Increased herd size may be taken care of by adding to the length of the barn. It is usually not practical to try to widen an existing barn. Be sure the remodeled barn is properly ventilated. Older barns seldom have the proper ventilation.

Free Stall Barns

A free stall barn is a loose housing system in which stalls are provided for the cows (Figure 45-5). The cows are not fastened in the stalls. They may enter and leave the stalls whenever they want to. The system usually has a resting and a feeding area. The stalls are located in the resting area.

Advantages of the free stall barn include:

- Requires less bedding than other loose housing systems
- Cows stay cleaner compared to loose housing

FIGURE 45-5 Free stalls in a dairy barn. *Photo by Michael Dzaman.*

- Fewer injuries to teats and udders
- Requires less space than for other loose housing systems
- Easier to use with a milking parlor
- Easier to use automatic feeding equipment
- Cow disposition is better
- Less disturbance from boss cows and cows in heat

Disadvantages of free stall barns include:

- Manure is usually fluid and therefore harder to handle.
- Some cows will not use the stalls and must be trained to enter them.
- Some systems require more daily labor for manure handling.

A number of different designs are used for free stall housing systems. An enclosed warm system is fully enclosed, insulated, and mechanically ventilated. An enclosed cold system is fully enclosed, not insulated, and uses natural ventilation. Some designs use an open front with a roof, no insulation, and natural ventilation. An additional paved area may be provided outside of the roofed area. Another design provides a building only for resting. The feeding area, which may or may not be roofed, is outside and paved.

A stall is provided for each cow. Current practice is to allow about 10 percent more cows in the area than there are stalls. Not all the cows lie down at the same time.

Recommended stall sizes for various size animals are shown in Table 45-2.

To force the cow to move back when standing up, place a neck board across the top of the stall about two feet (0.6 m) from the front of the stall. This causes more of the droppings to be deposited in the alley. The cows will stay cleaner.

Floors of stalls may be earth or concrete. Compacted clay is preferred to other types of earth fill. Deep holes may develop in earth floors. These can cause injury to the cow's hips and legs. Dirt from an earth floor may get into the liquid manure system and cause problems. Concrete adds to the cost but eliminates some of the problems of earth floors. Rubber mats may be used on concrete floors. Mattresses

TABLE 45-2 Stall Sizes for Free Stall Barns.

Type of Animal	Width	Length
Average Herd weight (cows)	(center to center)	(outside dimensions)
over 1,500 lb (680 kg)	4'-0'' (1.2 m)	7'-6'' to 8'-0'' (2.3 to 2.4 m)
1,400 lb (635 kg)	4'-0'' (1.2 m)	7'-0'' to 7'-6'' (2.1 to 2.3 m)
1,200 lb (544 kg)	3'-9'' (1.14 m)	6'-6'' to 7'-0'' (1.98 to 2.1 m)
1,000 lb (454 kg)	3'-6'' (1.07 m)	6'-6'' to 7'-0'' (1.98 to 2.1 m)
Heifers	3'-0'' to 3'-6'' (0.9 to 1.07 m)	5'-6'' to 6'-0'' (1.67 to 1.8 m)
Calves	2'-0'' to 2'-6'' (0.6 to 0.76 m)	4'-6'' to 5'-0'' (1.37 to 1.5 m)

made from a 22-oz, woven, nylon polyester material are durable and may be used in free stall housing. The mattresses are stuffed with shredded tires and don't shift under a cow's weight.

Several different kinds of materials may be used for bedding in free stalls. Sawdust and wood shavings make excellent bedding materials. Other materials such as chopped straw, chopped hay, shredded corn stalks, ground corn cobs, and peanut hulls are used for bedding.

Manure must be removed from the stalls each day. Use a rake to move the manure into the alley behind the stall.

Free stall barns usually have two, three, or four alleys. The width of the barn depends on the number of rows of stalls and the width of the alleys. A variety of plans and arrangements for free stall barns are available from the Cooperative Extension Service and agriculture engineers at state universities. Figure 45-6 shows a floor plan for a cold slat-floor free stall dairy barn.

Slat floors may be used in alleys as part of the manure handling system. The manure is worked through the slats by cow traffic. The manure is held in a pit beneath the floor until it is removed. Continuous ventilation must be provided in enclosed barns to prevent serious health hazards.

Scrapers or small tractors with blades are used to remove manure from solid floor alleys. The manure may be moved into a holding area

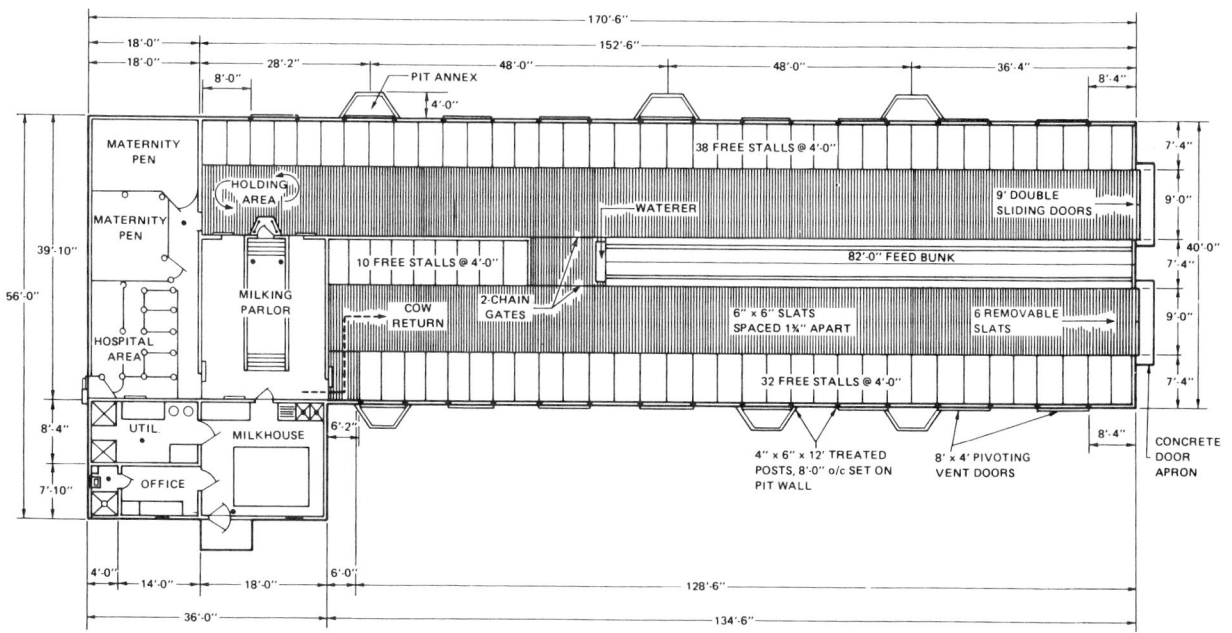

Source: Bates, D.W., *Free-Stall Housing For Dairy Cattle, M-132* (Revised 1975), Cooperative Extension Service, University of Minnesota.

FIGURE 45-6   Floor plan for a cold slat-floor free stall dairy barn. *Courtesy of Cooperative Extension Service, University of Minnesota.*

and later hauled to the field. Manure may also be disposed of in lagoons.

MILKING IN STALL BARNS

Three types of milking equipment are commonly used in stall barns—pail milkers, suspension milkers, and pipeline milkers.

Pail milkers are set on the floor beside the cow. The milker claw is attached to the udder and draws the milk into the pail. The pail must be emptied by hand after each cow is milked. The machine is operated by a vacuum pump with a vacuum line installed along the top of the stalls.

Suspension milkers are hung from a strap, called a surcingle, which is placed over the back of the cow. This type of milker is also operated by a vacuum pump and has a vacuum line along the top of the stalls.

Pipeline milkers consist of a vacuum line and a milk line that are installed along the top of the stalls. The milker claw is attached to the udder of the cow. An inlet for the milker claw is usually located between each set of two stalls. The milk is drawn through the milk line into the bulk tank located in the milk house.

Pipeline milkers are more expensive than pail or suspension milkers. More vacuum is needed to operate the system. However, less labor is needed in handling the milk. Milk can be kept cleaner when a pipeline milker is used.

MILKING PARLORS

A milking parlor consists of a separate area in which the cows are milked. Usually there is a pit in which the operator works. This eliminates the stooping that is necessary when milking in stall barns. Milking parlors are more commonly used with free stall housing. However, they may also be used when cows are kept in stall barns.

Advantages of milking parlors include:

- Reduces the labor and stooping needed to milk the cows
- Lower vacuum needed than in pipeline milkers in stall barns
- Less stainless steel pipe needed as compared to pipeline milkers in stall barns
- Can handle herd expansion easily
- Total cost of the system may be less because free stall housing is usually cheaper to build than stall barns
- Requires less milk handling than pail or suspension milkers

Disadvantages of milking parlors include:

- High investment cost is necessary to build the milking parlor. However, total cost of the system may be less than stall barns.

- When grain is fed in the milking parlor, high-producing cows may not have enough time to eat their ration of grain while being milked.
- Less individual attention can be given to each cow in the herd.

Four types of milking parlors are:

- Herringbone
- Side-opening
- Rotary or carousel
- Polygon

The most common milking parlor in current use is the *herringbone* (Figure 45-7). Common sizes range from the double-4 to the double-10. In the herringbone parlor, the cows enter and leave in groups. Cows stand at an angle to the operator pit, which is usually about 30 inches (76 cm) below the level at which the cows stand. A double-4 means that four cows stand on each side of the operator pit. Operator travel is reduced by having the cows stand at an angle to the operator pit.

Gates that allow the cows to enter and leave the parlor are controlled by the operator. Cows are kept in a holding area while waiting to be milked. A crowding gate is often used in the holding area to force the cows into the milking parlor. Holding areas should be paved to help keep the cows clean.

The pipeline milker may be along only one side of the parlor or along both sides. It is more expensive to have the pipeline along both sides. However, milking can be done faster with more units.

Cows may be fed grain in the milking parlor. The operator controls the amount of grain each cow receives. With the increased use of group feeding, it is becoming less common to feed grain in the milking parlor.

One disadvantage of the herringbone is that all the cows on one side must enter and leave at the same time. This means that a slow-milking cow will hold up an entire group. Separating slow-milking cows from the rest of the herd will help solve this problem.

Side-opening milking parlors are arranged with the cows standing parallel to the operator pit (Figure 45-8). Each cow enters and leaves the parlor individually. Most common sizes are two, three, or four milking stalls on each side of the operator's pit. Walking distances between udders are 8 to 10 feet (2.4–3 m) as compared to 3 to 4 feet (0.9–1.2 m) for herringbone parlors. One advantage of side-opening parlors is that slow-milking cows do not hold up an entire group of cows during milking.

Rotary milking parlors are arranged so that cows enter onto a turning platform that rotates slowly. Cows may be arranged to stand parallel to the operator pit or in a herringbone pattern next to the operator pit. A large number of cows can be milked in a small space with

FIGURE 45-7 A herringbone milking parlor. *Photo by Michael Dzaman.*

FIGURE 45-8 A side-opening milking parlor. *Photo by Michael Dzaman.*

the rotary parlor. It is a high-cost type of milking parlor. Two or more operators are usually needed to milk cows in a rotary parlor.

Each cow enters and leaves the rotary parlor by itself. The milker is attached and the cow is milked while the platform rotates. If the cow is not done milking by the time she reaches the exit she must go all the way around again.

The *polygon* milking parlor consists of a herringbone arrangement on four sides of the operator pit. Common sizes have four, five, six, eight, or ten cows on each side. The capacity of the polygon parlor, in terms of cows milked per hour, is about 25 percent greater than in herringbone parlors.

Generally, the polygon parlor is arranged in a diamond shape with corners not exceeding a 60-degree angle. Other shapes are possible but the diamond shape is the most common.

The recommended procedure is for the operator to move around the side of the pit preparing cows, attaching milkers, and removing milkers. It is not recommended that the operator cross from side to side in the pit to perform the milking operation.

MECHANIZATION IN THE MILKING PARLOR

There are a number of ways to increase mechanization in milking parlors. These include crowding gates, power gates, feedgates, prep stalls, and automatic detaching units.

Crowding gates are used to move cows forward in the holding pen area. Do not use electrically charged gates.

Power gates are used for entrance and exit of cows to and from the parlor. Pneumatic (air-operated) gates are commonly used. Controls are placed in several locations in the operator pit to reduce walking.

Feedgates are used to cover feed in the milking parlor. The feed is automatically covered when the cows are released. The purpose is to keep cows from stopping at the first milking stall as they come into the parlor. They also prevent cows from stopping to eat from stalls as they move out of the parlors. This helps control cow movement and cuts down on the amount of operator time needed to chase cows that do not want to leave the area.

Prep stalls are located ahead of the milking stall. A timed wash spray is used to preclean the cow and stimulate milk let-down.

Automatic detaching units are used with the milk claw or suspension cup milker. These units detect a decreased flow of milk and automatically shut off the vacuum. These units speed the milking and prevent overmilking.

HOLDING AREA

A paved holding area is recommended for confining the cows before they enter the milking parlor. A long, narrow area is best because this allows the use of a crowding gate to move the cows forward to the milking parlor. Provide 12 to 15 square feet (1.1–1.4 m^2) for each cow. Do not leave cows in the holding area more than two hours. Slope the holding area to provide good drainage. Clean the area after each milking.

MILK HOUSE

The milk house contains the equipment for filtering, cooling, and storing milk. Dairy utensils are also cleaned and stored in the milk house. The design and construction of the milk house is subject to approval by public health agencies. Be sure all plans are approved before beginning construction.

The size of the milk house depends on the size of the herd and the type of equipment to be used. Consider the possibility of herd expansion when planning the size of the milk house.

Locate the milk house on the side of the barn away from the cow pens or holding areas. Provide easy access for milk trucks to pick up milk.

Heat the milk house to prevent freezing and to keep the floors dry. Heat may come from the compressor, water heater, and a furnace. Insulate the milk house for the type of climate found locally.

Use a fan to ventilate the milk house. A filter is used to help keep dust out of the area. Screened air inlets are used in the milk house. Use screens on any milk house windows that may be opened.

Provide floor drains to remove water. Do not locate floor drains under the bulk tank or the milk outlet valve. Check local and state building codes for proper drain location.

Use concrete for the floor of the milk house. Slope the floor one-fourth inch per foot toward the drain. Extend the footing about six inches (15 cm) above the floor to protect the wall from excess water.

Use smooth, tight construction for the interior walls and ceiling. The material used must permit washing of the walls and ceiling.

Provide plenty of light in the milk house. Moisture-proof lighting fixtures should be used.

Use solid, tight fitting, and self-closing doors. Direct openings between the milk house and the barn or milking parlor must have doors. Use screen doors in addition to the solid doors on any outside openings.

Water under pressure must be available in the milk house. An automatic hot water heater is also needed. Clean-in-place pipelines may require extra-capacity hot water heaters. Properly vent any gas water heaters used. Two-compartment sanitary wash and rinse vats are needed. Provide facilities for hand washing.

Office facilities are included in some milk house designs. Toilet facilities are also sometimes included. The toilet facility may not open directly into the milkroom. Make sure there is proper waste disposal

FIGURE 45-9   Vacuum bulk tank. *Photo by Michael Dzaman.*

FIGURE 45-10   Milk pre-cooler on the milk line. *Photo by Michael Dzaman.*

from any toilet facilities. Do not run waste disposal from toilet facilities into the waste disposal system for the milkroom and milking parlor. Milkroom and milking parlor wastes may be disposed of through a common system. Because there is usually a large amount of fibrous material in the system, use a settling tank that can be easily cleaned.

BULK TANKS

Milk is cooled and stored in a stainless steel tank. A refrigerant is compressed by a compressor. Water or air may be used to cool the refrigerant. When water is used, the heat that is removed is used to heat water for use in the milk house. Direct expansion bulk tanks have the milk contact surface cooled directly by the refrigerant. Some bulk tanks use an ice bank that is located in the wall of the tank. The refrigerant circulates through the evaporator coils. This forms the ice bank. The milk is cooled by coming in contact with the walls of the tank in which the ice bank is located. Direct expansion systems require a larger compressor than ice bank systems. However, ice bank systems use more total energy.

Atmospheric tanks are not under pressure. They may be opened at any time. It is easy to reach all surfaces for cleaning. Vacuum tanks are under pressure. They are typically round with curved end plates (Figure 45-9). It is harder to clean vacuum tanks.

The size of the bulk tank needed depends on the amount of milk produced each day and the frequency of milk pickup. For everyday pickup, buy a tank large enough to hold three milkings plus 10 percent. For every-other-day pickup, use a tank large enough to hold five milkings plus 10 percent.

Consider future herd expansion when buying a bulk tank. It may be wise to buy a tank larger than those recommended in the previous paragraph if herd expansion is probable in the near future.

Bulk tanks must be capable of quickly cooling milk to meet local requirements for Grade A production. Pre-coolers may be located on the milk line before it reaches the bulk tank to remove some of the heat from the milk (Figure 45-10).

Locate the bulk tank at least two feet (0.6 m) from the wall on the side and rear of the tank. Provide at least three feet (0.9 m) between the wall and the outlet end of the tank.

SAVING ENERGY ON THE DAIRY FARM

The annual per cow energy use for milk production in the United States is 550 kilowatt-hours of electricity, 6 gallons (22.7 liters) of LP gas, and 10 gallons (37.8 liters) of gasoline equivalent. Careful management can save from 10 to 20 percent of this energy. Some energy-saving activities cost little money. Others require a large investment. Each energy-saving activity needs to be evaluated on the basis of its cost effectiveness. Detailed energy-saving ideas for dairy farms are available from the USDA and many state Cooperative Extension Services.

Areas that should be explored for energy savings include:

- Water heating
- Building ventilation and heating
- Milk cooling
- Vacuum pumps on the milking system
- Electric motors used in and around the dairy
- Lighting

Stray electrical current sometimes causes a problem in dairy facilities. A voltage difference as small as one-half volt can make cows nervous and cause them to withhold milk. Stray electrical current may originate off the farm or it may originate on the farm. The dairy farmer needs to work with the local utility company to identify the source of the stray current and take steps to correct the problem. Properly bonding all the metal parts of the facility that cows may contact will prevent stray currents because no voltage difference will exist between different metal objects. Permanent bonds between metal objects can be made by welding connections between them. A grounding wire (minimum size #8) should be connected between the metal stalls and the grounding bar in the electrical service panel. Make sure all connections meet electrical code requirements. It may also be necessary, especially in an old facility, to cover the floor with a non-conductive material. Asphalt, concrete, or epoxy materials may be used.

MILKING EQUIPMENT

The four parts of a milking system are the milking unit, the pulsation system, the vacuum supply system, and the milk flow system.

The *milking unit* is attached to the udder (Figure 45-11). The parts of the milking unit include the teat-cup assembly, the claw or suspension cup, and the connecting air and milk tubes. The teat-cup assembly consists of a steel shell with a liner. This fits over the teats. The liner is called an inflation. The inflation squeezes and relaxes on the teat as the pulsator operates. This causes the milk to flow into the system.

Many different types of inflations are on the market. They may be molded, one-piece stretch, or ring-type stretch. Both wide- and narrow-bore inflations are available. Inflations may be made of natural rubber, synthetic rubber, or a combination of the two.

Molded inflations are easier to use and usually last longer than other types. Narrow-bore (less than 3/4 inch; 1.9 cm) inflations cause less mastitis problems than wide-bore inflations. Narrow-bore shells and liners usually milk the cow faster than wide-bore shells and liners. Synthetic rubber inflations do not absorb milk particles as much as natural rubber inflations. They are easier to keep clean and do not tear

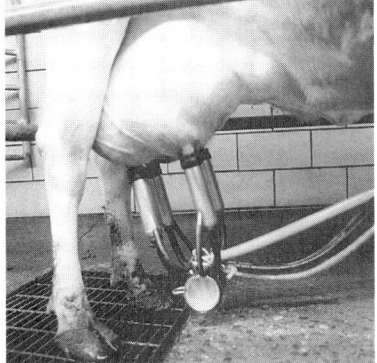

FIGURE 45-11   The milking unit is attached to the udder. *Photo by Michael Dzaman.*

as easily. The best molded inflations are made from synthetic rubber. Straight liners that are tightly stretched should be made from natural rubber.

The dairy farmer should try different types of inflations to see which work best for a specific herd. Make sure that the inflations match the type of shells in the system. Check the warranty on the machine to see if the use of a different type of inflation voids the warranty.

Narrow-bore inflations may be used for about 1,200 to 1,500 individual cow milkings. After this much use, they tend to stretch and bulge. Natural rubber inflations will last for about 800 individual cow milkings. The use of alternate sets of inflations each week make the inflations last longer. Always replace an entire set of four inflations at the same time rather than just one at a time. This makes it easier to keep track of the number of individual cow milkings per inflation.

Teat-cup shells are made of stainless steel. Several lengths and styles are manufactured. Make sure the inflation matches the type and style of shell used.

There are two types of milk receiving units used in the milking unit. One is the claw and the other is the suspension cup. The claw type is simpler than the suspension cup type. The suspension cup has a milk-receiving container on the bottom. It requires a support arm or surcingle.

An air bleeder vent in the claw increases the rate of milk flow away from the end of the teat. Make sure these air bleeder vents are kept clean. A plugged vent causes slower milking and problems with the teat cup dropping off the udder. A plugged vent can also cause milk flooding in the inflation, claw, or short hose leading from the teat cup. Milk flooding can cause a wide variation in the vacuum level at the end of the teat. Milk flooding in the inflation may also increase mastitis problems. The milk may contain bacteria that will enter the end of the teat.

Some newer types of inflations have an air inlet just below the shell. This reduces the problem of inflation flooding. More vacuum reserve is needed because this type lets more air into the system.

Air vents that are too large can cause too much foaming in the milk line and receiver jar. This can cause rancid milk.

The use of clear plastic in the construction of the claw or suspension cup permits the operator to see the milk flow from each quarter. This makes it possible to see when the quarter is milked out.

Synthetic rubber or clear plastic is used for milk hoses. It is easier to see the milk flow in clear plastic hoses. Rubber has more flexibility and lasts longer. The plastic hoses tend to crack at the end that is placed on the pipeline nipples. Do not use hoses that are too long. Looping long hoses increases vacuum changes in the system.

Air hoses are made of rubber. The air hose should be about the same length as the milk hose. A good seal around the pipe nipple is necessary for proper functioning.

The *pulsation system* controls the action of the inflation. This is done with a pulsator that alternately allows vacuum and atmospheric pressure into the space between the inflation and the metal teat-cup shell. Parts of the system include the pulsator, pipe, and hoses.

Pulsators may be vacuum or electrically operated. Air is used to move a plunger or slide valve in the vacuum-operated pulsator. This opens and closes the air passage to give the pulsation action. A needle valve controls the rate of pulsation. Temperature changes affect the operation of the vacuum-controlled pulsator. This type of pulsator must be kept at normal temperatures for best pulsation action. The electrically controlled pulsator is operated by a low-voltage electric current. The pulsation rate is constant. Temperature changes do not affect the operation of this type of pulsator.

The inside of the inflation is always under vacuum. When air is let into the space between the inflation and the shell, the inflation collapses against the teat. This is called the rest or massage phase of operation. The milk does not flow at this time. The teat is massaged in an upward direction. This lets the blood circulate out of the teat.

When a vacuum is created between the inflation and the shell, the inflation opens. This is called the milk phase of operation. The milk flows out of the teat and into the milking system.

The pulsator ratio is the ratio of time between the open and closed phases in the complete cycle. The range of ratios is 50:50 to 70:30 (milking time to resting time). A wider ratio gives faster milking. However, too wide a ratio may cause teat damage because of overmilking.

The pulsator rate is the number of times the inflation opens and closes per minute. Normal rates range from 45 to 70 per minute. The speed of milking is about the same within these rates. Too slow a pulsation rate slows milking. Too fast a pulsation rate allows too much air into the system, wears out inflations faster, does not give enough massage to the teat, and may slow the rate of milking. Follow the manufacturer's directions for the rate of pulsator operation.

A unit pulsator operates only one milking unit. A master pulsator operates two or more milking units at the same time. The unit pulsator is more commonly used.

Some milking systems use an alternating pulsation. That is, two quarters are milked while the other two are rested. Most machines milk all four quarters at the same time. There does not appear to be any advantage of one system over the other.

A pipe carries air from the pulsator to the vacuum pump. Follow the manufacturer's recommendations for the proper pipe size for the type of system used. When the pipe is installed, slope it down toward the vacuum pump.

A flexible rubber hose connects from the pipe to the milking unit. Sometimes the pulsator is at the stall cock, where the hose connects to the pipeline. Having the pulsator on the milking claw reduces the amount of pulsated hose.

The *vacuum supply system* includes the vacuum pump, vacuum lines, vacuum tank, vacuum controllers, and gauges. The functions of the vacuum supply system are to

- Cause the milk to flow from the end of the teat
- Massage the teat
- Cause the inflations to flex
- Move milk through the system

The vacuum produced in the system is about one-half atmosphere of pressure. It is important to maintain a steady rate of vacuum. The pump must be able to constantly remove the air at a rate equal to the amount of air entering the system.

Several types of vacuum pumps are used in milking systems. These include piston-type, rotary-vane, and centrifugal water displacement pumps.

The piston-type pumps stand up well under use. They operate at low to moderate speeds and are dependable.

Rotary-vane pumps may be high (1750 rpm or higher) or low (400 to 800 rpm) speed. High-speed pumps are noisier but have a higher capacity for moving air. Low-speed pumps are quieter, use less oil, and usually require less maintenance. Rotary pumps move more air at the same horsepower than do piston pumps.

Centrifugal water displacement pumps are quiet, reliable, and stand up well under use. They require a method of disposing of large volumes of water. Some recycle the water. This works well except where the water is quite hard.

The vacuum pump must be able to move the amount of air needed by the milking system. Vacuum pumps are rated in cfm (cubic feet per minute) at a given vacuum level and rpm (revolutions per minute). Use care when comparing pump ratings. A wide variation exists in pump ratings.

Two methods are used to rate the cfm of vacuum pumps. These are the ASME (American Society of Mechanical Engineers) Standard and the New Zealand Standard (NZ). The two rating systems use different pressure conditions to measure the volume of air moved. The ASME system uses normal atmospheric pressure. The NZ system uses 15 inches (38 cm) of mercury vacuum. Fifteen inches of vacuum is one-half normal atmospheric pressure. This means that the NZ Standard shows twice the volume of air as compared to the ASME Standard (1 cfm ASME = 2 cfm NZ).

Pump size needed for a particular dairy operation depends upon:

- Number of milking units in the system
- Size and length of pulsating lines
- Type of pulsator used
- Whether the system is a bucket or pipeline type

- Amount and type of other vacuum-operated equipment in the system
- Air leakage in the system

The pump needs to have enough capacity to prevent a drop of more than 1/2 inch (1.3 cm) of mercury under normal operating conditions. The vacuum level should be restored within five seconds.

Locate the vacuum pump as close as is practical to the milking area. Pipe the exhaust from the pump to the outside of the building. Use an exhaust pipe at least as large as the connection on the pump. The exhaust usually contains some oil fumes. Direct the exhaust pipe down and away from the building to keep oil off the side of the building.

Maintain the oil at the proper level in the pump. Keep the belt aligned and at the proper tension. Follow the manufacturer's directions for routine servicing.

Vacuum lines that are too small or partly clogged reduce the flow of air through the system. Follow the manufacturer's recommendations for proper size of vacuum lines to fit the system being used. Regular routine maintenance includes cleaning the vacuum lines.

Vacuum tanks are usually included in the system unless the vacuum lines are oversize. The vacuum tank prevents excessive vacuum fluctuation in the system. It is important to keep the vacuum level as even as possible.

A vacuum regulator is included in the system to maintain the right level of vacuum. Several types are used, including ball, poppet, and sliding sleeve valves. When the vacuum level reaches a preset amount, the vacuum regulator opens to let air into the system. The vacuum regulator is often located on the vacuum supply pipe just ahead of the sanitary trap or the vacuum tank. Make sure the regulator is kept clean and is functioning properly. Do not attempt to change the setting on the vacuum regulator. If the vacuum level is not normal, look for the cause and correct it.

A vacuum gauge is used to show the vacuum level in the system. Locate the gauge near the sanitary trap. Several gauges may be installed in the system to give an indication of the operating vacuum levels throughout the system.

A sanitary trap is used to separate the air line from the milk line in the system. This keeps liquid from moving from one part of the system to the other. Slope the connecting pipe from the milk receiver jar downward toward the sanitary trap. This prevents a reverse flow of liquid from the trap to the milk receiver jar.

The *milk flow system* includes the sanitary milk line, milk inlet valves, filters, milk receiver jar and releaser, and milk pump. The bulk tank is usually not considered to be a part of the milk flow system. When no receiver jar is used and a vacuum bulk tank is used, it is a part of the system.

TABLE 45-3 Milk Pipeline Size by Number of Milking Units.

Milk Line Size		Maximum Number of
Inches	cm	Units per Slope[a]
1.5	3.8	2
2	5.1	4
2.5	6.4	6
3	7.6	9

[a]Applies to systems in which the milk goes directly from the cow into the milk pipeline.

The purpose of the milk flow system is to move the milk from the milking unit to the bulk tank. A stable vacuum needs to be maintained from the sanitary trap to each milking unit.

Stainless steel or glass is used for the sanitary milk line. Stainless steel will not break and can be welded in place. Glass allows the operator to watch the flow of milk through the system. Sanitary couplings are used to connect sections of line that are not welded.

The milk line should not be more than 6.5 feet (1.98 m) above the cow platform. Low-level milk lines are installed at or below the level of the cow's udder. Low-level installations prevent vacuum fluctuation in the line.

Slope the milk line toward the receiver jar. Do not use any vertical risers in the line. Have as few sharp bends in the line as possible. Use the proper size of line for the number of units on the system (Table 45-3).

Locate milk inlet valves in the top half of the milk line. This prevents a reverse flow of milk in the milk hose when it is attached to the milk line. It also helps maintain the vacuum between the milk line and the milking unit.

A milk filter is used to remove sediment from the milk before it goes into the bulk tank. Milk must be filtered before it is cooled. Do not depend on the milk filter to detect mastitis in individual cows.

The four types of milk filters used are:

- *Unit filters* in the suspension cup or the milk hose between the claw and the milk line. This type clogs easily and must be watched closely. A clogged filter reduces the vacuum and slows the milking.
- *In-line suction filters* are commonly used with vacuum bulk tanks. They are placed in the line just ahead of the tank and must be watched closely to prevent clogging.
- *In-line pressure filters* are used with a milk pump. They do not affect the vacuum in the milk line. If they become clogged they may break and allow sediment to go through to the bulk tank. This is the most common type of filter used in closed milking systems that are under pressure.
- *Gravity filters* are not under pressure. They do not affect the milking vacuum and are less likely to break down.

Milk filters may be of woven or nonwoven material. Woven material can stand higher vacuum pressures. However, they do not filter as well as nonwoven material filters. Nonwoven material filters do a better job of taking small particles out of the milk.

The milk receiver-releaser collects the milk from the pipeline. It is made of either stainless steel or glass. Stainless steel will not break and gives more flexibility for placement of inlets. Stainless steel milk receivers are available in larger sizes than glass milk receivers. The glass receiver provides visibility of the milk flow.

The vacuum line and the milk line are connected to the milk receiver. The fittings are at or near the top of the receiver to maintain a stable vacuum. Electrodes are used to start and stop the milk pump.

The milk pump moves the milk from the milk receiver to the bulk tank. Most pumps operate under constant vacuum. Differential vacuum milk releasers are sometimes used. However, they create a demand for more vacuum in the system and give more trouble in operation.

Several methods are used to weigh the milk on test days. With modern computerized systems, most testing is done with the computer. Older systems include the use of milk meters, weigh jars, and buckets. Some milk meters are hard to clean in place. They also require additional vacuum to operate. Weigh jars slow the milking slightly. A regular milking machine bucket may be used to catch and weigh each cow's milk. They do not affect the vacuum in the system. However, the use of a bucket slows the milking operation.

HOUSING FOR DAIRY HERD REPLACEMENTS

Young calves up to two months of age may be housed in individual portable hutches (Figure 45-12) or in confinement calf barns. Either system works well with good management. Keep any type of housing clean, dry, well ventilated, and free from direct drafts.

Calf hutches are used successfully even in areas with cold winters. Hutches vary in size. A typical size is four feet by eight feet (1.2×2.4 m) for the shelter. Other sizes may be used depending on individual preference. The hutch is usually about four feet (1.2 m) high. An exercise area of similar size is provided in front of the shelter. Exterior grade plywood may be used to construct the hutch. Burlap, canvas, or plastic drop cloth may be used to close the hutch at night in severe weather.

The hutch may be bedded with wood shavings, sawdust, or straw. A manure pack will add warmth in the winter. The hutch is easily moved for cleaning. In wet areas, use a raised board floor to keep bedding dry.

Place the calf in the hutch as soon as it is dry after it is born. There is less shock to the calf than if it were to become used to warm conditions before being moved to the hutch.

Advantages of calf hutches include:

- Low cost
- Easy to design and build
- Easy to clean and disinfect
- Fewer disease problems
- Easy to move from contaminated areas
- Gets the calf used to existing weather conditions, therefore there is less shock when it is moved to minimal housing facility

FIGURE 45-12 Individual portable calf hutches. *Photo by Michael Dzaman.*

FIGURE 45-13   A group pen for calves. *Courtesy of William Grange.*

- Natural ventilation
- Calves grow as well in hutches as those housed in permanent barns

Disadvantages of calf hutches include:

- Takes more labor to feed and care for calves
- Takes more bedding
- Need another building to store feed
- More discomfort for person taking care of calves during bad weather

Confinement calf barns may be cold or warm. A cold calf barn is not heated or insulated and uses natural ventilation. Warm housing is heated, insulated, and uses a forced air ventilation system. Cold housing systems cost less than warm housing. Warm housing requires less space per calf, is easier to mechanize, is less likely to have its water freeze, is easier to use for treating sick animals, and provides greater operator comfort.

Use individual pens for young calves. This keeps the calves from sucking on each other after feeding. When the calves are weaned they may be kept in group pens (Figure 45-13).

Elevated pens or floor-level pens may be used for individual calves. Elevated pens do not require bedding but are more uncomfortable for the calves in cold weather. Individual elevated pens may be 2 × 4 feet (0.6 × 1.2 m). Floor-level pens may be 4 × 4 feet (1.2 × 1.2 m) in warm housing and 4 × 6 feet (1.2 × 1.8 m) in cold housing.

In group pens, provide 20 to 25 square feet (1.8 to 2.3 m²) of floor area for each calf up to eight months of age. Provide 25 square feet (2.3 m²) of bedded area per calf in open front loose housing. Do not put more than five to seven calves in a group pen.

Pole barn construction may be used for calf housing. Use concrete floors with gutters along the rear of the pens. Slope the floor toward the gutter to make manure handling easier.

A typical calf barn might have a row of individual pens along each side of a feed alley. Provide side alleys for moving calves in and out of the pens.

It is convenient to have a feed and supply storage room in one end of the calf barn. Include a hot water heater and a sink in the storage room.

Cold calf housing may be as simple as a row of pens open to the south. Each pen may be 4 × 8 feet (1.2 × 2.4 m). Provide an exercise area 8 feet (2.4 m) in front of each pen. A service alley with hay and bedding storage is provided on the interior back of the building.

Include a feed manger and watering facility for each pen regardless of the type of housing used.

Heifers from Two Months to Calving

Cold housing may be used for heifers after they are weaned. Heifers may be grouped in pens or free stall housing may be used. During warm weather, heifers may be kept on pasture. Provide shade and shelter for the animals on pasture.

Heifers in confinement housing should be grouped in pens by age and size. For heifers up to six months of age, provide 25 square feet (2.3 m²) per animal. From six months to nine months of age, provide 30 square feet (2.8 m²) per animal. From nine months to fifteen months of age, provide 35 square feet (3.2 m²) per animal. From fifteen months to calving provide 40 square feet (3.7 m²) per animal.

Solid floors may be used in pens or free stalls. A manure pack will provide added heat in cold weather. However, the manure pack system uses a lot of bedding. Less bedding is needed in free stalls but almost daily removal of the manure is required.

Slat floors may be used in free stall housing. Bedding needs are less and daily manure handling is eliminated.

Several stanchions should be included for holding heifers for artificial insemination. A head gate and squeeze chute will hold animals for treatment when required.

Do not feed and water animals in the resting area. Provide an outside lot for this purpose. Put pavement around feed bunks, waterers, and along the building. This will keep the animals out of the mud and make it easier to clean the area.

Cold housing is generally open on one side (Figure 45-14). Face the open side to the south or east. Make the building 30 feet (9.1 m) deep in moderate climates and 40 feet (12.2 m) deep in cold climates. Allow enough height for a 3 to 4 foot (0.9–1.2 m) manure pack. Use a door opening height of at least 10 feet (3 m). The floor of the building can be packed earth. Slope it toward the open side. Keep the floor about one foot (0.3 m) above the outside grade to prevent water and manure from running into the resting area.

Do not use a manure pack in the summer. If the building is used in the summer, a paved floor is recommended. Dry cows may be kept in open housing similar to that described for heifers. It is generally not good management to run the dry cows with the milking herd. Feed needs are different for dry cows.

MANURE HANDLING

Manure must be handled properly to maintain clean conditions for milk production. The best use for manure is as a fertilizer for crop production. The manure handling system must meet local health and milk market requirements. Handle manure in such a way that odors, runoff, seepage, and insects are controlled.

Manure may be handled in either a solid or liquid form. Liquid manure usually contains less than 15 percent solid material. A solid material content above 20 percent requires that the manure be handled as a solid.

FIGURE 45-14   An open front shed and hay bunks for dairy cattle. *Photo by Michael Dzaman.*

FIGURE 45-15 Gutter cleaner in a free stall barn. *Photo by Michael Dzaman.*

FIGURE 45-16 The manure spreader is loaded by this elevator portion of the gutter cleaner. *Courtesy of Badger Northland.*

Manure may be hauled every day as a solid and spread on crop ground. Another method is to haul the manure from storage as either a solid or liquid and spread it on crop ground. This is done when weather conditions are right, the ground is available, and the labor is available.

Manure mixed with bedding is handled as a solid. If the moisture content drains away or evaporates, the manure is also handled as a solid.

Barn cleaners located in the gutter are used in stall barns (Figures 45-15 and 45-16). A tractor scraper and a front-end loader may be used in free stall housing. A scrape-off ramp located at one end of the barn will permit the direct loading of the manure into the spreader.

The manure may be hauled daily or stored for later hauling. Daily hauling is not always convenient in bad weather or when crop land is not available on which to spread the manure. The cost is relatively low and odor problems are reduced with daily hauling.

Some type of stacking equipment is needed for storage of solid manure for a long period of time. Provide enough room for storage for up to six months. A manure loader on a tractor is needed to load the manure for hauling to the field.

Liquid manure may be stored in several ways, including:

- Under the barn with slat floors
- A below-ground storage tank outside the barn
- An outside basin with sloped earthen sides
- Above-ground silo storage

The size of the storage area depends on the number of cows in the herd and the length of time the manure is to be held. It is better to plan for extra capacity than to have too little storage volume. For example, Holsteins need about 2 cubic feet (0.06 m^3) per cow per day.

Manure may be moved to the storage area by pumps or by scraping. In slotted-floor facilities with a tank below, it drops directly into the storage area.

Automatic scrapers may be used. Tractors with scraper blades are sometimes used. The manure may be scraped through grates directly into a storage tank or into a holding tank to be pumped into the storage area.

Several types of pumps are used to move manure. Centrifugal pumps have a high capacity. Piston, auger, and diaphragm pumps have lower capacity. Large, hollow piston pumps work well with manure that has a low solids content.

The liquid manure in storage must be agitated just before unloading the storage tank. Centrifugal pumps with a bypass may be used. Recirculating pumps, paddle wheels, and inclined augers are also used for agitation in the storage tank.

Caution: Toxic gases are produced by liquid manure in storage tanks, especially during agitation. Make sure there is enough ventilation, especially in storage areas under the barn.

Liquid manure is hauled to the field in a tank holding up to 5,000 gallons (18,927 liters). Tanks may be loaded by gravity or with pumps. High-capacity centrifugal or vacuum pumps are used. The tank wagon is unloaded on the field by pressure or gravity.

Liquid runoff from the barnyard must be controlled to prevent contamination of rivers, lakes, or ponds. Check state Environmental Pollution Agency regulations for control of barnyard runoff.

FEED HANDLING

Many of the same kinds of feeding facilities are used for dairy cattle as are used for beef cattle. See Unit 18 for a discussion of silos, feed bunks, waterers, and other equipment.

SUMMARY

A high investment is required for buildings and equipment on the modern dairy farm. Dairy facilities must be planned for efficiency, economy, and convenience.

Two common types of housing used for the milking herd are stall (stanchion) barns and free stall barns. Large herds are more commonly housed in free stall barns.

Pail milkers, suspension milkers, and pipeline milkers are used in stall barns. Milking parlors with pipeline milkers are used with free stall housing. They may also be used for stall barn housing.

The most common type of milking parlor is the herringbone. Other types are side-opening, rotary or carousel, and polygon.

The use of crowding gates, power gates, feedgates, stimulating wash sprays, and automatic detaching units increase mechanization in the milking parlor.

Cows are held in a paved holding area while they are waiting to enter the milking parlor.

A milk house contains the equipment for filtering, cooling, and storing the milk. The milk house must be heated, insulated, ventilated, and kept clean.

Milk is stored in bulk tanks. These are large refrigeration units.

Dairying requires the use of large amounts of energy. Careful management can save 10 to 20 percent of the energy normally used on a dairy farm.

A milking system is made up of the milking unit, the pulsation system, the vacuum supply system, and the milk flow system. The milking unit is attached to the udder of the cow. The pulsation system controls the action of the milking unit. The vacuum supply system produces and distributes the vacuum necessary to operate the milking unit. The milk flow system moves the milk from the cow to the bulk tank.

Young calves may be housed in individual portable hutches or confinement barns. Either system works well with good management. Housing must be clean, dry, well ventilated, and free from drafts.

Older heifers may be grouped in pens or kept in free stall barns. Divide animals by age and size.

Manure may be handled as a solid or a liquid. Manure handling on the modern dairy farm is highly mechanized.

Student Learning Activities

1. Take field trips to local dairy farms in the area. Observe and report on the kinds of facilities used.
2. Prepare visuals on housing facilities and/or feed-handling facilities for a dairy farm and present an oral report to the class.
3. Prepare a bulletin board display of pictures of dairy housing and equipment.

Review

1. List the factors that should be considered when planning facilities for dairy cattle.
2. What are the two most common types of housing systems used for the milking herd?
3. What are the advantages and disadvantages of each system?
4. Briefly describe each system.
5. Briefly describe pail and suspension milkers.
6. What are the advantages and disadvantages of milking parlors?
7. Name the four types of milking parlors.
8. Briefly describe each type of milking parlor.
9. Describe some ways that mechanization in the milking parlor can be increased.
10. Describe the holding area needed for cows waiting to enter the milking parlor.
11. List the characteristics of a good milk house.
12. Describe the types of bulk tanks used to cool milk.
13. How much energy does it take to operate the average dairy farm in the United States?
14. Describe the parts of the milking unit.

15. Describe how the pulsation system works.

16. What are the functions of the vacuum supply system?

17. Describe the kinds of pumps used in the vacuum supply system.

18. Compare the two systems used to rate vacuum pumps.

19. List and briefly describe the other parts of the vacuum supply system.

20. Briefly describe the parts of the milk flow system.

21. Compare the use of individual hutches with confinement barns for housing dairy herd replacements.

22. Describe housing requirements for older heifers.

23. Compare handling manure as a solid and as a liquid.

Unit 46 *Marketing Milk*

The dairy farmer produces milk to sell at a profit. Careful management can help reduce the costs of producing the milk. A knowledge of the marketing structure for milk can help the dairy farmer make wise management decisions. Management decisions are influenced by the following factors:

- Price, supply, and demand trends for milk and dairy products
- Markets available for milk
- Pricing structure and regulation of milk marketing

PRICE OF MILK

Long-Term Trends

The average annual price received by farmers for all milk marketed in the United States rose steadily from 1967 to 1981; since then the price has fluctuated up and down in a narrower range (Figure 46-1). The milk-feed price ratio gives an indication of the true value of milk in terms of what it will buy. Figure 46-2 shows the value of one pound of a 16 percent dairy feed compared to the value of one pound of milk. The milk-feed price ratio shows a fairly wide variance during the period spanning 1967 to 1998.

Seasonal Trends

The amount of milk produced in the United States varies from month to month. The seasonal variation in production is less now than it

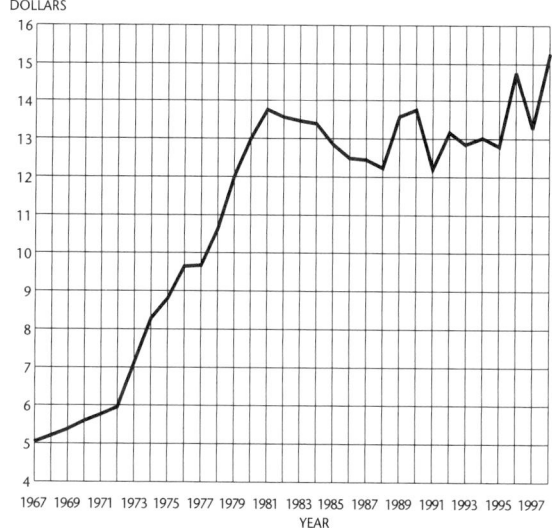

FIGURE 46-1 Average price received by farmers for all milk, 1967–1998. *Sources: Agricultural Statistics,* USDA, 1981, 1988, 1992, 1995, 1999; *Agricultural Prices,* USDA, December 1998.

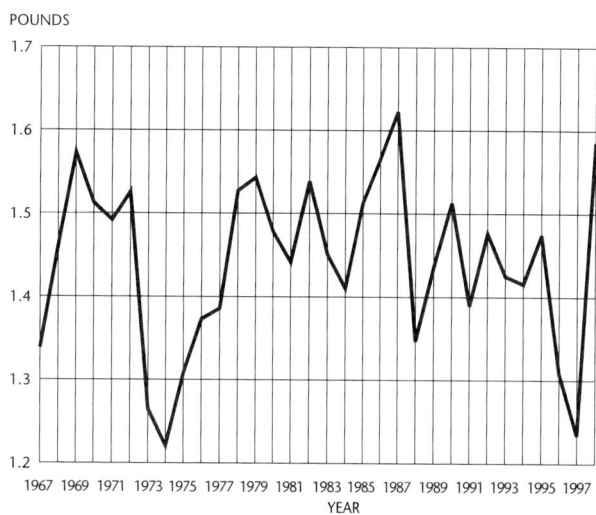

FIGURE 46-2 Milk-feed price ratio, 1967–1998. *Sources: Agricultural Statistics,* USDA, 1981, 1988, 1992, 1995, 1999; *Agricultural Outlook,* USDA, December 1998.

used to be. Currently, national average production is highest in May and lowest in November. Fluid milk consumption is at its lowest level in June.

The price for milk also varies seasonally (Figure 46-3). Lowest prices are paid in May and June. Highest prices are paid in October, November, and December.

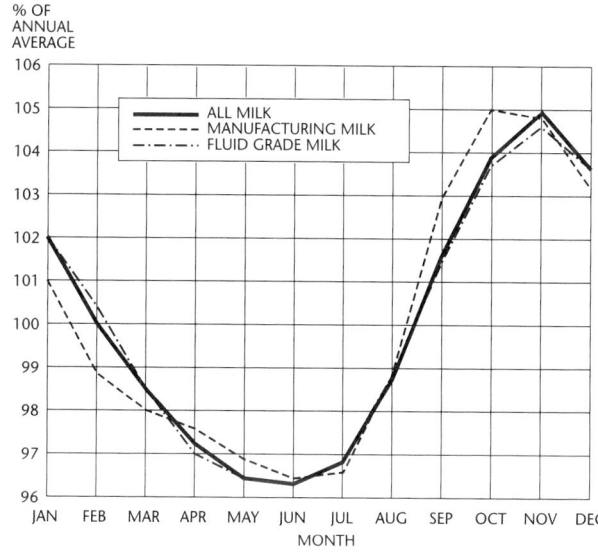

FIGURE 46-3 Monthly variation in prices received by farmers for all milk, manufacturing milk, and fluid grade milk as a percent of annual average. *Sources: Annual Price Summary,* USDA, June 1988 and July 1998.

DEMAND FOR DAIRY PRODUCTS

Products Made from Milk

The four major uses of milk are as fluid product, cheese, butter, or a frozen dairy product. These four types of product use about 96 percent of the milk processed in the United States.

Fluid milk products are mainly whole milk (3.25 percent milkfat), low-fat (from 0.5 to 2.0 percent milkfat), and skim milk (less than 0.5 milkfat). Coffee cream, whipping cream, half and half, and sour cream are also fluid milk products. About 36 percent of the milk produced in the United States is used for fluid milk products.

Many types of cheeses are produced from milk. About 30 percent of the milk produced in the United States is used for making cheese. About 67 percent of the cheese made in the United States is American-type, mainly cheddar and colby. The main Italian-type cheese made is mozzarella for pizza. Italian-type cheese makes up about 25 percent of the total cheese produced in the United States. Swiss cheese accounts for about 6 percent of the total cheese produced. Other types of cheese make up the rest of the cheese produced in the United States.

The production of butter uses about 20 percent of the milk produced in the United States. Only the milkfat is used in the production of butter. The solids-not-fat (SNF) is used to produce nonfat dry milk and condensed skim milk. More butter and nonfat dry milk are produced during the months when milk production is highest. The surplus milk is used for these products.

About 10 percent of the milk produced in the United States is used for frozen dairy products. Ice cream, ice milk, and sherbet are the major products.

About 4 percent of the milk produced in the United States goes for making other dairy products. These include evaporated and condensed milk, evaporated and condensed buttermilk, dry buttermilk, dry whole milk, dry skim milk, dry cream, dry whey, lactose, and yogurt.

Trends in the Consumption of Dairy Products

Figure 46-4 shows the percentage of change in per capita consumption of selected dairy products in the United States from 1977 to 1997. The consumption of cheese, non-fat dry milk, and dry whole milk increased during this period. The consumption of butter, fluid milk and cream, condensed and evaporated milk, and ice cream decreased since 1977.

Increased competition from imitation and substitute dairy products has caused a lower demand for real dairy products. Some of these products are margarine, imitation milk and cheese, nondairy creamers, and nondairy whipped products. Many of these products are sold at lower prices than real dairy products.

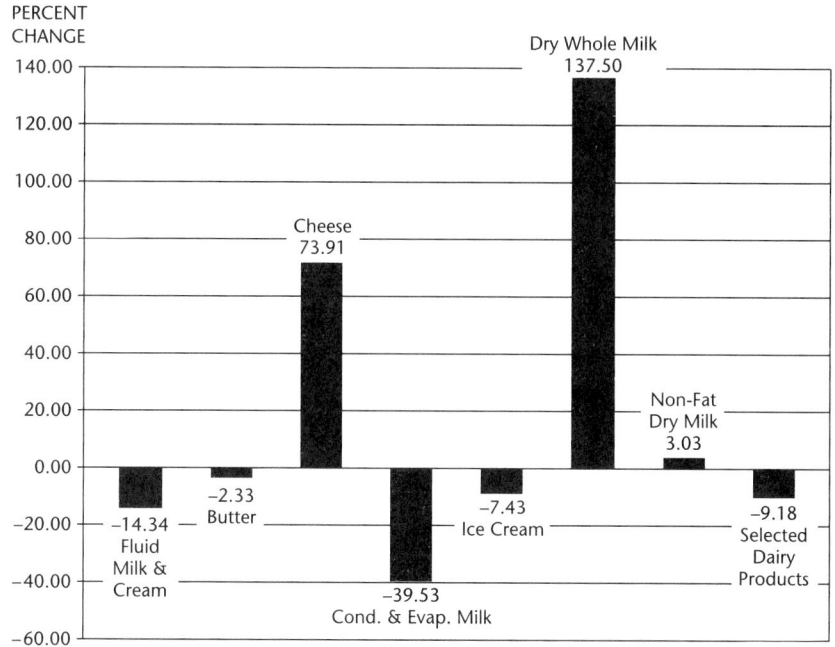

PERCENT
CHANGE

FIGURE 46-4   Percent change in per capita dairy product consumption, 1977–1997. *Sources: Agricultural Statistics,* USDA, 1992, 1999.

Research and development activities are needed to improve dairy products and develop new ones. The National Dairy Council and the United Dairy Industry Association have research and development programs for dairy products. These efforts must be maintained and expanded if recent trends toward a decrease in demand for dairy products are to be reversed.

Projections of demand for dairy products show an expected increase in use of fluid low-fat milk, fresh cream, cheese, ice cream, and butter. The trend toward lower demand for fluid whole milk, cottage cheese, and ice milk is expected to continue.

The demand for dairy products is affected by population, price of dairy products, price of competing products, purchasing power of consumers, and promotional advertising for dairy products.

Advertising Dairy Products

Before 1983 the generic or non-brand specific promotion of dairy products was funded primarily by voluntary contributions by dairy farmers through local and state dairy organizations. Congress established a mechanism for a dairy checkoff program with the passage of the Dairy and Tobacco Adjustment Act of 1983. This established the National Dairy Promotion and Research Board for the purpose of developing and administering promotion, research, and nutrition programs for the dairy industry.

Dairy farmers approved the continuation of the dairy checkoff program through a referendum held in 1993. In 1995, an entity called Dairy Management Inc. (DMI) was organized to coordinate local and national dairy promotion programs. This organization established the U.S. Dairy Export Council to help promote and market U.S. dairy products in international markets.

The promotion of fluid milk and cheese receives the highest priority for the use of the checkoff money. It is estimated that about 200 billion more pounds of milk has been sold than was projected by the USDA since the checkoff program began in 1984. Most of the surplus of dairy products that existed in 1984 has been sold, mainly because of promotion activities funded by the dairy checkoff. Promotion programs are funded by dairy farmers and do not use tax money. The USDA does exercise an oversight function in relation to all producer checkoff programs.

MARKETS FOR MILK

More than 21 billion dollars worth of milk is produced each year on dairy farms in the United States. About 86 percent of this is sold through farmer milk marketing cooperatives. The rest is sold to private firms, used on the farm, and sold directly to consumers.

Looking to the future, the number of dairy cooperatives is expected to decline. Cooperatives will become larger and provide more services to members.

Services Provided by Farmer Cooperatives

Cooperatives provide more services to milk producers than do private firms. The major services provided by cooperatives to milk producers include:

- Checking weights and tests
- Guaranteeing daily market for milk
- Providing marketing and outlook information
- Providing field services such as assisting with production problems
- Collecting and insuring payment from buyers
- Assisting with inspection problems
- Providing insurance programs
- Negotiating hauling rates
- Selling milking supplies and equipment

Cooperatives also provide marketwide services. The major marketwide services provided by cooperatives include:

- Maintaining quality control and related lab services
- Direct farm-to-market movement of milk
- Handling milk in excess of Class I use

FIGURE 46-5 Bulk milk truck loading milk on the farm. *Photos by Michael Dzaman.*

- Participating in federal order hearings
- Paying milk haulers
- Negotiating Class I prices and service charges
- Maintaining a full supply of milk
- Balancing milk supplies among processors to reduce reserve requirements
- Making out-of-market raw milk sales

HAULING MILK TO MARKET

Most of the milk marketed in the United States is hauled from the farm to the plant in bulk trucks (Figure 46-5). There has been a major shift from can to bulk handling of milk during the past forty years.

Fewer dairy farms, larger herds, and fewer dairy plants has made it necessary to haul milk farther. Larger bulk trucks are being used than a few years ago. Fuel and labor costs have increased sharply in recent years.

All of these factors have combined to increase the costs of hauling milk from the farm to the dairy plant.

A charge per hundredweight for hauling is usually deducted from the price the farmer receives for milk. Several studies have shown that the amount deducted does not cover the total cost of hauling. The amount not covered is absorbed by the plant buying the milk. This has the effect of lowering the price the plant is able to pay for milk.

Because of competition for milk suppliers, some plants use various methods of calculating the charge for hauling. In some cases the cost is based partially on the volume of milk picked up at the farm. In other cases a per stop charge is made in addition to the charge per hundredweight. It is expected that milk hauling charges will continue to increase in the future.

MILK GRADES

In 1924 the United States Public Health Service developed the *Standard Milk Ordinance* to help states and local governments prevent diseases that are spread through milk. The adoption of this ordinance is voluntary; however, it is widely used as the model for regulating the production and processing of Grade A fluid milk. The ordinance has been revised many times since 1924 and is now titled the *Grade "A" Pasteurized Milk Ordinance*. This ordinance is recognized by public health agencies and the milk industry as the national standard for milk sanitation. A copy of the complete ordinance may be obtained from the Public Health Service/Food and Drug Administration or may be downloaded from the Internet at the URL: http//vm.cfsan.fda.gov/list.html.

The ordinance sets forth standards for cleanliness of facilities, temperature for storing milk, bacterial count, somatic cell count, and chemical residue in milk, as well as other factors relating to the production and processing of milk for human consumption. Grade "A" raw milk must be cooled to 45°F (7°C) or less within two hours after milking. The blend temperature after the first and subsequent milkings cannot exceed 50°F (10°C). Each farmer's milk cannot exceed 100,000 bacterial count per milliliter before it is mixed with other producers' milk. The bacterial count of milk from several producers mixed together cannot exceed 300,000 bacterial count per milliliter before it is pasteurized. Each farmer's milk cannot exceed 750,000 somatic cells per milliliter. There can be no detectable antibiotics in the milk.

Inspections of dairy farms are made regularly to make sure there are no violations of the standards. The permit to sell Grade A milk may be suspended if violations are found that are not corrected.

Grade B milk is produced under standards that allow it to be used for manufacturing dairy products but not to be used for fluid milk consumption. The requirements for producing Grade B milk are not as strict as those for Grade A milk. However, the standards became stricter beginning July 1, 1980 for dairy plants that are inspected and approved by the USDA.

The standards are found in "Milk for Manufacturing Purposes and its Production and Processing, Recommended Requirements"; they became effective November 12, 1996. A complete copy of the standards may be obtained from the USDA, Agricultural Marketing Service, Dairy Programs or downloaded from the Internet at the URL: www.ams.usda.gov/dairy/index.htm. Many states have adopted these standards as the standard for Grade B milk sold to plants within the state. The adoption of the standard is voluntary. However, plants that do not meet the standards cannot sell products to the Commodity Credit Corporation under milk price support programs.

The Grade "B" milk standard addresses many of the same issues as the Grade "A" milk standard. If the bacterial count exceeds 500,000 per milliliter, the producer is given a warning. If two of the last four consecutive bacterial counts exceed 500,000 per milliliter, a written warning is given to the producer and the regulatory authority is notified. Another sample must be taken within 21 days after the written notice is issued. If the bacteria count still exceeds 500,000 per milliliter, the milk is not permitted on the market. The producer can be given a temporary permit to ship milk if a subsequent test shows a bacteria count below 500,000 per milliliter. Full reinstatement comes only after three out of four consecutive tests show bacteria counts below 500,000 per milliliter.

A somatic cell count above 750,000 per milliliter is considered excessive. The same testing procedures and time limits as described for bacteria counts also apply to somatic cell counts. No drug residues are permitted in the milk.

FEDERAL MILK MARKETING ORDER PROGRAM

The Federal Milk Marketing Order (FMMO) program began with the passage of the Agricultural Marketing Agreement Act of 1937. Since then, the U.S. Congress has passed a series of laws establishing price support programs for dairy products. These price support programs have been implemented through the purchase of surplus butter, cheese, and non-fat dry milk products by the Commodity Credit Corporation (CCC). Figure 46-6 shows milk solids removed from the market for the period 1978 to 1999. The number of Federal milk market orders has been declining for a number of years, dropping to 31 in 1997 (Table 46-1).

The Federal Agriculture Improvement and Reform Act of 1996 required the consolidation of the 31 federal milk marketing orders into no fewer than 10 nor more than 14 orders. This Act also eliminated the dairy price support operations of the CCC after December 31, 1999. It permitted the Secretary of Agriculture to review and revise related issues such as the method for determining the price of milk in the market orders.

Federal milk marketing orders are generally established at the request of the producers or through producer organizations. Two-thirds of the producers affected by the order must approve it by referendum. The order regulates the first buyers of milk and not the producers. The purposes of a federal order market include:

■ Establish and maintain orderly market conditions
■ Establish prices that are reasonable

BILLION POUNDS

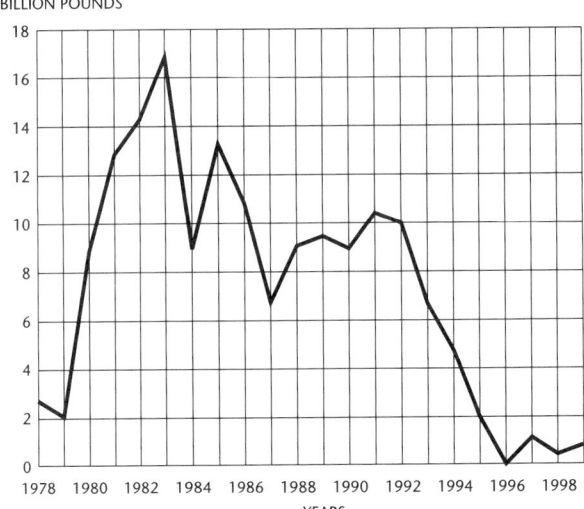

FIGURE 46-6 Net removal of milk from market by Commodity Credit Corporation programs, 1978–1999. *Source:* USDA.

- Insure a sufficient quantity of pure and wholesome milk
- Provide safeguards to protect producers from unfair and abusive trade practices

A milk order may be suspended or terminated by the Secretary of Agriculture. It may also be terminated at the request of more than 50 percent of the producers who supply more than one-half of the milk in the order. Federal milk market orders are administered by the USDA through a local administrator appointed by the Secretary of Agriculture.

In Federal order markets, Grade A milk is divided into classes, based on the final use of the milk, for the purpose of pricing. The federal order price is a minimum price; market conditions may cause the price paid for milk to be above the federal order price.

More than 95 percent of the milk sold to plants and dealers is Grade A. Class I milk is used for whole milk, skim milk, buttermilk, eggnog, and flavored milk drinks. Class II milk is used for ice cream, yogurt, cottage cheese, and cream. Class III milk is used for most cheeses. Class IV milk is used for butter and any milk product in dried form. Class I milk has the highest value and Class IV milk the lowest.

Prior to the reform of the federal milk market order system in 1999, the price of milk was determined by a basic formula price (BFP). This price was based on the price of Grade B milk in selected areas of Minnesota and Wisconsin. When a significant volume of milk sold was Grade B, it was believed that this represented the true competitive market value of milk. Grade A milk prices were then determined by setting differentials based on the class of milk. Because less than five

TABLE 46-1 Federal Milk Order Markets: Measures of Growth, 1966–1997.

Year	Markets[a] Number	Producers[b] Number	Milk Delivered Million lb	Milk Delivered Million kg	Milk Used in Class I Million lb	Milk Used in Class I Million kg	Percentage of Milk Used in Class I (%)	Receipts as Percentage of Milk Sold to Plants and Dealers Fluid grade (%)	Receipts as Percentage of Milk Sold to Plants and Dealers All milk (%)
1966	71	145,964	53,012	24,046	34,805	15,787	65.7	70	48
1967	74	140,657	53,761	24,386	34,412	15,609	64.0	71	49
1968	67	141,623	56,444	25,603	36,490	16,552	64.6	74	52
1969	67	144,275	61,026	27,681	39,219	17,789	64.3	77	56
1970	62	143,411	65,104	29,531	40,063	18,172	61.5	79	59
1971	62	141,347	67,872	30,786	40,268	18,265	59.3	80	60
1972	62	136,883	68,719	31,170	40,938	18,569	59.6	78	60
1973	61	131,565	66,229	30,041	40,519	18,379	61.2	78	60
1974	61	126,805	67,778	30,744	39,293	17,823	58.0	78	61
1975	56	123,855	69,149	31,365	40,106	18,192	58.0	78	63
1976	50	122,675	74,586	33,832	40,985	18,590	54.9	80	65
1977	47	122,755	77,947	35,356	41,125	18,654	52.8	80	66
1978	47	119,326	78,091	35,421	41,143	18,662	52.7	80	67
1979	47	116,447	79,436	36,032	41,011	18,602	51.6	80	66
1980	47	117,490	83,998	38,101	41,034	18,613	48.9	80	67
1981	48	119,323	87,989	39,911	40,746	18,482	46.3	80	68
1982	49	120,743	91,611	41,554	40,807	18,510	44.5	81	69
1983	46	121,052	95,757	43,435	41,091	18,639	42.9	82	70
1984	45	119,033	91,676	41,584	41,517	18,832	45.3	81	70
1985	44	116,765	97,764	44,345	42,201	19,142	43.2	80	70
1986	44	112,271	98,791	44,811	42,724	19,379	43.2	80	70
1987	43	105,896	98,163	44,526	42,897	19,458	43.7	80	70
1988	42	104,141	100,066	45,389	43,141	19,568	43.1	79	71
1989	41	100,291	95,871	43,486	43,367	19,671	45.2	75	68
1990	42	100,397	102,396	46,446	43,783	19,860	42.8	77	70
1991	40	100,267	103,252	46,834	45,033	20,427	43.6	76	71
1992	40	97,803	107,947	48,964	44,914	20,373	41.6	77	73
1993	38	92,934	103,979	47,164	44,805	20,323	43.1	73	69
1994	38	91,397	107,811	48,902	44,866	20,351	41.6	75	71
1995	33	88,717	108,548	49,237	45,004	20,413	41.5	75	71
1996	32	82,947	104,501	47,401	45,479	20,629	43.5	72	69
1997	31	78,590	105,221	47,727	44,916	20,374	42.7	70	68

[a]End of year (Date on which pricing provisions became effective)
[b]Average for year
[c]The decrease in these percentages from 1988 to date results from handlers electing, because of unusual price relationships and qualification circumstances, not to pool milk that normally would have been pooled under federal milk orders.

Sources: Agricultural Statistics, USDA, 1981, 1986, 1988, 1992, 1995, 1999.

percent of the milk currently sold to plants and dealers is Grade B, the price paid probably no longer represents a true competitive market value for milk. The FMMO reform of 1999 addressed this problem by establishing a different method for determining the price of Grade A milk.

This system of pricing uses multiple component pricing with differentials for various classes of milk. The base price for Class III and Class IV milk is determined by the value of protein, butterfat, and other nonfat solids used in manufactured dairy products. Monthly average commodity prices are used to determine the component values. Commodity prices tracked include block cheese, barrel cheese, butter, dry whey, and nonfat dry milk. Estimated manufacturing costs are deducted from the commodity price in order to arrive at a price based only on the component values. The USDA's National Agricultural Statistics Service surveys and reports on these prices. Class I milk in a given milk market order is then priced based on a differential for that market order. The Class II price of milk is set at 70 cents over the price of Class IV milk in all market orders.

Each producer in a milk market order receives a blend price for milk. This is a weighted average price determined from proceeds received for all the milk sold in all the classes in the milk market order. By using a blend price, all producers share both the higher price for Class I milk and the lower prices paid for milk used for manufactured products.

The FMMO reforms of 1999 also standardized a variety of provisions across all the milk market orders. These include definition of various terms used in the orders, regulatory standards for plants and handlers, uniform reporting of milk receipts and utilization, and uniform classification of milk. Prior to this reform, these provisions were not uniform across all milk market orders. Unique provisions that are necessary for specific milk market orders are still permitted.

After the reforms were approved by a majority of dairy producers, and were scheduled for implementation on October 1, 1999, there were some producers who were not satisfied with the revised pricing for their milk. These producers secured a court injunction delaying the implementation of the new milk pricing plan. The House of Representatives in the U.S. Congress passed a bill requiring the USDA to use a different method of determining the price differentials paid for milk in different parts of the United States. At the time the revision of this text was prepared, the U.S. Senate had not acted on the House bill and the issue remained unresolved. It is probable that some provisions of the FMMO reforms of 1999 will remain intact while others, especially the pricing of milk, will be revised. Current information regarding the pricing of milk may be secured from the USDA, the Cooperative Extension Service, or milk processors who buy milk directly from producers.

DAIRY IMPORTS AND EXPORTS

Imports and exports of dairy products have been at relatively low levels for a number of years. Imports have been generally less than two percent of total United States production. A number of dairy products are covered by specific import quotas. These products include dried milk, butter, and several types of cheeses.

Exports of dairy products have generally been below one percent of total U.S. production. Prices in the United States have generally been higher than in other parts of the world. This has tended to limit exports of dairy products.

SUMMARY

The price of milk in the United States has risen from 1967 to 1998. Milk production and prices vary seasonally. Highest production and lowest prices occur in May and June.

About 96 percent of the milk processed in the United States is used for fluid product, cheese, butter, or a frozen dairy product. Per capita consumption of cheese, dry whole milk, and non-fat dry milk has increased in recent years. Demand for other dairy products has decreased.

About 86 percent of the milk produced in the United States is marketed through cooperatives. Cooperatives provide many services to milk producers as well as to the entire market.

Most milk is hauled to market in bulk trucks. The cost of milk hauling has increased a great deal in recent years. A charge for hauling milk is deducted from the dairy farmer's milk check.

Milk is divided into classes based on use for pricing purposes. Class I milk has the highest price.

Federal milk marketing orders are established in many areas. Minimum prices for Grade A milk are set in these areas. The actual price paid is often above the minimum set in the federal order market.

Dairy imports and exports have generally been at a low level for a number of years.

Student Learning Activities

1. Prepare charts for the bulletin board that show price and production trends for milk.
2. Prepare and give an oral report to the class on any aspect of milk marketing.
3. Visit a local milk marketing cooperative and prepare a report for the class.

4. Have a representative from a milk marketing cooperative speak to the class.

Review

1. Describe the long-term trends in the price of milk.

2. Briefly explain the seasonal trends in the price of milk.

3. What kinds of products are made from milk?

4. How much of the milk produced in the United States is used for each general class of product?

5. Describe trends in the consumption of dairy products.

6. What methods are being used to try to increase the demand for dairy products?

7. How is most of the milk in the United States marketed?

8. List the kinds of services provided by marketing cooperatives to dairy producers.

9. List the marketwide services provided by marketing cooperatives.

10. How is most of the milk in the United States hauled to market?

11. How is the hauling paid for?

12. What is Grade A milk?

13. What is Grade B milk?

14. Describe the classes of milk.

15. Briefly explain federal milk marketing orders.

16. How is milk priced in federal milk marketing orders?

Rabbits

Unit 47 Rabbit Raising

CLASSIFICATION OF RABBITS

Rabbits belong to the order Lagomorpha, which is divided into two families: Leporidae, containing rabbits and hares, and Ochotonidae, containing the rock rabbit called pika. The three genera of Lagomorpha are *Lepus* (hares), *Sylvilagus* (American cottontail), and *Oryctolagus* (wild European and domestic rabbits). The wild European and domestic rabbits belong to the species *Oryctolagus cuniculus*. The various species of rabbits do not interbreed.

The ancestor of the domestic rabbit in the United States is the European wild rabbit. It is believed that the European wild rabbit developed in the area around the Mediterranean Sea. The modern breeds of domestic rabbit that are raised in the United States have been developed since the eighteenth century.

USES OF RABBITS

The primary use of rabbits in the United States is for meat production. Rabbits are also raised as pets, for use in research laboratories, and for wool production. There are about 200,000 producers of meat rabbits in the United States. Approximately 8 to 10 million pounds (3.6–4.5 million kg) of rabbit meat are consumed each year in the United States. About 6 to 8 million rabbits are raised each year in the United States for all purposes. The annual world consumption of Angora wool produced from rabbits is about 20 million pounds (9 million kg).

Rabbits produce a white meat that is palatable and nutritious, high in protein, and low in calories, fat, and cholesterol. Operations raising rabbits for meat range from small producers marketing just a few rabbits locally per year to large commercial farms that market many thousands of rabbits annually.

The skins from meat rabbits have some commercial value; to be commercially successful, however, a large volume is necessary. Skins are used for fur garments, slipper and glove linings, in toy making, and as felt. Remnants of flesh from dried skins are used in making glue.

Rabbits are also used for research purposes in medical schools, laboratories, and hospitals. Research relating to venereal disease, cardiac surgery, hypertension, virology, infectious diseases, toxins and antitoxins, and immunology is conducted utilizing rabbits. More than 600,000 rabbits are used annually for research purposes. Persons living near research facilities may find a market for rabbits raised for research purposes.

FIGURE 47-1   Checkered Giant. *Courtesy of American Rabbit Breeders Association.*

BREEDS

Forty-two breeds of rabbits are listed in the American Rabbit Breeders Association (ARBA) Standard of Perfection. Breed standards are based on color, type, shape, weights, fur, wool, and hair. Persons desiring complete information on these breeds of rabbits should contact the American Rabbit Breeders Association, P.O. Box 426, Bloomington, IL 61704. Information regarding rabbit raising, shows, judging, registration requirements, and sources of breeds is available from the ARBA.

The selection of a breed should be based on its expected use. Medium- and heavy-weight breeds are best adapted for meat production. Some breeds are used primarily for show and fur purposes. Persons wishing to market rabbits for research purposes should contact potential buyers in the area to determine the type, age, and size of rabbits preferred. Table 47-1 lists some of the common breeds of rabbits and their primary uses. Figures 47-1 through 47-6 show some typical breeds of rabbits.

At maturity small breeds weigh three to four pounds (1.4–1.8 kg); medium breeds weigh nine to 12 pounds (4–5.4 kg); and large breeds weigh 14 to 16 pounds (6.3–7.2 kg).

Breeding stock should be purchased from reliable breeders. Rabbits purchased for breeding purposes should be evaluated on the basis of their health, vigor, longevity, reproduction ability, and desirable type and conformation.

FIGURE 47-2   Dutch. *Courtesy of American Rabbit Breeders Association.*

FIGURE 47-3   English Spot. *Courtesy of American Rabbit Breeders Association.*

TABLE 47-1 Some Representative Breeds of Rabbits.

Breed	Color	Mature Weight (pounds)	Major Uses
Californian	Body white; nose, ears, feet, and tail colored.	8–10.5	Meat and show
Champagne d'Argent	Under fur dark slate blue; surface fur blue, white, or silver; liberal sprinkling of long black guard hairs.	9–12	Meat and show
Checkered Giant	Body white; black spots on cheeks, sides of body, and hindquarters. Wide spine stripe. Ears and nose black with black circles around the eye.	11 and over	Show and fur
Dutch	Body black, blue, chocolate, tortoise, steel gray, or gray. White saddle or band over the shoulder, under the neck, and over front legs and hind feet.	3.5–5.5	Show and laboratory
English Spot	Body white with black, blue, chocolate, tortoise, lilac, gray, or steel gray spots. Spots on nose, ears, cheeks; circles around eyes. Spine stripe from base of ears to end of tail. Side spots from base of ears to middle of hindquarters.	5–8	Meat, show, and laboratory
Flemish Giant	Body steel gray, light gray, sandy, blue, black, white, or fawn.	13 and over	Meat and show
Himalayan	Body white; nose, ears, feet, and tail colored.	2.5–5	Show and laboratory
New Zealand	Body white, red, or black.	9–12	Meat, show, and laboratory

FEEDING

Rabbits are simple-stomached animals and are herbivorous. *Herbivorous* means their diet comes mainly from plant sources. Rabbits have an enlarged cecum and therefore can use more forage in their diet than can other simple-stomached animals such as swine and poultry. The relative cost of feeds and local availability are two important factors to consider when selecting rabbit feeds.

Nutrient Requirements

The nutrient requirements for rabbits are shown in the Appendix of this text. Rabbits fed well-balanced rations are relatively efficient converters of feed into meat, having a conversion ratio of about 3:1 (three pounds of feed for each pound of meat produced).

Energy. Little research has been done on the energy needs of rabbits. It is believed that rabbits, like many other animals, adjust their energy intake to meet their needs.

Rabbits are efficient users of starch, which is found in cereal grains. Rabbits show a preference for barley or wheat over corn when given a choice of cereal grains. Diets that are based on corn have produced poorer growth rates as compared to barley- or oat-based diets. Oat-based diets appear to give the best results for lactation rations. Since the energy levels of oats, barley, and wheat are lower than corn, it appears that other factors such as palatability of the ration influence the intake level and thus the growth rate in rabbits.

The palatability of the diet may be improved with the addition of fat. Research has shown that rabbits prefer a diet with some fat added. Vegetable oils are a good source of fat in rabbit diets.

The energy requirement of rabbits is influenced by the environmental temperature. As temperatures decrease, the rabbit requires more energy to maintain normal body temperature. To compensate for this increased energy need, either the intake level of feed must be increased or the energy content of the ration must be increased. If the rabbit is already consuming feed at its maximum rate, then the energy content of the diet must be increased. The addition of fat to the ration is probably the best way to increase its energy level. Care must be taken not to have too high an energy level in the diet, because excess energy levels may increase the incidence of diarrhea and enteritis.

Fiber. Research data indicate that rabbits do not make as efficient use of plant fiber in the diet as do cattle, horses, and swine. However, the data indicate that plant fiber is necessary in rabbit diets for the normal functioning of the digestive tract. Crude fiber levels lower than 6 to 12 percent of the ration have been shown to increase the incidence of diarrhea and enteritis in rabbits.

High fiber levels in the ration lower its energy level; this may tend to reduce feed efficiency. Lowering the fiber level during cold weather will increase the energy level of the diet and help provide the needed energy intake for maintenance of body temperature.

Protein. Essential amino acids need to be included in the ration for rabbits. Lysine and methionine are usually the amino acids that are found to be deficient in rabbit rations. While there is some bacterial protein synthesis in the cecum, it is not enough to meet the essential amino acid requirements of rabbits. Non-protein nitrogen sources such as urea should not be used in rabbit diets.

Research has shown that soybean meal or fish meal promotes better growth rates than other protein supplements when the alternative supplements do not have essential amino acids added. When essential amino acids were added to protein supplements such as cottonseed meal, rapeseed meal, horsebeans, and peas, growth rates similar to

FIGURE 47-4   New Zealand White. *Courtesy of American Rabbit Breeders Association.*

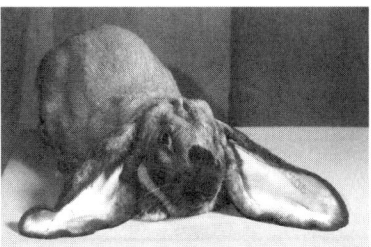

FIGURE 47-5   English Lop. *Courtesy of American Rabbit Breeders Association.*

FIGURE 47-6   English Angora. *Courtesy of American Rabbit Breeders Association.*

those achieved with soybean and fish meals were attained. The amino acid composition of the protein supplement has the greatest influence on its value when feeding rabbits.

The relative cost of protein supplements must be considered when formulating rations. Animal and fish meal supplements are generally more expensive than legume and oil seed supplements.

Rabbits do make efficient use of the protein in plants. This means that large amounts of alfalfa can be used in rabbit diets. Research has shown that adequate growth rates can be maintained when alfalfa protein concentrate is substituted for soybean meal in the ration.

Coprophagy. *Coprophagy* refers to the ingestion of fecal matter. Rabbits produce two kinds of feces, one hard and one soft; the soft feces is ingested directly from the anus as it is excreted, usually when the rabbit is unobserved. The practice of coprophagy is similar to rumination practiced by ruminants, in the sense that it provides a method of passing feedstuffs through the digestive tract a second time. It is sometimes referred to as pseudorumination. It is believed that coprophagy permits rabbits to synthesize the necessary B vitamins, contributes to better utilization of plant proteins, and provides some additional digestion of other nutrients. Rabbits begin the practice of coprophagy when they begin eating solid feed at about three to four weeks of age. The practice is normal in rabbits and does not indicate any nutritional deficiencies.

Minerals. A calcium/phosphorus ratio of 1:1 in rabbit diets will meet the need for these mineral elements. Rabbits can tolerate high levels of calcium in the diet without adverse effects. Levels of phosphorus above 1 percent of the diet reduce the palatability of the ration and may lower feed intake. Alfalfa and other legumes are good sources of calcium, while cereal grains are good sources of phosphorus; a combination of alfalfa and grain will generally supply the calcium and phosphorus needed in rabbit diets. Complete pelleted feeds will also supply the needed amounts of these minerals.

The use of iodized salt at the rate of 0.5 percent of the diet will supply the needed sodium, chlorine, and iodine for rabbits. Salt spools may be used, but they are more expensive, require more labor, and are corrosive to wire cages. Commercial rations for rabbits usually contain enough of these minerals; the use of salt spools, therefore, is generally not required.

Other required trace minerals in rabbit diets are usually provided through properly formulated commercial feeds. Deficiencies of trace minerals are rarely observed with rabbits.

Vitamins. While it is known that rabbits need fat-soluble vitamins in their diet, there is little available research information relating to quantities needed. Commercial feeds for rabbits are normally fortified with adequate amounts of the fat-soluble vitamins. Additionally, alfalfa is a good source of vitamin A. Vitamin K is synthesized in the intestinal tract of the rabbit; therefore, deficiencies rarely occur.

Several of the water-soluble vitamins, especially riboflavin, pantothenic acid, biotin, folic acid, and vitamin B_{12}, are synthesized by rabbits in amounts sufficient to meet their needs. No additional supplementation of the diet is usually needed for these vitamins. Some fortification of the ration with choline and niacin is suggested. Additional pyridoxine is usually not needed, as this vitamin is found in both cereal grains and forages, and is synthesized in the digestive tract as well. No additional vitamin C is needed in the ration.

Water. Rabbits need a good supply of clean, fresh water at all times. A commercial rabbitry should use an automated water system to meet the water needs of rabbits. Rabbits usually consume 2.5 to 3 times more water than dry matter. Water consumption increases with both hot and cold air temperatures. If the water supply is limited, feed intake is reduced. A meat doe, weighing 10 to 12 pounds (4.5–5.4 kg), and an eight-week-old litter of seven will consume about one gallon (3.8 liters) of water in each 24-hour period. Water intake is increased when the diet contains higher levels of protein and fiber.

Additives. The use of antibiotics on a regular basis in rabbit rations is not generally recommended. Because of the withdrawal period required before marketing, the use of antibiotics in the feed would require the use of a different feed formulation during the withdrawal period. Since there is little research to indicate that the regular use of antibiotics in rabbit feeding significantly improves rate of gain or feed efficiency, it is not considered to be economical to use them on a continuous basis.

Water treatment with drugs is recommended when it is required for the treatment of an outbreak of disease. Manufacturer's recommendations should be followed when using any drug treatment. Current regulations regarding drug treatments are available from the Federal Food and Drug Administration and must be followed.

FEEDS

Hay

Dry forages such as hay usually make up between 40 and 80 percent of a rabbit's diet. Legume hays, especially alfalfa, make good roughage feeds for rabbits. Grass hays are not as palatable for rabbits as legume hays, but may be used. The protein level of grass hays is lower than that of legume hays; therefore, protein supplementation of the ration is needed if they are used. Grass hays harvested before they are in bloom are more desirable than those cut after bloom. Hay provides bulk and fiber in the diet and its use reduces the incidence of fur chewing in rabbits.

Select hays that are leafy, fine stemmed, green, and well cured. Never feed hay containing mold or mildew. Coarse hay should be cut into 3- or 4-inch lengths to reduce waste.

Green Feeds/Root Crops

Fresh green feeds such as grasses, palatable weeds, cereal grains, and leafy vegetable crops may be fed. These feeds are high in vitamins, minerals, and proteins. They are of special value when feeding breeding animals.

Root crops such as carrots, sweet potatoes, turnips, mangels, beets, and Jerusalem artichokes may be fed to rabbits. They may be substituted for green feeds during the winter, when the green feeds are not available.

Green feeds and root crops are supplements to the concentrate in the diet. They will produce choice carcasses when fed to meat rabbits and may be used for maintenance for mature animals that are not in production.

Do not overfeed green feeds or root crops to rabbits that are not accustomed to them. If the feed is spoiled or contaminated, do not use it. Put the feed in a manger rather than on the floor and remove any that has not been eaten before feeding any additional amounts.

Commercial producers generally do not use fresh green feeds or root crops because of their high cost per unit of nutrient, the amount of labor required for their use, and their lower energy value; the low energy value reduces both the efficiency of meat production and their value for lactating females.

Grains

Oats, barley, wheat, buckwheat, grain sorghum, rye, and soft varieties of corn may be fed, either whole or milled. To reduce waste, flinty varieties of corn should be processed (ground or cracked) before feeding. One cereal grain may be substituted for another in the ration, although rabbits do show a preference for barley, wheat, or oats over corn.

Bran, middlings, shorts, and other cereal by-products may be used in mash mixtures or pellets for rabbits.

Protein Supplements

Soybean, linseed, sesame, and cottonseed oil meals are good protein supplements for use in mash or pelleted feeds. They should not be mixed with grains because they will settle out and be wasted. When feeding whole grains, feed the protein supplement in cake, flake, or pelleted form. Soybean oil meal is the most widely used protein supplement for rabbit rations.

Cottonseed meal must be treated to remove gossypol, which is toxic to rabbits. Cottonseed meal should be limited to no more than 5 to 7 percent of the total diet.

Animal protein supplements are generally not used in rabbit production because of their higher cost. Rabbits can meet their protein needs from plant protein sources.

PELLETING FEEDS

Rabbits prefer a pelleted ration to one in mash form. Both rate of gain and feed efficiency are improved when pelleted rations are used. Commercial producers of meat rabbits usually use complete pelleted rations. In some areas, the only commercial rabbit feeds available are in pelleted form.

There are two types of pelleted feeds available: all-grain pellets, which are designed to be used with hay, and complete pellets (green pellets). Complete pellets generally contain all the nutrients needed by rabbits.

The recommended pellet size is 1/8 to 3/16 inch in diameter and 1/8 to 1/4 inch in length. Rabbits will bite off only part of the pellet if it is too big and the rest of the pellet is dropped; the dropped portion is usually wasted.

STORING FEEDS

Feeds stored for over four weeks lose feed value and palatability. Keep rodents and insects out of stored feeds to avoid contamination of the feed. Do not allow cats or dogs in feed storage areas; cat or dog droppings may contain tapeworm eggs that could infest rabbits.

Store pellets in a dry area; excess humidity causes the pellets to soften and decreases the palatability of the feed.

FEEDING METHODS

Rabbits are generally hand-fed in small rabbitries. Feed is placed by hand in feed crocks or troughs. Regular feeding is more important than the number of times per day feed is provided. Rabbits generally eat more feed at night than during the day.

Rabbits may be self-fed or full-fed by means of a hopper or self-feeder. Rabbits on full-feed usually make more efficient use of feed; this results in more pounds of gain per pound of feed consumed. Gains are also generally faster on full-feed. Commercial meat producers usually utilize full-feeding with automated feeding equipment.

DIETS FOR COMMERCIAL PRODUCTION

Table 47-2 gives some examples of diets that are adequate for feeding rabbits in commercial production.

TABLE 47-2 Examples of Adequate Diets for Commercial Production.

Kind of Animal	Ingredients	Percent of Total Diet*
Growth, 0.5 to 4 kg	Alfalfa hay	50
	Corn, grain	23.5
	Barley, grain	11
	Wheat bran	5
	Soybean meal	19
	Salt	0.5
Maintenance, does, avg. 4.5 kg	Clover hay	70
	Oats, grain	29.5
	Salt	0.5
Pregnant does, avg. 4.5 kg	Alfalfa hay	50
	Oats, grain	45.5
	Soybean meal	4
	Salt	0.5
Lactating does, avg. 4.5 kg	Alfalfa hay	40
	Wheat, grain	25
	Sorghum, grain	22.5
	Soybean meal	12
	Salt	0.5

*Composition given on an as-fed basis.

Reprinted from "Nutrient Requirements of Rabbits, Second Revised Edition," 1977, with permission of the National Academy Press, Washington, D.C.

MAINTENANCE FEEDING

Junior does and bucks, mature dry does, and herd bucks that are not in service can be fed a maintenance ration. A fine-stemmed, leafy, legume hay will provide the nutrients needed for maintenance. Adding all-grain pellets or complete pellets to the maintenance ration is recommended if coarse legume hay or grass hay is fed. Medium-weight breeds should be fed about two ounces of pellets several times per week. Light-weight breeds should be fed less, and heavier-weight breeds should be fed more pellets along with the coarse legume or grass hay. Observe the rabbits closely to make sure they do not become too fat.

FEEDING HERD BUCKS IN SERVICE

Allow herd bucks free access to high-quality hay during the breeding season. Feed four to six ounces of a complete pellet if hay is not included in the ration. Do not allow bucks to become too fat, but feed them enough so they stay in good condition while in service.

FEEDING GROWING JUNIOR DOES AND BUCKS

A daily allowance of two to four ounces of all-grain or grain-protein pellets and free access to good-quality hay will provide the proper nutritional level for rabbits of the medium-weight breeds. A daily ration of four to six ounces of a complete pelleted feed is sufficient for proper growth of medium-weight breeds. Careful observation of the growing juniors is required to make sure they do not become too fat. Decrease the amount fed to light-weight breeds and increase the amount fed to heavier-weight breeds.

Alfalfa pellets containing 99 percent No. 2 or better-grade leafy alfalfa meal and 1 percent salt may be fed as the only ration to growing juniors from weaning until breeding. The pellets should contain 15 percent protein. A coarse alfalfa crumble or turkey alfalfa crumble may be substituted for alfalfa pellets if the latter are not locally available.

FEEDING PREGNANT AND NURSING DOES

After they are bred, continue feeding does the maintenance ration until it has been determined that they are pregnant. Pregnancy may be determined about 12 to 14 days after breeding by palpating, which is described further in the section on rabbit breeding. Good-quality hay or hay pellets may be fed. Limit the amount of feed when feeding an all-pellet ration to keep the does from becoming too fat.

After pregnancy is confirmed, does may be full-fed a complete pelleted feed. A ration of good-quality hay with a full feed of concentrates may also be fed. Grain plus protein pellets or all-grain pellets may be fed as the concentrate in the diet. Commercial rabbitries usually feed a complete pelleted ration. Pregnant females will normally eat six to eight ounces of feed per day.

Do not make sudden changes in the diet or the doe may go off feed; any change in the diet must be made gradually. Feed 1/4 of the new ration and 3/4 of the old ration for three or four days; then feed 50 percent of each for three or four days; complete the change by feeding 3/4 of the new ration and 1/4 of the old ration for another three or four days.

On the day of kindling, the does should be fed about one half the normal daily ration. After kindling, gradually increase the amount of daily feed until the does are back on full-feed at the end of one week.

Nursing does may be fed the same ration as pregnant does. Full-feed the does until the litter is weaned at about two months of age. Satisfactory diets include hay with an all-grain pellet, a grain-protein mixture, or a complete pelleted feed. Nursing does will eat six to eight ounces of feed per day until the litter is three weeks old. During the nursing period of three to eight weeks, does will eat from one to two pounds of feed per day.

RABBIT BREEDING

Does remain in heat for long periods of time during the breeding season. They do not show regular estrous cycles. Normally, ovulation occurs in the female about 10 hours after she is bred to the male; the sperm fertilize the eggs shortly after ovulation. The gestation period for rabbits is normally 30 to 32 days.

Litter size will vary with breeds and strains of rabbits. More prolific breeds will average about eight young per litter. Poor nutrition will lower the litter size.

The light-weight breeds of rabbits become sexually mature at an earlier age than the medium- and heavy-weight breeds. They may be bred when they are four to five months of age. The medium-weight breeds may be bred when they are five to six months of age and the heavy-weight breeds may be bred at eight to 10 months of age. Does generally reach sexual maturity earlier than bucks. Commercial rabbitries often allow bucks to grow a month longer than does before using them for breeding for the first time. There is, however, no research currently available that supports this practice.

Prolific does that are in good physical condition can be rebred six weeks after kindling (giving birth), even though they are still nursing their young. This makes it theoretically possible for a doe to produce five litters per year (assuming that there are no conception failures). Does that are rebred after weaning their litters at eight weeks of age can produce four litters per year. Commercial rabbitries often use breeding intervals of 21, 28, or 35 days to increase production. No research is currently available that measures the influence of accelerated breeding schedules on reproductive life, mortality, growth rate, feed conversion, or carcass quality. Does that are maintained in good physical condition will generally produce litters in commercial rabbitries for 2½ to three years.

Does that are full-fed a properly balanced ration can be rebred before weaning the current litter. Do not rebreed does that are in poor physical condition. Feed a ration that improves their physical condition before attempting to rebreed them. If the litter dies at birth or is small and the doe is in good physical condition, she may be rebred about 3 or 4 days after kindling.

In commercial rabbitries, a regular breeding schedule should be followed, regardless of whether or not the doe shows signs of being ready for mating; these signs include restlessness and nervousness, rubbing the chin on feeding and watering equipment, and attempting to join with rabbits in other cages.

The doe is taken to the buck for mating. Mating generally occurs within a few minutes of placing the doe with the buck. Return the doe to her cage as soon as mating is completed.

Records of matings that include the date and identification of the doe and buck may be kept. One buck should be kept for each 10 breeding does. Mature, healthy, vigorous bucks may be mated several times per day for short periods of time.

Does sometimes exhibit pseudopregnancy (false pregnancy); this may result from an infertile mating or one doe's riding another during a period of sexual excitement. The pseudopregnancy lasts for 17 days and the doe cannot be bred during this period. Separating does 18 days before mating will allow them to pass through any pseudopregnancy before breeding. The conception rate for rabbits is higher during the spring.

Research has shown that the conception rate in March and April may be as high as 85 percent, while it may be as low as 50 percent during September and October. There is a high level of individual variation in fertility among does and bucks; this factor should be considered when selecting breeding stock.

Artificial insemination is not commonly practiced in commercial rabbitries. It is sometimes used with rabbits used in research.

Pregnancy should be confirmed by palpating the doe 12 to 14 days after mating. Restrain the doe by holding the ears and a fold of skin over the shoulders in one hand. With the other hand, reach under the shoulder to the area between the hind legs and in front of the pelvis. Place the thumb on the right side and the fingers on the left side of the two uteri to palpate the fetuses. Move the hand gently back and forth, exerting a slight pressure. If the doe is pregnant, the fetuses may be felt as small marble-shaped forms, which will slip between the thumb and fingers. Do not exert too much pressure. Use caution when palpating to avoid bruising or tearing the tissue, which may cause abortion.

KINDLING

Prepare for kindling by providing a nesting box in the cage 27 days after the doe is bred. The nesting box should contain bedding materials. Suitable materials include clean straw or wood shavings. The doe will prepare a nest in the nesting box by pulling fur from her body to line the nest.

Feed intake is often reduced just before kindling. Adding some green feed to the diet may have a beneficial effect on the digestive system.

Kindling usually occurs at night. Does in good physical condition seldom have problems kindling. They are often nervous after kindling and should not be disturbed until they have quieted down.

If the doe fails to pull enough fur to cover the litter (or kindlings) on the hutch floor, warm the litter and add fur to keep them warm. Fur may be pulled from the doe's body for this purpose, or extra fur may be kept on hand to use.

The litter should be inspected no later than the day after kindling and any dead, deformed, or undersized young should be removed from the nest box. Take care not to disturb the doe when inspecting the litter. This inspection will not cause the doe to disown the litter. Nervous does may be quieted by placing feed in the hutch immediately after the inspection.

The litters of breeds used for meat production should contain seven to nine young. If the litter has more than nine young, the extra rabbits may be transferred to a smaller litter; make this transfer in the first three or four days after kindling. Place the young rabbits with a litter of approximately the same age. More uniform development and finish at weaning will result from keeping the litters at the desirable number.

Young rabbits usually open their eyes about 10 or 11 days after birth. If an infection prevents the eyes from opening normally, they should be treated promptly. Wash the eyes with warm water. After the tissue softens, the eyes may be opened using a gentle pressure. An antibiotic eye ointment should be used to treat the eyes if pus is present.

LOSSES AT KINDLING

Does sometimes kindle on the hutch floor rather than in the nest box. If this happens, the young will die of exposure unless they are warmed and placed in the nest box.

Frightened does may kindle prematurely. If a doe is frightened after kindling, she may jump into the nest box and injure or kill the young rabbits by stamping with her back feet. Take care to prevent does from becoming frightened just prior to or at kindling time.

If the doe fails to produce milk, the young will die in two or three days. Make sure the doe is producing milk and is feeding the young. If the doe is not producing milk, transfer the young rabbits to litters where they can nurse and be properly cared for.

While it is not common, a doe may eat her young. This may occur if the diet is not adequate to meet her needs. Does that are disturbed, nervous, or frightened after kindling may also eat their young. Providing a good diet and using care when handling the doe will usually prevent this problem. Does that continue to eat their young despite proper feeding and care should not be kept.

WEANING

Young rabbits will start to eat solid feed at about 19 to 20 days of age. They will come out of the nest box at this time for feeding.

Does usually nurse their young at night. If some of the young are out of the nest box and others are not, she will nurse only one group or the other, but not both. The doe will not carry the young rabbits back into the nest box to nurse; make sure all the young are nursing properly or losses will occur.

Rabbits may be weaned at eight weeks of age. Meat rabbits should weigh about four pounds (1.8 kg) and be ready for market at that time. Leaving the rabbits with the doe for nine or 10 weeks will produce fryers weighing about 4.5 to 5.5 pounds (2–2.5 kg).

Producers who use an accelerated breeding program, which breeds does at less than 35 days after kindling, should wean the young at five to seven weeks; this allows the doe to be in better physical condition for the next litter. It is a good practice to allow the doe to have a few days between weaning a litter and the birth of the next litter.

HANDLING RABBITS

Do not handle rabbits unless it is necessary. Young rabbits that are to be kept for breeding purposes may be handled occasionally to get them used to the moving done during breeding.

Do not lift rabbits by their ears or legs. To properly lift a rabbit, grasp the loose skin over the shoulders with one hand and support the weight of the rabbit with the other hand under the rump. Rabbits that weigh less than four pounds (1.8 kg) may be lifted and carried by the loin region. When carrying the rabbit in this manner, keep the heel of the hand toward the tail of the rabbit. Take care not to bruise or damage the skin of the animal.

SEXING RABBITS

Rabbits that are to be kept for breeding stock should be separated by sex at weaning. Hold the rabbit on its back, restraining it by holding the front legs up along the head. With the other hand, depress the tail back and down. Use the thumb to gently depress the area in front of the sex organs to expose the reddish mucous membrane. The organ protrudes as a rounded tip in the buck; it protrudes as a slit with a depression at the end next to the anus in the doe.

IDENTIFICATION OF BREEDING RABBITS

Tattoo breeding rabbits in the ear to mark them for identification. This provides a permanent mark that does not disfigure the ears. Do not use ear tags or clips for marking rabbits; these often tear out of the ear. An adjustable box may be used to restrain the rabbit while tattooing the ears. This makes it possible for one person to perform the operation. Biological and livestock supply houses have tattooing equipment available.

CASTRATION

Rabbits raised for meat purposes are generally not castrated. There is no research to indicate that castrating young bucks will improve feed conversion, rate of growth, or carcass quality.

CARE DURING HOT WEATHER

In hot weather, rabbits need to be kept cool. Good ventilation without drafts is necessary in the rabbitry. Pregnant does and newborn litters are especially susceptible to hot weather. Young rabbits become restless when the temperature is too high. Heat stress in older rabbits is indicated by rapid respiration, excessive moisture around the mouth, and sometimes slight bleeding around the nostrils. Rabbits that show symptoms of heat stress should be moved to a cooler area.

Automatic sprinkling equipment may be used in hot, dry areas to help cool rabbitries. Evaporative coolers also may be used to help keep the rabbitry cool. In areas where the humidity is high, use cooling fans in the rabbitry.

CARE DURING COLD WEATHER

Rabbits should be protected from direct exposure to rain, sleet, snow, and wind. Provide adequate ventilation in enclosed buildings to reduce excess moisture. Respiratory diseases are more common in rabbits exposed to drafts, high humidity, and cold. Mature rabbits are less susceptible to cold weather than younger rabbits if they are kept free of drafts. Make sure young rabbits are kept warm during cold weather.

Commercial rabbitries are increasingly turning to the use of controlled-environment housing for rabbits as a means of eliminating extremes in temperature. Seasonal variations in production are also reduced.

MAINTAINING HEALTHY RABBITS

Sanitation and Disease Control

Good sanitation in the rabbitry is the key to disease control. Remove manure and bedding on a regular basis. Protect water and feed from contamination from urine and feces. Contaminated feed must be removed daily. Store feed in rodent-proof areas. Clean watering and feeding equipment frequently by washing it in hot, soapy water. Rinse the equipment thoroughly and dry it in the sun or use rinse water with a disinfectant added and then rinse with clear water.

Do not overcrowd animals in the rabbitry; overcrowding makes it more difficult to follow good sanitation practices. When rabbits are overcrowded, their resistance to disease is lowered and diseases spread more rapidly through the herd.

After an outbreak of disease or parasites in the rabbitry, clean and disinfect hutches and equipment. Commercial rabbitries often use steam under pressure to clean facilities and equipment. Soak caked fecal matter before using a steam cleaner to facilitate removal of the feces. Make sure hutches and equipment are dry before putting rabbits in them. Nest boxes should be cleaned and disinfected after each use.

Movable equipment that has been thoroughly cleaned may be placed in the sun for a period of time to disinfect it. Dry heat from a flame may be used to disinfect equipment.

Caution: The use of a flame in the rabbitry is a fire hazard.

Maintaining Health

Good sanitation, proper management, adequate diet, and plenty of clean, fresh water will help the grower maintain healthy rabbits. Observe the herd closely for signs of disease outbreak. Isolate any animals that appear to be sick. Keep them isolated for at least two weeks. Treatment and caring for sick animals should be done only after completing work with the healthy rabbits. Practice good personal sanitation by washing hands and disinfecting boots after working with sick animals. Send some of the sick animals to a diagnostic laboratory if the cause of illness is not readily apparent. Destroy sick animals that cannot be successfully treated. Dead animals should be burned or buried. Do not use open pits for disposal of dead animals.

Preventing Disease Transmission

The most common health problems of rabbits are pasteurellosis, ear mange, and coccidiosis; these are often transmitted by contact with infected rabbits or by mechanical carriers.

Adult rabbits that have had a disease may appear healthy but will still spread the disease; infectious organisms may be present in the feces, urine, or moisture droplets exhaled with the breath of these animals. Pasteurellosis and liver coccidiosis are often spread by contact carriers. Isolate new animals or those that have returned from shows from the rest of herd for a period long enough to determine that they are free of disease. A minimum of two weeks of isolation is recommended. Tests to determine if the rabbit is a carrier of disease are often too expensive to be practical.

Sometimes the only solution to a disease problem in a rabbitry is to remove all the animals and clean and sanitize the facility. After the facility has been cleaned, use only disease-free animals to repopulate the rabbitry.

Disease organisms may be mechanically transmitted from one rabbitry to another. People such as feed salesmen, servicemen, buyers, and visitors who travel from place to place may carry disease organisms; growers who check or treat sick animals and then work with the healthy animals may carry an infection to the healthy rabbits. Keep dogs, cats, birds, and rodents out of the rabbitry; these animals may carry disease organisms. Control insects in the rabbitry to help prevent the spread of disease.

COMMON HEALTH PROBLEMS

Pasteurellosis

Pasteurellosis is the most serious health problem of rabbits. It appears in a number of forms, including snuffles, pneumonia, pyometra, orchitis, otitis media, conjunctivitis, subcutaneous abscesses, and septicemia. All of these diseases are associated with the bacteria *Pasteurella multocida.*

Snuffles, or cold, is caused by a bacterial infection of the nasal sinuses. Symptoms include sneezing, rubbing the nose, and a nasal discharge. Snuffles is usually a chronic condition that may develop into pneumonia. Stress contributes to the development of this disease. Antibiotics are used in treatment.

Pneumonia occurs when the bacterial infection develops in the lungs. Symptoms include labored breathing, depression, nasal discharge, elevated body temperature, and a bluish eye and ear color. This disease is the major single cause of death in mature rabbits. Antibiotics are used in treatment.

Pyometra is pus in the uterus of female rabbits. It occurs when Pasteurella bacteria get into the uterus during mating with a buck that has infected testicles. There is no treatment.

Orchitis is an infection of the testicles of the buck by the Pasteurella bacteria. Symptoms include enlarged testicles, which contain pus. Infected bucks will transmit the bacteria to does, causing pyometra. Generally, this condition is not treated. If the infection occurs in the membranes of the penis, swelling, reddening, and pus may be observed; antibiotics may be used to treat this condition.

Otitis media is an infection of the middle ear. If the infection spreads to the inner ear, the equilibrium of the rabbit is affected, and the animal will tilt its head. This condition is referred to as *wry neck.* Treatment of otitis media is usually not attempted because of the difficulty of getting antibiotics to the site of the infection. Wry neck may also be caused by some other bacterial infection, by parasitic infestation, or by injury.

Conjunctivitis occurs when the Pasteurella bacteria infect the membranes covering the eye and the inner part of the eyelid. The condition is sometimes called *weepy eye.* Symptoms include inflammation of the eyelids, discharge from the eye, and wet, matted fur around the eye. Antibiotic eye ointment is used in treatment.

Subcutaneous abscesses are soft swellings under the skin of the rabbit. The abscess is opened and drained and an antibiotic ointment is applied.

Septicemia, which is caused by Pasteurella bacteria, is an infection of the blood stream. It is usually of short duration and shows no visible symptoms. Because of the lack of symptoms and its short duration, treatment is usually not attempted.

Enteritis

Enteritis is a disease complex associated with bloating and diarrhea. This disease complex can be a serious health problem in rabbits. Death loss is often high. Enteritis takes several forms.

Mucoid enteritis is most commonly found in meat rabbits. Symptoms include loss of appetite, bloated appearance, grinding of teeth, and high death rate. The cause of this condition is not known. Reducing stress and feeding an adequate amount of fiber in the diet helps prevent the condition. Antibiotics may be used to treat secondary infections.

Tyzzer's disease is caused by the organism *Bacillus piliformis*. Symptoms include diarrhea, listlessness, loss of appetite, and dehydration. Affected animals usually die within 72 hours. Stress combined with exposure to fecal-contaminated feed and bedding contribute to the transmission of this disease. Antibiotic treatment of the entire herd has shown some success in reducing the effect of Tyzzer's disease.

Enterotoxemia is characterized by an acute diarrhea coupled with dehydration and death within 24 hours. No specific cause has been identified. The disease appears to be associated with a diet that is high in energy and low in fiber. Feeding a diet lower in energy and higher in fiber (18 percent) may help prevent this disease. Treating the drinking water with antibiotics may also help.

Other Health Problems

Table 47-3 describes some other health problems of rabbits.

KEEPING RECORDS

An important part of commercial rabbit raising is keeping good records. Records can provide information for culling unproductive animals from the herd, selecting breeding stock, determining feed efficiency, and determining costs and returns from the enterprise. Good records are also essential for tax purposes.

The doe performance record card should include identification data, date bred, buck used, date kindled, number of young born alive and dead, number of young retained, litter number, date weaned, number weaned, age at weaning, and weight at weaning; the buck performance card should include identification data, does bred, date of breeding, number of young kindled alive and dead, number weaned, and weaning weights. Annual summaries of the individual performance records may be made. Various record card forms are available from commercial firms that supply rabbitries.

Financial records should show investment in capital equipment, production expenses, and income records. The cost of production and returns over expenses may be determined from good records. Financial records also provide the information needed for income tax purposes.

TABLE 47-3 Some Common Health Problems of Rabbits.

Health Problem	Symptoms	Cause	Treatment/Control
Caked breasts	Breasts are swollen and firm; hard knots form at sides of nipples. May occur after kindling, when litter is small, when litter dies, or after weaning.	Excess supply of milk compared to rate of removal.	Reduce amount of concentrate in diet for several days. Reduce amount of feed before kindling, weaning, or if litter dies.
Coccidiosis, intestinal	Diarrhea containing blood. Slow or no weight gain, or loss of weight. May be pot-bellied after recovery.	Coccidia—protozoan parasite infects intestinal tract. (*Eimeria magna* and *E. irresidua*)	Treat with coccidiostat. Practice good sanitation. Prevent contamination of feed and water by feces.
Coccidiosis, liver	Diarrhea, loss of appetite, weight loss, pot belly may develop. Death may occur. White nodules on the liver.	Coccidia—protozoan parasite infects liver. (*Eimeria steidae*)	Same as above.
Ear mange (canker)	Scratching ears, shaking head. Scabs at base of inner ear.	Rabbit, goat, and cat ear mite.	Swab ear with medication. Isolation. Treat entire herd with medication once a month.
Fur block	Loss of appetite, loss of weight. Sometimes diarrhea. Fur is rough. Pneumonia may develop.	Small intestine blocked by fur. Blocks flow of feed through digestive system. Not enough fiber or roughage in diet.	Increase roughage in diet. Give mineral oil to affected animal.
Heat prostration	Respiration rate becomes rapid. Moisture around mouth and nose may contain blood. Ears and mouth become blue-tinged. Prostration followed by death.	High temperature often accompanied by high humidity.	Put affected animals in cool place. Keep animals cool with proper ventilation in buildings.
Mastitis (Blue breast)	Breasts become swollen, feverish, turn black and purple. Abscesses may form. Animal has fever.	Infection of breasts by bacteria (usually *Staphylococcus* or *Streptococcus*).	Treat with antibiotic. Sanitation and less concentrate in the diet helps prevent.
Paralyzed hindquarters	Drags hind legs, cannot stand, no control of bladder or bowel.	Injury that breaks back, displaces disk in spinal cord, or damages nerves in spinal cord.	No treatment: destroy affected animal. Prevent by careful handling. Protect from disturbances.
Pinworm	Mild infestation: no symptoms. Severe infestation: slow growth, weight loss, reduced disease resistance. Worms found in cecum and large intestine.	Pinworms (*Passalurus ambiguus*).	Treat with worm medicine. Prevent with good sanitation.

TABLE 47-3 Some Common Health Problems of Rabbits. *(Cont.)*

Health Problem	Symptoms	Cause	Treatment/Control
Ringworm (favus)	Hair loss in circular patches, scaly skin with red raised crust. Matted fur. May occur on any part of the body.	Fungus (*Trichophyton* and *Microsporum*). May be transmitted to humans: Use caution when handling animals.	Individual treatment with medication applied to skin. Herd treatment with oral or systemic medication. Disinfect cages and equipment.
Skin mange	Dry, scaly, irritated skin, scratching, loss of fur on head, ears, and neck.	Mange mites (rabbit fur mite and scabies or itch mite).	Consult veterinarian. Treat with miticide. Isolate herd, control rodents.
Sore hocks	Infected, inflamed, bruised, or abscessed bare areas on hind legs. Animal shifts weight to front legs. May affect front legs in severe cases. Secondary infections may occur.	Wet floors in hutch, unsanitary conditions, irritation from wire floors. Nervous stompers more likely to be affected.	Place animals on dry, solid surface. Good sanitation. Cull severe cases. Select breeding stock with well-furred feet.
Tapeworm larvae	Mild cases: no symptoms. Severe cases: diarrhea and loss of weight.	Larva from dog or cat tapeworm. Infestation occurs when rabbit ingests feed or water contaminated by eggs or tapeworm segments.	No practical treatment. Prevent by keeping dogs and cats out of rabbitry.
Urine-hutch burn	External genitals and anus become inflamed. Crusts may form, bleeding may occur. Pus may develop in severe cases.	Condition starts with wet, dirty hutches. Bacteria infect skin membranes.	Treat with antibiotic cream. Prevent with good sanitation practices.
Vent disease (Spirochetosis)	Blisters, lesions, scabs around genital organs. Lesions may occur around nose, ears, or mouth. (Sometimes confused with symptoms of urine-hutch burn.)	Spirochete (*Treponema cuniculi*). Usually transmitted by mating.	Treat with antibiotic. Do not use infected animals for breeding. Cull infected animals. Do not loan bucks to other breeders.

Microcomputers may be used for keeping the necessary records of the rabbitry. Commercial data base and general accounting programs may be used for keeping records. There are some software packages available specifically for the rabbit producer. These include pedigree, doe and buck performance record, and feed programs. There are a number of farm accounting programs available for microcomputers that can be adapted for use in rabbit production.

FACILITIES AND EQUIPMENT

Buildings

Climatic conditions, local building codes, and available capital are factors to consider when determining the kind of housing for rabbit hutches. Buildings of simple design that protect the rabbits from the weather and provide adequate ventilation are desirable.

In mild climates, hutches may be outside if shade is provided. Trees or a lath superstructure may be used for shade.

Where the weather is hot part of the year, some method of cooling the building is necessary. Overhead sprinklers or foggers inside the building may be used for cooling. In hot, humid areas, evaporation coolers with fans may be used to provide cooling in the building. High-volume commercial operations often use automated ventilation and cooling equipment. In areas where the weather is cold part of the year, the building must be designed to provide more protection (Figure 47-7).

Hutches

Hutches are provided for individual rabbits. These may be made of wood or wire construction. Hutches should be 2.5 feet (0.76 m) deep and 2 feet (0.6 m) high. Length varies with breed: Small breeds need 3 feet (0.9 m), medium breeds need 3 to 4 feet (0.9–1.2 m), and large breeds need 4 to 6 feet (1.2–1.8 m).

Hutches may be arranged in one-, two-, or three-tier configurations. If enough room is available, the one-tier arrangement is the most convenient for caring for the rabbits. Commercial rabbitries usually hang all-wire cages from eye hooks in the ceiling of the building.

Wire cages of various designs are available from commercial firms: Self-cleaning wire cages require no bedding and are easy to maintain. Wire mesh floors are usually used in self-cleaning hutches. Commercial rabbitries use self-cleaning cages because of the high labor requirement of solid-floor cages. Scraper blades are used to remove the manure from the building.

Feeding and Watering Equipment

Feeding equipment should be designed to prevent feed waste and contamination. Crocks, hoppers, and hay mangers with troughs may be used for feeding rabbits. Commercial rabbitries may find it desirable to use electric feed carts to reduce labor costs.

Water crocks may be used in the hutch to provide fresh, clean water for rabbits. However, the labor requirement is high when large numbers of rabbits are raised. Commercial rabbitries usually use auto-

FIGURE 47-7 Example of commercial rabbitry housing. *Courtesy of Safeguard.*

mated watering systems to reduce labor costs. In cold climates, automated watering systems must be protected against freezing; this may be done by heating the building or wrapping the pipes with heating cables.

Nest Boxes

Nest boxes are used to provide seclusion for the doe when kindling and to provide protection for the litter. The type used should be simple to clean.

Several types of nest boxes are available for use in rabbitries. Be sure the type used is large enough to prevent crowding, but small enough to keep the young rabbits together.

Nest boxes may be made of wood or metal; apple packing boxes or nail kegs may be used. Metal nest boxes may be cold, with water condensation being a problem in cold climates. Counterset nest boxes are recessed below the hutch floor; this type provides a more natural environment for the young rabbits. They are also easier to keep clean. Make sure nest boxes are well drained and ventilated. In cold weather, a well-insulated nest box will provide protection for the young rabbits at temperatures as low as $-15°$ to $-20°F$ ($-26°$ to $-29°C$).

MARKETING RABBITS

Commercial rabbitries usually sell fryers to processors who slaughter the rabbits and market the meat. Processors usually use trucks to pick up the live animals at the rabbitry. Rabbits with colored pelts bring lower prices; therefore, commercial operations usually raise only white-pelted rabbits.

The USDA defines a fryer as "a young domestic rabbit carcass weighing not less than $1\frac{1}{2}$ pounds and rarely more than $3\frac{1}{2}$ pounds processed from a rabbit usually less than 12 weeks of age." The rabbits are marketed when they reach fryer weight, which is a live weight of three to six pounds (1.36–2.7 kg). Live weights of 4 to $4\frac{3}{4}$ pounds (1.8–2.1 kg) produce the best carcasses, with a dressing percent ranging from 50 to 59 percent. Prime grade fryers have a yield of 57.7 percent; Choice grade fryers have a yield of 55.9 percent; and Commercial grade fryers have a yield of 52.2 percent.

The USDA defines a roaster as "a mature or old domestic rabbit carcass of any weight, but usually over 4 pounds processed from a rabbit usually 8 months of age or older." Roasters have a dressing percent of 55 to 65 percent. Culls from the breeding herd may be fattened and sold as roasters. It is usually not profitable to feed younger rabbits to roaster weights because of the higher feed requirements and the possibility of additional death loss.

Some small-volume producers slaughter, package, and market their own rabbits. Strict sanitation must be practiced when doing this.

Producers should determine local slaughtering regulations and restrictions when planning to market their rabbits in this manner.

SUMMARY

Modern breeds of domestic rabbits in the United States are descendants of the European wild rabbit and have been developed since the eighteenth century. The major use of domestic rabbits is for meat production. Rabbits produce a white meat that is high in protein and low in calories, fat, and cholesterol. Medium- and heavy-weight breeds are best suited for meat production. The most common breed used for this purpose is the New Zealand White.

Rabbits are simple-stomached, herbivorous animals. The major part of their diet comes from plants and plant products. Most commercial rabbitries use complete pelleted feeds for rabbit rations. A continuous supply of fresh, clean water should be available for rabbits. Feed additives are usually not used on a regular basis in rabbit feeding.

Maintenance rations may be fed to junior does and bucks, mature dry does, and herd bucks not in service. A fine-stemmed, leafy, legume hay will provide the nutrients needed for maintenance. Herd bucks in service should be fed four to six ounces of a complete pellet feed if hay is not included in the ration. Growing junior does and bucks may be fed four to six ounces of a complete pelleted ration. Pregnant does may be fed six to eight ounces of a complete pelleted feed, while nursing does will eat six to eight ounces during the first three weeks of lactation and one to two pounds of feed per day for the remainder of the nursing period.

The gestation period for rabbits is 30 to 32 days. Does in good physical condition can be rebred six weeks after kindling. The litter is normally weaned at eight weeks of age. On an accelerated breeding program, a doe can produce five litters per year, although four litters per year is more common. Meat rabbits are usually sold for market as fryers when they are weaned at eight weeks of age.

Good sanitation in the rabbitry is the key to disease control. Keep manure and soiled bedding cleaned out, isolate sick animals and new breeding stock coming into the herd, and prevent visitors from coming into contact with the rabbits. Keep dogs, cats, birds, rodents, and insects out the rabbitry, as they sometimes carry diseases or parasites that may be transmitted to the rabbits.

Rabbit producers need to keep good records. Performance records and financial records are important. A microcomputer may be used to keep the necessary records for the rabbitry.

Buildings for rabbits need to protect the animals from weather conditions. Commercial rabbitries generally use wire cages and utilize automated feeding and watering equipment. Nest boxes are provided for kindling and to protect the young rabbits during their first few weeks of life.

Commercial rabbitries usually sell live fryers to processors. The live weight of fryers ranges from three to six pounds (1.36–2.7 kg); the best carcasses dress out at 50 to 59 percent. Roasters are older rabbits that have been culled from the herd.

Student Learning Activities

1. Have a commercial rabbit raiser talk to the class and describe the business.
2. Present oral reports on commercial rabbit raising.
3. Prepare a bulletin board display on the breeds of rabbits.

Review

1. List the uses of rabbits in the United States. Which of these is the most important?
2. Describe and list the major uses of each of the following breeds of rabbits: (a) Californian, (b) Dutch, (c) New Zealand.
3. What are the typical mature weights of (a) small breeds, (b) medium breeds, (c) large breeds?
4. What are the major factors to evaluate when purchasing breeding stock?
5. Why can rabbits use more forage in their diet than other simple-stomached animals, such as swine and poultry?
6. When given a choice, which cereal grains do rabbits prefer?
7. How may the palatability of rabbits' rations be improved?
8. What effect does temperature have on the energy requirements of rabbits?
9. What percent of the ration for rabbits should consist of fiber? What effects do higher or lower levels of fiber have on rabbit health and growth?
10. Which two amino acids are most likely to be deficient in rabbit diets?
11. What is coprophagy? How does coprophagy affect rabbit nutrition?
12. What is the recommended ratio of calcium to phosphorus in rabbit diets?
13. Briefly explain the need for vitamins in rabbit diets.
14. Briefly explain the need for water in rabbit diets.
15. Briefly explain the use of feed additives in rabbit rations.
16. Why are legume hays preferred over grass hays in rabbit diets? Which legume hay is considered the best for rabbits?

17. Briefly explain the use of green feeds and root crops for rabbits.
18. Which protein supplement is most commonly used in rabbit feeds?
19. What is the recommended pellet size for rabbits?
20. Why do commercial rabbitries usually full-feed rabbits for meat production?
21. Briefly explain maintenance feeding for junior does and bucks, mature dry does, and herd bucks that are not in service.
22. What kind of a ration should be fed to herd bucks that are in service?
23. What kind of a ration should be fed to growing junior does and bucks?
24. Briefly explain the feeding of pregnant does.
25. Briefly explain the feeding of nursing does.
26. What is the gestation period for rabbits?
27. What are some common breeding intervals used by commercial rabbitries?
28. Describe the method of determining pregnancy by palpating the doe.
29. List several good management practices to follow at kindling time.
30. List several causes of losses at kindling and describe how to prevent these losses.
31. At what age are meat rabbits usually weaned?
32. Describe the method of determining the sex of young rabbits.
33. List and describe several good management practices in a disease control program in rabbitries.
34. List and describe several good management practices to prevent the transmission of disease in rabbitries.
35. What is the most serious health problem of rabbits? Discuss this problem.
36. Briefly explain the enteritis disease complex.
37. What kind of records should be kept for the rabbitry?
38. Briefly explain the kind of buildings needed for rabbits.
39. What kind of cages are most commonly used in commercial rabbitries?
40. What is the purpose of the nest box? Describe a nest box.
41. How do commercial rabbitries usually sell rabbits for meat?
42. Define: (a) fryers, (b) roasters.

Alternative Animals

Unit **48** *Alternative Animals*

Objectives

After studying this unit, the student should be able to

- describe the origin, history, and general characteristics of bison.
- describe the characteristics of the bison industry.
- discuss the management, feeding practices, health maintenance, facilities and equipment, and marketing of bison.
- discuss the characteristics and origin of ratites.
- discuss management practices used when caring for ratites.
- discuss facilities needed for ratites.
- discuss maintaining the health of ratites.
- discuss getting started with ratites.
- discuss the marketing of ratites and their products.
- discuss the history of and getting started in business with llamas and alpacas.
- describe the characteristics of llamas and alpacas.
- discuss the feeding and management of llamas and alpacas.

BISON

The information about bison in this unit was adapted from material furnished to the author by the American Bison Association (ABA), 4701 Marion St., Suite 100, Denver CO 80216 (telephone: 303-292-2833). The ABA was formed in 1975 to promote the production, marketing, and preservation of bison. Readers who desire more detailed information about raising bison should contact this association.

BACKGROUND

Origin

The American bison is a member of the Bovidae family, genus *Bison*, species *bison*. The Bovidae family includes cattle, sheep, and goats (see

Unit 1, Table 1-1). It is believed that the bison crossed the Bering Strait land bridge from Asia to North America approximately 20,000 to 30,000 years ago. The Bering Strait land bridge was a landmass that emerged during an ice age when a large amount of the earth's water was frozen in ice sheets and glaciers. When the ice melted at the end of the ice age, the seas rose in depth and the land bridge was again covered by water.

The American bison is related to the European bison, *Bison bonasus*, a species that is almost extinct. The European bison is known as the wisent; only a few specimens remain, mostly in parks and zoos.

The name "buffalo" is often used when referring to the bison. This is not a correct use of the word. Buffalo refers to the Cape Buffalo or Water Buffalo found in other parts of the world but not in the United States.

History of Bison

Bison were an important source of food, skins, bones (for implements), and fuel (dried dung) for the Plains Indians before European explorers reached the North American continent. It is estimated that there were between thirty and sixty million head of bison ranging the western part of the continent at that time. European settlers and explorers engaged in a methodical slaughter of bison during the 19th century. By the end of the century there were probably no more than 300 bison left on the North American continent. The bison had been hunted almost to extinction.

Through the efforts of a few conservationists and ranchers, the few remaining animals were saved from extinction. A slow rebuilding of bison numbers continued throughout the 20th century. A 1929 census reported 3,385 bison. By the late 1980s the number of bison in North America had grown to over 80,000. It is estimated that there are currently over 125,000 bison in North America. About 15 percent of the bison in the United States are currently maintained on public lands. The rest are in the hands of private owners. The American bison is no longer on the endangered species list.

Description and Characteristics

Bison have a hump over the front shoulders (Figure 48-1). The bull has larger horns than the cow; the horns curve outward and up from the head. The head is large and the body narrows down toward the hindquarters. Long, dark hair covers the head and forequarters; the hindquarters are covered with shorter hair. The bull has a black beard that is about twelve inches (30 cm) long. A mature bull may weigh from 1,500 to 2,000 pounds (680–907 kg) and stands about 6.5 feet (2 m) at the hump; the body ranges from 9 to 12 feet (2.7–3.7 m) in length. The mature cow is smaller than the bull, weighing about 1,000

FIGURE 48-1 Bison. *Courtesy of American Bison Association.*

FIGURE 48-2   Bison cow and calf. *Courtesy of American Bison Association.*

pounds (453 kg) (Figure 48-2). Bison have 14 pairs of ribs instead of the 13 pairs found in cattle.

With proper care and management, bison flourish in a wide range of environments. Their hair coat thickens in cold weather, providing them with protection from low temperatures. Bison can find grass under heavy snow cover and will eat snow to get the water they need. They can survive equally well in warm climates.

Bison are territorial animals, that is, they will defend their territory from intruders. They can successfully fight off most predators. People working with bison must exercise care. While generally not aggressive, they are wild animals; cows with calves and bulls in rutting (breeding) season can be dangerous.

CHARACTERISTICS OF THE BISON INDUSTRY

The bison industry in the United States is small compared to other livestock enterprises on farms and ranches. People interested in becoming involved in this industry have several options:

- absentee ownership
- hobby
- small producer
- medium-size ranches
- large ranches

Types of Bison Ownership

Absentee Ownership. People who want to invest in bison but do not have the land or knowledge to raise them might choose to contract with an existing bison ranch to own one or more animals. The extent of control the absentee owner has over the bison owned depends on the terms of the contract. Absentee ownership offers investment and tax shelter opportunities. Because of rising costs and other problems associated with this option, there are few absentee ownership programs currently being offered in the bison industry.

Hobby. A person with an interest in bison may keep only a few (usually fewer than 25) head as a hobby or attraction. This kind of ownership may be just for personal pleasure and interest or may be to serve as an attraction to some other type of existing business such as a restaurant, museum, recreational area, or tourist attraction. Care must be taken to protect the bison and the public in such operations.

Small producer. A small producer is a person with a small herd of bison (usually 25 to 100 head) who gets some income from the herd, usually from the sale of breeding stock. The small producer may have a goal of expanding into a larger operation.

Medium-size ranches. A person with a medium-size ranch has a herd of 100 to 250 head of bison. Sales from this kind of operation usually include breeding stock and excess bulls for a meat market.

Large ranches. The large producer is one with a herd of bison ranging in size from several hundred to several thousand head. The bison are the major source of income in this operation. Income comes from the sale of breeding stock and excess bulls sold or fed out for a meat market.

Demand for Breeding Stock

Because of a growing interest in raising bison, the demand for breeding stock has increased in recent years. There are a limited number of bison available for breeding purposes, therefore the price of breeding animals has risen. Prices for breeding stock at public actions of bison have risen from an average of $379 per animal in 1983 to $1,500 in 1993. In 1997 the average price per animal at the National Bison Association's Gold Trophy Show and Sale was almost $4,500. This indicates a continuing strong demand for bison breeding stock. In addition to private owners, surplus bison are being offered for sale at auctions from some public herds.

Demand for Meat

Excess bulls and animals not suitable for breeding stock are sold for meat. Bison produce a lean, red meat low in fat, calories, and cholesterol. It has a high protein content. It does not have a gamey or wild flavor. Hormones and subtherapeutic antibiotics are rarely used in the production of bison, making the meat desirable for those who have concerns about the presence of these substances in the food supply. It is expected that the demand for bison meat will grow as the public becomes more aware of its characteristics.

MANAGEMENT

Selecting Bison for the Herd

When starting out in the business of raising bison, it is usually best to begin by buying heifer calves. Calves adjust to a new environment easier than older animals. Yearlings from a good herd are also a good choice. If buying two- or three-year-old cows, be sure they are bred and tested for pregnancy.

To develop a good breeding herd, select animals from several herds that have developed superior breeding animals. The prospective buyer should learn as much as possible about the health program, the source of stock, and the calving rate of the herd from which the purchase is made. Look at the entire herd to get an idea of the overall condition of the animals. Determine the vaccination and worming schedule followed; ask about any other health problems. Inbreeding is a serious problem with bison because the current stock began with so few animals. Generally, buy from herds that avoid inbreeding. Good calving rates (90–95 percent) indicate a vigorous culling program with generally sound herd health practices.

When selecting animals, look for large, lengthy animals with well-developed hindquarters. An animal with good conformation has a flat back with a slight slope toward the tail. The top and back show a rounded, angular development. A large-frame body with adequate width and depth over its length is desirable. The parts of the bison are shown in Figure 48-3.

Avoid animals with a severe sloping of the back downward toward the tail and a severe taper from the hind legs toward the tail. This is referred to as a pencil-shaped rear end. Do not select females that show masculine traits or males that show feminine traits. Both conditions appear to correlate with sterility. Crooked legs and deformed body parts are also undesirable.

The presence or absence of horns, hair coat color, and length of hair on the forehead are generally matters of personal preference. They have little overall effect on the desirability of the animals as breeding stock.

Herd Management

Handling bison. Always use caution when working with bison. They are wild animals and have great strength. Serious injury to workers may result from carelessness around the bison herd.

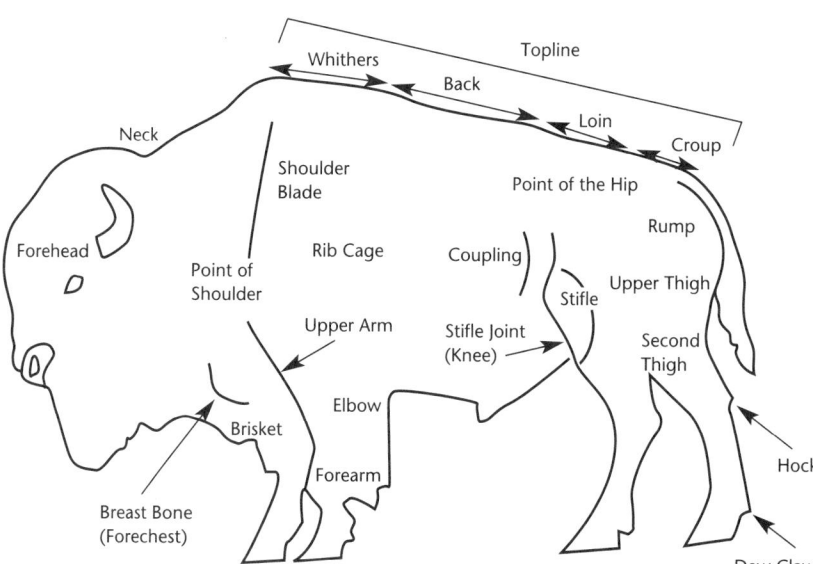

FIGURE 48-3 Parts of the bison. *Courtesy of American Bison Association.*

Do not attempt to work with bison during the breeding season and when cows are calving. Bulls become aggressive and dangerous during the breeding season. Cows that are calving will aggressively protect their calves, charging workers to drive them away. Young calves are not as dangerous as adult animals but should still be handled with care.

When working with bison, plan ahead and move slowly. Do not attempt to force the animals. Bison generally follow the herd leaders; therefore, if the herd leaders can be moved to the desired area, the rest of the herd will tend to follow. Using feed may help to get the bison to move to another area.

Use a closed truck or trailer with a top to transport bison. A stock trailer is recommended for use in hauling bison. Get the bison used to being confined by holding them in a closed corral for a time before transporting them. To prevent injuries, separate young animals from older ones in the trailer. Keep bulls in a compartment separate from other bison.

Bringing new animals into the herd. New animals being brought into a herd should be isolated in a separate corral or holding area for a period of two to four weeks before mixing them with the rest of the animals. This allows them to become acclimated to the new area and gives them time to settle down. There will be some pushing and shoving when the new animal is moved into the herd as a new dominance order is established.

Breeding. Bison reach maturity about one year later than cattle. Generally the females can be bred when they are two years old. The gestation period is 275 days. Bison generally breed in July and August, with the calves being born the following spring. It is usually a good practice to keep yearling heifers separated from the breeding herd during the breeding season because some may become pregnant at this age. Breeding bulls can be put into service when they are two years old. Breeding bulls may be semen tested to ensure that they are not sterile. A normal ratio is one mature bull to every ten cows or two younger bulls with ten cows. Some producers let the bulls run with the herd all year; others keep them separate except during the breeding season. Artificial insemination is not generally used with bison herds.

Inbreeding should be avoided; replace herd bulls about every two years to avoid breeding back to mothers or sisters in the herd. Inbreeding has the following undesirable effects:

- fertility is decreased in bulls that are the product of inbreeding.
- lower feed efficiency; it may take as much as two years longer to produce a desirable carcass.
- dished in, shortened faces can appear in the offspring.
- unsoundness in the legs; a condition called rabbit-legged bison appears.

Outbreeding (mating animals that are not closely related) develops individuals with more vigor, size, and hardiness and reduces the effects of recessive genes. Outbreeding is recommended in bison herds.

Weaning. Some producers regularly wean calves and others do not. Bison calves may be weaned when they are eight to nine months old. Weaned calves need to be started on supplements by creep feeding them for one to two months before weaning. Worm, ear-tag, and vaccinate calves at weaning time. Vaccinate heifer calves for Brucellosis. The weaned calves should be kept in a shelter with plenty of feed and water. The major reason for weaning calves is to reduce stress on the cows; they also appear to grow faster.

Culling. Bison herds should be culled to remove old, unproductive, or undesirable individuals. Cull animals that do not meet the criteria for productivity, health, vigor, temperament, and physical characteristics. Usually culling isn't necessary until several years after the herd has been established. Keep only the best-quality bulls for breeding purposes. The remainder should be removed, generally as two-year-olds and sold for meat. This will maintain herd size at a desirable level.

Branding and identification. Most states do not require bison to be branded for identification purposes because they are classified as wild animals. A herd owner needs to check state requirements. Some owners prefer not to brand bison because branding reduces the value of the hides. Most owners use ear tags as a method of identifying animals. A disadvantage of ear tags is that they may become lost.

Dehorning. The advantages of dehorning bison are:

- animals (especially cows) are not as dangerous.
- animals are more docile.
- less damage to hides and meat in the feedlot.
- reduces the space requirements in the feedlot.
- requires less feed bunk space.

Those who oppose dehorning cite the following reasons:

- detracts from the natural appearance of bison.
- proper handling can minimize the danger and economic losses without changing their appearance.

Owners who keep bison primarily for an attraction and do not handle the herd extensively probably will not want to dehorn the animals. Large herds that are kept for meat and where extensive handling is necessary probably should consider dehorning at least the females and the bulls that do not appear to have good trophy heads.

To prevent maggot infestation, dehorn at least six weeks before fly season begins or after a killing frost in the fall. Commercial dehorning instruments are available. Remove horns below the hairline. Cauterizing the wound with a hot iron helps prevent bleeding and the growth of spurs from missed horn cells.

Castration. Bison bulls are usually not castrated. The bulls are sold for meat before they reach breeding age, therefore there is no undesirable flavor in the meat from uncastrated animals. Castrated bison appear to grow more slowly than those that are not castrated.

Immobilization. Occasionally it may be necessary to immobilize an animal for treatment. There are several drugs that are used for this purpose. A veterinarian should always be present when it is necessary to immobilize an animal. Generally, do not use drugs to immobilize a bison unless it is absolutely necessary.

Orphan calves. In general, do not remove calves from their mothers unless it is necessary. Circumstances that may make it necessary to remove the calf from the mother include the following:

- the cow dies.
- the cow is unhealthy and not able to care for the calf.
- the cow is unable to nurse the calf.
- the cow is too wild to handle and must be removed from the herd.

When it is necessary to raise an orphan calf, it is important that the calf receive colostrum milk in the first few hours of life. Bison colostrum milk is usually not commercially available; dairy cow colostrum milk will work as well and is more often available. After the calf has received colostrum milk, goat's milk or a mixture of evaporated milk and water may be used to bottle feed the calf. Lamb milk replacer is a better choice than beef milk replacer if goat's milk or evaporated milk is not available. Lamb milk replacer is higher in protein and fat; it is more nearly like bison milk than beef milk replacer.

The calf should be fed one-half to one pint of milk four to six times per day for about two weeks. Gradually increase the amount of milk per feeding until the calf is being fed 1½–2 quarts (1.4–1.9 liters) of milk replacer twice a day. Watch carefully for signs of scours. Overfeeding or too rich a mixture may cause scours. Treat calves with scours the same way a beef calf is treated.

Provide access to water, hay, and grain free choice. The calf will start eating at a fairly young age. Provide a pen with shelter from the weather and an area for exercise. Wean calves from the bottle when they are about six to seven months old. Hay supplemented with grain can be fed after the calf is weaned.

Registering animals and bloodtyping. The American Bison Association founded the North American Bison Registry in 1980. Bloodtyping is used to identify individual animals and to determine parentage (which is possible if the sire and dam are also bloodtyped). The use of bloodtyping can determine that the animal is a purebred bison and not a crossbred animal. It is a good management tool for selecting breeding stock, especially herd bulls. Detailed information on registering animals and bloodtyping is available from the American Bison Association.

FEEDING AND NUTRITION

Feeding the Bison Herd

The nutrient requirements of bison are similar to those for beef cattle. Bison herds can use a variety of grasslands to meet their basic nutritional requirements (Figure 48-4). When the size of the rangeland available is limited in relation to the size of the herd, it may be wise to use a rotational grazing system. One method of rotational grazing is to move the herd to different pastures during the grazing season. Another method is to fence the pasture into several smaller areas and move the herd from one area to another. When using rotational grazing, the pasture should be allowed about 30 days for regrowth before moving the herd back onto it. Rotational grazing is recommended when the amount of pasture available is limited. It does require additional investment in fence building and time.

When the size of the pasture is large in relation to the size of the herd, it is possible to allow the bison to range freely over the entire area. If enough space is available, bison tend to move from one area to another each day. This method of grazing works best on large rangeland areas, where it can be used most easily and efficiently.

During the winter, when pasture is not available, the herd requires supplemental feeding. Hay or other forages may be used along with protein supplement, salt, and minerals. Range cubes are commercially available and are often used for supplemental winter feeding. The protein level of the ration should be from 11 to 13 percent. The amount of protein supplement needed is dependent upon the quality of the hay or other forage fed.

Be sure there is an adequate supply of fresh, clean water for the herd. Mineral requirements for bison are similar to those for beef cattle. Free choice feeding of an iodized salt mix along with a good commercial mineral mix will generally meet the requirements of the bison. Salt blocks and mineral blocks are commercially available for use on pasture.

FIGURE 48-4 Bison can use a variety of grasslands to meet their basic nutritional requirements. *Courtesy of American Bison Association.*

Finishing in the Feedlot

Extra young bulls from the herd are usually finished in the feedlot to be sold for meat. After weaning, bull calves that are to be fed out may be fed a roughage ration in drylot or carried on pasture until the next year when they are moved into the feedlot for finishing. The animals are then fed a finishing ration until they reach a liveweight of 1,000–1,100 pounds (453–499 kg), at which point they are marketed as meat. Bison will generally yield about 62 percent dressed meat.

Bison may be successfully started on feed with a ration of prairie and alfalfa hay, supplemented with some corn. After they become accustomed to eating corn, a free choice ration of alfalfa hay and corn may be used. The ration should contain 10–12 percent protein. Soybean oil meal may be used to add protein to the ration if necessary. A complete cubed ration containing roughage, grain, and supplement may be used. There is no apparent advantage to grinding, cracking, or rolling the corn in the ration when it is not fed in a complete ration. Consumption is about ten pounds (4.5 kg) of alfalfa and 7–10 pounds (3–4.5 kg) of corn per head per day. A mineral supplement should be available free choice. Make sure there is plenty of water at all times.

Bison do best in the feedlot if all are about the same age and size. Dehorned animals require less pen and feed bunk space. Self-feeders for the concentrate mix are recommended.

Average daily gain for well-managed bison in the feedlot can be as much as two to three pounds (0.9–1.4 kg). Because of genetic differences, some bison do not gain well in the feedlot. If they do not appear to be gaining adequately after three to four months on feed, they should be slaughtered. They will probably never make adequate gains no matter how long they are fed.

HERD HEALTH

Preventing health problems in the bison herd is better than trying to treat the problem after it occurs. Proper feeding and good sanitation will help maintain a herd in a healthy condition. Keep feedlots, feed bunks, and waterers clean. Bring only healthy animals into the herd. Worming and vaccinating new animals is generally recommended. Work with a veterinarian to determine the proper treatment of new breeding stock as well as the existing herd.

Bison are subject to many of the same health problems that affect other ruminants such as cattle and sheep. The following diseases are discussed in Units 17 or 29:

Anaplasmosis (bison are rarely affected but can be carriers—testing is recommended)

Anthrax (bison are affected—vaccinate if a problem exists)

Blackleg (vaccination is recommended)

Bloat (bison generally do not bloat)

Brucellosis (vaccinate female calves at 3–10 months to keep the herd clean)

Coccidiosis (bloody scours) (bison are affected)

Enterotoxemia (vaccination is recommended)

Hoof Rot (Foot Rot) (bison are affected)

Johne's Disease (rarely affects bison)

Leptospirosis (vaccination is recommended)

Malignant Edema (vaccination is recommended)

Neonatal Scours (Calf Enteritis) (rarely seen in bison—vaccinate if a problem exists)

Pinkeye (bison are generally resistant, but a few will be affected)

Shipping Fever (bison are affected—use preventative treatment)

Tetanus (rarely affects bison)

Other diseases (not discussed in Units 17 or 29) that may affect bison include malignant catarrhal fever and tuberculosis.

Malignant catarrhal fever is caused by a virus. Symptoms include a high fever and death. Bison may be infected by coming into contact with sheep that carry the virus but do not develop symptoms. There is no treatment. Prevent infection by keeping bison away from contact with sheep.

There are three forms of *tuberculosis*: human, bovine, and avian. The bacteria that cause tuberculosis, mycobacterium, only multiply in the host. The disease affects the lungs, causing a chronic weakening condition. In some animals it becomes acute and may cause death. Infection is spread from animal to animal by bodily discharges that get on the feed, in the water, or in the manure. Removal and destruction of infected animals is the only way to control this disease.

Using Medications

Medications that are approved for cattle are usually safe for use with bison. Medications may be administered through water or feed, by spraying or pouring on the animal, and by injection. Always follow the advice on the label, including withdrawal time before slaughter, when administering medications. It is usually a good management practice to consult a veterinarian when health problems arise.

Parasites

Bison are generally subject to the same external and internal parasites that affect cattle. A general discussion of these parasites is found in Unit 17.

Poisonous Plants

Bison are affected by the same poisonous plants that affect cattle and sheep. In some cases, the bison appear to be less susceptible to some poisonous plants than cattle. It is a good management practice to avoid exposing animals to poisonous plants and moldy feeds whenever possible.

FACILITIES AND EQUIPMENT

Fences

Bison need higher fences than cattle (Figure 48-5). Fences should be at least six feet (1.8 m) high. Bison can jump higher than some horses; if they can get their nose over the fence they may try to jump over it. Fences surrounding the pasture area need to be strong and high enough to keep the animals confined.

FIGURE 48-5 A combination of pipe and barbed wire fencing may be used for small pastures, feedlots, or corrals. *Courtesy of American Bison Association.*

Barbed wire can be used to fence bison; however, because of their thick hair and hide, these fences are not as effective as other types. Double-strand (one roll on top of another) woven wire fences will confine bison. Woven wire fences are more expensive and require more labor to install. High-tensile wire fencing works well with bison. This is a smooth wire with high tensile strength. High-tensile wire is less costly than woven wire and about the same cost as barbed wire fencing. Electric fence may be used if the bison are trained to respect it. Pipe or a combination of pipe and cable is especially useful for small pastures, feedlots, and corrals.

Temporary fencing may be used when moving bison from one area to another. Portable cattle panels may work for temporary fencing if they are tall enough. Black plastic has been used successfully to cover a weak area in an existing fence or to make a temporary holding pen. Use care when working with bison if temporary fencing is used. The animals may soon discover that the fencing is not substantial and will go through it.

Small pastures need higher and stronger fences than larger pastures. Corral type fencing should be used in small areas where bison are confined for long periods of time.

Gates

Pasture gates may be made of the same material as the fence. Locate pasture gates away from corners because bison do not like to be driven into a corner area. Corral gates should be constructed of heavier material than pasture gates. Solid gates are better for working with bison in the corral. Use rope-activated, spring-loaded gate latches if possible. These are safer and easier to use than other types of gate latches.

Cattle Guards

Ordinary cattle guards do not work well with bison. They have a tendency to jump over them. A cattle guard for use with bison should be at least 12–16 feet (3.6–4.9 m) long and 8 feet (2.4 m) wide. Keeping the

FIGURE 48-6   A running alley is used to move animals from the catch pens or holding pen into the crowding area. *Courtesy of American Bison Association.*

FIGURE 48-7   A half-circle crowding pen and alleyway leading to the squeeze chute. *Courtesy of American Bison Association.*

FIGURE 48-8   A catwalk along the alleyway and crowding pen makes it easier and safer to work with the animals. *Courtesy of American Bison Association.*

cattle guard hole cleaned out or using a light-colored material in the bottom of the hole may deter bison from attempting to jump over it.

Handling Facilities

Handling facilities must be built stronger than pasture facilities because bison tend to become excitable in confined areas. Bison are extremely large and strong animals; the safety of the workers must always be considered when designing handling facilities.

Catch pen.   A catch pen is used to move bison from the pasture to the corral. Use a funnel or triangular shape with an exit gate in the corner leading to the holding pen.

Holding pen.   A holding pen is used to sort animals before working with them. The pen should be 6–7 feet (1.8–2.1 m) in height.

Running alley.   The running alley is used to move the animals from the catch pen or holding pen into the crowding area (Figure 48-6).

Crowding pen.   The two basic types of crowding pens are funnel and circle or half-circle (Figure 48-7). Bison remain calmer in solid-sided pens. Catwalks built along the sides of the pen make it easier for workers to handle gates with greater safety (Figure 48-8). Gates in the crowding pen help move the bison into the working area. The alleyway leading from the crowding pen to the working area should be equipped with a series of gates to prevent the animals from turning back. The animals should be moved single file through this alleyway to the working area. An adjustable alleyway permits working with different sizes of animals more easily. A scale might be included in the area just before the squeeze chute.

Squeeze chute.   The squeeze chute is used to hold the animal for testing, vaccination, administering medications, or other work that may need to be done when the animal is confined (Figure 48-9 and Figure 48-10). The animal is held in a head catch and the sides can be squeezed together to provide further confinement. An ordinary cattle squeeze chute is generally not suitable for use because it is not tall enough or strong enough for a mature bison. Squeeze chutes designed for bison are commercially available.

MARKETING

Bison are marketed as breeding stock, for meat, and for by-products. Herd owners may choose to market their bison in any or all of these categories.

Selling Breeding Stock

Bison sold as breeding stock may be marketed by sale to private individuals or at auction. Personal reputation and advertising are keys to successful marketing of breeding stock. Certificates of health are generally needed, especially if the buyer is from a different state. Breeding stock should be guaranteed free of Brucellosis and tuberculosis. Testing is required for these two diseases. Some states require additional testing before bison can be brought into the state.

Careful selection of the best stock for sale as breeding animals will help maintain the reputation of the seller and improve the bison industry. Poorer-quality animals should only be marketed for meat.

A general livestock auction is not a good choice for selling bison. Well-advertised bison auctions will attract the kind and number of buyers the seller wants. These auctions are a better choice for the seller. Video auctions have been held for bison; these provide exposure of animals to a large number of potential buyers without having to transport the bison.

FIGURE 48-9   The squeeze chute and drop in gates as seen from the alleyway. *Courtesy of American Bison Association.*

Selling to Feedlot Operators

Feedlot operators need a consistent source of animals. The bison herd owner who does not want to feed bison for meat can sell the surplus animals to a feedlot operator.

Meat Marketing

Bison herd owners who want to market some of their animals as meat have two general types of markets to consider. The traditional bison market consists of consumers who want a low-fat red meat. Animals sold to this market are generally grass fed with little or no finishing in the feedlot. The gourmet market consists of consumers who want grain-fed meat. Animals sold to this market are finished on grain in a feedlot before being marketed.

Several channels are open to the producer for marketing bison meat:

FIGURE 48-10   The squeeze chute is used to hold the animal for testing, vaccination, or other work that may need to be done when the animal is confined. *Courtesy of American Bison Association.*

■ Distributors and meat brokers
■ Restaurants
■ Diet and health food stores
■ Direct to consumers

There are an increasing number of meat distributors and meat brokers that purchase bison meat for resale. When selling to restaurants, the producer must be able to provide delivery of the kinds of cuts the restaurant needs. Diet and health food stores are looking for wholesome products that are packaged to their specifications. Direct sales to consumers may involve selling entire carcasses, halves, or quarters. Some consumers will want only a small number of cuts packaged for their use. Direct sales may also be done through mail order. All customers want a quality product with good service.

Marketing By-products

There are a number of by-products from bison that the producer may want to market. These include:

- heads
- robes
- skulls

- leather goods
- wool
- bones and horns

Heads, robes, and skulls are often used for decoration. Some heads are mounted with the hump area; others do not include the hump. Winter robes are hides that have been tanned while covered with the winter hair coat. These may be used for wall hangings, furniture throws, rugs, pillow covers, seat cushion covers, and clothing. Leather is in demand for garments. Bison skulls may be bleached or used without bleaching. They are sometimes used by artists for paintings or are decorated with stones. Bison shed their hair in the spring of the year. This may be gathered and carded like lamb's wool. It can then be spun into yarn. Bison bones and horns are used in many decorative ways. By-products from bison are a specialty market that can add profit to the enterprise.

RATITES

Introduction

There is an interest in the United States in raising ratites as an alternative to other livestock production. Commercial production of ostriches, primarily for feathers, began in the 1800s in South Africa. After World War II, a market developed for the meat and leather from ostriches. South Africa has retained a virtual monopoly on the commercial production of ostriches until recently.

Most of the ratite industry in the United States involves ostriches, emus, and rheas. The care and management of ostriches, emus, and rheas are similar.

Ratites are a group of flightless birds that include ostrich, emu, rhea, cassowary, and kiwi. The breastbone of ratites is flat, lacking the keel that is characteristic of most birds that fly. The shape of the breastbone resembles a raft; the name ratite comes from the Latin words *ratis* (raft) and *ratitus* (shape of a raft).

The ostrich at one time was found in the Middle East and Africa; currently, its natural range is only in Africa. The natural range of the only surviving species of emu is in Australia. The rhea is from South America; the cassowary from Australia and the Malay Archipelago; and the kiwi is from New Zealand.

Description

Ostrich. The ostrich is the largest of the ratites and is also the largest bird in existence. Mature ostriches are seven to nine feet (2.1–2.7 m) in height and weigh from 200 to 350 pounds (91–159 kg); males are generally larger than females. The ostrich is capable of running at speeds of up to 30–50 miles per hour (48–80 km/h) for up to 30 minutes. The length of its stride is from 15 to over 20 feet (4.5–6 m), depending on its size. It has only two toes on its feet; the other ratites have three. The male ostrich has black feathers with white feathers on the wings and tail; the female is more drab and has light gray to grayish brown feathers. The feathers of the male become brighter during the mating season. Ostriches have long necks and small heads with large eyes and a short, flat bill.

The domestic ostrich is a result of crossbreeding several subspecies, all of which belong to the species *Struthio camelus*. The initial purpose of the crossbreeding was to improve feather quality. The emphasis on breeding today is selection for meat production.

Emu. The emu is the second largest bird in existence; mature adults are five to six feet (1.5–1.8 m) high and weigh from 125 to 150 pounds (57–68 kg). The female is dominant. The male is slightly smaller than the female. Emus are fast runners, being capable of speeds up to 30 miles per hour (48 km/h). The head and neck of mature birds are colored a grayish blue. There are black feathers on the head and mottled brown feathers on the rest of the body. The plumage is coarse and hairlike. The emu has three toes on its feet.

Rhea. The rhea is similar in appearance to the ostrich but is smaller and has three toes on its feet instead of two. The mature rhea is about five and one-half feet (1.7 m) high and weighs 44–55 pounds (20–25 kg). Like the ostrich, the rhea can run very rapidly. The feathers of the rhea are pale gray to brown; the head and neck are completely feathered and there are no tail feathers. The long body feathers droop over the posterior of the bird.

Cassowary. The mature cassowary is 4–6 feet (1.2–1.8 m) tall. There is no plumage on the head and neck, the skin of which is brightly colored with combinations of blue, red, and yellow. Some species have long brightly colored wattles. They have a large bony crest on the top of the head. The hairlike feathers on the body are brownish black in color. Cassowaries are capable of running at speeds of up to 30 miles per hour (48 km/h). They have long sharp toenails on their inner toes that they use for defense. In their natural habitat, cassowaries are found in the forest rather than on the open plain.

Kiwi. Kiwis are small birds, similar in size to chickens. There are three species of kiwi; the two larger species (brown kiwi and great spotted kiwi) are about 20 inches (50 cm) long and the smaller species (little spotted kiwi) is about 16 inches (40 cm) long. Kiwis are 2.75–9 pounds (1.2–4 kg) in weight and stand up to one foot (0.3 m) in height. The females of all three species are larger than the males. Kiwis have brown or gray hairlike feathers that may be streaked or barred. They have small heads, a long slender bill with the nostrils near the tip, no tail feathers, and short legs with three toes on the feet.

Breeding and Egg Production

Ostrich. The male ostrich matures at about two and one-half years of age, the female at about two years of age. Ostriches will mate under both monogamous and polygamous conditions; breeders in the United States typically mate bonded pairs. They engage in elaborate courtship behavior; the male vocalizes with a loud booming noise, inflates his neck area, sits on his hocks, moves his wings up and down and his head from side to side. The female indicates receptive behavior by bowing, ruffling her wings, sitting, and allowing the male to mount her from the back.

In the United States, mating and egg laying generally begins in March and may extend to September; under some conditions the breeding season may be shorter or longer. Breeding pairs should be established about two months before the breeding season begins. The female begins to lay eggs about five to ten days after the first mating. The male digs a shallow depression in the ground, forming a nest where the female lays the eggs. One egg is usually laid every other day; the total number of eggs laid during the season varies widely. When the eggs are left in the nest after they are laid, a total of about 15–20 eggs will be laid. More eggs will be laid if each egg is removed from the nest after it is laid. While estimates of total egg production range up to 90, the average number is closer to 40 to 50 eggs per hen. Ostrich eggs weigh from three to five pounds (1.4–2.3 kg). A diet that does not meet the needs of the hen will reduce the number of eggs laid. The male aggressively defends the female and the nest. Both the male and female sit on the nest, the male usually at night and the female during the day.

Emu. Most breeders pair emus in monogamous relationships; however, it is possible to group them in polygamous groups of one male to more than one female. Emu females are the dominant member of the pair; they choose the male for mating, choose the nesting site, and defend their territory. Sexual maturity is reached in one and one-half to three years. The mating season usually is from November to March. Vocalization occurs during courtship, with the male making a grunting sound and the female making a drumming sound. They both strut and display their neck feathers. Eggs are laid in a shallow depression in the ground. On average, 30 eggs are laid by one female during the egg laying season. Typically, one egg is laid about every three to four days. Emu eggs weigh about one and one-half pounds (0.7 kg). The male sits on the nest to incubate the eggs if they are not removed for artificial incubation.

Rhea. In their natural habitat, rheas are polygamous; a group usually consists of one male and six or more females. The females lay their eggs in the same nest and the male sits on the nest to incubate the eggs. The management of rhea breeding in domestic production is similar to practices followed for the ostrich.

Incubation and Hatching of Eggs

Most ratite producers in the United States use artificial rather than natural incubation for eggs. The incubation time for ostrich eggs ranges from 39 to 59 days with an average time of 42 days; for emu eggs the range is 46 to 56 days with an average time of 50 days. Commercial incubators and hatchers are available for use by ratite producers.

Collect the eggs from the nests each day and store them in a cool place at 55°–65°F (12.8°–18.3°C) at a humidity of 75 percent. Before placing the eggs in the incubator, store them with the large end up or on their side and turn them three times each day. To achieve maximum hatchability, the eggs should be placed in the incubator within seven days. Eggs that are dirty should be cleaned by sanding or washing (if extremely dirty) in water containing a sanitizing agent that is ten degrees warmer than the egg temperature. Do not use water that is too hot. The shell of the egg is porous and care must be taken not to allow water and bacteria to penetrate the egg.

Use care when handling eggs to prevent contamination. Some producers wear disposable latex gloves when collecting and handling the eggs. Contamination may also be prevented by thoroughly washing the hands before handling eggs.

The incubator temperature for ratite eggs is usually set in a range of 97.5°–99°F (36.4°–37.2°C). The producer should follow the directions for the specific brand of incubator being used. Humidity in the incubator is adjusted to achieve the desired moisture loss from the eggs. Typical relative humidity settings for ostrich eggs are in a range of 35 to 40 percent and, for emu eggs, 24 to 35 percent. Follow the manufacturers' directions for relative humidity setting. Eggs are weighed weekly to determine the amount of moisture loss that is occurring. This information is used to set the appropriate relative humidity in the incubator. The moisture loss should be about 12–16 percent over the incubation period.

Eggs need to be positioned properly so the air cell is at the top and turned while in the incubator. Ostrich eggs can be candled prior to placement in the incubator to determine the position of the air cell. Emu and other ratite eggs that are opaque should be placed on their side. Hand turn the eggs through a 90 degree angle several times daily. Some incubators automatically turn the eggs 12 to 24 times during a 24-hour period.

One to two days before hatching, the chick will break (pip) through the internal shell membrane into the air cell. At this time the egg is moved to the hatcher. Twelve to eighteen hours after internal piping occurs, the chick will begin to break through the shell of the egg. The chick will normally hatch within twelve hours after first breaking the egg shell. They are moved to the brooding area after they are dry.

Not all eggs set in the incubator will hatch. On average, about 70 percent of ostrich eggs set in the incubator will hatch; the average for emu eggs ranges from 50 to 80 percent. To determine the number of fertile eggs use this formula: (# Fertile eggs ÷ # Total eggs set) × 100 = Percent fertility. Fertile eggs are any that showed indications of fertility even if an embryo did not develop. Hatchability of fertile eggs is determined by this formula: (# Chicks hatched ÷ # Fertile eggs set) × 100 = Percent hatch of fertile eggs. If the percentage of fertile eggs is low, then a problem with the breeding program is indicated. If the percentage of hatched eggs is low, then a problem probably exists in the incubator.

It is important to inspect the material left in the hatcher tray after the birds have hatched. This material will give an indication of possible problems. Common problems found in the hatch waste breakout are indicated in Table 48-1.

Brooding

Ratite chicks need supplemental heat for several weeks after they hatch. The brooder area at floor level should be kept at about 90°F (32°C) for ten days to two weeks. After that the temperature can be reduced about five degrees every two weeks until supplemental heat is no longer needed. Do not allow chicks to become chilled or overheated. Provide enough space so they can move away from the heat source if necessary.

The brooding area needs to be clean, dry, and well ventilated. Provide protection from inclement weather if necessary. A variety of materials can be used for the floor area including wood shavings, straw, rice hulls, or clean sand. The material used should be easy to clean. Do not put any slick material such as newspaper, cardboard, or plastic on the floor. On a slick floor chicks may develop "spraddle leg," which may be fatal. Ratite chicks tend to eat the litter, which may result in impaction in the digestive tract. This may be prevented by covering the litter with burlap for the first two weeks. Stir the litter periodically to keep it dry and prevent caking.

Do not overcrowd the chicks in the brooding area, as this may result in their piling up and smothering. As they get older they may

TABLE 48-1 Common Symptoms of Problems Found in Hatch Waste Breakout.

Symptom	Check for
Smelly, brown-colored albumen	Contamination of eggs
Missing or small lower jaws	High/low selenium, low calcium
Chicks with head away from air cell	Proper positioning in incubator tray
Extremely small embryo or chick, no feathering seen (early death)	High incubator temperatures during beginning period of incubation
Little formation of air cell, failure to pip by embryo	High humidity in incubator
Low water loss within eggs, edema seen in chicks	High humidity in incubator, vitamin E/calcium/selenium deficiencies
Chicks with dry membranes stretched and surrounding chick	Low humidity in incubator and possibly within the hatching cabinet
Liquid seeping from eggs	Contamination of eggs
Chicks hatched with large amounts of yolk exposed	Late high temperatures in incubator and hatcher, low water loss during incubation
Spraddled legs, problems standing	Slick hatcher trays, giving improper footing. Low calcium in some cases
No embryo seen within egg, undeveloped germinal disk	Infertility, lack of breeding activity

Source: Martin, G.P., R.C. Fanguy, and J.S. Jeffrey, *The Incubation of Ratite Eggs.* College Station, TX: Texas A&M University.

be moved to larger pens. It is a good idea to provide outside exercise areas if the weather permits. This area should be free of trash, small rocks, or other material the birds may eat.

Chicks should be weighed and marked with numbered leg bands at the time they are placed in the brooder. Ratite chicks are not fed for one to several days after they have hatched so they will absorb the nutrients from the yolk sac. Feeding too soon can prevent the absorption of the yolk nutrients.

Sexing Young Birds

It is difficult to determine the sex of young ratite birds by observation alone. An examination of the sex organs is generally required. This is most easily done when the birds are one to two months of age. Because the sexual organs are small in immature birds, it requires some practice to accurately determine the sex of the individual bird. Both the penis and the clitoris are located in the ventral cloaca. To determine the sex of the bird, apply pressure on the cloaca so that it inverts, revealing the penis or clitoris. The penis is larger, curved, and has more cartilage than the clitoris.

Feeding

Only a limited amount of research has been done on the nutritional requirements of ratites. Commercial ratite starter, grower, and breeder rations in mash or pellet form are available from a number of feed companies. Most of these are based on nutritional research with chickens, turkeys, and game birds. The protein content of commercial feeds for ratites ranges from 16 to 22 percent.

After ratite chicks have absorbed the yolk nutrients (usually one to several days after hatching) they may be fed a starter ration. Commercial ratite, turkey, or game bird starter rations may be used. Do not use a ration that is too high in fiber, as this may cause intestinal blockage. The fiber content of the ration should range from five to fifteen percent. Some growers feed a vitamin/mineral supplement in addition to the starter ration. Never feed moldy feed.

Limited feeding for ratite chicks is generally recommended for a period of time up to three months. Typical feeding patterns are two or three times per day for about one hour for each feeding period. Limiting the feed intake during this time reduces problems with bacterial enteritis that may occur if continuous feeding is done.

A continuous supply of fresh water should be available. Some growers provide supplemental feed such as cut alfalfa, clover, or alfalfa pellets during this time. Grit may need to be supplied if fresh roughage is provided.

Impaction of the digestive system is sometimes a problem with young ratites. They have a habit of eating an excessive amount of sand, small stones, or fiber if these materials are available. Some growers add mineral oil to the ration about once a week to reduce the problem of impaction. Impaction problems are less severe with older birds, probably because of the larger size of their digestive system.

The birds may be switched to a grower ration at two to three months of age; the grower ration is fed until the birds reach maturity. Some growers continue to limit feed the birds at this time; others begin free choice unlimited feeding at ten to twelve weeks of age. Either system appears to be satisfactory. Supplemental feeding of alfalfa pellets or green forage may be done. Birds that have access to green forage should be provided with grit.

A breeder ration is fed when the birds are mature and beginning egg production. The ration needs to provide sufficient calcium and phosphorus for egg production. Forages such as alfalfa, oats, bermuda grass, or wheat may be used as part of the diet; a complete breeder ration should also be fed to birds that have access to forage.

While birds are not laying eggs, a maintenance ration may be used. This ration should be sufficient to prevent weight loss but not allow the birds to gain weight. About two months before the laying season begins, switch to a breeder ration that provides more protein.

Some sample starter, grower, and breeder ration mixes for ostriches are shown in Table 48-2.

TABLE 48-2 Ostrich Ration Mixes.

Ostrich Starter		Ostrich Grower	
Ingredient Name	*Amount (Pounds)*	*Ingredient Name*	*Amount (Pounds)*
Soybean Meal	26.50	Soybean Meal	26.00
Corn Grain Yellow	25.20	Corn Grain Yellow	22.90
Alfalfa (Dehydrated)	18.60	Alfalfa (Dehydrated)	15.00
Peanut Hulls	9.53	Sorghum	15.00
Sorghum	8.50	Wheat Midds	6.50
Corn Screening	5.50	Fish Meal (Menhaden)	5.00
Calcium Carbonate	1.84	Peanut Hulls	4.80
Dicalcium Phosphate	1.84	Calcium Carbonate	1.47
Nutra Blend Vitamins	1.25	Nutra Blend Vitamins	1.25
Fat (Animal)	0.80	Dicalcium Phosphate	1.23
Salt	0.60	Fat (Animal)	0.50
DL Methionine	0.07	Salt	0.30
Total Weight	100.00	DL Methionine	0.04
		Total Weight	100.00

Ostrich Breeder	
Ingredient Name	*Amount (Pounds)*
Soybean Meal	27.40
Corn Grain Yellow	21.00
Corn Screenings	10.20
Sorghum	10.00
Alfalfa (Dehydrated)	9.80
Meat and Bone Meal	8.18
Calcium Carbonate	4.90
Peanut Hulls	4.80
Nutra Blend Vitamins	1.25
Dicalcium Phosphate	1.20
Fat (Animal)	0.60
Salt	0.60
DL Methionine	0.06
Total Weight	100.00

Source: Berry, J., *Ostrich Production.* OSU Extension Facts No. 3988. Stillwater, OK: Oklahoma State University.

Handling and Transporting

Mature ratites can be dangerous to handle; they can kick with sufficient force to inflict serious injury to a human. They become nervous, jump about, and flail their wings when they are frightened or threatened. It is necessary to restrain a mature bird when handling it for medical treatment such as vaccination or examination.

Young birds can be picked up by the legs or by supporting the chest with one hand while holding the other hand on the back. Wrapping the young bird in a large towel makes it easier to handle and transport.

It may take several people to handle a mature bird. Because birds kick to the front or side, it is safer to approach them from the rear when preparing to restrain them. An opaque hood is often placed over the bird's head when handling is necessary. Usually the bird will calm down when hooded. Restrain the bird by grasping its wings and pushing down. An adult bird that is used to being handled can be moved by guiding it with one hand holding a wing and the other hand holding the tail. Another method of moving a bird involves two people, one behind the bird holding it by the wings and lifting on the rump while the other holds it by the neck or beak and guides it in the desired direction.

It is easier to move birds if alleys are constructed between pens. A solid or semi-solid panel may be used to push the bird down the alley. Birds are easier to handle under limited light conditions.

Transport ratites in enclosed trailers that are well ventilated and have a non-slip floor. Old rugs or carpets may be used to provide a non-slip surface on the floor of the trailer. Make sure there are no sharp projections in the trailer that might injure the birds. Separate birds of different sizes when transporting them in a trailer.

Identification

It is recommended that plastic leg bands be placed on the chicks when they hatch. Larger leg bands that can be adjusted in size may be used when the birds get older; these help identify the birds at a distance without having to restrain them for microchip reading. Microchips can be implanted, usually in the piping muscle just behind the head, at hatching time for identification purposes. Electronic identification is widely used in the ratite industry. Some insurance companies require microchip identification for insurance purposes. The microchip is implanted in young birds and can be read with a microchip reader to facilitate record keeping. Some breeders implant a second microchip in the tail head of older birds because it is easier to read when the bird is drinking or eating. Some companies offer DNA identification services. Tattooing has also been used for identification of birds.

Electronic identification provides proof of ownership, helps in breeding programs by keeping track of genetic bloodlines, and makes it easier to trace and recover missing birds. Most birds offered for sale have the microchip electronic identification devices already implanted.

Facilities

Young ratite chicks can be kept in small pens with adequate shade and protection from inclement weather. Generally they should be moved inside at night and kept in a heated room until they are two to three months of age. Provide larger pens for older birds; shelter will be needed if the temperature of the region is cold. Ground cover of alfalfa, clover, or grass is recommended.

The size of pen for breeding pairs of ratites may be from one-quarter to three acres. One-third to one-half acre per pair is adequate if space is limited. Do not mix other animals in the same area, as this may have a negative effect on their nesting and breeding. Provide an adequate area for exercise along with housing and shade. Woven wire, chain link, or smooth wire fencing may be used. Fences should be five feet (1.5 m) high and flush with the ground to keep other animals out of the pen.

Alleys built between the pens makes it easier to move birds from one area to another. Restraint facilities are useful for treating sick or injured birds. An isolation pen for sick birds is recommended. High solid walls should be used in crowding areas and chutes. This makes it easier to handle the birds.

Health

Ratites are subject to many of the same illnesses and parasites as other poultry. It is wise to establish a good working relationship with a veterinarian who is knowledgeable about ratites to plan a good health maintenance program. Always consult a veterinarian when treatment for an illness or injury is necessary. Birds should be examined by a qualified veterinarian before they are purchased.

Birds under six months of age have the highest mortality rate. Emu chicks appear to have a lower mortality rate than ostrich chicks. Yolk sac and umbilical infections are two common problems with young chicks. Careful management at hatching will help reduce these problems. Sometimes chicks become listless, stop eating and drinking, and die; this condition is called *malabsorption syndrome* or *fading chick syndrome*. The cause of this problem is not known. Another health problem of young chicks is impaction caused by ingesting foreign items such as sticks, fabric, sand, and grass. Impaction has been treated successfully with surgery, by adding mineral oil to the diet, or with the use of a laxative. Impaction is generally not a major health problem with older birds. Respiratory diseases are more common in young birds as compared to older birds. Antibiotics may be used to treat respiratory diseases.

Ratites are affected by both internal and external parasites including nematodes, ticks, lice, and mites. Several anthelmintic drugs and dusting powders may be used for treatment. Consultation with a veterinarian to determine the appropriate treatment is recommended.

A good disease prevention program is based upon the premise of reducing the opportunity for infectious organisms to enter the facilities where birds are kept. The three major components of a good program are isolation, traffic control, and sanitation. *Isolation* means to keep the birds in a facility that does not allow other animals to enter, keeping birds separated by age group, and isolating new birds until their health status is determined (2–4 weeks). *Traffic control* refers to not allowing people uncontrolled access to the facilities and birds. *Sanitation* refers to making sure all equipment is clean and disinfected, and that the facilities are properly cleaned and disinfected at regular intervals.

Getting Started in Business

Serious interest in raising ratites in the United States began to develop in the mid-1980s. During its first ten years, the ratite industry in the United States was primarily in a breeder phase; that is, birds were being raised to supply breeding stock to increase the total numbers. Prices for breeding stock were very high during this breeder phase. By the mid-1990s, the ratite industry began to move from a breeder phase to a slaughter and product production phase.

Current prices for ratite eggs and breeding stock are much lower than they were during the breeder phase of the industry. Fertile hatching eggs were being advertised on the Internet in late 1999 for $100 each. During the breeder phase hatching eggs were selling in the range of $800 to $2,000 each. Buying eggs and hatching them requires the lowest initial investment; however, there are several disadvantages to this method. The mortality rate is highest, incubation and brooding equipment is necessary, eggs are generally not insurable, too many chicks of one sex may be produced, and it is two to three years before any return on investment is realized. Generally, only people with experience in raising ratites should purchase eggs. Emu eggs are seldom sold because they are opaque and cannot be candled to determine fertility.

During the breeder phase, sexed chick pairs from one day to six months of age sold in a price range of $1,500 to $9,500 for a male-female pair depending on the age of the chicks. In December of 1999 ostrich chicks were advertised on the Internet in a price range from $150 to $250 each depending upon their size. Advantages of this method of getting started include knowing the sex of the birds, the purchase of an incubator can be delayed, conformation can be judged, and the initial investment is less than buying adults. The mortality rate is still relatively high and it is still two to three years before any return on investment is realized.

Yearling pairs of male-female birds may be purchased. The advantages of buying yearling birds include a lower mortality rate, birds are insurable, and conformation can be judged. Birds of this age are still one to two years from producing a return on investment.

During the breeder phase, two-year-old breeding pairs sold in a price range of $22,000 to $30,000; older, proven breeding pairs sold for as much as $80,000. Prices advertised on the Internet in December 1999 for breeding females ranged from $3,000 to $4,000 each depending on their age; mature breeding males are being advertised for $1,500 each. Ratites being raised for slaughter are currently valued in a range of $500 to $750 per bird. Two-year-old breeding pairs are closer to producing a return on investment; other advantages are the same as for buying yearling pairs. The major advantages of buying a proven breeding pair are that their reproductive potential is known and a quicker return on investment can be realized.

Some breeders are offering shares in proven breeding pairs. This is an option for a person who wants to invest some money in ratites but does not want to be directly involved in raising the birds. A minimum investment (typically $500) is usually required.

Investment in ratite raising is high; the potential returns are also high. People interested in this enterprise need to carefully research the costs, requirements for facilities, their ability to care for these birds, and the risks that are involved. They need to visit current operations and talk with those who are in the business. Ratites are currently being raised in the United States primarily for meat, hide, oil, and feathers. Some early investors in the industry have lost money and some have gotten out of the business. One of the major problems currently facing producers is the lack of facilities for slaughtering and processing ratites. The market for products from ratites is limited and retail prices are very high.

Purchases of breeding birds should be made from reputable breeders who are willing to guarantee the fertility and health of the birds. It is also important to purchase unrelated breeding stock. It is not possible to get unrelated breeding stock from a seller who has only one pair of breeding birds.

Before purchasing, be sure birds are healthy (get a health certificate from a licensed veterinarian), have been implanted with microchip identification or are DNA identified, and are guaranteed as to sex. Microchip numbers should be verified and checked against a current list of stolen birds. Get written records on the birds that include origin, family bloodline, and vaccinations. Get all warranties or guarantees in writing. Contact other customers of the breeder from whom the birds are being purchased, the Better Business Bureau, and the State Attorney General's Office to determine if there have been any problems with this breeder.

Products

Ratite products include meat, leather, oil, feathers, and other by-products. The meat is promoted as a low-fat, low-cholesterol red meat. Research shows that an ostrich has a dressing percentage of about 57 percent and an emu has a dressing percentage of about 54 percent. While not currently generally available in the United States, except in some restaurants, ratite meat is popular in many parts of Europe.

Ratite hides are used to produce leather products including shoes, boots, coats, belts, purses, wallets, and briefcases. Emu leather is thinner and finer textured than ostrich leather.

Emu oil is produced from the fat and is used in cosmetic and pharmaceutical products. One emu will produce five to six liters (1.3–1.6 gal.) of oil. Research is currently underway to determine the full potential of emu oil.

Ratite feathers are used in fashion, costuming, and feather dusters. Feathers can be harvested once a year. Eggshells and toenails are used in arts and crafts.

LLAMMAS AND ALPACAS

Introduction and History

Forty million years ago, the Camelidae family originated on the North American continent. About three million years ago, some members of this family migrated to the Asian and African continents and others migrated to South America. The Camelidae family became extinct on the North American continent about 10,000–12,000 years ago. The Dromedary camel and the Bactrian camel are the present-day descendants of those that migrated to Asia and Africa. The llama, alpaca, guanaco, and vicuna are the present-day descendants of those that migrated to South America.

The llama and alpaca were domesticated from the guanacos in Peru about 4,000–5,000 years ago. In South America, llamas have long been used as pack animals, and for their meat, milk, hides, and wool; alpacas are not usually used as pack animals but are valued for their wool. There are currently about 3.5 million llamas and 3.5 million alpacas in South American and over 70,000 llamas and over 5,000 alpacas in the United States and Canada.

Llamas were first imported into the United States in the late 1800s to be exhibited in zoos. Interest in raising llamas and alpacas increased in the 1970s with a primary emphasis on raising and selling breeding stock. In addition to breeding stock, llamas are used as pack animals, fiber production, guard animals for sheep herds, as show animals, and for pets. Alpacas are also used for breeding stock, fiber production, investments, and as pets; they are generally not used as pack animals or guard animals.

Initial investment costs for both llama and alpaca breeding stock vary widely in different parts of the United States. Prices for female llamas for breeding stock can range from $350 to $6,000 or more. Prices for males can range from $500 to $3,000 or more. Stud fees for breeding tend to be in the range of $700 to $1,500. Trained pack animals will cost about $2,500 to $3,000. Generally, alpaca prices are in the same ranges as the prices for llamas. Current prices for llamas and alpacas can be found by searching the Internet. Many producers advertise their animals for sale on their web pages.

Most llamas and alpacas in the United States are registered with the International Lama Registry (ILR). This is a closed registry, which means that both parents of an animal must be registered before the offspring can be registered. Blood typing with the ILR is commonly used to maintain and track genealogy records.

Persons interested in owning llamas or alpacas can get more information from the following organizations:

■ International Llama Association
P.O. Box 1891
Kalispell, MT 59903
■ Alpaca Owners and Breeders Association
c/o Hobert Office Services, Ltd.
P.O. Box 1992
Estes Park, CO 80517

For those with access to the Internet, there are many sites on the World Wide Web that provide information about llamas and alpacas. Entering "llama" or "alpaca" as keywords in a search will provide listings of current sites with their URL addresses.

Characteristics

Llama wool color is usually white but may be black; sometimes there are shades of beige, brown, red, or roan. The wool coat may be marked in a variety of patterns from solid to spotted. The average weight of mature llamas is 280–450 pounds (127–204 kg). Llamas have a long neck, and they stand 5.5–6 feet (1.7–1.8 m) at the head and approximately four feet (1.2 m) at the shoulder. The life span of the llama is 20–25 years.

Alpacas have 22 distinct natural colors including black, white, roan, pinto, brown, red, fawn, rose, and gray. There are two types of alpacas based on the type of fleece they produce. Huacaya fiber has a dense, crimped fiber; Suri fiber is usually white and is not crimped but hangs straight down and is curly. An adult alpaca produces about four pounds (1.8 kg) of fleece per year. The average weight of mature alpacas is 100–175 pounds (45–75 kg); overall, alpacas are about one-half the size of llamas. Alpacas are about 4.5 feet (1.4 m) at the head and approximately three feet (0.9 m) at the shoulders. The life span of the alpaca is 20–25 years.

Llamas and alpacas have six incisor teeth on the front bottom jaw and a hard dental pad (no teeth) on top. The back of their jaw contains five molar teeth on the top and bottom on both sides. Their upper lip is split in a manner that allows them to use it in grasping forage when grazing.

Their feet are two toed and have a broad, leathery pad on the bottom that helps give them excellent footing on many types of terrain. They have a scent gland located on the side of each rear leg and one between the toes.

Feeding

Llamas and alpacas are modified ruminants; the stomach has three compartments instead of four. They chew their cud just as other ruminants do and the stomach functions in essentially the same manner. They make more efficient use of feed than do cattle or sheep, converting more of it to the energy and protein they need to meet their requirements. Care must be taken not to overfeed these animals; it will cause them to become too fat. The feeding program is based on roughage with concentrates as a supplement to the diet. The total daily dry matter intake should be in the range of 1.8–2.0 percent of body weight. The crude fiber content of the diet should be about 25 percent.

Llamas will eat both forage and browse; alpacas are less likely to eat browse. Browse is the shoots, twigs, and leaves of brush plants found growing on rangeland. A moderate producing pasture will carry three to five adult animals per acre. If a good-quality forage is available, grain supplementation of the diet is usually not needed except during the last 2.5–3 months of pregnancy and during the first three months of lactation if the animal is not too fat. Working pack animals may also need some grain supplementation..

The protein requirement is generally low for llamas and alpacas. A maintenance ration should contain 8–10 percent crude protein; a 10–12 percent protein level is recommended for growing animals, and for pregnant and nursing females. Feed for an early weaned cria (young offspring) should contain 16 percent protein, with the amount being gradually reduced to 12 percent by the time it is six months of age.

Make sure the newborn cria receives the colostrum milk. In an emergency, cow or goat colostrum milk may be used. While it may be necessary to bottle feed a premature, weak, or orphaned cria to save its life, this practice should be held to a minimum. Cria that are bottle fed have a tendency to imprint on and become aggressive toward humans. If bottle feeding of colostrum milk is necessary, feed a total of about ten percent of the cria's weight in two to three feedings during the first day. After the first day, llama or goat milk, or a milk replacer, may be fed at a rate of ten percent of the cria's body weight per day. Follow the manufacturer's directions for preparing and using the milk replacer.

Cria will begin eating a small amount of forage when they are two or three weeks of age. They may be creep fed a limited amount of a 16 percent grain and forage mix starting at about three months of age and continuing until they are weaned. Feeding a free choice mineral supplement with the proper calcium-phosphorus ratio is recommended when creep feeding cria.

While llamas and alpacas have less need for water than most other farm animals, it is recommended that a fresh, clean supply of water be available at all times. Provide loose, trace-mineralized salt in containers to protect it from the weather. A calcium-phosphorus ratio of 1.2–2.0:1 is recommended. A free choice vitamin-mineral mix is used to provide the necessary vitamins and minerals in the diet.

Good quality hay that is not moldy and that has been analyzed for nutrient content should form the basis for any llama or alpaca feeding program. Care must be taken when feeding alfalfa hay as it may be too high in protein content. Brome grass, timothy, fescue, orchard grass, oat hay, and pea hay are good choices for feeding.

Many different kinds of pasture, including brome grass, timothy, orchard grass, bluegrass, white clover, and alfalfa, are suitable for llamas and alpacas. It may be necessary to limit access to pasture to prevent excessive dry matter intake.

Commercially prepared complete feeds in pelleted form are available for llamas and alpacas. While this is a convenient method of feeding, the diet may not contain enough fiber unless the animals are on pasture or being fed hay with the pelleted feed. Using a pelleted feed is more expensive than other methods of feeding. When using a pelleted feed, care must be taken not to overfeed the animals; also, some animals may choke on the pelleted feed.

Management

Llamas and alpacas are herd animals that desire the companionship of other animals. When kept in isolation, they do not thrive. If a person wants only one, it can be kept with other domesticate animals such as sheep. Some sheep ranchers use llamas as guard animals to protect the flock from predators.

When llamas or alpacas are kept on a breeding farm, there are several groupings that may be used:

■ *All the females and their young cria.* Generally, it is not wise to keep a stud with this group. Doing so makes it difficult to maintain good breeding records and makes it harder to work with the group because of the territorial behavior of the male. Some males experience a decline in libido when kept with a group of females.

■ *Pregnant females.* By keeping pregnant females in a separate group, it is easier to meet their nutritional needs and to more closely monitor them as they approach parturition. Do not keep any males with this group; they sometimes attempt to initiate breeding behavior as the females approach parturition.

- *Weanlings.* Females up to two years of age and males up to nine months of age may be kept in this group. It is easier to meet the nutritional needs of this age group when they are kept in a separate group. Males are removed from the group by the time they are nine months of age to prevent breeding problems.
- *Young males from nine to eighteen months of age.* By keeping this group separate, breeding problems with females are prevented. They should not be kept with older males because of problems with fighting.
- *Castrated (gelded) males eighteen months of age and older.* Many breeders castrate all the males they do not plan to use as studs.
- *Individual stud male.* When males that have not been castrated are kept together, they tend to fight. Some breeders do keep uncastrated males together successfully if they are not near a group of females.
- *New purchases in a quarantine area.* By quarantining all animals brought into the herd for two weeks, disease problems are reduced. The quarantine area must be located so contact with other animals is not possible.

Good record keeping is essential when managing breeding herds. The following kinds of records should be kept:

- Identification of sire and dam.
- Breeding dates.
- Parturition dates.
- Transfers among groups.
- Medical problems, treatments, and vaccinations.

Llamas and alpacas respond well to training and are relatively easy to handle; however, they do not like to be touched and do not respond well to physical punishment. They are intelligent and naturally curious about their surroundings. They have a strong herd instinct, tending to move together as a group. They may be moved in the pasture by utilizing the flight zone concept; approach them from the direction opposite to the direction of desired movement. A large group may require two or more handlers to keep the group moving in the desired direction. After they have been moved from pasture to a smaller space, an individual animal may be caught, when necessary, by moving it into a corner or smaller space. Approach the animal slowly to put a halter on its head. Feeding in a small pen makes it easier to catch individual animals; just shut the gate after they have entered for feeding and then proceed to isolate and catch the individual.

Occasionally, the toe nails of llamas and alpacas need to be trimmed. This is usually not necessary if they are kept on, or have intermittent access to, stony ground, roughened concrete, or sandy lots. If the nails grow too long the animal may become lame. Trimming is done with the same side-cutting nippers that are used on sheep. The animal will probably need to be restrained for nail trimming. Care must be taken not to cut into the live flesh around the nail.

At about two years of age, fighting teeth (canines) erupt in males. These teeth have sharp points and cutting edges and should be removed to prevent injury if the animals fight. The owner needs to recheck these teeth periodically because they will continue to erupt until the animal is four to five years of age. These teeth should be cut off when they are about one-quarter of an inch in length. A flexible braided cutting wire called an obstetrical wire is used for this procedure. The animal needs to be restrained when these teeth are removed. The wire is hooked behind the tooth and drawn back and forth in a sawing motion to cut the tooth. Soft tissue is not normally damaged by the obstetrical wire. Fighting teeth also are present in females but they are small and erupt when the animal is about 5–6 years of age; normally they are not removed.

In areas where temperatures go above 90°F (32°C), several methods of keeping the animals cool may be used. These include:

- Providing shade in an area with good air circulation.
- Shearing once a year.
- Providing an area filled with sand or pea gravel that can be wetted down.
- Using a sprinkler system or providing a shallow wading pool.
- Scheduling breeding so they do not give birth during hot weather.
- Using fans or evaporative coolers in barns.
- Limiting the feeding of forages and high-protein supplements. The digestion of these feeds produces a higher heat increment than does the digestion of concentrates. However, be sure the crude fiber content of the diet remains at about 25 percent.
- Feeding in the evening so digestion takes place during the cooler night hours.

Reproduction

Llamas and alpacas do not have estrus cycles; they are induced ovulators and can be bred at any time during the year. The act of mating stimulates the release of an egg for fertilization. Ovulation occurs 24–36 hours after copulation. Females can be bred at one year of age; however, the recommended age for breeding is one and one-half to two years of age. The gestation period is about 11–11.5 months. Parturition normally occurs during daylight hours; assistance at birth is rarely needed. The birth weight of llama crias is about 20–35 pounds (9–16 kg) and of alpaca crias about 10–20 pounds (4.5–9 kg). Crias are weaned at about six months of age. Llama males become capable of breeding at about seven to nine months of age; however, it is recommended they not be used for breeding until they are approximately three years old. Male alpacas are used for breeding at about two to three years of age.

Health

There are few disease problems with llamas and alpacas. It is generally wise to have a good working relationship with a veterinarian who is knowledgeable about them. A good preventative medicine program, including proper nutrition, vaccinations, parasite control, and sanitation of facilities, should be followed. Vaccines that are currently used for other domestic animals are used for llamas and alpacas; there are no vaccines currently available that have been developed specifically for these animals. Animals should be examined by a veterinarian before they are purchased and should be accompanied by a certificate of health.

The same internal and external parasites that affect cattle and sheep also affect llamas and alpacas. Control these parasites with the same methods used for cattle and sheep. The animals should be dewormed once or twice a year if internal parasites are a problem.

Digestive disorders are rare in llamas and alpacas. They rarely bloat on feed; however, overeating grain can cause severe, sometimes fatal, illness. Poisonous plants should be eliminated from pasture areas to prevent ingestion.

Housing and Fencing

Llamas and alpacas require little housing, especially in a mild climate. In areas where temperatures may be extremely cold or hot, some shelter should be provided. A closed barn will provide adequate shelter in cold weather; a three-sided shelter is usually sufficient in a milder climate. Orient the shelter to provide maximum protection from the prevailing wind. Good ventilation must be provided in closed barns. Many breeders provide an area where sick animals or a new mother with her cria can be kept for treatment. Heat lamps may be used to provide warmth for the cria in cold weather.

Providing hay mangers or feed bunks will prevent animals from picking up parasites from the ground. Feed can be protected by placing the mangers or bunks in a sheltered area.

Catch corrals make it easier to manage such activities as vaccinating, nail trimming, shearing, brushing, and training. A holding chute may be built into a corner of the catch corral. Holding pens, alleyways, and sorting gates help control the animals when it is necessary to work with them.

Fencing requirements vary with herd size, groupings used, pasture size, and management practices. Four-foot (1.2-m) fences are usually adequate if there is little pressure for the animals to want to leave the enclosure. Llamas, however, can jump a fence of this height. A fence height of 5.5–6.0 feet (1.7–1.8 m) may be necessary if the pressure to leave the enclosure is high. Higher pressure against a fence can be caused by separating cria from mothers at weaning time, keeping studs away from unbred females, and keeping stud males in separate enclosures. In some areas it may be necessary to erect fences sufficiently secure to keep out wild and domestic predators.

Fences may be constructed from a variety of suitable materials, including woven wire, chain link, smooth wire strand, cattle wire panels, wooden rails, or boards. Cost and durability of fencing are factors to be considered when constructing fencing. Electric fencing is effective once the animals learn what it is. Barbed wire should never be used because it may cause injury. Standard farm livestock gates work well and allow machinery access to pasture areas if necessary.

SUMMARY

The American bison once numbered in the millions on the North American continent and were an important resource for the Plains Indians. Hunted almost to extinction, their numbers are now on the increase. A small but growing number of people are raising bison for sale as breeding stock and for meat. Bison are large, strong animals and must be handled with care. Bison are similar to beef cattle in feed requirements, facilities needed, and health care. Breeding herds can be generally maintained on good grassland pasture. Animals being fed for meat are finished on roughage and concentrate rations.

There is some interest in the raising of ratites as an alternative to other livestock in the United States. Ratites are flightless birds and include ostrich, emu, rhea, cassowary, and kiwi. The ratite industry in the United States is mainly concerned with ostriches, emus, and rheas. The ostrich is the largest of the birds still in existence and the emu is the second largest. The ostrich hen will usually produce 40 to 50 eggs per year and the emu hen will produce about 30 eggs each year. Producers usually use artificial incubation to hatch the eggs. Only a limited amount of research has been done on the nutritional requirements of ratites. Some feed companies produce commercial ratite feeds. Mature ratites can be dangerous to handle. Electronic identification is widely used in the ratite industry in the United States. Producers need to follow a good disease prevention program. Prices for breeding stock are high. The meat of ratites is promoted as a low-fat, low-cholesterol red meat.

Llamas and alpacas were domesticated in Peru about 4,000–5,000 years ago. Llamas are used as pack animals as well as for meat, hides, and wool. Alpacas are not usually used as pack animals. There has been a growing interest in the United States in raising llamas and alpacas. The initial investment required for breeding animals is high. Llamas and alpacas are modified ruminants, having three stomachs. The feeding program is based on the use of roughages. Commercially prepared complete pelleted feeds are available for llamas and alpacas.

These animals respond well to training and are relatively easy to handle. They do not have estrus cycles; they are induced ovulators and can be bred at any time of the year. There are relatively few health problems but a producer should follow good health management practices. Housing needs for llamas and alpacas are generally simple.

Student Learning Activities

1. Visit an operation where bison are being raised.
2. Have someone who raises bison talk to the class.
3. Prepare and give an oral report to the class on any aspect of raising bison.
4. Visit a facility where ratites are being raised.
5. Have someone who raises ratites talk to the class.
6. Prepare and give an oral report to the class on any aspect of raising ratites.
7. Visit a facility where llamas or alpacas are being raised.
8. Have someone who raises llamas or alpacas talk to the class.
9. Prepare and give an oral report to the class on any aspect of raising llamas or alpacas.

Review

1. Briefly explain how the bison came to the North American continent.
2. Describe the appearance of the bison male and female.
3. Briefly explain the five options people have who want to become involved in the bison industry.
4. Why are people interested in bison meat for food?
5. What are the desirable and undesirable physical characteristics one should look for when selecting bison for the breeding herd?
6. When are bison most dangerous to handle?
7. At what age are bison heifers usually bred?
8. What months of the year are considered to be the breeding season for bison?
9. Why should the bison breeder avoid inbreeding?
10. Briefly explain procedures to be followed when weaning bison calves.
11. Briefly explain the advantages and disadvantages of dehorning bison.
12. Briefly explain how orphan calves can be cared for.
13. Briefly explain the recommended practices for feeding the bison herd on pasture.

14. Briefly explain the recommended practices for finishing bison for meat.

15. Name and describe the kinds of fencing that might be used for bison.

16. What kinds of facilities might be needed for a bison operation?

17. Briefly describe the important parts of the handling facilities for a bison herd.

18. List and briefly explain the methods of marketing bison.

19. What by-products may be marketed from a bison herd?

20. Describe the most common species of ratites used in the ratite industry in the United States.

21. Discuss the breeding of ratites.

22. How should ratite eggs be cared for?

23. Discuss the brooding of ratite chicks.

24. How is the sex of ratite chicks determined?

25. Discuss the feeding of ratite chicks and birds.

26. Why should care be taken when handling and transporting mature ratites?

27. Describe the proper method of handling and transporting mature ratites.

28. What methods may be used for identification of individual ratites and why are they used?

29. Describe facilities that might be used for raising ratites.

30. What health problems may occur with ratites?

31. How might a person get started in the ratite business and how expensive would it be?

32. What precautions should a person take when getting started in the ratite business?

33. What products are produced in a ratite business?

34. Discuss the background and history of llamas and alpacas.

35. What are llamas and alpacas used for in the United States?

36. Describe the characteristics of llamas and alpacas.

37. Discuss the feeding of llamas and alpacas.

38. Discuss the principles of managing llamas and alpacas.

39. How should llamas and alpacas be handled?

40. Discuss reproduction of llamas and alpacas.

41. Discuss health management of llamas and alpacas.

42. What kind of housing and fencing is needed for llamas and alpacas?

Glossary

abomasum—the fourth compartment, or true stomach, of the ruminant animal.

absorption—the process by which digested nutrients are taken into the bloodstream.

acid cleaner—cleaning compound made up of a combination of mild acids and wetting agents used to clean milkstone from milking equipment.

acid detergent fiber—that portion of a feed that contains mainly cellulose, lignin, and some silica.

additive—a drug or druglike material added to a basic feed mix.

additive gene affects—when many different genes are involved in the expression of a trait.

adenine—a purine base found in DNA and RNA; it normally pairs with thymine in DNA.

aflatoxin—a compound produced by a mold, especially *Aspergillus flavus,* that can contaminate stored animal feed.

afterbirth—the membranes expelled after the birth of the fetus.

agricultural biotechnology—the science of altering genetic and reproductive processes in animals and plants.

air-dry—See *as-fed basis.*

alkaline cleaners—group of general cleaners used to clean milking equipment; includes such substances as caustic soda (lye), soda ash, baking soda, and metasilicate of soda.

allele—one member of a pair or several genes that are located on a specific position on homologous chromosomes.

alveoli—the part of the udder that manufactures milk.

amino acid—a compound that contains carbon, hydrogen, oxygen, and nitrogen; certain amino acids are essential for growth and maintenance of cells.

ampicillin—a type of penicillin effective against gram negative and gram positive bacteria; used to treat infections of the intestinal, urinary, and respiratory tracts.

anaphase—one of the stages of mitosis and meiosis; at this stage the chromosomes move to opposite ends of the nuclear spindle.

Anas Boschas—wild mallard duck believed to be the ancestor of all domestic breeds of ducks.

androsterone—a steroid hormone that intensifies masculine characteristics.

animal protein—a protein supplement that comes from animals or animal by-products.

anthelmintic—a chemical compound used for treating internal worms in animals.

antibiotic—a chemical agent that prevents the growth of a germ or bacteria.

antimicrobial—a substance that can destroy or inhibit the growth of microorganisms, i.e., an antimicrobial drug.

anus—the opening at the end of the large intestine that is the termination of the digestive system and through which feces pass out of the body.

apparel wool—a fine wool used in making clothing.

arsenical—a drug containing arsenic.

artificial insemination—the placing of sperm in the female reproductive tract by other than natural means.

as-fed basis—data on feed composition or nutrient requirements calculated on the basis of the average amount of moisture found in the feed as it is used on the farm; sometimes referred to as *air-dry.*

Asiatic urial—a breed of wild sheep believed to be ancestors of some present-day domestic breeds.

ass—a four-footed animal of the genus *Equus* that is smaller than a horse, has longer ears, and a short, erect mane.

axial feather—in poultry, the feather that separates the primary feathers on the outer part of the wing from the secondary feathers on the inner part of the wing.

backcrossing—See *crisscrossing*.

backgrounding—the growing and feeding of calves from weaning until they are ready to enter the feedlot.

bacteria—one-celled microorganisms.

balanced ration—feed allowance for an animal during a 24-hour period that has all the nutrients the animal needs in the right proportions and amounts.

bantam—small chicken, usually weighing from 16 to 30 ounces (0.45–0.85 kg) as an adult.

barrow—a male swine castrated while young, in which physical traits of the boar have not developed.

basal metabolism—energy for the functioning of the heart, for breathing, and for other vital body processes.

best linear unbiased prediction (BLUP)—the system of determining the estimated breeding value or the expected progeny difference by taking into account all the known sources of information about the evaluated traits.

bile—a fluid produced by the liver that aids in the digestion of fats and fatty acids.

bit—the metal part of a bridle that goes in the horse's mouth.

bladder—storage compartment for liquid waste in the animal's body.

bloodtyping—a process whereby an individual's blood group is identified by a serological test on a sample of blood.

boar—a male swine that has not been castrated.

Bos indicus—humped cattle found in tropical countries.

Bos taurus—domestic cattle originating from either the Aurochs or the Celtic Shorthorn.

bovine somatotropin (BST)—a hormone produced naturally by the pituitary gland of the cow. (See *somatotropin*.)

bovine spongiform encephalopathy (BSE)—degenerative disease that affects the central nervous system of cattle.

breed crossing—in poultry, the crossing of different breeds to get the desired traits.

breeding value—the value of the sire or dam as a parent; based on differences that exist between a large number of offspring and the average performance of a trait within a population.

bridle—headgear used to control a horse.

broodiness—in poultry, when a hen stops laying eggs and wants to sit on a nest of eggs to hatch them.

browse—the shoots, twigs, and leaves of brush plants found growing on rangeland.

buck—a male of such species as goats, deer, rabbits, etc.

bull—a male bovine of any age that has not been castrated.

cabrito—the meat from young kids slaughtered at approximately 30 to 40 pounds (13.6–18 kg).

calf—a beef animal under one year of age.

candling—determining the interior quality of an egg with the use of a high-intensity light.

capon—a male chicken that has been surgically castrated.

carbohydrate—organic compound containing carbon, hydrogen, and oxygen.

cardia—a valve located at the end of the esophagus that prevents food in the stomach from coming back into the esophagus.

carpet wool—a coarse wool used for making carpets.

carrying capacity—the number of animals that can be grazed on a pasture during the grazing season.

castration—the removal of the testicles of a male animal.

catarrhal—inflammation of mucous membranes, especially of the nose and throat.

cattle—beef animals over one year of age.

cecum (caecum)—a blind pouch located at the point where the small intestine joins the large intestine.

cell—a microscopic mass of protoplasm that is the structural and functional unit of a plant or animal organism.

cell wall—membrane that surrounds the nucleus and cytoplasm of a cell.

centriole—a structure located just outside of the cell nucleus that organizes the spindle during mitosis and meosis.

centromere—that part of the chromosome where the spindle fiber attaches during mitosis.

cervix—in the female reproductive system, the neck of the uterus, which separates the uterus from the vagina.

chevon—meat from goats.

cholesterol—a steroid that occurs mainly in the bile, gallstones, brain, and blood cells. It is the most common of the animal sterols and is a precursor of a form of vitamin D.

chromatid—one of the two daughter strands of a duplicated chromosome, joined by a single centromere, that separate during cell division to form individual chromosomes.

chromosome—small rod-shaped body found in the nucleus of a cell.

chyme—partially digested feed that moves from the stomach to the small intestine.

clitoris—the sensory and erectile organ of the female reproductive system.

cloaca—in poultry, an enlarged part of the digestive tract where the large intestine joins the vent.

closebreeding—an intensive form of inbreeding in which the animals being mated are very closely related and have ancestory that can be traced back to more than one common ancestor.

cockerel—a young male chicken.

codominance—in heterozygotes the situation in which both alleles of a gene pair are fully expressed; neither being dominant or recessive to the other.

codon—a set of three nucleotides that make up the genetic code that specifies the specific amino acid to be inserted at a particular position in a polypeptide chain during protein systhesis.

color breed—in horses, a breed that registers horses on the basis of color; horses in the breed may have a variety of conformation types.

colostrum—the milk produced the first few days after parturition.

colt—generally, a male horse under three years of age; the Thoroughbred breed includes males that are four years old.

comb—in poultry, a fleshy crest on the top of the head.

concentrate—feed containing less than 18 percent crude fiber when dry; grains and protein supplements are concentrates.

conception rate—the percentage of a group of animals that become pregnant when bred.

contemporary group—a group of animals that have been raised under the same conditions, have the same genetic background, and are of the same sex.

coprophagy—the practice of ingesting feces.

copulation—the mating of a male and a female.

corpus luteum—a reddish-yellow mass that forms in a ruptured follicle in the ovary of mammals; the hormone progesterone is released by the corpus luteum.

cow—a female bovine that has had one or more calves; or an older female that has not had a calf but has matured.

Cow Index (CI)—in dairy cattle, a method of evaluating the genetic value of dairy cows.

cowper's gland—an accessory gland in the male reproductive system that produces a fluid which moves ahead of the seminal fluid, cleaning and neutralizing the urethra.

crimping—passing a feed crop through a set of corrugated rollers that are set close together.

crisscrossing—mating crossbred females to a sire belonging to one of the parent breeds of the female; also called *backcrossing.*

Critical Control Points (CCPs)—points at which identified hazards may be controlled. (See *Hazard Analysis/Critical Control Point.*)

crop—in poultry, an enlargement of the gullet that serves as a storage area for feed.

crossbreeding—mating animals of different breeds.

cross-mating—in poultry, the crossing of two or more strains within the same breed.

crossover—the formation of new chromosomes resulting from the splitting and rejoining of the original chromosomes.

crude protein—the amount of ammoniacal nitrogen in a feed multiplied by 6.25.

crutching—See *tagging.*

cryptorchidism—See *ridgeling.*

cubing—processing a feed by grinding and then forming it into a hard form called a cube; cubes are larger than *pellets.*

cud—in ruminants, a ball-like mass of feed that is brought up from the stomach to be rechewed.

culling—removing animals from a herd that are not as productive or desirable as the others in the herd.

cutability—the yield of closely trimmed, boneless retail cuts that come from the major wholesale cuts of an animal carcass.

cyst—a swelling containing a fluid or semisolid substance.

cytoplasm—material surrounding the nucleus of a cell.

cytosine—a pyrimidine base that is a part of DNA and RNA; pairs with guanine in DNA.

Dairy Herd Improvement Association (DHIA)—a group of dairy farmers in a local area that is formed to provide production testing services.

debeaking—cutting off part of the upper and lower beak of poultry.

demand—the amount of a product that buyers will purchase at a given time for a given price.

deoxyribose—the 5-carbon sugar found in DNA.

desnooding—the removal of the tubular fleshy appendage on the top of the head of some types of poultry.

dewlap—a hanging fold of skin under the neck of animals, especially some breeds of cattle and goats.

digestible energy (DE)—gross energy of a feed minus the energy remaining in the feces of the animal after the feed is digested.

digestible protein (DP)—that portion of the crude protein in a feed that can be utilized by an animal.

digestion—the process of breaking down feed into simple substances that can be absorbed by the body.

digestive system—the parts of the body involved in chewing and digesting feed.

diploid—an individual or cell having two sets of chromosomes.

DNA—deoxyribonucleic acid is a compound composed of deoxyribose (a sugar), phosphoric acid, and nitrogen. The DNA molecule contains genes on strands in the form of a double helix.

doe—a female of those species in which the male is called a buck; for example: goats, deer, rabbits, etc.

domesticate—to adapt the behavior of an animal to fit the needs of people.

dominant gene—one of a pair of genes that hides the effect of the other gene in the pair.

donkey—the common name for the *ass*.

draft animal—an animal used for pulling loads.

drake—a male duck.

drench—medicine in a liquid form administered to the animal through the mouth.

egg (ovum)—female gamete or reproductive cell.

electrolyte—a solution containing salts and energy sources used to feed young animals suffering from scours (diarrhea).

embryo—the early stage in the development of the fetus.

embryo transfer—moving an embryo from one animal to another.

endotoxin—poisonous substances produced by certain bacteria.

energy feed—livestock feed containing less than 20 percent crude protein; most grains are energy feeds.

environment—the total of the external conditions and influences that affect the life and development of living organisms.

enzootic—a disease that affects or is peculiar to animals in a specific geographic area.

enzyme—an organic catalyst that speeds up the digestive process without being used up in the process.

Eohippus—tiny (about one-foot high), four-toed ancestor of today's horse; native to the North American continent about 58 million years ago.

epididymis—a long, coiled tube that is connected to each testicle.

epithelium—in poultry, a thick, horny membranelike material that lines the muscular stomach or gizzard.

equitation—See *horsemanship*.

esophagus—the tubelike passage from the mouth to the stomach; sometimes called the *gullet*.

Estimated Breeding Value (EBV)—a method of combining performance data on a given trait about an individual animal and its close relatives to determine the individual's ability to transmit that performance to its offspring.

Estimated Relative Producing Ability (ERPA)—in dairy cattle, a predication of 305-day, two times per day milking, mature equivalent (305-2x-ME) production compared to other cows in the herd.

estray—a domestic animal of unknown ownership that is running at large.

estrogen—a hormone produced by the ovaries.

estrus—the time during which the female will accept the male for copulation; also referred to as being "in heat."

ewe—a female sheep or lamb.

expected progeny difference—an estimate of the genetic value of an animal in passing genetic traits to its offspring.

fagopyrin—a photosensitizing agent found in buckwheat that can cause rashes and itching when white pigs are exposed to sunlight.

fallopian tubes—See *oviduct*.

farrier—a person who works on horses' feet.

fat—organic compound composed of carbon, hydrogen, and oxygen; fats contain more carbon and hydrogen than do carbohydrates and are mainly glyceryl esters of certain acids that are soluble in ether but not in water.

feces—undigested material that is passed out of the digestive system through the anus.

Feed Additive Compendium—a publication that lists feed additives in current use and the regulations for their use.

feed composition table—a table showing the nutrients found in feeds.

feed conversion—See *feed efficiency*.

feed efficiency—the ratio of units of feed needed per one unit of production.

feeder calf—a weaned calf that is under one year of age and is sold to be fed for more growth.

feeding standard—a table of nutrient requirements for an animal.

fertilization—the union of a sperm cell with an egg cell.

fetus—unborn animal.

fiber—complex carbohydrates such as cellulose and lignin.

filly—generally, a female horse under three years of age; the Thoroughbred breed includes females that are four years old.

first cross (two-breed cross)—mating a sire from one breed to a female from another breed.

flaking—See *rolling*.

flushing—increasing the amount of feed fed to an animal for a short period of time, usually just prior to breeding.

foal—a young horse of either sex up to one year of age.

free-choice—making feed available to the animal at all times.

fullblood—an animal of unmixed breed.

full feed—giving an animal all it wants to eat.

gait—the movement of the horse's feet and legs when the horse is in motion.

Gallus gallus—a wild jungle fowl found in India and believed to be an early ancestor of most tame chickens.

gamete—a mature germ cell (sperm or egg) that is capable of initiating the formation of a new individual by fusion with another gamete.

gander—a male goose.

gastric juice—a fluid secreted by glands in the wall of the stomach, containing hydrochloric acid and the enzymes gastric lipase, pepsin, and rennin.

gelding—a male horse that has been castrated.

gene—a complex molecule located on the chromosomes that is involved in the transmission of inherited traits.

genetic engineering—a technology involving the removal, modification, or addition of genes to a DNA molecule.

genetics—the study of heredity, or the way in which traits of parents are passed on to offspring.

genotype—the kinds of gene pairs possessed by the animal.

gestation—the time during which the animal is pregnant.

gilt—a young female swine that has not farrowed and is not showing any signs of pregnancy.

gizzard—in poultry, the muscular stomach that crushes and grinds the feed and mixes it with digestive juices.

gossypol—a material found in some cottonseed meal that is toxic to swine and certain other simple-stomached animals.

Grade A milk—milk produced under high standards that make it acceptable for fluid use.

Grade B milk—milk produced under standards that allow it to be used for manufacturing dairy products but not for fluid consumption.

grading up—the mating of purebred sires to grade females.

grease—impurities present in a fleece.

grease mohair—a mohair fleece before it is cleaned.

grease wool—a wool fleece before it is cleaned.

Grecian ibex—species of wild goat believed to be ancestors of some of today's domestic breeds.

green geese—geese full fed for fast growth and marketed at 10 to 13 weeks of age when they weigh 10 to 12 pounds (4.5–5.4 kg); also called *junior geese*.

grinding—processing a feed by breaking it up into smaller particles.

grit—small particles of granite used in poultry rations to help in grinding the feed in the gizzard.

gross energy—total amount of heat released by completely burning a feed in a bomb calorimeter.

guanine—a pruine base that occurs in both RNA and DNA; pairs with cytosine in DNA.

gullet—See *esophagus*.

gymkhana—games on horseback.

halter—a rope or leather headgear used to tie or lead an animal.

hand—a unit of measurement used to describe the height of horses; one hand is 4 inches (10.2 cm).

haploid—a cell that has only a single set of chromosomes.

harness—the gear used to attach a horse or other draft animal to a load.

haylage—low-moisture grass silage.

Hazard Analysis/Critical Control Point (HACCP)—a system used to monitor the production of food (developed for NASA in 1971); currently used on a voluntary basis at many meat and poultry processing plants.

headgate—a device for restraining livestock.

heifer—a female bovine that has not had a calf or has not matured as a cow.

hen—a female of domestic poultry.

herbivorous—the practice of eating plants as the main part of the diet.

heritability—the amount of the difference between animals that is passed from the parent to the offspring.

heritability estimate—the likelihood of a trait being passed on from parent to offspring.

heterosis—improvement in the offspring resulting from favorable combinations of gene pairs; sometimes called *hybrid vigor*.

heterozygous gene pair—a gene pair that carries two unlike genes for a trait.

high-moisture storage—harvesting a feed crop when it has a high moisture content and storing it in a silo.

hinny—the offspring of a cross between a stallion and a jennet.

homozygous gene pair—a gene pair that carrier two like genes for a trait.

hormone—an organic material given off (secreted) by a body gland that helps to regulate body functions.

horsemanship—the art of riding a horse; also called *equitation*.

hothouse lambs—young sheep sold at 50 to 90 days of age, weighing 35 to 60 pounds (15–27 kg).

hutch—a pen or cage for rabbits.

hybrid vigor—See *heterosis*.

inbreeding—the mating of related animals.

incomplete dominance—a situation in which one gene does not completely hide or mask the effect of the other gene in a gene pair.

incubation—keeping eggs at the right temperature and humidity for hatching.

inflation—in reference to milking equipment, the liner used in the teat cup.

infundibulum—in the female reproductive system, the funnel-shaped end of the oviduct that is close to the ovary.

intestinal juice—fluid produced by the walls of the small intestine that contains peptidase, sucrase, maltase, and lactase.

intestine, large—the tube from the small intestine to the anus; shorter and larger in diameter than the small intestine.

intestine, small—the long, folded tube attached to the lower end of the stomach.

intranasal vaccination—in poultry, the placement of a vaccine directly into the nose opening.

intraocular vaccination—in poultry, the placement of a vaccine directly into the eye.

isthmus—in poultry, that part of the oviduct where the two shell membranes are added to the egg.

jack—a male ass.

jennet—a female ass.

junior geese—See *green geese.*

kemp—large, chalky white hairs found in the fleece of some breeds of goats, especially the Angora.

kid—a goat under one year of age.

kindle—to give birth; generally used when referring to rabbits.

lamb—a young sheep.

legume—a plant of the family *Leguminosae* that carries its seeds in a pod that splits along its seams; many legumes have nitrogen fixing bacteria in nodules on the roots that can transform nitrogen in the air into a form (NH_3) that can be used by the plant; peanuts, soybeans, clovers and alfalfa are common legumes used in agriculture.

limited feeding—See *restricted feeding.*

limit-fed—a method of feeding in which the amount of feed given the animal is controlled or limited to less than the animal would eat if given free access to the feed.

linebreeding—a form of inbreeding in which the animals being mated are more distantly related than in closebreeding and their ancestry can be traced back to one common ancestor.

linecrossing—mating animals from two different lines of breeding within a breed.

linkage—the tendency for genes that are located close together on the chromosome to stay together.

locus—the location of a given gene on a chromsome.

longeing—training a horse at the end of a 25- to 30-foot (7.6–9.1 m) line.

lysosome—that part of the cytoplasm in most cells that contain various hydrolytic enzymes involved in digestion.

magnum—in poultry, that part of the oviduct where the thick white of the egg is secreted.

marbling—the presence and distribution of fat and lean in a cut of meat.

mare—generally, a mature female horse, four years of age or older; in the Thoroughbred breed, five years of age or older.

markhors—species of wild goat believed to be ancestors of some of today's domestic breeds.

martingale—a strap attached to a horse's girth, passed through the forelegs, and fastened to the bit (standing martingale) or the reins (running martingale);

the purpose is to hold the horse's head down so it will not rear up.

meiosis—division of gamete in which chromosome pairs split, each individual chromosome going to one of the new gametes.

metabolizable energy (ME)—for ruminants, the gross energy in the feed eaten minus the energy found in the feces, the energy in the gaseous products of digestion, and the energy in the urine; for poultry and simple-stomached animals, the energy in the gaseous products of digestion is not considered when determining metabolizable energy.

metaphase—in mitosis and meiosis, the stage between prophase and anaphase when the chromosomes are aligned along the metaphase plate.

micronutrients—feed ingredients, such as minerals and vitamins, that are used in small amounts in the ration.

middlings—coarsely ground wheat mixed with bran.

milkstone—a combination of organic and inorganic materials that accumulate on improperly cleaned milking equipment.

mineral—inorganic substance needed in small amounts for proper nutrition.

mitosis—cell division that increases the number of total cells and results in growth.

mohair—the fleece of the Angora goat.

molting—in poultry, the process of losing the feathers from the body and wings.

moufflons—a breed of wild sheep believed to be ancestors of some present-day domestic breeds.

mouth—that part of the digestive system through which feed enters the animal's body.

mule—the offspring of a cross between a jack and a mare; usually sterile.

multiple farrowing—arranging the breeding program so that groups of sows farrow at regular intervals throughout the year.

mutation—the appearance of a new trait in the offspring that did not exist in the genetic makeup of the parents.

NE_g—net energy used for animal growth.
NE_l—net energy used for milk production (lactation).
NE_m—net energy used for animal maintenance.

neck-reining—controlling the response of a horse by the weight of the rein against the neck.

net energy (NE)—metabolizable energy minus the heat increment; used for maintenance, for production, or both.

neutral detergent fiber—the portion of a feed that is of lower digestibility, consisting of the more insoluble material found in the cell wall; mainly cellulose, lignin, silica, hemicellulose, and some protein.

nitrogen-free extract—simple carbohydrates such as sugar and starches.

nonruminant—an animal that has a simple, one-compartment stomach; for example: pigs, horses, poultry.

nucleotide—a combination of one of the nitrogenous bases, one phosphate, and one deoxyribose; forms the basic constituent of DNA and RNA.

nucleus—center of a cell.

nutrient—a chemical element or compound that aids in the support of life.

nutrient-sparing—a substance that allows animals to use available nutrients more effectively.

off flavor—any undesirable odor or flavor in milk.

oil—fat that is soluble at body temperature.

omasum—the third compartment of the ruminant stomach.

one hundred percent (100%) dry matter basis—data on feed composition or nutrient requirement calculated on the basis of all the moisture being removed from the feed.

oogenesis—formation, development, and maturation of an ovum.

outcrossing—the mating of animals of different families within the same breed.

ovary—organ in the female reproductive system that produces eggs.

overo—a color pattern in horses in which the head has variable color markings; the white usually does not cross the back between the withers and tail; one or more legs are dark colored; and the body markings are irregular and scattered.

oviduct—the tube in the female reproductive system that carries the eggs from the ovaries to the uterus; also called *fallopian tubes*.

ovulation—the release of the egg cell from the ovary.

ovum—See *egg*.

palatable—good tasting.

palpating—examining a doe to determine pregnancy.

pancreatic juice—fluid secreted by the pancreas that contains the enzymes trypsin, pancreatic amylase, pancreatic lipase, and maltase.

papillae—in poultry, the organ in the wall of the cloaca that deposits the sperm cells into the hen's reproductive system.

parturition—the act of giving birth.

pasang—species of wild goat believed to be ancestors of some of today's domestic breeds.

pedigree—the record of the ancestors of an animal.

pelleting—grinding a feed into small particles and then forming it into a small, hard form called a *pellet.*

penis—male reproductive organ that contains the urethra and deposits the sperm into the female reproductive tract.

performance stimulants—term sometimes applied to feed additives and hormone implants.

performance testing—a method of collecting records on an animal herd to be used for selecting the most productive animals.

phenotype—the physical appearance of an animal.

pinfeather—in poultry, a feather that is not fully developed.

placenta—in mammals, the structure by which the fetus is nourished in the uterus.

point source—a place from which pollutants may originate; for example, a large feedlot.

polled—not having horns.

pollutants—impurities in the environment, such as odors or waste from a feedlot.

polynucleotide—a linear sequence of a number of nucleotides.

preconditioning—the process of preparing calves for the stress of being moved into the feedlot.

predicated difference (PD)—in dairy cattle, the expected 305-day equivalent production of milk and/or fat of a bull's daughters as compared to genetic group herdmates.

production testing—measuring a brood female's production by the performance of her offspring.

progeny testing—evaluation of a male by the performance of a number of his offspring.

progesterone—hormone produced by the ovaries that maintains pregnancy in the animal.

prophase—the first stages of mitosis and meiosis.

prostate gland—an accessory gland of the male reproductive system that produces a fluid that is mixed with the seminal fluid.

protein—an organic compound made up of amino acids and containing carbon, hydrogen, oxygen, and nitrogen.

protein supplement—livestock feed that contains 20 percent or more protein.

protoplasm—the material that constitutes the living matter of cells; it includes the nucleus and cytoplasm.

protozoa—one-celled animals.

pseudopregnancy—false pregnancy.

puberty—the age at which sexual maturity is reached.

pullet—a female chicken.

pulsator—a control unit used with milking equipment to control the action of the inflation in the teat cup shell.

rabbitry—a place where rabbits are kept.

radura—the international symbol that must be placed on all foods treated by irradiation.

ram—a male sheep or lamp that has not been castrated.

ration—the total amount of feed that an animal is allowed during a 24-hour period.

recessive gene—one of a pair of genes, the effect of which is hidden by the other gene in the pair.

recombinant bovine somatotropin (rBST)—bovine somatotropin manufactured by genetic engineering. (See *genetic engineering, bovine somatotropin,* and *somatotropin.*)

recombinant DNA technology—See *genetic engineering.*

rectum—the last part of the large intestine.

relative feed value (RFV)—a number used to compare the quality of various legume and legume-grass forages as determined by their acid detergent fiber and neutral detergent fiber content.

repeatability—in dairy cattle, a measure of the confidence that can be placed on the Predicted Difference being a true measure of a bull's ability to transmit genetic characteristics.

reproduction—the production of offspring.

restricted feeding—feeding 75 to 80 percent of full feed; also referred to as *limited feeding.*

reticulum—the second compartment of the ruminant stomach.

retractor muscle—part of the male reproductive system that helps extend the penis from the sheath and draws it back after copulation.

ribonucleic acid—See *RNA.*

ribosome—a small particle composed of RNA and protein that is in the cytoplasm of cells and is active in protein synthesis.

ridgel—See *ridgeling.*

ridgeling—a male in which one or both testicles are held in the body cavity; also called *ridgel* or *cryptorchidism.*

RNA—regulates protein synthesis in animals; primary function is carrying the genetic message from DNA for the building of the polypeptide chains that begin the process of protein synthesis; contains the bases adenine, guanine, cytosine, and uracil bonded to the ribose.

rolling—processing grain through a set of smooth rollers that are set close together; sometimes called *flaking.*

rotavirus—wheel-shaped viruses that cause gastroenteritis, especially in newborn animals.

roughage—a feed containing more than 18 percent crude fiber when dry; examples: hay, silage, and pasture.

rumen—the first and largest compartment of the ruminant stomach.

rumen organisms—bacteria found in the rumen of cattle and other ruminant animals.

ruminant—an animal that has a stomach divided into several compartments; for example, cattle, sheep, goats.

rumination—in ruminants, the process of chewing the cud.

saliva—fluid secreted into the mouth by glands and containing enzymes that aid in digestion.

sanitizing—making a surface sanitary by using a sanitizing agent such as chlorine or Quaternary Ammonium compound.

scrotum—the saclike part of the male reproductive system outside the body cavity that contains the testicles and the epididymis.

selection—identification and use for breeding purposes of those animals with traits that are considered by the breeder to be desirable.

self-fed—a method of feeding in which the animal is given free access to all the feed it will eat.

semen—the mixture of the seminal and prostate fluid and the sperm.

seminal fluid—fluid in the *seminal vesicles* that mixes with the sperm cells.

seminal vesicles—an accessory gland in the male reproductive system that produces a fluid that protects and transports the sperm.

sexed chicks—young chickens sorted into groups of male and female.

sex-influenced gene—gene that is dominant in one sex but recessive in the other.

sex-limited gene—the phenotypic expression of some genes is determined by the presence or absence of one of the sex hormones; its expression is limited to one sex.

sex-linked genes—genes that are carried only on the sex chromosomes.

sheath—in mammal reproductive systems, a tubular fold of skin that covers the penis.

shigella—rod-shaped bacteria of the genus *Shigella,* including some species that cause dysentery.

shrinkage—the weight lost by an animal while it is being shipped to market.

signoid flexure—the part of the male reproductive system that helps to extend the penis from the sheath for the purpose of copulation.

skatole—a white crystalline organic compound, C_9H_9N, which has a strong fecal odor and is found in feces.

slaughter calves—calves that are between three months and one year of age at the time of slaughter.

somatotropin—a polypeptide hormone secreted by the anterior lobe of the pituitary gland that promotes growth of the body and influences the metabolism of proteins, carbohydrates, and lipids; also called *growth hormone* or *somatotropic hormone.*

sow—an older female swine that has farrowed or is showing signs of pregnancy.

specific pathogen free (SPF)—swine breeding stock that are surgically removed from the sow under antiseptic conditions and raised in disease-free conditions.

sperm—male gamete or reproductive cell.

spermatic cord—protective sheath around the vas deferens.

spermatid—any one of the four haploid cells that form during meiosis in the male that development into spermatozoa.

spermatogenesis—formation and development of speratozoa by meiosis.

spermatogonia—cells in male gonads that are progenitors of spermatocytes.

spinning count—the number of hanks of yarn that can be spun from one pound of wool top.

spray dried porcine plasma (SDPP)—a by-product of blood from pork slaughter plants; also called *plasma protein.*

spring lambs—young sheep that are three to seven months of age and usually weigh from 70 to 90 pounds (31.8–40.8 kg).

stag—a male animal, usually beef or swine, which was castrated after reaching sexual maturity and shows the physical traits of the uncastrated male.

stallion—generally, a mature male horse four years of age or older; in the Thoroughbred breed, five years or older.

staple—a fiber of materials such as wool, cotton, or flax, either in its natural state or after it has been carded or combed; also refers to the length, fineness, condition, or grade of the fiber.

steam flaking—heating grain with steam and putting it through a set of smooth rollers that are set close together.

steer—a male bovine animal that was castrated before reaching sexual maturity.

sterile—unable to reproduce.

stomach—the organ in the digestive system that receives the feed and adds chemicals that help in the digestive process.

straight breeding—mating animals of the same breed.

straight-run chicks—young chickens not sorted by sex.

subtherapeutic—feeding an antimicrobial drug below the dosage level used to treat diseases; for example, the subtherapeutic feeding of penicillin to livestock.

supply—the amount of a product that producers will offer for sale at a given price at a given time.

Sus scrofa—European wild boar.

Sus vittatus—East Indian wild pig.

swirl—hair growing in a circular pattern from the roots.

synovia—a transparent lubricating fluid secreted by the synovial membrane.

synovial membrane—a thin, pliable structure surrounding a joint.

tack—the equipment used for riding and showing horses.

tagging—shearing a ewe around the udder, between the legs, and around the dock; also called *crutching*.

tahrs—species of wild goat believed to be the ancestor of some of today's domestic goats.

tankage—animal tissues and bones from animal slaughter houses and rendering plants that are cooked, dried, and ground and used as a protein supplement.

teat—the outlet for milk produced in the udder.

teat cup shell—the metal part of the milking unit that contains the inflation and is attached to the teat during the milking operation.

telophase—final stage of mitosis or meiosis; in this stage the chromosomes are grouped in new nuclei.

testicles—male organs that produce the sperm cells.

testosterone—male hormone that controls the traits of the male animal.

three-breed rotation cross—mating crossbred females with a sire of a third breed.

thymine—pyrimidine base that is a part of DNA.

tobiano—a color pattern in horses in which the head is a solid color; the legs are white, at least below the knees and hocks; and there are regular spots on the body.

tom—a male turkey.

total digestible nutrients (TDN)—the total of the digestible protein, digestible nitrogen-free extract, digestible crude fiber, and 2.25 times the digestible fat.

toxin—a poisonous substance produced by the metabolism of plant or animal organisms.

trachea—the passage through which air passes to and from the lungs (windpipe).

transponder—a radio receiver activated for transmission by the reception of a predetermined signal.

trapnest—a nest equipped with a door that automatically closes when the bird enters the nest.

trenbolone acetate—a synthetic male hormone.

turbinate bone—a thin, bony plate on the wall of the nose.

two-breed cross—See *first cross*.

udder—the milk-producing gland of mammals such as cows.

ultrasonics—the use of high-frequency sound waves to measure fat thickness and loin-eye area.

uracil—pyrimidine base that is a part of RNA.

urea—a synthetic nitrogen source that is manufactured from air, water, and carbon.

urethra—the tube that carries urine from the bladder.

urine—liquid waste collected in the bladder.

uterus—part of the female reproductive system where the fetus grows; also called the *womb*.

vagina—in the female reproductive system, the passage between the cervix and the vulva.

vas deferens—the tube that connects the epididymus with the urethra, providing a passageway for the sperm cells.

veal—calves younger than three months of age sold for slaughter.

vegetable protein—protein supplement that comes from plant sources.

vent—the external opening of the lower end of the digestive system in poultry.

vertical integration—the linking together of two or more steps of production, marketing, and processing.

villi—small fingerlike projections that line the walls of the small intestine; they increase the absorption area of the small intestine.

viscera—the internal organs of an animal.

vitamin—an organic compound needed in small amounts for nutrition.

vulva—the external opening of the female reproductive and urinary systems.

wattle—a projection of skin hanging from the chin or throat, especially in poultry and some breeds of goats.

wether—a male sheep or goat that has been castrated before reaching sexual maturity.

wing web vaccination—in poultry, the process of injecting a vaccine into the skin on the underside of the wing web at the elbow.

withdrawal period—the length of time a feed additive must not be fed to an animal prior to slaughter.

womb—See *uterus*.

wool top—partially processed wool.

yardage—the fee charged for the use of stockyard facilities.

yearling feeders—cattle that are one to two years of age and are sold to be fed to finish for slaughter.

yield—the dressing percent (weight of the chilled carcass compared to the live weight) of the animal.

zoonoses—diseases and parasites that may be transmitted between man and animals.

zygote—fertilized egg cell formed by the union of two gametes.

Appendix

TABLE 1 Net Energy Requirements of Growing and Finishing Beef Cattle (Mcal/day).[a]

Body Weight, kg:	150	200	250	300	350	400	450	500	550	600
NE$_m$ Required:	3.30	4.10	4.84	5.55	6.24	6.89	7.52	8.14	8.75	9.33
Daily gain, kg	NE$_g$ Required									
Medium-frame steer calves										
0.2	0.41	0.50	0.60	0.69	0.77	0.85	0.93	1.01	1.08	
0.4	0.87	1.08	1.28	1.47	1.65	1.82	1.99	2.16	2.32	
0.6	1.36	1.69	2.00	2.29	2.57	2.84	3.11	3.36	3.61	
0.8	1.87	2.32	2.74	3.14	3.53	3.90	4.26	4.61	4.95	
1.0	2.39	2.96	3.50	4.02	4.51	4.98	5.44	5.89	6.23	
1.2	2.91	3.62	4.28	4.90	5.50	6.69	6.65	7.19	7.73	
Large-frame steers, compensating medium-frame yearling steers and medium-frame bulls										
0.2	0.36	0.45	0.53	0.61	0.68	0.75	0.82	0.89	0.96	1.02
0.4	0.77	0.96	1.13	1.30	1.46	1.61	1.76	1.91	2.05	2.19
0.6	1.21	1.50	1.77	2.03	2.28	2.52	2.75	2.98	3.20	3.41
0.8	1.65	2.06	2.43	2.78	3.12	3.45	3.77	4.08	4.38	4.68
1.0	2.11	2.62	3.10	3.55	3.99	4.41	4.81	5.21	5.60	5.98
1.2	2.58	3.20	3.78	4.34	4.87	5.38	5.88	6.37	6.84	7.30
1.4	3.06	3.79	4.48	5.14	5.77	6.38	6.97	7.54	8.10	8.64
1.6	3.53	4.39	5.19	5.95	6.68	7.38	8.07	8.73	9.38	10.01
Large-frame bull calves and compensating large-frame yearling steers										
0.2	0.32	0.40	0.47	0.54	0.60	0.67	0.73	0.79	0.85	0.91
0.4	0.69	0.85	1.01	1.15	1.29	1.43	1.56	1.69	1.82	1.94
0.6	1.07	1.33	1.57	1.80	2.02	2.23	2.44	2.64	2.83	3.02
0.8	1.47	1.82	2.15	2.47	2.77	3.06	3.34	3.62	3.88	4.15
1.0	1.87	2.32	2.75	3.15	3.54	3.91	4.27	4.62	4.96	5.30
1.2	2.29	2.84	3.36	3.85	4.32	4.77	5.21	5.64	6.06	6.47
1.4	2.71	3.36	3.97	4.56	5.11	5.65	6.18	6.68	7.18	7.66
1.6	3.14	3.89	4.60	5.28	5.92	6.55	7.15	7.74	8.31	8.87
1.8	3.56	4.43	5.23	6.00	6.74	7.45	8.13	8.80	9.46	10.10
Medium-frame heifer calves										
0.2	0.49	0.60	0.71	0.82	0.92	1.01	1.11	1.20	1.29	
0.4	1.05	1.31	1.55	1.77	1.99	2.20	2.40	2.60	2.79	
0.6	1.66	2.06	2.44	2.79	3.13	3.46	3.78	4.10	4.40	
0.8	2.29	2.84	3.36	3.85	4.32	4.78	5.22	5.65	6.07	
1.0	2.94	3.65	4.31	4.94	5.55	6.14	6.70	7.25	7.79	
Large-frame heifer calves and compensating medium-frame yearling heifers										
0.2	0.43	0.53	0.63	0.72	0.81	0.90	0.98	1.06	1.14	1.21
0.4	0.93	1.16	1.37	1.57	1.76	1.95	2.13	2.31	2.47	2.64
0.6	1.47	1.83	2.16	2.47	2.78	3.07	3.35	3.63	3.90	4.16
0.8	2.03	2.62	2.98	3.41	3.83	4.24	4.63	5.01	5.38	5.74
1.0	2.61	3.23	3.82	4.38	4.92	5.44	5.94	6.43	6.91	7.37
1.2	3.19	3.97	4.69	5.37	5.03	6.67	7.28	7.88	8.47	9.03

NOTE: Tables 1–6 (beef), Tables 13–19 (swine), Table 48 (beef and swine data), and Table 49 (swine data) are from older publications of the National Research Council. Those tables are retained here for those who may want to use them for instructional purposes. See Unit 8 for information regarding the new Nutrient Requirement publications and the computer programs they contain for determining nutrient requirements and feed composition for beef cattle and swine.

[a]Shrunk liveweight basis (weight after an overnight feed and water shrink—generally equivalent to about 96 percent of unshrunk weights taken in the early morning).

Source: Nutrient Requirements of Beef Cattle, sixth revised edition, National Academy of Sciences, National Academy Press, Washington, DC.

TABLE 2 Protein Requirements of Growing and Finishing Cattle (g/day).[a]

Body Weight, kg:	150	200	250	300	350	400	450	500	550	600
Medium-frame steer calves										
Daily gain, kg										
0.2	343	399	450	499	545	590	633	675	715	
0.4	428	482	532	580	625	668	710	751	790	
0.6	503	554	601	646	688	728	767	805	842	
0.8	575	621	664	704	743	780	815	849	883	
1.0	642	682	720	755	789	821	852	882	911	
1.2	702	735	766	794	822	848	873	897	921	
Large-frame steer calves and compensating medium-frame yearling steers										
0.2	361	421	476	529	579	627	673	719	762	805
0.4	441	499	552	603	651	697	742	785	827	867
0.6	522	576	628	676	722	766	809	850	890	930
0.8	598	650	698	743	786	828	867	906	944	980
1.0	671	718	762	804	843	881	918	953	988	1021
1.2	740	782	822	859	895	929	961	993	1023	1053
1.4	806	842	877	908	938	967	995	1022	1048	1073
1.6	863	892	919	943	967	989	1011	1031	1052	1071
Medium-frame bulls										
0.2	345	401	454	503	550	595	638	680	721	761
0.4	430	485	536	584	629	673	716	757	797	835
0.6	509	561	609	655	698	740	780	819	856	893
0.8	583	632	677	719	759	798	835	871	906	940
1.0	655	698	739	777	813	849	881	914	945	976
1.2	722	760	795	828	860	890	919	947	974	1001
1.4	782	813	841	868	893	917	941	963	985	1006
Large-frame bull calves and compensating large-frame yearling steers										
0.2	355	414	468	519	568	615	661	705	747	789
0.4	438	494	547	597	644	689	733	776	817	857
0.6	519	574	624	672	718	761	803	844	884	923
0.8	597	649	697	741	795	826	866	905	942	979
1.0	673	721	765	807	847	885	922	958	994	1027
1.2	745	789	830	868	904	939	973	1005	1037	1067
1.4	815	854	890	924	956	986	1016	1045	1072	1099
1.6	880	912	943	971	998	1024	1048	1072	1095	1117
1.8	922	942	962	980	997	1013	1028	1043	1057	1071
Medium-frame heifer calves										
0.2	323	374	421	465	508	549	588	626	662	
0.4	409	459	505	549	591	630	669	706	742	
0.6	477	522	563	602	638	674	708	741	773	
0.8	537	574	608	640	670	700	728	755	781	
1.0	562	583	603	621	638	654	670	685	700	
Large-frame heifer calves and compensating medium-frame yearling heifers										
0.2	342	397	449	497	543	588	631	672	712	751
0.4	426	480	530	577	622	665	707	747	787	825
0.6	500	549	596	639	681	721	759	796	832	867
0.8	568	613	654	693	730	765	799	833	865	896
1.0	630	668	703	735	767	797	826	854	881	907
1.2	680	708	734	758	781	803	824	844	864	883

[a]Shrunk liveweight basis (weight after an overnight feed and water shrink—generally equivalent to about 96 percent of unshrunk weights taken in the early morning).

Source: Nutrient Requirements of Beef Cattle, sixth revised edition, National Academy of Sciences, National Academy Press, Washington, DC.

TABLE 3 Calcium and Phosphorus Requirements of Growing and Finishing Cattle (g/day).[a]

Body Weight, kg	Mineral	150	200	250	300	350	400	450	500	550	600
Medium-frame steer calves											
Daily gain, kg											
0.2	Ca	11	12	13	14	15	16	17	19	20	
	P	7	9	10	12	13	15	16	18	19	
0.4	Ca	16	17	17	18	19	19	20	21	22	
	P	9	10	12	13	14	16	17	18	20	
0.6	Ca	21	21	21	22	22	22	22	23	23	
	P	11	12	13	14	15	17	18	19	20	
0.8	Ca	27	26	25	25	25	25	24	24	24	
	P	12	13	14	15	16	17	19	20	21	
1.0	Ca	32	31	29	29	28	27	26	26	25	
	P	14	15	16	16	17	18	19	20	21	
1.2	Ca	37	35	33	32	31	29	28	27	26	
	P	16	16	17	17	18	19	20	21	21	
1.4	Ca	42	39	37	35	33	32	30	29	27	
	P	17	18	18	19	19	20	20	21	22	
Large-frame steer calves, compensating medium-frame yearling steers, and medium-frame bulls											
0.2	Ca	11	12	13	14	16	17	18	19	20	22
	P	7	9	10	12	13	15	16	18	20	21
0.4	Ca	17	17	18	19	19	20	21	22	23	24
	P	9	10	12	13	15	16	17	19	20	22
0.6	Ca	22	22	23	23	23	24	24	24	25	25
	P	11	12	13	15	16	17	18	20	21	22
0.8	Ca	28	27	27	27	27	27	27	27	27	27
	P	13	14	15	16	17	18	19	20	22	23
1.0	Ca	33	32	31	31	30	30	29	29	29	28
	P	14	15	16	17	18	19	20	21	22	23
1.2	Ca	38	37	36	35	34	33	32	31	30	30
	P	16	17	18	18	19	20	21	22	23	24
1.4	Ca	44	42	40	38	37	36	34	33	32	31
	P	18	18	19	20	20	21	22	22	23	24
1.6	Ca	49	47	44	42	40	38	37	35	34	32
	P	20	20	20	21	21	22	22	23	24	24
Large-frame bull calves and compensating large-frame yearling steers											
0.2	Ca	11	12	13	15	16	17	18	20	21	22
	P	7	9	10	12	13	15	17	18	20	21
0.4	Ca	17	18	19	19	20	21	22	23	24	25
	P	9	11	12	13	15	16	18	19	21	22
0.6	Ca	23	23	23	24	24	25	25	26	27	27
	P	11	12	14	15	16	18	19	20	22	23
0.8	Ca	28	28	28	28	28	29	29	29	29	30
	P	13	14	15	16	18	19	20	21	22	24
1.0	Ca	34	34	33	33	32	32	32	32	32	32
	P	15	16	17	18	19	20	21	22	23	24
1.2	Ca	40	39	38	37	36	36	35	35	34	34
	P	17	17	18	19	20	21	22	23	24	25
1.4	Ca	45	44	42	41	40	39	38	37	36	36
	P	18	19	20	20	21	22	23	24	25	26
1.6	Ca	51	49	47	45	44	42	41	40	39	38
	P	20	21	21	22	23	23	24	25	25	26
1.8	Ca	56	54	51	49	47	45	44	42	41	39
	P	22	22	22	23	23	24	25	25	26	26

(Table 3 continues)

TABLE 3 Calcium and Phosphorus Requirements of Growing and Finishing Cattle (g/day).[a] *(Cont.)*

Body Weight, kg	Mineral	150	200	250	300	350	400	450	500	550	600
Medium-frame heifer calves											
0.2	Ca	10	11	12	13	14	16	17	18	19	
	P	7	9	10	11	13	14	16	17	19	
0.4	Ca	15	16	16	16	17	17	18	19	19	
	P	9	10	11	12	14	15	16	18	19	
0.6	Ca	20	20	19	19	19	19	19	19	19	
	P	10	11	12	13	14	16	17	18	19	
0.8	Ca	25	23	23	22	21	20	20	19	19	
	P	12	12	13	14	15	16	17	18	19	
1.0	Ca	29	27	26	24	23	22	20	19	19	
	P	13	14	14	15	16	16	17	18	19	
Large-frame heifer calves and compensating medium-frame yearling heifers											
0.2	Ca	11	12	13	14	15	16	17	18	20	21
	P	7	9	10	12	13	15	16	18	19	21
0.4	Ca	16	16	17	17	18	19	19	20	21	22
	P	9	10	11	13	14	15	17	18	20	21
0.6	Ca	21	21	21	21	21	21	21	21	22	22
	P	10	12	13	14	15	16	17	19	20	21
0.8	Ca	26	25	24	24	23	23	23	22	22	22
	P	12	13	14	15	16	17	18	19	20	21
1.0	Ca	31	29	28	27	26	25	24	23	23	22
	P	14	14	15	16	17	18	18	19	20	21
1.2	Ca	35	33	31	30	28	27	25	24	23	22
	P	15	16	16	17	17	18	19	20	20	21

[a]Shrunk liveweight basis (weight after an overnight feed and water shrink—generally equivalent to about 96 percent of unshrunk weights taken in the early morning).

Source: Nutrient Requirements of Beef Cattle, sixth revised edition, National Academy of Sciences, National Academy Press, Washington, DC.

TABLE 4 Nutrient Requirements of Breeding Cattle (metric).

| | | | Energy | | | | | | | | | Total Protein | | Calcium | | Phosphorus | | Vitamin A[d] |
| | | | Daily | | | | In Diet DM | | | | | | | | | | | |
Weight[a] (kg)	Daily Gain[b] (kg)	Daily DM[c] (kg)	ME (Mcal)	TDN (kg)	NEm (Mcal)	NEg (Mcal)	ME (Mcal/kg)	TDN (%)	NEm (Mcal/kg)	NEg (Mcal/kg)	Daily (g)	In Diet DM (%)	Daily (g)	In Diet DM (%)	Daily (g)	In Diet DM (%)	Daily (1000's IU)
Pregnant yearling heifers—Last third of pregnancy																	
325	0.4	7.1	14.2	3.9	8.04	NA[e]	2.00	55.2	1.15	NA[e]	591	8.4	19	0.27	14	0.20	20
325	0.6	7.3	15.7	4.3	8.04	0.77	2.15	59.3	1.29	0.72	649	8.9	23	0.32	15	0.21	20
325	0.8	7.3	17.2	4.8	8.04	1.67	2.35	64.9	1.47	0.88	697	9.5	27	0.37	16	0.22	20
350	0.4	7.5	14.8	4.1	8.38	NA	1.99	55.0	1.14	NA	616	8.3	20	0.27	15	0.21	21
350	0.6	7.7	16.5	4.6	8.38	0.81	2.14	59.1	1.28	0.71	674	8.8	24	0.32	16	0.21	22
350	0.8	7.8	18.1	5.0	8.38	1.76	2.34	64.6	1.46	0.88	720	9.3	27	0.35	17	0.22	22
375	0.4	7.8	15.5	4.3	8.71	NA	1.98	54.7	1.13	NA	641	8.2	21	0.27	15	0.19	22
375	0.6	8.1	17.2	4.8	8.71	0.86	2.13	58.8	1.27	0.70	697	8.6	25	0.31	17	0.21	23
375	0.8	8.2	19.0	5.2	8.71	1.86	2.32	64.1	1.45	0.86	743	9.1	27	0.33	18	0.22	23
400	0.4	8.2	16.1	4.5	9.04	NA	1.97	54.4	1.12	NA	664	8.1	22	0.27	16	0.20	23
400	0.6	8.5	18.0	5.0	9.04	0.90	2.12	58.6	1.26	0.69	721	8.5	25	0.30	18	0.21	24
400	0.8	8.6	19.8	5.5	9.04	1.95	2.31	63.8	1.44	0.85	764	8.9	28	0.33	18	0.20	24
425	0.4	8.6	16.8	4.6	9.36	NA	1.96	54.1	1.11	NA	687	8.0	23	0.27	17	0.20	24
425	0.6	8.9	18.7	5.2	9.36	0.94	2.11	58.3	1.25	0.69	743	8.4	26	0.30	18	0.20	25
425	0.8	9.0	20.7	5.7	9.36	2.04	2.30	63.5	1.43	0.84	786	8.8	28	0.31	19	0.21	25
450	0.4	8.9	17.3	4.8	9.67	NA	1.95	53.9	1.10	NA	710	8.0	23	0.26	18	0.20	25
450	0.6	9.2	19.4	5.4	9.67	0.98	2.10	58.0	1.25	0.68	765	8.3	26	0.29	19	0.21	26
450	0.8	9.4	21.5	5.9	9.67	2.13	2.29	63.3	1.42	0.84	807	8.6	28	0.30	20	0.21	26
Dry pregnant mature cows—Middle third of pregnancy																	
350	0.0	6.8	11.9	3.3	6.23	NA	1.76	48.6	0.92	NA	478	7.1	12	0.16	12	0.18	19
400	0.0	7.5	13.1	3.6	6.89	NA	1.76	48.6	0.92	NA	525	7.0	13	0.17	13	0.17	21
450	0.0	8.2	14.3	4.0	7.52	NA	1.76	48.6	0.92	NA	570	7.0	15	0.17	15	0.18	23
500	0.0	8.8	15.5	4.3	8.14	NA	1.76	48.6	0.92	NA	614	7.0	17	0.19	17	0.19	25
550	0.0	9.5	16.7	4.6	8.75	NA	1.76	48.6	0.92	NA	657	6.9	18	0.19	18	0.19	27
600	0.0	10.1	17.8	4.9	9.33	NA	1.76	48.6	0.92	NA	698	6.9	20	0.20	20	0.20	28
650	0.0	10.7	18.9	5.2	9.91	NA	1.76	48.6	0.92	NA	739	6.9	22	0.21	22	0.21	30
Dry pregnant mature cows—Last third of pregnancy																	
350	0.4	7.4	14.7	4.1	8.38	NA	1.98	54.7	1.13	NA	609	8.2	20	0.27	15	0.20	21
400	0.4	8.2	16.0	4.4	9.04	NA	1.96	54.1	1.11	NA	657	8.0	22	0.27	16	0.20	23
450	0.4	8.9	17.2	4.8	9.67	NA	1.94	53.6	1.10	NA	703	7.9	23	0.26	18	0.21	24
500	0.4	9.5	18.3	5.1	10.29	NA	1.92	53.1	1.08	NA	746	7.8	25	0.26	20	0.21	27
550	0.4	10.2	19.5	5.4	10.90	NA	1.91	52.8	1.07	NA	790	7.8	26	0.25	21	0.21	29
600	0.4	10.8	20.6	5.7	11.48	NA	1.90	52.5	1.06	NA	832	7.7	28	0.26	23	0.21	30
650	0.4	11.5	21.7	6.0	12.06	NA	1.89	52.2	1.05	NA	872	7.6	30	0.26	25	0.22	32
Two-year-old heifers nursing calves—First 3–4 months postpartum—5.0 kg milk/day																	
300	0.2	6.9	16.6	4.6	9.30[f]	0.72	2.41	66.6	1.53	0.93	814[g]	11.8	26	0.38	17	0.25	27
325	0.2	7.3	17.4	4.8	9.64[f]	0.77	2.37	65.5	1.49	0.90	841[g]	11.5	27	0.37	18	0.25	28
350	0.2	7.8	18.1	5.0	9.98[f]	0.81	2.34	64.6	1.46	0.88	866[g]	11.2	27	0.35	19	0.24	30
375	0.2	8.2	18.9	5.2	10.31[f]	0.86	2.31	63.8	1.44	0.85	892[g]	10.9	28	0.34	19	0.23	32
400	0.2	8.6	19.7	5.4	10.64[f]	0.90	2.29	63.3	1.42	0.84	916[g]	10.7	28	0.33	20	0.23	34
425	0.2	9.0	20.4	5.6	10.96[f]	0.94	2.27	62.7	1.40	0.82	939[g]	10.5	29	0.32	21	0.23	35
450	0.2	9.4	21.1	5.8	11.27[f]	0.98	2.25	62.2	1.38	0.80	963[g]	10.3	29	0.31	22	0.23	37

(Table 4 continues)

TABLE 4 Nutrient Requirements of Breeding Cattle (metric). (Cont.)

Weight[a] (kg)	Daily Gain[b] (kg)	Daily DM[c] (kg)	Energy Daily ME (Mcal)	Energy Daily NE$_m$ (Mcal)	Energy Daily NE$_g$ (Mcal)	Energy In Diet DM ME (Mcal/kg)	Energy In Diet DM TDN (%)	Energy In Diet DM NE$_m$ (Mcal/kg)	Energy In Diet DM NE$_g$ (Mcal/kg)	Total Protein Daily (g)	Total Protein In Diet DM (%)	Calcium Daily (g)	Calcium In Diet DM (%)	Phosphorus Daily (g)	Phosphorus In Diet DM (%)	Vitamin A[d] Daily (1000's IU)
Cows nursing calves—Average milking ability—First 3–4 months postpartum—5.0 kg milk/day																
350	0.0	7.7	16.6	9.98[f]	NA	2.15	59.4	1.29	NA	814[g]	10.6	23	0.30	18	0.23	30
400	0.0	8.5	17.9	10.64[f]	NA	2.11	58.3	1.25	NA	864[g]	10.2	25	0.29	19	0.22	33
450	0.0	9.2	19.1	11.27[f]	NA	2.08	57.5	1.23	NA	911[g]	9.9	26	0.28	21	0.23	36
500	0.0	9.9	20.3	11.89[f]	NA	2.05	56.6	1.20	NA	957[g]	9.7	28	0.28	22	0.22	39
550	0.0	10.6	21.5	12.50[f]	NA	2.03	56.1	1.18	NA	1001[g]	9.5	29	0.27	24	0.23	41
600	0.0	11.2	22.6	13.08[f]	NA	2.01	55.5	1.16	NA	1044[g]	9.3	31	0.28	26	0.23	44
650	0.0	11.9	23.9	13.66[f]	NA	2.00	55.3	1.15	NA	1086[g]	9.1	33	0.28	27	0.23	46
Cows nursing calves—Superior milking ability—First 3–4 months postpartum—10.0 kg milk/day																
350	0.0	6.2	18.5	13.73[f]	NA	3.00	82.9	2.03	NA	1009[g]	16.4	36	0.58	24	0.39	24
400	0.0	7.6	21.4	14.39[f]	NA	2.80	77.4	1.86	NA	1099[g]	14.4	37	0.49	25	0.33	30
450	0.0	9.1	23.2	15.02[f]	NA	2.56	70.7	1.66	NA	1186[g]	13.1	39	0.43	26	0.29	35
500	0.0	10.0	24.6	15.64[f]	NA	2.45	67.7	1.56	NA	1246[g]	12.4	40	0.40	28	0.28	39
550	0.0	10.9	25.8	16.25[f]	NA	2.38	65.8	1.50	NA	1299[g]	12.0	42	0.39	30	0.27	42
600	0.0	11.6	27.0	16.83[f]	NA	2.32	64.1	1.45	NA	1348[g]	11.6	43	0.37	31	0.27	45
650	0.0	12.4	28.2	17.41[f]	NA	2.28	63.0	1.41	NA	1394[g]	11.3	45	0.36	33	0.26	48
Bulls, maintenance and regaining body condition																
For growth and development use requirements for bulls in Tables 1, 2, and 3.																
<650																
650	0.4	12.3	24.3	6.7	2.06	1.98	54.8	1.13	0.57	904	7.4	25	0.20	23	0.19	48
650	0.6	12.6	26.7	7.4	3.21	2.11	58.4	1.25	0.69	957	7.6	27	0.21	24	0.19	49
650	0.8	12.8	28.7	7.9	4.40	2.24	62.0	1.37	0.79	998	7.8	29	0.23	25	0.20	50
700	0.4	13.0	25.7	7.1	2.18	1.98	54.8	1.13	0.57	942	7.3	26	0.20	25	0.20	51
700	0.6	13.4	28.2	7.8	3.40	2.11	58.4	1.25	0.69	994	7.4	29	0.22	26	0.20	52
700	0.8	13.5	30.3	8.4	4.66	2.24	62.0	1.37	0.79	1032	7.6	30	0.22	26	0.19	53
800	0.0	12.9	22.6	6.3	NA	1.75	48.4	0.91	NA	882	6.8	27	0.21	27	0.21	50
800	0.2	13.7	25.5	7.1	1.12	1.86	51.5	1.02	0.47	956	7.0	27	0.20	27	0.20	53
900	0.0	14.1	24.7	6.8	NA	1.75	48.4	0.91	NA	958	6.8	30	0.21	30	0.21	55
900	0.2	15.0	27.9	7.7	1.23	1.86	51.5	1.02	0.47	1031	6.9	31	0.21	31	0.21	58
1000	0.0	15.3	26.8	7.4	NA	1.75	48.4	0.91	NA	1032	6.8	33	0.22	33	0.22	60

[a]Average weight for a feeding period.
[b]Approximately 0.4 ± 0.1 kg of weight gain/day over the last third of pregnancy is accounted for by the products of conception. Daily 2.15 Mcal of NE$_m$ and 55 g of protein are provided for this requirement for a calf with a birth weight of 36 kg.
[c]Dry matter consumption should vary depending on the energy concentration of the diet and environmental conditions. These intakes are based on the energy concentration shown in the table and assuming a thermoneutral environment without snow or mud conditions. If the energy concentrations of the diet to be fed exceed the tabular value, limit feeding may be required.
[d]Vitamin A requirements per kilogram of diet of diet are 2800 IU for pregnant heifers and cows and 3900 IU for lactating cows and breeding bulls.
[e]Not applicable.
[f]Includes .75 Mcal NE$_m$/kg of milk produced.
[g]Includes 33.5 g protein/kg of milk produced.

Source: *Nutrient Requirements of Beef Cattle*, sixth revised edition, National Academy of Sciences, National Academy Press, Washington, DC.

TABLE 5 Nutrient Requirements for Growing and Finishing Cattle (Nutrient Concentration in Diet Dry Matter, avoirdupois system).[a,b,c]

Weight (lb)	Daily Gain (lb)	Dry Matter Intake (lb)	Protein Intake (lb)	Protein (%)	ME (Mcal/lb)	NE$_m$ (Mcal/lb)	NE$_g$ (Mcal/lb)	TDN (%)	CA (%)	P (%)
Medium-frame steer calves										
300	0.5	7.8	0.75	9.6	0.89	0.50	0.25	54.0	0.31	0.20
	1.0	8.4	0.95	11.4	0.96	0.57	0.31	58.5	0.45	0.24
	1.5	8.7	1.14	13.2	1.04	0.64	0.38	63.0	0.58	0.28
	2.0	8.9	1.32	14.8	1.11	0.70	0.44	67.5	0.72	0.32
	2.5	8.9	1.48	16.7	1.21	0.79	0.51	73.5	0.87	0.37
	3.0	8.0	1.60	19.9	1.39	0.95	0.64	85.0	1.13	0.47
400	0.5	9.7	0.87	8.9	0.89	0.50	0.25	54.0	0.27	0.18
	1.0	10.4	1.06	10.3	0.96	0.57	0.31	58.5	0.38	0.21
	1.5	10.8	1.24	11.5	1.04	0.64	0.38	63.0	0.47	0.25
	2.0	11.0	1.41	12.7	1.11	0.70	0.44	67.5	0.56	0.26
	2.5	11.0	1.56	14.2	1.21	`0.79	0.51	73.5	0.68	0.30
	3.0	10.0	1.65	16.6	1.39	0.95	0.64	85.0	0.86	0.37
500	0.5	11.5	0.98	8.5	0.89	0.50	0.25	54.0	0.25	0.17
	1.0	12.3	1.16	9.5	0.96	0.57	0.31	58.5	0.32	0.20
	1.5	12.8	1.33	10.5	1.04	0.64	0.38	63.0	0.40	0.22
	2.0	13.1	1.49	11.4	1.11	0.70	0.44	67.5	0.47	0.24
	2.5	13.0	1.63	12.5	1.21	0.79	0.51	73.5	0.56	0.27
	3.0	11.8	1.69	14.4	1.39	0.95	0.64	85.0	0.69	0.32
600	0.5	13.2	1.08	8.2	0.89	0.50	0.25	54.0	0.23	0.18
	1.0	14.1	1.26	9.0	0.96	0.57	0.31	58.5	0.28	0.19
	1.5	14.7	1.42	9.8	1.04	0.64	0.38	63.0	0.35	0.21
	2.0	15.0	1.57	10.5	1.11	0.70	0.44	67.5	0.40	0.22
	2.5	14.9	1.69	11.4	1.21	0.79	0.51	73.5	0.46	0.24
	3.0	13.5	1.73	12.9	1.39	0.95	0.64	85.0	0.57	0.29
700	0.5	14.8	1.18	7.9	0.89	0.50	0.25	54.0	0.22	0.18
	1.0	15.8	1.35	8.6	0.96	0.57	0.31	58.5	0.27	0.18
	1.5	16.5	1.50	9.2	1.04	0.64	0.38	63.0	0.31	0.20
	2.0	16.8	1.65	9.8	1.11	0.70	0.44	67.5	0.34	0.21
	2.5	16.7	1.75	10.5	1.21	0.79	0.51	73.5	0.40	0.22
	3.0	15.2	1.77	11.7	1.39	0.95	0.64	85.0	0.49	0.26
800	0.5	16.4	1.27	7.7	0.89	0.50	0.25	54.0	0.22	0.17
	1.0	17.5	1.44	8.3	0.96	0.57	0.31	58.5	0.24	0.19
	1.5	18.2	1.58	8.8	1.04	0.64	0.38	63.0	0.28	0.19
	2.0	18.6	1.72	9.2	1.11	0.70	0.44	67.5	0.31	0.20
	2.5	18.5	1.81	9.8	1.21	0.79	0.51	73.5	0.35	0.21
	3.0	16.8	1.81	10.8	1.39	0.95	0.64	85.0	0.42	0.25
900	0.5	17.9	1.36	7.6	0.89	0.50	0.25	54.0	0.21	0.18
	1.0	19.1	1.52	8.0	0.96	0.57	0.31	58.5	0.23	0.18
	1.5	19.9	1.66	8.4	1.04	0.64	0.38	63.0	0.25	0.19
	2.0	20.3	1.79	8.8	1.11	0.70	0.44	67.5	0.28	0.20
	2.5	20.2	1.87	9.3	1.21	0.79	0.51	73.5	0.31	0.20
	3.0	18.3	1.85	10.1	1.39	0.95	0.64	85.0	0.37	0.23
1000	0.5	19.3	1.45	7.5	0.89	0.50	0.25	54.0	0.21	0.18
	1.0	20.7	1.60	7.8	0.96	0.57	0.31	58.5	0.21	0.18
	1.5	21.5	1.74	8.1	1.04	0.64	0.38	63.0	0.24	0.18
	2.0	22.0	1.85	8.4	1.11	0.70	0.44	67.5	0.25	0.19
	2.5	21.9	1.92	8.8	1.21	0.79	0.51	73.5	0.27	0.19
	3.0	19.8	1.88	9.5	1.39	0.95	0.64	85.0	0.32	0.22
Large-frame steer calves and compensating medium-frame yearling steers										
300	0.5	8.2	0.77	9.5	0.86	0.48	0.23	52.5	0.30	0.19
	1.0	8.7	0.99	11.3	0.92	0.54	0.28	56.0	0.46	0.23
	1.5	9.1	1.19	12.9	0.98	0.59	0.33	59.5	0.58	0.27
	2.0	9.4	1.37	14.6	1.04	0.64	0.38	63.5	0.70	0.30
	2.5	9.6	1.55	16.3	1.11	0.70	0.44	67.5	0.85	0.34
	3.0	9.6	1.73	18.0	1.18	0.77	0.49	72.0	0.99	0.39
	3.5	9.3	1.88	20.3	1.29	0.86	0.57	78.5	1.16	0.45
400	0.5	10.1	0.89	8.9	0.86	0.48	0.23	52.5	0.26	0.17
	1.0	10.8	1.10	10.2	0.92	0.54	0.28	56.0	0.37	0.20
	1.5	11.3	1.30	11.4	0.98	0.59	0.33	59.5	0.47	0.23
	2.0	11.7	1.47	12.7	1.04	0.64	0.38	63.5	0.57	0.26
	2.5	11.9	1.64	13.9	1.11	0.70	0.44	67.5	0.65	0.30
	3.0	11.9	1.81	15.2	1.18	0.77	0.49	72.0	0.76	0.33
	3.5	11.5	1.94	16.9	1.29	0.86	0.57	78.5	0.90	0.36

(Table 5 continues)

TABLE 5 Nutrient Requirements for Growing and Finishing Cattle *(Cont.)*

Weight (lb)	Daily Gain (lb)	Dry Matter Intake (lb)	Protein Intake (lb)	Protein (%)	ME (Mcal/lb)	NE$_m$ (Mcal/lb)	NE$_g$ (Mcal/lb)	TDN (%)	CA (%)	P (%)
500	0.5	12.0	1.0	8.5	0.86	0.48	0.23	52.5	0.24	0.17
	1.0	12.8	1.21	9.5	0.92	0.54	0.28	56.0	0.33	0.19
	1.5	13.4	1.40	10.4	0.98	0.59	0.33	59.5	0.39	0.21
	2.0	13.8	1.57	11.4	1.04	0.64	0.38	63.5	0.46	0.24
	2.5	14.0	1.73	12.4	1.11	0.70	0.44	67.5	0.55	0.25
	3.0	14.0	1.88	13.4	1.18	0.77	0.49	72.0	0.63	0.28
	3.5	13.6	2.00	14.7	1.29	0.86	0.57	78.5	0.73	0.32
600	0.5	13.8	1.11	8.2	0.86	0.48	0.23	52.5	0.22	0.18
	1.0	14.6	1.31	9.0	0.92	0.54	0.28	56.0	0.29	0.18
	1.5	15.3	1.50	9.7	0.98	0.59	0.33	59.5	0.35	0.20
	2.0	15.8	1.66	10.5	1.04	0.64	0.38	63.5	0.40	0.22
	2.5	16.1	1.81	11.3	1.11	0.70	0.44	67.5	0.47	0.23
	3.0	16.1	1.95	12.1	1.18	0.77	0.49	72.0	0.52	0.26
	3.5	15.6	2.05	13.2	1.29	0.86	0.57	78.5	0.61	0.28
700	0.5	15.4	1.21	7.9	0.86	0.48	0.23	52.5	0.21	0.17
	1.0	16.4	1.41	8.6	0.92	0.54	0.28	56.0	0.27	0.19
	1.5	17.2	1.59	9.2	0.98	0.59	0.33	59.5	0.31	0.19
	2.0	17.8	1.74	9.8	1.04	0.64	0.38	63.5	0.36	0.21
	2.5	18.0	1.88	10.5	1.11	0.70	0.44	67.5	0.40	0.22
	3.0	18.0	2.01	11.1	1.18	0.77	0.49	72.0	0.45	0.23
	3.5	17.5	2.10	12.0	1.29	0.86	0.57	78.5	0.52	0.26
800	0.5	17.1	1.31	7.7	0.86	0.48	0.23	52.5	0.21	0.18
	1.0	18.2	1.51	8.3	0.92	0.54	0.28	56.0	0.24	0.18
	1.5	19.0	1.68	8.8	0.98	0.59	0.33	59.5	0.28	0.19
	2.0	19.6	1.82	9.3	1.04	0.64	0.38	63.5	0.32	0.20
	2.5	19.9	1.96	9.8	1.11	0.70	0.44	67.5	0.35	0.21
	3.0	19.9	2.07	10.4	1.18	0.77	0.49	72.0	0.40	0.22
	3.5	19.3	2.15	11.1	1.29	0.86	0.57	78.5	0.45	0.24
900	0.5	18.6	1.40	7.6	0.86	0.48	0.23	52.5	0.20	0.18
	1.0	19.8	1.60	8.0	0.92	0.54	0.28	56.0	0.23	0.18
	1.5	20.8	1.77	8.5	0.98	0.59	0.33	59.5	0.27	0.18
	2.0	21.4	1.90	8.9	1.04	0.64	0.38	63.5	0.29	0.20
	2.5	21.8	2.03	9.3	1.11	0.70	0.44	67.5	0.31	0.20
	3.0	21.7	2.13	9.8	1.18	0.77	0.49	72.0	0.36	0.21
	3.5	21.1	2.19	10.4	1.29	0.86	0.57	78.5	0.40	0.23
1000	0.5	20.2	1.49	7.5	0.86	0.48	0.23	52.5	0.20	0.17
	1.0	21.5	1.69	7.8	0.92	0.54	0.28	56.0	0.23	0.17
	1.5	22.5	1.85	8.2	0.98	0.59	0.33	59.5	0.25	0.18
	2.0	23.2	1.98	8.6	1.04	0.64	0.38	63.5	0.27	0.18
	2.5	23.6	2.09	8.9	1.11	0.70	0.44	67.5	0.29	0.19
	3.0	23.6	2.19	9.3	1.18	0.77	0.49	72.0	0.32	0.20
	3.5	22.8	2.24	9.8	1.29	0.86	0.57	78.5	0.35	0.21
1100	0.5	21.7	1.58	7.4	0.86	0.48	0.23	52.5	0.19	0.18
	1.0	23.1	1.77	7.7	0.92	0.54	0.28	56.0	0.21	0.18
	1.5	24.1	1.93	8.0	0.98	0.59	0.33	59.5	0.23	0.18
	2.0	24.9	2.05	8.3	1.04	0.64	0.38	63.5	0.25	0.18
	2.5	25.3	2.16	8.5	1.11	0.70	0.44	67.5	0.26	0.18
	3.0	25.3	2.25	8.9	1.18	0.77	0.49	72.0	0.29	0.19
	3.5	24.5	2.28	9.3	1.29	0.86	0.57	78.5	0.32	0.21
Medium-frame bulls										
300	0.5	7.8	0.76	9.7	0.88	0.49	0.24	53.5	0.31	0.20
	1.0	8.3	0.96	11.6	0.94	0.56	0.30	57.5	0.48	0.24
	1.5	8.6	1.15	13.4	1.01	0.62	0.35	61.5	0.62	0.28
	2.0	8.8	1.34	15.2	1.08	0.68	0.41	65.5	0.75	0.33
	2.5	8.9	1.52	17.0	1.15	0.74	0.47	70.0	0.92	0.37
	3.0	8.7	1.68	19.3	1.26	0.84	0.54	76.5	1.09	0.43
400	0.5	9.6	0.87	9.0	0.88	0.49	0.24	53.5	0.28	0.18
	1.0	10.3	1.07	10.4	0.94	0.56	0.30	57.5	0.39	0.21
	1.5	10.7	1.26	11.8	1.01	0.62	0.35	61.5	0.49	0.25
	2.0	11.0	1.44	13.1	1.08	0.68	0.41	65.5	0.60	0.28
	2.5	11.1	1.60	14.4	1.15	0.74	0.47	70.0	0.70	0.32
	3.0	10.8	1.74	16.1	1.26	0.84	0.54	76.5	0.84	0.37
500	0.5	11.4	0.98	8.6	0.88	0.49	0.24	53.5	0.25	0.17
	1.0	12.1	1.17	9.7	0.94	0.56	0.30	57.5	0.35	0.20
	1.5	12.7	1.35	10.7	1.01	0.62	0.35	61.5	0.42	0.23

(Table 5 continues)

TABLE 5 Nutrient Requirements for Growing and Finishing Cattle *(Cont.)*

Weight (lb)	Daily Gain (lb)	Dry Matter Intake (lb)	Protein Intake (lb)	Protein (%)	ME (Mcal/lb)	NE$_m$ (Mcal/lb)	NE$_g$ (Mcal/lb)	TDN (%)	CA (%)	P (%)
	2.0	13.0	1.52	11.7	1.08	0.68	0.41	65.5	0.49	0.25
	2.5	13.1	1.68	12.8	1.15	0.74	0.47	70.0	0.59	0.27
	3.0	12.8	1.81	14.1	1.26	0.84	0.54	76.5	0.69	0.31
600	0.5	13.1	1.08	8.3	0.88	0.49	0.24	53.5	0.24	0.19
	1.0	13.9	1.27	9.2	0.94	0.56	0.30	57.5	0.30	0.19
	1.5	14.5	1.44	10.0	1.01	0.62	0.35	61.5	0.36	0.21
	2.0	14.9	1.61	10.8	1.08	0.68	0.41	65.5	0.43	0.24
	2.5	15.0	1.75	11.6	1.15	0.74	0.47	70.0	0.50	0.25
	3.0	14.7	1.86	12.7	1.26	0.84	0.54	76.5	0.57	0.29
700	0.5	14.7	1.18	8.0	0.88	0.49	0.24	53.5	0.23	0.18
	1.0	15.6	1.37	8.8	0.94	0.56	0.30	57.5	0.28	0.20
	1.5	16.3	1.53	9.4	1.01	0.62	0.35	61.5	0.32	0.20
	2.0	16.7	1.69	10.1	1.08	0.68	0.41	65.5	0.38	0.22
	2.5	16.8	1.82	10.8	1.15	0.74	0.47	70.0	0.43	0.24
	3.0	16.5	1.92	11.7	1.26	0.84	0.54	76.5	0.49	0.25
800	0.5	16.2	1.27	7.8	0.88	0.49	0.24	53.5	0.22	0.19
	1.0	17.3	1.45	8.4	0.94	0.56	0.30	57.5	0.25	0.19
	1.5	18.0	1.61	9.0	1.01	0.62	0.35	61.5	0.29	0.20
	2.0	18.5	1.76	9.5	1.08	0.68	0.41	65.5	0.33	0.21
	2.5	18.6	1.89	10.1	1.15	0.74	0.47	70.0	0.38	0.23
	3.0	18.2	1.97	10.8	1.26	0.84	0.54	76.5	0.44	0.24
900	0.5	17.7	1.36	7.7	0.88	0.49	0.24	53.5	0.21	0.19
	1.0	18.9	1.54	8.2	0.94	0.56	0.30	57.5	0.25	0.19
	1.5	19.7	1.69	8.6	1.01	0.62	0.35	61.5	0.28	0.19
	2.0	20.2	1.83	9.1	1.08	0.68	0.41	65.5	0.31	0.21
	2.5	20.3	1.95	9.6	1.15	0.74	0.47	70.0	0.34	0.22
	3.0	19.9	2.02	10.2	1.26	0.84	0.54	76.5	0.39	0.23
1000	0.5	19.2	1.45	7.5	0.88	0.49	0.24	53.5	0.21	0.18
	1.0	20.4	1.62	8.0	0.94	0.56	0.30	57.5	0.24	0.18
	1.5	21.3	1.77	8.4	1.01	0.62	0.35	61.5	0.26	0.19
	2.0	21.8	1.90	8.7	1.08	0.68	0.41	65.5	0.28	0.19
	2.5	22.0	2.01	9.1	1.15	0.74	0.47	70.0	0.31	0.20
	3.0	21.5	2.07	9.6	1.26	0.84	0.54	76.5	0.35	0.22
1100	0.5	20.6	1.54	7.4	0.88	0.49	0.24	53.5	0.20	0.19
	1.0	21.9	1.70	7.8	0.94	0.56	0.30	57.5	0.22	0.19
	1.5	22.9	1.85	8.1	1.01	0.62	0.35	61.5	0.24	0.19
	2.0	23.4	1.97	8.4	1.08	0.68	0.41	65.5	0.26	0.19
	2.5	23.6	2.07	8.7	1.15	0.74	0.47	70.0	0.28	0.20
	3.0	23.1	2.11	9.2	1.26	0.84	0.54	76.5	0.32	0.21
Large-frame bull calves and compensating large-frame yearling steers										
300	0.5	7.9	0.77	9.7	0.86	0.48	0.23	52.5	0.31	0.20
	1.0	8.4	0.98	11.7	0.92	0.54	0.28	56.0	0.47	0.24
	1.5	8.8	1.18	13.5	0.98	0.59	0.33	59.5	0.63	0.28
	2.0	9.0	1.38	15.1	1.03	0.63	0.37	62.5	0.76	0.32
	2.5	9.2	1.56	17.0	1.09	0.69	0.42	66.5	0.91	0.36
	3.0	9.2	1.74	18.8	1.16	0.75	0.47	70.5	1.08	0.43
	3.5	9.1	1.91	20.9	1.24	0.82	0.53	75.5	1.24	0.48
	4.0	8.2	2.04	24.7	1.41	0.96	0.66	86.0	1.53	0.59
400	0.5	9.8	0.89	9.0	0.86	0.48	0.23	52.5	0.27	0.18
	1.0	10.4	1.09	10.5	0.92	0.54	0.28	56.0	0.40	0.21
	1.5	10.9	1.29	11.9	0.98	0.59	0.33	59.5	0.51	0.24
	2.0	11.2	1.48	13.1	1.03	0.63	0.37	62.5	0.61	0.28
	2.5	11.4	1.65	14.5	1.09	0.69	0.42	66.5	0.72	0.31
	3.0	11.5	1.82	15.9	1.16	0.75	0.47	70.5	0.82	0.35
	3.5	11.3	1.98	17.5	1.24	0.82	0.53	75.5	0.96	0.39
	4.0	10.2	2.08	20.3	1.41	0.96	0.66	86.0	1.19	0.48
500	0.5	11.6	1.00	8.6	0.86	0.48	0.23	52.5	0.25	0.19
	1.0	12.3	1.20	9.8	0.92	0.54	0.28	56.0	0.36	0.21
	1.5	12.9	1.39	10.9	0.98	0.59	0.33	59.5	0.43	0.22
	2.0	13.2	1.58	11.8	1.03	0.63	0.37	62.5	0.52	0.25
	2.5	13.5	1.74	12.9	1.09	0.69	0.42	66.5	0.59	0.28
	3.0	13.6	1.90	14.0	1.16	0.75	0.47	70.5	0.68	0.31
	3.5	13.4	2.05	15.3	1.24	0.82	0.53	75.5	0.77	0.35
	4.0	12.0	2.13	17.5	1.41	0.96	0.66	86.0	0.97	0.40
600	0.5	13.3	1.10	8.3	0.86	0.48	0.23	52.5	0.23	0.18
	1.0	14.1	1.30	9.2	0.92	0.54	0.28	56.0	0.31	0.20

(Table 5 continues)

TABLE 5 Nutrient Requirements for Growing and Finishing Cattle (Cont.)

Weight (lb)	Daily Gain (lb)	Dry Matter Intake (lb)	Protein Intake (lb)	Protein (%)	ME (Mcal/lb)	NE_m (Mcal/lb)	NE_g (Mcal/lb)	TDN (%)	CA (%)	P (%)
	1.5	14.8	1.48	10.1	0.98	0.59	0.33	59.5	0.37	0.21
	2.0	15.2	1.67	10.9	1.03	0.63	0.37	62.5	0.44	0.23
	2.5	15.5	1.82	11.8	1.09	0.69	0.42	66.5	0.51	0.26
	3.0	15.5	1.97	12.7	1.16	0.75	0.47	70.5	0.58	0.27
	3.5	15.3	2.11	13.7	1.24	0.82	0.53	75.5	0.66	0.30
	4.0	13.8	2.16	15.6	1.41	0.96	0.66	86.0	0.81	0.37
700	0.5	14.9	1.20	8.0	0.86	0.48	0.23	52.5	0.22	0.18
	1.0	15.9	1.40	8.8	0.92	0.54	0.28	56.0	0.29	0.19
	1.5	16.6	1.57	9.6	0.98	0.59	0.33	59.5	0.35	0.21
	2.0	17.0	1.75	10.2	1.03	0.63	0.37	62.5	0.39	0.22
	2.5	17.4	1.90	11.0	1.09	0.69	0.42	66.5	0.44	0.24
	3.0	17.5	2.04	11.7	1.16	0.75	0.47	70.5	0.50	0.25
	3.5	17.2	2.16	12.5	1.24	0.82	0.53	75.5	0.56	0.28
	4.0	15.5	2.20	14.1	1.41	0.96	0.66	86.0	0.70	0.33
800	0.5	16.5	1.30	7.9	0.86	0.48	0.23	52.5	0.21	0.19
	1.0	17.5	1.49	8.5	0.92	0.54	0.28	56.0	0.26	0.19
	1.5	18.3	1.66	9.1	0.98	0.59	0.33	59.5	0.31	0.20
	2.0	18.8	1.84	9.7	1.03	0.63	0.37	62.5	0.35	0.21
	2.5	19.2	1.97	10.3	1.09	0.69	0.42	66.5	0.40	0.23
	3.0	19.3	2.11	10.9	1.16	0.75	0.47	70.5	0.45	0.24
	3.5	19.0	2.22	11.6	1.24	0.82	0.53	75.5	0.50	0.26
	4.0	17.1	2.24	13.0	1.41	0.96	0.66	86.0	0.61	0.31
900	0.5	18.0	1.39	7.7	0.86	0.48	0.23	52.5	0.22	0.18
	1.0	19.2	1.58	8.3	0.92	0.54	0.28	56.0	0.25	0.18
	1.5	20.0	1.74	8.8	0.98	0.59	0.33	59.5	0.29	0.20
	2.0	20.6	1.92	9.2	1.03	0.63	0.37	62.5	0.32	0.20
	2.5	21.0	2.04	9.8	1.09	0.69	0.42	66.5	0.36	0.21
	3.0	21.1	2.17	10.3	1.16	0.75	0.47	70.5	0.40	0.23
	3.5	20.8	2.27	10.9	1.24	0.82	0.53	75.5	0.45	0.24
	4.0	18.7	2.27	12.1	1.41	0.96	0.66	86.0	0.53	0.28
1000	0.5	19.5	1.48	7.6	0.86	0.48	0.23	52.5	0.21	0.18
	1.0	20.7	1.66	8.1	0.92	0.54	0.28	56.0	0.25	0.19
	1.5	21.7	1.83	8.5	0.98	0.59	0.33	59.5	0.27	0.19
	2.0	22.3	1.99	8.9	1.03	0.63	0.37	62.5	0.30	0.20
	2.5	22.7	2.11	9.3	1.09	0.69	0.42	66.5	0.33	0.20
	3.0	22.8	2.23	9.7	1.16	0.75	0.47	70.5	0.36	0.21
	3.5	22.5	2.32	10.3	1.24	0.82	0.53	75.5	0.40	0.24
	4.0	20.2	2.30	11.3	1.41	0.96	0.66	86.0	0.48	0.27
1100	0.5	20.9	1.57	7.5	0.86	0.48	0.23	52.5	0.21	0.19
	1.0	22.3	1.75	7.9	0.92	0.54	0.28	56.0	0.23	0.19
	1.5	23.3	1.91	8.3	0.98	0.59	0.33	59.5	0.26	0.19
	2.0	23.9	2.07	8.6	1.03	0.63	0.37	62.5	0.28	0.19
	2.5	24.2	2.18	9.0	1.09	0.69	0.42	66.5	0.30	0.20
	3.0	24.5	2.29	9.3	1.16	0.75	0.47	70.5	0.32	0.21
	3.5	24.1	2.37	9.8	1.24	0.82	0.53	75.5	0.36	0.22
	4.0	21.7	2.33	10.7	1.41	0.96	0.66	86.0	0.43	0.25
Medium-frame heifer calves										
300	0.5	7.5	0.73	9.6	0.92	0.54	0.28	56.0	0.29	0.21
	1.0	8.0	0.91	11.4	1.02	0.63	0.36	62.0	0.44	0.22
	1.5	8.2	1.08	13.1	1.13	0.72	0.44	68.5	0.59	0.27
	2.0	8.0	1.22	15.1	1.26	0.84	0.55	77.0	0.74	0.33
400	0.5	9.3	0.84	8.9	0.92	0.54	0.28	56.0	0.26	0.19
	1.0	9.9	1.01	10.2	1.02	0.63	0.36	62.0	0.36	0.20
	1.5	10.2	1.17	11.4	1.13	0.72	0.44	68.5	0.45	0.24
	2.0	10.0	1.29	12.9	1.26	0.84	0.55	77.0	0.57	0.29
500	0.5	11.0	0.94	8.5	0.92	0.54	0.28	56.0	0.24	0.18
	1.0	11.8	1.11	9.4	1.02	0.63	0.36	62.0	0.30	0.21
	1.5	12.1	1.25	10.3	1.13	0.72	0.44	68.5	0.38	0.22
	2.0	11.8	1.35	11.4	1.26	0.84	0.55	77.0	0.45	0.24
600	0.5	12.6	1.04	8.1	0.92	0.54	0.28	56.0	0.23	0.18
	1.0	13.5	1.19	8.8	1.02	0.63	0.36	62.0	0.28	0.20
	1.5	13.8	1.32	9.5	1.13	0.72	0.44	68.5	0.32	0.21
	2.0	13.5	1.41	10.4	1.26	0.84	0.55	77.0	0.38	0.23
700	0.5	14.1	1.13	7.9	0.92	0.54	0.28	56.0	0.22	0.19
	1.0	15.1	1.28	8.4	1.02	0.63	0.36	62.0	0.25	0.19
	1.5	15.5	1.39	9.0	1.13	0.72	0.44	68.5	0.28	0.20
	2.0	15.2	1.46	9.6	1.26	0.84	0.55	77.0	0.32	0.22

(Table 5 continues)

TABLE 5 Nutrient Requirements for Growing and Finishing Cattle *(Cont.)*

Weight (lb)	Daily Gain (lb)	Dry Matter Intake (lb)	Protein Intake (lb)	Protein (%)	ME (Mcal/lb)	NE_m (Mcal/lb)	NE_g (Mcal/lb)	TDN (%)	CA (%)	P (%)
800	0.5	15.6	1.22	7.7	0.92	0.54	0.28	56.0	0.21	0.18
	1.0	16.7	1.36	8.1	1.02	0.63	0.36	62.0	0.22	0.18
	1.5	17.2	1.46	8.5	1.13	0.72	0.44	68.5	0.24	0.19
	2.0	16.8	1.51	9.0	1.26	0.84	0.55	77.0	0.28	0.20
900	0.5	17.1	1.31	7.5	0.92	0.54	0.28	56.0	0.21	0.18
	1.0	18.3	1.44	7.8	1.02	0.63	0.36	62.0	0.22	0.18
	1.5	18.8	1.53	8.1	1.13	0.72	0.44	68.5	0.22	0.19
	2.0	18.3	1.56	8.5	1.26	0.84	0.55	77.0	0.25	0.19
1000	0.5	18.5	1.39	7.4	0.92	0.54	0.28	56.0	0.20	0.19
	1.0	19.8	1.51	7.6	1.02	0.63	0.36	62.0	0.20	0.18
	1.5	20.3	1.59	7.8	1.13	0.72	0.44	68.5	0.21	0.18
	2.0	19.8	1.61	8.1	1.26	0.84	0.55	77.0	0.22	0.19
Large-frame heifer calves and compensating medium-frame yearling heifers										
300	0.5	7.8	0.76	9.5	0.89	0.50	0.25	54.0	0.31	0.20
	1.0	8.4	0.95	11.3	0.98	0.58	0.32	59.0	0.45	0.24
	1.5	8.8	1.13	13.0	1.05	0.65	0.39	64.0	0.58	0.25
	2.0	8.9	1.30	14.6	1.14	0.74	0.46	69.5	0.69	0.30
	2.5	8.7	1.45	16.7	1.26	0.84	0.55	77.0	0.86	0.35
400	0.5	9.7	0.87	8.9	0.89	0.50	0.25	54.0	0.27	0.18
	1.0	10.5	1.06	10.1	0.98	0.58	0.32	59.0	0.36	0.21
	1.5	10.9	1.23	11.3	1.05	0.65	0.39	64.0	0.45	0.22
	2.0	11.1	1.38	12.6	1.14	0.74	0.46	69.5	0.54	0.26
	2.5	10.8	1.51	14.1	1.26	0.84	0.55	77.0	0.65	0.31
500	0.5	11.5	0.98	8.4	0.89	0.50	0.25	54.0	0.23	0.17
	1.0	12.4	1.16	9.4	0.98	0.58	0.32	59.0	0.30	0.20
	1.5	12.9	1.32	10.3	1.05	0.65	0.39	64.0	0.38	0.20
	2.0	13.1	1.46	11.2	1.14	0.74	0.46	69.5	0.44	0.24
	2.5	12.8	1.57	12.4	1.26	0.84	0.55	77.0	0.53	0.26
600	0.5	13.2	1.08	8.1	0.89	0.50	0.25	54.0	0.22	0.18
	1.0	14.1	1.25	8.9	0.98	0.58	0.32	59.0	0.28	0.19
	1.5	14.8	1.41	9.6	1.05	0.65	0.39	64.0	0.33	0.19
	2.0	15.0	1.54	10.3	1.14	0.74	0.46	69.5	0.38	0.22
	2.5	14.6	1.63	11.2	1.26	0.84	0.55	77.0	0.44	0.24
700	0.5	14.8	1.18	7.9	0.89	0.50	0.25	54.0	0.21	0.18
	1.0	15.9	1.34	8.5	0.98	0.58	0.32	59.0	0.25	0.18
	1.5	16.6	1.49	9.0	1.05	0.65	0.39	64.0	0.29	0.19
	2.0	16.8	1.61	9.6	1.14	0.74	0.46	69.5	0.33	0.20
	2.5	16.4	1.68	10.3	1.26	0.84	0.55	77.0	0.38	0.22
800	0.5	16.4	1.27	7.7	0.89	0.50	0.25	54.0	0.20	0.17
	1.0	17.6	1.43	8.2	0.98	0.58	0.32	59.0	0.24	0.18
	1.5	18.3	1.57	8.6	1.05	0.65	0.39	64.0	0.25	0.18
	2.0	18.6	1.67	9.0	1.14	0.74	0.46	69.5	0.28	0.19
	2.5	18.1	1.74	9.6	1.26	0.84	0.55	77.0	0.33	0.21
900	0.5	17.8	1.36	7.5	0.89	0.50	0.25	54.0	0.20	0.18
	1.0	19.2	1.52	7.9	0.98	0.58	0.32	59.0	0.22	0.18
	1.5	20.0	1.64	8.2	1.05	0.65	0.39	64.0	0.23	0.18
	2.0	20.3	1.74	8.6	1.14	0.74	0.46	69.5	0.26	0.18
	2.5	19.8	1.78	9.0	1.26	0.84	0.55	77.0	0.29	0.20
1000	0.5	19.3	1.45	7.4	0.89	0.50	0.25	54.0	0.19	0.18
	1.0	20.8	1.60	7.7	0.98	0.58	0.32	59.0	0.21	0.18
	1.5	21.7	1.71	8.0	1.05	0.65	0.39	64.0	0.21	0.18
	2.0	22.0	1.80	8.2	1.14	0.74	0.46	69.5	0.23	0.18
	2.5	21.5	1.83	8.6	1.26	0.84	0.55	77.0	0.25	0.18
1100	0.5	20.8	1.54	7.3	0.89	0.50	0.25	54.0	0.19	0.18
	1.0	22.3	1.68	7.5	0.98	0.58	0.32	59.0	0.20	0.18
	1.5	23.3	1.78	7.7	1.05	0.65	0.39	64.0	0.20	0.18
	2.0	23.6	1.86	7.9	1.14	0.74	0.46	69.5	0.21	0.18
	2.5	23.1	1.88	8.2	1.26	0.84	0.55	77.0	0.22	0.18

[a]Shrunk liveweight basis (weight after an overnight feed and water shrink—generally equivalent to about 96 percent of unshrunk weights taken in the early morning).

[b]Vitamin A requirements are 1000 IU per pound of diet.

[c]This table gives reasonable examples of nutrient concentrations that should be suitable to formulate diets for specific management goals. It does not imply that diets with other nutrient concentrations when consumed in sufficient amounts would be inadequate to meet nutrient requirements.

Source: Nutrient Requirements of Beef Cattle, sixth revised edition, National Academy of Sciences, National Academy Press, Washington, DC.

TABLE 6 Nutrient Requirements of Breeding Cattle (avoirdupois system).

Weight[a] (lb)	Gain[b] (lb)	DM[c] (lb)	Daily ME (Mcal)	In Diet NEm (Mcal)	TDN (lb)	NEg (Mcal)	In Diet ME (Mcal/lb)	TDN (%)	In Diet NEm (Mcal/lb)	NEg (Mcal/lb)	Total Protein Daily (lb)	DM (%)	Calcium Daily (g)	DM (%)	Phosphorus Daily (g)	DM (%)	Vitamin A[d] Daily (1000's IU)
Pregnant yearling heifers—Last third of pregnancy																	
700	0.9	15.3	13.9	7.95	8.5	NA[e]	0.91	55.4	0.52	NA[e]	1.3	8.4	19	0.27	14	0.20	19
700	1.4	15.8	15.7	7.95	9.6	0.87	0.99	60.3	0.60	0.34	1.4	9.0	24	0.33	15	0.21	20
700	1.9	15.8	17.4	7.95	10.6	1.89	1.10	67.0	0.70	0.43	1.5	9.8	27	0.33	16	0.21	20
750	0.9	16.1	14.6	8.25	8.9	NA	0.90	55.1	0.52	NA	1.3	8.3	20	0.27	14	0.19	20
750	1.4	16.6	16.4	8.25	10.0	0.92	0.98	59.9	0.60	0.33	1.5	8.9	24	0.32	16	0.21	21
750	1.9	16.6	18.2	8.25	11.1	1.99	1.09	66.5	0.69	0.42	1.6	9.5	28	0.37	17	0.23	21
800	0.9	16.8	15.2	8.56	9.2	NA	0.90	54.8	0.51	NA	1.4	8.2	21	0.28	15	0.20	21
800	1.4	17.4	17.1	8.56	10.4	0.96	0.98	59.6	0.59	0.33	1.5	8.8	25	0.33	16	0.21	22
800	1.9	17.5	19.0	8.56	11.6	2.09	1.08	66.1	0.69	0.42	1.6	9.3	28	0.35	17	0.21	22
850	0.9	17.6	15.7	8.85	9.6	NA	0.89	54.5	0.51	NA	1.4	8.2	21	0.26	16	0.20	22
850	1.4	18.2	17.8	8.85	10.8	1.01	0.97	59.3	0.59	0.32	1.6	8.6	25	0.30	17	0.21	23
850	1.9	18.3	19.8	8.85	12.1	2.19	1.08	65.7	0.68	0.41	1.7	9.1	28	0.34	18	0.22	23
900	0.9	18.3	16.3	9.15	9.9	NA	0.89	54.3	0.51	NA	1.5	8.1	22	0.26	17	0.20	23
900	1.4	19.0	18.5	9.15	11.3	1.05	0.97	59.1	0.58	0.32	1.6	8.5	26	0.30	18	0.21	24
900	1.9	19.2	20.6	9.15	12.5	2.28	1.07	65.4	0.68	0.41	1.7	9.0	28	0.32	19	0.21	24
950	0.9	19.0	16.9	9.44	10.3	NA	0.89	54.1	0.50	NA	1.5	8.0	23	0.27	17	0.20	24
950	1.4	19.8	19.1	9.44	11.7	1.09	0.97	58.9	0.58	0.32	1.7	8.4	26	0.29	19	0.21	25
950	1.9	20.0	21.3	9.44	13.0	2.38	1.07	65.1	0.67	0.40	1.8	8.8	29	0.32	19	0.21	25
Dry pregnant mature cows—Middle third of pregnancy																	
800	0.0	15.3	12.3	6.41	7.5	NA	0.80	48.8	0.42	NA	1.1	7.1	12	0.17	12	0.17	19
900	0.0	16.7	13.4	7.00	8.2	NA	0.80	48.8	0.42	NA	1.2	7.0	14	0.18	14	0.18	21
1000	0.0	18.1	14.5	7.57	8.8	NA	0.80	48.8	0.42	NA	1.3	7.0	15	0.18	15	0.18	23
1100	0.0	19.5	15.6	8.13	9.5	NA	0.80	48.8	0.42	NA	1.4	7.0	17	0.19	17	0.19	25
1200	0.0	20.8	16.6	8.68	10.1	NA	0.80	48.8	0.42	NA	1.4	6.9	18	0.19	18	0.19	26
1300	0.0	22.0	17.7	9.22	10.8	NA	0.80	48.8	0.42	NA	1.5	6.9	20	0.20	20	0.20	28
1400	0.0	23.3	18.7	9.75	11.4	NA	0.80	48.8	0.42	NA	1.6	6.9	21	0.20	21	0.20	30
Dry pregnant mature cows—Last third of pregnancy																	
800	0.9	16.8	15.0	8.56	9.2	NA	0.89	54.5	0.51	NA	1.4	8.2	20	0.26	15	0.20	21
900	0.9	18.2	16.2	9.15	9.8	NA	0.89	54.0	0.50	NA	1.5	8.0	22	0.27	17	0.21	23
1000	0.9	19.6	17.3	9.72	10.5	NA	0.88	53.6	0.50	NA	1.6	7.9	23	0.26	18	0.20	25
1100	0.9	21.0	18.3	10.28	11.2	NA	0.87	53.2	0.49	NA	1.6	7.8	25	0.26	20	0.21	26
1200	0.9	22.3	19.4	10.83	11.8	NA	0.87	52.9	0.49	NA	1.7	7.8	26	0.26	21	0.21	28
1300	0.9	23.6	20.4	11.37	12.5	NA	0.87	52.7	0.48	NA	1.8	7.7	28	0.26	23	0.21	30
1400	0.9	24.9	21.5	11.90	13.1	NA	0.86	52.5	0.48	NA	1.9	7.6	29	0.26	24	0.21	32
Two-year-old heifers nursing calves—First 3–4 months postpartum—10 lb milk/day																	
700	0.5	15.9	17.0	9.20[f]	10.3	0.87	1.07	65.1	0.67	0.40	1.8[g]	11.3	26	0.36	17	0.24	28
750	0.5	16.7	17.7	9.51[f]	10.8	0.92	1.06	64.4	0.66	0.40	1.8[g]	11.0	26	0.34	18	0.24	30
800	0.5	17.6	18.4	9.81[f]	11.2	0.96	1.05	63.8	0.66	0.39	1.9[g]	10.8	27	0.34	19	0.24	31
850	0.5	18.4	19.1	10.11[f]	11.6	1.01	1.04	63.2	0.65	0.38	1.9[g]	10.6	27	0.33	19	0.23	33
900	0.5	19.2	19.8	10.40[f]	12.0	1.05	1.03	62.7	0.64	0.37	2.0[g]	10.4	28	0.32	20	0.23	34
950	0.5	20.0	20.5	10.69[f]	12.5	1.09	1.02	62.3	0.63	0.37	2.0[g]	10.2	28	0.31	21	0.23	35
1000	0.5	20.8	21.1	10.98[f]	12.9	1.14	1.02	61.9	0.62	0.36	2.1[g]	10.0	29	0.31	22	0.23	37

Cows nursing calves—Average milking ability—First 3–4 months postpartum—10 lb milk/day

800	0.0	17.3	16.6	10.1	9.81[f]	NA	0.96	58.2	0.57	NA	1.8[g]	10.2	23	0.30	17	0.22	31
900	0.0	18.8	17.7	10.8	10.40[f]	NA	0.94	57.3	0.55	NA	1.9[g]	9.9	24	0.28	19	0.22	33
1000	0.0	20.2	18.8	11.5	10.98[f]	NA	0.93	56.6	0.55	NA	2.0[g]	9.6	25	0.28	20	0.22	36
1100	0.0	21.6	19.9	12.1	11.54[f]	NA	0.92	56.0	0.54	NA	2.0[g]	9.4	27	0.27	22	0.22	38
1200	0.0	23.0	21.0	12.8	12.09[f]	NA	0.91	55.5	0.53	NA	2.1[g]	9.3	28	0.27	23	0.22	41
1300	0.0	24.3	22.0	13.4	12.63[f]	NA	0.90	55.1	0.52	NA	2.2[g]	9.1	30	0.27	25	0.22	43
1400	0.0	25.6	23.0	14.0	13.15[f]	NA	0.90	54.7	0.51	NA	2.3[g]	9.0	31	0.27	26	0.22	46

Cows nursing calves—Superior milking ability—First 3–4 months postpartum—20 lb milk/day

800	0.0	15.7	19.9	12.1	13.22[f]	NA	1.27	77.3	0.85	NA	2.2[g]	14.2	34	0.48	22	0.31	28
900	0.0	18.7	21.5	13.1	13.81[f]	NA	1.15	69.8	0.74	NA	2.4[g]	12.9	35	0.41	24	0.28	33
1000	0.0	20.6	22.7	13.8	14.38[f]	NA	1.10	67.0	0.70	NA	2.5[g]	12.3	36	0.39	25	0.27	37
1100	0.0	22.3	23.8	14.5	14.94[f]	NA	1.07	65.2	0.67	NA	2.6[g]	11.9	38	0.38	27	0.27	40
1200	0.0	23.8	24.9	15.2	15.49[f]	NA	1.05	63.7	0.65	NA	2.7[g]	11.5	39	0.36	28	0.26	42
1300	0.0	25.3	26.0	15.9	16.03[f]	NA	1.03	62.6	0.64	NA	2.8[g]	11.2	41	0.36	30	0.26	45
1400	0.0	26.7	27.1	16.5	16.56[f]	NA	1.01	61.7	0.62	NA	2.9[g]	11.0	42	0.35	31	0.26	47

Bulls, maintenance and slow rate of growth (regain body condition)

For growth and development use requirements for bulls in Tables 1, 2, and 3.

<1300																		
1300	1.0	25.4	23.3	14.2	9.22	2.20	0.92	55.8	0.53	0.28	1.9	7.6	25	0.22	22	0.19	45
1300	1.5	26.1	25.5	15.6	9.22	3.43	0.98	59.7	0.59	0.33	2.0	7.9	28	0.24	23	0.19	46
1300	2.0	26.2	27.6	16.8	9.22	4.71	1.05	64.0	0.65	0.39	2.2	8.2	31	0.26	24	0.20	46
1400	1.0	26.8	24.6	15.0	9.75	2.33	0.92	55.8	0.53	0.28	2.0	7.5	26	0.21	23	0.19	48
1400	1.5	27.6	27.0	16.5	9.75	3.63	0.98	59.7	0.59	0.33	2.1	7.7	29	0.23	24	0.19	49
1400	2.0	27.7	29.1	17.8	9.75	4.98	1.05	64.0	0.65	0.39	2.2	8.0	31	0.25	25	0.20	49
1500	0.0	25.2	20.0	12.2	10.26	NA	0.79	48.4	0.41	NA	1.7	6.9	23	0.20	23	0.20	45
1500	1.0	28.3	25.9	15.8	10.26	2.45	0.92	55.8	0.53	0.28	2.1	7.4	27	0.21	24	0.19	50
1500	1.5	29.0	28.4	17.3	10.26	3.82	0.98	59.7	0.59	0.33	2.2	7.6	29	0.22	25	0.19	51
1600	0.0	26.5	21.0	12.8	10.77	NA	0.79	48.4	0.41	NA	1.8	6.9	23	0.19	24	0.20	47
1600	1.0	29.7	27.2	16.6	10.77	2.57	0.92	55.8	0.53	0.28	2.2	7.3	29	0.22	26	0.19	53
1600	1.5	30.4	29.8	18.2	10.77	4.01	0.98	59.7	0.59	0.33	2.3	7.4	31	0.22	27	0.20	54
1700	0.0	27.7	22.0	13.4	11.28	NA	0.79	48.4	0.41	NA	1.9	6.8	26	0.21	26	0.21	49
1700	0.5	29.6	25.3	15.4	11.28	1.26	0.85	52.0	0.47	0.22	2.1	7.0	27	0.20	26	0.19	52
1800	0.0	28.9	23.0	14.0	11.77	NA	0.79	48.4	0.41	NA	2.0	6.8	27	0.21	27	0.21	51
1800	0.5	30.9	26.4	16.1	11.77	1.31	0.85	52.0	0.47	0.22	2.2	7.0	28	0.20	28	0.20	53
1900	0.0	30.1	23.9	14.6	12.26	NA	0.79	48.4	0.41	NA	2.0	6.8	28	0.21	29	0.21	53
1900	0.5	32.2	27.5	16.8	12.26	1.37	0.85	52.0	0.47	0.22	2.2	6.9	29	0.20	29	0.20	57
2000	0.0	31.3	24.9	15.2	12.74	NA	0.79	48.4	0.41	NA	2.1	6.8	30	0.21	30	0.21	55
2100	0.0	32.5	25.8	15.7	13.21	NA	0.79	48.4	0.41	NA	2.2	6.8	32	0.22	32	0.22	58
2200	0.0	33.6	26.7	16.3	13.68	NA	0.79	48.4	0.41	NA	2.3	6.8	33	0.22	33	0.22	60

[a] Average weight for a feeding period.

[b] Approximately 0.9 ±0.2 lb of weight gain/day over the last third of pregnancy is accounted for by the products of conception. Daily 2.15 Mcal of NE_m and 0.1 lb of protein are provided for this requirement for a calf with a birth weight of 80 lb.

[c] Dry matter consumption should vary depending on the energy concentration of the diet and environmental conditions. These intakes are based on the energy concentration shown in the table and assuming a thermoneutral environment without snow or mud conditions. If the energy concentrations of the diet to be fed exceed the tabular value, limit feeding may be required.

[d] Vitamin A requirements per pound of diet are 1273 IU for pregnant heifers and cows and 1773 for lactating cows and breeding bulls.

[e] Not applicable.

[f] Includes 0.34 Mcal NE_m/lb of milk produced.

[g] Includes 0.03 lb protein/lb of milk produced.

Source: Nutrient Requirements of Beef Cattle, sixth revised edition, National Academy of Sciences, National Academy Press, Washington, DC.

TABLE 7 Daily Nutrient Requirements of Sheep.

Body Weight (kg)	(lb)	Weight Change/Day (g)	(lb)	Dry Matter per Animal[a] (kg)	(lb)	(% body weight)	Energy[b] TDN (kg)	TDN (lb)	DE (Mcal)	ME (Mcal)	Nutrients per Animal — Crude protein (g)	(lb)	Ca (g)	P (g)	Vitamin A Activity (IU)	Vitamin E Activity (IU)
Ewes[c]																
Maintenance																
50	110	10	0.02	1.0	2.2	2.0	0.55	1.2	2.4	2.0	95	0.21	2.0	1.8	2,350	15
60	132	10	0.02	1.1	2.4	1.8	0.61	1.3	2.7	2.2	104	0.23	2.3	2.1	2,820	16
70	154	10	0.02	1.2	2.6	1.7	0.66	1.5	2.9	2.4	113	0.25	2.5	2.4	3,290	18
80	176	10	0.02	1.3	2.9	1.6	0.72	1.6	3.2	2.6	122	0.27	2.7	2.8	3,760	20
90	198	10	0.02	1.4	3.1	1.5	0.78	1.7	3.4	2.8	131	0.29	2.9	3.1	4,230	21
Flushing—2 weeks prebreeding and first 3 weeks of breeding																
50	110	100	0.22	1.6	3.5	3.2	0.94	2.1	4.1	3.4	150	0.33	5.3	2.6	2,350	24
60	132	100	0.22	1.7	3.7	2.8	1.00	2.2	4.4	3.6	157	0.34	5.5	2.9	2,820	26
70	154	100	0.22	1.8	4.0	2.6	1.06	2.3	4.7	3.8	164	0.36	5.7	3.2	3,290	27
80	176	100	0.22	1.9	4.2	2.4	1.12	2.5	4.9	4.0	171	0.38	5.9	3.6	3,760	28
90	198	100	0.22	2.0	4.4	2.2	1.18	2.6	5.1	4.2	177	0.39	6.1	3.9	4,230	30
Nonlactating—First 15 weeks gestation																
50	110	30	0.07	1.2	2.6	2.4	0.67	1.5	3.0	2.4	112	0.25	2.9	2.1	2,350	18
60	132	30	0.07	1.3	2.9	2.2	0.72	1.6	3.2	2.6	121	0.27	3.2	2.5	2,820	20
70	154	30	0.07	1.4	3.1	2.0	0.77	1.7	3.4	2.8	130	0.29	3.5	2.9	3,290	21
80	176	30	0.07	1.5	3.3	1.9	0.82	1.8	3.6	3.0	139	0.31	3.8	3.3	3,760	22
90	198	30	0.07	1.6	3.5	1.8	0.87	1.9	3.8	3.2	148	0.33	4.1	3.6	4,230	24
Last 4 weeks gestation (130–150% lambing rate expected) or last 4–6 weeks lactation suckling singles[d]																
50	110	180 (45)	0.40 (0.10)	1.6	3.5	3.2	0.94	2.1	4.1	3.4	175	0.38	5.9	4.8	4,250	24
60	132	180 (45)	0.40 (0.10)	1.7	3.7	2.8	1.00	2.2	4.4	3.6	184	0.40	6.0	5.2	5,100	26
70	154	180 (45)	0.40 (0.10)	1.8	4.0	2.6	1.06	2.3	4.7	3.8	193	0.42	6.2	5.6	5,950	27
80	176	180 (45)	0.40 (0.10)	1.9	4.2	2.4	1.12	2.4	4.9	4.0	202	0.44	6.3	6.1	6,800	28
90	198	180 (45)	0.40 (0.10)	2.0	4.4	2.2	1.18	2.5	5.1	4.2	212	0.47	6.4	6.5	7,650	30
Last 4 weeks gestation (180–225% lambing rate expected)																
50	110	225	0.50	1.7	3.7	3.4	1.10	2.4	4.8	4.0	196	0.43	6.2	3.4	4,250	26
60	132	225	0.50	1.8	4.0	3.0	1.17	2.6	5.1	4.2	205	0.45	6.9	4.0	5,100	27
70	154	225	0.50	1.9	4.2	2.7	1.24	2.8	5.4	4.4	214	0.47	7.6	4.5	5,950	28
80	176	225	0.50	2.0	4.4	2.5	1.30	2.9	5.7	4.7	223	0.49	8.3	5.1	6,800	30
90	198	225	0.50	2.1	4.6	2.3	1.37	3.0	6.0	5.0	232	0.51	8.9	5.7	7,650	32
First 6–8 weeks lactation suckling singles or last 4–6 weeks lactation suckling twins[d]																
50	110	−25 (90)	−0.06 (0.20)	2.1	4.6	4.2	1.36	3.0	6.0	4.9	304	0.67	8.9	6.1	4,250	32
60	132	−25 (90)	−0.06 (0.20)	2.3	5.1	3.8	1.50	3.3	6.6	5.4	319	0.70	9.1	6.6	5,100	34
70	154	−25 (90)	−0.06 (0.20)	2.5	5.5	3.6	1.63	3.6	7.2	5.9	334	0.73	9.3	7.0	5,950	38
80	176	−25 (90)	−0.06 (0.20)	2.6	5.7	3.2	1.69	3.7	7.4	6.1	344	0.76	9.5	7.4	6,800	39
90	198	−25 (90)	−0.06 (0.20)	2.7	5.9	3.0	1.75	3.8	7.6	6.3	353	0.78	9.6	7.8	7,650	40

First 6–8 weeks lactation suckling twins

50	110	−60	2.4	5.3	4.8	1.56	3.4	6.9	5.6	389	0.86	10.5	7.3	5,000	36
60	132	−60	2.6	5.7	4.3	1.69	3.7	7.4	6.1	405	0.89	10.7	7.7	6,000	39
70	154	−60	2.8	6.2	4.0	1.82	4.0	8.0	6.6	420	0.92	11.0	8.1	7,000	42
80	176	−60	3.0	6.6	3.8	1.95	4.3	8.6	7.0	435	0.96	11.2	8.6	8,000	45
90	198	−60	3.2	7.0	3.6	2.08	4.6	9.2	7.5	450	0.99	11.4	9.0	9,000	48

Ewe lambs

Nonlactating—First 15 weeks gestation

40	88	160	1.4	3.1	3.5	0.83	1.8	3.6	3.0	156	0.34	5.5	3.0	1,880	21
50	110	135	1.5	3.3	3.0	0.88	1.9	3.9	3.2	159	0.35	5.2	3.1	2,350	22
60	132	135	1.6	3.5	2.7	0.94	2.0	4.1	3.4	161	0.35	5.5	3.4	2,820	24
70	154	125	1.7	3.7	2.4	1.00	2.2	4.4	3.6	164	0.36	5.5	3.7	3,290	26

Last 4 weeks gestation (100–120% lambing rate expected)

40	88	180	1.5	3.3	3.8	0.94	2.1	4.1	3.4	187	0.41	6.4	3.1	3,400	22
50	110	160	1.6	3.5	3.2	1.00	2.2	4.4	3.6	189	0.42	6.3	3.4	4,250	24
60	132	160	1.7	3.7	2.8	1.07	2.4	4.7	3.9	192	0.42	6.6	3.8	5,100	26
70	154	150	1.8	4.0	2.6	1.14	2.5	5.0	4.1	194	0.43	6.8	4.2	5,950	27

Last 4 weeks gestation (130–175% lambing rate expected)

40	88	225	1.5	3.3	3.8	0.99	2.2	4.4	3.6	202	0.44	7.4	3.5	3,400	22
50	110	225	1.6	3.5	3.2	1.06	2.3	4.7	3.8	204	0.45	7.8	3.9	4,250	24
60	132	225	1.7	3.7	2.8	1.12	2.5	4.9	4.0	207	0.46	8.1	4.3	5,100	26
70	154	215	1.8	4.0	2.6	1.14	2.5	5.0	4.1	210	0.46	8.2	4.7	5,950	27

First 6–8 weeks lactation suckling singles (wean by 8 weeks)

40	88	−50	1.7	3.7	4.2	1.12	2.5	4.9	4.0	257	0.56	6.0	4.3	3,400	26
50	110	−50	2.1	4.6	4.2	1.39	3.1	6.1	5.0	282	0.62	6.5	4.7	4,250	32
60	132	−50	2.3	5.1	3.8	1.52	3.4	6.7	5.5	295	0.65	6.8	5.1	5,100	34
70	154	−50	2.5	5.5	3.6	1.65	3.6	7.3	6.0	301	0.68	7.1	5.6	5,450	38

First 6–8 weeks lactation suckling twins (wean by 8 weeks)

40	88	−100	2.1	4.6	5.2	1.45	3.2	6.4	5.2	306	0.67	8.4	5.6	4,000	32
50	110	−100	2.3	5.1	4.6	1.59	3.5	7.0	5.7	321	0.71	8.7	6.0	5,000	34
60	132	−100	2.5	5.5	4.2	1.72	3.8	7.6	6.2	336	0.74	9.0	6.4	6,000	38
70	154	−100	2.7	6.0	3.9	1.85	4.1	8.1	6.6	351	0.77	9.3	6.9	7,000	40

Replacement ewe lambs[e]

30	66	227	1.2	2.6	4.0	0.78	1.7	3.4	2.8	185	0.41	6.4	2.6	1,410	18
40	88	182	1.4	3.1	3.5	0.91	2.0	4.0	3.3	176	0.39	5.9	2.6	1,880	21
50	110	120	1.5	3.3	3.0	0.88	1.9	3.9	3.2	136	0.30	4.8	2.4	2,350	22
60	132	100	1.5	3.3	2.5	0.88	1.9	3.9	3.2	134	0.30	4.5	2.5	2,820	22
70	154	100	1.5	3.3	2.1	0.88	1.9	3.9	3.2	132	0.29	4.6	2.8	3,290	22

Replacement ram lambs[e]

40	88	330	1.8	4.0	4.5	1.1	2.5	5.0	4.1	243	0.54	7.8	3.7	1,880	24
60	132	320	2.4	5.3	4.0	1.5	3.4	6.7	5.5	263	0.58	8.4	4.2	2,820	26
80	176	290	2.8	6.2	3.5	1.8	3.9	7.8	6.4	268	0.59	8.5	4.6	3,760	28
100	220	250	3.0	6.6	3.0	1.9	4.2	8.4	6.9	264	0.58	8.2	4.8	4,700	30

(Table 7 continues)

TABLE 7 Daily Nutrient Requirements of Sheep. (Cont.)

Body Weight		Weight Change/Day		Dry Matter per Animal[a]			Energy[b]				Crude protein		Ca (g)	P (g)	Vitamin A Activity (IU)	Vitamin E Activity (IU)
(kg)	(lb)	(g)	(lb)	(kg)	(lb)	(% body weight)	TDN (kg)	TDN (lb)	DE (Mcal)	ME (Mcal)	(g)	(lb)				
Lambs finishing—4 to 7 months old[f]																
30	66	295	0.65	1.3	2.9	4.3	0.94	2.1	4.1	3.4	191	0.42	6.6	3.2	1,410	20
40	88	275	0.60	1.6	3.5	4.0	1.22	2.7	5.4	4.4	185	0.41	6.6	3.3	1,880	24
50	110	205	0.45	1.6	3.5	3.2	1.23	2.7	5.4	4.4	160	0.35	5.6	3.0	2,350	24
Early weaned lambs—Moderate growth potential[f]																
10	22	200	0.44	0.5	1.1	5.0	0.40	0.9	1.8	1.4	127	0.38	4.0	1.9	470	10
20	44	250	0.55	1.0	2.2	5.0	0.80	1.8	3.5	2.9	167	0.37	5.4	2.5	940	20
30	66	300	0.66	1.3	2.9	4.3	1.00	2.2	4.4	3.6	191	0.42	6.7	3.2	1,410	20
40	88	345	0.76	1.5	3.3	3.8	1.16	2.6	5.1	4.2	202	0.44	7.7	3.9	1,880	22
50	110	300	0.66	1.5	3.3	3.0	1.16	2.6	5.1	4.2	181	0.40	7.0	3.8	2,350	22
Early weaned lambs—Rapid growth potential[f]																
10	22	250	0.55	0.6	1.3	6.0	0.48	1.1	2.1	1.7	157	0.35	4.9	2.2	470	12
20	44	300	0.66	1.2	2.6	6.0	0.92	2.0	4.0	3.3	205	0.45	6.5	2.9	940	24
30	66	325	0.72	1.4	3.1	4.7	1.10	2.4	4.8	4.0	216	0.48	7.2	3.4	1,410	21
40	88	400	0.88	1.5	3.3	3.8	1.14	2.5	5.0	4.1	234	0.51	8.6	4.3	1,880	22
50	110	425	0.94	1.7	3.7	3.4	1.29	2.8	5.7	4.7	240	0.53	9.4	4.8	2,350	25
60	132	350	0.77	1.7	3.7	2.8	1.29	2.8	5.7	4.7	240	0.53	8.2	4.5	2,820	25

[a]To convert dry matter to an as-fed basis, divide dry matter values by the percentage of dry matter in the particular feed.

[b]One kilogram TDN (total digestible nutrients) = 4.4 Mcal DE (digestible energy); ME (metabolizable energy) = 82% of DE. Because of rounding errors, values in Table 7 and Table 8 may differ.

[c]Values are applicable for ewes in moderate condition. Fat ewes should be fed according to the next lower weight category and thin ewes at the next higher weight category. Once desired or moderate weight condition is attained, use that weight category through all production stages.

[d]Values in parentheses are for ewes suckling lambs the last 4–6 weeks of lactation.

[e]Lambs intended for breeding; thus, maximum weight gains and finish are of secondary importance.

[f]Maximum weight gains expected.

Source: Nutrient Requirements of Sheep, sixth revised edition, National Academy of Sciences, National Academy Press, Washington, DC.

TABLE 8 Nutrient Concentration in Diets for Sheep (expressed on 100 Percent Dry Matter Basis[a]).

Body Weight (kg)	(lb)	Weight Change/Day (g)	(lb)	Energy[b] TDN[c] (%)	DE (Mcal/kg)	ME (Mcal/kg)	Example Diet Proportions Concentrate %	Forage %	Crude Protein (%)	Calcium (%)	Phosphorus (%)	Vitamin A Activity (IU/kg)	Vitamin E Activity (IU/kg)
Ewes[d]													
Maintenance													
70	154	10	0.02	55	2.4	2.0	0	100	9.4	0.20	0.20	2,742	15
Flushing—2 weeks prebreeding and first 3 weeks of breeding													
70	154	100	0.22	59	2.6	2.1	15	85	9.1	0.32	0.18	1,828	15
Nonlactating—First 15 weeks gestation													
70	154	30	0.07	55	2.4	2.0	0	100	9.3	0.25	0.20	2,350	15
Last 4 weeks gestation (130–150% lambing rate expected) or last 4–6 weeks lactation suckling singles[e]													
70	154	180 (0.45)	0.40 (0.10)	59	2.6	2.1	15	85	10.7	0.35	0.23	3,306	15
Last 4 weeks gestation (180–225% lambing rate expected)													
70	154	225	0.50	65	2.9	2.3	35	65	11.3	0.40	0.24	3,132	15
First 6–8 weeks lactation suckling singles or last 4–6 weeks lactation suckling twins[e]													
70	154	−25 (90)	−0.06 (0.20)	65	2.9	2.4	35	65	13.4	0.32	0.26	2,380	15
First 6–8 weeks lactation suckling twins													
70	154	−60	−0.13	65	2.9	2.4	35	65	15.0	0.39	0.29	2,500	15
Ewe Lambs													
Nonlactating—First 15 weeks gestation													
55	121	135	0.30	59	2.6	2.1	15	85	10.6	0.35	0.22	1,668	15
Last 4 weeks gestation (100–120% lambing rate expected)													
55	121	160	0.35	63	2.8	2.3	30	70	11.8	0.39	0.22	2,833	15
Last 4 weeks gestation (130–175% lambing rate expected)													
55	121	225	0.50	66	2.9	2.4	40	60	12.8	0.48	0.25	2,833	15
First 6–8 weeks lactation suckling singles (wean by 8 weeks)													
55	121	−50	0.22	66	2.9	2.4	40	60	13.1	0.30	0.22	2,125	15
First 6–8 weeks lactation suckling twins (wean by 8 weeks)													
55	121	−100	−0.22	69	3.0	2.5	50	50	13.7	0.37	0.26	2,292	15
Replacement Ewe Lambs[f]													
30	66	227	0.50	65	2.9	2.4	35	65	12.8	0.53	0.22	1,175	15
40	88	182	0.40	65	2.9	2.4	35	65	10.2	0.42	0.18	1,343	15
50–70	110–154	115	0.25	59	2.6	2.1	15	85	9.1	0.31	0.17	1,567	15

(Table 8 continues)

TABLE 8 Nutrient Concentration in Diets for Sheep (expressed on 100 Percent Dry Matter Basis^a). (Cont.)

Body Weight		Weight Change/Day		Energy[b]			Example Diet Proportions		Crude Protein (%)	Calcium (%)	Phosphorus (%)	Vitamin A Activity (IU/kg)	Vitamin E Activity (IU/kg)
(kg)	(lb)	(g)	(lb)	TDN[c] (%)	DE (Mcal/kg)	ME (Mcal/kg)	Concentrate %	Forage %					
Replacement Ram Lambs[f]													
40	88	330	0.73	63	2.8	2.3	30	70	13.5	0.43	0.21	1,175	15
60	132	320	0.70	63	2.8	2.3	30	70	11.0	0.35	0.18	1,659	15
80–100	176–220	270	0.60	63	2.8	2.3	30	70	9.6	0.30	0.16	1,979	15
Lambs Finishing—4 to 7 months old[g]													
30	66	295	0.65	72	3.2	2.5	60	40	14.7	0.51	0.24	1,085	15
40	88	275	0.60	76	3.3	2.7	75	25	11.6	0.42	0.21	1,175	15
50	110	205	0.45	77	3.4	2.8	80	20	10.0	0.35	0.19	1,469	15
Early Weaned Lambs—Moderate and rapid growth potential[g]													
10	22	250	0.55	80	3.5	2.9	90	10	26.2	0.82	0.38	940	20
20	44	300	0.66	78	3.4	2.8	85	15	16.9	0.54	0.24	940	20
30	66	325	0.72	78	3.3	2.7	85	15	15.1	0.51	0.24	1,085	15
40–60	88–132	400	0.88	78	3.3	2.7	85	15	14.5	0.55	0.28	1,253	15

[a]Values in Table 8 are calculated from daily requirements in Table 7 divided by DM intake. The exception, vitamin E daily requirements/head, are calculated from vitamin E/kg diet × DM intake.

[b]One kilogram TDN = 4.4 Mcal DE (digestible energy); ME (metabolizable energy) = 82% of DE. Because of rounding errors, values in Table 7 and Table 8 may differ.

[c]TDN calculated on following basis: hay DM, 55% TDN and on as-fed basis 50% TDN; grain DM, 83% TDN and on as-fed basis 75% TDN.

[d]Values are for ewes in moderate condition. Fat ewes should be fed according to the next lower weight category and thin ewes at the next higher weight category. Once desired or moderate weight condition is attained, use that weight category through all production stages.

[e]Values in parentheses are for ewes suckling lambs the last 4–6 weeks of lactation.

[f]Lambs intended for breeding; thus, maximum weight gains and finish are of secondary importance.

[g]Maximum weight gains expected.

Source: Nutrient Requirements of Sheep, sixth revised edition, National Academy of Sciences, National Academy Press, Washington, DC.

TABLE 9 Net Energy Requirements for Lambs of Small, Medium, and Large Mature Weight Genotypes[a] (kcal/d).

Body Weight (kg)[b]:	10	20	25	30	35	40	45	50
NE$_m$ Requirements[c]:	315	530	626	718	806	891	973	1053
Daily Gain (g)[b]								
NE$_g$ Requirements								
Small mature weight lambs[d]								
100	178	300	354	406	456	504	551	596
150	267	450	532	610	684	756	826	894
200	357	600	708	812	912	1,008	1,102	1,192
250	446	750	886	1,016	1,140	1,261	1,377	1,490
300	535	900	1,064	1,219	1,368	1,513	1,652	1,788
Medium mature weight lambs[e]								
100	155	261	309	354	397	439	480	519
150	233	392	463	531	596	658	719	778
200	310	522	618	708	794	878	960	1,038
250	388	653	771	884	993	1,097	1,199	1,297
300	466	784	926	1,062	1,191	1,316	1,438	1,557
350	543	914	1,080	1,238	1,390	1,536	1,678	1,816
400	621	1,044	1,234	1,415	1,589	1,756	1,918	2,076
Large mature weight lambs[f]								
100	132	221	262	300	337	372	407	439
150	197	332	392	450	505	558	610	660
200	263	442	524	600	674	744	813	880
250	329	553	654	750	842	930	1,016	1,099
300	394	663	785	900	1,010	1,116	1,220	1,320
350	461	775	916	1,050	1,179	1,303	1,423	1,540
400	526	885	1,046	1,200	1,347	1,489	1,626	1,760
450	592	996	1,177	1,350	1,515	1,675	1,830	1,980

[a]Approximate mature ram weights of 95 kg, 115 kg, and 135 kg, respectively.
[b]Weights and gains include fill.
[c]NE$_m$ = 56 kcal · W$^{0.75}$ · d^{-1}.
[d]NE$_g$ = 317 kcal · W$^{0.75}$ · LWG, kg · d^{-1}.
[e]NE$_g$ = 276 kcal · W$^{0.75}$ · LWG, kg · d^{-1}.
[f]NE$_g$ = 234 kcal · W$^{0.75}$ · LWG, kg · d^{-1}.

Source: Nutrient Requirements of Sheep, sixth revised edition, National Academy of Sciences, National Academy Press, Washington, DC.

TABLE 10 NE$_{preg}$ (NE$_y$) Requirements of Ewes Carrying Different Numbers of Fetuses at Various Stages of Gestation.

Number of Fetuses Being Carried	Stage of Gestation (days)[a]					
	100	%[b]	120	%[b]	140	%[b]
	NE$_{preg}$ Required (kcal/day)					
1	70	100	145	100	260	100
2	125	178	265	183	440	169
3	170	243	345	238	570	219

[a]For gravid uterus (plus contents) and mammary gland development only.
[b]As a percentage of a single fetus's requirement.

Source: Nutrient Requirements of Sheep, sixth revised edition, National Academy of Sciences, National Academy Press, Washington, DC.

TABLE 11 Crude Protein Requirements for Lambs of Small, Medium, and Large Mature Weight Genotypes[a] (g/d).

Body Weight (kg)[b]:	10	20	25	30	35	40	45	50
Daily Gain (g)[b]								
Small mature weight lambs								
100	84	112	122	127	131	136	135	134
150	103	121	137	140	144	147	145	143
200	123	145	152	154	156	158	154	151
250	142	162	167	168	168	169	164	159
300	162	178	182	181	180	180	174	168
Medium mature weight lambs								
100	85	114	125	130	135	140	139	139
150	106	132	141	145	149	153	151	149
200	127	150	158	160	163	166	163	160
250	147	167	174	175	177	179	175	171
300	168	185	191	191	191	191	186	181
350	188	203	207	206	205	204	198	192
400	209	221	224	221	219	217	210	202
Large mature weight lambs								
100	94	128	134	139	145	144	150	156
150	115	147	152	156	160	159	164	169
200	136	166	170	173	176	174	178	182
250	157	186	188	190	192	189	192	195
300	179	205	206	207	208	204	206	208
350	200	224	224	224	224	219	220	221
400	221	243	242	241	240	234	234	234
450	242	262	260	256	256	249	248	248

[a]Approximate mature ram weights of 95 kg, 115 kg, and 135 kg, respectively.
[b]Weights and gains include fill.

Source: Nutrient Requirements of Sheep, sixth revised edition, National Academy of Sciences, National Academy Press, Washington, DC.

TABLE 12 Daily Nutrient Requirements of Goats.

Body Weight (kg)	TDN (g)	DE (Mcal)	ME (Mcal)	NE (Mcal)	TP (g)	DP (g)	Ca (g)	P (g)	Vitamin A (1000 IU)	Vitamin D IU	1 kg = 2.0 Mcal ME Total (kg)	1 kg = 2.0 Mcal ME % of kg BW	1 kg = 2.4 Mcal ME Total (kg)	1 kg = 2.4 Mcal ME % of kg BW
		Feed Energy			**Crude Protein**						**Dry Matter per Animal**			
Maintenance only (includes stable feeding conditions, minimal activity, and early pregnancy)														
10	159	0.70	0.57	0.32	22	15	1	0.7	0.4	84	0.28	2.8	0.24	2.4
20	267	1.18	0.96	0.54	38	26	1	0.7	0.7	144	0.48	2.4	0.40	2.0
30	362	1.59	1.30	0.73	51	35	2	1.4	0.9	195	0.65	2.2	0.54	1.8
40	448	1.98	1.61	0.91	63	43	2	1.4	1.2	243	0.81	2.0	0.67	1.7
50	530	2.34	1.91	1.08	75	51	3	2.1	1.4	285	0.95	1.9	0.79	1.6
60	608	2.68	2.19	1.23	86	59	3	2.1	1.6	327	1.09	1.8	0.91	1.5
70	682	3.01	2.45	1.38	96	66	4	2.8	1.8	369	1.23	1.8	1.02	1.5
80	754	3.32	2.71	1.53	106	73	4	2.8	2.0	408	1.36	1.7	1.13	1.4
90	824	3.63	2.96	1.67	116	80	4	2.8	2.2	444	1.48	1.6	1.23	1.4
100	891	3.93	3.21	1.81	126	86	5	3.5	2.4	480	1.60	1.6	1.34	1.3
Maintenance plus low activity (= 25% increment, intensive management, tropical range and early pregnancy)														
10	199	0.87	0.71	0.40	27	19	1	0.7	0.5	108	0.36	3.6	0.30	3.0
20	334	1.47	1.20	0.68	46	32	2	1.4	0.9	180	0.60	3.0	0.50	2.5
30	452	1.99	1.62	0.92	62	43	2	1.4	1.2	243	0.81	2.7	0.67	2.2
40	560	2.47	2.02	1.14	77	54	3	2.1	1.5	303	1.01	2.5	0.84	2.1
50	662	2.92	2.38	1.34	91	63	4	2.8	1.8	357	1.19	2.4	0.99	2.0
60	760	3.35	2.73	1.54	105	73	4	2.8	2.0	408	1.36	2.3	1.14	1.9
70	852	3.76	3.07	1.73	118	82	5	3.5	2.3	462	1.54	2.2	1.28	1.8
80	942	4.16	3.39	1.91	130	90	5	3.5	2.6	510	1.70	2.1	1.41	1.8
90	1030	4.54	3.70	2.09	142	99	6	4.2	2.8	555	1.85	2.1	1.54	1.7
100	1114	4.91	4.01	2.26	153	107	6	4.2	3.0	600	2.00	2.0	1.67	1.7
Maintenance plus medium activity (= 50% increment, semiarid rangeland, slightly hilly pastures, and early pregnancy)														
10	239	1.05	0.86	0.48	33	23	1	0.7	0.6	129	0.43	4.3	0.36	3.6
20	400	1.77	1.44	0.81	55	38	2	1.4	1.1	216	0.72	3.6	0.60	3.0
30	543	2.38	1.95	1.10	74	52	3	2.1	1.5	294	0.98	3.3	0.81	2.7
40	672	2.97	2.42	1.36	93	64	4	2.8	1.8	363	1.21	3.0	1.01	2.5
50	795	3.51	2.86	1.62	110	76	4	2.8	2.1	429	1.43	2.9	1.19	2.4
60	912	4.02	3.28	1.84	126	87	5	3.5	2.5	492	1.64	2.7	1.37	2.3
70	1023	4.52	3.68	2.07	141	98	6	4.2	2.8	552	1.84	2.6	1.53	2.2
80	1131	4.98	4.06	2.30	156	108	6	4.2	3.0	609	2.03	2.5	1.69	2.1
90	1236	5.44	4.44	2.50	170	118	7	4.9	3.3	666	2.22	2.5	1.85	2.0
100	1336	5.90	4.82	2.72	184	128	7	4.9	3.6	723	2.41	2.4	2.01	2.0
Maintenance plus high activity (= 75% increment, arid rangeland, sparse vegetation, mountainous pastures, and early pregnancy)														
10	278	1.22	1.00	0.56	38	26	2	1.4	0.8	150	0.50	5.0	0.42	4.2
20	467	2.06	1.68	0.94	64	45	2	1.4	1.3	252	0.84	4.2	0.70	3.5
30	634	2.78	2.28	1.28	87	60	3	2.1	1.7	342	1.14	3.8	0.95	3.2
40	784	3.46	2.82	1.59	108	75	4	2.8	2.1	423	1.41	3.5	1.18	3.0
50	928	4.10	3.34	1.89	128	89	5	3.5	2.5	501	1.67	3.3	1.39	2.7
60	1064	4.69	3.83	2.15	146	102	6	4.2	2.9	576	1.92	3.2	1.60	2.7
70	1194	5.27	4.29	2.42	165	114	6	4.2	3.2	642	2.14	3.0	1.79	2.6
80	1320	5.81	4.74	2.68	182	126	7	4.9	3.6	711	2.37	3.0	1.98	2.5
90	1442	6.35	5.18	2.92	198	138	8	5.6	3.9	777	2.59	2.9	2.16	2.4
100	1559	6.88	5.62	3.17	215	150	8	5.6	4.2	843	2.81	2.8	2.34	2.3
Additional requirements for late pregnancy (for all goat sizes)														
	397	1.74	1.42	0.80	82	57	2	1.4	1.1	213	0.71		0.59	
Additional requirements for growth—weight gain at 50 g per day (for all goat sizes)														
	100	0.44	0.36	0.20	14	10	1	0.7	0.3	54	0.18		0.15	
Additional requirements for growth—weight gain at 100 g per day (for all goat sizes)														
	200	0.88	0.72	0.40	28	20	1	0.7	0.5	108	0.36		0.30	
Additional requirements for growth—weight gain at 150 g per day (for all goat sizes)														
	300	1.32	1.08	0.60	42	30	2	1.4	0.8	162	0.54		0.45	

(Table 12 continues)

TABLE 12 Daily Nutrient Requirements of Goats. *(Cont.)*

	Feed Energy			Crude Protein					
TDN (g)	DE (Mcal)	ME (Mcal)	NE (Mcal)	TP (g)	DP (g)	Ca (g)	P (g)	Vitamin A (1000 IU)	Vitamin D IU

Additional requirements for milk production per kg at different fat percentages
(including requirements for nursing single, twin or triplet kids at the respective milk production level)

(% fat)

2.5	333	1.47	1.20	0.68	59	42	2	1.4	3.8	760
3.0	337	1.49	1.21	0.68	64	45	2	1.4	3.8	760
3.5	342	1.51	1.23	0.69	68	48	2	1.4	3.8	760
4.0	346	1.53	1.25	0.70	72	51	3	2.1	3.8	760
4.5	351	1.55	1.26	0.71	77	54	3	2.1	3.8	760
5.0	356	1.57	1.28	0.72	82	57	3	2.1	3.8	760
5.5	360	1.59	1.29	0.73	86	60	3	2.1	3.8	760
6.0	365	1.61	1.31	0.74	90	63	3	2.1	3.8	760

Additional requirements for mohair production by Angora at different production levels

Annual Fleece Yield (kg)

2	16	0.07	0.06	0.03	9	6
4	34	0.15	0.12	0.07	17	12
6	50	0.22	0.18	0.10	26	18
8	66	0.29	0.24	0.14	34	24

Source: Nutrient Requirements of Goats: Angora, Dairy, and Meat Goats in Temperate and Tropical Countries, National Academy of Science, National Academy Press, Washington, DC.

TABLE 13 Nutrient Requirements of Swine Allowed Feed Ad Libitum (90 percent dry matter).

Intake and Performance Levels	Swine Liveweight (kg)				
	1–5	5–10	10–20	20–50	50–110
Expected weight gain (g/day)	200	250	450	700	820
Expected feed intake (g/day)	250	460	950	1,900	3,110
Expected efficiency (gain/feed)	0.800	0.543	0.474	0.368	0.264
Expected efficiency (feed/gain)	1.25	1.84	2.11	2.71	3.79
Digestible energy intake (kcal/day)	850	1,560	3,230	6,460	10,570
Metabolizable energy intake (kcal/day)	805	1,490	3,090	6,200	10,185
Energy concentration (kcal ME/kg diet)	3,220	3,240	3,250	3,260	3,275
Protein (%)	24	20	18	15	13
	Requirement (% or amount/kg diet)[a]				
Nutrient					
Indispensable amino acids (%)					
Arginine	0.60	0.50	0.40	0.25	0.10
Histidine	0.36	0.31	0.25	0.22	0.18
Isoleucine	0.76	0.65	0.53	0.46	0.38
Leucine	1.00	0.85	0.70	0.60	0.50
Lysine	1.40	1.15	0.95	0.75	0.60
Methionine + cystine	0.68	0.58	0.48	0.41	0.34
Phenylalanine + tyrosine	1.10	0.94	0.77	0.66	0.55
Threonine	0.80	0.68	0.56	0.48	0.40
Tryptophan	0.20	0.17	0.14	0.12	0.10
Valine	0.80	0.68	0.56	0.48	0.40
Linoleic acid (%)	0.1	0.1	0.1	0.1	0.1
Mineral elements					
Calcium (%)	0.90	0.80	0.70	0.60	0.50
Phosphorus, total (%)	0.70	0.65	0.60	0.50	0.40
Phosphorus, available (%)	0.55	0.40	0.32	0.23	0.15
Sodium (%)	0.10	0.10	0.10	0.10	0.10
Chlorine (%)	0.08	0.08	0.08	0.08	0.08
Magnesium (%)	0.04	0.04	0.04	0.04	0.04
Potassium (%)	0.30	0.28	0.26	0.23	0.17
Copper (mg)	6.0	6.0	5.0	4.0	3.0
Iodine (mg)	0.14	0.14	0.14	0.14	0.14
Iron (mg)	100	100	80	60	40
Manganese (mg)	4.0	4.0	3.0	2.0	2.0
Selenium (mg)	0.30	0.30	0.25	0.15	0.10
Zinc (mg)	100	100	80	60	50
Vitamins					
Vitamin A (IU)	2,200	2,200	1,750	1,300	1,300
Vitamin D (IU)	220	220	200	150	150
Vitamin E (IU)	16	16	11	11	11
Vitamin K (menadione) (mg)	0.5	0.5	0.5	0.5	0.5
Biotin (mg)	0.08	0.05	0.05	0.05	0.05
Choline (g)	0.6	0.5	0.4	0.3	0.3
Folacin (mg)	0.3	0.3	0.3	0.3	0.3
Niacin, available (mg)	20.0	15.0	12.5	10.0	7.0
Pantothenic acid (mg)	12.0	10.0	9.0	8.0	7.0
Riboflavin (mg)	4.0	3.5	3.0	2.5	2.0
Thiamin (mg)	1.5	1.0	1.0	1.0	1.0
Vitamin B_6 (mg)	2.0	1.5	1.5	1.0	1.0
Vitamin B_{12} (μg)	20.0	17.5	15.0	10.0	5.0

NOTE: The requirements listed are based upon the principles and assumptions described in the text of this publication. Knowledge of nutritional constraints and limitations is important for the proper use of this table.

[a]These requirements are based upon the following types of pigs and diets: 1- to 5-kg pigs, a diet that includes 25 to 75 percent milk products; 5- to 10-kg pigs, a corn-soybean meal diet that includes 5 to 25 percent milk products; 10- to 110-kg pigs, a corn-soybean meal diet. In the corn-soybean meal diets, the corn contains 8.5 percent protein; the soybean meal contains 44 percent.

Source: Nutrient Requirements of Swine, ninth revised edition, © 1988 by the National Academy of Sciences, Washington, DC.

TABLE 14 Daily Nutrient Intakes and Requirements of Swine Allowed Feed Ad Libitum.

Intake and Performance Levels	Swine Liveweight (kg)				
	1–5	5–10	10–20	20–50	50–110
Expected weight gain (g/day)	200	250	450	700	820
Expected feed intake (g/day)	250	460	950	1,900	3,110
Expected efficiency (gain/feed)	0.800	0.543	0.474	0.368	0.264
Expected efficiency (feed/gain)	1.25	1.84	2.11	2.71	3.79
Digestible energy intake (kcal/day)	850	1,560	3,230	6,460	10,570
Metabolizable energy intake (kcal/day)	805	1,490	3,090	6,200	10,185
Energy concentration (kcal ME/kg diet)	3,220	3,240	3,250	3,260	3,275
Protein (g/day)	60	92	171	285	404
	Requirement (amount/day)				
Nutrient					
Indispensable amino acids (g)					
Arginine	1.5	2.3	3.8	4.8	3.1
Histidine	0.9	1.4	2.4	4.2	5.6
Isoleucine	1.9	3.0	5.0	8.7	11.8
Leucine	2.5	3.9	6.6	11.4	15.6
Lysine	3.5	5.3	9.0	14.3	18.7
Methionine + cystine	1.7	2.7	4.6	7.8	10.6
Phenylalanine + tyrosine	2.8	4.3	7.3	12.5	17.1
Threonine	2.0	3.1	5.3	9.1	12.4
Tryptophan	0.5	0.8	1.3	2.3	3.1
Valine	2.0	3.1	5.3	9.1	12.4
Linoleic acid (g)	0.3	0.5	1.0	1.9	3.1
Mineral elements					
Calcium (g)	2.2	3.7	6.6	11.4	15.6
Phosphorus, total (g)	1.8	3.0	5.7	9.5	12.4
Phosphorus, available (g)	1.4	1.8	3.0	4.4	4.7
Sodium (g)	0.2	0.5	1.0	1.9	3.1
Chlorine (g)	0.2	0.4	0.8	1.5	2.5
Magnesium (g)	0.1	0.2	0.4	0.8	1.2
Potassium (g)	0.8	1.3	2.5	4.4	5.3
Copper (mg)	1.50	2.76	4.75	7.60	9.33
Iodine (mg)	0.04	0.06	0.13	0.27	0.44
Iron (mg)	25	46	76	114	124
Manganese (mg)	1.00	1.84	2.85	3.80	6.22
Selenium (mg)	0.08	0.14	0.24	0.28	0.31
Zinc (mg)	25	46	76	114	155
Vitamins					
Vitamin A (IU)	550	1,012	1,662	2,470	4,043
Vitamin D (IU)	55	101	190	285	466
Vitamin E (IU)	4	7	10	21	34
Vitamin K (menadione) (mg)	0.02	0.02	0.05	0.10	0.16
Biotin (mg)	0.02	0.02	0.05	0.10	0.16
Choline (g)	0.15	0.23	0.38	0.57	0.93
Folacin (mg)	0.08	0.14	0.28	0.57	0.93
Niacin, available (mg)	5.00	6.90	11.88	19.00	21.77
Pantothenic acid (mg)	3.00	4.60	8.55	15.20	21.77
Riboflavin (mg)	1.00	1.61	2.85	4.75	6.22
Thiamin (mg)	0.38	0.46	0.95	1.90	3.11
Vitamin B_6 (mg)	0.50	0.69	1.42	1.90	3.11
Vitamin B_{12} (μg)	5.00	8.05	14.25	19.00	15.55

Source: *Nutrient Requirements of Swine,* ninth revised edition, © 1988 by the National Academy of Sciences, Washington, DC.

TABLE 15 Nutrient Requirements of Breeding Swine.

Intake Levels	Bred Gilts, Sows, and Adult Boars	Lactating Gilts and Sows
Digestible energy (kcal/kg diet)	3,340	3,340
Metabolizable energy (kcal/kg diet)	3,210	3,210
Crude protein (%)	12	13

Requirement (% or amount/kg diet)[a]

Nutrient		
Indispensable amino acids (%)		
Arginine	0.00	0.40
Histidine	0.15	0.25
Isoleucine	0.30	0.39
Leucine	0.30	0.48
Lysine	0.43	0.60
Methionine + cystine	0.23	0.36
Phenylalanine + tyrosine	0.45	0.70
Threonine	0.30	0.43
Tryptophan	0.09	0.12
Valine	0.32	0.60
Linoleic acid (%)	0.1	0.1
Mineral elements		
Calcium (%)	0.75	0.75
Phosphorus, total (%)	0.60	0.60
Phosphorus, available (%)	0.35	0.35
Sodium (%)	0.15	0.20
Chlorine (%)	0.12	0.16
Magnesium (%)	0.04	0.04
Potassium (%)	0.20	0.20
Copper (mg)	5.00	5.00
Iodine (mg)	0.14	0.14
Iron (mg)	80.00	80.00
Manganese (mg)	10.00	10.00
Selenium (mg)	0.15	0.15
Zinc (mg)	50.00	50.00
Vitamins		
Vitamin A (IU)	4,000	2,000
Vitamin D (IU)	200	200
Vitamin E (IU)	22	22
Vitamin K (menadione) (mg)	0.50	0.50
Biotin (mg)	0.20	0.20
Choline (g)	1.25	1.00
Folacin (mg)	0.30	0.30
Niacin, available (mg)	10.00	10.00
Pantothenic acid (mg)	12.00	12.00
Riboflavin (mg)	3.75	3.75
Thiamin (mg)	1.00	1.00
Vitamin B_6 (mg)	1.00	1.00
Vitamin B_{12} (μg)	15.00	15.00

NOTE: The requirements listed are based upon the principles and assumptions described in the text of this publication. Knowledge of nutritional constraints and limitations is important for the proper use of this table.

[a]These requirements are based upon corn-soybean meal diets, feed intakes, and performance levels listed in Tables 16, 18, and 19. In the corn-soybean meal diets, the corn contains 8.5 percent protein; the soybean meal contains 44 percent.

Source: Nutrient Requirements of Swine, ninth revised edition, © 1988 by the National Academy of Sciences, Washington, DC.

TABLE 16 Daily Nutrient Intake and Requirements of Intermediate-Weight Breeding Animals.

Intake and Performance Levels	Mean Gestation or Farrowing Weight (kg) of:	
	Bred Gilts, Sows, and Adult Boars	Lactating Gilts and Sows
	162.5	165.0
Daily feed intake (kg)	1.9	5.3
Digestible energy (Mcal/day)	6.3	17.7
Metabolizable energy (Mcal/day)	6.1	17.0
Crude protein (g/day)	228	689
Requirement (amount/day)		
Nutrients		
Indispensable amino acids (g)		
Arginine	0.0	21.2
Histidine	2.8	13.2
Isoleucine	5.7	20.7
Leucine	5.7	25.4
Lysine	8.2	31.8
Methionine + cystine	4.4	19.1
Phenylalanine + tyrosine	8.6	37.1
Threonine	5.7	22.8
Tryptophan	1.7	6.4
Valine	6.1	31.8
Linoleic acid (g)	1.9	5.3
Mineral elements		
Calcium (g)	14.2	39.8
Phosphorus, total (g)	11.4	31.8
Phosphorus, available (g)	6.6	18.6
Sodium (g)	2.8	10.6
Chlorine (g)	2.3	8.5
Magnesium (g)	0.8	2.1
Potassium (g)	3.8	10.6
Copper (mg)	9.5	26.5
Iodine (mg)	0.3	0.7
Iron (mg)	152	424
Manganese (mg)	19	53
Selenium (mg)	0.3	0.8
Zinc (mg)'	95	265
Vitamins		
Vitamin A (IU)	7,600	10,600
Vitamin D (IU)	380	1,060
Vitamin E (IU)	42	117
Vitamin K (menadione) (mg)	1.0	2.6
Biotin (mg)	0.4	1.1
Choline (g)	2.4	5.3
Folacin (mg)	0.6	1.6
Niacin, available (mg)	19.0	53.0
Pantothenic acid (mg)	22.8	63.6
Riboflavin (mg)	7.1	19.9
Thiamin (mg)	1.9	5.3
Vitamin B_6 (mg)	1.9	5.3
Vitamin B_{12} (μg)	28.5	79.5

Source: Nutrient Requirements of Swine, ninth revised edition, © 1988 by the National Academy of Sciences, Washington, DC.

TABLE 17 Requirements for Several Nutrients of Breeding Herd Replacements Allowed Feed Ad Libitum.

| | Weight (kg) of: | | | |
| | Developing Gilts | | Developing Boars | |
Intake Levels	20–50	50–110	20–50	50–110
Energy concentration (kcal ME/kg diet)	3,255	3,260	3,240	3,255
Crude protein (%)	16	15	18	16
Nutrient[a]				
Lysine (%)	0.80	0.70	0.90	0.75
Calcium (%)	0.65	0.55	0.70	0.60
Phosphorus, total (%)	0.55	0.45	0.60	0.50
Phosphorus, available (%)	0.28	0.20	0.33	0.25

[a]Sufficient data are not available to indicate that requirements for other nutrients are different from those in Table 13 for animals of these weights.

Source: Nutrient Requirements of Swine, ninth revised edition, © 1988 by the National Academy of Sciences, Washington, DC.

TABLE 18 Daily Energy and Feed Requirements of Pregnant Gilts and Sows.

| | Weight (kg) of Bred Gilts and Sows at Mating[a] | | |
Intake and Performance Levels	120	140	160
Mean gestation weight (kg)[b]	142.5	162.5	182.5
Energy required (Mcal DE/day)			
Maintenance[c]	4.53	5.00	5.47
Gestation weight gain[d]	1.29	1.29	1.29
Total	5.82	6.29	6.76
Feed required/day (kg)[e]	1.8	1.9	2.0

[a]Requirements are based on a 25-kg maternal weight gain plus 20-kg increase in weight due to the products of conception; the total weight gain is 45 kg.
[b]Mean gestation weight is weight at mating + (total weight gain/2).
[c]The animal's daily maintenance requirement is 110 kcal of DE/kg$^{0.75}$.
[d]The gestation weight gain is 1.10 Mcal of DE/day for maternal weight gain plus 0.19 Mcal of DE/day for conceptus gain.
[e]The feed required/day is based on a corn-soybean meal diet containing 3.34 Mcal of DE/kg.

Source: Nutrient Requirements of Swine, ninth revised edition, © 1988 by the National Academy of Sciences, Washington, DC.

TABLE 19 Daily Energy and Feed Requirements of Lactating Gilts and Sows.

| | Weight (kg) of Lactating Gilts and Sows at Postfarrowing | | |
Intake and Performance Levels	145	165	185
Milk yield (kg)	5.0	6.25	7.5
Energy required (Mcal DE/day)			
Maintenance[a]	4.5	5.0	5.5
Milk production[b]	10.0	12.5	15.0
Total	14.5	17.5	20.5
Feed required/day (kg)[c]	4.4	5.3	6.1

[a]The animal's daily maintenance requirement is 110 kcal of DE/kg$^{0.75}$.
[b]Milk production requires 2.0 Mcal of DE/kg of milk.
[c]The feed required/day is based on a corn-soybean meal diet containing 3.34 Mcal of DE/kg.

Source: Nutrient Requirements of Swine, ninth revised edition, © 1988 by the National Academy of Sciences, Washington, DC.

TABLE 20 Daily Nutrient Requirements of Ponies (200 kg mature weight).

Animal	Weight (kg)	Daily Gain (kg)	DE (Mcal)	Crude Protein (g)	Lysine (g)	Calcium (g)	Phosphorus (g)	Magnesium (g)	Potassium (g)	Vitamin A (10³ IU)
Mature horses										
Maintenance	200		7.4	296	10	8	6	3.0	10.0	6
Stallions (breeding season)	200		9.3	370	13	11	8	4.3	14.1	9
Pregnant mares										
9 months	200		8.2	361	13	16	12	3.9	13.1	12
10 months			8.4	368	13	16	12	4.0	13.4	12
11 months			8.9	391	14	17	13	4.3	14.2	12
Lactating mares										
Foaling to 3 months	200		13.7	688	24	27	18	4.8	21.2	12
3 months to weaning	200		12.2	528	18	18	11	3.7	14.8	12
Working horses										
Light work[a]	200		9.3	370	13	11	8	4.3	14.1	9
Moderate work[b]	200		11.1	444	16	14	10	5.1	16.9	9
Intense work[c]	200		14.8	592	21	18	13	6.8	22.5	9
Growing horses										
Weanling, 4 months	75	0.40	7.3	365	15	16	9	1.6	5.0	3
Weanling, 6 months										
Moderate growth	95	0.30	7.6	378	16	13	7	1.8	5.7	4
Rapid growth	95	0.40	8.7	433	18	17	9	1.9	6.0	4
Yearling, 12 months										
Moderate growth	140	0.20	8.7	392	17	12	7	2.4	7.6	6
Rapid growth	140	0.30	10.3	462	19	15	8	2.5	7.9	6
Long yearling, 18 months										
Not in training	170	0.10	8.3	375	16	10	6	2.7	8.8	8
In training	170	0.10	11.6	522	22	14	8	3.7	12.2	8
Two year old, 24 months										
Not in training	185	0.05	7.9	337	13	9	5	2.8	9.4	8
In training	185	0.05	11.4	485	19	13	7	4.1	13.5	8

NOTE: Mares should gain weight during late gestation to compensate for tissue deposition. However, nutrient requirements are based on maintenance body weight (all values are on a dry matter basis).

[a]Examples are horses used in Western and English pleasure, bridle path hack, equitation, and so on.

[b]Examples are horses used in ranch work, roping, cutting, barrel racing, jumping, and so on.

[c]Examples are horses in race training, polo, and so on.

Source: Nutrient Requirements of Horses, fifth revised edition, National Research Council, National Academy of Sciences, Washington, DC.

TABLE 21 Daily Nutrient Requirements of Horses (400 kg mature weight).

Animal	Weight (kg)	Daily Gain (kg)	DE (Mcal)	Crude Protein (g)	Lysine (g)	Calcium (g)	Phosphorus (g)	Magnesium (g)	Potassium (g)	Vitamin A (10³ IU)
Mature horses										
Maintenance	400		13.4	536	19	16	11	6.0	20.0	12
Stallions (breeding season)	400		16.8	670	23	20	15	7.7	25.5	18
Pregnant mares										
9 months	400		14.9	654	23	28	21	7.1	23.8	24
10 months			15.1	666	23	29	22	7.3	24.2	24
11 months			16.1	708	25	31	23	7.7	25.7	24
Lactating mares										
Foaling to 3 months	400		22.9	1,141	40	45	29	8.7	36.8	24
3 months to weaning	400		19.7	839	29	29	18	6.9	26.4	24
Working horses										
Light work[a]	400		16.8	670	23	20	15	7.7	25.5	18
Moderate work[b]	400		20.1	804	28	25	17	9.2	30.6	18
Intense work[c]	400		26.8	1,072	38	33	23	12.3	40.7	18
Growing horses										
Weanling, 4 months	145	0.85	13.5	675	28	33	18	3.2	9.8	7
Weanling, 6 months										
Moderate growth	180	0.55	12.9	643	27	25	14	3.4	10.7	8
Rapid growth	180	0.70	14.5	725	30	30	16	3.6	11.1	8
Yearling, 12 months										
Moderate growth	265	0.40	15.6	700	30	23	13	4.5	14.5	12
Rapid growth	265	0.50	17.1	770	33	27	15	4.6	14.8	12
Long yearling, 18 months										
Not in training	330	0.25	15.9	716	30	21	12	5.3	17.3	15
In training	330	0.25	21.6	970	41	29	16	7.1	23.4	15
Two year old, 24 months										
Not in training	365	0.15	15.3	650	26	19	11	5.7	18.7	16
In training	365	0.15	21.5	913	37	27	15	7.9	26.2	16

NOTE: Mares should gain weight during late gestation to compensate for tissue deposition. However, nutrient requirements are based on maintenance body weight (all values are on a dry matter basis).

[a]Examples are horses used in Western and English pleasure, bridle path hack, equitation, and so on.

[b]Examples are horses used in ranch work, roping, cutting, barrel racing, jumping, and so on.

[c]Examples are horses in race training, polo, and so on.

Source: Nutrient Requirements of Horses, fifth revised edition, National Research Council, National Academy of Sciences, Washington, DC.

TABLE 22 Daily Nutrient Requirements of Horses (500 kg mature weight).

Animal	Weight (kg)	Daily Gain (kg)	DE (Mcal)	Crude Protein (g)	Lysine (g)	Calcium (g)	Phosphorus (g)	Magnesium (g)	Potassium (g)	Vitamin A (10³ IU)
Mature horses										
Maintenance	500		16.4	656	23	20	14	7.5	25.0	15
Stallions (breeding season)	500		20.5	820	29	25	18	9.4	31.2	22
Pregnant mares										
9 months	500		18.2	801	28	35	26	8.7	29.1	30
10 months			18.5	815	29	35	27	8.9	29.7	30
11 months			19.7	866	30	37	28	9.4	31.5	30
Lactating mares										
Foaling to 3 months	500		28.3	1,427	50	56	36	10.9	46.0	30
3 months to weaning	500		24.3	1,048	37	36	22	8.6	33.0	30
Working horses										
Light work[a]	500		20.5	820	29	25	18	9.4	31.2	22
Moderate work[b]	500		24.6	984	34	30	21	11.3	37.4	22
Intense work[c]	500		32.8	1,312	46	40	29	15.1	49.9	22
Growing horses										
Weanling, 4 months	175	0.85	14.4	720	30	34	19	3.7	11.3	8
Weanling, 6 months										
Moderate growth	215	0.65	15.0	750	32	29	16	4.0	12.7	10
Rapid growth	215	0.85	17.2	860	36	36	20	4.3	13.3	10
Yearling, 12 months										
Moderate growth	325	0.50	18.9	851	36	29	16	5.5	17.8	15
Rapid growth	325	0.65	21.3	956	40	34	19	5.7	18.2	15
Long yearling, 18 months										
Not in training	400	0.35	19.8	893	38	27	15	6.4	21.1	18
In training	400	0.35	26.5	1,195	50	36	20	8.6	28.2	18
Two year old, 24 months										
Not in training	450	0.20	18.8	800	32	24	13	7.0	23.1	20
In training	450	0.20	26.3	1,117	45	34	19	9.8	32.2	20

NOTE: Mares should gain weight during late gestation to compensate for tissue deposition. However, nutrient requirements are based on maintenance body weight (all values are on a dry matter basis).

[a]Examples are horses used in Western and English pleasure, bridle path hack, equitation, and so on.

[b]Examples are horses used in ranch work, roping, cutting, barrel racing, jumping, and so on.

[c]Examples are horses in race training, polo, and so on.

Source: Nutrient Requirements of Horses, fifth revised edition, National Research Council, National Academy of Sciences, Washington, DC.

TABLE 23 Daily Nutrient Requirements of Horses (600 kg mature weight).

Animal	Weight (kg)	Daily Gain (kg)	DE (Mcal)	Crude Protein (g)	Lysine (g)	Calcium (g)	Phosphorus (g)	Magnesium (g)	Potassium (g)	Vitamin A (10³ IU)
Mature horses										
Maintenance	600		19.4	776	27	24	17	9.0	30.0	18
Stallions (breeding season)	600		24.3	970	34	30	21	11.2	36.9	27
Pregnant mares										
9 months	600		21.5	947	33	41	31	10.3	34.5	36
10 months			21.9	965	34	42	32	10.5	35.1	36
11 months			23.3	1,024	36	44	34	11.2	37.2	36
Lactating mares										
Foaling to 3 months	600		33.7	1,711	60	67	43	13.1	55.2	36
3 months to weaning	600		28.9	1,258	44	43	27	10.4	39.6	36
Working horses										
Light work[a]	600		24.3	970	34	30	21	11.2	36.9	27
Moderate work[b]	600		29.1	1,164	41	36	25	13.4	44.2	27
Intense work[c]	600		38.8	1,552	54	47	34	17.8	59.0	27
Growing horses										
Weanling, 4 months	200	1.00	16.5	825	35	40	22	4.3	13.0	9
Weanling, 6 months										
Moderate growth	245	0.75	17.0	850	36	34	19	4.6	14.5	11
Rapid growth	245	0.95	19.2	960	40	40	22	4.9	15.1	11
Yearling, 12 months										
Moderate growth	375	0.65	22.7	1,023	43	36	20	6.4	20.7	17
Raid growth	375	0.80	25.1	1,127	48	41	22	6.6	21.2	17
Long yearling, 18 months										
Not in training	475	0.45	23.9	1,077	45	33	18	7.7	25.1	21
In training	475	0.45	32.0	1,429	60	44	24	10.2	33.3	21
Two year old, 24 months										
Not in training	540	0.30	23.5	998	40	31	17	8.5	27.9	24
In training	540	0.30	32.3	1,372	55	43	24	11.6	38.4	24

NOTE: Mares should gain weight during late gestation to compensate for tissue deposition. However, nutrient requirements are based on maintenance body weight (all values are on a dry matter basis).

[a]Examples are horses used in Western and English pleasure, bridle path hack, equitation, and so on.

[b]Examples are horses used in ranch work, roping, cutting, barrel racing, jumping, and so on.

[c]Examples are horses in race training, polo, and so on.

Source: Nutrient Requirements of Horses, fifth revised edition, National Research Council, National Academy of Sciences, Washington, DC.

TABLE 24 Daily Nutrient Requirements of Horses (700 kg mature weight).

Animal	Weight (kg)	Daily Gain (kg)	DE (Mcal)	Crude Protein (g)	Lysine (g)	Calcium (g)	Phosphorus (g)	Magnesium (g)	Potassium (g)	Vitamin A (10³ IU)
Mature horses										
Maintenance	700		21.3	851	30	28	20	10.5	35.0	21
Stallions	700		26.6	1,064	37	32	23	12.2	40.4	32
(breeding season)										
Pregnant mares										
9 months	700		23.6	1,039	36	45	34	11.3	37.8	42
10 months			24.0	1,058	37	46	35	11.5	38.5	42
11 months			25.5	1,124	39	49	37	12.3	40.9	42
Lactating mares										
Foaling to 3 months	700		37.9	1,997	70	78	51	15.2	64.4	42
3 months to weaning	700		32.4	1,468	51	50	31	12.1	46.2	42
Working horses										
Light work[a]	700		26.6	1,064	37	32	23	12.2	40.4	32
Moderate work[b]	700		31.9	1,277	45	39	28	14.7	48.5	32
Intense work[c]	700		42.6	1,702	60	52	37	19.6	64.7	32
Growing horses										
Weanling, 4 months	225	1.10	19.7	986	41	44	25	4.8	14.6	10
Weanling, 6 months										
Moderate growth	275	0.80	20.0	1,001	42	37	20	5.1	16.2	12
Rapid growth	275	1.00	22.2	1,111	47	43	24	5.4	16.8	12
Yearling, 12 months										
Moderate growth	420	0.70	26.1	1,176	50	39	22	7.2	23.1	19
Rapid growth	420	0.85	28.5	1,281	54	44	24	7.4	23.6	19
Long yearling, 18 months										
Not in training	525	0.50	27.0	1,215	51	37	20	8.5	27.8	24
In training	525	0.50	36.0	1,615	68	49	27	11.3	36.9	24
Two year old, 24 months										
Not in training	600	0.35	26.3	1,117	45	35	19	9.4	31.1	27
In training	600	0.35	36.0	1,529	61	48	27	12.9	42.5	27

NOTE: Mares should gain weight during late gestation to compensate for tissue deposition. However, nutrient requirements are based on maintenance body weight (all values are on a dry matter basis).

[a]Examples are horses used in Western and English pleasure, bridle path hack, equitation, and so on.
[b]Examples are horses used in ranch work, roping, cutting, barrel racing, jumping, and so on.
[c]Examples are horses in race training, polo, and so on.

Source: Nutrient Requirements of Horses, fifth revised edition, National Research Council, National Academy of Sciences, Washington, DC.

TABLE 25 Daily Nutrient Requirements of Horses (800 kg mature weight).

Animal	Weight (kg)	Daily Gain (kg)	DE (Mcal)	Crude Protein (g)	Lysine (g)	Calcium (g)	Phosphorus (g)	Magnesium (g)	Potassium (g)	Vitamin A (10³ IU)
Mature horses										
Maintenance	800		22.9	914	32	32	22	12.0	40.0	24
Stallions (breeding season)	800		28.6	1,143	40	35	25	13.1	43.4	36
Pregnant mares										
9 months	800		25.4	1,116	39	48	37	12.2	40.6	48
10 months			25.8	1,137	40	49	37	12.4	41.3	48
11 months			27.4	1,207	42	52	40	13.2	43.9	48
Lactating mares										
Foaling to 3 months	800		41.9	2,282	81	90	58	17.4	73.6	48
3 months to weaning	800		35.5	1,678	60	58	36	13.8	52.8	48
Working horses										
Light work[a]	800		28.6	1,143	40	35	25	13.1	43.4	36
Moderate work[b]	800		34.3	1,372	48	42	30	15.8	52.1	36
Intense work[c]	800		45.7	1,829	64	56	40	21.0	69.5	36
Growing horses										
Weanling, 4 months	250	1.20	21.4	1,070	45	48	27	5.3	16.1	11
Weanling 6 months										
Moderate growth	305	0.90	22.0	1,100	46	41	23	5.7	18.0	14
Rapid growth	305	1.10	24.2	1,210	51	47	26	6.0	18.6	14
Yearling, 12 months										
Moderate growth	460	0.80	28.7	1,291	55	44	24	7.9	25.4	21
Rapid growth	460	0.95	31.0	1,396	59	49	27	8.1	25.9	21
Long yearling, 18 months										
Not in training	590	0.60	30.2	1,361	57	43	24	9.6	31.3	27
In training	590	0.60	39.8	1,793	76	56	31	12.6	41.2	27
Two year old, 24 months										
Not in training	675	0.40	28.7	1,220	49	40	22	10.6	35.0	30
In training	675	0.40	39.1	1,662	66	54	30	14.5	47.6	30

NOTE: Mares should gain weight during late gestation to compensate for tissue deposition. However, nutrient requirements are based on maintenance body weight (all values are on a dry matter basis).

[a]Examples are horses used in Western and English pleasure, bridle path hack, equitation, and so on.
[b]Examples are horses used in ranch work, roping, cutting, barrel racing, jumping, and so on.
[c]Examples are horses in race training, polo, and so on.

Source: Nutrient Requirements of Horses, fifth revised edition, National Research Council, National Academy of Sciences, Washington, DC.

1067

TABLE 26 Daily Nutrient Requirements of Horses (900 kg mature weight).

Animal	Weight (kg)	Daily Gain (kg)	DE (Mcal)	Crude Protein (g)	Lysine (g)	Calcium (g)	Phosphorus (g)	Magnesium (g)	Potassium (g)	Vitamin A (10³ IU)
Mature horses										
Maintenance	900		24.1	966	34	36	25	13.5	45.0	27
Stallions (breeding season)	900		30.2	1,207	42	37	26	13.9	45.9	40
Pregnant mares	900									
9 months			26.8	1,179	41	51	39	12.9	42.9	54
10 months			27.3	1,200	42	52	39	13.1	43.6	54
11 months			29.0	1,275	45	55	42	13.9	46.3	54
Lactating mares										
Foaling to 3 months	900		45.5	2,567	89	101	65	19.6	82.8	54
3 months to weaning	900		38.4	1,887	66	65	40	15.5	59.4	54
Working horses										
Light work[a]	900		30.2	1,207	42	37	26	13.9	45.9	40
Moderate work[b]	900		36.2	1,448	51	44	32	16.7	55.0	40
Intense work[c]	900		48.3	1,931	68	59	42	22.2	73.4	40
Growing horses										
Weanling, 4 months	275	1.30	23.1	1,154	48	53	29	5.8	17.7	12
Weanling, 6 months										
Moderate growth	335	0.95	23.4	1,171	49	44	24	6.2	19.6	15
Rapid growth	335	1.15	25.6	1,281	54	50	28	6.5	20.2	15
Yearling, 12 months										
Moderate growth	500	0.90	31.2	1,404	59	49	27	8.6	27.7	22
Rapid growth	500	1.05	33.5	1,509	64	54	30	8.8	28.2	22
Long yearling, 18 months										
Not in training	665	0.70	33.6	1,510	64	49	27	10.9	35.4	30
In training	665	0.70	43.9	1,975	83	64	35	14.2	46.2	30
Two year old, 24 months										
Not in training	760	0.45	31.1	1,322	53	45	25	12.0	39.4	34
In training	760	0.45	42.2	1,795	72	61	34	16.2	53.4	34

NOTE: Mares should gain weight during late gestation to compensate for tissue deposition. However, nutrient requirements are based on maintenance body weight (all values are on a dry matter basis).

[a]Examples are horses used in Western and English pleasure, bridle path hack, equitation, and so on.

[b]Examples are horses used in ranch work, roping, cutting, barrel racing, jumping, and so on.

[c]Examples are horses in race training, polo, and so on.

Source: Nutrient Requirements of Horses, fifth revised edition, National Research Council, National Academy of Sciences, Washington, DC.

TABLE 27 Nutrient Concentrations in Total Diets for Horses and Ponies (dry matter basis).

	Digestible Energy[a]		Diet Proportions		Crude Protein (%)	Lysine (%)	Calcium (%)	Phosphorus (%)	Magnesium (%)	Potassium (%)	Vitamin A	
	(Mcal/kg)	(Mcal/lb)	Conc. (%)	Hay (%)							(IU/kg)	(IU/lb)
Mature horses												
Maintenance	2.00	0.90	0	100	8.0	0.28	0.24	0.17	0.09	0.30	1830	830
Stallions	2.40	1.10	30	70	9.6	0.34	0.29	0.21	0.11	0.36	2640	1200
Pregnant mares												
9 months	2.25	1.00	20	80	10.0	0.35	0.43	0.32	0.10	0.35	3710	1680
10 months	2.25	1.00	20	80	10.0	0.35	0.43	0.32	0.10	0.36	3650	1660
11 months	2.40	1.10	30	70	10.6	0.37	0.45	0.34	0.11	0.38	3650	1660
Lactating mares												
Foaling to 3 months	2.60	1.20	50	50	13.2	0.46	0.52	0.34	0.10	0.42	2750	1250
3 months to weaning	2.45	1.15	35	65	11.0	0.37	0.36	0.22	0.09	0.33	3020	1370
Working horses												
Light work[b]	2.45	1.15	35	65	9.8	0.35	0.30	0.22	0.11	0.37	2690	1220
Moderate work[c]	2.65	1.20	50	50	10.4	0.37	0.31	0.23	0.12	0.39	2420	1100
Intense work[d]	2.85	1.30	65	35	11.4	0.40	0.35	0.25	0.13	0.43	1950	890
Growing horses												
Weanling, 4 months	2.90	1.40	70	30	14.5	0.60	0.68	0.38	0.08	0.30	1580	720
Weanling, 6 months												
Moderate growth	2.90	1.40	70	30	14.5	0.61	0.56	0.31	0.08	0.30	1870	850
Rapid growth	2.90	1.40	70	30	14.5	0.61	0.61	0.34	0.08	0.30	1630	740
Yearling, 12 months												
Moderate growth	2.80	1.30	60	40	12.6	0.53	0.43	0.24	0.08	0.30	2160	980
Rapid growth	2.80	1.30	60	40	12.6	0.53	0.45	0.25	0.08	0.30	1920	870
Long yearling, 18 months												
Not in training	2.50	1.15	45	55	11.3	0.48	0.34	0.19	0.08	0.30	2270	1030
In training	2.65	1.20	50	50	12.0	0.50	0.36	0.20	0.09	0.30	1800	820
Two year old, 24 months												
Not in training	2.45	1.15	35	65	10.4	0.42	0.31	0.17	0.09	0.30	2640	1200
In training	2.65	1.20	50	50	11.3	0.45	0.34	0.20	0.10	0.32	2040	930

[a]Values assume a concentrate feed containing 3.3 Mcal/kg and hay containing 2.00 Mcal/kg of dry matter.
[b]Examples are horses used in Western and English pleasure, bridle path hack, equitation, etc.
[c]Examples are horses used in ranch work, roping, cutting, barrel racing, jumping, etc.
[d]Examples are race training, polo, etc.

Source: *Nutrient Requirements of Horses*, fifth revised edition, National Research Council, National Academy of Sciences, Washington, DC.

TABLE 28 Nutrient Concentrations in Total Diets for Horses and Ponies (90% dry matter basis).

	Digestible Energy[a]		Diet Proportions		Crude Protein (%)	Lysine (%)	Calcium (%)	Phosphorus (%)	Magnesium (%)	Potassium (%)	Vitamin A	
	(Mcal/kg)	(Mcal/lb)	Conc. (%)	Hay (%)							(IU/kg)	(IU/lb)
Mature horses												
Maintenance	1.80	0.80	0	100	7.2	0.25	0.21	0.15	0.08	0.27	1650	750
Stallions	2.15	1.00	30	70	8.6	0.30	0.26	0.19	0.10	0.33	2370	1080
Pregnant mares												
9 months	2.00	0.90	20	80	8.9	0.31	0.39	0.29	0.10	0.32	3330	1510
10 months	2.00	0.90	20	80	9.0	0.32	0.39	0.30	0.10	0.33	3280	1490
11 months	2.15	1.00	30	70	9.5	0.33	0.41	0.31	0.10	0.35	3280	1490
Lactating mares												
Foaling to 3 months	2.35	1.10	50	50	12.0	0.41	0.47	0.30	0.09	0.38	2480	1130
3 months to weaning	2.20	1.05	35	65	10.0	0.34	0.33	0.20	0.08	0.30	2720	1240
Working horses												
Light work[b]	2.20	1.05	35	65	8.8	0.32	0.27	0.19	0.10	0.34	2420	1100
Moderate work[c]	2.40	1.10	50	50	9.4	0.35	0.28	0.22	0.11	0.36	2140	970
Intense work[d]	2.55	1.20	65	35	10.3	0.36	0.31	0.23	0.12	0.39	1760	800
Growing horses												
Weanling, 4 months	2.60	1.25	70	30	13.1	0.54	0.62	0.34	0.07	0.27	1420	650
Weanling, 6 months												
Moderate growth	2.60	1.25	70	30	13.0	0.55	0.50	0.28	0.07	0.27	1680	760
Rapid growth	2.60	1.25	70	30	13.1	0.55	0.55	0.30	0.07	0.27	1470	670
Yearling, 12 months												
Moderate growth	2.50	1.15	60	40	11.3	0.48	0.39	0.21	0.07	0.27	1950	890
Rapid growth	2.50	1.15	60	40	11.3	0.48	0.40	0.22	0.07	0.27	1730	790
Long yearling, 18 months												
Not in training	2.30	1.05	45	55	10.1	0.43	0.31	0.17	0.07	0.27	2050	930
In training	2.40	1.10	50	50	10.8	0.45	0.32	0.18	0.08	0.27	1620	740
Two year old, 24 months												
Not in training	2.20	1.00	35	65	9.4	0.38	0.28	0.15	0.08	0.27	2380	1080
In training	2.40	1.10	50	50	10.1	0.41	0.31	0.17	0.09	0.29	1840	840

[a]Values assume a concentrate feed containing 3.3 Mcal/kg and hay containing 2.00 Mcal/kg of dry matter.
[b]Examples are horses used in Western and English pleasure, bridle path hack, equitation, etc.
[c]Examples are horses used in ranch work, roping, cutting, barrel racing, jumping, etc.
[d]Examples are race training, polo, etc.

Source: Nutrient Requirements of Horses, fifth revised edition, National Research Council, National Academy of Sciences, Washington, DC.

TABLE 29 Other Minerals and Vitamins for Horses and Ponies (dry matter basis).

| | Adequate Concentrations in Total Rations | | | | |
	Maintenance	Pregnant and Lactating Mares	Growing Horses	Working Horses	Maximum Tolerance Levels
Minerals					
Sodium (%)	0.10	0.10	0.10	0.30	3[a]
Sulfur (%)	0.15	0.15	0.15	0.15	1.25
Iron (mg/kg)	40	50	50	40	1,000
Manganese (mg/kg)	40	40	40	40	1,000
Copper (mg/kg)	10	10	10	10	800
Zinc (mg/kg)	40	40	40	40	500
Selenium (mg/kg)	0.1	0.1	0.1	0.1	2.0
Iodine (mg/kg)	0.1	0.1	0.1	0.1	5.0
Cobalt (mg/kg)	0.1	0.1	0.1	0.1	10
Vitamins					
Vitamin A (IU/kg)	2,000	3,000	2,000	2,000	16,000
Vitamin D (IU/kg)[b]	300	600	800	300	2,200
Vitamin E (IU/kg)	50	80	80	80	1,000
Vitamin K (mg/kg)	[c]				
Thiamin (mg/kg)	3	3	3	5	3,000
Riboflavin (mg/kg)	2	2	2	2	
Niacin (mg/kg)					
Pantothenic acid (mg/kg)					
Pyridoxine (mg/kg)					
Biotin (mg/kg)					
Folacin (mg/kg)					
Vitamin B_{12} (μg/kg)					
Ascorbic acid (mg/kg)					
Choline (mg/kg)					

[a]As sodium chloride.
[b]Recommendations for horses not exposed to sunlight or to artificial light with an emission spectrum of 280–315 nm.
[c]Blank space indicates that data are insufficient to determine a requirement or maximum tolerable level.
Source: Nutrient Requirements of Horses, fifth revised edition, National Research Council, National Academy of Sciences, Washington, DC.

TABLE 30 Expected Feed Consumption by Horses (Percent of Body Weight).[a]

	Forage	Concentrate	Total
Mature horses			
Maintenance	1.5–2.0	0–0.5	1.5–2.0
Mares, late gestation	1.0–1.5	0.5–1.0	1.5–2.0
Mares, early lactation	1.0–2.0	1.0–2.0	2.0–3.0
Mares, late lactation	1.0–2.0	0.5–1.5	2.0–2.5
Working horses			
Light work	1.0–2.0	0.5–1.0	1.5–2.5
Moderate work	1.0–2.0	0.75–1.5	1.75–2.5
Intense work	0.75–1.5	1.0–2.0	2.0–3.0
Young horses			
Nursing foal, 3 months	0	1.0–2.0	2.5–3.5
Weanling foal, 6 months	0.5–1.0	1.5–3.0	2.0–3.5
Yearling foal, 12 months	1.0–1.5	1.0–2.0	2.0–3.0
Long yearling, 18 months	1.0–1.5	1.0–1.5	2.0–2.5
Two year old (24 months)	1.0–1.5	1.0–1.5	1.75–2.5

[a]Air-dry feed (about 90% DM).
Source: Nutrient Requirements of Horses, fifth revised edition, National Research Council, National Academy of Sciences, Washington, DC.

TABLE 31A Nutrient Requirements of Immature Leghorn-Type Chickens as Percentages or Units per Kilogram of Diet.

Nutrient	Unit	White-Egg Laying Strains			
		0 to 6 Weeks; 450 g[a] 2,850[b]	6 to 12 Weeks; 980 g[a] 2,850[b]	12 to 18 Weeks; 1,375 g[a] 2,900[b]	16 Weeks to First Egg; 1,475 g[a] 2,900[b]
Protein and amino acids					
Crude protein[c]	%	18.00	16.00	15.00	17.00
Arginine	%	1.00	0.83	0.67	0.75
Glycine + serine	%	0.70	0.58	*0.47*	*0.53*
Histidine	%	0.26	0.22	0.17	0.20
Isoleucine	%	0.60	0.50	0.40	0.45
Leucine	%	1.10	0.85	0.70	0.80
Lysine	%	0.85	0.60	0.45	0.52
Methionine	%	0.30	0.25	0.20	0.22
Methionine + cystine	%	0.62	0.52	0.42	0.47
Phenylalanine	%	0.54	0.45	0.36	0.40
Phenylalanine + tyrosine	%	1.00	0.83	0.67	0.75
Threonine	%	0.68	0.57	0.37	0.47
Tryptophan	%	0.17	0.14	0.11	0.12
Valine	%	0.62	0.52	0.41	0.46
Fat					
Linoleic acid	%	1.00	*1.00*	*1.00*	*1.00*
Macrominerals					
Calcium[d]	%	0.90	0.80	0.80	2.00
Nonphytate phosphorus	%	0.40	0.35	0.30	0.32
Potassium	%	*0.25*	*0.25*	*0.25*	*0.25*
Sodium	%	0.15	0.15	0.15	0.15
Chlorine	%	0.15	*0.12*	*0.12*	*0.15*
Magnesium	mg	600.0	*500.0*	*400.0*	*400.0*
Trace minerals					
Manganese	mg	60.0	*30.0*	*30.0*	*30.0*
Zinc	mg	40.0	*35.0*	*35.0*	*35.0*
Iron	mg	80.0	*60.0*	*60.0*	*60.0*
Copper	mg	*5.0*	*4.0*	*4.0*	*4.0*
Iodine	mg	0.35	*0.35*	*0.35*	*0.35*
Selenium	mg	0.15	*0.10*	*0.10*	*0.10*
Fat soluble vitamins					
A	IU	1,500.0	1,500.0	1,500.0	1,500.0
D$_3$	ICU	200.0	200.0	200.0	200.0
E	IU	*10.0*	*5.0*	*5.0*	*5.0*
K	mg	0.5	*0.5*	*0.5*	*0.5*
Water soluble vitamins					
Riboflavin	mg	3.6	1.8	1.8	2.2
Pantothenic acid	mg	*10.0*	*10.0*	*10.0*	*10.0*
Niacin	mg	27.0	11.0	11.0	11.0
B$_{12}$	mg	0.009	0.003	0.003	0.004
Choline	mg	1,300.0	900.0	500.0	500.0
Biotin	mg	*0.15*	*0.10*	*0.10*	*0.10*
Folic acid	mg	*0.55*	*0.25*	*0.25*	*0.25*
Thiamin	mg	*1.0*	*1.0*	*0.8*	*0.8*
Pyridoxine	mg	3.0	*3.0*	*3.0*	*3.0*

NOTE: Where experimental data are lacking, values in bold italic represent an estimate based on values obtained for other ages or related species.
[a]Final body weight.
[b]These are typical dietary energy concentrations for diets based mainly on corn and soybean meal, expressed in kcal ME$_n$/kg diet.
[c]Chickens do not have a requirement for crude protein per se. However, there should be sufficient crude protein to ensure an adequate nitrogen supply for synthesis of nonessential amino acids. Suggested requirements for crude protein are typical of those derived with corn-soybean meal diets, and levels can be reduced somewhat when synthetic amino acids are used.
[d]The calcium requirement may be increased when diets contain high levels of phytate phosphorus.

Source: Nutrient Requirements of Poultry, ninth revised edition, copyright 1994 by the National Academy of Sciences, National Academy Press, Washington, DC., 1994.

TABLE 31B Nutrient Requirements of Leghorn-Type Laying Hens as Percentages or Units per Kilogram of Diet (90 percent dry matter).

Nutrient	Unit	Dietary Concentrations Required by White-Egg Layers at Different Feed Intakes			Amount Required per Hen Daily (mg or IU)	
		80[a,b]	100[a,b]	120[a,b]	White-Egg Breeders at 100 g of Feed per Hen Daily[b]	White-Egg Layers at 100 g of Feed per Hen Daily
Protein and amino acids						
Crude protein[c]	%	18.8	15.0	12.5	15,000	15,000
Arginine[d]	%	*0.88*	*0.70*	*0.58*	*700*	*700*
Histidine	%	*0.21*	*0.17*	*0.14*	*170*	*170*
Isoleucine	%	0.81	0.65	0.54	650	650
Leucine	%	*1.03*	*0.82*	*0.68*	*820*	*820*
Lysine	%	0.86	0.69	0.58	690	690
Methionine	%	0.38	0.30	0.25	300	300
Methionine + cystine	%	0.73	0.58	0.48	580	580
Phenylalanine	%	*0.59*	*0.47*	*0.39*	*470*	*470*
Phenylalanine + tyrosine	%	*1.04*	*0.83*	*0.69*	*830*	*830*
Threonine	%	*0.59*	*0.47*	*0.39*	*470*	*470*
Tryptophan	%	0.20	0.16	0.13	160	160
Valine	%	*0.88*	*0.70*	*0.58*	*700*	*700*
Fat						
Linoleic acid	%	1.25	1.0	0.83	1,000	1,000
Macrominerals						
Calcium[e]	%	4.06	3.25	2.71	3,250	3,250
Chloride	%	0.16	0.13	0.11	130	130
Magnesium	mg	625	500	420	50	50
Nonphytate phosphorus[f]	%	0.31	0.25	0.21	250	250
Potassium	%	0.19	0.15	0.13	150	150
Sodium	%	0.19	0.15	0.13	150	150
Trace minerals						
Copper	mg	?	?	?	?	?
Iodine	mg	0.044	0.035	0.029	0.010	0.004
Iron	mg	56	45	38	6.0	4.5
Manganese	mg	25	20	17	2.0	2.0
Selenium	mg	0.08	0.06	0.05	0.006	0.006
Zinc	mg	44	35	29	4.5	3.5
Fat soluble vitamins						
A	IU	3,750	3,000	2,500	300	300
D$_3$	ICU	375	300	250	30	30
E	IU	*6*	*5*	*4*	*1.0*	0.5
K	mg	0.6	0.5	0.4	0.1	*0.05*
Water soluble vitamins						
B$_{12}$	mg	0.004	*0.004*	*0.004*	0.008	*0.0004*
Biotin	mg	*0.13*	*0.10*	*0.08*	*0.01*	*0.01*
Choline	mg	1,310	1,050	875	105	105
Folacin	mg	0.31	0.25	0.21	0.035	0.025
Niacin	mg	12.5	10.0	8.3	1.0	1.0
Pantothenic acid	mg	2.5	2.0	1.7	0.7	0.20
Pyridoxine	mg	3.1	2.5	2.1	0.45	0.25
Riboflavin	mg	3.1	2.5	2.1	0.36	0.25
Thiamin	mg	0.88	0.70	0.60	0.07	*0.07*

NOTE: Where experimental data are lacking, values in bold italic represent an estimate based on values obtained for other ages or related species.
[a]Grams feed intake per hen daily
[b]Based on dietary ME$_n$ concentrations of approximately 2,900 kcal/kg and an assumed rate of egg production of 90 percent (90 eggs per 100 hens daily).
[c]Laying hens do not have a requirement for crude protein per se. However, there should be sufficient crude protein to ensure an adequate supply of nonessential amino acids. Suggested requirements for crude protein are typical of those derived with corn-soybean meal diets, and levels can be reduced somewhat when synthetic amino acids are used.
[d]Italicized amino acid values for white-egg-laying chickens were estimated by using Model B (Hurwitz and Bornstein, 1973), assuming a body weight of 1,800 g and 47 g of egg mass per day.
[e]The requirement may be higher for maximum eggshell thickness.
[f]The requirement may be higher in very hot temperatures.

Source: Nutrient Requirements of Poultry, ninth revised edition, copyright 1994 by the National Academy of Sciences, National Academy Press, Washington, DC, 1994.

TABLE 32 Nutrient Requirements of Broilers as Percentages or Units per Kilogram of Diet (90 percent dry matter).

Nutrient	Unit	0 to 3 Weeks[a]; 3,200[b]	3 to 6 Weeks[a]; 3,200[b]	6 to 8 Weeks[a]; 3,200[b]
Protein and amino acids				
Crude protein[c]	%	23.00	20.00	18.00
Arginine	%	1.25	1.10	1.00
Glycine + serine	%	1.25	*1.14*	*0.97*
Histidine	%	0.35	*0.32*	*0.27*
Isoleucine	%	0.80	*0.73*	*0.62*
Leucine	%	1.20	*1.09*	*0.93*
Lysine	%	1.10	1.00	0.85
Methionine	%	0.50	0.38	0.32
Methionine + cystine	%	0.90	0.72	0.60
Phenylalanine	%	0.72	*0.65*	*0.56*
Phenylalanine + tyrosine	%	1.34	*1.22*	*1.04*
Proline	%	0.60	*0.55*	*0.46*
Threonine	%	0.80	0.74	0.68
Tryptophan	%	0.20	0.18	0.16
Valine	%	0.90	*0.82*	0.70
Fat				
Linoleic acid	%	1.00	1.00	1.00
Macrominerals				
Calcium[d]	%	1.00	0.90	0.80
Chlorine	%	0.20	0.15	0.12
Magnesium	mg	600	*600*	*600*
Nonphytate phosphorus	%	0.45	0.35	0.30
Potassium	%	0.30	0.30	0.30
Sodium	%	0.20	0.15	0.12
Trace minerals				
Copper	mg	8	*8*	*8*
Iodine	mg	*0.35*	0.35	0.35
Iron	mg	80	*80*	*80*
Manganese	mg	*60*	*60*	*60*
Selenium	mg	0.15	0.15	0.15
Zinc	mg	40	*40*	*40*
Fat soluble vitamins				
A	IU	*1,500*	*1,500*	*1,500*
D₃	ICU	200	200	200
E	IU	*10*	*10*	*10*
K	mg	0.50	0.50	0.50
Water soluble vitamins				
B₁₂	mg	0.01	*0.01*	*0.007*
Biotin	mg	0.15	0.15	*0.12*
Choline	mg	1,300	*1,000*	750
Folacin	mg	0.55	0.55	*0.50*
Niacin	mg	35	30	25
Pantothenic acid	mg	*10*	*10*	*10*
Pyridoxine	mg	3.5	3.5	*3.0*
Riboflavin	mg	3.6	3.6	3
Thiamin	mg	*1.80*	*1.80*	*1.80*

NOTE: Where experimental data are lacking, values in bold italic represent an estimate based on values obtained for other ages or related species.

[a]The 0- to 3-, 3- to 6-, and 6- to 8-week intervals for nutrient requirements are based on chronology for which research data were available; however, these nutrient requirements are often implemented at younger age intervals or on a weight-of-feed consumed basis.

[b]These are typical dietary energy concentrations, expressed in kcal ME$_n$/kg diet. Different energy values may be appropriate depending on local ingredient prices and availability.

[c]Broiler chickens do not have a requirement for crude protein per se. However, there should be sufficient crude protein to ensure an adequate nitrogen supply for synthesis of nonessential amino acids. Suggested requirements for crude protein are typical of those derived with corn-soybean meal diets, and levels can be reduced somewhat when synthetic amino acids are used.

[d]The calcium requirement may be increased when diets contain high levels of phytate phosphorus.

Source: Nutrient Requirements of Poultry, ninth revised edition, copyright 1994 by the National Academy of Sciences, National Academy Press, Washington, DC, 1994.

TABLE 33 Nutrient Requirements of Meat-Type Hens for Breeding Purposes as Units per Hen per Day (90 percent dry matter).

Nutrient	Unit	Requirements
Protein and amino acids		
Crude protein[a]	g	*19.5*
Arginine	mg	*1,110*
Histidine	mg	205
Isoleucine	mg	850
Leucine	mg	1,250
Lysine	mg	765
Methionine	mg	*450*
Methionine + cystine	mg	*700*
Phenylalanine	mg	610
Phenylalanine + tyrosine	mg	*1,112*
Threonine	mg	720
Tryptophan	mg	190
Valine	mg	750
Minerals		
Calcium	g	*4.0*
Chloride	mg	*185*
Nonphytate phosphorus	mg	*350*
Sodium	mg	150
Vitamin		
Biotin	µg	16

NOTE: These are requirements for hens at peak production. Broiler breeder hens are usually fed on a controlled basis to maintain body weight within breeder guidelines. Daily energy consumption varies with age, stage of production, and environmental temperature but usually ranges between 400 and 450 ME kcal per hen at peak production. For nutrients not listed, see requirements for egg-type breeders (Table 31B) as a guide. Where experimental data are lacking, values in bold italics represent an estimate based on values obtained for other ages or related species.
[a]Broiler chickens do not have a requirement for crude protein per se. However, there should be sufficient crude protein to ensure an adequate nitrogen supply for synthesis of nonessential amino acids. Suggested requirements for crude protein are typical of those derived with corn-soybean meal diets, and levels can be reduced somewhat when synthetic amino acids are used.
Source: Nutrient Requirements of Poultry, ninth revised edition, copyright 1994 by the National Academy of Sciences, National Academy Press, Washington, DC, 1994.

TABLE 34 Estimates of Metabolizable Energy Required per Hen per Day by Chickens in Relation to Body Weight and Egg Production (kcal).

Body Weight (kg)	Rate of Egg Production (%)					
	0	50	60	70	80	90
1.0	130	192	205	217	229	242
1.5	177	239	251	264	276	289
2.0	218	280	292	305	317	330
2.5	259	321	333	346	358	371
3.0	296	358	370	383	395	408

NOTE: A number of formulas have been suggested for prediction of the daily energy requirements of chickens. The formula used here was derived from that in *Effect of Environment on Nutrient Requirements of Domestic Animals* (National Research Council, 1981c):

$$\text{ME per hen daily} = W^{0.75}(173 - 1.95T) + 5.5\,\Delta W + 2.07EE$$

where W = body weight (kg), T = ambient temperature (°C), ΔW = change in body weight (g/day), and EE = daily egg mass (g). Temperature of 22°C, egg weight of 60 g, and no change in body weight were used in calculations.
Source: Nutrient Requirements of Poultry, ninth revised edition, copyright 1994 by the National Academy of Sciences, National Academy Press, Washington, DC, 1994.

TABLE 35 Nutrient Requirements of Turkeys as Percentages or Units per Kilogram of Diet (90 percent dry matter).

Nutrient	Unit	Growing Turkeys, Male and Female (Age in Weeks[a,b]) M: 0–4 / F: 0–4 / 2,800[c]	4–8 / 4–8 / 2,900[c]	8–12 / 8–11 / 3,000[c]	12–16 / 11–14 / 3,100[c]	16–20 / 14–17 / 3,200[c]	20–24 / 17–20 / 3,300[c]	Breeders Holding 2,900[c]	Laying Hens 2,900[c]
Protein and amino acids									
Protein[d]	%	28.0	26	22	19	16.5	14	12	14
Arginine	%	1.6	*1.4*	*1.1*	*0.9*	*0.75*	*0.6*	*0.5*	*0.6*
Glycine + serine	%	1.0	*0.9*	*0.8*	*0.7*	*0.6*	*0.5*	*0.4*	*0.5*
Histidine	%	0.58	*0.5*	*0.4*	*0.3*	*0.25*	*0.2*	*0.2*	*0.3*
Isoleucine	%	1.1	*1.0*	*0.8*	*0.6*	*0.5*	*0.45*	*0.4*	*0.5*
Leucine	%	1.9	*1.75*	*1.5*	*1.25*	*1.0*	*0.8*	*0.5*	*0.5*
Lysine	%	1.6	1.5	1.3	1.0	0.8	0.65	*0.5*	*0.6*
Methionine	%	0.55	0.45	0.4	0.35	0.25	*0.25*	*0.2*	*0.2*
Methionine + cystine	%	1.05	0.95	0.8	0.65	0.55	*0.45*	*0.4*	*0.4*
Phenylalanine	%	1.0	*0.9*	*0.8*	*0.7*	*0.6*	*0.5*	*0.4*	*0.4*
Phenylalanine + tyrosine	%	1.8	*1.6*	*1.2*	*1.0*	*0.9*	*0.9*	*0.8*	*1.0*
Threonine	%	1.0	*0.95*	*0.8*	*0.75*	*0.6*	*0.5*	*0.4*	*0.45*
Tryptophan	%	0.26	*0.24*	*0.2*	*0.18*	*0.15*	*0.13*	*0.1*	*0.13*
Valine	%	1.2	*1.1*	*0.9*	*0.8*	*0.7*	*0.6*	*0.5*	*0.58*
Fat									
Linoleic acid	%	1.0	*1.0*	*0.8*	*0.8*	*0.8*	*0.8*	*0.8*	1.1
Macrominerals									
Calcium[e]	%	1.2	1.0	0.85	0.75	0.65	0.55	0.5	2.25
Nonphytate phosphorus[f]	%	0.6	0.5	0.42	0.38	0.32	0.28	0.25	0.35
Potassium	%	0.7	*0.6*	*0.5*	*0.5*	*0.4*	*0.4*	*0.4*	*0.6*
Sodium	%	0.17	*0.15*	*0.12*	*0.12*	*0.12*	*0.12*	*0.12*	*0.12*
Chlorine	%	0.15	*0.14*	*0.14*	*0.12*	*0.12*	*0.12*	*0.12*	*0.12*
Magnesium	mg	500	*500*	*500*	*500*	*500*	*500*	*500*	*500*
Trace minerals									
Manganese	mg	60	*60*	*60*	*60*	*60*	*60*	*60*	*60*
Zinc	mg	70	*65*	*50*	*40*	*40*	*40*	*40*	*65*
Iron	mg	*80*	*60*	*60*	*60*	*50*	*50*	*50*	*60*
Copper	mg	*8*	*8*	*6*	*6*	*6*	*6*	*6*	*8*
Iodine	mg	*0.4*	*0.4*	*0.4*	*0.4*	*0.4*	*0.4*	*0.4*	*0.4*
Selenium	mg	0.2	*0.2*	*0.2*	*0.2*	*0.2*	*0.2*	*0.2*	*0.2*
Fat soluble vitamins									
A	IU	5,000	5,000	*5,000*	*5,000*	*5,000*	*5,000*	*5,000*	*5,000*
D_3[g]	ICU	1,100	*1,100*	*1,100*	*1,100*	*1,100*	*1,100*	*1,100*	1,100
E	IU	12	*12*	*10*	*10*	*10*	*10*	*10*	25
K	mg	1.75	*1.5*	*1.0*	*0.75*	*0.75*	*0.50*	*0.5*	*1.0*
Water soluble vitamins									
B_12	mg	0.003	0.003	*0.003*	*0.003*	*0.003*	*0.003*	*0.003*	*0.003*
Biotin[h]	mg	0.25	0.2	*0.125*	*0.125*	*0.100*	*0.100*	*0.100*	*0.20*
Choline	mg	1,600	1,400	*1,100*	*1,100*	*950*	*800*	*800*	*1,000*
Folacin	mg	1.0	1.0	*0.8*	*0.8*	*0.7*	*0.7*	*0.7*	1.0
Niacin	mg	60.0	60.0	*50.0*	*50.0*	*40.0*	*40.0*	*40.0*	*40.0*
Pantothenic acid	mg	10.0	9.0	*9.0*	*9.0*	*9.0*	*9.0*	*9.0*	16.0
Pyridoxine	mg	*4.5*	*4.5*	*3.5*	*3.5*	*3.0*	*3.0*	*3.0*	*4.0*
Riboflavin	mg	4.0	3.6	*3.0*	*3.0*	*2.5*	*2.5*	*2.5*	*4.0*
Thiamin	mg	2.0	2.0	*2.0*	*2.0*	*2.0*	*2.0*	*2.0*	*2.0*

NOTE: Where experimental data are lacking, values in bold italic represent an estimate based on values obtained for other ages or related species.

[a]The age intervals for nutrient requirements of males are based on actual chronology from previous research. Genetic improvements in body weight gain have led to an earlier implementation of these levels, at 0 to 3, 3 to 6, 6 to 9, 9 to 12, 12 to 15, and 15 to 18 weeks, respectively, by the industry at large.

[b]The age intervals for nutrient requirements of females are based on actual chronology from previous research. Genetic improvements in body weight gain have led to an earlier implementation of these levels, at 0 to 3, 3 to 6, 6 to 9, 9 to 12, 12 to 14, and 14 to 16 weeks, respectively, by the industry at large.

[c]These are approximate metabolizable energy (ME) values provided with typical corn-soybean-meal-based feeds, expressed in kcal ME_n/kg diet. Such energy, when accompanied by the nutrient levels suggested, is expected to provide near maximum growth, particularly with pelleted feed.

(Table 35 continues)

[d]Turkeys do not have a requirement for crude protein per se. However, there should be sufficient crude protein to ensure an adequate nitrogen supply for synthesis of nonessential amino acids. Suggested requirements for crude protein are typical of those derived with corn-soybean meal diets, and levels can be reduced when synthetic amino acids are used.
[e]The calcium requirement may be increased when diets contain high levels of phytate phosphorus.
[f]Organic phosphorus is generally considered to be associated with phytin and of limited availability.
[g]These concentrations of vitamin D are considered satisfactory when the associated calcium and phosphorus levels are used.
[h]Requirement may increase with wheat-based diets.

Source: Nutrient Requirements of Poultry, ninth revised edition, copyright 1994 by the National Academy of Sciences, National Academy Press, Washington, DC, 1994.

TABLE 36 Nutrient Requirements of Geese as Percentages or Units per Kilogram of Diet (90 percent dry matter).

Nutrient	Unit	0 to 4 Weeks 2,900[a]	After 4 Weeks 3,000[a]	Breeding 2,900[a]
Protein and amino acids				
Protein	%	*20*	*15*	*15*
Lysine	%	1.0	0.85	*0.6*
Methionine + cystine	%	0.60	0.50	*0.50*
Macrominerals				
Calcium	%	*0.65*	*0.60*	*2.25*
Nonphytate phosphorus	%	*0.30*	*0.3*	*0.3*
Fat soluble vitamins				
A	IU	*1,500*	*1,500*	*4,000*
D$_3$	IU	*200*	*200*	*20*
Water soluble vitamins				
Choline	mg	*1,500*	*1,000*	?
Niacin	mg	*65.0*	*35.0*	*20.0*
Pantothenic acid	mg	*15.0*	*10.0*	*10.0*
Riboflavin	mg	3.8	*2.5*	*4.0*

NOTE: For nutrients not listed or those for which no values are given, see requirements of chickens (Table 32) as a guide. Where experimental data are lacking, values in bold italic represent an estimate based on values obtained for other ages or related species.
[a]These are typical dietary energy concentrations expressed in kcal ME$_n$/kg diet.

Source: Nutrient Requirements of Poultry, ninth revised edition, copyright 1994 by the National Academy of Sciences, National Academy Press, Washington, DC, 1994.

TABLE 37 Nutrient Requirements of White Pekin Ducks as Percentages or Units per Kilogram of Diet (90 percent dry matter).

Nutrient	Unit	0 to 2 Weeks 2,900[a]	2 to 7 Weeks 3,000[a]	Breeding 2,900[a]
Protein and amino acids				
Protein	%	*22*	16	*15*
Arginine	%	*1.1*	*1.0*	
Isoleucine	%	*0.63*	*0.46*	*0.38*
Leucine	%	*1.26*	*0.91*	*0.76*
Lysine	%	*0.90*	*0.65*	*0.60*
Methionine	%	0.40	*0.30*	*0.27*
Methionine + cystine	%	0.70	0.55	*0.50*
Tryptophan	%	*0.23*	*0.17*	*0.14*
Valine	%	*0.78*	*0.56*	*0.47*
Macrominerals				
Calcium	%	*0.65*	*0.60*	*2.75*
Chloride	%	0.12	*0.12*	*0.12*
Magnesium	mg	*500*	*500*	*500*
Nonphytate phosphorus	%	*0.40*	*0.30*	
Sodium	%	0.15	*0.15*	*0.15*
Trace minerals				
Manganese	mg	*50*	?[b]	?
Selenium	mg	0.20	?	?
Zinc	mg	*60*	?	?
Fat soluble vitamins				
A	IU	*2,500*	*2,500*	*4,000*
D₃	IU	*400*	*400*	*900*
E	IU	*10*	*10*	*10*
K	mg	*0.5*	*0.5*	*0.5*
Water soluble vitamins				
Niacin	mg	*55*	*55*	*55*
Pantothenic acid	mg	*11.0*	*11.0*	*11.0*
Pyridoxine	mg	*2.5*	*2.5*	*3.0*
Riboflavin	mg	*4.0*	*4.0*	*4.0*

NOTE: For nutrients not listed or those for which no values are given, see requirements of chickens (Table 32) as a guide. Where experimental data are lacking, values in bold italic represent an estimate based on values obtained for other ages or related species.
[a]These are typical dietary energy concentrations expressed in kcal ME_n/kg diet.
[b]Question marks indicate that no estimates are available.

Source: Nutrient Requirements of Poultry, ninth revised edition, copyright 1994 by the National Academy of Sciences, National Academy Press, Washington, DC, 1994.

TABLE 38 Nutrient Requirements of Ring-Necked Pheasants and Bobwhite Quail as Percentages or Units per Kilogram of Diet (90 percent dry matter).

Nutrient	Unit	Ring-Necked Pheasants				Bobwhite Quail		
		0 to 4 Weeks; 2,800[a]	4 to 8 Weeks; 2,800[a]	9 to 17 Weeks; 2,700[a]	Breeding 2,800[a]	0 to 6 Weeks; 2,800[a]	After 6 Weeks; 2,800[a]	Breeding 2,800[a]
Protein and amino acids								
Protein	%	28	*24*	*18*	15	*26*	*20.0*	24.0
Glycine + serine	%	*1.8*	*1.55*	*1.0*	*0.50*	—	—	—
Lysine	%	*1.5*	*1.40*	*0.8*	*0.68*	—	—	—
Methionine	%	*0.50*	0.47	*0.30*	*0.30*	—	—	—
Methionine + cystine	%	*1.0*	*0.93*	*0.6*	*0.60*	*1.0*	*0.75*	*0.90*
Fat								
Linoleic acid	%	*1.0*	*1.0*	*1.0*	*1.0*	*1.0*	*1.0*	*1.0*
Macrominerals								
Calcium	%	1.0	*0.85*	*0.53*	*2.5*	0.65	*0.65*	*2.4*
Chlorine	%	*0.11*	0.11	*0.11*	*0.11*	*0.11*	0.11	*0.11*
Nonphytate phosphorus	%	0.55	*0.50*	*0.45*	*0.40*	*0.45*	0.30	*0.70*
Sodium	%	*0.15*	0.15	*0.15*	*0.15*	*0.15*	0.15	*0.15*
Trace minerals								
Iodine	mg	—	—	—	—	0.30	*0.30*	*0.30*
Manganese	mg	*70*	*70*	*60*	*60*	—	—	—
Zinc	mg	60	*60*	*60*	*60*	—	—	—
Water soluble vitamins								
Choline	mg	*1,430*	*1,300*	*1,000*	*1,000*	1,500	*1,500*	*1,000*
Niacin	mg	70.0	*70.0*	*40.0*	*30.0*	30.0	*30.0*	*20.0*
Pantothenic acid	mg	10.0	*10.0*	*10.0*	*16.0*	12.0	*9.0*	*15.0*
Riboflavin	mg	3.4	*3.4*	*3.0*	*4.0*	3.8	*3.0*	*4.0*

NOTE: Where experimental data are lacking, values in bold italic represent an estimate based on values obtained for other ages or related species. For nutrients not listed or those for which no values are given, see requirements of turkeys (Table 35) as a guide.
[a]These are typical dietary energy concentrations expressed in kcal ME$_n$/kg diet.

Source: Nutrient Requirements of Poultry, ninth revised edition, copyright 1994 by the National Academy of Sciences, National Academy Press, Washington, DC, 1994.

TABLE 39 Nutrient Requirements of Japanese Quail (*Coturnix*) as Percentages or Units per Kilogram of Diet (90 percent dry matter).

Nutrient	Unit	Starting and Growing; 2,900[a]	Breeding 2,900[a]
Protein and amino acids			
Protein	%	24.0	20.0
Arginine	%	1.25	*1.26*
Glycine + serine	%	1.15	*1.17*
Histidine	%	0.36	*0.42*
Isoleucine	%	0.98	*0.90*
Leucine	%	1.69	*1.42*
Lysine	%	1.30	*1.00*
Methionine	%	0.50	0.45
Methionine + cystine	%	0.75	0.70
Phenylalanine	%	0.96	*0.78*
Phenylalanine + tyrosine	%	1.80	*1.40*
Threonine	%	1.02	*0.74*
Tryptophan	%	0.22	*0.19*
Valine	%	0.95	*0.92*
Fat			
Linoleic acid	%	*1.0*	1.0
Macrominerals			
Calcium	%	0.8	2.5
Chlorine	%	*0.14*	*0.14*
Magnesium	mg	300	*500*
Nonphytate phosphorus	%	0.30	*0.35*
Potassium	%	*0.4*	0.4
Sodium	%	0.15	*0.15*
Trace minerals			
Copper	mg	*5*	*5*
Iodine	mg	*0.3*	*0.3*
Iron	mg	*120*	*60*
Manganese	mg	*60*	*60*
Selenium	mg	*0.2*	*0.2*
Zinc	mg	25	*50*
Fat soluble vitamins			
A	IU	1,650	*3,300*
D$_3$	ICU	750	*900*
E	IU	*12*	*25*
K	mg	*1*	*1*
Water soluble vitamins			
B$_{12}$	mg	*0.003*	*0.003*
Biotin	mg	*0.3*	*0.15*
Choline	mg	*2,000*	*1,500*
Folacin	mg	*1*	*1*
Niacin	mg	40	*20*
Pantothenic acid	mg	*10*	*15*
Pyridoxine	mg	*3*	*3*
Riboflavin	mg	*4*	*4*
Thiamin	mg	*2*	*2*

NOTE: Where experimental data are lacking, values in bold italic represent an estimate based on values obtained for other ages or related species. For nutrients not listed or those for which no values are given, see requirements for turkeys (Table 35) as a guide.
[a]These are typical dietary energy concentrations expressed in kcal ME$_n$/kg diet.

Source: *Nutrient Requirements of Poultry,* ninth revised edition, copyright 1994 by the National Academy of Sciences, National Academy Press, Washington, DC, 1994.

TABLE 40 Nutrient Requirements of Rabbits Fed Ad Libitum (Percentage or Amount per kg of Diet).

Nutrients[a]	Growth	Maintenance	Gestation	Lactation
Energy and protein				
Digestible energy (kcal)	2500	2100	2500	2500
TDN (%)	65	55	58	70
Crude fiber (%)	10–12[b]	14[b]	10–12[b]	10–12[b]
Fat (%)	2[b]	2[b]	2[b]	2[b]
Crude protein (%)	16	12	15	17
Inorganic nutrients				
Calcium (%)	0.4	—[c]	0.45[b]	0.75[b]
Phosphorus (%)	0.22	—[c]	0.37[b]	0.5
Magnesium (mg)	300–400	300–400	300–400	300–400
Potassium (%)	0.6	0.6	0.6	0.6
Sodium (%)	0.2[b,d]	0.2[b,d]	0.2[b,d]	0.2[b,d]
Chlorine (%)	0.3[b,d]	0.3[b,d]	0.3[b,d]	0.3[b,d]
Copper (mg)	3	3	3	3
Iodine (mg)	0.2[b]	0.2[b]	0.2[b]	0.2[b]
Iron	—[c]	—[c]	—[c]	—[c]
Manganese (mg)	8.5[e]	2.5[e]	2.5[e]	2.5[e]
Zinc	—[c]	—[c]	—[c]	—[c]
Vitamins				
Vitamin A (IU)	580	—[c]	>1160	—[c]
Vitamin A as carotene (mg)	0.83[b,e]	—[f]	0.83[b,e]	—[f]
Vitamin D	—[g]	—[g]	—[g]	—[g]
Vitamin E (mg)	40[h]	—[c]	40[h]	40[h]
Vitamin K (mg)	—[i]	—[i]	0.2[b]	—[i]
Niacin (mg)	180	—[i]	—[i]	—[i]
Pyridoxine (mg)	39	—[i]	—[i]	—[i]
Choline (g)	1.2[b]	—[i]	—[i]	—[i]
Amino acids (%)				
Lysine	0.65	—[g]	—[g]	—[g]
Methionine + cystine	0.6	—[g]	—[g]	—[g]
Arginine	0.6	—[g]	—[g]	—[g]
Histidine	0.3[b]	—[g]	—[g]	—[g]
Leucine	1.1[b]	—[g]	—[g]	—[g]
Isoleucine	0.6[b]	—[g]	—[g]	—[g]
Phenylalanine + tyrosine	1.1[b]	—[g]	—[g]	—[g]
Threonine	0.6[b]	—[g]	—[g]	—[g]
Tryptophan	0.2[b]	—[g]	—[g]	—[g]
Valine	0.7[b]	—[g]	—[g]	—[g]
Glycine	—[c]	—[g]	—[g]	—[g]

[a]Nutrients not listed indicate dietary need unknown or not demonstrated.
[b]May not be minimum but known to be adequate.
[c]Quantitative requirement not determined, but dietary need demonstrated.
[d]May be met with 0.5 percent NaCl.
[e]Converted from amount per rabbit per day using an air-dry feed intake of 60 g per day for a 1-kg rabbit.
[f]Quantitative requirement not determined.
[g]Probably required, amount unknown.
[h]Estimated.
[i]Intestinal synthesis probably adequate.
[j]Dietary need unknown.

Source: Reprinted from *Nutrient Requirements of Rabbits,* second revised edition, with permission of the National Academy Press, Washington, DC.

TABLE 41 Daily Nutrient Requirements of Growing Dairy Cattle and Mature Bulls.

Live Weight (kg)	Gain (g)	Dry Matter Intake^a (kg)	Energy NEM (Mcal)	NEG (Mcal)	ME (Mcal)	DE (Mcal)	TDN (kg)	Protein UIP (g)	DIP (g)	CP (g)	Minerals Ca (g)	P (g)	Vitamins A (1,000 IU)	D (1,000 IU)
*Growing Large-Breed Calves Fed Only Milk or Milk Replacer*														
40	200	0.48	1.37	0.41	2.54	2.73	0.62	–	–	105	7	4	1.70	0.26
45	300	0.54	1.49	0.56	2.86	3.07	0.70	–	–	120	8	5	1.94	0.30
*Growing Large-Breed Calves Fed Milk Plus Starter Mix*														
50	500	1.30	1.62	0.72	5.90	6.42	1.46	–	–	290	9	6	2.10	0.33
75	800	1.98	2.19	1.30	8.98	9.78	2.22	–	–	435	16	8	3.20	0.50
*Growing Small-Breed Calves Fed Only Milk or Milk Replacer*														
25	200	0.38	0.96	0.37	2.01	2.16	0.49	–	–	84	6	4	1.10	0.16
30	300	0.51	1.10	0.52	2.70	2.90	0.66	–	–	112	7	4	1.30	0.20
*Growing Small-Breed Calves Fed Milk Plus Starter Mix*														
50	500	1.43	1.62	0.72	6.49	7.06	1.60	–	–	315	10	6	2.10	0.33
75	600	1.76	2.19	0.96	7.98	8.69	1.97	–	–	387	14	8	3.20	0.50
*Growing Veal Calves Fed Only Milk or Milk Replacer*														
40	200	0.45	1.37	0.55	1.89	2.07	0.47	–	–	100	7	4	1.70	0.26
50	400	0.57	1.62	0.57	2.39	2.63	0.59	–	–	125	9	5	2.10	0.33
60	540	0.80	1.85	0.81	2.84	3.17	0.71	–	–	176	13	8	2.60	0.40
75	900	1.36	2.19	1.47	4.82	5.39	1.21	–	–	300	16	9	3.20	0.50
100	1,250	2.00	2.72	2.26	6.22	7.06	1.58	–	–	440	20	11	4.20	0.66
125	1,250	2.38	3.21	2.44	7.40	8.40	1.88	–	–	524	22	13	5.30	0.82
150	1,100	2.72	3.69	2.29	8.46	9.60	2.15	–	–	598	24	15	6.40	0.99
*Large-Breed Growing Females*														
100	600	2.63	2.72	1.22	7.03	8.13	1.84	317	57	421	17	9	4.24	0.66
100	700	2.82	2.72	1.44	7.54	8.72	1.98	346	75	452	18	9	4.24	0.66
100	800	3.02	2.72	1.66	8.06	9.32	2.11	374	92	483	18	10	4.24	0.66
150	600	3.51	3.69	1.45	9.14	10.61	2.41	283	150	562	19	11	6.36	0.99
150	700	3.75	3.69	1.71	9.76	11.33	2.57	307	173	600	19	12	6.36	0.99
150	800	3.99	3.69	1.97	10.39	12.07	2.74	331	196	639	20	12	6.36	0.99
200	600	4.39	4.57	1.65	11.14	12.99	2.95	254	239	631	20	14	8.48	1.32
200	700	4.68	4.57	1.95	11.87	13.84	3.14	274	267	686	21	14	8.48	1.32
200	800	4.97	4.57	2.25	12.62	14.71	3.34	294	295	741	22	15	8.48	1.32
250	600	5.31	5.41	1.84	13.10	15.33	3.48	229	326	637	22	16	10.60	1.65
250	700	5.65	5.41	2.18	13.94	16.32	3.70	246	359	678	23	17	10.60	1.65
250	800	5.99	5.41	2.51	14.79	17.32	3.93	263	393	726	24	17	10.60	1.65
300	600	6.26	6.20	2.02	15.05	17.69	4.01	209	413	752	23	17	12.72	1.98
300	700	6.66	6.20	2.39	16.00	18.81	4.27	223	452	799	24	18	12.72	1.98
300	800	7.06	6.20	2.77	16.97	19.95	4.52	236	490	848	25	19	12.72	1.98
350	600	7.29	6.96	2.20	17.01	20.09	4.56	193	501	874	24	18	14.84	2.31
350	700	7.75	6.96	2.60	18.09	21.36	4.84	204	545	930	25	19	14.84	2.31
350	800	8.21	6.96	3.01	19.18	22.64	5.14	214	590	985	26	20	14.84	2.31
400	600	8.39	7.69	2.37	19.03	22.58	5.12	182	592	1,007	25	19	16.96	2.64
400	700	8.92	7.69	2.80	20.23	24.00	5.44	190	641	1,070	26	20	16.96	2.64
400	800	9.46	7.69	3.24	21.44	25.44	5.77	198	692	1,135	26	21	16.96	2.64
450	600	9.59	8.40	2.53	21.12	25.18	5.71	176	686	1,151	28	19	19.08	2.97
450	700	10.20	8.40	2.99	22.46	26.78	6.07	182	742	1,224	28	20	19.08	2.97
450	800	10.82	8.40	3.46	23.81	28.40	6.44	187	799	1,298	29	21	19.08	2.97
500	600	10.93	9.09	2.69	23.32	27.96	6.34	175	785	1,311	28	20	21.20	3.30
500	700	11.63	9.09	3.18	24.81	29.74	6.75	179	848	1,395	28	20	21.20	3.30
500	800	12.33	9.09	3.68	26.32	31.55	7.16	182	913	1,480	29	21	21.20	3.30
550	600	12.42	9.77	2.84	25.67	30.95	7.02	180	891	1,490	28	20	23.32	3.63
550	700	13.22	9.77	3.37	27.33	32.95	7.47	183	963	1,587	28	20	23.32	3.63
550	800	14.04	9.77	3.90	29.02	34.99	7.94	185	1,035	1,685	29	21	23.32	3.63
600	600	14.11	10.43	3.00	28.23	34.24	7.77	193	1,007	1,694	28	20	25.44	3.96
600	700	15.05	10.43	3.55	30.09	36.50	8.28	194	1,088	1,805	28	21	25.44	3.96
600	800	15.99	10.43	4.11	31.98	38.79	8.80	195	1,170	1,919	29	21	25.44	3.96
*Small-Breed Growing Females*														
100	400	2.41	2.72	0.91	6.34	7.35	1.67	249	38	386	15	8	4.24	0.66
100	500	2.64	2.72	1.16	6.92	8.03	1.82	275	59	422	16	8	4.24	0.66
100	600	2.86	2.72	1.40	7.51	8.71	1.98	300	80	458	17	9	4.24	0.66
150	400	3.31	3.69	1.09	8.39	9.78	2.22	222	129	512	17	10	6.36	0.99

(Table 41 continues)

TABLE 41 Daily Nutrient Requirements of Growing Dairy Cattle and Mature Bulls. *(Cont.)*

Live Weight (kg)	Gain (g)	Dry Matter Intake[a] (kg)	Energy NEM (Mcal)	NEG (Mcal)	ME (Mcal)	DE (Mcal)	TDN (kg)	Protein UIP (g)	DIP (g)	CP (g)	Minerals Ca (g)	P (g)	Vitamins A (1,000 IU)	D (1,000 IU)
150	500	3.60	3.69	1.39	9.12	10.63	2.41	243	156	567	18	11	6.36	0.99
150	600	3.89	3.69	1.69	9.86	11.50	2.61	263	185	622	19	11	6.36	0.99
200	400	4.24	4.57	1.26	10.38	12.16	2.76	201	217	513	19	13	8.48	1.32
200	500	4.60	4.57	1.60	11.25	13.19	2.99	217	251	562	20	13	8.48	1.32
200	600	4.96	4.57	1.95	12.14	14.23	3.23	232	286	611	20	14	8.48	1.32
250	400	5.24	5.41	1.41	12.36	14.57	3.30	185	305	629	21	15	10.60	1.65
250	500	5.68	5.41	1.80	13.38	15.78	3.58	197	346	681	21	16	10.60	1.65
250	600	6.12	5.41	2.20	14.43	17.01	3.86	209	389	735	22	16	10.60	1.65
300	400	6.34	6.20	1.56	14.38	17.06	3.87	176	395	761	22	16	12.72	1.98
300	500	6.87	6.20	1.99	15.57	18.48	4.19	184	445	824	23	17	12.72	1.98
300	600	7.40	6.20	2.43	16.79	19.92	4.52	192	495	888	23	17	12.72	1.98
350	400	7.57	6.96	1.71	16.50	19.71	4.47	173	490	909	23	17	14.84	2.31
350	500	8.20	6.96	2.18	17.87	21.35	4.84	178	548	985	23	18	14.84	2.31
350	600	8.85	6.96	2.66	19.28	23.03	5.22	183	608	1,062	24	18	14.84	2.31
400	400	8.98	7.69	1.84	18.77	22.58	5.12	177	592	1,078	24	18	16.96	2.64
400	500	9.74	7.69	2.35	20.36	24.50	5.56	181	661	1,169	24	19	16.96	2.64
400	600	10.52	7.69	2.87	21.98	26.45	6.00	183	730	1,263	25	19	16.96	2.64
450	400	10.64	8.40	1.98	21.27	25.80	5.85	191	706	1,276	27	18	19.08	2.97
450	500	11.56	8.40	2.52	23.12	28.04	6.36	193	786	1,387	28	19	19.08	2.97
450	600	12.50	8.40	3.08	25.01	30.33	6.88	194	867	1,500	28	19	19.08	2.97
Large-Breed Growing Males														
100	800	2.80	2.72	1.42	7.48	8.66	1.96	401	65	448	18	10	4.24	0.66
100	900	2.97	2.72	1.60	7.92	9.16	2.08	433	79	475	19	10	4.24	0.66
100	1,000	3.13	2.72	1.79	8.36	9.67	2.19	465	93	501	20	11	4.24	0.66
150	800	3.60	3.69	1.64	9.52	11.03	2.50	364	155	576	20	12	6.36	0.99
150	900	3.80	3.69	1.85	10.03	11.63	2.64	393	172	607	21	13	6.36	0.99
150	1,000	3.99	3.69	2.07	10.55	12.22	2.77	422	190	639	22	13	6.36	0.99
200	800	4.43	4.57	1.84	11.48	13.34	3.03	333	241	709	22	15	8.48	1.32
200	900	4.66	4.57	2.08	12.06	14.02	3.18	359	262	745	23	15	8.48	1.32
200	1,000	4.89	4.57	2.33	12.66	14.71	3.34	385	284	782	24	16	8.48	1.32
250	800	5.27	5.41	2.03	13.37	15.58	3.53	305	325	778	24	17	10.60	1.65
250	900	5.53	5.41	2.30	14.03	16.35	3.71	329	350	837	25	18	10.60	1.65
250	1,000	5.80	5.41	2.57	14.70	17.13	3.89	352	375	897	26	18	10.60	1.65
300	800	6.13	6.20	2.21	15.22	17.80	4.04	281	408	771	25	19	12.72	1.98
300	900	6.43	6.20	2.51	15.96	18.66	4.23	302	436	827	25	19	12.72	1.98
300	1,000	6.73	6.20	2.80	16.70	19.53	4.43	323	464	884	26	20	12.72	1.98
350	800	7.02	6.96	2.38	17.06	20.02	4.54	261	490	843	26	20	14.84	2.31
350	900	7.36	6.96	2.70	17.88	20.98	4.76	280	522	883	26	20	14.84	2.31
350	1,000	7.70	6.96	3.02	18.70	21.94	4.98	298	554	924	27	21	14.84	2.31
400	800	7.96	7.69	2.55	18.91	22.27	5.05	244	572	955	26	21	16.96	2.64
400	900	8.34	7.69	2.89	19.80	23.32	5.29	260	608	1,001	27	21	16.96	2.64
400	1,000	8.72	7.69	3.24	20.71	24.39	5.53	277	644	1,046	28	22	16.96	2.64
450	800	8.95	8.40	2.71	20.78	24.56	5.57	230	656	1,074	29	21	19.08	2.97
450	900	9.37	8.40	3.08	21.76	25.72	5.83	245	696	1,125	29	22	19.08	2.97
450	1,000	9.80	8.40	3.44	22.75	26.89	6.10	259	736	1,176	29	23	19.08	2.97
500	800	10.00	9.09	2.87	22.69	26.92	6.11	220	742	1,201	29	21	21.20	3.30
500	900	10.48	9.09	3.25	23.76	28.19	6.39	233	786	1,257	29	22	21.20	3.30
500	1,000	10.95	9.09	3.64	24.84	29.47	6.68	246	830	1,314	29	23	21.20	3.30
550	800	11.14	9.77	3.02	24.66	29.38	6.66	213	831	1,336	29	21	23.32	3.63
550	900	11.66	9.77	3.43	25.82	30.76	6.98	225	879	1,399	29	22	23.32	3.63
550	1,000	12.19	9.77	3.84	27.00	32.16	7.29	236	927	1,463	30	23	23.32	3.63
600	800	12.36	10.43	3.17	26.71	31.95	7.25	211	923	1,483	29	21	25.44	3.96
600	900	12.95	10.43	3.60	27.97	33.47	7.59	221	976	1,554	29	22	25.44	3.96
600	1,000	13.54	10.43	4.03	29.25	34.99	7.94	231	1,029	1,624	30	23	25.44	3.96
650	800	13.69	11.07	3.32	28.86	34.67	7.86	212	1,020	1,643	29	21	27.56	4.29
650	900	14.35	11.07	3.77	30.24	36.33	8.24	222	1,078	1,722	29	22	27.56	4.29
650	1,000	15.01	11.07	4.22	31.63	38.00	8.62	230	1,137	1,801	30	23	27.56	4.29
700	800	15.16	11.70	3.46	31.14	37.59	8.52	219	1,124	1,820	29	22	29.68	4.62
700	900	15.90	11.70	3.93	32.64	39.40	8.94	227	1,187	1,907	29	22	29.68	4.62
700	1,000	16.63	11.70	4.40	34.16	41.23	9.35	235	1,252	1,996	30	23	29.68	4.62

(Table 41 continues)

TABLE 41 Daily Nutrient Requirements of Growing Dairy Cattle and Mature Bulls. *(Cont.)*

Live Weight (kg)	Gain (g)	Dry Matter Intake[a] (kg)	Energy					Protein			Minerals		Vitamins	
			NEM (Mcal)	NEG (Mcal)	ME (Mcal)	DE (Mcal)	TDN (kg)	UIP (g)	DIP (g)	CP (g)	Ca (g)	P (g)	A (1,000 IU)	D
750	800	16.79	12.33	3.60	33.59	40.73	9.24	232	1,235	2,015	29	22	31.80	4.95
750	900	17.62	12.33	4.09	35.23	42.73	9.69	239	1,305	2,114	29	23	31.80	4.95
750	1,000	18.45	12.33	4.58	36.89	44.74	10.15	246	1,376	2,213	30	23	31.80	4.95
800	800	17.56	12.94	3.74	35.12	42.59	9.66	216	1,303	2,107	29	22	33.92	5.28
800	900	18.41	12.94	4.25	36.83	44.67	10.13	221	1,377	2,210	29	23	33.92	5.28
800	1,000	19.28	12.94	4.76	38.55	46.76	10.61	227	1,451	2,313	30	23	33.92	5.28
Small-Breed Growing Males														
100	500	2.45	2.72	1.02	6.54	7.56	1.72	287	41	392	16	8	4.24	0.66
100	600	2.64	2.72	1.23	7.04	8.15	1.85	316	58	422	17	9	4.24	0.66
100	700	2.83	2.72	1.45	7.55	8.74	1.98	345	75	453	18	9	4.24	0.66
150	500	3.28	3.69	1.20	8.55	9.92	2.25	257	129	525	18	11	6.36	0.99
150	600	3.52	3.69	1.46	9.16	10.64	2.41	282	151	563	19	11	6.36	0.99
150	700	3.76	3.69	1.71	9.78	11.36	2.58	306	174	601	19	12	6.36	0.99
200	500	4.12	4.57	1.37	10.45	12.18	2.76	232	213	573	20	13	8.48	1.32
200	600	4.40	4.57	1.66	11.17	13.02	2.95	252	241	629	20	14	8.48	1.32
200	700	4.69	4.57	1.96	11.90	13.87	3.15	273	268	684	21	14	8.48	1.32
250	500	4.99	5.41	1.53	12.31	14.41	3.27	210	296	598	21	16	10.60	1.65
250	600	5.32	5.41	1.86	13.14	15.38	3.49	228	328	638	22	16	10.60	1.65
250	700	5.66	5.41	2.19	13.97	16.35	3.71	245	361	679	23	17	10.60	1.65
300	500	5.89	6.20	1.68	14.15	16.64	3.77	193	378	707	23	17	12.72	1.98
300	600	6.28	6.20	2.04	15.09	17.74	4.02	207	415	754	23	17	12.72	1.98
300	700	6.68	6.20	2.41	16.04	18.85	4.28	221	453	801	24	18	12.72	1.98
350	500	6.86	6.96	1.82	16.01	18.91	4.29	180	461	823	23	18	14.84	2.31
350	600	7.31	6.96	2.22	17.06	20.15	4.57	191	503	877	24	18	14.84	2.31
350	700	7.76	6.96	2.62	18.13	21.41	4.86	203	547	932	25	19	14.84	2.31
400	500	7.90	7.69	1.96	17.91	21.25	4.82	171	545	947	24	19	16.96	2.64
400	600	8.41	7.69	2.39	19.08	22.64	5.14	180	594	1,010	25	19	16.96	2.64
400	700	8.94	7.69	2.82	20.27	24.06	5.46	189	644	1,073	26	20	16.96	2.64
450	500	9.03	8.40	2.10	19.87	23.70	5.37	166	634	1,083	28	19	19.08	2.97
450	600	9.62	8.40	2.55	21.18	25.26	5.73	174	689	1,155	28	19	19.08	2.97
450	700	10.23	8.40	3.01	22.51	26.84	6.09	180	744	1,227	28	20	19.08	2.97
500	500	10.28	9.09	2.23	21.93	26.29	5.96	167	726	1,233	28	19	21.20	3.30
500	600	10.96	9.09	2.71	23.39	28.04	6.36	173	788	1,315	28	20	21.20	3.30
500	700	11.65	9.09	3.20	24.87	29.81	6.76	177	851	1,398	28	20	21.20	3.30
550	500	11.67	9.77	2.36	24.12	29.08	6.60	174	825	1,400	28	19	23.32	3.63
550	600	12.46	9.77	2.87	25.75	31.05	7.04	178	895	1,495	28	20	23.32	3.63
550	700	13.26	9.77	3.39	27.40	33.03	7.49	181	966	1,591	28	20	23.32	3.63
600	500	13.25	10.43	2.48	26.50	32.14	7.29	187	933	1,590	28	19	25.44	3.96
600	600	14.16	10.43	3.02	28.32	34.35	7.79	190	1,012	1,699	28	20	25.44	3.96
600	700	15.08	10.43	3.57	30.17	36.59	8.30	192	1,091	1,810	28	21	25.44	3.96
Maintenance of Mature Breeding Bulls														
500	–	7.89	9.09	–	15.79	19.15	4.34	161	472	789	20	12	21.20	3.30
600	–	9.05	10.43	–	18.10	21.95	4.98	155	573	905	24	15	25.44	3.96
700	–	10.16	11.70	–	20.32	24.64	5.59	148	670	1,016	28	18	29.68	4.62
800	–	11.23	12.94	–	22.46	27.24	6.18	142	764	1,123	32	20	33.92	5.28
900	–	12.27	14.13	–	24.53	29.76	6.75	135	854	1,227	36	22	38.16	5.94
1,000	–	13.28	15.29	–	26.55	32.20	7.30	129	943	1,328	41	25	42.40	6.60
1,100	–	14.26	16.43	–	28.52	34.59	7.85	122	1,029	1,426	45	28	46.64	7.26
1,200	–	15.22	17.53	–	30.44	36.92	8.37	115	1,113	1,522	49	30	50.88	7.92
1,300	–	16.16	18.62	–	32.32	39.21	8.89	108	1,196	1,616	53	32	55.12	8.58
1,400	–	17.09	19.68	–	34.17	41.45	9.40	102	1,277	1,709	57	35	59.36	9.24

NOTE: The following abbreviations were used: NEM, net energy for maintenance; NEG, net energy for gain; ME, metabolizable energy; DE, digestible energy; TDN, total digestible nutrients; UIP, undegraded intake protein; DIP, degraded intake protein; CP, crude protein.
[a]The data for DMI are not requirements per se, unlike the requirements for net energy maintenance, net energy gain, and absorbed protein. They are not intended to be estimates of voluntary intake but are consistent with the specified dietary energy concentrations. The use of diets with decreased energy concentrations will increase dry matter intake needs; metabolizable energy, digestible energy, and total digestible nutrient needs; and crude protein needs. The use of diets with increased energy concentrations will have opposite effects on these needs.
Source: Nutrient Requirements of Dairy Cattle, sixth revised edition, updated 1989, copyright © 1988 by the National Academy of Sciences.

TABLE 42 Daily Nutrient Requirements of Lactating and Pregnant Cows.

Live Weight (kg)	Energy				Total Crude Protein (g)	Minerals		Vitamins	
	NEL (Mcal)	ME (Mcal)	DE (Mcal)	TDN (kg)		Ca (g)	P (g)	A (1,000 IU)	D
Maintenance of Mature Lactating Cows[a]									
400	7.16	12.01	13.80	3.13	318	16	11	30	12
450	7.82	13.12	15.08	3.42	341	18	13	34	14
500	8.46	14.20	16.32	3.70	364	20	14	38	15
550	9.09	15.25	17.53	3.97	386	22	16	42	17
600	9.70	16.28	18.71	4.24	406	24	17	46	18
650	10.30	17.29	19.86	4.51	428	26	19	49	20
700	10.89	18.28	21.00	4.76	449	28	20	53	21
750	11.47	19.25	22.12	5.02	468	30	21	57	23
800	12.03	20.20	23.21	5.26	486	32	23	61	24
Maintenance Plus Last 2 Months of Gestation of Mature Dry Cows[b]									
400	9.30	15.26	18.23	4.15	890	26	16	30	12
450	10.16	16.66	19.91	4.53	973	30	18	34	14
500	11.00	18.04	21.55	4.90	1,053	33	20	38	15
550	11.81	19.37	23.14	5.27	1,131	36	22	42	17
600	12.61	20.68	24.71	5.62	1,207	39	24	46	18
650	13.39	21.96	26.23	5.97	1,281	43	26	49	20
700	14.15	23.21	27.73	6.31	1,355	46	28	53	21
750	14.90	24.44	29.21	6.65	1,427	49	30	57	23
800	15.64	25.66	30.65	6.98	1,497	53	32	61	24
Milk Production—Nutrients/kg of Milk of Different Fat Percentages									
(Fat %)									
3.0	0.64	1.07	1.23	0.280	78	2.73	1.68	–	–
3.5	0.69	1.15	1.33	0.301	84	2.97	1.83	–	–
4.0	0.74	1.24	1.42	0.322	90	3.21	1.98	–	–
4.5	0.78	1.32	1.51	0.343	96	3.45	2.13	–	–
5.0	0.83	1.40	1.61	0.364	101	3.69	2.28	–	–
5.5	0.88	1.48	1.70	0.385	107	3.93	2.43	–	–
Live Weight Change During Lactation—Nutrients/kg of Weight Change[c]									
Weight loss	–4.92	–8.25	–9.55	–2.17	–320	–	–	–	–
Weight gain	5.12	8.55	9.96	2.26	320	–	–	–	–

NOTE: The following abbreviations were used: NEL, net energy for lactation; ME, metabolizable energy; DE, digestible energy; TDN, total digestible nutrients.

[a]To allow for growth of young lactating cows, increase the maintenance allowances for all nutrients except vitamins A and D by 20 percent during the first lactation and 10 percent during the second lactation.

[b]Values for calcium assume that the cow is in calcium balance at the beginning of the last 2 months of gestation. If the cow is not in balance, then the calcium requirement can be increased from 25 to 33 percent.

[c]No allowance is made for mobilized calcium and phosphorus associated with live weight loss or with live weight gain. The maximum daily nitrogen available from weight loss is assumed to be 30 g or 234 g of crude protein.

Source: Nutrient Requirements of Dairy Cattle, sixth revised edition, updated 1989, copyright © 1988 by the National Academy of Sciences.

TABLE 43 Recommended Nutrient Content of Diets for Dairy Cattle.

Cow weight legend (milk-yield columns vary by cow weight):

Cow Wt (kg)	Fat (%)	Wt. Gain (kg/d)
400	5.0	0.220
500	4.5	0.275
600	4.0	0.330
700	3.5	0.385
800	3.5	0.440

Milk Yield (kg/d) for the five Lactating Cow Diet columns, stacked by cow weight (400/500/600/700/800):

- Column 1: 7 / 8 / 10 / 12 / 13
- Column 2: 13 / 17 / 20 / 24 / 27
- Column 3: 20 / 25 / 30 / 36 / 40
- Column 4: 26 / 33 / 40 / 48 / 53
- Column 5: 33 / 41 / 50 / 60 / 67

	Lactating Cow Diets — Milk Yield (kg/d)					Early Lactation (wks 0–3)	Dry, Pregnant Cows	Calf Milk Replacer	Calf Starter Mix	Growing Heifers and Bulls[a] 3–6 Mos	6–12 Mos	>12 Mos	Mature Bulls	Maximum Tolerable Levels[b,c]
	7–13	13–27	20–40	26–53	33–67									
Energy														
NEL, Mcal/kg	1.42	1.52	1.62	1.72	1.72	1.67	1.25	—	—	—	—	—	—	—
NEM, Mcal/kg	—	—	—	—	—	—	—	2.40	1.90	1.70	1.58	1.40	1.15	—
NEG, Mcal/kg	—	—	—	—	—	—	—	1.55	1.20	1.08	0.98	0.82	—	—
ME, Mcal/kg	2.35	2.53	2.71	2.89	2.89	2.80	2.04	3.78	3.11	2.60	2.47	2.27	2.00	—
DE, Mcal/kg	2.77	2.95	3.13	3.31	3.31	3.22	2.47	4.19	3.53	3.02	2.89	2.69	2.43	—
TDN, % of DM	63	67	71	75	75	73	56	95	80	69	66	61	55	—
Protein equivalent														
Crude protein, %	12	15	16	17	18	19	12	22	18	16	12	12	10	—
UIP, %	4.4	5.2	5.7	5.9	6.2	7.0	—	—	—	8.2	4.4	2.1	—	—
DIP, %	7.8	8.7	9.6	10.3	10.4	9.7	—	—	—	4.6	6.4	7.2	—	—
Fiber content (min.)[d]														
Crude fiber, %	17	17	15	15	17	17	22	—	—	13	15	15	15	—
Acid detergent fiber, %	21	21	19	19	21	21	27	—	—	16	19	19	19	—
Neutral detergent fiber, %	28	28	25	25	28	28	35	—	—	23	25	25	25	—
Ether extract (min.)	3	3	3	3	3	3	3	10	3	3	3	3	3	—
Minerals														
Calcium, %	0.43	0.51	0.58	0.64	0.66	0.77	0.39[e]	0.70	0.60	0.52	0.41	0.29	0.30	2.00
Phosphorus, %	0.28	0.33	0.37	0.41	0.41	0.48	0.24	0.60	0.40	0.31	0.30	0.23	0.19	1.00
Magnesium, %[f]	0.20	0.20	0.20	0.25	0.25	0.25	0.16	0.07	0.10	0.16	0.16	0.16	0.16	0.50
Potassium, %[g]	0.90	0.90	0.90	1.00	1.00	1.00	0.65	0.65	0.65	0.65	0.65	0.65	0.65	3.00
Sodium, %	0.18	0.18	0.18	0.18	0.18	0.18	0.10	0.10	0.10	0.10	0.10	0.10	0.10	—
Chlorine, %	0.25	0.25	0.25	0.25	0.25	0.25	0.20	0.20	0.20	0.20	0.20	0.20	0.20	—
Sulfur, %	0.20	0.20	0.20	0.20	0.20	0.25	0.16	0.29	0.20	0.16	0.16	0.16	0.16	0.40
Iron, ppm	50	50	50	50	50	50	50	100	50	50	50	50	50	1,000
Cobalt, ppm	0.10	0.10	0.10	0.10	0.10	0.10	0.10	0.10	0.10	0.10	0.10	0.10	0.10	10.00
Copper, ppm[h]	10	10	10	10	10	10	10	10	10	10	10	10	10	100
Manganese, ppm	40	40	40	40	40	40	40	40	40	40	40	40	40	1,000
Zinc, ppm	40	40	40	40	40	40	40	40	40	40	40	40	40	500
Iodine, ppm[i]	0.60	0.60	0.60	0.60	0.60	0.60	0.25	0.25	0.25	0.25	0.25	0.25	0.25	50.00[j]
Selenium, ppm	0.30	0.30	0.30	0.30	0.30	0.30	0.30	0.30	0.30	0.30	0.30	0.30	0.30	2.00
Vitamins[k]														
A, IU/kg	3,200	3,200	3,200	3,200	3,200	4,000	4,000	3,800	2,200	2,200	2,200	2,200	3,200	66,000
D, IU/kg	1,000	1,000	1,000	1,000	1,000	1,000	1,200	600	300	300	300	300	300	10,000
E, IU/kg	15	15	15	15	15	15	15	40	25	25	25	25	15	2,000

NOTE: The values presented in this table are intended as guidelines for the use of professionals in diet formulation. Because of the many factors affecting such values, they are not intended and should not be used as a legal or regulatory base.

[a]The approximate weight for growing heifers and bulls at 3–6 mos is 150 kg; at 6–12 mos, it is 250 kg; and at more than 12 mos, it is 400 kg. The approximate average daily gain is 700 g/day.

(Table 43 continues)

TABLE 43 Recommended Nutrient Content of Diets for Dairy Cattle. *(Cont.)*

[b]The maximum safe levels for many of the mineral elements are not well defined and may be substantially affected by specific feeding conditions. Additional information is available in *Mineral Tolerance of Domestic Animals* (NRC, 1980).
[c]Vitamin tolerances are discussed in detail in *Vitamin Tolerance of Animals* (NRC, 1987b).
[d]It is recommended that 75 percent of the NDF in lactating cow diets be provided as forage. If this recommendation is not followed, a depression in milk fat may occur.
[e]The value for calcium assumes that the cow is in calcium balance at the beginning of the dry period. If the cow is not in balance, then the dietary calcium requirement should be increased by 25 to 33 percent.
[f]Under conditions conducive to grass tetany (see text), magnesium should be increased to 0.25 or 0.30 percent.
[g]Under conditions of heat stress, potassium should be increased to 1.2 percent (see text).
[h]The cow's copper requirement is influenced by molybdenum and sulfur in the diet (see text).
[i]If the diet contains as much as 25 percent strongly goitrogenic feed on a dry basis, the iodine provided should be increased two times or more.
[j]Although cattle can tolerate this level of iodine, lower levels may be desirable to reduce the iodine content of milk.
[k]The following minimum quantities of B-complex vitamins are suggested per unit of milk replacer: niacin, 2.6 ppm; pantothenic acid, 13 ppm; riboflavin, 6.5 ppm; pyridoxine, 6.5 ppm; folic acid, 0.5 ppm; biotin, 0.1 ppm; vitamin B_{12}, 0.07 ppm; thiamin, 6.5 ppm; and choline, 0.26 percent. It appears that adequate amounts of these vitamins are furnished when calves have functional rumens (usually at 6 weeks of age) by a combination of rumen synthesis and natural feedstuffs.
Source: Nutrient Requirements of Dairy Cattle, sixth revised edition, updated 1989, copyright © 1988 by the National Academy of Sciences.

TABLE 44 Dry Matter Intake Requirements to Fulfill Nutrient Allowances for Maintenance, Milk Production, and Normal Live Weight Gain During Mid- and Late Lactation.

Live Weight (kg)	400	500	600	700	800
FCM (4%)[a] (kg)			– – % Live Wt[b,c] – –		
10	2.7	2.4	2.2	2.0	1.9
15	3.2	2.8	2.6	2.3	2.2
20	3.6	3.2	2.9	2.6	2.4
25	4.0	3.5	3.2	2.9	2.7
30	4.4	3.9	3.5	3.2	2.9
35	5.0	4.2	3.7	3.4	3.1
40	5.5	4.6	4.0	3.6	3.3
45	–	5.0	4.3	3.8	3.5
50	–	5.4	4.7	4.1	3.7
55	–	–	5.0	4.4	4.0
60	–	–	5.4	4.8	4.3

NOTE: The following assumptions were made in calculating the DMI requirements shown in Table 44:
   1. The basic or reference cow used for calculations weighed 600 kg and produced milk with 4 percent milk fat. Other live weights in the table and corresponding fat percentages were 400 kg and 5 percent fat; 500 kg and 4.5 percent fat; and 700 and 800 kg and 3.5 percent fat.
   2. The concentration of energy in the diet for the reference cow was 1.42 Mcal of NEL/kg of DM for milk yields equal to or less than 10 kg/day. It increased linearly to 1.72 Mcal of NEL/kg for milk yields equal to or greater than 40 kg/day.
   3. The energy concentrations of the diets for all other cows were assumed to change linearly as their energy requirements for milk production, relative to maintenance, changed in a manner identical to that of the 600-kg cow as she increased in milk yield from 10 to 40 kg/day.
   4. Enough DM to provide sufficient energy for cows to gain 0.055 percent of their body weight daily was also included in the total. If cows do not consume as much DM as they require, as calculated from Table 44, their energy intake will be less than their requirements. The result will be a loss of body weight, reduced milk yields, or both. If cows consume more DM than what is projected as required from Table 44, the energy concentration of their diet should be reduced or they may become overly fat.

[a]4% Fat-corrected milk (kg) = (0.4) (kg of milk) + (15) (kg of milk fat).
[b]The probable DMI may be up to 18 percent less in early lactation.
[c]DMI as a percentage of live weight may be 0.02 percent less per 1 percent increase in diet moisture content above 50 percent if fermented feeds constitute a major portion of the diet.

Source: Nutrient Requirements of Dairy Cattle, sixth revised edition, updated 1989, copyright © 1988 by the National Academy of Sciences.

TABLE 45 Daily Nutrient Requirements of Growing Dairy Cattle and Mature Bulls.

Live Weight (lb)	Gain (lb)	Dry Matter Intake[a] (lb)	Energy NEM (Mcal)	NEG (Mcal)	ME (Mcal)	DE (Mcal)	TDN (lb)	Protein UIP (lb)	DIP (lb)	CP (lb)	Minerals Ca (lb)	P (lb)	Vitamins A (1,000 IU)	D (1,000 IU)
			*Growing Large-Breed Calves Fed Only Milk or Milk Replacer*											
90	0.6	1.08	1.39	0.37	2.59	2.79	1.32	–	–	0.24	0.015	0.009	1.73	0.27
110	0.8	1.32	1.61	0.52	3.17	3.41	1.70	–	–	0.29	0.019	0.013	2.20	0.33
			Growing Large-Breed Calves Fed Milk Plus Starter Mix											
100	1.0	2.00	1.50	0.64	4.11	4.48	2.24	–	–	0.44	0.018	0.012	1.93	0.30
150	1.8	3.50	2.04	1.29	7.20	7.84	3.92	–	–	0.77	0.034	0.017	2.90	0.50
			Growing Small-Breed Calves Fed Only Milk or Milk Replacer											
60	0.4	0.80	1.02	0.23	1.92	2.06	1.03	–	–	0.18	0.013	0.008	1.16	0.18
75	0.5	1.20	1.21	0.36	2.88	3.10	1.55	–	–	0.26	0.016	0.009	1.44	0.22
			Growing Small-Breed Calves Fed Milk Plus Starter Mix											
100	1.1	2.00	1.50	0.70	4.11	4.48	2.24	–	–	0.44	0.021	0.013	1.93	0.30
150	1.3	3.50	2.04	0.92	7.20	7.84	3.92	–	–	0.77	0.030	0.017	2.90	0.45
			Growing Veal Calves Fed Only Milk or Milk Replacer											
85	0.5	0.80	1.33	0.30	1.92	2.06	1.03	–	–	0.18	0.015	0.009	1.64	0.26
100	0.8	1.20	1.50	0.57	2.88	3.10	1.55	–	–	0.26	0.018	0.010	1.93	0.30
125	1.0	1.50	0.67	1.37	3.60	3.87	1.94	–	–	1.33	0.030	0.018	2.41	0.38
150	1.8	2.50	2.04	1.29	4.76	5.23	2.60	–	–	0.55	0.032	0.019	2.90	0.45
200	2.6	4.00	2.53	2.06	7.62	8.36	4.16	–	–	0.88	0.041	0.023	3.85	0.60
250	2.8	5.00	2.99	2.42	9.52	10.45	5.20	–	–	1.16	0.046	0.027	4.82	0.75
300	2.6	5.60	3.43	2.37	10.67	11.71	5.82	–	–	1.23	0.050	0.031	5.78	0.90
			Large-Breed Growing Females											
200	1.30	5.43	2.53	1.16	6.58	7.61	3.81	0.71	0.08	0.87	0.036	0.018	3.85	0.60
200	1.50	5.80	2.53	1.35	7.03	8.13	4.07	0.77	0.12	0.93	0.038	0.019	3.85	0.60
200	1.70	6.17	2.53	1.54	7.48	8.65	4.33	0.83	0.15	0.99	0.039	0.020	3.85	0.60
300	1.30	7.15	3.43	1.36	8.50	9.86	4.93	0.64	0.27	1.14	0.040	0.023	5.77	0.90
300	1.50	7.60	3.43	1.59	9.04	10.49	5.24	0.69	0.31	1.22	0.041	0.024	5.77	0.90
300	1.70	8.06	3.43	1.81	9.58	11.12	5.56	0.74	0.36	1.29	0.043	0.025	5.77	0.90
400	1.30	8.90	4.25	1.55	10.33	12.03	6.02	0.58	0.45	1.39	0.043	0.028	7.69	1.20
400	1.50	9.44	4.25	1.80	10.96	12.76	6.38	0.62	0.50	1.50	0.045	0.029	7.69	1.20
400	1.70	9.98	4.25	2.06	11.59	13.50	6.75	0.66	0.55	1.60	0.047	0.030	7.69	1.20
500	1.30	10.68	5.03	1.72	12.11	14.15	7.08	0.52	0.62	1.37	0.047	0.033	9.62	1.50
500	1.50	11.31	5.03	2.01	12.83	14.99	7.49	0.56	0.68	1.48	0.049	0.034	9.62	1.50
500	1.70	11.95	5.03	2.30	13.55	15.84	7.92	0.60	0.75	1.58	0.050	0.035	9.62	1.50
600	1.30	12.54	5.76	1.89	13.87	16.27	8.13	0.48	0.80	1.50	0.050	0.037	11.54	1.80
600	1.50	13.27	5.76	2.20	14.68	17.21	8.61	0.51	0.87	1.59	0.051	0.038	11.54	1.80
600	1.70	14.00	5.76	2.52	15.50	18.17	9.09	0.54	0.94	1.68	0.053	0.039	11.54	1.80
700	1.30	14.49	6.47	2.05	15.63	18.40	9.20	0.44	0.97	1.74	0.052	0.039	13.46	2.10
700	1.50	15.32	6.47	2.39	16.53	19.47	9.73	0.47	1.05	1.84	0.053	0.040	13.46	2.10
700	1.70	16.17	6.47	2.74	17.44	20.54	10.27	0.49	1.13	1.94	0.055	0.041	13.46	2.10
800	1.30	16.56	7.15	2.20	17.41	20.59	10.30	0.42	1.15	1.99	0.054	0.041	15.39	2.40
800	1.50	17.52	7.15	2.57	18.41	21.77	10.89	0.44	1.24	2.10	0.055	0.042	15.39	2.40
800	1.70	18.48	7.15	2.95	19.43	22.97	11.49	0.46	1.33	2.22	0.056	0.043	15.39	2.40
900	1.30	18.79	7.81	2.35	19.24	22.85	11.43	0.40	1.33	2.26	0.062	0.042	17.31	2.69
900	1.50	19.88	7.81	2.75	20.35	24.16	12.08	0.41	1.43	2.39	0.062	0.044	17.31	2.69
900	1.70	20.97	7.81	3.15	21.47	25.49	12.75	0.43	1.53	2.52	0.063	0.046	17.31	2.69
1,000	1.30	21.22	8.45	2.49	21.13	25.21	12.61	0.39	1.51	2.55	0.062	0.043	19.23	2.99
1,000	1.50	22.44	8.45	2.92	22.35	26.67	13.33	0.40	1.63	2.69	0.062	0.044	19.23	2.99
1,000	1.70	23.68	8.45	3.34	23.59	28.14	14.07	0.41	1.74	2.84	0.063	0.046	19.23	2.99
1,100	1.30	23.87	9.08	2.64	23.12	27.71	13.86	0.38	1.71	2.86	0.062	0.043	21.16	3.29
1,100	1.50	25.26	9.08	3.08	24.46	29.33	14.66	0.39	1.84	3.03	0.062	0.044	21.16	3.29
1,100	1.70	26.67	9.08	3.53	25.83	30.96	15.48	0.40	1.97	3.20	0.063	0.046	21.16	3.29
1,200	1.30	26.81	9.69	2.77	25.23	30.39	15.20	0.39	1.92	3.22	0.062	0.043	23.08	3.59
1,200	1.50	28.39	9.69	3.24	26.71	32.18	16.09	0.40	2.06	3.41	0.063	0.045	23.08	3.59
1,200	1.70	29.99	9.69	3.72	28.22	34.00	17.00	0.40	2.20	3.60	0.063	0.046	23.08	3.59
1,300	1.30	30.10	10.29	2.91	27.49	33.30	16.65	0.42	2.15	3.61	0.062	0.043	25.00	3.89
1,300	1.50	31.90	10.29	3.40	29.14	35.30	17.65	0.42	2.30	3.83	0.063	0.045	25.00	3.89
1,300	1.70	33.73	10.29	3.90	30.81	37.32	18.66	0.42	2.46	4.05	0.063	0.047	25.00	3.89
			Small-Breed Growing Females											
200	0.90	4.99	2.53	0.89	5.98	6.93	3.47	0.57	0.05	0.80	0.033	0.016	3.85	0.60
200	1.10	5.42	2.53	1.11	6.49	7.52	3.76	0.62	0.09	0.87	0.034	0.017	3.85	0.60
200	1.30	5.84	2.53	1.32	7.00	8.11	4.06	0.67	0.13	0.93	0.036	0.018	3.85	0.60
300	0.90	6.79	3.43	1.07	7.89	9.18	4.59	0.51	0.23	1.09	0.036	0.021	5.77	0.90
300	1.10	7.33	3.43	1.32	8.51	9.91	4.95	0.55	0.29	1.17	0.038	0.022	5.77	0.90
300	1.30	7.87	3.43	1.58	9.15	10.65	5.32	0.60	0.34	1.26	0.040	0.023	5.77	0.90
400	0.90	8.63	4.25	1.22	9.71	11.36	5.68	0.46	0.41	1.14	0.040	0.026	7.69	1.20

(Table 45 continues)

TABLE 45 Daily Nutrient Requirements of Growing Dairy Cattle and Mature Bulls. *(Cont.)*

Live Weight (lb)	Gain (lb)	Dry Matter Intake[a] (lb)	Energy NEM (Mcal)	Energy NEG (Mcal)	Energy ME (Mcal)	Energy DE (Mcal)	Energy TDN (lb)	Protein UIP (lb)	Protein DIP (lb)	Protein CP (lb)	Minerals Ca (lb)	Minerals P (lb)	Vitamins A (1,000 IU)	Vitamins D
400	1.10	9.29	4.25	1.52	10.46	12.23	6.12	0.50	0.48	1.24	0.042	0.027	7.69	1.20
400	1.30	9.97	4.25	1.82	11.21	13.11	6.56	0.53	0.54	1.34	0.043	0.028	7.69	1.20
500	0.90	10.58	5.03	1.37	11.51	13.54	6.77	0.42	0.59	1.27	0.044	0.031	9.62	1.50
500	1.10	11.38	5.03	1.71	12.38	14.56	7.28	0.45	0.66	1.37	0.045	0.032	9.62	1.50
500	1.30	12.19	5.03	2.05	13.26	15.59	7.80	0.48	0.74	1.46	0.047	0.033	9.62	1.50
600	0.90	12.68	5.76	1.51	13.33	15.77	7.88	0.40	0.77	1.52	0.047	0.034	11.54	1.80
600	1.10	13.63	5.76	1.88	14.33	16.95	8.47	0.42	0.86	1.64	0.048	0.036	11.54	1.80
600	1.30	14.60	5.76	2.26	15.35	18.14	9.07	0.44	0.95	1.75	0.050	0.037	11.54	1.80
700	0.90	14.99	6.47	1.65	15.21	18.09	9.05	0.38	0.95	1.80	0.049	0.036	13.46	2.10
700	1.10	16.11	6.47	2.05	16.35	19.44	9.72	0.40	1.06	1.93	0.050	0.038	13.46	2.10
700	1.30	17.26	6.47	2.46	17.51	20.82	10.41	0.41	1.16	2.07	0.052	0.039	13.46	2.10
800	0.90	17.57	7.15	1.78	17.18	20.56	10.28	0.38	1.15	2.11	0.051	0.039	15.39	2.40
800	1.10	18.90	7.15	2.22	18.48	22.12	11.06	0.39	1.27	2.27	0.052	0.040	15.39	2.40
800	1.30	20.25	7.15	2.66	19.80	23.70	11.85	0.40	1.39	2.43	0.054	0.041	15.39	2.40
900	0.90	20.50	7.81	1.91	19.29	23.25	11.62	0.39	1.36	2.46	0.060	0.039	17.31	2.69
900	1.10	22.08	7.81	2.38	20.78	25.03	12.52	0.40	1.50	2.65	0.061	0.041	17.31	2.69
900	1.30	23.68	7.81	2.85	22.28	26.85	13.42	0.41	1.64	2.84	0.062	0.042	17.31	2.69
1,000	0.90	23.75	8.45	2.03	21.54	26.13	13.06	0.42	1.58	2.85	0.060	0.039	19.23	2.99
1,000	1.10	25.61	8.45	2.53	23.23	28.18	14.09	0.42	1.74	3.07	0.061	0.041	19.23	2.99
1,000	1.30	27.50	8.45	3.04	24.95	30.26	15.13	0.42	1.91	3.30	0.062	0.043	19.23	2.99
					Large-Breed Growing Males									
200	1.80	5.91	2.53	1.40	7.16	8.28	4.14	0.91	0.11	0.95	0.040	0.021	3.85	0.6
200	2.00	6.23	2.53	1.57	7.54	8.73	4.36	0.98	0.14	1.00	0.042	0.022	3.85	0.6
200	2.20	6.55	2.53	1.73	7.93	9.17	4.59	1.04	0.16	1.05	0.044	0.023	3.85	0.6
300	1.80	7.50	3.43	1.61	9.04	10.46	5.23	0.83	0.29	1.20	0.044	0.026	5.77	0.9
300	2.00	7.87	3.43	1.80	9.48	10.98	5.49	0.89	0.33	1.26	0.046	0.027	5.77	0.9
300	2.20	8.24	3.43	1.99	9.93	11.50	5.75	0.95	0.36	1.32	0.047	0.028	5.77	0.9
400	1.80	9.17	4.25	1.81	10.85	12.60	6.30	0.77	0.47	1.47	0.048	0.031	7.69	1.2
400	2.00	9.60	4.25	2.02	11.36	13.19	6.59	0.82	0.51	1.54	0.049	0.032	7.69	1.2
400	2.20	10.03	4.25	2.23	11.87	13.78	6.89	0.88	0.55	1.60	0.051	0.033	7.69	1.2
500	1.80	10.85	5.03	1.99	12.60	14.67	7.33	0.71	0.64	1.74	0.051	0.036	9.62	1.5
500	2.00	11.34	5.03	2.22	13.17	15.33	7.67	0.76	0.69	1.81	0.053	0.037	9.62	1.5
500	2.20	11.83	5.03	2.46	13.74	16.00	8.00	0.81	0.73	1.89	0.055	0.037	9.62	1.5
600	1.80	12.55	5.76	2.16	14.30	16.70	8.35	0.66	0.81	1.73	0.053	0.040	11.54	1.8
600	2.00	13.11	5.76	2.41	14.94	17.44	8.72	0.70	0.86	1.84	0.055	0.041	11.54	1.8
600	2.20	13.67	5.76	2.67	15.57	18.18	9.09	0.75	0.91	1.96	0.056	0.042	11.54	1.8
700	1.80	14.31	6.47	2.32	15.99	18.72	9.36	0.61	0.97	1.72	0.055	0.042	13.46	2.1
700	2.00	14.93	6.47	2.60	16.69	19.54	9.77	0.65	1.03	1.83	0.057	0.043	13.46	2.1
700	2.20	15.56	6.47	2.87	17.39	20.36	10.18	0.69	1.09	1.94	0.058	0.044	13.46	2.1
800	1.80	16.13	7.15	2.48	17.67	20.76	10.38	0.57	1.14	1.94	0.057	0.044	15.39	2.4
800	2.00	16.83	7.15	2.78	18.43	21.65	10.83	0.61	1.20	2.02	0.059	0.045	15.39	2.4
800	2.20	17.52	7.15	3.07	19.20	22.55	11.27	0.64	1.27	2.10	0.060	0.046	15.39	2.4
900	1.80	18.04	7.81	2.63	19.36	22.82	11.41	0.54	1.31	2.16	0.063	0.047	17.31	2.6
900	2.00	18.81	7.81	2.95	20.19	23.79	11.90	0.57	1.38	2.26	0.064	0.048	17.31	2.6
900	2.20	19.58	7.81	3.26	21.02	24.77	12.39	0.60	1.45	2.35	0.065	0.050	17.31	2.6
1,000	1.80	20.05	8.45	2.78	21.08	24.92	12.46	0.51	1.47	2.41	0.063	0.047	19.23	2.9
1,000	2.00	20.90	8.45	3.12	21.97	25.98	12.99	0.54	1.55	2.51	0.064	0.048	19.23	2.9
1,000	2.20	21.76	8.45	3.45	22.88	27.05	13.52	0.57	1.63	2.61	0.065	0.050	19.23	2.9
1,100	1.80	22.18	9.08	2.93	22.83	27.08	13.54	0.49	1.65	2.66	0.064	0.047	21.16	3.2
1,100	2.00	23.12	9.08	3.28	23.79	28.23	14.12	0.52	1.74	2.77	0.064	0.049	21.16	3.2
1,100	2.20	24.07	9.08	3.63	24.77	29.39	14.69	0.54	1.82	2.89	0.065	0.050	21.16	3.2
1,200	1.80	24.45	9.69	3.07	24.62	29.32	14.66	0.48	1.83	2.93	0.064	0.047	23.08	3.5
1,200	2.00	25.48	9.69	3.44	25.67	30.57	15.28	0.50	1.92	3.06	0.064	0.049	23.08	3.5
1,200	2.20	26.53	9.69	3.81	26.72	31.82	15.91	0.52	2.02	3.18	0.065	0.050	23.00	3.6
1,300	1.80	26.88	10.29	3.21	26.48	31.65	15.83	0.47	2.01	3.23	0.064	0.047	25.00	3.8
1,300	2.00	28.03	10.29	3.60	27.61	33.00	16.50	0.49	2.11	3.36	0.064	0.049	25.00	3.8
1,300	2.20	29.18	10.29	3.98	28.75	34.36	17.18	0.51	2.22	3.50	0.065	0.051	25.00	3.8
1,400	1.80	29.51	10.88	3.35	28.42	34.11	17.05	0.47	2.20	3.54	0.064	0.048	26.93	4.1
1,400	2.00	30.78	10.88	3.75	29.64	35.57	17.79	0.49	2.32	3.69	0.065	0.049	26.93	4.1
1,400	2.20	32.06	10.88	4.15	30.87	37.05	18.52	0.51	2.43	3.85	0.065	0.051	26.93	4.1
1,500	1.80	32.37	11.46	3.48	30.47	36.71	18.35	0.48	2.41	3.88	0.064	0.048	28.85	4.4
1,500	2.00	33.78	11.46	3.90	31.79	38.30	19.15	0.50	2.53	4.05	0.065	0.049	28.85	4.4
1,500	2.20	35.19	11.46	4.32	33.12	39.90	19.95	0.51	2.66	4.22	0.065	0.051	28.85	4.4
1,600	1.80	35.51	12.03	3.62	32.63	39.49	19.74	0.50	2.63	4.26	0.064	0.048	30.77	4.7
1,600	2.00	37.07	12.03	4.05	34.06	41.22	20.61	0.51	2.76	4.45	0.065	0.050	30.77	4.7
1,600	2.20	38.64	12.03	4.49	35.51	42.96	21.48	0.53	2.90	4.64	0.065	0.051	30.77	4.7
1,700	1.80	38.04	12.58	3.75	34.51	41.86	20.93	0.50	2.81	4.56	0.064	0.048	32.70	5.0

(Table 45 continues)

TABLE 45 Daily Nutrient Requirements of Growing Dairy Cattle and Mature Bulls. *(Cont.)*

Live Weight (lb)	Gain (lb)	Dry Matter Intake[a] (lb)	Energy NEM (Mcal)	NEG (Mcal)	ME (Mcal)	DE (Mcal)	TDN (lb)	Protein UIP (lb)	DIP (lb)	CP (lb)	Minerals Ca (lb)	P (lb)	Vitamins A (1,000 IU)	D
1,700	2.00	39.72	12.58	4.19	36.03	43.70	21.85	0.51	2.96	4.77	0.065	0.050	32.70	5.0
1,700	2.20	41.40	12.58	4.65	37.56	45.56	22.78	0.52	3.10	4.97	0.066	0.052	32.70	5.0
1,800	1.80	39.57	13.14	3.87	35.90	43.54	21.77	0.47	2.95	4.75	0.064	0.048	34.62	5.3
1,800	2.00	41.31	13.14	4.34	37.47	45.45	22.73	0.48	3.10	4.96	0.065	0.050	34.62	5.3
1,800	2.20	43.05	13.14	4.81	39.06	47.37	23.69	0.48	3.25	5.17	0.066	0.052	34.62	5.3
Small-Breed Growing Males														
200	1.10	5.08	2.53	0.98	6.15	7.12	3.56	0.65	0.05	0.81	0.034	0.017	3.85	0.60
200	1.30	5.45	2.53	1.17	6.60	7.63	3.82	0.71	0.09	0.87	0.036	0.018	3.85	0.60
200	1.50	5.81	2.53	1.35	7.04	8.15	4.07	0.77	0.12	0.93	0.038	0.019	3.85	0.60
300	1.10	6.72	3.43	1.15	7.99	9.27	4.64	0.58	0.23	1.08	0.038	0.022	5.77	0.90
300	1.30	7.17	3.43	1.37	8.52	9.89	4.94	0.64	0.27	1.15	0.040	0.023	5.77	0.90
300	1.50	7.62	3.43	1.59	9.06	10.51	5.25	0.69	0.31	1.22	0.041	0.024	5.77	0.90
400	1.10	8.39	4.25	1.31	9.74	11.34	5.67	0.53	0.40	1.27	0.042	0.027	7.69	1.20
400	1.30	8.92	4.25	1.56	10.36	12.06	6.03	0.57	0.45	1.38	0.043	0.028	7.69	1.20
400	1.50	9.46	4.25	1.81	10.98	12.79	6.39	0.62	0.50	1.50	0.045	0.029	7.69	1.20
500	1.10	10.09	5.03	1.45	11.44	13.37	6.68	0.48	0.57	1.26	0.045	0.032	9.62	1.50
500	1.30	10.71	5.03	1.74	12.15	14.19	7.10	0.52	0.63	1.37	0.047	0.033	9.62	1.50
500	1.50	11.34	5.03	2.02	12.86	15.02	7.51	0.56	0.69	1.47	0.049	0.034	9.62	1.50
600	1.10	11.85	5.76	1.59	13.12	15.38	7.69	0.44	0.73	1.42	0.048	0.036	11.54	1.80
600	1.30	12.57	5.76	1.90	13.91	16.31	8.16	0.48	0.80	1.51	0.050	0.037	11.54	1.80
600	1.50	13.30	5.76	2.22	14.71	17.25	8.63	0.51	0.87	1.60	0.051	0.038	11.54	1.80
700	1.10	13.71	6.47	1.73	14.79	17.41	8.71	0.41	0.90	1.65	0.050	0.038	13.46	2.10
700	1.30	14.53	6.47	2.06	15.68	18.46	9.23	0.44	0.97	1.74	0.052	0.039	13.46	2.10
700	1.50	15.36	6.47	2.41	16.57	19.51	9.76	0.47	1.05	1.84	0.053	0.040	13.46	2.10
800	1.10	15.68	7.15	1.86	16.48	19.49	9.74	0.39	1.06	1.88	0.052	0.040	15.39	2.40
800	1.30	16.61	7.15	2.22	17.47	20.65	10.33	0.41	1.15	1.99	0.054	0.041	15.39	2.40
800	1.50	17.56	7.15	2.59	18.46	21.82	10.91	0.43	1.24	2.11	0.055	0.042	15.39	2.40
900	1.10	17.79	7.81	1.98	18.21	21.63	10.81	0.37	1.23	2.13	0.061	0.041	17.31	2.69
900	1.30	18.85	7.81	2.37	19.30	22.92	11.46	0.39	1.33	2.26	0.062	0.042	17.31	2.69
900	1.50	19.92	7.81	2.76	20.40	24.22	12.11	0.41	1.43	2.39	0.062	0.044	17.31	2.69
1,000	1.10	20.08	8.45	2.10	20.00	23.86	11.93	0.37	1.41	2.41	0.061	0.041	19.23	2.99
1,000	1.30	21.28	8.45	2.52	21.20	25.29	12.65	0.38	1.52	2.55	0.062	0.043	19.23	2.99
1,000	1.50	22.50	8.45	2.94	22.41	26.73	13.37	0.39	1.63	2.70	0.062	0.044	19.23	2.99
1,100	1.10	22.58	9.08	2.22	21.87	26.22	13.11	0.37	1.60	2.71	0.061	0.041	21.16	3.29
1,100	1.30	23.94	9.08	2.66	23.19	27.80	13.90	0.38	1.72	2.87	0.062	0.043	21.16	3.29
1,100	1.50	25.32	9.08	3.10	24.53	29.40	14.70	0.39	1.84	3.04	0.062	0.044	21.16	3.29
1,200	1.10	25.34	9.69	2.34	23.85	28.73	14.37	0.38	1.79	3.04	0.061	0.041	23.08	3.59
1,200	1.30	26.89	9.69	2.80	25.31	30.49	15.25	0.39	1.93	3.23	0.062	0.043	23.08	3.59
1,200	1.50	28.46	9.69	3.27	26.78	32.27	16.13	0.40	2.07	3.42	0.063	0.045	23.08	3.59
1,300	1.10	28.43	10.29	2.45	25.97	31.46	15.73	0.41	2.00	3.41	0.061	0.042	25.00	3.89
1,300	1.30	30.20	10.29	2.94	27.58	33.41	16.71	0.41	2.16	3.62	0.062	0.043	25.00	3.89
1,300	1.50	31.98	10.29	3.43	29.22	35.39	17.70	0.42	2.31	3.84	0.063	0.045	25.00	3.89
Maintenance of Mature Breeding Bulls														
1,200	–	18.55	9.69	–	16.83	20.41	10.20	0.35	1.14	1.86	0.049	0.030	23.08	3.59
1,400	–	20.82	10.88	–	18.89	22.91	11.45	0.34	1.34	2.08	0.057	0.035	26.93	4.19
1,600	–	23.01	12.03	–	20.88	25.32	12.66	0.32	1.53	2.30	0.065	0.040	30.77	4.79
1,800	–	25.14	13.14	–	22.81	27.66	13.83	0.31	1.72	2.51	0.073	0.045	34.62	5.39
2,000	–	27.21	14.22	–	24.68	29.93	14.97	0.30	1.90	2.72	0.081	0.050	38.47	5.99
2,200	–	29.22	15.27	–	26.51	32.15	16.08	0.28	2.08	2.92	0.089	0.055	42.31	6.59
2,400	–	31.19	16.30	–	28.30	34.32	17.16	0.27	2.25	3.12	0.097	0.060	46.16	7.19
2,600	–	33.12	17.31	–	30.05	36.44	18.22	0.26	2.42	3.31	0.105	0.065	50.00	7.78
2,800	–	35.01	18.30	–	31.77	38.53	19.26	0.24	2.58	3.50	0.113	0.070	53.85	8.38
3,000	–	36.87	19.27	–	33.45	40.57	20.29	0.23	2.75	3.69	0.122	0.075	57.70	8.98

NOTE: The following abbreviations were used: NEM, net energy for maintenance; NEG, net energy for gain; ME, metabolizable energy; DE, digestible energy; TDN, total digestible nutrients; UIP, undegraded intake protein; DIP, degraded intake protein; CP, crude protein.
[a]The data for DMI are not requirements per se, unlike the requirements for net energy maintenance, net energy gain, and absorbed protein. They are not intended to be estimates of voluntary intake but are consistent with the specified dietary energy concentrations. The use of diets with decreased energy concentrations will increase dry matter intake needs; metabolizable energy, digestible energy, and total digestible nutrient needs; and crude protein needs. The use of diets with increased energy concentrations will have opposite effects on these needs.

Source: Nutrient Requirements of Dairy Cattle, sixth revised edition, updated 1989, copyright © 1988 by the National Academy of Sciences.

TABLE 46 Daily Nutrient Requirements of Lactating and Pregnant Cows.

Live Weight (lb)	Energy				Total Crude Protein (lb)	Minerals		Vitamins	
	NEL (Mcal)	ME (Mcal)	DE (Mcal)	TDN (lb)		Ca (lb)	P (lb)	A (1,000 IU)	D
Maintenance of Mature Lactating Cows[a]									
700	6.02	10.10	11.61	5.80	0.613	0.028	0.020	24	10
800	6.65	11.17	12.83	6.42	0.661	0.032	0.023	28	11
900	7.27	12.20	14.01	7.01	0.708	0.036	0.026	31	12
1,000	7.86	13.20	15.17	7.58	0.755	0.041	0.029	34	14
1,100	8.45	14.18	16.29	8.15	0.801	0.045	0.031	38	15
1,200	9.02	15.13	17.39	8.70	0.846	0.049	0.034	41	16
1,300	9.57	16.07	18.47	9.23	0.892	0.053	0.037	45	18
1,400	10.12	16.99	19.52	9.76	0.932	0.057	0.040	48	19
1,500	10.66	17.89	20.56	10.28	0.973	0.061	0.043	52	20
1,600	11.19	18.78	21.58	10.79	1.011	0.065	0.046	55	22
1,700	11.71	19.65	22.58	11.29	1.049	0.069	0.049	59	23
1,800	12.22	20.51	23.57	11.79	1.087	0.073	0.051	62	24
Maintenance Plus Last 2 Months of Gestation of Mature Dry Cows[b]									
700	7.82	12.96	15.20	7.63	1.651	0.046	0.028	24	10
800	8.65	14.33	16.81	8.43	1.825	0.053	0.032	28	11
900	9.45	15.65	18.36	9.21	1.993	0.059	0.036	31	12
1,000	10.22	16.94	19.87	9.97	2.157	0.066	0.040	34	14
1,100	10.98	18.19	21.34	10.71	2.317	0.072	0.044	38	15
1,200	11.72	19.42	22.78	11.43	2.473	0.079	0.048	41	16
1,300	12.45	20.62	24.19	12.14	2.626	0.086	0.052	45	18
1,400	13.16	21.80	25.57	12.83	2.776	0.092	0.056	48	19
1,500	13.86	22.98	26.93	13.51	2.924	0.099	0.060	52	20
1,600	14.54	24.10	28.26	14.18	3.069	0.105	0.064	55	22
1,700	15.22	25.22	29.58	14.84	3.211	0.112	0.068	59	23
1,800	15.89	26.32	30.87	15.49	3.352	0.118	0.072	62	24
Milk Production—Nutrients/lb of Milk of Different Fat Percentages									
(Fat %)									
3.0	0.29	0.49	0.56	0.280	0.078	0.0027	0.0017	–	–
3.5	0.31	0.52	0.60	0.301	0.084	0.0030	0.0018	–	–
4.0	0.33	0.56	0.64	0.322	0.090	0.0032	0.0020	–	–
4.5	0.36	0.60	0.69	0.343	0.096	0.0035	0.0021	–	–
5.0	0.38	0.63	0.73	0.364	0.101	0.0037	0.0023	–	–
5.5	0.40	0.67	0.77	0.385	0.107	0.0039	0.0024	–	–
Live Weight Change During Lactation—Nutrients/lb of Weight Change									
Weight loss	−2.23	−3.74	−4.33	−2.17	−0.320	–	–	–	–
Weight gain	2.32	3.88	4.52	2.26	0.320	–	–	–	–

NOTE: The following abbreviations were used: NEL, net energy for lactation; ME, metabolizable energy; DE, digestible energy; TDN, total digestible nutrients.

[a]To allow for growth of young lactating cows, increase the maintenance allowances for all nutrients except vitamins A and D by 20 percent during the first lactation and 10 percent during the second lactation.

[b]Values for calcium assume that the cow is in calcium balance at the beginning of the last 2 months of gestation. If the cow is not in balance, then the calcium requirement can be increased from 25 to 33 percent.

[c]No allowance is made for mobilized calcium and phosphorus associated with live weight loss or with live weight gain. The maximum daily nitrogen available from weight loss is assumed to be 0.066 lb or 0.515 lb of crude protein.

Source: Nutrient Requirements of Dairy Cattle, sixth revised edition, updated 1989, copyright © 1988 by the National Academy of Sciences.

TABLE 47 Recommended Nutrient Content of Diets for Dairy Cattle.

Cow Wt (lb)	Fat (%)	Wt. Gain (lb/d)
900	5.0	0.50
1,100	4.5	0.60
1,300	4.0	0.72
1,500	3.5	0.82
1,700	3.5	0.94

	Lactating Cow Diets (Milk Yield, lb/d)					Early Lactation (wks 0–3)	Dry, Pregnant Cows	Calf Milk Replacer	Calf Starter Mix	Growing Heifers and Bulls[a] — 3–6 Mos	Growing 6–12 Mos	Growing >12 Mos	Mature Bulls	Maximum Tolerable Levels[b,c]
Milk Yield (lb/d)	14	29	43	58	74									
	18	36	55	73	91									
	23	47	70	93	117									
	26	52	78	104	130									
	29	57	86	114	143									
Energy														
NEL, Mcal/lb	0.65	0.69	0.73	0.78	0.78	0.76	0.57	—	—	—	—	—	—	—
NEM, Mcal/lb	—	—	—	—	—	—	—	1.09	0.86	0.77	0.72	0.63	0.52	—
NEG, Mcal/lb	—	—	—	—	—	—	—	0.70	0.54	0.49	0.44	0.37	—	—
ME, Mcal/lb	1.07	1.16	1.25	1.31	1.31	1.27	0.93	1.71	1.41	1.18	1.12	1.03	0.91	—
DE, Mcal/lb	1.26	1.35	1.44	1.50	1.50	1.46	1.12	1.90	1.60	1.37	1.31	1.22	1.10	—
TDN, % of DM	63	67	71	75	75	73	56	95	80	69	66	61	55	—
Protein equivalent														
Crude protein, %	12	15	16	17	18	19	12	22	18	16	14	12	10	—
UIP, %	4.5	5.4	5.7	6.0	6.3	7.2	—	—	—	8.2	4.3	2.1		—
DIP, %	7.9	8.8	9.7	10.4	10.4	9.7	—	—	—	4.6	6.4	7.2		—
Fiber content (min.)[d]														
Crude fiber, %	17	17	17	15	15	17	22	—	—	13	15	15	15	—
Acid detergent fiber, %	21	21	21	19	19	21	27	—	—	16	19	19	19	—
Neutral detergent fiber, %	28	28	28	25	25	28	35	—	—	23	25	25	25	—
Ether extract (min.)	3	3	3	3	3	3	3	10	3	3	3	3	3	—
Minerals														
Calcium, %	0.43	0.53	0.60	0.65	0.66	0.77	0.39[e]	0.70	0.60	0.52	0.41	0.29	0.30	2.00
Phosphorus, %	0.28	0.34	0.38	0.42	0.41	0.49	0.24	0.60	0.40	0.31	0.30	0.23	0.19	1.00
Magnesium, %[f]	0.20	0.20	0.20	0.25	0.25	0.25	0.16	0.07	0.10	0.16	0.16	0.16	0.16	0.50
Potassium, %[g]	0.90	0.90	0.90	1.00	1.00	1.00	0.65	0.65	0.65	0.65	0.65	0.65	0.65	3.00
Sodium, %	0.18	0.18	0.18	0.18	0.18	0.18	0.10	0.10	0.10	0.10	0.10	0.10	0.10	—
Chlorine, %	0.25	0.25	0.25	0.25	0.25	0.25	0.20	0.20	0.20	0.20	0.20	0.20	0.20	—
Sulfur, %	0.20	0.20	0.20	0.20	0.20	0.25	0.16	0.29	0.20	0.16	0.16	0.16	0.16	0.40
Iron, ppm	50	50	50	50	50	50	50	100	50	50	50	50	50	1,000
Cobalt, ppm	0.10	0.10	0.10	0.10	0.10	0.10	0.10	0.10	0.10	0.10	0.10	0.10	0.10	10.00
Copper, ppm[h]	10	10	10	10	10	10	10	10	10	10	10	10	10	100
Manganese, ppm	40	40	40	40	40	40	40	40	40	40	40	40	40	1,000
Zinc, ppm	40	40	40	40	40	40	40	40	40	40	40	40	40	500
Iodine, ppm[i]	0.60	0.60	0.60	0.60	0.60	0.60	0.25	0.25	0.25	0.25	0.25	0.25	0.25	50.00[j]
Selenium, ppm	0.30	0.30	0.30	0.30	0.30	0.30	0.30	0.30	0.30	0.30	0.30	0.30	0.30	2.00
Vitamins[k]														
A, IU/lb	1,450	1,450	1,450	1,450	1,450	1,800	1,800	1,700	1,000	1,000	1,000	1,000	1,450	30,000
D, IU/lb	450	450	450	450	450	450	540	270	140	140	140	140	140	4,500
E, IU/lb	7	7	7	7	7	7	7	18	11	11	11	11	7	900

NOTE: The values presented in this table are intended as guidelines for the use of professionals in diet formulation. Because of the many factors affecting such values, they are not intended and should not be used as a legal or regulatory base.

(Table 47 continues)

TABLE 47 Recommended Nutrient Content of Diets for Dairy Cattle. *(Cont.)*

[a]The approximate weight for growing heifers and bulls at 3–6 mos is 331 lb; at 6–12 mos, it is 559 lb; and at more than 12 mos, it is 881 lb. The approximate average daily gain is 1.543 lb/day.

[b]The maximum safe levels for many of the mineral elements are not well defined and may be substantially affected by specific feeding conditions. Additional information is available in *Mineral Tolerance of Domestic Animals* (NRC, 1980).

[c]Vitamin tolerances are discussed in detail in *Vitamin Tolerance of Animals* (NRC, 1987b).

[d]It is recommended that 75 percent of the NDF in lactating cow diets be provided as forage. If this recommendation is not followed, a depression in milk fat may occur.

[e]The value for calcium assumes that the cow is in calcium balance at the beginning of the dry period. If the cow is not in balance, then the dietary calcium requirement should be increased by 25 to 33 percent.

[f]Under conditions conducive to grass tetany (see text), magnesium should be increased to 0.25 or 0.30 percent.

[g]Under conditions of heat stress, potassium should be increased to 1.2 percent (see text).

[h]The cow's copper requirement is influenced by molybdenum and sulfur in the diet (see text).

[i]If the diet contains as much as 25 percent strongly goitrogenic feed on a dry basis, the iodine provided should be increased two times or more.

[j]Although cattle can tolerate this level of iodine, lower levels may be desirable to reduce the iodine content of milk.

[k]The following minimum quantities of B-complex vitamins are suggested per unit of milk replacer: niacin, 2.6 ppm; pantothenic acid, 13 ppm; riboflavin, 6.5 ppm; pyridoxine, 6.5 ppm; folic acid, 0.5 ppm; biotin, 0.1 ppm; vitamin B_{12}, 0.07 ppm; thiamin, 6.5 ppm; and choline, 0.26 percent. It appears that adequate amounts of these vitamins are furnished when calves have functional rumens (usually at 6 weeks of age) by a combination of rumen synthesis and natural feedstuffs.

Source: Nutrient Requirements of Dairy Cattle, sixth revised edition, updated 1989, copyright © 1988 by the National Academy of Sciences.

TABLE 48 Composition of Some Common Feeds, Excluding Amino Acids.

Feed Name	International Reference Number	Dry Matter (%)	Crude Fiber[a] (%)	Crude Protein[a] (%)	Acid Detergent Fiber[a] (%)	BEEF CATTLE[a]				
						DE (Mcal/ kg)	ME (Mcal/ kg)	NE$_m$ (Mcal/ kg)	NE$_g$ (Mcal/ kg)	TDN (%)
ALFALFA *Medicago sativa*										
1. fresh	2-00-196	24	26.2	19.7	35	—	—	—	—	—
2. fresh, midbloom	2-00-185	24	28.0	18.3	35	2.56	2.10	1.24	0.68	58
3. hay, sun-cured, midbloom	1-00-063	90	26.0	17.0	35	2.56	2.10	1.24	0.68	58
4. meal, dehydrated, 17% protein	1-00-023	92	26.2	18.9	35	2.69	2.21	1.34	0.77	61
5. silage, late vegetative	3-00-204	21	34.6	20.0	47	—	—	—	—	—
6. silage, full bloom	3-00-207	26	34.9	17.5	38	—	—	—	—	—
7. silage	3-00-212	41	33.0	17.8	—	—	—	—	—	—
8. silage wilted, midbloom	3-00-217	38	30.0	15.5	35	2.56	2.10	1.24	0.68	58
BAHIAGRASS *Paspalum notatum*										
9. fresh	2-00-464	30	30.4	8.9	38	2.38	1.95	1.11	0.55	54
10. hay, sun-cured	1-00-462	91	32.0	8.2	41	2.25	1.84	1.00	0.45	51
BAKERY										
11. waste, dehydrated (dried bakery product)	4-00-466	92	1.3	10.7	13	3.92	3.22	2.21	1.52	89
BARLEY *Hordeum vulgare*										
12. grain	4-00-549	88	5.7	13.5	7	3.70	3.04	2.06	1.40	84
13. grain, pacific coast	4-07-939	89	7.1	10.8	9	3.79	3.11	2.12	1.45	86
14. hay, sun-cured	1-00-495	87	27.5	8.7	—	2.47	2.03	1.18	0.61	56
15. silage	3-00-512	31	30.0	10.3	—	2.25	1.84	1.00	0.45	51
BERMUDAGRASS, COASTAL *Cynodon dactylon*										
16. fresh	2-00-719	29	28.4	15.0	—	2.82	2.31	1.44	0.86	64
17. hay, sun-cured	1-00-716	90	30.7	6.0	38	2.16	1.77	0.93	0.39	49
BLOOD										
18. meal	5-00-380	92	1.1	87.2	—	—	—	—	—	—
19. meal spray dehydrated	5-00-381	93	0.6	88.9	—	—	—	—	—	—
BLUEGRASS, KENTUCKY *Poa pratensis*										
20. fresh, early vegetative	2-00-777	31	25.3	17.4	29	3.17	2.60	1.70	1.08	72
21. fresh, mature	2-00-784	42	32.2	9.5	40	2.47	2.03	1.18	0.61	56
22. hay, sun-cured	1-00-776	89	31.0	13.0	—	2.47	2.03	1.18	0.61	56
BONE MEAL										
23. feeding (more than 10% P)	6-00-397	94	—	24.5	—	—	—	—	—	—
BONE MEAL										
24. steamed	6-00-400	97	2.0	13.2	—	—	—	—	—	—
BREWERS										
25. grains, dehydrated	5-02-141	92	14.9	25.4	24	2.91	2.39	1.51	0.91	66
26. grains, wet	5-02-142	21	14.9	25.4	23	2.91	2.39	1.51	0.91	66
BROME *Bromus* spp.										
27. fresh, early vegetative	2-00-892	34	24.0	18.0	31	3.26	2.68	1.76	1.14	74
28. hay, sun-cured, late bloom	1-00-888	89	37.0	10.0	43	2.43	1.99	1.14	0.58	55
29. hay, sun-cured	1-00-890	91	33.3	9.7	—	—	—	—	—	—
BROME, SMOOTH *Bromus inermis*										
30. fresh, early vegetative	2-00-956	30	22.8	21.3	27	3.22	2.64	1.73	1.11	73
31. fresh, mature	2-08-364	55	34.8	6.0	—	2.34	1.92	1.07	0.52	53
32. hay, sun-cured, midbloom	1-05-633	90	31.8	14.6	37	2.47	2.03	1.18	0.61	56
CANARYGRASS, REED *Phalaris arundinacea*										
33. fresh	2-01-113	27	29.5	11.6	28	2.65	2.17	1.31	0.74	60
CATTLE *Bos taurus*										
34. manure, dehydrated, all forage	1-28-274	92	—	17.0	46	1.10	0.90	0.01	—	25
35. manure, dehy (high concentrate)	1-28-213	92	—	25.0	26	1.90	1.56	0.72	0.18	43
36. manure, dehy, forage and concentrate	1-28-214	92	31.4	17.0	34	1.32	1.09	0.22	—	30

(Table 48 continues)

TABLE 48 Composition of Some Common Feeds, Excluding Amino Acids. *(Cont.)*

SHEEP[a]						DAIRY CATTLE[a]					
								(Growing)		(Lactating)	
DE (Mcal/kg)	ME (Mcal/kg)	NE_m (Mcal/kg)	NE_g (Mcal/kg)	TDN (%)	Dig. Protein (%)	DE (Mcal/kg)	ME (Mcal/kg)	NE_m (Mcal/kg)	NE_g (Mcal/kg)	NE_l (Mcal/kg)	TDN (%)
2.56	2.10	1.24	0.68	58	14.6	—	—	—	—	—	—
—	—	—	—	—	—	—	—	—	—	—	—
2.47	2.03	1.18	0.61	56	12.9	2.56	2.13	1.24	0.68	1.30	58
2.65	2.17	1.34	0.77	60	12.7	2.69	2.27	1.34	0.77	1.38	61
—	—	—	—	—	—	—	—	—	—	—	—
2.60	2.13	1.28	0.71	59	13.0	—	—	—	—	—	—
—	—	—	—	—	—	—	—	—	—	—	—
—	—	—	—	—	—	2.38	1.95	1.11	0.55	1.20	54
—	—	—	—	—	—	—	—	—	—	—	—
—	—	—	—	—	—	3.92	3.51	2.20	1.52	2.06	89
3.79	3.11	2.12	1.45	86	11.1	3.70	3.29	2.06	1.40	1.94	84
3.88	3.18	2.18	1.50	88	7.1	3.79	3.38	2.12	1.45	1.99	86
2.47	2.03	1.18	0.61	56	4.7	2.47	2.04	1.18	0.62	1.25	56
—	—	—	—	—	—	—	—	—	—	—	—
2.38	1.95	1.11	0.55	54	3.8	—	—	—	—	—	—
2.95	2.42	1.54	0.94	67	71.0	—	—	—	—	—	—
—	—	—	—	—	—	—	—	—	—	—	—
2.87	2.35	1.47	0.88	65	13.2	3.17	2.76	1.60	1.08	1.64	72
—	—	—	—	—	—	—	—	—	—	—	—
—	—	—	—	—	—	—	—	—	—	—	—
—	—	—	—	—	—	—	—	—	—	—	—
—	—	—	—	—	—	—	—	—	—	—	—
3.09	2.53	1.63	1.03	70	21.5	2.91	2.49	1.51	0.91	1.50	66
—	—	—	—	—	—	2.91	2.49	1.51	0.91	1.50	66
3.53	2.89	1.94	1.30	80	14.8	3.26	2.85	1.75	1.13	1.69	74
—	—	—	—	—	—	2.60	2.18	1.27	0.70	1.33	59
2.43	1.99	1.14	0.58	55	5.3	—	—	—	—	—	—
—	—	—	—	—	—	—	—	—	—	—	—
—	—	—	—	—	—	—	—	—	—	—	—
—	—	—	—	—	—	—	—	—	—	—	—
2.47	2.03	1.18	0.61	56	1.7	—	—	—	—	—	—
—	—	—	—	—	—	—	—	—	—	—	—
—	—	—	—	—	—	—	—	—	—	—	—
—	—	—	—	—	—	—	—	—	—	—	—

(Table 48 continues)

TABLE 48 Composition of Some Common Feeds, Excluding Amino Acids *(Cont.)*.

| | HORSES[a] | | | GOATS[a] | | | | | | SWINE[b] | |
Feed Name	DE (Mcal/kg)	Neutral Detergent Fiber (%)	Crude Protein (%)	DE (Mcal/kg)	ME (Mcal/kg)	NE_m (Mcal/kg)	NE_g (Mcal/kg)	NE_l (Mcal/kg)	TDN (%)	DE (kcal/kg)	ME (kcal/kg)
ALFALFA *Medicago sativa*											
1. fresh	—	—	—	2.43[c]	1.99[c]	1.19[c]	0.47[c]	1.23[c]	55	—	—
2. fresh, midbloom	—	—	—	—	—	—	—	—	—	—	—
3. hay, sun-cured, midbloom	2.28	47.1	18.7	—	—	—	—	—	—	—	—
4. meal, dehydrated, 17% protein	2.36	45.0	18.9	2.69[c]	2.21[c]	1.33[c]	0.69[c]	1.38[c]	61[d]	1880	1705
5. silage, late vegetative	—	—	—	2.78[c]	2.28[c]	1.39[c]	0.75[c]	1.42[c]	63[d]	—	—
6. silage, full bloom	—	—	—	2.56[c]	2.10[c]	1.26[c]	0.58[c]	1.30[c]	58[d]	—	—
7. silage	—	—	—	—	—	—	—	—	—	—	—
8. silage wilted, midbloom	—	—	—	—	—	—	—	—	—	—	—
BAHIAGRASS *Paspalum notatum*											
9. fresh	2.03	73.1	12.6	—	—	—	—	—	—	—	—
10. hay, sun-cured	1.94	73.9	9.5	2.25[c]	1.84[c]	1.10[c]	0.32[c]	1.13[c]	51[d]	—	—
BAKERY											
11. waste, dehydrated (dried bakery product)	—	—	—	—	—	—	—	—	—	3983	3738
BARLEY *Hordeum vulgare*											
12. grain	3.68	19.0	13.2	3.70[c]	3.04[c]	2.00[c]	1.35[c]	1.94[c]	84	3120	3040
13. grain, pacific coast	3.58	21.0	11.0	—	—	—	—	—	—	3094	2918
14. hay, sun-cured	2.01	—	8.8	—	—	—	—	—	—	—	—
15. silage	—	—	—	—	—	—	—	—	—	—	—
BERMUDAGRASS, COASTAL *Cynodon dactylon*											
16. fresh	2.38	73.3	12.6	—	—	—	—	—	—	—	—
17. hay, sun-cured	—	—	—	—	—	—	—	—	—	—	—
BLOOD											
18. meal	—	—	—	—	—	—	—	—	—	—	—
19. meal spray dehydrated	—	—	—	4.01[c]	3.29[c]	2.22[c]	1.52[c]	2.11[c]	91[d]	2980	2330
BLUEGRASS, KENTUCKY *Poa pratensis*											
20. fresh, early vegetative	2.09	—	17.4	—	—	—	—	—	—	—	—
21. fresh, mature	—	—	—	—	—	—	—	—	—	—	—
22. hay, sun-cured	—	—	—	2.69[c]	2.21[c]	1.33[c]	0.69[c]	1.38[c]	61[d]	—	—
BONE MEAL											
23. feeding (more than 10% P)	—	—	—	—	—	—	—	—	—	—	—
BONE MEAL											
24. steamed	—	—	—	—	—	—	—	—	—	—	—
BREWERS											
25. grains, dehydrated	2.75	46.0	25.4	3.09[c]	2.53[c]	1.58[c]	0.97[c]	1.60[c]	70[d]	2090	1900
26. grains, wet	—	—	—	—	—	—	—	—	—	—	—
BROME *Bromus* spp.											
27. fresh, early vegetative	—	—	—	—	—	—	—	—	—	—	—
28. hay, sun-cured, late bloom	—	—	—	—	—	—	—	—	—	—	—
29. hay, sun-cured	—	—	—	—	—	—	—	—	—	—	—
BROME, SMOOTH *Bromus inermis*											
30. fresh, early vegetative	2.59	47.9	21.3	—	—	—	—	—	—	—	—
31. fresh, mature	1.62	64.5	6.1	—	—	—	—	—	—	—	—
32. hay, sun-cured, midbloom	2.13	57.7	14.4	—	—	—	—	—	—	—	—
CANARYGRASS, REED *Phalaris arundinacea*											
33. fresh	2.54	46.4	17.0	2.87[c]	2.35[c]	1.44[c]	0.82[c]	1.47[d]	65[d]	—	—
CATTLE *Bos taurus*											
34. manure, dehydrated, all forage	—	—	—	—	—	—	—	—	—	—	—
35. manure, dehy (high concentrate)	—	—	—	—	—	—	—	—	—	—	—
36. manure, dehy, forage and concentrate	—	—	—	—	—	—	—	—	—	—	—

(Table 48 continues)

TABLE 48 Composition of Some Common Feeds, Excluding Amino Acids *(Cont.)*.

| POULTRY[b] | RABBITS[b] | | | MINERALS[a] | | | VITAMINS[a] | | | |
ME$_n$ (kcal/ kg)	DE (kcal/ kg)	TDN (%)	Dig. Protein (%)	Calcium (%)	Phosphorous (%)	Potassium (%)	Carotene Vitamin A Activity (1000 IU/kg)	(mg/kg)	Vitamin E (mg/kg)	Vitamin D$_2$ (IU/g)
—	620	14	3.6	0.48	0.07	0.51	—	185	—	191
—	—	—	—	2.01	0.28	2.06	56.7	—	—	—
—	—	—	—	1.41	0.24	1.71	46.0	—	2	1,544
1200	2350	53	12.2	1.52	0.25	2.60	52.0	131	121	—
—	—	—	—	—	—	—	—	—	—	—
—	—	—	—	1.50	0.28	2.15	—	99	—	289
—	—	—	—	—	—	—	—	—	—	—
—	—	—	—	0.46	0.22	1.45	73.0	—	—	—
—	—	—	—	0.50	0.22	—	—	—	—	—
3862	4190	101	9.4	0.14	0.26	0.53	2.0	—	45	—
2640	3330	75	9.9	0.05	0.38	0.47	1.0	2	25	—
2620	3330	75	8.0	0.06	0.39	0.58	—	—	30	—
—	—	—	—	0.23	0.26	1.18	21.0	53	—	1,103
—	—	—	—	0.34	0.28	2.01	10.2	—	—	—
—	—	—	—	0.49	0.27	—	132.2	—	—	—
—	1770	40	5.7	0.43	0.20	1.61	41.8	105	—	—
—	—	—	—	0.32	0.26	0.10	—	—	—	—
3420	—	—	—	0.06	0.08	0.38	—	—	—	—
—	—	—	—	0.50	0.44	2.27	193.0	482	—	—
—	—	—	—	—	—	—	36.7	—	—	—
—	—	—	—	0.33	0.25	1.69	—	—	—	—
—	—	—	—	26.81	11.91	0.19	0.0	—	—	—
—	—	—	—	30.71	12.86	0.19	0.0	—	—	—
2080	—	—	—	0.33	0.55	0.09	—	—	29	—
—	—	—	—	0.33	0.55	0.09	—	—	29	—
—	—	—	—	0.50	0.30	2.30	184.0	459	—	—
—	—	—	—	0.30	0.35	2.32	15.0	—	—	—
—	—	—	—	0.35	0.19	1.93	—	34	—	1,407
—	—	—	—	0.55	0.45	3.16	233.2	—	—	—
—	—	—	—	0.26	0.16	—	—	—	—	—
—	—	—	—	0.29	0.28	1.99	—	—	—	—
—	—	—	—	0.41	0.35	3.64	—	—	—	—
—	—	—	—	—	—	—	—	—	—	—
—	—	—	—	—	—	—	—	—	—	—
—	—	—	—	—	—	—	—	—	—	—

(Table 48 continues)

TABLE 48 Composition of Some Common Feeds, Excluding Amino Acids. *(Cont.)*

Feed Name	International Reference Number	Dry Matter (%)	Crude Fiber[a] (%)	Crude Protein[a] (%)	Acid Detergent Fiber[a] (%)	BEEF CATTLE[a]				
						DE (Mcal/kg)	ME (Mcal/kg)	NE_m (Mcal/kg)	NE_g (Mcal/kg)	TDN (%)
CITRUS *Citrus* spp.										
37. pulp, silage	3-01-234	21	15.6	7.3	—	3.88	3.18	2.18	1.50	88
38. pulp, dried	4-01-237	91	12.7	6.7	22	3.62	2.97	2.00	1.35	82
CLOVER, ALSIKE *Trifolium hybridum*										
39. fresh, early vegetative	2-01-314	19	17.5	24.1	—	2.91	2.39	1.51	0.91	66
40. hay, sun-cured	1-01-313	88	30.1	14.9	—	2.56	2.10	1.24	0.68	58
CLOVER, CRIMSON *Trifolium incarnatum*										
41. fresh, early vegetative	2-20-890	18	28.0	17.0	—	2.78	2.28	1.41	0.83	63
42. hay, sun-cured	1-01-328	87	30.1	16.0	—	2.51	2.06	1.21	0.64	57
CLOVER, LADINO *Trifolium repens*										
43. fresh, early vegetative	2-01-380	19	14.0	24.7	—	3.00	2.46	1.57	0.97	68
44. hay, sun-cured	1-01-378	90	21.2	22.0	32	2.65	2.17	1.31	0.74	60
CLOVER, RED *Trifolium pratense*										
45. fresh, early bloom	2-01-428	20	23.2	19.4	31	3.04	2.50	1.60	1.00	69
46. hay, sun-cured	1-01-415	89	28.8	16.0	36	2.43	1.99	1.14	0.58	55
CORN, DENT YELLOW *Zea mays indentata*										
47. stover (straw)	1-28-233	85	34.4	5.9	39	2.21	1.81	0.97	0.42	50
48. cobs, ground	1-02-782	90	35.0	2.8	35	—	—	—	—	—
49. cobs, ground	1-28-234	90	36.2	3.2	35	2.21	1.81	0.97	0.42	50
50. distillers grains, dehydrated	5-02-842	92	13.0	29.5	—	—	—	—	—	—
51. distillers grains, dehydrated	5-28-235	94	12.1	23.0	17	3.79	3.11	2.12	1.45	86
52. distillers grains w/solubles, dehydrated	5-02-843	92	10.0	29.8	—	—	—	—	—	—
53. distillers grains w/solubles, dehydrated	5-28-236	92	9.9	25.0	18	3.88	3.18	2.18	1.50	88
54. distillers solubles, dehydrated	5-28-237	93	5.0	29.7	7	3.88	3.18	2.18	1.50	88
55. ears, ground (ground ear corn)	4-02-849	87	9.0	9.3	—	—	—	—	—	—
56. ears, ground (corn and cob meal)	4-28-238	87	9.4	9.0	11	3.66	3.00	2.03	1.37	83
57. ears with husks, silage	4-28-239	44	11.6	8.9	—	3.26	2.68	1.76	1.14	74
58. gluten, meal	5-28-241	91	4.8	46.8	9	3.79	3.11	2.12	1.45	86
59. gluten, meal 60% protein	5-28-242	90	2.2	67.2	5	3.92	3.22	2.21	1.52	89
60. gluten, meal 41%	5-12-354	91	3.8	42.1	—	—	—	—	—	—
61. gluten, w/bran (corn gluten feed)	5-28-243	90	9.7	25.6	12	3.66	3.00	2.03	1.37	83
62. grain, opaque 2 (high lysine)	4-28-253	90	3.7	11.3	—	—	—	—	—	—
63. grain	4-02-879	87	2.4	10.6	—	—	—	—	—	—
64. grain, grade 2, 69.5 kg/hl	4-02-931	88	2.2	10.1	—	3.97	3.25	2.24	1.55	90
65. grain, dent yellow	4-02-935	89	2.9	10.9	3	—	—	—	—	—
66. grain	4-02-985	88	2.0	10.9	—	—	—	—	—	—
67. grain, cracked, dent yellow	4-20-698	89	2.6	10	3	—	—	—	—	—
68. grain, ground, dent yellow	4-26-023	89	2.6	10	3	—	—	—	—	—
69. grain, flaked	4-28-244	89	2.6	10	3	4.19	3.44	2.38	1.67	95
70. grain, high moisture	4-20-770	77	2.6	10	3	4.10	3.36	2.33	1.62	93
71. grits by-product (hominy feed)	4-03-011	90	6.7	11.5	13	4.14	3.40	2.35	1.65	94
72. silage (stalkage) (stover)	3-28-251	31	31.3	5.9	43	2.43	1.99	1.14	0.58	55
73. silage, well eared	3-28-250	33	23.7	8.1	28	3.09	2.53	1.63	1.03	70
74. silage (ears with husks)	3-28-239	44	11.6	8.9	14	—	—	—	—	—
75. silage, dough stage	3-02-819	26	24.5	7.8	31	—	—	—	—	—
COTTON *Gossypium* spp.										
76. seeds, meal solvent extracted, 41% protein	5-01-621	91	13.3	45.2	17	3.25	2.75	1.82	1.19	76
77. seeds, meal solvent extracted	5-01-619	92	11.7	45.3	—	—	—	—	—	—
78. seeds, meal prepressed solvent extracted, 41% protein	5-07-872	91	14.1	45.6	19	—	—	—	—	—

(Table 48 continues)

TABLE 48 Composition of Some Common Feeds, Excluding Amino Acids. *(Cont.)*

		SHEEP[a]						DAIRY CATTLE[a]			
								(Growing)		(Lactating)	
DE (Mcal/kg)	ME (Mcal/kg)	NE$_m$ (Mcal/kg)	NE$_g$ (Mcal/kg)	TDN (%)	Dig. Protein (%)	DE (Mcal/kg)	ME (Mcal/kg)	NE$_m$ (Mcal/kg)	NE$_g$ (Mcal/kg)	NE$_l$ (Mcal/kg)	TDN (%)
3.88	3.18	2.18	1.50	88	3.7	3.44	3.02	1.88	1.24	1.79	78
3.70	3.04	2.06	1.40	84	3.4	3.40	2.98	1.86	1.22	1.77	77
—	—	—	—	—	—	2.91	2.49	1.51	0.92	1.50	66
2.56	2.10	1.24	0.68	58	9.9	2.56	2.13	1.25	0.68	1.30	58
—	—	—	—	—	—	2.78	2.36	1.41	0.83	1.42	63
2.43	1.99	1.14	0.58	55	12.7	2.51	2.09	1.21	0.64	1.28	57
—	—	—	—	—	—	3.00	2.58	1.57	0.97	1.55	68
2.91	2.39	1.51	0.91	66	16.7	2.87	2.45	1.47	0.88	1.47	65
3.00	2.46	1.57	0.47	68	15.0	3.04	2.62	1.60	1.00	1.57	69
2.65	2.17	1.31	0.74	60	10.1	2.43	2.00	1.14	0.58	1.23	55
2.60	2.13	1.28	0.71	59	2.9	2.21	1.78	0.97	0.42	1.11	50
—	—	—	—	—	—	2.21	1.78	0.97	0.42	1.11	50
—	—	—	—	—	—	—	—	—	—	—	—
3.84	3.15	2.12	1.48	87	16.7	3.79	3.38	2.12	1.45	1.99	86
—	—	—	—	—	—	—	—	—	—	—	—
3.84	3.15	2.15	1.48	87	12.3	3.88	3.47	2.18	1.50	2.04	88
3.75	3.07	2.09	1.43	85	23.5	3.88	3.47	2.18	1.50	2.04	88
—	—	—	—	—	—	—	—	—	—	—	—
3.66	3.00	2.03	1.37	83	5.5	3.66	3.25	2.03	1.37	1.91	83
3.26	2.68	1.76	1.14	74	4.8	—	—	—	—	—	—
3.88	3.18	2.18	1.50	88	39.8	3.79	3.38	2.12	1.45	1.99	86
—	—	—	—	—	—	3.92	3.51	2.20	1.52	2.06	89
—	—	—	—	—	—	—	—	—	—	—	—
3.66	3.00	2.03	1.37	83	—	3.66	3.25	2.03	1.37	1.91	83
—	—	—	—	—	—	3.92	3.51	2.20	1.52	2.06	89
—	—	—	—	—	—	—	—	—	—	—	—
3.84	3.15	2.15	1.48	87	6.5	—	—	—	—	—	—
—	—	—	—	—	—	—	—	—	—	—	—
—	—	—	—	—	—	3.53	3.12	1.94	1.30	1.84	80
—	—	—	—	—	—	3.75	3.34	2.10	1.43	1.96	85
—	—	—	—	—	—	3.88	3.47	2.18	1.50	2.04	88
—	—	—	—	—	—	3.88	3.47	2.18	1.50	2.04	88
—	—	—	—	—	—	3.84	3.42	2.16	1.48	2.01	87
2.34	1.92	1.07	0.52	53	1.7	2.43	2.00	1.14	0.58	1.23	55
3.09	2.53	1.63	1.03	70	3.6	3.09	2.67	1.63	1.03	1.60	70
—	—	—	—	—	—	3.26	2.85	1.75	1.13	1.69	74
—	—	—	—	—	—	—	—	—	—	—	—
3.13	2.57	1.67	1.06	71	—	—	—	—	—	—	—
—	—	—	—	—	—	—	—	—	—	—	—
—	—	—	—	—	—	3.35	2.93	1.82	1.19	1.74	76

(Table 48 continues)

TABLE 48 Composition of Some Common Feeds, Excluding Amino Acids *(Cont.)*.

Feed Name	HORSES[a]			GOATS[a]						SWINE[b]	
	DE (Mcal/kg)	Neutral Detergent Fiber (%)	Crude Protein (%)	DE (Mcal/kg)	ME (Mcal/kg)	NE_m (Mcal/kg)	NE_g (Mcal/kg)	NE_l (Mcal/kg)	TDN (%)	DE (kcal/kg)	ME (kcal/kg)
CITRUS *Citrus* spp.											
37. pulp, silage	—	—	—	—	—	—	—	—	—	—	—
38. pulp, dried	2.81	23.0	6.7	—	—	—	—	—	—	—	—
CLOVER, ALSIKE *Trifolium hybridum*											
39. fresh, early vegetative	—	—	—	—	—	—	—	—	—	—	—
40. hay, sun-cured	—	—	—	—	—	—	—	—	—	—	—
CLOVER, CRIMSON *Trifolium incarnatum*											
41. fresh, early vegetative	—	—	—	—	—	—	—	—	—	—	—
42. hay, sun-cured	—	—	—	—	—	—	—	—	—	—	—
CLOVER, LADINO *Trifolium repens*											
43. fresh, early vegetative	—	—	—	—	—	—	—	—	—	—	—
44. hay, sun-cured	2.20	36.0	22.4	—	—	—	—	—	—	—	—
CLOVER, RED *Trifolium pratense*											
45. fresh, early bloom	2.53	40.0	20.8	—	—	—	—	—	—	—	—
46. hay, sun-cured	2.22	46.9	15.0	2.43[c]	1.99[c]	1.19[c]	0.47[c]	1.23[c]	55[d]	—	—
CORN, DENT YELLOW *Zea mays indentata*											
47. stover (straw)	—	—	—	2.21[c]	1.81[c]	1.07[c]	0.28[c]	1.11[c]	50[d]	—	—
48. cobs, ground	—	—	—	—	—	—	—	—	—	—	—
49. cobs, ground	1.36	87.0	2.8	—	—	—	—	—	—	—	—
50. distillers grains, dehydrated	—	—	—	—	—	—	—	—	—	—	—
51. distillers grains, dehydrated	3.49	43.0	30.3	—	—	—	—	—	—	—	—
52. distillers grains w/solubles, dehydrated	—	—	—	—	—	—	—	—	—	—	—
53. distillers grains w/solubles, dehydrated	—	—	—	—	—	—	—	—	—	3640	3335
54. distillers solubles, dehydrated	—	—	—	—	—	—	—	—	—	3330	2945
55. ears, ground (ground ear corn)	—	—	—	3.66[c]	3.00[c]	1.97[c]	1.32[c]	1.91[c]	83[d]	—	—
56. ears, ground (corn and cob meal)	3.29	28.0	9.0	—	—	—	—	—	—	3127	2952
57. ears with husks, silage	—	—	—	—	—	—	—	—	—	—	—
58. gluten, meal	—	—	—	—	—	—	—	—	—	—	—
59. gluten, meal 60% protein	—	—	—	—	—	—	—	—	—	4065	3585
60. gluten, meal 41%	—	—	—	—	—	—	—	—	—	4290	3880
61. gluten, w/bran (corn gluten feed)	—	—	—	—	—	—	—	—	—	3155	2695
62. grain, opaque 2 (high lysine)	—	—	—	—	—	—	—	—	—	—	—
63. grain	—	—	—	3.92[c]	3.22[c]	2.15[c]	1.47[c]	2.06[c]	89	—	—
64. grain, grade 2, 69.5 kg/hl	—	—	—	—	—	—	—	—	—	—	—
65. grain, dent yellow	3.84	10.8	10.4	—	—	—	—	—	—	3530	3420
66. grain	—	—	—	—	—	—	—	—	—	—	—
67. grain, cracked, dent yellow	—	—	—	—	—	—	—	—	—	—	—
68. grain, ground, dent yellow	—	—	—	—	—	—	—	—	—	—	—
69. grain, flaked	—	—	—	—	—	—	—	—	—	—	—
70. grain, high moisture	—	—	—	—	—	—	—	—	—	—	—
71. grits by-product (hominy feed)	—	—	—	—	—	—	—	—	—	3495	3310
72. silage (stalkage) (stover)	—	—	—	—	—	—	—	—	—	—	—
73. silage, well eared	—	—	—	—	—	—	—	—	—	—	—
74. silage (ears with husks)	—	—	—	—	—	—	—	—	—	—	—
75. silage, dough stage	—	—	—	3.09[c]	2.53[c]	1.58[c]	0.97[c]	1.60[c]	70[d]	—	—
COTTON *Gossypium* spp.											
76. seeds, meal solvent extracted, 41% protein	3.01	27.9	45.4	3.35[c]	2.75[c]	1.76[c]	1.14[c]	1.74[c]	76[d]	—	—
77. seeds, meal solvent extracted	—	—	—	—	—	—	—	—	—	2670	2555
78. seeds, meal prepressed solvent extracted, 41% protein	—	—	—	—	—	—	—	—	—	—	—

(Table 48 continues)

TABLE 48 Composition of Some Common Feeds, Excluding Amino Acids *(Cont.)*.

POULTRY[b]	RABBITS[b]			MINERALS[a]			VITAMINS[a]			
ME_n (kcal/kg)	DE (kcal/kg)	TDN (%)	Dig. Protein (%)	Calcium (%)	Phosphorous (%)	Potassium (%)	Carotene Vitamin A Activity (1000 IU/kg)	(mg/kg)	Vitamin E (mg/kg)	Vitamin D_2 (IU/g)
—	—	—	—	2.04	0.15	0.62	—	—	—	—
—	—	—	—	1.84	0.12	0.79	—	—	—	—
—	—	—	—	1.29	0.26	2.46	154.0	—	—	—
—	—	—	—	1.29	0.26	2.46	75.0	187	—	—
—	—	—	—	1.40	0.22	2.40	95.0	—	—	—
—	—	—	—	1.40	0.22	2.40	9.0	23	—	—
—	—	—	—	1.35	0.31	2.62	141.0	—	—	—
—	—	—	—	1.35	0.31	2.62	33.0	83	—	—
—	—	—	—	2.26	0.38	2.49	99.0	248	—	—
—	2170	49	9.8	1.53	0.25	1.62	8.0	20	—	1914
—	—	—	—	0.57	0.10	1.45	2.0	4	—	1103
—	—	—	—	0.12	0.04	0.84	—	—	—	—
—	—	—	—	0.12	0.04	0.87	—	—	—	—
—	—	—	—	0.10	0.40	0.20	1.0	—	—	—
1972	—	—	—	0.11	0.43	0.18	1.0	3	—	—
—	—	—	—	0.16	0.79	0.50	2.0	—	40	—
2480	—	—	—	0.15	0.71	0.44	1.0	3	43	600
2930	—	—	—	0.35	1.37	1.80	—	—	49	—
—	—	—	—	0.05	0.26	0.56	3.0	—	—	—
—	—	—	—	0.07	0.27	0.53	2.0	4	20	—
—	—	—	—	0.10	0.29	0.49	3.1	8	—	—
—	—	—	—	0.16	0.50	0.03	7.0	18	34	—
3720	—	—	—	0.08	0.54	0.21	14.0	—	26	—
—	—	—	—	0.13	0.40	0.03	—	—	31	—
1750	—	—	—	0.36	0.82	0.64	3.0	7	14	—
—	—	—	—	0.03	0.22	0.39	2	—	—	—
—	—	—	—	0.05	0.28	1.10	—	—	—	—
—	—	—	—	0.02	0.35	0.37	0.8	2	25	—
3350	3790	83	7.3	0.03	0.29	0.37	—	3	25	—
—	—	—	—	0.05	0.60	0.35	—	—	—	—
—	—	—	—	0.03	0.29	0.37	1	—	25	—
—	—	—	—	0.03	0.29	0.37	1	—	25	—
—	—	—	—	0.03	0.29	0.37	1	—	25	—
—	—	—	—	0.02	0.32	0.35	1	—	25	—
2896	—	—	—	0.05	0.57	0.65	—	—	—	—
—	—	—	—	0.38	0.31	1.54	6	15	—	—
—	—	—	—	0.23	0.22	0.96	18	45	—	119
—	—	—	—	0.10	0.29	0.49	3	—	—	—
—	—	—	—	0.27	0.19	0.95	—	65.1	—	—
—	3090	67	34.5	0.18	1.21	1.52	—	—	17	—
—	—	—	—	—	—	—	—	—	—	—
2400	—	—	—	0.22	1.21	1.39	—	—	—	—

(Table 48 continues)

TABLE 48 Composition of Some Common Feeds, Excluding Amino Acids. *(Cont.)*

Feed Name	International Reference Number	Dry Matter (%)	Crude Fiber[a] (%)	Crude Protein[a] (%)	Acid Detergent Fiber[a] (%)	BEEF CATTLE[a]				
						DE (Mcal/ kg)	ME (Mcal/ kg)	NE_m (Mcal/ kg)	NE_g (Mcal/ kg)	TDN (%)
CUPGRASS, TEXAS *Eriochloa sericea*										
79. fresh, late vegetative	2-29-996	40	—	7.8	—	—	—	—	—	—
80. fresh, mature	2-30-059	83	—	5.0	—	—	—	—	—	—
CURLY MESQUITE *Hilaria belangeri*										
81. browse, fresh, early vegetative	2-01-723	23	25.6	17.2						
DALLISGRASS *Paspalum dilatatum*										
82. fresh, early vegetative	2-01-738	26	30.1	23.2	—	—	—	—	—	—
83. fresh, full bloom	2-01-739	30	32.2	7.1	—	—	—	—	—	—
84. DICALCIUM PHOSPHATE FATS AND OILS	6-01-080	96	—	—	—	—	—	—	—	—
85. fat, animal, hydrolized	4-00-376	99	—	—	—	7.80	6.40	4.75	3.51	177
86. fat, animal-poultry	4-00-409	99	—	—	—	7.80	6.40	4.75	3.51	177
87. oil, vegetable	4-05-077	100	—	—	—	7.80	6.40	4.75	3.51	177
FESCUE *Festuca* spp.										
88. hay, sun-cured, early vegetative	1-06-132	91	26.0	12.4	32	2.69	2.21	1.34	0.77	61
FESCUE, KENTUCKY 31 *Festuca arundinacea*										
89. hay, sun-cured	1-20-800	92	24.9	18.2	—	—	—	—	—	—
FESCUE, MEADOW *Festuca elatior*										
90. grazed	2-01-920	27	29	11.5	—	—	—	—	—	—
91. hay, sun-cured	1-01-912	88	33	10.5	43	—	—	—	—	—
FISH, MENHADEN *Brevoortia tyrannus*										
92. meal, mechanical extracted	5-02-009	92	1.0	66.7	—	—	—	—	—	—
FISH										
93. solubles, condensed	5-01-969	51	0.39	64.1	—	—	—	—	—	—
FLAX, COMMON *Linum usitatissimum*										
94. seeds, meal solvent extracted (linseed meal)	5-02-048	90	10.1	38.3	19	3.44	2.82	1.88	1.24	78
GRASS-LEGUME										
95. silage	3-02-303	29	31.8	11.3	—	—	—	—	—	—
GREASEWOOD, BLACK *Sarcobatus vermiculatus*										
96. browse, fresh	2-20-083	35	16.5	15.3	—	—	—	—	—	—
HACKBERRY, TETENDRA *Celtis tetendra*										
97. browse, fresh, mature	2-30-164	53	16.6	14.0	—	—	—	—	—	—
JUJUBE, COMMON *Ziziphus jujuba*										
98. browse, fresh	2-30-091	32	30.1	8.6	—	—	—	—	—	—
KUDZU *Pueraria*										
99. hay, sun-cured	1-02-478	91	39.1	14.3	—	—	—	—	—	—
LESPEDEZA, COMMON — LESPEDEZA, KOREAN										
Lespedeza striata—Lespedeza stipulacea										
100. fresh, early bloom	2-20-885	28	32	16.4	—	2.43	1.99	1.14	0.58	55
101. fresh, late vegetative	2-07-093	25	32	16.0	—	—	—	—	—	—
102. grazed	2-02-568	31	38	14.9	—	—	—	—	—	—
103. hay, sun-cured	1-08-591	91	32	13.9	—	—	—	—	—	—
104. hay, sun-cured, midbloom	1-26-026	93	30	14.5	—	2.21	1.81	0.97	0.42	50
105. hay, sun-cured, midbloom	1-02-554	92	28.8	15.1	—	—	—	—	—	—
106. hay, sun-cured, midbloom	1-21-021	93	30	14.5	—	—	—	—	—	—
107. hay, sun-cured, full bloom	1-20-887	89	30.7	14.3	—	—	—	—	—	—
LIMESTONE										
108. ground, mn 33% calcium	6-02-632	100	—	—	—	—	—	—	—	—

(Table 48 continues)

TABLE 48 Composition of Some Common Feeds, Excluding Amino Acids. *(Cont.)*

| | | SHEEP[a] | | | | | | DAIRY CATTLE[a] | | | |
| | | | | | | | | | (Growing) | | (Lactating) | |
DE (Mcal/kg)	ME (Mcal/kg)	NE_m (Mcal/kg)	NE_g (Mcal/kg)	TDN (%)	Dig. Protein (%)	DE (Mcal/kg)	ME (Mcal/kg)	NE_m (Mcal/kg)	NE_g (Mcal/kg)	NE_l (Mcal/kg)	TDN (%)
—	—	—	—	—	—	—	—	—	—	—	—
—	—	—	—	—	—	—	—	—	—	—	—
—	—	—	—	—	—	—	—	—	—	—	—
—	—	—	—	—	—	—	—	—	—	—	—
—	—	—	—	—	—	—	—	—	—	—	—
—	—	—	—	—	—	—	—	—	—	—	—
—	—	—	—	—	—	—	—	—	—	—	—
—	—	—	—	—	—	—	—	—	—	—	—
—	—	—	—	—	—	—	—	—	—	—	—
2.65	2.17	1.31	0.74	60	13.1	—	—	—	—	—	—
—	—	—	—	—	—	—	—	—	—	—	—
—	—	—	—	—	—	—	—	—	—	—	—
3.09	2.53	1.63	1.03	70	54.0	3.22	2.80	1.73	1.11	1.67	73
—	—	—	—	—	—	—	—	—	—	—	—
3.48	2.86	1.91	1.27	79	32.8	3.44	3.02	1.88	1.24	1.79	78
2.73	2.24	1.38	0.80	62	6.5	—	—	—	—	—	—
—	—	—	—	—	—	—	—	—	—	—	—
—	—	—	—	—	—	—	—	—	—	—	—
—	—	—	—	—	—	—	—	—	—	—	—
—	—	—	—	—	—	—	—	—	—	—	—
2.60	2.13	1.28	0.71	59	11.9	—	—	—	—	—	—
—	—	—	—	—	—	—	—	—	—	—	—
—	—	—	—	—	—	2.21	1.78	0.97	0.42	1.11	50
2.51	2.06	1.21	0.64	57	10.1	—	—	—	—	—	—
2.56	2.10	1.24	0.68	58	9.4	—	—	—	—	—	—
—	—	—	—	—	—	—	—	—	—	—	—

(Table 48 continues)

TABLE 48 Composition of Some Common Feeds, Excluding Amino Acids *(Cont.).*

	HORSES[a]			GOATS[a]						SWINE[b]	
Feed Name	DE (Mcal/ kg)	Neutral Detergent Fiber (%)	Crude Protein (%)	DE (Mcal/ kg)	ME (Mcal/ kg)	NE$_m$ (Mcal/ kg)	NE$_g$ (Mcal/ kg)	NE$_l$ (Mcal/ kg)	TDN (%)	DE (kcal/ kg)	ME (kcal/ kg)
CUPGRASS, TEXAS *Eriochloa sericea*											
79. fresh, late vegetative	—	—	—	1.68[c]	1.37[c]	0.87[c]	0.00[c]	0.81[c]	38[d]	—	—
80. fresh, mature	—	—	—	1.32[c]	1.09[c]	0.78[c]	0.00[c]	0.62[c]	30[d]	—	—
CURLY MESQUITE *Hilaria belangeri*											
81. browse, fresh, early vegetative	—	—	—	2.60[c]	2.13[c]	1.28[c]	0.62[c]	1.33[c]	59[d]	—	—
DALLISGRASS *Paspalum dilatatum*											
82. fresh, early vegetative	—	—	—	2.78[c]	2.28[c]	1.39[c]	0.75[c]	1.42[c]	63[d]	—	—
83. fresh, full bloom	—	—	—	2.56[c]	2.10[c]	1.26[c]	0.58[c]	1.30[c]	58[d]	—	—
84. DICALCIUM PHOSPHATE	—	—	—	—	—	—	—	—	—	—	—
FATS AND OILS											
85. fat, animal, hydrolized	—	—	—	—	—	—	—	—	—	—	—
86. fat, animal-poultry	—	—	—	—	—	—	—	—	—	—	—
87. oil, vegetable	—	—	—	—	—	—	—	—	—	—	—
FESCUE *Festuca* spp.											
88. hay, sun-cured, early vegetative	—	—	—	—	—	—	—	—	—	—	—
FESCUE, KENTUCKY 31 *Festuca arundinacea*											
89. hay, sun-cured	—	—	—	—	—	—	—	—	—	—	—
FESCUE, MEADOW *Festuca elatior*											
90. grazed	—	—	—	—	—	—	—	—	—	—	—
91. hay, sun-cured	—	—	—	—	—	—	—	—	—	—	—
FISH, MENHADEN *Brevoortia tyrannus*											
92. meal, mechanical extracted	3.20	—	67.9	3.22[c]	2.64[c]	1.67[c]	1.06[c]	1.67[c]	73[d]	3800	3300
FISH											
93. solubles, condensed	—	—	—	—	—	—	—	—	—	1909	1623
FLAX, COMMON *Linum usitatissimum*											
94. seeds, meal solvent extracted (linseed meal)	3.04	25.0	38.4	3.44[c]	2.82[c]	1.82[c]	1.19[c]	1.79[c]	78[d]	—	—
GRASS-LEGUME											
95. silage	—	—	—	—	—	—	—	—	—	—	—
GREASEWOOD, BLACK *Sarcobatus vermiculatus*											
96. browse, fresh	—	—	—	2.43[c]	1.99[c]	1.19[c]	0.47[c]	1.23[c]	55[d]	—	—
HACKBERRY, TETENDRA *Celtis tetendra*											
97. browse, fresh, mature	—	—	—	1.81[c]	1.48[c]	0.91[c]	0.00[c]	0.89[c]	41[d]	—	—
JUJUBE, COMMON *Ziziphus jujuba*											
98. browse, fresh	—	—	—	1.37[c]	1.12[c]	0.79[c]	0.00[c]	0.64[c]	31	—	—
KUDZU *Pueraria*											
99. hay, sun-cured	—	—	—	2.43[c]	1.99[c]	1.19[c]	0.47[c]	1.23[c]	55[d]	—	—
LESPEDEZA, COMMON — LESPEDEZA, KOREAN											
Lespedeza striata—Lespedeza stipulacea											
100. fresh, early bloom	—	—	—	—	—	—	—	—	—	—	—
101. fresh, late vegetative	2.20	—	16.4	—	—	—	—	—	—	—	—
102. grazed	2.13	—	12.6	—	—	—	—	—	—	—	—
103. hay, sun-cured	—	—	—	—	—	—	—	—	—	—	—
104. hay, sun-cured, midbloom	—	—	—	—	—	—	—	—	—	—	—
105. hay, sun-cured, midbloom	—	—	—	—	—	—	—	—	—	—	—
106. hay, sun-cured, midbloom	—	—	—	—	—	—	—	—	—	—	—
107. hay, sun-cured, full bloom	—	—	—	2.56[c]	2.10[c]	1.26[c]	0.58[c]	1.30[c]	58[d]	—	—
LIMESTONE											
108. ground, mn 33% calcium	—	—	—	—	—	—	—	—	—	—	—

(Table 48 continues)

TABLE 48 Composition of Some Common Feeds, Excluding Amino Acids (Cont.).

POULTRY[b]	RABBITS[b]			MINERALS[a]			VITAMINS[a]			
ME$_n$ (kcal/ D$_2$ kg)	DE (kcal/ kg)	TDN (%)	Dig. Protein (%)	Calcium (%)	Phosphorous (%)	Potassium (%)	Carotene Vitamin A Activity (1000 IU/kg)	(mg/kg)	Vitamin E (mg/kg)	Vitamin (IU/g)
—	—	—	—	—	0.14	—	—	—	—	—
—	—	—	—	—	0.05	—	—	—	—	—
—	—	—	—	1.04	0.26	0.79	—	—	—	—
—	—	—	—	0.65	0.42	—	—	426.6	—	—
—	—	—	—	—	—	—	—	—	—	—
—	0	0	0.0	23.70	18.84	0.04	—	—	—	—
—	—	—	—	—	—	—	—	—	—	—
—	—	—	—	—	—	0.23	—	—	8	—
—	—	—	—	—	—	—	—	—	—	—
—	—	—	—	0.51	0.36	2.30	—	—	—	—
—	—	—	—	0.44	0.40	2.80	—	—	—	—
—	—	—	—	0.60	0.43	2.34	—	—	—	—
—	—	—	—	0.57	0.37	1.74	—	—	—	—
2820	—	—	—	5.65	3.16	0.76	—	—	13	—
1460	—	—	—	0.58	0.59	3.41	—	—	—	—
—	3430	68	30.4	0.43	0.89	1.53	—	—	15	—
—	—	—	—	0.85	0.27	1.80	—	230	—	289
—	—	—	—	1.10	0.21	2.09	—	—	—	—
—	—	—	—	4.00	0.13	—	—	—	—	—
—	—	—	—	—	—	—	—	—	—	—
—	—	—	—	2.35	0.35	—	—	43.9	—	—
—	—	—	—	1.35	0.21	1.12	—	—	—	—
—	—	—	—	—	—	—	—	—	—	—
—	—	—	—	1.10	0.28	1.26	—	—	—	—
—	—	—	—	1.15	0.25	1.03	—	—	—	—
—	—	—	—	—	—	—	22.0	—	—	—
—	—	—	—	1.18	0.24	1.01	—	—	—	—
—	—	—	—	1.19	0.26	1.05	22	—	—	—
—	—	—	—	1.14	0.21	1.04	—	—	—	—
—	—	—	—	36.07	0.02	0.12	—	—	—	—

(Table 48 continues)

TABLE 48 Composition of Some Common Feeds, Excluding Amino Acids. *(Cont.)*

Feed Name	International Reference Number	Dry Matter (%)	Crude Fiber[a] (%)	Crude Protein[a] (%)	Acid Detergent Fiber[a] (%)	BEEF CATTLE[a] DE (Mcal/ kg)	ME (Mcal/ kg)	NE$_m$ (Mcal/ kg)	NE$_g$ (Mcal/ kg)	TDN (%)
MARGOSA *Azadirachta indica*										
109. browse, fresh	2-30-147	21	11.9	18.7	—	—	—	—	—	—
110. leaves, fresh	2-27-194	20	20.7	16.1	—	—	—	—	—	—
MEADOW PLANTS, INTERMOUNTAIN										
111. hay, sun-cured	1-03-181	95	32.3	8.7	—	2.56	2.10	1.24	0.68	58
MEAT										
112. meal rendered	5-00-385	94	2.8	54.8	—	—	—	—	—	—
113. with bone, meal rendered	5-00-388	93	2.4	54.1	—	—	—	—	—	—
114. meat and bone meal, 50%	5-09-322	94	2.4	50.9	—	—	—	—	—	—
115. meat meal, 55%	5-09-323	93	2.3	55.6	—	—	—	—	—	—
MESQUITE, HONEY *Prosopis glandulosa*										
116. browse, fresh late vegetative	2-29-999	48	—	16.2	—	—	—	—	—	—
MESQUITE, SPICIGERA *Prosopis spicigera*										
117. hay, sun-cured	1-30-160	86	22.1	14.2	—	—	—	—	—	—
MILK *Bos taurus*										
118. fresh (cattle)	5-01-168	12	—	26.7	—	—	—	—	—	—
119. skimmed, fresh (cattle)	5-01-170	10	0.2	31.2	—	—	—	—	—	—
120. skimmed, dehydrated	5-01-175	94	0.2	35.8	—	—	—	—	—	—
MILK *Ovis aries*										
121. fresh (sheep)	5-08-510	19	—	24.7	—	—	—	—	—	—
MILLET, PEARL *Pennisetum glaucum*										
122. grain	4-03-118	91	4.7	—	—	—	—	—	—	—
MILLET, PROSO *Panicum miliaceum*										
123. grain	4-03-120	90	6.8	12.9	17	3.70	3.04	2.06	1.40	84
MIMOSA *Mimosa* spp.										
124. browse, fresh, early vegetative	2-03-122	26	21.1	20.6	—	—	—	—	—	—
MOLASSES AND SYRUP *Beta vulgaris altissim*										
125. beet, sugar, molasses, more than 48% invert sugar, more than 79.5 degrees brix	4-00-668	78	—	8.5	—	3.48	2.86	1.91	1.27	79
MOLASSES AND SYRUP *Saccharum officinarum*										
126. sugarcane, molasses, more than 46% invert sugars, more than 79.5 degrees brix (black strap)	4-04-696	75	—	5.8	—	3.17	2.60	1.70	1.08	72
OAK, LIVE *Quercus virginiana*										
127. browse, fresh, early vegetative	2-29-815	36	—	20.3	—	—	—	—	—	—
128. leaves, fresh	2-29-859	50	—	10.2	—	—	—	—	—	—
OAK, SHIN *Quercus sinuata breviloba*										
129. browse, fresh, early vegetative	2-29-826	32	—	17.4	—	—	—	—	—	—
OATS *Avena sativa*										
130. grain	4-03-309	89	12.1	13.3	16	3.40	2.78	1.85	1.22	77
131. grain, Pacific coast	4-07-999	91	12.3	10.0	—	3.41	2.82	1.88	1.24	78
132. hay, sun-cured	1-03-280	91	30.4	9.3	36	2.43	1.99	1.14	0.58	55
133. silage, late vegetative	3-20-898	23	29.9	12.8	—	2.87	2.35	1.47	0.88	65
134. silage	3-03-298	31	31.5	9.6	—	—	—	—	—	—
ORCHARDGRASS *Dactylis glomerata*										
135. fresh, early vegetative	2-03-439	23	24.7	18.4	31	3.17	2.60	1.70	1.08	72
136. fresh, early bloom	2-03-442	25	30.0	16.0	—	—	—	—	—	—
137. fresh, midbloom	2-03-443	31	33.5	11.0	41	2.51	2.06	1.21	0.64	57
138. hay, sun-cured, early bloom	1-03-425	89	31.0	15.0	34	2.87	2.35	1.47	0.88	65
139. hay, sun-cured	1-03-438	91	35.1	11.2	—	—	—	—	—	—
OYSTER *Crassostrea* spp., *Ostrea* spp.										
140. shells, fine ground, min. 33% ca	6-03-481	100	—	1	—	—	—	—	—	—

(Table 48 continues)

TABLE 48 Composition of Some Common Feeds, Excluding Amino Acids. *(Cont.)*

| | | SHEEP[a] | | | | | | DAIRY CATTLE[a] | | | |
| | | | | | | | | | (Growing) | (Lactating) | |
DE (Mcal/kg)	ME (Mcal/kg)	NE_m (Mcal/kg)	NE_g (Mcal/kg)	TDN (%)	Dig. Protein (%)	DE (Mcal/kg)	ME (Mcal/kg)	NE_m (Mcal/kg)	NE_g (Mcal/kg)	NE_l (Mcal/kg)	TDN (%)
—	—	—	—	—	—	—	—	—	—	—	—
—	—	—	—	—	—	—	—	—	—	—	—
2.56	2.10	1.24	0.68	58	5.2	—	—	—	—	—	—
—	—	—	—	—	—	3.13	2.71	1.67	1.06	1.62	71
—	—	—	—	—	—	3.13	2.71	1.67	1.06	1.62	71
—	—	—	—	—	—	—	—	—	—	—	—
—	—	—	—	—	—	—	—	—	—	—	—
—	—	—	—	—	—	—	—	—	—	—	—
5.60	5.43	3.80	3.80	150	25.4	5.69	5.29	3.34	2.16	3.04	129
4.06	3.94	2.76	2.76	109	—	4.06	3.65	2.30	1.60	2.13	92
—	—	—	—	—	—	3.75	3.34	2.10	1.43	1.96	85
6.00	5.82	4.07	4.07	161	—	—	—	—	—	—	—
—	—	—	—	—	—	—	—	—	—	—	—
—	—	—	—	—	—	3.70	3.29	2.06	1.40	1.94	84
—	—	—	—	—	—	—	—	—	—	—	—
3.40	2.78	1.85	1.22	77	4.4	3.31	2.89	1.79	1.16	1.72	75
3.48	2.86	1.91	1.27	79	−1.7	3.17	2.76	1.69	1.08	1.64	72
—	—	—	—	—	—	—	—	—	—	—	—
—	—	—	—	—	—	—	—	—	—	—	—
—	—	—	—	—	—	—	—	—	—	—	—
3.40	2.78	1.85	1.22	77	10.4	3.40	2.98	1.86	1.22	1.77	77
3.44	2.82	1.88	1.24	78	6.4	3.44	3.02	1.88	1.24	1.79	78
2.34	1.92	1.14	0.58	53	5.7	—	—	—	—	—	—
—	—	—	—	—	—	—	—	—	—	—	—
2.73	2.24	1.38	0.80	62	4.9	—	—	—	—	—	—
2.95	2.42	1.54	0.94	67	13.3	3.17	2.76	1.69	1.08	1.64	72
2.69	2.21	1.34	0.77	61	8.9	—	—	—	—	—	—
2.60	2.13	1.28	0.71	59	8.9	—	—	—	—	—	—
2.29	1.88	1.04	0.49	52	8.2	2.87	2.45	1.47	0.88	1.47	65
2.56	2.10	1.24	0.68	58	7.1	—	—	—	—	—	—
—	—	—	—	—	—	—	—	—	—	—	—

(Table 48 continues)

TABLE 48 Composition of Some Common Feeds, Excluding Amino Acids (con't).

Feed Name	HORSES[a]			GOATS[a]						SWINE[b]	
	DE (Mcal/ kg)	Neutral Detergent Fiber (%)	Crude Protein (%)	DE (Mcal/ kg)	ME (Mcal/ kg)	NE$_m$ (Mcal/ kg)	NE$_g$ (Mcal/ kg)	NE$_l$ (Mcal/ kg)	TDN (%)	DE (kcal/ kg)	ME (kcal/ kg)
MARGOSA *Azadirachta indica*											
109. browse, fresh	—	—	—	2.95[c]	2.42[c]	1.50[c]	0.88[c]	1.52[c]	67	—	—
110. leaves, fresh	—	—	—	0.93[c]	0.76[c]	0.77[c]	0.00[c]	0.40[c]	21	—	—
MEADOW PLANTS, INTERMOUNTAIN											
111. hay, sun-cured	1.69	—	8.7	—	—	—	—	—	—	—	—
MEAT											
112. meal rendered	—	—	—	—	—	—	—	—	—	—	—
113. with bone, meal rendered	—	—	—	3.13[c]	2.57[c]	1.61[c]	1.00[c]	1.62[c]	71[d]	—	—
114. meat and bone meal, 50%	—	—	—	—	—	—	—	—	—	2540	2280
115. meat meal, 55%	—	—	—	—	—	—	—	—	—	2805	2415
MESQUITE, HONEY *Prosopis glandulosa*											
116. browse, fresh late vegetative	—	—	—	1.98[c]	1.63[c]	0.98[c]	0.06[c]	0.98[c]	45[d]	—	—
MESQUITE, SPICIGERA *Prosopis spicigera*											
117. hay, sun-cured	—	—	—	1.81[c]	1.48[c]	0.91[c]	0.00[c]	0.89[c]	41	—	—
MILK *Bos taurus*											
118. fresh (cattle)	—	—	—	—	—	—	—	—	—	—	—
119. skimmed, fresh (cattle)	—	—	—	—	—	—	—	—	—	—	—
120. skimmed, dehydrated	4.05	0.0	35.5	—	—	—	—	—	—	3845	3570
MILK *Ovis aries*											
121. fresh (sheep)	—	—	—	—	—	—	—	—	—	—	—
MILLET, PEARL *Pennisetum glaucum*											
122. grain	—	—	—	—	—	—	—	—	—	—	—
MILLET, PROSO *Panicum miliaceum*											
123. grain	—	—	—	—	—	—	—	—	—	3273	3057
MIMOSA *Mimosa* spp											
124. browse, fresh, early vegetative	—	—	—	3.04[c]	2.50[c]	1.55[c]	0.94[c]	1.57[c]	69[d]	—	—
MOLASSES AND SYRUP *Beta vulgaris altissim*											
125. beet, sugar, molasses, more than 48% invert sugar, more than 79.5 degrees brix	3.40	0.0	8.5	2.71[c]	2.22[c]	1.44[c]	0.95[c]	1.41[c]	61[d]	2510	2380
MOLASSES AND SYRUP *Saccharum officinarum*											
126. sugarcane, molasses, more than 46% invert sugars, more than 79.5 degrees brix (black strap)	3.50	—	5.8	3.17[c]	2.60[c]	1.64[c]	1.03[c]	1.64[c]	72[d]	2210	2005
OAK, LIVE *Quercus virginiana*											
127. browse, fresh, early vegetative	—	—	—	2.38[c]	1.95[c]	1.16[c]	0.43[c]	1.20[c]	54[d]	—	—
128. leaves, fresh	—	—	—	2.03[c]	1.66[c]	0.99[c]	0.10[c]	1.01[c]	46[d]	—	—
OAK, SHIN *Quercus sinuata breviloba*											
129. browse, fresh, early vegetative	—	—	—	3.17[c]	2.60[c]	1.64[c]	1.03[c]	1.64[c]	72[d]	—	—
OATS *Avena sativa*											
130. grain	3.20	27.3	13.3	3.40[c]	2.78[c]	1.79[c]	1.17[c]	1.77[c]	77[d]	2760	2735
131. grain, Pacific coast	3.20	—	10.0	—	—	—	—	—	—	—	—
132. hay, sun-cured	1.92	63.0	9.5	—	—	—	—	—	—	—	—
133. silage, late vegetative	—	—	—	—	—	—	—	—	—	—	—
134. silage	—	—	—	2.73[c]	2.24[c]	1.36[c]	0.72[c]	1.40[c]	62[d]	—	—
ORCHARDGRASS *Dactylis glomerata*											
135. fresh, early vegetative	—	—	—	—	—	—	—	—	—	—	—
136. fresh, early bloom	2.29	55.1	12.8	2.91[c]	2.39[c]	1.47[c]	0.85[c]	1.50[c]	66[d]	—	—
137. fresh, midbloom	2.20	57.6	10.1	—	—	—	—	—	—	—	—
138. hay, sun-cured, early bloom	2.17	59.6	12.8	—	—	—	—	—	—	—	—
139. hay, sun-cured	—	—	—	—	—	—	—	—	—	—	—
OYSTER *Crassostrea* spp., *Ostrea* spp.											
140. shells, fine ground, min. 33% ca	—	—	—	—	—	—	—	—	—	—	—

(Table 48 continues)

TABLE 48 Composition of Some Common Feeds, Excluding Amino Acids *(Cont.)*.

POULTRY[b]	RABBITS[b]			MINERALS[a]			VITAMINS[a]			
ME$_n$ (kcal/ kg)	DE (kcal/ kg)	TDN (%)	Dig. Protein (%)	Calcium (%)	Phosphorous (%)	Potassium (%)	Carotene Vitamin A Activity (1000 IU/kg)	(mg/kg)	Vitamin E (mg/kg)	Vitamin D$_2$ (IU/g)
—	—	—	—	3.80	0.19	—	—	—	—	—
—	—	—	—	—	—	—	—	—	—	—
—	—	—	—	0.61	0.18	1.58	13.4	—	—	—
2195	—	—	—	9.44	4.74	0.61	—	—	1	—
2150	—	—	—	11.06	5.48	1.43	—	—	1	—
—	—	—	—	10.00	4.87	1.53	—	—	1.2	—
—	—	—	—	8.89	4.41	0.59	—	—	1.1	—
—	—	—	—	—	0.08	—	—	—	—	—
—	—	—	—	1.80	0.16	—	—	—	—	—
—	—	—	—	0.95	0.76	1.12	—	—	—	—
—	—	—	—	1.31	1.04	1.90	—	—	10	—
—	—	—	32.9	1.36	1.09	1.70	—	—	—	400
—	—	—	—	—	—	—	—	—	—	—
2675	—	—	—	0.05	0.35	0.47	—	—	—	—
2898	—	—	—	0.03	0.34	0.48	—	—	—	—
—	—	—	—	2.38	0.17	1.14	—	—	—	—
—	—	—	—	0.17	0.03	6.07	—	—	5	—
—	—	—	2.0	1.00	0.11	3.84	—	—	7	—
—	—	—	—	—	0.38	—	—	—	—	—
—	—	—	—	—	0.11	—	—	—	—	—
—	—	—	—	—	0.31	—	—	—	—	—
2550	2950	65	9.2	0.07	0.38	0.44	—	—	15	—
2610	—	—	—	0.11	0.34	0.42	—	—	22	—
—	—	—	—	0.24	0.22	1.51	11.1	28	—	1544
—	—	—	—	—	0.10	2.44	65.0	—	—	—
—	—	—	—	0.34	0.24	2.74	—	45	—	—
—	—	—	—	0.58	0.54	3.58	193.0	482	—	—
—	—	—	—	0.25	0.39	3.38	—	—	—	—
—	—	—	—	0.23	0.23	—	—	—	—	—
—	—	—	—	0.27	0.34	2.91	15.0	38	—	—
—	—	—	—	0.39	0.35	3.36	—	22	191	—
—	—	—	—	38.22	0.07	0.10	0	—	—	—

(Table 48 continues)

TABLE 48 Composition of Some Common Feeds, Excluding Amino Acids. *(Cont.)*

Feed Name	International Reference Number	Dry Matter (%)	Crude Fiber[a] (%)	Crude Protein[a] (%)	Acid Detergent Fiber[a] (%)	BEEF CATTLE[a] DE (Mcal/kg)	ME (Mcal/kg)	NE$_m$ (Mcal/kg)	NE$_g$ (Mcal/kg)	TDN (%)
PANGOLAGRASS *Digitaria decumbens*										
141. fresh	2-03-493	21	30.5	10.3	38	2.43	1.99	1.14	0.58	55
142. hay, sun-cured	1-26-214	91	36.0	7.1	43	1.98	1.63	0.79	0.25	45
PEANUT *Arachis hypogaea*										
143. hay, sun-cured	1-03-619	91	33.2	10.8	—	2.43	1.99	1.14	0.58	55
144. kernels, meal mechanical extracted (peanut meal)	5-03-649	93	7.5	52.0	6	3.66	3.00	2.03	1.37	83
145. kernels, meal solvent extracted (peanut meal)	5-03-650	92	10.8	52.3	—	3.40	2.78	1.85	1.22	77
PERSIMMON, TEXAS *Diospyros texana*										
146. leaves, fresh	2-29-858	48	—	11.8	—	—	—	—	—	—
PHOSPHATE ROCK										
147. defluorinated, ground	6-01-780	100	—	—	—	—	—	—	—	—
POULTRY										
148. by-product, meal rendered	5-03-798	93	2.2	62.8	—	—	—	—	—	—
149. feathers, hydrolyzed	5-03-795	93	1.5	91.3	—	3.09	2.53	1.63	1.03	70
150. manure and litter, dehydrated	5-05-587	89	16.1	24.5	—	2.91	2.39	1.51	0.91	66
PRAIRIE PLANTS, MIDWEST										
151. hay, sun-cured	1-03-191	92	34.0	5.8	—	2.25	1.84	1.00	0.45	51
152. hay, sun-cured midbloom	1-07-956	95	32.1	7.0	—	—	—	—	—	—
PRICKLYPEAR *Opuntia* spp.										
153. fresh	2-01-061	17	13.5	4.8	—	—	—	—	—	—
154. fresh, mature	2-01-059	21	13.7	3.1	—	—	—	—	—	—
155. fruit, fresh, immature	4-30-020	26	—	6.8	—	—	—	—	—	—
RAPE *Brassica napus*										
156. grazed, early vegetative	2-03-865	18	13	16.4	—	—	—	—	—	—
157. fresh, early bloom	2-03-866	11	15.8	23.5	—	3.31	2.71	1.79	1.16	75
158. seeds, meal solvent extracted	5-03-871	91	13.2	40.6	—	3.04	2.50	1.60	1.00	69
REDTOP *Agrostis alba*										
159. fresh, midbloom	2-03-890	39	29.0	7.4	—	—	—	—	—	—
160. fresh, full bloom	2-03-891	26	30.0	8.1	—	—	—	—	—	—
161. fresh	2-03-897	29	26.7	11.6	—	2.78	2.28	1.41	0.83	63
162. hay, sun-cured, midbloom	1-03-886	94	30.7	11.7	—	2.51	2.06	1.21	0.64	57
RYE *Secale cereale*										
163. distillers grains dehydrated	5-04-023	92	13.4	23.5	—	2.69	2.21	1.34	0.77	61
164. grain	4-04-047	88	2.5	13.8	—	3.70	3.04	2.06	1.40	84
165. hay, sun-cured	1-04-004	93	33.3	8.5	—	—	—	—	—	—
RYEGRASS, ITALIAN *Lolium multiflorum*										
166. fresh	2-04-073	25	23.8	14.5	—	2.65	2.17	1.31	0.74	60
167. hay, sun-cured, early vegetative	1-04-064	89	19.7	15.2	38	—	—	—	—	—
168. hay, sun-cured, late vegetative	1-04-065	86	23.8	10.3	42	2.73	2.24	1.38	0.80	62
169. hay, sun-cured, early bloom	1-04-066	83	36.3	5.5	45	2.38	1.95	1.11	0.55	54
RYEGRASS, PERENNIAL *Lolium perenne*										
170. fresh	2-04-086	27	23.2	10.4	—	3.00	2.46	1.57	0.97	68
171. hay, sun-cured	1-04-077	86	24.6	8.6	30	2.65	2.17	1.31	0.74	60
SAGE, BLACK *Salvia mellifera*										
172. browse, fresh, stem-cured	2-05-564	65	—	8.5	—	2.16	1.77	0.93	0.39	49
SAGEBRUSH, BIG *Artemisia tridentata*										
173. browse, fresh, stem-cured	2-07-992	65	—	9.3	30	2.21	1.81	0.97	0.42	50
SAGEBRUSH, BUD *Artemisia spinescens*										
174. browse, fresh, early vegetative	2-07-991	23	—	17.3	—	2.25	1.84	1.00	0.45	51
175. browse, fresh, late vegetative	2-04-124	32	22.7	17.5	—	2.29	1.88	1.04	0.49	52

(Table 48 continues)

TABLE 48 Composition of Some Common Feeds, Excluding Amino Acids. *(Cont.)*

| | | SHEEP[a] | | | | | | DAIRY CATTLE[a] | | | |
| | | | | | | | | (Growing) | | (Lactating) | |
DE (Mcal/kg)	ME (Mcal/kg)	NE_m (Mcal/kg)	NE_g (Mcal/kg)	TDN (%)	Dig. Protein (%)	DE (Mcal/kg)	ME (Mcal/kg)	NE_m (Mcal/kg)	NE_g (Mcal/kg)	NE_l (Mcal/kg)	TDN (%)
—	—	—	—	—	—	—	—	—	—	—	—
—	—	—	—	—	—	1.98	1.55	0.78	0.25	0.98	45
—	—	—	—	—	—	2.43	2.00	1.14	0.58	1.23	55
4.14	3.40	2.35	1.65	94	47.3	3.66	3.25	2.03	1.37	1.91	83
3.40	2.78	1.85	1.22	77	—	3.40	2.98	1.86	1.22	1.77	77
—	—	—	—	—	—	—	—	—	—	—	—
—	—	—	—	—	—	—	—	—	—	—	—
—	—	—	—	—	—	—	—	—	—	—	—
—	—	—	—	—	—	—	—	—	—	—	—
—	—	—	—	—	—	—	—	—	—	—	—
2.38	1.95	1.11	0.55	54	2.8	—	—	—	—	—	—
2.51	2.06	1.21	0.64	57	2.1	—	—	—	—	—	—
—	—	—	—	—	—	—	—	—	—	—	—
—	—	—	—	—	—	—	—	—	—	—	—
—	—	—	—	—	—	3.57	3.16	1.97	1.32	1.87	81
3.31	2.71	1.79	1.16	75	20.2	—	—	—	—	—	—
3.26	2.68	1.76	1.14	74	—	3.04	2.62	1.60	1.00	1.57	69
2.60	2.13	1.28	0.71	59	4.5	—	—	—	—	—	—
2.56	2.10	1.24	0.68	58	4.5	—	—	—	—	—	—
—	—	—	—	—	—	2.78	2.36	1.41	0.83	1.42	63
2.47	2.03	1.18	0.61	56	7.3	2.51	2.09	1.21	0.64	1.28	57
2.82	2.31	1.44	0.86	64	14.1	—	—	—	—	—	—
3.75	3.07	2.09	1.43	85	10.9	3.70	3.29	2.06	1.40	1.94	84
—	—	—	—	—	—	—	—	—	—	—	—
2.60	2.13	1.28	0.71	59	6.2	—	—	—	—	—	—
2.65	2.17	1.31	0.74	60	10.2	3.00	2.58	1.57	0.97	1.55	68
—	—	—	—	—	—	2.73	2.31	1.38	0.80	1.40	62
2.51	2.06	1.21	0.64	57	8.1	2.38	1.96	1.11	0.55	1.20	54
—	—	—	—	—	—	—	—	—	—	—	—
—	—	—	—	—	—	2.82	2.40	1.44	0.85	1.45	64
2.16	1.77	0.93	0.39	49	4.5	—	—	—	—	—	—
2.21	1.81	0.97	0.42	50	4.9	—	—	—	—	—	—
2.25	1.84	1.00	0.45	51	13.7	—	—	—	—	—	—
2.47	2.03	1.18	0.61	56	13.3	—	—	—	—	—	—

(Table 48 continues)

TABLE 48 Composition of Some Common Feeds, Excluding Amino Acids *(Con't)*.

Feed Name	HORSES[a]			GOATS[a]						SWINE[b]	
	DE (Mcal/ kg)	Neutral Detergent Fiber (%)	Crude Protein (%)	DE (Mcal/ kg)	ME (Mcal/ kg)	NE$_m$ (Mcal/ kg)	NE$_g$ (Mcal/ kg)	NE$_l$ (Mcal/ kg)	TDN (%)	DE (kcal/ kg)	ME (kcal/ kg)
PANGOLAGRASS *Digitaria decumbens*											
141. fresh	1.95	—	9.1	—	—	—	—	—	—	—	—
142. hay, sun-cured	1.78	72.7	7.4	—	—	—	—	—	—	—	—
PEANUT *Arachis hypogaea*											
143. hay, sun-cured	1.91	—	10.9	—	—	—	—	—	—	—	—
144. kernels, meal mechanical extracted (peanut meal)	—	—	—	—	—	—	—	—	84	4055	3710
145. kernels, meal solvent extracted (peanut meal)	3.25	—	52.9	—	—	—	—	—	—	3140	2910
PERSIMMON, TEXAS *Diospyros texana*											
146. leaves, fresh	—	—	—	2.38[c]	1.95[c]	1.16[c]	0.43[c]	1.20[c]	54[d]	—	—
PHOSPHATE ROCK											
147. defluorinated, ground	—	—	—	—	—	—	—	—	—	—	—
POULTRY											
148. by-product, meal rendered	—	—	—	—	—	—	—	—	—	—	—
149. feathers, hydrolyzed	—	—	—	3.09[c]	2.53[c]	1.58[c]	0.97[c]	1.60[c]	70[d]	2729	2213
150. manure and litter, dehydrated	—	—	—	—	—	—	—	—	—	—	—
PRAIRIE PLANTS, MIDWEST											
151. hay, sun-cured	1.62	—	6.4	—	—	—	—	—	—	—	—
152. hay, sun-cured midbloom	—	—	—	—	—	—	—	—	—	—	—
PRICKLYPEAR *Opuntia* spp.											
153. fresh	—	—	—	—	—	—	—	—	—	—	—
154. fresh, mature	—	—	—	2.25[c]	1.84[c]	1.10[c]	0.32[c]	1.13[c]	51[d]	—	—
155. fruit, fresh, immature	—	—	—	1.50[c]	1.23[c]	0.82[c]	0.00[c]	0.71[c]	34[d]	—	—
RAPE *Brassica napus*											
156. grazed, early vegetative	—	—	—	—	—	—	—	—	—	—	—
157. fresh, early bloom	—	—	—	—	—	—	—	—	—	—	—
158. seeds, meal solvent extracted	—	—	—	3.04[c]	2.50[c]	1.55[c]	0.94[c]	1.57[c]	69[d]	—	—
REDTOP *Agrostis alba*											
159. fresh, midbloom	—	—	—	2.73[c]	2.24[c]	1.36[c]	0.72[c]	1.40[c]	62[d]	—	—
160. fresh, full bloom	—	—	—	—	—	—	—	—	—	—	—
161. fresh	—	—	—	—	—	—	—	—	—	—	—
162. hay, sun-cured, midbloom	1.97	—	12.0	—	—	—	—	—	—	—	—
RYE *Secale cereale*											
163. distillers grains, dehydrated	—	—	—	—	—	—	—	—	—	—	—
164. grain	3.84	18.6	13.7	3.70[c]	3.04[c]	2.00[c]	1.35[c]	1.94[c]	84[d]	3285	3005
165. hay, sun-cured	—	—	—	2.07[c]	1.70[c]	1.01[c]	0.15[c]	1.03[c]	47[d]	—	—
RYEGRASS, ITALIAN *Lolium multiflorum*											
166. fresh	2.20	61.0	17.9	—	—	—	—	—	—	—	—
167. hay, sun-cured, early vegetative	—	—	—	3.00[c]	2.46[c]	1.52[c]	0.91[c]	1.55[c]	68[d]	—	—
168. hay, sun-cured, late vegetative	1.84	64.0	10.3	—	—	—	—	—	—	—	—
169. hay, sun-cured, early bloom	—	—	—	—	—	—	—	—	—	—	—
RYEGRASS, PERENNIAL *Lolium perenne*											
170. fresh	—	—	—	—	—	—	—	—	—	—	—
171. hay, sun-cured	—	—	—	—	—	—	—	—	—	—	—
SAGE, BLACK *Salvia mellifera*											
172. browse, fresh, stem-cured	—	—	—	—	—	—	—	—	—	—	—
SAGEBRUSH, BIG *Artemisia tridentata*											
173. browse, fresh, stem-cured	—	—	—	—	—	—	—	—	—	—	—
SAGEBRUSH, BUD *Artemisia spinescens*											
174. browse, fresh, early vegetative	—	—	—	—	—	—	—	—	—	—	—
175. browse, fresh, late vegetative	—	—	—	—	—	—	—	—	—	—	—

(Table 48 continues)

TABLE 48 Composition of Some Common Feeds, Excluding Amino Acids (*Cont.*).

POULTRY[b]	RABBITS[b]			MINERALS[a]			VITAMINS[a]			
ME_n (kcal/ kg)	DE (kcal/ kg)	TDN (%)	Dig. Protein (%)	Calcium (%)	Phosphorous (%)	Potassium (%)	Carotene Vitamin A Activity (1000 IU/kg)	(mg/kg)	Vitamin E (mg/kg)	Vitamin D_2 (IU/g)
—	—	—	—	0.43	0.18	1.43	24.8	—	—	—
—	—	—	—	0.46	0.23	1.40	—	—	—	—
—	—	—	—	1.23	0.15	1.38	14.0	—	—	3600
2500	—	—	—	0.20	0.61	1.25	—	—	3	—
2200	4120	90	45.2	0.29	0.68	1.23	—	—	—	—
—	—	—	—	—	0.11	—	—	—	—	—
—	0	0	0.0	31.65	13.7	0.16	0	—	—	—
2950	—	—	—	3.22	1.83	0.32	—	—	2.2	—
2360	—	—	—	0.28	0.72	0.31	—	—	—	—
—	—	—	—	3.16	1.78	1.68	—	—	—	—
—	—	—	—	0.43	0.15	1.08	9.8	—	—	—
—	—	—	—	—	—	—	—	—	—	—
—	—	—	—	9.61	0.12	2.21	—	—	—	—
—	—	—	—	—	0.03	—	—	6	—	—
—	—	—	—	—	0.13	—	—	—	—	—
—	—	—	—	—	—	—	62	—	—	—
—	—	—	—	—	—	—	—	—	—	—
—	—	—	—	0.67	1.04	1.36	—	—	—	—
—	—	—	—	0.33	0.23	2.13	—	—	—	—
—	—	—	—	0.62	0.37	2.35	—	153	—	—
—	—	—	—	0.46	0.29	2.35	87	—	—	—
—	—	—	—	0.63	0.35	1.69	2.0	5	—	—
—	—	—	—	0.16	0.52	0.8	—	—	—	—
2626	3590	77	9.0	0.07	0.37	0.52	—	—	17	—
—	—	—	—	0.33	0.19	1.35	—	6.5	—	—
—	—	—	—	0.65	0.41	2.00	160.4	401	—	—
—	—	—	—	—	—	—	—	—	—	—
—	—	—	—	0.62	0.34	1.56	116.0	—	—	—
—	—	—	—	—	—	—	—	—	—	—
—	—	—	—	0.55	0.27	1.91	88.8	—	—	—
—	—	—	—	0.65	0.32	1.67	48.0	—	211	—
—	—	—	—	0.81	0.17	—	—	—	—	—
—	—	—	—	0.71	0.18	—	6.4	16	—	—
—	—	—	—	0.97	0.33	—	9.5	24	—	—
—	—	—	—	0.60	0.42	—	—	—	—	—

(Table 48 continues)

TABLE 48 Composition of Some Common Feeds, Excluding Amino Acids. *(Cont.)*

Feed Name	International Reference Number	Dry Matter (%)	Crude Fiber[a] (%)	Crude Protein[a] (%)	Acid Detergent Fiber[a] (%)	BEEF CATTLE[a] DE (Mcal/ kg)	ME (Mcal/ kg)	NE$_m$ (Mcal/ kg)	NE$_g$ (Mcal/ kg)	TDN (%)
SAGEBRUSH, FRINGED *Artemisia frigida*										
176. browse, fresh, midbloom	2-04-129	43	33.2	9.4	—	2.56	2.10	1.24	0.68	58
177. browse, fresh, mature	2-04-130	60	31.8	7.1	35	2.25	1.84	1.00	0.45	51
SAGEBRUSH, MEXICAN *Artemisia ludoviciana albula*										
178. browse, fresh, mature	2-30-052	44	—	10.2	—	—	—	—	—	—
SAGEBRUSH, SAND *Artemisia filifolia*										
179. browse, fresh, early vegetative	2-04-133	29	22.6	12.2	—	—	—	—	—	—
180. browse, fresh, mature	2-04-135	45	31.7	7.2	—	—	—	—	—	—
SALTBUSH, NUTTALL *Atriplex nuttallii*										
181. browse, fresh, stem-cured	2-07-993	55	—	7.2	—	1.59	1.30	0.45	—	36
SODIUM PHOSPHATE										
182. monobasic	6-04-228	87	—	—	—	—	—	—	—	—
183. SODIUM TRIPOLYPHOSPHATE	6-08-076	96	—	—	—	—	—	—	—	—
SORGHUM *Sorghum bicolor*										
184. aerial part, sun-cured, full bloom	1-04-371	90	23.8	6.4	0.62	—	—	—	—	—
185. distillers grains, dehydrated	5-04-374	94	12.7	34.4	—	3.66	3.00	2.03	1.37	83
186. grain	4-04-383	90	2.6	12.4	0.04	—	—	—	—	—
187. grain, 8–10% protein	4-20-893	87	2.0	9.7	9	3.70	3.04	2.06	1.40	84
188. grain, more than 10% protein	4-20-894	88	2.0	13.0	—	3.66	3.00	2.03	1.37	83
189. silage	3-04-323	30	27.9	7.5	38	2.65	2.17	1.31	0.74	60
SORGHUM, JOHNSONGRASS *Sorghum halepense*										
190. hay, sun-cured	1-04-407	89	33.5	9.5	—	2.34	1.92	1.07	0.52	53
SORGHUM, KAFIR *Sorghum bicolor caffrorum*										
191. grain	4-04-428	89	2.3	12.3	—	—	—	—	—	—
SORGHUM, MILO *Sorghum bicolor subglabrescens*										
192. grain	4-04-444	89	2.5	11.3	5	—	—	—	—	—
SORGHUM, SORGO *Sorghum bicolor saccharatum*										
193. silage, mature	3-04-467	27	27.6	6.6	—	—	—	—	—	—
194. silage	3-04-468	27	28.3	6.2	—	2.56	2.10	1.24	0.68	58
SORGHUM, SUDANGRASS *Sorghum bicolor sudanense*										
195. fresh, early vegetative	2-04-484	18	23.0	16.8	29	3.09	2.53	1.63	1.03	70
196. hay, sun-cured, late vegetative	1-04-474	88	31.9	13.9	33	—	—	—	—	—
197. hay, sun-cured	1-04-480	91	36.0	8.0	42	2.47	2.03	1.18	0.61	56
198. silage	3-04-499	28	33.1	10.8	42	2.43	1.99	1.14	0.58	55
SOYBEAN *Glycine max*										
199. hay, sun-cured, midbloom	1-04-538	94	29.8	17.8	40	2.34	1.92	1.07	0.52	53
200. hay, sun-cured	1-04-558	89	33.7	16.0	—	—	—	—	—	—
201. seeds, meal solvent extracted, 44% protein	5-20-637	89	7.0	49.9	—	3.70	3.04	2.06	1.40	84
202. seeds, meal solvent extracted	5-04-604	90	6.5	49.9	—	—	—	—	—	—
203. seeds without hulls, meal solvent extracted	5-04-612	90	3.7	55.1	6	3.84	3.15	2.15	1.48	87
SUGARCANE *Saccharum officinarum*										
204. bagasse, dehydrated	1-04-686	91	49.0	1.5	61	2.12	1.74	0.90	0.35	48
SUMAC, FRAGRANT *Rhus aromatica*										
205. browse, fresh, early vegetative	2-29-827	40	—	13.7	—	—	—	—	—	—
SWEETCLOVER, YELLOW *Melilotus officinalis*										
206. hay, sun-cured	1-04-754	87	33.4	15.7	—	2.38	1.95	1.11	0.55	54
TIMOTHY *Phleum pratense*										
207. fresh, midbloom	2-04-905	29	33.5	9.1	37	2.78	2.28	1.41	0.83	63
208. hay, sun-cured, late vegetative	1-04-881	89	27.0	17.0	29	2.73	2.24	1.41	0.83	62
209. hay, sun-cured, midbloom	1-04-883	89	31.0	9.1	36	2.51	2.06	1.21	0.64	57

(Table 48 continues)

TABLE 48 Composition of Some Common Feeds, Excluding Amino Acids. *(Cont.)*

| | | SHEEP[a] | | | | | | DAIRY CATTLE[a] | | | |
| | | | | | | | | (Growing) | | (Lactating) | |
DE (Mcal/kg)	ME (Mcal/kg)	NE_m (Mcal/kg)	NE_g (Mcal/kg)	TDN (%)	Dig. Protein (%)	DE (Mcal/kg)	ME (Mcal/kg)	NE_m (Mcal/kg)	NE_g (Mcal/kg)	NE_l (Mcal/kg)	TDN (%)
2.60	2.13	1.28	0.71	59	5.8	—	—	—	—	—	—
2.38	1.95	1.11	0.55	54	3.6	—	—	—	—	—	—
—	—	—	—	—	—	—	—	—	—	—	—
—	—	—	—	—	—	—	—	—	—	—	—
—	—	—	—	—	—	—	—	—	—	—	—
1.59	1.30	0.45	—	36	3.4	—	—	—	—	—	—
—	—	—	—	—	—	—	—	—	—	—	—
—	—	—	—	—	—	—	—	—	—	—	—
2.47	2.03	1.18	0.61	56	2.3	—	—	—	—	—	—
3.75	3.07	2.09	1.43	85	—	3.66	3.25	2.03	1.37	1.91	83
3.88	3.18	2.18	1.50	88	9.1	—	—	—	—	—	—
—	—	—	—	—	—	3.53	3.12	1.94	1.30	1.84	80
—	—	—	—	—	—	3.48	3.06	1.91	1.27	1.82	79
2.51	2.06	1.21	0.64	57	2.2	2.65	2.22	1.31	0.74	1.35	60
2.47	2.03	1.18	0.61	56	4.2	2.34	1.91	1.08	0.52	1.18	53
3.75	3.07	2.09	1.43	85	9.9	—	—	—	—	—	—
3.88	3.18	2.18	1.50	88	8.8	—	—	—	—	—	—
—	—	—	—	—	—	—	—	—	—	—	—
2.65	2.17	1.31	0.74	60	1.4	2.56	2.14	1.25	0.68	1.30	58
2.78	2.28	1.41	0.83	63	12.6	3.09	2.67	1.63	1.03	1.60	70
—	—	—	—	—	—	—	—	—	—	—	—
2.43	1.99	1.14	0.58	55	4.3	2.47	2.04	1.18	0.62	1.25	56
2.34	1.92	1.07	0.52	53	6.6	2.43	2.00	1.14	0.58	1.23	55
—	—	—	—	—	—	2.34	1.91	1.08	0.52	1.18	53
2.51	2.06	1.21	0.64	57	11.3	—	—	—	—	—	—
—	—	—	—	—	—	—	—	—	—	—	—
3.88	3.18	2.18	1.50	88	46.4	—	—	—	—	—	—
—	—	—	—	—	—	3.84	3.42	2.16	1.48	2.01	87
—	—	—	—	—	—	1.94	1.51	0.75	0.22	0.96	44
—	—	—	—	—	—	—	—	—	—	—	—
2.34	1.92	1.07	0.52	53	11.5	2.38	1.96	1.11	0.55	1.20	54
2.73	2.24	1.38	0.80	62	4.9	—	—	—	—	—	—
2.87	2.35	1.47	0.88	65	11.4	2.91	2.49	1.51	0.92	1.50	66
2.65	2.17	1.31	0.74	60	5.6	2.56	2.13	1.25	0.68	1.30	58

(Table 48 continues)

TABLE 48 Composition of Some Common Feeds, Excluding Amino Acids *(Con't)*.

Feed Name	HORSES[a]			GOATS[a]						SWINE[b]	
	DE (Mcal/kg)	Neutral Detergent Fiber (%)	Crude Protein (%)	DE (Mcal/kg)	ME (Mcal/kg)	NE_m (Mcal/kg)	NE_g (Mcal/kg)	NE_l (Mcal/kg)	TDN (%)	DE (kcal/kg)	ME (kcal/kg)
SAGEBRUSH, FRINGED *Artemisia frigida*											
176. browse, fresh, midbloom	—	—	—	—	—	—	—	—	—	—	—
177. browse, fresh, mature	—	—	—	—	—	—	—	—	—	—	—
SAGEBRUSH, MEXICAN *Artemisia ludoviciana albula*											
178. browse, fresh, mature			—	2.16[c]	1.77[c]	1.05[c]	0.23[c]	1.08[c]	49[d]		
SAGEBRUSH, SAND *Artemisia filifolia*											
179. browse, fresh, early vegetative	—	—	—	2.91[c]	2.39[c]	1.47[c]	0.85[c]	1.50[c]	66[d]	—	—
180. browse, fresh, mature	—	—	—	2.65[c]	2.17[c]	1.31[c]	0.65[c]	1.35[c]	60[d]	—	—
SALTBUSH, NUTTALL *Atriplex nuttallii*											
181. browse, fresh, stem-cured	—	—	—	—	—	—	—	—	—	—	—
SODIUM PHOSPHATE											
182. monobasic	—	—	—	—	—	—	—	—	—	—	—
183. SODIUM TRIPOLYPHOSPHATE	—	—	—	—	—	—	—	—	—	—	—
SORGHUM *Sorghum bicolor*											
184. aerial part, sun-cured, full bloom	—	—	—	2.91[c]	2.39[c]	1.47[c]	0.85[c]	1.50[c]	66	—	—
185. distillers grains, dehydrated	—	—	—	3.66[c]	3.00[c]	1.97[c]	1.32[c]	1.91[c]	83[d]	—	—
186. grain	3.56	23.0	12.7	—	—	—	—	—	—	—	—
187. grain, 8–10% protein	—	—	—	—	—	—	—	—	—	—	—
188. grain, more than 10% protein	—	—	—	—	—	—	—	—	—	—	—
189. silage	—	—	—	—	—	—	—	—	—	—	—
SORGHUM, JOHNSONGRASS *Sorghum halepense*											
190. hay, sun-cured	1.66	—	7.5	2.34[c]	1.92[c]	1.14[c]	0.40[c]	1.18[c]	53[d]	—	—
SORGHUM, KAFIR *Sorghum bicolor caffrorum*											
191. grain	—	—	—	—	—	—	—	—	—	—	—
SORGHUM, MILO *Sorghum bicolor subglabrescens*											
192. grain	—	—	—	3.88[c]	3.18[c]	2.12[c]	1.45[c]	2.04[c]	88[d]	3415	3280
SORGHUM, SORGO *Sorghum bicolor saccharatum*											
193. silage, mature	—	—	—	2.78[c]	2.28[c]	1.39[c]	0.75[c]	1.42[c]	63[d]	—	—
194. silage	—	—	—	—	—	—	—	—	—	—	—
SORGHUM, SUDANGRASS *Sorghum bicolor sudanense*											
195. fresh, early vegetative	—	—	—	—	—	—	—	—	—	—	—
196. hay, sun-cured, late vegetative	—	—	—	2.51[c]	2.06[c]	1.23[c]	0.55[c]	1.28[c]	57[d]	—	—
197. hay, sun-cured	—	—	—	—	—	—	—	—	—	—	—
198. silage	—	—	—	—	—	—	—	—	—	—	—
SOYBEAN *Glycine max*											
199. hay, sun-cured, midbloom	—	—	—	—	—	—	—	—	—	—	—
200. hay, sun-cured	—	—	—	—	—	—	—	—	—	—	—
201. seeds, meal solvent extracted, 44% protein	3.52	14.9	49.9	—	—	—	—	—	—	—	—
202. seeds, meal solvent extracted	—	—	—	3.88[c]	3.18[c]	2.12[c]	1.45[c]	2.04[c]	88[d]	3490	3220
203. seeds without hulls, meal solvent extracted	3.73	7.7	54.0	—	—	—	—	—	—	3680	3385
SUGARCANE *Saccharum officinarum*											
204. bagasse, dehydrated	—	—	—	—	—	—	—	—	—	—	—
SUMAC, FRAGRANT *Rhus aromatica*											
205. browse, fresh, early vegetative	—	—	—	3.40[c]	2.78[c]	1.79[c]	1.17[c]	1.77[c]	77[d]	—	—
SWEETCLOVER, YELLOW *Melilotus officinalis*											
206. hay, sun-cured	—	—	—	—	—	—	—	—	—	—	—
TIMOTHY *Phleum pratense*											
207. fresh, midbloom	2.00	—	9.1	—	—	—	—	—	—	—	—
208. hay, sun-cured, late vegetative	—	—	—	—	—	—	—	—	—	—	—
209. hay, sun-cured, midbloom	1.99	63.7	9.7	—	—	—	—	—	—	—	—

(Table 48 continues)

TABLE 48 Composition of Some Common Feeds, Excluding Amino Acids *(Cont.).*

POULTRY[b]	RABBITS[b]			MINERALS[a]			VITAMINS[a]			
							Carotene Vitamin A Activity		Vitamin E	Vitamin D$_2$
ME$_n$ (kcal/kg)	DE (kcal/kg)	TDN (%)	Dig. Protein (%)	Calcium (%)	Phosphorous (%)	Potassium (%)	(1000 IU/kg)	(mg/kg)	(mg/kg)	(IU/g)
—	—	—	—	—	—	—	—	—	—	—
—	—	—	—	—	—	—	—	—	—	—
—	—	—	—	—	0.15	—	—	—	—	—
—	—	—	—	—	—	—	—	—	—	—
—	—	—	—	0.48	0.12	—	—	—	—	—
—	—	—	—	2.21	0.12	—	7.6	19	—	—
—	—	—	—	—	25.80	—	0	—	—	—
—	—	—	—	—	25.98	—	0	—	—	—
—	—	—	—	0.62	0.19	1.24	—	—	—	—
—	—	—	—	0.16	0.74	0.38	—	—	—	—
—	3330	—	6.4	0.04	0.33	0.39	—	1	12	29
3288	—	—	—	0.04	0.34	0.40	—	—	12	—
3212	—	—	—	0.04	0.36	0.38	—	—	12	—
—	—	—	—	0.35	0.21	1.37	6.0	15	—	700
—	—	—	—	0.84	0.28	1.35	16.0	39	—	—
—	—	—	—	0.04	0.35	0.38	—	0	—	—
—	—	—	—	0.05	0.34	0.35	—	0	13	29
—	—	—	—	0.26	0.14	—	—	2.9	—	—
—	—	—	—	0.34	0.17	1.12	14.0	36	—	—
—	—	—	—	0.43	0.41	2.14	79.0	198	—	—
—	—	—	—	0.43	0.30	3.22	—	—	—	—
—	—	—	—	0.55	0.30	1.87	24.0	59	—	—
—	—	—	—	0.46	0.21	2.25	42.0	105	—	—
—	—	—	—	1.26	0.27	0.97	13.0	—	—	—
—	—	—	—	1.29	0.28	1.07	—	45	30	1059
—	—	—	—	0.33	0.71	2.14	—	—	—	—
2230	3770	82	41.4	0.34	0.70	2.20	—	—	3	—
2440	—	—	45.0	0.29	0.70	2.30	—	—	—	—
—	—	—	—	0.90	0.29	0.50	—	—	—	—
—	—	—	—	—	0.20	—	—	—	—	—
—	—	—	—	1.27	0.25	1.60	40.0	99	—	1874
—	—	—	—	0.38	0.30	2.06	78.0	195	—	—
—	—	—	—	0.66	0.34	1.68	50.0	125	—	—
—	1420	32	4.9	0.48	0.22	1.59	21.0	53	—	—

(Table 48 continues)

TABLE 48 Composition of Some Common Feeds, Excluding Amino Acids. *(Cont.)*

Feed Name	International Reference Number	Dry Matter (%)	Crude Fiber[a] (%)	Crude Protein[a] (%)	Acid Detergent Fiber[a] (%)	BEEF CATTLE[a] DE (Mcal/ kg)	ME (Mcal/ kg)	NE$_m$ (Mcal/ kg)	NE$_g$ (Mcal/ kg)	TDN (%)
TREFOIL, BIRDSFOOT *Lotus corniculatus*										
210. fresh	2-20-786	24	24.7	21.0	—	2.91	2.39	1.51	0.91	66
211. hay, sun-cured	1-05-044	92	30.7	16.3	36	2.60	2.13	1.28	0.71	59
212. hay, sun-cured, midbloom	1-20-790	91	—	14.5	38	—	—	—	—	—
TRITICALE *Triticale hexaploide*										
213. grain	4-20-362	90	4.4	17.6	8	3.70	3.04	2.06	1.40	84
UREA										
214. 45% nitrogen, 281% protein equivalent	5-05-070	99	—	281.0	—	—	—	—	—	—
VETCH *Vicia* spp.										
215. hay, sun-cured	1-05-106	89	30.6	20.8	33	2.51	2.06	1.21	0.64	57
WHEAT *Triticum aestivum*										
216. bran	4-05-190	89	11.3	17.1	15	3.09	2.53	1.63	1.03	70
217. grain	4-05-211	89	2.9	16.0	8	3.88	3.18	2.18	1.50	88
218. grain, hard red spring	4-05-258	88	2.9	17.2	13	3.92	3.22	2.21	1.52	89
219. grain, hard winter	4-05-268	88	2.8	14.4	4	3.88	3.18	2.18	1.50	88
220. grain, soft red winter	4-05-294	88	2.4	13.0	—	3.92	3.22	2.21	1.52	89
221. grain, soft white winter	4-05-337	89	2.6	11.3	4	3.92	3.22	2.21	1.52	89
222. flour-by-product, less than 9.5% fiber (wheat middlings)	4-05-205	89	8.2	18.4	10	3.04	2.50	1.60	1.00	69
223. flour by-product, less than 7% fiber (wheat shorts)	4-05-201	88	7.7	18.6	—	3.22	2.64	1.93	1.11	73
WHEAT, DURUM *Triticum durum*										
224. grain	4-05-224	88	2.5	15.9	—	3.75	3.07	2.09	1.43	85
WHEATGRASS, CRESTED *Agropyron desertorum*										
225. fresh, early vegetative	2-05-420	28	22.2	21.5	—	3.31	2.71	1.79	1.16	75
226. hay, sun-cured	1-05-418	93	32.9	12.4	36	2.34	1.92	1.07	2.52	53
WILDRYE, RUSSIAN *Elymus junceus*										
227. fresh	2-05-469	33	22.4	14.1	—	—	—	—	—	—
WHEY *Bos taurus*										
228. dehydrated (cattle)	4-01-182	93	0.2	14.2	0	3.57	2.93	1.97	1.32	81
229. fresh (cattle)	4-08-134	7	—	13.0	—	4.14	3.40	2.35	1.65	94
230. low lactose, dehydrated (dried whey product) (cattle)	4-01-186	93	0.2	17.7	—	3.48	2.86	1.91	1.27	79
YEAST, BREWERS *Saccharomyces cerevisiae*										
231. dehydrated	7-05-527	93	3.1	46.9	—	3.48	2.86	1.91	1.27	79
YEAST, IRRADIATED *Saccharomyces cerevisiae*										
232. dehydrated	7-05-529	94	6.6	51.2	—	3.35	2.75	1.82	1.19	76
YEAST, PRIMARY *Saccharomyces cerevisiae*										
233. dehydrated	7-05-533	93	3.3	51.8	—	3.40	2.78	1.85	1.22	77
YEAST, TORULA *Torulopsis utilis*										
234. dehydrated	7-05-534	93	2.4	52.7	—	3.44	2.82	1.88	1.24	78
YUCCA *Yucca* spp.										
235. flowers, fresh	2-05-536	15	13.3	19.7	—	—	—	—	—	—
236. leaves, fresh, immature	2-30-050	41	—	7.3	—	—	—	—	—	—

[a]100 percent dry matter basis.
[b]As fed basis.
[c]Calculated from data for cows.
[d]Data from sheep.
Reprinted from: "Nutrient Requirements of Beef Cattle," Sixth Revised Edition, 1984; "Nutrient Requirements of Dairy Cattle," Sixth Revised Edition, updated 1989, 1988; "Nutrient Requirements of Horses," Fifth Revised Edition, 1989; "Nutrient Requirements of Goats," 1981; "Nutrient Requirements of Poultry," Ninth Revised Edition, 1994; "Nutrient Requirements of Rabbits," Second Revised Edition, 1977; "Nutrient Requirements of Sheep," Sixth Revised Edition, 1985; "Nutrient Requirements of Swine," Ninth Revised Edition, 1988,
© 1988 by the National Academy of Sciences; with permission of the National Academy Press, Washington, DC.

(Table 48 continues)

TABLE 48 Composition of Some Common Feeds, Excluding Amino Acids. *(Cont.)*

		SHEEP[a]						(Growing)		(Lactating)	
DE (Mcal/kg)	ME (Mcal/kg)	NE_m (Mcal/kg)	NE_g (Mcal/kg)	TDN (%)	Dig. Protein (%)	DE (Mcal/kg)	ME (Mcal/kg)	NE_m (Mcal/kg)	NE_g (Mcal/kg)	NE_l (Mcal/kg)	TDN (%)
2.78	2.28	1.41	0.83	63	16.3	2.91	2.49	1.51	0.92	1.50	66
2.56	2.10	1.24	0.68	58	11.2	2.60	2.18	1.27	0.70	1.33	59
—	—	—	—	—	—	—	—	—	—	—	—
—	—	—	—	—	—	3.70	3.29	2.06	1.40	1.94	84
—	—	—	—	—	—	—	—	—	—	—	—
2.43	1.99	1.14	0.58	55	15.7	2.51	2.09	1.21	0.64	1.28	57
3.13	2.57	1.67	1.06	71	13.3	3.09	2.67	1.63	1.03	1.60	70
3.84	3.15	2.15	1.48	87	12.2	3.88	3.47	2.18	1.50	2.04	88
3.97	3.25	2.24	1.55	90	13.0	—	—	—	—	—	—
3.88	3.18	2.18	1.50	88	10.4	3.88	3.47	2.18	1.50	2.04	88
3.88	3.18	2.18	1.50	88	9.8	—	—	—	—	—	—
—	—	—	—	—	—	3.92	3.51	2.20	1.52	2.06	89
3.62	2.97	2.0	1.35	82	14.8	3.04	2.62	1.60	1.00	1.57	69
—	—	—	—	—	—	3.22	2.80	1.73	1.11	1.67	73
—	—	—	—	—	—	—	—	—	—	—	—
3.31	2.71	1.79	1.16	75	18.3	—	—	—	—	—	—
2.34	1.92	1.07	0.52	53	8.0	—	—	—	—	—	—
—	—	—	—	—	—	—	—	—	—	—	—
—	—	—	—	—	—	3.57	3.16	1.97	1.32	1.87	81
—	—	—	—	—	—	—	—	—	—	—	—
—	—	—	—	—	—	3.48	3.07	1.91	1.27	1.82	79
—	—	—	—	—	—	—	—	—	—	—	—
—	—	—	—	—	—	—	—	—	—	—	—
—	—	—	—	—	—	3.44	3.02	1.88	1.24	1.79	78
—	—	—	—	—	—	—	—	—	—	—	—
—	—	—	—	—	—	—	—	—	—	—	—

(Table 48 continues)

TABLE 48 Composition of Some Common Feeds, Excluding Amino Acids *Cont.*).

	HORSES[a]			GOATS[a]						SWINE[b]	
Feed Name	DE (Mcal/ kg)	Neutral Detergent Fiber (%)	Crude Protein (%)	DE (Mcal/ kg)	ME (Mcal/ kg)	NE$_m$ (Mcal/ kg)	NE$_g$ (Mcal/ kg)	NE$_l$ (Mcal/ kg)	TDN (%)	DE (kcal/ kg)	ME (kcal/ kg)
TREFOIL, BIRDSFOOT *Lotus corniculatus*											
210. fresh	2.18	46.7	20.6	—	—	—	—	—	—	—	—
211. hay, sun-cured	2.20	47.5	15.9	—	—	—	—	—	—	—	—
212. hay, sun-cured, midbloom	—	—	—	2.78[c]	2.28[c]	1.39[c]	0.75[c]	1.42[c]	63[d]	—	—
TRITICALE *Triticale hexaploide*											
213. grain	—	—	—	—	—	—	—	—	—	3299	3050
UREA											
214. 45% nitrogen, 281% protein equivalent	—	—	—	—	—	—	—	—	—	—	—
VETCH *Vicia* spp.											
215. hay, sun-cured	—	—	—	2.51[c]	2.06[c]	1.23[c]	0.55[c]	1.28[c]	57[d]	—	—
WHEAT *Triticum aestivum*											
216. bran	3.30	42.8	17.4	3.09[c]	2.53[c]	1.58[c]	0.97[c]	1.60[c]	70	2370	2155
217. grain	—	—	—	3.88[c]	3.18[c]	2.12[c]	1.45[c]	2.04[c]	88[d]	—	—
218. grain, hard red spring	—	—	—	—	—	—	—	—	—	—	—
219. grain, hard winter	3.86	11.7	14.6	—	—	—	—	—	—	3402	3300
220. grain, soft red winter	3.86	14.0	12.0	—	—	—	—	—	—	3402	3300
221. grain, soft white winter	3.92	9.7	11.8	—	—	—	—	—	—	—	—
222. flour-by-product, less than 9.5% fiber (wheat middlings)	3.42	35.0	18.5	—	—	—	—	—	—	3080	2965
223. flour by-product, less than 7% fiber (wheat shorts)	—	—	—	—	—	—	—	—	—	3060	2865
WHEAT, DURUM *Triticum durum*											
224. grain	—	—	—	—	—	—	—	—	—	—	—
WHEATGRASS, CRESTED *Agropyron desertorum*											
225. fresh, early vegetative	2.54	—	21.0	—	—	—	—	—	—	—	—
226. hay, sun-cured	—	—	—	—	—	—	—	—	—	—	—
WILDRYE, RUSSIAN *Elymus junceus*											
227. fresh	—	—	—	2.91[c]	2.39[c]	1.47[c]	0.85[c]	1.50[c]	66[d]	—	—
WHEY *Bos taurus*											
228. dehydrated (cattle)	4.06	0.0	14.0	3.57[c]	2.93[c]	1.91[c]	1.27[c]	1.87[c]	81	3215	3090
229. fresh (cattle)	—	—	—	—	—	—	—	—	—	—	—
230. low lactose, dehydrated (dried whey product) (cattle)	3.61	0.0	17.9	—	—	—	—	—	—	2934	2744
YEAST, BREWERS *Saccharomyces cerevisiae*											
231. dehydrated	3.30	—	46.6	—	—	—	—	—	—	3295	2865
YEAST, IRRADIATED *Saccharomyces cerevisiae*											
232. dehydrated	—	—	—	—	—	—	—	—	—	—	—
YEAST, PRIMARY *Saccharomyces cerevisiae*											
233. dehydrated	—	—	—	—	—	—	—	—	—	—	—
YEAST, TORULA *Torulopsis utilis*											
234. dehydrated	—	—	—	—	—	—	—	—	—	—	—
YUCCA *Yucca* spp.											
235. flowers, fresh	—	—	—	3.09[c]	2.53[c]	1.58[c]	0.97[c]	1.60[c]	70[d]	—	—
236. leaves, fresh, immature	—	—	—	1.90[c]	1.56[c]	0.94[c]	0.00[c]	0.93[c]	43[d]	—	—

[a]100 percent dry matter basis.
[b]As fed basis.
[c]Calculated from data for cows.
[d]Data from sheep.

Reprinted from: "Nutrient Requirements of Beef Cattle," Sixth Revised Edition, 1984; "Nutrient Requirements of Dairy Cattle," Sixth Revised Edition, updated 1989, 1988; "Nutrient Requirements of Horses," Fifth Revised Edition, 1989; "Nutrient Requirements of Goats," 1981; "Nutrient Requirements of Poultry," Ninth Revised Edition, 1994; "Nutrient Requirements of Rabbits," Second Revised Edition, 1977; "Nutrient Requirements of Sheep," Sixth Revised Edition, 1985; "Nutrient Requirements of Swine," Ninth Revised Edition, 1988, © 1988 by the National Academy of Sciences; with permission of the National Academy Press, Washington, DC.

(Table 48 continues)

TABLE 48 Composition of Some Common Feeds, Excluding Amino Acids *(Cont.)*.

POULTRY[b]	RABBITS[b]			MINERALS[a]			VITAMINS[a]			
ME_n (kcal/kg)	DE (kcal/kg)	TDN (%)	Dig. Protein (%)	Calcium (%)	Phosphorous (%)	Potassium (%)	Carotene Vitamin A Activity (1000 IU/kg)	(mg/kg)	Vitamin E (mg/kg)	Vitamin D_2 (IU/g)
—	—	—	—	1.91	0.22	1.99	—	—	—	—
—	—	—	—	1.70	0.27	1.92	75.0	188	—	1544
—	—	—	—	—	—	—	—	—	—	—
3163	—	—	—	0.06	0.33	0.40	—	—	—	—
—	—	—	—	—	—	—	—	—	—	—
—	—	—	—	1.18	0.32	2.32	184.0	461	—	—
1300	2610	57	12.8	0.13	1.38	1.56	1.0	3	21	—
—	—	—	—	0.04	0.42	0.42	—	—	17	—
—	3680	79	9.2	0.04	0.43	0.41	—	—	14	—
2900	3680	84	9.0	0.05	0.43	0.49	—	—	12	—
—	—	—	—	0.05	0.43	0.46	—	—	18	—
3120	3680	79	8.6	0.07	0.36	0.46	—	—	20	—
2000	—	—	—	0.13	0.99	1.13	—	—	—	—
2162	—	—	—	0.10	0.91	1.06	—	—	61	—
—	—	—	—	0.10	0.41	0.51	—	—	—	—
—	—	—	—	0.46	0.34	—	180.4	451	—	—
—	—	—	—	0.33	0.21	2.00	8.9	22	—	—
—	—	—	—	—	—	—	—	—	—	—
1900	—	—	—	0.92	0.82	1.23	—	—	—	—
—	—	—	—	0.73	0.65	2.75	—	—	—	—
2090	—	—	—	1.71	1.12	3.16	—	—	—	—
1990	—	—	—	0.13	1.49	1.79	—	—	2	—
—	—	—	—	0.83	1.51	2.28	—	—	—	—
—	—	—	—	0.39	1.86	—	—	—	—	—
2160	—	—	—	0.54	1.71	2.04	—	—	—	—
—	—	—	—	—	0.48	—	—	—	—	—
—	—	—	—	—	0.10	—	—	—	—	—

TABLE 49 Amino Acid Composition of Some Feeds Commonly Used for Swine, Poultry, and Rabbits (Data on as fed basis).

Line # Feed Name	IFN	Dry Matter (%)		Arginine (%)	Glycine (%)	Serine (%)	Histidine (%)	Isoleucine (%)
ALFALFA *Medicago sativa*								
1. hay, sun-cured, early bloom	1-00-059	89	R	0.80	—	—	0.24	0.72
2. meal, dehydrated, 17% protein	1-00-023	88	P	0.69	0.82	0.72	0.57	0.67
		92	S	0.77	—	—	0.33	0.81
		92	R	0.76	—	—	0.32	0.82
3. meal, dehydrated, 20% protein	1-00-024	92	P	0.92	0.97	0.89	0.34	0.88
			R	0.95	—	—	0.34	0.87
BAKERY								
4. waste, dehydrated (dried bakery product)	4-00-466	92	P	0.47	0.82	0.65	0.13	0.45
			S	0.47	—	—	0.13	0.45
			R	0.51	—	—	0.15	0.36
BARLEY *Hordeum vulgare*								
5. grain	4-00-549	89	P	0.52	0.44	0.46	0.27	0.37
			S	0.52	—	—	0.24	0.46
			R	0.55	—	—	0.26	0.51
6. grain, pacific coast	4-07-939	89	P	0.48	0.36	0.32	0.21	0.40
			S	0.44	—	—	0.21	0.40
			R	0.44	—	—	0.21	0.40
BLOOD								
7. meal spray dehydrated	5-00-381	93	P	3.62	3.95	4.25	5.33	0.98
			S	3.59	—	—	5.18	0.91
BREWERS								
8. grains, dehydrated	5-02-141	92	P	1.28	1.09	0.80	0.57	1.44
			S	1.27	—	—	0.53	1.57
CORN, DENT YELLOW *Zea mays indentata*								
9. distillers grains, dehydrated	5-28-235	94	P	0.97	0.49	0.70	0.62	0.99
10. distillers grains w/solubles, dehydrated	5-28-236	93	P	0.98	0.57	1.61	0.66	1.00
			S	0.96	—	—	0.64	1.38
11. distillers solubles, dehydrated	5-28-237	92	P	1.05	1.10	1.30	0.70	1.25
			S	0.96	—	—	0.66	1.30
12. gluten meal	5-02-900	91	R	1.42	—	—	0.96	2.21
13. corn and cob meal	4-28-238	87	S	0.37	—	—	0.17	0.35
14. gluten, meal 60% protein	5-28-242	88	P	1.82	1.67	2.96	1.20	2.45
		90	S	2.08	—	—	1.40	2.54
15. gluten, meal 41%	5-12-354	91	S	1.37	—	—	0.97	2.25
16. gluten w/bran (corn gluten feed)	5-28-243	90	P	1.01	0.99	0.80	0.71	0.65
			S	0.79	—	—	0.62	0.89
17. grain, dent yellow	4-02-935	88	P	0.38	0.33	0.37	0.23	0.29
		88	S	0.43	—	—	0.27	0.35
		89	R	0.52	—	—	0.20	0.40
18. grits by-product (hominy feed)	4-03-011	90	P	0.47	0.40	0.50	0.20	0.40
			S	0.50	—	—	0.21	0.40
COTTON *Gossypium* spp								
19. seeds, meal solvent extracted	5-01-621	90	R	4.36	—	—	1.02	1.20
20. seeds, meal solvent extracted	5-01-619	92	S	4.27	—	—	1.01	1.22
21. seeds, meal prepressed solvent extracted, 41% protein	5-07-872	90.4	P	4.66	1.69	1.78	1.10	1.33

(Table 49 continues)

TABLE 49 Amino Acid Composition of Some Feeds Commonly Used for Swine, Poultry, and Rabbits (Data on as fed basis). *(Cont.)*

Leucine (%)	Lysine (%)	Methionine (%)	Cystine (%)	Phenylalanine (%)	Tyrosine (%)	Threonine (%)	Tryptophan (%)	Valine (%)
1.13	0.89	0.16	0.32	0.71	0.48	0.63	0.24	0.72
1.19	0.73	0.24	0.19	0.81	0.84	0.69	0.23	0.84
1.28	0.85	0.27	0.29	0.80	0.54	0.71	0.34	0.88
1.27	0.90	0.21	0.30	0.79	0.56	0.68	0.35	0.83
1.30	0.87	0.31	0.25	0.85	0.59	0.76	0.33	0.97
1.37	0.90	0.32	0.31	0.86	0.61	0.74	0.43	0.97
0.73	0.31	0.17	0.17	0.40	0.41	0.49	0.10	0.42
0.73	0.31	0.17	0.17	0.40	0.41	0.49	0.10	0.42
0.70	0.32	0.17	0.18	0.40	0.30	0.44	0.10	0.40
0.76	0.40	0.18	0.24	0.56	0.35	0.37	0.14	0.52
0.75	0.40	0.16	0.21	0.58	0.35	0.36	0.15	0.57
0.82	0.46	0.18	0.21	0.61	0.36	0.38	0.16	0.62
0.60	0.29	0.13	0.18	0.48	0.31	0.30	0.12	0.46
0.60	0.26	0.14	0.19	0.47	0.31	0.31	0.12	0.46
0.60	0.27	0.14	0.20	0.48	—	0.29	0.12	0.46
11.32	7.88	1.09	1.03	5.85	2.63	3.92	1.35	7.53
10.97	7.44	1.05	1.03	5.89	2.26	3.63	1.05	7.52
2.48	0.90	0.57	0.39	1.45	1.19	0.98	0.34	1.66
2.53	0.88	0.46	0.35	1.46	1.16	0.93	0.37	1.58
3.01	0.78	0.40	0.24	0.94	0.84	0.49	0.20	1.18
2.20	0.75	0.60	0.40	1.20	0.74	0.92	0.19	1.30
2.21	0.70	0.49	0.29	1.47	0.69	0.92	0.17	1.48
2.11	0.90	0.50	0.40	1.30	0.95	1.00	0.30	1.39
2.31	0.91	0.55	0.44	1.46	0.86	1.00	0.24	1.52
7.33	0.83	1.07	0.66	2.78	1.00	1.43	0.21	2.24
0.87	0.17	0.14	0.12	0.39	0.33	0.29	0.07	0.31
10.04	1.03	1.49	1.10	3.56	3.07	2.00	0.36	2.78
10.44	1.03	1.78	1.01	4.04	3.31	2.25	0.30	3.11
6.00	0.78	1.07	0.66	2.84	1.01	1.42	0.21	2.22
1.89	0.63	0.45	0.51	0.77	0.58	0.89	0.10	0.05
2.21	0.64	0.37	0.43	0.82	0.74	0.79	0.15	1.11
1.00	0.26	0.18	0.18	0.38	0.30	0.29	0.06	0.40
1.19	0.25	0.18	0.22	0.46	0.38	0.36	0.09	0.48
1.10	0.20	0.13	0.13	0.50	0.44	0.40	0.09	0.38
0.84	0.40	0.13	0.13	0.35	0.49	0.40	0.10	0.49
0.87	0.36	0.15	0.13	0.33	0.47	0.40	0.13	0.51
2.17	1.59	0.49	0.67	2.00	1.06	1.21	0.48	1.64
2.21	1.70	0.49	0.57	2.05	1.04	1.23	0.48	1.67
2.41	1.76	0.51	0.62	2.23	1.14	1.34	0.52	1.82

(Table 49 continues)

TABLE 49 Amino Acid Composition of Some Feeds Commonly Used for Swine, Poultry, and Rabbits (Data on as fed basis). *(Cont.)*

Line # Feed Name	IFN	Dry Matter (%)		Arginine (%)	Glycine (%)	Serine (%)	Histidine (%)	Isoleucine (%)
FISH, MENHADEN *Brevoortia tyrannus*								
22. meal, mechanical extracted	5-02-009	92.1	P	3.68	4.46	2.37	1.42	2.28
		92	S	3.74	—	—	1.44	2.85
FISH								
23. solubles, condensed	5-01-969	51	P	1.61	3.41	0.83	1.56	1.06
			S	1.63	—	—	1.54	1.09
MEAT								
24. meal rendered	5-00-385	92	P	3.73	6.30	1.60	1.30	1.60
25. with bone, meal rendered	5-00-388	93.4	P	3.28	6.65	2.20	0.96	1.54
26. meat and bone meal, 50%	5-09-322	94	S	3.65	—	—	0.96	1.47
27. meat meal, 55%	5-09-323	93	S	3.79	—	—	1.04	1.84
MILK *Bos taurus*								
28. skimmed, dehydrated	5-01-175	94	S	1.16	—	—	0.86	2.18
			R	1.15	—	—	0.84	2.16
MILLET, PEARL *Pennisetum glaucum*								
29. grain	4-03-118	90	P	0.74	0.47	0.74	0.31	0.37
MILLET, PROSO *Panicum miliaceum*								
30. grain	4-03-120	87.5	P	0.35	0.31	0.40	0.22	0.35
		90	S	0.36	—	—	0.21	0.45
MOLASSES AND SYRUP *Beta vulgaris altissima*								
31. beet, sugar, molasses, more than 48% invert sugar, more than 79.5 degrees brix	4-00-668	78	S	—	—	—	—	—
MOLASSES AND SYRUP *Saccharum officinarum*								
32. sugarcane, molasses, more than 46% invert sugars, more than 79.5 degrees brix (black strap)	4-04-696	74	S	—	—	—	—	—
			R	—	—	—	—	—
OATS *Avena sativa*								
33. grain	4-03-309	89	P	0.79	0.50	0.40	0.24	0.52
			S	0.71	—	—	0.17	0.48
			R	0.57	—	—	0.09	0.53
34. grain, Pacific coast	4-07-999	91	P	0.60	0.40	0.30	0.10	0.40
PEANUT *Arachis hypogaea*								
35. kernels, meal mechanical extracted (peanut meal)	5-03-649	90	P	4.35	2.18	1.83	0.87	1.27
			S	5.08	—	—	1.03	1.78
36. kernels, meal solvent extracted (peanut meal)	5-03-650	91.9	P	5.33	2.67	2.25	1.07	1.55
		93	S	5.82	—	—	1.46	1.84
		92	R	5.90	—	—	1.20	2.00
POULTRY								
37. by-product, meal rendered	5-03-798	94.2	P	3.94	6.17	2.71	1.07	2.16
38. feathers, hydrolyzed	5-03-795	91	P	5.57	6.13	8.52	0.95	3.91
		93	S	5.33	—	—	0.47	3.51
RYE *Secale cereale*								
39. grain	4-04-047	88	P	0.53	0.49	0.52	0.26	0.47
			S	0.51	—	—	0.24	0.46
			R	0.53	—	—	0.26	0.50

(Table 49 continues)

TABLE 49 Amino Acid Composition of Some Feeds Commonly Used for Swine, Poultry, and Rabbits (Data on as fed basis). *(Cont.)*

Leucine (%)	Lysine (%)	Methionine (%)	Cystine (%)	Phenylalanine (%)	Tyrosine (%)	Threonine (%)	Tryptophan (%)	Valine (%)
4.16	4.51	1.63	0.57	2.21	1.80	2.46	0.49	2.77
4.48	4.74	1.75	0.58	2.46	1.93	2.51	0.65	3.19
1.86	1.73	0.50	0.30	0.93	0.40	0.86	0.31	1.16
1.94	1.85	0.70	0.39	1.07	0.50	0.90	0.33	1.26
3.32	3.00	0.75	0.66	1.70	0.84	1.74	0.36	2.30
3.28	2.61	0.69	0.69	1.81	1.20	1.74	0.27	2.36
3.02	2.89	0.68	0.46	1.65	0.79	1.60	0.28	2.14
3.51	3.09	0.73	0.68	1.91	0.96	1.78	0.38	2.61
3.30	2.54	0.90	0.45	1.57	1.14	1.57	0.43	2.29
3.23	2.48	0.90	0.45	1.58	1.13	1.58	0.43	2.30
1.14	0.45	0.25	0.24	0.56	0.35	0.48	0.08	0.49
1.14	0.21	0.16	0.17	0.47	0.34	0.29	0.08	0.44
1.15	0.26	0.29	—	0.57	—	0.40	0.17	0.58
—	—	—	—	—	—	—	—	—
—	—	—	—	—	—	—	—	—
—	—	—	—	—	—	—	—	—
0.89	0.50	0.18	0.22	0.59	0.53	0.43	0.16	0.68
0.87	0.40	0.18	0.19	0.57	0.45	0.38	0.15	0.62
0.09	0.34	0.18	0.15	0.60	1.07	0.40	0.12	0.70
0.30	0.40	0.13	0.17	0.44	0.20	0.20	0.12	0.51
2.42	1.26	0.45	0.52	1.97	1.47	1.01	0.39	1.53
3.13	1.69	0.50	0.75	2.38	1.59	1.27	0.46	2.29
2.97	1.54	0.54	0.64	2.41	1.80	1.24	0.48	1.87
3.27	1.45	0.44	0.73	2.12	—	1.37	0.48	2.16
3.70	2.30	0.40	—	2.70	1.80	1.50	0.50	2.80
3.99	3.10	0.99	0.98	2.29	1.68	2.17	0.37	2.87
6.94	2.28	0.57	4.34	3.94	2.48	3.81	0.55	5.93
6.42	1.67	0.54	3.21	3.59	2.35	3.63	0.52	5.85
0.70	0.42	0.17	0.19	0.56	0.26	0.36	0.11	0.56
0.67	0.41	0.17	0.19	0.58	0.22	0.35	0.11	0.56
0.69	0.42	0.17	0.19	0.57	0.26	0.36	0.13	0.60

(Table 49 continues)

TABLE 49 Amino Acid Composition of Some Feeds Commonly Used for Swine, Poultry, and Rabbits (Data on as fed basis). *(Cont.)*

Line # Feed Name	IFN	Dry Matter (%)		Arginine (%)	Glycine (%)	Serine (%)	Histidine (%)	Isoleucine (%)
SORGHUM *Sorghum bicolor*								
40. grain, 8–10% protein	4-20-893	87.5	P	0.35	0.31	0.40	0.22	0.35
41. grain, more than 10% protein	4-20-894	88	P	0.35	0.32	0.45	0.23	0.43
SORGHUM, MILO *Sorghum bicolor subolabrescens*								
42. grain	4-04-444	89	S	0.37	—	—	0.24	0.44
SOYBEAN *Glycine max*								
43. seeds, meal solvent extracted	5-04-604	88.2	P	3.14	1.90	2.29	1.17	1.96
		90	S	3.20	—	—	1.12	2.00
		89	R	3.25	—	—	1.14	2.44
44. seeds without hulls, meal solvent extracted	5-04-612	88.4	P	3.48	2.05	2.48	1.28	2.12
		90	S	3.67	—	—	1.20	2.13
		90	R	3.76	—	—	1.26	2.57
TRITICALE *Triticale hexaploide*								
45. grain	4-20-362	88	P	0.57	0.48	0.52	0.26	0.39
		90	S	0.86	—	—	0.40	0.61
WHEAT *Triticum aestivum*								
46. bran	4-05-190	88	P	1.02	0.81	0.67	0.46	0.47
		87	S	0.85	—	—	0.33	0.55
		89	R	0.96	—	—	0.35	0.58
47. grain, hard winter	4-05-268	88.1	P	0.60	0.59	0.59	0.31	0.44
		88	S	0.65	—	—	0.30	0.53
		89	R	0.57	—	—	0.22	0.57
48. grain, soft red winter	4-05-294	88	S	0.65	—	—	0.32	0.45
49. grain, soft white winter	4-05-337	89	P	0.40	0.49	0.55	0.20	0.42
			R	0.40	—	—	0.20	0.42
50. flour by-product, less than 9.5% fiber (wheat middlings)	4-05-205	88	P	1.15	0.63	0.75	0.37	0.58
			S	0.98	—	—	0.40	0.68
51. flour by-product, less than 7% fiber (wheat shorts)	4-05-201	88	P	1.18	0.96	0.77	0.45	0.58
			S	1.20	—	—	0.44	0.57
WHEY *Bos taurus*								
52. dehydrated (cattle)	4-01-182	93	P	0.34	0.30	0.32	0.18	0.82
			S	0.33	—	—	0.17	0.78
53. low lactose, dehydrated (dried whey product) (cattle)	4-01-186	91	P	0.67	1.04	0.76	0.25	0.90
			S	0.60	—	—	0.27	0.96
YEAST, BREWERS *Saccharomyces cerevisiae*								
54. dehydrated	7-05-527	93	P	2.19	2.09	—	1.07	2.14
			S	2.26	—	—	1.13	2.03
YEAST, TORULA *Torulopsis utilis*								
55. dehydrated	7-05-534	93	P	2.60	2.60	2.76	1.40	2.90

Species are indicated on data lines as: P = poultry; S = swine; R = rabbits

Reprinted from: "Nutrient Requirements of Poultry," Ninth Revised Edition, 1994
"Nutrient Requirements of Swine," Ninth Revised Edition, 1988, © 1988 by the National Academy of Sciences
"Nutrient Requirements of Rabbits," Second Revised Edition, 1977 with permission of the National Academy Press, Washington, DC.

(Table 49 continues)

TABLE 49 Amino Acid Composition of Some Feeds Commonly Used for Swine, Poultry, and Rabbits (Data on as fed basis). *(Cont.)*

Leucine (%)	Lysine (%)	Methionine (%)	Cystine (%)	Phenylalanine (%)	Tyrosine (%)	Threonine (%)	Tryptophan (%)	Valine (%)
1.14	0.21	0.16	0.17	0.47	0.34	0.29	0.08	0.44
1.37	0.22	0.15	0.11	0.52	0.17	0.33	0.09	0.54
1.32	0.23	0.16	0.13	0.49	0.37	0.27	0.10	0.53
3.39	2.69	0.62	0.66	2.16	1.91	1.72	0.74	2.07
3.37	2.90	0.52	0.66	2.10	1.50	1.70	0.64	2.02
3.49	2.92	0.60	0.67	2.26	1.24	1.78	0.65	2.36
3.74	2.96	0.67	0.72	2.34	1.95	1.87	0.74	2.22
3.63	3.12	0.71	0.70	2.36	1.71	1.90	0.69	2.47
3.82	3.22	0.72	0.77	2.57	2.01	1.92	0.69	2.72
0.76	0.39	0.26	0.26	0.49	0.32	0.36	0.14	0.51
1.18	0.52	0.21	0.29	0.80	0.51	0.57	0.18	0.84
0.96	0.61	0.23	0.32	0.61	0.46	0.50	0.23	0.70
0.89	0.56	0.17	0.26	0.52	0.38	0.41	0.25	0.67
0.90	0.58	0.17	0.25	0.50	0.40	0.43	0.33	0.73
0.89	0.37	0.21	0.30	0.60	0.43	0.39	0.16	0.57
0.87	0.40	0.22	0.30	0.71	0.46	0.37	0.17	0.58
0.87	0.38	0.23	0.25	0.65	0.50	0.39	0.18	0.56
0.90	0.36	0.22	0.36	0.64	0.37	0.39	0.27	0.58
0.59	0.31	0.15	0.22	0.45	0.39	0.32	0.12	0.44
0.59	0.32	0.17	0.27	0.45	0.39	0.32	0.12	0.44
1.07	0.69	0.21	0.32	0.64	0.45	0.49	0.20	0.71
1.11	0.68	0.19	0.22	0.66	0.43	0.57	0.19	0.80
1.09	0.79	0.27	0.36	0.67	0.47	0.60	0.21	0.83
1.07	0.80	0.28	0.38	0.67	0.47	0.60	0.23	0.82
1.19	0.97	0.19	0.30	0.33	0.25	0.89	0.19	0.68
1.18	0.94	0.19	0.30	0.35	0.25	0.89	0.18	0.67
1.35	1.47	0.57	0.57	0.50	0.35	0.85	0.23	0.83
1.54	1.40	0.41	0.43	0.55	0.46	0.95	0.27	0.87
3.19	3.23	0.70	0.50	1.81	1.49	2.06	0.49	2.32
3.19	3.23	0.66	0.52	1.60	1.47	2.06	0.51	2.25
3.50	3.80	0.80	0.60	3.00	2.10	2.60	0.50	2.90

TABLE 50 Conversion Factors

If the measure is given in this unit ×	Multiply by this conversion factor =	To obtain this unit
MASS		
μg	0.000001	g
μg	0.001	mg
μg/kg	0.453592	μg/lb
μg/lb	2.2046	μg/kg
g	0.0022046	lb
g	0.03527	oz
g	0.001	kg
g	1 000	mg
g	1 000 000	μg
g/kg	0.453592	g/lb
g/kg	0.1	%
g/lb	2.2046	g/kg
g/Mcal	0.0022046	lb/Mcal
ICU/kg	0.453592	ICU/lb
ICU/lb	2.2046	ICU/kg
IU/kg	0.453592	IU/lb
IU/lb	2.2046	IU/kg
kcal	0.001	Mcal
kcal/kg	0.453592	kcal/lb
kcal/lb	2.2046	kcal/kg
kg	2.204624	lb
kg	35.2734	oz
kg	0.0011	ton (short, 2,000lb)
kg	0.001	tonne
kg	1 000 000	mg
kg	1 000	g
kg/kg	0.453592	kg/lb
kg/lb	2.2046	kg/kg
lb	0.453592	kg
lb	453.592	g
lb	453592	mg
lb/Mcal	453.592	g/Mcal
Mcal	1 000	kcal
Mcal/kg	0.453592	Mcal/lb
Mcal/lb	2.2046	Mcal/kg
mg	0.0000022046	lb
mg	0.000001	kg
mg	0.001	g
mg	1 000	g
mg/g	453.592	mg/lb
mg/g	0.1	%
mg/kg	0.453592	mg/lb
mg/kg	0.0001	%
mg/lb	2.2046	mg/kg
mg/lb	0.0022046	mg/g
oz	0.02835	kg
oz	28.35	g
ppm	1.0	μg/g
ppm	1.0	mg/g
ppm	0.453592	mg/lb
ppm	0.0001	%
ton (short, 2,000lb)	907.185	kg
tonne	1 000	kg

If the measure is given in this unit ×	Multiply by this conversion factor =	To obtain this unit
LENGTH		
cm	0.01	m
cm	0.3937	in
ft	12	in
ft	0.3048	m
furlong	220	yd
furlong	201.1675	m
in	2.54	cm
in	0.0254	m
km	1 000	m
km	0.6214	mi
m	100	cm
m	0.001	km
m	39.37	in
m	3.2808	ft
m	1.0936	yd
m	0.1988	rd
m	0.00497	furlong
m	0.0006214	mi
mi	5,280	ft
mi	1,760	yd
mi	320	rd
mi	8	furlong
mi	1,609.34	m
mi	1.60934	km
rd	16.5	ft
rd	5.0292	m
yd	3	ft
yd	0.9144	m
AREA		
acre	4,046.86	m²
acre	0.404686	hectare
acre	43,560	ft²
acre	4,840	yd²
acre	160	rd²
cm²	0.1550	in²
cm²	0.001076	ft²
cm²	0.000 1	m²
ft²	929.03	cm²
ft²	0.092903	m²
ft²	144	in²
hectare	2.471	acre
hectare	10,000	m²
hectare	0.01	km²
in²	6.4516	cm²
in²	0.000645	m²
km²	0.3861	mi²
km²	1 000 000	m²
km²	100	hectare
m²	1,550.3875	in²
m²	10.764	ft²
m²	1.19599	yd²
m²	0.03954	rd²
m²	0.0002471	acre

(Table 50 continues)

TABLE 50 Conversion Factors *(Cont.)*

If the measure is given in this unit	×	Multiply by this conversion factor =	To obtain this unit		If the measure is given in this unit	×	Multiply by this conversion factor =	To obtain this unit
m^2		10 000	cm^2		liter		1.05669	quart
m^2		0.000 001	km^2		liter		0.26417	gallon
m^2		0.000 1	hectare		m		28.3776	bushel
mi^2		2.58998	km^2		m^3		113.636	peck
mi^2		640	acre		m^3		909	quart
rd^2		25.2928	m^2		m^3		264.2	gallon
rd^2		272.25	ft^2		m^3		1 000	dm^3
rd^2		30.25	yd^2		m^3		35.31	ft^3
yd^2		0.836127	m^2		m^3		1.3079	yd^3
yd^2		1,296	in^2		oz		1.8047	in^3
yd^2		9	ft^2		oz		6	teaspoon
					oz		29.5735	cm^3
					oz		29.5735	ml

VOLUME OR CAPACITY

bushel		4	peck		peck		8	quart
bushel		0.035239	m^3		peck		0.0088	m^3
cm^3		1 000	mm^3		pint		16	oz
cm^3		0.061	in^3		pint		2	cup
cup		8	oz		pint		0.473176	liter
dm^3		1 000	cm^3		quart		2	pint
ft^3		1,728	in^3		quart		0.946353	liter
ft^3		0.028317	m^3		quart		0.0011	m^3
gallon		4	quart		tablespoon		3	teaspoon
gallon		3.78541	liter		tablespoon		0.5	oz
gallon		0.003785	m^3		tablespoon		15	cm^3
in^3		16.387	cm^3		teaspoon		5	cm^3
liter		1 000	ml		yd^3		27	ft^3
liter		1 000	cm^3		yd^3		0.764555	m^3

TEMPERATURE

If the measure is given in this unit	Use this formula for conversion	To obtain this unit
Degrees Fahrenheit (F)	$\dfrac{(F-32)5}{9}$	Degrees Celsius (C)
Degrees Celsius (C)	$\left(\dfrac{C \times 9}{5}\right) + 32$	Degrees Fahrenheit (F)

TABLE 51 Abbreviations Used in Table 50.

cm	centimeter	in^2	square inch	m	meter	ml	milliliter
cm^2	square centimeter	in^3	cubic inch	m^2	square meter	oz	ounce
cm^3	cubic centimeter	IU	International Unit	m^3	cubic meter	ppm	parts per million
dm^3	cubic decimeter	kcal	kilocalorie	Mcal	megacalorie	rd	rod
ft^2	square foot	kg	kilogram	mg	milligram	rd^2	square rod
ft^3	cubic foot	km	kilometer	mi	mile	yd	yard
g	gram	km^2	square kilometer	mi^2	square mile	yd^2	square yard
ICU	International Chick Unit	lb	pound	μg	microgram	yd^3	cubic yard

TABLE 52 Dry Matter Intake Requirements to Fulfill Nutrient Allowances for Maintenance, Milk Production, and Normal Live Weight Gain During Mid- and Late Lactation.

Live Wt. (lb)	800	900	1,000	1,100	1,200	1,300	1,400	1,500	1,600	1,700	1,800
FCM (4%)[a] (lb)						% of Live Weight[b,c]					
20	2.8	2.6	2.5	2.3	2.2	2.1	2.1	2.0	1.9	1.9	1.8
30	3.2	3.0	2.9	2.7	2.6	2.5	2.4	2.3	2.2	2.1	2.1
40	3.6	3.4	3.2	3.1	2.9	2.8	2.7	2.5	2.4	2.4	2.3
50	4.0	3.8	3.6	3.4	3.2	3.1	3.0	2.8	2.7	2.6	2.5
60	4.4	4.1	3.9	3.7	3.5	3.4	3.2	3.1	3.0	2.9	2.7
70	4.8	4.6	4.3	4.0	3.8	3.6	3.5	3.3	3.2	3.1	2.9
80	5.4	5.1	4.7	4.3	4.1	3.8	3.7	3.5	3.4	3.2	3.1
90	—	5.5	5.1	4.7	4.4	4.1	3.9	3.7	3.6	3.4	3.3
100	—	—	5.5	5.0	4.7	4.4	4.2	3.9	3.8	3.6	3.5
110	—	—	—	5.4	5.1	4.8	4.5	4.2	4.0	3.8	3.7
120	—	—	—	—	5.4	5.0	4.8	4.5	4.3	4.1	3.9
130	—	—	—	—	—	5.4	5.1	4.8	4.6	4.4	4.2

NOTE: The following assumptions were made in calculating the dry matter intake requirements shown in Appendix Table 52.

    1. The basic or reference cow used for the calculations weighed 1,320 lb and produced milk with 4 percent milk fat. Other live weights in the table and corresponding fat percentages were 881 lb and 5 percent fat; 1,100 lb and 4.5 percent fat; and 1,540 and 1,760 lb and 3.5 percent fat.

    2. The concentration of energy in the diet for the reference cow was 0.65 Mcal of NEL/lb of DM for milk yields equal to or less than 22 lb/day. It increased linearly to 0.78 Mcal of NEL/lb for milk yields equal to or greater than 88 lb/day.

    3. The energy concentrations of the diets for all other cows were assumed to change linearly as their energy requirements for milk production, relative to maintenance, changed in a manner identical to that of the 1,320-lb cow as she increased in milk yield from 22 to 88 lb/day.

    4. Enough DM to provide sufficient energy for cows to gain 0.055 percent of their body weight daily was also included in the total. If cows do not consume as much DM as they require, as calculated from Appendix Table 52, their energy intake will be less than their requirements. The result will be a loss of body weight, reduced milk yields, or both. If cows consume more DM than what is projected as required from Appendix Table 52, the energy concentration of their diet should be reduced or they may become overly fat.

[a]4% Fat-corrected milk (lb) = (0.4) (lb of milk) + (15) (lb of milk fat).

[b]Probable DMI may be up to 18 percent less in early lactation.

[c]DMI as a percentage of live weight may be 0.02 percent less per 1 percent increase in diet moisture content above 50 percent if fermented feeds constitute a major portion of the diet.

Source: Nutrient Requirements of Dairy Cattle, sixth revised edition, updated 1989, copyright © 1988 by the National Academy of Sciences.

TABLE 53 Daily Nutrient Requirements of Lactating Cows Using Absorbable Protein.

Live Wt. (kg)	Fat (%)	Milk (kg)	Change (kg)	Live Weight Intake (kg)	Dry Matter NELDM (Mcal/kg)	NEL (Mcal)	TDN (kg)	UIP (g)	DIP (g)	Ca (g)	P (g)
					Energy			Protein		Minerals	
				Intake at 100% of the Requirement for Maintenance, Lactation, and Weight Gain							
400	4.5	8.0	0.220	10.14	1.43	14.55	6.44	511	753	44	28
400	4.5	14.0	0.220	12.66	1.52	19.26	8.48	710	1,052	65	41
400	4.5	20.0	0.220	14.91	1.61	23.96	10.51	880	1,355	85	54
400	4.5	26.0	0.220	16.94	1.69	28.67	12.54	1,026	1,662	106	67
400	4.5	32.0	0.220	19.41	1.72	33.37	14.58	1,220	1,962	127	80
400	5.0	8.0	0.220	10.36	1.44	14.94	6.60	525	778	46	30
400	5.0	14.0	0.220	13.00	1.53	19.93	8.77	730	1,096	68	43
400	5.0	20.0	0.220	15.35	1.62	24.93	10.93	902	1,419	90	57
400	5.0	26.0	0.220	17.44	1.72	29.92	13.07	1,048	1,745	112	71
400	5.0	32.0	0.220	20.30	1.72	34.91	15.25	1,277	2,061	134	84
400	5.5	8.0	0.220	10.57	1.45	15.32	6.77	538	803	48	31
400	5.5	14.0	0.220	13.33	1.55	20.61	9.07	748	1,140	71	45
400	5.5	20.0	0.220	15.77	1.64	25.89	11.34	923	1,483	95	60
400	5.5	26.0	0.220	18.13	1.72	31.17	13.62	1,091	1,826	118	75
400	5.5	32.0	0.220	21.20	1.72	36.45	15.92	1,334	2,160	142	89
500	4.0	9.0	0.275	11.59	1.42	16.49	7.30	540	883	49	32
500	4.0	17.0	0.275	14.78	1.51	22.38	9.86	797	1,257	75	48
500	4.0	25.0	0.275	17.62	1.61	28.27	12.40	1,015	1,635	101	64
500	4.0	33.0	0.275	20.14	1.70	34.15	14.93	1,201	2,018	126	80
500	4.0	41.0	0.275	23.29	1.72	40.04	17.49	1,453	2,392	152	95
500	4.5	9.0	0.275	11.84	1.43	16.92	7.49	556	911	51	33
500	4.5	17.0	0.275	15.20	1.53	23.20	10.21	821	1,310	79	50
500	4.5	25.0	0.275	18.16	1.62	29.47	12.92	1,043	1,715	107	68
500	4.5	33.0	0.275	20.79	1.72	35.74	15.61	1,230	2,124	134	85
500	4.5	41.0	0.275	24.44	1.72	42.02	18.35	1,526	2,519	162	102
500	5.0	9.0	0.275	12.08	1.44	17.36	7.68	571	939	53	35
500	5.0	17.0	0.275	15.60	1.54	24.01	10.57	844	1,364	83	53
500	5.0	25.0	0.275	18.68	1.64	30.67	13.44	1,069	1,795	113	71
500	5.0	33.0	0.275	21.71	1.72	37.33	16.31	1,289	2,226	142	89
500	5.0	41.0	0.275	25.58	1.72	43.99	19.21	1,599	2,646	172	108
600	3.0	10.0	0.330	12.52	1.42	17.79	7.87	533	974	52	34
600	3.0	20.0	0.330	16.20	1.49	24.18	10.67	845	1,375	79	51
600	3.0	30.0	0.330	19.37	1.58	30.58	13.43	1,102	1,784	106	68
600	3.0	40.0	0.330	22.21	1.67	36.98	16.19	1,323	2,198	133	84
600	3.0	50.0	0.330	25.23	1.72	43.38	18.95	1,565	2,608	161	101
600	3.5	10.0	0.330	12.86	1.42	18.27	8.08	557	1,004	54	35
600	3.5	20.0	0.330	16.70	1.51	25.15	11.08	874	1,438	84	54
600	3.5	30.0	0.330	20.04	1.60	32.03	14.06	1,137	1,879	113	72
600	3.5	40.0	0.330	23.00	1.69	38.90	17.01	1,360	2,326	143	90
600	3.5	50.0	0.330	26.63	1.72	45.78	20.00	1,654	2,763	173	109
600	4.0	10.0	0.330	13.20	1.42	18.75	8.30	581	1,034	56	37
600	4.0	20.0	0.330	17.19	1.52	26.11	11.50	902	1,501	89	57
600	4.0	30.0	0.330	20.69	1.62	33.47	14.68	1,170	1,975	121	77
600	4.0	40.0	0.330	23.78	1.72	40.83	17.84	1,395	2,454	153	96
600	4.0	50.0	0.330	28.03	1.72	48.19	21.05	1,744	2,918	185	116
700	3.0	12.0	0.385	14.46	1.42	20.54	9.09	607	1,154	61	40
700	3.0	24.0	0.385	18.75	1.50	28.21	12.44	968	1,638	94	60
700	3.0	36.0	0.385	22.48	1.60	35.89	15.76	1,269	2,129	127	81
700	3.0	48.0	0.385	25.80	1.69	43.57	19.05	1,525	2,627	159	101
700	3.0	60.0	0.385	29.81	1.72	51.25	22.39	1,857	3,114	192	121
700	3.5	12.0	0.385	14.86	1.42	21.11	9.34	636	1,190	64	42
700	3.5	24.0	0.385	19.34	1.52	29.37	12.94	1,002	1,713	100	64
700	3.5	36.0	0.385	23.26	1.62	37.62	16.50	1,309	2,244	135	86
700	3.5	48.0	0.385	26.72	1.72	45.88	20.04	1,567	2,781	171	108
700	3.5	60.0	0.385	31.48	1.72	54.13	23.65	1,964	3,300	207	130
700	4.0	12.0	0.385	15.20	1.43	21.69	9.60	658	1,227	67	44
700	4.0	24.0	0.385	19.92	1.53	30.52	13.44	1,035	1,789	105	68
700	4.0	36.0	0.385	24.02	1.64	39.35	17.25	1,347	2,359	144	91
700	4.0	48.0	0.385	28.03	1.72	48.19	21.05	1,648	2,930	182	115
700	4.0	60.0	0.385	33.16	1.72	57.02	24.91	2,071	3,485	221	139
800	3.0	14.0	0.440	16.36	1.42	23.24	10.29	682	1,331	71	46

(Table 53 continues)

TABLE 53 Daily Nutrient Requirements of Lactating Cows Using Absorbable Protein. *(Cont.)*

Live Wt. (kg)	Fat (%)	Milk (kg)	Change (kg)	Live Weight Intake (kg)	Dry Matter NELDM (Mcal/kg)	Energy NEL (Mcal)	TDN (kg)	Protein UIP (g)	DIP (g)	Minerals Ca (g)	P (g)
800	3.0	27.0	0.440	20.93	1.51	31.56	13.91	1,064	1,857	106	68
800	3.0	40.0	0.440	24.95	1.60	39.88	17.50	1,388	2,390	142	90
800	3.0	53.0	0.440	28.54	1.69	48.20	21.08	1,665	2,928	177	112
800	3.0	66.0	0.440	32.87	1.72	56.51	24.69	2,022	3,457	213	134
800	3.5	14.0	0.440	16.78	1.42	23.92	10.58	710	1,374	74	49
800	3.5	27.0	0.440	21.59	1.52	32.86	14.47	1,102	1,942	113	72
800	3.5	40.0	0.440	25.82	1.62	41.80	18.33	1,432	2,517	151	96
800	3.5	53.0	0.440	29.57	1.72	50.75	22.17	1,711	3,099	190	120
800	3.5	66.0	0.440	34.72	1.72	59.69	26.07	2,140	3,661	228	144
800	4.0	14.0	0.440	17.17	1.43	24.59	10.88	734	1,418	77	51
800	4.0	27.0	0.440	22.24	1.54	34.16	15.03	1,139	2,027	119	76
800	4.0	40.0	0.440	26.66	1.64	43.73	19.16	1,474	2,644	161	102
800	4.0	53.0	0.440	31.00	1.72	53.29	23.28	1,800	3,263	203	128
800	4.0	66.0	0.440	36.56	1.72	62.86	27.46	2,259	3,865	244	154
				Intake at 85% of the Requirement for Maintenance and Lactation							
400	4.5	20.0	−0.696	11.62	1.67	19.41	8.49	687	1,066	85	54
400	4.5	26.0	−0.840	14.02	1.67	23.41	10.24	931	1,310	106	67
400	4.5	32.0	−0.983	16.41	1.67	27.41	11.99	1,187	1,554	127	80
400	5.0	20.0	−0.726	12.11	1.67	20.23	8.85	720	1,118	90	57
400	5.0	26.0	−0.878	14.65	1.67	24.47	10.71	987	1,377	112	71
400	5.0	32.0	−1.030	17.20	1.67	28.72	12.56	1,255	1,635	134	84
400	5.5	20.0	−0.755	12.60	1.67	21.05	9.21	761	1,169	95	60
400	5.5	26.0	−0.916	15.29	1.67	25.54	11.17	1,042	1,443	118	75
400	5.5	32.0	−1.077	17.98	1.67	30.03	13.14	1,323	1,717	142	89
500	4.0	25.0	−0.819	13.67	1.67	22.83	9.99	810	1,286	101	64
500	4.0	33.0	−0.998	16.67	1.67	27.83	12.18	1,134	1,590	126	80
500	4.0	41.0	−1.178	19.66	1.67	32.84	14.37	1,458	1,894	152	95
500	4.5	25.0	−0.856	14.28	1.67	23.85	10.44	864	1,350	107	68
500	4.5	33.0	−1.047	17.48	1.67	29.18	12.77	1,205	1,674	134	85
500	4.5	41.0	−1.238	20.67	1.67	34.52	15.10	1,546	1,998	162	102
500	5.0	25.0	−0.892	14.89	1.67	24.87	10.88	917	1,414	113	71
500	5.0	33.0	−1.095	18.28	1.67	30.53	13.36	1,275	1,758	142	89
500	5.0	41.0	−1.298	21.67	1.67	36.19	15.83	1,633	2,103	172	108
600	3.0	30.0	−0.881	14.71	1.67	24.56	10.74	860	1,399	106	68
600	3.0	40.0	−1.076	17.96	1.67	30.00	13.12	1,223	1,728	133	84
600	3.0	50.0	−1.271	21.22	1.67	35.44	15.50	1,585	2,057	161	101
600	3.5	30.0	−0.925	15.44	1.67	25.79	11.28	924	1,476	113	72
600	3.5	40.0	−1.135	18.94	1.67	31.63	13.84	1,308	1,830	143	90
600	3.5	50.0	−1.344	22.44	1.67	37.48	16.40	1,692	2,184	173	109
600	4.0	30.0	−0.969	16.17	1.67	27.01	11.82	988	1,552	121	77
600	4.0	40.0	−1.193	19.92	1.67	33.27	14.55	1,393	1,932	153	96
600	4.0	50.0	−1.418	23.67	1.67	39.52	17.29	1,798	2,311	185	116
700	3.0	36.0	−1.034	17.26	1.67	28.83	12.61	1,054	1,669	127	81
700	3.0	48.0	−1.268	21.17	1.67	35.36	15.47	1,489	2,064	159	101
700	3.0	60.0	−1.502	25.08	1.67	41.88	18.32	1,924	2,458	192	121
700	3.5	36.0	−1.087	18.15	1.67	30.30	13.26	1,131	1,761	135	86
700	3.5	48.0	−1.339	22.35	1.67	37.32	16.33	1,591	2,186	171	108
700	3.5	60.0	−1.590	26.55	1.67	44.34	19.40	2,052	2,611	207	130
700	4.0	36.0	−1.140	19.03	1.67	31.78	13.90	1,208	1,853	144	91
700	4.0	48.0	−1.409	23.52	1.67	39.28	17.19	1,694	2,308	182	115
700	4.0	60.0	−1.678	28.02	1.67	46.79	20.47	2,180	2,764	221	139
800	3.0	40.0	−1.147	19.15	1.67	31.98	13.99	1,176	1,871	142	90
800	3.0	50.0	−1.342	22.41	1.67	37.42	16.37	1,538	2,200	169	107
800	3.0	60.0	−1.537	25.66	1.67	42.86	18.75	1,900	2,529	196	124
800	3.5	40.0	−1.206	20.13	1.67	33.62	14.71	1,261	1,973	151	96
800	3.5	50.0	−1.416	23.63	1.67	39.46	17.27	1,645	2,327	181	114
800	3.5	60.0	−1.625	27.13	1.67	45.31	19.82	2,028	2,682	211	133
800	4.0	40.0	−1.264	21.11	1.67	35.25	15.42	1,346	2,075	161	102
800	4.0	50.0	−1.489	24.86	1.67	41.51	18.16	1,751	2,455	193	122
800	4.0	60.0	−1.713	28.60	1.67	47.76	20.90	2,156	2,835	225	142

NOTE: The following abbreviations were used: NELDM, net energy for lactation/kg of dry matter; NEL, net energy for lactation; TDN, total digestible nutrients; UIP, undegraded intake protein; DIP, degraded intake protein.
Source: Nutrient Requirements of Dairy Cattle, sixth revised edition, updated 1989, copyright © 1988 by the National Academy of Sciences.

TABLE 54 Daily Nutrient Requirements of Lactating Cows Using Absorbable Protein.

Live Wt. (lb)	Fat (%)	Milk (lb)	Live Weight Change (lb)	Dry Matter Intake (lb)	Energy			Protein		Minerals	
					NELDM (Mcal/lb)	NEL (Mcal)	TDN (lb)	UIP (lb)	DIP (lb)	Ca (lb)	P (lb)

Intake at 100% of the Requirement for Maintenance, Lactation, and Weight Gain

Live Wt. (lb)	Fat (%)	Milk (lb)	Live Weight Change (lb)	Dry Matter Intake (lb)	NELDM (Mcal/lb)	NEL (Mcal)	TDN (lb)	UIP (lb)	DIP (lb)	Ca (lb)	P (lb)
900	4.5	14.0	0.495	20.79	0.64	13.39	13.07	0.98	1.50	0.085	0.056
900	4.5	29.0	0.495	27.42	0.68	18.73	18.20	1.51	2.25	0.137	0.087
900	4.5	43.0	0.495	32.76	0.72	23.71	22.94	1.92	2.95	0.185	0.117
900	4.5	58.0	0.495	37.86	0.77	29.04	28.00	2.29	3.72	0.237	0.149
900	4.5	74.0	0.495	44.54	0.78	34.73	33.45	2.81	4.52	0.292	0.183
900	5.0	14.0	0.495	21.26	0.64	13.70	13.37	1.01	1.54	0.088	0.058
900	5.0	29.0	0.495	28.13	0.69	19.36	18.80	1.55	2.34	0.144	0.092
900	5.0	43.0	0.495	33.70	0.73	24.65	23.84	1.97	3.09	0.195	0.124
900	5.0	58.0	0.495	38.98	0.78	30.31	29.19	2.34	3.90	0.251	0.158
900	5.0	74.0	0.495	46.61	0.78	36.35	35.00	2.94	4.74	0.310	0.194
900	5.5	14.0	0.495	21.73	0.64	14.01	13.66	1.05	1.59	0.092	0.060
900	5.5	29.0	0.495	28.84	0.69	20.00	19.41	1.59	2.43	0.151	0.096
900	5.5	43.0	0.495	34.62	0.74	25.59	24.73	2.01	3.23	0.206	0.130
900	5.5	58.0	0.495	40.49	0.78	31.57	30.41	2.43	4.08	0.265	0.166
900	5.5	74.0	0.495	48.68	0.78	37.96	36.56	3.08	4.97	0.328	0.205
1,100	4.0	18.0	0.605	24.62	0.64	15.86	15.47	1.11	1.88	0.102	0.067
1,100	4.0	36.0	0.605	32.01	0.68	21.87	21.25	1.71	2.70	0.160	0.103
1,100	4.0	55.0	0.605	38.77	0.73	28.21	27.29	2.23	3.60	0.221	0.140
1,100	4.0	73.0	0.605	44.45	0.77	34.22	32.98	2.65	4.46	0.279	0.176
1,100	4.5	91.0	0.605	54.13	0.78	42.22	40.65	3.39	5.58	0.359	0.225
1,100	4.5	18.0	0.605	25.23	0.64	16.25	15.86	1.16	1.91	0.107	0.070
1,100	4.5	36.0	0.605	32.89	0.69	22.65	22.00	1.76	2.81	0.169	0.108
1,100	4.5	55.0	0.605	39.95	0.74	29.41	28.43	2.29	3.77	0.234	0.149
1,100	4.5	73.0	0.605	45.92	0.78	35.81	34.49	2.72	4.69	0.297	0.187
1,100	4.5	91.0	0.605	54.13	0.78	42.22	40.65	3.39	5.58	0.359	0.225
1,100	5.0	18.0	0.605	25.74	0.65	16.65	16.24	1.19	1.97	0.111	0.072
1,100	5.0	36.0	0.605	33.76	0.69	23.44	22.75	1.81	2.92	0.177	0.113
1,100	5.0	55.0	0.605	41.11	0.74	30.61	29.57	2.35	3.95	0.248	0.157
1,100	5.0	73.0	0.605	47.97	0.78	37.41	36.02	2.85	4.92	0.314	0.198
1,100	5.0	91.0	0.605	56.68	0.78	44.20	42.56	3.55	5.86	0.381	0.239
1,300	3.0	23.0	0.715	27.80	0.64	17.91	17.47	1.21	2.16	0.115	0.076
1,300	3.0	47.0	0.715	36.40	0.68	24.87	24.17	1.94	3.13	0.181	0.116
1,300	3.0	70.0	0.715	43.53	0.72	31.55	30.53	2.51	4.07	0.244	0.155
1,300	3.0	93.0	0.715	49.90	0.77	38.22	36.85	3.01	5.02	0.306	0.194
1,300	3.0	117.0	0.715	57.94	0.78	45.19	43.52	3.67	6.00	0.372	0.234
1,300	3.5	23.0	0.715	28.58	0.64	18.41	17.96	1.27	2.23	0.121	0.079
1,300	3.5	47.0	0.715	37.55	0.69	25.90	25.15	2.00	3.27	0.192	0.123
1,300	3.5	70.0	0.715	45.05	0.73	33.08	31.98	2.59	4.29	0.261	0.165
1,300	3.5	93.0	0.715	51.70	0.78	40.25	38.77	3.09	5.32	0.329	0.208
1,300	3.5	117.0	0.715	61.21	0.78	47.74	45.97	3.88	6.36	0.400	0.251
1,300	4.0	23.0	0.715	29.25	0.65	18.91	18.45	1.31	2.30	0.127	0.083
1,300	4.0	47.0	0.715	38.68	0.70	26.92	26.13	2.07	3.42	0.204	0.130
1,300	4.0	70.0	0.715	46.53	0.74	34.60	33.43	2.67	4.52	0.277	0.176
1,300	4.0	93.0	0.715	54.22	0.78	42.28	40.72	3.24	5.61	0.351	0.221
1,300	4.0	117.0	0.715	64.49	0.78	50.29	48.43	4.09	6.72	0.428	0.269
1,500	3.0	26.0	0.825	31.23	0.64	20.12	19.63	1.32	2.48	0.132	0.087
1,500	3.0	52.0	0.825	40.51	0.68	27.66	26.88	2.10	3.53	0.203	0.130
1,500	3.0	78.0	0.825	48.58	0.73	35.21	34.07	2.75	4.60	0.274	0.174
1,500	3.0	104.0	0.825	55.76	0.77	42.76	41.22	3.31	5.67	0.345	0.218
1,500	3.0	130.0	0.825	64.50	0.78	50.30	48.44	4.03	6.73	0.416	0.262
1,500	3.5	26.0	0.825	32.11	0.64	20.69	20.18	1.39	2.56	0.138	0.091
1,500	3.5	52.0	0.825	41.79	0.69	28.80	27.97	2.18	3.69	0.215	0.138
1,500	3.5	78.0	0.825	50.27	0.73	36.91	35.69	2.84	4.84	0.292	0.186
1,500	3.5	104.0	0.825	57.77	0.78	45.02	43.36	3.40	6.01	0.370	0.233
1,500	3.5	130.0	0.825	68.14	0.78	53.14	51.17	4.26	7.13	0.447	0.281
1,500	4.0	26.0	0.825	32.83	0.65	21.25	20.73	1.43	2.64	0.144	0.094
1,500	4.0	52.0	0.825	43.04	0.70	29.93	29.05	2.25	3.86	0.228	0.146
1,500	4.0	78.0	0.825	51.92	0.74	38.61	37.30	2.92	5.09	0.311	0.197
1,500	4.0	104.0	0.825	60.65	0.78	47.29	45.55	3.58	6.33	0.395	0.249
1,500	4.0	130.0	0.825	71.77	0.78	55.97	53.90	4.50	7.53	0.478	0.300
1,700	3.0	29.0	0.935	34.60	0.64	22.29	21.75	1.43	2.80	0.148	0.097

(Table 54 continues)

TABLE 54 Daily Nutrient Requirements of Lactating Cows Using Absorbable Protein. *(Cont.)*

Live Wt. (lb)	Fat (%)	Milk (lb)	Live Weight Change (lb)	Dry Matter Intake (lb)	NELDM (Mcal/lb)	NEL (Mcal)	TDN (lb)	UIP (lb)	DIP (lb)	Ca (lb)	P (lb)
1,700	3.0	57.0	0.935	44.56	0.68	30.42	29.56	2.27	3.93	0.224	0.145
1,700	3.0	86.0	0.935	53.55	0.73	38.84	37.58	2.99	5.12	0.304	0.193
1,700	3.0	114.0	0.935	61.28	0.77	46.96	45.28	3.59	6.28	0.380	0.240
1,700	3.0	143.0	0.935	71.01	0.78	55.38	53.33	4.39	7.46	0.459	0.289
1,700	3.5	29.0	0.935	35.55	0.64	22.93	22.37	1.50	2.89	0.155	0.102
1,700	3.5	57.0	0.935	45.96	0.69	31.66	30.75	2.35	4.11	0.238	0.153
1,700	3.5	86.0	0.935	55.41	0.74	40.71	39.36	3.09	5.39	0.324	0.206
1,700	3.5	114.0	0.935	63.48	0.78	49.45	47.63	3.69	6.64	0.407	0.257
1,700	3.5	143.0	0.935	75.01	0.78	58.50	56.33	4.65	7.90	0.494	0.311
1,700	4.0	29.0	0.935	36.35	0.65	23.56	22.98	1.55	2.98	0.162	0.106
1,700	4.0	57.0	0.935	47.33	0.70	32.91	31.94	2.43	4.29	0.252	0.161
1,700	4.0	86.0	0.935	57.22	0.74	42.59	41.14	3.18	5.67	0.345	0.219
1,700	4.0	114.0	0.935	66.60	0.78	51.94	50.02	3.88	7.00	0.435	0.274
1,700	4.0	143.0	0.935	79.01	0.78	61.62	59.34	4.91	8.34	0.528	0.332
Intake at 85% of the Requirement for Maintenance and Lactation											
900	4.5	43.0	−1.516	25.31	0.76	19.18	18.49	1.47	2.32	0.185	0.117
900	4.5	58.0	−1.875	31.30	0.76	23.71	22.87	2.08	2.93	0.237	0.149
900	4.5	74.0	−2.257	37.69	0.76	28.55	27.53	2.76	3.58	0.292	0.183
900	5.0	43.0	−1.579	26.37	0.76	19.97	19.26	1.54	2.43	0.195	0.124
900	5.0	58.0	−1.960	32.72	0.76	24.79	23.91	2.20	3.08	0.251	0.158
900	5.0	74.0	−2.366	39.50	0.76	29.92	28.86	2.92	3.77	0.310	0.194
900	5.5	43.0	−1.642	27.42	0.76	20.77	20.03	1.62	2.54	0.206	0.130
900	5.5	58.0	−2.045	34.14	0.76	25.86	24.94	2.32	3.23	0.265	0.166
900	5.5	74.0	−2.474	41.31	0.76	31.29	30.18	3.08	3.96	0.328	0.205
1,100	4.0	55.0	−1.802	30.08	0.76	22.79	21.98	1.78	2.83	0.221	0.140
1,100	4.0	73.0	−2.206	36.82	0.76	27.89	26.90	2.51	3.51	0.279	0.176
1,100	4.5	91.0	−2.743	45.79	0.76	34.69	33.46	3.43	4.43	0.359	0.225
1,100	4.5	55.0	−1.882	31.43	0.76	23.81	22.96	1.90	2.97	0.234	0.149
1,100	4.5	73.0	−2.313	38.61	0.76	29.25	28.21	2.67	3.70	0.297	0.187
1,100	4.5	91.0	−2.743	45.79	0.76	34.69	33.46	3.43	4.43	0.359	0.225
1,100	5.0	55.0	−1.963	32.77	0.76	24.83	23.94	2.02	3.11	0.248	0.157
1,100	5.0	73.0	−2.420	40.40	0.76	30.60	29.52	2.82	3.88	0.314	0.198
1,100	5.0	91.0	−2.877	48.02	0.76	36.38	35.09	3.63	4.66	0.381	0.239
1,300	3.0	70.0	−2.009	33.54	0.76	25.40	24.50	2.04	3.19	0.244	0.155
1,300	3.0	93.0	−2.458	41.03	0.76	31.08	29.97	2.88	3.95	0.306	0.194
1,300	3.0	117.0	−2.926	48.84	0.76	37.00	35.68	3.75	4.74	0.372	0.234
1,300	3.5	70.0	−2.112	35.25	0.76	26.70	25.75	2.19	3.37	0.261	0.165
1,300	3.5	93.0	−2.594	43.30	0.76	32.80	31.64	3.07	4.19	0.329	0.208
1,300	3.5	117.0	−3.097	51.71	0.76	39.17	37.78	3.99	5.04	0.400	0.251
1,300	4.0	70.0	−2.214	36.96	0.76	28.00	27.01	2.34	3.55	0.277	0.176
1,300	4.0	93.0	−2.730	45.58	0.76	34.53	33.30	3.27	4.42	0.351	0.221
1,300	4.0	117.0	−3.269	54.57	0.76	41.34	39.87	4.24	5.34	0.428	0.269
1,500	3.0	78.0	−2.238	37.36	0.76	28.30	27.30	2.28	3.60	0.274	0.174
1,500	3.0	104.0	−2.745	45.83	0.76	34.71	33.48	3.23	4.46	0.345	0.218
1,500	3.0	130.0	−3.252	54.29	0.76	41.13	39.67	4.17	5.31	0.416	0.262
1,500	3.5	78.0	−2.352	39.27	0.76	29.75	28.69	2.45	3.80	0.292	0.186
1,500	3.5	104.0	−2.897	48.37	0.76	36.64	35.34	3.45	4.72	0.370	0.233
1,500	3.5	130.0	−3.443	57.48	0.76	43.54	41.99	4.45	5.64	0.447	0.281
1,500	4.0	78.0	−2.467	41.18	0.76	31.19	30.09	2.62	4.00	0.311	0.197
1,500	4.0	104.0	−3.050	50.92	0.76	38.57	37.20	3.67	4.99	0.395	0.249
1,500	4.0	130.0	−3.633	60.66	0.76	45.95	44.32	4.72	5.98	0.478	0.300
1,700	3.0	86.0	−2.464	41.14	0.76	31.17	30.06	2.53	4.01	0.304	0.193
1,700	3.0	114.0	−3.011	50.26	0.76	38.07	36.72	3.54	4.93	0.380	0.240
1,700	3.0	143.0	−3.576	59.70	0.76	45.23	43.62	4.59	5.88	0.459	0.289
1,700	3.5	86.0	−2.590	43.25	0.76	32.76	31.60	2.71	4.23	0.324	0.206
1,700	3.5	114.0	−3.178	53.05	0.76	40.19	38.76	3.79	5.22	0.407	0.257
1,700	3.5	143.0	−3.786	63.20	0.76	47.88	46.18	4.90	6.25	0.494	0.311
1,700	4.0	86.0	−2.717	45.35	0.76	34.36	33.14	2.89	4.45	0.345	0.219
1,700	4.0	114.0	−3.345	55.84	0.76	42.30	40.80	4.03	5.51	0.435	0.274
1,700	4.0	143.0	−3.996	66.71	0.76	50.53	48.74	5.20	6.61	0.528	0.332

NOTE: The following abbreviations were used: NELDM, net energy for lactation/lb of dry matter; NEL, net energy for lactation; TDN, total digestible nutrients; UIP, undegraded intake protein; DIP, degraded intake protein.

Source: Nutrient Requirements of Dairy Cattle, sixth revised edition, updated 1989, copyright © 1988 by the National Academy of Sciences.

TABLE 55 Age Computing Chart.[1]

Day	Jan.	Feb.	March	April	May	June	July	Aug.	Sept.	Oct.	Nov.	Dec.
1	1	32	60	91	121	152	182	213	244	274	305	335
2	2	33	61	92	122	153	183	214	245	275	306	336
3	3	34	62	93	123	154	184	215	246	276	307	337
4	4	35	63	94	124	155	185	216	247	277	308	338
5	5	36	64	95	125	156	186	217	248	278	309	339
6	6	37	65	96	126	157	187	218	249	279	310	340
7	7	38	66	97	127	158	188	219	250	280	311	341
8	8	39	67	98	128	159	189	220	251	281	312	342
9	9	40	68	99	129	160	190	221	252	282	313	343
10	10	41	69	100	130	161	191	222	253	283	314	344
11	11	42	70	101	131	162	192	223	254	284	315	345
12	12	43	71	102	132	163	193	224	255	285	316	346
13	13	44	72	103	133	164	194	225	256	286	317	347
14	14	45	73	104	134	165	195	226	257	287	318	348
15	15	46	74	105	135	166	196	227	258	288	319	349
16	16	47	75	106	136	167	197	228	259	289	320	350
17	17	48	76	107	137	168	198	229	260	290	321	351
18	18	49	77	108	138	169	199	230	261	291	322	352
19	19	50	78	109	139	170	200	231	262	292	323	353
20	20	51	79	110	140	171	201	232	263	293	324	354
21	21	52	80	111	141	172	202	233	264	294	325	355
22	22	53	81	112	142	173	203	234	265	295	326	356
23	23	54	82	113	143	174	204	235	266	296	327	357
24	24	55	83	114	144	175	205	236	267	297	328	358
25	25	56	84	115	145	176	206	237	268	298	329	359
26	26	57	85	116	146	177	207	238	269	299	330	360
27	27	58	86	117	147	178	208	239	270	300	331	361
28	28	59	87	118	148	179	209	240	271	301	332	362
29	29		88	119	149	180	210	241	272	302	333	363
30	30		89	120	150	181	211	242	273	303	334	364
31	31		90		151		212	243		304		365

[1]The numbers under the months represent the day of the year. In a leap year one (1) must be added to the numbers beginning on February 29. To determine the age of an animal: Follow down under the month until the birth date is found in the Day column; this will be the day of the year. Next, find the current date using the same method. Subtract the smaller number from the larger number to find the age of the animal. For example, an animal is born on February 17 (day of the year is 48); the current date is August 25 (day of the year is 237). Subtract 48 from 237 = 189. The animal is 189 days old.

Index